Functional Natural-Based Polymers

Functional Natural-Based Polymers

Editor

Arn Mignon

MDPI • Basel • Beijing • Wuhan • Barcelona • Belgrade • Manchester • Tokyo • Cluj • Tianjin

Editor
Arn Mignon
Engineering Technology
KU Leuven
Leuven
Belgium

Editorial Office
MDPI
St. Alban-Anlage 66
4052 Basel, Switzerland

This is a reprint of articles from the Special Issue published online in the open access journal *Polymers* (ISSN 2073-4360) (available at: www.mdpi.com/journal/polymers/special_issues/funct_natur_based_poly).

For citation purposes, cite each article independently as indicated on the article page online and as indicated below:

LastName, A.A.; LastName, B.B.; LastName, C.C. Article Title. *Journal Name* **Year**, *Volume Number*, Page Range.

ISBN 978-3-0365-5236-1 (Hbk)
ISBN 978-3-0365-5235-4 (PDF)

Contents

About the Editor

Arn Mignon

The research expertise of Prof. Arn Mignon is within the field of (smart) polymer science, with a strong emphasis on biomaterials. As a talented and highly-motivated early-career researcher, he is focusing on the interdisciplinarity of biomaterials research and development, with a special focus on the development of smart polymeric systems. During his PhD and the three years as a postdoctoral researcher, he has worked on an FWO project, INTERREG project and has received funding for an FWO postdoctoral mandate. In that period, he has gathered extensive expertise on polymer synthesis, processing (electrospinning) and characterization (physico-chemical, thermal, mechanical, visual) and gained hands-on experience with imaging techniques for in vitro characterization of biomaterials. The main focus of his research for the coming years lies in the synthesis of new smart polymers for biomedical applications (healing of burn wounds and diabetic ulcers, flexor tendon repair, dental implants, etc.) and the processing of these polymers through (co-axial) electrospinning.

polymers MDPI

Review

Synthetic, Natural, and Semisynthetic Polymer Carriers for Controlled Nitric Oxide Release in Dermal Applications: A Review

Carolina Gutierrez Cisneros [1], Veerle Bloemen [1,2] and Arn Mignon [1,*]

1 Surface and Interface Engineered Materials, Campus Group T, KU Leuven, Andreas Vesaliusstraat 13, 3000 Leuven, Belgium; Caro.Cisneros@KULeuven.be (C.G.C.); Veerle.Bloemen@KULeuven.be (V.B.)

2 Prometheus, Division of Skeletal Tissue Engineering, KU Leuven, Herestraat 49, 3000 Leuven, Belgium

* Correspondence: Arn.Mignon@KULeuven.be

Abstract: Nitric oxide (NO•) is a free radical gas, produced in the human body to regulate physiological processes, such as inflammatory and immune responses. It is required for skin health; therefore, a lack of NO• is known to cause or worsen skin conditions related to three biomedical applications—infection treatment, injury healing, and blood circulation. Therefore, research on its topical release has been increasing for the last two decades. The storage and delivery of nitric oxide in physiological conditions to compensate for its deficiency is achieved through pharmacological compounds called NO-donors. These are further incorporated into scaffolds to enhance therapeutic treatment. A wide range of polymeric scaffolds has been developed and tested for this purpose. Hence, this review aims to give a detailed overview of the natural, synthetic, and semisynthetic polymeric matrices that have been evaluated for antimicrobial, wound healing, and circulatory dermal applications. These matrices have already set a solid foundation in nitric oxide release and their future perspective is headed toward an enhanced controlled release by novel functionalized semisynthetic polymer carriers and co-delivery synergetic platforms. Finally, further clinical tests on patients with the targeted condition will hopefully enable the eventual commercialization of these systems.

Keywords: NO-donor; topical release; polymeric matrices; microbial infections; wound healing; blood circulation; semisynthetic polymers

Citation: Gutierrez Cisneros, C.; Bloemen, V.; Mignon, A. Synthetic, Natural, and Semisynthetic Polymer Carriers for Controlled Nitric Oxide Release in Dermal Applications: A Review. *Polymers* **2021**, *13*, 760. https://doi.org/10.3390/polym13050760

Academic Editor: Alexey Iordanskii

Received: 8 February 2021
Accepted: 25 February 2021
Published: 28 February 2021

1. Introduction

Nitric oxide is a free radical molecule that is produced endogenously in the human body. Its chemical properties make it suitable for the regulation of several physiological processes, including circulatory, immune, neurological, and antioxidant responses [1]. Given its widespread participation in biological systems, inadequate amounts of it (both deficiency and overproduction) can result in illness [2]. While an excess of nitric oxide is implicated in hypersensitive responses and neurodegenerative disorders, its lack results in cardiovascular and neurological complications, hypertension, sexual dysfunction, etc., especially for patients with a pre-existing condition such as diabetes [3,4].

This has brought special interest in the generation of biomedical strategies that carry and deliver nitric oxide, or its precursors, exogenously. Being a gaseous free radical, it has a short half-life in vivo; hence, the main challenges for its administration in the clinic are stable storage and controlled release. This has led to the blending of nitric oxide donors with carriers. Numerous of these platforms have been developed where polymeric scaffolds are often used [3]. Their physicochemical properties as well as their loading and release capacities have been studied and improved in the course of the past 20 years.

Polymers are a strong pillar of drug delivery technologies given that they can offer intrinsic therapeutic activity, be biodegradable to prevent accumulation or toxicity, and enhance release kinetics. Moreover, their molecular structure can be engineered for more biocompatibility and strict control over a wide range of drugs. This tunable release can be in constant doses over long periods, cycled, or condition-responsive, all of which can improve the therapeutic value of the system [5]. This is by the enhancement of molecule transport and because they can become active participants in the treatment. For instance, their biocompatibility can contribute to increased stability for cellular uptake [3]. The inherent properties of the materials vary according to their origin, while synthetic polymers such as poly(ethylene oxide) (PEO) and poly(vinyl alcohol) (PVA) are characterized by their reproducibility and desirable mechanical properties, natural polymers including chitosan, alginate, and gelatin, are valuable due to their biocompatibility, biodegradability, mucoadhesion and antibacterial capacity, for instance [6].

Considerable advances in polymer synthesis have contributed to the availability of a wide range of polymeric scaffolds with different architectures (multi-layered, compartmentalized, branched, etc.). These platforms have advanced regarding physical properties by particle entrapment, with the possibility to tune the capsule size and concerning chemical advances that allow for surface functionalization, all of which contribute to controlled release kinetics [7]. Further improvement in polymer carrier systems has been addressed by the introduction of semisynthetic polymers, where both natural and synthetic polymers are linked to each other to obtain a combination of their properties. Along with the evolution of polymers, the therapeutic effect of drug co-delivery by different release triggers is another increasing trend that has accomplished synergistic treatments.

Dermal (cutaneous) administration of NO-donors has been increasingly studied as a promising therapy for related skin conditions. The endogenous synthesis of nitric oxide in human skin can be either enzymatic or chemical and it is required for various processes such as pigmentation, blood flow dynamics, cutaneous tissue regeneration, and skin immune response [3,8]. The focus of this review consists of the topical NO-delivery by polymer matrices according to their application target, with a special highlight on the nature of the applied polymers. For a better understanding of these platforms, an overview of the biochemistry and the dermatological role of nitric oxide, and the resulting disorders is given in the upcoming sections (Sections 2–4). Afterward, nitric oxide donors and their possible carriage systems are explained (Sections 5 and 6) before reviewing the pharmaceutical dermal applications in detail (Section 7). Finally, remarks are made on the future perspectives of these technologies (Section 8).

2. Biochemistry of Nitric Oxide

Nitrogen monoxide, or nitric oxide (NO•), is a colorless, free-radical gas molecule. It is non-combustible but as a hydrogen bond acceptor, it can react with oxygen. The particular molecular characteristics and reactivity make this reactive oxygen species relevant in biological systems. In water, it has poor solubility and reactivity, but it has a high diffusion rate. Given that it complexes metals, it participates in corporal catalyzes by coordination with enzymatic units [9]. The bioabsorption of NO• is based on its high solubility in hydrophobic solvents as it can cross membranes, lipidic biological barriers, without the need for active transport. Given that such molecules with unpaired electrons are rare in the human metabolism, the reactivity of NO• is selective, which is a useful property to moderate biological systems [10–12].

Nitric oxide's reactivity with oxygen gas is due to an unpaired electron in the anti-bonding π molecular orbital between the nitrogen and oxygen atoms of the molecule, which needs another unpaired electron to stabilize. Oxygen gas is unpaired in the ground state because its electrons, despite being an even amount, occupy different molecular orbitals. This extensively studied reaction is still not fully understood, yet the currently accepted reaction scheme consists of the following two equations [10,13]:

$$2NO \leftrightarrow N_2O_2 \quad (1)$$

$$N_2O_2 + NO \rightarrow N_2O + NO_2 \quad (2)$$

Compared to other radical molecules, NO• is considered relatively stable because, in oxygen-free conditions, it can only dimerize at low temperatures or high pressures. The conditions for this are not fulfilled anatomically; thus, this reaction does not risk taking place in the human body. However, at high aerobic conditions, NO• is quickly oxidized into reactive nitrogen species, from which dinitrogen trioxide (N_2O_3) predominates in aqueous systems, such as that of the human body [14]. N_2O_3 is an unstable molecule, so it is quickly metabolized by hydrolysis [10,14,15].

The chemical properties of nitric oxide allow it to have a cell messenger role, which specifically contributes to the regulation of corporal responses by various signaling networks. The behavior of NO• under physiological conditions accounts for its participation in a wide variety of biological responses. This is further addressed in the following section [10,16].

3. Nitric Oxide in Human Physiology

NO• is present in the human body in a concentration range in the order of pM to nM and it is used for regulatory functions [10]. The metabolic synthesis of NO• results from the enzymatically controlled oxidation of L-arginine, an essential alpha-amino acid involved in protein synthesis. Nitric oxide synthases (NOS) are the enzymes that regulate this process; they have two redox regions—an electron generating carbon-terminal (reductase) and a nitrogen-terminal (oxidase). NOS synthetize one molecule of L-arginine into one molecule of NO• and stoichiometric amounts of L-citrulline as a byproduct [15,17,18].

There are two classes of NOS. The first types are constitutive, calcium-regulated, cNOS. When activated for a physiological function, they produce low amounts of nitric oxide. The activation is transient because it follows the signal's kinetics in a pulsed form. The second types are inducible NOS (iNOS), which are regulated genetically and transcribe for immune and inflammatory responses. They are continuously active once induced, for instance, by bacterial lipopolysaccharides or cytokines. There is a delay of hours for the production of NO• due to the time for protein synthesis, but the reached NO• concentration is higher and sustains for as long as the enzyme is present [10,15,19].

In addition to the enzymatic release, nitric oxide can be physiologically generated by the chemical reduction of nitrate (NO_3) or nitrite (NO_2). For metabolism, NO_3 is converted by commensal bacteria to NO_2 [20]. Sweat is a nitrate source for the body and given a higher amount of sudoriparous glands in the skin, higher NO• amounts are found there [17,21]. The present review focuses on the dermal application of nitric oxide; hence, the following subsection focuses on metabolic activity at a cutaneous level.

Dermatological Nitric Oxide

The skin comprises around 6% of the total body weight. Its outermost layer is the epidermis, a dense avascular epithelial structure that protects organs; it mainly consists of keratinocytes. The next skin layer is the dermis, which is a mucous polysaccharide matrix that contains encapsulated collagen and elastic fibrils, where metabolic exchange happens. Within this structure, different functions take place that are regulated by NO•, including the synthesis of connective tissue, the immune and inflammatory responses, and the pigmentation of the skin. The deepest skin layer is the hypodermis, a subcutaneous lipidic layer whose blood and lymphatic vessel networks respond to nitric oxygen radicals, for the formation of new blood vessels (angiogenesis). Finally, sudoriparous glands and hair follicles originate at a subcutaneous level and disperse in human skin. Sudoriparous glands produce sweat, which is an indirect source of NO• [21].

Both classes of NOS enzymes described earlier in Section 3 are needed for dermal homeostasis. There are three bioregulatory roles of NO• in human skin—vasodilation, cutaneous immune response, and tissue regeneration. The first is predominantly regulated by low cNOS synthesis. It consists of vascular smooth muscle relaxation following induction by endothelial cells of skin vessels. Its deficiency results in the contraction of vessels that narrows the arteries. When this happens, blood cannot flow properly and it causes circulatory conditions such as ischemia [4,22]. The second role is the immune response against environmental threats. Injury protection from UV-light or pathogens is provided by high NO• production. This can be either enzymatically activated by iNOS, or in an enzyme-independent manner from the metabolism of sweat's nitrates. Nitric oxide present on the skin is cytotoxic for a wide variety of pathogens; therefore, its chemical production on the body surface prevents and clears infections [22,23]. Finally, nitric oxide has an important role in skin regeneration by modulating inflammation, cell proliferation, extracellular matrix deposition, and angiogenesis. All three dermal NO-regulated processes contribute to wound healing [4,22–26].

Triggered for defense and repair, most skin cell types can synthetize NO• by one or more NOS for inflammatory, antimicrobial, and apoptotic responses. Another stimulant is exposure to UV, i.e., sunlight, which activates iNOS for skin pigmentation. Skin pigmentation occurs in the dermis and epidermis by melanocyte cells. Nitric oxide contributes to the regulation of this process as a response to UV-light exposure. Negative feedback is carried out indirectly by growth factors and directly by NO• molecules [8,10,14,21,27–29].

Under pathophysiological conditions, nitric oxide can be overproduced, due to increased immune and inflammatory responses; or underproduced by inhibition. The unsuitable presence of NO• in the body results in circulatory disorders regarding blood activity and vessels. Furthermore, some enzymatic antagonists of NOS have been proven to damage circulation locally [4,22].

4. Nitric Oxide Skin Disorders

Given that both excess and a lack of NO• lead to pathological states, fine control of NOS expression and NO• production is needed. Excessive production of nitric oxide promotes oxidative stress, causing degenerative diseases. Large amounts of NO• associate with undesirable infectious or inflammatory responses. NO• can induce toxic reactions against other host tissues, especially in particular types of inflammation, such as asthma. The role of NO• in allergen-induced skin inflammation is pro-inflammatory at low concentrations, while it induces inflammatory cell apoptosis at high concentrations. Elevated iNOS expression has been reported in skin conditions such as psoriasis, dermatitis, and atrophy. The extracellular matrix deposition, collagen synthesis, and cytokinetic promotion of angiogenesis, overstimulated by NO• may result in abnormal scarring by disproportionate (hypertrophy) or fibrous (keloid) tissue [14,24–26,28].

However, decreased skin NO• levels also cause impairment, of which the first consequence is endothelium dysfunction. This condition is related to atherosclerosis, high blood pressure, and other factors that may lead to cardiac events and disorders related to angiogenesis. Insulin resistance is associated with nitric oxide deactivation by dietary oxidation. At low availability of L-arginine, endothelial NOS can generate superoxide, which scavenges nitric oxide. This has been reported in pathological conditions such as diabetes and hypercholesterolemia. Furthermore, insufficiency of nitric oxide results in wound healing incapacity. This is especially problematic for patients that suffer from diabetes, malnutrition, or chronic steroid treatment. Sepsis patients present decreased NO• production, linked to a decreased and inadequate production of arginine. Undernourished infants also present arginine deficiency, which is similarly manifested by lowered nitric oxide production [8,10,15,26,30–33].

Because insufficient nitric oxide bioproduction results in a vast number of pathological conditions, at the end of the 19th century, sodium nitroprusside, glyceryl trinitrate, and isoamyl nitrite were used to treat the deficiency for the first time, although their physiological effects were not completely known. These pharmacologically active substances that carry nitric oxide are called NO-donors. Ideally, the donors stabilize the radical until the release is required [33,34].

5. Nitric Oxide Donors

Nitric oxide is an inconvenient molecule to handle in gas form due to long treatment time by impractical and expensive equipment for delivery (a gas tank with a diluting system). There is an additional concern about the presence of toxic concentrations for host cells [35,36]. Therefore, NO-donors have been designed as substances by which the radicals can be steadily delivered in situ [37]. Given that NO• is a radical of high therapeutic relevance, systems that improve their deliverability have been developed. Despite the simplicity of the molecule, its exogenous application is challenging due to the complexity and manifold of reactions, tissue specificity, and concentration dependence [8,34,37].

The biochemical pathways where nitric oxide is involved are the fundament for the design of NO-donors. There are eight molecules that can be metabolized into nitric oxide and that can be used as donors in biological systems (see Figure 1). These consist of arginine, metal nitrosyl complexes ($M^{(n-1)}$-NO^+), N-diazeniumdiolates (NONOates), nitrate ion (NO_3^-), nitrogen dioxide (NO_2), nitrite ion (dioxidonitrate(1-), NO_2^-), nitrosothiols (RSNO), organic nitrite (RNO_2) and organic nitrate (RNO_3) [33]. They can be classified as natural or chemical-occurring, according to their physiological presence.

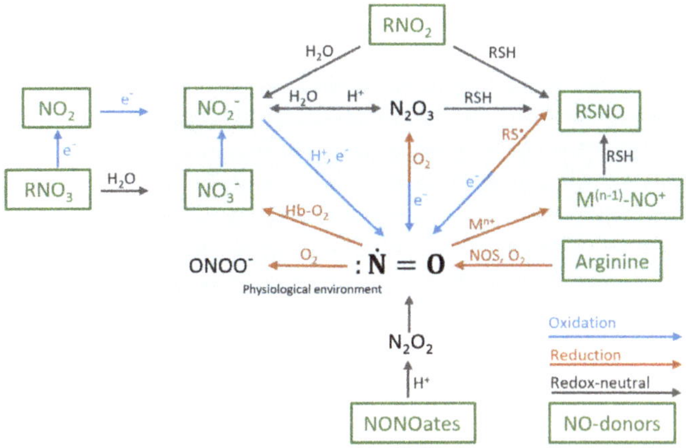

Figure 1. Physiological production of nitric oxide starting from different NO-donors. This figure was adapted from [33]. Abbreviations: hemoglobin (Hb).

Natural NO-donors occur biologically and include arginine, metallonitrosyl complexes ($M^{(n-1)}$-NO^+), NO_2^-, NO_3^- and RSNO. Assuming that NOS are present and active, L-arginine can be provided to increase NO• production. However, this can often not be affirmed; hence, enzyme-independent donors are needed. Nitric oxide binds to transition metals to form metallonitrosyl complexes. These are endogenously formed at a physiological level when NO• binds metallic centers of enzymatic units. Their usefulness as NO-donor suppliers relies on their photosensitivity to decompose to nitric oxide, which can be used for controlled release [38]. While NO_2^- can be directly metabolized to nitric oxide, NO_3^- has to be reduced to NO_2^- first. This conversion is performed by bacterial enzymes that are present in the human body. If not converted, NO_3^- remains stable to flow through the circulatory system. Finally, nitrosothiols require an electron to decompose into nitric oxide and thiolate (RS–).

Chemically occurring NO-donors do not exist physiologically, but they are viable donors because they can be metabolized by the human body. These are NO_2, RNO_3, RNO_2, and NONOates. Provision of nitric oxide by NO_2 requires its reduction to a nitrite ion. Organic nitrate can either be reduced to nitrogen dioxide or be hydrolyzed to nitrate ion. Organic nitrite is transnitrosated into a nitrite ion by water or into RSNO by a thiol (RSH). NONOates are one of the most frequently used NO-donors. Upon protonation to dinitrogen dioxide (N_2O_2), the dimerized form of nitric oxide (which does not occur in physiological conditions), provides two NO• radicals as shown in Scheme 1. NONOates are N-based analogs of diazeniumdiolates; their production consists of the reaction of nitric oxide gas with primary or secondary amines [33].

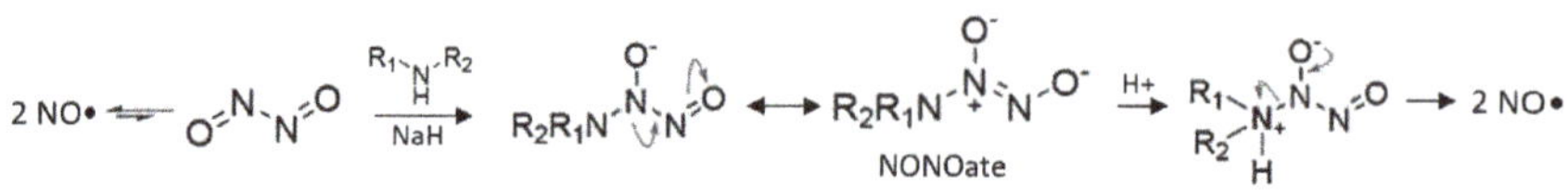

Scheme 1. Formation and degradation scheme of N-diazeniumdiolates (NONOates). This scheme was adapted from [33].

5.1. Controlled Release

Ideally, the nitric oxide delivery profile and release location must be precisely controlled, which can be achieved by a triggering substrate. As the first possibility, NONOates have been designed to respond to substrate modulation. Their spontaneous breakdown can be prevented, for instance, by an alkylation that can later be selectively removed by a particular molecule or enzyme. This technology is inspired by fluorescence biosensing chemistry [33].

The second possibility is to trigger NO• release photonically. This is convenient, given that light is adaptable, non-invasive, and adjustable. Sometimes, NO-donors are spontaneously photosensitive; therefore, unless this pathway is disabled, they cannot serve for controlled release. Usually, the alternative is to deactivate NONOates with a photo-labile group. To monitor the localization of the donors microscopically and to control the release of NO•, fluorescence emission has been used [33].

The variety of NO-donor structures available (summarized in Table 1) results in different reactivity and delivery kinetics. This can happen either spontaneously, enzymatically, or chemically. S-nitrosothiols and diazeniumdiolates, for instance, release it through an autolytic breakdown triggered by thermal or photonic factors. For an enzymatic source, either prodrugs or mimetics are used. Prodrugs are immobilized enzymes in biomaterials characteristic of a controlled bioconversion and mimetics consist of endogenous activation by non-enzymatic compounds [34,37].

Table 1. Representative NO-donors according to chemical class, adjusted from [33].

NO-donor.	Representative Examples		Trigger for NO• Generation	
	Structure	Name	Non-Enzymatic	Enzymatic
Organic nitrate	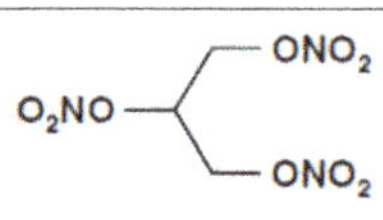	GTN [33]	Thiols	Cyt-P450, GST, etc.
Organic nitrite	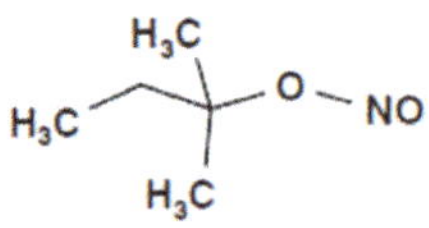	Tert-pentyl nitrite [33]	Hydrolysis, trans-nitrosation, thiols, light, heat	Xanthine oxidase
Metal–NO complex	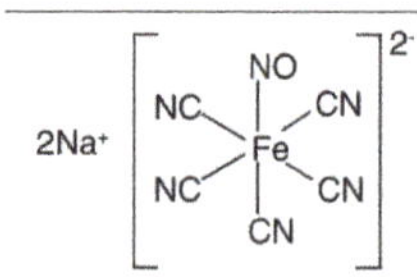	Sodium nitroprusside [38]	Light, thiols, reductants, nucleophiles	Membrane-bound enzyme

Table 1. *Cont.*

NO-donor.	Representative Examples Structure	Name	Trigger for NO• Generation Non-Enzymatic	Enzymatic
Nitrosothiol	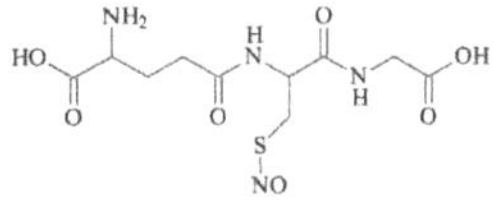	GSNO [39]	Spontaneous, enhanced by thiols, light, metal ions	Unknown enzymes
	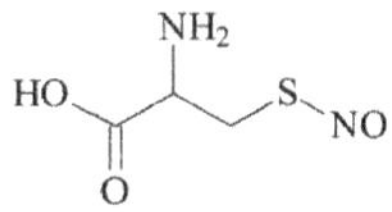	CySNO [39]		
	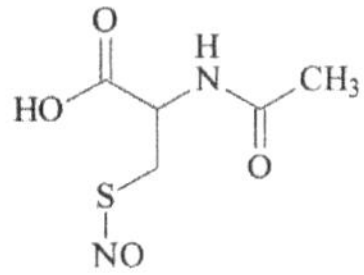	SNAC [39]		
NONOate	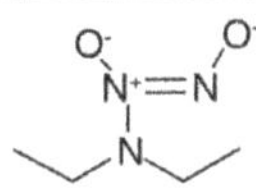	DEA/NO [40]	Spontaneous, engineered to be light or pH-responsive	Caged for control by β-galactosidase
	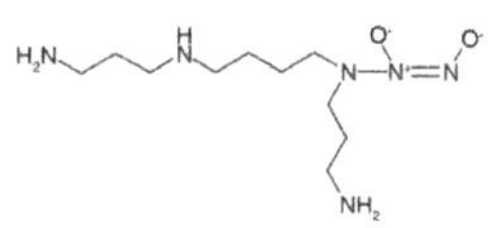	SPER/NO [40]		

Abbreviations: Nitroglycerin (GTN), S-nitrosoglutathione (GSNO), S-nitroso cysteine (CySNO), S-nitroso-N-acetylcysteine (SNAC), diethylamine NONOate (DEA/NO), spermine NONOate (SPER/NO).

5.2. Main NO-Donors

The most frequently studied and implemented NO-donors are S-nitrosothiols (RNSO) and NONOates [41]. The discovery of S-nitrosothiols (RSNO) as endogenous NO-donors encouraged the development of exogenous for the delivery of NO• [42]. On the other hand, diazeniumdiolates (NONOates) are the most widespread NO-donors given that they do not need to be redox-active for release because they imitate nitric oxide biosynthesis. They consist of structures with a functional NONO-group [1].

RSNO synthesis is carried-out in liquid solute by S-nitrosation of thiols. Some of the RSNO that result from this reaction are S-nitrosoglutathione (GSNO), S-nitroso cysteine (CySNO), S-nitroso-N-acetylcysteine (SNAC) (depicted in Table 1), S-nitrosoalbumin, and S-nitrosohemoglobin. GSNO is the most widely used RSNO in research because it can easily be purified by precipitation and drying, which makes it relatively stable for storage. On the contrary, the use of CysNO or SNAC implies their fresh preparation upon application. S-nitrosoalbumin and S-nitrosohemoglobin are less used in general, due to their instability in solution. RSNO are prone to photochemical decomposition so they require light protection. Another broadly studied synthetic RSNO, characterized by high thermal stability, is S-nitroso-N-acetylpenicillamine (SNAP). It is found commercially but cannot be integrated into hydrogels due to its low water solubility. Furthermore, it has been avoided in topical formulations to eliminate the possibility of undesired effects by excessive concentration in the hypodermis due to its high hydrophobicity [42].

NONOates are convenient NO-donors because of their biocompatibility, as they only release the initial amines from which they were made upon physiological decomposition to nitric oxide. Furthermore, they are stable in solid state and only decompose when dissolved in a biological environment [1]. The release is spontaneous, but it can be controlled by amine groups contained in a polymer matrix, which can also be convenient to avoid unwanted leakage [43]. Some common NONOates are diethylamine NONOate (DEA/NO), spermine NONOate (SPER/NO), and 1-(hydroxy-NNO-azoxy)-L-proline (PROLI/NO) (see Table 1) [40].

Despite the broad spectrum of NO-donors, only a few have been approved and are available commercially. This may be due to the amount of testing for approval and to the fact that these therapeutic applications have only been recently developed. Commercially used NO-donors are used for a particular, clinically proven application such as nitroglycerine for acute angina and Nitropress for heart failure and high blood pressure [33,34].

6. Carriers of Nitric Oxide Donors

For dermal applications, carriers are often used for localized release. There are several possibilities to contain NO-donors, such as in dendrimers, micelles, liposomes, polymer particles, and metalorganic crystals. An innovative approach to harness the beneficial properties of NO• is to attach a releasing moiety to an existing drug. Different hybrid compounds can offer various drug actions with synergistic effects, with reduced toxicity and side effects [34,37]. For dermatological application, in particular, there are three main classes of carriers—liposomes, metals, and polymers, which are discussed in further detail in the following subsections.

6.1. Liposomal Carriers

Liposomes are spherical vesicles made of bilayers of phospholipids and cholesterol; they are amphiphilic: hydrophobic at the outside while hydrophilic at the inside. Comparable to NO• biochemistry, liposomes can easily penetrate the epidermis due to their lipidic composition. This effect is also used to control the kinetics and physicochemical features together with the choice of the NO-donor [23,34].

The encapsulation of NO-donors inside liposomes results in sustained release because their decomposition is prevented by the lipidic bilayer coating. Some of the recent developments in liposomes as NO-donor platforms include the design of photoinduced NO-releasing liposomes by the incorporation of Ru salen, a metal photosensitive nitrosyl complex [44]. Attachment of fluorescent immunoassays on and inside liposomes has also been reported as it can be used for monitoring of NO• in time and location (spatiotemporal) [33,45,46]. Liposomes are biodegradable and biocompatible; therefore, research in this area is promising. However, they are hydrophobic, which limits their cutaneous application due to excessive accumulation in lipidic membranes [23,34].

6.2. Metallic Carriers

Metallic particles are highly tunable, and they have been studied extensively. Silver and gold are especially popular metals for skin application because of their antibacterial and anti-inflammatory properties, in addition to their functional modification possibilities. The capacity to prevent infection of NO• is enhanced by these metallic platforms [23].

A popular metallic carrier platform is nanoparticles. Their fabrication and surface functionalization is versatile for controlled release. The size and location of nanoparticles within the system control the kinetics by the probability of contact with water to release the NO-donors. Small particle size is directly proportional to the bactericidal effect, while the sphericity is inversely proportional. These nanoparticles range in size from 1 nm to 100 nm with convenient optical and electrical properties. For instance, functionalized gold nanoparticles with grafted NONOate-forming polymers have allowed for their proton-induced delivery [47]. Despite the steady progress of such therapeutic technology, the challenge for their medical use is still the possible toxic effect of bioaccumulation and the control over membrane crossing [46].

Another modern platform consists of metal–organic crystalline structures, they are metallic centers with bound organic linkers resulting in porous materials with a high surface area. Their properties are adjusted according to the combination of linker and metal. Because NO• can be adsorbed within the pores, they are pharmaceutically useful as NO-donor delivery systems. The release happens upon contact with water but it can be controlled. For instance, highly regulated, photon activated NO-donor delivery by nitroimidazole bound to zinc has been reported. The release varies according to the duration and intensity of the irradiation [48]. The photoactivation can also be monitored and induced by fluorescence [34]. This technology is based on covalent bond coordination with polymers, which is the following class of NO-donor carrier addressed.

6.3. Polymeric Carriers

The advantage of polymers for dermal-controlled delivery of NO• resides in the possibility of adjusting the physicochemical properties. These donors can be either synthetic or natural-based polymers or a combination of both. Synthetic polymers are well understood and convenient, especially for their mechanical properties. Synthetic poly(lactic-co-glycolic acid) (PLGA) nanoparticles, for example, are FDA approved for their use as drug carriers and they have been widely used due to their sustainable release properties. Interest in natural ones has increased in the last years because of the therapeutic properties of the carrier itself. An example is chitosan, displaying anti-inflammatory and antibacterial properties with sustainable, controlled drug release [23,49]. A combined type of polymer carriers has emerged in recent years with the intention of joining the advantages of synthetic and natural polymers, the so-called semisynthetic polymers. Different types of NO-donors have been encapsulated in these molecules via technologies such as emulsion polymerization, with control on their molecular mass and functional ends to adjust the profile of delivery [34,50].

The first described polymer carrier matrices were hydrogels; they are currently still used to entrap nanoparticles. This type of system is desirable due to optimizable polymer particle size, solubility, and release kinetics [51]. Alternative polymeric matrices consist of micelles and dendrimers, dissolved in a buffer or gel for topical applications. Self-assembled micelles are used to encapsulate NO-donors due to their amphiphilic nature with a hydrophobic core. Additionally, moieties can be adhered for a controlled release. Dendrimers are globular molecules produced by branching polymers with attached reactive groups. Their size and shape can be chemically tuned; they have been functionalized with NO• moieties to produce NONOates with a significant antimicrobial capacity [52].

Biomedical applications that make use of polymeric carriers exploit their specific properties being their molecular size, porosity, solubility, complexation, bioavailability, toxicity, cellular and molecular effects, etc. Currently, interdisciplinary research develops biocompatible and biodegradable drug delivery platforms made of synthetic, natural, and semisynthetic polymers to treat cutaneous inflammation, dysfunction, and infection. They seek to enhance the mechanical and physical characteristics of these materials regarding the efficiency of release [53]. There are multiple pharmacological factors that affect this, which result in a wide spectrum of properties. These factors include the polymer type, drug loading, polymer breakdown mechanisms, interactions between the drug and polymer, etc. [51].

Synthetic polymers have often superior mechanical properties compared to natural polymers, and they are abundantly present and are easier to process into suitable pore sizes and scaffold geometries than natural ones. Natural polymers do not have the robustness of the design of synthetic polymers, but their biocompatibility makes them stand out. Natural polymers are often more effective in physiological conditions than synthetic polymers because, in most cases, they are biodegradable in the human body. Important drawbacks they have are their limited purity and high molar mass dispersity, resulting in variability of the characteristics, which is unacceptable for the regulated environment of the pharmaceutical industry [54]. Therefore, laborious purification and analytical testing are needed to ensure high quality. Ideally, semisynthetic polymers succeed to preserve the desirable properties of synthetic polymers From a chemical point of view, semisynthetic polymers enrich the field of drug-carrying scaffolds by optimizing physicochemical properties for a targeted delivery [51].

The investigation and optimization of mechanical and physical material characteristics of novel polymer carriers is still the focus of drug delivery, as will be seen in the upcoming sections of this review. Strict regulation of the surface properties can prevent possible toxicity to human cells by controlling NO• delivery [34]. The uses of polymer matrices for cutaneous delivery of NO-donors are the object of this review and they are addressed according to their pharmaceutical applications in the following section.

Polymer Carrier Preparation

The carrier preparation technologies have advanced alongside the evolution of polymers. In the beginning, most materials were directly solvent cast with the simultaneous entrapment of NO-donors [55]. This allowed the loading of sufficient concentrations for therapeutic effects according to the NO-donor of choice (often in the millimolar range), but there was no control of the release given its spontaneous delivery in physiological conditions [56,57]. There was an initial burst in the first five minutes, where most of the loaded drug would be released [58]. Given that solvent casting is the simplest and the least controlled preparation, lyophilization and electrospinning were introduced to achieve higher quality and control of the material and their processing [6,52,59,60].

Simultaneously, polymer carrier nanoparticles began to be tested according to their size, porosity, and surface charges to achieve high NO-donor encapsulation efficiencies. Nitric oxide release from these nanoparticles depends on different mechanisms that have been measured and modeled, including desorption, diffusion, particle erosion, or a combination of them. Kinetic measurements have demonstrated sustained NO release for up to ca. 24 h after an initial peak [60–62].

In the case of modern polymers, highly controlled chemistry technologies were necessary for the preparation of nanoparticles, which were either directly tested or entrapped in hydrogels and electrospun mats. These implemented technologies were also necessary for applications where the NO-donor was grafted into the polymer, which is another possibility to extend and control the period of release. They include light-mediated, polymerization-induced self-assembly and reversible addition–fragmentation chain transfer [63,64].

7. Pharmaceutical Applications

The clinical applicability of NO-donors is impeded by a lack of control regarding the release rate and target location. Therefore, their delivery through carriers has enabled the proposal of various pharmaceutical applications. Research has focused on the employment against bacteria and biofilms, but they have also been increasingly investigated for their use in wound healing enhancement and in the treatment of circulatory conditions [34]. The evolution of each category is discussed based on an assemblage of pertinent publications.

7.1. *Antimicrobial Applications*

NO• is an antimicrobial agent whose activity is dependent on the concentration—low production by cNOS stimulates the immune system by regulation of the immune cell cycle, while iNOS, as part of the innate immune system, produce larger quantities that respond to bacterial polysaccharides and endotoxins, and pro-inflammatory cytokines [8]. When nitric oxide began to be used to treat skin infections in the late 1900s, it was mainly applied in a gaseous state or by nitrite provision to supply NO• as an alternative treatment to antibiotic-resistant microorganisms such as *C. albicans* bacteria and *Leishmania* parasites. It was also initially tested for particular skin conditions such as psoriasis [65–68]. However, the need for controlled and localized release and more convenient platforms shifted the development of nitric oxide supply to NO-donor carriers.

In the beginning, these platforms were tested using silane and zeolite. With the purpose to treat infections, silane particles were conjugated with NO-donors. Alternatively, these were carried by zeolites given that they are porous crystal minerals with high compatibility and storage capacity [69]. By 2000, interest in polymeric platforms began to rise significantly to achieve a steady prolonged release of NO•. Polymers are versatile materials, capable of delivering constant, sustained doses of drugs of which the dosage may be tuned and cycled. Advances in polymer technology have focused on the development of biocompatible degradable responsive systems with higher permeation and retention capacities [5]. In Table 2, a summary of publications on infection-treating NO-carrying polymeric formulations for dermal applications dating from 2003 until 2020 is presented (this is further discussed in more detail in the following paragraphs).

Table 2. Compilation of publications showing the evolution of NO-donor polymeric carriers intended to treat dermal infections.

Year	Origin	Carrier	Matrix	NO-Donor	Donor Coupling	Synergistic Additive	Release Trigger	Antimicrobial Test	Clinical Trial	Ref.
2020	N	Alg	np	Nitrosothiol	-	Silver np	-	Bacterial	-	[70]
2020	N	Alginate	np	Nitrosothiol	-	Silver np	-	Bacterial	-	[71]
2020	N	Cellulose	Hydrogel	NONOate	-	-	-	Fungal	Yes	[72]
2019	S	PEO	Micelles	N-nitrosamine	Yes	-	Light	Bacterial	-	[63]
2018	S	PEO	Hydrogel	NO• gas	Yes	-	-	Bacterial	-	[64]
2018	S	PEG/Disaccharide	np	Inorganic nitrite	-	Antifungal	-	Fungal	-	[73]
2018	SS	PEO/Cellulose	np	NONOate	-	-	-	Fungal	Yes	[74]
2018	SS	Chitosan/PL	np	Nitrosothiol	-		-	-	-	[2]
2018	SS	Chitosan/PL	Hydrogel	Nitrosothiol	-		-	Bacterial	-	[75]
2017	S	PEG/Disaccharide	Hydrogel	Inorganic nitrite	-	-	-	Fungal	-	[76]
2017	S	POEGMA	np	NONOate	Yes		-	Bacterial	-	[62]
2016	S	PU/PAMAM	Dendrimers	NONOate	Yes	-	-	Bacterial	-	[52]
2016	S	POEGMA	np	NONOate	-	Antibiotic	-	Bacterial	-	[61]
2014	N	Alginate/Chitosan	Capsules	Nitrosothiol	-	-	-	Bacterial	-	[60]
2013	N	Alginate/Chitosan	np	Nitrosothiol	-	-	-	-	-	[6]
2012	S	PEO	Chamber	NO• gas	-	-	Enzyme	Bacterial	-	[77]
2011	N	Alginate/Chitosan	np	Nitrosothiol	-	-	-	-	-	[59]
2010	S	PU	Hydrogel	Organic nitrite	-	-	-	Parasitic	Yes	[78]
2010	S	Acrylic polymer	Chamber	Inorganic nitrite	-	-	Bacteria	Bacterial, fungal	-	[79]
2009	S	PEG/Disaccharide	Hydrogel	Inorganic nitrite	-	-	-	Bacterial	-	[80]
2009	S	PU	Hydrogel	Metal-NO complex	-	H_2O_2, methyl-ene blue	Light	Bacterial	-	[58]
2008	S	PEG/Disaccharide	Hydrogel	Inorganic nitrite	-	-	-	-	-	[81]
2004	S	PEO	Hydrogel	Nitrosothiol	-	-	-	-	Yes	[56]
2004	S	PVA/PVP	Hydrogel	Nitrosothiol	-	-	-	-	-	[55]
2003	S	PL	Hydrogel	Nitrosothiol	-	-	Light	-	-	[57]

Abbreviations: Polymer origin: natural (N), semi synthetic (SS), synthetic (S). Polymer carriers: poly(ethylene oxide) (PEO), pol(ethylene glycol) (PEG), Pluronic F-127 (PL), poly(oligoethylene glycol) methyl ether methacrylate (POEGMA), poly(urethane) (PU), poly(amidoamine) (PAMAM), poly(vinyl alcohol) (PVA), poly(vinyl pyrrolidone) (PVP). Matrix: nanoparticles (np). NO-donor: diazeniumdiolate (NONOate).

7.1.1. Synthetic Polymer Carriers

Until 2010, only synthetic polymers were used to carry nitric oxide. These synthetic hydrogels have been made of materials including poly(vinyl alcohol) (PVA), poly(vinyl pyrrolidone), poly(urethane) (PU), Pluronic F-127 (PL), and (hydroxyethyl)methacrylate [55–58,80,81]. PU is used for its semi-permeability, i.e., impermeable to bacteria but permeable to vapor and air [82]. The main NO-donor used was GSNO, a well-known commercial nitrosothiol. Given that the microbicidal capacity of NO• was known, an early development was the incorporation of two different nitrosothiols—GSNO and SNAC into hydrogels made of Pluronic F-127. These were tested for NO-release by irradiation as potential topical therapeutics. They highlighted that a hydrogel matrix damps the initial rate of thermal NO release compared to aqueous solutions [57].

Other recurrent precursors of nitric oxide are organic and inorganic nitrites. Jones et al. reported the use of such a precursor in a nitric oxide-releasing chamber made of a gas-permeable acrylic polymer. Inside the chamber, the dismutation of inorganic sodium nitrite salt ($NaNO_2$) resulted in the sustained release of NO• during the fermentation of immobilized glucose by lactic acid probiotics (*L. fermentum*) [79]. In 2008, another delivery system based on $NaNO_2$ was published by Friedman et al. The platform consisted of glass-state disaccharide films (trehalose and sucrose) within which $NaNO_2$ was chemically reduced to generate and trap nitric oxide using lyophilization. The resulting material was incorporated in a composite hydrogel matrix with poly(ethylene glycol) (PEG). The NO-release profile varied in function of the PEG composition and they achieved a steady state level of nitric oxide that could be maintained for at least 24 h [81]. The system's release was sustained and tunable; therefore, it was tested against methicillin-resistant *Staphylococcus aureus* (MRSA) in 2009 by the optical density of bacterial cultures, histochemistry of cultured tissue, and histological processing of the skin of a mouse model [80]. In 2017 [54] and 2018 [73], the platform was tested again to treat reference strains of a dermatophytic fungus *Trichophyton rubrum*. These NO-releasing polymer matrices were assessed as a more effective alternative for current treatments, which face antifungal resistance, side effects, and insufficient penetration [76]. In all cases, the technology effectively reduced the microbial abscesses and proved promising for the intended therapeutic application. In 2018, the in vitro test consisted of the measurement of inhibition of 40 fungi compared to a standardized treatment. After 24 h of exposure, it resulted in 99.9% of fungal reduction [73].

Regarding the evaluation of the NO-releasing hydrogels, in the beginning, only the physicochemical properties and NO-release profile were studied, their bactericidal effect in vitro was not characterized yet. However, for the first time in 2004, GSNO-containing PEO hydrogels were tested on the skin of healthy human subjects and found to deliver nitric oxide sustainably [56]. Five years later, a successful antibacterial test against *P. aeruginosa* with a photoactive, NO-releasing platform was published by Halpenny et al. [58]. The system consisted of a PU hydrogel containing metal-NO complexes of Mn and Ru. This was also the first time that combined antimicrobial agents were tested. Additional to the NO-donor, two other active substances were included, i.e., hydrogen peroxide and methylene blue, a phenothiazinium dye with reported anti-MRSA activity. The light-controlled co-delivery proved more effective against the treated colonies than the separate effect of the agents. In the same year, Han et al. applied their NO-releasing hydrogel on cutaneous infections of a murine model to prove that the application of topical nitric oxide reduced the bacterial burden of MRSA [80].

In addition to testing the antibacterial efficacy, two publications in 2010 reported efficacy against other pathogens such as fungi and parasites. The first publication was the earlier explained probiotic chamber of Jones et al. [79], but in this system, it was tested for efficacy against bacteria (*E. coli*, *S. aureus*, *P. aeruginosa*, *A. baumannii*, and MRSA) and two pathogenic fungal strains (*T. mentagrophytes* and *T. rubrum*) [79]. In all cases, the eradication capacity was confirmed under sustained therapeutic levels of released nitric oxide. The second publication by López-Jaramillo et al. described a randomized clinical trial in which 178 patients suffering cutaneous leishmaniasis, a parasitic skin infection, were treated. The tested therapy consisted of acidified organic nitrite encapsulated during the electrospinning of thermoplastic PU hydrogels. Even though the results showed more than twice less effectiveness than the prevalent medication (meglumine antimoniate), the use of NO-donors remained an alternative for treatment given that adverse effects were greatly reduced [78]. At this point, the in vitro testing for antibacterial properties of NO-donors became a usual practice but this did not ensure the platforms' safety for their clinical use. For this purpose, in vivo experimentation and clinical trials were needed.

7.1.2. Natural Polymer Carriers

By 2011, the proof-of-concept of polymeric NO-donor systems was achieved, so efforts to improve the polymer carrier system were starting. Given the desire for suitable functional groups for medical applications and the need for toxicity reduction, natural polymers were introduced as NO-donor carriers. The manipulation and the mechanical versatility of synthetic polymers allow the production of materials that fit a particular biological application. However, they are not biodegradable, which means that the resulting byproducts may not be able to be discarded renally [83].

Therefore, natural polymers from hydroxyacids were implemented. They have molecular structures whose characteristics cannot be achieved in synthetic systems. The biodegradability, for instance, by enzymatic hydrolysis is an attractive property of natural polymers as toxic byproducts are avoided. These byproducts are already part of the human metabolic system. Natural carriers also have potential bioactivity, which could make them active contributors to the treatment [84]. The first natural-based polymer carrier for cutaneous NO-donor delivery consisted of chitosan and alginate combinations used to encapsulate GSNO in nanoparticles. They observed that NO• was released spontaneously in physiological conditions with an initial burst, which was tested for cytotoxicity against fibroblasts to ensure their safety for their application onto mammalian cells [59].

Two more publications using the same natural carriers were reported in 2013 and 2014. The first one consisted of a very similar nanoparticle system to the one discussed in the previous paragraph [59], with the difference that instead of direct GSNO encapsulation, glutathione, a GSNO precursor, was encapsulated in the chitosan/alginate mucoadhesive polymers where it was nitrosated to then release NO in a sustained manner. The delivery was carefully studied against a varying pH and temperature and the nanoparticles were tested with two cytotoxicity tests. The parameters were compared to the free release of GSNO. Its encapsulation in natural biodegradable, non-toxic nanoparticles resulted in a more stable release profile with complete removal of cytotoxicity [6]. The second publication consisted of the same natural polymers, where the encapsulation of a different nitrosothiol (S-nitroso-mercaptosuccinic acid) was tested against *S. aureus* and *E. coli* bacteria. The results showed a positive correlation between NO-delivery and its antibacterial effect [60].

Several years later, in 2020, three purely natural polymer-based platforms were reported. Two articles by Urzedo et al. reported the evaluation of alginate nanoparticles where the encapsulated NO-donor was S-nitroso-mercaptosuccinic acid. The antibacterial tests performed were against three bacterial strains—*E. coli*, *S. mutans*, and *S. aureus*. They compared the antibacterial effect of the pure nanoparticles to the co-release of biogenic-produced silver nanoparticles, which are antimicrobial as well. They reported that the biodegradable and biocompatible nanopolymers were highly bactericidal while non-toxic to mammalian cells. The co-delivery proved to have a synergistic effect by an increased antibacterial capacity [70,71].

Banerjee et al. tested a cellulose topical gel in a randomized clinical trial. The formulations contained different doses of the berdazimer sodium (NVN1000) NO-donor, a macromolecular polysiloxane polymer that stably stores NO• as a covalently bound N-diazeniumdiolate [85]. These formulations are commercially available under SB206 by Novan Inc. It was applied on 256 patients for the treatment of *M. contagiosum*, which is a viral infection [86]. The results revealed significant clearing of the lesions with minimal adverse effects after 12 weeks of daily application. The best performing dose was the highest (12% NVN1000) regarding the optimized balance between antiviral and side effects during the course of the study [72].

Natural polymers exhibit different chain sizes according to monomer ratios. This inherent variation is undesirable to obtain consistent materials. Another important shortcoming is compromised stability due to poor mechanical properties and high water solubility. Therefore, blending and crosslinking with other materials to enhance these properties has been researched [83]. Furthermore, although pure natural polymer delivery platforms have been often proven non-cytotoxic at high concentrations, the different combinations must also be carefully tested to ensure their safe application on biological systems. Since the introduction of natural polymers as nitric oxide carriers, they have been prepared by the ionic gelation method regardless of the presentation of the carrier [70,71].

7.1.3. Semisynthetic Polymer Carriers

By 2018, a new type of material was used as NO-donor carriers: semisynthetic polymers. They consist of the combination of natural polymers with synthetic polymers by blending, grafting, or crosslinking. The outcome is a material that combines the desirable mechanical and thermal strength of synthetic polymers, and they often are still biodegradable and as biocompatible as natural polymers. Thus, semisynthetic polymers can potentially overcome the limiting properties of purely synthetic and purely natural carriers [50]. Therefore, an increasing tendency in the last years to combine natural and synthetic polymers to optimize the particle properties is also observed in Table 2.

For instance, chitosan nanoparticles, containing S-nitrosothiol (RSNO) as NO-donor, incorporated into Pluronic F-127 hydrogels have been dermally tested in murine models. The incorporation of these systems has proven to profit the advantages of both polymers in terms of physicochemical and biological properties; hence, sustained NO• release for bactericidal concentrations while remaining harmless for mammalian cells [2,75].

The efficacy of additional semisynthetic polymer nanoparticles was reported in the same year by Stasko et al. The tested molecules made of hexylene glycol and cellulose, loaded with NONOate were applied to treat different cutaneous fungal infections, including *Trichophyton*, *Epidermophyton*, *Fusarium*, *Candida*, and *Malassezia* species. The gel formulation containing berdazimer sodium was characterized regarding NO• release, and it was reported to effectively clear the infected tissue from bacteria [74].

7.1.4. Further Enhancement of Polymer Carriers

In order to enhance the release profile and control of synthetic polymers, there has been growing interest in the direct functionalization of NO-donor into the polymer backbones. This has been achieved by highly controlled polymerization conditions and modern polymer chemistry technologies such as reversible addition–fragmentation chain transfer, a quasi-living radical polymerization. This approach is a very reliable one for tailor-made structures with reduced variability by decreasing the distribution range of molecular mass [83]. In 2016, NONOate-functionalized dendrimers were produced by the modification of poly(amidoamine) (PAMAM). PAMAM dendrimers are highly branched macromolecules synthetized into a particular structure. This is achieved by the addition of defined generations of methyl acrylate and ethylenediamine in an ethylenediamine core [87]. The utility of this system is the functionalization capacity of the superficial branches by the NONOate group. The dendrimers coupled with the NONOate were entrapped in electrospun PU fibers and they resulted effective in the inhibition of *P. aeruginosa*, *E. coli*, *S. aureus*, and MRSA, while not being cytotoxic in vitro [52].

In 2017, a second functionalized polymer carrier was published. Poly(oligoethylene) glycol methyl ether methacrylate (POEGMA) nanoparticles, produced by polymerization induced self-assembly, were functionalized with diazeniumdiolate to release NO• upon contact with water. The measured release was extended compared to the free NO-donors by limited access to water within the polymeric capsules [62]. In 2018, an oligoethylene glycol-based polymer was also functionalized with NONOate to release NO•gas; this was reported effective against *P. aeruginosa* biofilms. In this research, the polymer, an amphiphilic statistical ternary copolymer, was a carrier and also designed to be actively antimicrobial by cell membrane disruption [64]. Statistical copolymers are those in which the monomer distribution obeys established statistical laws. This can result from alternation, clustering, or no arrangement tendency [88]. One year later, Shen et al. reported polyethylene oxide micelles they functionalized with N-nitrosoamine to release NO• with a photoinduced control due to photon-cleaved N–NO bonds, and coumarin chromophore as a visual reporter for monitoring. These amphiphilic structures were loaded with the antibiotic ciprofloxacin in the core for more efficient bacterial eradication through synergistic co-delivery [63].

As mentioned earlier, co-delivery is another approach to enhance the therapeutic capacity of NO-releasing platforms. In the course of the last five years, developments on nanopolymeric NO-carrier synergistic systems have been made. An early publication of a combination of hydrogen peroxide and methylene blue into the NO-donor platform was reported in 2009 and detailed in Section 7.1.1. [58]. Another co-delivery was shown by

Nguyen et al. in 2016 and proven to be more effective than the separated treatments. Their POEGMA nanoparticles carrying acetonitrile NONOate and gentamycin antibiotic were tested against *P. aeruginosa* infection and a decrease of more than 90% of the viability was measured [61].

NO-donors have also been compared to available therapies separately. For instance, Sulemankhil et al. assessed the antimicrobial activity of NO• gas, released by enzymatic reaction of at a sustained rate during 18 h. The technology was proven to be more effective than other antimicrobial dressings containing iodine, silver, and antibiotics (chlorhexidine, gentamicin, and vancomycin) against three clinically relevant antibiotic-resistant microorganisms—*A. baumannii*, methicillin-resistant *S. aureus*, and *P. aeruginosa* [77].

Even if these composite and highly technological platforms hold a great potential application as NO-delivery platforms, exhaustive experimentation and clinical trials are required to ensure the safety and efficiency of the systems to eventually translate them to the medical field.

7.2. Wound Healing Applications

The role of NO• in wound healing includes its antimicrobial capacity, but it also has an effect on the stimulation of cytokine production, on the regulation of growth factor expression, and on the deposition of the extracellular matrix. Catalytic nitric oxide production is critical for wound healing, as the fluid in the lesion induces NOS in different cell types.

Wounds that fail to heal in form and type are called chronic. Chronic wounds are an ailment in multiple conditions, which could be related to decreased nitric oxide levels. In diabetes, for instance, insufficient production has been reported, related to a reduced inflammatory response. Malnutrition and radiotherapy-induced injury are additional conditions associated with inadequate healing. Finally, steroidal drugs, to treat conditions such as asthma and arthritis, strongly damage the healing process by altering arginine's metabolism. Both L-arginine supplementation and NO-donors can potentially restore such impairments [8,22,89].

On the contrary, in some conditions where the skin lesions are hypersensitive, such as in the case of psoriatic patients, the reduction of NO• production may be needed. One way to achieve this is by the suppression of enzymatic pathways to control the lesions' excessive angiogenesis [8,65].

The wound healing application of NO-donors has been increasingly tested once the research on the antimicrobial potential of similar systems had begun. This is probably explained by the fact that for wound healing applications, protection against microbes is only one of the functions of nitric oxide to be tested. Therefore, the platforms promoting wound healing correspond to more recent investigations, which are compiled chronologically in Table 3. In general, the target condition to be treated with the NO-donor was the improvement of the wound healing process in terms of time for regeneration and quality of the new tissue. However, additional complications have also been tested, including infection, diabetes, ischemia, and acne.

Table 3. Compilation of publications showing the evolution of NO-donor polymeric carriers intended to treat cutaneous wounds.

Year	Origin	Carrier	Matrix	NO-Donor	Coupled Donor	Synergistic Additive	Trigger for Release	Antibacterial Test	Trial	Extra Target Condition	Ref.
2020	S	PLGA	np	Nitrosothiol	Yes	-	-	Yes	in vivo	Infection	[90]
2020	N	Cellulose	Hydrogel	NO• gas	-	Plant extract	-	Yes	in vivo	Diabetic skin ulcer	[91]
2020	S	PEG	Hydrogel	Nitrosothiol	Yes	-	-	Yes	in vivo, in vitro	Infection	[92]
2020	SS	PU, gelatin, keratin	Hydrogel	Nitrosothiol	Yes	-	-	Yes	in vivo, in vitro	-	[93]
2020	N	GelMA	Hydrogel	N-nitrosamine	Yes	Hyaluronic acid	Photons	Yes	in vivo, in vitro	Infection	[94]
2020	N	Collagen	Sponge	Nitrosothiol	-	-	-	-	in vivo	-	[95]
2019	S	PEI	Hydrogel	NONOate	Yes	-	-	-	in vivo	-	[96]
2019	S	PAA	Hydrogel	Sildenafil citrate	-	-	-	-	in vivo, in vitro	-	[97]
2019	S	PEI, PLGA	np	NONOate	Yes	-	-	Yes	in vivo, ex vivo, in vitro	Diabetic infection	[98]
2018	S	PLGA	Emulsion	Nitrosothiol	-	-	-	Yes	in vivo, in vitro	Infection	[99]
2017	SS	PCL/chitosan	Hydrogel	NONOate	Yes	-	Enzyme	-	in vivo	Ischemia	[35]
2016	S	Poly-siloxane	Hydrogel	NONOate NONOate	Yes	-	-	Yes	Clinical	Acne lesion	[100]
2015	S	PVA/PL	Hydrogel	Nitrosothiol	Yes	-	-	-	in vivo	-	[101]
2015	S	PAN	Hydrogel	Acrylonitrile	Yes	-	-	-	in vivo	-	[102]
2015	N	Chitosan	Hydrogel	Nitrosothiol	-	-	-	Yes	in vivo, in vitro	-	[103]
2013	N	Nap-FFGGG	Hydrogel	NONOate	Yes	-	Enzyme	-	in vivo	-	[104]
2012	SS	Alginate/Acrylic polymer	Chamber	Inorganic nitrite	-	Probiotics	-	Yes	in vivo, in vitro	Ischemic, infection	[105]
2010	S	PL	Hydrogel	Nitrosothiol	-	-	-	-	in vivo	Ischemia	[106]
2009	S	PEG/Disaccharide	Hydrogel	Inorganic nitrite	-	-	-	Yes	in vivo, in vitro	Infection	[107]
2008	S	PL	Hydrogel	Nitrosothiol	-	-	-	-	in vivo	-	[108]
2007	S	PL	Hydrogel	Nitrosothiol	-	-	-	-	in vivo	-	[109]
2002	S	PVA	Hydrogel	NO• gas	Yes	-	-	-	in vivo	Diabetes	[110]

Abbreviations. Polymer origin: natural (N), semi synthetic (SS), synthetic (S). Polymer carriers: poly(lactic-co-glycolic acid) (PLGA), poly(ethylene glycol) (PEG), polyurethane (PU), gelatin methacrylamide (gelMA), poly(ethylene imine) (PEI), poly(acrylic acid) (PAA), poly(ε-caprolactone) (PCL), poly(vinyl alcohol) (PVA), Pluronic F-127 (PL), poly(acrylonitrile) (PAN), naphthalene-capped amino acids (phenylalanine/glycine 2:3) (Nap-FFGGG). Matrix: nanoparticles (np). NO-donor: diazeniumdiolate (NONOate).

Despite the encouraging results of the discussed platforms, the treatment of chronic wounds remains a challenge. For this reason, research has focused on improving the polymer carrier technologies in terms of stability, biodegradability, and controlled release to make them more suitable for skin repair. Additionally, NO-releasing wound dressings have been tested in combination with additional compounds such as anti-inflammatory agents to enhance the healing capacity.

7.2.1. Synthetic Polymer Carriers

In comparison to the evolution of platforms intended for microbial applications, the same trend with a delay can be seen in wound healing applications. For instance, initially, several purely synthetic polymer carriers were evaluated. Initially, Pluronic F-127 was a frequent polymer used to contain GSNO nitrosothiols in hydrogel matrices. Specifically, three publications based on the same platform were published in 2007 [109], 2008 [108], and 2010 [106]. The latter study focused particularly on impaired wound healing of patients with an ischemic condition. All three articles confirmed amelioration in the healing process by increased collagen deposition and tissue granulation, which resulted in earlier closure and re-epithelialization of the wound [106]. It was highlighted that the hydrogel application worked best when applied in the initial phases of re-epithelialization, i.e., proliferation and inflammation [108,109].

In 2009, the technology that was developed by Friedman et al. and published in 2008 (previously described in Section 7.1.1.), was used in a nanoparticle matrix to treat MRSA-infected wounds. Acceleration of the closure of infected wounds and antimicrobial properties confirmed that the system was effective in the dermal treatment of infected wounds [107]. Another unmodified synthetic polymer hydrogel matrix was published in 2018 by Hlaing et al. They tested for improvement of healing in wounds infected with MRSA through a PLGA emulsion with entrapped GSNO. They confirmed improvement in wound healing in vivo and confirmed the antibacterial capacity both in vitro and in vivo [99]. Finally, in 2020, a GSNO-incorporated PEG-gel was tested in vitro and in vivo on MRSA-infected wounds by Lee et al. [92].

Synthetic polymer carriers have performed outstandingly as carriers of NO-donors for wound healing applications. However, hydrogels made of commonly used non-hydrolyzable synthetic polymers, such as PVA and Pluronic-F127, demonstrate insufficient loading capacity and do not stimulate bioabsorption of the carried drug. This is undesirable for NO-delivery. Another shortcoming is the risk of harmful degradation products. To reduce these limitations, different natural polymers and polymer blends have been developed [41].

7.2.2. Natural Polymer Carriers

In the case of wound healing applications, the first natural-based carrier was published in 2013, while this was already reported by 2011 in antimicrobial applications. Polymer carriers of purely natural origin have been tested for wound healing since then as biocompatible and biodegradable alternatives. A stronger tendency to use polymers of proteic nature (i.e., gelatin and collagen) can be identified, in contrast to those intended to treat infections that are mostly based on polysaccharides. This can be explained by the fact that those natural polymers need to be actively synthetized during re-epithelialization (i.e., collagen deposition); therefore, they offer an additional benefit by actively helping to accelerate the healing process.

In 2013, an enzyme-controlled platform was developed by Gao et al. They prepared a naphthalene-capped short polymer gel composed of phenylalanine and glycine amino acids in a 2:3 ratio. β-galactoside-caged NO• was incorporated in the gel to regulate the release by the cleavage activity of β-galactosidase. Testing in vivo proved that controlled release of nitric oxide-induced angiogenesis and resulted in faster wound healing. Another natural-polymer platform was published in 2015, it consisted of a chitosan film with GSNO to treat infected wounds with two bacterial strains, *P. aeruginosa* and *S. aureus*, which were particularly selected for having different cell membrane compositions. Interestingly, the films showed an effective reduction of both bacterial strains, as opposed to the application of pure chitosan films, where a decrease in viability was measured due to the polymer's antibacterial properties. The results also proved an enhancement of healing by accelerated inflammation and tissue granulation [103].

Five years later, in 2020, three more platforms were published with a natural-based wound healing hydrogel as a carrier, namely by Nie et al. [91], Póvoa et al. [95], and Huang et al. [94]. The first system was designed to treat diabetic wounds based on a cellulose gel containing nitric oxide and asiaticoside, an anti-inflammatory plant extract. The results proved that the platform adequately adhered to the skin and promoted healing of diabetic skin ulcers by up-regulation of Wnt/β-Catenin signaling pathway, which activates iNOS [91]. The second matrix was a "self-expandable" sponge produced by lyophilization of collagen with GSNO. They reported accelerated healing by an effectively bioabsorbable platform [95].

The last system was developed to treat infected wounds with gelatin methacrylamide hydrogels. Hyaluronic acid was added due to its cell repulsive properties to avoid excessive adhesion. The NO-donor consisted of bis-N-nitroso compound 6 (BNN6, Bis(2-methoxyethyl)nitrosoamine). These hydrogels were light-responsive due to the presence of graphene oxide nanosheets functionalized with β-cyclodextrin and they were tested in two bacterial strains—*E. coli* and *S. aureus*—by in vitro and in vivo experiments [94].

The results demonstrated enhanced healing of wounds infected with bacteria [94]. However, lack of control and premature release of NO• has been measured due to instability of the NO-donors in heterogeneous, insufficiently strong natural carriers. Furthermore, certain natural polymers such as proteins promote cell adhesion, which in excess may be counterproductive as a support matrix during the treatment at the clinical level [83].

7.2.3. Semisynthetic Polymer Carriers

The combination of synthetic and natural polymers to enhance the properties of the separate carriers was reported earlier in wound healing applications (2012) than in antimicrobial applications (2018). This can be explained by the fact that in wound healing, the predominating matrices are biocompatible natural hydrogels, for which the enhancement of mechanical properties may be of higher priority than for the anti-microbial applications (where gel or spray administration alternatives are sufficient). In 2012, Jones et al., who also published the probiotic chamber in 2010 (detailed in Section 7.1.1. [79]), developed another chamber containing *L. fermentum* for the in-situ production of nitric oxide. In this study, the patch was made of a combination of alginate and a synthetic acrylic polymer and they proved that the healing of ischemic and infected dermal wounds in a rabbit model was accelerated [105].

Five years later, another semisynthetic platform was published by Zhou et al. [35]. They developed a poly(ε-caprolactone) (PCL)/chitosan electrospun hydrogel, where glycosylated diazeniumdiolates were grafted onto the side chain of chitosan. Given the formation of a CS–NO bond, the release of nitric oxide was enzymatically controlled by glycosidase. This technology was tested on ischemic chronic wounds and demonstrated enhanced healing by pro-angiogenesis, immunomodulation, and enhanced collagen synthesis. Finally, PU/gelatin/keratin electrospun hydrogels were produced in 2020 by Wan et al. [93]. In this development, further enhancement was achieved by N-nitrosation of keratin to form a coupled nitrosothiol NO-donor. Their composite system released nitric oxide with no cytotoxic effect to mammalian cells and it promoted skin repair by cell proliferation and a decreased inflammatory response. It also demonstrated an antibacterial effect against *E. coli*. These platforms feature functionalization, which is a strategy to improve polymer carriers, addressed in detail in the next section.

7.2.4. Further Enhancement of Polymer Carriers

In the attempt to improve the release properties of polymer carriers for the treatment of wounds by NO-donors, the active molecules have been functionally coupled into polymer chains in around 50% of the wound healing applications (Table 3). Additionally, synergistic additives have also been applied in order to enhance the tissue regeneration capacity of the treatment. In 2002 already, an exceptionally elaborate application was published by Bohl et al. They tested an aminated-PVA hydrogel functionalized by covalent attachment of NO• to the polymer backbone. The design was meant to treat chronic wounds of diabetic patients. Thus, it was tested in vivo in mice with induced diabetes. They clearly observed that the nitric oxide provided actively participated in the wound healing process by enhancement of the tissue quality, although the exact mechanisms of action were not yet known. They hypothesized it was related to the genetic regulation of growth factors and collagen synthesis [110].

NO-functionalized polymer hydrogels were reported in 2015 [101,102], 2016 [100], 2019 [96,98], and 2020 [90]. An acrylonitrile-functionalized vinyl terpolymer, with enhanced mechanical properties, was electrospun with a defined fiber-morphology, to control NO-release. The electrospun mats were tested in vivo [102]. Another system consisted of Pluronic F-127/PVA hydrogels, where PVA was chemically functionalized by esterification with mercaptosuccinic acid of SNO groups, with a subsequent S-nitrosation to release nitric oxide. The polymer combination was biologically tested in vivo only for the acceleration of wound healing [101]. Wound closure was also promoted in vivo by diazeniumdiolate-coupled propylene oxide mixed with poly(ethylene imine) (PEI) in a PEG-gel [96].

Particular wound conditions have also been targeted by research. For instance, NVN1000 covalent-functionalized polysiloxane was tested by Baldwin et al. in a clinical trial in 153 subjects as a therapy for acne vulgaris (lesions infected by *P. acnes*) [100]. It was found that the daily application successfully decreased the amount and size of wounds and it was well tolerated. Another tested condition was MRSA-infection of diabetic wounds; the system consisted of PEI/PLGA nanoparticles where PLGA was functionalized with NONOate. The results of in vivo, ex vivo, and in vitro testing were positively correlated with antibacterial and wound healing capacities [98]. PLGA nanoparticles were also conjugated to treat MRSA-infected wounds in 2020; however, the NO-donor, in this case, was GSNO [90].

In most research regarding wound healing, the experimental testing follows a general structure. The NO-delivering therapy is produced and tested for NO• release. In the case of a novel carrier, its physicochemical characterization is performed first. For instance, electrospun functionalized dressings have been characterized by techniques including attenuated total reflection-Fourier transform infrared spectroscopy for the verification of functional coupling of the NO-donor to the polymer carrier; this is verified with scanning electron microscopy by visualizing the morphology of the fibers [35,93]. In the case of hydrogel matrices, further physicochemical characterization has been often carried-out to obtain parameters that describe the swelling capacity, degradation rate, rheological properties, and compression strength [94]. When nanoparticles are incorporated into these platforms, they are carefully characterized separately. Relevant parameters for this form of carriers include size, polydispersity, sphericity, and electric potential, which are measured by dynamic light scattering [98,99].

In Table 3, it can be observed that almost all the wound healing research so far has been pre-clinical, i.e., in vivo testing in murine or rat models [101]. The wound assessment is quantified in terms of inflammation, wound discharge, granulation tissue, induration, skin maceration, color, etc. [102]. Additional complications of wound healing because of the treatment also require further testing. For instance, those studies focusing on infected wounds also performed in vitro tests to confirm the antibacterial capacity.

Finally, cytotoxicity for mammalian cells is also often tested in vitro for an initial safety test of the presented systems. Despite in vitro and in vivo testing, in addition to some clinical trials that have also been successfully conducted, it is not enough to achieve the introduction of these platforms to the clinic. Approval of molecular entities and platforms by the FDA is particularly strict; therefore, the extrapolation of in vitro and in vivo experimentation requires substantial funding and collaborations between the clinical and pharmacological research [111]. This translation takes time and NO-donor therapeutics are still a relatively new field.

7.3. Circulatory Applications

Given NO-donors' ability to activate hemodynamics by increased vasodilation and blood flow while reducing the aggregation risk, they stimulate research in therapeutic targets to treat related chronic and debilitating conditions [4]. This regulatory function of nitric oxide is less applicable at a cutaneous level (excluding its need during skin regeneration addressed in Section 7.2), but it is interesting for systemic delivery such as in the case of biomedical implants. The reason for this is that NO-donors for cutaneous applications (such as creams and gels) have low water solubility, while they are highly soluble in hydrophobic matrices. Accordingly, the released molecules can reach the smallest blood vessels and potentially lead to uncontrolled systemic vasodilation with undesirable side effects [11]. Because of this risk, safety is to be ensured in dermal NO-donors in circulatory applications by ensuring a localized, controlled release. In some studies, it has been assessed how the dermal release of NO• regulates circulation and these studies are addressed in the present section.

An early topical vasodilatory application was reported by Souto et al. in 2010. The unconventional application of a nitric oxide-releasing gel consisted in therapy for women with sexual disfunction by the enhancement of clitoral blood flow. The gel was made of Pluronic F-127 (PL) and it carried GSNO nitrosothiol as NO-donor. The measurement of clitoral arteries by Doppler ultrasound demonstrated that desired local vasodilation without systemic effects was achieved within 15 min of its application in 20 healthy, sexually active women [112]. Hence, this formulation can be of use to treat women's sexual disfunction if proven in a larger sample size, where subjects with the ongoing sexual disorder also participate.

Another topical NO-release platform for skin vasodilation was published by Vercelino et al. in 2013, as an alternative treatment of painful inflammation medically known as nociception. For the production of hydrogels, they made use of PL to carry GSNO, the same components as Souto et al. (see the previous paragraph). The hydrogels were tested in vivo in rats with induced hypernociception and healthy human volunteers. The platform resulted in a local increase of dermal blood flow as a potential analgesic (pain-relieving) dermal therapy of inflammation. Again, the limitation in this study was the sample size (n = 5) [11].

Circulatory applications at a cutaneous level were no longer reported for the following seven years, until 2020, when Soares et al. tested a commercial nitroglycerin patch for the treatment of microvascular dysfunction of ischemic patients [113]. Nitroglycerin patches date from 1985, and they are commercially available to treat chest pain, medically called angina pectoris [114]. Soares et al. applied Mylan-Nitro patches by Mylan Pharmaceuticals, consisting of silicone-coated polyester films in form of a pouch where nitroglycerin is contained [115]. The study was carried out to investigate if the placement of a NO-donor locally could serve as microvascular protector by dilation of brachial arteries. This was confirmed in 10 healthy men [113]. The same year, a similar application was recommended in a review published by Appleton et al., where they examined different trials on the potential capacity of stroke prevention by nitroglycerin NO-donor patches administered dermally. Their conclusion was to recommend the treatment given its safety proven by a meta-analysis. However, this was only ensured as a preventive mechanism. Thus, it is not yet safe to use in ultra-acute strokes [116].

A novel NO-donor hydrogel has been reported in 2020 as a microvascular therapeutic. The technology consists of a nitrite-functionalized PVA patch produced by reversible addition–fragmentation chain-transfer polymerization of nitrosothiols. Giglio et al. tested the platform in vivo in healthy volunteers as a proof-of-concept and determined a significant increase in blood circulation with potential application in microvascular diseases [117].

In most of the circulatory applications of NO-donor polymeric carrier systems healthy volunteers were invited for an initial stage, but in vivo tests still need to be performed on people suffering from each one of these diseases.

8. Future Perspectives

Plentiful polymer-based platforms have been designed, characterized, and tested as carriers of NO-donors for therapeutic applications. Nitric oxide is a radical gas species that participates in the regulation of several cutaneous processes. In topical applications, its half-life is brief. For an efficient and functional NO• delivery in biomedical applications, control of the release is needed. A recurrent problem of the available technologies is the unrestrained scattering of nitric oxide or leaching of byproducts produced during the release. Some platforms have aimed to solve these issues by the use of semisynthetic polymers, whose byproducts are often safe, or by the control of release through functionalization or caging the NO-donor. However, even if a more controlled release has been achieved, the developed systems tend to be complicated or not safe yet for clinical use, given a risk of interactions, cell toxicity, or other unknown side effects.

Novel NO-delivery carriers for topical therapies are still to be developed for targeted release. Some recent materials have been proposed as promising alternatives. For instance, PCL is a polymer that has been blended more recently for cutaneous application of nitric oxide as it is ductile and bioabsorbable. It has been reported to provide desirable mechanical properties and degradation kinetics [41]. Finally, a novel Laponite nanoclay has been proposed as a NO-donor carrier because it can potentially prevent leaching and release controlled amounts of NO• [118]. The development and implementation of these carriers could eventually offset the current challenges of NO-releasing platforms. This could stimulate their clinical testing for approval for administration to patients.

Since 1867, nitric oxide therapy has been the subject of pharmacological approaches because it was identified as a powerful cutaneous regulator of vasodilation, immune response, and skin regeneration. However, it is not absent of side effects such as generating tolerance or undesired oxidative stress [119]. Numerous conditions are related to the reduced bioavailability of NO•, but even if the administration of NO-donors has shown positive evidence from animal studies, more translational studies are required for their implementation in the clinical field.

Author Contributions: All authors contributed to this study. All authors have read and agreed to the published version of the manuscript.

Funding: This research received no external funding.

Institutional Review Board Statement: Not applicable.

Informed Consent Statement: Not applicable.

Data Availability Statement: Not applicable.

Conflicts of Interest: The authors declare no conflict of interest.

References

1. Li, B.; Ming, Y.; Liu, Y.; Xing, H.; Fu, R.; Li, Z.; Ni, R.; Li, L.; Duan, D.; Xu, J.; et al. Recent Developments in Pharmacological Effect, Mechanism and Application Prospect of Diazeniumdiolates. *Front. Pharmacol.* **2020**, *11*, 923. Available online: https://www.frontiersin.org/article/10.3389/fphar.2020.00923/full (accessed on 27 February 2021). [CrossRef]
2. Pelegrino, M.T.; de Araújo, D.R.; Seabra, A.B. S-nitrosoglutathione-containing chitosan nanoparticles dispersed in Pluronic F-127 hydrogel: Potential uses in topical applications. *J. Drug Deliv. Sci. Technol.* **2018**, *43*, 211–220. [CrossRef]
3. Seabra, A.B.; Justo, G.Z.; Haddad, P.S. State of the art, challenges and perspectives in the design of nitric oxide-releasing polymeric nanomaterials for biomedical applications. *Biotechnol. Adv.* **2015**, *33*, 1370–1379. [CrossRef] [PubMed]

4. Opländer, C.; Römer, A.; Paunel-Görgülü, A.; Fritsch, T.; van Faassen, E.E.; Mürtz, M.; Bozkurt, A.; Grieb, G.; Fuchs, P.; Pallua, N.; et al. Dermal Application of Nitric Oxide In Vivo: Kinetics, Biological Responses, and Therapeutic Potential in Humans. *Clin. Pharmacol. Ther.* **2012**, *91*, 1074–1082. Available online: http://doi.wiley.com/10.1038/clpt.2011.366 (accessed on 27 February 2021). [CrossRef]
5. Liechty, W.B.; Kryscio, D.R.; Slaughter, B.V.; Peppas, N.A. Polymers for Drug Delivery Systems. *Annu. Rev. Chem. Biomol. Eng.* **2010**, *1*, 149–173. Available online: https://pmc/articles/PMC3438887/?report=abstract (accessed on 27 February 2021). [CrossRef]
6. Marcato, P.D.; Adami, L.F.; de Melo, B.R.; Melo, P.S.; Ferreira, I.R.; de Paula, L.; Nelson, D.; Seabra, A.B. Development of a Sustained-release System for Nitric Oxide Delivery using Alginate/Chitosan Nanoparticles. *Curr. Nanosci.* **2013**, *9*, 1–7.
7. Kesharwani, P.; Jain, K.; Jain, N.K. Dendrimer as nanocarrier for drug delivery. *Prog. Polym. Sci.* **2014**, *39*, 268–307. [CrossRef]
8. Adler, B.L.; Friedman, A.J. Nitric oxide therapy for dermatologic disease. *Future Sci. OA* **2015**, *1*. Available online: https://pmc/articles/PMC5137922/?report=abstract (accessed on 27 February 2021). [CrossRef] [PubMed]
9. Franke, A.; Oszajca, M.; Brindell, M.; Stochel, G.; van Eldik, R. Metal-assisted activation of nitric oxide-mechanistic aspects of complex nitrosylation processes. In *Advances in Inorganic Chemistry*; van Eldik, R., Ford, P., Eds.; Elsevier Academic Press: San Diego, CA, USA, 2015; pp. 171–241.
10. Toledo, J.C.; Augusto, O. Connecting the Chemical and Biological Properties of Nitric Oxide. *Chem. Res. Toxicol.* **2012**, *25*, 975–989. Available online: https://pubs.acs.org/doi/full/10.1021/tx300042g (accessed on 27 February 2021). [CrossRef]
11. Vercelino, R.; Cunha, T.M.; Ferreira, E.S.; Cunha, F.Q.; Ferreira, S.H.; de Oliveira, M.G. Skin vasodilation and analgesic effect of a topical nitric oxide-releasing hydrogel. *J. Mater. Sci. Mater. Med.* **2013**, *24*, 2157–2169. Available online: https://link.springer.com/article/10.1007/s10856-013-4973-7 (accessed on 27 February 2021). [CrossRef] [PubMed]
12. Pub-Chem. Center for Biotechnology Information. Compound Summary for CID 145068, Nitric Oxide. Available online: https://pubchem.ncbi.nlm.nih.gov/compound/Nitric-oxide (accessed on 26 November 2020).
13. Lancaster, J.R. Nitric oxide: A brief overview of chemical and physical properties relevant to therapeutic applications. *Futur. Sci. OA* **2015**, *1*, FSO59. Available online: http://www.future-science.com/doi/10.4155/fso.15.59 (accessed on 27 February 2021). [CrossRef]
14. Hughes, M.N. Chemistry of Nitric Oxide and Related Species. In *Cellulases*; Elsevier BV: Amsterdam, The Netherlands, 2008; Volume 436, pp. 3–19.
15. Coleman, J.W. Nitric oxide in immunity and inflammation. *Int. Immunopharmacol.* **2001**, *1*, 1397–1406. [CrossRef]
16. Snyder, S.H.; Bredt, D.S. Biological Roles of Nitric Oxide. *Sci. Am.* **1992**, *266*, 68–77. Available online: https://www.jstor.org/stable/24939060 (accessed on 27 February 2021). [CrossRef] [PubMed]
17. Ignarro, L.J. Nitric oxide is not just blowing in the wind. *Br. J. Pharmacol.* **2019**, *176*, 131–134. Available online: http://onlinelibrary.wiley.com/doi/10.1111/bph.v176.2/issuetoc (accessed on 27 February 2021). [CrossRef]
18. PubChem. National Center for Biotechnology Information. PubChem Compound Summary for CID 6322, Arginine. Available online: https://pubchem.ncbi.nlm.nih.gov/compound/Arginine (accessed on 18 December 2020).
19. Ghavari, A.; Miller, C.C.; Mcmullin, B.; Ghahary, A. Potential application of gaseous nitric oxide as a topical antimicrobial agent. *Nitric Oxide* **2006**, *14*, 21–29. Available online: www.elsevier.com/locate/yniox (accessed on 27 February 2021). [CrossRef]
20. Ma, L.; Hu, L.; Feng, X.; Wang, S. Nitrate and nitrite in health and disease. *Aging Dis.* **2018**, *9*, 938–945. Available online: https://pmc/articles/PMC6147587/?report=abstract (accessed on 27 February 2021). [CrossRef] [PubMed]
21. Cracowski, J.; Roustit, M. Human Skin Microcirculation. In *Comprehensive Physiology*; Prakash, Y.S., Ed.; Wiley: Hoboken, NJ, USA, 2020; Volume 10, pp. 1105–1154. Available online: https://onlinelibrary.wiley.com/doi/10.1002/cphy.c190008 (accessed on 27 February 2021).
22. Bruch-Gerharz, D.; Ruzicka, T.; Kolb-Bachofen, V. Nitric oxide and its implications in skin homeostasis and disease—A review. *Arch. Dermatol. Res.* **1998**, *290*, 643–651. Available online: https://link.springer.com/article/10.1007/s004030050367 (accessed on 27 February 2021). [CrossRef] [PubMed]
23. Pieretti, J.C.; Seabra, A.B. Nitric Oxide-Releasing Nanomaterials and Skin Infections. In *Nanotechnology in Skin, Soft Tissue, and Bone Infections*; Mahendra, R., Ed.; Springer International Publishing: Cham, Switzerland, 2020; pp. 3–23. Available online: https://link.springer.com/chapter/10.1007/978-3-030-35147-2_1 (accessed on 27 February 2021).
24. Cobbold, C. The role of nitric oxide in the formation of keloid and hypertrophic lesions. *Med. Hypotheses* **2001**, *57*, 497–502. [CrossRef]
25. Limandjaja, G.C.; Niessen, F.B.; Scheper, R.J.; Gibbs, S. The Keloid Disorder: Heterogeneity, Histopathology, Mechanisms and Models. *Front. Cell Dev. Biol.* **2020**, *8*, 360. Available online: www.frontiersin.org (accessed on 27 February 2021). [CrossRef] [PubMed]
26. Luo, J.D.; Chen, A.F. Nitric oxide: A newly discovered function on wound healing. *Acta Pharmacol. Sin.* **2005**, *26*, 259–264. Available online: https://www.nature.com/articles/aps200541 (accessed on 27 February 2021). [CrossRef] [PubMed]
27. Holliman, G.; Lowe, D.; Cohen, H.; Felton, S.; Raj, K. Ultraviolet Radiation-Induced Production of Nitric Oxide: A multi-cell and multi-donor analysis. *Sci. Rep.* **2017**, *7*, 1–11. [CrossRef] [PubMed]
28. Ross, R.; Reske-Kunz, A.B. The role of NO in contact hypersensitivity. *Int. Immunopharmacol.* **2001**, *1*, 1469–1478. [CrossRef]

29. Furchgott, R.F.; Zawadzki, J.V. The obligatory role of endothelial cells in the relaxation of arterial smooth muscle by acetylcholine. *Nature* **1980**, *288*, 373–376. Available online: https://www.nature.com/articles/288373a0 (accessed on 27 February 2021). [CrossRef]
30. Luiking, Y.C.; Engelen, M.P.K.J.; Deutz, N.E.P. Regulation of nitric oxide production in health and disease. *Curr. Opin. Clin. Nutr. Metab. Care.* **2010**, *13*, 97–104. Available online: https://pmc/articles/PMC2953417/?report=abstract (accessed on 27 February 2021). [CrossRef] [PubMed]
31. Hadi, H.A.R.; Carr, C.S.; al Suwaidi, J. Endothelial dysfunction: Cardiovascular risk factors, therapy, and outcome. *Vasc. Health Risk Manag.* **2005**, *1*, 183–198. Available online: https://pmc/articles/PMC1993955/?report=abstract (accessed on 27 February 2021).
32. Hsu, Y.-C.; Hsiao, M.; Wang, L.-F.; Chien, Y.W.; Lee, W.-R. Nitric oxide produced by iNOS is associated with collagen synthesis in keloid scar formation. *Nitric Oxide* **2006**, *14*, 327–334. [CrossRef]
33. Zhang, Z.; Luo, X.; Yang, Y. From Spontaneous to Photo-triggered and Photo-calibrated Nitric Oxide Donors. *Isr. J. Chem.* **2020**, 202000084. Available online: https://onlinelibrary.wiley.com/doi/10.1002/ijch.202000084 (accessed on 27 February 2021). [CrossRef]
34. Yang, T.; Zelikin, A.N.; Chandrawati, R. Progress and Promise of Nitric Oxide-Releasing Platforms. *Adv. Sci.* **2018**, *5*, 1701043. Available online: http://doi.wiley.com/10.1002/advs.201701043 (accessed on 27 February 2021). [CrossRef]
35. Zhou, X.; Wang, H.; Zhang, J.; Li, X.; Wu, Y.; Wei, Y.; Ji, S.; Kong, D.; Zhao, Q. Functional poly(ε-caprolactone)/chitosan dressings with nitric oxide-releasing property improve wound healing. *Acta Biomater.* **2017**, *54*, 128–137. [CrossRef]
36. Kutner, A.; Friedman, A. Nitric oxide nanoparticles for wound healing: Future directions to overcome challenges. *Expert Rev. Dermatol.* **2013**, *8*, 451–461. Available online: https://www.tandfonline.com/doi/abs/10.1586/17469872.2013.837670 (accessed on 27 February 2021). [CrossRef]
37. Wang, P.G.; Xian, M.; Tang, X.; Wu, X.; Wen, Z.; Cai, T.; Janczuk, A.J. Nitric Oxide Donors: Chemical Activities and Biological Applications. *Chem. Rev.* **2002**, *102*, 1091–1134. Available online: https://pubs.acs.org/doi/full/10.1021/cr0000401 (accessed on 27 February 2021). [CrossRef]
38. Hayton, T.W.; Legzdins, A.P.; Sharp, W.B. Coordination and Organometallic Chemistry of Metal−NO Complexes. *Chem. Rev.* **2002**, *102*, 935–992. Available online: https://pubs.acs.org/doi/pdf/10.1021/cr000074t (accessed on 27 February 2021). [CrossRef] [PubMed]
39. Simplicio, F.I.; de Oliveira, M.G.; de Souza, G.F.P. In vitro inhibition of linoleic acid peroxidation by primary S-nitrosothiols. *J. Braz. Chem. Soc.* **2010**, *21*, 1885–1895. Available online: http://www.scielo.br/scielo.php?script=sci_arttext&pid=S0103-50532010001000013&lng=en&nrm=iso&tlng=en (accessed on 27 February 2021). [CrossRef]
40. Liang, H.; Nacharaju, P.; Friedman, A.; Friedman, J.M. Nitric oxide generating/releasing materials. *Future Sci. OA* **2015**, *1*. Available online: https://pmc/articles/PMC4739797/?report=abstract (accessed on 27 February 2021). [CrossRef] [PubMed]
41. Baldim, V.; de Oliveira, M.G. Poly-ε-caprolactone/polysulfhydrylated polyester blend: A platform for topical and degradable nitric oxide-releasing materials. *Eur. Polym. J.* **2018**, *109*, 143–152. [CrossRef]
42. De Oliveira, M.G. S-Nitrosothiols as Platforms for Topical Nitric Oxide Delivery. *Basic Clin. Pharmacol. Toxicol.* **2016**, *119*, 49–56. Available online: http://doi.wiley.com/10.1111/bcpt.12588 (accessed on 27 February 2021). [CrossRef]
43. Miller, M.R.; Megson, I.L. Recent developments in nitric oxide donor drugs. *Br. J. Pharmacol.* **2007**, *151*, 305–321. Available online: https://pmc/articles/PMC2013979/?report=abstract (accessed on 27 February 2021). [CrossRef]
44. Nakanishi, K.; Koshiyama, T.; Iba, S.; Ohba, M. Lipophilic ruthenium salen complexes: Incorporation into lipo-some bilayers and photoinduced release of nitric oxide. *Dalt. Trans.* **2015**, *44*, 14200–14203. Available online: https://pubmed.ncbi.nlm.nih.gov/26200295/ (accessed on 27 February 2021). [CrossRef]
45. Connelly, J.T.; Kondapalli, S.; Skoupi, M.; Parker, J.S.L.; Kirby, B.J.; Baeumner, A.J. Micro-total analysis system for virus detection: Microfluidic pre-concentration coupled to liposome-based detection. *Anal. Bioanal. Chem.* **2011**, *402*, 315–323. Available online: https://link.springer.com/article/10.1007/s00216-011-5381-9 (accessed on 27 February 2021). [CrossRef] [PubMed]
46. Chandrawati, R.; Städler, B.; Postma, A.; Connal, L.A.; Chong, S.F.; Zelikin, A.N.; Caruso, F. Cholesterol-mediated anchoring of enzyme-loaded liposomes within disulfide-stabilized polymer carrier capsules. *Biomaterials* **2009**, *30*, 5988–5998. [CrossRef] [PubMed]
47. Duong, H.T.T.; Adnan, N.N.M.; Barraud, N.; Basuki, J.S.; Kutty, S.K.; Jung, K.; Kumar, N.; Davis, T.P.; Boyer, C. Functional gold nanoparticles for the storage and controlled release of nitric oxide: Applications in biofilm dispersal and intracellular delivery. *J. Mater. Chem. B* **2014**, *2*, 5003–5011. Available online: https://pubs.rsc.org/en/content/articlehtml/2014/tb/c4tb00632a (accessed on 27 February 2021). [CrossRef] [PubMed]
48. Diring, S.; Wang, D.O.; Kim, C.; Kondo, M.; Chen, Y.; Kitagawa, S.; Kamei, K.-I.; Furukawa, S. Localized cell stimulation by nitric oxide using a photoactive porous coordination polymer platform. *Nat. Commun.* **2013**, *4*, 2684. Available online: https://www.nature.com/articles/ncomms3684 (accessed on 27 February 2021). [CrossRef] [PubMed]
49. Lu, B.; Lu, F.; Zou, Y.; Liu, J.; Rong, B.; Li, Z.; Dai, F.; Wu, D.; Lan, G. In situ reduction of silver nanoparticles by chitosan-l-glutamic acid/hyaluronic acid: Enhancing antimicrobial and wound-healing activity. *Carbohydr. Polym.* **2017**, *173*, 556–565. [CrossRef] [PubMed]

50. Sithole, M.N.; Choonara, Y.E.; du Toit, L.C.; Kumar, P.; Pillay, V. A review of semi-synthetic biopolymer complexes: Modified polysaccharide nano-carriers for enhancement of oral drug bioavailability. *Pharm. Dev. Technol.* **2016**, *22*, 283–295. Available online: https://www.tandfonline.com/doi/abs/10.1080/10837450.2016.1212882 (accessed on 27 February 2021). [CrossRef]
51. Dmour, I.; Taha, M.O. Natural and semisynthetic polymers in pharmaceutical nanotechnology. In *Organic Materials as Smart Nanocarriers for Drug Delivery*; Grumezescu, A.M., Ed.; Elsevier BV: Amsterdam, The Netherlands, 2018; pp. 35–100.
52. Worley, B.V.; Soto, R.J.; Kinsley, P.C.; Schoenfisch, M.H. Active Release of Nitric Oxide-Releasing Dendrimers from Electrospun Polyurethane Fibers. *ACS Biomater. Sci. Eng.* **2016**, *2*, 426–437. Available online: https://pubs.acs.org/doi/full/10.1021/acsbiomaterials.6b00032 (accessed on 27 February 2021). [CrossRef]
53. Dan Mogoşanu, G.; Grumezescu, A.M.; Bejenaru, L.E.; Bejenaru, C. Natural and synthetic polymers for drug delivery and targeting. In *Nanobiomaterials in Drug Delivery: Applications of Nanobiomaterials*; Grumezescu, A.M., Ed.; Elsevier: Amsterdam, The Netherlands, 2016; pp. 229–284.
54. Sikka, M.P.; Midha, V.K. The role of biopolymers and biodegradable polymeric dressings in managing chronic wounds. In *Advanced Textiles for Wound Care*; Rajendran, S., Ed.; Elsevier BV: Amsterdam, The Netherlands, 2019; pp. 463–488.
55. Seabra, A.B.; de Oliveira, M.G. Poly (vinyl alcohol) and poly (vinyl pyrrolidone) blended films for local nitric oxide release. *Biomaterials* **2004**, *25*, 3773–3782. [CrossRef]
56. Seabra, A.; Fitzpatrick, A.; Paul, J.; de Oliveira, M.; Weller, R. Topically applied S-nitrosothiol-containing hydrogels as experimental and pharmacological nitric oxide donors in human skin. *Br. J. Dermatol.* **2004**, *151*, 977–983. Available online: https://onlinelibrary.wiley.com/doi/full/10.1111/j.1365-2133.2004.06213.x (accessed on 27 February 2021). [CrossRef]
57. Shishido, S.M.; Seabra, A.B.; Loh, W.; de Oliveira, M.G. Thermal and photochemical nitric oxide release from S-nitrosothiols incorporated in Pluronic F127 gel: Potential uses for local and controlled nitric oxide release. *Biomaterials* **2003**, *24*, 3543–3553. [CrossRef]
58. Halpenny, G.M.; Steinhardt, R.C.; Okialda, K.A.; Mascharak, P.K. Characterization of pHEMA-based hydrogels that exhibit light-induced bactericidal effect via release of NO. *J. Mater. Sci. Mater. Med.* **2009**, *20*, 2353–2360. Available online: https://link.springer.com/article/10.1007/s10856-009-3795-0 (accessed on 27 February 2021). [CrossRef]
59. Marcato, P.D.; Adami, L.F.; Melo, P.S.; de Paula, L.B.; Durán, N.; Seabra, A.B. Glutathione and S-nitrosoglutathione in alginate/chitosan nanoparticles: Cytotoxicity. *J. Phys. Conf. Ser.* **2011**, *304*, 012045. Available online: https://iopscience.iop.org/article/10.1088/1742-6596/304/1/012045 (accessed on 27 February 2021). [CrossRef]
60. Cardozo, V.F.; Lancheros, C.A.C.; Narciso, A.M.; Valereto, E.C.S.; Kobayashi, R.K.T.; Seabra, A.B.; Nakazato, G. Evaluation of antibacterial activity of nitric oxide-releasing polymeric particles against *Staphylococcus aureus* and *Escherichia coli* from bovine mastitis. *Int. J. Pharm.* **2014**, *473*, 20–29. [CrossRef]
61. Nguyen, T.K.; Selvanayagam, R.; Ho, K.K.K.; Chen, R.; Kutty, S.K.; Rice, S.A.; Kumar, N.; Barraud, N.; Duong, H.T.T.; Boyer, C. Co-delivery of nitric oxide and antibiotic using polymeric nanoparticles. *Chem. Sci.* **2016**, *7*, 1016–1027. Available online: https://pubs.rsc.org/en/content/articlehtml/2016/sc/c5sc02769a (accessed on 27 February 2021). [CrossRef] [PubMed]
62. Sadrearhami, Z.; Yeow, J.; Nguyen, T.-K.; Ho, K.K.K.; Kumar, N.; Boyer, C. Biofilm dispersal using nitric oxide loaded nanoparticles fabricated by photo-PISA: Influence of morphology. *Chem. Commun.* **2017**, *53*, 12894–12897. Available online: https://pubs.rsc.org/en/content/articlehtml/2017/cc/c7cc07293g (accessed on 27 February 2021). [CrossRef]
63. Shen, Z.; He, K.; Ding, Z.; Zhang, M.; Yu, Y.; Hu, J. Visible-Light-Triggered Self-Reporting Release of Nitric Oxide (NO) for Bacterial Biofilm Dispersal. *Macromolecules* **2019**, *52*, 7668–7677. Available online: https://pubs.acs.org/doi/full/10.1021/acs.macromol.9b01252 (accessed on 27 February 2021). [CrossRef]
64. Namivandi-Zangeneh, R.; Sadrearhami, Z.; Bagheri, A.; Sauvage-Nguyen, M.; Ho, K.K.K.; Kumar, N.; Wong, E.H.H.; Boyer, C. Nitric Oxide-Loaded Antimicrobial Polymer for the Synergistic Eradication of Bacterial Biofilm. *ACS Macro Lett.* **2018**, *7*, 592–597. Available online: https://pubs.acs.org/doi/full/10.1021/acsmacrolett.8b00190 (accessed on 27 February 2021). [CrossRef]
65. Weller, R.; Dykhuizen, R.; Leifert, C.; Ormerod, A.; Christophers, E.; Henseler, T. Nitric oxide release accounts for the reduced incidence of cutaneous infections in psoriasis. *J. Am. Acad. Dermatol.* **1997**, *36*, 281–282. Available online: http://www.jaad.org/article/S0190962297703074/fulltext (accessed on 27 February 2021). [CrossRef]
66. Weller, R.; Price, R.; Ormerod, A.; Benjamin, N.; Leifert, C. Antimicrobial effect of acidified nitrite on dermatophyte fungi, *Candida* and bacterial skin pathogens. *J. Appl. Microbiol.* **2001**, *90*, 648–652. Available online: http://doi.wiley.com/10.1046/j.1365-2672.2001.01291.x (accessed on 27 February 2021). [CrossRef]
67. Li, J.; Hunter, C.A.; Farrell, J.P. Anti-TGF-β Treatment Promotes Rapid Healing of *Leishmania* major Infection in Mice by Enhancing In Vivo Nitric Oxide Production. *J. Immunol.* **1999**, *162*, 974–979. [PubMed]
68. López-Jaramillo, P.; Ruano, C.; Rivera, J.; Terán, E.; Salazar-Irigoyen, R.; Esplugues, J.V.; Moncada, S. Treatment of cutaneous leishmaniasis with nitric-oxide donor. *Lancet* **1998**, *351*, 1176–1177.
69. Seabra, A.B.; Duran, N. Nanotechnology Allied to Nitric Oxide Release Materials for Dermatological Applications. *Curr. Nanosci.* **2012**, *8*, 520–525. Available online: http://www.eurekaselect.com/openurl/content.php?genre=article&issn=1573-4137&volume=8&issue=4&spage=520 (accessed on 27 February 2021). [CrossRef]
70. Urzedo, A.L.; Gonçalves, M.C.; Nascimento, M.H.M.; Lombello, C.B.; Nakazato, G.; Seabra, A.B. Cytotoxicity and Antibacterial Activity of Alginate Hydrogel Containing Nitric Oxide Donor and Silver Nanoparticles for Topical Applications. *ACS Biomater. Sci. Eng.* **2020**, *6*, 2117–2134. Available online: https://dx.doi.org/10.1021/acsbiomaterials.9b01685 (accessed on 27 February 2021). [CrossRef]

71. Urzedo, A.L.; Gonçalves, M.C.; Nascimento, M.H.M.; Lombello, C.B.; Nakazato, G.; Seabra, A.B. Multifunctional alginate nanoparticles containing nitric oxide donor and silver nanoparticles for biomedical applications. *Mater. Sci. Eng. C* **2020**, *112*, 110933. [CrossRef]
72. Hebert, A.A.; Siegfried, E.C.; Durham, T.; de León, E.N.; Reams, T.; Messersmith, E.; Maeda-Chubachi, T. Efficacy and tolerability of an investigational nitric oxide–releasing topical gel in patients with molluscum contagiosum: A randomized clinical trial. *J. Am. Acad. Dermatol.* **2020**, *82*, 887–894. Available online: https://doi.org/10.1016/j.jaad.2019.09.064 (accessed on 27 February 2021). [CrossRef]
73. Costa-Orlandi, C.B.; Mordorski, B.; Baltazar, L.M.; Mendes-Giannini, M.J.S.; Friedman, J.M.; Nosanchuk, J.D.; Friedman, A.J. Nitric Oxide Releasing Nanoparticles as a Strategy to Improve Cur-rent Onychomycosis Treatments. *J. Drugs Dermatol.* **2018**, *7*, 717–720. Available online: https://pubmed.ncbi.nlm.nih.gov/30005092/ (accessed on 27 February 2021).
74. Stasko, N.; McHale, K.; Hollenbach, S.J.; Martin, M.; Doxey, R. Nitric Oxide-Releasing Macromolecule Exhibits Broad-Spectrum Antifungal Activity and Utility as a Topical Treatment for Superficial Fungal Infections. *Antimicrob. Agents Chemother.* **2018**, *62*. Available online: https://doi.org/10.1128/AAC.01026-17 (accessed on 27 February 2021). [CrossRef]
75. Pelegrino, M.T.; Lima, B.D.A.; Nascimento, M.H.M.D.; Lombello, C.B.; Brocchi, M.; Seabra, A.B. Biocompatible and Antibacterial Nitric Oxide-Releasing Pluronic F-127/Chitosan Hydrogel for Topical Applications. *Polymers* **2018**, *10*, 452. Available online: www.mdpi.com/journal/polymers (accessed on 27 February 2021). [CrossRef] [PubMed]
76. Mordorski, B.; Costa-Orlandi, C.B.; Baltazar, L.M.; Carreño, L.J.; Landriscina, A.; Rosen, J.; Navati, M.; Mendes-Giannini, M.J.S.; Friedman, J.M.; Nosanchuk, J.D.; et al. Topical nitric oxide releasing nanoparticles are effective in a murine model of dermal *Trichophyton rubrum* dermatophytosis. *Nanomed. Nanotechnol. Biol. Med.* **2017**, *13*, 2267–2270. [CrossRef] [PubMed]
77. Sulemankhil, I.; Ganopolsky, J.G.; Dieni, C.A.; Dan, A.F.; Jones, M.L.; Prakash, S. Prevention and Treatment of Virulent Bacterial Biofilms with an Enzymatic Nitric Oxide-Releasing Dressing. *Antimicrob. Agents Chemother.* **2012**, *56*, 6095–6103. Available online: https://pmc/articles/PMC3497171/?report=abstract (accessed on 27 February 2021). [CrossRef]
78. López-Jaramillo, P.; Rincón, M.Y.; García, R.G.; Silva, S.Y.; Smith, E.; Kampeerapappun, P.; García, C.; Smith, D.J.; López, M.; Vélez, I.D. A controlled, randomized-blinded clinical trial to assess the efficacy of a nitric oxide releasing patch in the treatment of cutaneous leishmaniasis by *Leishmania* (V.) panamensis. *Am. J. Trop. Med. Hyg.* **2010**, *83*, 97–101. Available online: https://www.ajtmh.org/view/journals/tpmd/83/1/article-p97.xml (accessed on 27 February 2021). [CrossRef] [PubMed]
79. Jones, M.L.; Ganopolsky, J.G.; Labbé, A.; Prakash, S. A novel nitric oxide producing probiotic patch and its antimicrobial efficacy: Preparation and in vitro analysis. *Appl. Microbiol. Biotechnol.* **2010**, *87*, 509–516. Available online: https://link.springer.com/article/10.1007/s00253-010-2490-x (accessed on 27 February 2021). [CrossRef] [PubMed]
80. Han, G.; Martinez, L.R.; Mihu, M.R.; Friedman, A.J.; Friedman, J.M.; Nosanchuk, J.D. Nitric Oxide Releasing Nanoparticles Are Therapeutic for *Staphylococcus aureus* Abscesses in a Murine Model of Infection. *PLoS ONE* **2009**, *4*, e7804. Available online: https://dx.plos.org/10.1371/journal.pone.0007804 (accessed on 27 February 2021). [CrossRef] [PubMed]
81. Friedman, A.J.; Han, G.; Navati, M.S.; Chacko, M.; Gunther, L.; Alfieri, A.; Friedman, J.M. Sustained release nitric oxide releas-ing nanoparticles: Characterization of a novel delivery platform based on nitrite containing hydrogel/glass composites. *Nitric Oxide* **2008**, *19*, 12–20. [CrossRef]
82. Mir, M.; Ali, M.N.; Barakullah, A.; Gulzar, A.; Arshad, M.; Fatima, S.; Asad, M. Synthetic polymeric biomaterials for wound healing: A review. *Prog. Biomater.* **2018**, *7*, 1–21. Available online: https://doi.org/10.1007/s40204-018-0083-4 (accessed on 27 February 2021). [CrossRef] [PubMed]
83. Coelho, J.F.; Ferreira, P.C.; Alves, P.; Cordeiro, R.; Fonseca, A.C.; Góis, J.R.; Gil, M.H. Drug delivery systems: Advanced technologies potentially applicable in personalized treatments. *EPMA J.* **2010**, *1*, 164–209. Available online: https://pmc/articles/PMC3405312/?report=abstract (accessed on 27 February 2021). [CrossRef]
84. Bhatia, S. Natural Polymers vs Synthetic Polymer. In *Natural Polymer Drug Delivery Systems*; Springer Science and Business Media LLC: Cham, Switzerland, 2016; pp. 95–118. Available online: https://link.springer.com/chapter/10.1007/978-3-319-41129-3_3 (accessed on 27 February 2021).
85. Banerjee, N.S.; Moore, D.W.; Wang, H.-K.; Broker, T.R.; Chow, L.T. NVN1000, a novel nitric oxide-releasing compound, inhibits HPV-18 virus production by interfering with E6 and E7 oncoprotein functions. *Antivir. Res.* **2019**, *170*, 104559. [CrossRef] [PubMed]
86. Maeda-Chubachi, T.; Messersmith, E.; Hebert, D.; de Leon, E.; Reams, T. LB1096 Results of phase 2 study evaluating the efficacy and safety of SB206, topical berdazimer sodium gel, in subjects with Molluscum Contagiosum. *J. Investig. Dermatol.* **2019**, *139*, B14. [CrossRef]
87. Abedi-Gaballu, F.; Dehghan, G.; Ghaffari, M.; Yekta, R.; Abbaspour-Ravasjani, S.; Baradaran, B.; Dolatabadie, J.E.N.; Hamblinfgh, M.R. PAMAM dendrimers as efficient drug and gene delivery nanosystems for cancer therapy. *Appl. Mater. Today* **2018**, *12*, 177–190. Available online: https://pmc/articles/PMC6269116/?report=abstract (accessed on 27 February 2021). [CrossRef] [PubMed]
88. Jenkins, A.D.; Loening, K.L. Nomenclature. In *Comprehensive Polymer Science and Supplements*; Allen, G., Bevington, J.C., Eds.; Elsevier: Amsterdam, The Netherlands, 1989; pp. 13–54. Available online: https://linkinghub.elsevier.com/retrieve/pii/B9780080967011000021 (accessed on 27 February 2021).
89. Witte, M.B.; Barbul, A. Role of nitric oxide in wound repair. *Am. J. Surg.* **2002**, *183*, 406–412. [CrossRef]

90. Lee, J.; Kwak, D.; Kim, H.; Kim, J.; Hlaing, S.P.; Hasan, N.; Cao, J.; Yoo, J.-W. Nitric Oxide-Releasing *S*-Nitrosoglutathione-Conjugated Poly (Lactic-*Co*-Glycolic Acid) Nanoparticles for the Treatment of MRSA-Infected Cutaneous Wounds. *Pharmaceutics* **2020**, *12*, 618. Available online: https://www.mdpi.com/1999-4923/12/7/618 (accessed on 27 February 2021). [CrossRef]
91. Nie, X.; Zhang, H.; Shi, X.; Zhao, J.; Chen, Y.; Wu, F.; Yang, J.; Li, X. Asiaticoside nitric oxide gel accelerates diabetic cutaneous ulcers healing by activating Wnt/β-catenin signaling pathway. *Int. Immunopharmacol.* **2020**, *79*, 106109. [CrossRef]
92. Lee, J.; Hlaing, S.P.; Cao, J.; Hasan, N.; Yoo, J.-W. In vitro and in vivo evaluation of a novel nitric oxide-releasing ointment for the treatment of methicillin-resistant *Staphylococcus aureus*-infected wounds. *J. Pharm. Investig.* **2020**, *50*, 505–512. Available online: https://doi.org/10.1007/s40005-020-00472-1 (accessed on 27 February 2021). [CrossRef]
93. Wan, X.; Liu, S.; Xin, X.; Li, P.; Dou, J.; Han, X.; Kang, I.-K.; Yuan, J.; Chi, B.; Shen, J. S-nitrosated keratin composite mats with NO release capacity for wound healing. *Chem. Eng. J.* **2020**, *400*, 125964. [CrossRef]
94. Huang, S.; Liu, H.; Liao, K.; Hu, Q.; Guo, R.; Deng, K. Functionalized GO Nanovehicles with Nitric Oxide Release and Photothermal Activity-Based Hydrogels for Bacteria-Infected Wound Healing. *ACS Appl. Mater. Interfaces* **2020**. Available online: https://pubs.acs.org/doi/full/10.1021/acsami.0c04080 (accessed on 27 February 2021). [CrossRef]
95. Póvoa, V.C.; dos Santos, G.J.; Picheth, G.F.; Jara, C.P.; da Silva, L.C.; de Araújo, E.P.; de Oliveira, M.G. Wound healing action of nitric oxide-releasing self-expandable collagen sponge. *J. Tissue Eng. Regen. Med.* **2020**, *14*, 807–818. Available online: https://onlinelibrary.wiley.com/doi/abs/10.1002/term.3046 (accessed on 27 February 2021). [CrossRef]
96. Zhang, Y.; Tang, K.; Chen, B.; Zhou, S.; Li, N.; Liu, C.; Yang, J.; Lin, R.; Zhang, T.; He, W. A polyethylenimine-based diazeniumdiolate nitric oxide donor accelerates wound healing. *Biomater. Sci.* **2019**, *7*, 1607–1616. Available online: https://pubmed.ncbi.nlm.nih.gov/30702089/ (accessed on 27 February 2021). [CrossRef] [PubMed]
97. Kulshrestha, S.; Chawla, R.; Alam, T.; Adhikari, J.; Basu, M. Efficacy and dermal toxicity analysis of Sildenafil citrate based topical hydrogel formulation against traumatic wounds. *Biomed. Pharmacother.* **2019**, *112*, 108571. [CrossRef] [PubMed]
98. Hasan, N.; Cao, J.; Lee, J.; Naeem, M.; Hlaing, S.P.; Kim, J.; Jung, Y.; Lee, B.-L.; Yoo, J.-W. PEI/NONOates-doped PLGA nanoparticles for eradicating methicillin-resistant *Staphylococcus aureus* biofilm in diabetic wounds via binding to the biofilm matrix. *Mater. Sci. Eng. C* **2019**, *103*, 109741. [CrossRef] [PubMed]
99. Hlaing, S.P.; Kim, J.; Lee, J.; Hasan, N.; Cao, J.; Naeem, M.; Lee, E.H.; Shin, J.H.; Jung, Y.; Lee, B.-L.; et al. S-Nitrosoglutathione loaded poly(lactic-co-glycolic acid) microparticles for prolonged nitric oxide release and enhanced healing of methicillin-resistant Staphy-lococcus aureus-infected wounds. *Eur. J. Pharm. Biopharm.* **2018**, *132*, 94–102. [CrossRef]
100. Baldwin, H.; Blanco, D.; McKeever, C.; Paz, N.; Vasquez, Y.N.; Quiring, J.; Enloe, C.; de León, E.; Stasko, N. Results of a phase 2 efficacy and safety study with SB204, an investigational topical nitric oxide-releasing drug for the treatment of *Acne vulgaris*. *J. Clin. Aesthetic Dermatol.* **2016**, *9*, 12. Available online: https://pmc/articles/PMC5022991/?report=abstract (accessed on 27 February 2021).
101. Schanuel, F.S.; Santos, K.S.R.; Monte-Alto-Costa, A.; de Oliveira, M.G. Combined nitric oxide-releasing poly (vinyl alcohol) film/F127 hydrogel for accelerating wound healing. *Colloids Surf. B Biointerfaces* **2015**, *130*, 182–191. [CrossRef] [PubMed]
102. Lowe, A.; Bills, J.; Verma, R.; Lavery, L.; Davis, K.; Balkus, K.J. Electrospun nitric oxide releasing bandage with enhanced wound healing. *Acta Biomater.* **2015**, *13*, 121–130. [CrossRef] [PubMed]
103. Kim, J.O.; Noh, J.K.; Thapa, R.K.; Hasan, N.; Choi, M.; Kim, J.H.; Lee, J.-H.; Ku, S.K.; Yoo, J.-W. Nitric oxide-releasing chitosan film for enhanced antibacterial and in vivo wound-healing efficacy. *Int. J. Biol. Macromol.* **2015**, *79*, 217–225. [CrossRef]
104. Gao, J.; Zheng, W.; Zhang, J.; Guan, D.; Yang, Z.; Kong, D.; Zhao, Q. Enzyme-controllable delivery of nitric oxide from a molecular hydrogel. *Chem. Commun.* **2013**, *49*, 9173–9175. Available online: https://pubmed.ncbi.nlm.nih.gov/23989671/ (accessed on 27 February 2021). [CrossRef] [PubMed]
105. Jones, M.; Ganopolsky, J.G.; Labbé, A.; Gilardino, M.; Wahl, C.; Martoni, C.; Prakash, S. Novel nitric oxide producing probiotic wound healing patch: Preparation and in vivo analysis in a New Zealand white rabbit model of ischaemic and infected wounds. *Int. Wound, J.* **2012**, *9*, 330–343. Available online: http://doi.wiley.com/10.1111/j.1742-481X.2011.00889.x (accessed on 27 February 2021). [CrossRef]
106. Georgii, J.L.; Amadeu, T.P.; Seabra, A.B.; de Oliveira, M.G.; Monte-Alto-Costa, A. Topical S-nitrosoglutathione-releasing hydrogel improves healing of rat ischaemic wounds. *J. Tissue Eng. Regen. Med.* **2010**, *5*, 612–619. Available online: http://doi.wiley.com/10.1002/term.353 (accessed on 27 February 2021). [CrossRef]
107. Martinez, L.R.; Han, G.; Chacko, M.; Mihu, M.R.; Jacobson, M.; Gialanella, P.; Friedman, A.J.; Nosanchuk, J.D.; Friedman, J.M. Antimicrobial and healing efficacy of sustained release nitric oxide nanoparticles against *Staphylococcus aureus* skin infection. *J. Investig. Dermatol.* **2009**, *129*, 2463–2469. Available online: www.jidonline.org (accessed on 27 February 2021). [CrossRef]
108. Amadeu, T.P.; Seabra, A.B.; de Oliveira, M.G.; Monte-Alto-Costa, A. Nitric Oxide Donor Improves Healing if Applied on Inflammatory and Proliferative Phase. *J. Surg. Res.* **2008**, *149*, 84–93. Available online: http://www.journalofsurgicalresearch.com/article/S0022480407006373/fulltext (accessed on 27 February 2021). [CrossRef]
109. Amadeu, T.P.; Seabra, A.B.; de Oliveira, M.G.; Monte-Alto-Costa, A. S-nitrosoglutathione-containing hydrogel accelerates rat cutaneous wound repair. *J. Eur. Acad. Dermatol. Venereol.* **2007**, *21*, 629–637. Available online: http://doi.wiley.com/10.1111/j.1468-3083.2006.02032.x (accessed on 27 February 2021). [CrossRef] [PubMed]
110. Masters, K.S.B.; Leibovich, S.J.; Belem, P.; West, J.L.; Poole-Warren, L.A. Effects of nitric oxide releasing poly (vinyl alcohol) hydrogel dressings on dermal wound healing in diabetic mice. *Wound Repair Regen.* **2002**, *10*, 286–294. Available online: https://onlinelibrary.wiley.com/doi/full/10.1046/j.1524-475X.2002.10503.x (accessed on 27 February 2021). [CrossRef]

111. Arora, N.; Maurya, P.K.; Kacker, P. Translational Research in Drug Discovery and Development. *Transl. Med. Res.* **2017**, 55–87. Available online: https://link.springer.com/chapter/10.1007/978-94-024-1045-7_3 (accessed on 27 February 2021).
112. Souto, S.; Palma, P.; Riccetto, C.; Seabra, A.; Oliveira, M.; Palma, T.; Capmartin, R. Impact of topic administration of nitric oxide donor gel in the clitoridian blood flow, assessed by Doppler ultra-sound. *Actas Urológicas Españolas* **2010**, *34*, 708–712. [CrossRef] [PubMed]
113. Soares, R.N.; Proctor, D.N.; de Oliveira, G.V.; Alvares, T.S.; Murias, J.M. Acute application of a transdermal nitroglycerin patch protects against prolonged forearm ischemia-induced microvascular dysfunction. *Microcirculation* **2019**, *27*, e12599. Available online: https://onlinelibrary.wiley.com/doi/abs/10.1111/micc.12599 (accessed on 27 February 2021). [CrossRef]
114. Pastore, M.N.; Kalia, Y.N.; Horstmann, M.; Roberts, M.S. Transdermal patches: History, development and pharmacology. *Br. J. Pharmacol.* **2015**, *172*, 2179–2209. Available online: https://pmc/articles/PMC4403087/?report=abstract (accessed on 27 February 2021). [CrossRef] [PubMed]
115. Kirkland, Q.C. *Product Monograph Including Patient Medication Information*; Mylan Pharmaceuticals ULC: Etobicoke, ON, USA, 2016.
116. Appleton, J.P.; Krishnan, K.; Bath, P.M. Transdermal delivery of glyceryl trinitrate: Clinical applications in acute stroke. *Expert Opin. Drug Deliv.* **2020**, *17*, 297–303. Available online: https://www.tandfonline.com/doi/abs/10.1080/17425247.2020.1716727 (accessed on 27 February 2021). [CrossRef]
117. Giglio, L.P.; Picheth, G.F.; Løvschall, K.B.; Zelikin, A.N.; de Oliveira, M.G. S-nitrosothiol-terminated poly (vinyl alcohol): Nitric oxide release and skin blood flow response. *Nitric Oxide* **2020**, *98*, 41–49. [CrossRef] [PubMed]
118. Park, K.; Dawson, J.I.; Oreffo, R.O.C.; Kim, Y.-H.; Hong, J. Nanoclay–Polyamine Composite Hydrogel for Topical Delivery of Nitric Oxide Gas via Innate Gelation Characteristics of Laponite. *Biomacromolecules* **2020**, *21*, 2096–2103. Available online: https://dx.doi.org/10.1021/acs.biomac.0c00086 (accessed on 27 February 2021). [CrossRef] [PubMed]
119. Gori, T. Exogenous no therapy for the treatment and prevention of atherosclerosis. *Int. J. Mol. Sci.* **2020**, *21*, 2703. Available online: www.mdpi.com/journal/ijms (accessed on 27 February 2021). [CrossRef] [PubMed]

polymers

MDPI

Review

Polymer-Based Constructs for Flexor Tendon Repair: A Review

Jef Brebels and **Arn Mignon** *

Surface and Interface Engineered Materials, Campus Group T, KU Leuven, Andreas Vesaliusstraat 13, 3000 Leuven, Belgium; jef.brebels@kuleuven.be
* Correspondence: arn.mignon@kuleuven.be

Abstract: A flexor tendon injury is acquired fast and is common for athletes, construction workers, and military personnel among others, treated in the emergency department. However, the healing of injured flexor tendons is stretched over a long period of up to 12 weeks, therefore, remaining a significant clinical problem. Postoperative complications, arising after traditional tendon repair strategies, include adhesion and tendon scar tissue formation, insufficient mechanical strength for early active mobilization, and infections. Various researchers have tried to develop innovative strategies for developing a polymer-based construct that minimalizes these postoperative complications, yet none are routinely used in clinical practice. Understanding the role such constructs play in tendon repair should enable a more targeted approach. This review mainly describes the polymer-based constructs that show promising results in solving these complications, in the hope that one day these will be used as a routine practice in flexor tendon repair, increasing the well-being of the patients. In addition, the review also focuses on the incorporation of active compounds in these constructs, to provide an enhanced healing environment for the flexor tendon.

Keywords: flexor tendon repair; anti-inflammatory; anti-adhesion; antimicrobial; polymer-based constructs

Citation: Brebels, J.; Mignon, A. Polymer-Based Constructs for Flexor Tendon Repair: A Review. *Polymers* **2022**, *14*, 867. https://doi.org/10.3390/polym14050867

Academic Editor: Cornelia Vasile

Received: 21 January 2022
Accepted: 20 February 2022
Published: 23 February 2022

1. Introduction

The flexor digitorum superficialis, or in short, the flexor tendon, is an irreplaceable part of the human body. Connecting muscle to bone provides strength and stability, the ability to withstand tension, transmit forces, and release stored energy. Since tendons are subjected to repeated motions and degeneration over time, they are vulnerable to acute and chronic injuries [1–3]. Hand tendon traumas comprise approximately 10% of all emergency department visits and up to 20% of all injuries treated [4,5]. Athletes, construction workers, military personnel, and others who make repetitive movements have a greater risk of injuring the flexor tendon by tearing or rupturing. A trauma impact directly to the hand could also lead to such an injury. Injured flexor tendons will exhibit a biological attempt to heal the inflicted damage. However, the speed at which this happens is greatly outpaced by the own capacity of accumulating further damage. Therefore, it should be noted that flexor tendons cannot undergo spontaneous healing and operational procedures are almost always required [6,7]. In addition, flexor tendons have an extended healing period of up to 12 weeks due to their limited blood flow and hypocellularity [8,9].

So far, multiple therapeutic reconstruction techniques such as suturing, auto-, allo-, and xenograft or replacement with a synthetic prosthesis have been used [10,11]. Unfortunately, none of these traditional techniques accomplish a long-term adequate solution for postoperative complications such as infection, wear, tendon scar tissue formation, mechanical failure, and excessive adhesion formation [12]. The success and effectiveness of these traditional repair techniques are mostly linked to the degree of undesired postoperative adhesion formation between surrounding tissue and the healing site [7]. It is important to note that the original mechanical properties are never fully restored after tendon repair due to scar tissue formation around the healing site. The scar tissue is inferior in mechanical properties due to the predominant presence of type III collagen, whereas healthy tendon tissue mainly consists of type I collagen. An excessive amount of type III collagen results in loosely organized fibrils [13]. These complications can be avoided by inducing a healing response that is faster than the rate of adhesion and scar tissue formation [14].

New treatment strategies have emerged to overcome these clinical challenges. Tissue engineering for flexor tendon repair by combining cells and growth factors on interactive scaffolds formed the next promising repair technique [15]. Tissue engineering has gained popularity in the field of regenerative medicine due to its bioactivity and biocompatibility. The scaffold is used as a biomaterial that enables critical functions such as cell adhesion, proliferation, cell-biomaterial responses, and cell differentiation in the body. Scaffold vascularization is often a problem occurring in tissue engineering. The supply of oxygen through the scaffold to the surrounding tissue is essential for maintaining cellular respiration. Additionally, cellular functions including proliferation and differentiation are only possible when essential nutrient exchange and removal of toxins and waste products from the scaffold is ensured [16]. The scaffold must be able to achieve these functions without inducing an immune reaction [17]. However, compared to the traditional solutions, similar problems, as described above, appear such as the lack of mechanical strength in vivo which is needed for flexor tendon repair since they support large mechanical stresses [18]. Therefore, an alternative route of current experimental research is more focused on producing a material-based mechanical construct that is placed around the damaged tendon area. The construct acts as a mechanical and physical barrier to minimize the formed adhesion, without compromising the diffusion of nutrients and by-products produced by the biodegradation of the construct [19]. The ideal construct should provide sufficient mechanical support as well as provide the tendon with a controlled environment to regrow and if needed reattach. This review article acts as a literature scan to identify the current state-of-the-art research for flexor tendon repair constructs, some of which have active compounds incorporated which induce anti-inflammatory and antimicrobial properties.

2. Flexor Tendon

2.1. Flexor Tendon and Function: Composition and Structure

The flexor tendon is composed of dense, fibrous, and highly organized tissue responsible for the connections between muscle and bone. Their function is to transmit the forces produced by muscle contraction to the skeletal bones, making body motion possible [1,2]. Other functionalities include the resistance against large tensions, storage, and release of energy, providing strength, and ultimately, stability, to the fingers [3]. Healthy tendons allow for great resistance to mechanical stress with minimal deformation due to their fibro-elastic texture [20,21].

Flexor tendon functionality is closely linked to the composition of the extracellular matrix (ECM). The major component of the flexor tendon is water (60–80%) whereas type I collagen makes up for approximately 70–80% of dry weight. Other types of collagens (III, V), proteoglycans, and glycoproteins such as mainly elastin account for the remainder of 20–30% of dry weight. Elastin in combination with type I collagen is responsible for flexibility and strength respectively [21–23].

The aggregation of type I collagen molecules forms a collagen fibril. These collagen fibrils bundle into fibers containing fibroblast cells, fibers further bundle into fascicles, and finally, fascicles bundle together to form the fascicular matrix. The endotenon occupies the small space between the fascicle bundles. The fascicular bundles can make small slip motions due to the endotenon containing limited nerves, blood, and lymphatic vessels. All fascicles group together and are surrounded by yet another connective tissue sheath, the epitenon [21,24–27]. The hierarchical structure of the tendon can be seen in Figure 1. Assemblies of type I collagen in fibrils, fibers, and fascicles exhibit a wave pattern, also known as crimp, that will align and straighten when a tensile loading is applied [28]. They also act as a shock absorber giving the tendon the possibility to absorb and transmit the applied tension forces, avoiding possible tissue damage.

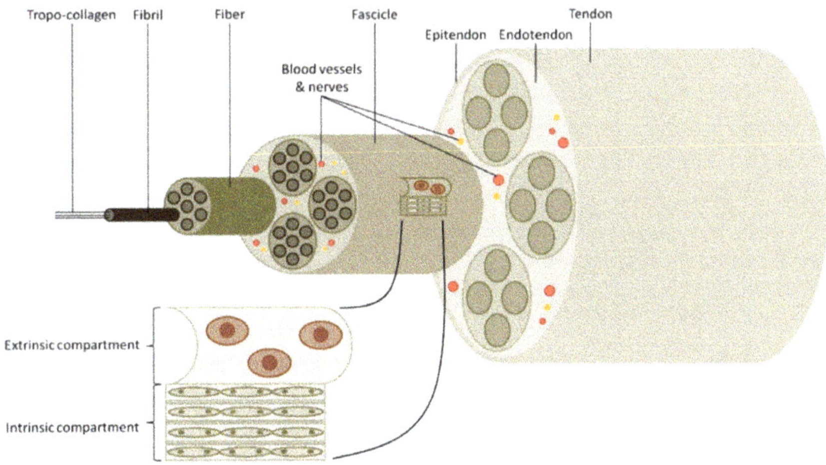

Figure 1. Hierarchical structure of the tendon. Collagen molecules are shown in the simplified model of the tendon structure to represent the forming complex arrangement from tropo-collagen up to tendon fascicles and the final tendon tissue. The intrinsic compartment is represented by the tendon fascicles as the basic unit. The extrinsic compartment is represented by the synovium-like tissue connecting the vascular, nervous, and immune systems. Reprinted with permission from ref. [12]. 2020 Angelo V. Vasiliadis.

As mentioned above, the functionality of flexor tendons is closely linked with the composition and structure of the ECM, responsible for a non-linear viscoelastic and anisotropic behavior, making tendons capable of withstanding high-tension loads [21]. This unique mechanical behavior can be reflected in four distinct regions of the stress-strain curve including the toe, linear, plastic, and finally the failure region [26] (See Figure 2). Tendons also exhibit a viscoelastic behavior, meaning the stress-strain response of the flexor tendon is strain rate dependent. Stretching of the tendon at higher strain rates results in less deformation and higher stiffness while maintaining the same regions during the damage process [29,30]. Therefore, tendons loaded at low strain rates will tend to absorb more energy, although they are less effective in withstanding mechanical stress. "Un-crimping" happens in the toe region for tensile loads up to 2% strain. The slope of the toe region is not constant, resulting in a non-linear stress-strain curve. Once all collagen fibrils have been aligned (2% strain), the slope becomes linear and Young's modulus can be determined. The collagen fibrils stretch physically due to intermolecular sliding of the collagen triple helices [31] (2–4% strain). Tendons are elastically deformed until this point and will still return to their original state and length once the tensile load is removed. Additional loading, thus for a strain >4%, causes microscopic failure to individual fibrils which are irreversible plastic deformation. The tendon is stretched beyond the physiological limit, initiating macroscopic failure followed by failure of the whole tendon when strains are reaching higher values than 8–10% [32–34]. Partly overloading the tendons before macroscopic failure or providing insufficient recovery time can cause permanent damage to the collagen fibrils, lowering the functionality of the tendon [2,35]. Fatigue loading also has a significant influence on the structure of the tendon, permanently altering the crimp characteristics [28].

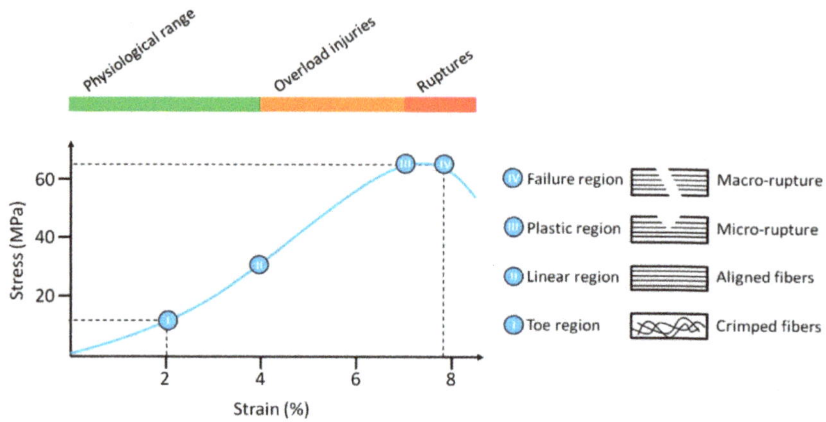

Figure 2. Typical stress-strain curve for a healthy tendon. The righthand side is a schematic representation of the mechanical behavior of the collagen fibers for the different regions. The physiological range (green) consists of the region with the normal use of the tendon and is followed by the overload injuries region (orange) where permanent damage occurs, starting with microscopic failure. Further strain of the tendon will lead to the failure region (red) where rupture of the tendon takes place. Reprinted with permission from ref. [12]. 2020 Angelo V. Vasiliadis.

2.2. *Flexor Tendon and Function: Causes and Impact*

Since tendons are subjected to repeated motions and degeneration over time, they are vulnerable to acute and chronic injuries, direct or indirect. Direct acute tendon injuries occur because of sports activities, blunt impact trauma, or lacerations by a sharp object, accompanied by a loss of tensile strength and disarrangement or possible rupture of the tendon bundles. Tensile overloading, repetitive microscopic failure (fatigue included) or intratendinous degeneration due to aging will lead to chronic indirect flexor tendon injuries. Most commonly, these indirect injuries will occur at the junctions (osteotendinous or musculotendinous) because a healthy tendon can withstand higher tensile stress in the central part of the tendon, compared to the junctions [36]. Tendinopathy, affecting approximately 100 million people worldwide annually, is a failed healing response of the tendon, characterized by swelling, hindered performance, and pain [2]. Interestingly, for decades, researchers believed tendinopathy to be non-inflammatory [37]. Nonetheless, more recent research on tendinopathy confirms the inrush of inflammatory cells, including macrophages and lymphocytes. In addition, it has been hypothesized that these inflammatory cells play a significant role in the early initiation of tendon pathologies. Various mediators of inflammation, pro-inflammatory cytokines, are proven to be expressed during tendon injury such as; Tumor Necrosis Factor-α (TNF-α), Interleukin-1 β (IL-1 β), and Interleukin-6 (IL-6) [38,39]. Even minor injuries may result in a significant high economic, social, and clinical impact on society and patients due to lost work time and high socioeconomic costs [40].

2.3. *Flexor Tendon and Function: Regeneration and Repair*

Tendons have a weak intrinsic healing ability due to their hypocellularity and hypovascularity structure. This problem, in combination with postoperative complications, makes for inefficient and prolonged healing [41]. Damaged flexor tendons tend to heal after injury, although original mechanical and biological properties are never completely restored due to the higher ratio of type III collagen to type I collagen in the collagen fibers after repair [42]. This increases the risk of degeneration and re-rupture at the repaired injury site along with functional impairment. Lower mechanical properties result in a decrease in grip strength of the fingers and hand. Literature [43] indicates a mean grip strength of 74.5% compared to the opposite uninjured hand.

The healing mechanism of tendons is divided into intrinsic and extrinsic healing operating either independently or in cooperation with one another. Tenocytes and fibrocytes at the injury site will result in intrinsic healing, while extrinsic healing is a result of the invasion and proliferation of fibroblasts and inflammatory cells from surrounding tissue and tendon sheath in order to produce a new collagen matrix. The degree of involvement of these external cells is dependent on the vascular perfusion to the injury site. Commonly, these external cells will predominate over the resident cells, causing the surrounding tissue to attach to the repair site, with cell adhesion as an adverse effect. Sufficient vascularity and nutrition by surrounding fluids, absence of cell adhesion, and resident cells proliferation are depending on factors of intrinsic healing [44–49].

Tendon healing after surgical repair follows three well-described, consecutive but overlapping phases: (i) inflammation, (ii) proliferation, and (iii) remodeling [50], represented in Figure 3. The inflammation phase is rather short, which lasts only approximately a week, followed by the proliferation phase, lasting up to 4 weeks. The last phase, remodeling, takes several months to complete [51]. During the inflammation phase, immediately after injury, vascular permeability increases, and inflammatory cells infiltrate the healing site on the tendon (forming a hematoma). These cells include macrophages, monocytes, neutrophils, and tendon stem/progenitor cells (TSPCs). These latter produce cytokines like IL-6 and IL-1 β, responsible for regulating the immune responses by attracting fibroblasts to the injury site [37].

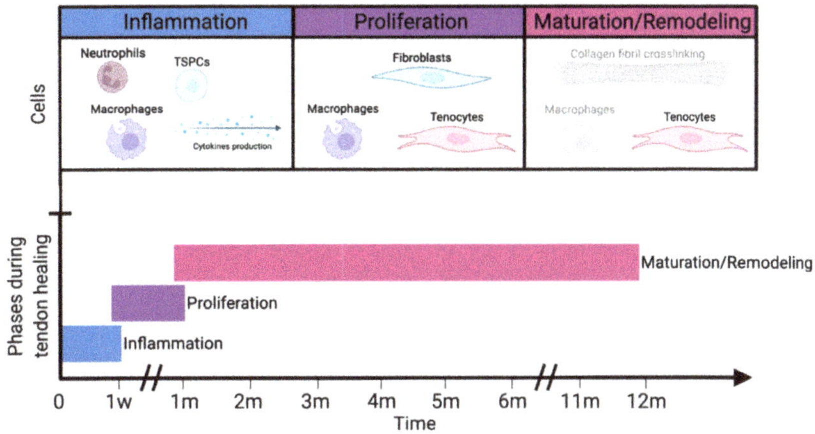

Figure 3. Overview of the process during tendon healing. Healing includes three phases, which overlap slightly. The duration of each phase is an estimate, as duration depends upon the location and severity of the tendon injury.

Neutrophils, released into the blood vessels, contribute to the healing process by removing foreign cells and strengthening the inflammatory response. Macrophages, in addition, are essential in both promoting and resolving the inflammatory response as well as both moderating and facilitating tissue repair [49]. They remove dead cells and toxic metabolites from the injury site by phagocytosis. TSPCs differentiate into tendon-like cells to play an essential role in the whole healing process. The inflammation phase is followed and partly overlapped, by the proliferation phase as can be seen in Figure 3 [36]. During proliferation, growth factors, including basic fibroblast growth factor (bFGF), bone morphogenetic proteins-12, -13, and -14 (BMPs) also known as growth and differentiation factors-5, -6, and -7 (GDFs) respectively, transforming growth factor-beta (TGF-β), insulin-like growth factor-1 (IGF-1), platelet-derived growth factor (PDGF) and vascular endothelial growth factor (VEGF) released by macrophages and endothelial cells which are involved during different phases of the healing and regeneration process [52]. Next, recruited fibroblasts (by cytokines) synthesize components of the ECM; mostly type III collagen, proteoglycans, and fibronectin. The newly formed ECM is randomly arranged and forms a bridge between the injured regions. Lastly, the remodeling phase (see Figure 3) can be divided into two, consolidation and maturation. Consolidation begins 6–8 weeks after the injury and can take up to 10 weeks to complete. Here, a decrease in cellularity and ECM production is observed. The tissue becomes more fibrous due to the replacement of randomly organized type III collagen with crimp-oriented type I collagen. The organization of type I collagen fibers to the longitudinal axis partly restores the tendon tensile strength and stiffness [45,53]. The maturation starts after approximately 10 to 12 weeks, characterized by an increase in collagen fibril crosslinking and further conversion of type III to type I collagen. An important point to note is that the tendon is considered "healed" after a maximum of 12 weeks. This means that the finger can be used again for normal movements. However, the maturation process will continue in the background to increase the mechanical properties of the tendon. Finally, more mature tendonous tissue is formed [36,54].

3. Traditional Strategies for the Repair of Flexor Tendon Injuries

Lacerated tendons do not have the ability to repair themselves without surgical intervention. Tension in the flexor tendon causes the injured end to retract up to several centimeters into the palm of the hand [55,56]. Ultrasound and magnetic resonance imaging can operate as a tool in order to locate the retracted flexor tendon end pre-operatively, although they are not routinely used [57,58]. Surgical repair is necessary to regain the finger mobility, lost after injury. In most cases, flexor tendon surgery is performed in the emergency department's operating room under general or local anesthesia. A flexor tendon injury is commonly accompanied by fracture and loss of the skin, nerve lacerations, and damage to the blood vessels [59]. All these injuries, along with the tendon injury, need to be repaired during the same operation. There are a variety of operative measures available for repairing the flexor tendon in humans, with suture and graft transplantation techniques being the two most commonly used.

3.1. Suture Techniques

Clinically, a lot of different suture techniques are used such as Lin locking, Kessler, Modified Kessler, Savage, Becker, Tajima, grasping, and epitendinous suture [36,60]. The following two are widely accepted: (i) the Modified Kessler suture utilizing 2 sutures [61] and (ii) the grasping suture [62] which uses knots located in the cross-section. Past research has proven that the number of core strands crossing the repair site is directly proportional to the strength of the tendon repair, whereas most ruptures occur at the knots [59,63–65]. However, an increased number of core strands will at the same time increase the complexity and tendon gliding resistance [11]. It is recommended to use a core suture purchase of 7 to 10 mm, as it provides higher gap strength and resistance [59].

In addition to the various possible techniques, several studies have been conducted on the influence of the suture material as well. Some of the commonly used non- adsorbable materials for sutures include monofilament nylon, braided polyester, and monofilament poly(propene). Nylon has the lowest ultimate tensile strength, followed by polypropylene and polyester respectively [66]. Both adsorbable and non-adsorbable materials are capable of early active rehabilitation along with similar rupture rates and mechanical outcomes [11,56,59,66]. Another material used in more recent research is monofilament stainless steel, characterized by its increased stiffness and strength compared to the other suture materials [67]. Reoperation is necessary when non- adsorbable threads are used because they are unaffected by the biological activities of the body tissues, which could increase the chance of infections and scar formation. Finally, the ideal suture should also respond minimal to surrounding tissue and be easily manipulated [68].

A meta-analysis recently performed revealed rates of 4% rupture, 6% re-operation, and 4% adhesions after operative suture repair. In addition, studies showed no difference in rupture rate when core strand or epitendinous suture technique was used. The same study also showed a decrease of 84% in re-operation when the epitendinous suture was used and adhesion formation decreases by 57% with the modified Kessler suture technique [69]. Rupture of the repaired tendon happens most commonly 1 or 2 weeks after repair due to the weakness of the repair site, overaggressive therapy, or non-compliance of the patient [70]. There is a lot of room for improvement despite the recent advances in clinical research. Reoperation is often required, and a significant adhesion formation is still present, these two being the main complications of the suture technique.

3.2. Graft Transplantation Techniques

The suture technique is seen as the primary repair method for flexor tendon injuries but is not always possible due to infections or when part of the tendon is non-viable. Grafting can be utilized in these cases to bridge the gap and restore the flexor tendon. It can be seen as a secondary reconstruction method when the primary suture technique has failed. Satisfactory results have been obtained by this procedure providing there was no damage to the digital sheath and the muscle was still intact [71].

Auto- (same species, own body) and allografts (same species, different body) are the two most used grafting techniques to repair flexor tendons. The palmaris longus, the plantaris, and sometimes the toe extensor include the most suited tendons used as grafts [72]. Allografts are limited in supply and may cause the transmission of diseases and increased inflammatory reactions to the donor cell tissue. In addition, a decrease in tensile properties has been observed for allografts that have been irradiated or chemically sterilized [73,74]. Autografts are less limited in supply although significant donor-site morbidity has been observed, causing tendonitis, pain, and muscle deterioration [75]. Unfortunately, identical to suture procedures, flexor tendon grafting procedures will also cause the formation of adhesion with all the negative consequences previously described. Although the mechanism of tendon graft repair is poorly understood, adhesion formation with autografts is believed to arise from intrinsic (tenocyte necrosis) and extrinsic fibrosis (influx of synovial and inflammatory cells), leading to excessive scarring [59,76,77]. Other research reported minimal formation of adhesion using acellular allografts on bovine flexor tendon repair [78,79] indicating that acellular allografts can heal without the intrinsic healing mechanism, often observed in autograft tissue.

In contrast to auto- and allografts, xenografts do not originate from the same species. Xenografts have higher immunogenicity than allografts, causing rejection by the host to be the most important limitation [10,80,81]. Similar to allografts, there is a risk of disease transmission from donor to host. The risk of disease transfer is higher when using xenografts due to the fact that there are several zoonotic diseases that are unknown and could pose a great risk to the patient and society [80,82]. A high postoperative infection percentage of 20.6% was observed in the literature where immunochemically modified porcine patellar tendon xenografts were used to reconstruct human anterior cruciate ligament (ACL) tissue [83]. Therefore, allo- and xenografts are often decellularized to reduce immunogenicity. Several processes have been used like cell rinsing, freeze-drying, sterilization, etc. These two types of grafts are as a result non-viable in nature and will act more as inert pieces in the body [84]. This causes the allo- and xenografts to incorporate significantly worse with the injured tissue resulting in rapid adsorption of the graft during healing [85].

James Hunter (1960s–1980s) was among the first to describe refined techniques of using artificial tendons as grafts, along with artificial gliding, and implants to reconstruct the tendon sheath [86,87]. Synthetic tendon grafts do not require graft harvesting from a donor or donor-site, shortening the surgical procedure. In addition, autografts often experience a necrotic stage before the prolonged healing stage, which is not the case for synthetic grafts. Finally, synthetic grafts have similar mechanical properties and dimensions with abundant availability compared to the other three earlier described graft types. Artificial tendon grafts clinically available include carbon (Johnson and Johnson, New Brunswick, NJ, USA), carbon and polyester (Surgicraft, Nairobi, Kenya), Dacron (Stryker-Meadox), Leeds-Keio polyester (Neoligaments, Yeadon, UK), and Gore-Tex polyctetrafluoroethylene)

(WL Gore, Hong Kong, China) [88]. The usage of these synthetic grafts has declined over the years due to many problems. Tissue reaction from debris particulates may cause pyogenic flexor tenosynovitis (PFT), characterized by inflammation of the fluid-filled sheath, after 4–5 years [89,90]. In addition, synthetic grafts often have a high failure rate and poorer outcomes. Finally, they also significantly increase the cost of the operation; artificial ligaments are costly [91].

4. New Strategies for the Repair of Flexor Tendon Injuries

The traditional strategies for flexor tendon repair can lead to serious complications, as mentioned above, despite the positive outcomes in the short term. The need of solving the shortcomings of traditional strategies for flexor tendon repair techniques has prompted the research of alternatives such as construct designs from polymeric materials that wrap around the injured tendon, not to be confused with synthetic grafts which replace the injured tendon.

4.1. Biochemical Solutions for Postoperative Complications

Research has shown that several factors affect flexor tendon healing and cell adhesion formation due to the invasion of external fibroblasts. The formation of post-surgical scar tissue and cell adhesion between surrounding tissue and tendon constricts tendon gliding and motion, causing a loss in functionality. In some situations, although they are rare, infections occur after flexor tendon repair [49]. These postoperative complications are still major clinical challenges to overcome.

4.1.1. Peritendinous Adhesion Formation

Adhesion formation can be minimized or even prevented by optimizing the intrinsic healing mechanism. Past researchers believed that flexor tendon healing was strongly dependent on extrinsic cellular ingrowth, which relies on adhesion formation at the site of injury. However, it was documented that flexor tendons should have the ability to heal by intrinsic healing mechanisms alone [48]. Intrinsic healing can be optimized by using biochemical factors to achieve scarless healing. Current intrinsic healing optimization methods include physical and mechanical barriers to prevent adhesion formation as well as chemical and molecular compound addition against scar tissue formation [92]. Ideally, physical barriers are combined with chemical/biological modulation to produce a superior biomaterial construct to prevent peritendinous adhesion.

Physical barriers form the first method for the prevention of peritendinous adhesion formation. Placing an anti-adhesive material, acting as a barrier, between the healing site and the surrounding tissue limits the contact between the tendon injury site and its sheath, diminishing the amount of surface available for adhesion formation. Hereby, the tendon is restricted to intrinsic healing, healing only to itself and not to surrounding tissue or the tendon sheath [14].

Other researchers have made use of the therapeutic properties of biological polymers in combination with synthetic polymers to form semi-synthetic materials as an attempt to minimize the formation of scar tissue and peritendinous adhesion. The ideal chemical compound should be limited to a single application, have no systemic side effects, and should target immediate inflammatory responses and the extrinsic healing mechanism. A decrease in inflammation is accompanied by a decrease of invasive external cells to the injured site of the tendon, minimizing the scar tissue and peritendinous adhesion. The anti-inflammation character of these chemical agents is thus responsible for the reduction of adhesion formation. Chemical compounds used in recent research include nonsteroidal anti-inflammatory drugs (NSAID's) like naproxen [93], ibuprofen [94–98], 5-fluorouracil [99], and celecoxib [100–102], steroids like dexamethasone [103] and other polymers such as hyaluronic acid (HA) [93,100,102,104–108] and chitosan [19,109] (Figure 3, see Section 4.3). The common function of these chemical agents is the reduction of the inflammatory response and consecutively the reduction of adhesion.

Another key aspect in the prevention of peritendinous adhesion after an operation is the early mobilization of the injured finger, as this will cause the tendon to glide, which will ensure adhesion formation is minimized. Historically, fingers were immobilized after flexor tendon repair using an external brace until the tendon healed sufficiently before allowing movement [110]. Hereby, adhesion formation could not be prevented leading to the decrease in functionality of the finger. New strategies of early passive postoperative mobilization have been developed in the late 1970s to promote tendon gliding. Early passive mobilization decreased adhesion formation and increased the tensile strength at the injured site [111], but might also cause deformation of the tendon [7]. Early active mobilization immediately after a repair has been recognized as an important treatment for enhancing flexor tendon healing [112]. Strong material-based constructs are needed for tolerating the forces of the early active movements, as the construct is solely responsible for the strength of the repair during the early days after operating [113]. However, this technique is also accompanied by an increase in ruptures of the load-bearing constructs due to the difficulty in controlling the loads from active movement [114,115].

4.1.2. Infections

Infection is most commonly caused by a significant degree of contamination during the initial injury [49]. Infection rates increase when the trauma was caused by for example maritime or agricultural activities. In addition, infections depend on the type of injury with increased infection rates for bite wounds, crush injuries, replantation, and injuries with accompanied fractures [116–118]. In 2018, a review paper [119] reported the most found microbial populations causing flexor tendon infections such as i.e., *Streptococcus pyogenes, Mycobacterium tuberculosis, Mycobacterium tuberculosis, Staphylococcus*, and many more with the latter being the most frequently isolated bacterium. Flexor tendon sheath infections can have a devastating effect leading to morbidity and even the loss of a finger. Therefore, it is important to address the presence of such contamination or early infection prior to the surgical procedure. Good results have been obtained using closed tendon sheath irrigation, antibiotics, and debridement [120]. However, patients with severe PFT are still at risk of morbidity. Leaving PFT untreated will result in rapid deterioration of the gliding mechanism and will cause adhesion formation. Fast observation and treatment of PFT are essential to prevent the disruption of finger and hand functionality. A physical examination is needed in order to identify Kanavel's four cardinal signs of infections as follows: (i) a finger held in slight flexion, (ii) fusiform swelling of the affected digit, (iii) tenderness along the flexor tendon sheath, and (iv) pain with passive extension of the digit [121,122]. Recent development indicates that it is possible to integrate an antimicrobial compound, such as silver nanoparticles (Ag NPs) [68,104,123–125], for example, into material-based constructs used for tendon repair in order to mitigate the infection risk.

4.2. Requirements of Polymeric Materials for Flexor Tendon Repair

Several polymeric (bio)materials have been explored as alternatives to the traditional repair strategies. Neighboring fields of cartilage and bone tissue engineering formed an inspiration for possible polymer-based constructs. In addition, some have been specifically developed to mimic the ECM and biomechanical structure of the flexor tendon. The ideal polymeric material should cover several requirements such as: (i) biodegradability; (ii) biocompatibility; (iii) processability and suitable structure architecture, and (iv) sufficient mechanical properties [42].

(i) Biodegradability

Polymeric constructs used for flexor tendon repair are not intended to permanently remain in the human body. Therefore, the construct must be preferably biodegradable. Biodegradation of such polymeric constructs releases by-products that should be non-toxic and able to be absorbed by, and ultimately exit, the body through metabolic pathways. These by-products also may not interfere with other organs in the body [12,42,126]. The construct cannot lose its mechanical properties during the biodegradation prior to the complete healing of the tendon. Generally, flexor tendon injuries can take 9 to even up to 12 weeks to heal (maturation takes even longer up to 12 months) [49], which must be matched with the biodegradation rate. The biodegradation process is greatly influenced by the external environment where it takes place. Polymer degradation is initiated when it comes into contact with surrounding fluids inside the body. This degradation process then leads to the formation of lower molecular weight polymers, oligomers, and eventually monomers by chain scission [127].

Chemical degradation via hydrolysis (which can be enzyme-catalyzed) is possible on every degradable polymer due to the presence of hydrolyzable bonds [128,129]. The molar mass and the degree of crystallinity are proven to have a significant effect on the degradation rate of polymers and are therefore important parameters to consider when designing a certain construct. An increase in molecular weight results in an increase of scissions needed to degrade the material [130]. Similar results can be observed for the degree of crystallinity where an increase resulted in a decrease in the degradation rate. This can be attributed to the amorphous sections of the polymer which will degrade first [131,132].

Research has observed that mechanical loads have a significant accelerating effect on the material biodegradation rate. Observations were made that the biodegradation rate of poly(D, L-lactic acid) (PDLLA) increased while applying a continuous tensile load in comparison to no applied tensile load. In addition, the combination of both tensile and compressive loads had an even further increased effect on the biodegradation rate [133,134]. The mechanical loading should always be considered when the biodegradation rate of a biodegradable polymer is regulated.

(ii) Biocompatibility

Biocompatibility is essential when polymeric materials are used for any biomedical application. This means that the construct should have an appropriate response to the host for flexor tendon repair [135,136]. Other definitions describe biocompatibility as non-immunogenic, non-toxic, non-thrombogenic, and non-carcinogenic. In most cases, a cytotoxicity test is conducted to determine the biocompatibility of a material. Hereby, the effect of toxic agents derived from the polymeric material on cell viability and cells can be determined [137]. A polymeric material is believed to be cytotoxic when cell viability is <70%, measured during in vitro cell seeding tests [138].

(iii) Processability and Structure Architecture

Processability is another important requirement to get a construct commercially and clinically viable. The processing technique used to create the construct should be easy to scale up and more importantly, be cost-efficient. Polymers can be processed into films/sheets or tubular, nanofibrous membranes by electrospinning or directly injected in vivo [139]. It is important that the construct enables nutrient, growth factor, and cytokine permeation during tendon healing, which is possible due to the porosity of the membrane. In addition, porosity will allow permeation of the degradation by-products through the membrane so they can be metabolized by the body, avoiding high concentrations of by-products (acidic when polyesters are used) leading to cytotoxicity [19].

In addition, sufficient porosity must be achieved to be able to promote the formation of blood vessels. Finally, the construct should also resemble the native ECM as much as possible to avoid rejection of the body, which is an advantage of using nanofibrous membranes.

(iv) Mechanical Properties

When developing a construct, utilized for flexor tendon repair and healing, having good mechanical properties is an essential factor to take into consideration. Identification of the mechanical strength can be achieved by measuring the impact resistance of the final construct to maintain its integrity during implantation [140]. Tensile tests include the most common mechanical tests to evaluate a construct. During tissue remodeling, each application requires a different working range for mechanical properties as it is desirable that the construction resembles the mechanical properties of the native organs or tissue. For example, the flexor tendon has ultimate tensile strength (UTS) values between 2.98–3.98 MPa, in combination with an elongation at a break between 10–12% (Figure 2) [141], which is needed for immediate active mobilization after repair, during the healing process [142,143]. Several research projects, focusing on polymer-based constructs for flexor tendon repair, have tried to achieve these optimum mechanical properties but often did not succeed, Chen et al. [98] for example achieved a UTS of 1.43 ± 0.13 MPa. However, last year (2021) Pien et al. [93] managed to achieve the required mechanical strength due to the innovative solution of working with a multi-layer construct, in which the middle layer of polyethylene acts as a Chinese finger trap, increasing the mechanical strength. The biomechanical stability also depends on other factors such as degradation rate, absorption at the interface, and elasticity, which should be considered.

4.3. *Materials for Flexor Tendon Scaffold and Construct Designs*

Polymeric materials used for flexor tendon repair can either be biological, synthetic (Figure 4), or semi-synthetic. Biological polymers have gained interest because of the therapeutic properties of the construct itself. Chitosan for example has both an anti-inflammatory as well as an antimicrobial response [19]. Problems such as toxicity and chronic immunological reactions, sometimes occurring with synthetic polymers, are frequently avoided using these biological materials [18,144]. However, synthetic polymers are widely understood in the field of biomedical applications, mainly for their mechanical properties and ease of processing into porous structures [145], compared to biological materials. Biological polymers do not have this wide variety of possible processing techniques, but their ability to mimic the native ECM and often better biocompatibility compared to synthetic polymers makes them very interesting [146]. Despite these advantages, biological polymers often have inferior mechanical properties [15] and (too) short biodegradation times compared to synthetic polymers. Finally, other major drawbacks of biological polymers are their high polydispersity, limited purity, and batch-to-batch molar mass variations [147] causing varying characteristics which is unacceptable from a biomedical point of view. Both polymer types have the ability to be functionalized by chemical and biological compounds, controlling their chemical, biological and physical properties [148].

Evaluation of recent literature shows that a combination of both polymer types can form a solution for current clinical complications in flexor tendon repair such as adhesion formation, lack of mechanical properties, and risk of infection. These are the so-called semi-synthetic materials, formed by the combination of biological and synthetic polymers using copolymerization, grafting, crosslinking, or blending. However, the incorporation of biological polymers in synthetic materials is also done in the literature. These are still classified as synthetic materials in this review paper as the biological compounds are solely released as drugs and are not added to strongly influence the properties of the construct (i.e., mechanical properties, biodegradation rate, etc.). In the case of semi-synthetic materials, the combination of both polymer types results in a superior polymer-based construct combining the desirable mechanical strength of synthetic polymers with the therapeutic and biocompatible properties of biological polymers [149], while maintaining a desired degradation rate.

This section is divided into three subsections based on the nature of the polymers. Each section also has an overview table of past research arranged according to the year of acceptance. This makes it possible to assess the evolution of these polymeric materials and to evaluate the advances in state-of-the-art development of flexor tendon repair using material-based constructs.

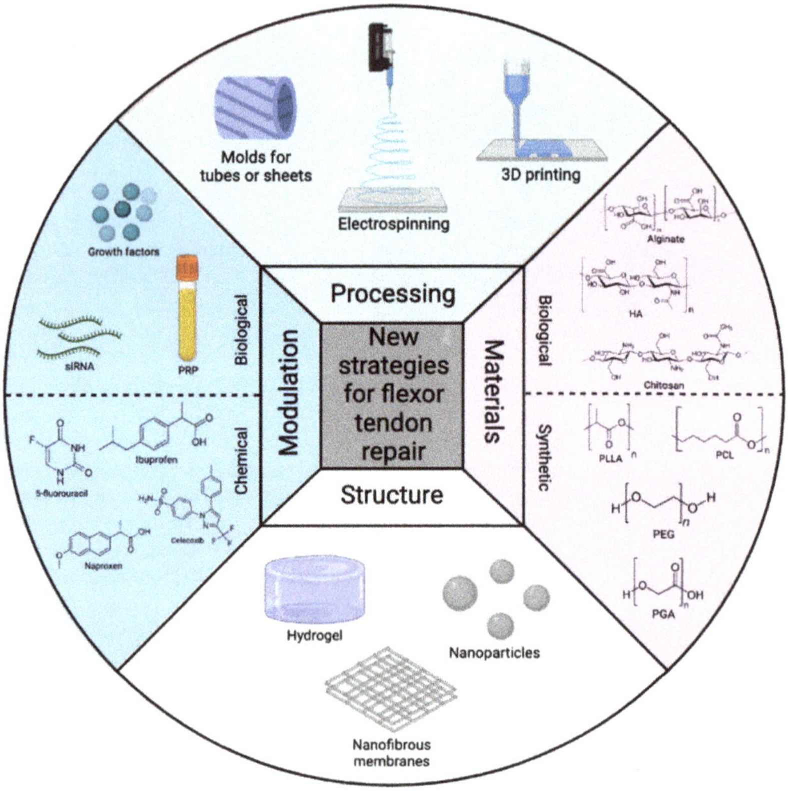

Figure 4. Overview of the most commonly used processing techniques, polymeric materials, structures, and modulations for flexor tendon repair in this review paper.

4.3.1. Biological Polymer Constructs

Polymeric constructs of purely biological origin have been studied extensively. In the case of flexor tendon repair applications (Table 1), there is a stronger tendency to use protein-based materials, such as collagen, as it is predominately present in the native tendon tissue. Collagen was therefore one of the first biological polymers to be used for tendon recovery [147]. Silks can also be defined as protein polymers, consisting of fibroin and sericin [150]. These polymers are mostly derived from animals, although they are often in need of crosslinking in order to improve the mechanical strength, remove foreign antigens and possible diseases and decrease the degradation rate [151]. This is necessary, since a construct may not biodegrade or lose mechanical properties prior to 9–12 weeks of healing [49]. The increased use of these fibrous proteins can be attributed to their highly repetitive primary sequence, leading to an improved homogeneity in the secondary structure, i.e., β-sheets for most silks and triple helices in the case of collagen. Because of these repetitive structures, this family of protein polymers provides better mechanical properties compared to other biological materials, in combination with their biocompatibility and ability to tailor sequence by genetic control provides a basis to exploit these biological polymers for flexor tendon repair construct design [152].

Between 2013 and 2015, as seen in Table 1, research [153,154] was published where collagen membranes were derived from animals followed by sterilization, freeze-drying, and crosslinking to wrap around the injured tendon site after suturing. An increase in maximal tensile loading was observed compared to the traditional suturing technique. In addition, the membrane acted as a physical and mechanical barrier to significantly prevent adhesion formation. In 2015, collagen was crosslinked with glycosaminoglycan (GAG) into a resorbable matrix, which further reduced the adhesion formation. Despite these promising results, it was proven in 2016 by Wichelhaus et al. [155] that the use of collagen, in combination with elastin, cannot be advised for biological constructs as it enhances both extracellular and cellular inflammation. In addition, increased tendon gapping was observed as a negative result. These mixed results, in combination with a fast in vivo degradation directly accompanied by loss of mechanical strength [151], are believed to be the reason why the use of pure collagen was never researched further as a suitable primary material for tendon repair. However, a few more recent studies [156–159] reported the use of amniotic membranes in combination with protein-based polymers like collagen and silk for construct design, suppressing inflammation, and neovascularization.

The use of polysaccharides has recently increased for tendon repair, explained by the fact that these biological polymers often have therapeutic properties, actively helping the healing process. HA can be used for tendon repair with an additional anti-inflammatory effect by suppressing pro-inflammatory cytokines and chemokines and creating improved cellular proliferation response at the injury site, preventing adhesion formation as described in Table 1. It is the largest component of GAGs and is known to be non-immunogenic [160]. Nevertheless, HA has a very short in vivo degradation time of around 4 days by hyaluronidase [161]. Therefore, several chemical modifications and crosslinking options have been explored to decrease the degradation rate and improve its processability. Past research from Yilmaz et al. [162] and Isık et al. [163] described a platform where a HA membrane, combined with carboxymethylcellulose (CMC), was created in order to achieve a slow degradation in the body. CMC acts as a chemical carrier, used in various drug and food preparations [157]. Despite these promising results, the biodegradation time of HA is still too short for it to be a primary material for construct design. Embedding HA in a mildly cross-linked alginate hydrogel [164] increased the biodegradation time of HA from 4 to 11 days in vivo, which for the current application is still too short. Alginate is commonly derived from brown seaweed or bacteria and shows low toxicity and excellent biocompatibility [165]. In 2006, the use of a high concentration alginate solution was tested as a coating over the injured site [166]. $CaCl_2$ was purposely not used as a hard gel would form, which would no longer be injectable. Animal studies on rabbits were performed, which were treated with the alginate solution, whereafter the adhesion formation was histologically and biomechanically tested. A higher flexion region and decrease in scar tissue formation were observed for rabbits treated with the alginate solution.

Another polymer in the list of polysaccharides with therapeutic properties is chitosan. These properties include cell proliferation, fibroblast growth inhibition, anti-microbial response, and induced collagen production [167–169]. Chitosan, which is cationic and commonly derived from shrimp shells [170], is often combined with alginate or HA due to their opposite charges to form poly-ionic complexes. Chitosan can be produced into an injectable gel that is capable of reducing adhesion formation [169]. The effect of chitosan on the sirtuin (SIRT)1 protein expression, responsible for the inhibition of inflammation in human tenocytes [171], was also investigated and proved to be elevated when chitosan was injected. In addition, a higher maximum tensile loading was observed when the flexor tendons were treated with chitosan injections. The use of a thermo-responsive in-situ forming hydrogel, based on poly(N-isopropylacrylamide) containing chitosan and HA, was also reported in 2017 [172]. A sol-gel transition of the injectable hydrogel was observed at a lower critical solution temperature (LCST) of 31.4 °C. One year later, Yousefi et al. [173] incorporated zinc oxide nanoparticles into a novel tubular construct made from a chitosan solution in a mold after freeze-drying (see Figure 5). Literature proved that zinc has an antimicrobial response and is an essential micronutrient for metabolism, catalyzing more than one hundred enzymes [174–176].

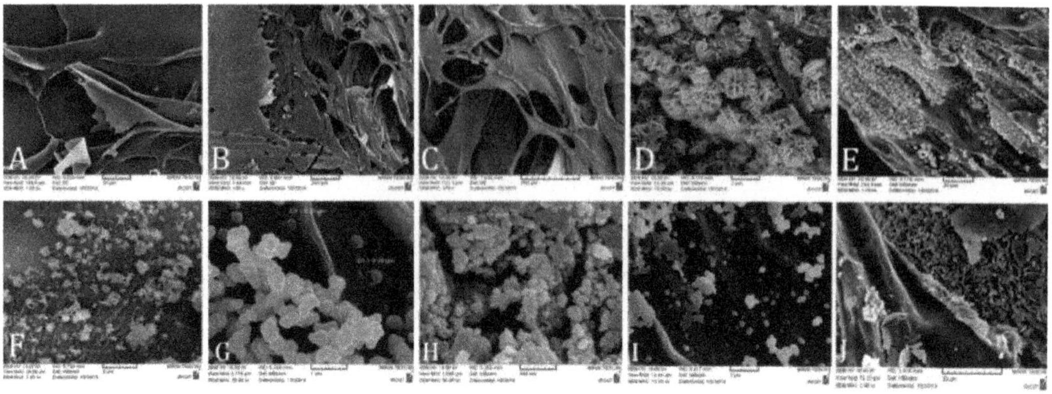

Figure 5. SEM ultra-micrographs of a tubular construct made from chitosan. (**A–C**): pure chitosan, (**D–F**): ZnO coated chitosan and (**G–J**): ZnO nanoparticles coated chitosan. Reprinted with permission from ref. [173]. 2018 A. Yousefi.

The degradation time of biological polymers is strongly dependent on their molar mass, which is no exception for chitosan. Commonly, chitosan shows fast partial degradation within the first 10 days whereafter it slows down and is fully degraded after 1–2 months [177]. For this reason, similar to HA, it is not possible to use chitosan as the main component of a polymeric construct for tendon repair.

These polysaccharides are often processed into hydrogels sheets or membranes (Figure 4) using various techniques such as crosslinking, grafting, ionic interactions, hydrogen bonding, etc. The use of hydrogels also allows for easy injection at the injured site to act as a barrier against adhesion formation combined with an increased residence stability time of the therapeutic agent in the body [178]. However, a major drawback of hydrogels for flexor tendon repair is their ability to absorb large amounts of water, causing the construct to strongly swell which puts pressure on surrounding tissue and decreases mechanical properties, therefore decreasing the healing capability [179]. For this reason, polysaccharides like HA, alginate, and chitosan have found more applications in wound healing and as drug delivery systems when loading as an active compound inside another polymer material due to their proven anti-inflammatory effect [14,19,93,106,180]. In addition, polysaccharides are commonly combined with synthetic polymers into semi-synthetic polymers.

4.3.2. Synthetic Polymer Constructs

In comparison to the evolution of platforms using biological polymers for flexor tendon construct design, the same trend, being that initially several constructs were evaluated made from one pure polymer, was observed already earlier in the use of synthetic polymers as biomedical construct designs as shown in Table 2. Each of these individual polymers has certain shortcomings whereafter it was observed that the mechanical properties, in addition to the degradation kinetics and their interactions with the surrounding tissue, could be tailored by blending or copolymerization [126].

Table 1. Compilation of publications showing the evolution of biological polymer-based constructs intended to treat flexor tendon injuries.

Year	Material and Structure	Material Processing Technique	Mechanical Response [1]	Biological Response [1]	Ref.
1999	Membrane of HA	Crosslinking	-	Decrease in adhesion formation	[163]
2006	Coating of ALG solution	Sodium ALG derived from Lassonia nigrescens, non-crosslinked and sterilized	Higher flexion region	Decrease in scar tissue formation Tendon healing with longitudinal remodeling compared to random remodeling in control	[166]
2010	Membrane of HA	-	Increased UTS	Decrease in adhesion formation	[162]
2012	Hydrogel of mildly crosslinked ALG embedded with HA (HA/mcALG)	Crosslinking	Sustained release of HA	Slower degradation rate of HA/mcALG compared to the pure HA hydrogel Decrease in adhesion formation for both HA as well as HA/mcALG constructs	[164]
2013	Membrane of collagen	Bovine derived, freeze-dried, and sterilized	Increased UTS after 8 weeks	Decrease in adhesion formation	[153]
2015	Membrane of collagen	Porcine derived, sterilized, and rehydrated	-	Decrease in adhesion formation for both the collagen membrane and collagen-GAG matrix sheet	[154]
	Porous matrix sheet of Collagen-GAG	Crosslinked and rehydrated			
2015	Hydrogel of CHI	-	Increased UTS	Decrease in adhesion formation CHI-induced SIRT1 protein expression	[169]
2016	Membrane of collagen/elastin	Matriderm®	Increased gapping	Decrease in adhesion formation	[155]
2016	Amnionic membrane combined with silk	Sericin removed by solution, soaked in 1% collagen/HA, and dried, freeze-dried, and sterilized. Amnionic membranes harvested from human placenta	Increased UTS	No immigration of inflammatory cells or fibroblast-like cells Formation of new blood vessels	[156]
2017	Thermoresponsive hydrogel of HA-CHI-PNIPAm	Copolymer by grafting CHI on PNIPAm whereafter dissolved in HA solution	Sol-gel in-situ transition at an LCST of 31.4 °C	Decrease in adhesion formation Decrease in fibroblast migration Limited cytotoxicity	[172]
2018	Tubular construct of CHI with zinc oxide nanoparticles	Mold with CHI solution containing ZnO NP followed by freeze-drying	Complete biodegradation in 8 weeks	Decrease in adhesion Improved gliding Improved collagen synthesis due to Zn	[173]
2021	Hydrogel of HA embedded with rhynchophylline	Crosslinking	Sustained release of rhynchophylline Increased healing strength	Decrease in adhesion formation Increased expression of type I and III collagen Increased gliding excursion	[178]

[1] Compared to traditional suture technique control. Abbreviations: Polymers: hyaluronic acid (HA), alginate (ALG), glycosaminoglycan (GAG), chitosan (CHI), poly(N-isopropylacrylamide) (PNIPAm). Other: ultimate tensile strength (UTS), sirtuin 1 (SIRT1), lower critical solution temperature (LCST) zinc oxide nanoparticles (ZnO NP).

Aliphatic polyesters including, poly(ε-caprolactone) (PCL), poly(L-lactic acid) (PLLA), and poly(glycolic acid) (PGA) (Figure 4) have been attractive for flexor tendon repair constructs due to their biodegradability by mostly non-enzymatic hydrolysis of the ester backbone into acidic by-products. Notice that these could cause an immunological reaction with the surrounding tissue at the repair site [12]. Synthesis of these aliphatic polyesters is commonly performed by ring-opening polymerization starting from the cyclic dimers [181]. These polyesters are commonly processed into fibers by i.e., electrospinning [97,105,108], melt electrowriting [182,183], or 3D printing [184,185] (Figure 4) to mimic the native ECM of the flexor tendon and are among the few synthetic polymers which have been approved by the FDA in several clinical applications since the 1970s [186].

Up until 2013, mostly biological polymers were used for flexor tendon repair constructs. A shift to more synthetic polymer-based constructs can be observed in the last decade and is described in Table 2. Materials like PLLA and PCL were among the first to be processed into nanofibrous membranes by electrospinning due to their relatively high toughness and mechanical strength [131]. In addition, PLLA and PCL are hydrophobic, meaning the swelling of the construct is minimal. The first paper describing the use of the synthetic polymer PLLA in construct design for flexor tendon repair was published in 2013 [187]. In this research, dextran glassy nanoparticles (DGNs) were used as a carrier for loading bFGFs. Hereafter, the loaded DGNs were blended with PLLA, and a fibrous membrane was produced by electrospinning, capable of sustainably releasing bFGFs. They highlighted that the use of the PLLA membrane resulted in enhanced cell proliferation and intrinsic tendon healing.

In the same year, the first papers were published that loaded ibuprofen into a nanofibrous membrane constructs based on synthetic polymers [94–96]. The membranes achieved improved biological outcomes (Table 2) compared to similar research due to the use of ibuprofen, which is a non-steroidal anti-inflammatory drug capable of reducing pain and improving the healing process. Additionally, ibuprofen is proven to have an antipyretic response in vivo. As opposed to its beneficial properties, higher concentrations of ibuprofen can lead to increased cytotoxicity, as was proven by Shalumon et al. [123]. From 2015 to 2021, four papers have been published capable of solving the risk of a cytotoxic ibuprofen burst release, of which two contained synthetic materials [97,188] (Table 2) and two semi-synthetic materials [98,123] (see further Section 4.3.3).

PLLA has a relatively long degradation period of approximately 6 months depending on the environment, accompanied by the acidic biodegradation by-products, which may be responsible for a prolonged risk of inflammation. The degradation period can be shortened by copolymerization with poly(ethylene glycol) (PEG) [95]. It is important to notice that PEG is hydrophilic, increasing the overall hydrophilicity of the construct, making it more prone to swelling. Jiang et al. [101] believed in 2014 that the peritendinous adhesion formation may be due to the proliferation of fibroblasts and excessive collagen synthesis. Therefore, they looked more into the Extracellular Signal-Regulated Kinases 1 and 2 (ERK1/2) and Small Mothers Against Decapentaplegic 2 and 3 (SMAD2/3) pathways responsible for this. A hypothesis was made that the proliferation of fibroblasts and excessive collagen synthesis could be inhibited by downregulating ERK1/2 and SMAD2/3 phosphorylation of exogenous fibroblasts by using celecoxib. Therefore, celecoxib was incorporated into a PELA fibrous membrane, processed by the commonly used electrospinning technique. Identical to the use of ibuprofen, an excellent continuous celecoxib release pattern was observed. Their hypothesis was confirmed by the reduced type I and type III collagen expression. In 2021, celecoxib was also used in a PCL electrospun nanofibrous membrane for the treatment of failed back surgery syndrome with similar results of anti-inflammatory response [189].

Regarding the evaluation of the complications occurring at flexor tendon operative measures, in the beginning, only adhesion formation was researched. Solutions for infections by microbials were at that time not yet incorporated into the constructs. As mentioned in Section 4.1.2, infections are caused by bacteria entering the repair site during operations or healing. In the literature, Ag NPs are commonly used as clinically approved material for inducing an antimicrobial response in the body [68,94,104,108,123–125,190]. They are highly effective against pathogens responsible for an infection such as *S. aureus* and *E. coli*, with the absence of inducing drug resistance. Their ability to inactivate enzymes and proteins located in the cell membrane blocks cellular functionally. In addition, internal wound and skin healing, often accompanied by tendon injuries, is promoted by reducing the activity of matrix metalloproteinases. Finally, Ag NPs are responsible for the decrease in K^+ ions, which results in the damaging of bacterial cell membranes and ultimately cellular death [191,192]. However, there still are some concerns about the use of Ag NPs, which may lead to liver damage and kidney failure. The group from Lui et al. [94] combined both the anti-inflammatory effect of ibuprofen and the antimicrobial effect of the Ag NPs into a PLLA fibrous membrane construct. The construct was capable of reducing both, most impacting, complications occurring during flexor tendon healing. However, the antimicrobial effect was only tested in vitro, therefore further in vivo tests are still required.

In 2015, a paper was published that described the use of PCL instead of PLLA as a construction material [193]. PCL, a semi-crystalline polymer, also widely used in biomedical applications, is compatible with both soft and hard tissue applications [194]. It is always present in its rubbery state due to the low glass-transition temperature of ~ -60 °C, with a degradation time of up to 4 years. Therefore, constructs produced solely from PCL have limited applications. PCL, similar to PLLA, has a very slow in vivo degradation rate, much lower compared to other synthetic polyesters, and is thus commonly combined with PGA or PEG to increase the degradation rate and decrease its brittle behavior. Therefore, a blend of PCL with different weight fractions of PEG was processed into a monolayer nanofibrous membrane using the same techniques mentioned above, where altering the PEG wt% influenced several mechanical and biological properties [193]. The most important observations made were that an increase in PEG wt% resulted in a lower UTS, faster biodegradation, and a reduced adhesion formation. Therefore, it is important to compare these properties when opting to choose possible blending combinations. Another study from 2020 combined the advantages of a PCL nanofibrous membrane with the slow release of bFGFs inside an amnionic membrane into a multi-layered construct [195]. The adhesion and proliferation of tenocytes and fibroblasts and collagen synthesis were enhanced by up-regulating the phosphorylation of ERK1/2 and SMAD2/3, which was contradictory to previous research [101]. In addition, the multi-layer nanofibrous membrane effectively isolated the exogenous adhesion tissue and promoted endogenous tendon healing.

Synthetic polymers, used for flexor tendon repair, are not always processed by electrospinning. Wu et al. [103] and Yuan et al. [99] developed and processed a block copolymer of PCL and PEG loaded with dexamethasone micelles and a block copolymer of PLGA and PEG loaded with 5-fluorouracil respectively into thermo-responsive injectable hydrogels. These hydrogels showed promising results in reducing the adhesion formation through a controlled release of dexamethasone as anti-inflammatory glucocorticoids [196] and 5-fluorouracil as a non-steroidal anti-inflammatory drug. Despite the promising results of utilizing hydrogels for flexor tendon repair, they commonly biodegrade in less than 28 days, which is too short for a complete recovery of the flexor tendon. Additionally, as mentioned before, they are able of absorbing large amounts of water, which causes swelling of the construct. Hydrogels are mostly injected at the healing site that has been repaired by a traditional technique such as suture, to achieve a more superior and controlled healing process.

Table 2. Compilation of publications showing the evolution of synthetic polymer-based constructs intended to treat flexor tendon injuries.

Year	Material and Structure	Chemical/Biological Modulation and Concentration	Material Processing Technique	Mechanical Response [1]	Biological Response [1]	Ref.
2013	NFM of PLLA	bFGFs loaded in DGNs	Electrospinning	Blending of DGNs decreased UTS and maximum elongation	Improved cell proliferation Improved intrinsic tendon healing	[187]
2013	NFM of PLLA-b-PELA	IBU	Electrospinning	-	Idem [187] Decrease in inflammation and adhesion formation	[95]
2013	NFM of PLLA–MMS	IBU	Electrospinning	Controlled release of IBU, without initial burst	Idem [95]	[96]
2014	NFM of PLLA-c-PELA	Celecoxib	Electrospinning	Controlled release of celecoxib	Idem [95] Decreased expression of type I and type III collagen	[101]
2014	NFM of PLLA	IBU + Ag NPs	Electrospinning	Controlled release of IBU, without initial burst	Idem [95] Antimicrobial response	[94]
2015	NFM of PCL/PEG blend (0, 25, 50 and 75 wt% PEG)	-	Electrospinning	Increasing wt% PEG decreased UTS and maximum elongation	Decrease in adhesion formation Higher PEG wt% led to decreased fibroblast attachment Good permeability for nutrients, growth factors, and cytokines	[193]
2015	Hydrogel of PLGA-PEG-PLGA	5-Fluorouracil	-	Sol-gel phase transition depending on temperature	Decrease in adhesion formation Full in vivo degradation in 28 days	[99]
2015	Hydrogel of PEG-b-PLC-b-PEG	DEX micelles	PEG-b-PLC-b-PEG dissolved in saline at 4 °C	Sol-gel phase transition depending on temperature	Decrease in adhesion formation Low cytotoxicity of hydrogel and micelles Full in vivo degradation in 20 days	[103]
2017	NFM of PLGA	IBU	Electrospinning of PCL Amniotic membrane was freeze-dried and decellularized	Increased UTS Sustainable release of IBU	Decrease in adhesion formation Inhibition of fibrosis via the COX2 pathway Reduced pain and neurological deficits	[188]
2020	MNFM of an amniotic membrane between two layers of PCL	bFGFs	Electrospinning	Increased work of flexion Decreased UTS	Increased phosphorylation of ERK1/2 and SMAD2/3 Enhanced collagen synthesis	[195]
2021	MNFM of PCL/AUP outer layer and inner layer, with braided monofilament PE as middle layer	Naproxen and HA loaded in PCL/AUP	Electrospinning	Significant increase in UTS No degradation prior to 9 weeks	Decrease in adhesion formation naproxen introduced no cytotoxicity	[93,105]

[1] Compared to traditional suture technique control. Abbreviations: Structures: nanofibrous membrane (NFM), multi-layer nanofibrous membrane (MNFM). Polymers: poly(L-lactic acid) (PLLA), poly(L-lactic acid)-co-poly(ethylene glycol) (PELA), poly(ε-caprolactone) (PCL), poly(ethylene glycol) (PEG), poly(lactic-co-glycolic acid (PLGA). Modulations: basic fibroblast growth factors (bFGFs), ibuprofen (IBU), silver nanoparticles (Ag NPs), dexamethasone (DEX). Other: dextran glassy nanoparticles (DGNs), modified mesoporous silica (MMS), ultimate tensile strength (UTS), cyclooxygenase-2 (COX2), acrylate endcapped urethane-based precursor (AUP).

Pien et al. [93,105] published the most recent research papers (2021) on processing synthetic polymer-based materials for flexor tendon repair into a multi-layered tubular construct (Figure 6) instead of sheath membranes. A novel acrylate endcapped urethane-based precursor (AUP) with a PCL backbone was used in combination with HA as anti-adhesion and naproxen as an anti-inflammatory compound respectively to create an outer and inner layer, by electrospinning. The construct is classified as synthetic because HA was added for the sole purpose of a drug and did not strongly influence the mechanical properties of the construct. Naproxen has a similar mechanism in the body compared to ibuprofen by reducing inflammation and pain due to reducing the responsible hormones. However, naproxen has a longer-lasting effect as proven by literature [197–199]. Often, the mechanical properties of the construct did not suffice for optimal healing of the tendon by active mobilization. Therefore, the middle layer was composed of a polypropylene braided structure, based on the mechanism of a Chinese finger trap, acting as a mechanical support and significantly enhancing the mechanical properties measured by uniaxial tensile tests.

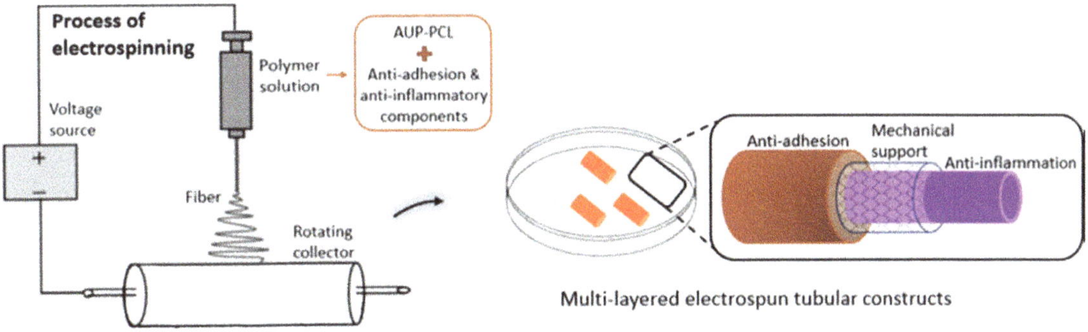

Figure 6. A Multi-layered tubular construct from a novel material, acrylate endcapped urethane-based precursor (AUP) with a PCL backbone as the outer and the inner layer, in combination with HA as anti-adhesion and naproxen as an anti-inflammatory. The polypropylene braided structure in the middle acts as mechanical support, based on the Chinese finger trap mechanism. Reprinted with permission from ref. [89]. 2021 N. Pien.

Additionally, the construct had been functionalized with methyl acrylate groups, which gives the AUP the possibility to form strong covalent bonds due to UV-crosslinking after electrospinning, therefore creating extra interactions between the inner and surrounding layers. In a further study [105], the tubular construct was evaluated using a rabbit model, proving its anti-adhesion response and sufficient mechanical properties for tendon repair and healing. A next step could still be to include antimicrobial compounds to counter infections. In recent years, apart from the constructs mentioned above, there has been a transition from synthetic, more towards semi-synthetic polymer-based constructs as explained in Section 4.3.3.

4.3.3. Semi-Synthetic Polymer Constructs

The use of semi-synthetic materials, combining biological and synthetic polymers, to design a polymer-based construct in flexor tendon repair has become more popular. A clear transition from synthetic to semi-synthetic polymer-based constructs is observed in the literature, starting from 2014 shown in Table 3. This can be explained by the fact that the therapeutic benefits of biological polymers were further researched around this time.

A study is mentioned in Section 4.3.2 details the preparation of a synthetic nanofibrous membrane made from electrospun ibuprofen-loaded poly(l-lactic acid)-poly(ethylene glycol) diblock copolymer. Indeed, literature expects that membranes made from polymer blends or copolymers enhance the anti-adhesion response compared to homopolymer membranes. However, the production of such copolymers includes complex steps. Additionally, extreme care is required in controlling the electrospinning conditions. An alternative for blends or copolymerization is to produce a surface-grafted membrane. As mentioned in Section 4.3, the synthesis of copolymers by grafting is one of the techniques for producing semi-synthetic materials. The grafting takes place after electrospinning, which allows for more flexibility in choosing the grafting material and in the electrospinning conditions. Often polysaccharides such as chitosan and HA, due to their therapeutic properties, are surface-grafted onto a polyester backbone [19,107,180]. Surface grafting of the anti-inflammatory polymers could provide a sustained release over several weeks during biodegradation. In 2014, Chen et al. [107] was the first in the literature to report the use of surface grafting chitosan to an electrospun PCL membrane using plasma. Surface grafting of chitosan on PCL had no effect on the fiber diameter, morphology, microstructure, and permeability, while at the same time improving its mechanical properties and reducing fibroblast attachment. Literature also reports the surface grafting of another polysaccharide such as HA onto PCL, which proved to have the same results compared to the chitosan grafting without the antimicrobial effect [107]. An increase in UTS was observed when HA was grafted on PCL which was a promising result as no trade-off had to be considered between the anti-inflammatory response and the UTS compared to other research in the same year [106] where a core-shell nanofibrous membrane of HA/PCL was produced. Hereby, the conclusion can be made that surface grafting of the chitosan and HA to PCL allows for a more sustainable and accurate anti-inflammation response at the injured flexor tendon site. Both membranes showed a significant decrease in peritendinous adhesion formation. In 2018, a paper was published describing the same plasma technique for surface grafting of HA, however, poly(lactide-co-ε-caprolactone) (PLCL) was used as the backbone instead of pure PCL [180]. The biodegradation period was slightly decreased due to the copolymerization, although the overall anti-inflammatory response remained similar.

In the last decade, another platform to produce a dual functional semi-synthetic polymer-based construct has been investigated. The platform describes the use of core-shell nanofibers by co-axial electrospinning. Several papers, starting from 2012 up until 2021 have been published using this core-shell co-axial electrospinning technique [100,104,106,108,123,200]. The core commonly consists of a therapeutic agent such as HA and the shell is made from polyester such as PCL, PLA, PELA, etc. Such a construction provides both the strong mechanical properties of the polyester and the slow release of HA for an anti-inflammation effect and improved tendon gliding. However, this processing technique requires much more complex electrospinning conditions compared to the previous surface-grafting technique. In addition, the HA core material is responsible for a decreased UTS compared to the pure polyester fibers. A significant advantage of the co-axial electrospinning technique is the possibility to load a different chemical/biological compound in the outer shell, compared to the inner core. Several papers report embedding Ag NPs into the outer shell of the fiber, therefore considering tackling the risk of infection during the operation or healing period [104,108,123]. The Ag NPs showed no cytotoxicity for all constructs. Another study, published in 2016, used the same platform and embedded celecoxib into the core which was responsible for a decreased expression of type I and type III collagen and cell proliferation [100]. A paper published in 2021 [108] investigated the difference between a thick and thin shell. The release rate of HA from the core could be controlled by varying the size of the shell to achieve the optimal anti-adhesion effect. The thin shell fibers demonstrated the optimal outcomes with limited cytotoxicity of the Ag NPs. The same year, another interesting state-of-the-art study [200] using core-shell fibers investigated the influence of the combined effect of fiber orientation and platelet-rich plasma (PRP) on tendon healing. The core of the fibers was loaded with PRP, which is a blood plasma fraction with platelet-rich cellular components and plays a key role during tendon healing and regeneration [201,202]. The sustained release of growth factors from PRP provided a biochemical stimulus for tendon healing. Another study described the function of aligned fibers (mimicking the physiological environment of the tendon). This alignment of the fibers improved cell proliferation compared to randomly oriented fibers.

Table 3. Compilation of publications showing the evolution of semi-synthetic polymer-based constructs intended to treat flexor tendon injuries.

Year	Material and Structure	Chemical/Biological Modulation and Concentration	Material Processing Technique	Mechanical Response [1]	Biological Response [1]	Ref.
2012	CSNFM of PCL shell with a HA/PCL core	HA	Sequential and microgel electrospinning	Increasing wt% HA decreased UTS	Decrease in adhesion formation Increased wt% HA resulted in higher cell viability and cell proliferation after 1 day in culture	[106]
2014	NFM of PCL-sg-CHI	CHI	Electrospinning	Increased UTS for healed flexor tendons treated with the PCL-g-CHI membrane compared to Seprafilm and PCL membrane	Decrease in adhesion formation CHI showed no cytotoxicity	[19]
2014	NFM PCL-sg-HA	HA	Electrospinning	Increased UTS, maximum elongation, and Young's modulus for PCL-g-HA compared to PCL	Decrease in adhesion formation Unaffected cell proliferation	[107]
2015	CSNFM of PCL shell with a HA core	Ag NPs + HA	Co-axial electrospinning	Decreased pull-out force for HA/PCL+Ag NPs Increased UTS for healed tendons treated with HA/PCL+Ag NPs compared to PCL membrane	Decrease in adhesion formation Ag NPs as an antimicrobial effect without significant cytotoxicity	[104]
2015	MNFM of PELA outer layer, HA middle layer, and PELA inner layer	HA + Celecoxib	Sequential electrospinning	Lower work of flexion	Decrease in adhesion formation Good permeability for nutrients, growth factors, and cytokines	[102]
2016	NFM of PLLA blended with collagen/CHI hydrogel, coated with ALG	CHI + ALG	Microgel electrospinning and solution coating	-	Decrease in adhesion formation Promotion of tendon gliding	[109]
2016	CSNFM with celecoxib loaded PELA shell with a HA/PELA core	Celecoxib + HA	Microgel and sequential electrospinning	No decrease in mechanical properties due to the use of celecoxib	Decrease in adhesion formation Promotion of tendon gliding Decreased expression of type I and type III collagen Decrease in cell proliferation	[100]
2018	CSNFM of PCL/PEG shell with HA core	HA embedded with IBU and PCL/PEG loaded with Ag NPs	Co-axial electrospinning	Elongation at break decreased and Young's modulus and UTS increased for higher IBU wt% IBU showed fast release during first 8 h, but slows down over time	Decrease in adhesion formation Reduced fibroblast attachment and proliferation Higher IBU concentrations lead to substantial cytotoxicity in vitro and in vivo Promotion of tendon gliding Ag NPs as an antimicrobial effect without additional cytotoxicity	[123]

Table 3. *Cont.*

Year	Material and Structure	Chemical/Biological Modulation and Concentration	Material Processing Technique	Mechanical Response [1]	Biological Response [1]	Ref.
2018	NFM of PLCL-sg-HA	HA	Electrospinning	Decreased UTS Release rate of HA was controlled by sheath thickness	Decrease in adhesion formation Prevention of penetrating fibroblasts	[180]
2019	NFM of HA	IBU	Electrospinning followed by covalently crosslinking to BDDE and ionic crosslinking to FeCl3	Higher IBU loading increased UTS and Young's modulus	Decrease in adhesion formation Prevention of cell attachment and penetration	[98]
2019	NFM of PLLA-HA	PDA loaded with ERK2-siRNA + HA	Electrospinning	Cumulative release of 80% ERK2-siRNA in 30 days	Decreased expression of type I and type III collagen Decrease in cell proliferation	[203]
2021	CSNFM of PLA shell with HA core	PLA loaded with Ag NPs + HA	Co-axial electrospinning	Sustainable release of HA	Best decrease in adhesion formation observed for thin shell fibers Ag NPs as an antimicrobial effect	[108]
2021	CSNFM of PCL shell with HA core	HA loaded with PRP	Co-axial electrospinning	Controlled release of proteins from PRP	Aligned fibers provided optimal cell proliferation Aligned fibers provided increase in type I and decrease in type III collagen Ag NPs as an antimicrobial effect without additional cytotoxicity	[200]
2021	NFM of PLGA coated with PEG-PLV hydrogel	IBU + bFGFs	Electrospinning followed by hydrogel coating	No obvious effect on tendon mechanical properties	Decrease in adhesion formation Increased expression of type I and III collagen	[97]

[1] Compared to traditional suture technique control. Abbreviations: Structures: core-shell nanofibrous membrane (CSNFM), nanofibrous membrane (NFM), multi-layer nanofibrous membrane (MNFM), surface graft (sg). Polymers: poly(ε-caprolactone) (PCL), hyaluronic acid (HA), chitosan (CHI), poly(L-lactic acid)-poly(ethylene glycol) (PELA), poly(L-lactic acid) (PLLA), alginate (ALG), poly(ethylene glycol) (PEG), poly(L-valine) (PLV). Modulations: silver nanoparticles (Ag NPs), ibuprofen (IBU), 2,6-pyridinedicarboxaldehyde-polyethylenimine (PDA), extracellular signal-regulated kinase 2 (ERK2), small interfering RNA (siRNA), 1,4-butanediol diglyceryl ether (BDDE), platelet-rich plasma (PRP). Other: ultimate tensile strength (UTS).

As mentioned in Section 4.3.2, a multi-layer tubular construct was produced by Pien et al. [93,105] in 2021. However, this was not the first published paper reporting the use of a multi-layered construct for flexor tendon repair. In 2015, a multi-layer semi-synthetic polymer-based construct was produced with celecoxib-loaded PELA electrospun fibrous membrane as the outer layer, HA hydrogel as the middle layer, and PELA electrospun fibrous membrane as the inner layer by sequential electrospinning [102]. The inner layer of PELA and HA was proven to mimic the biological response of HA secretion in order to enhance tendon gliding and tendon healing, while the outer celecoxib-loaded layer reduced the adhesion formation. A similar approach was taken by Deepthi et al. [109] one year later using electrospun PLLA, to mimic the aligned collagen fibers, blended with collagen-chitosan hydrogel in order to mimic the glycosaminoglycans of sheath ECM for tendon regeneration. Thereafter, the fibers were coated with alginate for a superior healing process. The construct achieved the same promising results without the suppression of collagen expression by inhibition of ERK1/2 and SMAD2/3 phosphorylation by celecoxib. A double layer composite membrane with electrospun nanofibrous poly(lactic-co-glycolic acid) (PLGA) inner layer and a hydrogel of poly(ethylene glycol)-block-poly(L-valine) as outer layer was produced in 2021 [97] and is shown in Figure 7. The addition of the outer hydrogel layer was responsible for a sustained release of ibuprofen over a period of 34 days in vitro without a burst release. Therefore, higher concentrations of ibuprofen could be loaded, increasing the anti-inflammatory effect of the construct.

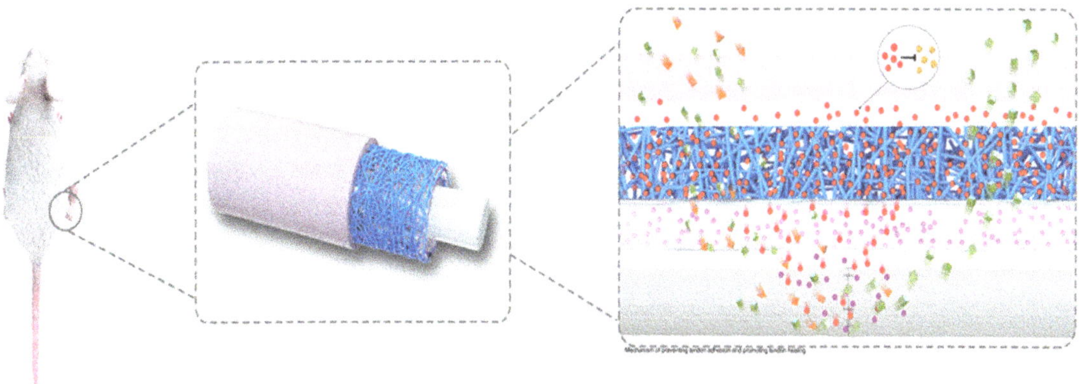

Figure 7. Double layer composite membrane with electrospun nanofibrous poly(lactic-co-glycolic) ibuprofen-loaded inner layer and a hydrogel of poly(ethylene glycol)-block-poly(L-valine) as an outer layer for preventing tendon adhesion and promoting tendon healing. The hydrogel coating prolonged the ibuprofen release in vivo. Reprinted with permission from ref. [97]. 2021 Z. Yan.

In 2019, different materials were used [98] in comparison to the commonly used aliphatic polyesters and PEG. HA loaded with ibuprofen was processed into a nanofibrous membrane by electrospinning, which was similar to previous research, but was followed by dual ionic crosslinking with $FeCl_3$ and covalently crosslinked with 1,4-butanediol diglyceryl ether (BDDE). The study aimed to use iron ions (Fe^{3+}) and BDDE as a crosslinking agent to achieve a prolonged release of ibuprofen. Higher concentrations of ibuprofen could be used without inducing cytotoxicity and improving the mechanical properties. However, ibuprofen loading above 30% induced cytotoxicity, tested by in vivo rabbit models. An innovating approach was published in [203] in the same year. They developed a promising platform using gene silencing in reducing the risk of peritendinous adhesion formation. In recent years, gene silencing via the delivery of small interfering RNA (siRNA) has experienced rapid advancement in the downregulation of specific genes [204]. As such, SiRNA therapies have been developed in the most recent years using viral or non-viral vectors, mainly involving peptide conjugation [205], cationic polymers [206], and liposomes [207].

Unfortunately, a bottleneck has gradually emerged in the siRNA technology around the transient silencing effect [208]. Therefore, an urgent need was created for a sustained release of siRNA via carriers, followed by a consequent increase in bioavailability. In this study, a cationic polymer, 2,6-pyridinedicarboxaldehyde-polyethylenimine (PDA), was used as a delivery system for ERK2-siRNA in combination with electrospun PLLA/HA fibers. Silencing ERK2 can indirectly inhibit type I and III collagen production and directly inhibit fibroblast proliferation as mentioned before [101]. A sustainable, cumulative release of 80% ERK2-siRNA in 30 days was achieved. Inhibition of cell proliferation showed that the membrane could retain the bioavailability of ERK2-siRNA, mostly due to PDA.

5. Conclusions and Perspectives

Flexor tendon injuries are common and mostly treated operationally. Hereby, several postoperative complications may occur such as infection, wear, tendon scar tissue formation, mechanical failure, and excessive adhesion formation. Multiple therapeutic reconstruction techniques, such as grafting or suturing, have been used in the past, but they don't offer an adequate long-term solution. Current strategies for preventing the formation of peritendinous adhesion rely mainly on the inhibition of inflammation, using physical/mechanical barriers for averting wound contact with surrounding tissue, downregulating the ERK1/2 and SMAD2/3 phosphorylation, and inhibiting the excessive type I/III collagen and proliferation of fibroblasts. The use of polymer-based constructs can play a variety of important roles in preventing the formation of adhesion, not only by forming a physical barrier. Most current researchers focusing on flexor tendon repair develop a construct that is equipped with both anti-inflammatory as well as antimicrobial agents, solving the most severe postoperative complications occurring during tendon healing. The use of only synthetic or biological polymers is not sufficient for creating a multi-functional construct, therefore functional materials and combinations of several polymer types (semi-synthetic) have been engineered in combination with drugs, which are as a result highly effective in preventing these postoperative complications. This review provided a comprehensive summary of the mechanism of tendon injuries and healing. In addition, it contains a compilation of constructs for tendon repair found in the literature for each different polymer type (biological, synthetic, and semi-synthetic). Besides the material type, several techniques, structures, and even the incorporation of chemical and biological drugs have attracted more attention recently and were included in the review. These new approaches in the development of new constructs are likely to result in an enhanced healing treatment for flexor tendon injuries. At present, a lot of these constructs already show great in vivo results in animal models, mostly mice or rabbits, although only limited studies have been performed on humans. The fast pace of development in the field will undoubtedly lead to the use of smart materials and multi-functional constructs, used in clinical practice. Although excellent results were obtained from past research, precise engineered constructs and the development of new drugs are not yet at their peak performance and the field of flexor tendon repair still requires new approaches and techniques.

Constructs can contain multiple drugs or biological compounds and fulfill combination treatment where the active payload release could be controlled in the future depending on the healing stage of the tendon. Hence, structures can heal the damaged or lacerated tendon at the first stage and later help in the maturation period of tendon healing. Active control of the drug release could also avoid commonly observed burst releases and avoid the accompanying possible side effects. Nano-based drug delivery has already proven to be successful in other medical fields to eradicate the problem of a burst release [209]. Future work may include the incorporation of nanoparticles into the nanofibrous constructs, i.e., electrospinning. The controlled active drug release can be accomplished in the future by engineering a smart polymeric construct that could have a trigger based on temperature, pH, electrical signals, magnetic field, and many more. Although such smart triggers have already been incorporated into other biomedical applications such as tumor immunotherapy [210], temperature-dependent drug and gene delivery [211], neural tissue engineering [212], skin wound healing [213], and many more, it is not yet developed for flexor tendon repair constructs.

Author Contributions: All authors contributed to this study. All authors have read and agreed to the published version of the manuscript.

Funding: This research received no external funding.

Institutional Review Board Statement: Not applicable.

Informed Consent Statement: Not applicable.

Data Availability Statement: Not applicable.

Conflicts of Interest: The authors declare no conflict of interest.

References

1. Kirkendall, D.T.; Garrett, W.E. Function and biomechanics of tendons. *Scand. J. Med. Sci. Sport.* **1997**, *7*, 62–66. [CrossRef] [PubMed]
2. Snedeker, J.G.; Foolen, J. Tendon injury and repair—A perspective on the basic mechanisms of tendon disease and future clinical therapy. *Acta Biomater.* **2017**, *63*, 18–36. [CrossRef] [PubMed]
3. Banik, B.L.; Lewis, G.S.; Brown, J.L. Multiscale Poly-(ϵ-caprolactone) Scaffold Mimicking Non-linearity in Tendon Tissue Mechanics. *Regen. Eng. Transl. Med.* **2016**, *2*, 1–9. [CrossRef] [PubMed]
4. Ghiya, M.N.; Murty, S.; Shetty, N.; D'Cunha, R. A descriptive study of hand injuries presenting to the adult emergency department of a tertiary care center in urban India. *J. Emerg. Trauma. Shock* **2017**, *10*, 19–25. [CrossRef]
5. Clark, D.P.; Scott, R.N.; Anderson, I.W. Hand problems in an accident and emergency department. *J. Hand Surg. Br.* **1985**, *10*, 297–299. [CrossRef]
6. Citeroni, M.R.; Ciardulli, M.C.; Russo, V.; Della Porta, G.; Mauro, A.; El Khatib, M.; Di Mattia, M.; Galesso, D.; Barbera, C.; Forsyth, N.R.; et al. Review in vitro innovation of tendon tissue engineering strategies. *Int. J. Mol. Sci.* **2020**, *21*, 1–78. [CrossRef]
7. Titan, A.L.; Foster, D.S.; Chang, J.; Longaker, M.T. Flexor Tendon: Development, Healing, Adhesion Formation, and Contributing Growth Factors. *Plast. Reconstr. Surg.* **2019**, *144*, 639e–647e. [CrossRef]
8. Legrand, A.; Kaufman, Y.; Long, C.; Fox, P.M. Molecular Biology of Flexor Tendon Healing in Relation to Reduction of Tendon Adhesions. *J. Hand Surg. Am.* **2017**, *42*, 722–726. [CrossRef]
9. Woo, S.L.Y.; Gelberman, R.H.; Cobb, N.G.; Amiel, D.; Lothringer, K.; Akeson, W.H. the importance of controlled passive mobilization on flexor tendon healing: A biomechanical study. *Acta Orthop.* **1981**, *52*, 615–622. [CrossRef]
10. Buschmann, J.; Meier Bürgisser, G. Autograft, allograft, and xenograft scaffolds for tendon and ligament repair. In *Biomechanics of Tendons and Ligaments*; Woodhead Publishing: Sawston, UK, 2017; ISBN 9780081004890.
11. Rawson, S.; Cartmell, S.; Wong, J. Suture techniques for tendon repair; a comparative review. *Muscles Ligaments Tendons J.* **2013**, *3*, 220–228. [CrossRef]
12. Vasiliadis, A.V.; Katakalos, K. The role of scaffolds in tendon tissue engineering. *J. Funct. Biomater.* **2020**, *11*, 78. [CrossRef] [PubMed]
13. Sigal, I.R.; Grande, D.A.; Dines, D.M.; Dines, J.; Drakos, M. Biologic and Tissue Engineering Strategies for Tendon Repair. *Regen. Eng. Transl. Med.* **2016**, *2*, 107–125. [CrossRef]
14. Zhou, H.; Lu, H. Advances in the Development of Anti-Adhesive Biomaterials for Tendon Repair Treatment. *Tissue Eng. Regen. Med.* **2021**, *18*, 1–14. [CrossRef] [PubMed]
15. Beldjilali-Labro, M.; Garcia, A.G.; Farhat, F.; Bedoui, F.; Grosset, J.F.; Dufresne, M.; Legallais, C. Biomaterials in tendon and skeletal muscle tissue engineering: Current trends and challenges. *Materials* **2018**, *11*, 1116. [CrossRef]
16. Rademakers, T.; Horvath, J.M.; van Blitterswijk, C.A.; LaPointe, V.L.S. Oxygen and nutrient delivery in tissue engineering: Approaches to graft vascularization. *J. Tissue Eng. Regen. Med.* **2019**, *13*, 1815–1829. [CrossRef]
17. Cravedi, P.; Farouk, S.; Angeletti, A.; Edgar, L.; Tamburrini, R.; Duisit, J.; Perin, L.; Orlando, G. Regenerative immunology: The immunological reaction to biomaterials. *Transpl. Int.* **2017**, *30*, 1199–1208. [CrossRef]
18. No, Y.J.; Castilho, M.; Ramaswamy, Y.; Zreiqat, H. Role of Biomaterials and Controlled Architecture on Tendon/Ligament Repair and Regeneration. *Adv. Mater.* **2020**, *32*, 1904511. [CrossRef]
19. Chen, S.H.; Chen, C.H.; Fong, Y.T.; Chen, J.P. Prevention of peritendinous adhesions with electrospun chitosan-grafted polycaprolactone nanofibrous membranes. *Acta Biomater.* **2014**, *10*, 4971–4982. [CrossRef]
20. Kannus, P. Structure of the tendon connective tissue. *Scand. J. Med. Sci. Sports* **2000**, *10*, 312–320. [CrossRef]
21. Benjamin, M.; Ralphs, J.R. The cell and developmental biology of tendons and ligaments. *Int. Rev. Cytol.* **2000**, *196*, 85–130. [CrossRef]
22. Wang, J.H.C.; Guo, Q.; Li, B. Tendon biomechanics and mechanobiology—A minireview of basic concepts and recent advancements. *J. Hand Ther.* **2012**, *25*, 133–141. [CrossRef] [PubMed]

23. Walden, G.; Liao, X.; Donell, S.; Raxworthy, M.J.; Riley, G.P.; Saeed, A. A Clinical, Biological, and Biomaterials Perspective into Tendon Injuries and Regeneration. *Tissue Eng. Part B Rev.* **2017**, *23*, 44–58. [CrossRef] [PubMed]
24. Steinmann, S.; Pfeifer, C.G.; Brochhausen, C.; Docheva, D. Spectrum of tendon pathologies: Triggers, trails and end-state. *Int. J. Mol. Sci.* **2020**, *21*, 844. [CrossRef] [PubMed]
25. Silver, F.H.; Freeman, J.W.; Seehra, G.P. Collagen self-assembly and the development of tendon mechanical properties. *J. Biomech.* **2003**, *36*, 1529–1553. [CrossRef]
26. Tempfer, H.; Lehner, C.; Grütz, M.; Gehwolf, R.; Traweger, A. Biological Augmentation for Tendon Repair: Lessons to be Learned from Development, Disease, and Tendon Stem Cell Research. In *Cell Engineering and Regeneration. Reference Series in Biomedical Engineering*; Gimble, J., Marolt Presen, D., Oreffo, R., Wolbank, S., Redl, H., Eds.; Springer: Cham, Switzerland, 2017; ISBN 9783319370767.
27. Canty, E.G.; Kadler, K.E. Collagen fibril biosynthesis in tendon: A review and recent insights. *Comp. Biochem. Physiol. A Mol. Integr. Physiol.* **2002**, *133*, 979–985. [CrossRef]
28. Zuskov, A.; Freedman, B.R.; Gordon, J.A.; Sarver, J.J.; Buckley, M.R.; Soslowsky, L.J. Tendon Biomechanics and Crimp Properties Following Fatigue Loading Are Influenced by Tendon Type and Age in Mice. *J. Orthop. Res.* **2020**, *38*, 36–42. [CrossRef]
29. Rosario, M.V.; Roberts, T.J. Loading Rate Has Little Influence on Tendon Fascicle Mechanics. *Front. Physiol.* **2020**, *11*, 255. [CrossRef]
30. Silver, F.H.; Christiansen, D.L.; Snowhill, P.B.; Chen, Y. Role of Storage on Changes in the Mechanical Properties of Tendon and Self-Assembled Collagen Fibers. *Connect. Tissue Res.* **2000**, *41*, 155–164. [CrossRef]
31. Mccormick, R.J. Extracellular modifications to muscle collagen: Implications for meat quality. *Poult. Sci.* **1999**, *78*, 785–791. [CrossRef]
32. Franchi, M.; Trirè, A.; Quaranta, M.; Orsini, E.; Ottani, V. Collagen structure of tendon relates to function. *ScientificWorldJournal* **2007**, *7*, 404–420. [CrossRef]
33. Johnson, G.A.; Tramaglini, D.M.; Levine, R.E.; Ohno, K.; Choi, N.-Y.; Woo, S.L.-Y. Tensile and viscoelastic properties of human patellar tendon. *J. Orthop. Res.* **1994**, *12*, 796–803. [CrossRef] [PubMed]
34. Martin, R.B.; Burr, D.B.; Sharkey, N.A.; Fyhrie, D.P. *Mechanical Properties of Ligament and Tendon In Skeletal Tissue Mechanics*; Springer: New York, NY, USA, 2015.
35. Zhang, C.; Zhu, J.; Zhou, Y.; Thampatty, B.P.; Wang, J.H.C. Tendon Stem/Progenitor Cells and Their Interactions with Extracellular Matrix and Mechanical Loading. *Stem Cells Int.* **2019**, *2019*, 3674647. [CrossRef] [PubMed]
36. Lin, T.W.; Cardenas, L.; Soslowsky, L.J. Biomechanics of tendon injury and repair. *J. Biomech.* **2004**, *37*, 865–877. [CrossRef] [PubMed]
37. Millar, N.L.; Murrell, G.A.C.; Mcinnes, I.B. Inflammatory mechanisms in tendinopathy—Towards translation. *Nat. Rev. Rheumatol.* **2017**, *13*, 110–122. [CrossRef] [PubMed]
38. Fatemi, M.J.; Shirani, S.; Sobhani, R.; Lebaschi, A.H.; Gharegozlou, M.J.; Bagheri, T.; Pedram, M.; Saberi, M.; Araghi, S.; Fatemi, M.A. Prevention of Peritendinous Adhesion Formation After the Flexor Tendon Surgery in Rabbits: A Comparative Study Between Use of Local Interferon-α, Interferon-β, and 5-Fluorouracil. *Ann. Plast. Surg.* **2018**, *80*, 171–175. [CrossRef] [PubMed]
39. D'Addona, A.; Maffulli, N.; Formisano, S.; Rosa, D. Inflammation in tendinopathy. *Surgeon* **2017**, *15*, 297–302. [CrossRef]
40. de Jong, J.P.; Nguyen, J.T.; Sonnema, A.J.M.; Nguyen, E.C.; Amadio, P.C.; Moran, S.L. The incidence of acute traumatic tendon injuries in the hand and wrist: A 10-year population-based study. *Clin. Orthop. Surg.* **2014**, *6*, 196–202. [CrossRef]
41. Sharma, P.; Maffulli, N. Tendon Injury and Tendinopathy. *J. Bone Jt. Surg.* **2005**, *87*, 187–202. [CrossRef]
42. Yang, G.; Rothrauff, B.B.; Tuan, R.S. Tendon and ligament regeneration and repair: Clinical relevance and developmental paradigm. *Birth Defects Res. Part C Embryo Today Rev.* **2013**, *99*, 203–222. [CrossRef]
43. Gault, D.T. Reduction of grip strength, finger flexion pressure, finger pinch pressure and key pinch following flexor tendon repair. *J. Hand Surg. Br.* **1987**, *12*, 182–184. [CrossRef]
44. Rodrigues, M.T.; Reis, R.L.; Gomes, M.E. Engineering tendon and ligament tissues: Present developments towards successful clinical products. *J. Tissue Eng. Regen. Med.* **2013**, *7*, 673–686. [CrossRef] [PubMed]
45. Docheva, D.; Müller, S.A.; Majewski, M.; Evans, C.H. Biologics for tendon repair. *Adv. Drug Deliv. Rev.* **2015**, *84*, 222–239. [CrossRef] [PubMed]
46. Strickland, J.W. Flexor Tendon Injuries: I. Foundations of Treatment. *J. Am. Acad. Orthop. Surg.* **1995**, *3*, 44–54. [CrossRef]
47. Gelberman, R.H.; Manske, P.R.; de Van Berg, J.S.; Lesker, P.A.; Akeson, W.H. Flexor tendon repair in vitro: A comparative histologic study of the rabbit, chicken, dog, and monkey. *J. Orthop. Res.* **1984**, *2*, 39–48. [CrossRef] [PubMed]
48. Beredjiklian, P.K. Biologic aspects of flexor tendon laceration and repair. *J. Bone Jt. Surg. Ser. A* **2003**, *85*, 539–550. [CrossRef]
49. Pearce, O.; Brown, M.T.; Fraser, K.; Lancerotto, L. Flexor tendon injuries: Repair & Rehabilitation. *Injury* **2021**, *52*, 2053–2067. [CrossRef]
50. Thomopoulos, S.; Parks, W.C.; Rifkin, D.B.; Derwin, K.A. Mechanisms of tendon injury and repair. *J. Orthop. Res.* **2015**, *33*, 832–839. [CrossRef]
51. Leong, N.L.; Kator, J.L.; Clemens, T.L.; James, A.; Enamoto-Iwamoto, M.; Jiang, J. Tendon and Ligament Healing and Current Approaches to Tendon and Ligament Regeneration. *J. Orthop. Res.* **2020**, *38*, 7–12. [CrossRef]
52. Galvez, M.G.; Crowe, C.; Farnebo, S.; Chang, J. Tissue engineering in flexor tendon surgery: Current state and future advances. *J. Hand Surg. Eur. Vol.* **2014**, *39*, 71–78. [CrossRef]

53. Voleti, P.B.; Buckley, M.R.; Soslowsky, L.J. Tendon healing: Repair and regeneration. *Annu. Rev. Biomed. Eng.* **2012**, *14*, 47–71. [CrossRef]
54. Jung, H.-J.; Fisher, M.B.; Woo, S.L.-Y. Role of biomechanics in the understanding of normal, injured, and healing ligaments and tendons. *BMC Sports Sci. Med. Rehabil.* **2009**, *1*, 9. [CrossRef] [PubMed]
55. Drapé, J.L.; Tardif-Chastenet De Gery, S.; Silbermann-Hoffman, O.; Chevrot, A.; Houvet, P.; Alnot, J.Y.; Benacerraf, R. Closed ruptures of the flexor digitorum tendons: MRI evaluation. *Skelet. Radiol.* **1998**, *27*, 617–624. [CrossRef] [PubMed]
56. Ivanova, N.; Gugleva, V.; Dobreva, M.; Pehlivanov, I.; Stefanov, S.; Andonova, V. Management of Flexor Tendon Injuries in Hand. In *Intech*; IntechOpen: London, UK, 2016; Volume i, p. 13.
57. Lee, D.H.; Robbin, M.L.; Galliott, R.; Graveman, V.A. Ultrasound evaluation of flexor tendon lacerations. *J. Hand Surg. Am.* **2000**, *25*, 236–241. [CrossRef] [PubMed]
58. Kumar, B.A.; Tolat, A.R.; Threepuraneni, G.; Jones, B. The role of magnetic resonance imaging in late presentation of isolated injuries of the flexor digitorum profundus tendon in the finger. *J. Hand Surg. Br.* **2000**, *25*, 95–97. [CrossRef]
59. Sandvall, B.K.; Kuhlman-Wood, K.; Recor, C.; Friedrich, J.B. Flexor tendon repair, rehabilitation, and reconstruction. *Plast. Reconstr. Surg.* **2013**, *132*, 1493–1503. [CrossRef]
60. Güntürk, Ö.B.; Kayalar, M.; Kaplan, İ.; Uludağ, A.; Özaksar, K.; Keleşoğlu, B. Results of 4-strand modified Kessler core suture and epitendinous interlocking suture followed by modified Kleinert protocol for flexor tendon repairs in Zone 2. *Acta Orthop. Traumatol. Turc.* **2018**, *52*, 382–386. [CrossRef]
61. Sebastin, S.J.; Ho, A.; Karjalainen, T.; Chung, K.C. History and evolution of the Kessler repair. *J. Hand Surg. Am.* **2013**, *38*, 552–561. [CrossRef]
62. Wada, A.; Kubota, H.; Hatanaka, H.; Hotokezaka, S.; Miura, H.; Iwamoto, Y. The mechanical properties of locking and grasping suture loop configurations in four-strand core suture techniques. *J. Hand Surg. Br.* **2000**, *25*, 548–551. [CrossRef]
63. Thurman, R.T.; Trumble, T.E.; Hanel, D.P.; Tencer, A.F.; Kiser, P.K. Two-, four-, and six-strand zone II flexor tendon repairs: An in situ biomechanical comparison using a cadaver model. *J. Hand Surg. Am.* **1998**, *23*, 261–265. [CrossRef]
64. Viinikainen, A.; Göransson, H.; Huovinen, K.; Kellomäki, M.; Rokkanen, P. A comparative analysis of the biomechanical behaviour of five flexor tendon core sutures. *J. Hand Surg. Br.* **2004**, *29*, 536–543. [CrossRef]
65. Winters, S.C.; Gelberman, R.H.; Woo, S.L.; Chan, S.S.; Grewal, R.; Seiler, J.G., 3rd. The effects of multiple-strand suture methods on the strength and excursion of repaired intrasynovial flexor tendons: A biomechanical study in dogs. *J. Hand Surg. Am.* **1998**, *23*, 97–104. [CrossRef]
66. Lawrence, T.M.; Davis, T.R.C. A biomechanical analysis of suture materials and their influence on a four-strand flexor tendon repair. *J. Hand Surg. Am.* **2005**, *30*, 836–841. [CrossRef]
67. McDonald, E.; Gordon, J.A.; Buckley, J.M.; Gordon, L. Comparison of a multifilament stainless steel suture with FiberWire for flexor tendon repairs–an in vitro biomechanical study. *J. Hand Surg. Eur. Vol.* **2013**, *38*, 418–423. [CrossRef] [PubMed]
68. Trail, I.A.; Powell, E.S.; Noble, J. An evaluation of suture materials used in tendon surgery. *J. Hand Surg. Br.* **1989**, *14*, 422–427. [CrossRef]
69. Dy, C.J.; Hernandez-Soria, A.; Ma, Y.; Roberts, T.R.; Daluiski, A. Complications after flexor tendon repair: A systematic review and meta-analysis. *J. Hand Surg. Am.* **2012**, *37*, 543–552. [CrossRef] [PubMed]
70. Zhao, C.; Amadio, P.C.; Zobitz, M.E.; Momose, T.; Couvreur, P.; An, K.-N. Effect of synergistic motion on flexor digitorum profundus tendon excursion. *Clin. Orthop. Relat. Res.* **2002**, 223–230. [CrossRef] [PubMed]
71. Stark, H.H.; Anderson, D.R.; Zemel, N.P.; Boyes, J.H.; Ashworth, C.R.; Rickard, T.A. Bridge flexor tendon grafts. *Clin. Orthop. Relat. Res.* **1989**, 51–59. [CrossRef]
72. Jakubietz, M.; Jakubietz, D.; Gruenert, J.G.P.; Zahn, R.; Meffert, R.; Jakubietz, R. Adequacy of Palmaris Longus and Plantaris Tendons for Tendon Grafting. *J. Hand Surg. Am.* **2011**, *36*, 695–698. [CrossRef] [PubMed]
73. Jackson, D.W.; Corsetti, J.; Simon, T.M. Biologic Incorporation of Allograft Anterior Cruciate Ligament Replacements. *Clin. Orthop. Relat. Res.* **1996**, *324*, 126–133. [CrossRef]
74. Robertson, A.; Nutton, R.W.; Keating, J.F. Current trends in the use of tendon allografts in orthopaedic surgery. *J. Bone Jt. Surg. Br.* **2006**, *88*, 988–992. [CrossRef]
75. Poehling, G.G.; Curl, W.W.; Lee, C.A.; Ginn, T.A.; Rushing, J.T.; Naughton, M.J.; Holden, M.B.; Martin, D.F.; Smith, B.P. Analysis of Outcomes of Anterior Cruciate Ligament Repair With 5-Year Follow-up: Allograft Versus Autograft. *Arthrosc. J. Arthrosc. Relat. Surg.* **2005**, *21*, 774.e1–774.e15. [CrossRef] [PubMed]
76. Taras, J.S.; Lamb, M.J. Treatment of flexor tendon injuries: Surgeons' perspective. *J. Hand Ther. Off. J. Am. Soc. Hand Ther.* **1999**, *12*, 141–148. [CrossRef]
77. Derby, B.M.; Wilhelmi, B.J.; Zook, E.G.; Neumeister, M.W. Flexor tendon reconstruction. *Clin. Plast. Surg.* **2011**, *38*, 607–619. [CrossRef] [PubMed]
78. Ramesh, R.; Kumar, N.; Sharma, A.K.; Maiti, S.K.; Singh, G.R. Acellular and glutaraldehyde-preserved tendon allografts for reconstruction of superficial digital flexor tendon in bovines: Part I—Clinical, radiological and angiographical observations. *J. Vet. Med. A Physiol. Pathol. Clin. Med.* **2003**, *50*, 511–519. [CrossRef]
79. Zhang, C.-H.; Jiang, Y.-L.; Ning, L.-J.; Li, Q.; Fu, W.-L.; Zhang, Y.-J.; Zhang, Y.-J.; Xia, C.-C.; Li, J.; Qin, T.-W. Evaluation of Decellularized Bovine Tendon Sheets for Achilles Tendon Defect Reconstruction in a Rabbit Model. *Am. J. Sports Med.* **2018**, *46*, 2687–2699. [CrossRef]

80. Moshiri, A.; Oryan, A. Role of tissue engineering in tendon reconstructive surgery and regenerative medicine: Current concepts, approaches and concerns. *Hard Tissue* **2012**, *1*, 11. [CrossRef]
81. Allman, A.J.; McPherson, T.B.; Badylak, S.F.; Merrill, L.C.; Kallakury, B.; Sheehan, C.; Raeder, R.H.; Metzger, D.W. Xenogeneic extracellular matrix grafts elicit a TH2-restricted immune response. *Transplantation* **2001**, *71*, 1631–1640. [CrossRef] [PubMed]
82. Oryan, A.; Moshiri, A.; Meimandi-Parizi, A. Graft selection in ACL reconstructive surgery: Past, present, and future. *Curr. Orthop. Pract.* **2013**, *24*, 321–333. [CrossRef]
83. Van Der Merwe, W.; Lind, M.; Faunø, P.; Van Egmond, K.; Zaffagnini, S.; Marcacci, M.; Cugat, R.; Verdonk, R.; Ibañez, E.; Guillen, P.; et al. Xenograft for anterior cruciate ligament reconstruction was associated with high graft processing infection. *J. Exp. Orthop.* **2020**, *7*, 79. [CrossRef]
84. Pridgen, B.C.; Woon, C.Y.L.; Kim, M.; Thorfinn, J.; Lindsey, D.; Pham, H.; Chang, J. Flexor tendon tissue engineering: Acellularization of human flexor tendons with preservation of biomechanical properties and biocompatibility. *Tissue Eng. Part C Methods* **2011**, *17*, 819–828. [CrossRef]
85. Veillette, C.J.; Cunningham, K.D.; Hart, D.A.; Fritzler, M.J.; Frank, C.B. Localization and characterization of porcine patellar tendon xenograft antigens in a rabbit model of medial collateral ligament replacement. *Transplantation* **1998**, *65*, 486–493. [CrossRef] [PubMed]
86. Hunter, J. ARTIFICIAL TENDONS. EARLY DEVELOPMENT AND APPLICATION. *Am. J. Surg.* **1965**, *109*, 325–338. [CrossRef]
87. Hunter, J.M.; Salisbury, R.E. Use of gliding artificial implants to produce tendon sheaths. Techniques and results in children. *Plast. Reconstr. Surg.* **1970**, *45*, 564–572. [CrossRef] [PubMed]
88. Abdullah, S. Usage of synthetic tendons in tendon reconstruction. *BMC Proc.* **2015**, *9*, A68. [CrossRef]
89. Murray, A.W.; MacNicol, M.F. 10–16 year results of Leeds-Keio anterior cruciate ligament reconstruction. *Knee* **2004**, *11*, 9–14. [CrossRef]
90. Bashaireh, K.M.; Audat, Z.; Radaideh, A.M.; Aleshawi, A.J. The Effectiveness of Autograft Used in Anterior Cruciate Ligament Reconstruction of the Knee: Surgical Records for the New Generations of Orthopedic Surgeons and Synthetic Graft Revisit. *Orthop. Res. Rev.* **2020**, *12*, 61–67. [CrossRef]
91. Satora, W.; Królikowska, A.; Czamara, A.; Reichert, P. Synthetic grafts in the treatment of ruptured anterior cruciate ligament of the knee joint. *Polim. Med.* **2017**, *47*, 55–59. [CrossRef]
92. McLaughlin, R.M. Complications After Treatment of Flexor Tendon Injuries. *Small Anim. Surg. Secrets Second Ed.* **2004**, *14*, 327–330. [CrossRef]
93. Pien, N.; Peeters, I.; Deconinck, L.; Van Damme, L.; De Wilde, L.; Martens, A.; Van Vlierberghe, S.; Dubruel, P.; Mignon, A. Design and development of a reinforced tubular electrospun construct for the repair of ruptures of deep flexor tendons. *Mater. Sci. Eng. C* **2021**, *119*, 111504. [CrossRef]
94. Chen, S.; Wang, G.; Wu, T.; Zhao, X.; Liu, S.; Li, G.; Cui, W.; Fan, C. Silver nanoparticles/ibuprofen-loaded poly(L-lactide) fibrous membrane: Anti-infection and anti-adhesion effects. *Int. J. Mol. Sci.* **2014**, *15*, 14014–14025. [CrossRef]
95. Liu, S.; Hu, C.; Li, F.; Li, X.J.; Cui, W.; Fan, C. Prevention of peritendinous adhesions with electrospun ibuprofen-loaded poly(l-Lactic Acid)-polyethylene glycol fibrous membranes. *Tissue Eng. Part A* **2013**, *19*, 529–537. [CrossRef] [PubMed]
96. Hu, C.; Liu, S.; Zhang, Y.; Li, B.; Yang, H.; Fan, C.; Cui, W. Long-term drug release from electrospun fibers for in vivo inflammation prevention in the prevention of peritendinous adhesions. *Acta Biomater.* **2013**, *9*, 7381–7388. [CrossRef] [PubMed]
97. Yan, Z.; Meng, X.; Su, Y.; Chen, Y.; Zhang, L.; Xiao, J. Double layer composite membrane for preventing tendon adhesion and promoting tendon healing. *Mater. Sci. Eng. C* **2021**, *123*, 111941. [CrossRef] [PubMed]
98. Chen, C.T.; Chen, C.H.; Sheu, C.; Chen, J.P. Ibuprofen-loaded hyaluronic acid nanofibrous membranes for prevention of postoperative tendon adhesion through reduction of inflammation. *Int. J. Mol. Sci.* **2019**, *20*, 5038. [CrossRef] [PubMed]
99. Yuan, B.; He, C.; Dong, X.; Wang, J.; Gao, Z.; Wang, Q.; Tian, H.; Chen, X. 5-Fluorouracil loaded thermosensitive PLGA-PEG-PLGA hydrogels for the prevention of postoperative tendon adhesion. *RSC Adv.* **2015**, *5*, 25295–25303. [CrossRef]
100. Li, L.; Zheng, X.; Fan, D.; Yu, S.; Wu, D.; Fan, C.; Cui, W.; Ruan, H. Release of celecoxib from a bi-layer biomimetic tendon sheath to prevent tissue adhesion. *Mater. Sci. Eng. C* **2016**, *61*, 220–226. [CrossRef]
101. Jiang, S.; Zhao, X.; Chen, S.; Pan, G.; Song, J.; He, N.; Li, F.; Cui, W.; Fan, C. Down-regulating ERK1/2 and SMAD2/3 phosphorylation by physical barrier of celecoxib-loaded electrospun fibrous membranes prevents tendon adhesions. *Biomaterials* **2014**, *35*, 9920–9929. [CrossRef]
102. Jiang, S.; Yan, H.; Fan, D.; Song, J.; Fan, C. Multi-layer electrospun membrane mimicking tendon sheath for prevention of tendon adhesions. *Int. J. Mol. Sci.* **2015**, *16*, 6932–6944. [CrossRef]
103. Wu, Q.; Wang, N.; He, T.; Shang, J.; Li, L.; Song, L.; Yang, X.; Li, X.; Luo, N.; Zhang, W.; et al. Thermosensitive hydrogel containing dexamethasone micelles for preventing postsurgical adhesion in a repeated-injury model. *Sci. Rep.* **2015**, *5*, 13553. [CrossRef]
104. Chen, C.H.; Chen, S.H.; Shalumon, K.T.; Chen, J.P. Dual functional core-sheath electrospun hyaluronic acid/polycaprolactone nanofibrous membranes embedded with silver nanoparticles for prevention of peritendinous adhesion. *Acta Biomater.* **2015**, *26*, 225–235. [CrossRef]
105. Peeters, I.; Pien, N.; Mignon, A.; Van Damme, L.; Dubruel, P.; Van Vlierberghe, S.; Mantovani, D.; Vermeulen, V.; Creytens, D.; Van Tongel, A.; et al. Flexor tendon repair using a reinforced tubular, medicated electrospun construct. *J. Orthop. Res.* **2021**, *40*, 750–760. [CrossRef]

106. Liu, S.; Zhao, J.; Ruan, H.; Tang, T.; Liu, G.; Yu, D.; Cui, W.; Fan, C. Biomimetic sheath membrane via electrospinning for antiadhesion of repaired tendon. *Biomacromolecules* **2012**, *13*, 3611–3619. [CrossRef] [PubMed]
107. Chen, S.H.; Chen, C.H.; Shalumon, K.T.; Chen, J.P. Preparation and characterization of antiadhesion barrier film from hyaluronic acid-grafted electrospun poly(caprolactone) nanofibrous membranes for prevention of flexor tendon postoperative peritendinous adhesion. *Int. J. Nanomed.* **2014**, *9*, 4079–4092. [CrossRef] [PubMed]
108. Chen, C.H.; Cheng, Y.H.; Chen, S.H.; Chuang, A.D.C.; Chen, J.P. Functional hyaluronic acid-polylactic acid/silver nanoparticles core-sheath nanofiber membranes for prevention of post-operative tendon adhesion. *Int. J. Mol. Sci.* **2021**, *22*, 8781. [CrossRef] [PubMed]
109. Deepthi, S.; Nivedhitha Sundaram, M.; Deepti Kadavan, J.; Jayakumar, R. Layered chitosan-collagen hydrogel/aligned PLLA nanofiber construct for flexor tendon regeneration. *Carbohydr. Polym.* **2016**, *153*, 492–500. [CrossRef] [PubMed]
110. Woythal, L.; Hølmer, P.; Brorson, S. Splints, with or without wrist immobilization, following surgical repair of flexor tendon lesions of the hand: A systematic review. *Hand Surg. Rehabil.* **2019**, *38*, 217–222. [CrossRef] [PubMed]
111. Neiduski, R.L.; Powell, R.K. Flexor tendon rehabilitation in the 21st century: A systematic review. *J. Hand Ther.* **2019**, *32*, 165–174. [CrossRef]
112. Pan, Z.J.; Xu, Y.F.; Pan, L.; Chen, J. Zone 2 flexor tendon repairs using a tensioned strong core suture, sparse peripheral stitches and early active motion: Results in 60 fingers. *J. Hand Surg. (Eur. Vol.)* **2019**, *44*, 361–366. [CrossRef]
113. Silfverskiöld, K.L.; May, E.J. Flexor tendon repair in zone II with a new suture technique and an early mobilization program combining passive and active flexion. *J. Hand Surg. Am.* **1994**, *19*, 53–60. [CrossRef]
114. Silfverskiöld, K.L.; Andersson, C.H. Two new methods of tendon repair: An in vitro evaluation of tensile strength and gap formation. *J. Hand Surg. Am.* **1993**, *18*, 58–65. [CrossRef]
115. Fathy Sadek, A.; Azmy, M.M.; Nady Saleh Elsaid, A.; Zein, A.M.N.; Yehya Hasan, M. Repair of flexor digitorum profundus avulsions including the palmar plate: A retrospective comparative study of 56 cases. *J. Hand Surg. (Eur. Vol.)* **2022**, 17531934221074514. [CrossRef] [PubMed]
116. Kheiran, A.; Palial, V.; Rollett, R.; Wildin, C.J.; Chatterji, U.; Singh, H.P. Cat bite: An injury not to underestimate. *J. Plast. Surg. Hand Surg.* **2019**, *53*, 341–346. [CrossRef]
117. Malizos, K.N.; Papadopoulou, Z.K.; Ziogkou, A.N.; Rigopoulos, N.; Athanaselis, E.D.; Varitimidis, S.E.; Dailiana, Z.C. Infections of Deep Hand and Wrist Compartments. *Microorganisms* **2020**, *8*, 838. [CrossRef] [PubMed]
118. Pickrell, B.B.; Eberlin, K.R. Secondary Surgery Following Replantation and Revascularization. *Hand Clin.* **2019**, *35*, 231–240. [CrossRef] [PubMed]
119. Mamane, W.; Lippmann, S.; Israel, D.; Ramdhian-Wihlm, R.; Temam, M.; Mas, V.; Pierrart, J.; Masmejean, E.H. Infectious flexor hand tenosynovitis: State of knowledge. A study of 120 cases. *J. Orthop.* **2018**, *15*, 701–706. [CrossRef] [PubMed]
120. Stone, J.F.; Davidson, J.S. The role of antibiotics and timing of repair in flexor tendon injuries of the hand. *Ann. Plast. Surg.* **1998**, *40*, 7–13. [CrossRef]
121. Chapman, T.; Ilyas, A.M. Pyogenic Flexor Tenosynovitis: Evaluation and Treatment Strategies. *J. Hand Microsurg.* **2019**, *11*, 121–126.
122. Chan, E.; Robertson, B.F.; Johnson, S.M. Kanavel signs of flexor sheath infection: A cautionary tale. *Br. J. Gen. Pract. J. R. Coll. Gen. Pract.* **2019**, *69*, 315–316. [CrossRef]
123. Shalumon, K.T.; Sheu, C.; Chen, C.H.; Chen, S.H.; Jose, G.; Kuo, C.Y.; Chen, J.P. Multi-functional electrospun antibacterial core-shell nanofibrous membranes for prolonged prevention of post-surgical tendon adhesion and inflammation. *Acta Biomater.* **2018**, *72*, 121–136. [CrossRef]
124. Bilal, M.; Rasheed, T.; Iqbal, H.M.N.; Li, C.; Hu, H.; Zhang, X. Development of silver nanoparticles loaded chitosan-alginate constructs with biomedical potentialities. *Int. J. Biol. Macromol.* **2017**, *105*, 393–400. [CrossRef]
125. Humayun, A.; Luo, Y.; Elumalai, A.; Mills, D.K. 3D printed antimicrobial PLA constructs functionalised with zinc- coated halloysite nanotubes-Ag-chitosan oligosaccharide lactate. *Mater. Technol.* **2022**, *37*, 28–35. [CrossRef]
126. Khan, F.; Tanaka, M. Designing smart biomaterials for tissue engineering. *Int. J. Mol. Sci.* **2018**, *19*, 17. [CrossRef] [PubMed]
127. Silva, M.; Ferreira, F.N.; Alves, N.M.; Paiva, M.C. Biodegradable polymer nanocomposites for ligament/tendon tissue engineering. *J. Nanobiotechnol.* **2020**, *18*, 23. [CrossRef] [PubMed]
128. Engineer, C.; Parikh, J.K.; Raval, A. Hydrolytic Degradation Behavior of Biodegradable Polymers from Controlled Drug Delivery System. *Trends Biomater. Artif. Organs* **2011**, *25*, 79–85.
129. Vroman, I.; Tighzert, L. Biodegradable polymers. *Materials* **2009**, *2*, 307–344. [CrossRef]
130. Migliaresi, C.; Fambri, L.; Cohn, D. A study on the in vitro degradation of poly(lactic acid). *J. Biomater. Sci. Polym. Ed.* **1994**, *5*, 591–606. [CrossRef]
131. Duek, E.A.R.; Zavaglia, C.A.C.; Belangero, W.D. In vitro study of poly(lactic acid) pin degradation. *Polymer* **1999**, *40*, 6465–6473. [CrossRef]
132. Jenkins, M.J.; Harrison, K.L. The effect of crystalline morphology on the degradation of polycaprolactone in a solution of phosphate buffer and lipase. *Polym. Adv. Technol.* **2008**, *19*, 1901–1906. [CrossRef]
133. Guo, M.; Chu, Z.; Yao, J.; Feng, W.; Wang, Y.; Wang, L.; Fan, Y. The effects of tensile stress on degradation of biodegradable PLGA membranes: A quantitative study. *Polym. Degrad. Stab.* **2016**, *124*, 95–100. [CrossRef]

134. Fan, Y.B.; Li, P.; Zeng, L.; Huang, X.J. Effects of mechanical load on the degradation of poly(d,l-lactic acid) foam. *Polym. Degrad. Stab.* **2008**, *93*, 677–683. [CrossRef]
135. Cooper, J.A.; Lu, H.H.; Ko, F.K.; Freeman, J.W.; Laurencin, C.T. Fiber-based tissue-engineered scaffold for ligament replacement: Design considerations and in vitro evaluation. *Biomaterials* **2005**, *26*, 1523–1532. [CrossRef] [PubMed]
136. Goh, J.C.; Sahoo, S. Scaffolds for tendon and ligament tissue engineering. In *Regenerative Medicine and Biomaterials for the Repair of Connective Tissues*; Woodhead Publishing: Sawston, UK, 2010.
137. Groth, T.; Falck, P.; Miethke, R.-R. Cytotoxicity of Biomaterials—Basic Mechanisms and In Vitro Test Methods: A Review. *Altern. Lab. Anim.* **1995**, *23*, 790–799. [CrossRef]
138. Cannella, V.; Altomare, R.; Leonardi, V.; Russotto, L.; Di Bella, S.; Mira, F.; Guercio, A. In Vitro Biocompatibility Evaluation of Nine Dermal Fillers on L929 Cell Line. *Biomed Res. Int.* **2020**, *2020*, 8676343. [CrossRef] [PubMed]
139. Yang, S.; Leong, K.F.; Du, Z.; Chua, K. The Design of Scaffolds for Use in Tissue Engineering. Part I. Traditional Factors. *Tissue Eng.* **2002**, *7*, 679–689. [CrossRef] [PubMed]
140. Song, A.; Rane, A.A.; Christman, K.L. Antibacterial and cell-adhesive polypeptide and poly(ethylene glycol) hydrogel as a potential scaffold for wound healing. *Acta Biomater.* **2012**, *8*, 41–50. [CrossRef]
141. Pring, D.J.; Amis, A.A.; Coombs, R.R.H. The mechanical properties of human flexor tendons in relation to artificial tendons. *J. Hand Surg. Br. Eur. Vol.* **1985**, *10*, 331–336. [CrossRef]
142. Verdan, C.; Potenza, A.D. Tendon Surgery of the Hand. *Plast. Reconstr. Surg.* **1980**, *66*, 1493–1503. [CrossRef]
143. Tang, J.B.; Gu, Y.T.; Rice, K.; Chen, F.; Pan, C.Z. Evaluation of four methods of flexor tendon repair for postoperative active mobilization. *Plast. Reconstr. Surg.* **2001**, *107*, 742–749. [CrossRef]
144. Mano, J.F.; Silva, G.A.; Azevedo, H.S.; Malafaya, P.B.; Sousa, R.A.; Silva, S.S.; Boesel, L.F.; Oliveira, J.M.; Santos, T.C.; Marques, A.P.; et al. Natural origin biodegradable systems in tissue engineering and regenerative medicine: Present status and some moving trends. *J. R. Soc. Interface* **2007**, *4*, 999–1030. [CrossRef]
145. Sensini, A.; Cristofolini, L. Biofabrication of electrospun scaffolds for the regeneration of tendons and ligaments. *Materials* **2018**, *11*, 1963. [CrossRef]
146. Narayanan, N.; Kuang, L.; Del Ponte, M.; Chain, C.; Deng, M. 1—Design and fabrication of nanocomposites for musculoskeletal tissue regeneration. In *Nanocomposites for Musculoskeletal Tissue Regeneration*; Liu, H., Ed.; Woodhead Publishing: Oxford, UK, 2016; pp. 3–29. ISBN 978-1-78242-452-9.
147. Smith, B.D.; Grande, D.A. The current state of scaffolds for musculoskeletal regenerative applications. *Nat. Rev. Rheumatol.* **2015**, *11*, 213–222. [CrossRef] [PubMed]
148. Langer, R.; Vacanti, J.P. Tissue Engineering. *Science* **1993**, *260*, 920–926. [CrossRef]
149. Sithole, M.N.; Choonara, Y.E.; du Toit, L.C.; Kumar, P.; Pillay, V. A review of semi-synthetic biopolymer complexes: Modified polysaccharide nano-carriers for enhancement of oral drug bioavailability. *Pharm. Dev. Technol.* **2017**, *22*, 283–295. [CrossRef] [PubMed]
150. Monti, P.; Freddi, G.; Bertoluzza, A.; Kasai, N.; Tsukada, M. Raman spectroscopic studies of silk fibroin from Bombyx mori. *J. Raman Spectrosc.* **1998**, *29*, 297–304. [CrossRef]
151. Vieira, A.C.; Guedes, R.M.; Marques, A.T. Development of ligament tissue biodegradable devices: A review. *J. Biomech.* **2009**, *42*, 2421–2430. [CrossRef]
152. Meinel, L.; Betz, O.; Fajardo, R.; Hofmann, S.; Nazarian, A.; Cory, E.; Hilbe, M.; McCool, J.; Langer, R.; Vunjak-Novakovic, G.; et al. Silk based biomaterials to heal critical sized femur defects. *Bone* **2006**, *39*, 922–931. [CrossRef]
153. Zhao, H.; Guan, H.G.; Gu, J.; Luo, Z.P.; Zhang, W.; Chen, B.; Gu, Q.L.; Yang, H.L.; Shi, Q. Collagen membrane alleviates peritendinous adhesion in the rat Achilles tendon injury model. *Chin. Med. J.* **2013**, *126*, 729–733. [CrossRef]
154. Turner, J.B.; Corazzini, R.L.; Butler, T.J.; Garlick, D.S.; Rinker, B.D. Evaluating adhesion reduction efficacy of type I/III collagen membrane and collagen-GAG resorbable matrix in primary flexor tendon repair in a chicken model. *Hand* **2015**, *10*, 482–488. [CrossRef]
155. Wichelhaus, D.A.; Beyersdoerfer, S.T.; Gierer, P.; Vollmar, B.; Mittlmeier, T. The effect of a collagen-elastin matrix on adhesion formation after flexor tendon repair in a rabbit model. *Arch. Orthop. Trauma Surg.* **2016**, *136*, 1021–1029. [CrossRef]
156. Seo, Y.K.; Kim, J.H.; Eo, S.R. Co-effect of silk and amniotic membrane for tendon repair. *J. Biomater. Sci. Polym. Ed.* **2016**, *27*, 1232–1247. [CrossRef]
157. Hortensius, R.A.; Ebens, J.H.; Dewey, M.J.; Harley, B.A.C. Incorporation of the Amniotic Membrane as an Immunomodulatory Design Element in Collagen Scaffolds for Tendon Repair. *ACS Biomater. Sci. Eng.* **2018**, *4*, 4367–4377. [CrossRef] [PubMed]
158. Zukawa, M.; Okabe, M.; Osada, R.; Makino, H.; Nogami, M.; Seki, S.; Yoshida, T.; Kimura, T.; Kawaguchi, Y. Effect of hyperdry amniotic membrane in preventing tendon adhesion in a rabbit model. *J. Orthop. Sci.* 2021, in press. [CrossRef] [PubMed]
159. Prakash, S.; Kalra, P.; Dhal, A. Flexor tendon repair with amniotic membrane. *Int. Orthop.* **2020**, *44*, 2037–2045. [CrossRef] [PubMed]
160. Cristino, S.; Grassi, F.; Toneguzzi, S.; Piacentini, A.; Grigolo, B.; Santi, S.; Riccio, M.; Tognana, E.; Facchini, A.; Lisignoli, G. Analysis of mesenchymal stem cells grown on a three-dimensional HYAFF 11-based prototype ligament scaffold. *J. Biomed. Mater. Res. A* **2005**, *73*, 275–283. [CrossRef] [PubMed]
161. Patterson, J.; Siew, R.; Herring, S.W.; Lin, A.S.P.; Guldberg, R.; Stayton, P.S. Hyaluronic acid hydrogels with controlled degradation properties for oriented bone regeneration. *Biomaterials* **2010**, *31*, 6772–6781. [CrossRef]

162. Yilmaz, E.; Avci, M.; Bulut, M.; Kelestimur, H.; Karakurt, L.; Ozercan, I. The Effect of Seprafilm on Adhesion Formation and Tendon Healing After Flexor Tendon Repair in Chicken. *Orthopedics* **2010**, *33*, 164–170. [CrossRef] [PubMed]
163. Işik, S.; Öztürk, S.; Gürses, S.; Yetmez, M.; Güler, M.M.; Selmanpakoglu, N.; Günhan, Ö. Prevention of restrictive adhesions in primary tendon repair by HA-membrane: Experimental research in chickens. *Br. J. Plast. Surg.* **1999**, *52*, 373–379. [CrossRef]
164. Na, S.Y.; Oh, S.H.; Song, K.S.; Lee, J.H. Hyaluronic acid/mildly crosslinked alginate hydrogel as an injectable tissue adhesion barrier. *J. Mater. Sci. Mater. Med.* **2012**, *23*, 2303–2313. [CrossRef]
165. Sun, J.; Tan, H. Alginate-Based Biomaterials for Regenerative Medicine Applications. *Materials* **2013**, *6*, 1285–1309. [CrossRef]
166. Namba, J.; Shimada, K.; Saito, M.; Murase, T.; Yamada, H.; Yoshikawa, H. Modulation of peritendinous adhesion formation by alginate solution in a rabbit flexor tendon model. *J. Biomed. Mater. Res. Part B Appl. Biomater.* **2007**, *80*, 273–279. [CrossRef]
167. Shao, H.J.; Lee, Y.T.; Chen, C.S.; Wang, J.H.; Young, T.H. Modulation of gene expression and collagen production of anterior cruciate ligament cells through cell shape changes on polycaprolactone/chitosan blends. *Biomaterials* **2010**, *31*, 4695–4705. [CrossRef] [PubMed]
168. Xia, C.-S.; Hong, G.-X.; Dou, R.-R.; Yang, X.-Y. Effects of chitosan on cell proliferation and collagen production of tendon sheath fibroblasts, epitenon tenocytes, and endotenon tenocytes. *Chin. J. Traumatol. (Zhonghua Chuang Shang Za Zhi)* **2005**, *8*, 369–374. [PubMed]
169. Chen, Q.; Lu, H.U.I.; Yang, H.U. Chitosan prevents adhesion during rabbit flexor tendon repair via the sirtuin 1 signaling pathway. *Mol. Med. Rep.* **2015**, *12*, 4598–4603. [CrossRef] [PubMed]
170. Radwan, M.A.; Farrag, S.A.A.; Abu-Elamayem, M.M.; Ahmed, N.S. Extraction, characterization, and nematicidal activity of chitin and chitosan derived from shrimp shell wastes. *Biol. Fertil. Soils* **2012**, *48*, 463–468. [CrossRef]
171. Busch, F.; Mobasheri, A.; Shayan, P.; Stahlmann, R.; Shakibaei, M. Sirt-1 is required for the inhibition of apoptosis and inflammatory responses in human tenocytes. *J. Biol. Chem.* **2012**, *287*, 25770–25781. [CrossRef]
172. Chen, C.H.; Chen, S.H.; Mao, S.H.; Tsai, M.J.; Chou, P.Y.; Liao, C.H.; Chen, J.P. Injectable thermosensitive hydrogel containing hyaluronic acid and chitosan as a barrier for prevention of postoperative peritoneal adhesion. *Carbohydr. Polym.* **2017**, *173*, 721–731. [CrossRef]
173. Yousefi, A.; Sarrafzadeh-Rezaei, F.; Asri-Rezaei, S.; Farshid, A.-A.; Behfar, M. Fabrication of novel tubular scaffold for tendon repair from chitosan in combination with zinc oxide nanoparticles. *Vet. Res. Forum* **2018**, *9*, 105–111. [CrossRef]
174. Jayasuriya, A.C.; Aryaei, A.; Jayatissa, A.H. ZnO nanoparticles induced effects on nanomechanical behavior and cell viability of chitosan films. *Mater. Sci. Eng. C Mater. Biol. Appl.* **2013**, *33*, 3688–3696. [CrossRef]
175. Saper, R.B.; Rash, R. Zinc: An essential micronutrient. *Am. Fam. Physician* **2009**, *79*, 768–772.
176. Sudheesh Kumar, P.T.; Lakshmanan, V.-K.; Anilkumar, T.V.; Ramya, C.; Reshmi, P.; Unnikrishnan, A.G.; Nair, S.V.; Jayakumar, R. Flexible and Microporous Chitosan Hydrogel/Nano ZnO Composite Bandages for Wound Dressing: In Vitro and In Vivo Evaluation. *ACS Appl. Mater. Interfaces* **2012**, *4*, 2618–2629. [CrossRef]
177. Ganji, F.; Abdekhodaie, M.J.; Ramazani, A. Gelation time and degradation rate of chitosan-based injectable hydrogel. *J. Sol-Gel Sci. Technol.* **2007**, *42*, 47–53. [CrossRef]
178. Yang, Q.Q.; Zhang, L.; Ju, F.; Zhou, Y.L. Sustained-Release Hydrogel-Based Rhynchophylline Delivery System Improved Injured Tendon Repair. *Colloids Surf. B BioInterfaces* **2021**, *205*, 111876. [CrossRef] [PubMed]
179. Ahmed, E.M. Hydrogel: Preparation, characterization, and applications: A review. *J. Adv. Res.* **2015**, *6*, 105–121. [CrossRef] [PubMed]
180. Liao, J.C.Y.; He, M.; Gan, A.W.T.; Wen, F.; Tan, L.P.; Chong, A.K.S. The effects of bi-functional anti-adhesion scaffolds on flexor tendon healing in a rabbit model. *J. Biomed. Mater. Res. Part B Appl. Biomater.* **2018**, *106*, 2605–2614. [CrossRef]
181. Nifant'ev, I.; Ivchenko, P. Coordination ring-opening polymerization of cyclic esters: A critical overview of DFT modeling and visualization of the reaction mechanisms. *Molecules* **2019**, *24*, 4117. [CrossRef]
182. Kade, J.C.; Dalton, P.D. Polymers for Melt Electrowriting. *Adv. Healthc. Mater.* **2021**, *10*, 2001232. [CrossRef]
183. King, W.E.; Bowlin, G.L. Near-Field Electrospinning and Melt Electrowriting of Biomedical Polymers—Progress and Limitations. *Polymers* **2021**, *13*, 1097. [CrossRef]
184. Chiulan, I.; Frone, A.N.; Brandabur, C.; Panaitescu, D.M. Recent Advances in 3D Printing of Aliphatic Polyesters. *Bioengineering* **2018**, *5*, 2. [CrossRef]
185. Jain, S.; Fuoco, T.; Yassin, M.A.; Mustafa, K.; Finne-Wistrand, A. Printability and Critical Insight into Polymer Properties during Direct-Extrusion Based 3D Printing of Medical Grade Polylactide and Copolyesters. *Biomacromolecules* **2020**, *21*, 388–396. [CrossRef]
186. Vieira, A.F.C.; Vieira, J.F.C.; Marques, A.T.; Guedes, R.M. Biomechanics of biomaterials used in soft tissue regenerative medicine. *Biomecánica* **2009**, *17*, 7–14. [CrossRef]
187. Liu, S.; Qin, M.; Hu, C.; Wu, F.; Cui, W.; Jin, T.; Fan, C. Tendon healing and anti-adhesion properties of electrospun fibrous membranes containing bFGF loaded nanoparticles. *Biomaterials* **2013**, *34*, 4690–4701. [CrossRef] [PubMed]
188. Liu, S.; Pan, G.; Liu, G.; das Neves, J.; Song, S.; Chen, S.; Cheng, B.; Sun, Z.; Sarmento, B.; Cui, W.; et al. Electrospun fibrous membranes featuring sustained release of ibuprofen reduce adhesion and improve neurological function following lumbar laminectomy. *J. Control. Release* **2017**, *264*, 1–13. [CrossRef] [PubMed]

189. Wang, W.; Wang, Y.; Lou, T.; Ding, M.; Li, J.; Xiong, H.; Yao, Z.; Ma, Y.; Chen, H.; Liu, S. Celecoxib-Loaded Electrospun Fibrous Antiadhesion Membranes Reduce COX-2/PGE_2 Induced Inflammation and Epidural Fibrosis in a Rat Failed Back Surgery Syndrome Model. *Neural Plast.* **2021**, *2021*, 6684176. [CrossRef]
190. Gunasekaran, T.; Nigusse, T.; Dhanaraju, M.D. Silver Nanoparticles as Real Topical Bullets for Wound Healing. *J. Am. Coll. Clin. Wound Spec.* **2011**, *3*, 82–96. [CrossRef] [PubMed]
191. Ge, L.; Li, Q.; Wang, M.; Ouyang, J.; Li, X.; Xing, M.M.Q. Nanosilver particles in medical applications: Synthesis, performance, and toxicity. *Int. J. Nanomed.* **2014**, *9*, 2399–2407. [CrossRef]
192. Wong, K.K.Y.; Liu, X. Silver nanoparticles—The real "silver bullet" in clinical medicine? *Med. Chem. Commun.* **2010**, *1*, 125–131. [CrossRef]
193. Chen, C.H.; Chen, S.H.; Shalumon, K.T.; Chen, J.P. Prevention of peritendinous adhesions with electrospun polyethylene glycol/polycaprolactone nanofibrous membranes. *Colloids Surf. B Biointerfaces* **2015**, *133*, 221–230. [CrossRef]
194. Malikmammadov, E.; Tanir, T.E.; Kiziltay, A.; Hasirci, V.; Hasirci, N. PCL and PCL-based materials in biomedical applications. *J. Biomater. Sci. Polym. Ed.* **2018**, *29*, 863–893. [CrossRef]
195. Liu, C.; Tian, S.; Bai, J.; Yu, K.; Liu, L.; Liu, G.; Dong, R.; Tian, D. Regulation of ERK1/2 and SMAD2/3 pathways by using multi-layered electrospun PCL–amnion nanofibrous membranes for the prevention of post-surgical tendon adhesion. *Int. J. Nanomed.* **2020**, *15*, 927–942. [CrossRef]
196. Yu, H.; Mei, S.; Zhao, L.; Zhao, M.; Wang, Y.; Zhu, H.; Wang, Y.; Wu, J.; Cui, C.; Xu, W.; et al. RGD-peptides modifying dexamethasone: To enhance the anti-inflammatory efficacy and limit the risk of osteoporosis. *Medchemcomm* **2015**, *6*, 1345–1351. [CrossRef]
197. The Manchester General Practitioner Group. A study of naproxen and ibuprofen in patients with osteoarthritis seen in general practice. *Curr. Med. Res. Opin.* **1984**, *9*, 41–46. [CrossRef] [PubMed]
198. Cooper, S.A.; Desjardins, P.; Brain, P.; Paredes-Diaz, A.; Troullos, E.; Centofanti, R.; An, B. Longer analgesic effect with naproxen sodium than ibuprofen in post-surgical dental pain: A randomized, double-blind, placebo-controlled, single-dose trial. *Curr. Med. Res. Opin.* **2019**, *35*, 2149–2158. [CrossRef] [PubMed]
199. Seager, J.M.; Cullen, D.J.; Pearson, G.; Holmes, S.; Doherty, M.; Wilson, J.V.; Garrud, P.; Garner, S.; Maynard, A.; Logan, R.F.; et al. Ibuprofen versus other non-steroidal anti-inflammatory drugs: Use in general practice and patient perception. *Aliment. Pharmacol. Ther.* **2000**, *14*, 187–191. [CrossRef] [PubMed]
200. Chen, C.H.; Li, D.L.; Chuang, A.D.C.; Dash, B.S.; Chen, J.P. Tension stimulation of tenocytes in aligned hyaluronic acid/platelet-rich plasma-polycaprolactone core-sheath nanofiber membrane scaffold for tendon tissue engineering. *Int. J. Mol. Sci.* **2021**, *22*, 11215. [CrossRef]
201. Yoshimi, R.; Yamada, Y.; Ito, K.; Nakamura, S.; Abe, A.; Nagasaka, T.; Okabe, K.; Kohgo, T.; Baba, S.; Ueda, M. Self-Assembling Peptide Nanofiber Scaffolds, Platelet-Rich Plasma, and Mesenchymal Stem Cells for Injectable Bone Regeneration with Tissue Engineering. *J. Craniofac. Surg.* **2009**, *20*, 1523–1530. [CrossRef]
202. Wang, X.; Qiu, Y.; Triffitt, J.; Carr, A.; Xia, Z.; Sabokbar, A. Proliferation and differentiation of human tenocytes in response to platelet rich plasma: An in vitro and in vivo study. *J. Orthop. Res. Off. Publ. Orthop. Res. Soc.* **2012**, *30*, 982–990. [CrossRef]
203. Liu, S.; Wu, F.; Gu, S.; Wu, T.; Chen, S.; Chen, S.; Wang, C.; Huang, G.; Jin, T.; Cui, W.; et al. Gene Silencing via PDA/ERK2-siRNA-Mediated Electrospun Fibers for Peritendinous Antiadhesion. *Adv. Sci.* **2019**, *6*, 1801217. [CrossRef]
204. Heidenreich, O. Targeting oncogenes with siRNAs. *Methods Mol. Biol.* **2009**, *487*, 221–242. [CrossRef]
205. Kurrikoff, K.; Gestin, M.; Langel, Ü. Recent in vivo advances in cell-penetrating peptide-assisted drug delivery. *Expert Opin. Drug Deliv.* **2016**, *13*, 373–387. [CrossRef]
206. Gwak, S.-J.; Nice, J.; Zhang, J.; Green, B.; Macks, C.; Bae, S.; Webb, K.; Lee, J.S. Cationic, amphiphilic copolymer micelles as nucleic acid carriers for enhanced transfection in rat spinal cord. *Acta Biomater.* **2016**, *35*, 98–108. [CrossRef]
207. Xiang, B.; Jia, X.-L.; Qi, J.-L.; Yang, L.-P.; Sun, W.-H.; Yan, X.; Yang, S.-K.; Cao, D.-Y.; Du, Q.; Qi, X.-R. Enhancing siRNA-based cancer therapy using a new pH-responsive activatable cell-penetrating peptide-modified liposomal system. *Int. J. Nanomed.* **2017**, *12*, 2385–2405. [CrossRef] [PubMed]
208. Hinterleitner, R.; Gruber, T.; Pfeifhofer-Obermair, C.; Lutz-Nicoladoni, C.; Tzankov, A.; Schuster, M.; Penninger, J.M.; Loibner, H.; Lametschwandtner, G.; Wolf, D.; et al. Adoptive transfer of siRNA Cblb-silenced CD8+ T lymphocytes augments tumor vaccine efficacy in a B16 melanoma model. *PLoS ONE* **2012**, *7*, e44295. [CrossRef] [PubMed]
209. Yeh, Y.-C.; Huang, T.-H.; Yang, S.-C.; Chen, C.-C.; Fang, J.-Y. Nano-Based Drug Delivery or Targeting to Eradicate Bacteria for Infection Mitigation: A Review of Recent Advances. *Front. Chem.* **2020**, *8*, 286. [CrossRef] [PubMed]
210. Yoshizaki, Y.; Yuba, E.; Komatsu, T.; Udaka, K.; Harada, A.; Kono, K. Improvement of Peptide-Based Tumor Immunotherapy Using pH-Sensitive Fusogenic Polymer-Modified Liposomes. *Molecules* **2016**, *21*, 1284. [CrossRef]
211. Gandhi, A.; Paul, A.; Sen, S.O.; Sen, K.K. Studies on thermoresponsive polymers: Phase behaviour, drug delivery and biomedical applications. *Asian J. Pharm. Sci.* **2015**, *10*, 99–107. [CrossRef]
212. Dong, M.; Shi, B.; Liu, D.; Liu, J.-H.; Zhao, D.; Yu, Z.-H.; Shen, X.-Q.; Gan, J.-M.; Shi, B.; Qiu, Y.; et al. Conductive Hydrogel for a Photothermal-Responsive Stretchable Artificial Nerve and Coalescing with a Damaged Peripheral Nerve. *ACS Nano* **2020**, *14*, 16565–16575. [CrossRef]
213. Zhang, K.; Lv, H.; Zheng, Y.; Yao, Y.; Li, X.; Yu, J.; Ding, B. Nanofibrous hydrogels embedded with phase-change materials: Temperature-responsive dressings for accelerating skin wound healing. *Compos. Commun.* **2021**, *25*, 100752. [CrossRef]

polymers

MDPI

Article

Bacterial Cellulose Membranes as Carriers for Nisin: Incorporation, Antimicrobial Activity, Cytotoxicity and Morphology

Gabriela Ribeiro dos Santos [1], Victória Soares Soeiro [1], Carolina Fernanda Talarico [1], Janaína Artem Ataide [2], André Moreni Lopes [2], Priscila Gava Mazzola [2], Thais Jardim Oliveira [1,3], José Martins Oliveira Junior [4], Denise Grotto [4] and Angela F. Jozala [1,*]

1 LAMINFE—Laboratory of Industrial Microbiology and Fermentation Process, University of Sorocaba, Sorocaba 18023-000, SP, Brazil
2 Faculty of Pharmaceutical Science, University of Campinas (Unicamp), Campinas 13083-871, SP, Brazil
3 LAFINAU—Laboratory of Nuclear Physics, University of Sorocaba, Sorocaba 18023-000, SP, Brazil
4 LAPETOX—Laboratory of Toxicological Research, University of Sorocaba, Sorocaba 18023-000, SP, Brazil
* Correspondence: angela.jozala@prof.uniso.br

Abstract: Based on the previous study, in which nisin and bacterial cellulose were utilized, this new experiment loads nisin into bacterial cellulose (N–BC) and evaluates the morphological characteristics, cytotoxicity, antimicrobial activity and stability of the developed system. The load efficiency of nisin in BC was evaluated by an agar diffusion assay, utilizing *Lactobacillus sakei*, and total proteins. After having found the ideal time and concentration for the loading process, the system stability was evaluated for 100 days at 4, 25 and 37 °C against *Staphylococcus aureus* and *L. sakei*. Thus, in this study, there is a system that proves to be efficient, once BC has enhanced the antimicrobial activity of nisin, acting as a selective barrier for other compounds present in the standard solution and protecting the peptide. After 4 h, with 45% of proteins, this activity was almost 2 $\log_{10}$ higher than that of the initial solution. Once the nisin solution was not pure, it is possible to suggest that the BC may have acted as a filter. This barrier enhanced the nisin activity and, as a consequence of the nisin loading, a stable N–BC system formed. The N–BC could create meaningful material for pharmaceutical and food applications.

Keywords: antimicrobial activity; bacterial cellulose; cytotoxicity; nisin; stability

Citation: dos Santos, G.R.; Soeiro, V.S.; Talarico, C.F.; Ataide, J.A.; Lopes, A.M.; Mazzola, P.G.; Oliveira, T.J.; Oliveira Junior, J.M.; Grotto, D.; Jozala, A.F. Bacterial Cellulose Membranes as Carriers for Nisin: Incorporation, Antimicrobial Activity, Cytotoxicity and Morphology. *Polymers* **2022**, *14*, 3497. https://doi.org/10.3390/polym14173497

Academic Editor: Arn Mignon

Received: 16 February 2022
Accepted: 20 July 2022
Published: 26 August 2022

1. Introduction

Antimicrobial peptides are synthesized by several microorganisms. Nisin is a lantibiotic peptide, with 34 amino acids, and is secreted by Lactococcus [1–4]. The antimicrobial property of nisin is attributed to pore formation in the cell membrane of microorganisms, with specific binding to the lipidic precursor of the cell wall, attached to the membrane [4,5].

Nisin is considered safe by the World Health Organization (WHO) and the Food and Drug Administration (FDA-United States), being used initially as a food additive [6,7]. Further, its antimicrobial action encourages the clinical use of nisin, whether in topical or systemic therapies, due to its broad-spectrum activity and lower probability of developing microbial resistance [6–8]. The application of nisin extends to several medical areas, from mastitis to oral and gastrointestinal diseases [5,9,10].

Bacterial cellulose (BC) is a polysaccharide that is extracellularly secreted by several microorganisms, such as *Agrobacterium, Rhizobium, Escherichia, Sarcina*, and *Acetobacter* [11–13]. Namely, *Komagataeibacter xylinus* is a non-pathogenic Gram-negative bacterium that can produce significant amounts of cellulose [14]. BC is a linear glucose polymer, formed by a matrix of nanofibers, giving it porous characteristics in a three-dimensional network structure. Although its structure resembles vegetable cellulose, bacterial cellulose presents a high degree of purity, crystallinity, tensile strength, and high-water absorption [15].

Due to its biocompatibility and non-toxicity, the BC application has been directed to medical devices and tissue engineering [16]. Additionally, owing to its structural model, BC has been proposed as an ideal dermal substitute that is capable of inducing the direction of cells for repair in tissue reconstruction [17,18]. BC has been used as wound dressing because it forms a physical barrier against infections. It also allows gas exchange, absorbs exudates, keeps the wound moist to favor tissue reconstitution and it can be easily removed [19].

The incorporation of biomolecules into BC has been studied both by increasing its antibacterial or enzymatic properties and by providing a control release system [20,21]. For instance, nisin immobilization in solid matrices such as bacterial cellulose membranes could control its release [22]. Due to these properties, our research group previously evaluated the antimicrobial and antioxidant activity of nisin loaded into BC [23]. Since the work yielded satisfactory results, our group has provided a supplementary study. For this reason, this study has evaluated BC membranes as carriers for nisin regarding the morphological characteristics, cytotoxicity, and stability of the developed system.

2. Materials and Methods

2.1. Materials

The standard nisin and the bicinchoninic acid kit were purchased from Sigma-Aldrich (São Paulo, Brazil). All other reagents were of analytical grade.

2.2. BC Production and Purification

BC was produced by *Komagataeibacter xylinus* ATCC 53582, using 24-well plates with Hestrin and Schramm medium. Each well was filled with 1 mL and incubated at 30 °C in static conditions for seven days. After that, membranes were immersed in a 2% sulfate dodecyl sodium (SDS) solution under stirring overnight and washed in running water. The bleaching process was carried out with 4% NaOH for one hour until reaching 60 °C. The BC membranes were washed in order to remove NaOH and sterilized at 121 °C for 15 min [23,24].

2.3. Total Proteins Quantification

Protein concentration was determined by a bicinchoninic acid assay [25]. Bovine serum albumin (BSA) with different concentrations from 0.1 to 1.0 mg/mL was used as the standard protein. Absorbance reading was performed on 96-well microplates with a wavelength of 562 nm, by spectroscopy (Infinite M200 PRO, RCHISTO, Barueri, Brazil). Analyses were performed in triplicate, and the mean of these absorbance values was used to determine protein concentrations.

2.4. Nisin Standard Curve by Agar Diffusion Assay

The standard nisin (Sigma-Aldrich, St. Louis, MO, USA—containing 2.5% nisin, with 1,000,000 IU/g in its composition) solution was prepared by dissolving 1 g of nisin in 10 mL of phosphate buffer solution (PBS-pH 7.0), as in Table 1. The solution was centrifuged at 13,201 g for 10 min at 10 °C and the supernatant was collected and filtered in a 0.22 μm membrane (Millipore, Burlington, MA, USA). The nisin standard curve was evaluated by an agar diffusion assay. The nisin bioindicator *Lactobacillus sakei* ATCC 15521 was used for the agar diffusion assay [6,23].

Table 1. Relation of concentration of standard nisin product, containing 2.5% of nisin, to its activity in AU and Log_{10}AU.

Standard Nisin g/mL	0.1	0.01	0.001	0.0001
Standard Nisin Considering 2.5%	2500	250	25	2.5
Nisin Activity AU/mL	100,000	10,000	1000	100
Nisin Activity Log_{10} AU/mL	5	4	3	2

The concentrations of standard nisin were related by the diameter of the inhibition halo (H, mm), and the activity of nisin was determined and expressed in arbitrary units per mL (AU/mL). The activity of nisin was based on the dilution of the standard nisin calibration curves. The correlation between AU/mL and international units per mL (IU/mL) was 1.09 ± 0.17 AU to 1.0 IU (40 IU = 1 μg of pure nisin A) [14,26].

2.5. Evaluation of Nisin Loaded into BC by Time

A dose of 250 μg/mL of nisin was employed based on previous studies from our research group [23]. The BC membranes were arranged in 24-well plates and 1.0 mL of nisin solution (250 μg/mL) was added in each well. The plate was placed on a shaker (NT 715, Nova Tecnica) at 25 °C, under 100 rpm for 4, 8, 12, 18 and 24 h. After each period, residual samples were collected in order to determine the nisin load efficiency by the concentration of proteins and nisin activity. The BC membranes were collected and evaluated by the agar diffusion assay for the measurement of the nisin activity.

The residual samples were evaluated by the agar diffusion assay and the total protein method. The load efficiency (LE) of nisin was evaluated by the total protein method and calculated by the following Equation (1), which was elaborated by the authors:

$$\%LE = \frac{\text{Protein in the initial solution} - \text{Protein in the Residual Solution}}{\text{Protein in the initial solution}} \times 100 \quad (1)$$

2.6. Stability Test by Agar Difusion Assay

In this test, *Staphylococcus aureus* ATCC 10390 and *Lactobacillus sakei* ATCC 15521 were utilized. *S. aureus* was grown in Tryptic Soy Broth (TSB) supplied by Becton, Dickinson and Company (BD Difco™, Franklin Lakes, NJ, USA) for 24 h at 37 °C, whereas *L. sakei* was cultivated in De Man, Rogosa and Sharpe broth (MRS) supplied by Sigma-Aldrich Corporation (Sigma-Aldrich, St. Louis, MO, USA) for 24 h at 30 °C. All the media were prepared in distilled water and autoclaved at 121 °C for 30 min.

After the growth, the microorganisms were counted by the Pour Plate technique and the sample with 10^6 UFC/mL was selected to carry out the agar diffusion assay. For *S. aureus*, an amount of 20 mL of TSB agar was placed in a Petri plate. After the solidification, 1 mL was placed on the agar surface. For *L. sakei*, an amount of 20 mL of MRS agar was placed in a Petri plate. After the solidification, 1 mL (10^6 UFC/mL) was also placed on the agar surface.

The BC membranes loaded with nisin (N–BC) were kept in Petri plates and incubated at different temperatures (4, 25, 37 °C)—pre-set temperature conditions for stability evaluation—for 100 days. The samples were collected each day and the nisin activity was evaluated by the agar diffusion assay, adapted from ISO 20645, with *L. sakei* and *S. aureus*.

The BC was placed on an agar surface and the possible clear zones were observed after its incubation at 37 °C for 24 h. The presence of a clear halo that formed around BC was measured and indicated the antimicrobial activity [1].

2.7. Cytotoxicity Assay

The standard nisin and N–BC at the best time and with different concentrations of 7, 15, 31, 62, 125, 250 μg/mL were arranged in 24-well plates with 500 μL of DMEM (Dulbecco's modified eagle's medium, Low Glucose, Sigma®), with the addition of 10% fetal bovine serum (FBS, Sigma®) and 1% antibiotic (PEN/STREP, Sigma®) in each well. In each well, there were 75,000 L929 fibroblast cells. The plates were incubated at 37 °C in an atmosphere of 5% CO2 for 24 and 48 h. The assay was performed in triplicate and for positive control, PBS and DMEM were used.

After 24 and 48 h, cell viability was determined by the mitochondrial activity of the cell culture by the colorimetric assay MTS [3-(4,5-dimethylthiazol-2-yl)-5-(3-carboxymethoxyphenyl)-2-(4-sulfophenyl)-2H-tetrazolium]. Absorbance readings were taken at 490 nm, after 1 h of incubation of the cells with the reagent [27,28]. The results are expressed as mean percentage ± standard deviation (SD) of viable cells in relation to the positive control.

2.8. Morphology

2.8.1. Scanning Electron Microscopy (SEM)

The BC and N–BC (with 250 μ/mL of nisin) were stored at −80 °C for 24 h and lyophilized for 48 h. BC and N–BC were set with carbon tape and metallized for 2 min (DH-29010SCTR Smart Coater). SEM images were observed on the scanning electron microscopy equipment (JEOL, Tokyo, Japan, Model IT200) and obtained using an accelerating voltage of 20 kV.

2.8.2. X-ray Microtomography (μCT)

This equipment is responsible for analyzing pore size, porosity (%), and the interconnectivity of the pores in the membranes [29]. The membrane pictures were captured by X-ray microtomography (Bruckermicro CT-SkyScan 1174, Kontich, Belgium). The X-ray source was 29 kV and 661 mA. Here, 3D virtual models, representative of various sections of membranes, were built, and the data were mathematically managed by CT Analyzer software, v. 1.13.5.2.2.8.

2.9. Statistical Analyses

Differences among nisin concentrations are presented as mean ± standard deviation. ANOVA (analysis of variance) was used for parametric data, followed by Duncan or Scheffe tests for multiple comparisons. A p value < 0.05 was accepted as statistically significant. The data were analyzed using Statistica®8.0 (Statsoft software, Tulsa, OK, USA).

3. Results and Discussion

3.1. Nisin Standard Curve

For the nisin standard curve, zones of inhibition showed different sizes of halo ranging from 14 to 36.30 mm. Each halo corresponded to one concentration of nisin ranging from 2.5 μg/mL to 2500 μg/mL, as seen in Figure 1.

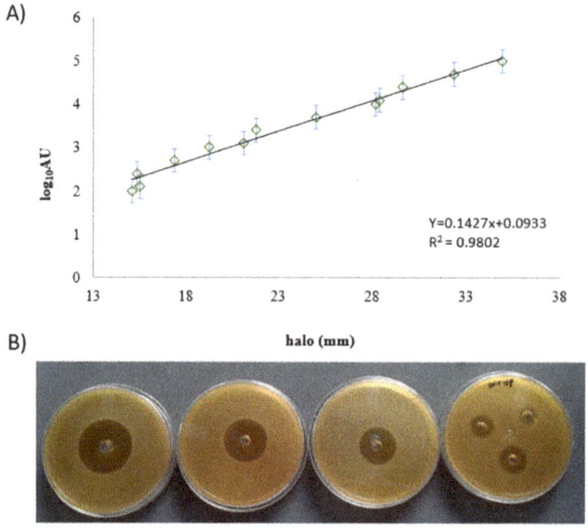

Figure 1. (**A**). Nisin standard curve by agar diffusion assay. (**B**) Agar diffusion assay image to represent the diameter of the inhibition halo.

The standard curve demonstrates the sensibility of the bioindicator *L. sakei* even when it was exposed to low nisin concentration. This finding is important in establishing the relationship between the amount of nisin and the method that was used [30]. Nisin is a soluble peptide and its low molecular weight facilitates its diffusion in other matrices, supplementing the antimicrobial activity [31,32].

3.2. Evaluation of Nisin Loading in BC by Time

The nisin concentration was set at 250 μg/mL and we were able to observe the loading efficiency (%LE) by amount of protein (mg/mL) and nisin activity ($\log_{10}$AU) at different periods (Table 2). The protein concentration was not directly proportional to AU. Despite different incubation periods, there was little variation in nisin titers as observed in Table 2. These results are key to the definition of the best time for the load efficiency. After 4 h, the N–BC with 45% proteins was almost 2 $\log_{10}$ higher than the initial solution. Once the nisin solution was not pure, it is possible to suggest that the BC may have acted as a filter, separating the nisin from the other components and maybe causing overly strong adsorption kinetics in 4 h [33].

Table 2. The load efficiency (LE) of the nisin in bacterial cellulose by concentration of proteins compared to the nisin activity in different periods.

Time (h)	% LE	Log_{10}AU
0	0	4.0 ± 0.00
4	43.7 ± 5.0	5.9 ± 0.12
8	47.6 ± 3.0	5.5 ± 0.35
12	42.9 ± 1.0	5.4 ± 0.08
18	40.0 ± 1.0	5.2 ± 0.12
24	17.8 * ± 3.0	5.5 ± 0.19

* $p < 0.05$ compared to each period (ANOVA test, followed by Duncan test).

The same observation was reported by Ataide and colleagues [20]; they evaluated the enzyme bromelain loaded into the BC membranes, and its release. They noticed a threefold increase in the specific bromelain activity when compared to the initial solution, with 31% of proteins when bromelain was loaded into BC, showing the selective behavior of the BC membrane.

Jorge et al. [34] studied a system based on BC for the stabilization of insulin. The results indicate that BC was able to enhance permeation. Nguyen et al. [8] also evaluated the nisin loaded in the BC membranes. However, they reported 6 h as the best period for nisin incorporation in the BC membranes, two hours longer than in our study. Moreover, Nguyen and colleagues used 600 IU/mL in the initial solution, and we experimented with 10,000 AU/mL (4 $\log_{10}$AU).

According to Moniri et al. [35], BC has been extensively studied for controlled drug delivery as its structure is suitable for the dispensing of biomolecules.

This structure, which consists of nanofibers, as reported in the SEM section, probably explains the enhancement of nisin's activity when loaded into BC. The BC nanofibers could act as a filter, retaining big molecules of dirt from the nisin solution and selectively allowing the absorption of the low-weight molecules, such as nisin.

3.3. Stability Test by Agar Diffusion Assay

Figure 2A shows the qualitative agar diffusion test of N–BC against *L. sakei* (Figure 2(Aa)) and *S. aureus* (Figure 2(Ab)) after 7 days of storage at room temperature. Additionally, the quantitative tests of stability against *L. sakei* and *S. aureus* are shown in Figure 2B,C, respectively.

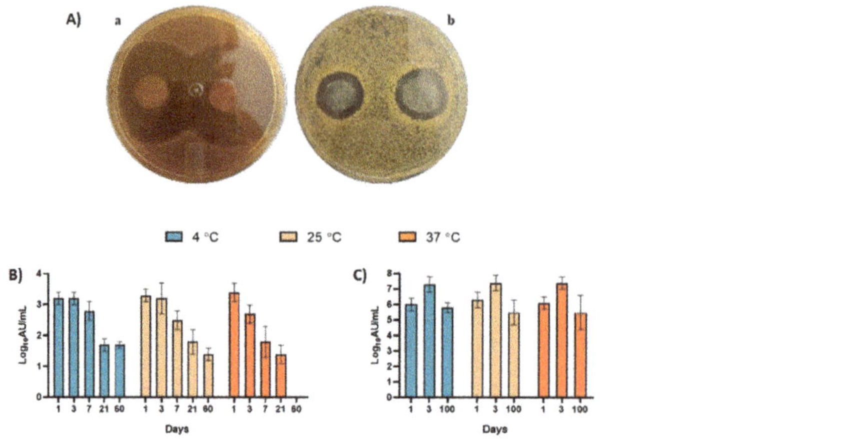

Figure 2. (**A**) Agar diffusion assay of the nisin loaded in BC (N–BC) against *L. sakei* (**a**) and *S. aureus* (**b**). (**B**) N–BC stability after storage at 4, 25 and 37 °C evaluated by agar diffusion assay against *S. aureus* ATCC 10390. (**C**) N–BC stability after storage at 4, 25 and 37 °C evaluated by agar diffusion assay against *L. sakei* ATCC 15521.

In general, the N–BC showed antimicrobial activity against *S. aureus* for 60 days at 4 and 25 °C. No antimicrobial activity was observed for 60 days at 37 °C. Despite having antimicrobial activity, N–BC was statistically stable for up to 7 days at 4 and 25 °C. At 37 °C, N–BC was stable for up to 3 days. After these periods of time, a significant decrease in activity was revealed.

On the other hand, N–BC was stable for up to 100 days against *L. sakei* at all incubation temperatures (Figure 2C), and no significant difference was seen.

By observing these results, we could hypothesize that if the N–BC system was used as wound dressing in tissues contaminated with *S. aureus*, it should be replaced—or even reapplied—after 7 days. Furthermore, if the N–BC system was used in food packaging, it should be considered for fresh food wrapping.

The process of nisin purification, described in the literature, shows that the activity of nisin can be reduced by fat clusters, salt concentrations and the aggregation of soluble proteins. It also shows that, after purification, nisin's antimicrobial activity was increased [26].

Staphylococci are bacteria that live on the epithelial surfaces of humans and animals, being responsible for food poisoning and infections. With respect to bacterial resistance, nisin has been proven to be useful for limiting the development of antimicrobial resistance [36,37]. *S. aureus* also appears among the main bacteria responsible for secondary infections in burns [38,39].

Nisin's antimicrobial activity against *S. aureus* generated a significant decrease in viable cells in a period of 1 day, and even inhibited biofilm formation [40]. Nisin encapsulation in nanofiber polymer provides stability to nisin for a period of up to seven days [39]. However, longer periods produced an inhibition zone inferior to what is presented in our study.

3.4. Cytotoxicity Assay

Figure 3 shows the cytotoxicity data of N–BC at 24 and 48 h, at different concentrations. The fibroblast cells did not exhibit any significant change after 24 h of treatment. On the other hand, after 48 h, proliferation rates were 25% higher than those produced after 24 h, indicating the N–BC was able to stimulate cell proliferation, and no toxicity was reported.

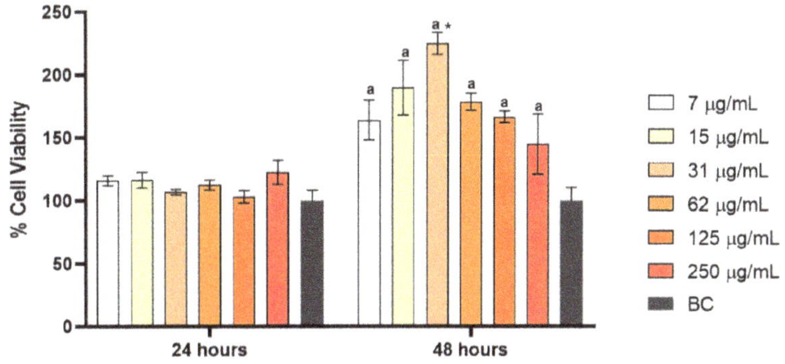

Figure 3. Cytotoxicity of nisin loaded in bacterial cellulose (N–BC) at concentrations of 7, 15, 31, 62, 125 and 250 µg/mL.; [a] $p < 0.05$ if compared to the respective concentrations at 24 h. * $p < 0.05$ if compared to the concentration of 7 µg/mL at 48 h.

Nisin is a safe molecule for both FDA and WHO [7–37], and BC is also considered a non-cytotoxic material [41,42]. Corroborating the Gao et al. [43] study, a cytotoxicity test of N–BC was performed in a co-culture and no cytotoxicity was observed. Proliferation in fibroblasts is important to accelerate the healing process.

The BC membranes loaded with nisin were used in large animals to treat surgical dehorn wounds [44]. The authors observed that the treatment with the BC membrane accelerated the healing process. Wound healing involves a dynamic set of tissue changes, important for maintaining the integrity of an organism. The authors also noted that BC may have created a barrier between the wound and the environment, preventing contamination.

In a previous work, dos Santos et al. [23] analyzed the antioxidant activity of nisin. The antioxidant activity may be related to the proliferation of fibroblasts since nisin has the potential to neutralize free radicals, facilitating cell proliferation.

3.5. Morphology

3.5.1. Scanning Electron Microscopy (SEM)

Figure 4A,B shows the BC without and with nisin, respectively. Comparing the morphology, it is possible to identify a similarity between the fibers on a micrometric and nanometric scale for both forms of BC. However, in N–BC, the fibers were more swollen than in BC, suggesting nisin incorporation.

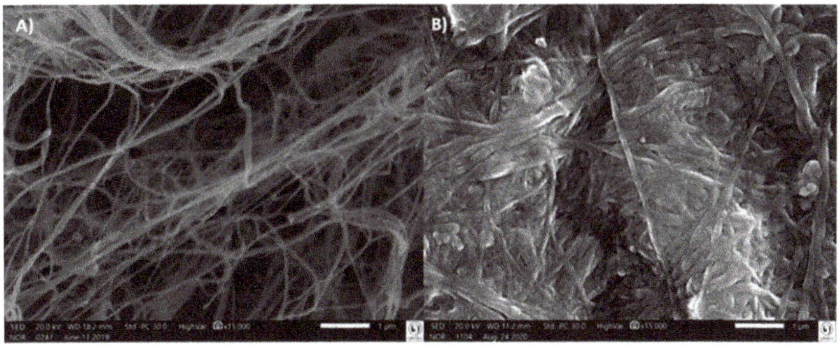

Figure 4. Scanning electron microscopy of the bacterial cellulose (BC) (**A**) and the bacterial cellulose loaded with nisin (N–BC) (**B**), at 15,000× magnification.

The fiber network accounts for a large surface area and the porous structure of the BC facilitates the immobilization of nisin. In addition, this structure allows the diffusion of water, providing a functional stabilization of the biomolecules.

3.5.2. X-ray Microtomography

The quantitative results of the structure were confirmed by X-ray microtomography assay, as presented in Table 3. The BC membranes presented more connectivity than the N–BC ones. Additionally, the presence of nisin decreased the porosity of membranes by 2%, supporting the SEM images. The pores were interconnected and distributed throughout the sample. The BC structure had a lower degree of anisotropy than the N–BC membrane.

Table 3. X-ray microtomography parameters of the bacterial cellulose (BC) and the bacterial cellulose with nisin (N–BC) membranes.

Parameter	BC	N–BC
Connectivity	22.421 ± 1.121	1.100 ± 0.55
Degree of anisotropy	0.692 ± 0.035	0.769 ± 0.038
Total porosity (%)	79.15 ± 3.96	77.95 ± 3.89
Open porosity (%)	79.17 ± 3.91	77.93 ± 3.89

Note: data were mathematically managed by CT Analyzer software, v. 1.13.5.2.2.8.

Regarding membrane morphology, Gedarawatte et al. [45] performed the SEM test in nanocrystal BC containing nisin, and the presence of the antimicrobial was identified in the swelling of samples. The same behavior was observed in BC and nisin samples examined in co-culture by Gao et al. [43], reporting a large presence of nisin in the morphology of swelling fibers. The increase in fiber size indicated a high amount of antibacterial substance loaded in the BC membrane. Gao et al. [43] additionally observed that nisin can affect the BC fibers, resulting in a slight decrease in crystallinity and improving the antimicrobial property.

4. Conclusions

Here, we have proven the possibility of loading nisin into the BC membranes in 4 h, and also improving the antimicrobial activity with less protein, which suggests the BC membrane served as a selective barrier and it separated nisin from other components. This barrier enhanced nisin's activity, and as a consequence of the nisin loading, a stable N–BC system was created and considered ideal for nisin delivery, with no toxicity and capacity to control the growth of microorganisms.

Thus, N–BC has the potential to be used as a dressing for skin wounds since nisin may prevent infections upon its release, requiring only a replacement at a point between 3 and 7 days.

Moreover, N–BC can be applied in in food packaging and other fields because its antimicrobial activity will probably reduce food deterioration and improve shelf-life.

Author Contributions: Conceptualization, G.R.d.S., V.S.S. and A.F.J.; methodology C.F.T., G.R.d.S., J.M.O.J., A.F.J. and D.G.; software, V.S.S. and T.J.O.; writing—original draft preparation V.S.S., T.J.O. and A.F.J.; writing—review and editing, P.G.M., J.A.A., A.M.L., D.G. and A.F.J.; supervision, A.F.J. All authors have read and agreed to the published version of the manuscript.

Funding: This research was funded by UNISO. Coordination for Higher Level Graduate Improvements (CAPES/Brazil, finance code 001), National Council for Scientific and Technological Development (CNPq/Brazil Process #428751/2016-4), and the State of São Paulo Research Foundation (FAPESP/Brazil, processes #2016/05930-4).

Informed Consent Statement: Not applicable.

Data Availability Statement: The data presented in this study are available on request from the corresponding author.

Acknowledgments: The authors acknowledge financial support from University of Sorocaba. Coordination for Higher Level Graduate Improvements (CAPES/Brazil, finance code 001), National Council for Scientific and Technological Development (CNPq/Brazil), and the State of São Paulo Research Foundation (FAPESP/Brazil).

Conflicts of Interest: The authors declare no conflict of interest.

References

1. Aldarhami, A.; Felek, A.; Sharma, V.; Upton, M. Purification and characterization of nisin P produced by a strain of *Streptococcus gallolyticus*. *J. Med. Microbiol.* **2020**, *69*, 605–616. [CrossRef] [PubMed]
2. Dosler, S.; Gerceker, A.A. In vitro activities of antimicrobial cationic peptides; melittin and nisin, alone or in combination with antibiotics against Gram-positive bacteria. *J. Chemother.* **2012**, *24*, 137–143. [CrossRef]
3. Krivorotova, T.; Cirkovas, A.; Maciulyte, S.; Staneviciene, R.; Budriene, S.; Serviene, E.; Sereikaite, J. Nisin-loaded pectin nanoparticles for food preservation. *Food Hydrocoll.* **2016**, *54*, 49–56. [CrossRef]
4. Peng, X.; Zhu, L.; Wang, Z.; Zhan, X. Enhanced stability of the bactericidal activity of nisin through conjugation with gellan gum. *Int. J. Biol. Macromol.* **2020**, *148*, 525–532. [CrossRef] [PubMed]
5. Wiedemann, I.; Breukink, E.; Kraaij, V.; Kuipers, O.P.; de Kruijff, B.; Chem, J.B.; Wiedemann, I.; Breukink, E.; van Kraaij, C.; Kuipers, O.P.; et al. Lipids and lipoproteins: Specific Binding of Nisin to the Peptidoglycan Precursor Lipid II Combines Pore Formation and Inhibition of Cell Wall Biosynthesis for Potent Antibiotic Activity Specific Binding of Nisin to the Peptidoglycan Precursor Lipid II C. *J. Biol. Chem.* **2001**, *276*, 1772–1779. [CrossRef] [PubMed]
6. de Arauz, L.J.; Jozala, A.F.; Baruque-Ramos, J.; Mazzola, P.G.; Júnior, A.P.; Penna, T.C.V. Culture medium of diluted skimmed milk for the production of nisin in batch cultivations. *Ann. Microbiol.* **2012**, *62*, 419–426. [CrossRef]
7. de Arauz, L.J.; Jozala, A.F.; Mazzola, P.G.; Vessoni Penna, T.C. Nisin biotechnological production and application: A review. *Trends Food Sci. Technol.* **2009**, *20*, 146–154. [CrossRef]
8. Nguyen, T.; Brody, H.; Lin, G.H.; Rangé, H.; Kuraji, R.; Ye, C.; Kamarajan, P.; Radaic, A.; Gao, L.; Kapila, Y. Probiotics, including nisin-based probiotics, improve clinical and microbial outcomes relevant to oral and systemic diseases. *Periodontology 2000* **2020**, *82*, 173–185. [CrossRef]
9. Kitazaki, K.; Koga, S.; Nagatoshi, K.; Kuwano, K.; Zendo, T.; Nakayama, J.; Sonomoto, K.; Ano, H.; Katamoto, H. In vitro synergistic activities of cefazolin and nisin a against mastitis pathogens. *J. Vet. Med. Sci.* **2017**, *79*, 1472–1479. [CrossRef]
10. Sandiford, S.K. Current developments in lantibiotic discovery for treating Clostridium difficile infection. *Expert Opin. Drug Discov.* **2019**, *14*, 71–79. [CrossRef]
11. Gromovykh, T.I.; Pigaleva, M.A.; Gallyamov, M.O.; Ivanenko, I.P.; Ozerova, K.E.; Kharitonova, E.P.; Bahman, M.; Feldman, N.B.; Lutsenko, S.V.; Kiselyova, O. IStructural organization of bacterial cellulose: The origin of anisotropy and layered structures. *Carbohydr. Polym.* **2020**, *237*, 116140. [CrossRef] [PubMed]
12. Hassan, E.; Abdelhady, H.; Abd l-Salam, S.; Abdullah, S. The Characterization of Bacterial Cellulose Produced by Acetobacter xylinum and Komgataeibacter saccharovorans under Optimized Fermentation Conditions. *Br. Microbiol. Res. J.* **2015**, *9*, 1–13. [CrossRef]
13. Huang, Y.; Zhu, C.; Yang, J.; Nie, Y.; Chen, C.; Sun, D. Recent advances in bacterial cellulose. *Cellulose* **2014**, *21*, 1–30. [CrossRef]
14. Wang, J.; Tavakoli, J.; Tang, Y. Bacterial cellulose production, properties and applications with different culture methods—A review. *Carbohydr. Polym.* **2019**, *219*, 63–76. [CrossRef]
15. Costa, A.F.S.; Almeida, F.C.G.; Vinhas, G.M.; Sarubbo, L.A. Production of bacterial cellulose by Gluconacetobacter hansenii using corn steep liquor as nutrient sources. *Front. Microbiol.* **2017**, *8*, 2027. [CrossRef]
16. Zhang, C.; Cao, J.; Zhao, S.; Luo, H.; Yang, Z.; Gama, M.; Zhang, Q.; Su, D.; Wan, Y. Biocompatibility evaluation of bacterial cellulose as a scaffold material for tissue-engineered corneal stroma. *Cellulose* **2020**, *27*, 2775–2784. [CrossRef]
17. de Oliveira Barud, H.G.; da Silva, R.R.; da Silva Barud, H.; Tercjak, A.; Gutierrez, J.; Lustri, W.R.; de Oliveira, O.B.; Ribeiro, S.J.L. A multipurpose natural and renewable polymer in medical applications: Bacterial cellulose. *Carbohydr. Polym.* **2016**, *153*, 406–420. [CrossRef]
18. Numata, Y.; Mazzarino, L.; Borsali, R. A slow-release system of bacterial cellulose gel and nanoparticles for hydrophobic active ingredients. *Int. J. Pharm.* **2015**, *486*, 217–225. [CrossRef]
19. Anton-Sales, I.; Beekmann, U.; Laromaine, A.; Roig, A.; Kralisch, D. Opportunities of Bacterial Cellulose to Treat Epithelial Tissues. *Curr. Drug Targets* **2019**, *20*, 808–822. [CrossRef]
20. Ataide, J.A.; De Carvalho, N.M.; Rebelo, M.D.A.; Chaud, M.V.; Grotto, D.; Gerenutti, M.; Rai, M.; Mazzola, P.G.; Jozala, A.F. Bacterial Nanocellulose Loaded with Bromelain: Assessment of Antimicrobial, Antioxidant and Physical-Chemical Properties. *Sci. Rep.* **2017**, *7*, 18031. [CrossRef]
21. Malheiros, P.S.; Jozala, A.F.; Pessoa-Jr, A.; Vila, M.M.D.C.; Balcão, V.M.; Franco, B.D.G.M. Immobilization of antimicrobial peptides from Lactobacillus sakei subsp. sakei 2a in bacterial cellulose: Structural and functional stabilization. *Food Packag. Shelf Life* **2018**, *17*, 25–29. [CrossRef]
22. Bayazidi, P.; Almasi, H.; Asl, A.K. Immobilization of lysozyme on bacterial cellulose nanofibers: Characteristics, antimicrobial activity and morphological properties. *Int. J. Biol. Macromol.* **2018**, *107*, 2544–2551. [CrossRef] [PubMed]

23. dos Santos, C.A.; dos Santos, G.R.; Soeiro, V.S.; dos Santos, J.R.; de Rebelo, M.A.; Chaud, M.V.; Gerenutti, M.; Grotto, D.; Pandit, R.; Rai, M.; et al. Bacterial nanocellulose membranes combined with nisin: A strategy to prevent microbial growth. *Cellulose* **2018**, *25*, 6681–6689. [CrossRef]
24. Jozala, A.F.; Pértile, R.A.N.; dos Santos, C.A.; de Carvalho Santos-Ebinuma, V.; Seckler, M.M.; Gama, F.M.; Pessoa, A. Bacterial cellulose production by Gluconacetobacter xylinus by employing alternative culture media. *Appl. Microbiol. Biotechnol.* **2014**, *99*, 1181–1190. [CrossRef] [PubMed]
25. Smith, P.K.; Krohn, R.I.; Hermanson, G.T.; Mallia, A.K.; Gartner, F.H.; Provenzano, M.D.; Fujimoto, E.K.; Goeke, N.M.; Olson, B.J.; Klenk, D.C. Measurement of protein using bicinchoninic acid. *Anal. Biochem.* **1985**, *150*, 76–85. [CrossRef]
26. Jozala, A.F.; De Lencastre Novaes, L.C.; Mazzola, P.G.; Oliveira-Nascimento, L.; Vessoni Penna, T.C.; Teixeira, J.A.; Passarinha, L.A.; Queiroz, J.A.; Pessoa, A. Low-cost purification of nisin from milk whey to a highly active product. *Food Bioprod. Process.* **2015**, *93*, 115–121. [CrossRef]
27. Aslantürk, Ö.S. In Vitro Cytotoxicity and Cell Viability Assays: Principles, Advantages, and Disadvantages. In *Genotoxicity—A Predictable Risk to Our Actual World*; InTech: London, UK, 2018; pp. 64–80. [CrossRef]
28. Riss, T.L.; Moravec, R.A.; Niles, A.L.; Duellman, S.; Benink, H.A.; Worzella, T.J.; Minor, L. Cell Viability Assays. In *Assay Guidance Manual*. 2004. Available online: http://www.ncbi.nlm.nih.gov/pubmed/23805433. (accessed on 1 July 2022).
29. Araújo, L.C.P.; de Oliveira, J.M., Jr.; Aranha, N. Síntese e caracterização de scaffolds de fibroína. *Matéria* **2018**, *23*, 4. [CrossRef]
30. Goudarzi, F.; Asadi, A.; Afsharpour, M.; Jamadi, R.H. In Vitro Characterization and Evaluation of the Cytotoxicity Effects of Nisin and Nisin-Loaded PLA-PEG-PLA Nanoparticles on Gastrointestinal (AGS and KYSE-30), Hepatic (HepG2) and Blood (K562) Cancer Cell Lines. *AAPS PharmSciTech* **2018**, *19*, 1554–1566. [CrossRef]
31. Azhar, N.S.; Zin, N.H.; Haziyamin, T.; Abdul, T. Lactococcus Lactis Strain A5 Producing Nisin-like Bacteriocin Active against Gram Positive and Negative Bacteria Driven by the increasing demand for animal protein and food security, aquaculture is a fast-growing industry that becomes a vital economy in. *Trop. Life Sci. Res.* **2017**, *28*, 107–118. [CrossRef]
32. Niaz, T.; Imran, M. Diffusion kinetics of nisin from composite coatings reinforced with nano-rhamnosomes. *J. Food Eng.* **2021**, *288*, 110143. [CrossRef]
33. Rabe, M.; Verdes, D.; Seeger, S. Understanding protein adsorption phenomena at solid surfaces. *Adv. Colloid Interface Sci.* **2011**, *162*, 87–106. [CrossRef] [PubMed]
34. Jorge, L.; Harada, L.; Silva, E.; Campos, W.; Oliveira, J., Jr.; Vila, M.; Tubino, M.; Balcão, V. Bacterial Nanocellulose biomembrane as a support for human insulin aiming at transdermal permeation. *Química Nova* **2020**, *42*, 572–578. [CrossRef]
35. Moniri, M.; Moghaddam, A.B.; Azizi, S.; Rahim, R.A.; Ariff, A.B.; Saad, W.Z.; Navaderi, M.; Mohamad, R. Production and status of bacterial cellulose in biomedical engineering. *Nanomaterials* **2017**, *7*, 257. [CrossRef] [PubMed]
36. Field, D.; O'Connor, R.; Cotter, P.D.; Ross, R.P.; Hill, C. In vitro activities of nisin and nisin derivatives alone and in combination with antibiotics against Staphylococcus biofilms. *Front. Microbiol.* **2016**, *7*, 508. [CrossRef]
37. Han, D.; Sherman, S.; Filocamo, S.; Steckl, A.J. Long-term antimicrobial effect of nisin released from electrospun triaxial fiber membranes. *Acta Biomater.* **2017**, *53*, 242–249. [CrossRef]
38. Macedo, J.L.S.J. Complicações infecciosas em pacientes queimados. *Rev. Bras. Cir. Plástica* **2001**, *21*, 108–111.
39. Oliveira, F.L.; do Serra, M.C.V.F. Infecções em queimaduras: Revisão. *Rev. Bras. Queimaduras* **2011**, *10*, 96–99.
40. de Oliveira, A.A., Jr.; Silva de Araújo Couto, H.G.; Barbosa, A.A.T.; Carnelossi, M.A.G.; de Moura, T.R. Stability, antimicrobial activity, and effect of nisin on the physico-chemical properties of fruit juices. *Int. J. Food Microbiol.* **2015**, *211*, 38–43. [CrossRef]
41. Chen, Y.M.; Xi, T.; Zheng, Y.; Guo, T.; Hou, J.; Wan, Y.; Gao, C. In Vitro Cytotoxicity of Bacterial Cellulose Scaffolds Used for Tissue-engineered Bone. *J. Bioact. Compat. Polym.* **2009**, *24* (Suppl. 1), 137–145. [CrossRef]
42. Frone, A.N.; Panaitescu, D.M.; Nicolae, C.A.; Gabor, A.R.; Trusca, R.; Casarica, A.; Stanescu, P.O.; Baciu, D.D.; Salageanu, A. Bacterial cellulose sponges obtained with green cross-linkers fortissue engineering. *Mater. Sci. Eng. C* **2020**, *110*, 110740. [CrossRef]
43. Gao, G.; Fan, H.; Zhang, Y.; Cao, Y.; Li, T.; Qiao, W.; Wu, M.; Ma, T.; Li, G. Production of nisin-containing bacterial cellulose nanomaterials with antimicrobial properties through co-culturing Enterobacter sp. FY-07 and Lactococcus lactis N8. *Carbohydr. Polym.* **2021**, *251*, 117131. [CrossRef] [PubMed]
44. Custódio, F.A.F.; de Castro, L.M.; Unterkircher, E.; Porto, A.C.R.C.; Braga, I.S.; Hataka, A.; Jozala, A.F.; Grotto, D. Evaluation of Bacterial Nanocellulose Membranes Loaded or Not with Nisin as a Complementary Treatment in Surgical Dehorning Wounds in Bovines. *Pharmaceutics* **2021**, *13*, 688. [CrossRef] [PubMed]
45. Gedarawatte, S.T.G.; Ravensdale, J.T.; Al-Salami, H.; Dykes, G.A.; Coorey, R. Antimicrobial efficacy of nisin-loaded bacterial cellulose nanocrystals against selected meat spoilage lactic acid bacteria. *Carbohydr. Polym.* **2021**, *251*, 117096. [CrossRef] [PubMed]

polymers

MDPI

Article

Extraction Optimization of Mucilage from Seeds of *Mimosa pudica* by Response Surface Methodology

Syed Nasir Abbas Bukhari [1,*], Arshad Ali [2], Muhammad Ajaz Hussain [2,*], Muhammad Tayyab [3], Nasser F. Alotaibi [4], Mervat A. Elsherif [4], Kashaf Junaid [5] and Hasan Ejaz [5]

1 Department of Pharmaceutical Chemistry, College of Pharmacy, Jouf University, Sakaka 72388, Saudi Arabia
2 Institute of Chemistry, University of Sargodha, Sargodha 40100, Pakistan; arshadali04@yahoo.com
3 Department of Pharmacy, Quaid-i-Azam University, Islamabad 45320, Pakistan; ph.tayyab@gmail.com
4 Chemistry Department, College of Science, Jouf University, Sakaka 72388, Saudi Arabia; nfalotaibi@ju.edu.sa (N.F.A.); maelsherif@ju.edu.sa (M.A.E.)
5 Department of Clinical Laboratory Sciences, College of Applied Medical Sciences, Jouf University, Sakaka 72388, Saudi Arabia; kjunaid@ju.edu.sa (K.J.); hetariq@ju.edu.sa (H.E.)
* Correspondence: sbukhari@ju.edu.sa (S.N.A.B.); majaz172@yahoo.com (M.A.H.); Tel.: +966-565738896 (S.N.A.B.); Tel: +92-3458614959 (M.A.H.)

Citation: Bukhari, S.N.A.; Ali, A.; Hussain, M.A.; Tayyab, M.; Alotaibi, N.F.; Elsherif, M.A.; Junaid, K.; Ejaz, H. Extraction Optimization of Mucilage from Seeds of *Mimosa pudica* by Response Surface Methodology. *Polymers* **2022**, *14*, 1904. https://doi.org/10.3390/polym14091904

Academic Editor: Arn Mignon

Received: 17 April 2022
Accepted: 5 May 2022
Published: 6 May 2022

Abstract: *Mimosa pudica* seed mucilage (MPM) is composed of glucuronoxylan, which is a swellable, pH-responsive and non-toxic biomaterial. Herein, we aimed to extract MPM from *M. pudica* seeds (MP seeds) to ascertain optimization of extraction conditions to get highest yield by response surface methodology, via Box-Behnken design (RSM-BBD). MPM was extracted from MP seeds by a hot water extraction method. The effects of four different parameters on the extraction yield of MPM were evaluated: pH of the extraction medium (1–10), seed/water contact time (1–12 h), the temperature of extraction medium (30–90 °C), and seed/water ratio (1:5–1:35 *w*/*v*). The maximum yield of MPM obtained by Design-Expert software was 10.66% (10.66 g/100 g) at pH 7, seed/water contact time of 6 h, extraction temperature of 50 °C, and seed/water ratio of 1:20 *w*/*v*. The *p* values of ANOVA were found to be less than 0.0001, which indicated that the extraction yield of MPM was significantly affected by all the study parameters. The results revealed that pH and extraction temperature were the most significant factors affecting the yield of MPM. MPM in compressed tablet form showed pH-responsive on–off switching behavior at pH 7.4 and 1.2 in a reversible manner. MPM in compressed tablet form sustained the release of itopride for 16 h following a super case-II transport mechanism and zero-order release kinetics.

Keywords: *Mimosa pudica* mucilage; extraction optimization; Box-Behnken design; response surface methodology; pH-responsive on–off switching; zero-order release

1. Introduction

Mucilage from plant seeds has been extensively used in food systems as additives, emulsifiers, stabilizers, gelling agents, and texture modifiers, due to the safety profile, ease of availability, low cost, and biodegradable nature of mucilage [1–4]. Mucilaginous materials are polysaccharides in nature and have been extensively used in every sphere of life [5–9].

Mimosa pudica (Family, *Mimosaceae*) is a native Brazilian plant. Its height is 1 to 2 cm. The plant as a whole is enriched with phytochemicals. Its seeds (MP seeds) are brown and 2.5 mm in length. MP seeds have been utilized for the treatment of ulcers, piles, constipation, snake bites, depression, and smallpox [10]. MP seeds release mucilage (MPM) upon soaking in water. MPM contains glucuronoxylan, which is composed of D-xylose and D-glucuronic acid [11,12]. The glucuronoxylan from MP seeds is a water-swellable, pH-responsive, bio-compatible, non-toxic biomaterial [13,14]. MP seeds are commercially available, and many studies have been reported on their utilization as pharmaceutical and industrial excipients. However, no study is available in the literature so far for the optimization of extraction conditions at which maximum yield of MPM can be obtained to further broaden its spectrum of applications.

The optimization of the mucilage extraction yield by a single extraction factor is a costly and time-consuming process. Moreover, where there are multiple independent factors involved in the extraction process of mucilage from plant seeds, the quadratic and interaction terms of independent factors also affect the mucilage yield [15]. Therefore, due to the increase in research activities on plant seed mucilage, their extraction needs design of experiment (DOE) to optimize the ideal extraction conditions and to save time and cost. Response surface methodology (RSM) is a collection of mathematical and statistical tools for designing experiments to get ideal extraction conditions and elucidation of the combined effect of independent factors on extraction yield [16]. Many researchers have effectively used RSM for optimizing extraction conditions to get the maximum yield of mucilage, i.e., polysaccharides, from different parts of plants [17–22].

Herein, we aimed to extract mucilage from MP seeds and optimize the extraction conditions using RSM-BBD. The effects of four independent factors, i.e., pH of the extraction medium (A: 1–10), seed/water contact time (B: 1–12 h), extraction temperature (C: 30–90 °C), and seed/water ratio (D: 1:5 to 1:35 w/v) on the extraction yield of MPM were evaluated. The combined effect of the studied parameters, in terms of linear, quadratic, and interaction effects on the extraction yield of MPM, were also evaluated. Furthermore, we can report that MPM is a pH-responsive biomaterial for successful zero-order release of the standard drug itopride.

2. Materials and Methods

2.1. Materials

MP seeds were procured from the local market of District Sargodha, Pakistan. MP seeds were first cleaned manually and then sieved to eradicate any unwanted material and stored in an air-tight jar. Analytic grade reagents and solvents were purchased from Sigma-Aldrich Co., (St. Louis, MO, USA) and used as such, without further purification, to perform experiments. Buffers of pH 1.2, 6.8, and 7.4 were prepared according to the protocol given in the United States Pharmacopeia (USP 34-NP 29). Itopride was used as a standard drug and received from Dyson Research Laboratories, Lahore, as a gift sample. Deionized water (DI) was used for washing and preparation of necessary solutions and dilutions.

2.2. Methods

2.2.1. Extraction of Mucilage

The mucilage (MPM) was extracted from MP seeds using the hot water extraction method. MP seeds were soaked in DI at 50 °C and allowed to swell for 6.0 h. The mucilage, i.e., the MPM released from the seed coats of swollen MP seeds, was isolated by using a clean cloth, and washed with n-hexane to remove non-polar/lipophilic substances. The MPM was further washed with ethanol to ensure it was free from polar impurities. Later, it was dried in a vacuum oven at 50 °C for 24 h. The completely dried MPM was ground and powdered by passing through a sieve no. 60, and stored in a vacuum desiccator.

2.2.2. Calculation of Yield

The actual yield of MPM in percentage was calculated using Equation (1) [23].

$$Extraction\ yield\ of\ \mathrm{MPM}\ (\%) = \frac{weight\ of\ extracted\ \mathrm{MPM}\ after\ drying}{weight\ of\ \mathrm{MP}\ seeds\ taken\ for\ extraction\ of\ \mathrm{MPM}} \times 100 \quad (1)$$

2.2.3. Experimental Design and Statistical Analysis

By following the aforesaid extraction procedure, the effect of four different parameters on the extraction yield of MPM were studied: pH of the extraction medium (*A*: 1–10), seed/water contact time (*B*: 1–12 h), the temperature of the extraction medium (*C*: 30–90 °C), and seed/water ratio (*D*: 1:5–1:35 *w*/*v*). These studies provided the effect of a single parameter on the extraction yield of MPM and were considered preliminarily experimental studies. The responses (% yield) obtained from these preliminary studies were further utilized for designing the model, according to RSM-BBD.

Three different levels were selected from the responses of individual parameters and were assigned as low (−), moderate (0), and high (+). For every individual parameter, the levels were: pH (low = 6, moderate = 7, high = 8), seed/water contact time (low = 4 h, moderate = 6 h, high = 8 h), extraction temperature (low = 30 °C, moderate = 50 °C, high = 70 °C), and seed/water ratio (low = 1:10 *w*/*v*, moderate = 1:20 *w*/*v*, high = 1:30 *w*/*v*). RSM-BBD was applied to the extracted yield data using the statistical package, Design-Expert version 12.0.3.0 (Stat-Ease Inc., Minneapolis, MN, USA). RSM-BBD provided the statistical treatment of the tested parameters on the yield of MPM in terms of regression (analysis of variance, ANOVA) and graphically (two dimensional, i.e., 2D response surface and three dimensional, i.e., 3D contours plots) analyses.

The ANOVA offers linear, quadratic, and composite interactions, through which their effects on the yield of MPM were also studied by noting *p*-values. The *F*-values and coefficients of regression, including (R^2), adjusted-R^2, and predicted-R^2, were noted to check the significance of ANOVA. Also, the fitness level of the RSM-BBD model to the extraction yield data was predicted in terms of sum square error (SSE), standard error (DE), mean, coefficient of variance (% CV), adequate precision (ADP), and lack of fit.

The 3D response surface and 2D contour plots were acquired from RSM-BBD to obtain the points of optimum yield and optimum conditions. Also, from these plots, the combined effect of studied parameters, co-relationships between the parameters, and model desirability were studied. Moreover, the extraction yield data of MPM was further put into a second-order polynomial equation to assess the statistical significance of the RSM-BBD. A general linearized form of a second-order polynomial Equation (2) is presented below.

$$Y = \alpha_o + \alpha_1 A + \alpha_2 B + \alpha_3 C + \alpha_4 D + \alpha_{11} A^2 + \alpha_{22} B^2 + \alpha_{33} C^2 + \alpha_{44} D^2 + \alpha_1\alpha_2 AB + \alpha_1\alpha_3 AC + \alpha_1\alpha_4 AD + \alpha_2\alpha_3 BC + \alpha_2\alpha_4 BD + \alpha_3\alpha_4 CD + E \quad (2)$$

where, Y is the yield response (%). The A, B, C, and D represent the pH of the extraction medium, seed/water contact time, the temperature of the extraction medium, and seed/water ratio, respectively. The α_o is the intercept. α_1, α_2, α_3, and α_4 are the coefficient of linearity. α_{11}, α_{22}, α_{33}, and α_{44} are the quadratic coefficients. $\alpha_1\alpha_2$, $\alpha_1\alpha_3$, $\alpha_1\alpha_4$, $\alpha_2\alpha_3$, $\alpha_2\alpha_4$, and $\alpha_3\alpha_4$ are the coefficient of interactions. E is the error function of the model.

2.2.4. Evaluation of MPM as a Sustained Release Material

Preparation of MPM-Based Tablets

Itopride was used as a model drug to assess the potential of MPM as a pH-responsive and sustained release material. Three different oral tablet formulations (F1, F2, and F3), based on MPM and itopride, were prepared using the wet granulation method [13]. The composition of the tablets is given in Table 1. The mixture of MPM and itopride was prepared in a pestle and mortar and homogenized using a 5% solution of polyvinyl pyrrolidone (PVP) prepared in isopropyl alcohol. The mixture was sieved, granulated, dried in an oven at 50 °C under vacuum, lubricated using 5 mg of magnesium stearate, and pressed on a rotary press having a 9 mm flat surface punch. The hardness of every tablet was maintained between 6–9 kg/cm^2.

Table 1. Composition of MPM-based oral tablet formulations in mg/tablet.

Constituents of the Tablets	F1	F2	F3
MPM	75	100	125
Itopride	100	100	100
Microcrystalline cellulose	120	95	70
Magnesium stearate	5	5	5
Total weight	300	300	300

pH-Responsive On–Off Switching Studies

The pH-responsive on–off switching (swelling and deswelling) properties of the oral tablet formulation F3 (having high concentration of MPM) were studied by alternatively immersing the tablet in beakers having 100 mL of each buffer of pH 7.4 (swelling medium) and pH 1.2 (deswelling medium), after enclosing the tablet in a pre-weighed tea bag [13]. The tea bag was removed from the swollen medium after pre-defined intervals of time and the swelling and deswelling responses were recorded using Equation (3). These experiments were conducted to record three cycles of swelling and deswelling studies. The swelling and deswelling experiments were performed three times and the mean of the values calculated as below.

$$Swelling\ capacity\ (g/g) = \frac{W_s - W_o - W_e}{W_o} \tag{3}$$

where, W_s, W_o, and W_e represent the weight of a wet tea bag having swollen tablet of MPM formulation F3, the weight of the dry tablet of MPM formulation F3, and the weight of an empty wet tea bag, respectively.

In Vitro Release Study of Itopride

The in vitro release of itopride from oral tablet formulation F3 was studied in a buffer of pH 6.8 (900 mL) for 16 h on USP Dissolution Apparatus II. The dissolution apparatus was run by maintaining the temperature of the dissolution medium at 37 ± 0.5 °C and the rotation speed of the paddles at 50 rpm. The sample (10 mL) was taken out from the dissolution medium after a pre-defined interval of time, filtered using a 0.45 μm nylon filter, necessarily diluted, and run on a UV-Vis spectrophotometer (UV-1600 Shimadzu, Germany) at 258 nm to record the absorbance. The dissolution medium was restored by adding the same amount of freshly prepared buffer of pH 6.8. The release study was also performed at pH 1.2 for 2 h and at pH 6.8 for the next 14 h to study the release behavior that mimics the pH and transit time of the gastrointestinal tract.

Drug Release Kinetics and Mechanism

The zero-order kinetic model (Equation (4)) [24] was used to study the rate of itopride release from F3 and the Korsmeyer-Peppas model (Equation (5)) [25] was applied to the release data. The mechanism of itopride release from F3 was evaluated.

$$Q_t = K_0 \tag{4}$$

where, Q_t and K_0 shows quantity of itopride release at time t and rate constant for zero-order kinetic model, respectively.

$$\frac{M_t}{M_\infty} = k_p t^n \tag{5}$$

where, M_t/M_∞, k_p, and n represent the quantity of itopride released at any time t, rate constant for the Korsmeyer-Peppas model, and diffusion exponent, respectively. The value of n describes the mechanism of drug release from oral tablet formulation. The value of n can be ≤0.45 for Fickian diffusion, from 0.45-0.89 for non-Fickian diffusion, =0.89 for case-II transport, and >0.89 for super case-II transport [26,27].

3. Results and Discussion

3.1. Extraction of Mucilage

The MPM was extracted from MP seeds by the hot water extraction method. MP seeds were soaked in DI. Water penetrated the microscopic pores of the MP seeds because of the swellable nature of glucuronoxylan present in MPM. MPM appeared as a colorless powdery material in dry form. Its yield was optimized by RSM-BBD.

3.2. Preliminary Studies for the Extraction Optimization of MPM Yield

3.2.1. Effect of pH

The selectivity of any material for pharmaceutical applications, particularly regarding drug delivery, is based on its pH-responsive nature. Therefore, it is essential to know the effect of pH on the extraction yield of the material. The effect of pH was evaluated by changing the pH of the extraction medium from 1 to 10. The rest of the extraction conditions, such as the seed/water contact time (6 h), extraction temperature (50 °C), and seed/water ratio (1:20 *w*/*v*), were kept constant. A negligible, i.e., less than 0.5%, yield of MPM was found at the lower pH ranging from 1–4 (not reported here). However, beyond pH 4.0 the yield of MPM increased and reached a maximum of 11.26% (9.0 g/100 g) at pH 7 = DI. On further increase in the pH, the yield of MPM dropped to 6.11% at pH 10.0. Moreover, it was noted that the yield of MPM was higher at an alkaline pH than that of an acidic pH (Figure 1a). This may be due to the fact that in an alkaline medium, due to possible hydrolysis, some of the insoluble polysaccharide fractions get converted into soluble ones [28]. A similar trend has previously been reported by Somboonpanyakul et al., Balke and Diosady, and Esteves et al. [29–31]. Hence, MPM extracted from MP seeds is a pH-responsive polysaccharide and could be an ideal candidate for pharmaceutical applications.

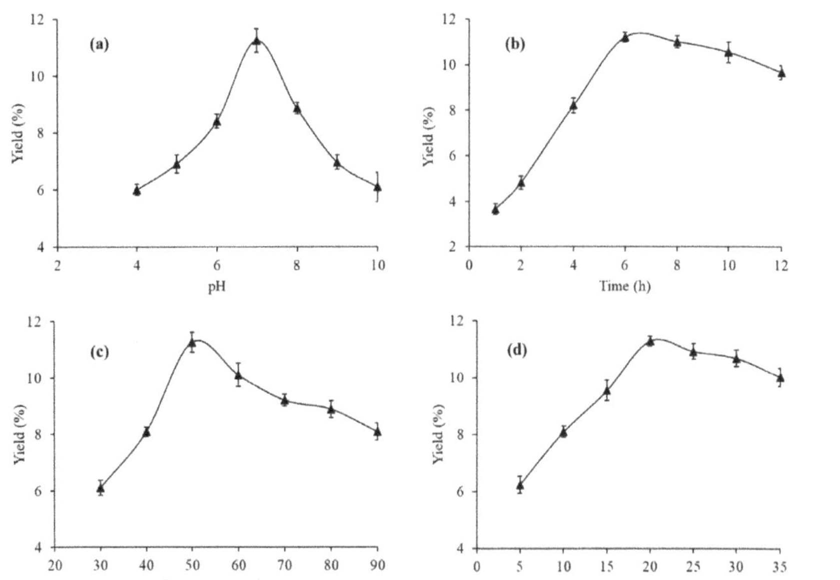

Figure 1. Effect of the pH of extraction medium (**a**), seed/water contact time (**b**), the temperature of the extraction medium (**c**), and seed/water ratio (**d**) on the yield of MPM extracted from MP seeds.

3.2.2. Effect of Seed/Water Contact Time

To study the effect of seed/water contact time on the yield of MPM, this contact time was varied from 1–12 h. The other extraction conditions were adjusted at their optimum levels. It was revealed that as seed/water contact time increased from 1 to 6 h, the yield of MPM also increased abruptly, and after 6 h the yield of MPM decreased by a small extent (Figure 1b). As far as the extraction medium and seeds are in contact with each other, the exposure of water to seed increases, due to which more and more water enters into the pores of the seeds and acts as a driving force to remove mucilage from seeds [15,32]. Reports on the mucilage extraction from *Malva sylvestris* [29], cress seeds [28], and *Arctic chlorella* sp., [33] also favor greater yield in percentage at a longer extraction time.

3.2.3. Effect of Temperature

The effect of extraction temperature on the yield of MPM was studied within temperatures ranging from 30 to 90 °C. The conditions of the pH 7, seed/water contact time of 6 h, and seed/water ratio 1:20 *w*/*v* were maintained. It was evaluated that initially at low temperature, i.e., at 30 °C, the yield of MPM was low and equal to 6.11%. However, with increase in temperature from 30 to 50 °C, the yield of MPM also increased from 6.11 to 11.27%. After 50 °C, the yield of MPM decreased to 8.09% at 90 °C (Figure 1c). As the maximum yield of MPM was achieved at 50 °C, this temperature was considered to be the optimum extraction temperature. The reason underlying this trend is that with increase in the temperature of the extraction medium the polysaccharide gets solubilized in it, due to which the value of the diffusion coefficient of the polysaccharide increases. At a high value of diffusion coefficient, a greater mass of polysaccharides is extracted from the plant seeds and this leads to an increase in the yield [34]. Another possible reason for this trend is that at high extraction temperatures, the seeds become less sticky and, consequently, release a high amount of mucilage. However, after the optimum temperature, the polysaccharide may degrade and release less mucilage [2]. Many other studies have reported a decrease in mucilage yield on increasing temperature of the extraction medium. which may be due to thermal degradation of the polysaccharides in the mucilage at high temperature [35]. Similar findings have been demonstrated for extraction optimization of chia seed mucilage [1] and *Alyssum homolocarpum* seed mucilage [36].

3.2.4. Effect of Seed/Water Ratio

At pH 7, extraction temperature 50 °C, and contact time 6 h, the influence of seed/water ratio on the yield of MPM was studied by varying seed/water ratio from 1:5 to 1:35 *w*/*v*. The results obtained are incorporated in Figure 1d. On increasing seed/water ratio from 1:5 to 1:20, the yield of MPM also increased. At 1:20 seed/water ratio the maximum yield of MPM, i.e., 11.29% was achieved, and after that, no more significant increase or decrease in the yield was observed, due to the attainment of a state of dynamic equilibrium. At a high seed/water ratio, the water molecule of the extraction medium exerts more driving force on the seeds and pushes the mucilage to come out from the seeds [37,38]. Hence, the highest yield of MPM was achieved at seed/water ratio 1:20 *w*/*v*. The yield of mucilage extracted from *Dioscorea nipponica* [39] and *Phoenix dactylifera* [40] also reported similar results.

Conclusively, the above discussion shows that all the tested parameters have a significant effect on the yield of MPM extracted from MP seeds. The maximum yield, i.e., 11.29% (11.29 g/100 g), of MPM was obtained at pH 7.0, contact time 6 h, extraction temperature 50 °C, and seed/water ratio 1:20 *w*/*v*. Therefore, based on these preliminary experiments, three different levels, such as low (−), moderate (0), and high (+) for each parameter were selected and RSM-BBD containing 29 different sets of experimental runs were constructed (Table 2). At every set of experimental runs, MPM was extracted and its yield in percentage was tabulated and validated statistically by applying ANOVA.

3.3. Fitting of Model

Findings of 29 different experimental runs, according to RSM-BBD, are incorporated in Table 2, which includes the design, actual, and predicted yields of MPM. Both the actual and predicted yields were found very close to each other. The actual yield of MPM was found in the range of 3.68 to 11.25%. The yield data of MPM was fitted to the second-order polynomial response surface model and the following quadratic Equation (6) was obtained.

$$\text{Actual yield (\%) of MPM} = 10.66 + 0.8858A + 0.7092B + 0.7992C + 0.61587D - 2.31A^2 - 1.578B^2 - 2.03C^2 - 1.09D^2 - 0.5925AB - 0.9700AC - 1.32AD + 0.4950BC + 0.4800BD + 1.64CD \quad (6)$$

Table 2. RSM-BBD experimental design and actual vs. predicted yield of MPM (%).

Run Nos.	Independent Variables				Yield (%)	
	pH A	Seed/Water Contact Time (h) B	Extraction Temperature (°C) C	Seed/Water Ratio (*w*/*v*) D	Actual Y	Predicted Z
1	7	6	30	10	8.44	7.76
2	6	6	30	20	4.16	3.66
3	8	6	70	20	7.23	7.03
4	7	4	30	20	5.32	6.05
5	7	6	50	20	11.25	10.66
6	7	4	70	20	6.26	6.65
7	7	8	70	20	9.15	9.06
8	7	6	70	10	6.75	6.08
9	8	6	50	10	8.34	8.85
10	7	8	50	30	10.25	9.81
11	7	8	30	20	6.23	6.47
12	8	6	50	30	7.55	7.44
13	8	6	30	20	7.89	7.37
14	7	8	50	10	7.28	7.62
15	8	8	50	20	7.68	7.78
16	7	4	50	10	7.41	7.16
17	6	6	50	30	8.19	8.31
18	7	4	50	30	8.46	7.43
19	6	6	70	20	7.38	7.20
20	7	6	70	30	9.85	10.59
21	8	4	50	20	7.34	7.55
22	7	6	30	30	4.99	5.72
23	7	6	50	20	10.36	10.66
24	7	6	50	20	11.17	10.66
25	6	4	50	20	4.64	4.59
26	7	6	50	20	10.31	10.66
27	6	8	50	20	7.35	7.20
28	6	6	50	10	3.68	4.43
29	7	6	50	20	10.21	10.66

Table 3 shows the results of ANOVA regarding the RSM-BBD. As the p-value of ANOVA is significant, i.e., <0.001, it means that the RSM-BBD model is satisfactorily applied to the yield data for optimization of extraction conditions. The p-values in the case of linear independent parameters were found to be significant, i.e., $p < 0.001$, which shows that the yield of MPM was purely dependent on them. Among all the independent parameters, the pH of the extraction medium has a pronounced effect on the yield of MPM followed by extraction temperature, seed/water contact time, and seed/water ratio. Moreover, $p < 0.0001$ also indicated that the extraction yield of MPM related linearly to the studied parameters.

The yield of MPM was also found to be dependent on quadratic (A^2, B^2, C^2, D^2) and interaction (AB, AC, AD, BC, BD, and CD) terms. All the quadratic terms have a significant effect on the yield of MPM in the sequence of $A^2 > C^2 > B^2 > D^2$ by considering p-values. The interaction between pH vs. extraction temperature (AC), pH vs. seed/water ratio (AD), and extraction temperature vs. seed/water ratio (CD) was found highly significant because the p-values in these cases are less than 0.01. The interaction terms pH vs. contact time (AB), contact time vs. extraction temperature (BC), and contact time vs. seed/water ratio (BD) were found non-significant because the p-values in these cases are greater than 0.05. The effect of interaction terms on the yield of MPM was found in the order of CD > AD > AC > AB > BC > BD.

Table 3. ANOVA of the experimental results of the RSM-BBD.

Source	Sum of Squares	DF [a]	Mean	F-Value	p-Value [b,c]
Model	110.10	14	7.86	16.39	<0.0001 **
Linear	9.42	1	9.42	19.62	0.0006 **
A-pH	6.04	1	6.04	12.57	0.0032 **
B—Contact time (h)	7.66	1	7.66	15.97	0.0013 **
C—Temperature (°C)	4.55	1	4.55	9.48	0.0082 **
D—Seed/water ratio	110.10	14	7.86	16.39	<0.0001 **
Quadratic					
A^2	34.63	1	34.63	72.15	<0.0001 **
B^2	15.95	1	15.95	33.23	<0.0001 **
C^2	26.81	1	26.81	55.86	<0.0001 **
D^2	7.71	1	7.71	16.07	0.0013 **
Interaction					
AB	1.40	1	1.40	2.93	ns
AC	3.76	1	3.76	7.84	0.0142 *
AD	7.02	1	7.02	14.63	0.0019 **
BC	0.9801	1	0.9801	2.04	ns
BD	0.9216	1	0.9216	1.92	ns
CD	10.73	1	10.73	22.35	0.0003 **
Residual	6.72	14	0.4799		
Lack of Fit	5.70	10	0.5696	2.23	ns
Pure Error	1.02	4	0.2558		
Cor Total	116.82	28			
R^2 = 0.94255; Adjusted-R^2 = 0.8850; Predicted-R^2 = 0.7055; CV = 8.92%; Mean = 7.76; ADP = 14.0462					

[a] DF: Degree of freedom; [b] Significant (* $p < 0.05$); [c] Highly significant (** $p < 0.01$); ns: Not significant.

The adequacy of the RSM-BBD model for yield data was further assessed by comparing the values of R^2, adjusted-R^2, and predicted-R^2. The difference between R^2 (0.9425) and adjusted-R^2 (0.8850) was found to be 0.0575 and the difference between R^2 (0.9425) and predicted-R^2 (0.7055) was found to be 0.237. This minute difference, and high value of R^2, indicated that RSM-BBD is an adequate model for yield data of MPM [41].

The % CV describes the variation in the mean value and interprets the adequacy of the model for the studied response with great accuracy, high precision, and reliability of experiment. If the value of % CV is less than 10%, then the variation in the mean value is low and hence the response model will develop with great satisfaction and vice versa. In this research, the % CV value was found to be 8.92%. This low value of % CV indicated that the RSM-BBD model was adequate and desirable to identify the optimum extraction conditions for the highest yield of MPM from MP seeds. Moreover, a low % CV value is testament to the fact that the extraction experiments had high precision and reliability [42].

The value of ADP helps in predicting the signal-to-noise ratio as well as model desirability. Generally, the greater the value of ADP, the lower will be the signal-to-noise ratio and the higher will be the model desirability. The normal value of ADP is 4.0. This means that if the ADP for any designed model is greater than 4.0, then it is desirable for optimizing the extraction conditions. The ADp value was found to be 14.0462 which is greater than a normal value [43]. Hence, RSM-BBD is a desirable model for optimizing the yield of MPM.

Lack of fit is another vital factor to determine whether the RSM-BBD model successfully fits, or fails to fit, in representing the extraction yield data in the experimental domain. A significant lack of fit ($p < 0.05$) means that the model is not satisfactorily applied to the yield data and the corresponding response factors should not be included in the regression. Whereas, a non-significant lack of fit ($p > 0.05$) means that the model is satisfactorily applied to the yield data and the corresponding response factors should be included in the regression. This study found a non-significant lack of fit, i.e., 0.0760, for all independent factors at a 95% confidence interval. This means that all models can equally predict the corresponding responses accurately [38].

Therefore, it can be concluded that the second-order quadratic was the most significant model to identify the optimum conditions for the extraction of MPM from MP seeds.

3.4. Interpretation of Response Surface Plots

The 3D response surface and 2D contour plots were obtained from RSM-BBD to evaluate the combined effect of all the studied parameters on the extraction yield of MPM.

The 3D response surface plot and 2D contour plots showing the combined effect of pH and extraction time on the yield of MPM at a constant temperature of 50 °C and seed/water ratio of 1:20 *w*/*v* are presented in Figures 2a and 3a. It can be seen that, as the pH and seed/water ratio increased, the yield of MPM also increased and reached a maximum of 10.71% at pH 6.77 and contact time 7.26 h. However, after these threshold values, the yield of MPM decreased up to 7.93% at pH 7.95 and contact time 7.98 h.

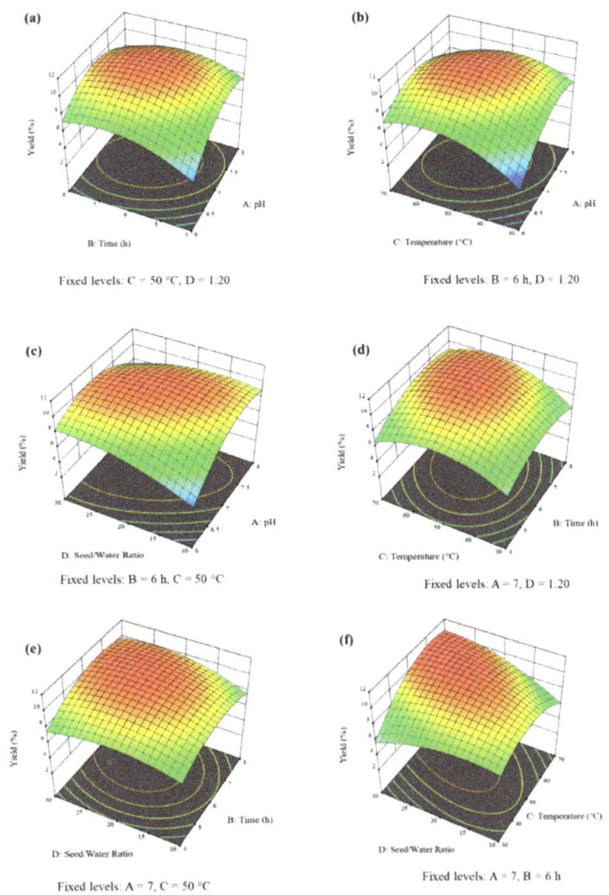

Figure 2. 3D response surface plots pH vs. contact time (**a**), pH vs. extraction temperature (**b**), pH vs. seed/water ratio (**c**), contact time vs. extraction temperature (**d**), contact time vs. seed/water ratio (**e**), and extraction temperature vs. seed/water ratio (**f**), showing significant interaction effect on the yield of MPM (%) from MP seeds.

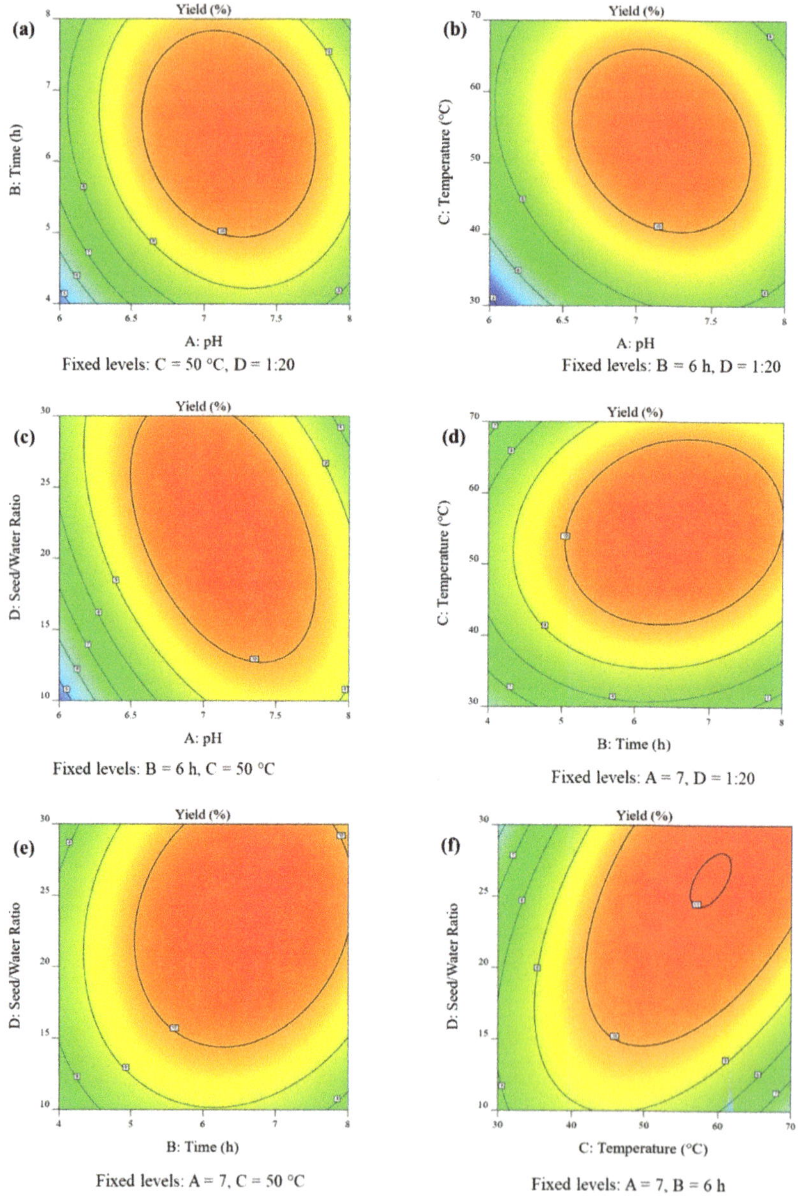

Figure 3. 2D contour plots pH vs. contact time (**a**), pH vs. extraction temperature (**b**), pH vs. seed/water ratio (**c**), contact time vs. extraction temperature (**d**), contact time vs. seed/water ratio (**e**), and extraction temperature vs. seed/water ratio (**f**), showing significant interaction effect on the yield of MPM (%) from MP seeds.

The combined effect of pH and temperature on the yield of MPM is shown in Figures 2b and 3b at a constant contact time of 6 h and seed/water ratio of 1:20 w/v. The yield of MPM increased from 3.71–10.77% by increasing pH from 6.0–7.21 and temperature from 30–54.77 °C. Beyond, these levels the MPM yield decreased to 7.36% at pH 7.96 and a temperature of 69.23 °C.

At constant contact time, 6 h, and extraction temperature of 50 °C, the finding of the effect of pH and seed/water ratio on the yield of MPM is shown in Figures 2c and 3c. At low pH 6.0 and seed/water ratio 1:10.13 *w*/*v*, the yield of MPM was poor, i.e., 4.5%. However, by increasing pH and seed/water ratio, the MPM yield also increased significantly and reached a maximum of 10.75% at pH 7.1 and seed/water ratio 1:23.42 *w*/*v*. Later, the MPM yield decreased and achieved a minimum point of 7.57% at pH 7.98 and a seed/water ratio of 29.77 *w*/*v*.

The different extraction temperatures and contact also had significant effects on the yield of MPM. At an extraction temperature of 30 °C and contact time of 4 h, the yield of MPM was quite low, i.e., 6.0%. On increasing both of these parameters the yield of MPM increased to 10.83% at an extraction temperature of 55.47 °C and contact time of 6.47 h. Afterwards, the MPM yield decreased to 9.15% at an extraction temperature of 69.48 °C and a contact time of 7.98 h (Figure 2d or Figure 3d).

The quadratic effect of contact time and seed/water on the yield of MPM is graphically presented in Figures 2e and 3e. The conditions of pH and extraction temperature were fixed at their optimum levels. It can be seen that as the contact time and seed/water ratio increased from 4.02–6.61 h and 1:10.09–1:22.74 *w*/*v*, respectively, the MPM yield also increased from 7.19–10.85%. The maximum yield of MPM, i.e., 10.85% was achieved at contact time 6.51 h and seed/water ratio 1:22.74 *w*/*v*. After these plateau regions, the MPM yield decreased to 9.88% at a contact time of 7.94 h and seed/water ratio 1:29.77 *w*/*v*.

The 3D response surface plot (Figure 2f) and 2D contour plot (Figure 3f) were recorded to study the dependency of the yield of MPM on extraction temperature and seed/water ratio at a fixed value of pH 7 and contact time of 6 h. The maximum yield of MPM, i.e., 10.99%, was obtained at an extraction temperature of 62.84 °C and seed/water ratio of 1:26.63 *w*/*v*. However, afterwards, a slight decrease in the MPM yield was observed to 10.61% at extraction temperature 69.58 °C and seed/water ratio 1:29.63 *w*/*v*.

In 3D response surface plots (Figure 2a–f), the areas that bulged out near 11% yield of MPM show the optimal regions of maximum yield. In 2D contour plots (Figure 3a–f), the regions demarcated by clear circular lines are showing the optimal regions for maximum yield of MPM.

3.5. Model Adequacy

The RSM-BBD model adequacy was checked by plotting a graph between the actual (experimental) vs. predicted (theoretical) extraction yield of MPM (Figure 4; Table 2). The straight line in Figure 4 shows the actual yield of MPM whereas randomly displayed scattered points indicate the predicted yield. It is obvious that the values of the actual yield of MPM are very close to the predicted yield of MPM and are comparable. Hence, this good agreement between the actual and predicted yields illustrated that the second-order quadratic regression model satisfactorily described the yield of MPM by RSM-BBD.

3.6. Optimization of Extraction Yield and Checking of Model Desirability

The graphical and numerical optimizations were recorded (Figures 2a–f and 3a–f) and the maximum yield of MPM 10.66% was obtained at pH 7.0, contact time 6 h, extraction temperature 50 °C, and seed/water ratio 1:20 *w*/*v* by Design-Expert software. This optimized yield of MPM is close to experimental ones, i.e., 11.25% under similar extraction conditions (Table 2, run 5). The same is also evident from scattered plots between actual and predicted (Figure 4). Furthermore, the RSM-BBD model desirability for the extraction of MPM from MP seeds was found at 1.000 which also supported the ideal nature of the aforesaid conditions for the highest extraction yield of MPM from MP seeds according to RSM-BBD (Figure 5).

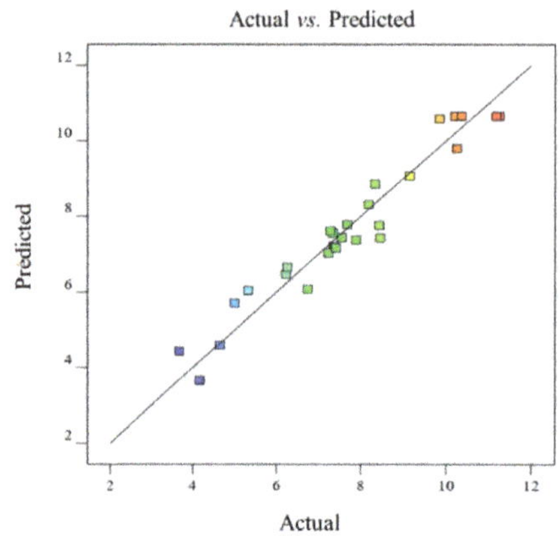

Figure 4. Comparison between actual vs. predicted yield of MPM obtained from MP seeds.

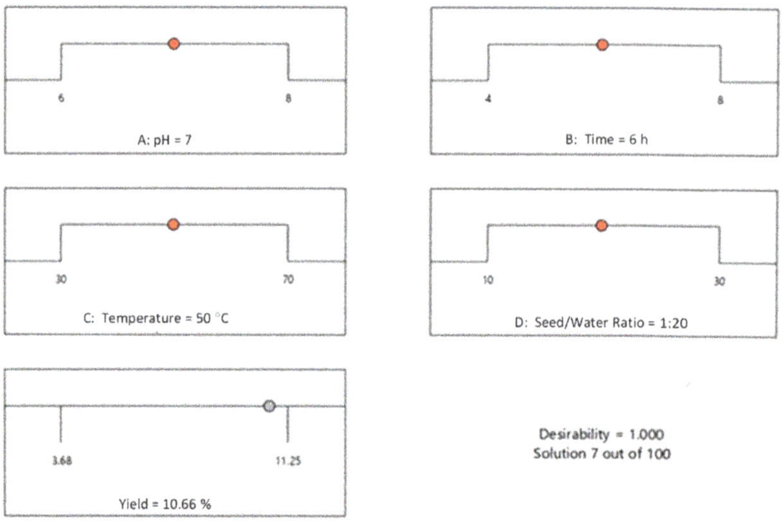

Figure 5. Desirability plot for the extraction optimization of MPM from MP seeds.

3.7. *Comparison of Extraction Yield of MPM with Already Reported Mucilages*

The extraction yield of MPM (11.25%) was found to be greater than mucilage from durian seeds (1.2%) [44], cress seed (6.46%) [45], chia seeds (6.96%) [3], kanocha seeds (7.35%) [46], and flax seeds (7.9%) [15]. Therefore, due to its high yield, MPM has the potential to become an important commercial gum.

3.8. *pH-Responsive On–Off Switching Studies*

The MPM-based oral tablet formulation F3 showed swelling at pH 7.4 and deswelling at pH 1.2 (Figure 6a). At pH 7.4, the carboxylic group (−COOH) present in the polymeric chains of MPM loses its protons and gets ionized to a carboxylate anion (COO^-) which consequently offers anion-anion repulsions. Due to these repulsions, the adjacent polymeric chains repel each other and allow the swelling medium, i.e., the buffer of pH 7.4. to penetrate the polymeric matrix of MPM, which results in rapid swelling [47]. After record swelling at pH 7.4 for an hour, and upon shifting the tablet F3 to the deswelling medium, the COO^- ions accept protons from the deswelling medium, i.e., the buffer of pH 1.2, and become protonated to COOH [48–50]. Consequently, at pH 1.2, deswelling of F3 was observed. The on–off switching properties of F3 were observed up to three consecutive cycles and found to be reproducible.

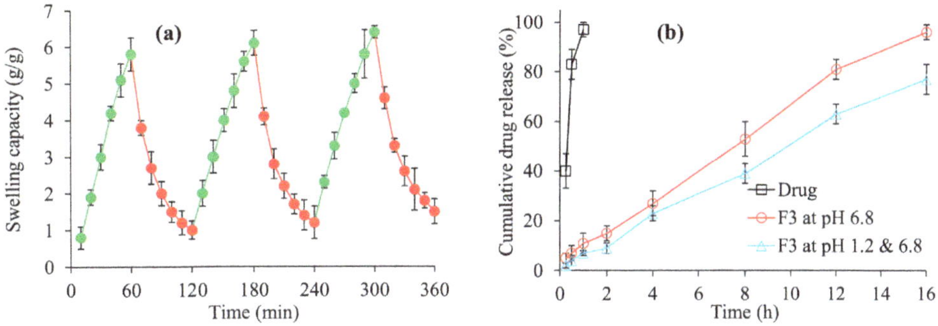

Figure 6. On–off switching (swelling and deswelling) studies of MPM-based oral tablet formulation F3 (**a**) and drug (itopride) release from MPM-based oral tablet formulation F3 at pH 6.8 and pH 1.2 (**b**).

3.9. In Vitro Release Study of Itopride

It was evaluated that MPM sustained the release of itopride for 16 h at pH 6.8. The net release of itopride was 96.0% after 16 h at pH 6.8 (Figure 6b). The release of itopride followed a zero-order kinetic pattern and super case-II transport mechanism by considering the values of R^2 which is 0.9907, and n which is 0.915, respectively [25] (Table 4). The standard drug itopride was found completely dissolved at pH 6.8 within just 1.5 h (Figure 6a). Furthermore, in gastrointestinal tract mimicking conditions, a negligible amount of the itopride, i.e., <9.0% was released in the first 2 h at acidic pH, i.e., pH 1.2, whereas up to 77.0% of itopride was released after 16 h study at basic pH, i.e., pH 6.8. The less release of itopride at pH 1.2 is due to the inability of MPM to swell.

Table 4. Mathematical data of zero-order kinetics and Korsmeyer-Peppas model.

Formulation Code	Zero-Order Kinetic Model		Korsmeyer-Peppas Model		
	R^2	K_0	R^2	K_{KP}	n
F3	0.9907	6.758	0.9933	8.178	0.915

3.10. Applications and Future Research Perspectives

Extraction optimization of MPM from MP seeds via RSM clearly appeared to be a tool to reduce extraction time and economically enhance yield of mucilage. Hence, the findings of the presented research are valuable for future pharmacists and material chemists to develop new materials as Inactive Pharmaceutical Ingredients. Such materials, offering pH-sensitive swelling and deswelling responses are highly prosperous for the development of novel smart materials for intelligent drug delivery, due to their stimuli-responsive nature. Another perspective regarding mucilage, which is under discussion, is that it (MPM) could be a novel smart material for sustained/delayed/targeted drug delivery applications.

4. Conclusions

The RSM-BBD appeared to be an effective tool for the optimization of extraction conditions to get the highest yield of MPM from MP seeds. Results revealed that the pH of the extraction has a significant effect on the yield of MPM, followed by extraction temperature, contact time, and seeds/water ratio. A second-order quadratic model was obtained to predict the extraction yield of MPM. The RSM-BBD exhibited a high value of R^2 (0.94255) and a non-significant lack of fit. The maximum yields of MPM 10.66% were obtained at pH 7, seed/water contact time of 6 h, extraction temperature of 50 °C, and seed/water ratio of 1:20 w/v by Design-Expert software, which is very close to experimental yield, i.e., 11.25%, under the same extraction conditions. Our results demonstrated that MPM is a material of superb choice for the development of a zero-order sustained release drug delivery system. The preliminary results of drug release studies appear promising and it is expected that MPM may also deliver different non-steroidal anti-inflammatory drugs and antibiotics to the colon after bypassing the stomach.

Author Contributions: Conceptualization, M.A.H.; Data curation, A.A. and K.J.; Formal analysis, A.A., M.A.E. and K.J.; Funding acquisition, S.N.A.B.; Investigation, M.T.; Methodology, A.A., M.T. and H.E.; Resources, M.T., N.F.A. and H.E.; Software, N.F.A.; Supervision, M.A.H.; Writing—original draft, M.A.E.; Writing—review & editing, S.N.A.B. and M.A.H. All authors have read and agreed to the published version of the manuscript.

Funding: The authors' work was supported through grant number "375213500" from the Deputyship for Research and Innovation, Ministry of Education in Saudi Arabia.

Acknowledgments: The authors extend their appreciation to the Deputyship for Research and Innovation, Ministry of Education in Saudi Arabia, and the central laboratory at Jouf University for supporting this study.

Conflicts of Interest: The authors declare that they have no known competing financial interests or personal relationships that could have appeared to influence the work reported in this paper.

References

1. Junior, F.A.L.; Conceição, M.C.; de Resende, J.V.; Junqueira, L.A.; Pereira, C.G.; Prado, M.E.T. Response surface methodology for optimization of the mucilage extraction process from *Pereskia aculeata* Miller. *Food Hydrocoll.* **2013**, *33*, 38–47. [CrossRef]
2. Chen, Z.; Cheng, L.; He, Y.; Wei, X. Extraction, characterization, utilization as wound dressing and drug delivery of *Bletilla striata* polysaccharide: A review. *Int. J. Biol. Macromol.* **2018**, *120*, 2076–2085. [CrossRef] [PubMed]
3. Campos, B.E.; Dias Ruivo, T.; da Silva Scapim, M.R.; Madrona, G.S.; de Bergamasco, R.C. Optimization of the mucilage extraction process from chia seeds and application in ice cream as a stabilizer and emulsifier. *LWT Food Sci. Technol.* **2016**, *65*, 874–883. [CrossRef]
4. Keshani-Dokht, S.; Emam-Djomeh, Z.; Yarmand, M.S.; Fathi, M. Extraction, chemical composition, rheological behavior, antioxidant activity and functional properties of *Cordia myxa* mucilage. *Int. J. Biol. Macromol.* **2018**, *118*, 485–493. [CrossRef] [PubMed]
5. You, J.; Liu, C.; Feng, X.; Lu, B.; Xia, L.; Zhuang, X. In situ synthesis of ZnS nanoparticles onto cellulose/chitosan sponge for adsorption-photocatalytic removal of Congo red. *Carbohydr. Polym.* **2022**, *288*, 119332. [CrossRef]
6. Yu, Z.F.; Song, S.; Xu, X.L.; Ma, Q.; Lu, Y. Sources, migration, accumulation and influence of microplastics in terrestrial plant communities. *Environ. Exp. Bot.* **2021**, *192*, 104635. [CrossRef]
7. Irfan, J.; Hussain, M.A.; Haseeb, M.T.; Ali, A.; Farid-ul-Haq, M.; Tabassum, T.; Hussain, S.Z.; Hussain, I.; Naeem-ul-Hassan, M. A pH-sensitive, stimuli-responsive, superabsorbent, smart hydrogel from psyllium (*Plantago ovata*) for intelligent drug delivery. *RSC Adv.* **2021**, *11*, 19755–19767. [CrossRef]
8. Soltani, M.D.; Meftahizadeh, H.; Barani, M.; Rahdar, A.; Hosseinikhah, S.M.; Hatami, M.; Ghorbanpour, M. Guar (*Cyamopsis tetragonoloba* L.) plant gum: From biological applications to advanced nanomedicine. *Int. J. Biol. Macromol.* **2021**, *193*, 1972–1985. [CrossRef]
9. Qing, W.; Xinmin, W.; Shuo, P. The three-dimensional molecular structure model of Fushun oil shale kerogen, China. *J. Mol. Struct.* **2022**, *1255*, 132380. [CrossRef]
10. Muhammad, G.; Hussain, M.A.; Jantan, I.; Bukhari, S.N.A. *Mimosa pudica* L., a high-value medicinal plant as a source of bioactives for pharmaceuticals. *Compr. Rev. Food Sci. Food Saf.* **2016**, *15*, 303–315. [CrossRef]
11. Saraswat, R.; Pokharkar, R. GCMS studies of *Mimosa pudica*. *Int. J. PharmTech Res.* **2012**, *4*, 93–98.
12. Rezagholi, F.; Hashemi, S.M.B.; Gholamhosseinpour, A.; Sherahi, M.H.; Hesarinejad, M.A.; Ale, M.T. Characterizations and rheological study of the purified polysaccharide extracted from quince seeds. *J. Sci. Food Agric.* **2019**, *99*, 143–151. [CrossRef] [PubMed]
13. Muhammad, G.; Hussain, M.A.; Ashraf, M.U.; Haseeb, M.T.; Hussain, S.Z.; Hussain, I. Polysaccharide based superabsorbent hydrogel from *Mimosa pudica*: Swelling-deswelling and drug release. *RSC Adv.* **2016**, *6*, 23310–23317. [CrossRef]
14. Muhammad, G.; Ashraf, M.U.; Naeem-ul-Hassan, M.; Bukhari, S.N.A. Appraisal of acute oral toxicity of glucuronoxylan hydrogel from *Mimosa pudica* seeds. *Braz. J. Pharm. Sci.* **2018**, *54*, e17579. [CrossRef]
15. Cui, W.; Mazza, G.; Oomah, B.D.; Biliaderis, C.G. Optimization of an aqueous extraction process for flaxseed gum by response surface methodology. *LWT Food Sci. Technol.* **1994**, *27*, 363–369. [CrossRef]
16. Wu, Y.; Cui, S.W.; Tang, J.; Gu, X. Optimization of extraction process of crude polysaccharides from boat-fruited sterculia seeds by response surface methodology. *Food Chem.* **2007**, *105*, 1599–1605. [CrossRef]

17. Cheng, S.; He, F.; Fu, L.; Zhang, Y. Polysaccharide from rubescens: Extraction, optimization, characterization and antioxidant activities. *RSC Adv.* **2021**, *11*, 18974–18983. [CrossRef]
18. Zhong, K.; Wang, Q. Optimization of ultrasonic extraction of polysaccharides from dried longan pulp using response surface methodology. *Carbohydr. Polym.* **2010**, *80*, 19–25. [CrossRef]
19. Wang, B.; Xu, Y.; Chen, L.; Zhao, G.; Mi, Z.; Lv, D.; Niu, J. Optimizing the extraction of polysaccharides from *Bletilla ochracea* Schltr. using response surface methodology (RSM) and evaluating their antioxidant activity. *Processes* **2020**, *8*, 341. [CrossRef]
20. Tahmouzi, S. Optimization of polysaccharides from Zagros oak leaf using RSM: Antioxidant and antimicrobial activities. *Carobohydr. Polym.* **2014**, *106*, 238–246. [CrossRef]
21. Sui, Z.; Li, L.; Liu, B.; Gu, T.; Zhao, Z.; Liu, C.; Shi, C.; Yang, R. Optimum conditions for *Radix Rehmanniae* polysaccharides by RSM and its antioxidant and immunity activity in UVB mice. *Carbohydr. Polym.* **2013**, *92*, 283–288. [CrossRef] [PubMed]
22. Liu, Z.; Dang, J.; Wang, Q.; Yu, M.; Jiang, L.; Mei, L.; Shao, Y.; Tao, Y. Optimization of polysaccharides from *Lycium ruthenicum* fruit using RSM and its anti-oxidant activity. *Int. J. Biol. Macromol.* **2013**, *61*, 127–134. [CrossRef] [PubMed]
23. Samavati, V. Polysaccharide extraction from *Abelmoschus esculentus*: Optimization by response surface methodology. *Carbohydr. Polym.* **2013**, *95*, 588–597. [CrossRef] [PubMed]
24. Gibaldi, M.; Feldman, S. Establishment of sink conditions in dissolution rate determinations- theoretical considerations and application to non-disintegrating dosage forms. *J. Pharm. Sci.* **1967**, *56*, 1238–1242. [CrossRef] [PubMed]
25. Korsmeyer, R.W.; Gurny, R.; Doelker, E.; Buri, P.; Peppas, N.A. Mechanisms of solute release from porous hydrophilic polymers. *Int. J. Pharm.* **1983**, *15*, 25–35. [CrossRef]
26. Ritger, P.L.; Peppas, N.A. A simple equation for description of solute release II. Fickian and anomalous release from swellable devices. *J. Control. Release* **1987**, *5*, 37–42. [CrossRef]
27. Siepmann, J.; Peppas, N.A. Modeling of drug release from delivery systems based on hydroxypropylmethylcellulose. *Adv. Drug Deliv. Rev.* **2001**, *48*, 139–157. [CrossRef]
28. Karazhiyan, H.; Razavi, S.M.; Phillips, G.O. Extraction optimization of a hydrocolloid extract from cress seed (*Lepidium sativum*) using response surface methodology. *Food Hydrocoll.* **2011**, *25*, 915–920. [CrossRef]
29. Somboonpanyakul, P.; Wang, Q.; Cui, W.; Barbut, S.; Jantawat, P. Malva nut gum. (Part I): Extraction and physicochemical characterization. *Carbohydr. Polym.* **2006**, *64*, 247–253. [CrossRef]
30. Balke, D.T.; Diosady, L.L. Rapid aqueous extraction of mucilage from whole white mustard seed. *Food Res. Int.* **2000**, *33*, 347–356. [CrossRef]
31. Estévez, A.M.; Saenz, C.; Hurtado, M.L.; Escobar, B.; Espinoza, S.; Suárez, C. Extraction methods and some physical properties of mesquite (*Prosopis chilensis* (Mol) Stuntz) seed gum. *J. Sci. Food Agric.* **2004**, *84*, 1487–1492. [CrossRef]
32. Nazir, S.; Wani, I.A.; Masoodi, F.A. Extraction optimization of mucilage from Basil (*Ocimum basilicum* L.) seeds using response surface methodology. *J. Adv. Res.* **2017**, *8*, 235–244. [CrossRef] [PubMed]
33. Song, H.; He, M.; Gu, C.; Wei, D.; Liang, Y.; Yan, J.; Wang, C. Extraction optimization, purification, antioxidant activity, and preliminary structural characterization of crude polysaccharide from an Arctic *Chlorella* sp. *Polymers* **2018**, *10*, 292. [CrossRef] [PubMed]
34. Ye, C.L.; Jiang, C.J. Optimization of extraction process of crude polysaccharides from *Plantago asiatica* L. by response surface methodology. *Carbohydr. Polym.* **2011**, *84*, 495–502. [CrossRef]
35. Singh, B.; Oberoi, D.P.S.; Wani, I.A.; Sogi, D.S. Effect of temperature, salt concentration, pH and time on thermal degradation of pumpkin (*Cucurbita pepo*) puree. *Adv. Food Sci.* **2009**, *31*, 96–101.
36. Koocheki, A.; Mortazavi, S.A.; Shahidi, F.; Razavi, S.M.A.; Taherian, A.R. Rheological properties of mucilage extracted from *Alyssum homolocarpum* seed as a new source of thickening agent. *J. Food Eng.* **2009**, *91*, 490–496. [CrossRef]
37. Koocheki, A.; Taherian, A.R.; Razavi, S.M.; Bostan, A. Response surface methodology for optimization of extraction yield, viscosity, hue and emulsion stability of mucilage extracted from *Lepidium perfoliatum* seeds. *Food Hydrocoll.* **2009**, *23*, 2369–2379. [CrossRef]
38. Jouki, M.; Mortazavi, S.A.; Yazdi, F.T.; Koocheki, A. Optimization of extraction, antioxidant activity and functional properties of quince seed mucilage by RSM. *Int. J. Biol. Macromol.* **2014**, *66*, 113–124. [CrossRef]
39. El Batal, H.; Hasib, A. Optimization of extraction process of carob bean gum purified from carob seeds by response surface methodology. *Chem. Process Eng. Res.* **2013**, *12*, 1–8.
40. Bendahou, A.; Dufresne, A.; Kaddami, H.; Habibi, Y. Isolation and structural characterization of hemicelluloses from palm of *Phoenix dactylifera* L. *Carbohydr. Polym.* **2007**, *68*, 601–608. [CrossRef]
41. Joglekar, A.M.; May, A.T. Optimization of multigrain premix for high protein and dietary fibre biscuits using response surface methodology (RSM). *Cereal Food World* **1987**, *32*, 857–868.
42. Baş, D.; Boyacı, I.H. Modeling and optimization I: Usability of response surface methodology. *J. Food Eng.* **2007**, *78*, 836–845. [CrossRef]
43. Myers, R.H.; Montgomery, R.C. *Response Surface Methodology, Process and Product Optimization Using Design Experiment*; Wiley: New York, NY, USA, 2002.
44. Amin, M.A.; Ahmad, A.S.; Yinyin, Y.; Yahya, N.; Ibrahim, N. Extraction, purification and characterization of durian (*Durio zibethinus*) seed gum. *Food Hydrocoll.* **2007**, *21*, 273–279. [CrossRef]

45. Koocheki, A.; Taherian, A.R.; Bostan, A. Studies on the steady shear flow behavior and functional properties of *Lepidium perfoliatum* seed gum. *Food Res. Int.* **2012**, *50*, 446–456. [CrossRef]
46. Ali, A.; Hussain, M.A.; Haseeb, M.T.; Bukhari, S.N.A.; Tabassum, T.; Farid-ul-Haq, M.; Sheikh, F.A. A pH-responsive, biocompatible, and non-toxic citric acid cross-linked polysaccharide-based hydrogel from *Salvia spinosa* L. offering zero-order drug release. *J. Drug Deliv. Sci. Technol.* **2022**, *69*, 103144. [CrossRef]
47. Pooresmaeil, M.; Namazi, H. Facile preparation of pH-sensitive chitosan microspheres for delivery of curcumin; Characterization, drug release kinetics and evaluation of anticancer activity. *Int. J. Biol. Macromol.* **2020**, *162*, 501–511. [CrossRef]
48. Cok, M.; Viola, M.; Vecchies, F.; Sacco, P.; Furlani, F.; Marsich, E.; Donati, I. N-isopropyl chitosan. A pH- and thermo-responsive polysaccharide for gel formation. *Carbohydr. Polym.* **2020**, *230*, 115641. [CrossRef]
49. Lodhi, B.A.; Hussain, M.A.; Sher, M.; Haseeb, M.T.; Ashraf, M.U.; Hussain, S.Z.; Hussain, I.; Bukhari, S.N.A. Polysaccharide-based superporous, superabsorbent, and stimuli responsive hydrogel from sweet basil: A novel material for sustained drug release. *Adv. Polym. Technol.* **2019**, *2019*, 9583516. [CrossRef]
50. Lodhi, B.A.; Hussain, M.A.; Ashraf, M.U.; Haseeb, M.T.; Muhammad, G.; Farid-ul-Haq, M.; Naeem-ul-Hassan, M. Basil (*Ocimum basilicum* L.) seeds engender a smart material for intelligent drug delivery: On-off switching and real-time swelling, in vivo transit detection, and mechanistic studies. *Ind. Crop. Prod.* **2020**, *155*, 112780. [CrossRef]

polymers

Article

Free Radical Copolymerization of Diallylamine and Itaconic Acid for the Synthesis of Chitosan Base Superabsorbent

Wafa Al-Mughrabi [1,2], Abeer O. Al-dossary [1,2] and Abir Abdel-Naby [1,2,*]

1 Department of Chemistry, College of Science, Imam Abdulrahman Bin Faisal University, P.O. Box 1982, Dammam 31441, Saudi Arabia; walmagribi@iau.edu.sa (W.A.-M.); aodossary@iau.edu.sa (A.O.A.-d.)
2 Water Treatment Unit, Basic & Applied Scientific Research Center (BASRC), Imam Abdulrahman Bin Faisal University, P.O. Box 1982, Dammam 31441, Saudi Arabia
* Correspondence: aabdelnaby@iau.edu.sa

Abstract: Copolymerization of diallylamine (DAA) and itaconic acid (IA) was synthesized using benzoyl peroxide as a free radical initiator, in dioxane as the solvent. The composition of the copolymer was determined by the nitrogen content using Edx. The solubility of the copolymer was also investigated. The water solubility of the synthesized copolymer depends on the comonomers' ratio. The structure of the copolymer was confirmed by ^{13}C-NMR spectroscopy. To increase the water insolubility of the copolymers, and keep their hydrophilicity, the copolymer was allowed to react with chitosan to form a superabsorbent polymeric material (SP). The structure of the synthesized superabsorbent was confirmed using ^{13}C-NMR spectroscopy. The thermal property of the (SP) was also investigated by TGA. The investigation of the chitosan-based superabsorbent, as water-retaining agents, was studied. The results revealed that the superabsorbent polymers exhibited a good swelling ability and salt tolerance.

Keywords: free radical polymerization; superabsorbent; water-retaining agent; thermal properties

Citation: Al-Mughrabi, W.; Al-dossary, A.O.; Abdel-Naby, A. Free Radical Copolymerization of Diallylamine and Itaconic Acid for the Synthesis of Chitosan Base Superabsorbent. *Polymers* **2022**, *14*, 1707. https://doi.org/10.3390/polym14091707

Academic Editor: Yung-Chung Chen

Received: 24 March 2022
Accepted: 20 April 2022
Published: 22 April 2022

1. Introduction

Water is considered an essential need for living creatures, humans, animals, and plants. It is also important for various activities such as agriculture, especially in countries exhibiting a desert nature [1–3]. Superabsorbent polymers (SA) are hydrophilic polymers, that can absorb and retain water for short time but do not dissolve in water [4–6]. The (SA) polymers are widely used in various applications, such as agriculture, industry, drug delivery, and personal care [7–9]. In agriculture, the (SA) was used to reduce the frequency of irrigation, as it improves soil water retaining which enhances the plant survival rates, especially in the desert [10–14]. As an environmental requirement, naturally based superabsorbents are always used, such as starch, cellulose, proteins, and chitosan [15–18].

Chitosan is one of the most abundant natural polymers. It is used in various applications because of its compatibility, degradability, and nontoxicity [19]. Despite these advantages, chitosan suffers from low thermal stability at high temperatures. Additionally, the high percentage of hydrogen bonding limits its adjustment to various applications without modification [20,21]. Chitosan was known to undergo graft copolymerization with vinyl monomers, such as acrylic acid, acrylamide, and acrylonitrile [22–24].

Diallylamine (DAA) is known to undergo copolymerization by free radical mechanism, forming pyrrolidine rings [25]. The pyrrolidine ring is considered a proton adsorption site. Itaconic acid exhibits two carboxylic groups and an ethene bond, which enable it to undergo both copolymerization and polycondensation. IA and its producing special hydrogels for water decontamination, targeted drug delivery as well as smart nanohydrogels in food applications, coatings, and elastomers [26].

In the present work, copolymerization of DAA and IA with chitosan will be synthetized to produce a superabsorbent chitosan base polymeric material for the performance of hydrogels.

2. Experimental Section

2.1. Materials

Chitosan (CS), degree of acetylation 80%, average molecular weight 50,000, was purchased from Sigma-Aldrich. Itaconic acid, sodium bisulfite, and benzoyl peroxide were obtained from Loba Chemie. Potassium persulphate was obtained from Winlab limited. Ethanol, acetic acid, and DAA were obtained from Sigma-Aldrich.

All chemical reagents are of analytical grade.

2.2. Copolymerization of Diallylamine and Itaconic Acid

Various in-feed concentrations of the two comonomers were allowed to undergo free-radical copolymerization, in dioxane, using 0.02 M benzoyl peroxide as the initiator. The reaction was carried out under a nitrogen atmosphere for certain intervals of time, at 60 °C, in an ultrasonic bath of power 300 watts. The copolymer was filtered with diethyl ether and then washed with ethanol using Soxhlet system to remove any homopolymer. The copolymer was dried, weighted, and the found composition of the copolymer was determined by deducing the nitrogen content of the prepared copolymer.

2.3. Synthesis of the Superabsorbent Polymers

Synthesis of the superabsorbent polymer was carried out in two-neck round-bottom flask. A pure (2 g) chitosan was dissolved in 100 mL of 1% acetic acid solution. Proper concentrations of initiators (0.01 M, 0.015 M, 0.02 M) of Sodium bisulfite and potassium persulfate were added at temperatures (30 °C, 40 °C, 50 °C, 60 °C), in ultrasonic bath of 300 watts, under nitrogen atmosphere. After 15 min the comonomers were successively added. The reaction was carried out for a given interval of time (1–6 h). The Soxhlet extraction system was used to remove the homopolymers. After drying, the graft copolymer was weighted and the graft percentage (G%) was calculated according to the following equation:

$$G\% = \frac{w - w_o}{w_o} \times 100$$

w_o = original weight of chitosan, w = weight of the graft copolymer.

2.4. Spectroscopic Analysis

The structure of the copolymer and the superabsorbent polymer was confirmed using solid-state NMR Bruker Avance III spectrometer, operating at 400 MHz.

2.5. SEM and Energy Dispersive Spectrometer (EDS)

A VEGA 3 TESCAN scanning electron microscope (Tescan, Czech Republic) with a detector of secondary electron (SE). In addition, detector and energy-dispersive spectrometer (EDS) were used to determine the nitrogen content of the copolymer formed. The analysis was carried out at voltage of 15 KeV with a working distance of 10 mm between the specimen and the detector.

2.6. Swelling Measurements

The superabsorbent powder (0.05 g) was dispersed in distilled water (500 mL) for 4 h, at room temperature to reach the swelling equilibrium. The residual water was removed by filtration using a 100 mesh stainless steel screen until water ceased to drip.

The water absorbency was calculated according to the following Equation (1)

$$Q = \frac{(m_1 - m_0)}{m_0} \tag{1}$$

where Q is the water absorbency (w/w), m_0 is the weight of the dry superabsorbent polymer, and m_1 is the swollen superabsorbent polymer.

2.7. Swelling in Salt Solutions

The absorbency of the superabsorbent polymer was evaluated in variable NaCl solutions (from 0.1 to 1.0 $w/w\%$), using the above method described for the swelling measurements in distilled water.

2.8. Water Retention of the Superabsorbent Polymers

The water retention of the superabsorbent polymer (SA) was tested using the following method.

A specific amount of (SA) was allowed to swell to saturation in distilled water. The superabsorbent polymers were filtered using a 100 mesh screen and placed in Petri dishes, at room temperature. The weight of SA was determined after regular time intervals. The process was continued until saturation (no change in weight) was detected.

The water retention was obtained by applying the following equation:

$$water\ retention(\%) = \frac{(wt - wd)}{(wi - wd)} \times 100 \tag{2}$$

where, wt is the weight of SA, at time (t), wd is the weight of the dry superabsorbent polymer, and wi is the initial weight of swollen superabsorbent polymer.

3. Results and Discussion

3.1. Copolymerization of DAA and IA

The composition of the copolymer was determined according to the nitrogen content, as the nitrogen atom is a direct confirmation of the DAA moieties present in the synthetized copolymer (Table 1).

Table 1. Water solubility of various synthesized (DAA-IA) copolymers.

Sample No.	In-Feed Composition (M) (DAA, IA)	Found Composition (DAA, IA)	Water Solubility
1	(0.8, 0.2)	(0.95, 0.05)	Insoluble
2	(0.6, 0.4)	(0.85, 0.15)	Insoluble
3	(0.5, 0.5)	(0.8, 0.2)	soluble
4	(0.4, 0.9)	(0.7, 0.3)	Soluble

The structure of the copolymer formed in sample 2 was confirmed using ^{13}C-NMR (Figure 1).

The results revealed that the DAA cyclo-polymerized to form pyrrolidine rings [25], while the IA formed a condensation product with the pyrrolidine -NH group. Moreover, the presence of the ethylenic carbons accounted for the low composition of IA in the synthetized copolymers (Table 2).

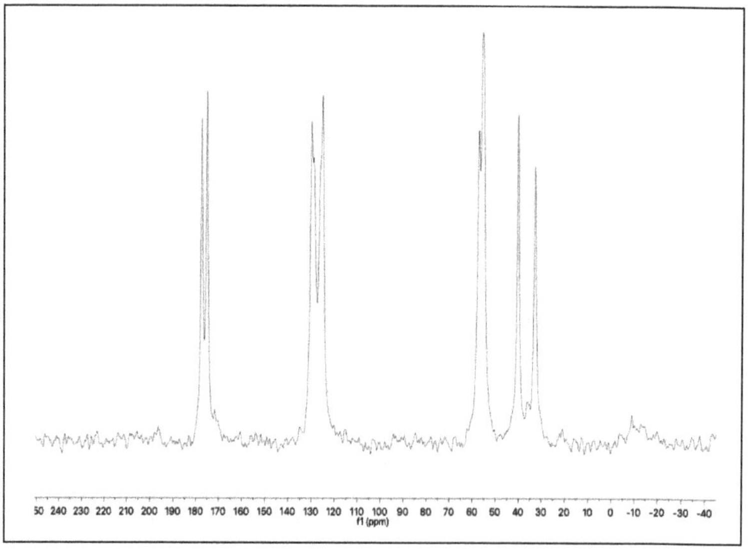

Figure 1. ^{13}C-NMR spectrum of (DAA:IA) copolymer (sample 2).

Table 2. ^{13}C-NMR spectral data of copolymerization of IA and DAA.

Structure	Carbon Atoms	^{13}C-NMR (δ ppm)
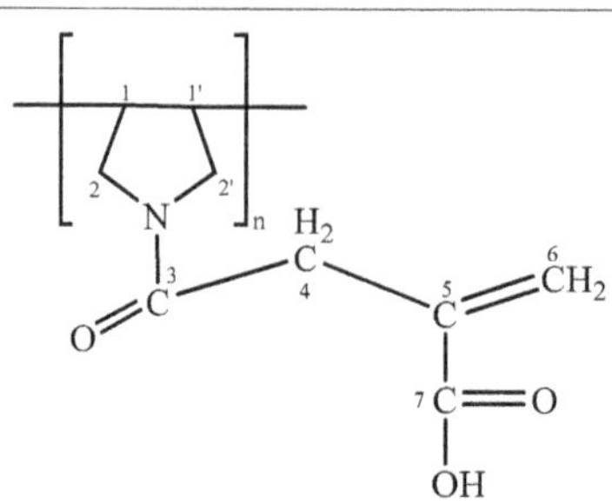	C1, C1′	40
	C2, C2′	55
	C3	173
	C4	31
	C5	133
	C6	123
	C7	177

3.2. Synthesis of the Superabsorbent Polymer

Although IA comonomer exhibits two carboxylic groups, which enable it to absorb either water or salty water, it also increases its water solubility. To profit from the advantage of the presence of carboxylic groups and overcome the water solubility, graft copolymer of both comonomers with chitosan (CS) occurred using redox polymerization.

3.2.1. Effect of Various Parameters on the Percentage of Graft

The following equation was used to calculate the grafting percentage

$$G = \frac{W - W_0}{W_0} \times 100 \tag{3}$$

To synthetize insoluble superabsorbent, the comonomers concentrations to the chitosan were kept (CS = 2 g, [DAA] = 0.4 M, [IA] = 0.15 M).

Effect of Time

The effect of various intervals of time on the grafting copolymerization of comonomers onto CS is listed in Table 3. The results revealed that the percentage of the graft increased gradually with time. This might be attributed to the increase in the chain length of the grafting branches until it reached a maximum value. Afterward, the increase in the reaction time led to a steady state for the G%, due to the consumption of monomer units in the formation of branches.

Table 3. Effect of time on the graft percentage.

Grafting Time (h)	Grafting (%) (CS = 2 g, [DAA] = 0.4 M, [IA] = 0.15 M, [I] = 0.02 M, T = 60 °C)
0	0
1	13.7
2	25.85
3	38.85
4	38.75
5	38.5
6	38.45

Effect of Temperature

To investigate the effect of reaction temperature on the graft copolymerization reaction, the temperature was increased from room temperature to 60 °C. The maximum percentage of the graft was achieved at 40 °C. Afterward, a gradual decrease in the percentage of the graft was observed with the increase in temperature, which could be attributed to the achievement of ceiling temperature of the polymeric branches as shown in Table 4.

Table 4. Effect of temperature on the graft percentage.

Temperature (°C)	Grafting (%) (CS = 2 g, [DAA] = 0.4 M, [IA] = 0.15 M, [I] = 0.02 M, Grafting Time = 3 h)
25	0
30	33.85
40	72.05
50	51.45
60	38.75

Effect of Initiators Concentration

The effect of initiator concentration on the percentage of graft onto Chitosan is listed in Table 5. A gradual increase in G% was observed with the increase in initiator concentration from 0.015 M to 0.02 M. Afterward, a decrease in the G% was observed by increasing the initiator concentration, which could be ascribed to the increase in the probability of chain transfer to initiator reactions.

Table 5. Effect of initiator concentration on the percentage of graft.

Concentration of Initiators (M)	Grafting (%) (CS = 2 g, [DAA] = 0.4 M, [IA] = 0.15 M, Grafting time = 3 h, T = 40 °C)
0.015	60.65
0.02	72.05
0.025	49.65

Thus, the optimum conditions for the synthesis of the superabsorbent were (CS = 2 g, [DAA] = 0.4 M, [IA] = 0.15 M, [I] = 0.02 M, and T = 40 °C).

3.2.1.4. ^{13}C-NMR Spectroscopic Analyses

The CS-g-(DAA-IA) ^{13}C-NMR spectrum confirmed the structure of the superabsorbent polymer (Figure 2). The results showed the condensation of the IA carboxylic group with an amino group of chitosan ($-\overset{O}{\overset{\|}{C}}-$ = 174 ppm) as well as the condensation of the IA carboxylic group with -NH groups of pyrrolidine ring ($-\overset{O}{\overset{\|}{C}}-$ = 175 ppm). In addition to another two types of carbonyl groups of IA moieties at 176.5 ppm and that of chitosan amino group at 178 ppm. The ethylenic bond was confirmed by the peaks at 129 ppm, 136 ppm.

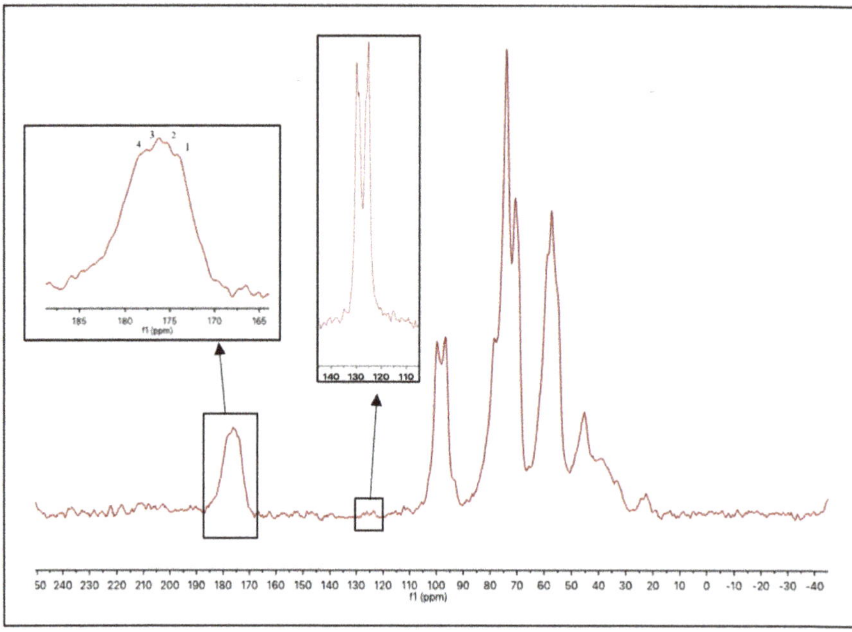

Figure 2. ^{13}C-NMR spectrum of SA.

From the above data, one can determine the structure of SA as follows:

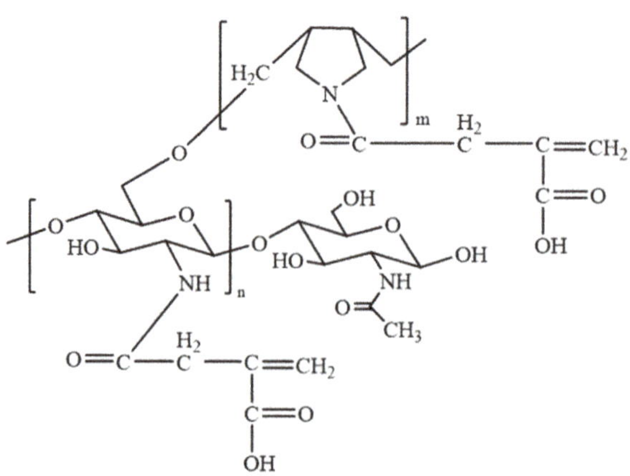

The Structure of SA according to its NMR spectrum

Another confirmation of the modification of chitosan is the surface morphology of the superabsorbent as compared to parent chitosan (Figure 3).

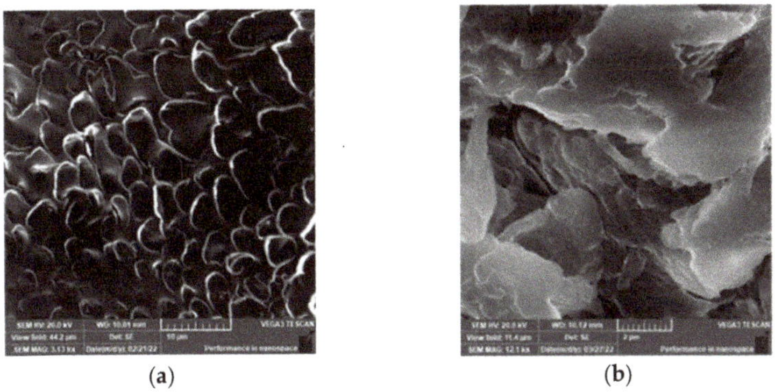

Figure 3. SEM Morphology of the superabsorbent (**b**) as compared to unmodified chitosan (**a**).

Figure 3b shows the branches built as extra layers on the top of the main chitosan chains. In addition, the chitosan shows smooth morphology, while the superabsorbent showed a rough surface due to the branches formed by the pyrrolidine moieties. The surface roughness is known to enhance the water permeability [27].

3.3. Thermal Properties of the Superabsorbent

To adjust the synthetized superabsorbent to any application, its thermal behavior should be examined. Figure 4 shows the TGA curves of various graft copolymers as compared to that of the native chitosan.

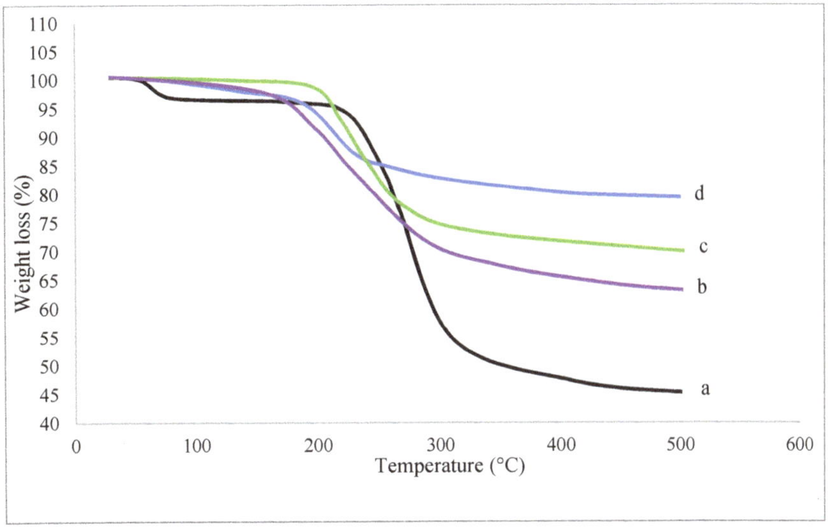

Figure 4. TGA of various CS-g-(DAA-IA) copolymers (%G: b = 49.1%, c = 60.65%, d = 72.05%) as compared to CS (a).

The results revealed that the graft copolymerization affected the initial decomposition temperature (T_o), the temperature at which the polymer starts to lose part of its polymeric matrix. This is attributed to the decrease in the matrix crystallinity as a consequence of the formation of branches onto the chitosan main chains. Despite the decrease in T_o values, the thermal stability of the graft copolymers could be shown by the decrease in weight loss percentages at high temperatures as compared to the native chitosan, which lost almost half of its weight at 500 °C. Thus, the increase in the percentage of the graft gave the copolymer its extra thermal stability.

3.4. *Water Absorbency Measurements*

3.4.1. Effect of the Percentage of Graft

The effect of comonomers contents on the water absorption of the superabsorbent is shown in Figure 5.

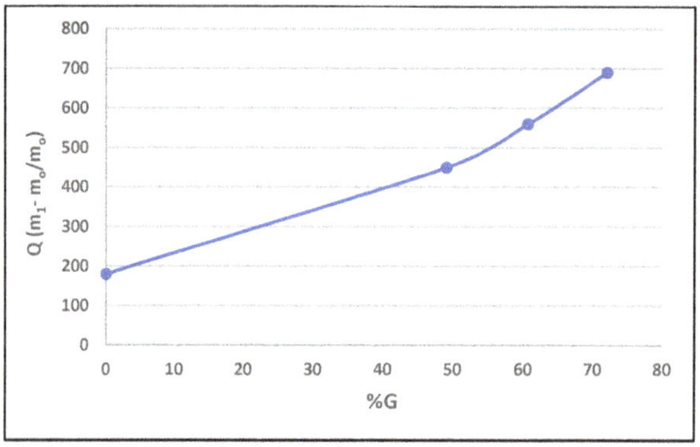

Figure 5. Water absorption capacity as a function of the percentage of graft.

The results revealed that the increase in the graft percentage led to an increase in the water absorbency. This is attributed to the increase in the number of carboxylic groups present in the superabsorbent polymer as itaconic acid moieties.

3.4.2. Swelling in Salt Solutions

The absorbency of the superabsorbent polymer in different salt concentration solutions is illustrated in Figure 6.

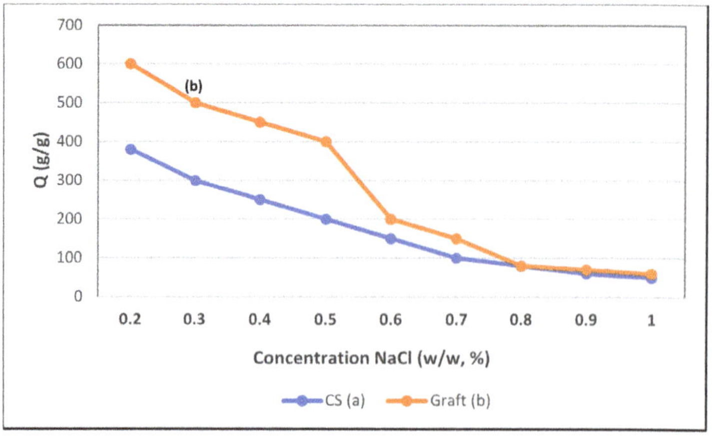

Figure 6. Water absorption capacity in different concentration salt solutions of superabsorbent (b) (%G = 60.65), as compared to CS (a).

The superabsorbent polymer swelling in different salt concentration solutions is illustrated in Figure 6. The results revealed that the water absorption capacity decreased with the increase in salt concentration. This is attributed to a charge screening effect of additional cations, causing anion–anion electrostatic repulsion [27].

The absorbency of SA in the salt solution is higher than that of CS. This is attributed to the basicity of the pyrrolidine ring to attract the protons of carboxylic groups of IA and the chlorine anion of NaCl to form quaternary ammonium salt.

From the above-mentioned data, the swelling efficiency of the superabsorbent polymer depends on its chemical structure and the medium.

3.5. Water Retention

The water retention property of SA was investigated. The results are illustrated in Figure 7.

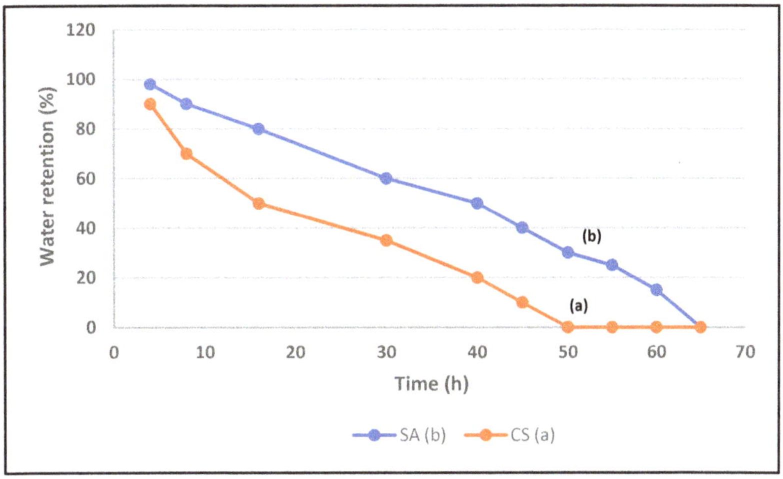

Figure 7. Water retention of SA (%G = 60.65) (b) as compared to that of CS (a).

The results revealed that the water retention of SA decreased with time. The superabsorbent polymer (%G = 60.65) exhibited higher water retention efficiency as compared to CS, as it retained up to 50% of water after 40 h. This is attributed to the ability of the SA polymer to undergo hydrogen bonding with water molecules [28].

4. Conclusions

A novel superabsorbent copolymer was successfully synthesized by the free radical polymerization of chitosan with diallylamine and itaconic acid comonomers. The reaction conditions were optimized. The structure of the superabsorbent was confirmed by ^{13}C NMR. The diallylamine was found to form polymeric branches as pyrrolidine rings while itaconic acid formed condensation products with the amino group of chitosan and the -NH group of pyrrolidine rings. The superabsorbent exhibited better thermal stability as compared to the native chitosan. The results of the effect of comonomers contents on the water absorption of the superabsorbent revealed that the increase in the graft percentage led to the increase in the water absorbency. This is attributed to the increase in the number of carboxylic groups present in the superabsorbent matrix as itaconic acid moieties. The absorbency of SA in the salt solution is higher than that of CS, this is attributed to the basicity of the pyrrolidine ring to attract the protons of carboxylic groups of IA moieties and the chlorine anions of NaCl salts to form quaternary ammonium salt.

The superabsorbent polymer (%G = 60.65) exhibited higher water retention efficiency, as compared to CS, as it retained up to 50% of water after 40 h. This is attributed to the ability of the SA polymer to undergo hydrogen bonding with water molecules.

The new superabsorbent is recommended to become an ideal soil water retention agent.

Author Contributions: Conceptualization, A.A.-N.; Data curation, W.A.-M., A.O.A.-d. and A.A.-N.; Formal analysis, W.A.-M., A.O.A.-d. and A.A.-N.; Methodology, W.A.-M., A.O.A.-d. and A.A.-N.; Supervision, A.A.-N.; Writing—original draft, A.O.A.-d. and A.A.-N.; Writing—review & editing, A.O.A.-d. and A.A.-N. All authors have read and agreed to the published version of the manuscript.

Funding: No grant is applicable.

Institutional Review Board Statement: Not applicable.

Informed Consent Statement: Not applicable.

Data Availability Statement: Not applicable.

Acknowledgments: The authors greatly acknowledge the facilities offered by the water treatment unit, at the Basic and Applied Research Center, College of Science, Imam Abdulrahman Bin Faisal University, and the institutional funding from the Ministry of Education, KSA. Project No IF 20-2020. The Authors also Acknowledge the Work Done by Zahra Sadiq (E mail: zahrasadiq43@gmail.com) for proof reading this manuscript, as a native British speaker.

Conflicts of Interest: The authors declare no conflict of interest.

References

1. Xu, Q.; Song, W.; Zhang, Y. Forecast and Optimal Allocation of Production, Living and Ecology Water Consumption in Zhangye, China. *Phys. Chem. Earth Parts A/B/C* **2016**, *96*, 16–25. [CrossRef]
2. Zhang, Y.; Wu, F.; Liu, L.; Yao, J. Synthesis and Urea Sustained-Release Behavior of an Eco-Friendly Superabsorbent Based on Flax Yarn Wastes. *Carbohydr. Polym.* **2013**, *91*, 277–283. [CrossRef] [PubMed]
3. Pathak, V.M.; Kumar, N. Dataset on the Superabsorbent Hydrogel Synthesis with SiO_2 Nanoparticle and Role in Water Restoration Capability of Agriculture Soil. *Data Br.* **2017**, *13*, 291–294. [CrossRef] [PubMed]
4. Feng, E.; Ma, G.; Wu, Y.; Wang, H.; Lei, Z. Preparation and Properties of Organic-Inorganic Composite Superabsorbent Based on Xanthan Gum and Loess. *Carbohydr. Polym.* **2014**, *111*, 463–468. [CrossRef] [PubMed]
5. Zhang, M.; Cheng, Z.; Zhao, T.; Liu, M.; Hu, M.; Li, J. Synthesis, Characterization, and Swelling Behaviors of Salt-Sensitive Maize Bran-Poly(Acrylic Acid) Superabsorbent Hydrogel. *J. Agric. Food Chem.* **2014**, *62*, 8867–8874. [CrossRef] [PubMed]
6. Zohuriaan-Mehr, M.J.; Omidian, H.; Doroudiani, S.; Kabiri, K. Advances in Non-Hygienic Applications of Superabsorbent Hydrogel Materials. *J. Mater. Sci.* **2010**, *45*, 5711–5735. [CrossRef]
7. Liu, J.; Wang, Q.; Wang, A. Synthesis and Characterization of Chitosan-g-Poly(Acrylic Acid)/Sodium Humate Superabsorbent. *Carbohydr. Polym.* **2007**, *70*, 166–173. [CrossRef]
8. Liu, T.; Qian, L.; Li, B.; Li, J.; Zhu, K.; Deng, H.; Yang, X.; Wang, X. Homogeneous Synthesis of Chitin-Based Acrylate Superabsorbents in NaOH/Urea Solution. *Carbohydr. Polym.* **2013**, *94*, 261–271. [CrossRef]
9. Hussien, R.A.; Donia, A.M.; Atia, A.A.; El-Sedfy, O.F.; El-Hamid, A.R.A.; Rashad, R.T. Studying Some Hydro-Physical Properties of Two Soils Amended with Kaolinite-Modified Cross-Linked Poly-Acrylamides. *Catena* **2012**, *92*, 172–178. [CrossRef]
10. Wilske, B.; Bai, M.; Lindenstruth, B.; Bach, M.; Rezaie, Z.; Frede, H.G.; Breuer, L. Biodegradability of a Polyacrylate Superabsorbent in Agricultural Soil. *Environ. Sci. Pollut. Res. Int.* **2014**, *21*, 9453–9460. [CrossRef]
11. Feng, D.; Bai, B.; Wang, H.; Suo, Y. Novel Fabrication of Biodegradable Superabsorbent Microspheres with Diffusion Barrier through Thermo-Chemical Modification and Their Potential Agriculture Applications for Water Holding and Sustained Release of Fertilizer. *J. Agric. Food Chem.* **2017**, *65*, 5896–5907. [CrossRef]
12. Wu, L.; Liu, M.; Liang, R. Preparation and Properties of a Double-Coated Slow-Release NPK Compound Fertilizer with Superabsorbent and Water-Retention. *Bioresour. Technol.* **2008**, *99*, 547–554. [CrossRef] [PubMed]
13. Yang, Y.; Wang, H.; Huang, L.; Zhang, S.; He, Y.; Gao, Q.; Ye, Q. Effects of Superabsorbent Polymers on the Fate of Fungicidal Carbendazim in Soils. *J. Hazard. Mater.* **2017**, *328*, 70–79. [CrossRef] [PubMed]
14. Chen, Y.; Tang, H.; Liu, Y.; Tan, H. Preparation and Study on the Volume Phase Transition Properties of Novel Carboxymethyl Chitosan Grafted Polyampholyte Superabsorbent Polymers. *J. Taiwan Inst. Chem. Eng.* **2016**, *59*, 569–577. [CrossRef]
15. Zou, W.; Yu, L.; Liu, X.; Chen, L.; Zhang, X.; Qiao, D.; Zhang, R. Effects of Amylose/Amylopectin Ratio on Starch-Based Superabsorbent Polymers. *Carbohydr. Polym.* **2012**, *87*, 1583–1588. [CrossRef]
16. Sawut, A.; Yimit, M.; Sun, W.; Nurulla, I. Photopolymerisation and Characterization of Maleylatedcellulose-g-Poly(Acrylic Acid) Superabsorbent Polymer. *Carbohydr. Polym.* **2014**, *101*, 231–239. [CrossRef]
17. Li, Q.; Liu, J.; Su, Y.; Yue, Q.; Gao, B. Synthesis and Swelling Behaviors of Semi-IPNs Superabsorbent Resin Based on Chicken Feather Protein. *J. Appl. Polym. Sci.* **2014**, *131*, 39748. [CrossRef]
18. Chen, Y.; Liu, Y.; Tan, H.; Jiang, J. Synthesis and Characterization of a Novel Superabsorbent Polymer of *N,O*-Carboxymethyl Chitosan Graft Copolymerized with Vinyl Monomers. *Carbohydr. Polym.* **2009**, *75*, 287–292. [CrossRef]
19. Dash, M.; Chiellini, F.; Ottenbrite, R.M.; Chiellini, E. Chitosan—A Versatile Semi-Synthetic Polymer in Biomedical Applications. *Prog. Polym. Sci.* **2011**, *36*, 981–1014. [CrossRef]

20. Abdel-Naby, A.S.; Nabil, S.; Aldulaijan, S.; Ababutain, I.M.; Alghamdi, A.I.; Almubayedh, S.; Khalil, K.D. Synthesis, Characterization of Chitosan-Aluminum Oxide Nanocomposite for Green Synthesis of Annulated Imidazopyrazol Thione Derivatives. *Polymers* **2021**, *13*, 1160. [CrossRef]
21. Jiang, J.; Hua, D.; Tang, J. One-Pot Synthesis of PH- and Thermo-Sensitive Chitosan-Based Nanoparticles by the Polymerization of Acrylic Acid/Chitosan with Macro-RAFT Agent. *Int. J. Biol. Macromol.* **2010**, *46*, 126–130. [CrossRef] [PubMed]
22. Anitha, A.; Sowmya, S.; Kumar, P.T.S.; Deepthi, S.; Chennazhi, K.P.; Ehrlich, H.; Tsurkan, M.; Jayakumar, R. Chitin and Chitosan in Selected Biomedical Applications. *Prog. Polym. Sci.* **2014**, *39*, 1644–1667. [CrossRef]
23. Dutkiewicz, J.K. Superabsorbent Materials from Shellfish Waste—A Review. *J. Biomed. Mater. Res.* **2002**, *63*, 373–381. [CrossRef] [PubMed]
24. Kong, M.; Chen, X.G.; Xing, K.; Park, H.J. Antimicrobial Properties of Chitosan and Mode of Action: A State of the Art Review. *Int. J. Food Microbiol.* **2010**, *144*, 51–63. [CrossRef]
25. Abdel-Naby, A.S.; Alharthi, S.N. Dyeability and Mechanical Properties of Acrylonitrile—Diallylamine Salts Copolymers. *Am. J. Appl. Sci.* **2013**, *10*, 525–535. [CrossRef]
26. Rezania, H.; Vatanpour, V.; Faghani, S. Poly(Itaconic Acid)-Assisted Ultrafiltration of Heavy Metal Ions' Removal from Wastewater. *Iran. Polym. J.* **2019**, *28*, 1069–1077. [CrossRef]
27. Fang, S.; Wang, G.; Li, P.; Xing, R.; Liu, S.; Qin, Y.; Yu, H.; Chen, X.; Li, K. Synthesis of chitosan derivative graft acrylic acid superabsorbent polymers and its application as water retaining agent. *Int. J. Biol. Macromol.* **2018**, *115*, 754–761. [CrossRef]
28. Wu, F.; Zhang, Y.; Liu, L.; Yao, J. Synthesis and Characterization of a Novel Cellulose-g-Poly(Acrylic Acid-Co-Acrylamide) Superabsorbent Composite Based on Flax Yarn Waste. *Carbohydr. Polym.* **2012**, *87*, 2519–2525. [CrossRef]

polymers

Review

Recent Advances in Chemically Modified Cellulose and Its Derivatives for Food Packaging Applications: A Review

Zhuolun Jiang and To Ngai *

Department of Chemistry, The Chinese University of Hong Kong, Shatin, N.T., Hong Kong, China; 1155140454@link.cuhk.edu.hk
* Correspondence: tongai@cuhk.edu.hk

Abstract: The application of cellulose in the food packaging field has gained increasing attention in recent years, driven by the desire for sustainable products. Cellulose can replace petroleum-based plastics because it can be converted to biodegradable and nontoxic polymers from sustainable natural resources. These products have increasingly been used as coatings, self-standing films, and paperboards in food packaging, owing to their promising mechanical and barrier properties. However, their utilization is limited because of the high hydrophilicity of cellulose. With the presence of a large quantity of functionalities within pristine cellulose and its derivatives, these building blocks provide a unique platform for chemical modification via covalent functionalization to introduce stable and permanent functionalities to cellulose. A primary aim of chemical attachment is to reduce the probability of component leaching in wet and softened conditions and to improve the aqueous, oil, water vapor, and oxygen barriers, thereby extending its specific use in the food packaging field. However, chemical modification may affect the desirable mechanical, thermal stabilities and biodegradability exhibited by pristine cellulose. This review exhaustively reports the research progress on cellulose chemical modification techniques and prospective applications of chemically modified cellulose for use in food packaging, including active packaging.

Keywords: cellulose; biodegradable polymers; chemical modification; food packaging

Citation: Jiang, Z.; Ngai, T. Recent Advances in Chemically Modified Cellulose and Its Derivatives for Food Packaging Applications: A Review. *Polymers* **2022**, *14*, 1533. https://doi.org/10.3390/polym14081533

Academic Editor: Arn Mignon

Received: 8 March 2022
Accepted: 6 April 2022
Published: 10 April 2022

1. Introduction

Plastics and common polymers made from fossil feedstocks have helped to build the modern world. However, ubiquitous discarded plastic contaminates the environment and has caused a serious microplastics problem, threatening the health of marine life, with associated risks for ecosystems and ultimately for human health, owing to plastic's long durability over centuries in terrestrial and water environments [1,2]. Efforts to reuse and reduce plastic have proven insufficient, and most plastic waste accumulates in landfills or is released into the environment [3,4]. Food packaging materials account for approximately 26% of all plastic produced worldwide and play a vital role in preserving food throughout the distribution chain, owing to plastic's water barrier capability, low cost, and light weight. However, approximately 90% of food packaging becomes waste after only one use [5,6]. With increased environmental concerns regarding sustainability and end-of-life disposal challenges, governments have proposed regulations such as limits or bans on single-use plastic bags and straws. Thus, there is an urgent need to transition from using plastic packaging materials to sustainable, biodegradable or compostable materials for food packaging [2,7–13]. Moreover, the ongoing COVID-19 pandemic has caused factory and transport disruptions worldwide, and much more food packaging material needs to be produced to meet the need for food preservation.

Cellulose refers to a class of natural carbohydrate polymers that are found in a virtually inexhaustible source of raw materials, such as plants, agricultural residues, shells of marine organisms, and microorganisms (Figure 1a) [14,15]. The plant cell wall of cellulose fibers contains different ratios of cellulose embedded as microfibrils in a matrix of hemicellulose, lignin, pectin, ash, and other extractives, depending on the origin of plants [14,16,17]. Cellulose is a class of linear, stereoregular, semicrystalline polysaccharides composed of repeated β-1,4-linked D-anhydroglucose units (AGU) with three hydroxyl groups (–OHs) per unit (Figure 1b) [18,19]. Compared with conventional petroleum-based polymers, such as polyethylene (PE), polypropylene (PP), and polystyrene (PS), cellulose-based materials have higher thermal stability, better film-forming capability, and better application qualities, such as a lower thermal expansion coefficient and improved mechanical strength and thermal stability at extreme temperatures [20–22]. Its intrinsic renewability and biodegradability make this material highly promising for use in various practical fields, e.g., packaging, straws, foams, flexible electronics, and tissue engineering materials, for the realization of sustainable product solutions and achieving a low carbon footprint (Figure 1c) [13,22–33].

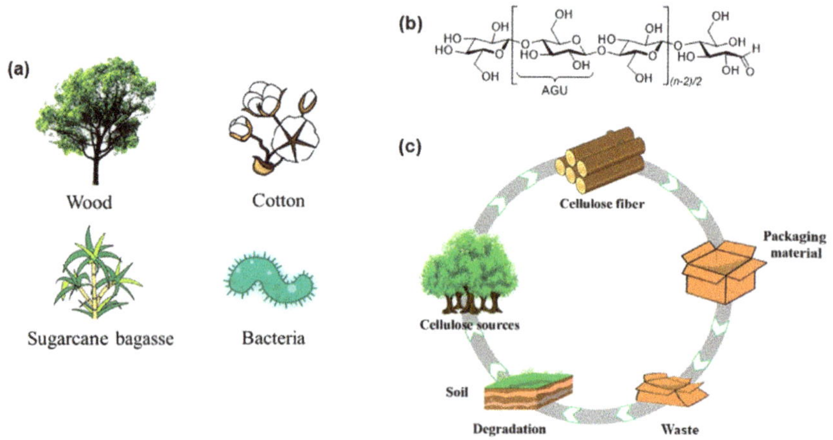

Figure 1. (**a**) Common sources of cellulose; (**b**) Chemical structure of cellulose; (**c**) Life cycle of cellulosic food packaging [15].

Desirable food packaging should provide a sufficient barrier against grease, water vapor, and oxygen, maintain good mechanical strength, and continually release antimicrobial agents during food storage [34–40]. These properties will extend the shelf-life of packaged food by inhibiting microbiological growth and reducing gas and moisture exchange between the food and the surrounding environment to slow chemical and physical changes in food [35,41–46]. Cellulose fibers can be manufactured into wrapping films and containers with various shapes, owing to the multiple hydrogen bonds among fibrils [6,13,47,48]. Nanosized cellulosic fibers have a higher specific surface area than cellulose fibers, and their high hydrogen bonding formation ability allows the material to create a strong and dense network, which makes it difficult for molecules to pass through [49–51]. This property is beneficial for barrier applications, especially to prevent the passage of oxygen, which is highly useful for the food packaging industry [52–54]. The water vapor transmission rate (WVTR) and oxygen transmission rate (OTR) of cellulose-based packaging are two critical indices that should be reduced as much as possible to achieve sufficient food packaging barrier performance, and should be comparable with those of commercial plastic products, such as low-density polyethylene (LDPE) [15,55,56]. It is essential to increase the hydrophobicity of cellulosic packaging to maintain its strong barrier ability and to enable its use in extended applications, even under humid conditions or when in the presence of moist foods [57]. The inherent hydrophilic character of cellulose fiber results in moisture absorption and the swelling of the polymer, leading to a more porous structure and thus a reduction in the packaging barrier to moisture and oxygen and degraded mechanical properties caused by interfibrillar slippage [17]. However, current methods of reducing the high hydrophilicity of cellulose-based food packaging materials are usually not good for the environment, such as coating the surface of the cellulose-based material with a thin layer of wax, plastic (e.g., PE), or aluminum. These coatings, which are prevalent in the production of milk cartons and paper cups, hinder the biodegradability of the packaging, resulting in potential environmental contamination [17].

This review begins with a discussion of fundamental information regarding the natural origin of cellulose, fiber extraction methods, and the scale, crystallinity, and functionality of fibers. It then systematically describes and provides up-to-date information on the chemical modification techniques of cellulose and its derivatives, along with giving details about the degree of reaction, degree of substitution (DS), and packaging preparation methods. Moreover, the influence of chemical modification on cellulose properties is discussed, with a focus on the barrier properties (surface hydrophobicity, oil/water vapor/oxygen barrier) and mechanical and thermal properties of cellulose-based packaging. Routes for improving the packaging performance to mitigate the intrinsic drawbacks of certain chemical modifications that can reduce the mechanical strength and water vapor barrier are outlined. Safety and biodegradability issues are also briefly considered. To highlight value-added applications in food preservation, this review also gives several examples in which various categories of additives (plasticizers, antioxidants, antimicrobial agents) are loaded. Approaches to overcoming the gap between industrially required extrusion–melting processing and the highly crystalline nature of cellulose are also examined, with the anticipation that a new generation of high-performance processable cellulose bioplastics is achievable.

Given the abundance of -OHs and derived functionalities of cellulose, chemical modifications such as (trans)esterification, amidation, silylation, urethanization, polymer grafting and crosslinking have been investigated intensively in the past decades to enhance the hydrophobicity of cellulose materials [16,58–64]. This review focuses on esterification, transesterification, and crosslinking techniques. The modification techniques can be further divided into two categories depending on whether the reaction occurs on the -OHs of neat cellulose or elsewhere on functional groups of cellulose derivatives (Figure 2). Theoretically, the extent of cellulose covalent functionalization is represented by a DS of 0–3, corresponding to the average number of -OHs per cellulosic AGU.

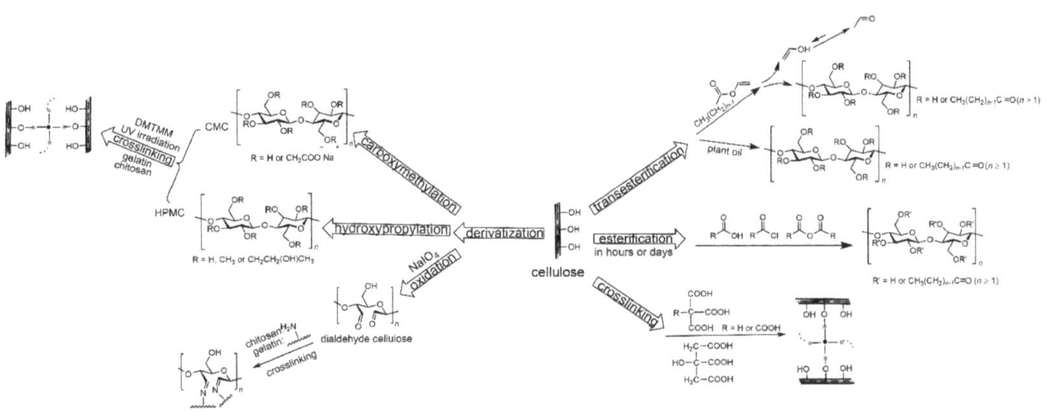

Figure 2. (Trans)esterification and crosslinking reaction schemes of cellulose and cellulose derivatives.

2. Cellulose Sources, Extraction Methods, and Size, Crystallinity and Functionality of Fibers

Approximately 36 cellulose chains come together to form a basic fibrillar unit, known as an elementary fibril, which has a characteristic lateral dimension of 1.5–3.5 nm and a length of up to 100 nm. These nanosized cellulose chains are bound together into larger cellulose microfibrils (CMF, 10–100 nm with a length up to 50 μm) through both intra- and intermolecular hydrogen bonds. CMF further associates to form plant fibers with diameters of 10–50 μm and lengths of several micrometers, depending on the cellulose source. The hierarchical microstructure of wood cellulose fibers is shown in Figure 3a [65,66]. The three-dimensional arrangement of the cellulose chains leads to the coexistence of crystalline and amorphous regions within the cellulose fibers [67,68]. Different mechanical and chemical treatments for cellulose plants can produce various micro- and nanosized cellulosic products through transverse dissociation in the amorphous region, namely, CMF, cellulose microcrystals (CMC), cellulose nanofibrils (CNF), and cellulose nanocrystals (CNC) (Figure 3b) [16,17,19]. These treatments lead to cellulose fibers with differing crystallinities, fiber sizes, and functionalities [16,69,70]. The traditional mechanical production of cellulose fiber mainly includes refining and high-pressure homogenization, microfluidization, and grinding [18,67,71]. CNC, also known as cellulose whiskers or rodlike cellulose (3–10 nm in diameter and 100–250 nm in length), is usually obtained via sulfuric acid hydrolysis of the amorphous regions of cellulose [72,73]. However, the introduction of charged sulfate groups compromises the thermostability of CNC, which can be a serious drawback for food packaging applications [74]. Bacterial nanofibrils (BNF) are composed of very pure cellulose nanofiber networks produced from low-molecular-weight sugars and alcohols through bacterial synthesis and have ultra-long nanofibers with diameter of 20–100 nm [26,75]. CMF, CMC, CNF, CNC, and BNF have emerged as key components for the production of cellulose-based food packaging. The first section of this review describes several studies in the last 10 years that have investigated fundamental aspects of the raw sources of cellulose, the production methods of micro/nanosized cellulose fibers, fiber size and crystallinity, and corresponding functionalities of the cellulose anhydroglucosic backbone (Table 1).

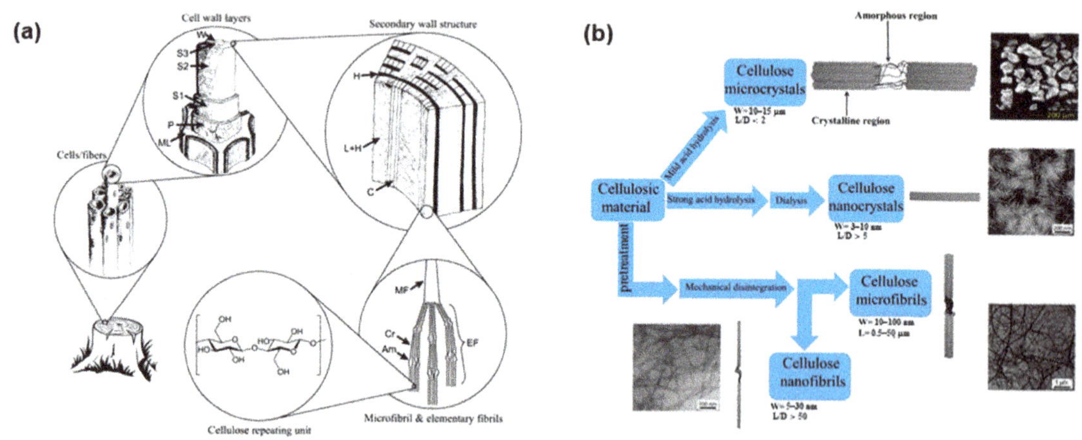

Figure 3. (**a**) Hierarchical structure of wood, showing: the middle lamella (ML), the primary wall (P), the outer (S1), middle (S2), and inner (S3) layers of secondary wall, the warty layer (W), cellulose (C), hemicellulose (H), lignin (L), microfibril (MF), elementary fibril (EF), crystalline domain (Cr) and amorphous domain (Am) [65]; (**b**) Methods to manufacture cellulose fibers with various sizes [17].

Table 1. Extraction methods and properties of cellulose from diverse natural sources.

Cellulose Source	Extraction Method	Fiber Size	Crystallinity	Functionality	Reference
kenaf bast fiber	disintegration, refining, cryo-crushing and cylinder homogenization	CNF, diameter: 10–30 nm	81%	–OH	[76]
Gluconacetobacter xylinus NRRL B-42	blender homogenization	BNF, fiber size: n.d. [a]	91.8%	–OH	[77]
tunicin cellulose	TEMPO [b]-mediated oxidation, blender homogenization, sonication	CMF, diameter: 10–20 nm	n.d.	–OH and sodium carboxylate groups (0.31 mmol/g)	[78]
wood pulp filter paper	blender homogenization, refining, freeze-drying	CNC, fiber size: n.d.	85%	–OH and sulfate groups	[79]
commercial never-dried CNC suspension in water	-	CNC, length: 64 nm, width: 7 nm	72%	–OH and sulfate groups	[80]
wheat straw CMF	blender homogenization	CNF, diameter: 10–40 nm	89%	–OH	[62]
softwood pulp dissolved with sulfite	carboxymethylation and cylinder homogenization	CMF, diameter: 5–15 nm	n.d.	–OH and sodium carboxylate groups (586 μ-equiv./g)	[81]
softwood and hardwood bleached kraft pulp	TEMPO-mediated oxidation, blender homogenization, sonication	TOCN [c], length: several μm, diameter: 3–4 nm	75%	-OH and sodium carboxylate groups	[82]
spruce/pine ($w/w = 7/3$) bleached softwood pulp	enzymatic treatment and cylinder homogenization	CNF, diameter: 20 nm	n.d.	-OH and sulfate groups	[83]
CMC from cotton linters	CNC: acid hydrolysis using sulfuric acid and freeze-dried, regenerated cellulose: treated by *N*-Methylmorpholine *N*-oxide under heating	CNC (length: 300 nm, diameter: 10 nm), regenerated cellulose (length: 100 μm)	80%	CNC: –OH and sulfate groups (70 mmol kg^{-1})	[84]

[a] Not detected; [b] 2,2,6,6-Tetramethyl-1-piperidinyloxy; [c] TEMPO-oxidized CNF.

3. Esterification

The reactivity of acylation agents (acylants) has provided a broad spectrum of strategies for producing cellulose esters. The esterification of cellulose using carboxylic acids as acylants has low reactivity without a catalyst but can be activated by adding a strong acid such as sulfuric acid. However, during acidic catalysis, cellulose hydrolysis generally occurs simultaneously, which causes the degradation of glucosidic bonds. A good acylant alternative is to replace carboxylic acid with anhydride or acyl chloride, which possesses activated acyl moieties, in combination with a tertiary base (e.g., pyridine, triethylamine, 4-dimethylaminopyridine [DMAP]) [19,85]. Employing acyl chlorides and anhydrides for esterification leads to the formation of hydrochloric acid or a carboxylic acid, respectively, which are involved in the acidic hydrolysis of cellulose chains; this process can be alleviated by adding a base to the reaction medium to neutralize the progressive formation of acid products. The added base also has the advantage of being a catalyst, as bases can form reactive intermediates that drive the reaction forward [19,86].

3.1. Esterification with Acylants Bearing a Short Substituent Chain (C2–C6)

Cellulose esters have been used in commercial processes for decades. These materials mainly consist of cellulose with substituent lengths of no more than four carbon atoms.

Such cellulose esters with short substitution chains (CESs) are generally produced by esterification using anhydrides in the presence of sulfuric acid [19,87]. Cellulose acetate (CA) is a commonly used CES and is described as the first organic cellulose ester, synthesized more than 150 years ago from wood pulp by Paul Schutzenberger [15]. Other CESs, including cellulose butyrate (CB), cellulose acetate propionate, and cellulose acetate butyrate (CAB), are also common commercial cellulose derivatives and are widely applied as components of coatings, paints, inks, membranes, and filters. These coatings are highly resistant to heat, UV, and moisture [88].

3.1.1. Synthesis of CESs with and without Solvents

CESs can be synthesized using different types of acylants with and without organic solvents in the reaction system. Jonoobi et al. adopted a solventless method to synthesize CA using acetic anhydride (or acetic acid) as the acylant and dispersion medium with pyridine (or sulfuric acid) as a catalyst to modify kenaf CNF, CNC, and wood filter pulp. Under optimal reaction conditions obtained by adjusting the temperature and reaction period, the DS of CA was relatively low, and the water contact angle (WCA) of CA was increased to 113°, demonstrating the effectiveness of the hydrophobic modification [76,79]. Ramírez et al. also performed solventless esterification between BNF and acetic or propionic acid (as dispersant) at 120 °C for 1–8 h using nontoxic biobased tartaric acid as a catalyst. The acetylated BNF (DS = 0.45) and propionized BNF (DS = 0.23) samples did not absorb water and remained floating on the water surface, while the pristine BNF sample absorbed water and immediately sank to the vial bottom. The decomposition onset temperature ($T_{d\text{-onset}}$, weight loss of approximately 5% upon heating) of the esterified samples was found to increase with the DS. The acetylated and propionized BNF showed higher thermal stability ($T_{d\text{-onset}}$ = 275–330 °C) than that of pure BNF (225 °C) [77]. With toluene as a solvent, Singh et al. conducted esterification between CMF and propionic anhydride at 60–100 °C for 0.45–5 h with pyridine as a catalyst, obtaining CMF propionate with a DS of 0.34–2.56 and a WCA of up to 121° [62].

Many studies have used large quantities of anhydride, acyl chloride, or organic solvents (e.g., dimethylacetamide [DMAc], chloroform, toluene) for extraction or in various steps of the reaction. However, the reliance on these organic solvents is undesirable because they are toxic, flammable, and/or volatile, cause air pollution, are hazardous to human health, and lead to difficult recoveries. To avoid these noxious substances, an effective water-based method was developed to endow the cellulose surface with a high carboxyl content through the esterification of cellulose with oxalic acid. The obtained oxalic acid modified CMF (OCMF) aqueous solution was then deposited onto filter paper through vacuum filtration to prepare the composite paper. The $T_{d\text{-onset}}$ of the OCMF sample decreased by approximately 40 °C compared with that of pristine CMF. Notably, while the OCMF deposited on the filter paper was still hydrophilic (WCA = 63°), its WVTR and grease-stained area markedly decreased from 694 to 123 g/m^2 24 h and from 91% to 0%, respectively, compared with pristine filter paper [89]. Swatlosk et al. reported a novel type of "green" solvents called ionic liquids, which show very low vapor pressure and the possibility for recycling and reuse after the reaction [90]. Ionic liquids have attracted considerable attention over the past two decades for the homogeneous modification of cellulose, despite their high cost [80,83,91,92]. Missoum et al. used acetic, butyric, iso-butyric, and hexanoic anhydrides to esterify CNF in an ionic liquid 1-butyl-3-methylimidazolium hexafluorophosphate [bmim][PF_6] at 100 °C for 2 h. The WCAs of CNF acetate, CNF butylate, CNF iso-butylate, and CNF hexanoate films were found to be 56°, 99°, 100°, and 105°, respectively. The ionic liquid was successfully recycled without any impurities or by-products formed during the reaction through two cycles of washing in sodium hydroxide solution and further purification through two cycles of extraction with diethyl ether [83].

3.1.2. CESs in Food Packaging Applications

In the past 10 years, CES-coated/coupled papers and CES self-standing films have become commonly used as food packaging materials. CA film displayed a higher WCA (83°) and lower WVTR (167 g/m^2 24 h) than an unmodified CMF film (WCA = 41, WVTR = 234 g/m^2 24 h), and the OTR of the CA film (7.5 mL m^{-2} day^{-1}) was higher than that of the original CMF film (4.2 mL m^{-2} day^{-1}), demonstrating that acetylation could weaken the oxygen barrier of the cellulose film. Furthermore, the tensile strength (σ) of CA films was slightly reduced compared with the original CMF film [93]. CA can be composited with paperboard to produce a material that can be made into food trays. When the active filler-layered double hydroxide (LDH) intercalated with antimicrobial 4-hydroxybenzoate anion (LDH-HB) was embedded in food-grade resin and coated onto CA-coupled cardboard, the material showed a high antibacterial effect and was suitable for food contact. This packaging was found to effectively preserve cooked tomato pasta for up to 30 days at 4 °C [5]. Transparent and flexible butyrated cellulose nanocrystal (Bu-CNC) films and coatings (DS = 2.1) showed an enhanced WCA of 92°. The Bu-CNC finally formed a dry, white, spongy material on the water surface when CNC formed a clear transparent suspension in water. The $T_{d\text{-onset}}$ of Bu-CNC was nearly 50 °C higher than that of the original CNC (257 °C). When Bu-CNC was coated onto mung bean seeds, there was no sprout formation, while CNC-coated seeds showed obvious sprouting, demonstrating Bu-CNC as a promising waterproofing coating for food [94]. Ioelovich et al. prepared various CESs with high DS values of 2.7–3, including CA, cellulose propionate (CP), CB, and CAB. The CESs were made into films and coatings for paperboard. The hydrophobicity of the cellulose esters increased with substituent chain lengths from C2 to C4. CAB showed high hydrophobicity that was comparable to that of CP. The 10 wt% CES-coated paper became completely resistant to water and oil, while the original paper had no effective barrier against water or oil. The water absorption (WA) of the CB- and CAB-coated paper was reduced to 0.1 g/m^2 and 0.2 g/m^2, respectively, compared with 98 g/m^2 for pristine paper. These CES films displayed a high elongation at break (ε) (43–48%). When immersed in deionized (DI) water for 24 h, the σ of the CAB and CB films (~33 MPa) remained nearly unchanged compared with their dry counterparts [95]. Deng et al. reported an efficient and easily industrialized method for the synthesis of cellulose hexanoate (CH) through the acylation reaction between CMF and hexanoyl chloride in dimethyl formamide (DMF) activated by mechanical ball milling at room temperature (RT). The CH films had good transparency compared with the white and opaque pristine CMF film. At the optimal milling time and acylant dose, the CH films (DS = 0.25–0.45) displayed lower WA (4%), a higher WCA (113°), and a lower WVTR (101 g/m^2 24 h) than the pristine CMF film (70%, 20°, 607 g/m^2 24 h). Furthermore, the acylation reaction (3 mL of acylant) combined with ball milling treatment (6 h) endowed the CH film with a markedly increased σ and ε (140 MPa, 21.3%) compared with the control CMF film (7 MPa, 4.4%). However, the $T_{d\text{-onset}}$ of CH markedly decreased with increasing DS, but remained above 280 °C [96]. Esterification of cellulose using C2–C6 acylants, reaction degree, and the associated packaging forming methods are listed in Table 2.

Table 2. Esterification of cellulose using C2–C6 acylants, reaction degree, and the associated packaging forming methods.

Cellulose	Acylation Process	Packaging Type and Its Formation	Reaction Degree	Ref.
softwood cellulose pulp; CMF	acetic acid or acetic anhydride as acylant with or without sulfuric acid as catalyst, reacted at 60–70 °C for 0.5–4 h with or without toluene as solvent	CA coating or film; CA solution was coated on paper via the hand lay-up technique or solvent-casted in air	DS = 0.21–0.32 for C2-CMF CES	[93,97]
CMF	oxalic acid as esterifying agent, reacted at 90 °C for 0.25–4 h in DI water	C2-CES coating; C2-CES aqueous suspension was deposited on filter paper through vacuum filtration, then oven-dried	carboxyl group content of 0.21–0.43 mmol/g fibrils	[89]
CNC extracted from bamboo waste pulp	butyric anhydride as acylant and iodine as catalyst, reacted at 105–110 °C for 30 min without solvent	C4-CES film or coating; C4-CES solution was solvent-casted or coated on mung bean seeds	DS = 2.1	[94]
cotton cellulose	acetic, propionic or butyric anhydride as acylant with trifluoroacetic acid as solvent/catalyst, reacted at 50 °C for 1 h	C2, C3, C4 (or their mixture)-CES films or coatings; CES solution was solvent-casted or coated on white paper with a bar coater	DS = 2.7–3	[95]
CMF from wood pulp	hexanoyl chloride as acylant and activated by mechanical ball milling, reacted at RT for 1–12 h in DMF	C6-CES film; C6-CES solution was solvent-casted, dried in an air-circulating oven at 60 °C and treated at 60 °C under vacuum for another 8 h	DS = 0.25–0.45 (3 h of milling time, 1–4 mL of acylant)	[96]

3.1.3. CES-Based Food Packaging Containing Various Additives

Additives such as plasticizers, antioxidants, and antimicrobials are often incorporated to improve the specific properties of CES films. Cinnamaldehyde (CIN) or green-synthesized silver nanoparticles (G-AgNPs) were incorporated into CA coatings or self-standing films, greatly increasing the antimicrobial performance of the CA film or paper substrate. The composite packaging showed low cytotoxicity and high antioxidation ability, and beneficial properties were promoted, such as an increased barrier to oil, water, water vapor, oxygen, and increased mechanical strength. In addition, the food shelf-life of beef was prolonged by 4–5 d at 4 °C [37,97]. As pure CA films are very brittle, plasticizers are generally incorporated into the CA film to make it flexible by increasing the mobility of the polymer chains, which greatly influences the Young's modulus (E), σ and ε of the CA packaging. Poly(ethylene glycol) (PEG) and carotenoids have been shown to be effective plasticizers for CA films due to their ability to enhance film elongation. The addition of cetrimonium bromide (CTAB)-modified montmorillonite (CTAB-MMT) to PEG-CA film led to good antimicrobial performance with no cytotoxicity. With the addition of 3 wt% CTAB-MMT, the nanocomposite film exhibited the most improved WVTR (5.84 g/m^2 24 h) and σ (40.9 MPa) compared with the PEG-CA film, as the incorporation of clay layers into the polymer matrix creates a tortuous path that decreases water vapor diffusion through the polymer matrix [98]. The incorporation of carotenoids (lycopene, norbixin and zeaxanthin) into CA films protected sunflower oil and vitamin B_2 from light oxidation. The films containing norbixin showed the highest barrier to UV-Vis light. The σ was increased from 65.3 for the CA film to 82.6, 105.6, and 87.4 MPa for norbixin-, lycopene-, and zeaxanthin-loaded films, respectively. A higher concentration of norbixin contributed to a higher water vapor permeability (WVP) (0.1 wt%: 0.035 g $\times$ mm m^{-2} h^{-1} kPa^{-1}), which was associated with its hydrophilic character, while hydrophobic zeaxanthin and lycopene led to a decreasing trend in the WVP (0.1 wt%: 0.023 and 0.022 g $\times$ mm·m^{-2} h^{-1} kPa^{-1}, respectively) compared with 0.032 g $\times$ mm·m^{-2} h^{-1} kPa^{-1} for the original CA film. The film containing 0.1%

zeaxanthin showed decreased thermal stability ($T_{d\text{-onset}}$ = 200 °C), which was approximately 50 °C lower than that of the pure CA film and the films blended with the other antioxidants [99,100]. C2–C6 CES materials used in food packaging applications are listed in Table 3.

Table 3. Overview of C2–C6 CES materials in food packaging applications.

Cellulose Packaging	Additives	Packaging Properties			Food Packaging Application	Ref.
		Barrier Properties	**Mechanical Properties**	**Other Properties**		
kraft paper (CA as coating)	2–8 mL of CIN (*v*/*v* in CA solution)	oil resistance: kit number 12; WA, WVP and OTR of the CIN-CA-coated paper markedly decreased by 96.2%, 76.8%, and million times, respectively	dry and wet σ was increased from 55.8 and 2.3 MPa to 88.2 and 12.9 MPa, respectively, compared with kraft paper	good cytocompatibility, high antioxidation with 8% CIN, excellent antibacterial performance with 6% CIN	extend beef's shelf-life by 4–5 d at 4 °C	[97]
CA-coupled cellulose cardboard (food-grade resin as coating)	10% (*w*/*w* to resin) LDH-HB	n.d. [a]	n.d.	good cytocompatibility in the release test, excellent antibacterial performance	preserved cooked pasta for up to 30 days at 4 °C	[5]
CA film	10–50 wt% PEG, 1–5 wt% MMTCTAB-MMT or 0.05 wt% G-AgNPs	3 wt% CTAB-MMT-incorporated CA film showed the lowest WVTR of 5.84 g/m^2 24 h; 0.05% G-AgNP-incorporated CA film showed an increased degree of water swelling from 0.28 to 0.44–0.62	ε of 20 wt% PEG-incorporated CA film (CAP20) was increased from 3.8% to 31.0%, while σ decreased from 43.3 MPa to 32.6 MPa compared with CA film	CAP20 film incorporated with CTAB-MMT showed slightly increased thermal stability, good antimicrobial properties, and no cytotoxicity; G-AgNP-incorporated CA film showed strong antibacterial activity and no cytotoxicity	n.d.	[37,98]
CA film	0.1–1% of carotenoids (lycopene, norbixin and zeaxanthin) (*w*/*w* to CA)	0.1 wt% carotenoids: WVP = 0.035 g × mm m^{-2} h^{-1} kPa^{-1} for norbixin, 0.023 g × m m^{-2} h^{-1} kPa^{-1} for zeaxanthin and 0.022 g × mm m^{-2} h^{-1} kPa^{-1} for lycopene	0.1 wt% lycopene or zeaxanthin: ε increased from 3.9% to 15%, while ε of 0.1 wt% norbixin-CA film remained unchanged; σ increased from 65.3 MPa to 84, 104 and 86 MPa, respectively, for norbixin-, lycopene- and zeaxanthin-loaded films	films with norbixin or lycopene displayed better light protection for sunflower oil; films with norbixin showed the best UV protection for vitamin B_2; film with 0.1% zeaxanthin showed 50 °C lower $T_{d\text{-onset}}$ (200 °C) compared with CA film and the films with other carotenoids	n.d.	[99,100]

[a] Not detected.

3.2. *Esterification with Acylants Bearing Medium (C8–C12) or Long Substituent Chains (>C12)*

Grafting long substituent chains onto the cellulose surface greatly affects the hydrophobicity, barrier, mechanical strength, and thermal properties of cellulose. Generally, surface hydrophobicity (indicated by the WCA and WA), water vapor resistance (indicated by the WVTR), and elasticity (indicated by the ε) increase with the length of the substituent chain. The oxygen barrier (indicated by the OTR), σ, E and thermal stability (indicated by $T_{d\text{-onset}}$) decrease with the length of the substituent chain.

3.2.1. Influence of Substituent Length and DS on Surface Hydrophobicity and WVTR of Cellulose Esters with Medium (CEMs) and Long Substitution Chains (CELs)

Grafting medium or long substituent alkyl chains onto cellulose can effectively enhance the WCA and lower the water vapor transport of cellulose packaging (Figure 4). One study modified CNF using a polymer anhydride (maleated styrene block copolymers, M_n = 230,000) as an acylant, conferring CEL with a WCA of up to 130° [101]. Another study revealed the effectiveness of long alkyl chains for lowering the WVTR of cellulose film. Tall oil fatty acid (TOFA) is a side-stream of the pulping industry from coniferous trees. TOFA-based C18-CEL (DS = 2.5–2.9) film showed a strongly decreased WVTR of 22–43 g/m^2 24 h, which was similar to 20 g/m^2 24 h for LDPE [102]. In fact, under the same reaction period and temperature conditions, acylants with higher substituent lengths had lower DS values for cellulose esters. For example, with increasing substituent lengths of acylants from C2 to C12, the cellulose esters had a decreasing trend in the DS; from 1.23 to 0.64 for C2-CES to C12-CEM [103].

A higher DS of cellulose esters always results in a greater WCA and water vapor resistance. Zhang et al. synthesized C18-CEL0.3 (DS = 0.3) and C18-CEL3 (DS = 3) through esterification between CMC and stearoyl chloride. The C18-CEL3 film did not show significant WA, and its WCA increased to 110°, slightly higher than 102° for the C18-CEL0.3 film. The WVP of the C18-CEL3 film (5×10^{-12} g m^{-1} s^{-1} Pa^{-1}) was much lower than that of the C18-CEL0.3 film (75×10^{-12} g·m^{-1}·s^{-1} Pa^{-1}), and both were considerably lower than 158×10^{-12} g m^{-1} s^{-1} Pa^{-1} for the neat CMC film [104]. Another report found that C6-CES film with a low DS (1.3) had a low WCA of 66°, with higher WCAs of 95–118° at higher DS values (1.6–2.9). This trend was similar for C16-CEL films, which showed a low WCA of 87° at DS = 0.8 and higher WCAs of 100–123° at DS = 1.6–2.6 [105,106]. At similar DS values, cellulose esters with longer substituent lengths have higher WCAs and water vapor barriers in the resulting cellulose ester packaging. C16-CEL film with a low DS of 0.9 showed a lower WVP (1.6 cc × mm m^{-2} d^{-1} kPa^{-1}) compared with 6 cc × mm m^{-2} d^{-1} kPa^{-1} for a C6-CES film with a higher DS (1.3) [106]. The WVP of fully substituted cellulose esters (DS = 2.9–3) films was reported to decline from 32 to 4.5×10^{-12} g m^{-1} s^{-1} Pa^{-1} with increasing substituent length from C8 to C18 [107]. The WVTR (100–300 g/m^2 24 h) of C16-CEL (DS = 1.6–2.6)-coated paperboard was found to be considerably lower than that of C6-CES (DS = 1.6–2.8)-coated paperboard (400–1020 g/m^2 24 h) [105].

Self-standing films and paperboards are typically made through hot-pressing [107] or solution casting [102,106,107], and some studies have investigated the effects of coating techniques, such as spray-coating [108], bar-coating [105,109], or brush-coating [110] on the WCA and WVTR of cellulose ester-coated paperboards. A spray-coated ethanol suspension of C11-CEM (DS = 0.62) on filter paper conferred the coated paper with superhydrophobicity (WCA = 152°) due to the formation of claw-like bulges on the surface. However, the WVTR showed no change in comparison with that of uncoated filter paper because the micro/nanosized C11-CEM particles could not totally cover the pores of the underlying paper [108,109]. When a C11-CEM (DS = 2.75) solution immobilized of the antibacterial poly(hexamethylene guanidine hydrochloride) (PHGH) and 3-mercaptopropionic acid (MPA) was bar-coated on paper, the WCA and WVTR of the coated paper gradually decreased from 117° to 101° and from 441 to 192 g/m^2 24 h, respectively, with increasing coating thickness due to the markedly increased surface smoothness [109]. The number of coating sides also has a significant effect on the water vapor barrier of cellulose ester-coated paper. Balasubramaniam et al. investigated the effect of whole-paper immersion acylation and one-sided surface acylation on the hydrophobicity and water vapor barrier properties of premade CNF films. Lauroyl chloride, palmitoyl chloride, and stearoyl chloride were used as the acylants. For the one-sided acylation method, acyl chlorides were applied with a brush onto one surface of pyridine (100 °C)-swollen CNF film, and the films were allowed to stand overnight at RT. For the whole-paper immersion acylation method, CNF films were immersed in pyridine at 100 °C for 30 min, after which fatty acid chlorides were added and acylated for 90 min; the films were then dried. Immersion modification resulted in a higher DS (0.91–1.8) than the one-sided modification (0.37–0.55). The WCA of the one-sided-acylated CNF film was improved to 105–121°, and the WCA of the immersion-acylated CNF film was improved to 112–114°. The films formed through immersion modification showed a strongly enhanced water vapor barrier property (WVP = 0.006–0.021 ng s^{-1} m^{-1} Pa^{-1}), while the one-sided modified films did not show any decrease in WVP compared with that of unmodified CNF films (WVP = 0.057 ng s^{-1} m^{-1} Pa^{-1}) [110]. However, the WVTR values of the above cellulose ester films and coated paperboard were still much higher than that of commercial LDPE (20 g/m^2 24 h) and LDPE-coated paperboard (11 g/m^2 24 h) [102,105], although their surface hydrophobicity and water barrier property were considerably enhanced compared with unmodified cellulose film and paper.

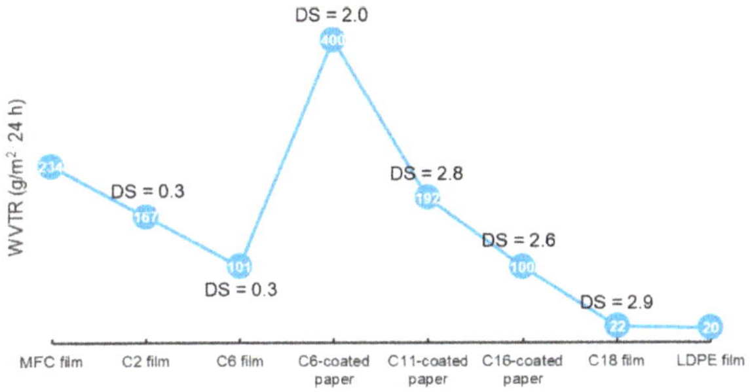

Figure 4. WVTR of CMF film, cellulose ester packaging and LPDE film [5,96,102,105,109].

3.2.2. Influence of Substituent Length on OTR of CEMs and CELs

The oxygen barrier capacity of cellulose ester packaging can be seriously compromised by the disruption of the crystalline region, especially for CEL packaging, resulting in the acceleration of microbial activity and food spoilage [111]. Modified TOFA-acylated cellulose films (DS = 2.5–2.9) showed greatly decreased WVTRs of 22–43 g/m^2 24 h. However, these films could not be used as an oxygen barrier, because the concentration of oxygen passing through the films was unacceptably high, indicating that CELs function well as water vapor barriers but are not effective as oxygen barriers [102]. Films made from fully substituted cellulose esters (C8–C18) (DS = 2.9–3) showed a declining WVP from 32 to 4.5×10^{-12} g m^{-1} s^{-1} Pa^{-1} with an increasing length of the substituent chain. However, their oxygen permeability (OP) increased from 0.24 to 1.3 cm^3 m^{-1} d^{-1} bar^{-1} with increasing substituent chain length, revealing that longer substituent chains caused poorer oxygen resistance [106,107].

3.2.3. Influence of Substituent Length and DS on the Mechanical Strength of CEMs and CELs

Cellulose films have inferior mechanical strength compared with conventional plastics such as LDPE, which shows σ = 7–16 MPa, E = 102–240 MPa, and ε = 100%–800%. Grafting long substituent alkyl chains onto cellulose can further decrease the mechanical strength of the packaging, as hydroxyl bonding among cellulose fibrils is destroyed (Figure 5a,b).

At similar DS, cellulose ester packaging with increasing substituent lengths has decreased mechanical strength. The C8-CEM (DS = 1.3) film exhibited a similar E (380 MPa), higher σ, and markedly higher ε (19 MPa, 90%) compared with C18-CEL (DS = 1.3) films (E = 397 MPa, σ = 10.4 MPa, ε = 10.7%) [104,106]. The films of fully substituted C8–C16 cellulose esters (DS = 2.8) were ductile, except for the C18-CEL film obtained from stearoyl chloride, which was extremely brittle. The ε of C8–C16 cellulose ester films decreased significantly from 100% to 40% with increasing substituent chain length, while all had a low σ, ranging from 4.8 to 7.1 MPa [112]. A higher DS in cellulose ester films results in lower mechanical strength. C18-CEL (DS = 0.3, 3) films had a lower mechanical strength than that of unmodified CMC film (σ = 169.2 MPa, E = 7230 MPa, ε = 15.9%), but C18-CEL0.3 films showed a much better mechanical strength (σ = 28.5 MPa, E = 1118 MPa, ε = 12.7%) than C18-CEL3 films (σ = 5.5 MPa, E = 286 MPa, ε = 2.5%) [104]. C12-CEM (DS = 0.9) film had a significantly higher σ (35 MPa) than that of higher substituted C12-CEM films (DS = 1.9–2.8) (σ = 1.8–2.5 MPa) [106,113].

To overcome the serious degradation in mechanical strength of CEM and CEL self-standing films and paperboard, coating technology has been frequently adopted to maintain the original mechanical strength of the neat cellulose packaging (Figure 6). One study reported that C11 CEM-coated film and paperboard displayed unchanged mechanical strength, as revealed by a slightly increased ε and nonsignificant change in tensile index [109]. Another study reported that, when solutions of C6-CES and C16-CEL with DS values of 1.6–2.6 were bar-coated on paperboard, the σ of the coated paper was slightly enhanced to 12.5 MPa, compared with 11.8 MPa for the uncoated paperboard [105]. Balasubramaniam et al. investigated the effect of chemical coating method (immersion acylation or one-sided surface acylation) on the mechanical properties of premade CNF films. The mechanical strength of the one-sided modified C12–C18 CNF film (specific σ = 51–72 MPa g^{-1} cm^3, specific E = 2.4–2.8 GPa g^{-1} cm^3) was comparable to that of the original CNF film, while those of the C12–C18 films prepared from the immersion method were much weaker (10–23 MPa g^{-1} cm^3, 0.4–1.6 GPa g^{-1} cm^3) [110]. Sehaqui et al. immersed CNF wet cake into linear or cyclic C2–C16 anhydride liquid for 2 d, followed by hot pressing treatment. The C2-CES nanopaper (DS = 0.38) had the highest storage modulus, and the C16-CEL nanopaper (DS = 0.1) displayed the lowest storage modulus. However, the wet stability of the cellulose ester nanopapers progressively increased with the length of grafted moieties from C2 to C16. The C16-CEL nanopaper had the highest wet strength ratio (32%) of its initial dry strength, which constituted a 7-fold improvement compared with the reference CNF nanopaper [114]. Esterification of cellulose using C2–C20 acylants, reaction degree, and the associated packaging forming methods are listed in Table 4. C2–C18 cellulose ester materials used in food packaging applications are listed in Table 5. And the influence of packaging manufacturing technologies on properties of cellulose ester-based packaging (DS = 0.2–2.8) are listed in Table 6.

Table 4. Esterification of cellulose using C2–C20 acylants, reaction degree, and the associated packaging forming methods.

Cellulose	Acylation Process	Packaging Type and Its Formation	Reaction Degree	Ref.
unbleached eucalyptus CNF; wheat bran and maize bran residue cellulose; cellulose sheet; softwood cellulose; CMC; α-cellulose	C6–C20 acyl chlorides as acylants, pyridine, sulfuric acid or DMAP as catalyst, reacted at 50–130 °C in DMAc, DMAc/LiCl or cosolvent of toluene and pyridine for hours	C6–C20 films; solvent casting or vacuum dried	DS = 0.19–3 (DS of cellulose esters decreased with increasing substituent chain length from C6 to C18)	[87,102,104,106, 107,112,113,115]
BNF, α-cellulose	C2–C12 carboxylic acids or C8–C18 acyl chlorides as acylants, reacted at 50–130 °C for 2 h in pyridine or pyridine/tosyl chloride	C2–C12 cellulose ester papers or C8–C18 cellulose ester films; wet cellulose cakes or films were hot-pressed at 90–110 °C	DS = 0.64–3 (DS of cellulose esters decreased with increasing substituent chain length from C2 to C12)	[103,107]
α-cellulose, bleached bagasse pulp, CMC	C6–C16 acyl chlorides as acylants and pyridine as catalyst, reacted in DMAc, DMAc/LiCl or pyridine at RT–100 °C for hours or days	C6–C16 cellulose ester-coated paper; cellulose ester suspension or solution was spray- or bar-coated on paperboard via air brush or bar coater	DS = 0.62–2.9	[105,108,109]
premade CNF film	lauroyl, palmitoyl or stearoyl chloride as acylant, reacted in pyridine at 100 °C	C12-, C16- and C18-cellulose ester films; one-sided acylation using a brush and reacted at 100 °C; immersion acylation at 100 °C for 90 min	immersion modification resulted in a higher DS (0.91–1.8) than one-sided modification (0.37–0.55)	[110]
oat straw CNF	acetic, butyric, hexanoic or 2-dodecen-1-yl-succinnic anhydride as acylant, reacted at 80 °C in an oven for 2 h with a 10 kg weight on top	C2–C16 cellulose ester nanopapers; CNF wet cake was immersed in acylant liquid for 2 d, then put in an oven at 80 °C for 2 h under hot pressing	DS of cellulose esters decreased from 0.38 to 0.1 with increasing substituent chain length from C2 to C16	[114]

Table 5. Overview of cellulose ester (C2–C18) materials in food packaging applications.

Cellulose Packaging	Packaging Properties			Ref.
	Barrier Properties	Mechanical Properties	Other Properties	
C11-CEM film (DS = 0.19)	higher WCA (101°) and lower WA (6%) than original CNF film (54°, 95%), lower WVP at 3.4×10^{-9} g·m^{-1} s^{-1} Pa^{-1} than pristine CNF film (9.0×10^{-9} g·m^{-1} s^{-1} Pa^{-1})	decreased σ and E (47 MPa, 2075 MPa) compared with that of neat CNF film (57, 3847 MPa), while the ε was slightly increased (6.2% *vs.* 4.5%)	$T_{d\text{-}onset}$ was increased slightly to 350 °C compared with that of CNF (343 °C)	[115]
C18-CEL film (DS = 2.53–2.86)	WVTRs of isostearic-, oleic- and modified TOFA- cellulose ester films were markedly reduced to 21.7, 22.4 and 43.4 g/m^2 24 h, poor oxygen resistance (too high for the sensor)	isostearic CEL film showed the highest ε of 101% (twice that of the original cellulose film), while oleic CEL film and modified TOFA CEL film had lower εs (57% and 45%)	$T_{d\text{-}onset}$ of C18-CELs (327~340 °C) was 7~22 °C higher than that of unmodified cellulose (320 °C)	[102]
C18-CEL film (DS = 0.3–3)	C18-CEL (DS = 0.3) film absorbed less water (13.9%) than pristine CMC film (28.9%), while C18-CEL (DS = 3) did not show significant WA, C18-CEL3 film had higher WCA (110°) than C18-CEL0.3 film (102°), WVPs of C18-CEL3 and C18-CEL0.3 films were decreased to 5×10^{-12} and 75×10^{-12} g m^{-1} s^{-1} Pa^{-1}, respectively, compared with 158×10^{-12} g m^{-1} s^{-1} Pa^{-1} for neat CMC film	C18-CEL0.3 film had higher mechanical strength (σ = 28.5 MPa, E = 1118 MPa, ε = 12.7%) than C18-CEL3 film (σ = 5.5 MPa, E = 286 MPa, ε = 2.5%), but both substituted films had decreased mechanical strength compared with unmodified CMC film (σ = 169.2 MPa, E = 7230 MPa, ε = 15.9%)	C18-CEL0.3 film exhibited moisture-responsiveness, while C18-CEL3 film showed thermal responsiveness, the C18-CEL3 film displayed a distinct T_m at 55 °C, while no T_m was found for the C18-CEL0.3 film, reversible changes in the C18-CEL3 film volumes were observed when varying the temperature	[104]
C6-, C8-, C10-, C12-, C14- and C18-cellulose ester films (DS = 0.8–1.3)	WCA was increased from 66° to 90°, WVP was decreased from 6 to 1.6 cc × mm m^{-2} d^{-1} kPa^{-1} for C6 to C18, which was markedly lower than 20–25 cc × mm m^{-2} d^{-1} kPa^{-1} for pristine CNF film, C6–C18 cellulose ester films all exhibited poor oxygen resistance	C6-CES, C10-CEM and C12-CEM films had higher E (550–600 MPa) compared with other cellulose ester films, C8-CEM (DS = 1.3) film had the highest ε (90%), C12-CEM (DS = 0.9) film had the highest σ (35 MPa); the other cellulose ester films had lower σ (20–24 MPa)	cellulose ester films were transparent, flexible and heat-sealable, melted at 170–225 °C, and were able to be squeezed through a 2-mm rod die	[106]
C11-CEM-coated paper (DS = 2.75)	WCA decreased from 117° to 101° with coating grammage increasing from 0.97 to 6.25 g m^{-2}, WVTR decreased from 441 to 192 g/m^2 24 h with increasing coating grammage, compared with 622 g/m^2 24 h of uncoated paper	slightly increased ε an nonsignificant change in tensile index	PHGH and MPA were attached to C11-CEM-coated paper, giving the paper desirable antimicrobial performance	[109]

Table 5. *Cont.*

Cellulose Packaging	Packaging Properties			Ref.
	Barrier Properties	Mechanical Properties	Other Properties	
C6- and C16-cellulose ester-coated paper (DS = 1.6–2.9)	WCA was enhanced to 95–123°, WVTR of C16-CEL-coated paper (100–300 g/m^2 24 h) was considerably lower than that of C6-CES-coated paper (400–1020 g/m^2 24 h)	cellulose ester-coated paper showed slightly enhanced σ (12.4–12.7 MPa) compared with uncoated paper (11.8 MPa)	C16-CEL-coated paper showed higher $T_{d\text{-onset}}$ (350 °C) than C6-CES-coated paper (320 °C). C6-CES and C16-CEL powder became liquid at 160 °C and 220 °C, respectively, in an oven	[105]
C12-, C16- and C18-cellulose ester films, DS (immersion method) = 0.9–1.8, DS (one-sided method) = 0.37–0.55	WCA of one-sided acylated film: 105–121°, WCA of immersion-acylated film: 113°, WVP of one-sided acylated film did not show any decrease compared with that of unmodified CNF films (WVP = 0.057 ng s^{-1} m^{-1} Pa^{-1}), WVP of immersion-acylated film decreased to 0.006–0.021 ng s^{-1} m^{-1} Pa^{-1}	mechanical strength of one-side-acylated film was comparable to that of the original CNF film, immersion-acylated film showed much weaker mechanical strength	thermal stability was increased for immersion-acylated films, whereas that of one-side-acylated films was similar to pristine CNF film ($T_{d\text{-onset}}$ = 245 °C)	[110]
C2-, C4-, C6- and C16-cellulose ester nanopapers (DS = 0.1–0.38)	WCA was enhanced from 32° to 119° from C2 to C16 and was 24° for the neat CNF nanopaper, C16-CEL paper floated on the water surface for several weeks, while pristine CNF paper sank	C2-CES nanopaper had the highest E, and C16-CEL nanopaper had the lowest E, C16-CEL had the highest wet strength, which was 7- fold greater than that of the reference CNF nanopaper	n.d. [a]	[114]

[a] Not detected.

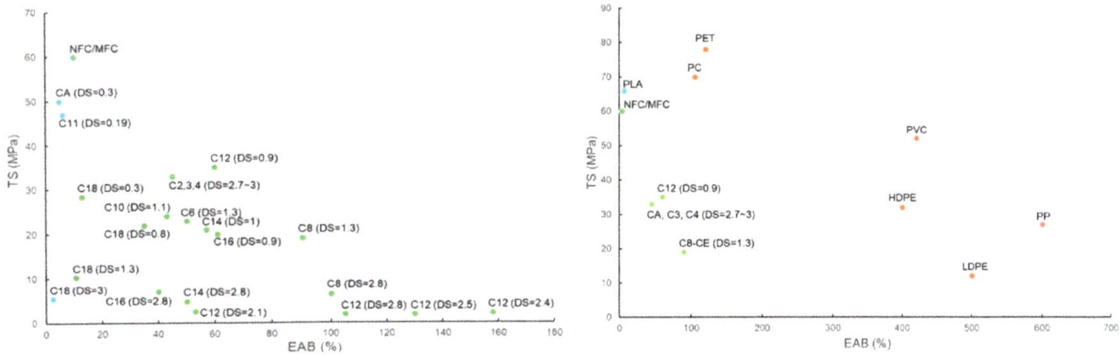

(a) (b)

Figure 5. (**a**) Mechanical strength (ε *vs.* σ) of cellulose and cellulose esters with various chain lengths and DS values; (**b**) mechanical strength (ε *vs.* σ) of cellulose and cellulose esters compared with poly(lactic acid) (PLA) and petroleum-derived plastics [116].

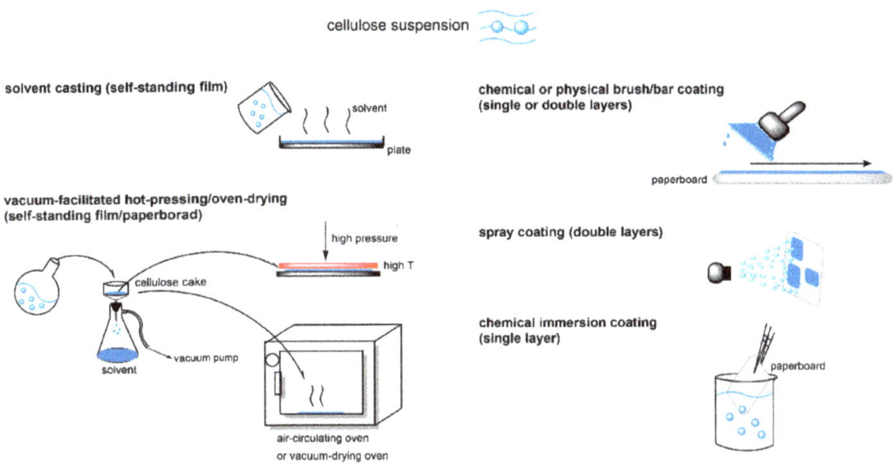

Figure 6. Strategies for cellulose ester-based packaging manufacture.

Table 6. Influence of packaging manufacturing technologies on properties of cellulose ester-based packaging (DS = 0.2–2.8).

Technology	Smoothness	WCA [a]	WVP	TS	Efficiency	Ref.
self-standing film	'4–5' [b]	84–107°, '2–3'	reduced 62–90%, '4–5'	decreased 18%, '4'	'2'	[103,106,115]
physical bar coating on paperboard	'3–4'	100–124°, '3–4'	reduced 29~69%, '2–4'	no change, '5'	'1'	[105,109]
chemical brush coating on film	'4'	109°, '3'	no change, '1'	decreased 32%, '3'	'4'	[110]
spray coating on paperboard	'1'	152°, '5'	no change, '1'	n.d. [c]	'3'	[108,109]
chemical immersion coating on film	'3'	114°, '3'	reduced 63~90%, '4–5'	decreased 68~80%, '1–2'	'5'	[110]

[a] WCA of neat cellulose film or paperboard is 15–57°; [b] With increasing number, cellulose ester-based packaging displays poorest ('1') or best ('5') performance for each parameter, and '2–4' is the extent between them; [c] Not detected.

3.2.4. Thermoplasticity of CELs

Cellulose has not yet achieved wide industrial acceptability for thermal processing; for example, it cannot undergo extrusion melting. The lack of thermal processing has led to increased production costs and times for cellulose-based packaging. One prerequisite for a polymer material to be suitable for melt processing is a broad temperature window, which means that the polymer's melt temperature (T_m) needs to be far lower than its $T_{d\text{-onset}}$ [105]. The thermoplasticity of cellulose esters with substituent chains from C6–C16 has been reported in several studies, and is enabled by the internal plasticization effect of the long alkyl chains [84,103,110]. These studies have found that C6–C16 cellulose esters with low or moderate DS melt at 160–230 °C, but it has been difficult to measure their exact $T_{d\text{-onset}}$ [105,106].

Generally, the $T_{d\text{-onset}}$ of cellulose esters decreases with increasing substituent length [103]. The $T_{d\text{-onset}}$ of fully substituted cellulose esters (C2–C18) has been reported to be 250–330 °C [102,117,118]. Fully substituted C8–C18 cellulose esters (DS = 2.7–3) are a class of thermosensitive polymers with a distinct T_m lower than 100 °C, but exhibiting a typical melting phenomenon (a thermoplastic feature) at relatively low heat, conferring the cellulose ester packaging with useful thermal responsiveness and processibility, greatly broadening its value-added application potential [104–106]. When the substituent length reaches C12 or higher, cellulose tri-esters are able to crystallize and form ordered regions via their side-chains. The T_m and heat of fusion of side-chain crystals increase with increasing side length from C12 to C20, which is attributable to an increase in side-chain crystal thickness as more methylene units participate in crystallization at longer side-chain lengths [87]. Sealey et al. (1996) found that fully substituted C12-CEM and C14–C20 CELs (DS =2.8–2.9) showed linearly increasing T_m values from −15 °C for C12-CEM to 55 °C for C20-CEL [87,113].

Zhang et al. found that fully substituted C18-CEL film displayed a distinct T_m at 55 °C, while no T_m was identified for the C18-CEL0.3 film. Reversible changes in C18-CEL3 film volume were observed when varying the temperature; these changes were caused by the construction and destruction of crystalline regions consisting of stearoyl moieties [104]. Geissler et al. prepared a transparent and thermally responsive cellulose film to controllably release rhodamine B molecules. C18-CEL (DS = 2.95) was synthesized through esterification between CMC and stearoyl chloride in pyridine. By dip-coating hydrophilic or hydrophobic silicon wafers into a C18-CEL toluene solution, ultrathin films were formed with thicknesses of 8–360 nm. The C18-CEL film was thermally responsive at temperatures above or below its T_m (56 °C). When the temperature was higher than 56 °C, the film melted and released the incorporated active agents (i.e., rhodamine B molecules), and the release of active molecules could be stopped by decreasing the temperature [119]. Crépy et al. synthesized fully substituted cellulose esters (DS = 2.8) using acyl chlorides from C8 to C18. The T_m values of the cellulose esters slightly decreased with increasing side-chain length from 58 °C for C8-CEM to 44 °C for C18-CEL. The results indicated that the cellulose esters organized into a layered structure in which the cellulosic backbones were arranged in a plane, and the flexible side-chains were fully extended and perpendicular to the cellulosic backbone. Furthermore, when the fatty chain length exceeded 12 carbon atoms, a portion of the alkyl chains crystallized into a hexagonal packing [112].

4. Transesterification

4.1. *Transesterification with Vinyl Esters as Acylants*

Many types of vinyl esters can be easily synthesized from the corresponding carboxylic acid. Transesterification is a relatively mild acetylation method, which uses acetate ester as an innocuous acyl donor [91]. Applying vinyl esters as transesterification agents in the presence of a catalyst has been commonly used for the formation of cellulose esters in the past 10 years [88,91,92,120]. The acylation approach based on the transesterification of vinyl esters is particularly attractive because it proceeds with high efficiency and without releasing acidic by-products. Even if the transesterification reaction is reversible, the vinyl alcohol leaving product is unstable and immediately tautomerizes to acetaldehyde, which is volatile and can be easily removed from the reaction system; this removal drives the reaction toward the formation of the expected esters, although the small-molecule aldehydes are toxic [88,120].

Brand et al. performed transesterification between CNC and vinyl acetate in dimethyl sulfoxide (DMSO) or DMF under microwave heating at 80–100 °C for 15–240 min, with K_2CO_3 as a catalyst. The fraction of accessible -OHs (N_{OHs}) on CNC was calculated to be 16.7 mol%. The impacts of solvent, temperature, and the ratio of vinyl acetate to N_{OHs} on the reaction efficiency were investigated. The acetyl content as a percentage of total -OHs on CNC reached 14.4% in DMF and 12.8% in DMSO. The reaction was more efficient in DMSO than in DMF, owing to the better dispersion of CNC in DMSO. Increasing the molar ratio of vinyl acetate to N_{OHs} from 1.5 to 3 also significantly increased the reaction efficiency [120]. Cao et al. synthesized CA, CP, and CB, with high DS values of 2.14–2.34 through the transesterification between CMC and vinyl acetate, propionate, or butyrate in various solvents at 100 °C for 5 min with LiOH, NaOH, or KOH as a catalyst. Almost all reactants were soluble in DMSO, while considerable crystal precipitation was observed in DMF, DMAc, pyridine, and *N*-methylpyrrolidone. DMSO was thus identified as the most suitable solvent for this reaction. NaOH and KOH showed high catalytic activity in this transesterification reaction, and nearly all of the crystals disappeared within 5 min. In contrast, LiOH did not show sufficient catalytic activity. Therefore, DMSO, NaOH, and KOH were found to be suitable for the synthesis of CESs in this study. No T_m could be identified from the differential scanning calorimetry curve of CA, but T_m values could be identified from the curves of CP (234 °C) and CB (182 °C) [88]. Chen et al. used vinyl acetate as a transesterification agent to prepare CA from six cellulose raw materials (CMC, cotton linter pulp, wheat straw pulp, bamboo pulp, bleached softwood sulfite-dissolved pulp, and bleached hardwood kraft pulp (HP)) in DMSO at 100 °C for only 15 min, with NaOH as catalyst. The CA fibers prepared from the cellulose raw materials with the highest viscosity-average degree of polymerization (DPv) exhibited the lowest DS. CA from CMC had the highest DS (2.94) due to CMC having the lowest DPv, while CA from HP had the lowest DS (2.55), as a result of HP having the highest DPv, indicating that cellulose with longer chains has lower transesterification reactivity. The mechanical strength of CA increased with increasing DPv, and the thermal stability of CA decreased with the increasing DPv of cellulose. CA fibers showed higher thermal stability ($T_{d\text{-onset}}$ = 290–320 °C) than unmodified cellulose ($T_{d\text{-onset}}$ = 280 °C) [91]. Cao et al. performed transesterification between CMC and C2–12 vinyl esters in DMSO at 100 °C for 5 min with NaOH as catalyst. The DS values of the obtained C2–12 cellulose esters showed a decreasing trend from 2.02 to 0.76 with increasing substituent chain length, indicating that vinyl esters with longer aliphatic chains have lower transesterification reactivity. The $T_{d\text{-onset}}$ values of the cellulose esters (320–333 °C) were higher than that of pristine CMC ($T_{d\text{-onset}}$ = 314.5 °C), except for C2-CMC (312 °C) [121]. Wen et al. obtained C12-CEM through transesterification between CMC and vinyl laurate in the cosolvent of ionic liquid 1-allyl-3-methylimidazolium chloride [amim]Cl and DMSO at 70–120 °C for 1–6 h, with 1,8-Diazabicyclo[5.4.0]undec-7-ene (DBU) as a catalyst. The DS of C12-CEM increased from 1.47 to 2.41 when the reaction temperature was increased from 70 °C to 120 °C. At 120 °C, prolonging the reaction time from 1 to 6 h increased the DS from 1.97 to 2.74. The DS increased from 1.80 to 2.62 when the molar ratio of vinyl laurate to AGU was increased from 3:1 to 9:1, and the WCA of C12-CEM was increased to 96–121°. The E decreased from 73.28 MPa to 33.65 MPa and the ε increased from 50% to 116% upon increasing the DS from 1.97 to 2.74, suggesting that the cellulose films became more ductile after the introduction of more aliphatic chains. Increasing the DS of C12-CMC film from 1.97 to 2.41 resulted in an improved σ from 4.2 to 5.9 MPa, while further increasing the DS to 2.74 led to a decrease in σ to 5.1 MPa. C12-CEM displayed higher thermal stability ($T_{d\text{-onset}}$ = 314–340°) than the original CMC ($T_{d\text{-onset}}$ = 295°) [118].

4.2. Transesterification with Plant Oils as Acylants

Effective cellulose hydrophobization strategies have been demonstrated, based on applying activated acid derivatives (i.e., acyl chlorides or anhydrides) in stoichiometric amounts or excessive vinyl esters [91]. These treatments lead to cellulose with increased hydrophobicity. However, the reactive acyl chloride and anhydrides or vinyl esters are obtained from non-renewable sources, and previous investigations have required the utilization of anhydrous conditions and non-eco-friendly (even toxic) organic solvents, making the reaction difficult and time-consuming due to the difficulty of reducing the high water content of the reaction medium. Furthermore, these acylants are too expensive for use in large-scale industrial applications. In addition, the concurrent release of hydrohalic or carboxylic acid by-products from acyl chlorides and anhydrides can adversely impact the cellulose structure and/or contaminate the fiber surface [104,120]. As an efficient and environmentally friendly approach, plant-derived vinyl esters (plant oils) have been used in place of petroleum compounds. This strategy has attracted substantial attention in recent years because the chemicals are easier to handle, less toxic, noncorrosive, and produce less waste throughout the reaction [92].

Plant oils as transesterification agents are triglyceride fatty acid esters derived from abundant and renewable resources. These compounds have been mainly used in food applications as more sustainable reagents for synthesizing CELs. The first study utilizing plant oils directly for cellulose modification was by Dankovich et al., who investigated increasing the water repellency of cellulosic materials such as CMC powder, cellulose filter paper, and cotton fabrics. The authors modified the cellulosic materials with several plant oils (soybean, rapeseed, olive, coconut) in acetone, ethanol, or DI water emulsion combined with a surfactant at 110 °C for 1 h. All of the oils, except for coconut, produced a hydrophobic cellulose surface with less water absorption capacity. The most hydrophobic surfaces were obtained with 1% soybean oil in acetone. The WCA was improved to 81° and the WA was decreased to 0.82 mL/mg. The maximum decomposition temperature of the cellulose esters was increased by 21 °C compared with the original CMC [122]. Dong et al. carried out transesterification between CMC and soybean oil. The CMC and soybean oil were first dispersed in ethanol and sonicated for 1 min. The samples were then dried at RT and placed in an oven at 110 °C for 15–120 min. After the modification, the CMC displayed a higher affinity for low-polar solvents; additionally, the hydrophobicity of the cellulose esters could be adjusted by controlling the heating time [123]. In another study, CMC was uniformly distributed in an emulsion mixture of rice bran oil and ethanol by sonicating for 15 min, followed by RT drying and reacting in an oven at 110 °C for 30, 60, or 90 min without the addition of a catalyst. The percentage of C18-CEL acylation was 15.4% at 3.3 wt% rice bran oil and 90 min of treatment time. The WCA of C18-CEL was enhanced to 92°, and WA was reduced to 0.9 μL/mg after modification compared with 2.5 μL/mg for unmodified CMC. As the treatment time increased, the $T_{d\text{-onset}}$ of C18-CEL decreased to 325, 311, and 306 °C, respectively, for 30, 60, and 90 min of treatment time, while the $T_{d\text{-onset}}$ of the original CMC was 306 °C [124]. Onwukamike et al. carried out transesterification between high-oleic sunflower oil and CMC (M_W = 89.0 kDa), cellulose filter paper (FP, M_W = 190.0 kDa), and cellulose pulp (M_W = 132.0 kDa) in a DBU-DMSO-CO_2 switchable solvent system at 115 °C for 6–24 h with DBU as a catalyst. The DS was increased from 0.34 to 1.59, as the reaction time was prolonged from 6 to 24 h. Among the investigated cellulose samples, the C18-CEL film from cellulose pulp (C18-CEL pulp) showed the highest E (478 MPa), followed by the C18-CEL FP film (458 MPa) and C18-CEL CMC film (376 MPa). Higher E values were obtained for films with lower DS values because the increasing presence of aliphatic side groups from the fatty acids increased the flexibility. All films showed an increased $T_{d\text{-onset}}$ after modification. C18-CEL CMC had a $T_{d\text{-onset}}$ of 320 °C (DS = 1.59) (compared with 280 °C for pristine CMC). C18-CEL FP had a $T_{d\text{-onset}}$ of 329 °C (DS = 1.48) (compared with 298 °C for pristine FP). C18-CEL pulp had a $T_{d\text{-onset}}$ of 327 °C (DS = 1.40) (compared with 261 °C for pristine pulp) [92].

The abovementioned transesterification processes required the use of organic solvents. The development of a greener process in water could therefore be beneficial from a sustainability point of view and be advantageous for industrial applications. Yoo et al. adopted a green procedure in aqueous lactic acid syrup for the acylation of CNC with plant oils (tung oil, linseed oil) with zinc acetate dihydrate and dibutyltin dilaurate as catalysts. The reactive solvent and intermediate product (CNC-*g*-PLA) system allowed for an in situ solvent exchange from DI water to lactic acid without prior drying of the CNC and facilitated the subsequent efficient acylation of CNC with plant oils. Approximately one third of the available -OHs on the CNC surface were substituted with PLA oligomers and plant oil. Furthermore, the side products derived from lactic acid can be recycled and reused, providing a simple, ecofriendly, and industrially amenable strategy for the hydrophobic modification of cellulose [80].

5. Crosslinking

Crosslinking can enhance the water stability, thermal stability, and mechanical properties of cellulose. In this strategy, crosslinked covalent linkages are formed between crosslinking agents and -OHs or functional groups on cellulose and cellulose derivatives (Table 7) [70].

Table 7. Influence of crosslinking on water sensitivity and mechanical strength of cellulose packaging (after vs. before).

Parameters / Cellulose	Crosslinker	Water Solubility (WS)/WA	WVTR/WVP	σ, ε	Ref.
sorbitol plasticized nanocellulose-coated filter paper	CAC	n.d. [a]	WVP: reduced by 88%	no change in σ; ε: 3.7% *vs.* 2.5%	[125]
HPMC	CAC	WS: reduced by 74%	WVP: reduced by 43%	n.d. [a]	[126]
CMCS	UV irradiation	WS: reduced by 21%	WVTR: reduced by 99%	σ: 46.8 MPa *vs.* 14.2 MPa; ε: 9.1% *vs.* 19.9%	[127]
DACNF	gelatin	WA: 44% *vs.* 167%	WVP: reduced by 99.9%	wet σ: 15.4 MPa *vs.* 9.9 MPa; wet ε: 23.2% *vs.* 16.5%	[57]
CMCS	DMTMM	WS: 18% *vs.* 100%	WVP: reduced by 40%	σ: 55 MPa *vs.* 32 MPa; ε: 25% *vs.* 30%	[128]

[a] Not detected.

5.1. Crosslinking on Hydroxyl Groups of Cellulose

In one study, the chemical crosslinking of softwood cellulose papers was achieved through dipping papers in aqueous solutions containing the polyfunctional carboxylic acids 1,2,3,4-butanetetra-carboxylic acid (BTCA), tricarballylic acid (TCA), or succinic acid (SA) using NaH_2PO_4 as a catalyst for 30 min. These treated cellulose papers were then cured in an oven at 150 °C for 5–30 min. BTCA showed better crosslinking capability than TCA. Compared with pristine paper, the wet σ of the cellulose papers increased with an increasing number of crosslinking sites. However, papers treated with SA showed very little wet strength enhancement because cellulose modification by SA resulted in the attachment of cellulosic units with a single pendant carboxylic group, which has low reactivity with the cellulosic -Ohs, and thus yields little crosslinking [129]. Citric acid (CAC) is non-toxic, and is widely used in the food industry as a safe natural additive. He et al. used CAC as a crosslinker to increase the hydrophobicity of CNC. An aqueous colloidal suspension of CNC and CAC was reacted in an oven at 130 °C for 4 h with NaH_2PO_4 as a catalyst. The WCA of CAC crosslinked CNC (CAC-CNC) was increased to only 55° (compared with 41° for CNC). However, the CAC-CNC had a higher thermal stability ($T_{d\text{-}onset}$ = 298 °C) than unmodified CNC ($T_{d\text{-}onset}$ = 249°) [130]. In another study, sorbitol plasticized nanocellulose-coated filter paper was crosslinked in a CAC aqueous solution and then cured in an oven at 150 °C for 5 min, using NaH_2PO_4 as a catalyst. When three layers of CAC crosslinked sorbitol-NC coating were applied, the resulting plasticized, crosslinked and coated filter paper exhibited a decreased WVP and OP of 0.5 g mm kPa^{-1} m^{-2} day^{-1} and 2 mL μm m^{-2} day^{-1} kPa^{-1}, compared with 4.0 g mm kPa^{-1} m^{-2} day^{-1} and 197 mL μm m^{-2} day^{-1} kPa^{-1} for unmodified filter paper, respectively. The modified paper did not show obvious changes in σ, but its E decreased from 570 to 310 MPa and the ε slightly increased from 2.5% to 3.7% compared with the control paper. The $T_{d\text{-onset}}$ of the modified paper decreased from 311 °C to 288 °C in comparison with that of the non-modified paper [125].

5.2. Crosslinking with Functional Groups of Cellulose Derivatives

Cellulose derivatives bear various functional groups (e.g., hydroxy, propyl, methyl, carboxymethyl, and formyl groups). When reacted at 190 °C for 15 min with PEG400 as a plasticizer and NaH_2PO_4 as a catalyst, 5 wt% CAC as a crosslinker for hydroxy propyl methyl cellulose (HPMC) decreased the WS of CAC crosslinked HPMC (CAC-HPMC) films by 74% compared with pristine HPMC film, and 15 wt% CAC resulted in the highest reduction in WVTR by 47% for CAC-HPMC film compared with neat HPMC film [126]. Another study modified carboxymethyl cellulose (CMCS) in two ways. One method used photo-crosslinking via UV irradiation at RT for 30–180 min with sodium benzoate (SB) as a photo-initiator. The other method involved chemical crosslinking using saturated glutaraldehyde (GLA) vapor plus gelatin as a synergistic crosslinker at 80 °C for 30–180 min. The photo-crosslinked film treated with 20 wt% SB with an irradiation time of 180 min and the chemically crosslinked film modified by adding 0.2 g of gelatin and exposed to GLA vapor for 90 min were identified as having the best crosslinking degrees. The photo-crosslinking treatment was more effective than the chemical crosslinking process in terms of WCA, WVP, and σ. WS of the CMCS/SB/UV and CMCS/GLA/gelatin films were reduced to 57.1% and 50.1%, respectively, compared with 78.2% for the pristine CMCS film. The CMCS/SB/UV film had a slightly higher WCA (78°) than the CMCS/GLA/gelatin film (67°). WVP of the CMCS/SB/UV film was greatly decreased to 9×10^{-7} g m^{-1} s^{-1} Pa^{-1} (compared with 819×10^{-7} g m^{-1} s^{-1} Pa^{-1} for the neat CMCS film), while the CMCS/GLA/gelatin film showed a smaller decrease in WVP (to 50×10^{-7} g m^{-1} s^{-1} Pa^{-1}). σ of the CMCS/SB/UV and CMCS/GLA/gelatin films was considerably enhanced to 46.8 MPa and 33.7 MPa, respectively, compared with 14.2 MPa for the unmodified CMCS film. However, ε of the CMCS films decreased from 19.9% to 9.1% and 13.8% after crosslinked with SB/UV and GLA/gelatin, respectively. The photo-crosslinked and chemically-crosslinked films showed no cytotoxicity [127]. In another study, softwood CNF was periodate-oxidized to dialdehyde CNF (DACNF) by sodium periodate with 6% of AGU on cellulose oxidized. Through the Maillard reaction, gelatin was crosslinked to DACNF at 60 °C for 3 h. The gelatin-crosslinked DACNF (G-DACNF) film (crosslinking degree = 57%) showed markedly decreased WA when immersed in DI water for 1 h, resulting in a weight increase of only 44%, while the neat CNF film weight increased by 167%. The G-DACNF film also exhibited a much higher wet mechanical strength (wet ε: 23.2%, wet σ: 15.4 MPa) than unmodified CNF film (6.7%, 0.9 MPa) and DACNF (16.5%, 9.9 MPa) [57]. In another study, chitosan was used as the crosslinker for DACNF. There was a slight increase in the WVP of the crosslinked films compared with that of the raw cellulose film. The OP of chitosan crosslinked films was greatly reduced to $4.3\text{–}8.9 \times 10^{-15}$ cm^3 cm/cm^2 s Pa, while the OP of the pure cellulose film was beyond the detection limit of the OP analyzer. In addition, σ of the crosslinked films was much higher than that of the unmodified film [131]. Furthermore, 4-(4,6-Dimethoxy-1,3,5-triazin-2-yl)-4-methyl-morpholinium chloride (DMTMM) was found to be an efficient crosslinking agent that promoted the reaction between carboxylic groups and -OHs under mild reaction conditions to form esters. With gelatin as a plasticizer, Beghetto et al. employed DMTMM as a crosslinking agent for CMCS. The reaction was carried out in DI water at RT for 2 h, resulting in a transparent DMTMM crosslinked CMCS film (DMTMM-CMCS) with an optical transmittance of 80–90%. By-products formed from DMTMM during the reaction were 2,4-dimethoxy-6-hydroxy-1,3,5-triazine (DMTOH) and *N*-methyl morpholinium hydrochloride, which are nontoxic and can easily be removed by water washing. A further advantage is that DMTOH can be recovered and recycled to regenerate DMTMM. Furthermore, 5 wt% DMTMM-CMCS film showed the lowest WS (18.1%), while CMCS completely dissolved in DI water within 4 h. The 10 wt% DMTMM-CMCS film had a decreased WVP and oil resistance ability (0.68×10^{-7} g m^{-1} h^{-1} Pa^{-1} and 0.29%), which was considerably lower than that of the unmodified CMCS film (1.14×10^{-7} g m^{-1} h^{-1} Pa^{-1}, 1.89%). Increasing amounts of DMTMM led to significantly improved σ, although the ε gradually decreased. With the addition of 5 wt% DMTMM, the σ was enhanced from 32 MPa to 54.9 MPa, while the ε decreased from 30.1% to 25.4% compared with the neat CMCS film. Adding 50 wt% glycerol as a plasticizer to the 5% DMTMM-CMCS film achieved optimum σ and ε (52.3 MPa, 37.3%), satisfying the need for a high σ and a high ε for food packaging applications. The crosslinked CMCS film also showed higher thermal stability than the unmodified CMCS film [128]. Crosslinking methods of cellulose derivatives, reaction degree, and the associated packaging forming methods are listed in Table 8. And the crosslinked cellulose materials used in food packaging applications are listed in Table 9.

Table 8. Crosslinking of cellulose derivatives, reaction degree, and the associated packaging forming methods.

Cellulose	Crosslinking Process	Packaging Type and Its Formation	Reaction Degree	Ref.
PEG400 plasticized HPMC	CAC as crosslinker and NaH_2PO_4 as catalyst, reacted in a mixture of DI water and ethanol with homogenization for 15 min, dried at 60 °C for 60 min and cured at 190 °C for 15 min	CAC-HPMC film; cured at high temperature	crosslinking rate ranged between 0% and 65% with a CAC content of 0–15% (*w*/*w* to HPMC)	[126]
CMCS	photo-crosslinking via UV irradiation at RT for 30–180 min with SB as photo-initiator or chemical crosslinking with saturated GLA vapor plus gelatin as synergistic crosslinkers at 80 °C for 30–180 min	CMCS/SB/UV or CMCS/GLA/gelatin film; films were prepared by the casting method at 45 °C for 18 h	photo-crosslinked film treated with 20 wt% SB and irradiated for 180 min, or chemically-crosslinked film modified with 0.2 g gelatin and exposed to GLA vapor for 90 min were found to have optimized crosslinking degrees	[127]
DACNF	gelatin as crosslinker, reacted in DI water at 60 °C for 3 h	G-DACNF film; vacuum filtration and solvent casting	crosslinking degree = 57%	[57]
gelatin plasticized CMCS	DMTMM as crosslinker for CMCS, reacted in DI water at RT for 2 h	DMTMM-CMCS film; films were prepared by the casting method at 40 °C overnight	optimum crosslinking degree was achieved in the presence of 5 wt% DMTMM and 50 wt% glycerol	[128]

Table 9. Overview of crosslinked cellulose materials in food packaging applications.

Cellulose Packaging	Additives	Packaging Properties			Ref.
		Barrier Properties	Mechanical Properties	Other Properties	
BTCA, TCA, SA crosslinked paper	-	n.d. [a]	wet tensile index of BTCA crosslinked (27 mN/g) and TCA crosslinked paper (16.5 mN/g) were markedly enhanced compared with 1.3 mN/g of pristine paper, while papers treated with SA showed little wet strength enhancement	n.d.	[129]
CAC-CNC-coated filter paper (three layers of coating)	sorbitol	modified paper showed increased WA (37%) compared with uncoated filter paper (29%), WVP and OP of modified paper was considerably decreased to 0.5 g mm kPa^{-1} m^{-2} day^{-1} and 2 mL μm m^{-2} day^{-1} kPa^{-1} compared with 4 g mm kPa^{-1} m^{-2} day^{-1} and 197 mL μm m^{-2} day^{-1} kPa^{-1} of uncoated filter paper	modified paper had a nearly unchanged σ, a reduced E (from 570 to 310 MPa) and increased ε (from 2.5% to 3.7%) compared with the control filter paper	$T_{d\text{-}onset}$ decreased from 311 °C to 288 °C	[125]
CAC-HPMC film	PEG400	increasing CAC content decreased WS of films, with an optimum CAC content of approximately 14%; 5 wt% CAC (*w*/*w* to HPMC) reduced WS of CAC-HPMC films by 74%; 15% CAC loading resulted in the highest reduction of WVTR by 47% for CAC-HPMC film (168 g/m^2 24 h) compared with neat HPMC film (316 g/m^2 24 h)	n.d.	CAC-HPMC films were transparent	[126]
CMCS/SB/UV or CMCS/GLA/gelatin film	-	WS of CMCS/SB/UV and CMCS/GLA/gelatin films was reduced to 57.1% and 50.1%, respectively, compared with 78.2% for pristine CMCS film, WVP of CMCS/SB/UV film was markedly decreased to 9×10^{-7} g m^{-1} s^{-1} Pa^{-1} compared with 8.19×10^{-5} g m^{-1} s^{-1} Pa^{-1} for neat CMCS film, while CMCS/GLA/gelatin showed a smaller WVP decrease (to 50×10^{-7} g m^{-1} s^{-1} Pa^{-1})	σ of CMCS/SB/UV and CMCS/GLA/gelatin films were considerably enhanced to 46.8 MPa and 33.7 MPa, respectively, compared with 14.2 MPa of unmodified CMCS film. However, the ε of the CMCS films decreased from 19.9% to 9.1% and 13.8% after crosslinking with SB/UV or GLA/gelatin, respectively	both crosslinked films were noncytotoxic	[127]
G-DACNF film	-	G-DACNF displayed greatly decreased WA when immersed in DI water for 1 h, resulting in a weight increase of only 44%, while the neat CNF film weight increased by 167%	G-DACNF film exhibited a much higher wet mechanical strength (ε: 23.2%, wet σ: 15.4 MPa, wet E: 94 MPa) compared with unmodified CNF film (6.7%, 0.9 MPa, 26 MPa)	G-DACNF films were transparent	[57]
DMTMM- CMCS film	gelatin (added or not)	5 wt% DMTMM-CMCS showed the lowest WS (18.1%); while CMCS completely dissolved in water within 4 h, the 10 wt% DMTMM-CMCS film showed a decreased WVP and oil absorption (0.68×10^{-7} g m^{-1} h^{-1} Pa^{-1} and 0.29%); however, the addition of glycerol increased the water sensitivity of the DMTMM-CMCS films	σ was enhanced from 32 MPa to 54.9 MPa, while the ε decreased from 30.1% to 25.4% with the addition of 5 wt% DMTMM compared with neat CMCS film. With the addition of glycerol, the optimum σ and ε values (52.3 MPa, 37.3%) were achieved in the presence of 5 wt% DMTMM and 50 wt% glycerol	DMTMM-CMCS films were transparent with optical transmittance of 80–90%	[128]

[a] Not detected.

6. Degradability of Esterified, Cross-Linked Cellulose and Its Derivatives

Products featuring cellulosic materials often advertise their complete biodegradability, which is in contrast with traditional petroleum-based plastics. According to the 94/62 EC Directive, a material can be defined as biodegradable if 90% of the material decomposes naturally within 6 months when used as compost [132]. The enzymatic biodegradation of cellulose has been reported to proceed completely within 60 d, during which the microorganisms break down the cellulose backbones and transform the glucose units into CO_2, water, and CH_4 [133]. Increasing the production of cellulose packaging to meet demand requires the hydrophobic modification of cellulose to improve the structural stability of the cellulose material in high humidity environments. The chemical modification of cellulose through (trans)esterification or crosslinking has been proven feasible.

The literature shows that the key architectural properties for biodegradability are the molecular weight (M_w) and crystallinity. In general, there is an inverse relationship between the mechanical performance and the biodegradability of cellulose and petroleum-based plastic packaging. Increasing the M_w and crystallinity is generally associated with better mechanical performance but decreased biodegradability, while lower crystallinity corresponds to looser chain packing, facilitating enzyme access [116]. For covalently modified cellulose, the biodegradability depends on both the DS and the nature of the chemical linkages (e.g., ether, ester). Studies have found that cellulose functionalized with a range of esters, ethers, or crosslinked moieties exhibits decreased rates and extents of biodegradation compared with unmodified cellulose, with etherified cellulose exhibiting the highest recalcitrance. One study reported that CNF ethers became nonbiodegradable at low surface DS values ($\approx$0.1), while the biodegradability of CNF esters at similar DS values was less affected [133]. The removal of functional groups (e.g., through the hydrolysis of ester groups) before attacking the -OHs present in native cellulose is the major rate-limiting step in the biodegradation of functionalized cellulose.

While native cellulose is readily and fully biodegraded, esterification has the potential to interfere with its biodegradability. Bare CA film partially degraded in soil with a 65% weight loss after 58 d, while pristine cellulose film completely degraded in soil within 60 d. When CA films were incorporated with additives such as sodium alginate or carrageenan, the composite films degraded at a slower rate. CA-sodium alginate and CA-carrageenan films showed 55% and 50% weight loss after 58 d, respectively [134]. One study found that the moderately substituted hexyl-esterified CNF (DS = 1.19) exhibited comparable biodegradation to unmodified CNF, while a highly functionalized C6-CES sample (DS = 2.43) showed 70% biodegradation compared with unmodified CNF. Dodecyl-esterified CNF (DS = 2.46) displayed considerably lower biodegradation of only 37% compared with unmodified CNF, indicating that cellulose esters with similar DS values but increasing substituent chain lengths biodegrade to a much lesser extent [133]. Fully substituted cellulose esters (C12–C20) were found to display little biodegradability [87].

The use of crosslinking to reinforce cellulose is thought to also significantly reduce the overall biodegradability, but there are few studies on this topic. One study found that DMTMM as a crosslinker contributed to a more compact and resistant structure, thereby decreasing the biodegradability rate of cellulose-based films [128]. The films crosslinked with 5 wt% DMTMM only began to biodegrade on day 7 and did not completely degrade in soil until day 15, while neat CMCS films degraded entirely by day 7. However, the addition of highly hydrophilic glycerol to cellulose film increased its water retention and consequently accelerated its biodegradability.

7. Conclusions

The overuse of petroleum-based plastics in the past few decades has led to growing concerns about environmental pollution. In recent years, research and innovation on cellulose-based food packaging have provided solutions to help reduce our dependency on fossil fuel-based packaging films. The advantages of cellulose films are the abundant natural origin of cellulose and its complete biodegradability. However, the ability of chemical modification to mitigate the hydrophilic nature of cellulose and to ensure that the mechanical and barrier performance of cellulose-based packaging is comparable to that of petroleum-based counterparts, while maintaining the biodegradability of the films, must be exploited and evaluated. Several mechanisms have been used to achieve mechanical and barrier performances that are equivalent to, and even exceed those of, corresponding petroleum-based plastics such as LDPE. Methods for producing active cellulose packaging (e.g., antioxidant, antimicrobial packaging) have also been developed. Blending cellulosic material with additives such as plasticizers and active agents has been demonstrated to improve the mechanical features and prolong the food storage ability of cellulose-based packaging without greatly reducing its biodegradability.

With respect to industrial-scale applicability, chemically modified hydrophobic cellulose packaging has not yet been adequately explored to establish its capacities for current and future applications. Given the range and diversity of options available for chemical hydrophobization strategies, there are excellent prospects for extending the application range of cellulose-based plastics on a scale comparable to that of fossil fuel-based thermoplastics and beyond. Highly hydrophobic cellulosic packaging, without the drawbacks of resource depletion, pollution, and recalcitrance, has the potential to act as a potent alternative in the food packaging market, and could contribute to future prosperity for the planet and its inhabitants.

Author Contributions: Conceptualization, Z.J. and T.N.; writing original draft preparation, Z.J.; writing—review and editing, T.N.; supervision, T.N. All authors have read and agreed to the published version of the manuscript.

Funding: We thank the Research Matching Grant Scheme of CUHK (8601309) for financial support.

Institutional Review Board Statement: Not applicable.

Informed Consent Statement: Not applicable.

Data Availability Statement: All supporting data are reported in the manuscript.

Conflicts of Interest: The authors declare no conflict of interest.

Abbreviation

Abbr.	Full Name
σ	tensile strength
ε	elongation at break
AGU	β-1,4-linked D-anhydroglucose units
BNF	bacterial nanofibrils
BTCA	1,2,3,4-butanetetra-carboxylic acid
CA	cellulose acetate
CAB	cellulose acetate butyrate
CAC	citric acid
CB	cellulose butyrate
CESs	cellulose esters with short substitution chains (C2–C6)
CEMs	cellulose esters with medium substitution chains (C8–C12)
CELs	cellulose esters with long substitution chains (> C12)
CH	cellulose hexanoate
CIN	cinnamaldehyde
CMC	cellulose microcrystals
CMCS	carboxymethyl cellulose
CMF	cellulose microfibrils
CNC	cellulose nanocrystals
CNF	cellulose nanofibrils
CP	cellulose propionate

DACNF	dialdehyde CNF
DI water	deionized water
DMAP	4-dimethylaminopyridine
DMTMM	4-(4,6-Dimethoxy-1,3,5-triazin-2-yl)-4-methyl-morpholinium chloride
DS	degree of substitution
E	Young's modulus
G-AgNPs	green-synthesized silver nanoparticles
G-DACNF	gelatin-crosslinked DACNF
GLA	glutaraldehyde
HPMC	hydroxy propyl methyl cellulose
OP	oxygen permeability
OTR	oxygen transmission rate
RT	room temperature
SA	succinic acid
SB	sodium benzoate
TCA	tricarballylic acid
$T_{d\text{-}onset}$	decomposition onset temperature
T_m	melting temperature
TOFA	tall oil fatty acid
WA	water absorption
WCA	water contact angle
WS	water solubility
WVP	water vapor permeability
WVTR	water vapor transmission rate

References

1. Wang, J.; Emmerich, L.; Wu, J.; Vana, P.; Zhang, K. Hydroplastic polymers as eco-friendly hydrosetting plastics. *Nat. Sustain.* **2021**, *4*, 877–883. [CrossRef]
2. Guan, Q.F.; Yang, H.B.; Han, Z.M.; Ling, Z.C.; Yang, K.P.; Yin, C.H.; Yu, S.H. Plant Cellulose Nanofiber-Derived Structural Material with High-Density Reversible Interaction Networks for Plastic Substitute. *Nano Lett.* **2021**, *21*, 8999–9004. [CrossRef]
3. Geyer, R.; Jambeck, J.R.; Law, K.L. Production, use, and fate of all plastics ever made. *Sci. Adv.* **2017**, *3*, e1700782. [CrossRef]
4. Shieh, P.; Zhang, W.; Husted, K.E.L.; Kristufek, S.L.; Xiong, B.; Lundberg, D.J.; Lem, J.; Veysset, D.; Sun, Y.; Nelson, K.A.; et al. Cleavable comonomers enable degradable, recyclable thermoset plastics. *Nature* **2020**, *583*, 542–547. [CrossRef]
5. Bugatti, V.; Viscusi, G.; Gorrasi, G. Formulation of a Bio-Packaging Based on Pure Cellulose Coupled with Cellulose Acetate Treated with Active Coating: Evaluation of Shelf Life of Pasta Ready to Eat. *Foods* **2020**, *9*, 1414. [CrossRef]
6. Liu, C.; Luan, P.; Li, Q.; Cheng, Z.; Sun, X.; Cao, D.; Zhu, H. Biodegradable, Hygienic, and Compostable Tableware from Hybrid Sugarcane and Bamboo Fibers as Plastic Alternative. *Matter* **2020**, *3*, 2066–2079. [CrossRef]
7. Lu, Y.; Weng, L.; Zhang, L. Morphology and Properties of Soy Protein Isolate Thermoplastics Reinforced with Chitin Whiskers. *Biomacromolecules* **2004**, *5*, 1046–1051. [CrossRef]
8. Ma, X.; Chang, P.R.; Yu, J.; Stumborg, M. Properties of biodegradable citric acid-modified granular starch/thermoplastic pea starch composites. *Carbohyd. Polym.* **2009**, *75*, 1–8. [CrossRef]
9. Chen, M.-J.; Li, R.-M.; Zhang, X.-Q.; Feng, J.; Feng, J.; Liu, C.-F.; Shi, Q.-S. Homogeneous Transesterification of Sugar Cane Bagasse toward Sustainable Plastics. *ACS Sustain. Chem. Eng.* **2016**, *5*, 360–366. [CrossRef]
10. He, Y.; Tang, H.; Chen, Y.; Zhang, S. Facile Strategy to Construct Metal–Organic Coordination Thermoplastic Starch with High Hydrophobicity, Glass-Transition Temperature, and Improved Shape Recovery. *ACS Sustain. Chem. Eng.* **2020**, *8*, 8655–8663. [CrossRef]
11. Ortega-Toro, R.; Morey, I.; Talens, P.; Chiralt, A. Active bilayer films of thermoplastic starch and polycaprolactone obtained by compression molding. *Carbohyd. Polym.* **2015**, *127*, 282–290. [CrossRef] [PubMed]
12. Blilid, S.; Kędzierska, M.; Miłowska, K.; Wrońska, N.; El Achaby, M.; Katir, N.; Belamie, E.; Alonso, B.; Lisowska, K.; Lahcini, M.; et al. Phosphorylated Micro- and Nanocellulose-Filled Chitosan Nanocomposites as Fully Sustainable, Biologically Active Bioplastics. *ACS Sustain. Chem. Eng.* **2020**, *8*, 18354–18365. [CrossRef]
13. Guan, Q.F.; Yang, H.B.; Han, Z.M.; Ling, Z.C.; Yu, S.H. An all-natural bioinspired structural material for plastic replacement. *Nat. Commun.* **2020**, *11*, 5401. [CrossRef] [PubMed]
14. Li, T.; Chen, C.; Brozena, A.H.; Zhu, J.Y.; Xu, L.; Driemeier, C.; Dai, J.; Rojas, O.J.; Isogai, A.; Wagberg, L.; et al. Developing fibrillated cellulose as a sustainable technological material. *Nature* **2021**, *590*, 47–56. [CrossRef]
15. Liu, Y.; Ahmed, S.; Sameen, D.E.; Wang, Y.; Lu, R.; Dai, J.; Li, S.; Qin, W. A review of cellulose and its derivatives in biopolymer-based for food packaging application. *Trends Food Sci. Technol.* **2021**, *112*, 532–546. [CrossRef]
16. Rol, F.; Belgacem, M.N.; Gandini, A.; Bras, J. Recent advances in surface-modified cellulose nanofibrils. *Prog. Polym. Sci.* **2019**, *88*, 241–264. [CrossRef]

17. Saedi, S.; Garcia, C.V.; Kim, J.T.; Shin, G.H. Physical and chemical modifications of cellulose fibers for food packaging applications. *Cellulose* **2021**, *28*, 8877–8897. [CrossRef]
18. Ventura-Cruz, S.; Tecante, A. Nanocellulose and microcrystalline cellulose from agricultural waste: Review on isolation and application as reinforcement in polymeric matrices. *Food Hydrocoll.* **2021**, *118*, 106771. [CrossRef]
19. Wang, Y.; Wang, X.; Xie, Y.; Zhang, K. Functional nanomaterials through esterification of cellulose: A review of chemistry and application. *Cellulose* **2018**, *25*, 3703–3731. [CrossRef]
20. Song, L.; Wang, Z.; Lamm, M.E.; Yuan, L.; Tang, C. Supramolecular Polymer Nanocomposites Derived from Plant Oils and Cellulose Nanocrystals. *Macromolecules* **2017**, *50*, 7475–7483. [CrossRef]
21. Farooq, M.; Zou, T.; Riviere, G.; Sipponen, M.H.; Osterberg, M. Strong, Ductile, and Waterproof Cellulose Nanofibril Composite Films with Colloidal Lignin Particles. *Biomacromolecules* **2019**, *20*, 693–704. [CrossRef]
22. Guan, Q.F.; Yang, H.B.; Han, Z.M.; Zhou, L.C.; Yu, S.H. Lightweight, tough, and sustainable cellulose nanofiber-derived bulk structural materials with low thermal expansion coefficient. *Sci. Adv.* **2020**, *6*, eaaz1114. [CrossRef] [PubMed]
23. Fei, Y.; Chen, F.; Fang, W.; Xu, L.; Ruan, S.; Liu, X.; Zhong, M.; Kuang, T. High-strength, flexible and cycling-stable piezo-resistive polymeric foams derived from thermoplastic polyurethane and multi-wall carbon nanotubes. *Compos. Part B* **2020**, *199*, 108279. [CrossRef]
24. Fei, Y.; Chen, F.; Fang, W.; Hejna, A.; Xu, L.; Liu, T.; Zhong, M.; Yang, J.; Kuang, T. Conductive thermoplastic polyurethane nanocomposite foams derived from a cellulose/MWCNTs aerogel framework: Simultaneous enhancement of piezoresistance, strength, and endurance. *J. Mater. Chem. C* **2021**, *9*, 13103–13114. [CrossRef]
25. Cui, X.; Lee, J.J.L.; Chen, W.N. Eco-friendly and biodegradable cellulose hydrogels produced from low cost okara: Towards non-toxic flexible electronics. *Sci. Rep.* **2019**, *9*, 18166. [CrossRef] [PubMed]
26. Yang, H.B.; Liu, Z.X.; Yin, C.H.; Han, Z.M.; Guan, Q.F.; Zhao, Y.X.; Ling, Z.C.; Liu, H.C.; Yang, K.P.; Sun, W.B.; et al. Edible, Ultrastrong, and Microplastic-Free Bacterial Cellulose-Based Straws by Biosynthesis. *Adv. Funct. Mater.* **2021**, 2111713. [CrossRef]
27. Jiang, B.; Chen, C.; Liang, Z.; He, S.; Kuang, Y.; Song, J.; Mi, R.; Chen, G.; Jiao, M.; Hu, L. Lignin as a Wood-Inspired Binder Enabled Strong, Water Stable, and Biodegradable Paper for Plastic Replacement. *Adv. Funct. Mater.* **2019**, *30*, 1906307. [CrossRef]
28. Shu, L.; Zhang, X.-F.; Wang, Z.; Yao, J. Structure reorganization of cellulose hydrogel by green solvent exchange for potential plastic replacement. *Carbohyd. Polym.* **2022**, *275*, 118695. [CrossRef]
29. Basu, A.; Lindh, J.; Alander, E.; Stromme, M.; Ferraz, N. On the use of ion-crosslinked nanocellulose hydrogels for wound healing solutions: Physicochemical properties and application-oriented biocompatibility studies. *Carbohyd. Polym.* **2017**, *174*, 299–308. [CrossRef]
30. Aditya, T.; Allain, J.P.; Jaramillo, C.; Restrepo, A.M. Surface Modification of Bacterial Cellulose for Biomedical Applications. *Int. J. Mol. Sci.* **2022**, *23*, 610. [CrossRef]
31. Sun, Y.; Chen, D.; Li, Y.; Sun, S.; Zheng, J.; Cui, J.; Wang, G.; Zheng, L.; Wang, Y.; Zhou, H. High-performance green electronic substrate employing flexible and transparent cellulose films. *Carbohydr. Polym.* **2021**, *270*, 118359. [CrossRef] [PubMed]
32. Ferreira, E.S.; Cranston, E.D.; Rezende, C.A. Naturally Hydrophobic Foams from Lignocellulosic Fibers Prepared by Oven-Drying. *ACS Sustain. Chem. Eng.* **2020**, *8*, 8267–8278. [CrossRef]
33. Ajdary, R.; Tardy, B.L.; Mattos, B.D.; Bai, L.; Rojas, O.J. Plant Nanomaterials and Inspiration from Nature: Water Interactions and Hierarchically Structured Hydrogels. *Adv. Mater.* **2020**, *33*, 2001085. [CrossRef] [PubMed]
34. Zaitoon, A.; Lim, L.-T.; Scott-Dupree, C. Activated release of ethyl formate vapor from its precursor encapsulated in ethyl Cellulose/Poly(Ethylene oxide) electrospun nonwovens intended for active packaging of fresh produce. *Food Hydrocoll.* **2021**, *112*, 106313. [CrossRef]
35. Xia, C.; Wang, W.; Wang, L.; Liu, H.; Xiao, J. Multilayer zein/gelatin films with tunable water barrier property and prolonged antioxidant activity. *Food Packag. Shelf Life* **2019**, *19*, 76–85. [CrossRef]
36. Kasaai, M.R.; Moosavi, A. Treatment of Kraft paper with citrus wastes for food packaging applications: Water and oxygen barrier properties improvement. *Food Packag. Shelf Life* **2017**, *12*, 59–65. [CrossRef]
37. Marrez, D.A.; Abdelhamid, A.E.; Darwesh, O.M. Eco-friendly cellulose acetate green synthesized silver nano-composite as antibacterial packaging system for food safety. *Food Packag. Shelf Life* **2019**, *20*, 100302. [CrossRef]
38. Wu, J.; Chen, S.; Ge, S.; Miao, J.; Li, J.; Zhang, Q. Preparation, properties and antioxidant activity of an active film from silver carp (*Hypophthalmichthys molitrix*) skin gelatin incorporated with green tea extract. *Food Hydrocoll.* **2013**, *32*, 42–51. [CrossRef]
39. Wang, W.; Xiao, J.; Chen, X.; Luo, M.; Liu, H.; Shao, P. Fabrication and characterization of multilayered kafirin/gelatin film with one-way water barrier property. *Food Hydrocoll.* **2018**, *81*, 159–168. [CrossRef]
40. He, X.; Li, M.; Gong, X.; Niu, B.; Li, W. Biodegradable and antimicrobial CSC films containing cinnamon essential oil for preservation applications. *Food Packag. Shelf Life* **2021**, *29*, 100697. [CrossRef]
41. Wan, Z.; Wang, L.; Yang, X.; Guo, J.; Yin, S. Enhanced water resistance properties of bacterial cellulose multilayer films by incorporating interlayers of electrospun zein fibers. *Food Hydrocoll.* **2016**, *61*, 269–276. [CrossRef]
42. Wen, Y.; Liu, J.; Jiang, L.; Zhu, Z.; He, S.; He, S.; Shao, W. Development of intelligent/active food packaging film based on TEMPO-oxidized bacterial cellulose containing thymol and anthocyanin-rich purple potato extract for shelf life extension of shrimp. *Food Packag. Shelf Life* **2021**, *29*, 100709. [CrossRef]

43. Yang, J.; Li, M.; Wang, Y.; Wu, H.; Zhen, T.; Xiong, L.; Sun, Q. Double Cross-Linked Chitosan Composite Films Developed with Oxidized Tannic Acid and Ferric Ions Exhibit High Strength and Excellent Water Resistance. *Biomacromolecules* **2019**, *20*, 801–812. [CrossRef] [PubMed]
44. Zhang, C.W.; Nair, S.S.; Chen, H.; Yan, N.; Farnood, R.; Li, F.Y. Thermally stable, enhanced water barrier, high strength starch bio-composite reinforced with lignin containing cellulose nanofibrils. *Carbohydr. Polym.* **2020**, *230*, 115626. [CrossRef] [PubMed]
45. Xu, Y.; Liu, X.; Jiang, Q.; Yu, D.; Xu, Y.; Wang, B.; Xia, W. Development and properties of bacterial cellulose, curcumin, and chitosan composite biodegradable films for active packaging materials. *Carbohydr. Polym.* **2021**, *260*, 117778. [CrossRef] [PubMed]
46. Zhou, X.; Cheng, R.; Wang, B.; Zeng, J.; Xu, J.; Li, J.; Kang, L.; Cheng, Z.; Gao, W.; Chen, K. Biodegradable sandwich-architectured films derived from pea starch and polylactic acid with enhanced shelf-life for fruit preservation. *Carbohydr. Polym.* **2021**, *251*, 117117. [CrossRef]
47. Xia, Q.; Chen, C.; Yao, Y.; Li, J.; He, S.; Zhou, Y.; Li, T.; Pan, X.; Yao, Y.; Hu, L. A strong, biodegradable and recyclable lignocellulosic bioplastic. *Nat. Sustain.* **2021**, *4*, 627–635. [CrossRef]
48. Guan, Q.-F.; Ling, Z.-C.; Han, Z.-M.; Yang, H.-B.; Yu, S.-H. Ultra-Strong, Ultra-Tough, Transparent, and Sustainable Nanocomposite Films for Plastic Substitute. *Matter* **2020**, *3*, 1308–1317. [CrossRef]
49. Shimizu, M.; Saito, T.; Isogai, A. Water-resistant and high oxygen-barrier nanocellulose films with interfibrillar cross-linkages formed through multivalent metal ions. *J. Membr. Sci.* **2016**, *500*, 1–7. [CrossRef]
50. Wang, Q.; Du, H.; Zhang, F.; Zhang, Y.; Wu, M.; Yu, G.; Liu, C.; Li, B.; Peng, H. Flexible cellulose nanopaper with high wet tensile strength, high toughness and tunable ultraviolet blocking ability fabricated from tobacco stalk via a sustainable method. *J. Mater. Chem. A* **2018**, *6*, 13021–13030. [CrossRef]
51. Zhou, J.; Fang, Z.; Cui, J.; Zhang, X.; Qian, Y.; Liu, W.; Yang, D.; Qiu, X. Wood-inspired strategy to toughen transparent cellulose nanofibril films. *Carbohydr. Polym.* **2021**, *259*, 117759. [CrossRef] [PubMed]
52. Plappert, S.F.; Quraishi, S.; Pircher, N.; Mikkonen, K.S.; Veigel, S.; Klinger, K.M.; Potthast, A.; Rosenau, T.; Liebner, F.W. Transparent, Flexible, and Strong 2,3-Dialdehyde Cellulose Films with High Oxygen Barrier Properties. *Biomacromolecules* **2018**, *19*, 2969–2978. [CrossRef] [PubMed]
53. Cheng, S.; Zhang, Y.; Cha, R.; Yang, J.; Jiang, X. Water-soluble nanocrystalline cellulose films with highly transparent and oxygen barrier properties. *Nanoscale* **2016**, *8*, 973–978. [CrossRef] [PubMed]
54. Isogai, A.; Saito, T.; Fukuzumi, H. TEMPO-oxidized cellulose nanofibers. *Nanoscale* **2011**, *3*, 71–85. [CrossRef]
55. Song, Z.; Xiao, H.; Zhao, Y. Hydrophobic-modified nano-cellulose fiber/PLA biodegradable composites for lowering water vapor transmission rate (WVTR) of paper. *Carbohydr. Polym.* **2014**, *111*, 442–448. [CrossRef]
56. Nguyen, H.L.; Tran, T.H.; Hao, L.T.; Jeon, H.; Koo, J.M.; Shin, G.; Hwang, D.S.; Hwang, S.Y.; Park, J.; Oh, D.X. Biorenewable, transparent, and oxygen/moisture barrier nanocellulose/nanochitin-based coating on polypropylene for food packaging applications. *Carbohydr. Polym.* **2021**, *271*, 118421. [CrossRef]
57. Kriechbaum, K.; Bergstrom, L. Antioxidant and UV-Blocking Leather-Inspired Nanocellulose-Based Films with High Wet Strength. *Biomacromolecules* **2020**, *21*, 1720–1728. [CrossRef]
58. Gaiolas, C.; Belgacem, M.N.; Silva, L.; Thielemans, W.; Costa, A.P.; Nunes, M.; Silva, M.J. Green chemicals and process to graft cellulose fibers. *J. Colloid Interface Sci.* **2009**, *330*, 298–302. [CrossRef]
59. Bayer, I.S.; Fragouli, D.; Attanasio, A.; Sorce, B.; Bertoni, G.; Brescia, R.; Di Corato, R.; Pellegrino, T.; Kalyva, M.; Sabella, S.; et al. Water-repellent cellulose fiber networks with multifunctional properties. *ACS Appl. Mater. Interfaces* **2011**, *3*, 4024–4031. [CrossRef]
60. Eyley, S.; Thielemans, W. Surface modification of cellulose nanocrystals. *Nanoscale* **2014**, *6*, 7764–7779. [CrossRef]
61. Habibi, Y. Key advances in the chemical modification of nanocelluloses. *Chem. Soc. Rev.* **2014**, *43*, 1519–1542. [CrossRef] [PubMed]
62. Singh, M.; Kaushik, A.; Ahuja, D. Surface functionalization of nanofibrillated cellulose extracted from wheat straw: Effect of process parameters. *Carbohyd. Polym.* **2016**, *150*, 48–56. [CrossRef] [PubMed]
63. Avila Ramirez, J.A.; Gomez Hoyos, C.; Arroyo, S.; Cerrutti, P.; Foresti, M.L. Acetylation of bacterial cellulose catalyzed by citric acid: Use of reaction conditions for tailoring the esterification extent. *Carbohyd. Polym.* **2016**, *153*, 686–695. [CrossRef] [PubMed]
64. Ali, H.T.; Tajvidi, M. Sustainable Barrier System via Self-Assembly of Colloidal Montmorillonite and Cross-linking Resins on Nanocellulose Interfaces. *ACS Appl. Mater. Interfaces* **2019**, *11*, 1604–1615.
65. Nechyporchuk, O.; Belgacem, M.N.; Bras, J. Production of cellulose nanofibrils: A review of recent advances. *Ind. Crops Prod.* **2016**, *93*, 2–25. [CrossRef]
66. Lavoine, N.; Desloges, I.; Dufresne, A.; Bras, J. Microfibrillated cellulose—Its barrier properties and applications in cellulosic materials: A review. *Carbohydr. Polym.* **2012**, *90*, 735–764. [CrossRef]
67. Abdul Khalil, H.P.S.; Davoudpour, Y.; Saurabh, C.K.; Hossain, M.S.; Adnan, A.S.; Dungani, R.; Paridah, M.T.; Islam Sarker, M.Z.; Fazita, M.R.N.; Syakir, M.I.; et al. A review on nanocellulosic fibres as new material for sustainable packaging: Process and applications. *Renew. Sustain. Energy Rev.* **2016**, *64*, 823–836. [CrossRef]
68. Xiong Chang, X.; Mujawar Mubarak, N.; Ali Mazari, S.; Sattar Jatoi, A.; Ahmad, A.; Khalid, M.; Walvekar, R.; Abdullah, E.C.; Karri, R.R.; Siddiqui, M.T.H.; et al. A review on the properties and applications of chitosan, cellulose and deep eutectic solvent in green chemistry. *J. Ind. Eng. Chem.* **2021**, *104*, 362–380. [CrossRef]
69. Ferrer, A.; Pal, L.; Hubbe, M. Nanocellulose in packaging: Advances in barrier layer technologies. *Ind. Crops Prod.* **2017**, *95*, 574–582. [CrossRef]

70. Liang, L.; Bhagia, S.; Li, M.; Huang, C.; Ragauskas, A.J. Cross-Linked Nanocellulosic Materials and Their Applications. *ChemSusChem* **2020**, *13*, 78–87. [CrossRef]
71. Li, F.; Mascheroni, E.; Piergiovanni, L. The Potential of NanoCellulose in the Packaging Field: A Review. *Packag. Technol. Sci.* **2015**, *28*, 475–508. [CrossRef]
72. Ben Shalom, T.; Belsey, S.; Chasnitsky, M.; Shoseyov, O. Cellulose Nanocrystals and Corn Zein Oxygen and Water Vapor Barrier Biocomposite Films. *Nanomaterials* **2021**, *11*, 247. [CrossRef] [PubMed]
73. Wang, B.; Zhou, J.; Wang, Z.; Mu, S.; Wu, R.; Wang, Z. Cellulose nanocrystal/plant oil polymer composites with hydrophobicity, humidity-sensitivity, and high wet strength. *Carbohydr. Polym.* **2020**, *231*, 115739. [CrossRef]
74. Nagalakshmaiah, M.; El Kissi, N.; Dufresne, A. Ionic Compatibilization of Cellulose Nanocrystals with Quaternary Ammonium Salt and Their Melt Extrusion with Polypropylene. *ACS Appl. Mater. Interfaces* **2016**, *8*, 8755–8764. [CrossRef]
75. Wang, S.; Jiang, F.; Xu, X.; Kuang, Y.; Fu, K.; Hitz, E.; Hu, L. Super-Strong, Super-Stiff Macrofibers with Aligned, Long Bacterial Cellulose Nanofibers. *Adv. Mater.* **2017**, *29*, 1702498. [CrossRef] [PubMed]
76. Jonoobi, M.; Harun, J.; Mathew, A.P.; Hussein, M.Z.B.; Oksman, K. Preparation of cellulose nanofibers with hydrophobic surface characteristics. *Cellulose* **2010**, *17*, 299–307. [CrossRef]
77. Avila Ramirez, J.A.; Suriano, C.J.; Cerrutti, P.; Foresti, M.L. Surface esterification of cellulose nanofibers by a simple organocatalytic methodology. *Carbohydr. Polym.* **2014**, *114*, 416–423. [CrossRef]
78. Saito, T.; Nishiyama, Y.; Putaux, J.-L.; Vignon, M.; Isogai, A. Homogeneous Suspensions of Individualized Microfibrils from TEMPO-Catalyzed Oxidation of Native Cellulose. *Biomacromolecules* **2006**, *7*, 1687–1691. [CrossRef]
79. Tang, L.; Huang, B.; Lu, Q.; Wang, S.; Ou, W.; Lin, W.; Chen, X. Ultrasonication-assisted manufacture of cellulose nanocrystals esterified with acetic acid. *Bioresour. Technol.* **2013**, *127*, 100–105. [CrossRef]
80. Yoo, Y.; Youngblood, J.P. Green One-Pot Synthesis of Surface Hydrophobized Cellulose Nanocrystals in Aqueous Medium. *ACS Sustain. Chem. Eng.* **2016**, *4*, 3927–3938. [CrossRef]
81. Minelli, M.; Baschetti, M.G.; Doghieri, F.; Ankerfors, M.; Lindström, T.; Siró, I.; Plackett, D. Investigation of mass transport properties of microfibrillated cellulose (MFC) films. *J. Membr. Sci.* **2010**, *358*, 67–75. [CrossRef]
82. Fukuzumi, H.; Saito, T.; Iwata, T.; Kumamoto, Y.; Isogai, A. Transparent and High Gas Barrier Films of Cellulose Nanofibers Prepared by TEMPO-Mediated Oxidation. *Biomacromolecules* **2009**, *10*, 162–165. [CrossRef] [PubMed]
83. Missoum, K.; Belgacem, M.N.; Barnes, J.-P.; Brochier-Salon, M.-C.; Bras, J. Nanofibrillated cellulose surface grafting in ionic liquid. *Soft Matter* **2012**, *8*, 8338. [CrossRef]
84. Uschanov, P.; Johansson, L.-S.; Maunu, S.L.; Laine, J. Heterogeneous modification of various celluloses with fatty acids. *Cellulose* **2011**, *18*, 393–404. [CrossRef]
85. Cunha, A.G.; Freire, C.S.; Silvestre, A.J.; Neto, C.P.; Gandini, A. Reversible hydrophobization and lipophobization of cellulose fibers via trifluoroacetylation. *J. Colloid Interface Sci.* **2006**, *301*, 333–336. [CrossRef]
86. Crepy, L.; Chaveriat, L.; Banoub, J.; Martin, P.; Joly, N. Synthesis of cellulose fatty esters as plastics-influence of the degree of substitution and the fatty chain length on mechanical properties. *ChemSusChem* **2009**, *2*, 165–170. [CrossRef]
87. Sealey, J.E.; Samaranayake, G.; Todd, J.G.; Glasser, W.G. Novel Cellulose Derivatives. Preparation and Thermal Analysis of Waxy Esters of Cellulose. *J. Polym. Sci. Part B Polym. Phys.* **1996**, *34*, 1613–1620. [CrossRef]
88. Cao, X.; Sun, S.; Peng, X.; Zhong, L.; Sun, R.; Jiang, D. Rapid synthesis of cellulose esters by transesterification of cellulose with vinyl esters under the catalysis of NaOH or KOH in DMSO. *J. Agric. Food Chem.* **2013**, *61*, 2489–2495. [CrossRef]
89. Wang, W.; Gu, F.; Deng, Z.; Zhu, Y.; Zhu, J.; Guo, T.; Song, J.; Xiao, H. Multilayer surface construction for enhancing barrier properties of cellulose-based packaging. *Carbohydr. Polym.* **2021**, *255*, 117431. [CrossRef]
90. Swatloski, R.P.; Spear, S.K.; Holbrey, J.D.; Rogers, R.D. Dissolution of Cellose with Ionic Liquids. *J. Am. Chem. Soc.* **2002**, *124*, 4974–4975. [CrossRef]
91. Chen, J.; Xu, J.; Wang, K.; Cao, X.; Sun, R. Cellulose acetate fibers prepared from different raw materials with rapid synthesis method. *Carbohydr. Polym.* **2016**, *137*, 685–692. [CrossRef] [PubMed]
92. Onwukamike, K.N.; Grelier, S.; Grau, E.; Cramail, H.; Meier, M.A.R. Sustainable Transesterification of Cellulose with High Oleic Sunflower Oil in a DBU-CO_2 Switchable Solvent. *ACS Sustain. Chem. Eng.* **2018**, *6*, 8826–8835. [CrossRef]
93. Rodionova, G.; Lenes, M.; Eriksen, Ø.; Gregersen, Ø. Surface chemical modification of microfibrillated cellulose: Improvement of barrier properties for packaging applications. *Cellulose* **2011**, *18*, 127–134. [CrossRef]
94. Abraham, E.; Nevo, Y.; Slattegard, R.; Attias, N.; Sharon, S.; Lapidot, S.; Shoseyov, O. Highly Hydrophobic Thermally Stable Liquid Crystalline Cellulosic Nanomaterials. *ACS Sustain. Chem. Eng.* **2016**, *4*, 1338–1346. [CrossRef]
95. Ioelovich, M. Adjustment of Hydrophobic Properties of Cellulose Materials. *Polymers* **2021**, *13*, 1241. [CrossRef]
96. Deng, S.; Huang, R.; Zhou, M.; Chen, F.; Fu, Q. Hydrophobic cellulose films with excellent strength and toughness via ball milling activated acylation of microfibrillated cellulose. *Carbohydr. Polym.* **2016**, *154*, 129–138. [CrossRef]
97. Zhang, J.; Guo, Z.; Chen, S.; Dong, H.; Zhang, X.; Qin, Y.; Yao, C.; Xu, F. High-barrier, strong, and antibacterial paper fabricated by coating acetylated cellulose and cinnamaldehyde for food packaging. *Cellulose* **2021**, *28*, 4371–4384. [CrossRef]
98. Saha, N.R.; Sarkar, G.; Roy, I.; Bhattacharyya, A.; Rana, D.; Dhanarajan, G.; Banerjee, R.; Sen, R.; Mishra, R.; Chattopadhyay, D. Nanocomposite films based on cellulose acetate/polyethylene glycol/modified montmorillonite as nontoxic active packaging material. *RSC Adv.* **2016**, *6*, 92569–92578. [CrossRef]

99. Assis, R.Q.; Rios, P.D.A.; Rios, A.D.O.; Olivera, F.C. Biodegradable packaging of cellulose acetate incorporated with norbixin, lycopene or zeaxanthin. *Ind. Crops Prod.* **2020**, *147*, 112212. [CrossRef]
100. Assis, R.Q.; Pagno, C.H.; Stoll, L.; Rios, P.D.; Rios, A.O.; Olivera, F.C. Active food packaging of cellulose acetate: Storage stability, protective effect on oxidation of riboflavin and release in food simulants. *Food Chem.* **2021**, *349*, 129140. [CrossRef]
101. Mulyadi, A.; Deng, Y. Surface modification of cellulose nanofibrils by maleated styrene block copolymer and their composite reinforcement application. *Cellulose* **2016**, *23*, 519–528. [CrossRef]
102. Kulomaa, T.; Matikainen, J.; Karhunen, P.; Heikkilä, M.; Fiskari, J.; Kilpeläinen, I. Cellulose fatty acid esters as sustainable film materials—Effect of side chain structure on barrier and mechanical properties. *RSC Adv.* **2015**, *5*, 80702–80708. [CrossRef]
103. Lee, K.Y.; Quero, F.; Blaker, J.J.; Hill, C.A.; Eichhorn, S.J.; Bismarck, A. Surface only modification of bacterial cellulose nanofibers with organic acids. *Cellulose* **2011**, *18*, 595–605. [CrossRef]
104. Zhang, K.; Geissler, A.; Standhardt, M.; Mehlhase, S.; Gallei, M.; Chen, L.; Marie Thiele, C. Moisture-responsive films of cellulose stearoyl esters showing reversible shape transitions. *Sci. Rep.* **2015**, *5*, 11011. [CrossRef]
105. Havimo, M.; Jalomäki, J.; Granström, M.; Rissanen, A.; Iivanainen, T.; Kemell, M.; Heikkilä, M.; Sipi, M.; Kilpeläinen, I. Mechanical strength and water resistance of paperboard coated with long chain cellulose esters. *Packag. Technol. Sci.* **2011**, *24*, 249–258. [CrossRef]
106. Willberg-Keyriläinen, P.; Vartiainen, J.; Harlin, A.; Ropponen, J. The effect of side-chain length of cellulose fatty acid esters on their thermal, barrier and mechanical properties. *Cellulose* **2017**, *24*, 505–517. [CrossRef]
107. Bras, J.; Vaca-Garcia, C.; Borredon, M.-E.; Glasser, W. Oxygen and water vapor permeability of fully substituted long chain cellulose esters (LCCE). *Cellulose* **2007**, *14*, 367–374. [CrossRef]
108. Zhang, S.; Li, W.; Wang, W.; Wang, S.; Qin, C. Reactive superhydrophobic paper from one-step spray-coating of cellulose-based derivative. *Appl. Surf. Sci.* **2019**, *497*, 143816. [CrossRef]
109. Qin, C.; Wang, W.; Li, W.; Zhang, S. Reactive Water Vapor Barrier Coatings Derived from Cellulose Undecenoyl Esters for Paper Packaging. *Coatings* **2020**, *10*, 1032. [CrossRef]
110. Balasubramaniam, S.L.; Patel, A.S.; Nayak, B. Surface modification of cellulose nanofiber film with fatty acids for developing renewable hydrophobic food packaging. *Food Packag. Shelf Life* **2020**, *26*, 100587. [CrossRef]
111. Sun, B.; Wang, W.; Zhang, M.; Sain, M. Biomass-based edible film with enhanced mass barrier capacity and gas permeable selectivity. *Cellulose* **2018**, *25*, 5919–5937. [CrossRef]
112. Crépy, L.; Miri, V.; Joly, N.; Martin, P.; Lefebvre, J.-M. Effect of side chain length on structure and thermomechanical properties of fully substituted cellulose fatty esters. *Carbohydr. Polym.* **2011**, *83*, 1812–1820. [CrossRef]
113. Chauvelon, G.; Gergaud, N.; Saulniera, L.; Lourdin, D.; Buleon, A.; Thibault, J.-F.; Krausz, P. Esterification of cellulose-enriched agricultural by-products and characterization of mechanical properties of cellulosic films. *Carbohydr. Polym.* **2000**, *42*, 385–392. [CrossRef]
114. Sehaqui, H.; Zimmermann, T.; Tingaut, P. Hydrophobic cellulose nanopaper through a mild esterification procedure. *Cellulose* **2014**, *21*, 367–382. [CrossRef]
115. Li, W.; Wang, S.; Wang, W.; Qin, C.; Wu, M. Facile preparation of reactive hydrophobic cellulose nanofibril film for reducing water vapor permeability (WVP) in packaging applications. *Cellulose* **2019**, *26*, 3271–3284. [CrossRef]
116. Attallah, O.A.; Mojicevic, M.; Garcia, E.L.; Azeem, M.; Chen, Y.; Asmawi, S.; Brenan Fournet, M. Macro and Micro Routes to High Performance Bioplastics: Bioplastic Biodegradability and Mechanical and Barrier Properties. *Polymers* **2021**, *13*, 2155. [CrossRef] [PubMed]
117. Ratanakamnuan, U.; Atong, D.; Aht-Ong, D. Cellulose esters from waste cotton fabric via conventional and microwave heating. *Carbohydr. Polym.* **2012**, *87*, 84–94. [CrossRef]
118. Wen, X.; Wang, H.; Wei, Y.; Wang, X.; Liu, C. Preparation and characterization of cellulose laurate ester by catalyzed transesterification. *Carbohydr. Polym.* **2017**, *168*, 247–254. [CrossRef]
119. Geissler, A.; Bonaccurso, E.; Heim, L.-O.; Heinze, T.; Zhang, K. Temperature-Responsive Thin Films from Cellulose Stearoyl Triester. *J. Phys. Chem. C* **2014**, *118*, 2408–2417. [CrossRef]
120. Brand, J.; Pecastaings, G.; Sebe, G. A versatile method for the surface tailoring of cellulose nanocrystal building blocks by acylation with functional vinyl esters. *Carbohydr. Polym.* **2017**, *169*, 189–197. [CrossRef]
121. Cao, X.; Peng, X.; Zhong, L.; Sun, S.; Yang, D.; Zhang, X.; Sun, R. A novel transesterification system to rapidly synthesize cellulose aliphatic esters. *Cellulose* **2014**, *21*, 581–594. [CrossRef]
122. Dankovich, T.A.; Hsieh, Y.-L. Surface modification of cellulose with plant triglycerides for hydrophobicity. *Cellulose* **2007**, *14*, 469–480. [CrossRef]
123. Dong, X.; Dong, Y.; Jiang, M.; Wang, L.; Tong, J.; Zhou, J. Modification of microcrystalline cellulose by using soybean oil for surface hydrophobization. *Ind. Crops Prod.* **2013**, *46*, 301–303. [CrossRef]
124. Gorade, V.G.; Kotwal, A.; Chaudhary, B.U.; Kale, R.D. Surface modification of microcrystalline cellulose using rice bran oil: A bio-based approach to achieve water repellency. *J. Polym. Res.* **2019**, *26*, 217. [CrossRef]
125. Herrera, M.A.; Mathew, A.P.; Oksman, K. Barrier and mechanical properties of plasticized and cross-linked nanocellulose coatings for paper packaging applications. *Cellulose* **2017**, *24*, 3969–3980. [CrossRef]
126. Coma, V.; Sebti, I.; Pardon, P.; Pichavant, F.H.; Deschamps, A. Film properties from crosslinking of cellulosic derivatives with a polyfunctional carboxylic acid. *Carbohydr. Polym.* **2003**, *51*, 265–271. [CrossRef]

127. Shahbazi, M.; Ahmadi, S.J.; Seif, A.; Rajabzadeh, G. Carboxymethyl cellulose film modification through surface photo-crosslinking and chemical crosslinking for food packaging applications. *Food Hydrocoll.* **2016**, *61*, 378–389. [CrossRef]
128. Beghetto, V.; Gatto, V.; Conca, S.; Bardella, N.; Buranello, C.; Gasparetto, G.; Sole, R. Development of 4-(4,6-dimethoxy-1,3,5-triazin-2-yl)-4-methyl-morpholinium chloride cross-linked carboxymethyl cellulose films. *Carbohydr. Polym.* **2020**, *249*, 116810. [CrossRef]
129. Zhou, Y.J.; Luner, P.; Caluwe, P. Mechanism of Crosslinking of Papers with Polyfunctional Carboxylic Acids. *Appl. Polym. Sci.* **1995**, *58*, 1523–1534. [CrossRef]
130. He, X.; Luzi, F.; Yang, W.; Xiao, Z.; Torre, L.; Xie, Y.; Puglia, D. Citric Acid as Green Modifier for Tuned Hydrophilicity of Surface Modified Cellulose and Lignin Nanoparticles. *ACS Sustain. Chem. Eng.* **2018**, *6*, 9966–9978. [CrossRef]
131. Wu, Y.; Luo, X.; Li, W.; Song, R.; Li, J.; Li, Y.; Li, B.; Liu, S. Green and biodegradable composite films with novel antimicrobial performance based on cellulose. *Food Chem.* **2016**, *197*, 250–256. [CrossRef]
132. Andersson, C. New ways to enhance the functionality of paperboard by surface treatment—A review. *Packag. Technol. Sci.* **2008**, *21*, 339–373. [CrossRef]
133. Frank, B.P.; Smith, C.; Caudill, E.R.; Lankone, R.S.; Carlin, K.; Benware, S.; Pedersen, J.A.; Fairbrother, D.H. Biodegradation of Functionalized Nanocellulose. *Environ. Sci. Technol.* **2021**, *55*, 10744–10757. [CrossRef] [PubMed]
134. Rajeswari, A.; Christy, E.J.S.; Swathi, E.; Pius, A. Fabrication of improved cellulose acetate-based biodegradable films for food packaging applications. *Environ. Chem. Ecotoxicol.* **2020**, *2*, 107–114. [CrossRef]

polymers

Article

Heavy Metal Ions Removal from Aqueous Solutions by Treated Ajwa Date Pits: Kinetic, Isotherm, and Thermodynamic Approach

Mohammad Azam *, Saikh Mohammad Wabaidur, Mohammad Rizwan Khan, Saud I. Al-Resayes and Mohammad Shahidul Islam

Department of Chemistry, College of Science, King Saud University, P.O. Box 2455, Riyadh 11451, Saudi Arabia; swabaidur@KSU.EDU.SA (S.M.W.); mrkhan@ksu.edu.sa (M.R.K.); sresayes@ksu.edu.sa (S.I.A.-R.); mislam@KSU.EDU.SA (M.S.I.)

* Correspondence: mhashim@ksu.edu.sa

Abstract: In the current study we prepared cost-effective adsorbents based on ajwa date pits to remove Cu(II) ions from aqueous medium. Adsorbents were studied using scanning electron microscopy (SEM), FT-IR, and Brunauer-Emmett-Teller (BET) methods to characterize the surface functionalities, morphology, pore size, and particle size. The concentration of Cu(II) ions in the studied samples was determined by atomic adsorption spectrometry technique (AAS). Adsorption method was performed sequentially in a batch system followed by optimization by studying the numerous conditions, for instance the initial amounts of Cu(II) ions, dosages of the adsorbent, contact time, and pH of the solution. The ideal pH observed for maximum adsorption capacity was ~6.5. Langmuir and Freundlich isotherm models correctly predicted the investigation results, with the maximum monolayer adsorption capacities for Cu(II) ions at 328 K being 1428.57 mg/g (treated ajwa date pits, TADP) and 1111.1 mg/g for as produced ajwa date pits (ADP). It was revealed that TADP possess greater adsorption capability than ADP. Recovery investigations revealed that the saturated adsorbents eluted the maximum metal with 0.1 M HCl. Cu(II) ions adsorption was observed to be reduced by 80–89% after the second regeneration cycle. For the raw and chemically processed ajwa date pits adsorbent, the Langmuir model performed significantly better than the Freundlich model. The results demonstrated that the adsorbent made from ajwa date pits could be an economical and environmentally friendly alternative for removing Cu(II) ion pollutant from aqueous media.

Keywords: biosorbent; copper; adsorption; model studies; aqueous medium

Citation: Azam, M.; Wabaidur, S.M.; Khan, M.R.; Al-Resayes, S.I.; Islam, M.S. Heavy Metal Ions Removal from Aqueous Solutions by Treated Ajwa Date Pits: Kinetic, Isotherm, and Thermodynamic Approach. *Polymers* **2022**, *14*, 914. https://doi.org/10.3390/polym14050914

Academic Editors: Stefano Vecchio Ciprioti and Arn Mignon

Received: 21 November 2021
Accepted: 9 February 2022
Published: 25 February 2022

1. Introduction

Heavy metals hazardous waste has currently become one of the most serious environmental issues [1,2]. Heavy metal contamination accumulates in the environment and is not naturally decomposable, posing a concern to human health [3,4]. Cu(II) ion is a heavy metal ion that is widely used in a variety of manufacturing industries, including fertilizer, paints, plating baths, and paper goods [5–8]. The discharges from these sources contain a significant amount of Cu(II) ions, which are discharged into the environment via watercourses and earth [7,8]. It eventually combines with the food chain to cause a variety of severe human ailments [9]. Long-term copper exposure is harmful to health and can cause several diseases such as diarrhoea, vomiting, central nervous system and, mucosal irritation, hepatic, gastrointestinal distress and, kidney and liver damage [10–14]. In addition, if a high quantity of copper salts is consumed, it forms severe copper toxicity in individuals, most likely because of redox cycling and the production of reactive oxygen species that harm deoxyribonucleic acid [15,16]. Regarding its exposure to human, the US Occupational Safety and Health Administration has labelled an allowable exposure limit for copper fumes and dusts in the workplace as a time-weighted average of 1 mg/m^3 [17]. The

National Institute for Occupational Safety and Health has fixed a suggested exposure limit of 1 mg/m^3 as the time-weighted average. The immediate risk to life and health concentration is 100 mg/m^3 [17]. Recently, the National Primary Drinking Water Regulations of the USEPA has also regulated the levels of Cu(II) in drinking water with MCL of 1.3 mg/L [14]. The Cu(II) ion has also been regulated by the National Primary Drinking Water Regulations of the United States Environmental Protection Agency (USEPA) in drinking water with a Maximum Contaminant Level (MCL) of 1.3 mg/L.

Several traditional techniques such as ion-exchange [18], oxidation [19], electrocoagulation [20], solvent extraction [21], chemical precipitation [22], ultrafiltration [23], adsorption [24], activated carbon [25], membrane separation [26], and evaporation [27] have been applied to eliminate heavy metal ions. Nonetheless, most of these techniques have some drawbacks such as high production cost, consumption of high energy, high volumes of waste, and challenging treatment systems. Although, the adsorbent based on activated carbon has great microporous features, including high adsorption capacity and high surface area, it has demonstrated its capability as an adsorbent for the heavy metals removal from wastewater [28]. However, the application of activated carbon is very costly, with moderately high operational overheads. In addition, there is a requirement of regeneration prior to each adsorption phase [25]. Among these procedures, adsorption has been the most extensively accepted process because of its simplicity, versatility, ecofriendly, and being highly cost effective [24]. However, for the present, there is a rising demand to discover effortlessly obtainable, highly efficient, and very cost effective adsorbents for the adsorption of Cu(II) ions from different matrices. Cashew nut shell [5], saw dust [29], spent grain [30], wheat straw [31], pine cone powder [32], sawdust [33], herbaceous peat [34], peanut hull [6] sugar beet pulp [35], and others [36–39] are examples of effective adsorbents to remove copper ions. Due to the great obtainability and low-price of these adsorbents a complex regeneration method is not needed, and this cheaper adsorption system has involved numerous researchers. Thus, in order to substitute the current marketable materials, we have chosen ajwa date pits as the adsorbent, because Saudi Arabia is the world's third main producer of dates, and date pits are easily available and free of cost. In addition, the occurrence of numerous functional groups, for instance amino acid, carboxylic acid, phenols, ester, hydroxyl, and carboxylate can assist them as possible adsorbents for heavy metals removal from aqueous medium [40]. Therefore, we have developed adsorbents to remove Cu(II) ions from aqueous medium that are both cost effective and high-yielding, and are derived from a common waste product—ajwa date pits. Hydrogen peroxide (H_2O_2) treatment enhanced the surface functionalities (–OH) of the synthesized ADP powder. The raw and treated materials were characterized by different methods, for instance SEM, FT-IR and BET, certifying the excellent particle size, surface morphology, and adsorption capabilities. Several investigational parameters, including solution contact time, pH, starting concentration, and Cu(II) ion temperature, were also studied. Langmuir and Freundlich adsorption isotherms were applied to study the equilibrium modelling [36]. Thermodynamic and kinetic characteristics have been applied to assess the nature of the sorption method. In this study, we are using ajwa date pits as adsorbents for the first time to remove Cu(II) heavy metal ion from an aqueous solution.

2. Materials and Methods

2.1. Chemicals and Reagents

All the chemicals and reagents namely, copper(II) chloride ($CuCl_2$), sodium hydroxide (NaOH), hydrochloric acid (HCl), sodium hydroxide (NaOH), and H_2O_2, 30% v/v were of analytical grade, and were obtained from Sigma-Aldrich (Steinheim, Germany). A stock solution of Cu(II) ion (1000 µg/mL) prepared in ultrapure water was purified using Milli–Q water purification system (Millipore Corporation, Bedford, NH, USA). For complete mixing, the solution was agitated for 5 min. The required Cu(II) ion test solutions were prepared by suitable successive dilutions of the stock solution. For calibration purposes the, concentration range of the Cu(II) ion ranged from 1 to 10 μg/mL. The pH of solutions was adjusted to the optimum value by NaOH/HCl.

2.2. Instrumentation

The amount of Cu(II) ion was identified by high-throughput atomic absorption spectrometry (AAS) (Thermo Scientific, Waltham, MA, USA). The point of zero charge was measured by means of the salt addition process (pHPZC). The surface morphology of the synthesized and treated date pit materials was determined using SEM (Jeol JSM 5400 L V, Tokyo, Japan). The surface functionalities present on the produced biosorbents were determined using FT-IR (Thermo Scientific Nicolet 6700, Waltham, MA, USA) with 1:100 wt ratio of KBr dilution equivalent to the 400 cm^{-1} and 4000 cm^{-1} wave numbers, obtained at 32 scans average value with a ± 4 cm^{-1} resolution applied. Multifunctional X-ray diffractometer (XRD, Ultima IV, Rigaku, Charlestown, MA, USA) was used to determine the surface crystallinity and average particle size of the samples. The surface area and pore size of the samples were measured by BET surface area analyzer (Micromeritics-Gemini VII 2390 V1.03, Norcross City, GA, USA).

2.3. Biosorbent Preparation

The ajwa dates (5 kg) were purchased from Kingdom Dates, Riyadh Saudi Arabia. The date pits were removed and thoroughly rinsed with MilliQ water to exclude any remnants of edible portion and dust materials, then sun dried for many days, and then dried in an oven for 2 h at 80 °C. The date pits were mechanically crushed and processed in a ball mill, homogenized, and sieved at 120 μm (Figure 1) [39,41].

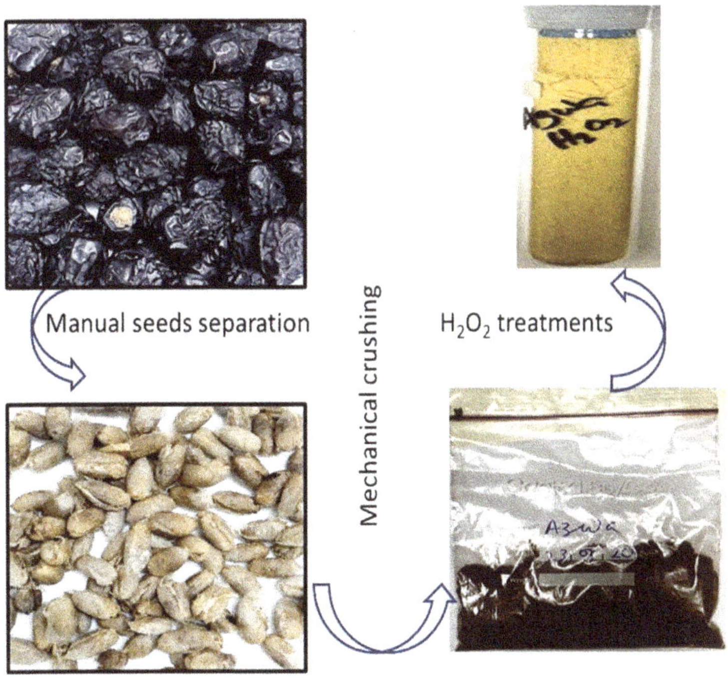

Figure 1. Ajwa dates, pits, and treated and untreated dates powders.

To establish the negative functionalities (-OH) onto the surface of the ADP adsorbent, the prepared ADP adsorbents were treated with 200 mL of 30% H_2O_2 (v/v). Treatment with H_2O_2 also offers advantages to decompose organic parts, reduction of biomass resistance, and to avert the ADP cellulose degradation [41]. To perform the treatment procedure, ajwa date powder (10 g) was taken out in a beaker (100 mL) followed by the addition of the H_2O_2 solution (100 mL). The solution was then thoroughly mixed overnight using a magnetic stirrer. The Buckner funnel was used to filter the sample suspension with a vacuum pump, which was then washed with MilliQ water to eliminate any remaining content of hydrogen peroxide, and dried in an oven overnight. The samples were then allowed to cool to room temperature and bottled in transparent glass vials for further analysis.

2.4. Adsorption and Desorption Studies

Batch adsorption tests were carried out to eliminate Cu(II) ions from aqueous media. ADP powder (0.05 g) was weighed in a cleaned Erlenmeyer flask (100 mL) and Cu(II) (100 mg/L, C0) solution was added to the flask, which was shaken overnight at 100 rpm on a shaker. The solution mixture was separated and kept in a cold, dry environment. The AAS method was employed to quantify the amounts of Cu(II) ion, C_e after the solution reached equilibrium.

At the time of equilibration time (q_e) and arbitrary time t (q_t), the adsorption (%) and Cu(II) ion removing property of the adsorbents were measured, as described in a previous study [41]:

$$q\,(\%) = \frac{(C_o - C_e)}{C_o} \times 100, \tag{1}$$

where q is the adsorption, C_o is the initial concentration, and C_e is the equilibrium concentration.

$$q_e = (C_o - C_e) \times \frac{V}{m}, \tag{2}$$

where q_e is the adsorption capacity at equilibrium, C_o is the initial concentration, C_e is the equilibrium concentration, V is the volume (L), and m is the mass of adsorbent (g).

$$q_t = (C_o - C_t) \times \frac{V}{m}, \tag{3}$$

where q_t is the adsorption capacity at time t concentration (μg/mL) of Cu(II) solution, C_o is the initial concentration, C_t is the concentration at time t, and V is the volume (L), m is the mass of adsorbent (g). The adsorption capacities of (q_e) and (q_t) were measured in milligrams per gram.

For the pH investigations, at a concentration of 25 mg/L, the initial pH (pHi) of Cu(II) ions was changed from 2 to 10. At concentration 20–100 mg/L and temperatures 293–323 K, the effects of Cu(II) ion initial concentration (C_o) on adsorption were tested.

Contact time (t) of Cu(II) ion adsorption was studied at C_o: 25 mg/L for periods ranging from 2 min to 24 h.

Cu(II) ion solutions (50 mL; 25 mg/L) were saturated for 24 h with the adsorbents (0.05 g) for desorption studies. The saturated adsorbents were moderately washed with MilliQ water to eliminate the un-adsorbed remnants of treated Cu(II) ions. The Cu(II) ion was then eluted from the saturated adsorbent using a variety of eluents, including CH_3COOH, H_2SO_4, HNO_3, NaOH, and HCl solutions (50 mL, 0.1 mol/L). The percentage of desorption was calculated using the following amount of Cu(II) ions in the eluent:

% Desorption = (Desorbed amount of Cu(II) ions/adsorbed amount of Cu(II) ions on adsorbent) × 100

2.5. Thermodynamic Studies of Cu(II) Ions Adsorption

The thermodynamic conditions, which define adsorption spontaneity, are extremely important in adsorption investigations. The spontaneity of the adsorption process is demonstrated by a negative Gibb's free energy change (ΔG°) at a known temperature. Equation (4) was used to calculate ΔG°

$$\Delta G^{\circ} = -\,RT\,\ln \mathrm{Ka}. \tag{4}$$

Equation (5) was used to calculate the values for enthalpy change (ΔH°) and entropy change (ΔS°).

$$\Delta G^\circ = \Delta H^\circ - T\Delta S^\circ, \quad (5)$$

where R, T, and Ka represent the gas constant (8.314 J/mol K), temperature (K), and Langmuir constant [42], respectively.

3. Results and Discussion

3.1. Characterization

3.1.1. Surface and Pore Size Analysis of Adsorbents

The morphology and physical properties of the adsorbent surfaces of the prepared ADP and TADP adsorbents were evaluated using SEM. The SEM images of ADP and TADP are shown in Figure 2B,C. There are numerous fragmented particles that resemble spots, which can be seen in the SEM images of ADP (Figure 2A–C), showing the porous character of the adsorbents appropriate for target metal adsorption [39]. However, Scanning Electron Microscopic images of TADP discovered that the TADP causes the formation of a larger porosity surface, when compared to the raw material (Figure 2D–F).

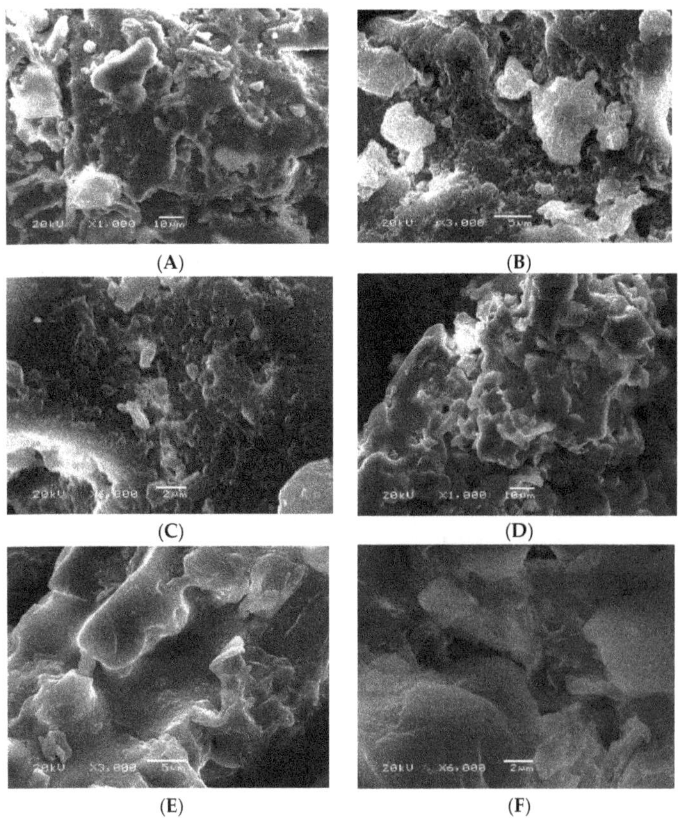

Figure 2. SEM images of ADP (**A–C**) and TADP (**D–F**) at various magnification: 1000× and 10 μm; 3000× and 5 μm: and 6000× magnification with 2 μm diameter.

The EDX analysis of the adsorbents was investigated to evaluate the composition of the active components presents in the adsorbents. The results indicate that the adsorbents are comprised of carbon as the major element (42 to 45%) and other negative heteroatoms such as oxygen (O_2) and nitrogen (N_2) with a relatively lower percentage (21 to 35%)

compared to carbon (Table 1). It is obvious that the H_2O_2 treatment of ADP adsorbent shows higher contents of O_2 as expected (35.8%), which indicates successful modification of the ADP surface into negatively charged hydroxyl functionalities (Table 1).

Table 1. EDX analysis of the as-prepared ADP and TADP samples.

Element	(keV)	Mass%			
		ADP	Error	TADP	Error
C K	0.277	45.37	0.15	42.96	0.21
N K	0.392	25.60	0.68	21.18	0.45
O K	0.525	29.03	0.18	35.86	0.16
Total		100		100	

The powder X-ray diffraction analysis for the ADP and treated adsorbents (TADP) was carried out and displayed in Figure 3. The sharp peaks in the diffractograms indicates the crystalline nature of the prepared adsorbents, which were converted to wide peaks for Cu(II) ions saturated materials. This is likely due to the adsorption of metal ions on the surface of the materials, which caused the materials to change from crystalline to amorphous. The average crystallite size was also calculated by Scherrer's equation and was determined to be 27 nm.

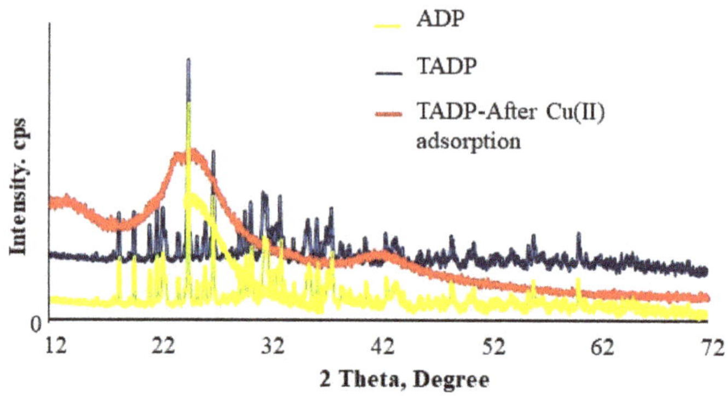

Figure 3. XRD analysis of ADP, TADP, and TADP after Cu(II) adsorption.

The BET surface areas for both ADP and TADP were determined to find the pore size of the materials (Figure 4). The surface area of TADP was found to be amplified from 2.47 to 8.48 m^2/g. The changes in pore areas show that during the H_2O_2 treatment of the ADP materials, increased porosity develops, resulting in higher adsorption efficiencies [41]. The adsorption average pore widths were also determined for ADP and TADP and were found to be in the range of 6–12 nm.

3.1.2. FTIR

FTIR analysis was carried out to identify the surface functionalities and adsorption mechanism for ADP and TADP adsorbents before and after the adsorption Cu(II) ions (Figure 5). FTIR analysis helps to identify the functional groups as well as the bindings of the analyte with the adsorbents. The adsorbents materials are the lignocellulosic and are made up of lignin, cellulose, hemicellulose, and protein. Hydroxyl, ether, and carbonyl types oxygen-rich functionalities are abundant in both cellulosic and hemicellulose materials, which are mainly responsible for binding the metal ion with the adsorbent materials [41]. A broad peak in the FTIR spectra of both adsorbents at 3238–3570 cm^{-1} range, is attributed to the presence of the characteristic lignocellulosic peaks of -OH, -NH, or both -OH and NH_2 [35].

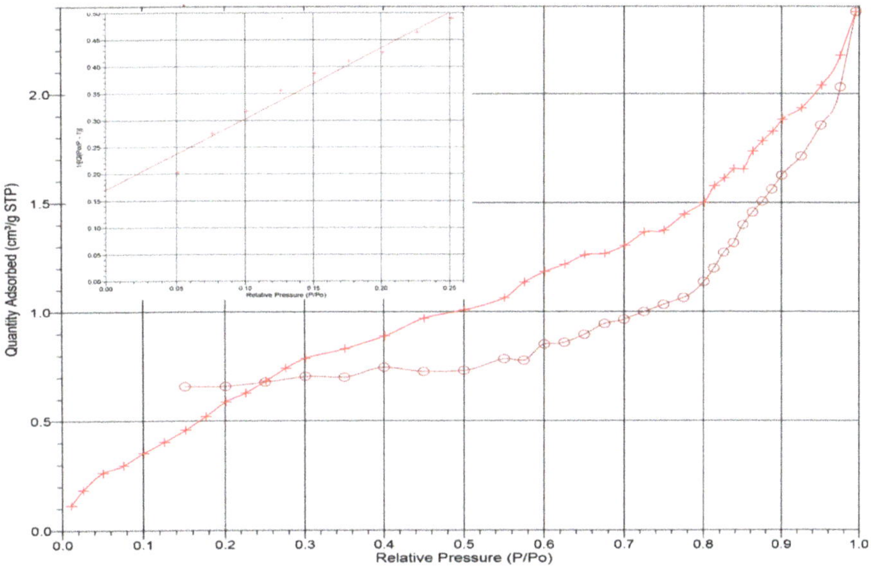

Figure 4. ADP linear isotherm plot with nitrogen adsorption plot (inset).

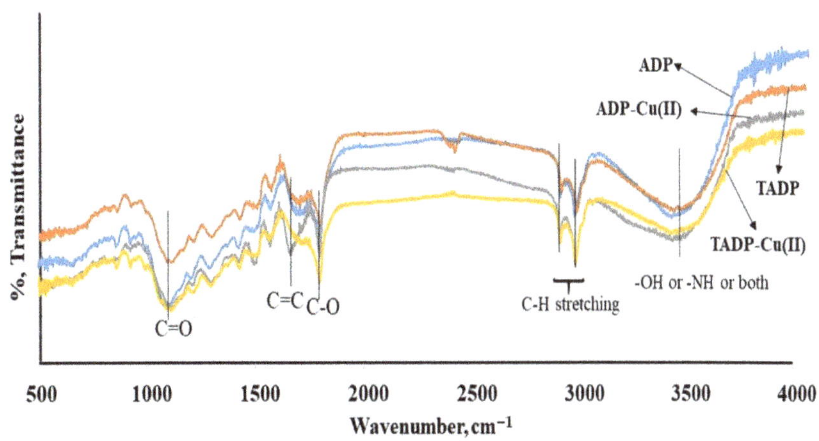

Figure 5. FT-IR analysis of ADP, ADP-Cu(II) ions, TADP, and TADP-Cu(II) ions (experimental conditions: initial conc. 25 mg/L; adsorbent weight 0.005 g; volume 0.05 L; temperature 25 °C; contact time 5 h; shaker speed: 100 rpm).

The stretching vibrations of aliphatic C-H functionalities are responsible for the absorbance peaks of 2927 and 2850 cm^{-1}. The occurrence of unconjugated carbonyl (C=O), imine (C=C) and C-O functionalities is indicated by vibrations at 1745, 1609, and 1045 cm^{-1}, respectively. The presence of small peaks in the spectral area between the C-O and C=C, corresponds to bending peaks of -CH_3 group [34,36]. The band at 514–708 cm^{-1} proved the existence of –OH polysaccharide groups. The number of small bands between the region 1410–1452 cm^{-1} appeared, owing to the presence of stretching vibrations of carboxylates (C–O and phenolic). Carboxylic acid and C-O-C vibrations were assigned to the bands appearing between 1210 and 1360 cm^{-1} and 1008 and 1159 cm^{-1}, respectively.

The FTIR spectra of Cu(II) adsorbed ADP and TADP were also analyzed, and a minor shift in peaks (to higher wavenumbers) was observed at 3246–3580 cm^{-1} range, 1745 cm^{-1}, and 1045 cm^{-1} as well as slight increases in those peak intensities, indicating the Cu(II) adsorption on adsorbents. The participation of the appropriate moieties (-OH, -NH, C-O, and C=O) in the adsorption of target metal ions is confirmed by small shifts and increases in peak intensities of such peaks. Furthermore, these findings reveal that the increased electronegativity of heteroatoms and the lone pair of electrons of the C=O, C-O, and –OH or NH functional groups are mostly liable for the coordination binding and electrostatic interactions that bind Cu(II) ions to the adsorbent surface [43].

3.2. Adsorption Properties

3.2.1. Effect of pH of the Solution

Cu(II) adsorption on ADP and TADP was studied for initial pH (pH_i) values ranging from 2.7 to 10.5. pH values higher than this range was omitted to avoid metal precipitation [41]. Because the Cu(II) metal ion was discovered to be cationic in the solutions phase, adsorption competition between the proton (H+) and metal ions is more likely at lower pH. This may be responsible for lower adsorption of the metal at highly acidic environments (Figure 6). Cu(II) ion adsorption capabilities rapidly increased after initial slowing, with maximum adsorption capacities of 131 mg/g and 177 mg/g for ADP and TADP at pH 6.5, respectively. The adsorption capacities of both adsorbents steadily decreased upon increasing the pH of the adsorption media (Figure 6).

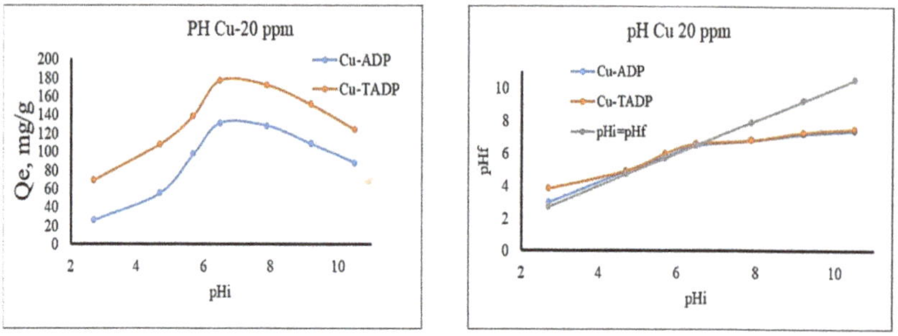

Figure 6. Adsorption capacity (qe) as a function of initial pH (pHi) and pHi versus pHf (final pH) plot for Cu(II) adsorption onto ADP and TADP. (Optimal conditions: initial concentration 25 mg/L; adsorbent weight 0.005 g; volume 0.05 L; temperature 25 °C; contact time 5 h; shaker speed 100 rpm).

Cu(II) ion adsorption was highest at pH 6.5, which was higher than the point of zero charges (pH_{PZC}), and pH 6.4 for both ADP and TADP (Figure 6). The pH_{PZC} value becomes zero at pH 6.4, confirming the neutral surface of the adsorbent materials. The surfaces of the adsorbents become positively and negatively charged, respectively, below and above this pHpzc, suggesting that the surfaces of TMDP and TSDP are more likely too protonated under acidic conditions to prevent metal ion adsorption.

Furthermore, as the pH increases, protonation decreases, resulting in a decrease in surface positivity and an enhancement in Cu(II) ion binding over the ADP and TADP surfaces. Above pH 6.5, a drop in the pH graph was observed (Figure 6), indicating that the adsorbents surface was neutralized in combination with the adsorption equilibrium.

3.2.2. Effect of Initial Concentration and Temperature

To determine the corresponding equilibrium isotherms, a different concentration of Cu(II) adsorption ranging from 10 to 50 mg/L on both ADP and TADP was investigated at temperatures of 25, 35, 45, and 55 °C. The results are shown in Figure 7. The vertical slopes of the plots for both adsorbents indicate the speedy enhancement in solid phase concentrations of Cu(II) with the increase of liquid phase concentrations from 10 to 50 mg/L. Although, when compared to untreated adsorbents (max 290 mg/g), the H_2O_2 treated adsorbents showed better adsorption capabilities (max 360 mg/g). For the graph of initial concentration (C_e) vs. adsorption capacities (q_e) of Cu(II) ions at different temperatures, the correlation coefficient (R^2) ranged from 0.9816 to 0.9926. Higher adsorbate concentration gradients may act as a driving factor for mass transfer between the solution and solid phases to overcome the resistance barrier [39].

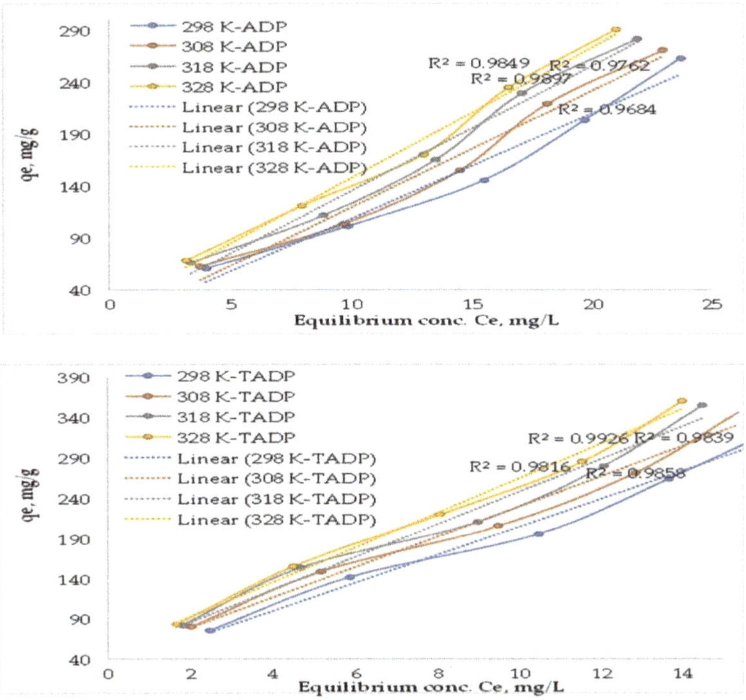

Figure 7. Effect of initial metal concentration and temperature (experimental conditions: initial conc: 25 mg/L; adsorbent weight: 0.005 g; volume: 0.05 L; temperature: 25 °C; contact time: 5 h; shaker speed: 100 rpm).

The slopes of the plots eventually fell and became nearly parallel to the x-axis, suggesting that the adsorbent surfaces were saturated with Cu(II) ions. The rise in Cu(II) ions adsorption capacity on both ADP and TADP with temperature supports the endothermic adsorption process, and the results are consistent with the literature on metal ion adsorption [44].

Cu(II) ions adsorption capacities on ADP were 60.0–2263.0 mg/g at 298 K, while TADP adsorption capacities (qe) were found to be in the range of 75.3–333.0 mg/g for concentrations ranging from 10 to 50 mg/L. The qe values were increased with increasing temperature and were found to be highest at 318 K for all analyzed concentrations, and the solid phase concentration (q_e) of Cu(II) ions on ADP were 68.2–290.0 mg/g, while the ranges at temperature 298 K were 60.0–2263.0 mg/g. Similarly, the solid phase concentration on TADP was found to be in the range of 83.2–360.0 mg/g, while the qe were in the ranges of 75.3–333.0 mg/g at temperature 318, and 298 K, respectively. It is obvious that a better adsorption of the target metals was achieved for the treated adsorbent material.

3.2.3. Effects of Contact Time

Cu(II) ions adsorption on ADP and TADP as a function of contact duration was investigated for Cu(II) concentrations ranging from 5 to 1440 min. Figure 8 shows how contact time (up to 360 min) affects Cu(II) adsorption.

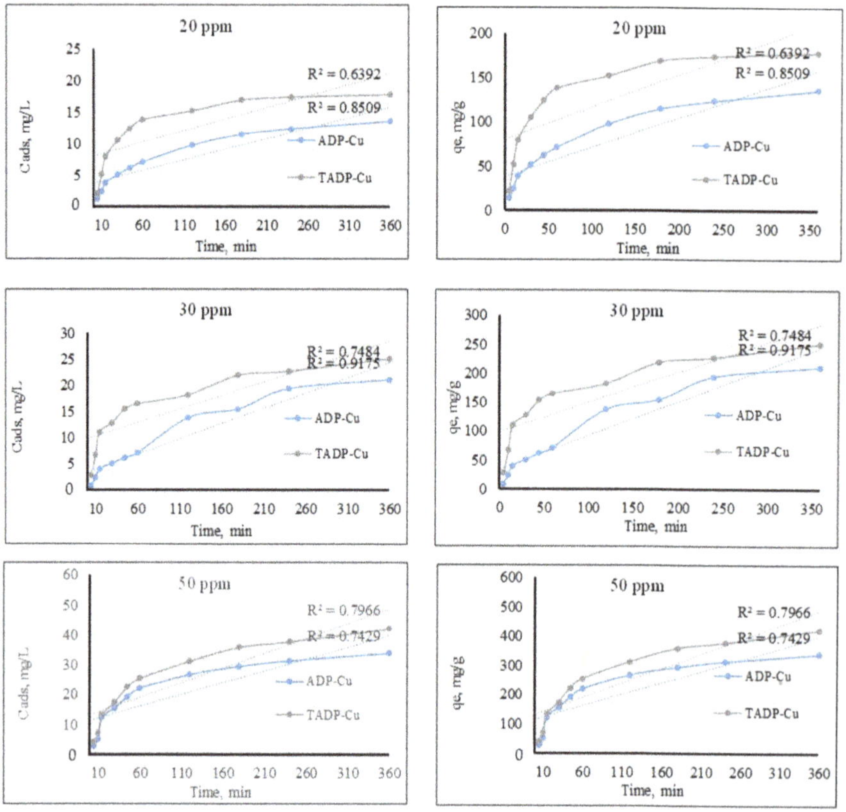

Figure 8. Effect of contact time at concentrations 20, 30, and 50 mg/L of the adsorbates (experimental conditions: Co: 25 mg/L; m: 0.005 g; V: 0.05 L; T: 25 °C; agitation speed: 100 rpm).

The higher contact time was neglected due to a very minor increase in the adsorption capacities after 360 min. Cu(II) adsorption was detected on ADP and TADP at a slow rate, reaching equilibrium adsorption in 360 min (6 h). For Cu(II) ions adsorption, the correlation coefficient (R^2) for the graph of time (min) vs. adsorbed concentration (Cads) and Time vs. Qe (mg/g) at concentration levels 20, 30, and 50 mg/L was found to be in the range of 0.6392 to 0.9175. The equilibrium adsorption capacity (Qe) of Cu(II) ions on ADP was 136, 211, and 339 mg/g, for concentrations of 20, 30, and 50 ppm, respectively, and 178, 251, and 421 mg/g for concentrations of 20, 30, and 50 ppm, respectively. The results show that ADP as produced materials have lower adsorption capabilities than TADP.

3.2.4. Adsorption Modeling

Equilibrium Isotherm

During the isotherm study, linear isothermal models such as the Langmuir and Freundlich [45] isotherms were applied.

The non-linearized and linearized Langmuir isotherm models are written as follows:

$$q_e = \frac{q_m K_L C_e}{1 + K_L C_e}, \tag{6}$$

$$\frac{C_e}{q_e} = \frac{1}{K_L q_m} + \frac{1}{q_m} \times C_e, \tag{7}$$

where q_m (mg/g) symbolizes maximum monolayer adsorption capacity and K_L (L/mg) denotes the process heat of adsorption.

The Freundlich isotherm, on the other hand, is stated as in [45] in both linear and non-linear models:

$$q_e = K_F \times C_e{}^{1/n}, \tag{8}$$

$$\log q_e = \log K_F + \frac{1}{n}\log C_e. \tag{9}$$

The Freundlich constants for bonding energy and deviation from linearity in adsorption are K_F ((mg/g) (L/mg)$^{(1/n)}$) and n, respectively. When the value of n in the equation is equal to 1, <1 or >1, the adsorption process is said to be linear, chemical, or physical, respectively.

The adsorption of Cu(II) on ADP and TADP adsorbents was carried out at varied temperatures and the corresponding parameters for adsorption are given in Table 2. The regression coefficient (R^2) of Langmuir isotherm model for Cu(II) ions adsorption on both ADP and TADP materials at various temperatures was found to be closer to unity than the Freundlich model, which is consistent with previously reported results of metal ion removal [39]. Furthermore, the appropriateness of the Langmuir model shows that monolayer metal ions cover both ADP and TADP adsorbents at various temperatures. The maximum adsorption capacity (q_m) for Cu(II) was found to be increased from 666.67 mg/g (666.67/63.546 = 10.491 mmol/g) to 1428.57 mg/g (1428.57/63.546 = 22.481 mmol/g) for TADP for changing the temperature from 298 K to 328 K. On the other hand, for the same temperature changes the q_m of ADP was changed from 348.0 mg/g (348.0/63.546 = 5.476 mmol/g) to 1111.1 mg/g (1111.1/63.546 = 17.485 mmol/g). The adsorbents and Cu(II) ions may have collided more frequently as the temperature increased, enhancing the adsorption capability of the materials. As the temperature rises, the surface bindings of the adsorbents are ruptured, exposing the active site to a greater extent and therefore increasing the adsorption of the metal ions.

Table 2. Isotherm parameters for the adsorption of Cu(II) on ADP and TADP.

Isotherm	**Temperature, K**			
	298	**308**	**318**	**328**
TADP				
Langmuir				
q_m (mg/g)	666.67	909.09	1111.11	1428.57
K_L (L/mg)	0.057915	0.01449	0.01289	0.01107
R^2	0.9889	0.9972	0.9986	0.9932
Freundlich				
KF (mg/g) (L/mg)$^{1/n}$	19.1134	37.6320	53.6320	57.0176
n	1.2449	1.2358	1.2858	1.4550
R^2	0.9096	0.9253	0.9170	0.9250
ADP				
Langmuir				
q_m (mg/g)	348.0	625.0	769.2	1111.1
K_L (L/mg)	0.03275	0.07048	0.01289	0.03904
R^2	0.9853	0.9901	0.9932	0.9891
Freundlich				
KF (mg/g) (L/mg)$^{1/n}$	18.52	23.43	23.69	48.56
n	1.0132	1.2168	1.6211	1.7621
R^2	0.7192	0.8586	0.8914	0.8979

The separation factor (K_L) for Cu(II) ions adsorption on ADP and TADP was in the range of 0.01 to 0.08 at the investigated temperature ranges (298 to 328 K), indicating an adsorption process. The K_F values enhanced as the temperature increased due to the increased contact between metal ions adsorbents, and n > 1 for both adsorbents confirm the physical adsorption process (Table 2).

3.3. Adsorption Kinetics

Using the adsorption kinetic data and the models, the Pseudo-first [45] and Pseudo-second-order [45] kinetics models in their linear form were investigated to determine the kinetic parameters. The linearized kinetic models are expressed as follows:

$$\log(q_{e1} - q_t) = \log q_{e1} - \frac{k_1}{2.303} \times t, \quad (10)$$

$$\frac{t}{q_t} = \frac{1}{k_2 q_{e2}{}^2} + \frac{1}{q_{e2}} \times \text{t}, \quad (11)$$

where q_{e1}, q_{e2}, and q_t signify the pseudo-first order model, the pseudo-second order model at equilibriums, and the pseudo-third order model at time t, respectively. Furthermore, k_1 and k_2 are the rate constants for pseudo-first-order and pseudo-second-order kinetics, respectively.

The obtained kinetic parameters for Cu(II) adsorption data on both ADP and TADP adsorbents are depicted in Table 3. Based on higher R2 values, the Cu(II) ions adsorption data in all tested adsorbents supports a kinetic model with a pseudo-second order (closer to unity). The close agreement between experimental and calculated adsorption capabilities ($q_{e,exp}$) further confirmed the findings, showing that Cu(II) ion adsorption on ADP and TADP was a chemical adsorption process rather than a conventional mass transport phenomenon [46]. Similar findings were described by Trikkaliotis et al. for Cu(II)) adsorption onto chitin based materials [47].

Table 3. Kinetic parameters calculated for the adsorption of Cu(II) on ADP and TADP adsorbents (initial concentration, C_0 for Cu(II) was 20 mg/L).

Kinetics Model	ADP	TADP
	Cu(II) 20 mg/L	
$q_{e,exp}$ (mg/g)	136	178
Pseudo-first-order		
$q_{e1,cal}$ (mg/g)	118.38	126.88
K_1(1/min)	0.0097	0.0145
R^2	0.991	0.978
Pseudo-second-order		
$q_{e2,cal}$ (mg/g)	145.25	189.07
k_2(g/mg-min)	0.00005	0.00003
R^2	0.996	0.995

3.4. Adsorption Thermodynamics

Van't Hoff plots [48] are used to estimate the thermodynamic parameters related to Cu(II) ions adsorption on ADP and TADP, such as standard free entropy changes ($\Delta S°$), enthalpy changes ($\Delta H°$), and free energy changes ($\Delta G°$), and the results are shown in Table 4. Cu(II) ions adsorption on both ADP and TADP was endothermic, as demonstrated by positive $\Delta H°$ values at all concentrations tested. The $\Delta S°$ values for both adsorbents were all positive, suggesting a random adsorption process due to energy redistribution between Cu(II) ions and the adsorbent [49]. Moreover, the $\Delta G°$ values for Cu(II) adsorption were negative for all types of adsorbents, suggesting the adsorption process to be spontaneous. The negative $\Delta G°$ values rose with an increase in temperature, indicating that the adsorption process favored spontaneity [41].

Table 4. Thermodynamic parameters for the adsorption of Cu(II) on ADP and TADP adsorbents (initial concentration, C_0 was 20 mg/L).

Adsorbents	Concentration mg/L	ΔH° (kJ/mol)	ΔS° (J/mol-K)	ΔG° (kJ/mol) 298 K	308 K	318 K	328 K
ADP	20	2608.34	86.29	−156.34	−566.44	−951.43	−1020.48
	30	2122.10	69.92	−165.79	−341.35	−708.64	−916.92
	40	1810.70	60.54	−419.21	−479.90	−562.61	−821.86
	50	1762.03	59.16	−64.62	−92.97	−423.11	−634.68
TADP	20	1017.01	41.52	−2167.69	−2537.27	−2934.88	−3064.55
	30	704.95	28.99	−1533.09	−1904.93	−2098.57	−2178.40
	40	854.00	34.71	−1786.02	−2073.66	−2380.42	−2550.16
	50	896.01	30.83	−257.90	−409.23	−899.72	−1004.94

3.5. Elution and Regeneration Studies

The elution efficiency of sufficiently Cu(II) ions saturated ADP and TADP adsorbents with various eluents was tested (Figure 9). Metal ion elution from both saturated adsorbents was discovered to be modest (3.0–3.2%) when eluted with 0.1 mol/L NaOH, however 0.1 M HCl caused a substantial increase in metal ion desorption for both adsorbents (84–98%).

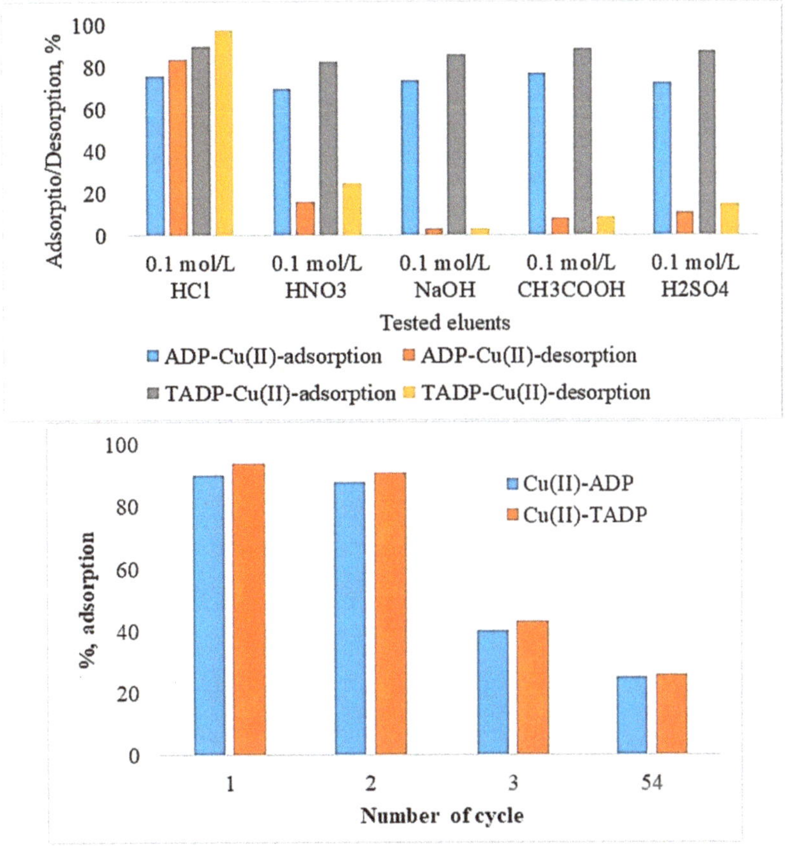

Figure 9. Cu(II) ion elution from saturated adsorbents (experimental conditions: Co: 25 mg/L; m: 0.05 g; V: 0.05 L; T: 25 °C; contact time: 5 h; agitation speed: 100 rpm).

The order of elution efficacy with the treated eluents was as follows: 0.1 mol/L NaOH < 0.1 mol/L CH_3COOH < 0.1 mol/L H_2SO_4 < 0.1 mol/L HNO_3 < 0.1 mol/L HCl. The data shows that the highest Cu(II) ion elution with HCl (0.1 mol/L) was 98% for TADP and 84% for ADP. The increased desorption efficacy of Cu(II) ions with relatively strong acid can be attributed to ion-exchange adsorption, which is mediated by metal ions binding on adsorbent surfaces.

After optimizing the eluent (HCl-0.1 mol/L), the regeneration studies for the adsorbents were also carried out to estimate their reusing efficiency. A drastic decrease in Cu(II) (80–89%) adsorption on both ADP and TADP were noticed after the second regeneration investigation. Repeated adsorption-regeneration studies may cause surface morphological deformation, which could be one of the reasons for a decrease in metal ion adsorption with increased regeneration experiments. As a result, ADP and TADP may be repurposed for Cu(II) adsorption with minimal adsorption efficiency loss during the second regeneration.

4. Conclusions

The optimized ADP and TADP revealed the excellent adsorption potential of Cu(II) in its respective adsorption methods. The biosorbent TADP showed higher adsorption capabilities compared to ADP. The R^2 values for C_e v.s q_e graph at various temperatures for Cu(II) adsorption were found to be 0.9816 to 0.9926, where the R^2 for time (min) vs. Qe (mg/g) graph at variable concentrations ranged from 0.6392 to 0.9175. The metal ions adsorptions by ADP and TADP were endothermic, indicating that a pseudo-second-order kinetics model would be more appropriate in all circumstances. The Langmuir isotherm model produced somewhat better findings in comparison to the Freundlich model for both ADP and TADP, indicating monolayer coverage. Desorption experiments with 0.1 mol/L HCl yielded the highest Cu(II) ions recovery. The findings of the adsorption and desorption outcomes reveal that the synthesized materials have the potential to be useful for removing heavy metal ions and for other contaminants as well from a variety of media while maintaining a safe environment.

Author Contributions: Conceptualization, M.A. and S.I.A.-R.; methodology, S.M.W. and M.R.K.; validation, M.A. and S.M.W.; formal analysis, S.M.W.; investigation, M.R.K. and M.S.I.; data curation, M.A.; writing—original draft preparation, S.M.W. and M.R.K. writing—review and editing, M.A. and S.I.A.-R.; supervision, M.A.; project administration, M.A.; funding acquisition, M.A. All authors have read and agreed to the published version of the manuscript.

Funding: This Project was funded by the National Plan for Science, Technology, and Innovation (MAARIFAH), King Abdulaziz City for Science and Technology (KACST), Kingdom of Saudi Arabia, Award Number (14-ENV2458-02).

Institutional Review Board Statement: Not applicable.

Informed Consent Statement: Not applicable.

Acknowledgments: This Project was funded by the National Plan for Science, Technology, and Innovation (MAARIFAH), King Abdulaziz City for Science and Technology (KACST), Kingdom of Saudi Arabia, Award Number (14-ENV2458-02).

Conflicts of Interest: The authors declare no conflict of interest.

References

1. Xu, D.-M.; Fu, R.-B.; Tong, Y.-H.; Shen, D.-L.; Guo, X.-P. The potential environmental risk implications of heavy metals based on their geochemical and mineralogical characteristics in the size-segregated zinc smelting slags. *J. Clean. Prod.* **2021**, *315*, 128199. [CrossRef]
2. Zhang, Y.; Wu, C.; Liu, H.; Khan, M.R.; Zhao, Z.; He, G.; Luo, A.; Zhang, J.; Deng, R.; He, Q. Label-free DNAzyme assays for dually amplified and one-pot detection of lead pollution. *J. Hazard. Mater.* **2021**, *406*, 124790. [CrossRef]
3. Khan, M.R.; Ahmad, N.; Ouladsmane, M.; Azam, M. Heavy Metals in Acrylic Color Paints Intended for the School Children Use: A Potential Threat to the Children of Early Age. *Molecules* **2021**, *26*, 2375. [CrossRef]
4. Mahurpawar, M. Effects of heavy metals on human health. *Int. J. Res. Granthaalayah* **2015**, *3*, 1–7. [CrossRef]

5. SenthilKumar, P.; Ramalingam, S.; Sathyaselvabala, V.; Kirupha, S.D.; Sivanesan, S. Removal of copper (II) ions from aqueous solution by adsorption using cashew nut shell. *Desalination* **2011**, *266*, 63–71. [CrossRef]
6. Zhu, C.S.; Wang, L.P.; Chen, W.B. Removal of Cu (II) from aqueous solution by agricultural by-product: Peanut hull. *J. Hazard. Mater.* **2009**, *168*, 739–746. [CrossRef] [PubMed]
7. Wu, W.; Yang, Y.; Zhou, H.; Ye, T.; Huang, Z.; Liu, R.; Kuang, Y. Highly efficient removal of Cu (II) from aqueous solution by using graphene oxide. *Water Air Soil Pollut.* **2013**, *224*, 1–8. [CrossRef]
8. Singha, B.; Das, S.K. Adsorptive removal of Cu (II) from aqueous solution and industrial effluent using natural/agricultural wastes. *Colloids Surf. B Biointerfaces* **2013**, *107*, 97–106. [CrossRef]
9. Edding, M.; Tala, F. Copper transfer and influence on a marine food chain. *Bull. Environ. Contam. Toxicol.* **1996**, *57*, 617–624. [CrossRef]
10. Stenhammar, L. Diarrhoea following contamination of drinking water with copper. *Eur. J. Med. Res.* **1999**, *4*, 217–218.
11. Ross, A. Vomiting and Diarrhoea due to Copper in Stewed Apples. *Lancet* **1955**, 87–88. [CrossRef]
12. Bergeron, C.; Petrunka, C.; Weyer, L. Copper/zinc superoxide dismutase expression in the human central nervous system. Correlation with selective neuronal vulnerability. *Am. J. Pathol.* **1996**, *148*, 273. [PubMed]
13. Araya, M.; Peña, C.; Pizarro, F.; Olivares, M. Gastric response to acute copper exposure. *Sci. Total Environ.* **2003**, *303*, 253–257. [CrossRef]
14. Acrylamide, O. National Primary Drinking Water Regulations. *Kidney* **2009**, *2*, 1–7.
15. Li, Y.; Trush, M.A.; Yager, J.D. DNA damage caused by reactive oxygen species originating from a copper-dependent oxidation of the 2-hydroxy catechol of estradiol. *Carcinogenesis* **1994**, *15*, 1421–1427. [CrossRef]
16. Starkebaum, G.; Harlan, J.M. Endothelial cell injury due to copper-catalyzed hydrogen peroxide generation from homocysteine. *J. Clin. Investig.* **1986**, *77*, 1370–1376. [CrossRef] [PubMed]
17. Renfew, M.M. *NIOSH Pocket Guide to Chemical Hazards*; ACS Publications; US Department of Health and Human Services—National Institute for Occupational Safety and Health: Washington, DC, USA, 1991.
18. Bashir, A.; Malik, L.A.; Ahad, S.; Manzoor, T.; Bhat, M.A.; Dar, G.; Pandith, A.H. Removal of heavy metal ions from aqueous system by ion-exchange and biosorption methods. *Environ. Chem. Lett.* **2019**, *17*, 729–754. [CrossRef]
19. Zhang, Q.; Hou, Q.; Huang, G.; Fan, Q. Removal of heavy metals in aquatic environment by graphene oxide composites: A review. *Environ. Sci. Pollut. Res.* **2020**, *27*, 190–209. [CrossRef] [PubMed]
20. Un, U.T.; Ocal, S.E. Removal of heavy metals (Cd, Cu, Ni) by electrocoagulation. *Int. J. Environ. Sci. Dev.* **2015**, *6*, 425. [CrossRef]
21. Li, Y.; Yang, L.; Xu, Z.; Sun, Q. *Separation and Recovery of Heavy Metals from Waste Water Using Synergistic Solvent Extraction*; IOP Conference Series: Materials Science and Engineering; IOP Publishing: Bristol, UK, 2017.
22. BrbootI, M.M.; AbiD, B.A.; Al-ShuwaikI, N.M. Removal of heavy metals using chemicals precipitation. *Eng. Technol. J* **2011**, *29*, 595–612.
23. Huang, Y.; Wu, D.; Wang, X.; Huang, W.; Lawless, D.; Feng, X. Removal of heavy metals from water using polyvinylamine by polymer-enhanced ultrafiltration and flocculation. *Sep. Purif. Technol.* **2016**, *158*, 124–136. [CrossRef]
24. Lakherwal, D. Adsorption of heavy metals: A review. *Int. J. Environ. Res. Dev.* **2014**, *4*, 41–48.
25. Shahrokhi-Shahraki, R.; Benally, C.; El-Din, M.G.; Park, J. High efficiency removal of heavy metals using tire-derived activated carbon vs commercial activated carbon: Insights into the adsorption mechanisms. *Chemosphere* **2021**, *264*, 128455. [CrossRef]
26. Sunil, K.; Karunakaran, G.; Yadav, S.; Padaki, M.; Zadorozhnyy, V.; Pai, R.K. Al-Ti2O6 a mixed metal oxide based composite membrane: A unique membrane for removal of heavy metals. *Chem. Eng. J.* **2018**, *348*, 678–684. [CrossRef]
27. Wu, S.; Xu, Y.; Sun, J.; Cao, Z.; Zhou, J.; Pan, Y.; Qian, G. Inhibiting evaporation of heavy metal by controlling its chemical speciation in MSWI fly ash. *Fuel* **2015**, *158*, 764–769. [CrossRef]
28. Karnib, M.; Kabbani, A.; Holail, H.; Olama, Z. Heavy metals removal using activated carbon, silica and silica activated carbon composite. *Energy Procedia* **2014**, *50*, 113–120. [CrossRef]
29. Yu, B.; Zhang, Y.; Shukla, A.; Shukla, S.S.; Dorris, K.L. The removal of heavy metal from aqueous solutions by sawdust adsorption—removal of copper. *J. Hazard. Mater.* **2000**, *80*, 33–42. [CrossRef]
30. Lu, S.; Gibb, S.W. Copper removal from wastewater using spent-grain as biosorbent. *Bioresour. Technol.* **2008**, *99*, 1509–1517. [CrossRef] [PubMed]
31. Wu, Y.; Zhang, L.; Gao, C.; Ma, J.; Ma, X.; Han, R. Adsorption of copper ions and methylene blue in a single and binary system on wheat straw. *J. Chem. Eng. Data* **2009**, *54*, 3229–3234. [CrossRef]
32. Ofomaja, A.; Naidoo, E.; Modise, S. Removal of copper (II) from aqueous solution by pine and base modified pine cone powder as biosorbent. *J. Hazard. Mater.* **2009**, *168*, 909–917. [CrossRef] [PubMed]
33. Larous, S.; Meniai, A. Removal of copper (II) from aqueous solution by agricultural by-products-sawdust. *Energy Procedia* **2012**, *18*, 915–923.
34. Gündoğan, R.; Acemioğlu, B.; Alma, M.H. Copper (II) adsorption from aqueous solution by herbaceous peat. *J. Colloid Interface Sci.* **2004**, *269*, 303–309. [CrossRef]
35. Aksu, Z.; İşoğlu, İ.A. Removal of copper (II) ions from aqueous solution by biosorption onto agricultural waste sugar beet pulp. *Process Biochem.* **2005**, *40*, 3031–3044. [CrossRef]

36. Jawad, A.H.; Abdulhameed, A.S.; Hanafiah, M.; ALOthman, Z.A.; Khan, M.R.; Surip, S. Numerical desirability function for adsorption of methylene blue dye by sulfonated pomegranate peel biochar: Modeling, kinetic, isotherm, thermodynamic, and mechanism study. *Korean J. Chem. Eng.* **2021**, *38*, 1–11. [CrossRef]
37. Jawad, A.H.; Bardhan, M.; Islam, M.A.; Islam, M.A.; Syed-Hassan, S.S.A.; Surip, S.; ALOthman, Z.A.; Khan, M.R. Insights into the modeling, characterization and adsorption performance of mesoporous activated carbon from corn cob residue via microwave-assisted H3PO4 activation. *Surf. Interfaces* **2020**, *21*, 100688. [CrossRef]
38. Jawad, A.H.; Abdulhameed, A.S.; Wilson, L.D.; Syed-Hassan, S.S.A.; ALOthman, Z.A.; Khan, M.R. High surface area and mesoporous activated carbon from KOH-activated Dragon fruit peels for methylene blue dye adsorption: Optimization and mechanism study. *Chin. J. Chem. Eng.* **2021**, *32*, 281–290. [CrossRef]
39. Azam, M.; Wabaidur, S.M.; Khan, M.R.; Al-Resayes, S.I.; Islam, M.S. Removal of Chromium (III) and Cadmium (II) Heavy Metal Ions from Aqueous Solutions Using Treated Date Seeds: An Eco-Friendly Method. *Molecules* **2021**, *26*, 3718. [CrossRef]
40. Idowu, A.T.; Igiehon, O.O.; Adekoya, A.E.; Idowu, S. Dates palm fruits: A review of their nutritional components, bioactivities and functional food applications. *AIMS Agric. Food* **2020**, *5*, 734–755. [CrossRef]
41. Khan, M.A.; Otero, M.; Kazi, M.; Alqadami, A.A.; Wabaidur, S.M.; Siddiqui, M.R.; Alothman, Z.A.; Sumbul, S. Unary and binary adsorption studies of lead and malachite green onto a nanomagnetic copper ferrite/drumstick pod biomass composite. *J. Hazard. Mater.* **2019**, *365*, 759–770. [CrossRef]
42. Tran, H.N.; You, S.-J.; Chao, H.-P. Thermodynamic parameters of cadmium adsorption onto orange peel calculated from various methods: A comparison study. *J. Environ. Chem. Eng.* **2016**, *4*, 2671–2682. [CrossRef]
43. Reddy, D.H.K.; Seshaiah, K.; Reddy, A.; Lee, S. Optimization of Cd (II), Cu (II) and Ni (II) biosorption by chemically modified Moringa oleifera leaves powder. *Carbohydr. Polym.* **2012**, *88*, 1077–1086. [CrossRef]
44. Alqadami, A.A.; Khan, M.A.; Siddiqui, M.R.; Alothman, Z.A. Development of citric anhydride anchored mesoporous MOF through post synthesis modification to sequester potentially toxic lead (II) from water. *Microporous Mesoporous Mater.* **2018**, *261*, 198–206. [CrossRef]
45. Wabaidur, S.M.; Khan, M.A.; Siddiqui, M.R.; Otero, M.; Jeon, B.-H.; Alothman, Z.A.; Hakami, A.A.H. Oxygenated functionalities enriched MWCNTs decorated with silica coated spinel ferrite–A nanocomposite for potentially rapid and efficient de-colorization of aquatic environment. *J. Mol. Liq.* **2020**, *317*, 113916. [CrossRef]
46. Khan, M.A.; Alqadami, A.A.; Otero, M.; Siddiqui, M.R.; Alothman, Z.A.; Alsohaimi, I.; Rafatullah, M.; Hamedelniel, A.E. Heteroatom-doped magnetic hydrochar to remove post-transition and transition metals from water: Synthesis, characterization, and adsorption studies. *Chemosphere* **2019**, *218*, 1089–1099. [CrossRef]
47. Trikkaliotis, D.G.; Christoforidis, A.K.; Mitropoulos, A.C.; Kyzas, G.Z. Adsorption of copper ions onto chitosan/poly (vinyl alcohol) beads functionalized with poly (ethylene glycol). *Carbohydr. Polym.* **2020**, *234*, 115890. [CrossRef] [PubMed]
48. Lima, É.C.; Adebayo, M.A.; Machado, F.M. Kinetic and equilibrium models of adsorption. In *Carbon Nanomaterials as Adsorbents for Environmental and Biological Applications*; Springer: Berlin/Heidelberg, Germany, 2015; pp. 33–69.
49. Argun, M.E.; Dursun, S.; Ozdemir, C.; Karatas, M. Heavy metal adsorption by modified oak sawdust: Thermodynamics and kinetics. *J. Hazard. Mater.* **2007**, *141*, 77–85. [CrossRef] [PubMed]

polymers

MDPI

Article

Antimicrobial Activity of Cellulose Based Materials

Nicoleta Sorina Nemeş [1], Cristina Ardean [2], Corneliu Mircea Davidescu [1,*], Adina Negrea [2], Mihaela Ciopec [2], Narcis Duţeanu [2,*], Petru Negrea [2], Cristina Paul [2], Daniel Duda-Seiman [3] and Delia Muntean [4]

1 Renewable Energy Research Institute-ICER, Politehnica University of Timisoara, 138 Gavril Musicescu Street, 300501 Timisoara, Romania; nicoleta.nemes@upt.ro

2 Faculty of Industrial Chemistry and Environmental Engineering, Politehnica University of Timişoara, 2 Piaţa Victoriei, 300006 Timisoara, Romania; cristina.ardean@student.upt.ro (C.A.); adina.negrea@upt.ro (A.N.); mihaela.ciopec@upt.ro (M.C.); petru.negrea@upt.ro (P.N.); cristina.paul@upt.ro (C.P.)

3 Department of Cardiology, "Victor Babeş" University of Medicine and Pharmacy Timişoara, 2 Piata Eftimie Murgu, 300041 Timisoara, Romania; dannymduda@gmail.com

4 Multidisciplinary Research Center on Antimicrobial Resistance, Department of Microbiology, "Victor Babeş" University of Medicine and Pharmacy, 2 Eftimie Murgu Square, 300041 Timisoara, Romania; muntean.delia@umft.ro

* Correspondence: corneliu.davidescu@upt.ro (C.M.D.); narcis.duteanu@upt.ro (N.D.)

Abstract: Biomaterials available for a wide range of applications are generally polysaccharides. They may have inherent antimicrobial activity in the case of chitosan. However, in order to have specific functionalities, bioactive compounds must be immobilized or incorporated into the polymer matrix, as in the case of cellulose. We studied materials obtained by functionalizing cellulose with quaternary ammonium salts: dodecyl-trimethyl-ammonium bromide (DDTMABr), tetradecyl-trimethyl-ammonium bromide (TDTMABr), hexadecyl-trimethyl ammonium chloride (HDTMACl), some phosphonium salts: dodecyl-triphenyl phosphonium bromide (DDTPPBr) and tri n-butyl-hexadecyl phosphonium bromide (HDTBPBr) and extractants containing sulphur: 2-mercaptobenzothiazole (MBT) and thiourea (THIO). Cel-TDTMABr material, whose alkyl substituent chain conformation was shortest, showed the best antimicrobial activity for which, even at the lowest functionalization ratio, 1:0.012 (w:w), the microbial inhibition rate is 100% for *Staphylococcus aureus, Escherichia coli,* and *Candida albicans*. Among the materials obtained by phosphonium salt functionalization, Cel-DDTPPBr showed a significant bactericidal effect compared to Cel-HDTBPBr. For instance, to the same functionalization ratio = 1:0.1, the inhibition microbial growth rate is maximum in the case of Cel-DDTPPBr for *Staphylococcus aureus, Escherichia coli,* and *Candida albicans*. At the same time, for the Cel-HDTBPBr material, the total bactericidal effect is not reached even at the functionalization ratio 1:0.5. This behavior is based on the hydrophobicity difference between the two extractants, DDTPPBr and HDTBPBr. Cel-MBT material has a maximum antimicrobial effect upon *Staphylococcus aureus, Escherichia coli,* and *Candida albicans* at functionalized ratio = 1:0.5. Cel-THIO material showed a bacteriostatic and fungistatic effect, the inhibition of microbial growth being a maximum of 76% for *Staphylococcus aureus* at the functionalized ratio = 1:0.5. From this perspective, biomaterials obtained by SIR impregnation of cellulose can be considered a benefit to be used to obtain biomass-derived materials having superior antimicrobial properties versus the non-functional support.

Keywords: cellulose; antimicrobial activities; functionalized materials; cellulose derivatives

Citation: Nemeş, N.S.; Ardean, C.; Davidescu, C.M.; Negrea, A.; Ciopec, M.; Duţeanu, N.; Negrea, P.; Paul, C.; Duda-Seiman, D.; Muntean, D. Antimicrobial Activity of Cellulose Based Materials. *Polymers* **2022**, *14*, 735. https://doi.org/10.3390/polym14040735

Academic Editor: Luis Alves

Received: 20 January 2022
Accepted: 13 February 2022
Published: 14 February 2022

1. Introduction

Cellulose is one of the major components of plant cell walls. Cellulose is mainly obtained from natural resources such as plant materials: grass (bamboo), straw (wheat, rice, barley), leaves (pineapple, palm), fibers (flax, jute, hemp, sugar cane, sisal) [1,2], wood and wood residues such as wood chips or branches [1,3–6], marine animals [6,7], fungi, algae and bacteria [6,8–12].

Regardless of the source [13,14], cellulose can be characterized as a linear homopolysaccharide with high molecular weight, consisting of monomer units arranged alternately with an angle of 180°C between them, linked by β-1,4 bonds [15,16]. The repeating segment is a glucose dimer known as cellobiose [6,14,17]. Each monomer unit, identified as an anhydrous-glucose unit, has three hydroxyl groups, giving cellulose the most important properties and determining its microfibrillated structure [2,9,14] and the hierarchical organization in crystalline and amorphous regions and the extremely cohesive structure of the molecule [7,18].

Each cellulose chain has a directional asymmetry of its molecular axis with respect to the ends. One end has reduced chemical functionality due to the hemiacetal unit being a COC-shaped etheric bridge. The other end has "pendant" type hydroxyl groups with no reduction properties [14]. However, hydroxyl groups play a major role in forming covalent bonds with various reactants [19], which increase the chemical activity of cellulose and thus the antibacterial effect.

Due to the biocompatibility and biodegradability of cellulose [20], but also through the production of a wide range of cellulose derivatives [21], this biopolymer is used in various technological fields: in the food industry, it is used as a gelling or stabilizing agent [20,22], while in cosmetics it is used as a scrub or water retention agent [16,23,24], emulsifier [22,25], stabilizer of suspensions in paste and cream [24]. Other applications are in the paper industry [23], textile industry [16,23], and leather [26] or as combustible [27]. Various materials made from bacterial cellulose and polymeric compounds were tested to obtain artificial bones, cartilage, or other medical devices [23,28]. Of all the known types, the most studied form of cellulose for obtaining materials with antibacterial properties was cellulose with Ag nanoparticles, considering the antibacterial potential of Ag nanoparticles [11,29–31].

Microcrystalline cellulose (like the cellulose used in this study, Avicel PH 101) is a partially depolymerized pure cellulose synthesized from the α-cellulose precursor (type Iβ). This type of cellulose is considered a remarkable excipient. It remains the most widely used and the strongest dry binder, disintegrating tablet coating, absorbent, lubricant, and non-stick material [32]. Cellulose has good water purifying effects [16,33] because it has many free -OH groups on the chain, allowing the efficient removal of metal ions and organic matter from water [33] having an excellent chelating effect. However, due to its low water solubility and relatively low chemical reactivity, the use of unmodified cellulose as a flocculant is limited [34]. Modified, water-soluble (hydrophilic) cellulose plays a very important role as a potential substitute for oil-based flocculants [35]. Some cellulose derivatives have already been successfully tested to remove suspended solids. There is also a growing interest in developing low-cost biomass (cellulosic) absorbents for the treatment of colour-contaminated wastewater (agricultural, industrial, municipal waste). Thus, anionic sodium carboxymethylcellulose, prepared from palm agricultural waste, has been tested as an environmentally friendly flocculant coupled with aluminum sulfate as a coagulant to remove turbidity in drinking water treatment [34,36].

As was said, natively, cellulose has a relatively low reactivity [5]. Therefore, introducing new functional groups on the cellulose structure can increase the polarity and hydrophilicity of its surface [23], which can increase the adsorption of polar compounds and the selectivity of cellulose for the targeted pollutant. However, the chemical modification of this natural polymer is slightly difficult due to its relatively low reactivity. This is also influenced by the many hydrogen bonds that decrease the potential solubility in most common solvents [23]. Typically, the modification of cellulose fibers is achieved by heterogeneous synthesis [23,37], which can lead to some unexpected by-products [37] or the decomposition of cellulose. Although obtaining materials with antibacterial properties based on cellulose by simply mixing with antibacterial agents could be a way to induce antibacterial activity to the biocellulose, antibacterial agents being leached during use and thus bioactivity will decrease. That is why it is preferable to obtain cellulose derivatives by grafting chains with antibacterial activity on the cellulose skeleton, as we will show in this paper. Quaternary ammonium salts or polymers containing quaternary ammonium groups are well known for their increased antibacterial activity [5,21,38]. On the other hand, by impregnating cellulose with various extractants containing active groups (SIR Method) [39–42], the use of harsh reaction conditions is avoided, which is in accordance with the principles of green chemistry.

2. Materials and Methods

2.1. Obtaining and Characterizing Materials

The cellulose used to obtain the materials used in the present study (powder, Avicel PH–101, Sigma-Aldrich, Merck, KGaA, Darmstadt, Germany) was used as a support, and dodecyl-trimethyl-ammonium bromide (DDTMABr—Acros Organics, Waltham, MA, USA), tetradecyl-trimethyl-ammonium bromide (TDTMABr—ThermoFisher, Waltham, MA, USA, purity 98%), hexadecyl-trimethyl ammonium chloride (HDTMACl—ThermoFisher, Germany, purity 98%), dodecyl-triphenyl phosphonium bromide (DDTPPBr—Thermo Fisher, Germany purity 98%), tri n-butyl-hexadecyl phosphonium bromide (HDTBPBr—ThermoFisher, Germany, purity 98%), 2-mercaptobenzothiazole (MBT—Janssen Chemistry, Bucharest, Romania), thiourea (THIO—Fluka AG, Charlotte, NC, USA). The method used to obtain the materials was the dry method SIR (Solvent Impregnated Resin), by which the support is functionalized by impregnation [41]. The support and the extractant (dissolved in distilled water) are in contact for 24 h at 323 K. Then, the obtained sample is filtered, and the material obtained is dried in an oven (POLENKO SLW53, Poland). Materials were obtained in which the support:extractant (w/w) ratio of is different, namely: 1:0.012; 1:0.025; 1:0.050; 1:0.075; 1:0.1; 1:0.2; 1:0.3; 1:0.4; 1:0.5. The obtained material was characterized by scanning electron microscopy, SEM, and X-ray energy dispersion (EDX), using the X-ray energy dispersion spectrometer and the FEI Quanta FEG 250 instrument (FEI, Hillsboro, OR, USA).

2.2. Bacterial Culture Preparation

(a) Microbial cultures preparation, in order to establish the optimal cellulose: extractant ratio

Bacterial cultures were obtained by inoculating a non-selective solid nutrient medium into which the test material was incorporated, then poured into Petri dishes, with a bacterial suspension of approx. 1×10^8 UFC/mL (0,5 Mc Farland). Completely dehydrated culture medium was used for all seedings, Plate Count Agar-produced by Merck (Peptone from Casein 5.0 g/L; Yeast Extract 2.5 g/L; D (+)-Glucose 1.0 g/L; Agar 14.0 g/L).

In each set of experiments, a Petri dish containing the control sample (M0-control sample) was seeded, consisting of culture medium and 1 mL of bacterial inoculum, and a Petri dish containing culture medium, bacterial inoculum, and non-functionalized support material, in solid-state (M1-cellulose (Cel)).

Each functionalized material (containing both supporting and extracting material) to be tested was analysed in all functionalization ratios: 1:0.012; 1:0.025; 1:0.050; 1:0.075; 1:0.1; 1:0.2; 1:0.3; 1:0.4; 1:0.5. An amount of non-functionalized or functionalized solid material of about 0.2 g was used in each Petri dish, which was distributed as evenly as possible in the culture medium and 1 mL of bacterial inoculum (approx. 1×10^8 UFC/mL), in an amount of culture medium of about 15 mL.

All inoculations were performed with 3 repetitions. The plates were incubated for 48 h at 310 K. Subsequently, the colonies developed on the surface of the Petri dishes were counted using an automatic colony counter produced by YUL (Flash&Go, YUL Instruments SA, Barcelona, Spain).

(b) Microbial cultures preparation for further tests of antimicrobial effect of the synthesized materials on the reference strains

In order to test the antimicrobial effect of the synthesized materials on the reference microbial strains, both the optimal support:extractant ratios determined in the first stage and other support:extractant ratios were considered so that the efficiency of the functionalization ratio can be monitored. In order to establish the efficacy of the functionalized material, microbiological control tests were performed on 1×10^5 CFU/mL of reference microbial strains (Thermo Scientific, Waltham, MA, USA). To highlight these effects on Gram-negative bacteria, strains of *Escherichia coli ATCC 25922* and *Pseudomonas aeruginosa ATCC 27853* were studied, while the effect of the materials on Gram-positive bacteria was studied on the strain of *Staphylococcus aureus ATCC 25923. Candida albicans* strain *ATCC 10231* was used to observe the antifungal activity. All determinations were performed with 3 repetitions. The plates were incubated for 48 h at 310 K. The number of colonies developed on the surface of the work plates was determined using the automatic colony counter from YUL instruments (YUL Instruments SA, Barcelona, Spain). The efficiency of the functionalized material on the reference strains was expressed as the rate of inhibition of microbial growth [43–45], calculated as the ratio between the number of colony-forming units on functionalized and non-functionalized material, expressed as a percentage ratio, according to the equation:

$$inhibition\ rate = \frac{[UFC_{control} - UFC_{test}]}{UFC_{control}} \times 100 \quad (1)$$

$UFC_{control}$ = number of colonies on the control plate
UFC_{test} = the number of colonies on the test plate

Using this method, the optimal support:extractant ratio was established, representing the optimum ratio at which no microbial growth was observed or the case when the rate of inhibition of microbial growth was the highest.

Total bactericidal effect was considered where the rate of inhibition of bacterial growth was 100%.

3. Results and Discussion

3.1. Characterization of Materials Obtained by Cellulose Functionalizing

Obtained materials were characterized by scanning electron microscopy, SEM, X-ray energy dispersion, EDX and infrared spectroscopy with Fourier transform, FT-IR to highlight the presence of pending groups N, P, and S of the extractant (DDTMABr, TDTMABr, HDTMACl, DDTPPBr, HDTBPBr, MBT and THIO) on the support surface (Cel).

a. Scanning electron microscopy, SEM

To highlight the morphology of the support surface (cellulose), as well as of the materials obtained by impregnation by functionalization with the extractants DDTMABr, TDTMABr, HDTMACl, DDTPPBr, HDTBPBr, MBT, and THIO, the probes were analyzed by scanning electron microscopy operated at 10 kV and a magnification of 2000× (obtained micrograph being presented in Figure 1). The assays were performed in a low vacuum to prevent the sample's destruction.

It can be seen from Figure 1 that for all obtained materials, cellulose fibers are more agglomerated due to the functionalization. It was observed that the obtained materials have more slender fibers. However, the major structural difference between control and extractants-treated cellulose fibers was not clearly observed [46].

This morphological change is attributed to the swelling and partial disintegration of the cellulosic fibers during the surface modification process, followed by the insertion of amine moieties that cover the surface of the fibers [47,48]. Nevertheless, analogous micro-sized fibers with a rough and rod-shaped surface microstructure are revealed for samples based on microcrystalline cellulose, demonstrating the homogenous surface functionalization without much variation of surface morphology.

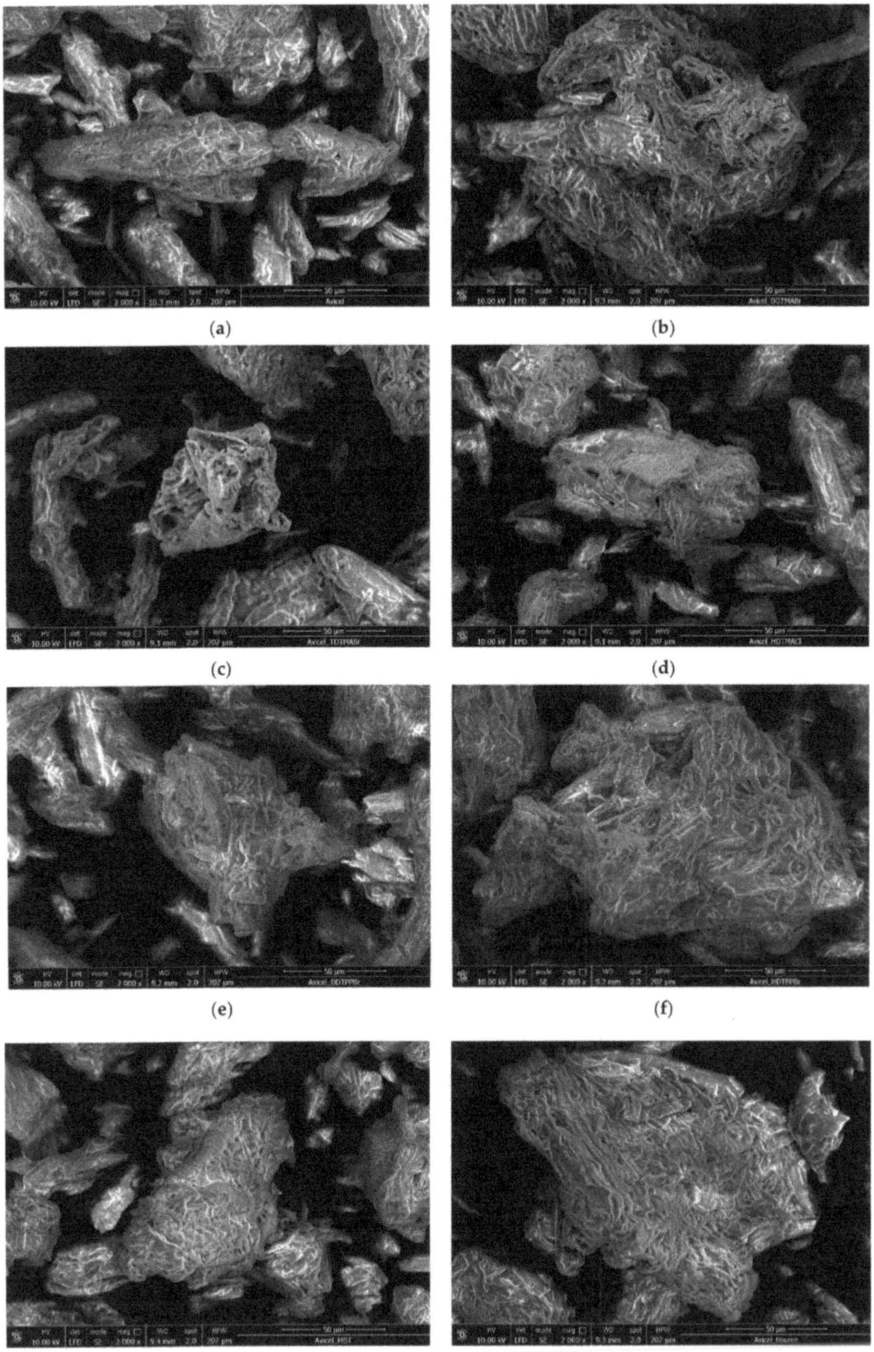

Figure 1. Scanning electron microscopy, SEM. (**a**). Cellulose (Cel); (**b**). Cel-DDTMABr; (**c**). Cel-TDTMABr; (**d**). Cel-HDTMACl; (**e**). Cel-DDTPPBr; (**f**). Cel-HDTBPBr; (**g**). Cel-MBT; (**h**). Cel-THIO.

b. X-ray energy dispersion, EDX

The EDX spectra for support, Cel, and the obtained materials are presented in Figure 2. Figure 2a shows the peaks specific to the composition of cellulose, namely C and O. Figure 2b–h shows the specific EDX spectra representing the specific peaks of the elements present in the composition of cellulose and the extractants. Thus, in Figure 2b,c there are drawn the specific peaks of the DDTMABr and TDTMABr extractants, associated with the presence of Br and N atoms, and in Figure 2d, in addition to the specific peak of N, there is also the specific peak in case of Cl. Figure 2e,f show specific P and Br peaks, elements that are present in the DDTPPBr and HDTBPBr extractants. In Figure 2g,h. specific S and N peaks appear, elements that are present in the MBT and THIO extractants. These observations reveal a successful functionalization of the natural polymer with thiourea extractant [49].

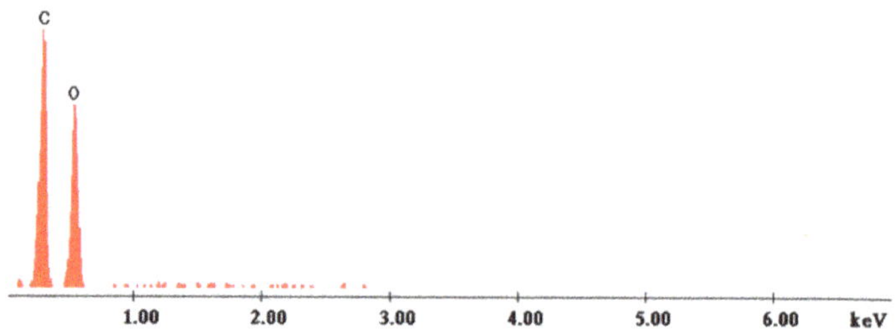

(a)

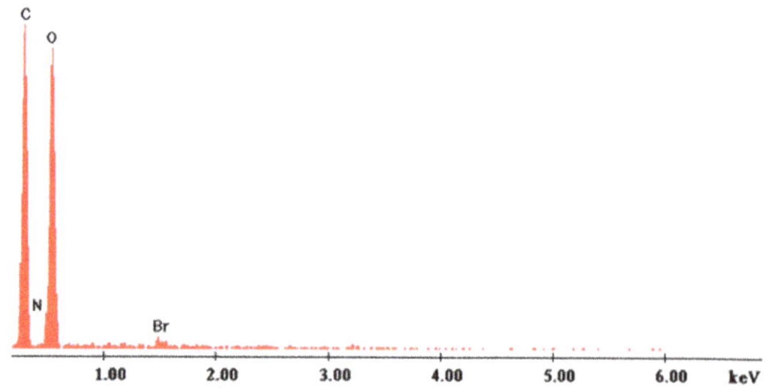

(b)

Figure 2. *Cont.*

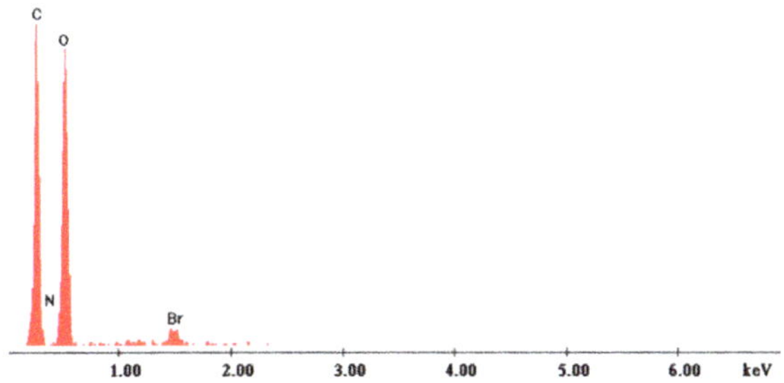

(**c**)

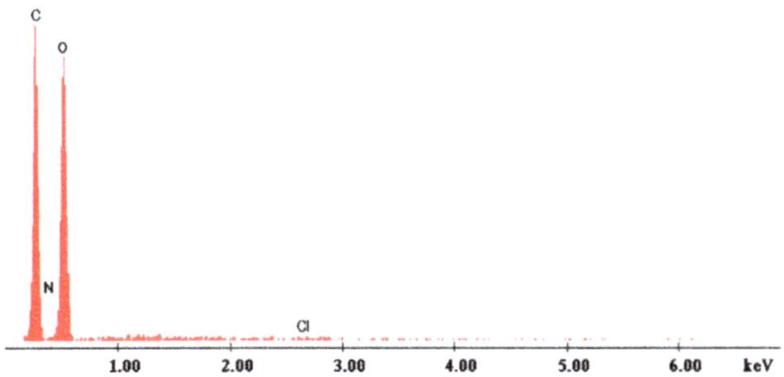

(**d**)

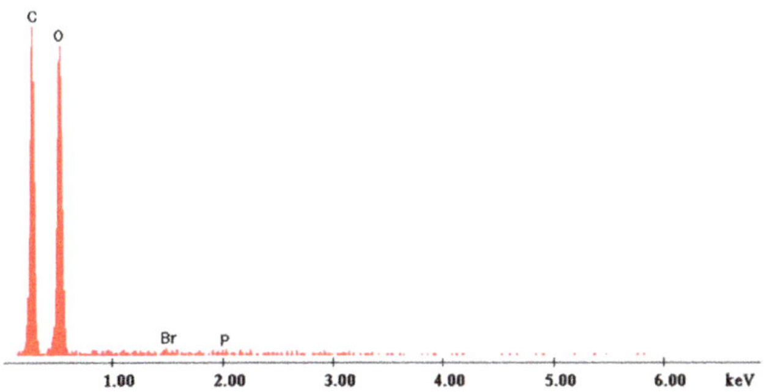

(**e**)

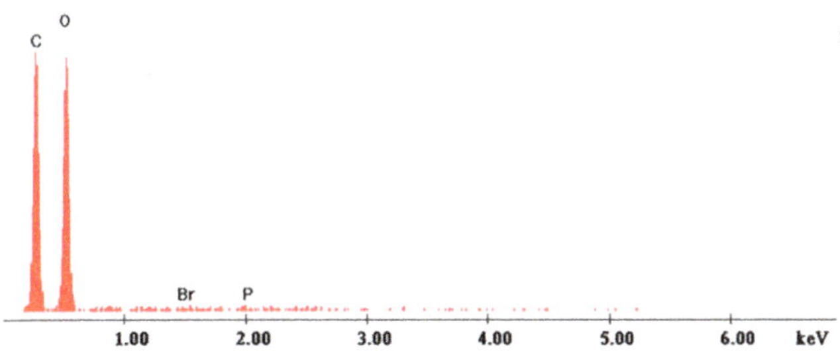

(**f**)

Figure 2. *Cont.*

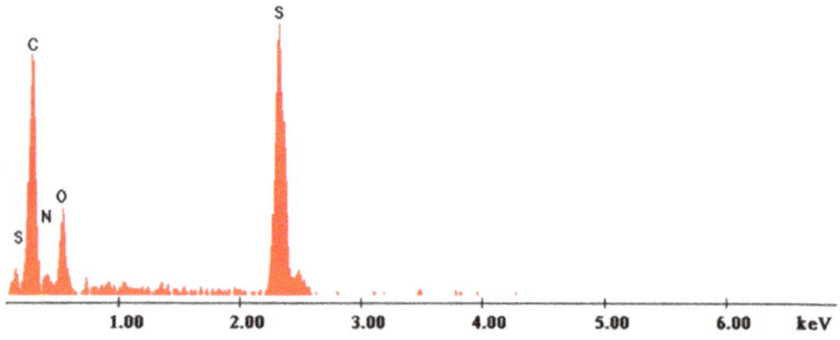

(g)

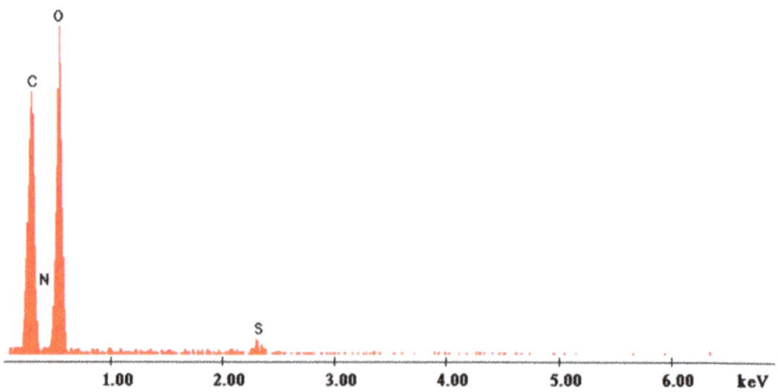

(h)

Figure 2. X-ray energy dispersion spectra recorded for pure cellulose and for prepared materials. (**a**). Cellulose (Cel); (**b**). Cel-DDTMABr; (**c**). Cel-TDTMABr; (**d**). Cel-HDTMACl; (**e**). Cel-DDTPPBr; (**f**). Cel-HDTBPBr; (**g**). Cel-MBT; (**h**). Cel-THIO.

c. Infrared spectroscopy with Fourier transform, FT-IR

In order to identify those functional groups specific to the extractants used for cellulose functionalization, cellulose and newly obtained materials FT-IR spectra were performed. These spectra are shown in Figure 3.

Based on the presented data, it can be stated that cellulose was successfully functionalized with the seven studied extractants. Small displacements of hydroxylic groups stretching vibrations observed in FT-IR spectra recorded for functionalized cellulose were associated with the formation of hydrogen bonds (physical intermolecular bonds) between used extractants and cellulosic free hydroxyl groups. This observation was in concordance with the low reactivity of cellulose in used working conditions, when only the cellulose swelling is taking place.

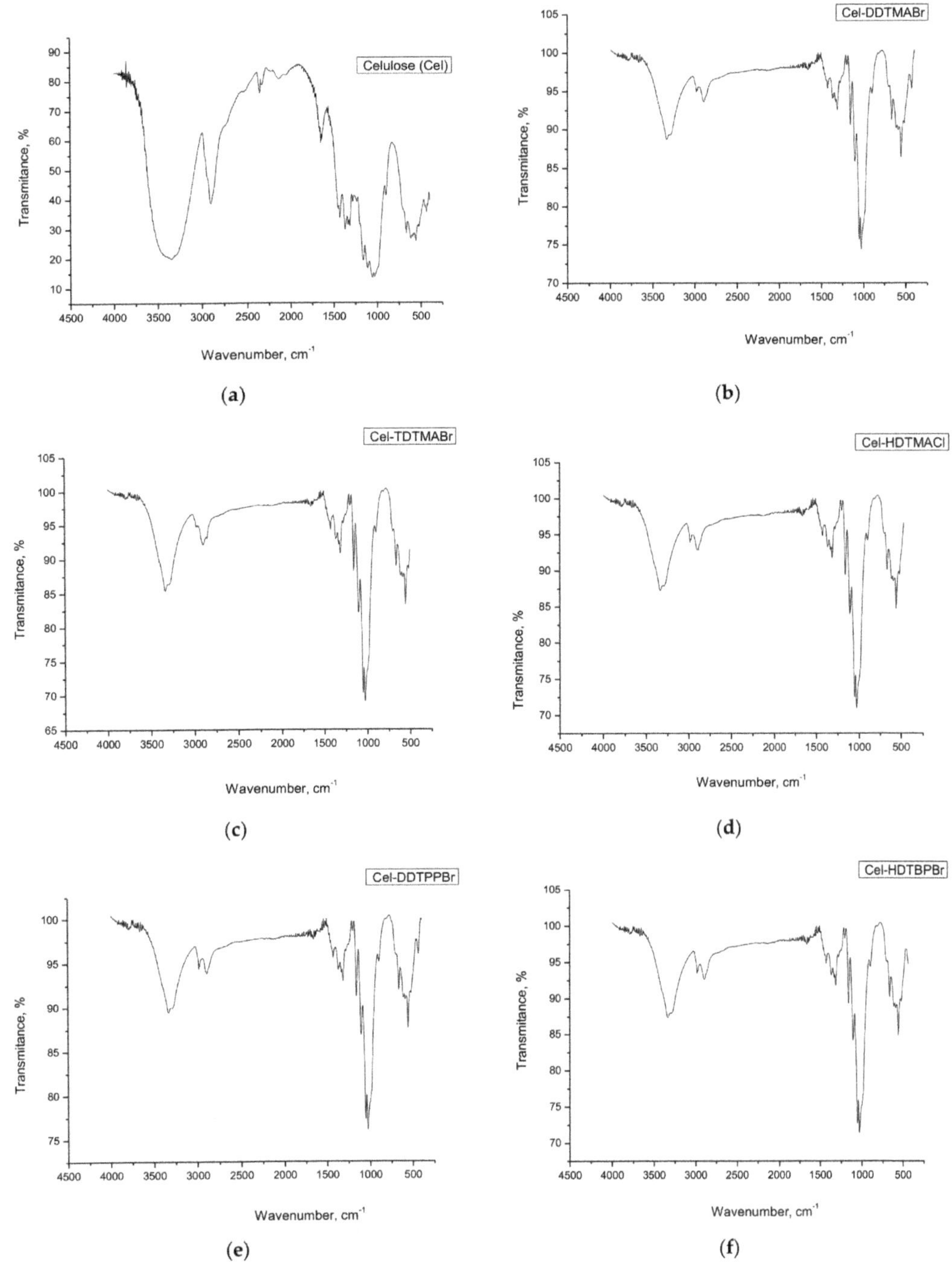

Figure 3. *Cont.*

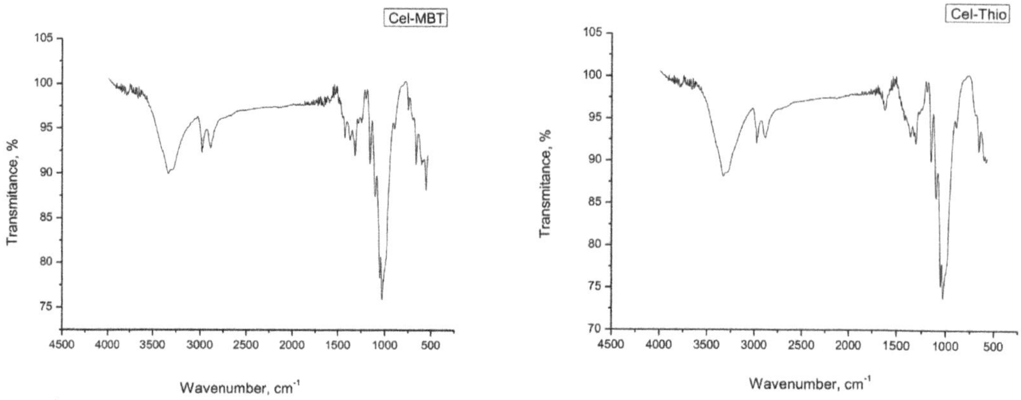

(g) (h)

Figure 3. FT-IR spectra recorded for cellulose and for produced materials. (**a**) Cellulose (Cel); (**b**) Cel-DDTMABr; (**c**) Cel-TDTMABr; (**d**) Cel-HDTMACl; (**e**) Cel-DDTPPBr; (**f**) Cel-HDTBPBr; (**g**) Cel-MBT; (**h**) Cel-THIO.

There are small differences due to the presence of active groups in the extractants. Table 1 shows the bands specific to the cellulose and extractant groups.

Table 1. FT-IR specific bands in cellulose and extractants [49,50].

Group	FT-IR Bands (cm^{-1})	Observations
	Cellulose (Cel)	
O-H C-H	3660 2893	Large band Small plateau; stretching vibrations in polysaccharides
CH_2 C-O O-C-O	1428; 1367 1334; 1158 1104; 1027 897	Vibrations specific to the crystalline structure of cellulose
OH_2	1600–900 1633	Amorphous region in cellulose Water molecules vibrations
	Cel-DDTMABr	
>N-CH_2	2700–2800	specific link e$^-$ nonparticipants from N
	Cel-TDTMABr	
>N-CH_2	2700–2800	specific link e$^-$ nonparticipants from N
	Cel-HDTMACl	
>N-CH_2	2700–2800	specific link e$^-$ nonparticipants from N
	Cel-DDTPPBr	
P-O-Aril C-O (fenil) O-H	1190–1240 1200 3500–3200	
	Cel-HDTBPBr	
P-O-Alchil	1150–1180; 1080	

Table 1. *Cont.*

Group	FT-IR Bands (cm^{-1})	Observations
	Cel-MBT	
S-C-S C-N C-H; N-H	568–600 1030–1074 1250–1320 750	Aromatic ring -torsion Stretching vibration
	Cel-THIO	
$-NH_2$	3395 1627	Asymmetric vibration; symmetric vibration Elongation vibration
N-H C=S	3179; 1520 1425;1395 1380; 1290 1074	Vibrations strongly influenced by intramolecular effects

3.2. Studies on the Antimicrobial Activity of Materials

A. Materials obtained by functionalizing cellulose with quaternary ammonium salts

A.1. Studies to determine the optimal cellulose:extractant ratio

The bacterial activity on chitosan can be considered intrinsic due to the presence of the amino group in its basic structure. The antibacterial activity of pure cellulose (avicel) can be highlighted only in its derivatives, obtained by various chemical functionalization with antibacterial groups or by combination with natural or synthetic bioactive components or polymers and with metal nanoparticles and metal oxides [21].

In order to establish the lowest cellulose:extractant ratio, which presents a bactericidal effect on the bacterial inoculum, each material was inoculated after incorporation into the culture medium and incubated, after which the total number of colonies was determined, according to the corresponding M1 non-functionalized cellulose (Cel) control sample, respectively the M0 control sample corresponding to the microbial inoculum. This ratio, for which the complete absence of microbial growth was observed, was considered the optimal, minimum required ratio, producing a total bactericidal effect. However, where microbial growth was observed onto the culture media, the significant antimicrobial effect was considered the highest calculated inhibition rate.

Analyzing the data presented in Figure 4, the non-functionalized cellulose sample (M1—yellow) has an insignificant antimicrobial effect compared to the control sample (M0—red).

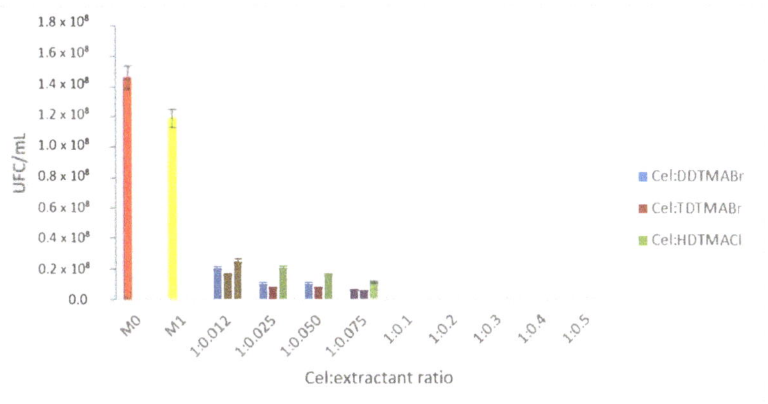

Figure 4. Comparison of the control sample (M0)—cellulose (M1)—cellulose:extractant (quaternary ammonium salts).

Regarding the three ammonium salts used as extractants, the best antimicrobial activity was highlighted by TDTMABr, in which case the ratio Cel:TDTMABr = 1:0.1 expresses the fact that the inhibition rate of bacterial growth was 100%. The DDTMABr extractant showed a total bactericidal effect when the ratio Cel:DDTMABr was 1:0.1 also. However, the inhibition rate of bacterial growth in the ratio of less than 1: 0.01 was lower for DDTMABr than for TDTMABr. For example, when cellulose:extractant ratio = 1:0.075 has been used, DDTMABr had an inhibition rate of 40%, while it was 95.8% for TDTMABr, respectively. HDTMACl extractant also showed total bactericidal effect at the functional ratio Cel:HDTMACl = 1:0.01. At the same time, the bacterial growth inhibition rate calculated for the HDTMACl extractant, at all functional ratios of less than 1:0.1, was lower compared to that calculated for TDTMABr and DDTMABr, respectively.

Consequently, the optimum functional ratio to express the total bactericidal effect in the case of quaternary ammonium salt extractants was considered when the cellulose:extractant ratio was 1:0.1.

A.2. Antimicrobial effect of ammonium materials on reference microbial strains

The most accessible groups present on the cellulose surface that contribute to the formation of new materials with antibacterial properties are the hydroxyl groups at the non-reducing end of the cellulose chain. Thus, the—OH groups on the crystalline cellulose were grafted with quaternary ammonium compounds through an adsorption process by SIR impregnation to allow a high concentration of quaternary ammonium groups on the cellulose surface. The effects of the obtained materials on four reference strains were tested. Table 2 shows the inhibition rate of microbial growth, calculated for each tested material.

Table 2. Antimicrobial growth inhibition rate when using ammonium salts materials.

Material	Cel:Extractant Ratio	Inhibition Rate (%)				OBSERVATIONS
		Staphylococcus aureus ATCC 25923	*Pseudomonas aeruginosa ATCC 27853*	*Escherichia coli ATCC 25922*	*Candida albicans ATCC 10231*	
Cellulose (Cel)	-	2.8	7.7	20	12.2	Slightly better bactericidal effect on Gram-negative bacteria and *C. albicans* strain
Cel:DDTMABr	1:0.012	100.0	40.8	100.0	100.0	Maximum bactericidal effect on Gram-positive bacteria and *C. albicans* strain
	1:0.1	100.0	52.8	100.0	100.0	
Cel:TDTMABr	1:0.012	100.0	49.9	100.0	100.0	
	1:0.1	100.0	54.8	100.0	100.0	
Cel:HDTMACl	1:0.012	100.0	17.0	53.4	100.0	
	1:0.05	100.0	34.3	58.1	100.0	
	1:0.1	100.0	40.8	67.5	100.0	

Non-functional cellulose showed a slightly better bactericidal effect on Gram-negative bacteria compared to Gram-positive ones. All cellulose derivatives based on quaternary ammonium salts used in this study (DDTMABr, TDTMABr, HDTMACl) showed very good antibacterial activity on *S. aureus* and *C. albicans*, regardless of the functionalization ratio of the support material. This effect may be due to the similar behavior of the cell wall of *S. aureus* to the fungal cell of *C. albicans* [51,52]. The extractant strongly adheres to the bacterial cell wall and the fungal cell membrane, damaging their structure and preventing the exchange of nutrients necessary for microbial development, which subsequently leads to cell death, meaning the accomplishment of the antimicrobial effect. Another important role in the permeability of the cell membrane is the number of fatty acids present in its structure, which subsequently correlates with the degree of penetration of the toxic element into the cell, as shown in numerous studies [51,53].

Moreover, on the *E. coli* strain, the bactericidal effect was maximum when using DDTMABr and TDTMABr. At the same time, for HDTMACl, the highest inhibition rate was 67.5%, corresponding to the functionalization ratio Cel:HDTMACl = 1:0.1. A possible explanation would be that a particularly important role for antibacterial effect has the hydrophobic component given by the alkyl radical of the quaternary salt. On the other hand, there is the distance from the positive charge given by the quaternary nitrogen to the basic skeleton of the cellulose, respectively, the conformation of the alkyl substituent that forms the chain of the extractant and distances the hanging group. The length of the alkyl substituent, although differing only by two carbon atoms, in the order of dodecyl, tetradecyl, hexadecyl, tends to adopt different conformations [38], and implicitly the distance determined by the alkyl chain from the cellulose skeleton increases in the order of TDTMABr, DDTMABr, HDTMACl. Therefore, the strongest antibacterial effect has the TDTMABr extractant, regardless of the studied functionalization ratio. This statement is in line with the data in the literature, which claims that the structure of the cell wall is destroyed and cell viability is disrupted by the action of hydrophobic compounds [53–57].

Regarding the effect on *P. aeruginosa*, all tested ammonium salt extractants showed a bactericidal effect, correlated with the amount of extractant used for functionalization. The maximum inhibition rate was not reached at any of the studied ratios, and a much higher amount of biocide is likely to be required. The outer membrane of the bacterial cell in *P. aeruginosa* is an important selective barrier, which results in reduced adsorption of the biocide in the cell due to the reduced susceptibility of this species to biocidal agents [58].

Previous research results on quaternary ammonium cellulose salts have shown that they can be used successfully in wastewater treatment as flocculants [21,59,60]. According to this study, its use is also recommended in terms of the effectiveness of its antibacterial effect on wastewater.

B. Materials obtained by functionalization of cellulose with extractants containing phosphonium salts

B.1. Studies to determine the optimal cellulose:extractant ratio

Regardless of the phosphonium salt used for functionalization, the obtained material's antimicrobial activity is directly dependent on the increase of the functionalization ratio (Figure 5).

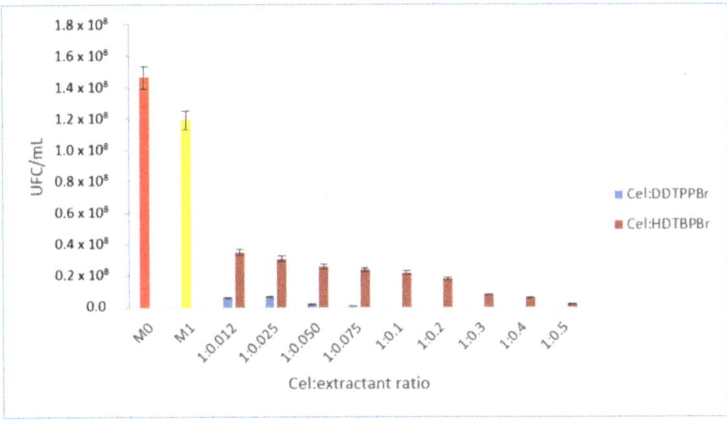

Figure 5. Comparison of the control sample (M0)—cellulose (M1)—cellulose:extractant (quaternary phosphonium salts).

When phosphonium salts were used as extractants, DDTPPBr showed a total antimicrobial effect at the functionalization ratio Cel:DDTPPBr = 1:0.1, while HDTBPBr showed a very good antimicrobial effect starting with the functional ratio Cel:HDTBPBr = 1:0.3, but did not show total bactericidal effect even at the ratio Cel:HDTBPBr = 1:0.5. Consequently, the optimal functionalization ratio for DDTPPBr is considered Cel:DDTPPBr = 1:0.1, and for HDTBPBr the optimal ratio is Cel:HDTBPBr = 1:0.5.

B.2. Antimicrobial effect of phosphonium materials on reference microbial strains

In the case of phosphonium salt extractants, they differ both in the length of the chain of the alkyl substituent grafted on the basic structure of the cellulose (dodecyl and hexadecyl) and in the different quaternization substituent (triphenyl and tributyl, respectively). The calculated microbial inhibition rates for each material studied are shown in Table 3.

Table 3. Inhibition rate of microbial growth when using phosphonium materials.

Material	Cel:Extractant Ratio	Inhibition Rate (%)			
		Staphylococcus aureus ATCC 25923	*Pseudomonas aeruginosa* ATCC 27853	*Escherichia coli* ATCC 25922	*Candida albicans* ATCC 10231
Cel:DDTPPBr	1:0.012	100.0	32.0	100.0	100.0
	1:0.1	100.0	39.8	100.0	100.0
Cel:HDTPPBr	1:0.012	30.7	8.4	42.7	65.3
	1:0.05	58.7	20.7	44.9	67.0
	1:0.1	69.7	21.6	52.6	77.9
	1:0.3	100.0	26.6	87.3	100.0
	1:0.5	100.0	42.6	89.1	100.0

In the case of DDTPPBr, the antibacterial and antifungal effect was manifested with an inhibition rate of 100%, regardless of the functionalization ratio, except for the plates inoculated with *P. aeruginosa*. In the case of *P. aeruginosa*, the 10-fold increase in functional ratio did not lead to a substantial increase in the rate of bacterial inhibition, which increased only from 32% to 39.8%. This aspect can be explained by the resistance of *P. aeruginosa* species [58] to numerous agents with bactericidal effects and the need to use a quantity of extractant for functionalization that is not economically justified.

When using materials with HDTBPBr, the total bactericidal effect was obtained for *S. aureus* strain at the functionalization ratio Cel:HDTBPBr = 1:0.3. The total antifungal effect was obtained in the case of this material, also at the functionalization ratio Cel:HDTBPBr = 1:0.3. This could be explained by the resemblance of the cell walls of Gram-positive bacteria to the fungal wall, at least in terms of the wall structure of *C. albicans* species [51,52]. It was also found that the rate of inhibition of bacterial and fungal growth, respectively, is directly proportional to the functionalization ratio. In both Gram-negative bacteria tested, HDTBPBr failed to reach the maximum inhibition rate. On *P. aeruginosa*, even if the Cel:HDTBPBr functionalization ratio was increased 10-fold, from 0.01 to 0.1, the inhibition rate increased from 6% to 21.6%. Respectively, as the functional ratio increased from 0.1 to 0.5, the inhibition rate increased from 21.6% to 42.6%. The same effect was observed on the *E. coli* strain when the increase of the functionalization ratio by 50 times (from 0.01 to 0.5) determined only a doubling of the increase of the inhibition rate, from 42.7% to 89.1%.

DDTPPBr is the extractant whose hydrophobicity is much higher than that of HDTBPBr because even if the alkyl substituent is shorter in the case of DDTPPBr (dodecyl versus hexadecyl), the phenyl substituent with which this extractant is quaternized has a high degree of hydrophobicity compared to the butyl substituent from HDTBPBr. On the other hand, being shorter, the alkyl segment causes a smaller distance of the pendant group from the basic structure of cellulose, which again increases the antimicrobial effect. All these findings are supported by data from the literature that emphasize the role of the hydrophilic-hydrophobic balance of the bacterial cell wall and the hydrophobicity of the antibacterial agent in the manifestation of the bactericidal effect [53,55,56,61]. Moreover, the electrostatic attraction between the functionalized cellulose and the bacterial cell membrane plays an essential role in the bactericidal effect, as the strong electrostatic attraction forces in the adsorbed polyelectrolyte and the lipid molecules in the bacterial cell membrane ultimately destroy the bacterial cell [61,62].

C. Materials obtained by functionalizing cellulose with sulfur compounds

C.1. Studies to determine the optimal cellulose:extractant ratio

Compared to the analysis of the M1 control sample, when non-functionalized cellulose allowed the growth of bacteria on the surface of the culture medium, it can be concluded that by functionalizing cellulose with MBT and THIO, increasing functionalization ratio, bactericidal capacity is positively influenced (Figure 6).

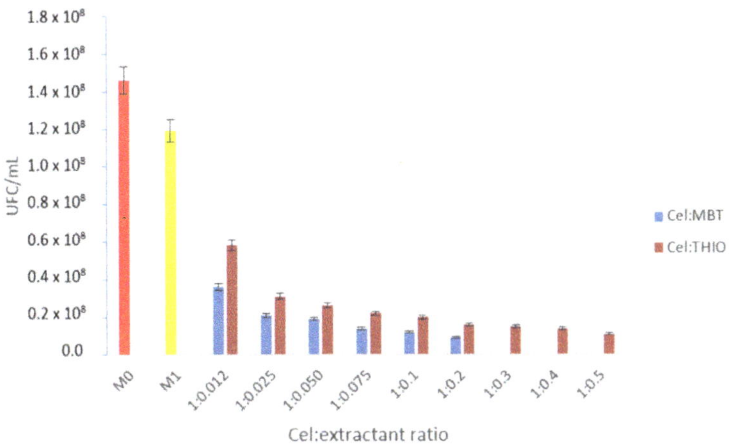

Figure 6. Comparison of the control sample (M0)—cellulose (M1)—cellulose:extractant (compounds containing sulfur).

After examining the series of microbial cultures in which MBT was used as an extractant, it was found that there was no microbial growth at the ratio Cel:MBT = 1:0.3; there was no microbial growth. Thus, the ratio that showed total bactericidal effect and is considered as optimal working ratio is 1:0.3.

For the THIO extractant, regardless of the studied functionalization ratio, no total bactericidal effect was obtained, proven by the maximum calculated bactericidal inhibition rate of 90.8% for the functionalization ratio of 1:0.5. In order to establish the optimal working ratio, given that we cannot talk about the total bactericidal effect, we considered the optimal ratio Cel:THIO = 1:0.5.

C.2. Antimicrobial effect of sulfur materials on reference microbial strains

The efficiency of the materials obtained by functionalizing cellulose with sulfur-containing extractants is expressed, in the form of the rate of microbial inhibition, in Table 4.

The functionalization of cellulose with MBT or THIO has led to the improved antibacterial and antifungal activity of native cellulose, proven by obtaining a higher inhibition rate for Cel-MBT and Cel-THIO materials than native cellulose, even for the minimum functionalization ratio Cel:extractant = 1:0.012.

We compared the efficiency of the functionalized material with MBT on the tested species. We find that the antimicrobial effect is better in the case of Gram-positive bacterial and fungal species, proven by the 100% inhibition rate achieved for a functionalization ratio Cel:MBT = 1:0.1 for both *S. aureus* and *C. albicans*. The similar manifestation of the antimicrobial effect exerted by the two species under the action of the MBT-based materials tested is suggestive, as the inhibition rate calculated for the functional ratio 1:0.05 is approximately equal (62.9% for *S. aureus* and 63.2% for *C. albicans*). When the effect of these sulfur-containing materials on Gram-negative bacteria was studied, their efficacy was found to be directly related to the increase in the amount of extractant and to the increase in functionalization ratio, but the maximum bacterial inhibition rate was obtained only for *E. coli* at a ratio Cel:MBT = 1:0.5. The functionalization of cellulose with THIO, even at the highest functionalization ratio, had no total antimicrobial effect. We could rather speak of the bacteriostatic or fungistatic effect in the case of materials functionalized with thiourea. Suppose we consider sufficient the manifestation of the bacteriostatic or fungistatic effect. In that case, it is found that the newly synthesized biomaterial prevented the growth of bacteria without necessarily killing them, observing an inactivation of microbial strain growth, as shown by the analysis of inhibition rates presented in Table 4.

Table 4. Microbial growth inhibition rate when using sulfur materials.

Material	Cel:Extractant Ratio	Inhibition Rate (%)			
		Staphylococcus aureus ATCC 25923	*Pseudomonas aeruginosa* ATCC 27853	*Escherichia coli* ATCC 25922	*Candida albicans* ATCC 10231
Cel:MBT	1:0.012	25.6	10.0	20.5	54.0
	1:0.05	62.9	13.7	22.1	63.2
	1:0.1	100.0	19.5	50.4	100.0
	1:0.3	100.0	36.2	65.4	100.0
	1:0.5	100.0	41.0	100.0	100.0
Cel:THIO	1:0.012	20.7	12.9	28.9	52.5
	1:0.05	27.7	20.7	49.6	55.8
	1:0.1	39.3	39.6	51.0	54.5
	1:0.3	75.3	44.5	52.4	60.1
	1:0.5	76.0	45.1	55.5	60.9

In Table 5, we synthesized a comparison between the antimicrobial effect (regarding the inhibition of microbial growth) of the materials synthesized in the present paper and other studies.

Table 5. Comparison between cellulose derivative and antimicrobial effect.

Material	Antimicrobial Effect (Inhibition Rate %) upon Microbial Strain	References
Cellulose with octadecyldimethyl(3-trimethoxysilylpropyl) ammonium chloride	96–99% on *E. coli* >99% on *S. aureus*	[63]
Silver–cellulose fiber sheets	91.58–99.46% on *S. aureus* 93.6–99.78% on *E. coli* 97.5–100% on *C. albicans*	[43]
Propane sulfonated chitosan	78.8% on *Phomopsis asparagi* 80.2% on *Fusarium oxysporum*	[44]
Dipropane sulfonated chitosan	82.2% on *Phomopsis asparagi* 94% on *Fusarium oxysporum*	[44]
Gold nanoparticles	5.4–20.0% on *E. coli* 4.6–16.3% on *S. aureus*	[45]
Gold nanocapsules	23.7–40.0% on *E. coli* 18.6–34.9% on *S. aureus*	[45]

Table 5. *Cont.*

Material	Antimicrobial Effect (Inhibition Rate %) upon Microbial Strain	References
Chitosan derivatives based upon quaternary ammonium salt	100% on *S. aureus* * 13.1–60.3% on *P. aeruginosa* * 25.8–100.0% on *E. coli* * 76.0–100.0% on *C. parapsilosis* *	[38]
Chitosan derivatives based upon phosphonium salt	100% on *S. aureus* * 28.2–48.6% on *P. aeruginosa* * 20.2–100.0 on *E. coli* * 100.0% on *C. parapsilosis* *	[38]
Chitosan derivatives based upon sulfur compound	38.6–100% on *S. aureus* * 22.8–46.7% on *P. aeruginosa* * 25.8–72.3% on *E. coli* * 80.6–100.0% on *C. parapsilosis* *	[38]
Cellulose derivatives based upon quaternary ammonium salt	100.0% on *S. aureus* * 17.0–54.8% on *P. aeruginosa* * 53.4–100.0% on *E. coli* * 100% on *C. albicans* *	present study
Cellulose derivatives based upon phosphonium salt	30.7–100.0% on *S. aureus* * 8.4–42.6% on *P. aeruginosa* * 42.7–100.0% on *E. coli* * 65.3–100.0% on *C. albicans* *	present study
Cellulose derivatives based upon sulfur compound	20.7–100.0% on *S. aureus* * 10.0–45.1% on *P. aeruginosa* * 20.5–100.0% on *E. coli* * 52.5–100.0% on *C. albicans* *	present study

* inhibition rate value is depending on the functionalization ratio and the tested material.

4. Conclusions

In the context of the resistance of microbial strains to most bactericidal agents known to date, the development of new antibacterial materials or at least with bacteriostatic properties, starting from biomaterials, has opened new opportunities to inhibit microbial adhesion and limit their transmission. As the polyolic nature of cellulose allows it to bind a series of compounds with chemical functions that can modulate the hydrophilic/lipophilic balance so that it can provide improved properties, we are determined to consider this biopolymer as a good support material applicable in many fields. For example, with regard to the use of cellulose derivatives in food preservation technology, the identification of new biomaterials with antibacterial properties is one of the most promising trends. It is even desirable to obtain such antibacterial materials that are biodegradable, which is why the study of cellulose is desirable. Moreover, the antimicrobial potential of cellulose functionalized in wastewater treatment should not be neglected, especially in terms of water disinfection, as evidenced by the results obtained in this study.

The impregnation technique used in this paper for the functionalization of cellulose is an easy way to obtain cellulose derivatives without involving the production of reaction by-products. On the other hand, in the case of all cellulose derivatives obtained by impregnation, increased antimicrobial properties were obtained compared to native cellulose, which leads to the conclusion that this impregnation technique can be used successfully to obtain new materials with very good antimicrobial effect.

Growth inhibition of *S. aureus*, *E. coli*, and *C. albicans* strains was determined by using materials with DDTMABr, TDTMABr, and DDTPPBr, regardless of the functionalization ratio. Materials obtained by functionalizing cellulose with quaternary ammonium and phosphonium salts showed substantial antibacterial capacity against *S. aureus*, *E. coli*, *C. albicans* (up to 100%), and *P. aeruginosa* (maximum 54.8%) even at very low concentrations of surface-immobilized antimicrobial agent. In the case of materials obtained by functionalization with quaternary ammonium salts, the antibacterial activity was dependent on the alkyl chain's length, which distances the quaternary salt's pendant group from the basic structure of the cellulose. In this sense, the antimicrobial effect was proved in the order of TDTMABr > DDTMABr > HDTMACl, respecting at the same time the increase of the ratio between the support material and extractant.

In the case of materials obtained by the functionalization of cellulose with phosphonium quaternary salts, the antibacterial activity was strongly influenced by the hydrophobic character of the pendant group (phenyl compared to butyl), in addition to the distance given by the alkyl substituent of the quaternary salt (dodecyl compared to hexadecyl) was added. The best antimicrobial effect was observed when it was used DDTPPBr like extractant, even at lower support:extractant ratio = 1:0.012.

Materials obtained by functionalizing cellulose with sulfur extractants have generally shown a bacteriostatic or fungistatic effect, with a maximum of 76% inhibition rate upon *S. aureus* at the highest functionalized support:extractant ratio = 1:0.5. The exception was in the case of Cel-MBT when the maximum inhibition rate on *S. aureus* and *C. albicans* was reached at the functionalization ratio Cel:MBT = 1:0.3.

The abundance of polysaccharides can effectively prepare antibacterial materials either by enhancing or modulating natural antibacterial properties (such as chitosan) or by de novo imparting these properties to cellulose or other biopolymers makes these biomaterials increasingly often studied, and their applicability increasingly varied. The combination of antibacterial activity with other well-known properties of these polysaccharides opens up good perspectives for the future development of such innovative functional materials.

Author Contributions: N.S.N., C.A., M.C., C.P., D.D.-S. and D.M. carried out experiments, P.N. and C.M.D. analyzed the results, N.D., A.N. and N.S.N. wrote the paper. All authors have read and agreed to the published version of the manuscript.

Funding: Not applicable.

Institutional Review Board Statement: Not applicable.

Informed Consent Statement: Not applicable.

Data Availability Statement: Not applicable.

Conflicts of Interest: The authors declare no conflict of interest.

References

1. Shanks, R.A. 2—Chemistry and structure of cellulosic fibres as reinforcements in natural fibre composites. In *Natural Fibre Composites*; Hodzic, A., Shanks, R., Eds.; Woodhead Publishing: Sawston, UK, 2014; pp. 66–83.
2. Malmström, E.; Carlmark, A. Controlled grafting of cellulose fibres—An outlook beyond paper and cardboard. *Polym. Chem.* **2012**, *3*, 1702–1713. [CrossRef]
3. Moriana, R.; Vilaplana, F.; Ek, M. Cellulose Nanocrystals from Forest Residues as Reinforcing Agents for Composites: A Study from Macro- to Nano-Dimensions. *Carbohydr. Polym.* **2016**, *139*, 139–149. [CrossRef] [PubMed]
4. Malucelli, L.C.; Lacerda, L.G.; Dziedzic, M.; da Silva Carvalho Filho, M.A. Preparation, properties and futureperspectives of nanocrystals from agro-industrial residues: A review of recentresearch. *Rev. Environ. Sci. Bio/Technol.* **2017**, *16*, 131–145. [CrossRef]
5. Khattak, S.; Wahid, F.; Liu, L.-P.; Jia, S.-R.; Chu, L.-Q.; Xie, Y.-Y.; Li, Z.-X.; Zhong, C. Applications of cellulose and chitin/chitosan derivatives and composites as antibacterial materials: Current state and perspectives. *Appl. Microbiol. Biotechnol.* **2019**, *103*, 1989–2006. [CrossRef] [PubMed]
6. Kupnik, K.; Primožič, M.; Kokol, V.; Leitgeb, M. Nanocellulose in Drug Delivery and Antimicrobially Active Materials. *Polymers* **2020**, *12*, 2825. [CrossRef] [PubMed]
7. Lavoine, N.; Desloges, I.; Dufresne, A.; Bras, J. Microfibrillated cellulose—Its barrier properties and applications in cellulosic materials: A review. *Carbohydr. Polym.* **2012**, *90*, 735–764. [CrossRef] [PubMed]
8. Wu, J.; Zheng, Y.; Song, W.; Luan, J.; Wen, X.; Wu, Z.; Chen, X.; Wang, Q.; Guo, S. In situ synthesis of silver-nanoparticles/bacterial cellulose composites for slow-released antimicrobial wound dressing. *Carbohydr. Polym.* **2014**, *102*, 762–771. [CrossRef]
9. Jonoobi, M.; Oladi, R.; Davoudpour, Y.; Oksman, K.; Dufresne, A.; Hamzeh, Y.; Davoodi, R. Different preparation methods and properties of nanostructured cellulose from various natural resources and residues: A review. *Cellulose* **2015**, *22*, 935–969. [CrossRef]

10. Kafle, K.; Shin, H.; Lee, C.M.; Park, S.; Kim, S.H. Progressive structural changes of Avicel, bleached softwood and bacterial cellulose during enzymatic hydrolysis. *Sci. Rep.* **2015**, *5*, 15102. [CrossRef]
11. Pal, S.; Nisi, R.; Stoppa, M.; Licciulli, A. Silver-Functionalized Bacterial Cellulose as Antibacterial Membrane for Wound-Healing Applications. *ACS Omega* **2017**, *2*, 3632–3639. [CrossRef]
12. Zheng, L.; Li, S.; Luo, J.; Wang, X. Latest Advances on Bacterial Cellulose-Based Antibacterial Materials as Wound Dressings. *Front. Bioeng. Biotechnol.* **2020**, *8*, 1334. [CrossRef] [PubMed]
13. Bano, S.; Negi, Y.S. Studies on cellulose nanocrystals isolated from groundnut shells. *Carbohydr. Polym.* **2017**, *157*, 1041–1049. [CrossRef] [PubMed]
14. Habibi, Y.; Lucia, L.A.; Rojas, O.J. Cellulose Nanocrystals: Chemistry, Self-Assembly, and Applications. *Chem. Rev.* **2010**, *110*, 3479–3500. [CrossRef] [PubMed]
15. George, J.; Sabapathi, S.N. Cellulose nanocrystals: Synthesis, functional properties, and applications. *Nanotechnol. Sci. Appl.* **2015**, *8*, 45. [CrossRef]
16. Wang, S.; Lu, A.; Zhang, L. Recent advances in regenerated cellulose materials. *Prog. Polym. Sci.* **2016**, *53*, 169–206. [CrossRef]
17. Siqueira, G.; Bras, J.; Dufresne, A. Cellulosic Bionanocomposites: A Review of Preparation, Properties and Applications. *Polymers* **2010**, *2*, 728. [CrossRef]
18. Bruel, C.; Tavares, J.R.; Carreau, P.J.; Heuzey, M.-C. The structural amphiphilicity of cellulose nanocrystals characterized from their cohesion parameters. *Carbohydr. Polym.* **2019**, *205*, 184–191. [CrossRef]
19. Sperandeo, P.; Bosco, F.; Clerici, F.; Polissi, A.; Gelmi, M.L.; Romanelli, A. Covalent Grafting of Antimicrobial Peptides onto Microcrystalline Cellulose. *ACS Appl. Bio Mater.* **2020**, *3*, 4895–4901. [CrossRef]
20. Baghaei, B.; Skrifvars, M. All-Cellulose Composites: A Review of Recent Studies on Structure, Properties and Applications. *Molecules* **2020**, *25*, 2836. [CrossRef]
21. Coma, V.; Freire, V.; Silvestre, A. *Recent Advances on the Development of Antibacterial Polysaccharide-Based Materials*; Springer: Cham, Switzerland, 2015; pp. 1751–1803. [CrossRef]
22. Kar, M.; Chourasiya, Y.; Maheshwari, R.; Tekade, R.K. Chapter 2—Current Developments in Excipient Science: Implication of Quantitative Selection of Each Excipient in Product Development. In *Basic Fundamentals of Drug Delivery*; Tekade, R.K., Ed.; Academic Press: Cambridge, MA, USA, 2019; pp. 29–83.
23. Klemm, D.; Heublein, B.; Fink, H.-P.; Bohn, A. Cellulose: Fascinating Biopolymer and Sustainable Raw Material. *Angew. Chem. Int. Ed.* **2005**, *44*, 3358–3393. [CrossRef]
24. Li, Q.; Wang, Y.; Wu, Y.; He, K.; Li, Y.; Luo, X.; Li, B.; Wang, C.; Liu, S. Flexible cellulose nanofibrils as novel pickering stabilizers: The emulsifying property and packing behavior. *Food Hydrocoll.* **2019**, *88*, 180–189. [CrossRef]
25. Huang, S.; Liu, X.; Chang, C.; Wang, Y. Recent developments and prospective food-related applications of cellulose nanocrystals: A review. *Cellulose* **2020**, *27*, 2991–3011. [CrossRef]
26. Tamilselvi, A.; Jayakumar, G.C.; Sri Charan, K.; Sahu, B.; Deepa, P.R.; Kanth, S.V.; Kanagaraj, J. Extraction of cellulose from renewable resources and its application in leather finishing. *J. Clean. Prod.* **2019**, *230*, 694–699. [CrossRef]
27. Salmeia, K.; Jovic, M.; Ragaišienė, A.; Rukuižienė, Ž.; Milašius, R.; Mikučionienė, D.; Gaan, S. Flammability of Cellulose-Based Fibers and the Effect of Structure of Phosphorus Compounds on Their Flame Retardancy. *Polymers* **2016**, *8*, 293. [CrossRef]
28. Ullah, H.; Wahid, F.; Santos, H.A.; Khan, T. Advances in biomedical and pharmaceutical applications of functional bacterial cellulose-based nanocomposites. *Carbohydr. Polym.* **2016**, *150*, 330–352. [CrossRef]
29. Ayala Valencia, G.; Cristina de Oliveira Vercik, L.; Ferrari, R.; Vercik, A. Synthesis and characterization of silver nanoparticles using water-soluble starch and its antibacterial activity on *Staphylococcus aureus*. *Starch—Stärke* **2013**, *65*, 931–937. [CrossRef]
30. Kalwar, K.; Bhutto, M.A.; Li, D.L.; Shan, D. Cellulose based nanofabrication; immobilization of silver nanoparticales and its size effect against *Escherichia coli*. *Mater. Res. Express* **2017**, *4*, 6. [CrossRef]
31. Wang, J.; Liu, W.; Li, H.; Huili, W.; Wang, Z.; Zhou, W.; Liu, H. Preparation of cellulose fiber–TiO_2 nanobelt–silver nanoparticle hierarchically structured hybrid paper and its photocatalytic and antibacterial properties. *Chem. Eng. J.* **2013**, *228*, 272–280. [CrossRef]
32. Bolhuis, G.K.; Armstrong, N.A. Excipients for direct compaction—An update. *Pharm. Dev. Technol.* **2006**, *11*, 111–124. [CrossRef]
33. Liu, S.; Zeng, J.; Tao, D.; Na, N. Microfiltration performance of regenerated cellulose membrane prepared at low temperature for wastewater treatment. *Cellulose* **2010**, *17*, 1159–1169. [CrossRef]
34. Fauzani, D.; Notodarmojo, S.; Handajani, M.; Helmy, Q.; Kardiansyah, T. Cellulose in natural flocculant applications: A review. *J. Phys. Conf. Ser.* **2021**, *2047*, 012030. [CrossRef]
35. Suopajärvi, T.; Liimatainen, H.; Hormi, O.; Niinimäki, J. Coagulation–flocculation treatment of municipal wastewater based on anionized nanocelluloses. *Chem. Eng. J.* **2013**, *231*, 59–67. [CrossRef]
36. Khiari, R.; Dridi-Dhaouadi, S.; Aguir, C.; Mhenni, M.F. Experimental evaluation of eco-friendly flocculants prepared from date palm rachis. *J. Environ. Sci.* **2010**, *22*, 1539–1543. [CrossRef]
37. Onwukamike, K.N.; Grelier, S.; Grau, E.; Cramail, H.; Meier, M.A.R. Critical Review on Sustainable Homogeneous Cellulose Modification: Why Renewability Is Not Enough. *ACS Sustain. Chem. Eng.* **2019**, *7*, 1826–1840. [CrossRef]
38. Ardean, C.; Davidescu, C.M.; Nemeş, N.S.; Negrea, A.; Ciopec, M.; Duteanu, N.; Negrea, P.; Duda-Seiman, D.; Muntean, D. Antimicrobial Activities of Chitosan Derivatives. *Pharmaceutics* **2021**, *13*, 1639. [CrossRef]

39. Cortina, J.L.; Miralles, N.; Sastre, A.; Aguilar, M.; Profumo, A.; Pesavento, M. Solvent-impregnated resins containing di-(2,4,4-trimethylpentyl) phosphonic acid I. Comparative study of di-(2,4,4-trimethylpentyl)phosphinic acid adsorbed into Amberlite XAD-2 and dissolved in toluene. *React. Polym.* **1993**, *21*, 89–101. [CrossRef]
40. Cortina, J.L.; Miralles, N.; Sastre, A.M.; Aguilar, M.; Profumo, A.; Pesavento, M. Solvent-impregnated resins containing di-(2,4,4-trimethylpentyl) phosphinic acid II. Study of the distribution equilibria of Zn(II), Cu(II) and Cd(II). *React. Polym.* **1993**, *21*, 103–116. [CrossRef]
41. Cortina, J.L.; Miralles, N.; Aguilar, M.; Sastre, M. Solvent impregnated resins containing di(2-ethylhexyl)phosphoric acid. I. Preparation and study of the retention and distribution of the extractant on the resin. *Solvent Extr. Ion Exch.* **1994**, *12*, 349–369. [CrossRef]
42. Warshawsky, A. Extraction with Solvent-Impregnated Resins. In *Ion Exchange and Solvent Extraction;* Marinsky, J.A., Marcus, Y., Eds.; Marcel Dekker: New York, NY, USA, 1981; Volume 8, pp. 229–310.
43. Csóka, L.; Božanić, D.K.; Božanić, D.K.; Nagy, V.; Dimitrijević-Branković, S.; Luyt, A.S.; Grozdits, G.; Djoković, V. Viscoelastic properties and antimicrobial activity of cellulose fiber sheets impregnated with Ag nanoparticles. *Carbohydr. Polym.* **2012**, *90*, 1139–1146. [CrossRef]
44. Luan, F.; Wei, L.; Zhang, J.; Mi, Y.; Dong, F.; Li, Q.; Guo, Z. Antioxidant Activity and Antifungal Activity of Chitosan Derivatives with Propane Sulfonate Groups. *Polymers* **2018**, *10*, 395. [CrossRef]
45. Belbekhouche, S.; Bousserrhine, N.; Alphonse, V.; Le Floch, F.; Mechiche, Y.C.; Menidjel, I.; Carbonnier, B. Chitosan based self-assembled nanocapsules as antibacterial agent. *Colloids Surf. B Biointerfaces* **2019**, *181*, 158–165. [CrossRef] [PubMed]
46. Syamani, F.A.; Suryani, A. Changes in Oil Palm Frond Fiber Morphology, Cellulose Crystallinity and Chemical Functional Groups during Cellulose Extraction Phases. *Chem. Mater. Res.* **2015**, *7*, 113–121.
47. Nazir, F.; Iqbal, M. Synthesis, Characterization and Cytotoxicity Studies of Aminated Microcrystalline Cellulose Derivatives against Melanoma and Breast Cancer Cell Lines. *Polymers* **2020**, *12*, 2634. [CrossRef] [PubMed]
48. Tarchoun, A.F.; Trache, D.; Klapötke, T.M.; Derradji, M.; Bessa, W. Ecofriendly isolation and characterization of microcrystalline cellulose from giant reed using various acidic media. *Cellulose* **2019**, *26*, 7635–7651. [CrossRef]
49. Negrea, A.; Gabor, A.; Davidescu, C.M.; Ciopec, M.; Negrea, P.; Duteanu, N.; Barbulescu, A. Rare Earth Elements Removal from Water Using Natural Polymers. *Sci. Rep.* **2018**, *8*, 316. [CrossRef] [PubMed]
50. Hospodarova, V.; Singovszka, E.; Stevulova, N. Characterization of Cellulosic Fibers by FTIR Spectroscopy for Their Further Implementation to Building Materials. *Am. J. Anal. Chem.* **2018**, *09*, 303–310. [CrossRef]
51. Brown, L.; Wolf, J.M.; Prados-Rosales, R.; Casadevall, A. Through the wall: Extracellular vesicles in Gram-positive bacteria, mycobacteria and fungi. *Nat. Rev. Microbiol.* **2015**, *13*, 620–630. [CrossRef]
52. Carolus, H.; Van Dyck, K.; Van Dijck, P. *Candida albicans* and *Staphylococcus* Species: A Threatening Twosome. *Front. Microbiol.* **2019**, *10*, 2162. [CrossRef]
53. Kumariya, R.; Sood, S.K.; Rajput, Y.S.; Saini, N.; Garsa, A.K. Increased membrane surface positive charge and altered membrane fluidity leads to cationic antimicrobial peptide resistance in *Enterococcus faecalis*. *Biochim. Biophys. Acta (BBA)—Biomembr.* **2015**, *1848*, 1367–1375. [CrossRef]
54. Russell, A.D. Similarities and differences in the responses of microorganisms to biocides. *J. Antimicrob. Chemother.* **2003**, *52*, 750–763. [CrossRef]
55. Krasowska, A.; Sigler, K. How microorganisms use hydrophobicity and what does this mean for human needs? *Front. Cell. Infect. Microbiol.* **2014**, *4*, 112. [CrossRef] [PubMed]
56. Palmer, J.; Flint, S.; Brooks, J. Bacterial cell attachment, the beginning of a biofilm. *J. Ind. Microbiol. Biotechnol.* **2007**, *34*, 577–588. [CrossRef] [PubMed]
57. Bassetti, M.; Baguneid, M.; Bouza, E.; Dryden, M.; Nathwani, D.; Wilcox, M. European perspective and update on the management of complicated skin and soft tissue infections due to methicillin-resistant *Staphylococcus aureus* after more than 10 years of experience with linezolid. *Clin. Microbiol. Infect. Off. Publ. Eur. Soc. Clin. Microbiol. Infect. Dis.* **2014**, *20* (Suppl. S4), 3–18. [CrossRef] [PubMed]
58. Ciofu, O.; Tolker-Nielsen, T. Tolerance and Resistance of *Pseudomonas aeruginosa* Biofilms to Antimicrobial Agents-How *P. aeruginosa* Can Escape Antibiotics. *Front. Microbiol.* **2019**, *10*, 913. [CrossRef] [PubMed]
59. Koshani, R.; Tavakolian, M.; van de Ven, T.G.M. Cellulose-based dispersants and flocculants. *J. Mater. Chem. B* **2020**, *8*, 10502–10526. [CrossRef] [PubMed]
60. Song, Y.; Zhang, J.; Gan, W.; Zhou, J.; Zhang, L. Flocculation Properties and Antimicrobial Activities of Quaternized Celluloses Synthesized in NaOH/Urea Aqueous Solution. *Ind. Eng. Chem. Res.* **2010**, *49*, 1242–1246. [CrossRef]
61. Thewes, N.; Loskill, P.; Jung, P.; Peisker, H.; Bischoff, M.; Herrmann, M.; Jacobs, K. Hydrophobic interaction governs unspecific adhesion of staphylococci: A single cell force spectroscopy study. *Beilstein J. Nanotechnol.* **2014**, *5*, 1501–1512. [CrossRef]
62. Chen, C.; Petterson, T.; Illergård, J.; Ek, M.; Wågberg, L. Influence of Cellulose Charge on Bacteria Adhesion and Viability to PVAm/CNF/PVAm-Modified Cellulose Model Surfaces. *Biomacromolecules* **2019**, *20*, 2075–2083. [CrossRef]
63. Andresen, M.; Stenstad, P.; Møretrø, T.; Langsrud, S.; Syverud, K.; Johansson, L.S.; Stenius, P. Nonleaching antimicrobial films prepared from surface-modified microfibrillated cellulose. *Biomacromolecules* **2007**, *8*, 2149–2155. [CrossRef]

polymers

MDPI

Article

pH-Induced 3D Printable Chitosan Hydrogels for Soft Actuation

Sheila Maiz-Fernández [1,2], Leyre Pérez-Álvarez [1,2,*], Unai Silván [1,3], José Luis Vilas-Vilela [1,2] and Senentxu Lanceros-Méndez [1,3]

1 BCMaterials (Basque Center for Materials), Applications and Nanostructures, UPV/EHU Science Park, 48940 Leioa, Spain; sheila.maiz@bcmaterials.net (S.M.-F.); unai.silvan@bcmaterials.net (U.S.); joseluis.vilas@ehu.eus (J.L.V.-V.); senentxu.lanceros@bcmaterials.net (S.L.-M.)

2 Macromolecular Chemistry Group (LABQUIMAC), Department of Physical Chemistry, Faculty of Science and Technology, University of the Basque Country, UPV/EHU, Barrio Sarriena, s/n, 48940 Leioa, Spain

3 Ikerbasque, Basque Foundation for Science, 48009 Bilbao, Spain

* Correspondence: leyre.perez@ehu.eus

Abstract: Three-dimensional (3D) printing represents a suitable technology for the development of biomimetic scaffolds for biomedical and tissue engineering applications. However, hydrogel-based inks' printability remains a challenge due to their restricted print accuracy, mechanical properties, swelling or even cytotoxicity. Chitosan is a natural-derived polysaccharide that has arisen as a promising bioink due to its biodegradability, biocompatibility, sustainability and antibacterial properties, among others, as well as its ability to form hydrogels under the influence of a wide variety of mechanisms (thermal, ionic, pH, covalent, etc.). Its poor solubility at physiological pH, which has traditionally restricted its use, represents, on the contrary, the simplest way to induce chitosan gelation. Accordingly, herein a NaOH strong base was employed as gelling media for the direct 3D printing of chitosan structures. The obtained hydrogels were characterized in terms of morphology, chemical interactions, swelling and mechanical and rheological properties in order to evaluate the influence of the gelling solution's ionic strength on the hydrogel characteristics. Further, the influence of printing parameters, such as extrusion speed (300, 600 and 800 mm/min) and pressure (20–35 kPa) and the cytocompatibility were also analyzed. In addition, printed gels show an electro-induced motion due to their polycationic nature, which highlights their potential as soft actuators and active scaffolds.

Keywords: chitosan; 3D printing; hydrogels

Citation: Maiz-Fernández, S.; Pérez-Álvarez, L.; Silván, U.; Vilas-Vilela, J.L.; Lanceros-Méndez, S. pH-Induced 3D Printable Chitosan Hydrogels for Soft Actuation. *Polymers* **2022**, *14*, 650. https://doi.org/10.3390/polym14030650

Academic Editor: Arn Mignon

Received: 20 January 2022
Accepted: 4 February 2022
Published: 8 February 2022

1. Introduction

Three-dimensional (3D) printing has experienced a fast development in the last decade and has become a promising manufacturing method in the biomedical and tissue engineering fields [1]. This technology represents a great advance in comparison with traditional methods, such as electrospinning or solvent casting, which typically provide uncontrolled geometries, restrict biomaterial functionality and lead to uncontrolled cell distributions, which consequently limit their use in biological applications [2]. On the contrary, 3D printing allows the production of complex and precise three-dimensional shapes and structures that are of great interest in the development of medical devices and tissue engineering scaffolds [3].

Polymeric hydrogels, which are three-dimensional structures formed by the chemical or physical crosslinking of polymeric chains [4], are able to maintain a large amount of water inside their structure without dissolving. This confers on them the ability to imitate natural tissues, and consequently, hydrogel scaffolds are considered suitable biomaterials for soft tissue regeneration [5].

Hydrogels can be based on natural or synthetic polymers; however, the use of natural polymers, such as polysaccharides, is preferred over traditional biosynthetic ones (poly-L-lactide or polyethyleneglycol, among others) due to their enhanced biological response [6], as well as their renewable nature, which is required in the scope of sustainable economy policies [7].

Although the 3D printing of polysaccharides has made steady progress over the past decade, it remains a challenge due to the high water content of hydrogels inks. Indeed, in order to be printable, polysaccharide inks still have to overcome their low mechanical properties and poor printing accuracy [8].

In this regard, one of the most important considerations in improving printability is that a fast in situ sol–gel transition must occur after the printing of the ink. Thus, before being deposited on the printing surface the bioink is presented in a liquid state, and once it is printed, as a consequence of applying an external change, this becomes a gel by an in situ-promoted physically or covalently induced crosslinking. In addition, printed soft structures must be long-lived and robust enough to self-support and accurately reproduce the designed geometry. Finally, an adequate ink must also exhibit at the same time structural integrity and controlled biodegradation at physiological conditions [9]. Consequently, research into new printing processes for polysaccharide-based inks that meet these stringent requirements is still necessary in order to make progress in this scope.

Chitosan, a biopolymer mainly obtained from chitin, which is the second-most-abundant natural polymer after cellulose, is a polysaccharide that has been studied in recent years as ink because of its biocompatibility, biodegradability, mucoadhesion, ease of functionalization and suitability for injection [10]. This polysaccharide is a weak polycation with a pKa value of 6.5 [11], and it is formed by two units (*N*-acetyl-glucosamine and glucosamine) which are randomly ordered. Since chitosan is not ionized in a pH above its pKa values, it is able to create strong inter- and intramolecular interactions, causing the precipitation of the polymer. This precipitation is related with the neutralization of the amino groups present in its chemical structure. Neutralization leads to the disappearance of the repulsive electrostatic forces that keep the polymers in the solution at acidic pH, resulting in the formation of intramolecular–intermolecular hydrogen bonds and hydrophobic interactions responsible for the precipitation of the polysaccharide [12].

Therefore, chitosan is not soluble in most solvents and conditions, including aqueous solutions or physiological pH. This fact has restricted its applicability as bioink, although some comparative studies on hydrogels' cellular proliferation and differentiation properties have highlighted the interest of chitosan instead of alginate, which is one of the most employed polysaccharide hydrogel inks [13].

Specifically, chitosan is an advantageous polysaccharide ink because it undergoes sol–gel transitions that can be induced by different gelation processes, which confer on it versatility as a bioink [14]. In this line, chitosan can be ionotropically crosslinked with a large variety of compounds, such as citrates, and phosphates such as tripolyphosphate (TPP) are the most employed ones [15]. Complexation with other polyelectrolytes and concretely polyanions, such as alginate or hyaluronic acid, has also been demonstrated to be an efficient way to produce chitosan gelation and printable polycomplex hydrogels [16].

Indeed, the ease of functionalization of chitosan offers this polysaccharide great versatility for the development of new bioinks based on specific gelation mechanisms driven by different chitosan derivatives. Examples of this can be found in the large number of publications that report the UV crosslinking of chitosan vinyl derivatives developed for 3D printing applications [17].

Further, the poor solubility properties of chitosan at physiological conditions can also be exploited to provide gelation opportunities by means of a coagulation-mediated gelation process. As explained above, chitosan in an acidic solution rapidly precipitates at pH values above its pKa, leading to the fast formation of a hydrogel as a consequence of the neutralization of its amino groups and the promotion of hydrogen bonding and hydrophobic interactions. Nevertheless, this simple approach for chitosan hydrogel printing has been poorly addressed. Almeida et al. [18] employed 3D-printed chitosan scaffolds obtained from the gelation of a high-concentrated polymer solution by contact with a mixture of

NaOH solution (8%) in ethanol (70%) to study the modulation of macrophage responses in the inflammation process. Wu et al. [19] characterized the printing process of chitosan in which in situ gelation took place by citrate action, although they observed also the positive effect of NaOH (1M) used in posterior neutralization steps on both the printability and the mechanical stability of printed complex structures [19]. Prolonged contact with NaCl solutions has been also employed as a post-printing step of chitosan bioinks in order to achieve strong mechanical properties [1]. Zhang et al. [20] also employed chitosan gelation induced by contact with a NaOH solution (3 M) and demonstrated the improvement of printed hydrogels' mechanical properties and the restriction on post-shrinkage derived from precipitation, but in this case promoted by the addition of silk particles. Bergonzni et al. [12] investigated the combination of cryogenic printing, in which the polymer is instantaneously frozen after being extruded in a cool surface, with the gelation promoted with different agents, such as ammonia gas, a strong base KOH with variable concentrations and weak bases such as carbonate or phosphate salts. Neutralization times were determined in order to guarantee the appropriate conditions (gelation agent and concentration) to reproduce and maintain 3D-printed shapes, showing, among the solutions, that KOH and Na_2CO_3 lead to the lowest gelation times [12].

In the light of these works that demonstrate the effective printability of chitosan inks without chemical modification by contact with bases, the present work reports the room temperature direct printing of chitosan by simple contact with concentrated NaOH solutions. Besides, the effect of NaOH concentration on the printing and on the physicochemical properties of printed structures was also evaluated. Printability, rheological, mechanical and swelling properties, the in vitro degradation profile and cytotoxicity were studied at different neutralization solution concentrations, optimizing the printing conditions that lead to an optimized performance of the printed biomaterial. On the other hand, the characteristic pH sensitivity of chitosan provides the hydrogels with the inherent ability to bend in response to an electrical stimulus, which additionally makes them suitable candidates for use as soft actuators and active scaffolds [21].

2. Materials and Methods

2.1. Materials

Chitosan from crab shells (Sigma-Aldrich, St. Louis, MO, USA, highly viscous, $8.7 \times 10^5 \pm 4.0 \times 10^4$ g/mol (PDI = 1.037), deacetylation degree of 85% determined by ^{1}H-NMR) was used for the synthesis of hydrogels. The average molecular weight (M_w) was measured by gel permeation chromatography equipped with refractive index (RID) and light-scattering (LS15 and LS90) detectors (HPLC Agilent Technologies, Agilent Technologies, Santa Clara, CA, USA). The employed column was a PolySep-GFC-P Linear 300 $\times$ 7.8 mm Phenomenex and acetic acid 0.15 M with 1 mL/min flow, and 20 μL injection volume was employed as eluent, followed a detector calibration against a poly(ethylene oxide) narrow standard (1.5 MDa). Acetic acid (for analysis, 99.8%) and lysozyme enzyme (from chicken egg white, ~70,000 U/mg), sodium hydroxide (pure, pharma grade) and sodium phosphate monobasic (BioXtra, $\geq$ 99.0%) were purchased from Sigma-Aldrich, St. Louis, MO, USA.

2.2. Methods

2.2.1. Hydrogel Synthesis and Printability

To prepare bulk hydrogels 1.5% (w/w) solution of chitosan was prepared adding 0.75 g of chitosan to 50 mL of 0.5% (v/v) acetic acid solution with homogeneous and controlled stirring overnight. Chitosan-based bulk hydrogels were prepared by mixing 3 g of chitosan solution with NaOH solutions of different concentrations (1, 3 and 5M) under vigorous mechanical stirring. Once chitosan solution was prepared, it was loaded into the printer syringe and printed onto a Petri dish, which had previously been covered by sodium hydroxide solution (1, 3 and 5M).

Using the CellInk HeartWare designer software, a strand of 4 cm and a square scaffold with 4 × 4 pores per side were designed (pore dimensions of 5 × 5 mm). For both strands and square scaffolds, hydrogel was printed (Cellink, INKREDIBLE+, Brighton, UK) from chitosan 1.5% (*w*/*w*) solution with a viscosity of 6092 ± 110 cP. The effect of printing speed (300, 600 and 800 mm/min), pressure (25, 30, 35 and 40 kPa) and nozzle's diameter (22 and 25 G, corresponding to 0.41 and 0.254 mm, respectively) (Adhesive Dispensing, Ltd., Milton Keynes, UK) was evaluated on the definition of the printed structures by the determination of the uniformity factor (U), expansion ratio (α) and, in the case of the square scaffolds, the size accuracy and squareness parameters.

Uniformity factor (Equation (1)) is defined as the difference between the length of the printed structure (l) and the theoretical length (L):

$$\text{Uniformity factor (U)} = l/L \tag{1}$$

Further, the expansion ratio (Equation (2)) indicates the ability of the bioink to spread over the printed surface. It is defined as the relation between the diameter of the printed filament (d) and the theoretical diameter of the nozzle (D):

$$\text{Expansion ratio } (\alpha) = d/D \tag{2}$$

The size accuracy (Equation (3)) quantifies the printing fidelity by comparing the theoretical area of the pores of the designed structure with the real area of the printed squares. This parameter is calculated taking into consideration the definition given by Di Giuseppe et al. [22]:

$$\text{Size accuracy} = 1 - (A_t - A)/A_t \tag{3}$$

where A_t is the theoretical pore area (25 mm^2), and A is the pore area of each printed square scaffold.

On the other hand, the squareness of printed scaffolds was determined by the equation defined by Ouyang et al. [23] (Equation (4)):

$$\text{Squareness} = L^2/16A \tag{4}$$

where L is the perimeter of the pore, and A is the area of the printed square scaffolds.

Each structure was printed in triplicate under the same conditions, and the estimation of the different parameters was determined by the analysis of the images of the printed structures using Fiji software.

Gelation Time

Gelation time of hydrogels is a key factor for the materials used as bioinks since this parameter might compromise printing fidelity. It was determined following the so-called inverted tube test [24]. In short, gelation time is determined as the moment when the solution stops flowing after inverting the tube. Gelation times of the hydrogels were determined at different NaOH concentrations (1, 3 and 5 M).

2.2.2. Physico-Chemical Characterization

Fourier-Transform Infrared Spectroscopy (FTIR)

Nicolet Nexus FTIR spectrometer (Thermo Scientific, Loughborough, UK) was employed to corroborate the formation of chitosan-based hydrogels. For this, previously dried hydrogels were analyzed by ATR, while in the case of pristine chitosan the analysis was carried out by KBr pellets. In both cases, the experiments were carried out at a resolution of 4 cm^{-1} and 32 scans per spectrum.

Morphological Characterization

Morphology and pore size of chitosan-based hydrogels were analyzed by scanning electron microscope (Hitachi S-4800, 150 s, 20 mA, 15 kV, zoom at ×30,000, Tokyo, Japan).

Chitosan hydrogels were lyophilized (−50 °C, 0.1 mBar) and coated with a thin gold layer before SEM characterization. Finally, the average pore size was determined using ImageJ software [25], and at least 35 pores were analyzed in each sample. The error of the distribution is represented as the mean +/− standard deviation.

In Vitro Swelling

The swelling behaviour of hydrogels was evaluated by immersing freeze-dried chitosan hydrogels (−50 °C and 0.1 mBar) in phosphate-buffered solution (PBS) (pH = 7.4) at 37 °C in order to imitate physiological media. The swelling kinetic was evaluated by measuring the absorption of water into the hydrogels gravimetrically over time, and then the swelling ratio was measured following Equation (5):

$$\text{Swelling factor} = (W_s - W_d)/W_d \quad (5)$$

where, W_s and W_d are the weights of the swollen and dried hydrogels, respectively.

Rheology

To evaluate the dynamic rheological behaviour of prepared hydrogels oscillatory rheometry was employed using, for this aim, a rheometric scientific advanced rheometric expansion system (ARES, New Castle, PA, USA) equipped with a parallel plate geometry (25 mm of diameter). The study was carried out, firstly by a shear strain sweep to determine the linear viscoelastic region of the materials. Subsequently, frequency sweep measurements were carried out from 0.1 to 500 rad/s at a fixed strain of 1% and at a gap distance of 1.5 mm at 25 °C. Both strain and frequency sweeps were measured in triplicate.

Compressive Stress–Strain Tests

Compression tests were carried out at room temperature using a texture analyzer instrument (Metrotec FTM-50, Lezo, Spain) equipped with a 20 N load cell. Measurements were carried out under progressive compression at a rate of 1 mm/min until breaking. Compression moduli were calculated from the slope of the linear portion (40–60% strain range) of the stress–strain plot.

2.2.3. Functional Characterization

In Vitro Hydrolytic and Enzymatic Biodegradation

The degradation of freshly prepared chitosan hydrogels was evaluated by immersing hydrogels in PBS solution in the presence and in the absence of lysozyme (1 mg/mL in PBS solution) at pH 7.4 and 37 °C. The quantification of the in vitro biodegradation (Equation (6)) was determined by weighing the mass loss at different time intervals:

$$\text{Mass loss } (\%) = (100 - (W_0 - W_t)/W_0) \times 100 \quad (6)$$

where W_0 is the weight of the swollen hydrogel at initial time and W_t at t_∞. Three samples were evaluated for each data point.

In Vitro Cytotoxicity Assay

Biocompatibility of the hydrogels was estimated using a live–dead assay. A total of 24 h before the assay, mouse embryonic fibroblasts (MEFs) were seeded in a 24-well plate at a density of 2.105 cells per well and cultured under standard conditions (37 °C and 5% CO_2) in complete medium (DMEM containing 10% fetal bovine serum and 1% penicillin). Chitosan hydrogels (5–10 mg) formed with NaOH solutions of increasing molarity were washed thoroughly with D-PBS, dried and sterilized under UV light for 1 h before being added to the wells containing MEFs. After 24 h, hydrogels were discarded, and cell cultures were washed with PBS and stained with Calcein-AM (2 μM), ethidium homodimer (EthD-1, 4 μM) and NucBlue (Thermo Fisher (Waltham, MA, USA, R37605, 2 drops/well). Fluorescence images of the blue (NucBlue, cell nuclei), green (calcein, cytoplasm of live cells) and red (EthD-1, nuclei of dead cells) nuclei acquired using a Leica DMi8 fluorescence microscope, and the collected data were quantified using Fiji software [25]. Cell viability was calculated as the ratio between red-stained nuclei and the total number of cells (blue-stained nuclei) in at least five images for three independent samples.

Actuator Bending Response

Actuator bending response of prepared chitosan-based hydrogels was evaluated under electrical stimuli. For this, printed chitosan hydrogel strands were placed in a 0.1 M NaCl electrolytic solution between two platinum electrodes separated with a gap distance of 2 cm. Subsequently, a potential difference of 15 V was applied with a Hewlett-Packard E3615A DC power supply (0–20 V), and images were acquired using a digital camera.

3. Results

3.1. Synthesis and Swelling of Chitosan-Based Hydrogels

Chitosan is not soluble in water or physiological media; however, a chitosan solution can be obtained in acid media when the solvent's pH is below its pKa value ($pKa_{chitosan}$~6.5) [26]. In acidic aqueous media, the protonation of the primary amine groups present in the polysaccharide backbone takes place, leading to positively charged polymer chains and favouring electrostatic repulsion between chitosan chains. However, when a strong base (NaOH) is added to the media, chitosan remains stable in the solution until the pH becomes higher than its pKa value, which leads to the formation of hydrated gel-like precipitate [27]. This gelling mechanism is driven by the neutralization of primary amine groups, which promotes extensive hydrophobic interactions and hydrogen bonds among different moieties, such as amines, hydroxides and carbonyl groups present in the polysaccharide backbone (Figure 1). Thus, chitosan hydrogels can be formed by controlling chitosan concentration as well as the pH of the precipitating solution [28].

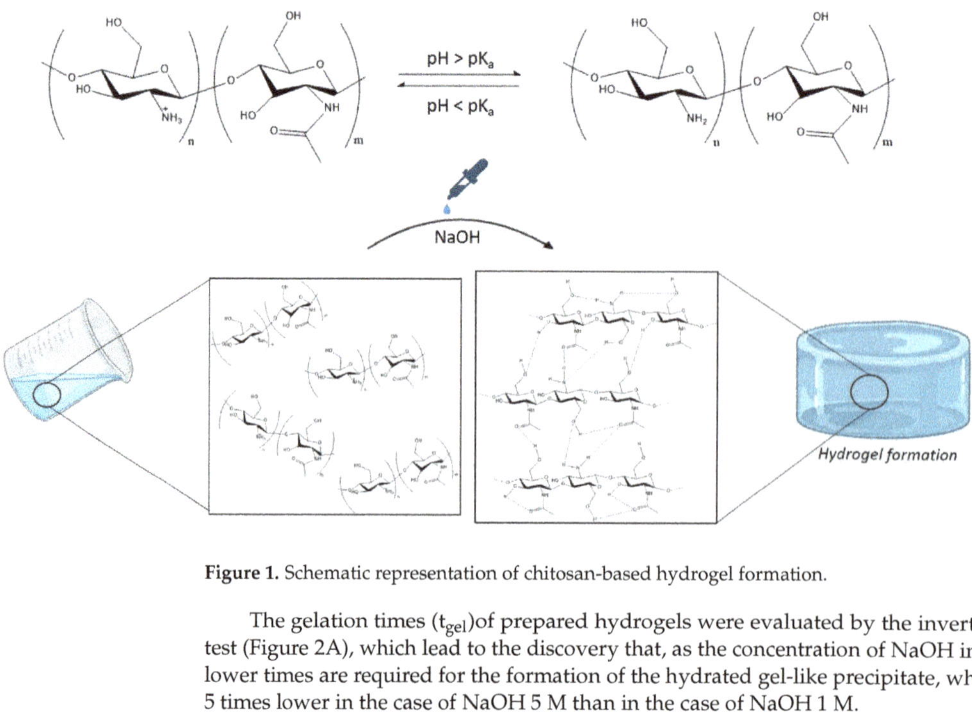

Figure 1. Schematic representation of chitosan-based hydrogel formation.

The gelation times (t_{gel})of prepared hydrogels were evaluated by the inverted tube test (Figure 2A), which lead to the discovery that, as the concentration of NaOH increases, lower times are required for the formation of the hydrated gel-like precipitate, which was 5 times lower in the case of NaOH 5 M than in the case of NaOH 1 M.

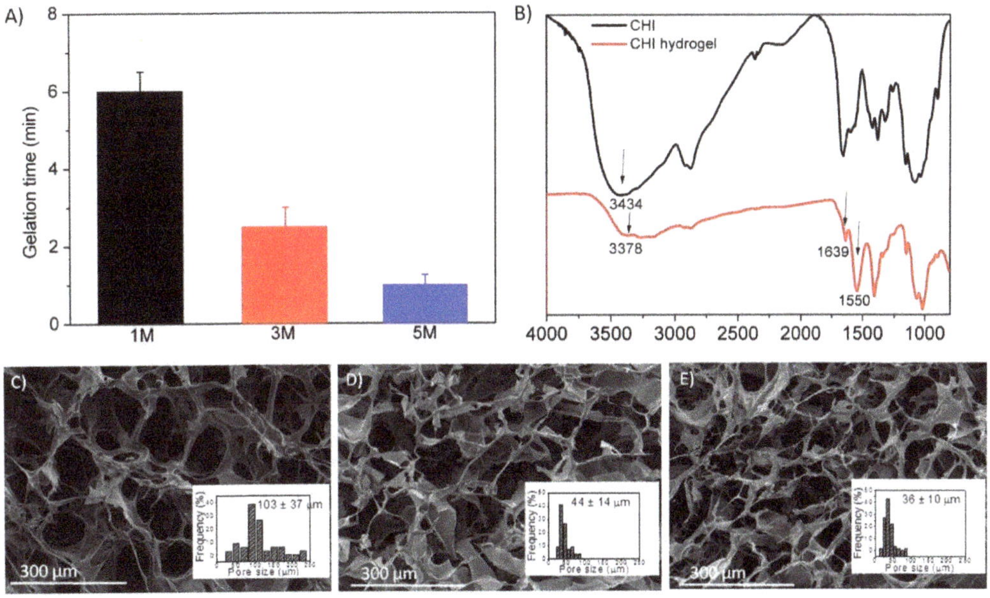

Figure 2. (**A**) Coagulation time of hydrogels prepared with different ionic strength, (**B**) FTIR spectra of pristine chitosan (black) and chitosan-based precipitate like gel (red) and SEM micrographs of hydrogels prepared by coagulation with NaOH (**C**) 1 M, (**D**) 3 M and (**E**) 5 M.

FTIR spectra of both pristine chitosan and hydrated gel-like precipitate that are compared in Figure 2B corroborated the above-described gelling mechanism. On the one hand, a slight shift to lower frequencies was observed in the band at 3434 cm^{-1} of the gelled chitosan that was assigned to O−H stretching vibrations, which points out the reinforcement of hydrogen bonding. On the other hand, an increased intensity in the bands at 1639 and 1550 cm^{-1} (assigned to N−H deformation vibrations), attributed to hydrophobic interactions and molecular chain entanglement, can be also appreciated in chitosan hydrogels [1].

Further, hydrogel formation was also evaluated by SEM micrographs, shown in Figure 2C–E (for NaOH 1, 3 and 5 M, respectively), in which it is confirmed that the interactions established between polysaccharide chains lead to interconnected porous three-dimensional structures. These results are in agreement with the literature on this type of physical hydrogels [29,30].

In addition, a slight dependence of the pore size of the hydrogels and the concentration of the NaOH solution can be observed in Figure 2C–E. Although a low concentration (1 M) of a strong base such as NaOH caused the precipitation of chitosan hydrogels, it seems to have led to a decrease in hydrogen bonds or hydrophobic interactions within the polymer, with respect to higher concentrations of NaOH (3 M and 5 M), resulting in more open networks with remarkable larger pore dimensions (~103 µm vs. ~40 µm).

Furthermore, the influence of the NaOH solution concentration and coagulation time on the swelling behaviour of the hydrogels was also studied, and collected data are shown in Figure 3A,B.

Figure 3A shows a clear dependence of hydrogel swelling on the NaOH concentration. Certainly, the high pH values of the medium (NaOH 3 or 5 M) promoted chitosan intramolecular–intermolecular interactions, resulting in a higher crosslinked network with smaller pore size, as shown in the SEM analyses, which limited water absorption and, therefore, the swelling capacity of the hydrogels. Indeed, significantly lower swelling factors were obtained for NaOH 3 M (swelling factor ~5) or NaOH 5 M (swelling factor ~3) in comparison with NaOH 1 M (swelling factor ~7). On the other hand, the influence of the coagulation time for a fixed concentration of sodium hydroxide (3 M) on the swelling capacity of the prepared materials was also studied (Figure 3B). Higher coagulation times favoured the establishment of hydrogen bonds and/or hydrophobic interactions, increasing crosslinking density and, consequently, leading to networks that are more compact with a lower swelling ability. This influence reveals that despite the fact that gel formation occurs in situ (t_{gel}), the establishment of interactions continues over time (t_{∞}~3 h).

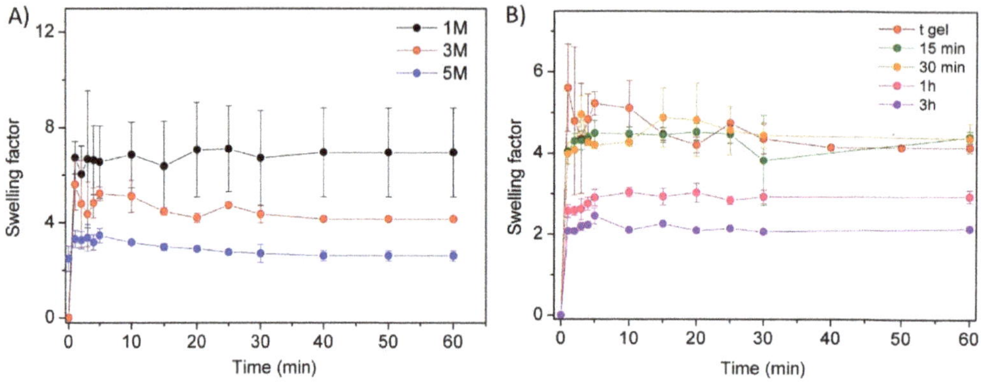

Figure 3. Swelling response of CHI/NaOH hydrogels. Influence of (**A**) NaOH ionic strength and (**B**) crosslinking time for CHI/NaOH 3 M hydrogels.

3.2. Mechanical and Rheological Properties

The unconfined compression tests of freshly prepared chitosan-based hydrated gel-like precipitates are shown in Figure 4A,B. These data reveal the influence of the concentration of NaOH solution and the exposure time to this solution on the mechanical stability of the studied materials. In both figures the typical non-linear mechanical response for this type of physical hydrogels when they are subjected to compression tests can be observed. However, chitosan-based hydrogels present a significantly different ability to deform depending on the concentration of the medium and the exposure time.

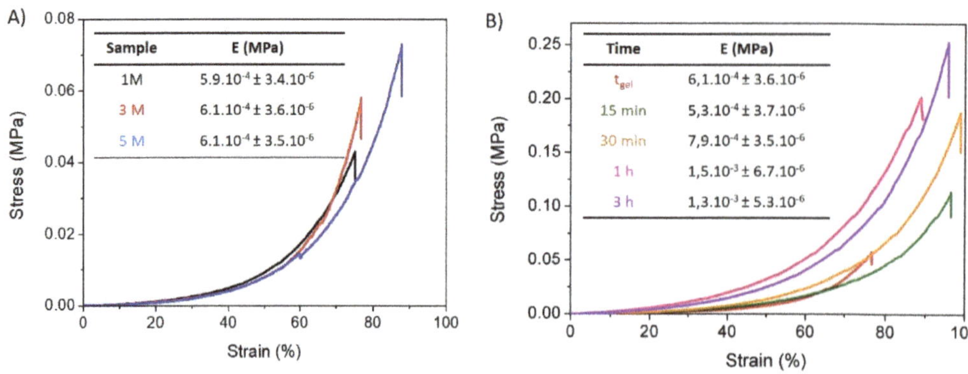

Figure 4. Mechanical compression stress–strain tests. Influence of (**A**) ionic strength of the gelling medium and (**B**) of the gelling time for CHI/NaOH 3 M hydrogels.

As is presented in Figure 4A, all hydrogels displayed a similar compression modulus (~$6.1.10^{-4}$ MPa); however, an increasing stress was required to reach their maximum deformation as NaOH solution concentration increased. In this sense, the gel formed by 5 M NaOH showed the largest resistance to deformation and required the highest stress to reach its maximum deformation (89%). This may be a consequence of the higher interactivity between polymer chains derived from a more effective coagulation, leading to higher crosslinking density. These results are in good agreement with morphological and swelling data. In contrast, when the concentration of NaOH decreased to 1 M and 3 M, lower deformations at breakage and lower stress were needed for maximum deformation. As previously shown, the use of a lower ionic strength results in less interconnected three-dimensional networks, leading to softer hydrogels.

The influence of the crosslinking time on the mechanical stability was also evaluated (Figure 4B) for hydrogels formed by the addition of NaOH 3 M. Higher complexation times led to a higher force needed for breakage, and higher Young moduli were measured for the hydrogels with the longest crosslinking times. These effects can be ascribed to the higher number of interactions established between chitosan chains, resulting in higher crosslinking density, which is in agreement with above-presented results.

The rheological behaviour of the hydrogels was evaluated by frequency sweep tests at room temperature in order to study the effect of sodium hydroxide solution concentration (Figure 5A) and coagulation time (Figure 5B). It was noticeable that all samples behaved like a hydrogel once the storage modulus (G′) was higher than the loss modulus (G″) in all frequency ranges. Furthermore, as the concentration of NaOH increased, both G′ and G″ increased, leading to an improvement of the rheological properties, especially for those hydrogels prepared with NaOH 3 M and 5 M, which presented similar behaviour. Additionally, hydrogels formed by the addition of NaOH 3 M showed a slight improvement in their rheological properties with increasing coagulation time. This behaviour, as was also revealed in the mechanical properties, can be ascribed to the higher number of interactions established between chitosan chains over time, which leads to the formation of elastic auto-stable materials, i.e., strong hydrogels [31].

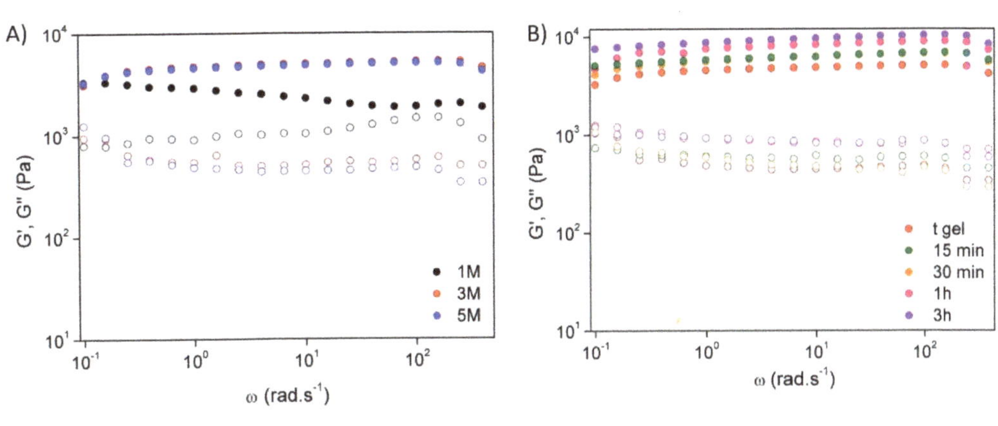

Figure 5. Rheology measurements. Storage modulus (G′) and loss modulus (G″) of the CHI-based hydrogels as a function of frequency: influence of the (**A**) ionic strength and (**B**) crosslinking time for hydrogels prepared with NaOH 3 M.

3.3. In Vitro Hydrolytic and Enzymatic Biodegradation

The evaluation of the degradation kinetics of the hydrogels is an essential factor since it determines their stability and, as a consequence, their applicability. Thus, the degradation profiles of chitosan-based hydrogels prepared with different NaOH concentrations (1, 3 and 5 M) were studied in vitro at 37 °C and pH 7, as presented in Figure 6. The study was carried out in the presence and in the absence of lysozyme. This enzyme was specifically used to catalyze the biodegradation of chitosan since it favours the hydrolysis of β-(1,4) glycosidic linkages.

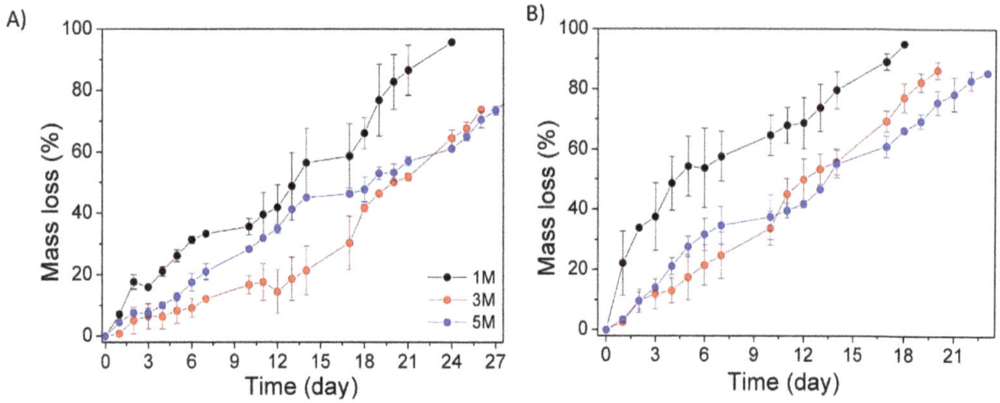

Figure 6. Degradation kinetics of the hydrogels prepared with different NaOH concentrations and different degradation media. (**A**) Hydrolytic media and (**B**) enzymatic media (lysozyme, 1 mg/mL).

In general, it can be observed that hydrogels with a remarkably higher swelling ability, such as those coagulated with 1 M NaOH solution, experienced faster and larger mass loss regardless of the presence of the enzyme as a consequence of their more open networks. In addition, the presence of the enzyme accelerated (~30%) the mass loss of the hydrogels.

It can also be observed in Figure 6 that no major differences were observed between the different hydrogels prepared with variable ionic strengths, as in all cases degradation was achieved, which was almost complete after 20 days, in agreement with the literature for similar hydrogels [16,32].

3.4. In Vitro Cytotoxicity Assay

To test the biocompatibility of chitosan gels obtained by NaOH precipitation, embryonic mouse fibroblasts (MEFs) were cultured in the presence of small fragments of hydrogels (5–10 mg) prepared with increasing molarities of NaOH (1, 3 and 5 M). After 24 h, cell viability was assessed using a fluorescent staining of the cytoplasm of live cells (calcein, green channel), the nuclei of all cells (NucBlue, blue channel) and the nuclei of dead cells (EthD-1, red channel) (Figure 7A–C). Cell viability was calculated as the ratio between red-stained and blue-stained nuclei (Figure 7D) and did not reveal any statistical differences between the control cells cultured in absence of hydrogels and the experimental conditions. This result exposes the excellent biocompatibility of the chitosan scaffolds independently of the molarity of the NaOH solution used to produce them.

3.5. Hydrogel Printability

3D bioprinting technology has become a promising technique to fabricate scaffolds with high accuracy and precision, allowing the development of precise biomimetic 3D structures. Thus, hydrogels that can be used as bioinks for 3D printing technique represent a critical component in the preparation of hydrogel-based scaffolds for biomedical applications. The printability of easily gelled chitosan hydrogels was evaluated (Figure 8A–F) in order to optimize the printing parameters, such as speed and pressure, as well as to analyze the influence of the coagulating solution (NaOH 1, 3 and 5 M).

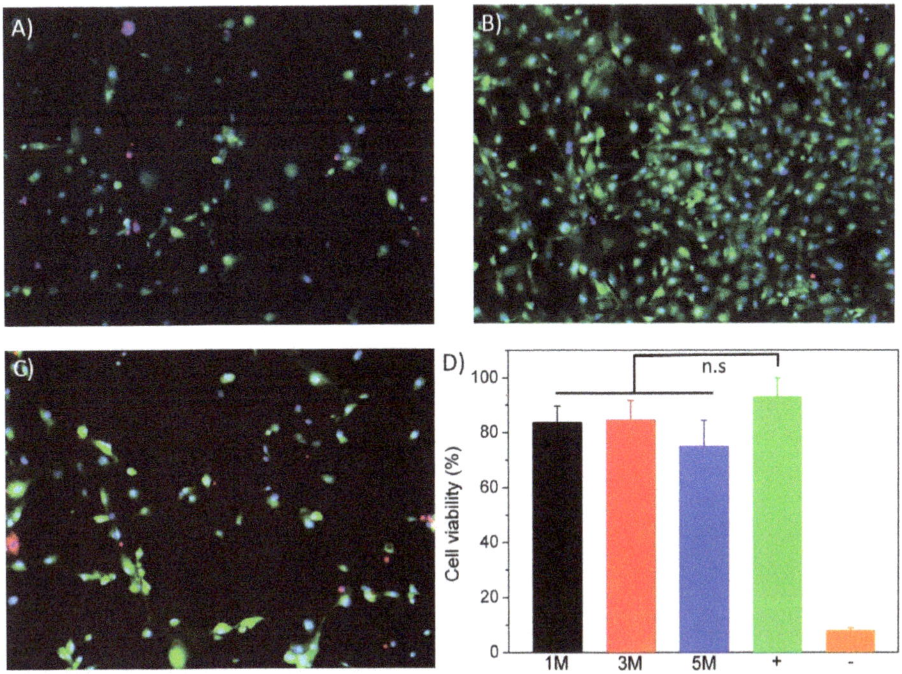

Figure 7. (**A–C**) Fluorescent images corresponding to CHI-based hydrogels coagulated with (**A**) NaOH 1 M, (**B**) NaOH 3 M and (**C**) NaOH 5 M. (**D**) Normalized cell viability results. In the positive control conditions, cell membranes were permeabilized by incubation in ethanol. Cells in the negative control conditions were cultured in absence of hydrogels. One-way ANOVA test with Tukey's multiple comparison test was used for the statistical analysis ($p < 0.05$). (n.s) No significant differences.

In all cases and regardless the ionic strength of the NaOH coagulating solution it was observed that a decrease in printing speeds and an increase in extrusion pressure led to higher values of both the expansion and uniformity factors. Indeed, intermediate pressures (30 kPa) only led to the optimal expansion rate for gels prepared with 1 and 3 M gelling solutions. It was also observed that at slower speeds the material had more time to spread over the surface, leading to extremely high expansion ratio values for high pressures. However, one of the main advantages of printing at low speeds, as observed in Figure 8, is that the uniformity factors were strongly improved, reaching values close to 1 for high pressures (40 kPa). These results are in agreement with recent published results on related materials [33]. Thus, there is a compromise between expansion and uniformity when intermediate pressures are employed, and higher accuracy is promoted by intermediate pressures and slow speed rates. With respect to ionic strength, it was observed that as the NaOH solution concentration increased, ink spread decreased as its precipitation occurred more rapidly, as 3 M is the optimum concentration, leading to expansion ratio and uniformity factor values closer to 1 for intermediate pressures. The 5 M concentration promoted fast gelling that leds to a decrease in the printing accuracy. Thus, an intermediate ionic strength (3 M NaOH) was selected as the optimal ionic strength for printing pure chitosan (1.5% w/w)-based inks. The digital images of the strands printed in the different conditions with 3 M NaOH and a 25 G nozzle are shown in Figure 8G. Accordingly, under these conditions chitosan scaffolds with square geometry were printed (Figure 9).

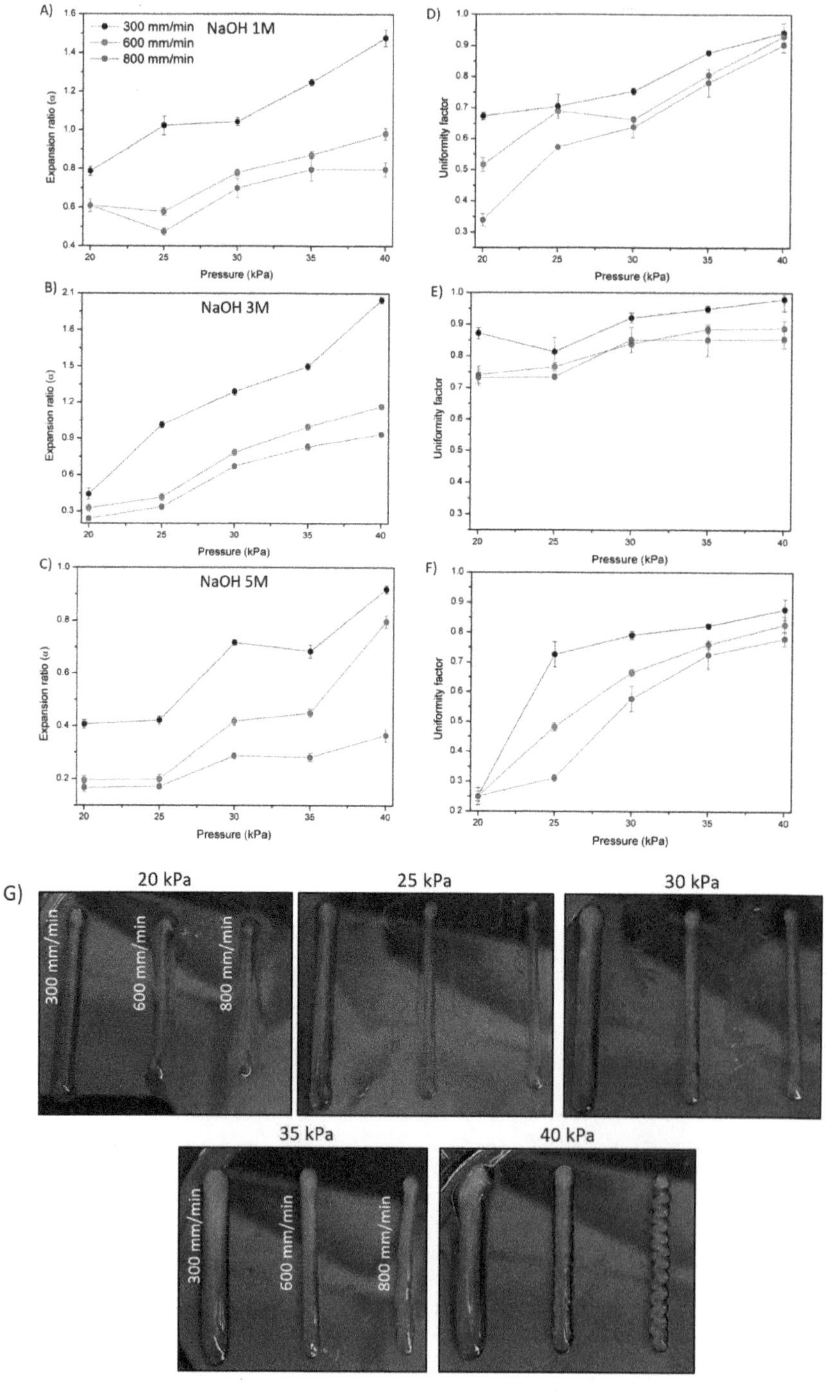

Figure 8. Chitosan hydrogel based on 1.5% w/w. solution with different NaOH concentrations. (**A–C**) expansion factors and (**D–F**) uniformity factor. (**G**) Digital photographs of CHI hydrogel coagulated with 3 M at different speeds and pressures with a nozzle of 25 G.

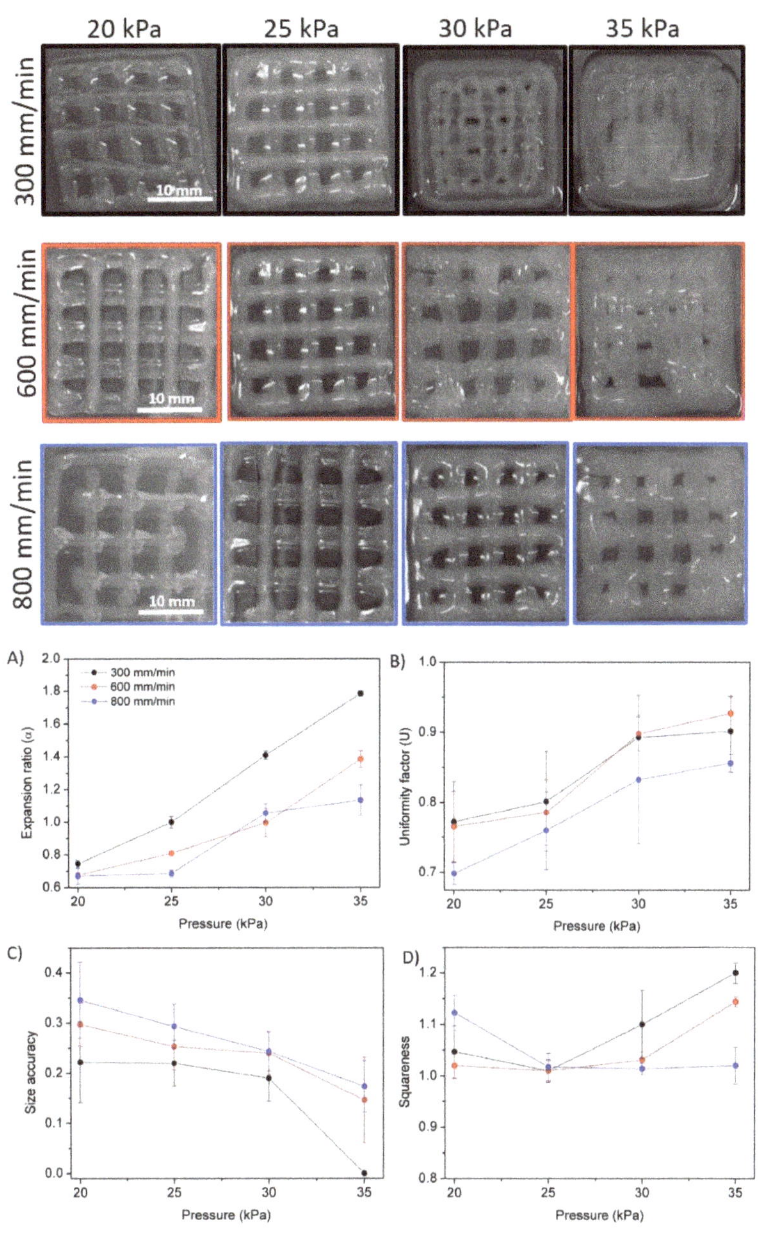

Figure 9. Digital images of printed square scaffolds of CHI/NaOH 3M hydrogel at different pressures and extrusion speeds with a 25 G nozzle. (**A–D**) Evaluation of the quality of printed structures: (**A**) expansion ratio, (**B**) uniformity factor, (**C**) size accuracy and (**D**) squareness.

Figure 9 shows that the appearance and definition of the square scaffolds varied significantly depending on the pressure and the speed at which they were printed. The combination of intermediate pressures with slow printing speeds favoured the printing of scaffolds with improved resolution and precision since less material was deposited at a time in the printing surface, which in turn led to a reduced pooling or collapse of the strands and vertices of the printed scaffold. Intermediate pressure (25 kPa) was found to promote optimum squareness, while no significant changes for variable speed rates were measured. However, it was observed that size accuracy differs significantly between the different printed scaffolds. At higher pressures and lower speeds, due to a larger flow of material and more time to spread, shape fidelity decreased sharply. This deterioration is reflected in the fact that the cavities of the printed pores did not maintain the theoretical area and instead began to fill with the bioink, which decreased their size and even filled the entire cavity. Accordingly, conditions of 30 kPa and 600 mm/min were shown to be the optimal printing conditions, that is, those that best promote expansion ratio, size accuracy, uniformity factor and squareness values closer to 1, leading to the highest quality of the printed chitosan scaffolds.

3.6. Bending Actuator

Hydrogels based on polyelectrolytes constitute interesting materials in the area of electroactive hydrogels due to their ability to locally swell and shrink in the opposite side under electric field, promoting their electromechanical bending response. Under an electric field, materials are capable of transforming electrical energy into mechanical energy due to the migration of ions in solutions, which leads to the swelling of one side of the hydrogel and the contraction of the opposite side [34,35]. Chitosan is a weak base with numerous ionizable primary amine groups along its structure that are able to induce this reversible electrical response.

Due to this, the electrically induced bending of the chitosan hydrogels was observed (Video S1). As is represented in Figure 10 printed chitosan strands bent towards the anode when a potential difference of 15 V was applied through two platinum electrodes. It is important to highlight that this bending behaviour is reversible; when the polarity of the electric field was inverted, chitosan strands bente back to the anode.

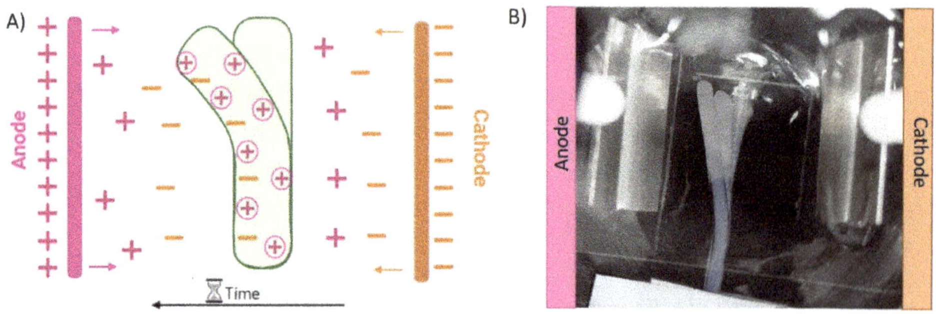

Figure 10. (**A**) Illustrative representation of the migration of ions through the hydrogel and electrolytic solution. (**B**) Bending behaviour of the CHI/NaOH 3 M strand (15 v).

There are different theories to explain the behaviour of electroactive hydrogels when an electric field is applied. Flory's theory of osmotic pressure is typically used for the explanation of polyelectrolyte-based hydrogels' bending behaviour [36]. In this case, when an electric field is applied through the material, the free ions present in solution (Na^+ and Cl, and in this case possibly OH^- residues) move toward their counter electrodes, leading to an osmotic pressure difference ($\Delta\pi$) along the direction of the potential difference inside and outside the hydrogel, due to the differences in concentration of these ions [21]. Consequently, in polycationic hydrogels ($\Delta\pi < 0$) the hydrogel swells on the cathode side while it shrinks in the anode, therefore bending to the anode side. In addition, the application of an electric field produces the electrolysis of the NaCl solution of the electrolytic media. In the negative pole (cathode), hydroxyl (OH^-) groups are created, and hydrogen gas is produced ($2H_2O + 2e^- \rightarrow 2H_2(g) + OH^-$). On the other hand, in the positive pole (anode), chlorine gas is formed due to the consumption of chloride ions ($2Cl^- \rightarrow Cl_2(g) + 2e^-$). In this way, the produced hydroxyl groups move towards the anode while the free sodium cations that are present in the electrolytic solutions travel up to the cathode [34]. Within chitosan-printed strands, ionized amine groups (positively charged) are balanced by the negatively charged ions (OH^-) that are moving from the cathode to the anode and, at the same time, they are repelled from ions in the cathode (Na^+) (Figure 10A). This fact makes the hydrogel swell on the side that is closest to the cathode and shrink on the anode side, thus bending toward the positive electrode [21,34]. Taking this electric response into consideration, the hydrogels showed a bending response suitable for the development of bioinspired soft actuators for soft robotics, allowing the further development of dynamic scaffold-based microenvironments for cell regeneration in biomedical applications [37].

4. Conclusions

Chitosan-based hydrogels were successfully prepared by coagulation using a strong base, such as NaOH, with different ionic strengths. The CHI/NaOH gel-like precipitates exhibited ionic strength dependence since their properties, such as morphology, swelling, mechanical stability and rheology, varied significantly when the ionic strength was changed. The printing process was optimized as a function of the disposal pressure, speed rate and gelling solution concentration. An intermediate concentration (3 M), pressure (30 KPa) and speed (600 mm/min) were shown to lead to the highest printing accuracy of square scaffolds. The study of the biodegradation process in the presence and in the absence of the enzyme showed no influence on the composition of the hydrogel; however, the presence of lysozyme in the degradation media accelerated the hydrolysis of O-glucosidic bonds present in chitosan backbone. In addition, it is important to highlight the ability of the hydrogel to deflect toward the anode when an electric field of 15 V is applied. Finally, the cytotoxicity tests showed the high biocompatibility of the gels, especially those coagulated with NaOH 1 and 3 M, which reveals their potential for the development of three-dimensional biomimetic scaffolds and as robotic actuators in medical applications.

Supplementary Materials: The following supporting information can be downloaded at: https://www.mdpi.com/article/10.3390/polym14030650/s1. Video S1: CHI/NaOH 3M hydrogel strand electrically induced bending.

Author Contributions: Conceptualization, S.L.-M. and L.P.-Á.; methodology, S.M.-F. and L.P.-Á.; formal analysis, U.S., L.P.-Á. and J.L.V.-V.; investigation, S.M.-F.; writing—original draft preparation, S.M.-F. and L.P.-Á.; writing—review and editing, U.S., J.L.V.-V. and S.L.-M.; supervision, L.P.-Á. and S.L.-M.; funding acquisition, S.L.-M. and J.L.V.-V. All authors have read and agreed to the published version of the manuscript.

Funding: This research was funded by Spanish State Research Agency (AEI) and the European Regional Development Fund (ERFD) through the project PID2019-106099RB-C43/AEI/10.13039/501100011033, as well as the Basque Government Industry Department under the ELKARTEK programme (KK-2021/00040 and KK-2021/00082).

Institutional Review Board Statement: Not applicable.

Informed Consent Statement: Not applicable.

Data Availability Statement: The data presented in this study are available on request from the corresponding author.

Acknowledgments: The authors thank Cristina Eguizabal for giving them access to the group of cell therapy, stem cells and tissues linked to the Basque Center for Transfusion and Human Tissues at Galdakao Hospital. The technical and human support provided by SGIker (UPV/EHU, MICINN, GV/EJ, EGEF and ESF) is gratefully acknowledged.

Conflicts of Interest: The authors declare no conflict of interest.

References

1. Wang, X.; Wei, C.; Cao, B.; Jiang, L.; Hou, Y.; Chang, J. Fabrication of Multiple-Layered Hydrogel Scaffolds with Elaborate Structure and Good Mechanical Properties via 3D Printing and Ionic Reinforcement. *ACS Appl. Mater. Interfaces* **2018**, *10*, 18338–18350. [CrossRef]
2. Do, A.V.; Khorsand, B.; Geary, S.M.; Salem, A.K. 3D Printing of Scaffolds for Tissue Regeneration Applications. *Adv. Healthc. Mater.* **2015**, *4*, 1742–1762. [CrossRef]
3. Yan, Q.; Dong, H.; Su, J.; Han, J.; Song, B.; Wei, Q.; Shi, Y. A Review of 3D Printing Technology for Medical Applications. *Engineering* **2018**, *4*, 729–742. [CrossRef]
4. Gutowska, A.; Jeong, B.; Jasionowski, M. Injectable gels for tissue engineering. *Anat. Rec.* **2001**, *263*, 342–349. [CrossRef] [PubMed]
5. Van Vlierberghe, S.; Dubruel, P.; Schacht, E. Biopolymer-Based Hydrogels as Scaffolds for Tissue Engineering Applications: A Review. *Biomacromolecules* **2011**, *12*, 1387–1408. [CrossRef]
6. Hinderer, S.; Layland, S.L.; Schenke-Layland, K. ECM and ECM-like materials—Biomaterials for applications in regenerative medicine and cancer therapy. *Adv. Drug Deliv. Rev.* **2016**, *97*, 260–269. [CrossRef]
7. Zhu, Y.; Romain, C.; Williams, C.K. Sustainable polymers from renewable resources. *Nature* **2016**, *540*, 354–362. [CrossRef]
8. Champeau, M.; Heinze, D.A.; Viana, T.N.; de Souza, E.R.; Chinellato, A.C.; Titotto, S. 4D Printing of Hydrogels: A Review. *Adv. Funct. Mater.* **2020**, *30*, 1910606. [CrossRef]
9. Gopinathan, J.; Noh, I. Recent trends in bioinks for 3D printing. *Biomater. Res.* **2018**, *22*, 11. [CrossRef]
10. Collado-González, M.; Espinosa, Y.G.; Goycoolea, F.M. Interaction between Chitosan and Mucin: Fundamentals and applications. *Biomimetics* **2019**, *4*, 32. [CrossRef]
11. Nilsen-Nygaard, J.; Strand, S.P.; Vårum, K.M.; Draget, K.I.; Nordgård, C.T. Chitosan: Gels and interfacial properties. *Polymers* **2015**, *7*, 552–579. [CrossRef]
12. Bergonzi, C.; Di Natale, A.; Zimetti, F.; Marchi, C.; Bianchera, A.; Bernini, F.; Silvestri, M.; Bettini, R.; Elviri, L. Study of 3D-printed chitosan scaffold features after different post-printing gelation processes. *Sci. Rep.* **2019**, *9*, 362. [CrossRef]
13. Kean, T.J.; Thanou, M. Utility of Chitosan for 3D Printing and Bioprinting. In *Sustainable Agriculture Reviews*, 1st ed.; Crini, G., Lichtfouse, E., Eds.; Springer: Cham, Switzerland, 2019; Volume 3, pp. 154–196. ISBN 9783319267760.
14. Rajabi, M.; McConnell, M.; Cabral, J.; Ali, M.A. Chitosan hydrogels in 3D printing for biomedical applications. *Carbohydr. Polym.* **2021**, *260*, 117768. [CrossRef]
15. Shu, X.Z.; Zhu, K.J. The influence of multivalent phosphate structure on the properties of ionically cross-linked chitosan films for controlled drug release. *Eur. J. Pharm. Biopharm.* **2002**, *54*, 235–243. [CrossRef]
16. Maiz-Fernández, S.; Barroso, N.; Pérez-Álvarez, L.; Silván, U.; Vilas-Vilela, J.L.; Lanceros-Mendez, S. 3D printable self-healing hyaluronic acid/chitosan polycomplex hydrogels with drug release capability. *Int. J. Biol. Macromol.* **2021**, *188*, 820–832. [CrossRef]
17. Chia, H.N.; Wu, B.M. Recent advances in 3D printing of biomaterials. *J. Biol. Eng.* **2015**, *9*, 4. [CrossRef]
18. Almeida, C.R.; Serra, T.; Oliveira, M.I.; Planell, J.A.; Barbosa, M.A.; Navarro, M. Impact of 3-D printed PLA- and chitosan-based scaffolds on human monocyte/macrophage responses: Unraveling the effect of 3-D structures on inflammation. *Acta Biomater.* **2014**, *10*, 613–622. [CrossRef]
19. Wu, Q.; Therriault, D.; Heuzey, M.C. Processing and Properties of Chitosan Inks for 3D Printing of Hydrogel Microstructures. *ACS Biomater. Sci. Eng.* **2018**, *4*, 2643–2652. [CrossRef]
20. Zhang, J.; Allardyce, B.J.; Rajkhowa, R.; Zhao, Y.; Dilley, R.J.; Redmond, S.L.; Wang, X.; Liu, X. 3D Printing of Silk Particle-Reinforced Chitosan Hydrogel Structures and Their Properties. *ACS Biomater. Sci. Eng.* **2018**, *4*, 3036–3046. [CrossRef]
21. Zolfagharian, A.; Kaynak, A.; Khoo, S.Y.; Kouzani, A.Z. Polyelectrolyte Soft Actuators: 3D Printed Chitosan and Cast Gelatin. *3D Print. Addit. Manuf.* **2018**, *5*, 138–150. [CrossRef]
22. Di Giuseppe, M.; Law, N.; Webb, B.; Macrae, R.A.; Liew, L.J.; Sercombe, T.B.; Dilley, R.J.; Doyle, B.J. Mechanical behaviour of alginate-gelatin hydrogels for 3D bioprinting. *J. Mech. Behav. Biomed. Mater.* **2018**, *79*, 150–157. [CrossRef] [PubMed]
23. Ouyang, L.; Yao, R.; Zhao, Y.; Sun, W. Effect of bioink properties on printability and cell viability for 3D bioplotting of embryonic stem cells. *Biofabrication* **2016**, *8*, 035020. [CrossRef] [PubMed]
24. Chung, Y.M.; Simmons, K.L.; Gutowska, A.; Jeong, B. Sol-Gel transition temperature of PLGA-g-PEG aqueous solutions. *Biomacromolecules* **2002**, *3*, 511–516. [CrossRef]
25. Schindelin, J.; Arganda-Carreras, I.; Verena Kaynig, E.F.; Longair, M.; Pietzsch, T.; Preibisch, S.; Rueden, C.; Saalfeld, S.; Schmid, B.; Cardona, A. Fiji: An open-source platform for biological-image analysis. *Nat. Methods* **2012**, *9*, 676–682. [CrossRef]
26. Mohammed, M.A.; Syeda, J.T.M.; Wasan, K.M.; Wasan, E.K. An overview of chitosan nanoparticles and its application in non-parenteral drug delivery. *Pharmaceutics* **2017**, *9*, 53. [CrossRef]
27. Mora Boza, A.; Wlodarczyk-Biegun, M.K.; Del Campo, A.; Vázquez-Lasal, B.; San Román, J. Chitosan-based inks: 3D printing and bioprinting strategies to improve shape fidelity, mechanical properties, and biocompatibility of 3D scaffolds. *Biomecánica* **2019**, *27*, 7–16. [CrossRef]
28. Berger, J.; Reist, M.; Mayer, J.M.; Felt, O.; Peppas, N.A.; Gurny, R. Structure and interactions in covalently and ionically crosslinked chitosan hydrogels for biomedical applications. *Eur. J. Pharm. Biopharm.* **2004**, *57*, 19–34. [CrossRef]
29. Rastogi, S.K.; Anderson, H.E.; Lamas, J.; Barret, S.; Cantu, T.; Zauscher, S.; Brittain, W.J.; Betancourt, T. Enhanced Release of Molecules upon Ultraviolet (UV) Light Irradiation from Photoresponsive Hydrogels Prepared from Bifunctional Azobenzene and Four-Arm Poly(ethylene glycol). *ACS Appl. Mater. Interfaces* **2018**, *10*, 30071–30080. [CrossRef]

30. Gao, L.; Gan, H.; Meng, Z.; Gu, R.; Wu, Z.; Zhang, L.; Zhu, X.; Sun, W.; Li, J.; Zheng, Y.; et al. Effects of genipin cross-linking of chitosan hydrogels on cellular adhesion and viability. *Colloids Surf. B Biointerfaces* **2014**, *117*, 398–405. [CrossRef]
31. Lalevée, G.; Sudre, G.; Montembault, A.; Meadows, J.; Malaise, S.; Crépet, A.; David, L.; Delair, T. Polyelectrolyte complexes via desalting mixtures of hyaluronic acid and chitosan—Physicochemical study and structural analysis. *Carbohydr. Polym.* **2016**, *154*, 86–95. [CrossRef]
32. Ganji, F.; Abdekhodaie, M.J.; Ramazani, S.A.A. Gelation time and degradation rate of chitosan-based injectable hydrogel. *J. Sol-Gel Sci. Technol.* **2007**, *42*, 47–53. [CrossRef]
33. Dutta, S.D.; Hexiu, J.; Patel, D.K.; Ganguly, K.; Lim, K.T. 3D-printed bioactive and biodegradable hydrogel scaffolds of alginate/gelatin/cellulose nanocrystals for tissue engineering. *Int. J. Biol. Macromol.* **2021**, *167*, 644–658. [CrossRef] [PubMed]
34. Fu, J.; Yang, F.; Guo, Z. The chitosan hydrogels: From structure to function. *New J. Chem.* **2018**, *42*, 17162–17180. [CrossRef]
35. Kim, S.J.; Shin, S.R.; Lee, J.H.; Lee, S.H.; Kim, S.I. Electrical response characterization of chitosan/polyacrylonitrile hydrogel in NaCl solutions. *J. Appl. Polym. Sci.* **2003**, *90*, 91–96. [CrossRef]
36. Shang, J.; Shao, Z.; Chen, X. Chitosan-based electroactive hydrogel. *Polymer* **2008**, *49*, 5520–5525. [CrossRef]
37. Cui, H.; Zhao, Q.; Zhang, L.; Du, X. Intelligent Polymer-Based Bioinspired Actuators: From Monofunction to Multifunction. *Adv. Intell. Syst.* **2020**, *2*, 2000138. [CrossRef]

polymers

MDPI

Article

Revealing the Effect of MnO_2, Activated Carbon and MnO_2/Activated Carbon on Chitosan Polymer Host Fabricated Co NPs: Antibacterial Performance and Degradation of Organic Compounds

Hani S. H. Mohammed Ali [1,2], Sumiya [4], Yasir Anwar [1,2], Youssef O. Al-Ghamdi [6], Muhammad Fakieh [1] and Shahid Ali Khan [3,4,5,*]

1 Department of Biological Sciences, Faculty of Science, King Abdulaziz University, P.O. Box 80203, Jeddah 21589, Saudi Arabia; haniolfat@gmail.com (H.S.H.M.A.); yasirpcsir2006@gmail.com (Y.A.); mf1881@yahoo.com (M.F.)
2 Princess Dr. Najla Bint Saud Al-Saud Center for Excellence Research in Biotechnology, King Abdulaziz University, P.O. Box 80203, Jeddah 21589, Saudi Arabia
3 Center of Excellence for Advanced Materials Research (CEAMR), King Abdulaziz University, P.O. Box 80203, Jeddah 21589, Saudi Arabia
4 Department of Chemistry, University of Swabi, Anbar, Swabi 23561, Pakistan; ssumiya208@gmail.com
5 Department of Chemistry, School of Natural Sciences, National University of Science and Technology (NUST), Islamabad 44000, Pakistan
6 Department of Chemistry, College of Science Al-Zulfi, Majmaah University, Al-Majmaah 11952, Saudi Arabia; Yo.alghamdi@mu.edu.sa
* Correspondence: skhan@uoswabi.edu.pk or shahid.ali@sns.nust.edu.pk

Citation: Mohammed Ali, H.S.H.; Sumiya; Anwar, Y.; Al-Ghamdi, Y.O.; Fakieh, M.; Khan, S.A. Revealing the Effect of MnO_2, Activated Carbon and MnO_2/Activated Carbon on Chitosan Polymer Host Fabricated Co NPs: Antibacterial Performance and Degradation of Organic Compounds. *Polymers* **2022**, *14*, 627. https://doi.org/10.3390/polym14030627

Academic Editor: Alina Sionkowska

Received: 23 December 2021
Accepted: 29 January 2022
Published: 6 February 2022

Abstract: MnO_2 and MnO_2 blended with 1 and 2 weight percent of activated carbon (AC), MnO_2/AC1 and MnO_2/AC2 were synthesized through the sol–gel method. The pure chitosan (CS) films were cast in the form of films. Similarly, 5 weight% of each MnO_2, AC, MnO_2/AC1 and MnO_2/AC2 was intermingled with the CS to produce different films, such as CS-AC, CS-MnO_2, CS-MnO_2/AC1 and CS-MnO_2/AC2. Zero-valent Co NPs were then supported on these films through the chemical reduction method and expressed as CS@Co, CS-AC@Co, CS-MnO_2@Co, CS-MnO_2/AC1@Co and CS-MnO_2/AC2@Co NPs. All the catalysts were characterized by field emission scanning electron microscopy (FESEM), energy-dispersive spectroscopy (EDS) and X-ray diffraction (XRD) techniques. The synthesized catalysts were used as a dip catalyst against the hydrogenation of 4-nitrophenol (4NP), and for the degradation of methyl orange (MO) and Congo red (CR) dyes. The k_{app} and R^2 values were deduced from pseudo-first-order kinetics for 4NP and MO and zero-order kinetics for CR dye. The k_{app} values of CS-AC@Co and CS-MnO_2/AC1@Co NPs for 4NP hydrogenation were higher than those for any other member of the series, at 1.14×10^{-1} and 1.56×10^{-1} min^{-1} respectively. Similarly, the rate of CR degradation was highest with CS-AC@Co. The R^2 values for 4NP, MO and CR dyes were above 0.9, which indicated that the application of pseudo-first- and zero-order models were appropriate for this study. Furthermore, the antibacterial activity of all the catalysts was evaluated against *Pseudomonas aeruginosa* and *Escherichia coli*. The CS-AC@Co NPs exhibited the highest zone of inhibition compared to other catalysts against *P. aeruginosa*, while all the catalysts were inactive against *E. coli*. This study reveals that the catalyst can be used for the degradation of other pollutants and for microbial inhibition.

Keywords: chitosan; activated carbon; MnO_2; Co NPs; antibacterial activity; degradation

1. Introduction

A dramatic increase in the anthropogenic activities, such as industrialization, modernization and urbanization, has contaminated water bodies to a serious extent. These activities are indeed exerting adverse effects on flora and fauna in ecosystems [1]. Nowadays, innumerable industries discharge their effluents directly into water bodies without any treatment, hence posing potential threats to aquatic biota [2]. The effluents from textile, paper, cosmetics, pharmaceuticals and food processing units are heavily loaded with dyes and other chemicals, which decrease the transparency of the water and stimulate their toxicity [3]. Dyes and dyestuff significantly block light penetration into water bodies, thereby impeding the vital activity of photosynthesis, hence rendering the entire ecosystem deprived of oxygen [4]. Reports have revealed that dyes and their degradation products are genotoxic and cytotoxic to living organisms [5,6]. Moreover, dyes accelerate immunological reactions, induce hyperactivity in children [7] and cause nasal congestion, urticaria, asthmatic disorders and rhinitis in men [8]. Besides the degradation of azo dyes, this study also aimed to decontaminate the water from hazardous chemicals such as 4-nitrophenol (4NP), which has been proven to be highly toxic to human beings. Indeed, 4NP is primarily discharged as an effluent by the pharmaceutical, pesticide, petrochemical and dye industries. Moreover, 4NP is a well-known potential carcinogen and mutagen; 4NP has been categorized as a priority contaminant by the United States (US) Environmental Protection Agency (EPA) [9]. As a consequence of insights into the hazardous effects caused by dyes and PNP, they should be removed from water bodies at the earliest opportunity.

In the last two decades, scientists and researchers have been attempting to remove as much dyestuff from water as possible [10]. In this regard, they have incorporated several water purification techniques, including adsorption, filtration [11] membrane technologies, Fenton's oxidation processes, advanced oxidation processes, electrochemical oxidation, biological methods and the use of zero-valent metal (ZVM) nanocatalysis, which are documented in appropriate detail in the literature [12]. Biological methods for wastewater treatment represent the most promising techniques, owing to their simplicity, ease of applicability, versatility and low costs [13]. However, most of the synthetic dyes are non-biodegradable under their complex benzenoid frameworks; thus, biological treatments cannot be extended to a broader spectral domain of dyes [14]. Chemical treatment of wastewater, such as oxidation via ozone, ClO_2 and chlorine, has also been proven to be very effective due to its remarkably high efficiency and reproducibility, but they generate chlorinated hydrocarbons as byproducts, which are well-established carcinogens [15]. Nowadays, advanced oxidation processes (AOPs) such as Fenton's reagent [16], electro-Fenton's, photo-electro Fenton's, photocatalytic and photo-ozonation techniques, are the main areas of interest for environmental chemists worldwide for the decontamination of waste water [17], but their rising prices and operational difficulties limit the applicability of these techniques.

The introduction of catalytic degradation techniques for pollution control has gained the attention of researchers globally, to a greater extent than any other technique. Over the past few decades, among the catalytic techniques, the incorporation of zero-valent metal (ZVM) nanocatalytic techniques for water pollution abatement constitutes a newly emerging and highly promising tool due to its elevated catalytic potential [18–22]. The catalytic potential of ZVM and various metal oxide nanoparticles is attributed to their remarkably high surface to volume ratio, large surface energy and tiny quantum size effect, diversity, versatility and ease of application over a broad spectral regime of pollutants in water [23,24]. ZVM is synthesized in an eco-friendly way by treating a pristine or a supported metal ion with a suitable reducing agent such as $NaBH_4$; however, ZVM NPs are always vulnerable to agglomeration and aggregation, which consequently leads to a loss of catalytic activity, but this problem can be easily avoided by using a suitable supporting matrix [25]. The supported catalyst after use can be conveniently recovered from the reaction mixture and can be reused as a dip catalyst. The employment of polymer-coated nanoscale composites for ZVM is gaining popularity in academic and industrial sectors due to their intrinsic features and significantly enhanced surface area. Some serious issues are always associated with the use of naked ZVM NPs for pollution abatement, such as the percolation of ZVM in the reaction mixture, and its cyclization, mass transport and effective isolation from the reaction mixture after use; however, these problems can be minimized by anchoring the ZVM over an appropriate polymeric support. Various polymer supports have been mentioned in the literature for the stabilization of zero-valent metal nanoparticles [10,26]. Chitosan is one of the emerging natural polymers that is used for NP stabilization due to their various inherent functional groups. Chitosan is a natural biopolymer with a linear structure and is derived from chitin. Chitosan possesses several important characteristics, such as biodegradability, non-toxic nature and biocompatibility; therefore, it is largely applied in various biomedical sectors [27,28]. All these characteristics make chitosan unique compared to other polymers; however, several drawbacks are associated with chitosan polymers, such as their low antimicrobial characteristics. To increase the antimicrobial characteristics and facilitate the anchoring of the NPs on their surface, several modifications have been adopted—for instance, the incorporation of inorganic fillers, which not only increases various characteristics of the polymer, but is also reported to increase the bioactivity of the polymer [1,29]. Therefore, the current study involved the stabilization of ZVM Co NPs over a variety of supports, such as pristine chitosan (CS), CS-blended composites with activated carbon (CS-AC), CS blended with MnO_2 (CS-MnO_2), CS-blended MnO_2/AC1(CS-MnO_2/AC1) and CS-blended MnO_2/AC1(CS-MnO_2/AC1). The above-mentioned composites were templates in the form of films and were employed as a solid supportive matrix for Co NPs' stabilization; they were evaluated in terms of the inhibition of pathogenic bacteria. These stabilized Co NPs were utilized to address the degradation/reduction of target contaminants such as 4NP and different dyes to evaluate the comparative role of the matrix.

2. Experimental

2.1. Reagents and Materials

Chitosan polymer with 85% degree of acetylation was procured from BDH Company (London, England), and Congo red and methyl orange dyes, *p*-nitrophenol (4NP), $NaBH_4$ and acetic acid were obtained from Dae-Jung Company (Sasang-gu, Busan, Korea). Moreover, $CoCl_2$, $MnCl_2$ salts and NaOH were purchased from Sigma Aldrich (Kawasaki, Kanagawa, Japan). Distilled water was obtained from the distillation plant of the Department of Chemistry, University of Swabi, KPK, Pakistan.

2.2. Instrumentation

The EDS and FESEM were performed on JSM-7600F, Tokyo, Japan, and FESEM JEOL (JSM-7600F, Japan). The X-ray diffraction (XRD) technique was performed on a PAN JDX-3532 JEOL Tokyo, JAPAN analytical diffractometer with a Cu Kα source of 1.5418 Å (40 kV, 30 mA, monochromatic). A UV–visible spectrophotometer was obtained from PerkinElmer Company (Lambda 365) and scanned from 190 to 800 nm (Waltham, MA, USA).

2.3. Synthesis of Materials and Composite Films

2.3.1. Synthesis and Activation of Activated Carbon

The peanut shell was ground and heated at 400 °C for 6 h and then sieved with a 20 µm pore size sieve. After this, it was treated with concentrated HNO_3, washed with distilled water and dried. The as-prepared activated carbon (AC) was used for the synthesis of other materials.

2.3.2. Synthesis of MnO_2

For the synthesis of MnO_2 NPs, 1 M $MnCl_2$ solution was prepared in 200 mL distilled water and basified with NaOH solution till pH 11, which was monitored on a pH meter. The solution was placed on a hot plate at 80 °C for 6 h. After the reaction's completion, the supernatant was decanted and the precipitate was washed with a 7:3 ethanol–water mixture, and then the solution was centrifuged three times for the separation of the precipitate based on mass gravity. The precipitate was heated at 80 °C in an oven overnight and then we calcined the MnO_2 NPs at 400 °C in a furnace for 6 h.

2.3.3. Synthesis of $MnO_2/AC1$ and $MnO_2/AC2$

First, 1 M $MnCl_2$ was mixed with 200 mL distilled water. After this, 1 and 2 weight% of AC was separately added to a beaker containing 100 mL of $MnCl_2$ solution to prepare $MnO_2/AC1$ and $MnO_2/AC2$, respectively. Each solution was basified with a dilute solution of NaOH till pH 11, which was monitored by a pH meter. The mixture was placed on the hot plate at 80 °C for 6 h and then washed with ethanol and water mixture (7:3) and was centrifuged three times to separate the precipitate. The precipitate was dried at 80 °C overnight, and calcined at 400 °C for 6 h in a furnace.

2.3.4. Synthesis of Pure CS Films

The solution of CS was prepared by mixing 2 g CS in a 7:3 (v/v) acetic acid and water mixture. The mixture was stirred at room temperature until a clear paste formed. The CS paste was cast in a petri dish and uniformly dispersed in the form of sheets, and it was kept in an open environment. The solvent and water molecules were evaporated from the sheets; as a result, one face of the sheet was porous, while the other was smooth. The dried sheets were dipped in a concentrated solution of NaOH for a few minutes and then washed with distilled water to remove acid/base content.

2.3.5. Synthesis of CS Hybrid Sheets

Various sheets were prepared by blending the synthesized nanocomposite with CS host polymer.

All the films, such as CS-AC, CS-MnO_2, CS-$MnO_2/AC1$ and CS-$MnO_2/AC2$, were prepared by adding 5 weight% of AC, MnO_2, $MnO_2/AC1$ and $MnO_2/AC2$, respectively, to the CS polymer and cast in a petri dish in the form of sheets. The remaining procedure for the sheet synthesis was the same as discussed above for the CS sheet preparation.

2.3.6. Stabilization of Co NPs on CS and CS Hybrid Sheets

The CS and CS hybrid sheets were dipped in 1 M Co salt solution for 6 h. All the sheets that adsorbed Co^{2+} ions were washed with distilled water to remove the un-adsorbed ions, and treated with freshly prepared $NaBH_4$ solution, which changed the color of the sheets from pink to black. The black color was an indication of Co^0 NPs synthesis. The synthesized Co NPs were used to treat discoloration caused by MO and CR dyes, as well as for the hydrogenation of 4NP. A generalized schematic representation is given in Scheme 1, including the preparation of CS-$MnO_2/AC1$ and the synthesis of CS-$MnO_2/AC1$@Co NPs.

2.4. Antibacterial Characteristics

The antibacterial potential of all the catalyst-supported Co NPs was assessed against *P. aeruginosa* and *E. Coli* on Mueller–Hinton agar using the Kirby–Bauer disk diffusion method [30]. The plates were sterilized, and then the culture of *P. aeruginosa* and *E. coli* was spread uniformly via a sterilized spreader across the whole plate. Different plates were used for *P. aeruginosa* and *E. coli*. After bacterial culture, the catalyst was cut into appropriate dimensions and then positioned in the plates containing bacterial colonies. These plates were incubated at 37 °C for 24 h and the zone of inhibition was measured through a scale in cm. The zone of inhibition was calculated through the mean value around the nutrient agar disk.

2.5. Evaluation of Catalyst Activity in 4NP Reduction and Dye Degradation

All the catalysts' activities were assessed for the hydrogenation of 4NP and degradation of MO and CR dyes. Briefly, 0.2 mM 4NP solution was prepared in 100 mL of distilled water and then 2.5 mL of it was placed in a cuvette and the absorption noted. After this, 0.5 mL of $NaBH_4$ (0.5 mM) was added to the same cuvette and then 20 mg of the catalyst was added. The hydrogenation of 4NP or degradation of dyes was recorded using a UV–vis spectrophotometer. Similarly, 0.1 mM solutions of CR and MO were prepared in 100 mL of distilled water and then 2.5 mL of each dye was placed in a cuvette, along with 0.5 mL

$NaBH_4$ (0.5 mM) solution and 20 mg of the catalyst, and their absorbance was recorded on the UV–vis spectrophotometer (Waltham, MA, USA).

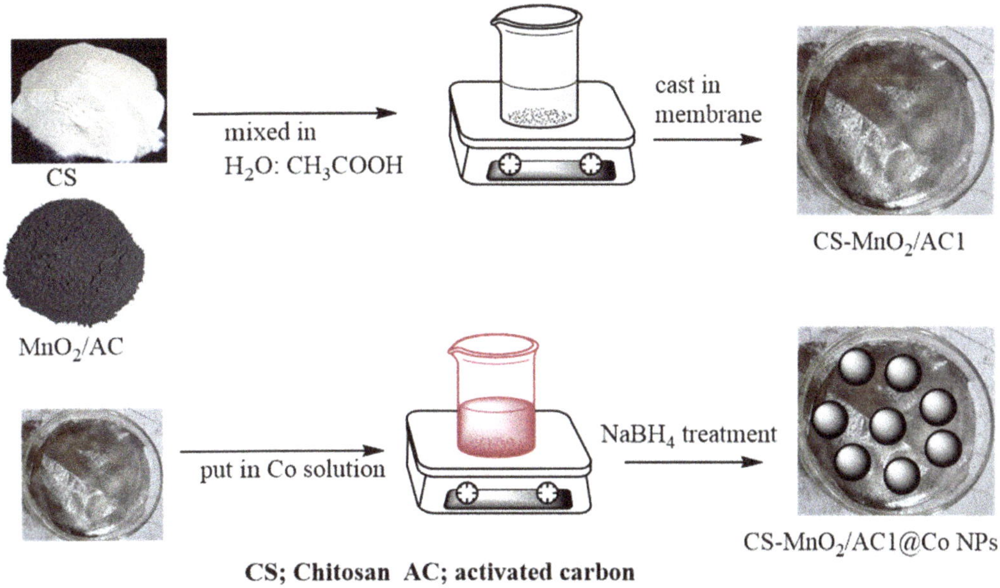

Scheme 1. General depiction of the synthesis of CS-MnO_2/AC1 and CS-MnO_2/AC1@Co NPs.

3. Results and Discussion

3.1. FESEM and EDS

The FESEM images and EDS spectrum are provided in the inset of Figure 1. The left- and right-hand sides of Figure 1 present the FESEM images and EDS spectrum. The FESEM image of CS@Co NPs shows a smooth surface with small pores (Figure 1a). The EDS spectrum and elemental window indicate peaks for C, O and N with 9.99, 31.68 and 2.75 weight%. The C, O and N atoms arise from the chitosan skeleton. Other peaks arise for Co and Cu elements. The Co and Cu elements are present in 24.08 and 16.18 by weight% (Figure 1b). Peaks for Cu are observed throughout the EDS spectrum, which is due to the Cu sputtering. Cu sputtering was performed before the FESEM and EDS analyses. Figure 1c indicates that AC covered the CS polymer and there were voids in their morphology. The EDS spectrum indicated C, O, N and Co elements with 10.31, 31.51, 3.01 and 23.76 weight percent in the CS-AC@Co catalyst (Figure 1d). Similarly, the FESEM spectrum of CS-MnO_2 indicated a rough surface of the polymer sheet, with small sphere-shaped embedded Co NPs (Figure 1e). The elemental window indicated peaks for C, O, N, Mn and Co elements, which were present in 10.65, 37.54, 2.42, 0.16 and 18.11 by weight percent in the CS-MnO_2@Co NPs (Figure 1f). The FESEM image of CS-MnO_2/AC1 (Figure 1g) indicated a protruding surface, while the EDS indicated peaks for C, O, N and Co elements with 27.74, 36.39, 5.87, 0.23 and 11.64 weight percent in the CS-MnO_2/AC1@Co catalyst (Figure 1h). The CS-MnO_2/AC2@Co catalyst showed a flat surface of the films, with numerous white spots on them (Figure 1i). The EDS and elemental window indicated peaks for C, N, O, Mn and Co with 31.55, 8.29, 32.93, 0.13 and 5.47 weight%, respectively (Figure 1j).

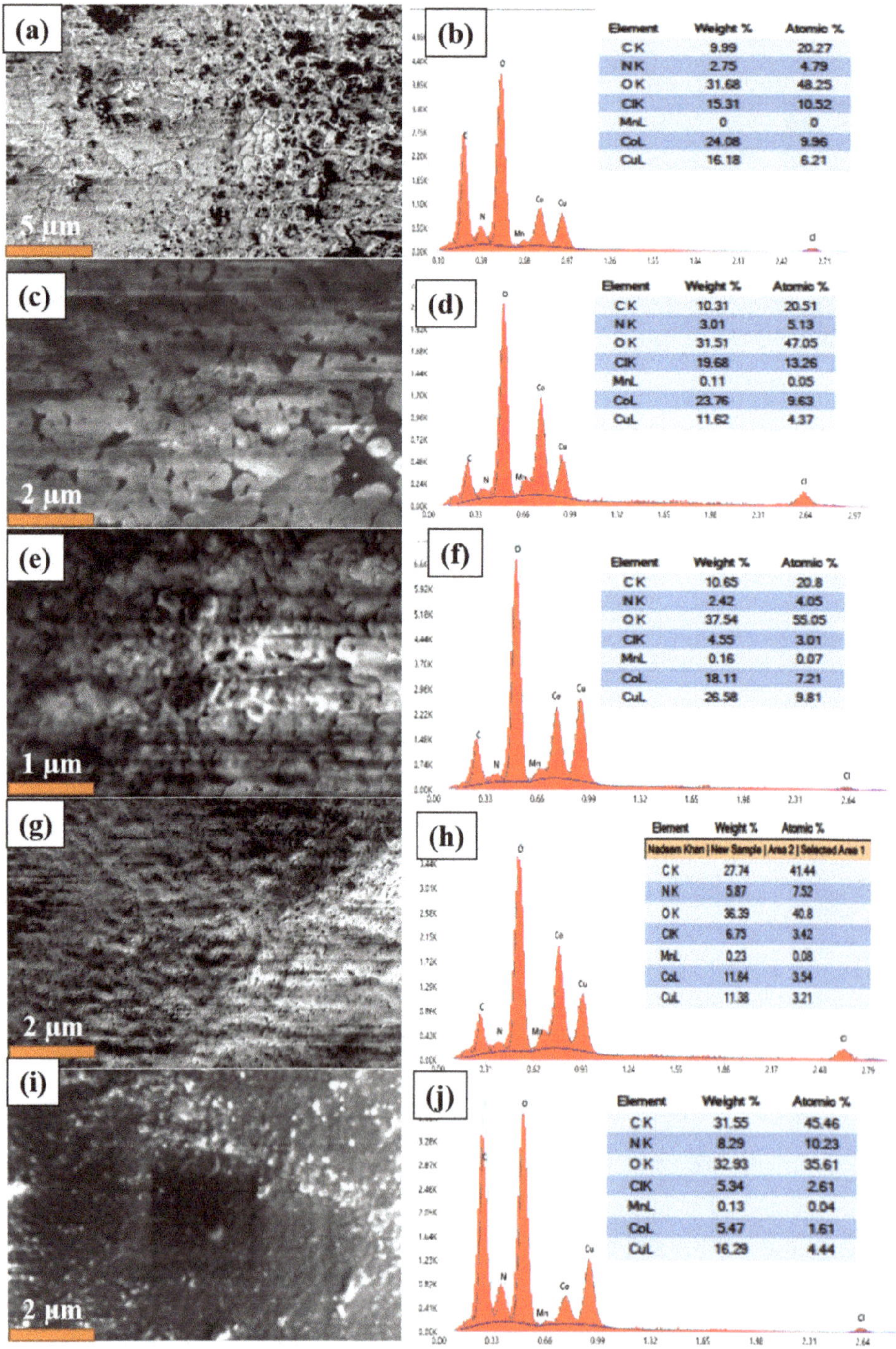

Figure 1. The FESEM and EDS images of CS@Co NPs (**a**,**b**), CS-AC@Co NPs (**c**,**d**), CS-MnO_2@Co NPs, (**e**,**f**), CS-MnO_2/AC1@Co NPs (**g**,**h**) and CS-MnO_2/AC2@Co NPs (**i**,**j**).

3.2. XRD

The XRD spectrum of the synthesized catalysts is shown in Figure 2. The XRD spectra of CS@Co, CS-AC@Co, CS-MnO_2/AC1@Co and CS-MnO_2/AC2@CoNPs indicated an amorphous peak at 2θ = 22.5°, while CS-MnO_2@Co exhibited an amorphous peak at 27.8°. These amorphous peaks reveal that Co NPs grew in an amorphous nature during their fabrication. The literature also includes similar reports for Co and Cu NPs, respectively [31,32]. Therefore, it is suggested that Co NPs with a larger particle size are formed on CS and CS hybrid catalysts during their synthesis.

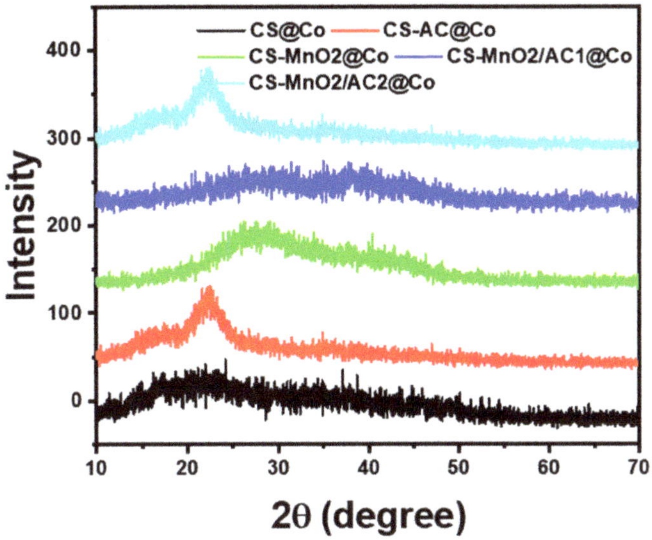

Figure 2. XRD spectrum of the synthesized catalysts.

3.3. Antibacterial Activity

Antibacterial characteristics are important for applications in the biomedical field [33,34]. All the catalysts were tested against two bacterial species, *P. aeruginosa* and *E. coli*, for 24 h of incubation time. The upper row of Figure 3 presents the disk diffusion results for *P. aeruginosa*, while the lower row show the results for *E. coli*. The zones of inhibition of all the catalysts were measured in cm. In the current study, CS-AC@Co NPs exhibited the strongest antibacterial activity, inhibiting a 2.2 ± 0.1 cm zone of *P. aeruginosa*, while the CS@Co NPs showed a 1.5 ± 0.1 cm zone of inhibition. Similarly, CS-MnO_2@Co and CS-MnO_2/AC1@Co NPs also showed an approximately 0.3 ± 0.1 cm zone of inhibition, while CS-MnO_2/AC2@Co NPs was inactive against *P. aeruginosa*. Furthermore, all the synthesized catalysts was inactive against *E. coli*.

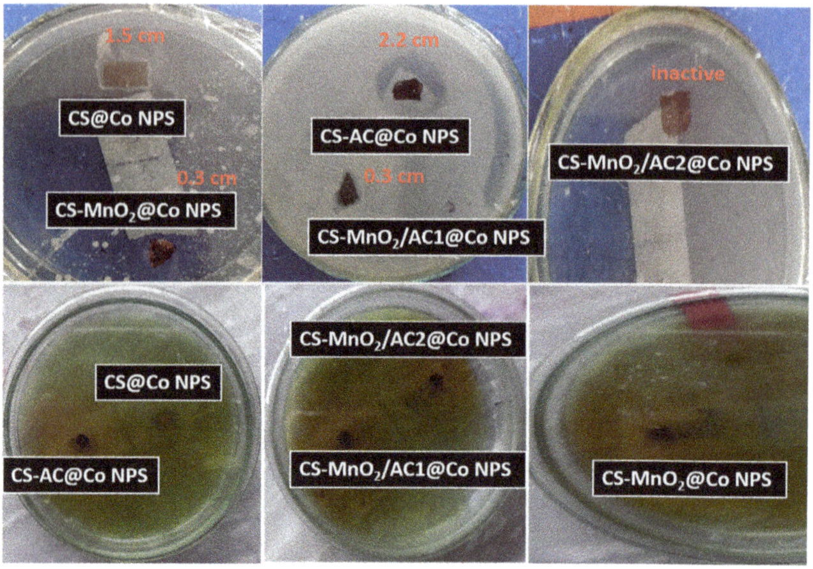

Figure 3. Antibacterial activity of all the catalysts against *P. aeruginosa* (**upper row**) and *E. coli* (**lower row**).

3.4. Catalyst Activity

3.4.1. Hydrogenation of 4NP

The hydrogenation of the $-NO_2$ group of 4NP was carried out by using $NaBH_4$ as a reductant to assess the catalyst activity of CS@Co, CS-AC@Co, CS-MnO_2@Co, CS-MnO_2/AC1@Co and CS-MnO_2/AC2@Co NPs. Firstly, 4NP exhibited a peak at 315 nm in the absorbance spectrum, with a light yellow color; however, soon after the addition of $NaBH_4$, a red shift was observed and the λ_{max} changed to 400 nm, with a deep yellow color [35]. The increase in the wavelength of 4NP was due to the increase in conjugation in the nitrophenolate anion. Numerous studies are available indicating that a reducing agent has a negligible effect on the degradation of 4NP, and thus the hydrogenation of the $-NO_2$ group in the presence of a reducing agent is considered a kinetically unfavorable reaction [36–38]. However, many catalyst systems have been described by various researchers for the hydrogenation of the $-NO_2$ group [1,39–41]; nevertheless, a more efficient and retrievable catalyst system is required. The hydrogenation of 4NP is considered an atom economic reaction where 4-aminophenol (4AmP) is the only major product. Furthermore, this reaction can be easily recorded on a UV–vis spectrophotometer. The decrease in the absorbance of 4-nitrophenolate was recorded at 400 nm as the reaction progressed, where a peak at 290 nm was attributed to the formation of the 4AmP product.

The decline in the absorbance at 400 nm was documented after the addition of CS@Co, CS-AC@Co, CS-MnO_2@Co, CS-MnO_2/AC1@Co and CS-MnO_2/AC2@Co NPs. As depicted in Figure 4a, the CS@Co NPs reduced 92.62% of 4NP to 4AmP in 37 min, 81.67% by CS-AC@Co in 16 min (Figure 4b), 85.55% by CS-MnO_2@Co in 54 min (Figure 4c) and 89.63% by CS-MnO_2/AC1@Co in 12 min (Figure 4d), while CS-MnO_2/AC2@Co NPs reduces 81.60% of 4NP to 4AmP in 45 min (Figure 4e). The percent reduction of the $-NO_2$ group of 4NP to $-NH_2$ of 4AmP is presented in Figure 4f. The rate constant data were extracted from the pseudo-first-order kinetics, which indicated the highest value for CS-MnO_2/AC1@Co NPs, which was 1.56×10^{-1} min^{-1}, and the slowest for CS-MnO_2@Co NPs. The rate constant value and the % reduction of the $-NO_2$ group to $-NH_2$ indicated that the addition of either AC or MnO_2/AC composite had a significant effect on the chitosan polymer and enhanced the rate of this hydrogenation reaction. However, amongst all the catalysts, the

CS-MnO_2/AC1@Co NPs showed superior catalyst performance compared to other catalysts in this study. However, as is obvious from Table 1, all the hybrid films indicated stronger activity as compared to the CS, except CS-MnO_2@Co NPs; therefore, we can conclude that the addition of AC or MnO_2/AC played a prominent role in the improvement of the CS polymer host. Furthermore, the value of R^2 indicates that the experimental data are in good agreement and the pseudo-first-order kinetic model is appropriate.

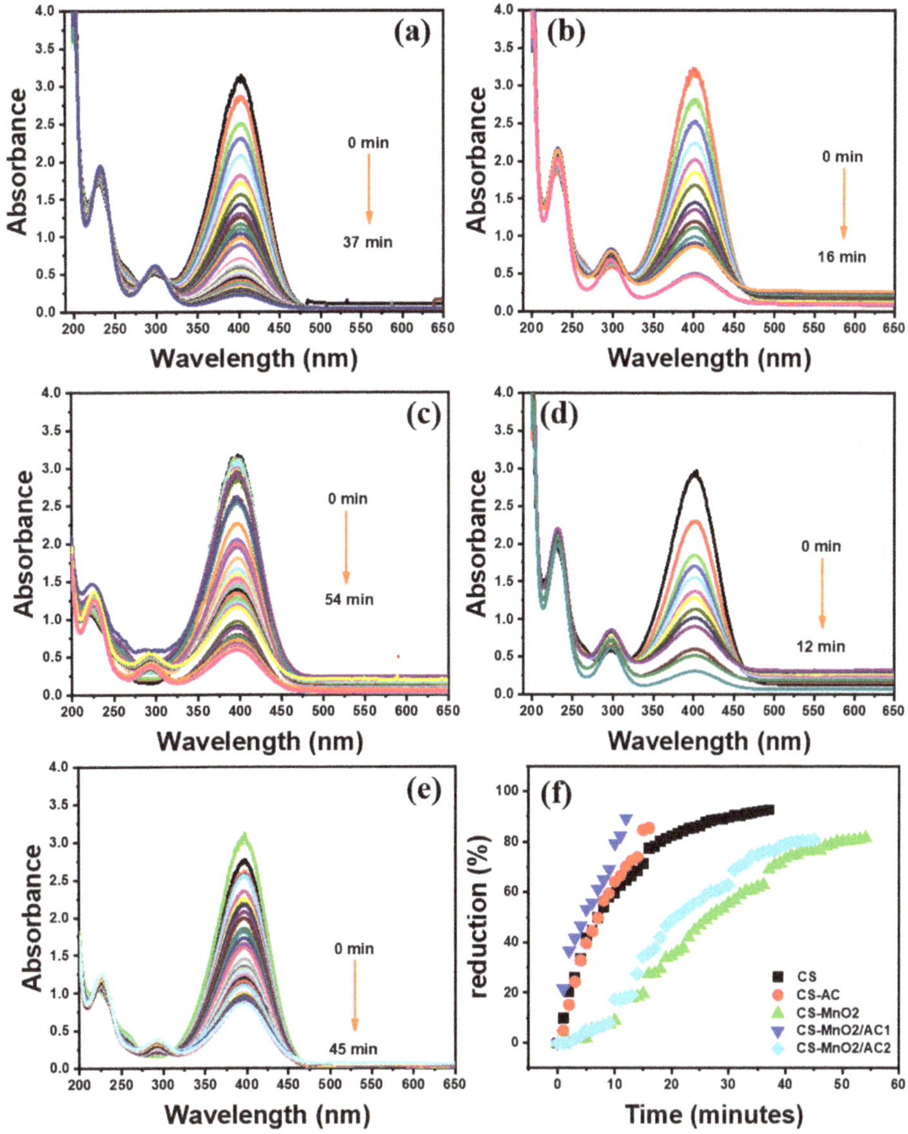

Figure 4. Absorbance spectrum of 4NP reduction in the presence of $NaBH_4$ by using CS@Co (**a**), CS-AC@Co (**b**), CS-MnO_2@Co (**c**), CS-MnO_2/AC1@Co (**d**), CS-MnO_2/AC2@Co NPs (**e**) and percentage reduction as a function of time (**f**). Experimental conditions: 2.5 mL of 0.2 mM 4NP solution; 0.5 mL of 0.5 mM $NaBH_4$ solution and 20 mg of each catalyst.

Table 1. Rate constant, R^2 values and % reduction or degradation values of 4NP and MO and CR dyes by using Co NPs supported on different polymer sheets.

Catalyst	Analyte	R^2	k_{app}	% Degradation
CS@Co		0.9835	7.01×10^{-2}	92.62
CS-AC@Co		0.9698	1.14×10^{-1}	81.67
CS-MnO_2@Co	4NP	0.9821	3.59×10^{-2}	85.55
CS-MnO_2/AC1@Co		0.9371	1.56×10^{-1}	89.63
CS-MnO_2/AC2@Co		0.9842	4.28×10^{-2}	81.60
CS@Co		0.9792	1.43×10^{-1}	93.28
CS-AC@Co		0.9622	2.25×10^{-1}	89.49
CS-MnO_2@Co	CR	0.9171	1.44×10^{-1}	93.53
CS-MnO_2/AC1@Co		0.9410	1.33×10^{-1}	86.31
CS-MnO_2/AC2@Co		0.7940	6.45×10^{-2}	83.44
CS@Co		0.8543	1.50×10^{-2}	71.19
CS-AC@Co		0.9558	5.81×10^{-2}	86.64
CS-MnO_2@Co	MO	0.9543	1.05×10^{-2}	48.61
CS-MnO_2/AC1@Co		0.9977	4.36×10^{-2}	81.42
CS-MnO_2/AC2@Co		0.9119	6.75×10^{-2}	87.97

The k_{app} and R^2 values of 4NP and MO are based on pseudo-first-order kinetics, while the k_{app} and R^2 values for CR discoloration were deduced from zero-order kinetics.

3.4.2. Discoloration of MO Dye

All the catalysts were applied to treat the discoloration caused by MO dye in the presence of $NaBH_4$. MO dye is a mono azo dye with a –N = N– functional group. A peak at 465 nm appeared in the absorbance spectrum of the MO dye; however, the azo group of MO dye reduces to the –NH_2-NH_2– group after the addition of $NaBH_4$. It is noted in the literature that the MO dye is not or is less degraded in the presence of $NaBH_4$, and thus it is kinetically an unfavorable process. The degradation of MO dye at 465 nm was recorded using all the catalysts in the presence of $NaBH_4$; as a result, a peak at 250 nm was observed. The absorbance spectrum of CS@Co NPs is presented in Figure 5a, which shows that approximately 71.19% of the MO dye was decolorized in 60 min. The CS-AC@Co NPs degraded 86.64% of MO dye in 33 min (Figure 5b), while CS-MnO_2@Co NPs degraded only 48.61% in 54 min (Figure 5c). Similarly, CS-MnO_2/AC1 and CS-MnO_2/AC2 decolorized 81.42 and 87.97% of MO dye in 60 and 29 min (Figure 5d,f), respectively. The % degradation of MO dye with all the catalysts is presented in Figure 5f. Based on the data, it is inferred that MnO_2 further diminished the catalyst activity of CS, while the AC, MnO_2/AC1 and MnO_2/AC2 enhanced the catalyst activity compared to CS. The pseudo-first-order model was used to assess the kinetic rate of all the catalysts against the degradation of MO dye. The experimental data fitted well in the pseudo-first-order model; therefore, this model was applied to MO degradation to deduce the k_{app} and R^2 values. As manifested in Table 1, the highest catalyst activity was displayed by CS-MnO_2/AC2, with a rate constant value of 6.75×10^{-2} min^{-1}, followed by CS-AC@Co NPs (5.81×10^{-2} min^{-1}), while the lowest MO degradation rate was displayed by CS-MnO_2@Co NPs (1.05×10^{-2} min^{-1}). The linearity of all the catalysts except CS@Co NPs was above 0.9, which indicates that the experimental data fitted well in the pseudo-first-order model.

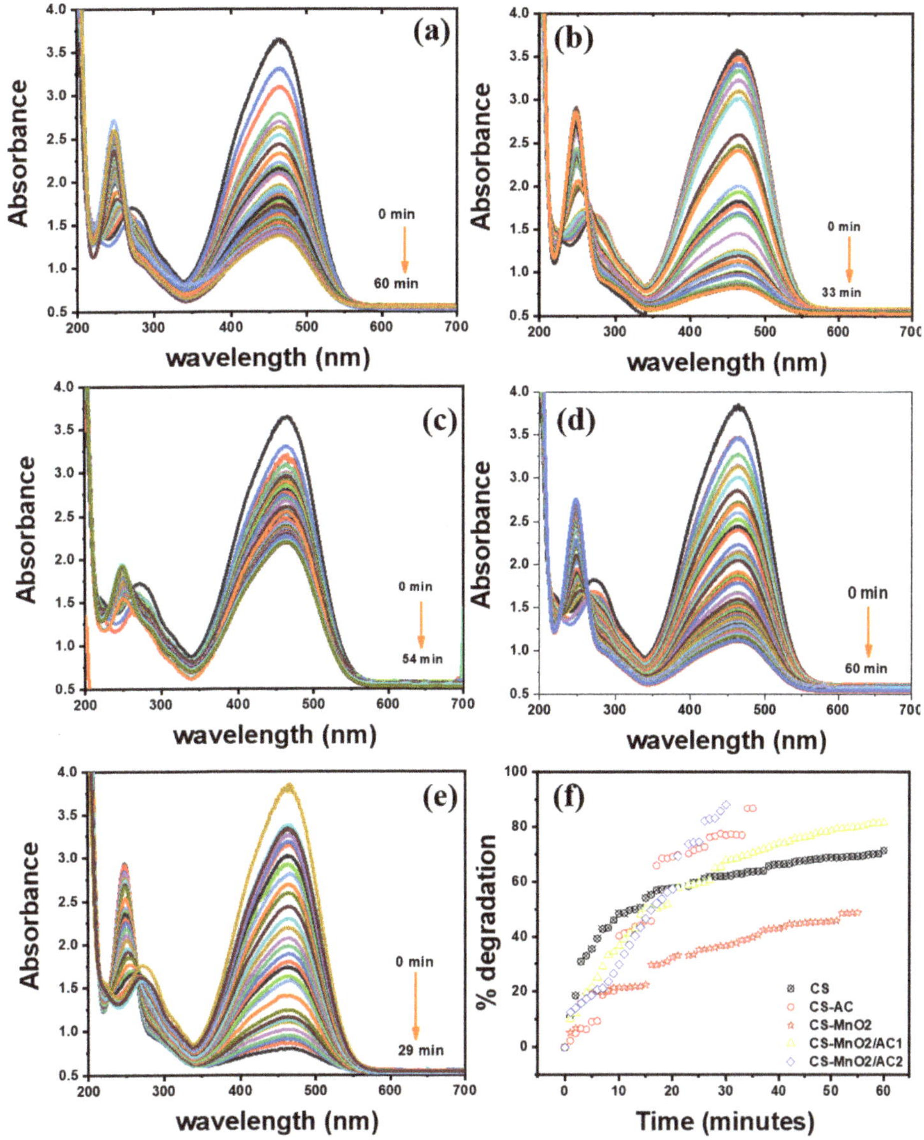

Figure 5. Absorbance spectrum of MO degradation in the presence of $NaBH_4$ by using CS@Co (**a**), CS-AC@Co (**b**), CS-MnO_2@Co (**c**), CS-MnO_2/AC1@Co (**d**) and CS-MnO_2/AC2@Co NPs (**e**) and % degradation of MO dye (**f**). Experimental conditions: 2.5 mL of 0.1 mM MO solution; 0.5 mL of 0.5 mM $NaBH_4$ solution and 20 mg of each catalyst.

3.4.3. Discoloration of CR Dye

CR is a diazo dye and its λ_{max} appeared at 495 nm in the UV–vis spectrum. The azo group is converted to the hydrazine group after $NaBH_4$ treatment [38,42]. As with 4NP and MO dye, the CR dye is also kinetically an unfavorable process with $NaBH_4$ treatment [40,43]. A decrease in the absorbance of the CR dye at 495 nm was recorded during the UV–vis spectroscopy and a few new peaks were noted at 248 and 280 nm due to the formation of amine-containing products. The CS@Co NPs degraded 93.28% of CR dye (see Figure 6a for the absorbance spectrum) in 18 min. Similarly, CS-AC@Co NPs (Figure 6b) decolorized 89.49% of CR dye in 12 min, while CS-MnO_2@Co NPs (Figure 6c) decolorized 93.53% in 19 min. The CS-MnO_2/AC1@Co (Figure 6d) and CS-MnO_2/AC2@Co NPs (Figure 6e) decolorized 86.31% and 83.44% CR dye in 18 and 27 min, respectively. The % discoloration of CR dye is depicted in Figure 6f. Unlike the 4NP and MO dye, the experimental data of the CR dye are fixed in the zero-order kinetics because the linearity of the zero-order kinetics for CR degradation is greater than that of the pseudo-first-order kinetics model. The rate constant with CS-AC@Co NPs was the highest, at 2.25×10^{-1} min^{-1}, and the lowest was observed with CS-MnO_2/AC2, at 6.75×10^{-2} min^{-1}. The R^2 values of all the catalysts except CS-MnO_2/AC2 were above 0.9, which suggests the agreement of the data in the zero-order kinetics (Table 1).

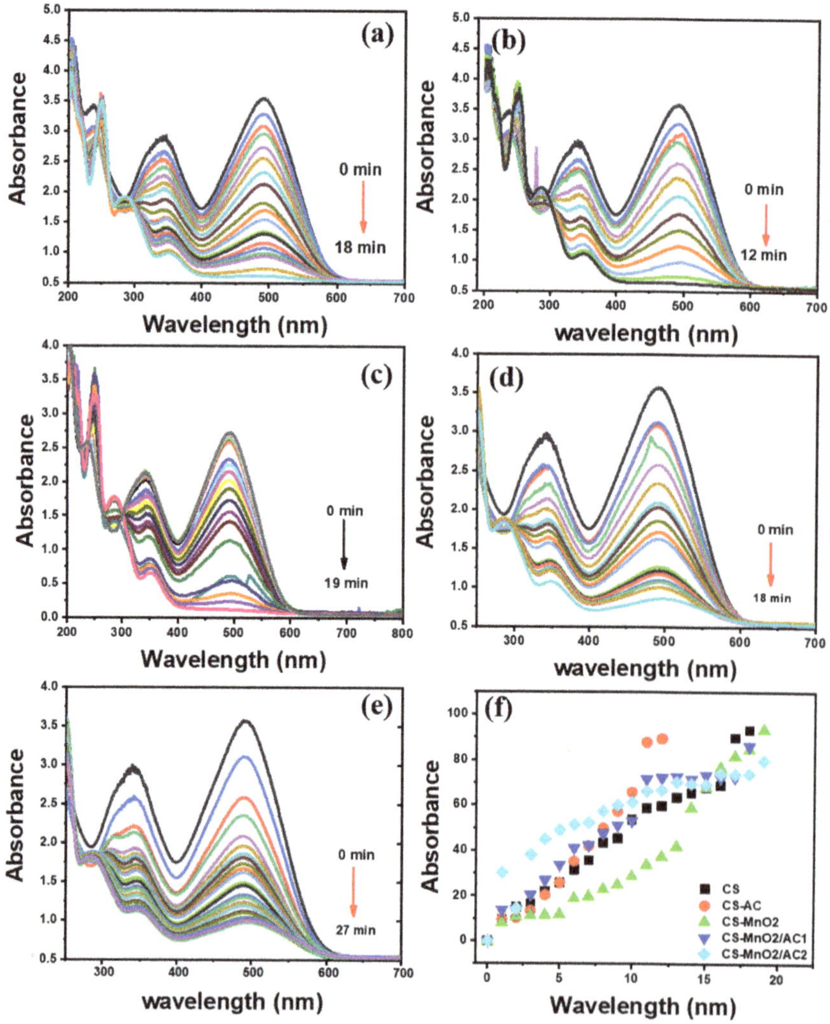

Figure 6. Absorbance spectra of CR degradation in the presence of $NaBH_4$ by using CS@Co (**a**), CS-AC@Co (**b**), CS-MnO_2@Co (**c**), CS-MnO_2/AC1@Co (**d**) and CS-MnO_2/AC2@Co NPs (**e**), and % degradation of CR dye (**f**). Experimental conditions: 2.5 mL of 0.1 mM CR solution; 0.5 mL of 0.5 mM $NaBH_4$ solution and 20 mg of each catalyst.

3.4.4. Recyclability of the Catalyst

The recyclability of the CS-AC catalyst was evaluated against the hydrogenation reaction of the 4NP reaction. First, 0.13 mM solution of the 4NP was treated with 0.5 mM solution of $NaBH_4$ and the reaction's progress was recorded using the UV–vis spectrophotometer. The reaction was completed in 11 min, and afterwards, the catalyst was fed to the second cycle for 11 min and similarly for the third cycle under the same experimental conditions. As shown in Figure 7a, the CS-AC@Co NPs reduced approximately 85% of 4NP to AmP in 11 min; however, the second and third (see Figure 7b,c for UV–vis spectrum) cycles approximately reduced 40% of 4NP to 4AmP in 11 min. As is clear from Figure 7, the catalyst activity was lost after the first cycle; however, no further loss in activity was observed in the second and third cycles. This can be explained by the low availability of the catalyst's active sites after the first cycle. Figure 7d presents a bar graph of the % reduction of 4NP and the inset in Figure 7d shows the % reduction of 4NP as a function of reaction time.

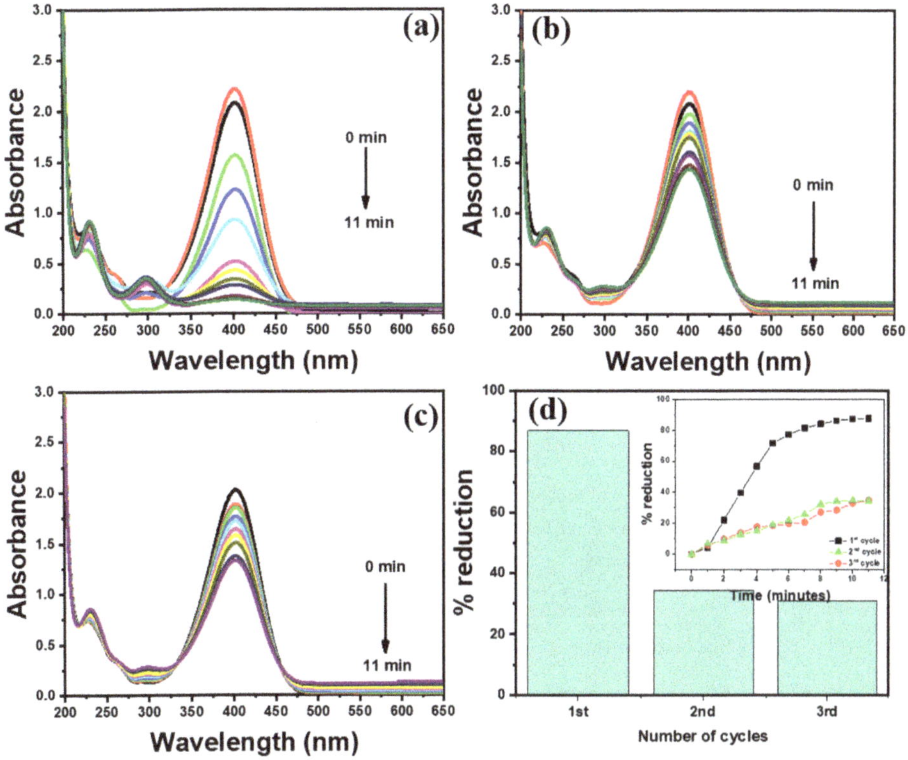

Figure 7. Recyclability of the CS-AC catalyst against 4NP reduction. The absorbance spectrum of 4NP reduction for the 1st cycle (**a**), 2nd cycle (**b**) and 3rd cycle (**c**), and the % reduction of 4NP against the number of cycles (**d**). The bar graph shows the % reduction of 4NP as a function of time in Figure 7d. Experimental conditions: 4NP (2.5 mL of 0.13 mM); $NaBH_4$ (0.5 mL of 0.5 mM); 20 mg of CS-AC catalyst.

4. Conclusions

In the current study, various nanocomposites, including MnO_2, MnO_2/AC1 and MnO_2/AC2, were synthesized through the sol–gel process. The effect of these nanocomposites along with activated carbon (AC) was revealed on chitosan polymer (CS), and 5 weight percent of each nanocomposite and AC was added as a filler in the CS polymer and cast in the form of films. All these films were fabricated with Co^{+2} ions and then converted to their corresponding Co NPs by treatment with $NaBH_4$. Among all the catalysts, CS-AC@Co NPs indicated good antibacterial activity against *P. aeruginosa*, while all the catalysts were inactive against *E. coli*. The synthesized Co NPs on the CS and CS hybrid film templates were used as a dip catalyst for the hydrogenation of 4NP and degradation of MO and CR dyes. The rate constant value of 4NP and MO dye was extracted from the pseudo-first-order model, while it was based on zero-order kinetics for CR dye degradation. The rate constant value for 4NP hydrogenation was higher for CS-MnO_2/AC1@Co and CS-AC@Co NPs (1.56×10^{-1} and 1.14×10^{-1} min^{-1}, respectively). Furthermore, the degradation of MO and CR dye was also well executed using CS-AC@Co, CS-MnO_2/AC1@Co and CS-MnO_2/AC1@Co NPs. The CS-AC@Co NPs also showed good recyclability for the hydrogenation of 4NP. In the first cycle, approximately 84% of 4NP was reduced to 4AmP in 11 min, which was reduced to 38% in the second and third cycles, respectively. However, a negligible loss of catalyst activity was observed in the second and third cycles.

This work applies generally to the removal of azo dyes and reduction of 4NP, as well as the inhibition/killing of *P. aeruginosa* and *E. coli*. However, these catalysts may be used for the degradation of different dyes, reduction of nitroaromatic compounds and inhibition/killing of different microbes.

Author Contributions: Conceptualization H.S.H.M.A., S., Y.A., Y.O.A.-G., M.F. and S.A.K.; methodology, S. and S.A.K.; software, S.A.K.; validation, H.S.H.M.A., S., Y.A., Y.O.A.-G., M.F. and S.A.K.; formal analysis, S. and S.A.K.; investigation, S. and S.A.K.; resources, H.S.H.M.A., Y.A., Y.O.A.-G. and S.A.K.; data curation, H.S.H.M.A., S., Y.A., Y.O.A.-G., M.F. and S.A.K.; writing—original draft preparation, S.A.K.; writing—review and editing, H.S.H.M.A., S., Y.A., Y.O.A.-G., M.F. and S.A.K.; visualization, S.A.K.; supervision, S.A.K.; project administration, H.S.H.M.A., Y.A. and S.A.K.; funding acquisition, H.S.H.M.A., Y.A. and S.A.K. All authors have read and agreed to the published version of the manuscript.

Funding: This research was funded by the Deputyship for Research & Innovation, Ministry of Education in Saudi Arabia and King Abdulaziz University, DSR, Jeddah, Saudi Arabia, under grant number IFPHI-001-130-2020.

Institutional Review Board Statement: Not applicable.

Informed Consent Statement: Not applicable.

Data Availability Statement: The data presented in this study are available on request from the corresponding author.

Acknowledgments: The authors extend their appreciation to the Deputyship for Research & Innovation, Ministry of Education, in Saudi Arabia for funding this research work through project number IFPHI-001-130-2020 and King Abdulaziz University, DSR, Jeddah, Saudi Arabia.

Conflicts of Interest: The authors declare no conflict of interest.

References

1. Khan, S.A.; Khan, S.B.; Kamal, T.; Yasir, M.; Asiri, A.M. Antibacterial nanocomposites based on chitosan/Co-MCM as a selective and efficient adsorbent for organic dyes. *Int. J. Biol. Macromol.* **2016**, *91*, 744–751. [CrossRef] [PubMed]
2. Chanderia, K.; Kumar, S.; Sharma, J.; Ameta, R.; Punjabi, P.B. Degradation of Sunset Yellow FCF using copper loaded bentonite and H_2O_2 as photo-Fenton like reagent. *Arab. J. Chem.* **2017**, *10*, S205–S211. [CrossRef]
3. Kus, E.; Eroglu, H.E. Genotoxic and cytotoxic effects of Sunset Yellow and Brilliant Blue, colorant food additives, on human blood lymphocytes. *Pak. J. Pharm. Sci.* **2015**, *28*, 28.
4. Ghoneim, M.M.; El-Desoky, H.S.; Zidan, N.M. Electro-Fenton oxidation of Sunset Yellow FCF azo-dye in aqueous solutions. *Desalination* **2011**, *274*, 22–30. [CrossRef]
5. Gosetti, F.; Gianotti, V.; Polati, S.; Gennaro, M.C. HPLC-MS degradation study of E110 Sunset Yellow FCF in a commercial beverage. *J. Chromatogr. A* **2005**, *1090*, 107–115. [CrossRef] [PubMed]
6. Kanarek, R.B. Artificial food dyes and attention deficit hyperactivity disorder. *Nutr. Rev.* **2011**, *69*, 385–391. [CrossRef]
7. Gomes, K.M.S.; Oliveira, M.V.G.A.d.; Carvalho, F.R.d.S.; Menezes, C.C.; Peron, A.P. Citotoxicity of food dyes sunset yellow (E-110), bordeaux red (E-123), and tatrazine yellow (E-102) on *Allium cepa* L. root meristematic cells. *Food Sci. Technol.* **2013**, *33*, 218–223. [CrossRef]

8. Rai, H.S.; Bhattacharyya, M.S.; Singh, J.; Bansal, T.; Vats, P.; Banerjee, U. Removal of dyes from the effluent of textile and dyestuff manufacturing industry: A review of emerging techniques with reference to biological treatment. *Crit. Rev. Environ. Sci. Technol.* **2005**, *35*, 219–238. [CrossRef]
9. Liu, S.; Lai, C.; Li, B.; Zhang, C.; Zhang, M.; Huang, D.; Qin, L.; Yi, H.; Liu, X.; Huang, F. Role of radical and non-radical pathway in activating persulfate for degradation of p-nitrophenol by sulfur-doped ordered mesoporous carbon. *Chem. Eng. J.* **2020**, *384*, 123304. [CrossRef]
10. Sohni, S.; Khan, S.A.; Akhtar, K.; Khan, S.B.; Asiri, A.M.; Hashim, R.; Omar, A.K.M. Room temperature preparation of lignocellulosic biomass supported heterostructure (Cu+Co@OPF) as highly efficient multifunctional nanocatalyst using wetness co-impregnation. *Colloids Surf. A Physicochem. Eng. Asp.* **2018**, *549*, 184–195. [CrossRef]
11. Erkurt, E.A.; Erkurt, H.A.; Unyayar, A. Decolorization of azo dyes by white rot fungi. In *Biodegradation of Azo Dyes*; Springer: Berlin/Heidelberg, Germany, 2010; pp. 157–167.
12. Akhtar, K.; Khan, S.A.; Khan, S.B.; Asiri, A.M. *Nanomaterials and Environmental Remediation: A Fundamental Overview*; Bentham Science Publishers: Sharjah, United Arab Emirates, 2018; Volume 2, pp. 1–36.
13. Aoudjit, L.; Martins, P.; Madjene, F.; Petrovykh, D.; Lanceros-Mendez, S. Photocatalytic reusable membranes for the effective degradation of tartrazine with a solar photoreactor. *J. Hazard. Mater.* **2018**, *344*, 408–416. [CrossRef] [PubMed]
14. Moreira, F.C.; Boaventura, R.A.; Brillas, E.; Vilar, V.J. Electrochemical advanced oxidation processes: A review on their application to synthetic and real wastewaters. *Appl. Catal. B Environ.* **2017**, *202*, 217–261. [CrossRef]
15. Martínez-Huitle, C.A.; Brillas, E. Decontamination of wastewaters containing synthetic organic dyes by electrochemical methods: A general review. *Appl. Catal. B Environ.* **2009**, *87*, 105–145. [CrossRef]
16. Vilar, V.J.; Amorim, C.C.; Brillas, E.; Puma, G.L.; Malato, S.; Dionysiou, D.D. AOPs: Recent advances to overcome barriers in the treatment of water, wastewater and air. *Environ. Sci. Pollut. Res.* **2017**, *24*, 5987–5990. [CrossRef]
17. Hussain, S.; Steter, J.R.; Gul, S.; Motheo, A.J. Photo-assisted electrochemical degradation of sulfamethoxazole using a $Ti/Ru_{0.3}Ti_{0.7}O_2$ anode: Mechanistic and kinetic features of the process. *J. Environ. Manag.* **2017**, *201*, 153–162. [CrossRef]
18. Khan, S.A.; Mohammed, S.A.; Bakhsh, E.M.; Al-Ghamdi, Y.O.; Rauf, A.; Akhtar, K.; Begum, A.; Khan, S.B. Reduction of nitrophenol isomers and degradation of azo dyes through zero-valent Ni nanoparticles anchored on cellulose acetate coated Ce/Zr composite. *J. Water Process Eng.* **2021**, *44*, 102383. [CrossRef]
19. Anwar, Y.; Mohammed Ali, H.S.; Rehman, W.U.; Hemeg, H.A.; Khan, S.A. Antibacterial Films of Alginate-CoNi-Coated Cellulose Paper Stabilized Co NPs for Dyes and Nitrophenol Degradation. *Polymers* **2021**, *13*, 4122. [CrossRef]
20. Khan, S.B.; Khan, M.S.J.; Kamal, T.; Asiri, A.M.; Bakhsh, E.M. Polymer supported metallic nanoparticles as a solid catalyst for the removal of organic pollutants. *Cellulose* **2020**, *27*, 5907–5921. [CrossRef]
21. Liu, W.; Yang, X.; Huang, W. Catalytic properties of carboxylic acid functionalized-polymer microsphere-stabilized gold metallic colloids. *J. Colloid Interface Sci.* **2006**, *304*, 160–165. [CrossRef]
22. Kuroda, K.; Ishida, T.; Haruta, M. Reduction of 4-nitrophenol to 4-aminophenol over Au nanoparticles deposited on PMMA. *J. Mol. Catal. A Chem.* **2009**, *298*, 7–11. [CrossRef]
23. Panimalar, S.; Logambal, S.; Thambidurai, R.; Inmozhi, C.; Uthrakumar, R.; Muthukumaran, A.; Rasheed, R.A.; Gatasheh, M.K.; Raja, A.; Kennedy, J. Effect of Ag doped MnO_2 nanostructures suitable for wastewater treatment and other environmental pollutant applications. *Environ. Res.* **2022**, *205*, 112560. [CrossRef] [PubMed]
24. Shoba, A.; Kavitha, B.; Aswathaman, H.; Ganesan, H.; Kumar, N.S. Synthesis, characterization and electrochemical sensors application of MnO_2 nanoparticles. *Mater. Today Proc.* **2022**, *48*, 521–526. [CrossRef]
25. Khan, S.A.; Bakhsh, E.M.; Asiri, A.M.; Khan, S.B. Chitosan coated NiAl layered double hydroxide microsphere templated zero-valent metal NPs for environmental remediation. *J. Clean. Prod.* **2021**, *285*, 124830. [CrossRef]
26. Khan, S.A.; Baksh, E.M.; Akhtar, K.; Khan, S.B. A template of cellulose acetate polymer-ZnAl layered double hydroxide composite fabricated with Ni NPs: Applications in the hydrogenation of nitrophenols and dyes degradation. *Spectrochim. Acta Part A Mol. Biomol. Spectrosc.* **2020**, *241*, 118671. [CrossRef]
27. Ashrafizade, M.; Delfi, M.; Hashemi, F.; Zabolian, A.; Saleki, H.; Bagherian, M.; Azami, N.; Farahani, M.V.; Sharifzadeh, S.O.; Hamzehlou, S. Biomedical application of chitosan-based nanoscale delivery systems: Potential usefulness in siRNA delivery for cancer therapy. *Carbohydr. Polym.* **2021**, *260*, 117809. [CrossRef] [PubMed]
28. Bagheri, M.; Validi, M.; Gholipour, A.; Makvandi, P.; Sharifi, E. Chitosan nanofiber biocomposites for potential wound healing applications: Antioxidant activity with synergic antibacterial effect. *Bioeng. Transl. Med.* **2021**, *7*, e10254. [CrossRef] [PubMed]
29. Hussain, S.; Kamran, M.; Khan, S.A.; Shaheen, K.; Shah, Z.; Suo, H.; Khan, Q.; Shah, A.B.; Rehman, W.U.; Al-Ghamdi, Y.O. Adsorption, kinetics and thermodynamics studies of methyl orange dye sequestration through chitosan composites films. *Int. J. Biol. Macromol.* **2021**, *168*, 383–394. [CrossRef] [PubMed]
30. Hudzicki, J. Kirby-Bauer disk diffusion susceptibility test protocol. *Am. Soc. Microbiol.* **2009**, *66*, 208.
31. Kamal, T.; Khan, S.B.; Haider, S.; Alghamdi, Y.G.; Asiri, A.M. Thin layer chitosan-coated cellulose filter paper as substrate for immobilization of catalytic cobalt nanoparticles. *Int. J. Biol. Macromol.* **2017**, *104*, 56–62. [CrossRef]
32. Zhu, J.; Hu, G.; Zhang, J. Preparation of Sn-Cu-graphene nanocomposites with superior reversible lithium ion storage. *Mater. Lett.* **2016**, *185*, 565–568. [CrossRef]

33. Tamer, T.M.; Sabet, M.M.; Omer, A.M.; Abbas, E.; Eid, A.I.; Mohy-Eldin, M.S.; Hassan, M.A. Hemostatic and antibacterial PVA/Kaolin composite sponges loaded with penicillin–streptomycin for wound dressing applications. *Sci. Rep.* **2021**, *11*, 3428. [CrossRef] [PubMed]
34. Tamer, T.M.; Alsehli, M.H.; Omer, A.M.; Afifi, T.H.; Sabet, M.M.; Mohy-Eldin, M.S.; Hassan, M.A. Development of Polyvinyl Alcohol/Kaolin Sponges Stimulated by Marjoram as Hemostatic, Antibacterial, and Antioxidant Dressings for Wound Healing Promotion. *Int. J. Mol. Sci.* **2021**, *22*, 13050. [CrossRef] [PubMed]
35. Aljohny, B.O.; Ahmad, Z.; Shah, S.A.; Anwar, Y.; Khan, S.A. Cellulose acetate composite films fabricated with zero-valent iron nanoparticles and its use in the degradation of persistent organic pollutants. *Appl. Organomet. Chem.* **2020**, *34*, e5892. [CrossRef]
36. Aditya, T.; Pal, A.; Pal, T. Nitroarene reduction: A trusted model reaction to test nanoparticle catalysts. *Chem. Commun.* **2015**, *51*, 9410–9431. [CrossRef]
37. Bello, B.A.; Khan, S.A.; Khan, J.A.; Syed, F.Q.; Mirza, M.B.; Shah, L.; Khan, S.B. Anticancer, antibacterial and pollutant degradation potential of silver nanoparticles from Hyphaene thebaica. *Biochem. Biophys. Res. Commun.* **2017**, *490*, 889–894. [CrossRef] [PubMed]
38. Khan, S.A.; Khan, N.; Irum, U.; Farooq, A.; Asiri, A.M.; Bakhsh, E.M.; Khan, S.B. Cellulose acetate-Ce/Zr@Cu^0 catalyst for the degradation of organic pollutant. *Int. J. Biol. Macromol.* **2020**, *153*, 806–816. [CrossRef]
39. Abazari, R.; Heshmatpour, F.; Balalaie, S. Pt/Pd/Fe trimetallic nanoparticle produced via reverse micelle technique: Synthesis, characterization, and its use as an efficient catalyst for reductive hydrodehalogenation of aryl and aliphatic halides under mild conditions. *ACS Catal.* **2013**, *3*, 139–149. [CrossRef]
40. Khan, S.A.; Bello, B.A.; Khan, J.A.; Anwar, Y.; Mirza, M.B.; Qadri, F.; Farooq, A.; Adam, I.K.; Asiri, A.M.; Khan, S.B. Albizia chevalier based Ag nanoparticles: Anti-proliferation, bactericidal and pollutants degradation performance. *J. Photochem. Photobiol. B Biol.* **2018**, *182*, 62–70. [CrossRef]
41. Ni, W.; Wang, B.; Cheng, J.; Li, X.; Guan, Q.; Gu, G.; Huang, L. Hierarchical foam of exposed ultrathin nickel nanosheets supported on chainlike Ni-nanowires and the derivative chalcogenide for enhanced pseudocapacitance. *Nanoscale* **2014**, *6*, 2618–2623. [CrossRef]
42. Ali, H.S.H.M.; Khan, S.A. Stabilization of Various Zero-Valent Metal Nanoparticles on a Superabsorbent Polymer for the Removal of Dyes, Nitrophenol, and Pathogenic Bacteria. *ACS Omega* **2020**, *5*, 7379–7391. [CrossRef]
43. Khan, S.A.; Khan, S.B.; Farooq, A.; Asiri, A.M. A facile synthesis of CuAg nanoparticles on highly porous ZnO/carbon black-cellulose acetate sheets for nitroarene and azo dyes reduction/degradation. *Int. J. Biol. Macromol.* **2019**, *130*, 288–299. [CrossRef] [PubMed]

polymers MDPI

Article

Effects of Extrusion and Irradiation on the Mechanical Properties of a Water–Collagen Solution

Hynek Chlup [1], Jan Skočilas [2,*], Jaromír Štancl [2], Milan Houška [3] and Rudolf Žitný [2]

1 Department of Mechanics, Biomechanics and Mechatronics, Faculty of Mechanical Engineering, Czech Technical University in Prague, Technická 4, 160 00 Prague, Czech Republic; hynek.chlup@fs.cvut.cz
2 Department of Process Engineering, Faculty of Mechanical Engineering, Czech Technical University in Prague, Technická 4, 160 00 Prague, Czech Republic; jaromir.stancl@fs.cvut.cz (J.Š.); rudolf.zitny@fs.cvut.cz (R.Ž.)
3 Food Research Institute Prague, Radiová 1285/7, 102 00 Prague, Czech Republic; milan.houska@vupp.cz
* Correspondence: jan.skocilas@fs.cvut.cz; Tel.: +42-02-2435-2719

Abstract: This article describes 1D extension tests on bovine collagen samples (8% collagen in water). At such a high collagen concentration, the mechanical properties of semi-solid samples can be approximated by hyperelastic models (two-parametric HGO and Misof models were used), or simply by Hooke's law and the modulus of elasticity E. The experiments confirm a significant increase in the E-modulus of the samples irradiated with high-energy electrons. The modulus E ~ 9 kPa of non-irradiated samples increases monotonically up to E ~ 250 kPa for samples absorbing an e-beam dose of ~3300 Gy. This amplification is attributed to the formation of cross-links by irradiation. However, E-modulus can be increased not only by irradiation but also by exposure to a high strain rate. For example, soft isotropic collagen extruded through a 200 mm long capillary increases the modulus of elasticity from 9 kPa to 30 kPa, and the increase is almost isotropic. This stiffening occurs when the corrugated collagen fibers are straightened and are aligned in the flow direction. It seems that the permanent structural changes caused by extrusion mitigate the effects of the ex post applied irradiation. Irradiation of extruded samples by 3300 Gy increases the modulus of E-elasticity only three times (from 30 kPa to approximately 90 kPa). Extruded and ex post irradiated samples show slight anisotropy (the stiffness in the longitudinal direction is on an average greater than the transverse stiffness).

Keywords: collagen; anisotropy; electron irradiation; extrusion; tensile test

Citation: Chlup, H.; Skočilas, J.; Štancl, J.; Houška, M.; Žitný, R. Effects of Extrusion and Irradiation on the Mechanical Properties of a Water–Collagen Solution. *Polymers* **2022**, *14*, 578. https://doi.org/10.3390/polym14030578

Academic Editor: Arn Mignon

Received: 5 December 2021
Accepted: 28 January 2022
Published: 31 January 2022

1. Introduction

Collagen-based materials are used, for example, in manufacturing vascular grafts in biomedicine [1], and in making sausage casings in the food industry [2]. However, there is little information regarding the mechanical properties of these collagenous systems. Skocilas et al. [3] used an extrusion rheometer on a fluid-like water suspension of native collagen and described the "shear" behavior at high shear rates, using a power law model. Semisolid highly elastic collagenous systems (for example tendons) are usually tested using extension machines, which provide information about the elastic modulus, stress relaxation, and the ultimate tensile strength [4]. During or in combination with uniaxial loading tests, structural information regarding the arrangement of collagen fibers can be determined using atomic force microscopy, electron microscopy, orcein staining, or X-ray scattering. This information enables entropic structural models describing the "toe" and "heel" regions of the stress–strain relationship to be identified [5,6].

The mechanical, thermodynamic, and biochemical properties of collagen can be modified through the addition of suitable additives (e.g., glutaraldehyde or polyvinyl alcohol) or through irradiation, which affects the crosslinking between long collagen macromolecules; however, in some cases the result is fragmentation of the irradiated dried collagen macromolecules. Electron irradiation [7], UV irradiation [8] and gamma irradiation [9] have all been used on various collagen materials; these tests were usually performed on soft tissues, e.g., tendons [6], or pericardium [4]. Only a small number of experiments have focused on the effects of radiation on the extracellular matrix or on water-collagen mixtures [10,11]. Riedel et al. [12] found out that, for collagen matrices (rat tail collagen, 2 mg/mL) irradiated by a dose from 0–100 kGy, the crosslinking density increases with the irradiation dose.

Water-soluble polymers can be converted into hydrogels with the use of irradiation [12]. The properties of the irradiated product depend on the concentration of the polymer and the radiation dose. In general, the crosslink density increases with the polymer concentration and the radiation dose [13]. Usually, high power of the dose is used for crosslinking enhancement from 10 kGy up to 0.5 MGy. Manaila et al. [14] applied high power of the dose of electron beam irradiation of 150, 300 and 450 kGy to investigate the mechanical properties and the degradation process of a composite of natural rubber and plasticized starch. They found that the tensile strength and also the cross-link density increase slightly at 150 kGy and then decrease with the increase in the irradiation dose.

Jamal et al. [15] investigated the effect of irradiation treatment on the mechanical properties of samples made from high-density polyethylene (HDPE, 70 vol%), ethylene propylene diene monomer rubber (EPDM 30 vol%) and nanoparticles of organophilic montmorillonite clay (OMMT, 2–8 vol%). An irradiation dose of 50, 100, 150 and 200 kGy was applied. The mechanical properties (the tensile strength) of the mixture increase with an irradiation dose of up to 100 kGy, then decrease with the irradiation dose. The best improvement of the mechanical properties was achieved with clay (4 vol%) and with a 100 kGy irradiation dose. The same results were obtained by the authors of [16] for EPDM polymer using gamma irradiation treatment (up to 1.4 MGy), and by [17] using gamma irradiation treatment (up to 250 kGy).

A similar phenomenon was observed after irradiation of chitosan powder for membrane production, see [18]. The molecular weight of a chitosan solution and its mechanical properties (Young's modulus, tensile strength and elongation at break) decreased with the increase in the irradiation dose (12.5, 100 and 250 kGy). This behavior can be attributed to a reduction in molecular weight and degradation of the molecular chain.

The authors of [19] used e-beam treatment of sulfonated polyether ketone (CSPEEK) membranes and investigated the mechanical properties of the membranes. They used doses in the range from 25 to 200 kGy, and they found that both the stress at break and the strain at break increased gradually with the increase in the irradiation dose, but no significant changes were observed from the yield points. The results showed that the CSPEEK membranes became harder and the degree of crosslinking increased with the increasing irradiation dose.

Similar trends [20] of dependence of mechanical properties on irradiation dose were observed for chitosan blends with polyvinyl alcohol PVA within the range of 0–75 kGy by [21,22].

Similar results were obtained for other polymer materials by [23] for silica-filled silicone elastomer (up to 200 kGy gamma irradiation treatment), by [24] for fluorocarbon elastomer (up to 200 kGy), and by [25] for styrene/butadiene-based linear triblock copolymer (up to 200 kGy).

Cui and Zhang [26] also used electron beam irradiation to modify the mechanical and thermal properties of ternary polyamide copolymer (TPA) crosslinked by various agents. They also found the maximum tensile strength of samples irradiated by doses between 100–250 kGy with respect to the crosslinking agent that was used. The range of the irradiation dose was from 50–300 kGy. The trend of the dependency of the mechanical properties on the irradiation dose is the same. There is an optimum dose corresponding to the maximum tensile strength for each polymer and for each mixture of polymer with additives. Above this optimum dose, the mechanical properties of the treated sample decrease. In order to improve or optimize the mechanical properties of a polymer product by irradiation treatment, the maximum dose has to be found, e.g., for a further engineering application [27,28].

The aim of this paper is to describe the changes in the mechanical properties of a bovine collagen solution after electron irradiation at various energy levels, which is an interesting and available method for changing the mechanical properties of biopolymer products. An even more interesting conclusion concerns the significant difference in mechanical properties in the highly concentrated isotropic collagen mass after passing through the narrow slit of the extruder may be due to structural changes in the material.

2. Materials and Methods

Extension tests were performed on an aqueous solution of natural bovine collagen (type I) with a collagen content of 8.0% by weight. Bovine collagen matter is made in tanneries from cattle hides according to established techniques [29]. The hypodermis is separated from the epidermis, the fat and the hair by soaking, liming, neutralization and bleaching, and the final material is washed, cut, milled and mixed with water to the required water/collagen content. No additives were used for the material investigated here. Further information on the composition and other properties of non-irradiated collagen material has been published previously [30]. Namely, the existence of very long collagen fibers (24% of the longest fraction 780 kDa) was identified with the use of size exclusion chromatography and UV detection, see [30]. Viscoelastic behavior of the material was detected by a cone and plate oscillating rheometer, modules of G' 15–18 kPa and G'' of 18–21 kPa were measured, see [30]. Measurements of the thermal properties of the collagen matter by DSC showed thermal capacity values in the range 3700–4400 J/(kg·K) and the transition temperature into gelatin in the range between 33–43 °C, see [31]. The parameters of the power-law model of the shear flow properties were identified as N = 0.28 and K = 700 Pa·s^n for the investigated collagen matter, see [3].

Isotropic samples of collagen material were irradiated at doses of 858 Gy and 3380 Gy using a Microtrone MT 25 cyclic electron accelerator; see [32]. In our experiments, an approximately constant intensity of electron irradiation was used. The distance of the irradiated samples from the e-beam emitter (500 W) was constant, and only the irradiation time changed. The cake sample was irradiated in a plastic bag, was then cut into strips (typical size 40 × 4 × 10 mm), and was clamped in the jaws of a Zwick Roell biaxial extension tester, see Figure 1. The sampling frequency of the force and displacement sensing was 20 Hz. The exact geometry and the different thicknesses of each sample were measured (optically), and the most uniform cross-section was selected for targeting the Zwick Roell optical extensometer and for evaluating the Cauchy stress. The dimensions of the isotropic samples are presented in Table 1.

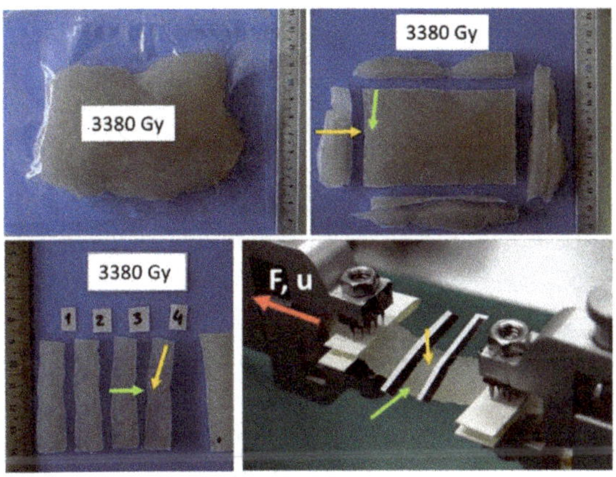

Figure 1. Sample storage, preparation, and attachment to a Zwick Roell biaxial extension tester. Non-extruded collagen irradiated by dose 3380 Gy. The arrows indicate the direction of sample cutting and attachment for isotropy analysis. F—force, u—displacement.

Table 1. Dimensions of non-extruded isotropic samples. *H*—sample thickness, *B*—sample width, *A*—sample cross-section. Values with 95% confidence bounds.

Dose (Gy)	*H* (mm)	*B* (mm)	*A* (mm^2)
0	8.27 ± 1.40	10.70 ± 1.46	89.10 ± 21.58
858	6.12 ± 0.62	11.23 ± 0.65	68.79 ± 8.33
3380	3.47 ± 0.40	10.44 ± 0.44	35.34 ± 4.63

The same raw collagen mass (isotropic collagen from the same batch) was used to analyze planar samples formed by extrusion of isotropic collagen through a narrow slit (cross-section 2×20 mm, capillary length 200 mm) at a very high strain rate of 2000 s^{-1}, see [31]. The dissipated mechanical energy in the material pushed through the capillary was 2.07 ± 0.31 kJ/kg for the geometry, the process conditions (wall shear stress 11.3 kPa) and the rheology parameters (flow index 0.25 and coefficient of consistency 1700 $Pa{\cdot}s^n$) that were used, see [31]. The extruded strips were slightly anisotropic (when clamping the strips, it was necessary to distinguish between the longitudinal orientation and the transverse orientation of the filaments), and the orientation according to the flow direction had to be respected, see Figure 2. Otherwise, the experimental procedure was the same as in the isotropic case. The dimensions of the extruded samples are presented in Table 2.

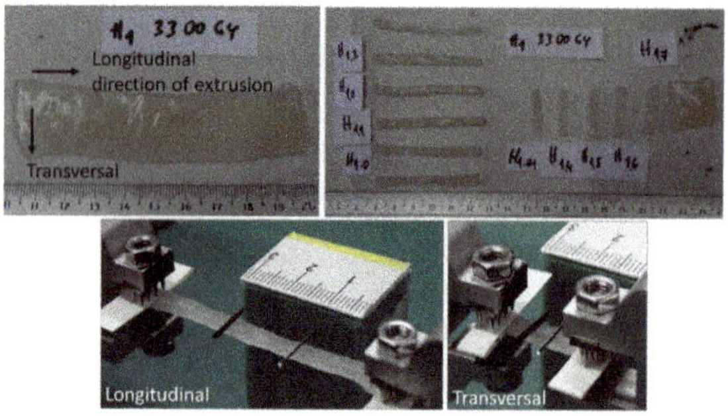

Figure 2. Sample preparation and attachment to a Zwick Roell biaxial extension tester. Extruded collagen irradiated by dose 3300 Gy. The arrows indicate the direction of sample orientation during extrusion.

Table 2. Dimensions of the extruded samples. *H*—sample thickness, *B*—sample width, *A*—sample cross section. Values with 95% confidence bounds.

Dose (Gy)	*H* (mm)	*B* (mm)	*A* (mm^2)
0	2.18 ± 0.12	6.25 ± 0.37	13.61 ± 0.97
250	2.10 ± 0.06	4.60 ± 0.28	9.66 ± 0.59
850	2.17 ± 0.11	3.98 ± 0.19	8.63 ± 0.42
3300	2.20 ± 0.12	4.08 ± 0.24	8.98 ± 0.54

Simple tension tests were performed. The jaw displacement was controlled during the experiment. The samples were loaded by deformation. The jaw velocity was 0.1 mm/s. The average value of the strain rate for all samples was 0.005 1/s. The force was measured using HBM U9B force sensors with a range of 50 N, accuracy class 0.5 (tensile/compressive force transducer, nominal forces 50 N, accuracy class 0.5). The recorded data were recalculated to the relationship between the deformation ε and the true Cauchy stress σ, using Matlab. Matlab functions were used for identifying the hyperelastic models and for calculating the associated statistics. The measured data are characterized by the time-independent stress stiffening and can be described using almost any hyperelastic model, e.g., three-parameter HGO [33].

For uniaxial loading and for deformation ε less than 0.3, the HGO model can be simplified to a two-parameter model

$$\sigma = E\varepsilon + 18k_1\varepsilon^3. \tag{1}$$

The same data can also be approximated using a two-parameter structural model [6]

$$\sigma = \frac{E\varepsilon}{1 - \frac{\varepsilon}{\varepsilon_l}} \tag{2}$$

where ε_l is the limiting deformation. For a very small deformation ε (or for $k_1 = 0$ or for $\varepsilon_l \to \infty$) both models reduce to Hooke's model with Young's modulus E.

3. Results and Discussion

A total of 31 samples were processed with three different doses of irradiation during our study of isotropic randomly-oriented collagen. The stress–strain relationship depends slightly on the elongation rate. However, there were no discernible trends, and therefore the viscous effects were neglected during the analysis. Measured data for three levels of irradiation (10 non-irradiated samples, 7 samples exposed to 858 Gy, and 14 samples exposed to 3380 Gy) and the corresponding predictions of the hyperelastic models are shown in Figure 3 separately for each dose. For a clear comparison, the predictions of the models are shown together in one graph, see Figure 4.

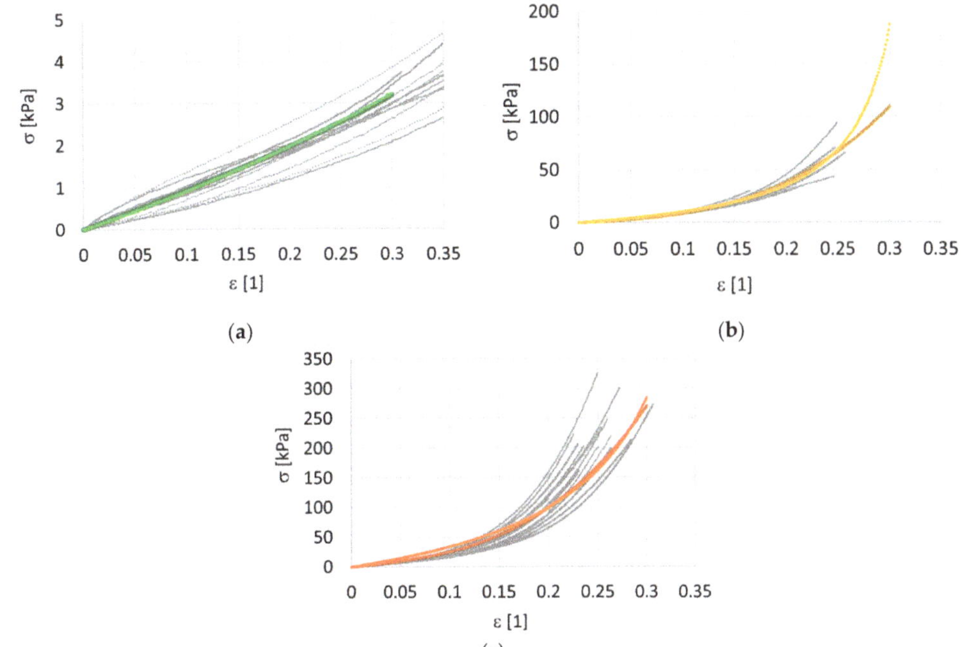

Figure 3. Cauchy stress—deformation data for isotropic collagen mass. Modified Holzapfel–Gasser–Ogden (HGO) model and Misof model. The grey points represent data, the colored lines represent models. Two doses of irradiation were used, 858 Gy and 3380 Gy. (**a**) non-irradiated 0 Gy, models are overlapping, (**b**) dose 858 Gy, the light line represents the HGO model, and the dark line represents the Misof model, (**c**) dose 3380 Gy, models are overlapping.

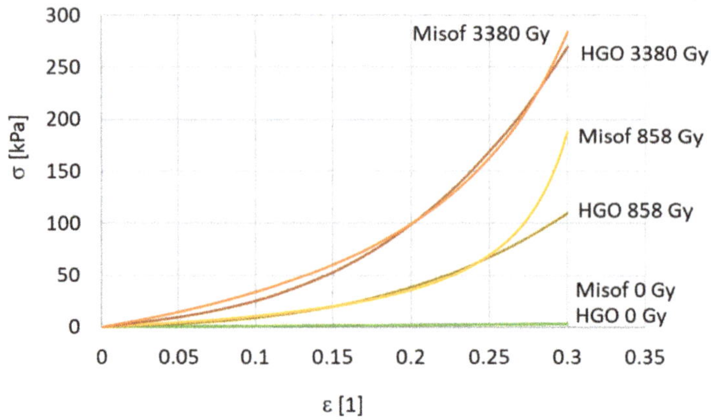

Figure 4. A comparison of the models for different irradiation doses, isotropic collagen mass.

The non-irradiated samples had the worst repeatability. For the non-irradiated samples, it was difficult to prepare a good-quality sample and the measurements were also difficult to make. The non-irradiated samples were sticky and were very pliable with minimal flexural stiffness. They were difficult to handle. To provide a clear comparison, the true (Cauchy) stress was 10 kPa for the non-irradiated samples, 110 kPa for the samples exposed to 858 Gy, and more than 270 kPa for the samples irradiated at the maximum dose of 3380 Gy, at the same 0.3 deformation. The 10 non-irradiated samples were very soft, with very high limiting deformation $\varepsilon_l > 1$. The identified parameters of the simplified HGO model, Equation (1), are presented in Table 3, while the parameters of the Misof model, Equation (2), are presented in Table 4. Note that elastic moduli of the order of 10 kPa are quite small for non-irradiated samples, and the maximum of 170 kPa is exhibited by samples irradiated by 3380 Gy. At first glance, it seems that the HGO model produced a better approximation of the averaged data, when the standard deviations between the measured and predicted stresses are taken into consideration.

Table 3. Isotropic samples. Parameters of the simplified HGO model. Values with 95% confidence bounds.

Dose (Gy)	E (kPa)	k_1 (kPa)	R-Square (-)
0	9.24 ± 0.10	0.84 ± 0.02	0.94
858	52.55 ± 1.96	193.20 ± 2.70	0.96
3300	169.9 ± 6.90	450.9 ± 7.70	0.89

Table 4. Isotropic samples. Parameters of the Misof model. Values with 95% confidence bounds.

Dose (Gy)	E (kPa)	ε_l (-)	R-Square (-)
0	8.49 ± 0.08	1.439 ± 0.022	0.94
858	73.55 ± 0.95	0.337 ± 0.002	0.95
3380	254.4 ± 3.60	0.414 ± 0.003	0.87

The second group of experiments investigated the effects of flow and anisotropy. Samples in the form of extruded bands were analyzed separately in the longitudinal direction (20 samples) and in the transversal direction (20 samples) at four different levels of dosing: 0, 250 Gy, 850 Gy and 3300 Gy. For each radiation dose in each direction, 5 valid measurements were evaluated. The identified model parameters are presented in Table 5—HGO longitudinal, in Table 6—transversal, in Table 7—Misof longitudinal, and in Table 8—transversal. The measured data for longitudinal deformation are shown in Figure 5, and the transversal deformation is shown in Figure 6. For a clear comparison, the predictions of the models are shown together in a single graph, see Figure 7a for the longitudinal direction and Figure 7b for the transversal direction. For the transversal samples, the data are thinner because the samples were shorter, but the loading rate was the same as for the longitudinal samples.

Table 5. Extruded samples. Parameters of the simplified HGO model. Longitudinal direction. Values with 95% confidence bounds.

Dose (Gy)	E (kPa)	k_1 (kPa)	R-Square (-)
0	30.96 ± 0.83	11.58 ± 0.74	0.95
250	46.00 ± 3.65	27.31 ± 7.75	0.79
850	90.16 ± 3.50	59.85 ± 10.02	0.90
3300	56.88 ± 2.00	57.35 ± 3.51	0.96

Table 6. Extruded samples. Parameters of the simplified HGO model. Transversal direction. Values with 95% confidence bounds.

Dose (Gy)	E (kPa)	k_1 (kPa)	R-Square (-)
0	27.37 ± 0.80	1.89 ± 0.93	0.90
250	23.31 ± 2.09	6.10 ± 1.48	0.89
850	40.39 ± 1.64	15.90 ± 4.46	0.97
3300	99.82 ± 5.89	11.85 ± 3.45	0.90

Table 7. Extruded samples. Parameters of the Misof model. Longitudinal direction. Values with 95% confidence bounds.

Dose (Gy)	E (kPa)	ε_l (-)	R-Square (-)
0	28.64 ± 0.73	0.718 ± 0.031	0.95
250	41.95 ± 3.78	0.559 ± 0.121	0.79
850	82.78 ± 3.72	0.538 ± 0.067	0.90
3300	54.30 ± 2.03	0.452 ± 0.016	0.96

Table 8. Extruded samples. Parameters of the Misof model. Transversal direction. Values with 95% confidence bounds.

Dose (Gy)	E (kPa)	ε_l (-)	R-Square (-)
0	25.96 ± 1.11	2.050 ± 0.774	0.90
250	21.09 ± 1.92	0.824 ± 0.136	0.89
850	37.80 ± 2.06	0.771 ± 0.187	0.97
3300	91.85 ± 6.56	1.286 ± 0.289	0.90

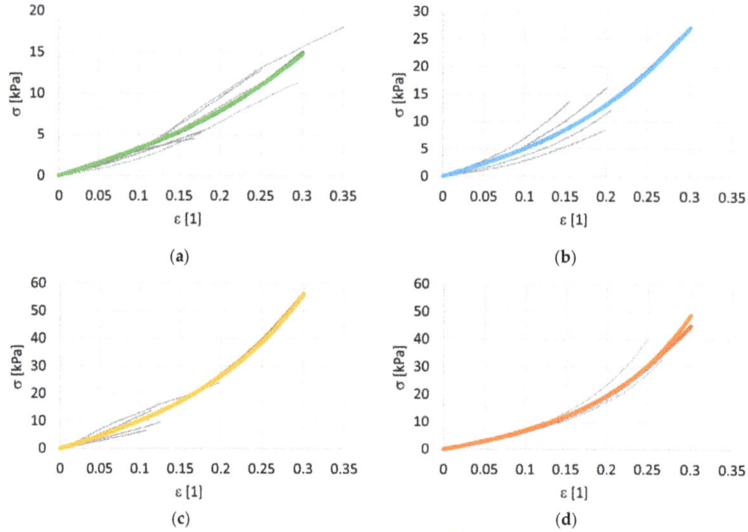

Figure 5. Cauchy stress—deformation for extruded collagen mass. The modified Holzapfel–Gasser–Ogden (HGO) model and the Misof model. Longitudinal direction. Three doses of irradiation were used: 250 Gy, 850 Gy and 3300 Gy. The grey points represent data, and the colored lines represent models. (**a**) non-irradiated 0 Gy, models are overlapping, (**b**) dose 250 Gy, models are overlapping, (**c**) dose 850 Gy, models are overlapping, (**d**) dose 3380 Gy, the light line represents the HGO model, and the dark line represents the Misof model.

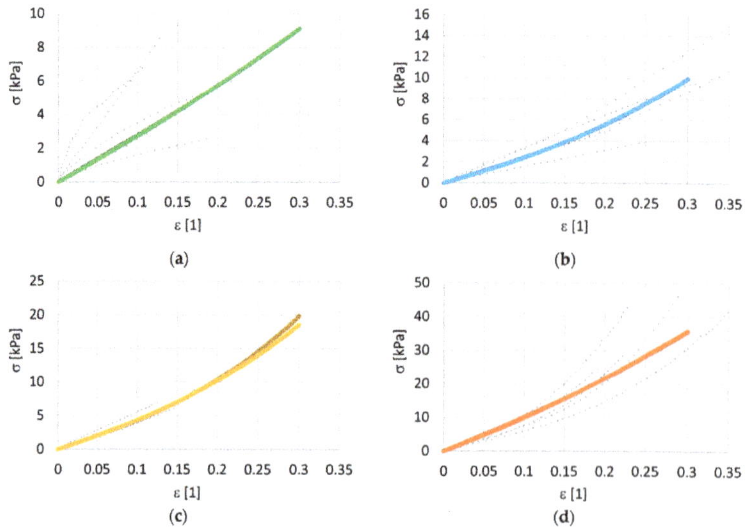

Figure 6. Cauchy stress—deformation for extruded collagen mass. The modified Holzapfel–Gasser–Ogden (HGO) model and the Misof model. Transversal direction. Three doses of irradiation were used: 250 Gy, 850 Gy and 3300 Gy. The grey points represent data, and the colored lines represent models. (**a**) non-irradiated 0 Gy, models are overlapping, (**b**) dose 250 Gy, models are overlapping, (**c**) dose 850 Gy, the light line represents the HGO model, and the dark line represents the Misof model, (**d**) dose 3380 Gy, models are overlapping.

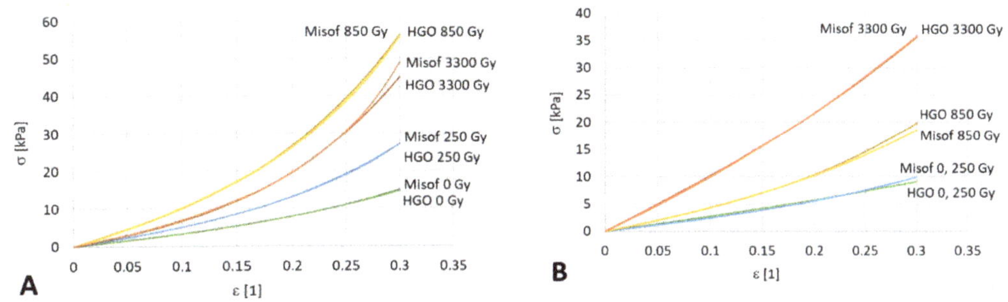

Figure 7. A comparison of the HGO and Misof model trends for extruded collagen mass. (**A**)—Longitudinal direction, (**B**)—Transversal direction.

Tables 5–8 and associated Figure 7 demonstrate a significant change of the mechanical properties of collagen after soft isotropic collagen matter has passed through a narrow gap. Extruding fresh collagen through a 200 mm long capillary increases the modulus of elasticity E approximately threefold (from 9 kPa to 30 kPa), and the increase is almost isotropic. The increase in stiffness due to extrusion can be explained by stretching of the wavy collagen fibers flowing through a narrow slit at a very high rate of deformation. Even greater stiffening can be achieved by e-beam irradiation of extruded samples (one order of magnitude can be accomplished by a radiation dose of the order of kGy). However, the supposedly irreversible structural changes caused by extrusion mitigate the effect of ex post applied irradiation (compare the increase of E at the same dose of 3300 Gy for an isotropic sample and for an extruded sample). The increases in stiffness when e-bean irradiation is applied in the longitudinal direction and also in the transversal direction can be explained by the assumption of the possible formation of crosslinks in the native collagen structure and also in the irreversibly modified collagen structure. The observed slight decrease in longitudinal stiffness for a very high dose (3300 Gy) can be explained by breaking of the collagen fibers, manifested by a decrease in stiffness and in molecular mass at an extremely high intensity of radiation. The same phenomenon of stiffness decreasing with an increasing irradiation dose was observed at 10 kGy [14] and at 12.5 kGy [18], respectively. After reaching the turning point of the irradiation dose, the mechanical properties of the sample decrease due to the reduction in molecular weight and degradation of its molecular chain [18]. For a bovine collagen solution, this critical irradiation dose together with pretreatment by extrusion is approximately 3 kGy, while for untreated collagen solution the critical dose is much more than 3 kGy.

4. Conclusions

The stiffening of a highly concentrated collagen solution can be explained as a manifestation of significant structural changes (not measured), probably the formation of the cross-links initiated by e-beam irradiation and irreversible straightening of the collagen fibers in the flow of isotropic collagen through a narrow gap. The extruded material with fibers oriented in the direction of flow is non-isotropic, and is stiffer than untreated isotropic collagen. The results of our study have confirmed that irradiating a water-collagen suspension with an electron beam significantly increases the final stiffness. The experimentally determined deformation-stress relationship has been approximated using two parametric hyperelastic models (HGO and Misof). A significantly increased modulus of elasticity (E, k_1) was found after electron irradiation (doses of 0, 250, 850 and 3300 Gy were tested). Further experiments should be carried out (on the role of concentration, critical loading and the duration of structural changes). Our results show that modification of the mechanical properties of collagen by irradiation has potential, for example, in the field of tissue engineering in addition to the food and pharmaceutical industries.

Author Contributions: Conceptualization, H.C., J.S. and R.Ž.; methodology, H.C., J.S. and R.Ž.; software, H.C.; validation, H.C., J.S. and R.Ž.; investigation and experiments, H.C., J.S. and J.Š.; writing—original draft preparation, H.C. and R.Ž.; writing—review and editing all authors.; supervision, M.H. All authors have read and agreed to the published version of the manuscript.

Funding: This research was funded by The Czech Science Foundation grant number 21-07851S.

Institutional Review Board Statement: Not applicable.

Informed Consent Statement: Not applicable.

Data Availability Statement: Exclude this statement. All important data are reported within article.

Conflicts of Interest: The authors declare no conflict of interest. The founding sponsors had no role in the design of the study; in the collection, analyses, or interpretation of data; in the writing of the manuscript; or in the decision to publish the results.

References

1. Kumar, V.; Caves, J.M.; Haller, C.A.; Dai, E.; Liu, L.; Grainger, S.; Chaikof, E.L. Acellular vascular grafts generated from collagen and elastin analogs. *Acta Biomater.* **2013**, *9*, 8067–8074. [CrossRef] [PubMed]
2. Deiber, J.A.; Peirotti, M.B.; Ottone, M.L. Rheological characterization of edible films made from collagen colloidal particle suspensions. *Food Hydrocoll.* **2011**, *25*, 1382–1392. [CrossRef]
3. Skočilas, J.; Žitný, R.; Štancl, J.; Dostál, M.; Landfeld, A.; Houška, M. Rheological properties of collagen matter predicted using an extrusion rheometer. *J. Texture Stud.* **2016**, *47*, 514–522. [CrossRef]
4. Daar, E.; Kaabar, W.; Woods, E.; Lei, C.; Nisbet, A.; Bradley, D. Atomic force microscopy and mechanical testing of bovine pericardium irradiated to radiotherapy doses. *Radiat. Phys. Chem.* **2014**, *96*, 176–180. [CrossRef]
5. Puxkandl, R.; Zizak, I.; Paris, O.; Keckes, J.; Tesch, W.; Bernstorff, S.; Purslow, P.; Fratzl, P. Viscoelastic properties of collagen: Synchrotron radiation investigations and structural model. *Philos. Trans. R. Soc. B Biol. Sci.* **2002**, *357*, 191–197. [CrossRef] [PubMed]
6. Misof, K.; Rapp, G.; Fratzl, P. A new molecular model for collagen elasticity based on synchrotron X-ray scattering evidence. *Biophys. J.* **1997**, *72*, 1376–1381. [CrossRef]
7. Gotoh, T.; Itoh, M.; Okuda, T. The changes of electrochemical and mechanical properties through gamma irradiation of a collagen aqueous solution. *Int. J. Appl. Radiat. Isot.* **1977**, *28*, 933–937. [CrossRef]
8. Sionkowska, A.; Wisniewski, M.; Skopinska, J.; Poggi, G.; Marsano, E.; Maxwell, C.; Wess, T. Thermal and mechanical properties of UV irradiated collagen/chitosan thin films. *Polym. Degrad. Stab.* **2006**, *91*, 3026–3032. [CrossRef]
9. Liu, B.; Harrell, R.; Davis, R.H.; Dresden, M.H.; Spira, M. The effect of gamma irradiation on injectable human amnion collagen. *J. Biomed. Mater. Res.* **1989**, *23*, 833–844. [CrossRef]
10. Mohamed, F.; Bradley, D.A.; Winlove, C.P. Effect of ionizing radiation on extracellular matrix. *Nucl. Instrum. Methods Phys. Res. Sect. A Accel. Spectrometers Detect. Assoc. Equip.* **2007**, *580*, 566–569. [CrossRef]
11. Giammona, G.; Pitarresi, G.; Cavallaro, G.; Spadaro, G. New biodegradable hydrogels based on an acryloylated polyaspartamide cross-linked by gamma irradiation. *J. Biomater. Sci. Polym. Ed.* **1999**, *10*, 969–987. [CrossRef]
12. Riedel, S.; Hietschold, P.; Krömmelbein, C.; Kunschmann, T.; Konieczny, R.; Knolle, W.; Mierke, C.T.; Zink, M.; Mayr, S.G. Design of biomimetic collagen matrices by reagent-free electron beam induced crosslinking: Structure-property relationships and cellular response. *Mater. Des.* **2019**, *168*, 168. [CrossRef]
13. Hennink, W.; van Nostrum, C. Novel crosslinking methods to design hydrogels. *Adv. Drug Deliv. Rev.* **2012**, *64*, 223–236. [CrossRef]
14. Manaila, E.; Craciun, G.; Ighigeanu, D.; Lungu, I.; Dumitru, M.; Stelescu, M. Electron Beam Irradiation: A Method for Degradation of Composites Based on Natural Rubber and Plasticized Starch. *Polymers* **2021**, *13*, 1950. [CrossRef] [PubMed]
15. Jamal, N.A.; Anuar, H.; Bahri, A.R.S. Enhancing the Mechanical Properties of Cross-Linked Rubber-Toughened Nanocomposites via Electron Beam Irradiation. *J. Nanotechnol.* **2011**, *2011*, 1–8. [CrossRef]
16. Šarac, T.; Quiévy, N.; Gusarov, A.; Konstantinović, M. Influence of γ -irradiation and temperature on the mechanical properties of EPDM cable insulation. *Radiat. Phys. Chem.* **2016**, *125*, 151–155. [CrossRef]
17. Zeid, M.A.; Rabie, S.; Nada, A.; Khalil, A.; Hilal, R. Effect of gamma irradiation on ethylene propylene diene terpolymer rubber composites. *Nucl. Instrum. Methods Phys. Res. Sect. B Beam Interact. Mater. At.* **2007**, *266*, 111–116. [CrossRef]
18. Baroudi, A.; García-Payo, C.; Khayet, M. Structural, Mechanical, and Transport Properties of Electron Beam-Irradiated Chitosan Membranes at Different Doses. *Polymers* **2018**, *10*, 117. [CrossRef]
19. Song, J.-M.; Shin, D.W.; Sohn, J.-Y.; Nho, Y.C.; Lee, Y.M.; Shin, J. The effects of EB-irradiation doses on the properties of crosslinked SPEEK membranes. *J. Membr. Sci.* **2013**, *430*, 87–95. [CrossRef]
20. Žitný, R.; Landfeld, A.; Skočilas, J.; Stancl, J.; Flegl, V.; Zachariášová, M.; Jírů, M.; Houška, M. Hydraulic characteristic of collagen. *Czech J. Food Sci.* **2016**, *33*, 479–485. [CrossRef]
21. Bano, I.; Ghauri, M.A.; Yasin, T.; Huang, Q.; Palaparthi, A.D. Characterization and potential applications of gamma irradiated chitosan and its blends with poly(vinyl alcohol). *Int. J. Biol. Macromol.* **2014**, *65*, 81–88. [CrossRef]

22. García, M.A.; Pérez, L.; de la Paz, N.; González, J.; Rapado, M.; Casariego, A. Effect of molecular weight reduction by gamma irradiation on chitosan film properties. *Mater. Sci. Eng. C* **2015**, *55*, 174–180. [CrossRef] [PubMed]
23. Labouriau, A.; Cady, C.; Gill, J.; Stull, J.; Ortiz-Acosta, D.; Henderson, K.; Hartung, V.; Quintana, A.; Celina, M. Gamma irradiation and oxidative degradation of a silica-filled silicone elastomer. *Polym. Degrad. Stab.* **2015**, *116*, 62–74. [CrossRef]
24. Banik, I.; Bhowmick, A.K. Influence of electron beam irradiation on the mechanical properties and crosslinking of fluorocarbon elastomer. *Radiat. Phys. Chem.* **1999**, *54*, 135–142. [CrossRef]
25. Khatiwada, S.P.; Gohs, U.; Janke, A.; Jehnichen, D.; Heinrich, G.; Adhikari, R. Influence of electron beam irradiation on the morphology and mechanical properties of styrene/butadiene triblock copolymers. *Radiat. Phys. Chem.* **2018**, *152*, 56–62. [CrossRef]
26. Cui, C.; Zhang, Y. Effect of Electron Beam Irradiation on the Mechanical and Thermal Properties of Ternary Polyamide Copolymer. *Macromol. Res.* **2018**, *26*, 359–364. [CrossRef]
27. Chmielewski, A.; Haji-Saeid, M.; Ahmed, S. Progress in radiation processing of polymers. *Nucl. Instrum. Methods Phys. Res. Sect. B Beam Interact. Mater. Atoms* **2005**, *236*, 44–54. [CrossRef]
28. Makuuchi, K.; Cheng, S. *Radiation Processing of Polymer Materials and Its Industrial Applications;* John Wiley & Sons: Hoboken, NJ, USA, 2012.
29. Klüver, E.; Meyer, M. Preparation, processing, and rheology of thermoplastic collagen. *J. Appl. Polym. Sci.* **2013**, *128*, 4201–4211. [CrossRef]
30. Landfeld, A.; Houška, M.; Skočilas, J.; Žitny, R.; Novotna, P.; Štangl, J.; Dostal, M.; Chvatil, D. The effect of irradiation on Rheological and electrical properties of collagen. *Appl. Rheol.* **2016**, *26*, 35–41. [CrossRef]
31. Štípek, J.; Skočilas, J.; Štancl, J.; Žitný, R. Extrusion rheometry of collagen dough. *Czech J. Food Sci.* **2021**, *39*, 384–392. [CrossRef]
32. Krist, P.; Bíla, J. Microtrone Modelling and Control, in Nuclear Physics Methods and Accelerators in Biology and Medicine. *Amer. Inst. Phys. (AIP) Conf. Proc.* **2009**, *1204*, 183–185. [CrossRef]
33. Holzapfel, G.A.; Gasser, T.C.; Ogden, R. Comparison of a Multi-Layer Structural Model for Arterial Walls with a Fung-Type Model, and Issues of Material Stability. *J. Biomech. Eng.* **2004**, *126*, 264–275. [CrossRef] [PubMed]

polymers

MDPI

Article

Highly Stretchable Bacterial Cellulose Produced by *Komagataeibacter hansenii* SI1

Izabela Cielecka [1,*], Małgorzata Ryngajłło [1], Waldemar Maniukiewicz [2] and Stanisław Bielecki [1]

1 Institute of Molecular and Industrial Biotechnology, Lodz University of Technology, 90-573 Łódź, Poland; malgorzata.ryngajllo@p.lodz.pl (M.R.); stanislaw.bielecki@p.lodz.pl (S.B.)

2 Institute of General and Ecological Chemistry, Lodz University of Technology, 90-924 Łódź, Poland; waldemar.maniukiewicz@p.lodz.pl

* Correspondence: izabela.cielecka1@gmail.com; Tel.: +48-42-631-3367

Abstract: A new strain of bacteria producing cellulose was isolated from Kombucha and identified as *Komagataeibacter hansenii*, named SI1. In static conditions, the strain synthesises bacterial nanocellulose with an improved ability to stretch. In this study, utilisation of various carbon and nitrogen sources and the impact of initial pH was assessed in terms of bacterial nanocellulose yield and properties. *K. hansenii* SI1 produces cellulose efficiently in glycerol medium at pH 5.0–6.0 with a yield of 3.20–3.60 g/L. Glucose medium led to the synthesis of membrane characterised by a strain of 77%, which is a higher value than in the case of another *Komagataeibacter* species. Supplementation of medium with vitamin C results in an enhanced porosity and improves the ability of bacterial nanocellulose to stretch (up to 123%). The properties of modified membranes were studied by scanning electron microscopy, Fourier transform infrared spectroscopy, X-ray diffraction and mechanical tests. The results show that bacterial nanocellulose produced in SH medium and vitamin C-supplemented medium has unique properties (porosity, tensile strength and strain) without changing the chemical composition of cellulose. The method of production BNC with altered properties was the issue of Polish patent application no. P.431265.

Keywords: *Komagataeibacter*; stretchable bacterial cellulose; enhanced strain; vitamin C

Citation: Cielecka, I.; Ryngajłło, M.; Maniukiewicz, W.; Bielecki, S. Highly Stretchable Bacterial Cellulose Produced by *Komagataeibacter hansenii* SI1. *Polymers* **2021**, *13*, 4455. https://doi.org/10.3390/polym13244455

Academic Editors: Arn Mignon and Antonio Pizzi

Received: 1 November 2021
Accepted: 16 December 2021
Published: 19 December 2021

1. Introduction

Biopolymers produced by bacteria, e.g., polyamides, polyesters, polysaccharides and extracellular proteins, are the part of the bacteria inherent physiology. They outperform properties of polymers extracted from natural origin because microbial biopolymers can be modified to specific application by biotechnology tools [1]. Bacterial nanocellulose (BNC) is an increasingly used natural polymer in constructing new smart biomaterials that can be applied in many fields [2]. It is produced in nanostructured membranes by many bacterial strains, such as *Komagataeibacter*, *Agrobacterium*, *Sarcina* or *Rhizobium* [3]. The most studied cellulose producers are the *Komagataeibacter* species, synthesising BNC in a pellicle at the air/liquid interface [4]. The chemical structure of BNC is identical to that of the plant cellulose, but it does not contain impurities such as hemicelluloses, lignin or dyes. Bacteria can produce cellulose from different carbon sources, including monosaccharides (glucose, fructose, galactose), disaccharides (sucrose, maltose) and sugar alcohols (glycerol, mannitol) [5]. Much research has recently focused on screening new strains producing BNC efficiently or with unique properties. *Komagataeibacter* strain could be isolated from Kombucha [6], vinegar [7,8], rotten green grapes [9], coconut milk [10] or fruits [11].

The structure of BNC consists of nanofibrils assembled into ultrathin fibres with a width of approximately 8 nm and next into ribbons of 70–150 nm width. Those ribbons are organised into the three-dimensional network [12]. The structure of BNC determines unique properties of this biopolymer, such as mechanical stability, high porosity, high water holding capacity (up to 99%) and crystallinity over 74% [13]. Moreover, the biocompatibility of BNC is well documented, which makes it a suitable material for biomedical and tissue engineering applications [14,15]. Physicochemical properties of bacterial cellulose has aroused the interest of a wide range of industries, where it can be useful as a part of biosensors [16], food and cosmetics stabilisers [17], packaging [18] and drug or enzymes delivery matrix [19,20]. The scope of BNC applications is highly dependent on biomaterial structure, defined properties and production costs. In the reported research, two main trends can be observed. The first of them regards to the economical aspects of BNC production. There are sought new, efficient strains, the culture conditions are optimised, and cheaper carbon sources are tested [21]. The second approach is attributed to in situ and ex situ methods of BNC properties modification. Those efforts are focused on the achievement of biomaterial with desirable features. Specific parameters of BNC membrane depend on fibres organisation in the three-dimensional structure, amount of branching points, the width of fibres, crystallinity and water content [22]. Those properties can be influenced by biological, engineering and material sciences approaches. Physico-chemical parameters of BNC can be affected by the producing strain, culture conditions, culture method and a wide range of additives. Cellulose produced in static conditions is formed as a flat pellicle with a dense three-dimensional structure. In contrast, agitated fermentation generates BNC in a spherical shape and is characterized by an enhanced porosity and looser arrangement of fibres in the structure [23]. On the other hand, it is known that different strains can produce cellulose with various crystallinity [24]. *Gluconacetobacter xylinus* ATCC 10,245 is able to produce BNC with crystallinity at 88% [25], while the crystallinity of cellulose synthesised by *G. xylinus* PTCC 1734 is only 63% [26]. In our previous study, the impact of culture conditions (such as culture time, glucose concentration, pH of culture medium and air-flow ratio) on mechanical strength has been shown [27].

In the presented research, a new cellulose producing strain was isolated and identified as *Komagataeibacter hansenii*, named SI1. This strain was microbiologically characterised, and kinetic growth, BNC structure, chemical composition, and BNC accumulation have been studied. Next, various culture conditions (carbon source, nitrogen source, pH, culture additives) were investigated for BNC production by *K. hansenii* SI1. Bacterial cellulose produced in the glucose medium exhibited unique mechanical properties and high porosity. Which were further improved by supplementation of vitamin C. Afterwards, the metabolism of *K. hansenii* SI1 growing in both control conditions and vitamin C-supplemented medium was assessed and compared during a 7-day culture. Next, modified cellulose membranes were analysed in the context of structure arrangement by scanning electron microscopy, chemical composition, mechanical properties, crystallinity and crystallite size.

2. Materials and Methods

2.1. Isolation and Identification of Microorganism

The bacterial cellulose (BNC) producing bacteria were isolated from commercial Kombucha beverages available in the Polish market. A 1 mL sample was inoculated into 10 mL of the Schramm–Hestrin (SH) medium consisting of 20 g/L glucose, 5 g/L yeast extract, 5 g/L peptone, 2.7 g/L Na_2HPO_4, 1.15 g/L citric acid and 0.5 g/L $MgSO_4$ with pH 5.7 adjusted with 0.1 M acetic acid. The culture was incubated at 30 °C for three days in static conditions until a pellicle on the culture surface appeared. Afterwards, 0.1 mL of those cultures were spread in a solid SH-agar (2% agar) medium containing cyclohexamide (0.1 g/L) and incubated at 30 °C for three days. Potential cellulose-producing bacteria were separately transferred into 5 mL of SH medium and incubated at 30 °C for three days. Next, from each culture with a pellicle, 0.1 mL was spread on an SH-agar medium and incubated. The procedure was repeated until single colonies were obtained. After three days, pure colonies were cultivated in 5 mL of SH medium and transferred into 100 mL of fresh SH medium. Cultures were incubated at 30 °C for seven days, and the microorganism for genetic identification was chosen based on cellulose characteristics. The BNC producer was preserved under freezing (−80 °C) using 20% glycerol as a cell cryoprotectant.

The genomic DNA of Kombucha was isolated as described by Ryngajłło et al. [28]. Genome sequencing was performed by BioNanoPark Łódź, Poland. Briefly, NGS libraries were prepared using NEBNext DNA Library Preparation Kit (New England Biolabs, Ipswich, MA, USA). Genome sequencing was performed using the Illumina MiSeq platform (Illumina, San Diego, CA, USA) in 2 × 250 bp paired-end reads mode. The sequencing reads were assembled de novo using SPAdes (v. 3.6.2, [29]). The genome sequence of Kombucha strain has been deposited into the NCBI database under the BioProject number: PRJNA751727.

Comparisons of genomic similarity are currently considered a standard for species classification. In particular, the in silico comparisons involving whole genome sequence allows for obtaining exact results. One of the most accurate genome-wide similarity statistics is average nucleotide identity (ANI) [30]. ANI-based phylogenetics of the newly isolated cellulose producers is now possible due to the availability of the type strain genome sequences for the species of the *Komagataeibacter* genus [31]. ANI analysis was performed using PYANI (v. 0.2.9; [32]) python program employing BLAST+ program [33]. The UPGM tree based on ANI-1 values was calculated using the phangorn R package [34]. Genome sequences of 16 type *Komagataeibacter* strains and *Glouconacetobacter entanii* LTH 4560 strain were downloaded from NCBI [31].

2.2. Culture Conditions

2.2.1. Inoculum Preparation

Bacteria from frozen stock were activated by spreading on SH-agar plate and incubation at 30 °C for three days. Next, a single colony was transferred into 5 mL of SH medium and incubated at 30 °C for three days. Then, cultures were transferred into 100 mL of SH medium and incubated at 30 °C. After three days, culture was used as an inoculum.

2.2.2. Culture Medium

In this study, Schramm–Hestrin medium, containing 20 g/L glucose (POCh, Gliwice, Poland), 5 g/L yeast extract (BTL, Łódź, Poland), 5 g/L bacterial peptone (BTL, Łódź, Poland), 2.7 g/L sodium phosphate dibasic (Chempur, Piekary Śląskie, Poland), 1.15 g/L citric acid (Chempur, Piekary Śląskie, Poland) and 0.5 g/L magnesium sulfate (Chempur, Piekary Śląskie, Poland), was used as the basal medium. The initial pH of the SH medium was adjusted to 5.7 using 0.1 M acetic acid.

2.3. Primary Characterisation of Bacterial Cellulose Production by the Komagataeibacter hansenii SI1 Strain

In our study, we assessed the kinetics of BNC production in standard conditions (SH medium, 30 °C, seven days) and BNC properties. Colonies were observed by using a microscope after four days. A time-course experiment determined growth rate, glucose consumption, BNC accumulation, and pH for seven days. The structure was visualised by a macro-photograph and scanning electron microscopy (SEM). Fibres width was determined based on SEM images as described in Section 2.7.2.

2.4. Time Course of BNC Biosynthesis

The metabolism of *K. hansenii* SI1 was studied for SH medium (primary characterisation) and SH medium supplemented with ascorbic acid at 0.5% and 1.0% concentration. The *K. hansenii* SI1 strain was cultivated in 5 mL of medium using 10 mL test tubes. Each replicate was inoculated with a single colony and incubated at 30 °C for seven days. The kinetic study was carried out on each day. The BNC yield was calculated as the dry weight of purified membranes per 1 L of culture medium. Bacterial growth as CFU was evaluated using the serial dilution method after degradation of cellulose membrane with cellulase (100 µL/culture, dilution of 5:3 in SH medium, Ultraflo Max, Novozymes, Kalundborg, Denmark). After serial dilution, 100 µL of the cell suspension was spread on a solid SH medium. The colony-forming units were counted after three days of incubation and expressed as a log(CFU). According to the manufacturer's protocol, the glucose concentration was assessed using a GLUCOSE test (BioMaxima, Lublin, Poland). The concentration of residual ascorbic acid was determined using a K-ASCO assay kit (Megazyme, Bray, Ireland). Each test was performed at least in three replicates.

2.5. Impact of Culture Conditions on Bacterial Cellulose Yield

In this study, the impact of carbon and nitrogen source was evaluated. Thus, glucose was substituted with one of the following carbon sources: fructose (Chempur, Piekary Śląskie, Poland), galactose (Chempur, Piekary Śląskie, Poland), glycerol (Chempur, Piekary Śląskie, Poland), lactose (Chempur, Piekary Śląskie, Poland), maltose (Chempur, Piekary Śląskie, Poland), mannitol (Chempur, Piekary Śląskie, Poland) and sucrose (Chempur, Piekary Śląskie, Poland). The concentration of carbon source in each variant was always 20 g/L. Standard nitrogen sources (yeast extract and peptone; BTL, Łódź, Poland) were substituted with another, added individually to SH medium with the concentration of 5 g/L, namely ammonium sulphate (Chempur, Piekary Śląskie, Poland), sodium nitrate (Chempur, Piekary Śląskie, Poland), urea (Chempur, Piekary Śląskie, Poland), corn steep liquor (Merck, Darmstadt, Germany), peptone (BTL, Łódź, Poland) or yeast extract (BTL, Łódź, Poland). The influence of BNC enhancers was studied for ethanol (1%) (Chempur, Piekary Śląskie, Poland), lactic acid (0.6%) (Merck KGaA, Darmstadt, Germany) and vitamin C (0.5% and 1.0%) (Stanlab, Lublin, Poland). The control was a culture grown in the SH medium.

Each culture was prepared in cuboid bioreactors filled with 200 mL of the modified SH medium. The medium was inoculated with 5% inoculum and cultivated for seven days at 30 °C. Afterwards, membranes were collected and purified according to Cielecka et al. [35]. The BNC membranes were then dried at 90 °C in a gel drier (Bio-Rad Laboratories, Hercules, CA, USA) until a constant weight was achieved [36]. The biosynthesis yield was expressed as the dry weight of membrane obtained from 1 L of culture medium. The samples were cultured in triplicate for each variant of the medium.

2.6. Evaluation of the Impact of Ascorbic Acid on Bacterial Cellulose Production and Properties

The influence of vitamin C on BNC biosynthesis was assessed in a time-course experiment described in the time course experiment section. The experiment was performed using two concentrations of vitamin C, namely 0.5% and 1.0%. Before adding to the culture medium, vitamin C was dissolved in SH medium (10 g/L) and sterilised using a syringe filter (0.22 μm). Next, an appropriate volume was added to each medium sample (5 mL) placed in the test tube. The supplemented medium was inoculated with a single colony. Cultures were incubated at 30 °C for seven days. Moreover, we evaluated the arrangement of fibres in the three-dimensional structure, chemical composition by FTIR analysis, crystallinity and average crystallite size and mechanical properties.

2.7. Analytical Methods

2.7.1. Mechanical Strength

The membranes were examined for tensile strength using a universal testing machine (Zwick/Roell Z1.0, Ulm, Germany), according to the method described by Cielecka et al. [35], with slight changes. The BNC was pressed before the tensile tests until 1 mm thickness was achieved to remove any excess water attached to membranes. Next, membranes were carefully cut into rectangular samples (20 mm × 45 mm) with a scalpel. The thickness of a strip was measured using a digital calliper. The samples were placed between two clamps (the gauge length was 15 mm) and subjected to deformation at a rate of 10 mm/min, while the pre-load was 0.1 N. The maximum stress and elongation at break were estimated using TestXpert@II software. Stress (MPa) was calculated as F/A, where F is the loading force expressed in Newtons (N), and A is the cross-section area of a sample. Strain (%) was calculated as $\Delta L/L_0 \times 100\%$, where L_0 is the initial length, and ΔL is the exerted extension from starting point. Values for Young's modulus (YM) under tension were calculated from the stress/strain relationship in the first linear region of the graph. The measurements were performed in at least nine replicates.

2.7.2. Scanning Electron Microscopy

Before testing, the samples were freeze-dried and then sputter-coated with a gold layer. A FEI QUANTA 250 FEG microscope (Thermo Fischer Scientific, Waltham, MA, USA) was used to visualise BNC structure with scanning parameters: HV = 2 kV and magnification × 40,000, for each sample. Fibre thickness was evaluated using Makroaufmassprogramm software (open source software by Jens Rüdig, http://ruedig.de/tmp/messprogramm.html, accessed on 3 April 2017).

2.7.3. Fourier Transform Infrared Spectrometry in Attenuated Total Reflectance Mode Analysis

Before testing, samples were freeze-dried in an ALPHA 1–2/LD freeze dryer (Martin Christ GmbH, Osterode am Harz, Germany). Chemical analysis of variations in the structure of BNC produced under different conditions was performed by FT-IR with attenuated total reflectance mode (ATR). The spectra were recorded at a resolution of 8 cm^{-1}, in the range of 4000 to 650 cm^{-1} using a Nicolet 6700 FT-IR (Thermo Fischer Scientific, Waltham, MA, USA). For each sample, 200 scans were taken.

2.7.4. X-ray Diffractometry

Room temperature powder X-ray diffraction patterns were collected using a PANalytical X'Pert Pro MPD diffractometer (Malvern Panalytical Ltd., Malvern, UK) in the Bragg–Brentano reflection geometry and the graphite monochromated Cu-Kα radiation. The PANalytical X'Celerator detector was used. All data were collected in the 2θ range 5–60° with a step of 0.0167° and an exposure per step of 30 s. The samples were spun during data collection to minimise preferred orientation effects. A PANalytical X'Celerator detector based on the Real-Time Multiple Strip technology and simultaneously measuring intensities in the 2θ range of 2.122° was used. The WAXFIT program was used to resolve the X-ray diffraction patterns [37]. The crystallinity index was calculated as the ratio of the integral intensity under all crystalline peaks to the sum of integral intensity under the crystalline peaks and amorphous halo [38]. The initial positions of crystalline peaks were assumed in accordance with the literature data [39]. According to the hybrid optimisation procedure, each diffraction curve was analysed by creating a theoretical function best fitted to the experimental curve, which combines a genetic algorithm with a modified Rosenbrock optimisation method. A linear combination of the Gauss and Cauchy profiles were used to construct the theoretical function approximating crystalline peaks and an amorphous halo. As a result of fitting, all the parameters of the component functions were determined.

The average crystallite size (ACS) was calculated according to the Scherrer equation [40]. The broadening of diffraction peaks due to crystallite size can be expressed as:

$$\text{Crystallite size (average)} = K\,\lambda/(Bs\cos\theta) \quad (1)$$

where: Bs is broadening due solely to crystallite size, K is a constant, the value of which depends on the particle shape (taken as 0.9 in this case), θ is the Bragg's angle, and λ is the wavelength of the incident X-ray beam.

3. Results and Discussion

3.1. Strain Classification

To analyse the phylogenetic relationship of the newly isolated strain with *Komagataeibacter* species, its genome was sequenced (manuscript describing the genome is in preparation). The whole-genome sequence comparison with type *Komagataeibacter* strains revealed that the strain isolated from Kombucha clustered with *K. hansenii* JCM 7643 strain (Figure 1). Based on these results, the new strain was classified as *K. hansenii* SI1.

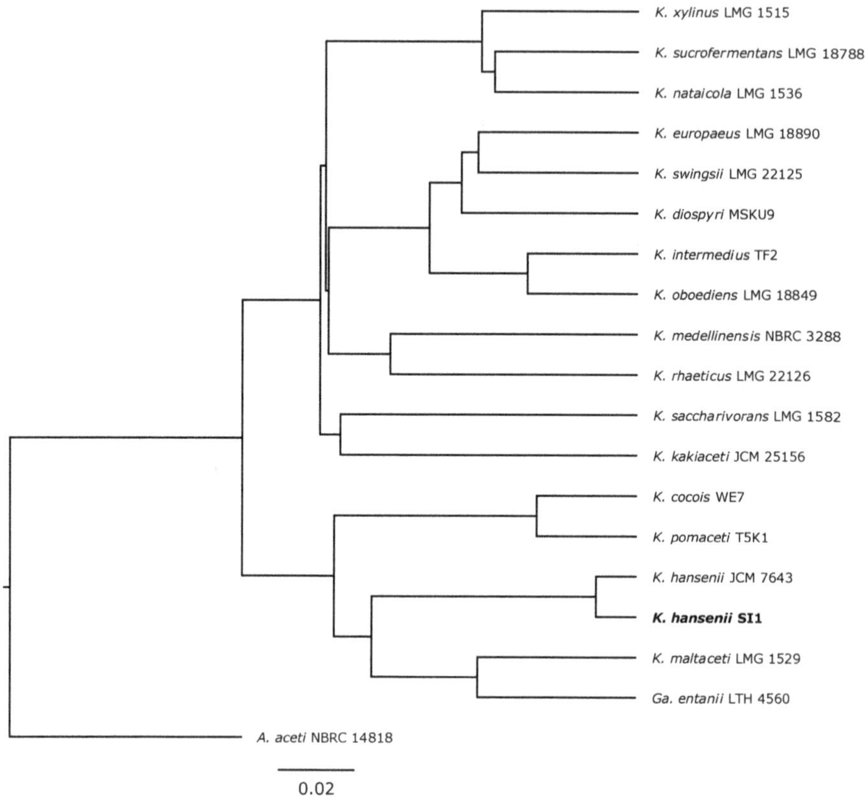

Figure 1. Phylogenetic relationship of the newly isolated Kombucha strain and type *Komagataeibacter* strains. The phylogenetic tree was constructed on ANI-1 values calculated based on whole-genome alignment. The *A. aceti* NBRC 14,818 was used as an outgroup. The tree was drawn in the FigTree program (v. 1.4.4). The scale bar represents the sequence divergence.

3.2. Characterisation of Strain K. hansenii SI1, BNC Properties and Kinetics of Biosynthesis in Standard Conditions

This study isolated a new cellulose-producing strain from a commercial Kombucha beverage. The isolate produced smooth, pale yellow colonies with a circular shape and a diameter of approximately 0.72 mm (Figure 2a). Bacteria were characterised as Gram-negative, aerobic microorganisms which did not form spores. Bacteria were rod-shaped, occurring singly or in short chains. We found that it metabolised carbohydrates and sugar alcohols and grew in the presence of ethanol and lactic acid. In liquid media, bacteria produced a pellicle at the surface of the medium (Figure 2b), which has a three-dimensional structure (Figure 2c) consisting of fibres with a diameter of 10–150 nm (Figure 2d).

The growth profile of *K. hansenii* SI1 growing on SH medium over seven days was examined. The yield of BNC, glucose concentration, amount of viable cells and pH were assessed each day (Figure 3). Bacterial cells grew rapidly until the second day, and afterwards, a stationary phase was observed. The maximum value of cells number reached 8.93 logCFU/mL on the third day. BNC synthesis started from the first day and linearly increased to the fourth day, where the plateau was reached. Glucose was consumed during the whole cultivation time, and the final concentration was 6.80 g/L. Thus, the medium was still rich in carbon source after culture, but cellulose production was diminished after the fourth day. The pH value decreased rapidly to 4.25 during the first three days, and subsequent, the pH value slowly increased to the final pH of 4.48. Changes in pH value can be ascribed to the fluctuations of gluconic acid concentration (Supplementary Figure S1), which is produced until the third day (2.99 g/L) and next consumed (final concentration at 2.31 g/L).

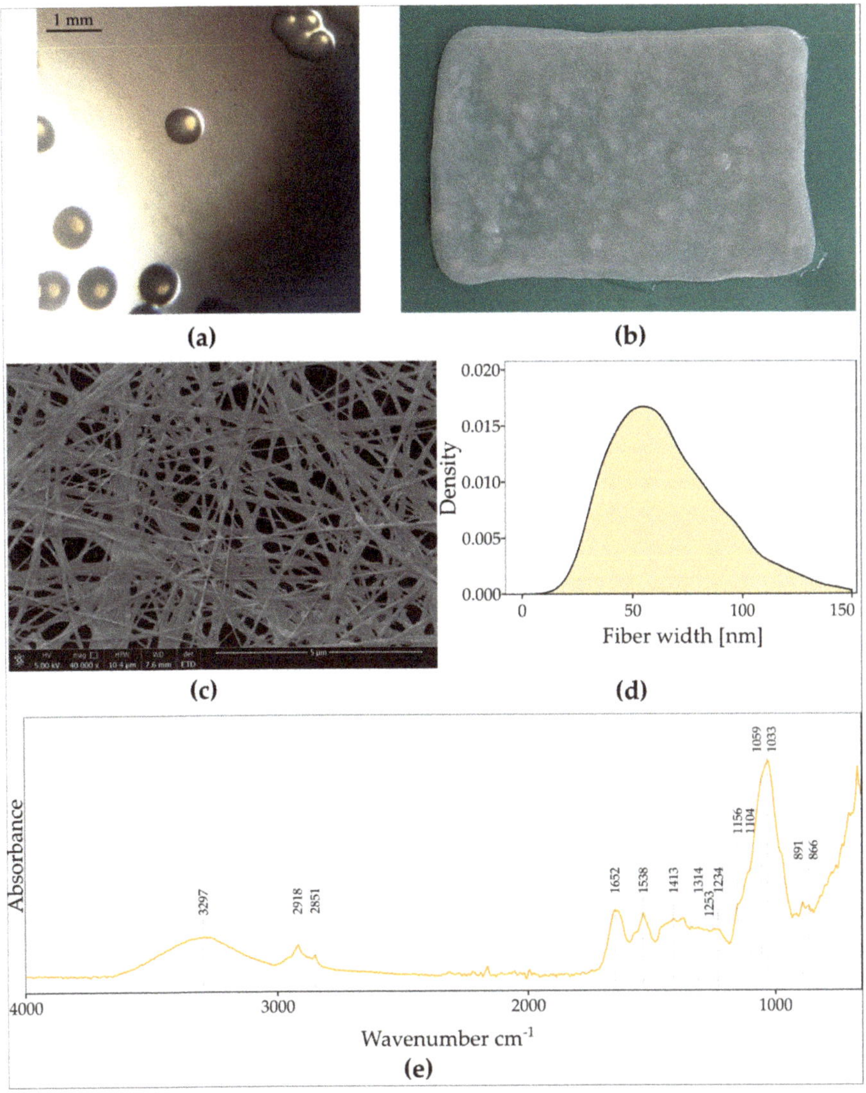

Figure 2. (**a**) Three-days colonies of *K. hansenii* SI1, (**b**) BNC membrane produced in SH medium, (**c**) scanning electron microscope image of BNC membrane produced in SH medium; the image was recorded at a magnification of 40,000× (bar—5 μm), (**d**) the probability density function of fibre width in BNC produced in SH medium and (**e**) FTIR-ATR spectrum of BNC produced in SH medium.

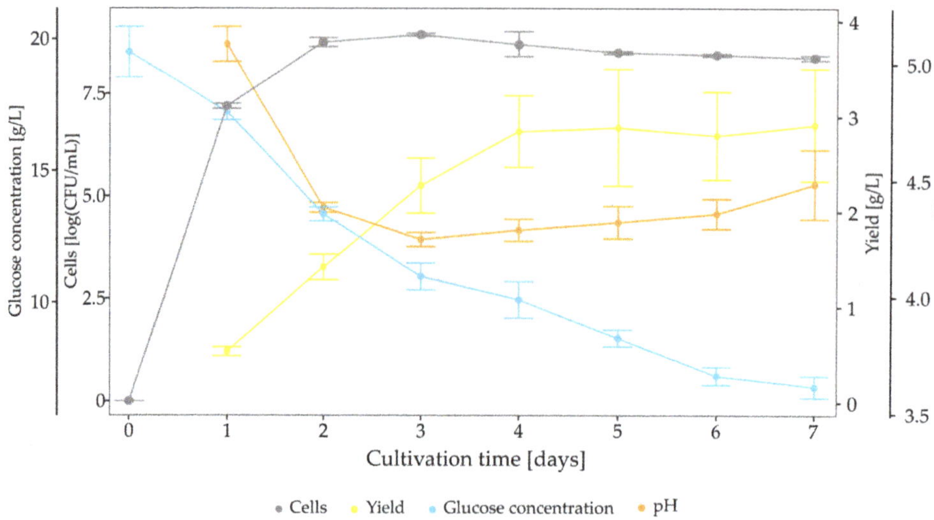

Figure 3. Time course of *K. hansenii* SI1 growth in SH medium.

The FTIR-ATR study was conducted to confirm that a pellicle is composed of cellulose. The FTIR spectrum of a biomaterial produced by *K. hansenii* SI1 is presented in Figure 2e. The absorption bands are a fingerprint that can confirm the structure (as cellulose) and slightly vary between cellulose from different origins [41]. Two main peaks can be observed, namely at 3297 cm^{-1} and in the range of 1200–1000 cm^{-1}. The first one corresponds to the –OH stretching. At the same time, a series of bands are assigned to the stretching of C-O-C of sugar rings and C-O stretching vibrations of the primary (C6) and the secondary hydroxyl (C2, C3) groups [42,43]. A peak at 2918 cm^{-1} is related to stretching vibrations of C-H groups, and 1652 cm^{-1} indicates deformational vibrations of –OH groups originated from bound water [44]. The absorption band at 891 cm^{-1} is assigned to the β-glucosidic linkage [45]. The FTIR spectrum does not differ from the reported spectra of BNC produced by other bacteria of the *Komagataeibacter* genus. According to Fuller et al. we do not observe impurities such as protein, lipids, nucleic acids or bacterial cells [46]. Thus, we conclude that *K. hansenii* SI1 produces chemically pure cellulose in standard culture conditions.

Although cellulose produced by *K. hansenii* SI1 does not differ chemically from other reported BNCs, we noticed that, in this case, the pellicle has an unexpectedly high ability to stretch. It could be stretched in all directions and maintain the formed shape. What is more, the pellicle can be easily manually shaped without using excessive tension. The BNC yield was not as high as other reported strains [47], but the newly isolated strain produces BNC with unique properties. These properties may find future application in medicine as, e.g., a shapable and transparent dressing (Supplementary Figure S2) or as a scaffold for three-dimensional cultures in tissue engineering by enabling shape and dimensions adjustment during culture.

3.3. The Impact of Culture Conditions on BNC Biosynthesis

Bacteria from the *Komagataeibacter* genus can synthesise cellulose from a wide range of carbon sources, including monosaccharides, disaccharides, oligosaccharides and sugar alcohols [5]. We examined various carbon substrates as a single carbon source for BNC production. Figure 4 shows the yield and final pH form SH medium modified with 2% carbon sources in static conditions after seven days of cultivation. Cellulose synthesis was observed in all tested conditions, but the highest yield (3.85 g/L) was noted in the glycerol medium. BNC yield in the fructose- and the glucose-medium were slightly lower (2.09 g/L and 2.29 g/L, respectively). The moderate yield was recorded for mannitol, which is transformed into fructose and afterwards converted into UDP-glucose [48]. For all of the studied disaccharides and galactose, low effectiveness was observed. The highest yield in the glycerol medium may be ascribed to the altered and more efficient glycerol metabolism compared to glucose [49,50]. In this way, cellulose is synthesised without gluconic acid generation, which could explain higher BNC biosynthesis from glycerol-medium than SH medium. The high yield of BNC production in the glycerol-medium was shown for other strains, e.g., *Acetobacter* sp. V6 [49], *Ga. xylinus* CGMCC2955 [51], *K. xylinus* E26 [28] and *Gluconacetobacter* sp. RKY5 [52]. On the other hand, Abdelhady et al. who used *K. saccharivorans* PE5 and *A. xylinum* ATCC 10245, did not report satisfactory yield results from glycerol-medium in comparison to other carbon sources, e.g., mannitol, starch or sucrose [53]. Glucose and fructose gave similar results for BNC yield, although in the glucose medium, gluconic acid was generated, and the final pH was significantly lower. By taking into account pH changes during culturing, it can be seen that in the glucose medium, pH did not drop below pH 4.2. It is considered that cellulose is produced efficiently in the pH range of 4.0–7.0 [54]. Sucrose as a single carbon source was not suitable for *K. hansenii* SI1. Similar observations were reported for *K. xylinus* E25, *K. xylinus* E26, *K. hansenii* ATCC 53,582 [28] and *G. hansenii* ATCC 23,769 [55]. Mikkelsen et al. observed that cellulose production starts later (after 84 h) when sucrose is used as a carbon source [5]. On the other hand, Mohammadkazemi et al. [56] and Santos et al. [57] reported a relatively high yield of BNC in sucrose medium. It has been suggested that sucrose, which consists of glucose and fructose, is not transported through the cell membrane, but has to be previously hydrolysed into monosugars [58]. Thus, the yield of BNC in this condition is affected by the strain's ability to the production of β-fructofuranosidase. Maltose appeared to be the least suitable carbon source for *K. hansenii* SI1, followed by galactose and lactose. Our data are in accordance with the yields obtained by Hungund and Gupta [59], Castro et al. [60], Wang et al. [50] and Rani and Appaiah [61]. Among reported strains, the most suitable carbon sources are glucose, fructose, glycerol and mannitol [25,62]. Therefore, our findings are in line with the published data.

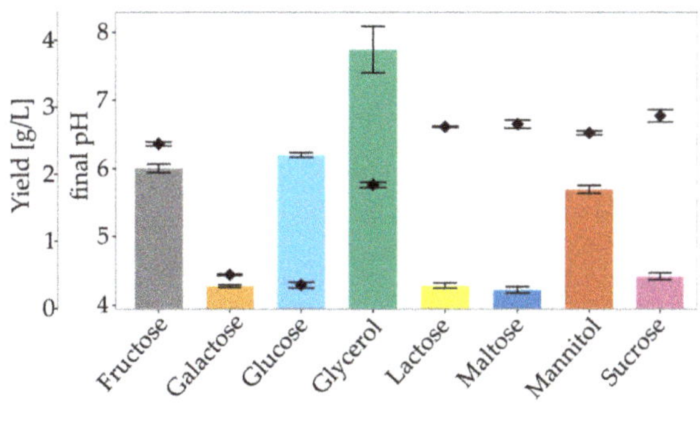

Figure 4. The yield of BNC and final pH in media modified with different carbon sources. Bars represent yield and points pH. Error bars represent standard deviation.

Nitrogen sources do not contribute directly to BNC synthesis, but they are vital for bacterial growth and survival [48]. Thus, proper nitrogen supplementation indirectly affects BNC yield. We investigated the impact of various organic and inorganic nitrogen sources of BNC synthesis in SH medium. Standard nitrogen sources (yeast extract and peptone) were substituted with another, added individually to SH medium with the concentration of 5 g/L. It was considered that organic nitrogen sources gave higher BNC yields than inorganic nitrogen sources [63]. As shown in Figure 5, medium containing yeast extract was the most suitable for BNC production, followed by medium containing peptone (1.69 g/L and 1.09 g/L, respectively). Yeast extract is rich in vitamins, especially B complex, amino acids, and trace elements that stimulate bacteria's growth [64]. Although some authors reported the stimulatory effect of CSL on BNC yield [65–67], we observed a decrease in the yield compared to other organic nitrogen sources. This difference can be explained by different culture conditions, such as carbon source, additional components of medium like lactate, methionine [67]. Ramana et al. observed that the most optimal nitrogen source varies for different carbon sources [68]. In sucrose-medium, the highest BNC yield was noted for casein hydrolysate, while in the case of glucose medium, the most suitable were sodium glutamate and $(NH_4)_2SO_4$. El-Saied et al. reported that medium containing CSL and treated molasses maximises BNC production [67]. The authors also studied the impact of CSL concentration on the BNC yield, and the optimal concentration was found to be 8%. Jang and Jeong observed a similar relationship for *G. persimmonis* KJ145 in apple juice-medium supplemented with ethanol [69].

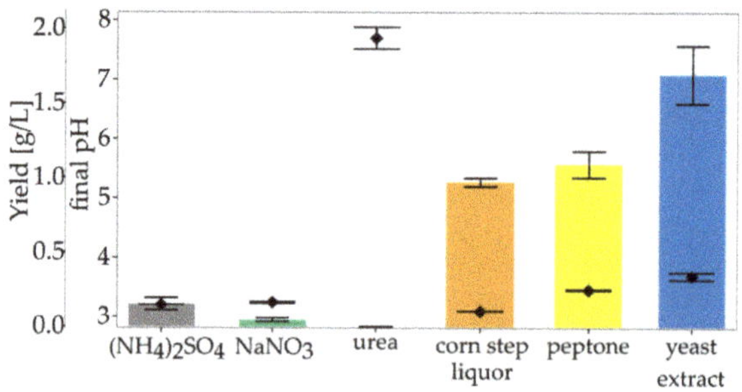

Figure 5. The yield of BNC and final pH in media modified with different nitrogen sources. Bars represent yield and points pH. Error bars represent standard deviation.

In these conditions, we found 5 g/L to be a more economical concentration of CSL. In our study, we used 0.5% of each carbon source, and it is possible that the concentration of CSL in glucose-medium was too low to achieve a satisfying yield. Although our data showed the superiority of organic nitrogen sources, the yield was significantly lower than for standard SH medium containing both yeast extract and peptone. Our finding is supported by a study conducted for *A. xylinum* ATCC 10,245 [53]. The authors also reported that yeast extract gave the highest yield as a single nitrogen source. Still, BNC was produced the most efficiently in the presence of complex nitrogen sources (yeast extract combined with peptone or with tryptone). Santos et al. noticed that *G. sucrofermentas* CECT 7291 produces BNC with the highest yield only in the presence of yeast extract and peptone, or yeast extract and CSL added together [57]. *Komagataeibacter* strains require nutrients for growth and BNC biosynthesis [49]. Thus, it can be concluded that a complex, organic nitrogen source is also obligatory for efficient BNC biosynthesis by *K. hansenii* SI1.

The pH of the culture medium is one of the crucial parameters in terms of optimisation of BNC yield. The pH value determines the activity of enzymes correlated with bacteria growth and BNC biosynthesis [70]. For this purpose, we studied the impact of pH in an SH medium containing glucose or glycerol. Glucose medium led to the production of BNC with unique mechanical properties. At the same time, glycerol-medium resulted in the highest yield in standard conditions. *K. hansenii* SI1 was cultured for its ability to produce cellulose over the range of initial pH from 3 to 7. Above pH 7 and below pH 3, we did not observe the growth of bacteria (data not shown). From Figure 6a,b, it can be seen that the highest BNC production was obtained at pH 7.0 in glucose medium (4.46 g/L), while the most suitable pH in glycerol medium was in the range of 5–6 (3.20–3.60 g/L). At pH 5.0–6.0 in the glucose medium, the yield was diminished to 2.25 g/L–2.49 g/L. The lower pH of both media resulted in a significant decrease in the yield. The difference between optimum initial pH values for the studied carbon source can be ascribed to gluconic acid generation in the glucose medium, which is not generated in the glycerol medium. The final pH glucose medium is dropping from 7.0 to 5.7. Similar final pH (5.4–5.6) was observed for glycerol medium but initial pH at 5.0–6.0. In the case of glycerol medium, the low variation between initial and final pH results in stable conditions during culture and solves pH regulation problems. Our results agree with another research reported by Thorat and Dastager [62]. The authors studied the impact of pH in glycerol medium on BNC production by *K. rhaeticus* PG2. Similar to our results, the authors noticed an optimum pH range from 5.0 to 6.0 and the lowest yield at pH 7.0. Analogous observations in glycerol medium were reported for *Acetobacter* sp. V6 [49]. Generally, it is considered that most strains produce BNC with the highest yield in the pH range of 5.0–6.0 [27,71,72]. Still, it slightly differs between strains and dependents on culture conditions. For *K. hansenii* SI1, the 50% decrease in the yield can be observed at pH 7 in glycerol medium, while in glucose medium, it was optimal initial pH. *K. xylinus* ATCC 700,178 produced BNC with a high field in the medium with pH from 4.5 to 6.2 [73] without a specific optimum point, while for *Acetobacter* sp. A9 and *K. hansenii* AS.5, the optimum was 6.5 [74] and 5.5 [75], respectively. There are also known low- and high-pH resistant strains. Castro et al. described *K. medellinensis* strain from Colombian vinegar, producing BNC efficiently in pH at 3.5 [60]. On the other hand, Pourramezan et al. [76] and Raghunathan [77] reported that *Acetobacter* sp. 4B-2 and *Acetobacter* sp. DR-1, respectively, the most suitable pH level was 7.0, similar to our findings in the glucose medium.

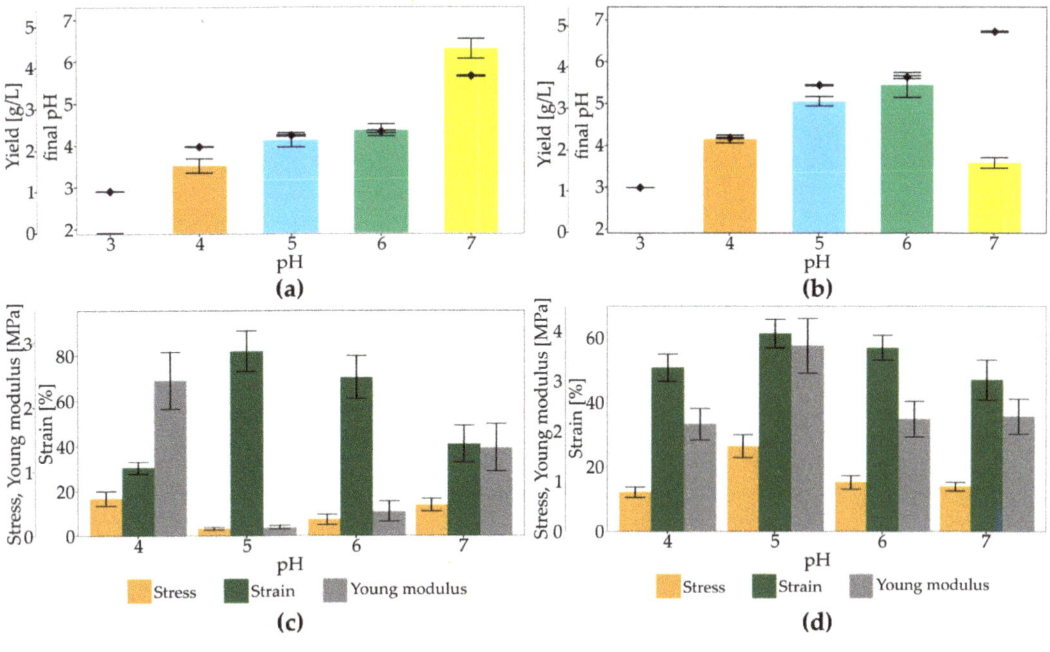

Figure 6. The yield (bars) of BNC and final pH (diamonds) in (**a**) SH medium and (**b**) SH medium modified glycerol at different initial pH, and mechanical properties of BNC membranes produced in (**c**) SH medium and (**d**) SH medium modified glycerol at different initial pH; error bars represent standard deviation.

Mechanical strength varies between samples from media with different initial pH and carbon sources (Figure 6c,d). In the glucose medium, high elongation was observed for membranes produced in pH range 5.0–6.0 (82% and 70%, respectively). Those pellicles were characterised by the lowest stress (0.12 MPa and 0.26 MPa, respectively) and Young modulus (0.14 MPa and 0.39 MPa, respectively). For extreme values of studied initial pH in glucose medium, the strain is decreased over two times. Although the BNC yield differed significantly for media with pH 4.0 and 7.0, the stress values were similar. Membranes produced in glycerol medium were less stretchable than BNC produced in glucose medium at pH 5.0–6.0, but this parameter did not vary highly between pH variants (from 47% to 61%).

Moreover, values of stress and Young modulus were significantly higher than for BNC membranes from the glucose medium. Pellicles produced at pH 5.0 in the glycerol medium had stress and Young modulus of 1.70 MPa and 3.70 MPa, respectively, while those parameters were 14 times and 26 times lower in the case of glucose medium, respectively. The difference in tensile properties between BNC membranes produced in different carbon sources can be ascribed to the crystallisation process [51]. In glycerol medium, bacteria produce cellulose with the lowest porosity [78] and the relative amorphous regions in the structure are reduced [49,62]. During culturing without high pH variations, bacteria synthesise larger cellulose microfibrils, which results in a compact structure [51]. These conclusions are in accordance with our findings. In the case of glycerol medium, membranes were rigid and compact irrespective of the initial pH of the culture medium. Denser arrangement of fibres in BNC form glycerol medium than from glucose medium was also reported for *K. rhaeticus* PG2 [62].

On the other hand, an unstable pH environment led to BNC production with higher porosity [6], which is more susceptible to stretching. We observed the highest difference between initial and final pH in case of glucose medium. Achieved pellicles were highly stretchable and porous (Figure 2c). We conclude that BNC characteristics produced in the glycerol medium do not expand the unique mechanical properties, such as the high ability to stretch, which is not observed for other *Komagataeibacter* strains. Thus, the next experiments will focus on modifying only glucose medium with standard initial pH (5.7), although glycerol as a single carbon source results in an improved yield.

3.4. The Influence of Culture Additives on BNC Biosynthesis and Mechanical Properties

Secondary substrates, such as organic acids, alcohols, vitamins and amino acids, have been proven as suitable enhancers of BNC biosynthesis [49,52,53,79,80]. Thus, we investigated the impact of ethanol, lactic acid vitamin C on BNC yield, structure and mechanical properties. Adding ethanol into the culture medium results in changes in global gene expression and the metabolic profile [81]. Ethanol also affects the enzymes from the BNC biosynthesis pathway [56]. Those changes cause an increase in BNC production. Moreover, ethanol diminishes the number of Cel (-) forms of bacteria, which do not produce bacterial cellulose [82]. On the other hand, lactate could stimulate BNC biosynthesis other than ethanol. The initial stage switches carbon flux to the TCA cycle and promotes cell growth [66]. Lactate functions as an additional energy source, generated during its oxidation into pyruvate [65]. Although vitamin C has been shown as a good enhancer for BNC biosynthesis for several *Komagataeibacter* strains [83], the direct impact of this vitamin reminds unknown. In Figure 7a, the BNC yields in the presence of culture supplements are shown. It can be seen that only ethanol improved the BNC production by *K. hansenii* SI1, but only by 18%. Reported studies show that ethanol highly increases the BNC yield. Our previous study [84] studied the effect of different ethanol concentrations on BNC production by *K. xylinus* E25. The yield was improved by 380% in the case of a 1% concentration of ethanol. Volova et al. reported a 2.2 times higher yield for *K. xylinus* B-12068 cultured in the presence of 3% of ethanol [28]. Additionally, Son et al. observed that *Acetobacter* sp. V6 in ethanol-supplemented medium produced 3.1 times more cellulose than in the medium without ethanol [64]. On the other way, El-Saied et al. noticed only a 6% improvement in the yield in the case of *Gluconacetobacter* subsp. *xylinus* ATCC 10,245 [67], which is closer to our results.

The yield in the lactic acid-supplemented medium did not differ from the yield in the SH medium. These data contradict the findings of Jang and Jeong [69] and Matsuoka et al. [66], who reported the enhanced cellulose production in lactate-medium. According to Matsuoka et al. adding 0.15% lactate into fructose-medium caused increased yield from 0.7 g/L to 3.2 g/L [66]. Jang and Jeong, who used apple juice as a carbon source, reported that at 1% concentration, lactate improved the yield 6.17-fold [69]. However, specific authors reported a lower impact of lactate on BNC biosynthesis. Jung et al. observed only a 32% increase in BNC yield in *Acetobacter* sp. V6 cultured in glycerol-medium [49]. Differences in response to the presence of lactate in the culture medium can be ascribed to varied growth rates and dynamics of BNC production by various *Komagataeibacter* strains and diverse cultivation strategies.

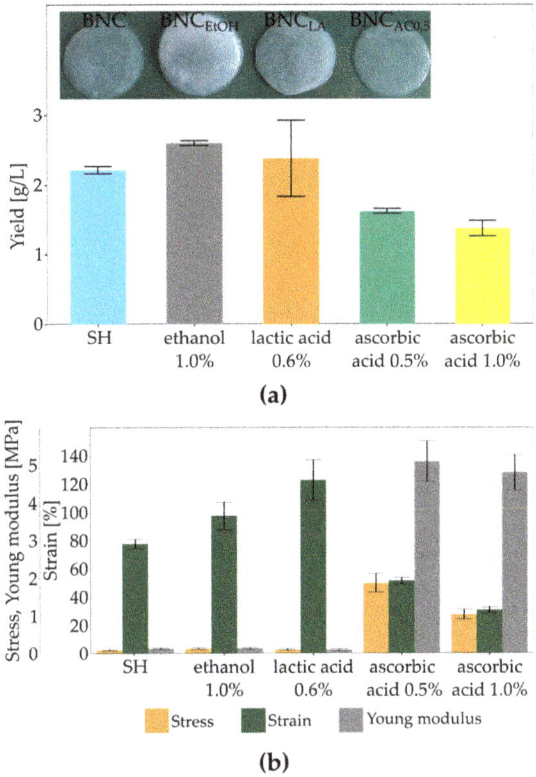

Figure 7. Impact of different culture enhancers on (**a**) BNC yield and (**b**) mechanical properties; error bars represent standard deviation.

Although Keshk and Atykyan et al. reported stimulatory impact of vitamin C on the BNC yield in *Gluconacetobacter xylinus* (ATCC 10,245, IFO 13,693, 13,772 and 13,773) [83] and *Gluconacetobacter sucrofermentans* VKPM B-11267 [85], we observed a reverse effect for *K. hansenii* SI1. The higher was vitamin C concentration, the lower the amount of achieved cellulose. Other authors [86] reported data similar to our findings. They studied lower concentrations of vitamin C, namely 0.01% and 0.04%, and noticed a concentration-dependent decrease in BNC production by *K. xylinus* PTCC1734. As the mechanism of vitamin C action on *Komagataeibacter* metabolism is not yet established, the diverse response of various strains cannot be discussed in detail. Pandit et al. described the inhibitory effect of sodium ascorbate on growth rate and biofilm formation by *Bacillus subtilis* by reducing EPS production [87]. The authors concluded that vitamin C inhibits bacterial quorum sensing and other regulatory mechanisms related to biofilm formation. We observe a difference in soluble EPS (Table S1; extraction according to Fang and Catchmark [88]) between BNC from SH-medium and vitamin C supplemented medium (78 mg/L, 53 mg/L and 27 mg/L for SH medium, SH medium supplemented with 0.5% vitamin C and SH medium supplemented with 1.0% vitamin C, respectively). There were no differences in hard to extract EPS levels between culture variants (approx. 290 mg/L). One might suspect that the metabolism of vitamin C and its effects on cellulose and other EPS is similar to the mechanism described for *B. subtilis* [87]. Still, the molecular aspects of vitamin C metabolism for the genus *Komagataeibacter* require further study.

The pellicles from ethanol- and lactic acid-supplemented media were rigid and not easily stretched. SEM images show (Figure 8) that the three-dimensional structure was changed in both cases. The fibres formed a dense structure with diminished porosity, unlike the loose arrangement in the control sample (Figure 2c). As a result, the tensile parameters of both variants were different (Figure 7b). Maximum stress at break was improved by 19.7 times in the case of BNC produced in the presence of lactic acid and by 10.9 times for BNC from ethanol-medium. The rigid structure of pellicles resulted in decreased strain from 77% to 31% and 51%, respectively. The most important change was observed for Young Modulus, 37–39 times higher than control BNC (from SH medium) due to strain and highly increased stress. Increased Young modulus for pellicles from ethanol and lactic acid supplemented media can be ascribed to the rigid structure and decreased porosity. Similar observations were reported for *K. xylinus* E25 cultured in SH medium with the addition of lactic acid, ethanol or both supplements [84]. In the case of this strain, both additives affected the structure of BNC and its mechanical properties. Membranes were compact with low porosity and were characterised by improved tensile strength (from 1.42 MPa to 4.0 MPa and 5.9 MPa for ethanol and lactic acid medium, respectively) and Young modulus (from 6.7 MPa to 20.7 MPa and 23.4 MPa for ethanol and lactic acid medium, respectively).

(a) **(b)**

Figure 8. Scanning electron microscope images of BNC synthesised by *K. hansenii* SI1 in SH medium supplemented with (**a**) ethanol and (**b**) lactic acid; the images were recorded at a magnification of 40,000× (bar—5 µm).

In contrast, vitamin C improved the porosity of membranes (Figure 9e,f) and positively affected the ability of pellicles to stretch. BNC membranes could easily manually spread on a flat surface (Supplementary Video S1). Raiszadeh-Jahromi et al. did not observe a significant difference in surface morphology caused by vitamin C, but they noticed lower compactness of cellulose layers in cross-section [86]. The authors did not mention the higher plasticity of membranes, which is observed for BNC produced by *K. hansenii* SI1. The surface of spread pellicles increased with the increasing concentration of vitamin C in the culture medium (Figure 9g,i). Fibres in the structure were orientated parallel to the direction of stretching (Figure 9k,l). The behaviour of BNC produced in vitamin C supplemented medium during hand-spreading is in accordance with mechanical properties. Both stress and Young modulus are much lower than the reported data. Costa et al. who used *G. hansenii* UCP1619, determined the Young modulus of BNC in wet state at approx. 10 MPa [89]. *K. xylinus* E25 produced membranes in SH medium with Young modulus equal to 9.0 MPa [35]. In this study, we achieved Young modulus at 130 kPa for membrane produced by *K. hansenii* SI1 in SH medium and at 142 kPa and 100 kPa from vitamin C supplemented medium (0.5% and 1%, respectively). Du et al. found that stress for BNC produced by *Ga. xylinus* isolated from Chinese persimmon vinegar was 15.2 MPa [90], similar to Kwak et al. who reported stress at 12.1 MPa for BNC produced by *Acetobacter* sp. A10 [91]. The strain (elongation) was found at 13–36% for pure, never-dried cellulose membrane [89,91–93]. As shown in Figure 7b, the strain of BNC membranes produced by *K. hansenii* SI1 is higher (77%). It increases with the addition of vitamin C to the culture medium (97% and 123% for 0.5% and 1% of vitamin C, respectively). To the best of our knowledge, such a high level of strain has not been yet reported.

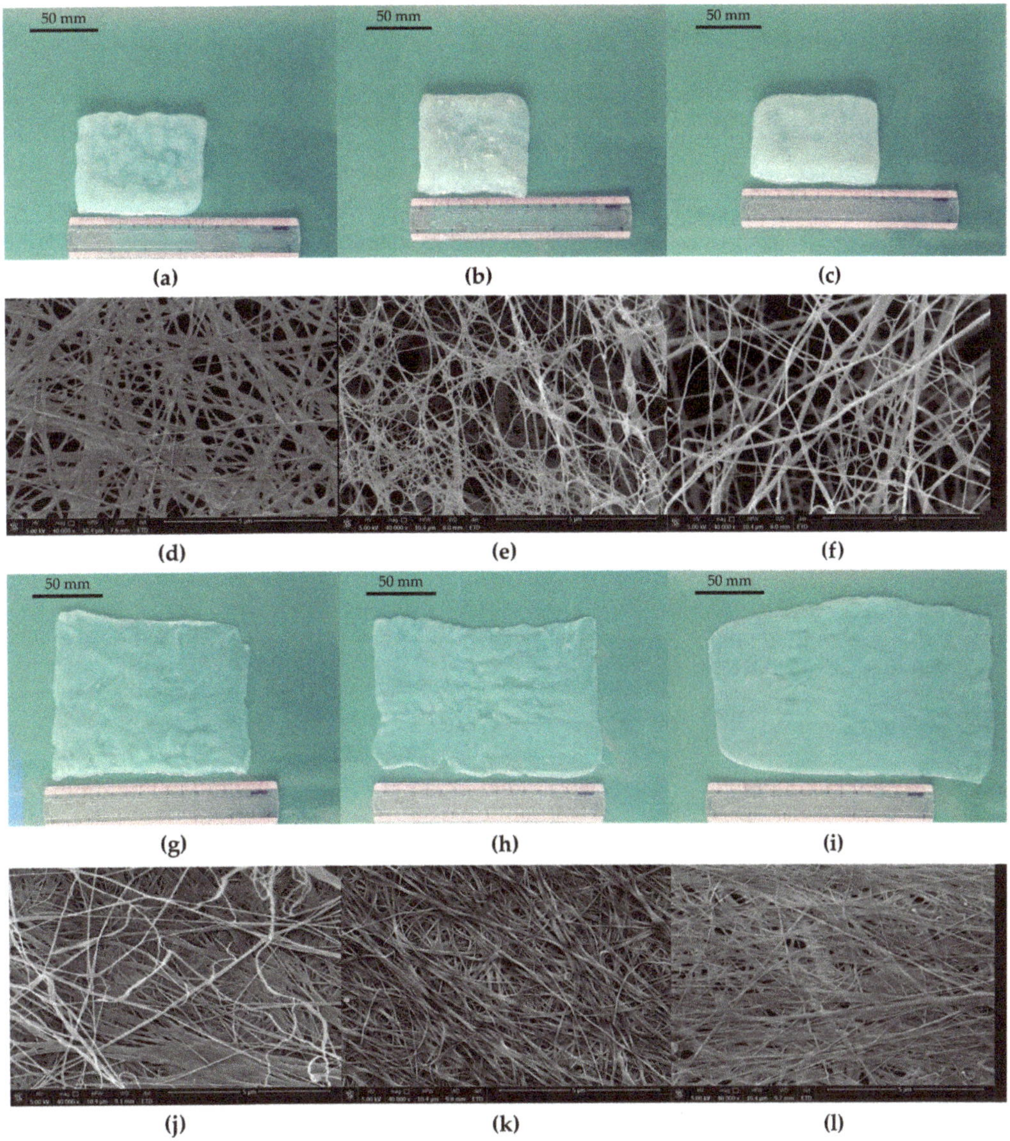

Figure 9. Photographs and SEM images of BNC synthesised by *K. hansenii* SI1 in SH medium (**a**,**d**) before and (**g**,**j**) after stretching; photographs and SEM images of BNC synthesised by *K. hansenii* SI1 in SH medium supplemented with 0.5% vitamin C (**b**,**e**) before and (**h**,**k**) after stretching; photographs and SEM images of BNC synthesised by *K. hansenii* SI1 in SH medium supplemented with 1.0% vitamin C (**c**,**f**) before and (**i**,**l**) after stretching; the SEM images were recorded at a magnification of 40,000× (bar—5 µm).

3.5. The Impact of Vitamin C on BNC Biosynthesis and Chemical Structure

As stated above, *K. hansenii* SI1 produces BNC in the vitamin C-supplemented medium with a lower yield, but unique mechanical properties characterise the membrane. Thus, we evaluated the impact of vitamin C on membranes' biosynthesis kinetic and chemical composition. In this study, bacteria were cultured in the presence of 0.5% and 1.0% vitamin C for seven days. The time course of cultures is presented in Figure 10. In comparison to SH medium (Figure 3), the growth rate of cells was diminished in the case of both modified media. Although rapid exponential growth can be observed within the first two days for all studied variants, the number of cells reached 8.50 logCFU/mL and 8.42 logCFU/mL (0.5% and 1.0% vitamin C, respectively), which is significantly lower than in the case of the control conditions. This difference could be one of the reasons for decreased BNC yield. The glucose consumption rate was almost unchanged, while the pH level was affected by adding vitamin C. In the case of supplemented media, the initial pH was 4.35 and 3.98 for vitamin C concentrations at 0.5% and 1.0%, respectively. It is worth mentioning that the BNC yield produced in 200 mL of medium containing 1% of vitamin C was lower (Figure 7a) than the yield in SH medium with the initial pH at 4.0 in analogous culture conditions (Figure 6a). During the time course study, bacteria were cultured in test tubes, which could affect the modified medium's yield [94]. The difference between initial and final pH was also higher in modified media than in a medium with pH 4.0, e.g., ΔpH = 0.52 in case of 0.5% of vitamin C, while in case of a medium with initial pH 4.0, ΔpH was 0.01. Vitamin C was metabolised linearly during culture in 18% (for 0.5% concentration) and 13% (for 1.0% concentration). Although supplement was not fully metabolised in both cases, the yield and properties of membranes were changed in a concentration-dependent manner.

The chemical composition of BNC produced in the presence of vitamin C was evaluated by FTIR and XRD analysis. The FTIR spectra presented in Figure 11 did not vary from the FTIR spectrum of the control sample (Figure 2e), except for the region between 1200 cm^{-1} and 1000 cm^{-1}. With the increase of vitamin C concentration, peaks in this range are sharper and less overlapped. The bands from 1200 cm^{-1} and 1000 cm^{-1} can be ascribed to the C-O group in the primary and secondary alcohols [42,43]. Fijałkowski et al. correlated the intensity of those peaks with the progressive aggregation of cellulose microfibrils and the formation of hydrogen bonds [95]. According to the authors, differentiation of molecular structure results in enhanced crystallinity and decreased fraction of amorphous regions. Thus, based on the FTIR spectra analysis, an improved crystallinity of BNC produced in vitamin C supplemented media can be observed. This conclusion is in agreement with the XRD study.

The X-ray diffraction patterns (Figure 12) of the studied samples showed five characteristic peaks at the Bragg angle 2θ of 14.7, 17.0, 20.4, 22.7 and 34.7°, corresponding with the crystal planes (100), (010), (11-2), (110) and (11-4), respectively (Table 1) [96,97]. These peaks corresponded to the structure of cellulose I [38]. A higher peak intensity 2θ could easily identify the cellulose Iα at 14.7° compared to peak intensity at 17.0° [98].

The diffractograms calculated the crystallinity index (CI), the interplanar crystal distance (i.e., d-spacing), and the average crystallite size. The results are presented in Table 1. The interplanar crystal distances of the recorded peaks were similar for all BNC samples. The average crystallite size (ASC) slightly varied between the samples, but without any trend. Although the interplanar crystal distances were nearly the same, the CI of samples varied significantly. BNC produced in SH medium had the lowest crystallinity (58%), and with increasing concentration of vitamin C in the medium, CI was also increasing.

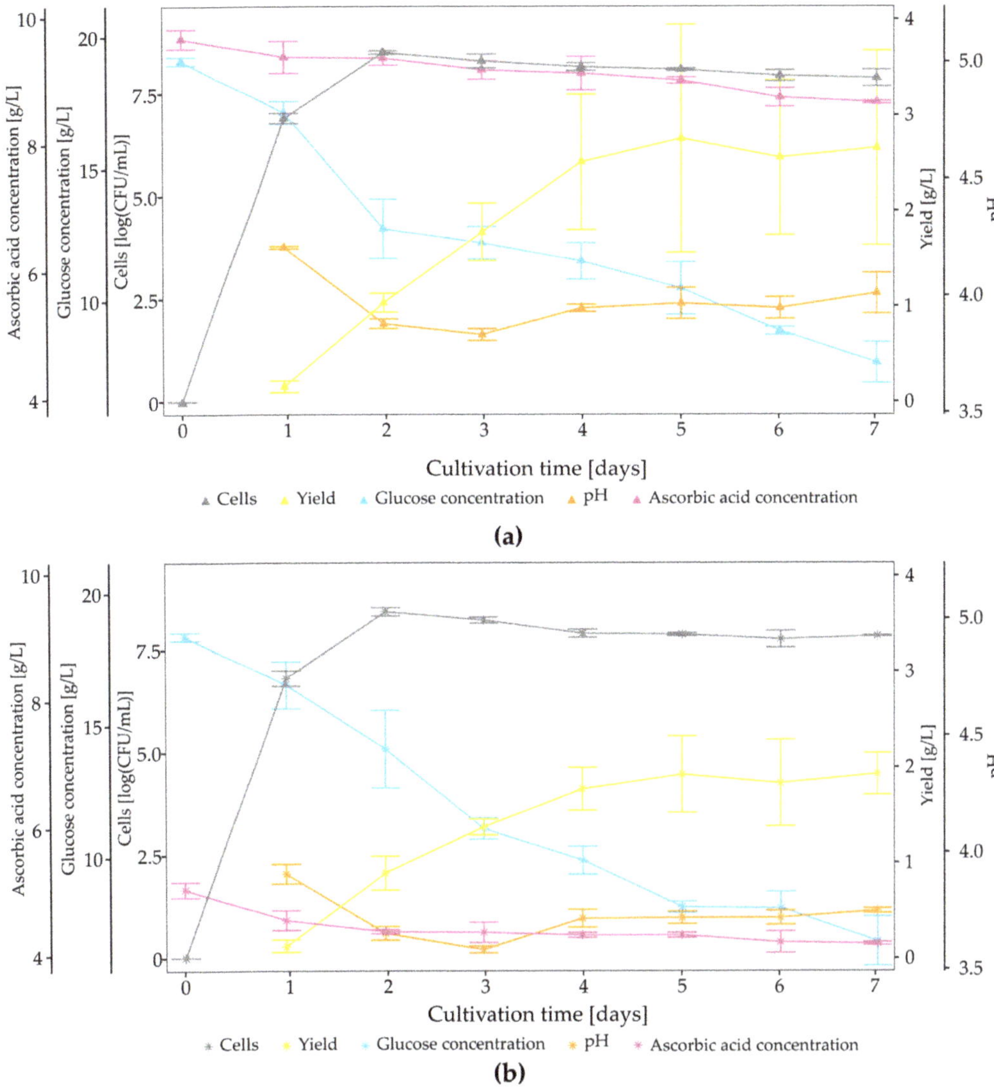

Figure 10. Time course of *K. hansenii* SI1 growth in (**a**) SH medium supplemented with 0.5% vitamin C and (**b**) SH medium supplemented with 1.0% vitamin C. Error bars represent standard deviation.

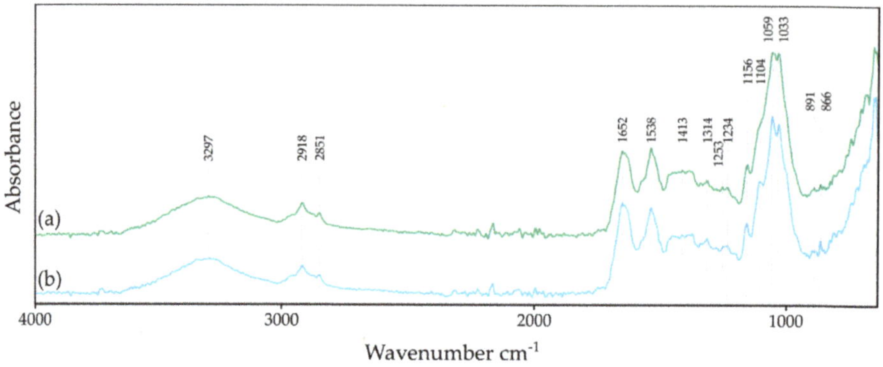

Figure 11. The FTIR-ATR spectra of BNC were produced in (**a**) SH medium supplemented with 0.5% vitamin C and (**b**) SH medium supplemented with 1.0% vitamin C.

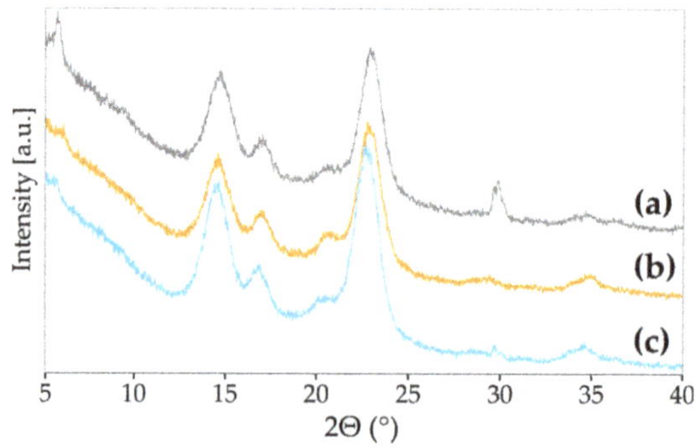

Figure 12. The XRD diffraction patterns of BNC were produced in (**a**) SH medium, (**b**) SH medium supplemented with 0.5% vitamin C, and (**c**) SH medium supplemented with 1.0% vitamin C.

Table 1. Crystallinity index, the interplanar crystal distance and the average crystallite size of BNC produced in SH medium and SH medium supplemented with 0.5% or 1.0% vitamin C.

Sample	CI [%]	Peak (101)		Peak (10-1)		Peak (012)		Peak (002)		Peak (040)	
		$d_{(hkl)}$	ACS [nm]	$d_{(hkl)}$	ACS [nm]	$d_{(hkl)}$	ACS [nm]	$d_{(hkl)}$	ACS [nm]	$d_{(hkl)}$	ACS [nm]
SH	58	6.02	5	5.21	6	4.27	6	3.87	6	2.55	4
0.5% vitamin C	77	6.07	5	5.23	6	4.28	4	3.88	6	2.57	5
1.0% vitamin C	87	6.11	6	5.25	5	4.30	5	3.90	5	2.59	5

On the other hand, Keshk reported a contradictory impact of vitamin C on crystallinity for BNC produced by four different strains [83]. The author suggested that vitamin C decreases the amount of hydrogen bonds between cellulose chains, which opposes our data obtained from FTIR analysis. Nevertheless, a positive impact of vitamin C supplementation on CI was reported for *K. xylinus* PTCC 1734 cultured in the medium containing cheese whey and date syrup as a nutrient source [86]. In the studies mentioned above, it was proposed that the excess vitamin C in the culture medium influences the cellulose chains' orientation, resulting in highly ordered structure and compact cellulose domains. The mechanism of vitamin C action was hypothesised to be the nucleation or cross-linking activity [86].

4. Conclusions

This study isolated a bacterial cellulose-producing strain from Kombucha and identified it as a *K. hansenii*, named SI1, based on the whole-genome sequencing. The isolated strain produces bacterial nanocellulose with unique properties such as enhanced mechanical properties, as compared to the other *Komagataeibacter* strains. To the best of our knowledge, such a high level of cellulose pellicle elongation has not yet been reported.

The production of BNC in various culture conditions was evaluated. As a single carbon source, glycerol was found to produce cellulose with the highest yield, while glucose led to the synthesis of stretchable cellulose membranes. Although glucose medium with initial pH at 7.0 stimulated high BNC production, membranes lost their unique mechanical properties. Supplementation of SH medium with common culture enhancers did not result in a significant increase in the yield.

What is the most significant, the addition of vitamin C to the culture increased porosity and improved mechanical properties of BNC, especially the strain. Moreover, those membranes were characterised by higher crystallinity than BNC produced in control conditions and were free of impurities. Based on the reported experiments, the production method of a novel type, highly stretchable bacterial cellulose was proposed. The obtained results of our study show the increased the potential of BNC in various fields, e.g., in tissue engineering as a material used for preparation of shapeable scaffold or tissue replenishment. The new type of BNC could also be a source of highly ordered cellulose nanocrystallites. The direct impact of vitamin C on *Komagataeibacer hansenii* SI1 metabolism will be a topic of a further detailed transcriptomic study which will allow us to explain changes more thoroughly in physical properties of BNC membranes.

5. Patents

The described method of production of BNC with enhanced porosity and high ability to stretching was subjected to an issue of Polish patent application no P.431265.

Supplementary Materials: The following are available online at https://www.mdpi.com/article/10.3390/polym13244455/s1, Figure S1: Gluconic acid concentration during 7-day culture of *K. hansenii* SI1 in SH medium (grey circles), SH medium supplemented with 0.5% vitamin C (orange triangles) and SH medium supplemented with 1.0% vitamin C (blue stars). Error bars represent standard deviation. Figure S2: Hand covered with transparent stretched BNC membrane produced by *K. hansenii* SI1. Table S1: Concentration of soluble EPS and HE-EPS, Video S1: Stretching of cellulose membrane produced in the presence of 0.5% vitamin C.

Author Contributions: Conceptualisation, I.C.; methodology, I.C. and M.R.; formal analysis, I.C., W.M. and M.R.; investigation, I.C. and M.R.; writing—original draft preparation, I.C.; writing—review and editing, I.C., M.R. and S.B.; visualisation, I.C. and M.R.; supervision, S.B. All authors have read and agreed to the published version of the manuscript.

Funding: This research received no external funding.

Institutional Review Board Statement: Not applicable.

Informed Consent Statement: Not applicable.

Data Availability Statement: The authors confirm that the data supporting the findings of this study are available within the article.

Acknowledgments: The authors would like to thank Marcin Rosowski for his contribution to recording the SEM images. We want to thank BioNanoPark Łódź, Poland, for performing the genome sequencing, especially Justyna Łukasiak and Katarzyna Kubiak. We thank Jolanta Płoszyńska for technical support. The authors thank Benoit Panissie for his careful reading of this manuscript and English corrections.

Conflicts of Interest: The authors declare no conflict of interest.

References

1. Hernandez-Arriaga, A.M.; Campano, C.; Rivero-Buceta, V.; Prieto, M.A. When microbial biotechnology meets material engineering. *Microb. Biotechnol.* **2021**, *00*, 1–15. [CrossRef]
2. Ludwicka, K.; Jedrzejczak-Krzepkowska, M.; Kubiak, K.; Kolodziejczyk, M.; Pankiewicz, T.; Bielecki, S. Chapter 9—Medical and Cosmetic Applications of Bacterial NanoCellulose. In *Bacterial Nanocellulose: From Biotechnology to Bio-Economy*; Gama, M., Dourado, F., Bielecki, S., Eds.; Elsevier: Amsterdam, The Netherlands, 2016; pp. 145–162.
3. Ross, P.; Mayer, R.; Benziman, M. Cellulose biosynthesis and function in bacteria. *Microbiol. Rev.* **1991**, *55*, 35–58. [CrossRef]
4. Jonas, R.; Farah, L.F. Production and application of microbial cellulose. *Polym. Degrad. Stab.* **1998**, *59*, 101–106. [CrossRef]
5. Mikkelsen, D.; Flanagan, B.M.; Dykes, G.A.; Gidley, M.J. Influence of different carbon sources on bacterial cellulose production by *Gluconacetobacter xylinus* strain ATCC 53524. *J. Appl. Microbiol.* **2009**, *107*, 576–583. [CrossRef]
6. Zhang, W.; Wang, X.; Qi, X.; Ren, L.; Qiang, T. Isolation and identification of a bacterial cellulose synthesizing strain from kombucha in different conditions: *Gluconacetobacter xylinus* ZHCJ618. *Food Sci. Biotechnol.* **2018**, *27*, 705–713. [CrossRef]
7. Lavasani, P.S.; Motevaseli, E.; Shirzad, M.; Modarressi, M.H. Isolation and identification of *Komagataeibacter xylinus* from Iranian traditional vinegars and molecular aspects. *Iran J. Micribiol.* **2017**, *9*, 338–347.
8. Fernandez, J.; Gala Morena, A.; Valenzuela, S.V.; Pastor, F.I.J.; Diaz, P.; Martinez, J. Microbial cellulose from *Komagataeibacter intermedius* strain isolated from commercial wine vinegar. *J. Polym. Environ.* **2019**, *27*, 956–967. [CrossRef]
9. Dubey, S.; Sharma, R.K.; Agarwal, P.; Singh, J.; Sinha, N.; Singh, R.P. From rotten grapes to industrial exploitation: *Komagataeibacter europaeus* SGP37, a micro-factory for macroscale production of bacterial nanocellulose. *Int. J. Biol. Macromol.* **2017**, *96*, 52–60. [CrossRef]
10. Liu, L.X.; Liu, S.X.; Wang, Y.M.; Bi, J.C.; Chen, H.M.; Deng, J.; Zhang, C.; Hu, Q.S.; Li, C.F. *Komagataeibacter cocois* sp. nov., a novel cellulose-producing strain isolated from coconut milk. *Int. J. Syst. Evol. Microbiol.* **2018**, *68*, 3125–3131. [CrossRef]
11. Park, M.S.; Jung, Y.H.; Oh, S.Y.; Kim, M.J.; Bang, W.Y.; Lim, Y.W. Cellulosic Nanomaterial Production via Fermentation by *Komagataeibacter* sp. SFCB22-18 Isolated from Ripened Persimmons. *J. Microbiol. Biotechnol.* **2019**, *29*, 617–624. [CrossRef]
12. Bielecki, S.; Krystynowicz, A.; Turkiewicz, M.; Kalinowska, H. Bacterial Cellulose. In *Biopolymers*; Steinbuchel, A., Ed.; Wiley-VCH: Weinheim, Germany, 2002; pp. 37–90.
13. Wang, J.; Tavakoli, J.; Tang, Y. Bacterial cellulose production, properties and applications with different culture methods—A review. *Carbohydr. Polym.* **2019**, *219*, 63–76. [CrossRef] [PubMed]
14. De Oliveira Barud, H.G.; da Silva, R.R.; da Silva Barud, H.; Tercjak, A.; Gutierrez, J.; Lustri, W.R.; de Oliveira, O.B.; Ribeiro, S.J.L. A multipurpose natural and renewable polymer in medical applications: Bacterial cellulose. *Carbohydr. Polym.* **2016**, *153*, 406–420. [CrossRef] [PubMed]
15. Torres, F.G.; Commeaux, S.; Troncoso, O.P. Biocompatibility of bacterial cellulose based biomaterials. *J. Funct. Biomater.* **2012**, *3*, 864–878. [CrossRef]
16. Gomes, N.O.; Carrilho, E.; Machado, S.A.S.; Sgobbi, L.F. Bacterial cellulose-based electrochemical sensing platform: A smart material for miniaturized biosensors. *Electrochim. Acta* **2020**, *349*, 136341. [CrossRef]
17. Ullah, H.; Santos, H.A.; Khan, T. Applications of bacterial cellulose in food, cosmetics and drug delivery. *Cellulose* **2016**, *23*, 2291–2314. [CrossRef]
18. Ludwicka, K.; Kaczmarek, M.; Białkowska, A. Bacterial nanocellulose—A biobased polymer for active and intelligent food packaging applications: Recent advances and developments. *Polymers* **2020**, *12*, 2209. [CrossRef]
19. Abeer, M.M.; Mohd Amin, M.C.; Martin, C. A review of bacterial cellulose-based drug delivery systems: Their biochemistry, current approaches and future prospects. *J. Pharm. Pharmacol.* **2014**, *66*, 1047–1061. [CrossRef] [PubMed]
20. Wu, S.-C.; Wu, S.-M.; Su, F.-M. Novel process for immobilizing an enzyme on a bacterial cellulose membrane through repeated absorption. *J. Chem. Technol. Biotechnol.* **2017**, *92*, 109–114. [CrossRef]
21. Volova, T.G.; Prudnikova, S.V.; Sukovatyi, A.G.; Shishatskaya, E.I. Production and properties of bacterial cellulose by the strain *Komagataeibacter xylinus* B-12068. *Appl. Microbiol. Biotechnol.* **2018**, *102*, 7417–7428. [CrossRef] [PubMed]
22. Stumpf, T.R.; Yang, X.; Zhang, J.; Cao, X. in situ and ex situ modifications of bacterial cellulose for applications in tissue engineering. *Mater. Sci. Eng. C* **2018**, *82*, 372–383. [CrossRef] [PubMed]
23. Singhsa, P.; Narain, R.; Manuspiya, H. Physical structure variations of bacterial cellulose produced by different *Komagataeibacter xylinus* strains and carbon sources in static and agitated conditions. *Cellulose* **2018**, *25*, 1571–1581. [CrossRef]
24. Van Zyl, E.M.; Coburn, J.M. Hierarchical structure of bacterial-derived cellulose and its impact on biomedical applications. *Curr. Opin. Chem. Eng.* **2019**, *24*, 122–130. [CrossRef]
25. Keshk, S.M.A.S.; Sameshima, K. Evaluation of different carbon sources for bacterial cellulose production. *Afr. J. Biotechnol.* **2005**, *4*, 478–482.
26. Mohammadkazemi, F.; Azin, M.; Ashori, A. Production of bacterial cellulose using different carbon sources and culture media. *Carbohydr. Polym.* **2015**, *117*, 518–523. [CrossRef]
27. Cielecka, I.; Ryngajłło, M.; Bielecki, S. BNC biosynthesis with increased productivity in a newly designed surface air-flow bioreactor. *Appl. Sci.* **2020**, *10*, 3850. [CrossRef]
28. Ryngajłło, M.; Kubiak, K.; Jędrzejczak-Krzepkowska, M.; Jacek, P.; Bielecki, S. Comparative genomics of the *Komagataeibacter* strains—Efficient bionanocellulose producers. *Mycrobiol. Open* **2018**, *8*, e731.

29. Nurk, S.; Bankevich, A.; Antipov, D.; Gurevich, A.; Korobeynikov, A.; Lapidus, A.; Prjibelsky, A.; Pyshkin, A.; Sirotkin, A.; Sirotkin, Y.; et al. Assembling single-cell genomes and mini-metagenomes from chimeric MDA products. *J. Comput. Biol.* **2013**, *20*, 714–737. [CrossRef]
30. Goris, J.; Konstantinidis, K.T.; Klappenbach, J.A.; Coenye, T.; Vandamme, P.; Tiedje, J.M. DNA-DNA hybridization values and their relationship to whole-genome sequence similarities. *Int. J. Syst. Evol. Microbiol.* **2007**, *57*, 81–91. [CrossRef] [PubMed]
31. Ryngajłło, M.; Jędrzejczak-Krzepkowska, M.; Kubiak, K.; Ludwicka, K.; Bielecki, S. Towards control of cellulose biosynthesis by the *Komagataeibacter* using systems-level and strain engineering strategies: Current progress and perspectives. *Appl. Microbiol. Biotechnol.* **2020**, *104*, 6565–6585. [CrossRef]
32. Pritchard, L.; Glover, R.H.; Humphris, S.; Elphinstone, J.G.; Toth, I.K. Genomics and taxonomy in diagnostics for food security: Soft-rotting enterobacterial plant pathogens. *Anal. Methods* **2016**, *8*, 12–24. [CrossRef]
33. Camacho, C.; Coulouris, G.; Avagyan, V.; Ma, N.; Papadopoulos, J.; Bealer, K.; Madden, T.L. BLAST+: Architecture and applications. *BMC Bioinform.* **2009**, *10*, 421. [CrossRef] [PubMed]
34. Schliep, K.P. Phangorn: Phylogenetic analysis in R. *Bioinformatics* **2011**, *27*, 592–593. [CrossRef]
35. Cielecka, I.; Szustak, M.; Gendaszewska-Darmach, E.; Kalinowska, H.; Ryngajłło, M.; Maniukiewicz, W.; Bielecki, S. Novel bionanocellulose/κ-carrageenan composites for tissue engineering. *Appl. Sci.* **2018**, *8*, 1352. [CrossRef]
36. Tiboni, M.; Grzybowski, A.; Passos, M.; Barison, A.; Liao, L.M.; Campos, F.R.; Pontarolo, R.; Fontana, J.D. The use of dyed bacterial cellulose to monitor cellulase complex activity. *Cellulose* **2012**, *19*, 1867–1877. [CrossRef]
37. Rabiej, M. A hybrid immune-evolutionary strategy algorithm for the analysis of the wide-angle X-ray diffraction curves of semicrystalline polymers. *J. Appl. Crystallogr.* **2014**, *47*, 1502–1511. [CrossRef]
38. Park, S.; Baker, J.O.; Himmel, M.E.; Parilla, P.A.; Johnson, D.K. Cellulose crystallinity index: Measurement techniques and their impact on interpreting cellulose performance. *Biotechnol. Biofuels* **2010**, *3*, 3–10. [CrossRef]
39. Roman, M.; Winter, W.T. Effect of Sulfate Groups from Sulfuric Acid Hydrolysis on the Thermal Degradation Behavior of Bacterial Cellulose. *Biomacromolecules* **2004**, *5*, 1671–1677. [CrossRef]
40. Klug, H.P.; Alexander, L.E. *X-ray Diffraction Procedures: For Polycrystalline and Amorphous Materials*, 2nd ed.; Wiley: New York, NY, USA, 1974.
41. Halib, N.; Amin, M.C.I.M.; Ahmad, I. Physicochemical properties and characterization of nata de coco from local food industries as a source of cellulose. *Sains Malays.* **2012**, *41*, 205–211.
42. Wang, S.S.; Han, Y.H.; Ye, Y.X.; Shi, X.X.; Xiang, P.; Chen, D.L.; Li, M. Physicochemical characterization of high-quality bacterial cellulose produced by *Komagataeibacter* sp. strain W1 and identification of the associated genes in bacterial cellulose production. *RSC Adv.* **2017**, *7*, 45145–45155. [CrossRef]
43. Liang, C.Y.; Marchessault, R.H. Infrared spectra of crystalline polysaccharides. II. Native celluloses in the region from 640 to 1700 cm^{-1}. *J. Polym. Sci.* **1959**, *39*, 269–278. [CrossRef]
44. Gorgieva, S.; Kokol, V. Synthesis and application of new temperature-responsive hydrogels based on carboxymethyl and hydroxyethyl cellulose derivatives for the functional finishing of cotton knitwear. *Carbohydr. Polym.* **2011**, *85*, 664–673. [CrossRef]
45. Sugiyama, J.; Persson, J.; Chanzy, H. Combined infrared and electron diffraction study of the polymorphism of native celluloses. *Macromolecules* **1991**, *24*, 2461–2466. [CrossRef]
46. Fuller, M.E.; Andaya, C.; McClay, K. Evaluation of ATR-FTIR for analysis of bacterial cellulose impurities. *J. Microbiol. Methods* **2018**, *144*, 145–151. [CrossRef]
47. Fernandes, I.A.A.; Pedro, A.C.; Ribeiro, V.R.; Bortolini, D.G.; Ozaki, M.S.C.; Maciel, G.M.; Haminiuk, C.W.I. Bacterial cellulose: From production optimization to new applications. *Int. J. Biol. Macromol.* **2020**, *164*, 2598–2611. [CrossRef] [PubMed]
48. Gopu, G.; Govindan, S. Production of bacterial cellulose from *Komagataeibacter saccharivorans* strain BC1 isolated from rotten green grapes. *Prep. Biochem. Biotechnol.* **2018**, *48*, 842–852. [CrossRef]
49. Jung, H.-I.; Jeong, J.-H.; Lee, O.-M.; Park, G.-T.; Kim, K.-K.; Park, H.-C.; Lee, S.-M.; Kim, Y.-G.; Son, H.-J. Influence of glycerol on production and structural–physical properties of cellulose from *Acetobacter* sp. V6 cultured in shake flasks. *Bioresour. Technol.* **2010**, *101*, 3602–3608. [CrossRef] [PubMed]
50. Wang, S.S.; Han, Y.H.; Chen, J.L.; Zhang, D.C.; Shi, X.X.; Ye, Y.X.; Chen, D.L.; Li, M. Insights into Bacterial Cellulose Biosynthesis from Different Carbon Sources and the Associated Biochemical Transformation Pathways in *Komagataeibacter* sp. W1. *Polymers* **2018**, *10*, 963. [CrossRef] [PubMed]
51. Zhong, C.; Zhang, G.-C.; Liu, M.; Zheng, X.-T.; Han, P.-P.; Jia, S.-R. Metabolic flux analysis of *Gluconacetobacter xylinus* for bacterial cellulose production. *Appl. Microbiol. Biotechnol.* **2013**, *97*, 6189–6199. [CrossRef]
52. Kim, S.Y.; Kim, J.N.; Wee, Y.J.; Park, D.H.; Ryu, H.W. Production of bacterial cellulose by *Gluconacetobacter* sp. RKY5 isolated from persimmon vinegar. *Appl. Biochem. Biotechnol.* **2006**, *131*, 705–715. [CrossRef]
53. Abdelhady, H.M.; Hassan, E.A.; El-Salam, S.S.A.; Abdullah, S.M. Bacterial cellulose production as affected by bacterial strains and some fermentation conditions. *Nat. Sci.* **2015**, *13*, 30–40.
54. Galas, E.; Krystynowicz, A.; Tarabasz-Szymanska, L.; Pankiewicz, T.; Rzyska, M. Optimization of the production of bacterial cellulose using multivariable linear regression analysis. *Acta Biotechnol.* **1999**, *19*, 251–260. [CrossRef]
55. Hodel, K.V.S.; Fonseca, L.M.D.S.; Santos, I.M.D.S.; Cerqueira, J.C.; Santos-Júnior, R.E.D.; Nunes, S.B.; Barbosa, J.D.V.; Machado, B.A.S. Evaluation of Different Methods for Cultivating *Gluconacetobacter hansenii* for Bacterial Cellulose and Montmorillonite Biocomposite Production: Wound-Dressing Applications. *Polymers* **2020**, *12*, 267. [CrossRef] [PubMed]

56. Mohammadkazemi, F.; Doosthoseini, K.; Azin, M. Effect of ethanol and medium on bacterial cellulose (BC) production from *Gluconacetobacter xylinus* PTCC 1734. *Cellul. Chem. Technol.* **2015**, *49*, 5–6.
57. Santos, S.M.; Carbajo, J.M.; Villar, J.C. The Effect of Carbon and Nitrogen Sources on Bacterial Cellulose Production and Properties from *Gluconacetobacter sucrofermentans* CECT 7291 Focused on Its Use in Degraded Paper Restoration. *BioResources* **2013**, *8*, 3630–3645. [CrossRef]
58. Velasco-Bedrán, H.; López-Isunza, F. The unified metabolism of *Gluconacetobacter entanii* in continuous and batch processes. *Process. Biochem.* **2007**, *42*, 1180–1190. [CrossRef]
59. Hungund, B.S.; Gupta, S.G. Improved production of bacterial cellulose from *Gluconacetobacter persimmonis* GH-2. *J. Microb. Biochem. Technol.* **2010**, *2*, 127–133. [CrossRef]
60. Castro, C.; Zuluaga, R.; Álvarez, C.; Putaux, J.L.; Caro, G.; Rojas, O.J.; Mondragon, I.; Gañán, P. Bacterial Cellulose Produced by a New Acid-Resistant Strain of *Gluconacetobacter* Genus. *Carbohydr. Polym.* **2012**, *89*, 1033–1037. [CrossRef] [PubMed]
61. Rani, M.U.; Appaiah, A. Optimization of culture conditions for bacterial cellulose production from *Gluconacetobacter hansenii* UAC09. *Ann. Microbiol.* **2011**, *61*, 781–787. [CrossRef]
62. Thorat, M.; Dastager, S. High yield production of cellulose by a *Komagataeibacter rhaeticus* PG2 strain isolated from pomegranate as a new host. *RSC Adv.* **2018**, *8*, 29797–29805. [CrossRef]
63. Embuscado, M.E.; Marks, J.S.; BeMiller, J.N. Bacterial cellulose. I. Factors affecting the production of celulose by *Acetobacter xylinum*. *Food Hydrocoll* **1994**, *8*, 407–418. [CrossRef]
64. Son, H.J.; Kim, H.G.; Kim, K.K.; Kim, H.S.; Kim, Y.G.; Lee, S.J. Increased production of bacterial cellulose by *Acetobacter* sp. V6 in synthetic media under shaking culture conditions. *Bioresour. Technol.* **2003**, *86*, 215–219. [CrossRef]
65. Naritomi, T.; Kouda, T.; Yano, H.; Yoshinaga, F. Effect of ethanol on bacterial cellulose production from fructose in continuous culture. *J. Ferment. Bioeng.* **1998**, *85*, 598–603. [CrossRef]
66. Matsuoka, M.; Tsuchida, T.; Matsushita, K.; Adachi, O.; Yoshinaga, F. A Synthetic Medium for Bacterial Cellulose Production by *Acetobacter xylinum* subsp. *sucrofermentans*. *Biosci. Biotechnol. Biochem.* **1996**, *60*, 575–579. [CrossRef]
67. El-Saied, H.; Basta, A.H.; Gobran, R.H. Research Progress in Friendly Environmental Technology for the Production of Cellulose Products (Bacterial Cellulose and Its Application). *Polym. Plast. Technol. Eng.* **2004**, *43*, 797–820. [CrossRef]
68. Ramana, K.V.; Tomar, A.; Singh, L. Effect of various carbon and nitrogen sources on cellulose synthesis by *Acetobacter xylinum*. *World J. Microbiol. Biotechnol.* **2000**, *16*, 245–248. [CrossRef]
69. Jang, S.Y.; Jeong, Y.J. Effect of lactate and corn steep liquor on the production of bacterial cellulose by *Gluconobacter perimmonis* KJ145T. *Food Sci. Biotechnol.* **2005**, *5*, 561–565.
70. Rahman, S.S.A.; Vaishnavi, T.; Vidyasri, G.S.; Sathya, K.; Priyanka, P.; Venkatachalam, P.; Karuppiah, S. Production of bacterial cellulose using *Gluconacetobacter kombuchae* immobilized on Luffa aegyptiaca support. *Sci. Rep.* **2021**, *11*, 2912. [CrossRef]
71. Lima, H.L.S.; Nascimento, E.S.; Andrade, F.K.; Brígida, A.I.S.; Borges, M.F.; Cassales, A.R.; Muniz, C.R.; Filho, M.D.S.M.S.; Morais, J.P.S.; Rosa, M.D.F. Bacterial cellulose production by *Komagataeibacter hansenii* ATCC 23769 using sisal juice—an agroindustry waste. *Braz. J. Chem. Eng.* **2017**, *34*, 671–680. [CrossRef]
72. Rangaswamy, B.E.; Vanitha, K.P.; Hungund, B.S. Microbial Cellulose Production from Bacteria Isolated from Rotten Fruit. *Int. J. Polym. Sci.* **2015**, *2015*, 1–8. [CrossRef]
73. Bilgi, E.; Bayir, E.; Sendemir-Urkmez, A.; Hames, E.E. Optimization of bacterial cellulose production by *Gluconacetobacter xylinus* using carob and haricot bean. *Int. J. Biol. Macromol.* **2016**, *90*, 2–10. [CrossRef]
74. Son, H.-J.; Heo, M.-S.; Kim, Y.-G.; Lee, S.-J. Optimization of fermentation conditions for the production of bacterial cellulose by a newly isolated *Acetobacter* sp. A9 in shaking cultures. *Biotechnol. Appl. Biochem.* **2001**, *33*, 1–5. [CrossRef] [PubMed]
75. Saleh, A.K.; Soliman, N.A.; Farrag, A.A.; Ibrahim, M.M.; El-Shinnawy, N.A.; Abdel-Fattah, Y.R. Statistical optimization and characterization of a biocellulose produced by local Egyptian isolate *Komagataeibacter hansenii* AS.5. *Int. J. Biol. Macromol.* **2020**, *144*, 198–207. [CrossRef] [PubMed]
76. Pourramezan, G.Z.; Roayaei, A.M.; Qezelbash, Q.R. Optimization of culture conditions for bacterial cellulose production by *Acetobacter* sp. 4B-2. *Biotechnology* **2009**, *8*, 150–154. [CrossRef]
77. Raghunathan, D. Production of Microbial Cellulose from the New Bacterial Strain Isolated from Temple Wash Waters. *Microbiol. Appl. Sci.* **2013**, *2*, 275–290.
78. Tang, W.; Jia, S.; Jia, Y.; Yang, H. The influence of fermentation conditions and post-treatment methods on porosity of bacterial cellulose membrane. *World J. Microbiol. Biotechnol.* **2010**, *26*, 125. [CrossRef]
79. Lu, Z.; Zhang, Y.; Chi, Y.; Xu, N.; Yao, W.; Sun, B. Effects of alcohols on bacterial cellulose production by *Acetobacter xylinum* 186. *World J. Microbiol. Biotechnol.* **2011**, *27*, 2281–2285. [CrossRef]
80. Premjet, S.; Premjet, D.; Ohtani, Y. The effect of ingredients of sugar cane molasses on bacterial cellulose production by *Acetobacter xylinum* ATCC 10245. *Sen'i Gakkaishi* **2007**, *63*, 193–199. [CrossRef]
81. Ryngajłło, M.; Jacek, P.; Cielecka, I.; Kalinowska, H.; Bielecki, S. Effect of ethanol supplementation on the transcriptional landscape of bionanocellulose producer *Komagataeibacter xylinus* E25. *Appl. Microbiol. Biotechnol.* **2019**, *103*, 6673–6688. [CrossRef] [PubMed]
82. Chawla, P.R.; Bajaj, I.B.; Survase, S.A.; Singhal, R.S. Microbial Cellulose: Fermentative Production and Applications. *Food Technol. Biotechnol.* **2009**, *47*, 107–124.
83. Keshk, S.M.A.S. Vitamin C enhances bacterial cellulose production in *Gluconacetobacter xylinus*. *Carbohydr. Polym.* **2014**, *99*, 98–100. [CrossRef] [PubMed]

84. Cielecka, I.; Ryngajłło, M.; Maniukiewicz, W.; Bielecki, S. Response surface methodology-based improvement of the yield and differentiation of properties of bacterial cellulose by metabolic enhancers. *Int. J. Biol. Macromol.* **2021**, *187*, 584–593. [CrossRef] [PubMed]
85. Atykyan, N.; Revin, V.; Shutova, V. Raman and FT-IR Spectroscopy investigation the cellulose structural differences from bacteria *Gluconacetobacter sucrofermentans* during the different regimes of cultivation on a molasses media. *AMB Express* **2020**, *10*, 1–11. [CrossRef]
86. Raiszadeh-Jahromi, Y.; Rezazadeh-Bari, M.; Almasi, H.; Amir, S. Optimization of bacterial cellulose production by *Komagataeibacter xylinus* PTCC 1734 in a low-cost medium using optimal combined design. *J. Food Sci. Technol.* **2020**, *57*, 2524–2533. [CrossRef] [PubMed]
87. Pandit, S.; Ravikumar, V.; Adbel-Haleem, A.M.; Derouiche, A.; Mokkapati, V.R.S.S.; Shibom, C.; Mineta, K.; Gojobori, T.; Gao, X.; Westerlund, F.; et al. Low concentration of vitamin C reduce the synthesis of extracellular polymers and destabilize bacterial biofilms. *Front. Microbiol.* **2017**, *8*, 2599. [CrossRef] [PubMed]
88. Fang, L.; Catchmark, J.M. Characterization of cellulose and other exopolysaccharides produced from *Gluconacetobacter* strains. *Carbohydr. Polym.* **2015**, *115*, 663–669. [CrossRef] [PubMed]
89. Costa, A.F.S.; Almeida, F.C.G.; Vinhas, G.M.; Sarubbo, L.A. Production of bacterial cellulose by *Gluconacetobacter hansenii* using corn steep liquor as nutrient sources. *Front. Microbiol.* **2017**, *8*, 2027. [CrossRef] [PubMed]
90. Du, R.; Zhao, F.; Peng, Q.; Zhou, Z.; Han, Y. Production and characterization of bacterial cellulose produced by *Gluconacetobacter xylinus* isolated from Chinese persimmon vinegar. *Carbohydr. Polym.* **2018**, *194*, 200–207. [CrossRef] [PubMed]
91. Kwak, M.H.; Kim, J.E.; Go, J.; Koh, E.K.; Song, S.H.; Son, H.J.; Kim, H.S.; Yun, Y.H.; Jung, Y.J.; Hwang, D.Y. Bacterial cellulose membrane produced by *Acetobacter* sp. A10 for burn wound dressing applications. *Carbohydr. Polym.* **2015**, *122*, 387–398. [CrossRef]
92. Lin, W.-C.; Lien, C.-C.; Yeh, H.-J.; Yu, C.-M.; Hsu, S.-H. Bacterial cellulose and bacterial cellulose-chitosan membranes for wound dressing applications. *Carbohydr. Polym.* **2013**, *94*, 603–611. [CrossRef]
93. Cielecka, I.; Szustak, M.; Kalinowska, H.; Gendaszewska-Darmach, E.; Ryngajłło, M.; Maniukiewicz, W.; Bielecki, S. Glycerol-plasticized bacterial nanocellulose-based composites with enhanced flexibility and liquid sorption capacity. *Cellulose* **2019**, *26*, 5409–5426. [CrossRef]
94. Ruka, D.R.; Simon, G.P.; Dean, K.M. Altering the growth conditions of *Gluconacetobacter xylinus* to maximize the yield of bacterial cellulose. *Carbohydr. Polym.* **2012**, *89*, 613–622. [CrossRef] [PubMed]
95. Fijałkowski, K.; Żywicka, A.; Drozd, R.; Kordas, M.; Rakoczy, R. Effect of *Gluconacetobacter xylinus* cultivation conditions on the selected properties of bacterial cellulose. *Pol. J. Chem. Technol.* **2016**, *18*, 116–122. [CrossRef]
96. French, A.D. Idealized powder diffraction patterns for cellulose polymorphs. *Cellulose* **2014**, *21*, 885–896. [CrossRef]
97. Nishiyama, Y.; Sugiyama, J.; Chanzy, H.; Langan, P. Crystal structure and hydrogen bonding system in cellulose Iα, from synchrotron X-ray and neutron fiber diffraction. *J. Am. Chem. Soc.* **2003**, *125*, 14300–14306. [CrossRef] [PubMed]
98. Ahvenainen, P.; Kontro, I.; Svedström, K. Comparison of sample crystallinity determination methods by X-ray diffraction for challenging cellulose I materials. *Cellulose* **2016**, *23*, 1073–1086. [CrossRef]

polymers

MDPI

Article

Characterization of Birch Wood Residue after 2-Furaldehyde Obtaining, for Further Integration in Biorefinery Processing

Maris Puke [1,*], Daniela Godina [1,2], Mikelis Kirpluks [1], Prans Brazdausks [1] and Janis Rizikovs [1]

1 Latvian State Institu of Wood Chemistry, Dzerbenes 27, LV-1006 Riga, Latvia; daniela.godina@kki.lv (D.G.); mikelis.kirpluks@kki.lv (M.K.); prans.brazdausks@kki.lv (P.B.); janis.rizikovs@kki.lv (J.R.)
2 Department of Chemsitry, Latvia University of Latvia, Jelgavas 1, LV-1004 Riga, Latvia
* Correspondence: maris.puke@kki.lv; Tel.: +371-29874322

Abstract: Latvia is a large manufacturer of plywood in Eastern Europe, with an annual production of 250,000 m^3. In Latvia's climatic conditions, birch (*Betula pendula*) is the main tree species that is mainly used for plywood production. A significant part of the processed wood makes up residues like veneer shorts, cores, and cut-offs (up to 30%), which have a high potential for value-added products. The aim of this research was to comprehensively characterize lignocellulosic (LC) biomass that was obtained after 2-furaldehyde production in terms of further valorization of this resource. The polymeric cellulose-enriched material can be used in the new biorefinery concept for the production of 2-furaldehyde, acetic acid, cellulose pulp, thermomechanical (TMP) and an alkaline peroxide mechanical (APMP) pulping process. In addition, we experimentally developed the best 2-furaldehyde production conditions to optimize the purity and usability of cellulose in the leftovers of the LC material. The best experimental results in terms of both 2-furaldehyde yield and the purity of residual lignocellulose were obtained if the catalyst concentration was 70%, the catalyst amount was 4 wt.%, the reaction temperature was 175 °C,and the treatment time was 60 min. After process optimization with DesignExpert11, we concluded that the best conditions for maximal glucose content (as cellulose fibers) was a catalyst concentration of 85%, a catalyst amount of 5 wt.%, a temperature of 164 °C, and a treatment time of 52 min.

Keywords: birch wood; pre-treatment; process parameter; lignocellulose; cellulose; 2-furaldehyde

Citation: Puke, M.; Godina, D.; Kirpluks, M.; Brazdausks, P.; Rizikovs, J. Characterization of Birch Wood Residue after 2-Furaldehyde Obtaining, for Further Integration in Biorefinery Processing. *Polymers* **2021**, *13*, 4366. https://doi.org/10.3390/polym13244366

Academic Editor: Antonio Pizzi

Received: 25 October 2021
Accepted: 11 December 2021
Published: 13 December 2021

1. Introduction

Nowadays, one of the most important challenges that mankind faces is climate change, which is caused by ever-increasing greenhouse gases (GHG) in the atmosphere produced by the burning and processing of fossil resources. To combat this problem, the European Union (EU) has long pursued a leading role in policies to tackle climate change. In December 2019, the EU made the energy transition one of its main goals and announced that it would pursue a "European Green deal" [1]. The Green Deal can be conceptualized as a roadmap of key policies for the EU's climate agenda, based on which the Commission has started and will continue to develop legislative proposals and strategies from 2020 onwards. EU climate and energy governance are structured around three main headline targets, concerning: (1) a GHG reduction from 1990 levels, (2) the share of renewable energy in final energy consumption, and (3) an improvement in energy efficiency [2]. To decrease industrial dependence on fossil energy resources, there is an ever-more increasing interest in the integrated biorefinery approach development, in which fossil resources are substituted with renewable, biomass-based resources in a closed-loop system [3]. A renewable and sustainable source of organic feedstock with zero carbon emission is one of the most promising pathways to produce green products and replace fossil resources that are in limited and dwindling supply.

From biomass-based feedstock, LC biomass is the most commonly available renewable resource [4] that do not compete with food supplies. Moreover, LC biomass is commonly derived in large quantities from forestry and agricultural waste products. Taking this into account, LC biomass has great importance as the main source of biofuels and other high value-added products such as 2-furaldehyde, lactic acid, or monomeric phenols, which can be utilized as building blocks for the production of polymeric materials [5,6]. In addition, LC biomass is a potential source for chemical resource products, such as ethanol, reducing sugars, and 2-furaldehyde, using hydrolysis processing either via an acid-catalyzed or enzymatic route. All pentose-containing material could, in theory, be used as a raw material for 2-furaldehyde production. However, for an economically viable process, 2-furaldehyde industrial production requires a minimum content of around 15% to 20% of pentoses in biomass [7]. Only about one-third of the pentosanes in raw materials can be converted into 2-furaldehyde through the existing production processes. 2-furaldehyde is an important chemical because it is a selective solvent for separating saturated and unsaturated compounds in petroleum refining, gas, oil, diesel fuel and for the high demand of its derivatives, especially 2-furaldehyde alcohol, used mainly in the production of furan resins for foundry sand binders, which is considered the major market for 2-furaldehyde.

In addition to this, 2-furaldehyde can be used as a starting material to produce a wide range of chemicals, and as such, it is a natural precursor to a range of furan-based chemicals and solvents, including methylfuran, furfuryl alcohol, tetrahydrofurfuryl alcohol, tetrahydrofuran, methyltetrahydrofuran, dihydropyran, and furoic acid. Hydrogenation of the aldehyde group or furan ring remains the most versatile reaction to upgrade furanic components and can be employed to synthesize hydrocarbon fuels directly from furan derivatives. Cleavage of the furan ring by hydrogenolysis can produce alcohols such as 1,5-pentanediol. To synthesize longer-chain hydrocarbons from furfural, adduct formation by aldol condensation and dimerization followed by hydrodeoxygenation can produce C8 to C13 and longer alkanes [8–10].

There is no synthetic route available for 2-furaldehyde production in the chemical industry. So far, all of the 2-furaldehyde is exclusively produced by acid hydrolysis and dehydration of pentoses (mainly xylose) contained in LC biomass resources [11–13]. In Latvia, the most promising LC feedstock for 2-furaldehyde obtaining is birch chips due to its industrial-scale availability in plywood production as well as its carbohydrate content suitable for effective 2-furaldehyde obtaining [14–16].

There have been different publications on 2-furaldehyde production by using varying process conditions, such as by using catalyst or starting feedstock, both from our research group as well as others, which focuses on maximizing obtained 2-furaldehyde yields and obtaining various other side-stream products [17–21]. 2-furaldehyde is a useful bio-based chemical that can be used as a starting block for the production of various value-added chemicals, such as solvents, plastics and various additives for the agricultural industry. 2-furaldehyde is only being produced from biological resources, and it has been identified as one of the top 30 most important biomass-based chemicals [22,23].

Although 2-furaldehyde obtaining is widely studied and used in industry, for successful process integration into the biorefinery processing chain, it is necessary to effectively utilize all parts of used feedstock to obtain value-added products. Carbohydrates extracted from the glucose-enriched residues obtained from furfural production could be used to produce 5-HMF, levulinic acid, or bioalcohol as well as used as a source of cellulosic fiber. The remaining lignin from this LC residue could be used for manufacturing various platform chemicals, such as aromatics, olefins, dibasic acids, lignin-based epoxydes, and lignin-derived polyurethanes [24–26].

To be in line with this, more attention is being spent on characterizing LC residue leftover after 2-furaldehyde obtaining. One of the ways for utilizing this residue could be as a pulp and lignin source by using it in various delignification processes such as TMP or sulphate pulping.

The main goal of this paper is to present information regarding obtained LC residue after 2-furaldehyde obtaining, its composition, and to show its potential as a feedstock in various pulping processes. The goal of the study was to obtain 2-furaldehyde at more than 60% of the theoretical amount and the cellulose degradation step at no more than 10%.

2. Materials and Methods

2.1. Materials and Chemicals

Orthophosphoric acid (H_3PO_4) (85%), sulfuric acid (95–97%), D-(+)-cellobiose (≥99%), D-(+)-glucose, (≥95%), D-(+)-xylose (≥99%), L-(+)-arabinose (≥99%), D-(+)-galactose (≥99%), D-(+)-mannose (≥99%), 2-furaldehyde (≥99%), acetic acid (≥99%), 5-hydroxymethylfurfural (5-HMF) (≥99%), levulinic acid (≥98%), and formic acid (≥95%) were purchased from Merck (Darmstadt, Germany) and used without further purification.

2.2. Samples

Birch wood chips (BWCs) were supplied by the A/S Latvijas Finieris company "Lignums", focusing on the production of plywood and processed wood chips. The company supplies BWCs to pulp producers in Scandinavia. We are using standard wood chips, which are used in pulp mills for cellulose production. The fractional distribution is shown in Figure 1.

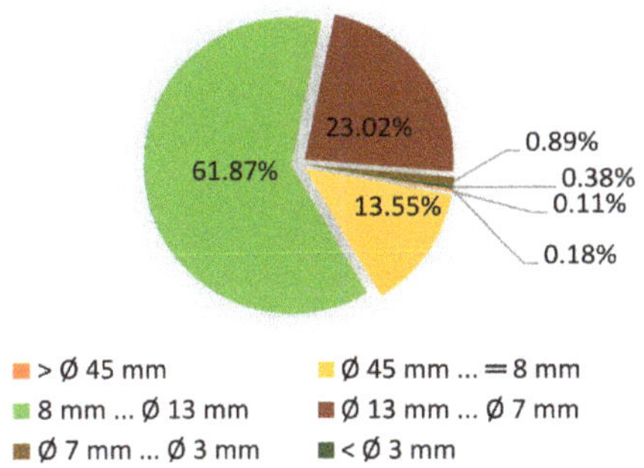

Figure 1. Birch wood chip fractional distribution.

After obtaining, BWCs were air-dried and stored at 15–20 °C to prevent degradation prior to use. The relative humidity in the laboratory was 25–35%. BWCs were of particle size between 45 and 47 mm.

2.3. Catalyzed Pre-Treatment of BWCs

BWCs (particle size 45–47 mm and moisture content W_{rel} = 40.43%) were mixed in a catalyst solution in a blade-type mixer of a special design. Orthophosphoric acid solution of a varied concentration (55–85%) was used as a catalyst. After mixing the chips with a defined amount of the catalyst, the obtained material was treated with a continuous superheated steam flow in an original pilot plant. These value-added products were obtained: condensate containing formic acid, acetic acid, levulinic acid, 5-HMF and 2-furaldehyde, and carbohydrates-enriched LC residue. The diameter of the main reactor camera was 110 mm, its height was 1450 mm, it had a volume of 13.7 L, and a max pressure of 1.2 MPa (bench-scale reactor is shown in our previous publication [21]).

The reactor had two heat insulation systems with automatic equipment to ensure a constant temperature in the reaction zone during the whole process time and with different process parameters. The steam leaving the reactor, which contained mainly a water solution of 2-furaldehyde and acetic acid, was condensed, and samples were taken every 10 min. The steam-treated wood chips lignocellulose (LC) was discharged from the reactor. The chemical composition of the birch chips was determined by the wet chemistry analytical standard methods as described in the Technical Association of the Pulp and Paper Industry (TAPPI) standards [27,28]. All yields of the products and catalyst amounts were calculated on the oven-dried mass of the initial feedstock. For each sample, three parallel experiments were carried out, and the obtained results are shown as the average, with the relative standard deviation (RSD) for all experiments being less than 5%.

2.4. Experimental Design

Based on the data described in our previous scientific publication [21] and information available in studies [10,29,30], the following process parameters were set, and after its implementation, it was possible to judge the direction in which to continue the experimental work (Table 1). It is important to note that the experimental design and process parameters depend on the used type of biomass, the catalyst, and the type of hydrolysis reactor. In turn, the constant factors were the moisture of the raw material (w)—40%, and the steam flow rate (v)—120 mL·min^{-1}.

Table 1. The process parameters.

No.	Catalyst Conc., (c) %	Temperature, (T) °C	Catalyst Amo., (m) wt.%	Treatment Time, (τ) min
1	70	175	4	10
2	70	165	5	10
3	85	165	4	60
4	85	165	3	35
5	85	155	4	35
6	70	155	4	10
7	55	165	3	35
8	70	165	3	60
9	85	175	4	35
10	55	165	5	35
11	70	155	5	35
12	55	155	4	35
13	70	175	3	35
14	70	155	4	60
15	70	165	3	10
16	70	165	4	35
17	85	165	5	35
18	70	175	4	60
19	70	155	3	35
20	55	165	4	10
21	55	175	4	35
22	70	175	5	35
23	70	165	5	60
24	85	165	4	10
25	55	165	4	60

In our previous work [21], using the computer program DesignExpert11, the initial full factorial experimental plan was developed, and after its implementation, it was possible to decide in which direction to continue the experimental studies and find the optimal process parameters for the pre-treatment process. Experimental work was performed on the bench equipment for 25 different experiments, continuing the sample listing from our previous experiments [21]. During these twenty-five experiments, in total ninety-four samples with two parallel samples were obtained. Samples were collected in different time increments. All of the condensate samples containing 2-furaldehyde, acetic acid and other compounds were obtained and analyzed by HPLC (Shimadzu 20AD). Furthermore, samples of LC residue were obtained, and their chemical composition was determined. Obtained data are an average value of the 2 parallel experiments of the 94 condensate samples.

2.5. HPLC Analysis

The contents of monosaccharides, 2-furaldehyde, 5-HMF, and organic acids in the obtained hydrolysates were determined using a Shimadzu LC-20A HPLC (Shimadzu,

Tokyo, Japan) with a refractive index detector. Cellobiose, glucose, xylose, arabinose, galactose, mannose, 2-furaldehyde, acetic acid, 5-HMF, levulinic acid and formic acid (Merck, Darmstadt, Germany) with purity ≥99.0% were used as reference standards. For the cellobiose, glucose, 2-furaldehyde, acetic acid, 5-HMF, levulinic acid and formic acid, we used a Shodex Sugar SH1821 column at 60 °C, with eluent 0.008 M H_2SO_4 at a flow rate of 0.6 mL·min^{-1}. For the carbohydrate analysis, we used a Shodex Sugar SP0810 column at 80 °C, with deionized water as the mobile phase under a flow rate of 0.6 mL·min^{-1}. Samples were neutralized to pH 5–7 with $BaCO_3$ and filtered through a 0.2 µm membrane filter before injection. All samples were tested three times.

For each analyzed standard, the equations of the calibration curves are given in our previous publication [21].

3. Results and Discussion

3.1. Analysis of the Raw Material

The chemical composition of the BWCs was determined, and characteristics are given in Table 2.

Table 2. Chemical composition of BWCs.

Compound	Amount (% of o.d.m.)
Extractives (ethanol-benzene)	4.24 ± 0.06
Extractives (hot water)	1.57 ± 0.44
Glucan	37.84 ± 0.05
Xylan	21.96 ± 0.06
Galactan	0.83 ± 0.05
Arabinan	0.66 ± 0.06
Mannan	1.56 ± 0.50
Acid-insoluble lignin	19.42 ± 0.04
Acid-soluble lignin	3.71 ± 0.06
Ash	0.60 ± 0.01
Acetyl group amount	4.80 ± 0.30
Other unidentified compounds	1.32 ± 0.05

The chemical composition of the used material is comparable to information found in the literature [16]. Characteristics of feedstock show that birch chips have the potential to be used in 2-furaldehyde obtaining due to its content of glucose and xylose. Xylose is a raw material for the production of 2-furaldehyde, while glucose-enriched LC residue can be further utilized as a raw material in pulping processes to obtain cellulosic fibers. From the obtained results, it is possible to calculate that the maximum theoretical amount of 2-furaldehyde obtainable from the BWCs is 16.45% on the oven-dried mass (o.d.m.).

3.2. Selection of the Initial Pre-Treatment Process Parameters for the Experimental Plan

3.2.1. Condensate Chemical Composition after Hydrolysis

Experimental conditions of these experiments are given in Table 1. After the obtained data (Table 3), it can be seen that the best experimental conditions for 2-furaldehyde and acetic acid production is achieved in experiment No. 18, where the catalyst concentration was 70%, the catalyst amount was 4 wt.%, the reaction temperature was 175 °C, and the treatment time was 60 min. The lowest 2-furaldehyde and acetic acid amount (% of o.d.m.) are for experiment No. 6, where the catalyst concentration was 70%, the catalyst amount was 4 wt.%, the reaction temperature was 155 °C, and the treatment time was 10 min. The total formic acid, levulinic acid, and 5-HMF amount for all of the experiments are below 1%.

Table 3. The chemical composition of condensate after hydrolysis.

No. of Experiments	Amount, %.o.d.m.				
	Formic Acid	Acetic Acid	Levulinic Acid	5-HMF	2-Furaldehyde
1	0.22 ± 0.01	1.68 ± 0.01	0.05 ± 0.04	<0.01	1.42 ± 0.10
2	0.14 ± 0.02	0.91 ± 0.03	<0.01	<0.01	0.51 ± 0.04
3	0.59 ± 0.02	4.86 ± 0.01	0.05 ± 0.01	0.03 ± 0.01	7.05 ± 0.11
4	0.35 ± 0.02	3.53 ± 0.01	0.04 ± 0.01	<0.01	2.78 ± 0.04
5	0.29 ± 0.11	3.22 ± 0.03	0.03 ± 0.01	<0.01	1.55 ± 0.19
6	0.09 ± 0.05	0.52 ± 0.04	<0.01	<0.01	0.17 ± 0.09
7	0.38 ± 0.03	2.47 ± 0.11	0.04 ± 0.01	<0.01	2.45 ± 0.07
8	0.52 ± 0.01	4.71 ± 0.01	0.05 ± 0.01	0.02 ± 0.01	5.60 ± 0.02
9	0.53 ± 0.06	4.60 ± 0.01	0.05 ± 0.01	0.02 ± 0.01	6.62 ± 0.09
10	0.40 ± 0.01	4.36 ± 0.02	0.03 ± 0.01	0.01 ± 0.01	4.26 ± 0.02
11	0.31 ± 0.04	3.42 ± 0.02	0.03 ± 0.01	<0.01	1.78 ± 0.03
12	0.28 ± 0.11	3.03 ± 0.03	0.03 ± 0.01	<0.01	1.34 ± 0.04
13	0.48 ± 0.06	4.25 ± 0.02	0.06 ± 0.01	0.02 ± 0.01	5.79 ± 0.05
14	0.40 ± 0.01	4.20 ± 0.01	0.04 ± 0.01	0.01 ± 0.01	2.91 ± 0.20
15	0.13 ± 0.03	0.71 ± 0.03	0.01 ± 0.01	<0.01	0.38 ± 0.01
16	0.40 ± 0.07	4.04 ± 0.01	0.03 ± 0.01	<0.01	3.58 ± 0.01
17	0.43 ± 0.01	4.16 ± 0.01	0.04 ± 0.01	0.01 ± 0.01	3.94 ± 0.03
18	0.67 ± 0.04	5.3 ± 0.02	0.08 ± 0.02	0.06 ± 0.01	10.04 ± 0.02
19	0.27 ± 0.07	2.57 ± 0.01	0.03 ± 0.01	<0.01	1.05 ± 0.14
20	0.15 ± 0.07	0.96 ± 0.01	<0.01	<0.01	0.52 ± 0.08
21	0.47 ± 0.03	4.24 ± 0.01	0.05 ± 0.01	0.02 ± 0.01	5.48 ± 0.04
22	0.54 ± 0.03	4.63 ± 0.01	0.05 ± 0.01	0.02 ± 0.01	6.85 ± 0.04
23	0.62 ± 0.01	5.15 ± 0.01	0.05 ± 0.02	0.03 ± 0.01	7.19 ± 0.01
24	0.15 ± 0.04	0.95 ± 0.01	0.02 ± 0.01	<0.01	0.52 ± 0.14
25	0.52 ± 0.08	4.50 ± 0.02	0.02 ± 0.01	0.02 ± 0.01	5.82 ± 0.03

After Figure 2, it can be observed that the lowest lignocellulose amount was for experiment No. 18, where maximal treatment parameters were used. The highest lignocellulose amount was obtained for experiments No. 15 (catalyst concentration 70%, catalyst amount 3 wt.%, reaction temperature 165 °C and treatment time 10 min) and No. 24 (catalyst concentration 85%, catalyst amount 4 wt.%, reaction temperature 165 °C and treatment time 10 min).

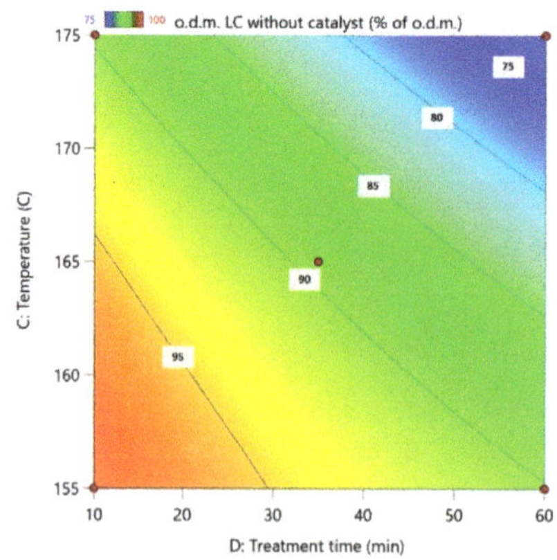

Figure 2. Influence of pre-treatment process parameters on the amount of lignocellulose without catalyst after hydrolysis; the actual factors for the regression model are A = 70 and B = 4.

For this regression model, the standard deviation is 1.57 with an R^2 of 0.9638.

Based on the process results, the following process parameter Equation (1) was obtained:

$$\text{LC without catalyst} = -319.83337 + 0.214028(c) - 2.24448(m) + 5.32509(T) + 1.72771(\tau) - 0.006273(c\cdot\tau) - 0.009337(T\cdot\tau) - 0.017144\,(T^2) \quad (1)$$

3.2.2. LC Residue Chemical Composition after Hydrolysis

Experimental conditions of these experiments are given in Table 1. The highest acid-insoluble lignin amount (%) is for experiment No. 18, and the lowest is for experiments No. 6 and No. 19 (Figure 3). The unifying element of these experiments are the catalyst concentration (70%) and the reaction temperature (155 °C). The highest yield of glucan is for experiments No. 3 (catalyst concentration 85%, catalyst amount 4 wt.%, reaction temperature 165 °C and treatment time 60 min) and No. 18 (Table 4). The yields of arabinan for all experiments are below 1%, but the yields of mannan and galactan are below 5%.

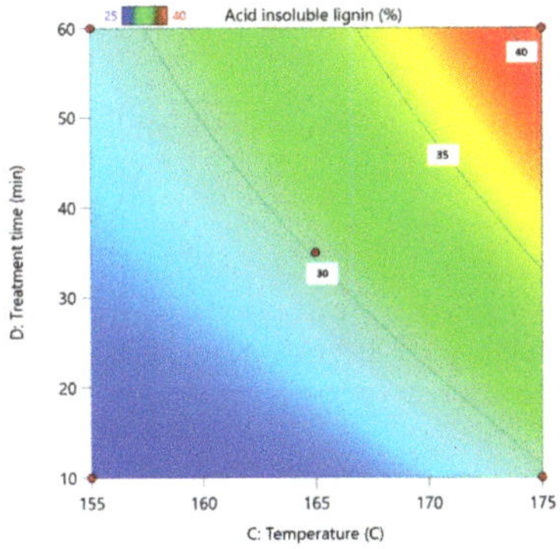

Figure 3. Influence of pre-treatment process parameters on the amount of acid-insoluble lignin in obtained LC residue; the actual factors for the regression model are A = 70 and B = 4.

Table 4. The yield of carbohydrates in lignocellulose after hydrolysis.

No. of Experiments	Amount, %.o.d.m.				
	Glucan	Xylan	Arabinan	Galactan	Mannan
	40.10 ± 4.00	21.90 ± 0.51	0.51 ± 0.21	2.09 ± 0.17	1.14 ± 0.08
1	40.12 ± 5.11	22.31 ± 0.30	0.82 ± 0.54	1.87 ± 0.12	1.30 ± 0.16
2	46.12 ± 6.00	15.51 ± 1.71	0.42 ± 0.23	1.86 ± 0.18	0.96 ± 0.05
3	40.13 ± 6.12	20.63 ± 0.20	0.51 ± 0.31	1.90 ± 0.12	1.06 ± 0.01
4	39.00 ± 6.00	21.71 ± 0.21	0.53 ± 0.22	1.79 ± 0.15	0.88 ± 0.05
5	36.12 ± 5.10	22.81 ± 0.08	0.41 ± 0.20	1.86 ± 0.32	0.99 ± 0.04
6	41.10 ± 5.11	21.60 ± 0.50	0.31 ± 0.10	2.26 ± 0.19	0.74 ± 0.03
7	43.13 ± 5.10	17.20 ± 0.40	0.27 ± 0.10	2.32 ± 0.14	0.76 ± 0.05
8	44.12 ± 5.00	13.51 ± 0.51	0.29 ± 0.11	1.75 ± 0.01	0.59 ± 0.04
9	42.11 ± 5.10	19.86 ± 0.04	0.38 ± 0.11	2.35 ± 0.03	0.80 ± 0.04
10	39.11 ± 4.13	22.14 ± 0.06	0.37 ± 0.11	2.08 ± 0.02	0.99 ± 0.05
11	38.00 ± 4.10	22.11 ± 0.06	0.47 ± 0.03	2.37 ± 0.01	0.97 ± 0.05
12	45.11 ± 5.12	14.21 ± 0.41	0.30 ± 0.20	0.74 ± 0.19	0.97 ± 0.24
13	40.11 ± 5.00	19.12 ± 0.71	0.60 ± 0.40	0.69 ± 0.10	1.11 ± 0.12
14	37.13 ± 5.00	22.50 ± 0.06	0.61 ± 0.22	0.84 ± 0.01	1.08 ± 0.04
15	42.14 ± 6.00	18.83 ± 0.24	0.38 ± 0.05	0.82 ± 0.07	0.72 ± 0.02
16	43.10 ± 6.11	18.47 ± 0.11	0.38 ± 0.10	0.56 ± 0.13	0.72 ± 0.08
17	46.10 ± 4.12	17.82 ± 0.10	0.17 ± 0.11	0.90 ± 0.01	0.80 ± 0.03
18	39.10 ± 3.13	22.86 ± 0.14	0.29 ± 0.07	2.60 ± 0.04	0.71 ± 0.05
19	37.11 ± 4.00	23.21 ± 0.51	0.33 ± 0.07	2.69 ± 0.02	0.83 ± 0.08
20	43.13 ± 5.00	15.20 ± 0.11	0.22 ± 0.03	1.98 ± 0.01	0.69 ± 0.05
21	45.12 ± 6.13	12.91 ± 0.21	0.22 ± 0.06	1.82 ± 0.18	0.53 ± 0.02
22	44.12 ± 5.10	16.43 ± 0.43	0.23 ± 0.03	1.59 ± 0.08	0.56 ± 0.03
23	37.00 ± 4.00	22.66 ± 0.02	0.37 ± 0.06	3.08 ± 0.02	0.84 ± 0.05
24	44.12 ± 3.00	15.21 ± 0.71	0.26 ± 0.02	2.45 ± 0.14	0.67 ± 0.01

The admixture content was lowest in experiment No. 18, being 1.78 ± 0.02% and 0.30 ± 0.02% for 2-furaldehyde and acetic acid, respectively (Table 5). The conditions for this experiment are as follows: the catalyst concentration was 70%, the catalyst amount was 4 wt.%, the reaction temperature was 175 °C, and the treatment time was 60 min.

Table 5. The yield of admixtures in condensate left in lignocellulose residue after hydrolysis.

No. of Experiments	Amount, %.o.d.m.				
	Formic Acid	Acetic Acid	Levulinic Acid	5-HMF	2-Furaldehyde
	0.53 ± 0.01	1.26 ± 0.01	0.70 ± 0.04	0.40 ± 0.01	3.03 ± 0.10
1	0.44 ± 0.02	1.87 ± 0.03	0.63 ± 0.03	0.37 ± 0.01	3.02 ± 0.04
2	0.51 ± 0.02	0.46 ± 0.01	0.82 ± 0.03	0.43 ± 0.05	2.31 ± 0.11
3	0.51 ± 0.02	1.10 ± 0.01	0.63 ± 0.02	0.36 ± 0.01	2.87 ± 0.04
4	0.46 ± 0.11	1.22 ± 0.03	0.55 ± 0.04	0.35 ± 0.01	2.82 ± 0.19
5	0.40 ± 0.05	2.58 ± 0.04	0.48 ± 0.01	0.34 ± 0.02	2.93 ± 0.09
6	0.37 ± 0.03	1.07 ± 0.11	0.46 ± 0.12	0.40 ± 0.01	2.68 ± 0.07
7	0.35 ± 0.01	0.59 ± 0.01	0.62 ± 0.01	0.39 ± 0.01	2.39 ± 0.02
8	0.48 ± 0.06	0.42 ± 0.01	0.77 ± 0.04	0.42 ± 0.01	2.17 ± 0.09
9	0.44 ± 0.01	0.69 ± 0.02	0.62 ± 0.02	0.39 ± 0.01	2.72 ± 0.02
10	0.39 ± 0.04	1.12 ± 0.02	0.53 ± 0.04	0.37 ± 0.01	2.72 ± 0.03
11	0.43 ± 0.11	1.40 ± 0.03	0.50 ± 0.03	0.37 ± 0.01	2.93 ± 0.04
12	0.50 ± 0.06	0.57 ± 0.02	0.74 ± 0.02	0.40 ± 0.01	2.33 ± 0.05
13	0.39 ± 0.01	0.84 ± 0.01	0.54 ± 0.01	0.42 ± 0.01	2.61 ± 0.21
14	0.43 ± 0.03	2.52 ± 0.03	0.49 ± 0.04	0.40 ± 0.01	3.06 ± 0.01
15	0.41 ± 0.07	0.71 ± 0.01	0.61 ± 0.03	0.42 ± 0.01	2.83 ± 0.01
16	0.43 ± 0.01	0.72 ± 0.01	0.66 ± 0.06	0.43 ± 0.01	2.82 ± 0.03
17	0.56 ± 0.04	0.30 ± 0.02	1.09 ± 0.06	0.44 ± 0.03	1.78 ± 0.02
18	0.51 ± 0.07	1.65 ± 0.01	0.60 ± 0.02	0.37 ± 0.01	2.77 ± 0.14
19	0.45 ± 0.07	2.06 ± 0.01	0.58 ± 0.01	0.37 ± 0.01	3.18 ± 0.08
20	0.63 ± 0.03	0.57 ± 0.01	0.84 ± 0.01	0.43 ± 0.01	2.38 ± 0.04
21	0.71 ± 0.03	0.44 ± 0.01	0.92 ± 0.04	0.42 ± 0.01	2.22 ± 0.04
22	0.68 ± 0.01	0.48 ± 0.01	0.84 ± 0.01	0.43 ± 0.01	2.28 ± 0.01
23	0.50 ± 0.04	2.05 ± 0.01	0.62 ± 0.04	0.40 ± 0.03	3.15 ± 0.14
24	0.51 ± 0.08	0.44 ± 0.01	0.68 ± 0.04	0.48 ± 0.01	2.05 ± 0.03

For this regression model, the standard deviation is 0.38 with an R^2 of 0.994.

Based on the process results, the following process parameter Equation (2) was obtained:

$$\text{Acid insoluble lignin} = 351.58226 + 0.019073(c) + 1.09748(m) - 4.14765(T) - 1.19312(\tau) + 0.008137(c \cdot \tau) + 0.012927(T^2) \quad (2)$$

3.3. Changes in the Average Degree of Cellulose Polymerization after the 2-Furaldehyde Obtaining Process

It is very important that the average degree of the polymerization of the cellulose in the LC residue during the 2-furaldehyde obtaining process is decreased to 200. This means that the structure of the wood cell wall is affected, and less energy will be required for the thermomechanical pulping process.

The changes of the cellulose degree of polymerization in separated LC residuals after the 2-furaldehyde obtaining process indicate the availability of this resource for further processing (Figure 4). For the most promising experiments (No. 18), the degree of polymerization is decreased to 200. This indicates that for this sample, the cell-wall structure is significantly altered, and further pulping of this resource can be implemented at milder conditions, preserving energy resources as well as the quality of the obtained pulp. It could be assumed that a lower degree of polymerization in the obtained LC residuals can further facilitate the pulping process performance, and obtained cellulose fibers in the pulp will be shorter in comparison to the pulping process of the unmodified raw material.

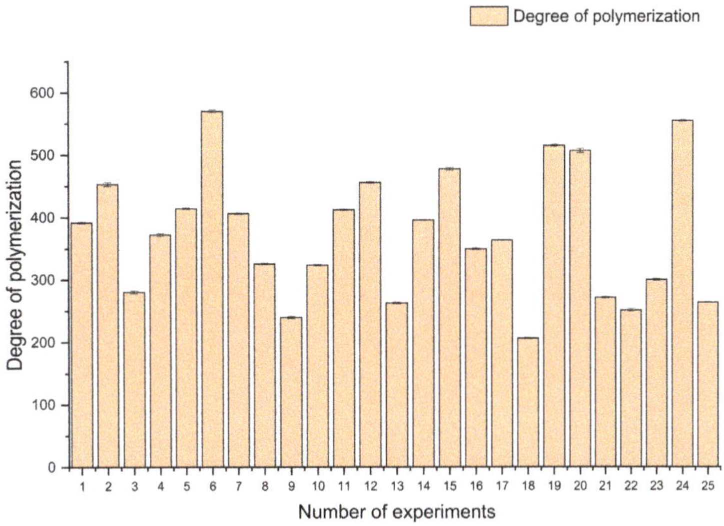

Figure 4. Changes in the degree of cellulose polymerization during the 2-furaldehyde obtaining process.

3.4. Experimental Design Modulation Using DesignExpert11

To characterize obtained results in terms of process efficiency, results were processed using DesignExpert11. This allows for the obtaining of mathematical equations that predict obtained yields of 2-furaldehyde, acetic acid, and glucose-enriched LC residue by varied input parameters. Obtained parity plots reveal how the predicted results correlate with the obtained actual results, and the obtained surface plots show how process parameters influenced the obtained results.

Based on the process results, the following process parameter Equation (3) was obtained:

$$\begin{aligned}\text{2-furaldehyde} = {} & 83.58384 - 0.014482(c) + 0.025299(m) - 1.03242(T) - 0.971688(\tau) + 0.000825(c\cdot\tau) \\ & + 0.014727(m\cdot\tau) + 0.005891(T\cdot\tau) + 0.003196(T^2)\end{aligned} \quad (3)$$

For this regression model, the standard deviation is 0.29 with an R^2 of 0.993, which indicates that the prepared model correlates closely with the actual obtained data. The surface graph for 2-furaldehyde shows that to obtain the maximum amount of 2-furaldehyde, process conditions have to be optimized using the highest possible temperature as well as the longest process duration, which leads to a higher amount of LC material conversion to 2-furaldehyde (Figure 5). This, of course, does not take into account any unwanted side reactions and degradation that can occur in such harsh conditions.

Based on the process results, the following process parameter Equation (4) was obtained:

$$\text{Acetic acid} = -19.31076 + 0.093366(c) + 1.82872(m) + 0.064675(T) + 0.173131(\tau) - 0.020892(c\cdot m) - 0.001378(\tau^2) \quad (4)$$

Similar to results obtained for 2-furaldehyde, the acetic acid prediction model corresponds nicely with obtained data with the obtained process equation as shown in Equation (2).

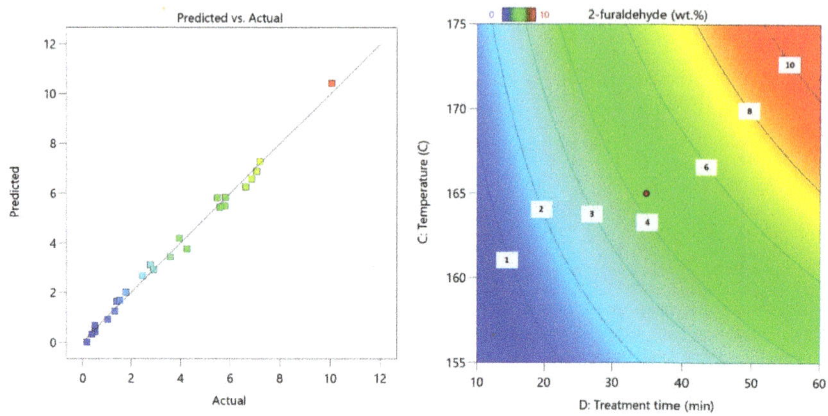

Figure 5. Parity plot of 2-furaldehyde yield (**left**); influence of pre-treatment process parameters on 2-furaldehyde yield (**right**); the actual factors for the regression model are A = 85 and B = 5.

For this prediction model, the standard deviation is 0.23 with an R^2 of 0.9832 (Figure 6).

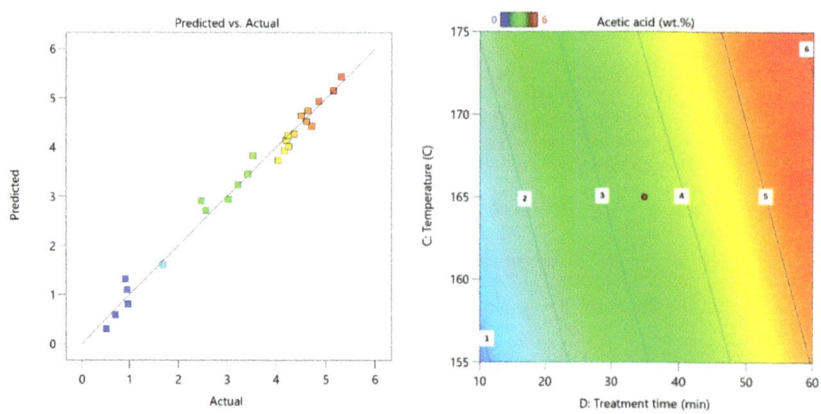

Figure 6. Parity plot of acetic acid (**left**); influence of pre-treatment process parameters on acetic acid yield (**right**); the actual factors for the regression model are A = 85 and B = 5.

Next was optimization of the process parameters for maximal glucose content in leftover LC residue after 2-furaldehyde obtaining. Based on the process results, the following process parameter Equation (5) was obtained:

$$\begin{aligned}\text{Glucose} = {} & 387.26156 + 0.074569(c) + 10.46853(m) + 4.93048(T) + 0.241925(\tau) - 0.001571(c\cdot\tau) - 0.065291(m\cdot T) \\ & - 0.014377(T^2) - 0.001696(\tau^2)\end{aligned} \quad (5)$$

For this prediction model, the standard deviation is 0.58 with an R^2 of 0.837.

Here, the surface plot reveals that to ensure maximal cellulose fiber conservation in the leftover residual fraction, it is vital that the 2-furaldehyde obtaining temperature is around 162 °C and the process length is around 40 min (Figure 7). Deviation from these conditions leads to either incomplete hemicellulose transformation in milder condition, which leads to a lower relative amount of cellulose, or in the case of a harsher-than-optimal condition, it leads to unnecessary degradation of cellulose fiber in addition to unwanted degradation/condensation reactions, which also lowers the amount of cellulosic fiber in the LC residue.

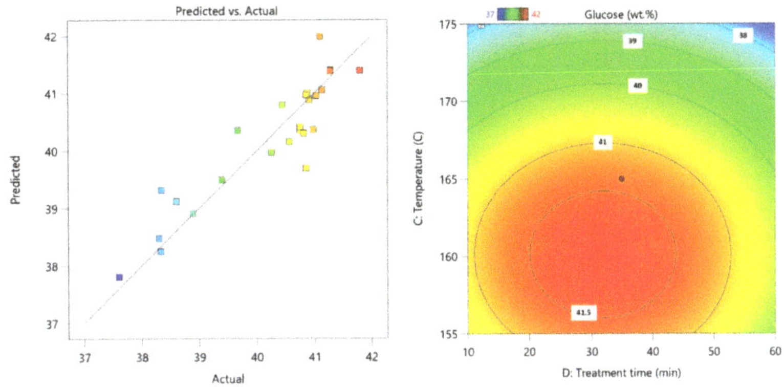

Figure 7. Parity plot of glucose yield (**left**); influence of pre-treatment process parameters on glucose yield (**right**); the actual factors for the regression model are A = 85 and B = 5.

3.5. Experimental Design Optimization Using DesignExpert11

In DesignExpert11, process parameters were set to obtain the best possible outcome in terms of glucose content in the leftover LC residue (Table 6). Parameter limits were set based on previously obtained results and experimental design [21].

Table 6. Set parameters and their limits for experimental model optimization.

Name	Goal	Lower Limit	Upper Limit	Lower Weight	Upper Weight	Importance
A:Catalyst conc.	Is in range	55	85	1	1	1
B:Catalyst amo.	Is in range	3	5	1	1	1
C:Temperature	minimize	155	175	1	1	1
D:Treatment time	is in range	10	60	1	1	5
2-furaldehyde	maximize	0	11	1	1	5
Acetic acid	none	0.52	5.31	1	1	3
Glucose	maximize	37.61	41.80	1	1	5
o.d.m. LC without catalyst	none	75.21	98.83	1	1	3
Acid insoluble lignin	none	25.50	41.71	1	1	3

Optimized process conditions are shown in Table 7, and the influence of the process time and temperature on the total desirability process is shown in Figure 8. Utilizing these conditions, it is possible to obtain 2-furaldehyde and acetic acid from biomass with yields of 1.6% and 2.29% respectively, while obtaining lignocellulose residue, which is enriched with cellulose. Glucose content in the obtained residue was 41.8%, which is the maximum possible amount based on the starting feedstock.

Table 7. Obtained optimal process conditions.

No.	Catalyst Conc. (c), %	Catalyst Amo. (m), wt.%	Temperature (T), °C	Treatment Time (τ), min	2-Furaldehyde	Glucose	Desirability
1	85	5	164	52	6	41	0.7

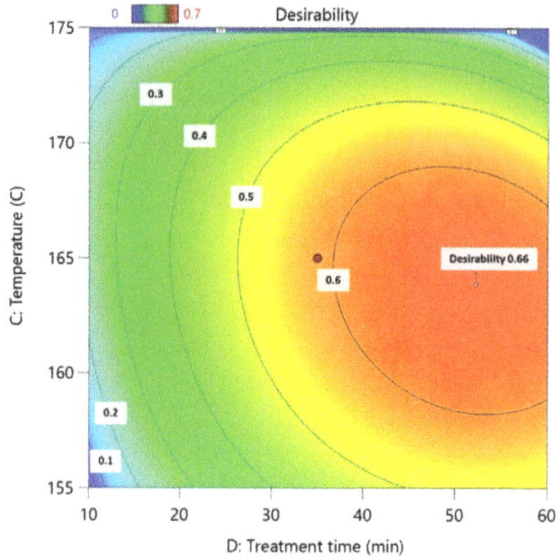

Figure 8. Influence of process time and temperature on total desirability of process; the actual factors for the regression model are A = 85 and B = 5.

4. Conclusions

Birch wood chips are a resource that can be successfully integrated into a biorefinery-based processing pathway whilst simultaneously obtaining 2-furaldehyde, acetic acid, and cellulose fiber-enriched LC residue. The best results in terms of both 2-furaldehyde yield and purity of residual lignocellulose were obtained at a catalyst concentration of 70%, a catalyst amount of 4 wt.%, a reaction temperature of 175 °C, and a treatment time of 60 min (No. 18). Obtained residue at these conditions is a promising feedstock for further use in pulping processes to obtain high-quality and purity of cellulosic fibers. After gathering the DesignExpert11 data, the optimized process conditions were as follows: a catalyst concentration of 85%, a catalyst amount of 5 wt.%, a temperature of 164 °C, and a treatment time of 52 min.

Author Contributions: Conceptualization, M.P.; methodology, M.P. and P.B.; formal analysis, M.P.; writing—original draft preparation, D.G. and M.P.; writing—review and editing, J.R., D.G. and M.P.; data analysis and visualization, M.K. and D.G.; supervision, J.R.; project administration, M.P. All authors have read and agreed to the published version of the manuscript.

Funding: This research was funded by ERDF project No. 1.1.1.2/VIAA/2/19/392 "Study of a novel bio-refining method for obtaining of 2-furaldehyde, acetic acid and pulp from birch wood".

Institutional Review Board Statement: Not applicable.

Informed Consent Statement: Not applicable.

Data Availability Statement: Not applicable.

Conflicts of Interest: The authors declare no conflict of interest.

References

1. Siddi, M. The European Green Deal: Assessing Its Current State and Future Implementation; FIIA Working Paper; Finish Institute of Insternational Affairs, Helsinki, Finland, 2020; p. 14. FIIA Working Paper; Finish Institute of Insternational Affairs: Helsinki, Finland, 2020; p. 14.
2. Oberthür, S. Hard or Soft Governance? The EU's Climate and Energy Policy Framework for 2030. *Politics Gov.* **2019**, *7*, 17–27. [CrossRef]
3. Clark, J.; Deswarte, F. The Biorefinery Concept. In *Introduction to Chemicals from Biomass*; Clark, J., Deswarte, F., Eds.; Wiley: Hoboken, NJ, USA, 2015; pp. 1–29, ISBN 9781118714485.

4. Balan, V.; Bals, B.; Chundawat, S.P.S.; Marshall, D.; Dale, B.E. Lignocellulosic Biomass Pretreatment Using AFEX. *Methods Mol. Biol. (Clifton N. J.)* **2009**, *581*, 61–77. [CrossRef]
5. Adsul, M.G.; Singhvi, M.S.; Gaikaiwari, S.A.; Gokhale, D.V. Development of Biocatalysts for Production of Commodity Chemicals from Lignocellulosic Biomass. *Bioresour. Technol.* **2011**, *102*, 4304–4312. [CrossRef]
6. Isikgor, F.H.; Becer, C.R. Lignocellulosic Biomass: A Sustainable Platform for the Production of Bio-Based Chemicals and Polymers. *Polym. Chem.* **2015**, *6*, 4497–4559. [CrossRef]
7. Yahyazadeh, A. Extraction and Investigation of Furfural in Tea Leaves and Comparing with Furfural in Rice Hull. *J. Pharm. Res.* **2011**, *4*, 4338–4339.
8. Huber, G.W.; Dumesic, J.A. An Overview of Aqueous-Phase Catalytic Processes for Production of Hydrogen and Alkanes in a Biorefinery. *Catal. Today* **2006**, *111*, 119–132. [CrossRef]
9. Wojcik, B.H. Catalytic Hydrogenation of Furan Compounds. *Ind. Eng. Chem.* **1948**, *40*, 210–216. [CrossRef]
10. Cai, C.M.; Zhang, T.; Kumar, R.; Wyman, C.E. Integrated Furfural Production as a Renewable Fuel and Chemical Platform from Lignocellulosic Biomass. *J. Chem. Technol. Biotechnol.* **2014**, *89*, 2–10. [CrossRef]
11. Yemiş, O.; Mazza, G. Acid-Catalyzed Conversion of Xylose, Xylan and Straw into Furfural by Microwave-Assisted Reaction. *Bioresour. Technol.* **2011**, *102*, 7371–7378. [CrossRef]
12. Riansa-Ngawong, W.; Prasertsan, P. Optimization of Furfural Production from Hemicellulose Extracted from Delignified Palm Pressed Fiber Using a Two-Stage Process. *Carbohydr. Res.* **2011**, *346*, 103–110. [CrossRef]
13. Vázquez, M.; Oliva, M.; Téllez-Luis, S.J.; Ramírez, J.A. Hydrolysis of Sorghum Straw Using Phosphoric Acid: Evaluation of Furfural Production. *Bioresour. Technol.* **2007**, *98*, 3053–3060. [CrossRef] [PubMed]
14. Demirbas, A. Competitive Liquid Biofuels from Biomass. *Appl. Energy* **2011**, *88*, 17–28. [CrossRef]
15. Gomez, L.D.; Steele-King, C.G.; McQueen-Mason, S.J. Sustainable Liquid Biofuels from Biomass: The Writing's on the Walls. *New Phytol.* **2008**, *178*, 473–485. [CrossRef] [PubMed]
16. Lachowicz, H.; Wróblewska, H.; Sajdak, M.; Komorowicz, M.; Wojtan, R. The Chemical Composition of Silver Birch (Betula Pendula Roth.) Wood in Poland Depending on Forest Stand Location and Forest Habitat Type. *Cellulose* **2019**, *26*, 3047–3067. [CrossRef]
17. Brazdausks, P.; Puke, M.; Vedernikovs, N.; Kruma, I. Influence of Biomass Pretreatment Process Time on Furfural Extraction from Birch Wood. *Rigas Teh. Univ. Zinat. Raksti* **2013**, *11*, 5. [CrossRef]
18. Yang, W.; Li, P.; Bo, D.; Chang, H. The Optimization of Formic Acid Hydrolysis of Xylose in Furfural Production. *Carbohydr. Res.* **2012**, *357*, 53–61. [CrossRef]
19. Anthonia, E.E.; Philip, H.S. An Overview of the Applications of Furfural and Its Derivatives. *Int. J. Adv. Chem.* **2015**, *3*, 42–47.
20. Vedernikov, N.; Kruma, I.; Puke, M. Furfural and Bioethanol Production from Hardwood and Agricultural Waste. In Proceedings of the UEAA General Assembly and the Associated Workshop: Renewable Energy Resources, Production and Technologies, 5, Riga, Latvia, 28–31 May 2008.
21. Puke, M.; Godina, D.; Kirpluks, M.; Rizikovs, J.; Brazdausks, P. Residual Birch Wood Lignocellulose after 2-Furaldehyde Production as a Potential Feedstock for Obtaining Fiber. *Polymers* **2021**, *13*, 1816. [CrossRef] [PubMed]
22. Bhaumik, P.; Dhepe, P.L. Exceptionally High Yields of Furfural from Assorted Raw Biomass over Solid Acids. *RSC Adv.* **2014**, *4*, 26215–26221. [CrossRef]
23. Campos Molina, M.J.; Mariscal, R.; Ojeda, M.; López Granados, M. Cyclopentyl Methyl Ether: A Green Co-Solvent for the Selective Dehydration of Lignocellulosic Pentoses to Furfural. *Bioresour. Technol.* **2012**, *126*, 321–327. [CrossRef]
24. Baker, D.A.; Gallego, N.C.; Baker, F.S. On the Characterization and Spinning of an Organic-Purified Lignin toward the Manufacture of Low-Cost Carbon Fiber. *J. Appl. Polym. Sci.* **2012**, *124*, 227–234. [CrossRef]
25. Demirbas, M.F. Current Technologies for Biomass Conversion into Chemicals and Fuels. *Energy Sources Part A Recovery Util. Environ. Eff.* **2006**, *28*, 1181–1188. [CrossRef]
26. Hüttermann, A.; Mai, C.; Kharazipour, A. Modification of Lignin for the Production of New Compounded Materials. *Appl. Microbiol. Biotechnol.* **2001**, *55*, 384–387. [CrossRef]
27. Technical Association of Pulp and Paper Industry. *T204 Solvent Extractives of Wood and Pulp*; Lin, T., Ed.; InTech: London, United Kingdom, 2011.
28. Technical Association of Pulp and Paper Industry. *Ash in Wood, Pulp, Paper and Paperboard: Combustion at 525 °C. TAPPI Test Methods*; Lin, T., Ed.; InTech: London, UK, 2009; Volume 1995, pp. 1–173.
29. Padilla-Rascón, C.; Romero-García, J.M.; Ruiz, E.; Castro, E. Optimization with Response Surface Methodology of Microwave-Assisted Conversion of Xylose to Furfural. *Molecules* **2020**, *25*, 3574. [CrossRef]
30. Mittal, A.; Black, S.K.; Vinzant, T.B.; O'Brien, M.; Tucker, M.P.; Johnson, D.K. Production of Furfural from Process-Relevant Biomass-Derived Pentoses in a Biphasic Reaction System. *ACS Sustain. Chem. Eng.* **2017**, *5*, 5694–5701. [CrossRef]

polymers

MDPI

Article

Malva parviflora Leaves Mucilage: An Eco-Friendly and Sustainable Biopolymer with Antioxidant Properties

Ans Munir [1,†], Fadia S. Youssef [2,†], Saiqa Ishtiaq [1,*], Sairah H. Kamran [3], Alaa Sirwi [4], Safwat A. Ahmed [5], Mohamed L. Ashour [2,6,*] and Sameh S. Elhady [4]

1 Department of Phrmacognosy, College of Pharmacy, University of the Punjab, Lahore 54000, Pakistan; ansmunir92@gmail.com
2 Department of Pharmacognosy, Faculty of Pharmacy, Ain-Shams University, Cairo 11566, Egypt; fadiayoussef@pharma.asu.edu.eg
3 Institute of Pharmacy, Faculty of Pharmaceutical Sciences and Allied Health Sciences, Lahore College for Women University, Lahore 54000, Pakistan; sairah.hafeez@lcwu.edu.pk
4 Department of Natural Products, Faculty of Pharmacy, King Abdulaziz University, Jeddah 21589, Saudi Arabia; asirwi@kau.edu.sa (A.S.); ssahmed@kau.edu.sa (S.S.E.)
5 Department of Pharmacognosy, Faculty of Pharmacy, Suez Canal University, Ismailia 41522, Egypt; safwat_ahmed@pharm.suez.edu.eg
6 Pharmacy Program, Department of Pharmaceutical Sciences, Batterjee Medical College, Jeddah 21442, Saudi Arabia
* Correspondence: saiqa.pharmacy@pu.edu.pk (S.I.); ashour@pharma.asu.edu.eg (M.L.A.)
† These authors are equally contributed to work.

Citation: Munir, A.; Youssef, F.S.; Ishtiaq, S.; Kamran, S.H.; Sirwi, A.; Ahmed, S.A.; Ashour, M.L.; Elhady, S.S. *Malva parviflora* Leaves Mucilage: An Eco-Friendly and Sustainable Biopolymer with Antioxidant Properties. *Polymers* **2021**, *13*, 4251. https://doi.org/10.3390/polym13234251

Academic Editor: Arn Mignon

Received: 28 October 2021
Accepted: 30 November 2021
Published: 3 December 2021

Publisher's Note: MDPI stays neutral with regard to jurisdictional claims in published maps and institutional affiliations.

Abstract: *Malva parviflora* L. is an edible and medicinal herb containing mucilaginous cells in its leaves. Mucilage obtained from *M. parviflora* leaves (MLM) was extracted in distilled water (1:10 *w*/*v*) at 70 °C followed by precipitation with alcohol. Preliminary phytochemical tests were performed to assess the purity of the extracted mucilage. Results showed that the yield of mucilage was 7.50%, and it was free from starch, alkaloids, glycosides, saponins, steroids, lipids and heavy metals. MLM had 16.19% carbohydrates, 13.55% proteins and 4.76% amino acids, which indicate its high nutritional value. Physicochemical investigations showed that MLM is neutral and water-soluble, having 5.84% moisture content, 15.60% ash content, 12.33 swelling index, 2.57 g/g water-holding capacity and 2.03 g/g oil-binding capacity. The functional properties, including emulsion capacity, emulsion stability, foaming capacity and stability increased with increased concentrations. Micromeritic properties, such as bulk density, tapped density, Carr's index, Hausner ratio, and angle of repose, were found to be 0.69 g/cm^3, 0.84 g/cm^3, 17.86%, 1.22 and 28.5, respectively. Scanning electron microscopy (SEM) showed that MLM is an amorphous powder possessing particles of varying size and shape; meanwhile, rheological studies revealed the pseudoplastic behavior of MLM. The thermal transition process of MLM revealed by a differential scanning calorimetry (DSC) thermogram, occurring at a reasonable enthalpy change (ΔH), reflects its good thermal stability. The presence of functional groups characteristic of polysaccharides was ascertained by the infrared (IR) and gas chromatography/mass spectrometry (GC/MS) analyses. GC revealed the presence of five neutral monosaccharides; namely, galactose, rhamnose, arabinose, glucose and mannose, showing 51.09, 10.24, 8.90, 1.80 and 0.90 mg/g of MLM, respectively. Meanwhile, galacturonic acid is the only detected acidic monosaccharide, forming 15.06 mg/g of MLM. It showed noticeable antioxidant activity against the DPPH (1,1-diphenyl-2-picrylhydrazyl) radical with an IC_{50} value of 154.27 μg/mL. It also prevented oxidative damage to DNA caused by the Fenton reagent, as visualized in gel documentation system. The sun protection factor was found to be 10.93 ± 0.15 at 400 μg/mL. Thus, MLM can be used in food, cosmetic and pharmaceutical industry and as a therapeutic agent due to its unique properties.

Keywords: emulsion capacity; FTIR; *Malva parviflora*; natural polymers; physicochemical properties; rheology

1. Introduction

Mucilage is a plant hydrocolloid that, upon hydrolysis, yields a mixture of sugars and uronic acids [1]. Structurally, mucilage is a complex of polymeric polysaccharides; in addition, they also contain glycoproteins and various bioactive components, represented by tannins, alkaloids and steroids. Plant-derived mucilage can be obtained from the special mucilage cells of different plant parts. Today, due to bio-incompatibility, toxicity and carcinogenicity of synthetic polymers, greater attention was given to plant-based, naturally derived biopolymers (mucilage, gums, etc.) as key ingredients in the formulation of sustainable, ecofriendly and cost effective products. Moreover, a greater interest was given to natural polymers aiming to encourage "green" materials from "green" chemistry and technologies. Their unique biological properties as well as their chemical flexibility make them the material of choice for efficient drug delivery [2].

Chemically, mucilages are polysaccharides extensively used in pharmaceutical and medicinal applications, due to their diverse physicochemical properties [3]. Pharmaceutically, they are employed as tablet binders and disintegrants, and as suspending, emulsifying, thickening, gelling, stabilizing and sustaining agents in tablets, in addition to their utilization as protective colloids in suspension [4]. Mucilages obtained from *Abelmoschus, Aloe, Asario, Bavchi, Fenugreek, Hibiscus, Ispagol, Ocimum* seed, *Satavari* and *Cactus* plants are examples of pharmaceutically important mucilages [4]. Traditionally, they are used to treat diabetes, enhance wound healing, boost immunity and prevent cancer. They are also used as antitussive, antispasmodic, antitoxin and angiotensin-converting enzyme inhibitory agents [5,6].

Mucilages possess antioxidant potential, which lower the aging process and hence can be used in cosmetics. Mucilage has high demand for water purification, due to high water absorption capacity. In the food industry, there is a growing interest in processed foods that contain mucilage, such as thickening, gel formation and emulsifying agents. It is also used as a stabilizer in the dairy industry, such as in ice cream, yogurt and flavored milk. In addition, mucilage-based hydrocolloids have been used in the food industry to offer textural functionality, such as fruit filling, water binding in meat products, milk-based beverages, desserts and jams. Mucilages have excessive use in ink, glue and adhesive industries [2,7].

Nowadays, intensive research is ongoing on plant-derived mucilage-based coating and films. In addition, mucilage is used to develop nanofibers that are used in textile industry fabrication. In coming years, there will be a significant rise in mucilage demand in food, cosmetics and pharmaceutical industries. However, a significant amount of mucilage for industries is not available and statistical consumption of mucilage is not reported. Thus, it was felt mandatory to start the research and commercialization of more natural polymers for drug delivery systems. Therefore, polymer and pharmaceutical scientists are interested in characterizing more natural polymers to be transferred to the market, where the study and consumption of natural polymer for drug delivery (naturapolyceutics) is essential for the present and the future [8,9].

Malva parviflora L. is a highly popular species in the genus *Malva* family Malvaceae, commonly known as Cheeseweed, or small whorl mallow, and named in Pakistan by Sonchal. Its leaves possess a pleasant taste and are used as vegetables and salad. Decoction of the whole plant is used to treat cold, fever and cough in folk medicine; meanwhile, a poultice made from the leaves is also used to treat wounds, swellings, abdominal pain, gastritis, constipation, diarrhea, anthelmintic, hair loss, scorpion sting, cataplasm, bladder ulcer, renal inflammation, diuretic, ocular disease and profuse menstruation [10,11].

Furthermore, *M. parviflora* was reported to possess many pharmacological activities comprising antibacterial, antifungal, antidiabetic, antioxidant, anti-inflammatory, neuroprotective, hepatoprotective, anti-irritant, wound healing, analgesic and antiulcerogenic activities [12]. These pronounced activities are mainly attributed to its richness with secondary metabolites, such as flavonoids and phenolic acids [13]. It is noteworthy to highlight that mucilage is one of the major constituents prevailing in genus *Malva*, owing to the presence of mucilaginous cells, which play a crucial role in its therapeutic activities [14–16].

Although mucilaginous cells were previously reported in the leaves of *M. parviflora*, nothing was found in the literature regarding its extraction and potential use as a source of eco-friendly and sustainable biopolymer with multiple medicinal values. Thus, in this study, we aimed to isolate mucilage from *M. parviflora* leaves and discover its pharmaceutical and biological properties. Mucilage was extracted in distilled water (1:10 *w*/*v*) at 70 °C followed by precipitation with alcohol. Preliminary phytochemical tests were performed to assess the purity of the extracted mucilage. Various physicochemical, functional and micromeritic properties of powdered mucilage were also investigated. Infrared and gas chromatography, coupled with mass spectrometry analyses, total phenolic and total flavonoid content were also performed. In vitro antioxidant activity was studied by using the 1,1-diphenyl-2-picrylhydrazyl (DPPH*) radical scavenging capacity assay; furthermore, in vitro DNA damage protection and in vitro solar protection factor (SPF) were also determined.

2. Materials and Methods

2.1. Plant Material

Malva parviflora L. (Malvaceae) was collected from Begum Kot near Shahdra, Lahore, Punjab, Pakistan, in February 2018. Authentication was kindly performed by Dr. Zaheer-ud-din Khan, Plant Taxonomist, Department of Botany, Government College University, Lahore. A specimen with voucher number GC.Herb.Bot.3533 was deposited in the Herbarium of the Department of Botany, GC University, Lahore, Pakistan. Leaves were separated and washed with tap water in order to remove any adhering materials. They were shade-dried and powdered by a grinder that was consequently exposed to extraction.

2.2. Extraction of Mucilage

Mucilage was extracted by the method previously described by Malviya [17], with slight modifications. Briefly, dried powdered leaves were soaked in warm distilled water in a ratio of 1:10 (*w*/*v*) for 4 h. Then, heating was performed in a water bath with continuous stirring at 70 °C for 1 h and cooled at room temperature for 1 h. The thick mixture was squeezed in a muslin cloth to remove the residue from the filtrate. The viscous filtrate was concentrated under reduced pressure at 60–70 °C using a rotary evaporator (Laborota 4002, Heidolph, Schwabach, Germany). Mucilage was precipitated from the viscous concentrate by the addition of two volumes of absolute alcohol. Precipitated mucilage was re-dissolved in distilled water, centrifuged and the supernatant was collected. Alcohol was added in supernatant to re-precipitate the mucilage. The precipitation step was repeated to improve the quality of mucilage. Precipitated mucilage was separated, washed with absolute alcohol and dried at 45 °C, and powdered and passed through sieve no. 80. The powdered mucilage was labeled as MLM and stored in a desiccator until further use.

2.3. Preliminary Confirmative Tests for Mucilage

The presence of mucilage was confirmed by performing Molisch's test and ruthenium red test; meanwhile, the iodine test was done to check the presence of starch [1].

2.4. Determination of the Yield Percentage

The percentage of mucilage yield was calculated by comparing the weight of dried extracted mucilage with the weight of the initial sample by using the following equation [18].

$$\%Yield = \frac{wt.\ of\ dried\ mucilage\ obtained}{wt.\ of\ sample\ taken} \times 100$$

2.5. Determination of the Purity of Mucilage

Qualitative phytochemical tests were performed to check the presence of impurities, such as reducing sugars, proteins, amino acids, lipids, tannins, alkaloids, flavonoids, steroids, glycosides, fats and oils [19].

2.6. Elemental Analysis Using Atomic Absorption Spectroscopy

MLM was digested using dry ash technique followed by the analysis of Cu, Pb, Ni, Cr, Fe, Zn and Cd using the atomic absorption flame spectrophotometer (AA7000, Shimadzu, Kyoto, Japan) [20].

2.7. Determination of the Nutritional Value of Mucilage

2.7.1. Estimation of Total Carbohydrates

Total carbohydrate content was estimated by the anthrone method, where various glucose concentrations were used as a standard for constructing the calibration curve [21].

2.7.2. Estimation of Total Proteins

Total proteins were determined employing the method previously reported by Lowry et al., in which bovine serum albumin (BSA) was used as a standard to establish the calibration curve [22].

2.7.3. Estimation of Total Free Amino Acids

Total free amino acids were estimated employing the method previously described by Sadasivam, where L-leucine was used as a standard to form a calibration curve [21]. In this method, 80% ethanol (10 mL) was added into a known amount of MLM (100 mg) and exposed to vortex for 1 min. The mixture was transferred to centrifuge tube, centrifuged at 3000 rpm and the supernatant was collected. A total of 0.1 mL of the supernatant was taken and made up the volume up to 1 mL with distilled water. The reagent blank was prepared in the same way by taking 0.1 mL of 80% ethanol. A total of 1 mL of ninhydrin reagent was added in all tubes and heated in a boiling water bath for 20 min. A total of 5 mL of distilled water were added in all test tubes and the absorbance was measured at 570 nm using a spectrophotometer.

2.8. Determination of the Physicochemical Properties of the Mucilage

2.8.1. Evaluation of the Organoleptic Properties of the Mucilage

Organoleptic properties, such as color, odor, taste, shape, touch and texture were evaluated using sense organs comprising eye, nose, tongue, skin, and ear [23].

2.8.2. Evaluation of the Solubility of the Mucilage

The solubility of extracted mucilage was checked in different solvents, such as water, hot water, 1 M NaOH, 1 M HCl and 1 M citric acid, and organic solvents, such as methanol, ethanol, acetone, petroleum ether and carbon tetrachloride [24].

2.8.3. Determination of the pH of the Mucilage

A pH of 1% w/v aqueous mucilage solution was determined using a pH meter [25].

2.8.4. Determination of the Moisture Content of the Mucilage

The moisture content of the mucilage was determined using a moisture analyzer (MA35, Sartorius, Gottingen, Germany) [24].

2.8.5. Determination of the Ash Content of the Mucilage

Total ash content, water-soluble and acid insoluble ash was determined by a previously reported procedure by Schuck et al. [26].

2.8.6. Determination of the Swelling Index of the Mucilage

The swelling index was determined using the method previously described by Killedar et al. [27]. Briefly, 1 g of dried powdered mucilage was taken in a 100 mL stopper graduated cylinder. The initial bulk volume was measured, and then 1 mL of alcohol was added, followed by a sufficient quantity of distilled water to form 100 mL of uniform dispersion. Consequently, the cylinder was agitated every 10 min for the first 1 h and kept aside for 3 h. The agitation was repeated at 3 h intervals, and finally, the volume was measured after 24 h. Three readings were taken simultaneously, and the average of the results was reported [27].

2.8.7. Determination of the Water-Holding Capacity of the Mucilage

Water-holding capacity (WHC) was determined using a method previously reported by Segura-Campos et al. [28]. An empty tube of 15 mL was weighed, then 1 g of powdered mucilage was put in the falcon tube and re-weighed, followed by the addition of 10 mL of distilled water, where the mixture was exposed to the vortex (Seoul in Biosciences, SL V-6, Seoul, Korea) for 5 min. Centrifugation was performed at 3000 rpm (2-16KC, Sigma, Taufkirchen, Germany) for 30 min. Then, the supernatant was removed carefully. Finally, the falcon tube was re-weighed along with the residue. The water-holding capacity of mucilage was calculated using the following equation:

$$WHC = \frac{weight\ of\ wet\ sample - weight\ of\ dry\ sample}{weight\ of\ the\ dry\ sample}.$$

2.8.8. Determination of the Oil Binding Capacity of the Mucilage

Oil binding capacity was determined using the method previously reported by Segura-Campos et al. [28]. An empty tube of 15 mL was weighed, then 1 g of powdered mucilage was put in the falcon tube and re-weighed, followed by the addition of 10 mL of sunflower oil where the mixture was exposed to the vortex (SL V-6, Seoul in Biosciences, Seongnam, Korea) for 5 min. Centrifugation was performed at 3000 rpm (2-16KC, Sigma, Taufkirchen, Germany) for 30 min. Then, the supernatant was removed carefully. Finally, the falcon tube was re-weighed along with the residue. The water-holding capacity of mucilage was calculated using the following equation:

$$Oil\ binding\ capacity = \frac{weight\ of\ wet\ sample - weight\ of\ dry\ sample}{weight\ of\ the\ dry\ sample}.$$

2.9. Determination of the Rheological Properties of the Mucilage

M. parviflora leaves mucilage (MLM) powder was dispersed in distilled water to prepare 2.5% (w/v). Then it was placed for 12 h at 4 °C for complete hydration. The rheological property of this dispersion was observed using a rheometer (AR 1500ex, TA instruments, New Castle, DE, USA). The rheograms of this dispersion were taken at room temperature (25 °C) using a rheometer (AR 1500ex, TA instruments, New Castle, DE, USA) with cone and plate geometry. The shear rate was uniformly increased from 0–1000 s^{-1} over 1 min and then decreased to zero over 1 min at room temperature [29].

2.10. Evaluation of the Functional Properties of the Mucilage

2.10.1. Determination of Emulsion Capacity

The sample was accurately weighed and dissolved in 10 mL of distilled water, and then 10 mL of refined commercial sunflower oil was added. In dispersion, the concentration (w/v) of mucilage was 0.20, 0.40, 0.60, 0.80 and 1%. Each sample was placed on a magnetic stirrer for 30 min. After that, all the emulsions were subjected to centrifugation at 800 rpm for 15 min. The emulsion capacity was calculated using the following equation [24]:

$$Emulsion\ capacity = \frac{initial\ emulsion\ volume}{total\ volume} \times 100.$$

2.10.2. Determination of Emulsion Stability

Emulsion stability of the prepared mucilage emulsion was determined by heating in a water bath at 80 °C for 30 min. After that, all the emulsions were subjected to centrifugation at 800 rpm for 15 min. Emulsion stability was calculated using the following equation [24]:

$$Emulsion\ stability = \frac{final\ emulsion\ volume}{initial\ emulsion\ volume} \times 100.$$

2.10.3. Determination of the Foaming Capacity

Foaming capacity was determined by the method previously described by Gemede et al. [30] with slight modifications. Briefly, 50 mL solution of mucilage in distilled water at concentrations of 0.2, 0.4, 0.6, 0.8 and 1% (w/v) were blended in a blender and transferred into a graduated cylinder. The volume was noted, and the foaming capacity was calculated using the following equation:

$$Foaming\ capacity = \frac{FV - IV}{IV} \times 100.$$

FV: final-volume; IV: initial-volume.

2.10.4. Determination of the Foaming Stability

Foaming stability was determined by the method previously described by Gemede et al. [30] with slight modifications. Briefly, 50 mL solution of mucilage in distilled water at concentrations of 0.2, 0.4, 0.6, 0.8 and 1% (w/v) were blended in a blender and transferred into a graduated cylinder. The volume was noted after 30 min and the foaming stability was calculated using the following equation:

$$Foaming\ stability = \frac{FV - IV}{IV} \times 100$$

FV: final foaming volume; IV: initial foaming volume.

2.11. Determination of the Micromeritic Properties of the Mucilage

2.11.1. Determination of the Bulk Density

The bulk density of mucilage is measured by transferring an accurately weighed powder to a graduated cylinder, and then the bulk volume is noted. The bulk density was calculated by dividing the weight of powder by its bulk volume [31]:

$$Bulk\ density = \frac{weight\ of\ mucilage}{bulk\ volume}$$

2.11.2. Determination of the Tapped Density

Tapped density was measured by mechanically tapping a graduated cylinder containing accurately weight mucilage powder until the constant volume was observed. The tapped density was calculated by dividing the weight of powder by its tapped volume [31]:

$$Tapped\ density = \frac{weight\ of\ mucilage}{tapped\ volume}$$

2.11.3. Determination of Carr's Index and Hausner Ratio

Carr's or compressibility index and Hausner ratio were calculated using the following formula [32]:

$$Carr's\ index = \frac{tapped\ density - bulk\ density}{tapped\ density} \times 100.$$

$$Hausner\ ratio = \frac{tapped\ density}{bulk\ density}.$$

2.11.4. Determination of Angle of Repose

The angle of repose is used to evaluate the flow property of powder, where a funnel with the end of the stem cut perpendicular to its axis of symmetry was fixed at a height above the graph paper and placed on a flat horizontal surface. The powder was carefully poured through the funnel until the apex of the conical pile just touched the tip of the funnel. The radius (r) of the base of the pile and height (h) of the pile was determined. The angle of repose (θ) was calculated using the following equation [32]:

$$\theta = \tan^{-1}\frac{h}{r}.$$

2.12. Morphological Analysis Using Scanning Electron Microscopy

Morphological analysis of powdered MLM surface was performed at different magnifications using a scanning electron microscope (SEM) (JSM-6480LV, JEOL, Tokyo, Japan). Samples were mounted on aluminum stubs, coated with a thin layer of gold and subjected to SEM analysis. The morphological analysis was performed to investigate the particle size, shape and form of the particles [33].

2.13. Thermal Analysis Using Differential Scanning Calorimetry

Differential scanning calorimetry (DSC) was used to study the thermal stability of mucilage. A total of 14 mg of MLM were placed into an aluminum pan and sealed. The measurements were made using DSC Q2000 v24.11 build 124 differential scanning calorimeter, where heating occurs at a rate of 20 °C/min with a temperature range from 0 °C to 300 °C, under a nitrogen atmosphere. Nitrogen gas purging was maintained at 20 mL/min. Meanwhile, an empty aluminum pan was taken as a reference during the analysis [34].

2.14. Phytochemical Analysis of the Mucilage

2.14.1. FTIR Spectroscopy Analysis

Attenuated total reflectance (ATR) FTIR spectrophotometric analysis was used to identify the functional group present in MLM powder. An ATR accessory equipped with diamond crystal (Cary 630, Agilent Technologies, Santa Clara, CA, USA) was used for sampling. The data were collected in absorbance (A) mode at frequency regions of 4000–650 cm^{-1} at a resolution of 4 cm^{-1} [35].

2.14.2. Gas Chromatography Coupled with Mass Spectrometry Analysis

Monosaccharide composition of MLM was determined by using a modified GC-MS analytical procedure developed by Xia et al. [36], which is based upon trimethylsilyl dithioacetal (TMSD) derivatization. Briefly, 6 mg of MLM powder were hydrolyzed using 3 mL of 2 M TFA at 110 °C for 2 h in a sealed reaction vessel and nitrogen atmosphere. A mixture of ethanethiol and TFA (2:1, 20 µL) was added in standards or samples and then the reaction vessels were tightly closed with a screw cap. The residue in vessels was dissolved by softly swirling and the resulting solution was kept for 10 min at 25 °C. For TMSD derivatization pyridine (100 µL), 68 µL of hexamethyl disilazane (HMDS) and 22 µL of trimethylchlorsilane (TMCS) were added in natural dried residue. The resulting mixture was heated in water bath at 70 °C for 30 min. Then, suitable amounts of water and chloroform were added for liquid–liquid extraction. The chloroform fraction was filtered through membrane filter (0.22 µm) for GC MS analysis. A total of 1 µL of the sample was injected to the Agilent 7890A gas chromatography coupled to the Agilent 5975C mass spectrometer and the analysis was performed on a column HP-5MS using the following temperature range: 80 °C for 0 min, 80–190 °C at 2.5 °C/min, 190–252 °C at 2 °C/min, 252–300 °C at 25 °C/min and 300–310 °C at 25 °C/min, and held for 15 min. Mass spectra was recorded using a total ion chromatogram (TIC) mode and interpreted using NIST 5 software.

2.14.3. Estimation of Total Flavonoid Content

Total flavonoid content was determined using the method previously described by Marinova et al. [37] with slight modifications. Briefly, the MLM sample was prepared in such a way that each sample contains 100 mg/mL. About 1 mL of MLM was mixed with 4 mL of distilled water and 0.3 mL of 5% $NaNO_2$. After 5 min, 0.6 mL of 10% $AlCl_3$ was added and the mixture incubated at room temperature for five minutes, followed by the addition of 2 mL of 1 M NaOH. The volume of mixture was made up to 10 mL by distilled water. The absorbance was taken at 510 nm against the blank. For the calibration curve, rutin was used with a concentration range of 10–100 µg/mL (Absorbance = 0.0079 rutin µg − 0.0267, R^2 = 0.996).

2.14.4. Estimation of Total Phenolic Content

Total phenolic content was determined using the method previously described by Saeed et al. [38] with slight modifications. Briefly, the MLM aqueous solution was prepared in such a way that it contained 1 mg/mL. The sample solution (1 mL) and Folin–Ciocalteu's phenol reagent (Central drug house, New Delhi, India) (1 mL) were mixed in 25 mL volumetric flask. After 5 min, 10 mL of 7% Na_2CO_3 solution were added to the mixture and the volume was made up to the mark in each flask by adding distilled water. The mixture was incubated for 90 min at room temperature. After incubation, the absorbance was determined against blank at 750 nm with UV-visible spectrophotometer. Gallic acid was used to construct the calibration curve used with a concentration range of 10–100 µg/mL (Absorbance = 0.0041 gallic acid µg + 0.0212, R^2 = 0.995).

2.15. In Vitro Biological Evaluation of the Mucilage

2.15.1. In Vitro Antioxidant Activity Assessment Using 1,1-Diphenyl-2-picrylhydrazyl (DPPH*) Radical Scavenging Capacity Assay

The mucilage was dissolved in distilled water using different concentrations from 10–1000 µg/mL. DPPH solution (0.1 mM) was prepared using ethanol as a solvent, then, 1 mL of the sample solution was mixed with 1 mL of DPPH. After incubation for 30 min in the dark at room temperature, the absorbance was measured at 517 nm. The scavenging activity of DPPH radical was calculated using the following equation [39–41]:

$$Scavenging\ ability\ (\%) = \left(1 - (\frac{A_1}{A_0})\right) \times 100$$

A_1 and A_0 are the absorbance of the sample and control (without sample), respectively.

2.15.2. In Vitro Determination of DNA Damage Protective Potential of the Mucilage

The DNA damage protective activity of MLM was evaluated at three different concentrations (50, 100 and 200 µg/mL), using the previously reported method by Kaur et al. with slight modifications [42]. Briefly, DNA was isolated from human blood using a DNA isolation kit (DNA mini kit 50, Qiagen, Venlo, The Netherland) and was then quantified using nanodrop technology (Nanodrop 2000m Thermo Fisher, Waltham, MA, USA). DNA damage was induced by using Fenton reagent, which is freshly prepared by mixing equal volumes of 80 mM $FeCl_3$ (BDH Chemicals Limited, Poole, UK), 50 mM ascorbic acid (Sigma–Aldrich Chemie GmbH, Taufkirchen, Germany) and 30 mM hydrogen peroxide (Merck AG, Darmstadt, Germany). Samples were prepared in such a way that the total volume of the mixture was 20 µL. Six groups were prepared as follows: Group A: 4 µL DNA + 16 µL deionized water (DW); Group B: 3 µL Fenton reagent (FR) +17 µL DW; Group C: 4 µL DNA + 3 µL FR + 13 µL DW; Group D: 4 µL DNA + 4 µL MLM (50 µg/mL) + 3 µL FR + 9 µL DW; Group E: 4 µL DNA + 4 µL MLM (100 µg/mL) + 3 µL FR + 9 µL DW and Group F: 4 µL DNA + 4 µL MLM (200 µg/mL) + 3 µL FR + 9 µL DW. All the above groups were incubated at 37 °C for 30 min. After incubation, 5 µL of bromophenol dye was added to each sample. For gel electrophoresis 1X TAE (40 mM Tris-acetate, 1 mM EDTA) buffer was used as a running buffer. A total of 10 µL sample from each group was loaded in each well of agarose gel (1%), and then the gel was run at 100 V for 1 h. Finally, the gel was observed in the gel documentation system (Gene genius, Syngene, Cambridge, UK) to get the final results.

2.15.3. In Vitro Evaluation of Sun Protection Factor of the Mucilage

Sun protection factor (SPF) was determined using a previously reported method by Suva [43] with slight modifications. MLM was dissolved in distilled water to get solutions with concentrations of 200 and 400 μg/mL. Then, spectrophotometer readings of these solutions were measured at wavelengths ranging from 290 to 320 nm at 5 nm intervals, where SPF for mucilage was calculated from the following equation:

$$SPF = CF \times \sum_{290}^{320} EE \times I \times Abs$$

EE (I) Erythemal effect spectrum; *I* (I)—solar intensity spectrum; *Abs*—Absorbance of sunscreen product; *CF*—correction factor (=10); the values of $EE \times I$ are constant and are predetermined.

2.16. Statistical Analysis

Statistical analysis was conducted using Microsoft excel (2010). All the experiments were conducted as triplicates and expressed as mean ± standard deviation.

3. Results and Discussion

Various methods were adopted to extract mucilage, comprising the precipitation method, ultrasound-assisted extraction, microwave-assisted extraction and enzyme-assisted extraction. MLM was extracted using the alcohol precipitation method, as it is the most common, inexpensive and easy method for extracting mucilage [44].

3.1. Determination of the Yield Percentage and Purity of Mucilage

M. parviflora leaves mucilage (MLM) yields 7.50 ± 0.20% (*w*/*w*); meanwhile, it stained pink with ruthenium red and gave a positive result with Molisch's test. Furthermore, it was free from starch, alkaloids, glycosides, saponins, lipids, steroids and heavy metals. Heavy metals have highly toxic and mutagenic effects even at a low concentrations, where lead, cadmium and chromium are the major cause of metallic toxicity [45]. Thus, MLM was subjected to elemental analysis using the atomic absorption spectroscopy technique. The results revealed that it was almost free from chromium, lead, cadmium and nickel and, in particular, did not have copper, zinc and iron.

3.2. Determination of the Nutritional Value of Mucilage

Results obtained from the nutritional value analysis of MLM revealed that carbohydrates, proteins and amino acids were abundantly present in *M. parviflora leaves mucilage*. This was evidenced by the presence of 16.19 ± 0.70% carbohydrates, 13.55 ± 0.44% proteins and 4.76 ± 0.56% amino acids, ensuring the nutritional value. This analysis was repeated in triplicates, where the results are expressed as mean ± standard deviation.

3.3. Determination of the Physicochemical Properties of the Mucilage

Mucilage, when used as an excipient, can change the color, odor and taste of pharmaceutical formulations; therefore, organoleptic evaluation of mucilages was carried out to assess the effect of MLM on the color, odor and taste on pharmaceutical products. Organoleptic examination of MLM revealed that it exhibited brown color, characteristic odor and bland taste with irregular shape, with hard and a rough to touch texture. Pharmaceutical applications of mucilage highly rely upon its physicochemical properties, illustrated in Table 1 for MLM. Solubility is a key factor to sustain the efficiency in biopolymer development, where high solubility improves the appearance and texture of the formulation. MLM solubility behavior was almost following those previously reported for already extracted mucilages [23,46,47]. It displayed solubility in cold and hot water and was less soluble in 1 M NaOH, HCl, citric acid and NaCl, and insoluble in all other organic solvents. Concerning the pH value of MLM, it is considered neutral, displaying a pH value of 6.94. It is noteworthy to mention that pH plays an important role in determining the suitability of the excipient in preparation, where the physiological activity and stability of the formulation depend on the pH of the excipient. In addition, pH is also crucial in the process of flocculation and coagulation. Neutral polymers are ideal for water treatment, with high salinity and water hardness [48,49], and additionally, it does not irritate the mucous membrane and gastrointestinal tract when ingested orally. Thus, MLM, being neutral, can be used safely as an excipient in pharmaceutical formulation.

Table 1. Physicochemical properties of *M. parviflora mucilage* obtained from the leaves (MLM).

Physicochemical Parameter	MLM
Solubility	Soluble in cold and hot water, less soluble in 1 M NaOH, HCl, citric acid and NaCl, and insoluble in all other organic solvents
pH (1% solution)	6.94 ± 0.02
Moisture content (%)	5.84 ± 0.03
Ash content (%)	15.60 ± 0.75
Water-soluble ash (%)	12.03 ± 0.55
Acid-insoluble ash (%)	0.87 ± 0.15
Swelling index	12.33 ± 2.51
Water-holding capacity (g/g)	2.57 ± 0.60
Oil-binding capacity (g/g)	2.03 ± 0.15

Results are expressed as Mean ± Standard deviation.

Regarding the moisture content of MLM, it showed 5.84% moisture, which perfectly lies within the pharmacopoeial limit (≤15%) for natural gums and mucilages [50]. Physical, chemical and microbiological properties of excipient and active pharmaceutical ingredients mainly depend on moisture content. Higher moisture content may cause various problems, such as poor powder flow, increased microbial contamination and instability of formulation. Furthermore, total ash content determines the quality and nutritional value of food products. It determines the whole residual material that remains after burning, comprising physiological and non-physiological ash, including extraneous matters, represented by sand and soil [50,51].

Meanwhile, acid-insoluble ash defines the remaining insoluble matter formed after boiling the total ash using dilute hydrochloric acid and followed by ignition. The acid-insoluble ash determines the quantity of silica present, especially sand and siliceous earth. It is often of higher value than the total ash in detecting earthy matter that adheres to the drug. The total ash content for MLM was found to be 15.60%, which is higher than commercially available gums, comprising xanthan gum (1.5%), gum Arabica (1.2%) and Guar gum (11.9%) [52]. Meanwhile, it showed 12.03% water-soluble ash and low acid-insoluble ash to be estimated by 0.87%, which in turn, verifies its purity and high nutritional value. Additionally, MLM displayed a considerable swelling index estimated by 12.33, where a good swelling index is an indication that the polymer can be used as a binder, disintegrant and an in-control release in the drug delivery system. It is also a prediction of the presence of the hydroxyl group, galactose moiety, and ascertains the purity of mucilage [44]. In addition, MLM displayed high water-holding capacity and oil-binding capacity, estimated by 2.57% and 2.03% (g/g), respectively. The water-holding capacity and oil-binding capacity play an important role in maintaining the food texture. High-oil binding capacity reduces the loss of oil and flavor from the food [53].

3.4. Determination of the Rheological Properties of the Mucilage

Rheological studies are usually performed to determine the behavior of polymer under stress, where the pseudoplastic behavior of polymers finds considerable application in the food industry. This makes them easily swallowed while eating. In addition, the pseudoplastic behavior of polymers helps in industrial operation comprising mixing and pumping [54]. Rheological studies showed that MLM had shear-thinning behavior or pseudoplastic behavior, a non-Newtonian attitude of fluids, where the viscosity decreases by increasing the shear stress. The rheogram showed that as the shear rate increased, the molecules in the polymers arranged themselves in the direction of flow, and thus the attraction between adjacent polymers decreased, and viscosity also decreased (Figure 1).

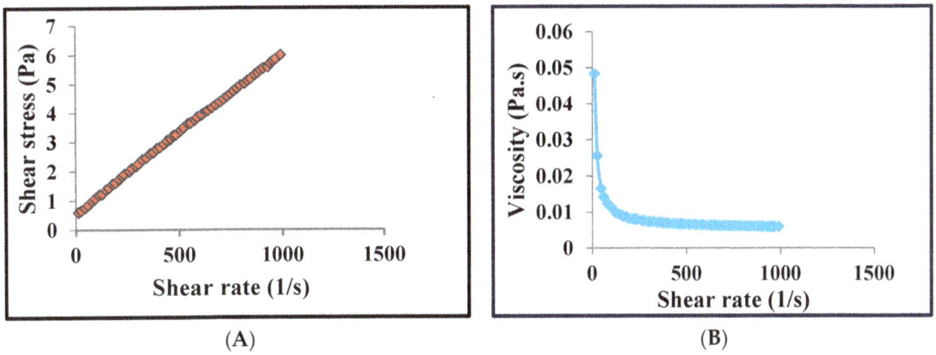

Figure 1. Rheogram of M. *parviflora mucilage* obtained from the leaves (MLM) showing the relationship between shear rate and shear stress (**A**) and the relationship between shear rate and viscosity (**B**). *Pa = Pascal—unit of shear stress whereas Pa. s—Pascal-second unit of viscosity.*

3.5. Determination of the Functional Properties of the Mucilage

Functional properties of mucilage relied upon its emulsion capacity, emulsion stability, foaming capacity and foaming stability. The mucilage's interfacial properties reduce the surface tension of water, where high emulsion capacity and emulsion stability indicate that the polymer can be used as a stabilizing agent in emulsions. Emulsion capacity and emulsion stability of MLM were found to be concentration-dependent, as shown in Figure 2A. As the concentration of mucilage increased, the entrapment of oil globules increased, and emulsion capacity and emulsion stability increased. The results of this study were comparable with previously reported values of some commercially available gums and mucilages. Hence it can be used as a stabilizer in oil in water emulsion [17]. The foaming properties are due to the ability of mucilage to increase viscosity and interfacial behavior [55]. The foaming capacity and foaming stability of MLM were also concentration-dependent, as shown in Figure 2B.

Meanwhile, MLM showed acceptable values for bulk density, tapped density, Carr's index, Hausner ratio and angle of repose, as displayed in Table 2. Biomaterials with low bulk density show better disintegration as the particles will absorb more water by capillary action through the pores. Packing characteristics, such as flowability, compressibility and cohesiveness depend upon the Carr's index and Hausner ratio, where MLM displayed Carr's index and Hausner ratio values of 17.86% and 1.22%, respectively. Packing characteristics are good when Carr's index and Hausner ratio are 12–18% and <1.25, respectively [24]. These findings suggest that MLM will show good packaging and disintegration properties when used in drug formulation.

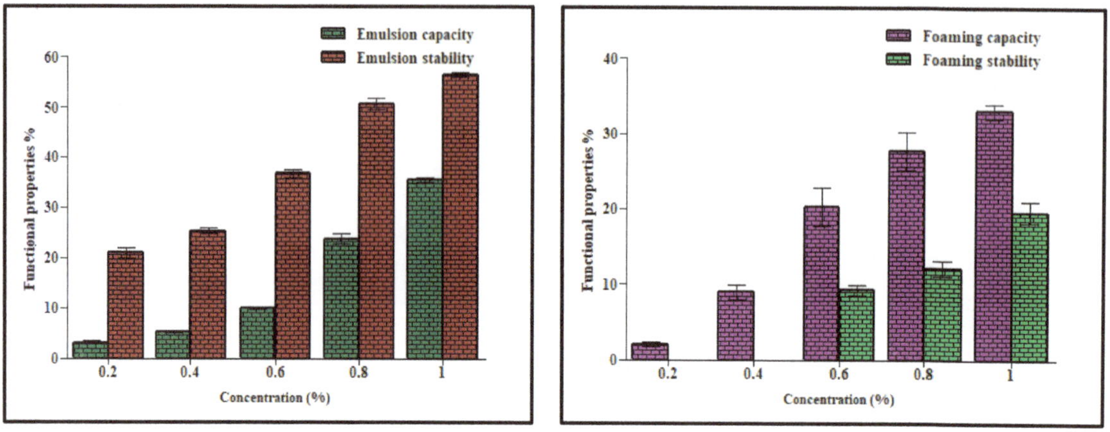

(A) (B)

Figure 2. Comparison between emulsion capacity (%), emulsion stability (%) (**A**), foaming capacity (%) and foaming stability (%) (**B**) of *M. parviflora mucilage* obtained from the leaves (MLM) at different concentrations.

Table 2. Micromeritic properties of *M. parviflora mucilage* obtained from the leaves (MLM).

Property	MLM
Bulk density (g/cm^3)	0.69 ± 0.06
Tapped density (g/cm^3)	0.84 ± 0.05
Carr's index (%)	17.86 ± 0.12
Hausner ratio	1.22 ± 0.12
Angle of repose	28.5 ± 0.5

Results are expressed as Mean ± Standard deviation.

3.6. Morphological Analysis Using Scanning Electron Microscopy

The scanning electron microscope is an important instrument for powder characterization, i.e., size, shape, surface topography, texture, porosity, microstructure and agglomeration tendencies. Scanning the electron micrograph of MLM showed that particles are irregular, as they do not have fixed dimensions, and some of them are in the form of clusters; this, in turn, confirms the amorphous nature of MLM. Studies showed that the surface properties are greatly affected by the extraction, purification and preparation [18] (Figure 3).

3.7. Thermal Analysis Using Differential Scanning Calorimetry

The DSC thermogram of MLM, displayed in Figure 4, represents a typical plot for polysaccharides. MLM displayed an early endothermic peak, appearing at 129 °C, that can be interpreted as losing the water that exists in the compound [56]. The endothermic peak below 150 °C makes natural polymers suitable to a wide class of therapeutic drugs [57]. The onset temperature was observed at 75 °C; meanwhile, the end state temperature was recorded at 185 °C. A pronounced melting peak was reported at around 129 °C, possessing an enthalpy value (ΔH) of 129.6 J/g. The onset, the glass transition temperature and the melting peak are mainly attributed to the structural stability of the mucilages [58]. The lower branching manner within the structure leads to greater binding energy among the monosaccharide backbones, resulting in a higher value of (ΔH) with a concomitant good thermal stability. From Figure 4, it is obvious that the thermal transition process of MLM occurs at a good enthalpy change (ΔH), which in turn reflects the good thermal stability of MLM.

Figure 3. Scanning electron microscopy micrographs of *M. parviflora mucilage* obtained from the leaves (MLM) at different magnifications (2700×, 1000×, 500×, 100×).

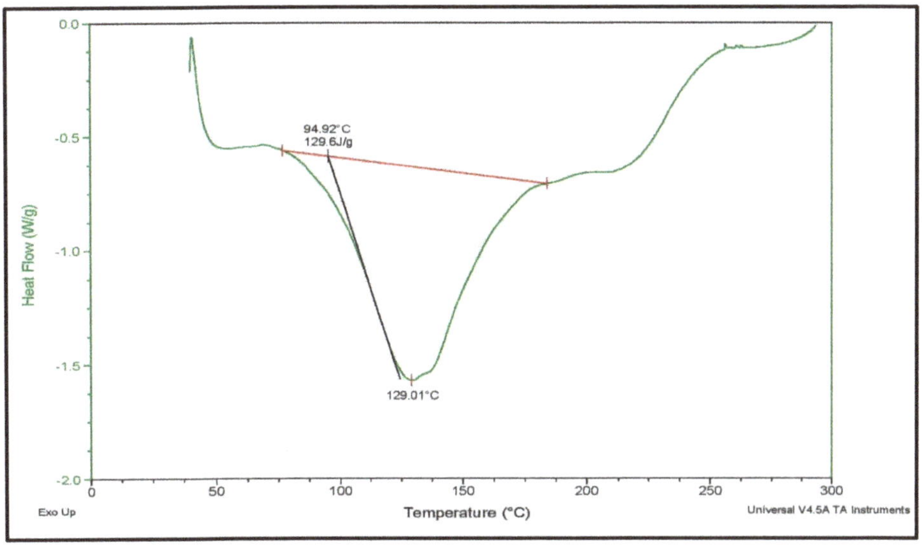

Figure 4. Differential scanning calorimetry (DSC) thermogram of *M. parviflora mucilage,* obtained from the leaves (MLM).

3.8. Phytochemical Analysis of the Mucilage

The IR spectrum of MLM represented in Figure 5 revealed absorption bands at 3250 (cm^{-1}) and 2850–2950 (cm^{-1}), attributed to the stretching of the O-H and C-H bond, respectively. O-H stretching vibrations occur within a broad range of wavenumbers between 3500 and 3000 cm^{-1} and demonstrate several different features, i.e., the stretching bonds of free hydroxyl group and bonded O-H bands of carboxylic acid. The wavenumber range of 3000–2800 cm^{-1} could be assigned to C-H stretching, symmetric and asymmetric of free sugars. These peaks may also be due to the double overlapping with O-H [59]. These are the bands for the alkyl and hydroxyl functionality of carbohydrates. Peaks at 2117 (cm^{-1}) indicate mono-substituted alkynes. FTIR spectra from the wavenumbers 1700–1500 cm^{-1} can be used to determine the structural properties of protein. Protein absorbance over these wavenumbers gives two absorption bands, amide I (1700–1600 cm^{-1}) and amide II (1600–1500 cm^{-1}). The amide II band is due to C-N stretching vibrations in combination with N-H bending [60]. The peak at 1576 (cm^{-1}) represents the amide II band as well as the amine groups, where the amide II band is characteristic of proteins. Meanwhile, the peak at 1397 (cm^{-1}) is mainly attributed to the presence of O-H bending and showed the existence of carboxylic acid moiety, where these peaks also represent the carboxylate ion stretches from uronic acid [61]. Peaks at 1259 (cm^{-1}), 1095 (cm^{-1}), and 1017 (cm^{-1}) are due to aryl-O stretch or C-O stretching, showing the presence of aromatic ethers or alkyl-substituted ethers, respectively [62]. These findings suggest that MLM has hydroxyl, methyl, carboxyl groups and glycosidic bonds, which are the characteristic moieties of polysaccharides. Spectra obtained from MLM closely resembles already studied plant-derived mucilages [2].

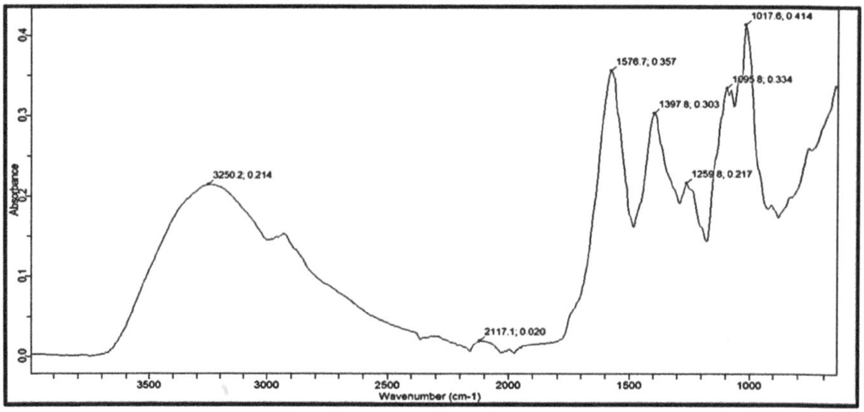

Figure 5. FTIR spectrum of M. *parviflora mucilage* obtained from the leaves (MLM).

Moreover, GC/MS analysis of MLM revealed the presence of five neutral monosaccharides; namely, galactose, rhamnose, arabinose, glucose and mannose showing 51.09, 10.24, 8.90, 1.80 and 0.90 mg/g of MLM, respectively. Meanwhile galacturonic acid is the only detected acidic monosaccharide, forming 15.06 mg/g of MLM. The results of the GC/MS analysis of MLM are displayed in Figure 6, whereas GC/MS chromatograms for the standard neutral and acidic monosaccharides were displayed in the Supplementary Data (Figure S1). Additionally, the total phenolic content (TPC) of MLM was estimated to be 46.91 ± 1.12 µg equivalent of the gallic acid/mL of the mucilage sample containing 1 mg of MLM. Meanwhile, the total flavonoid content (TFC) of MLM was found to be 413.71 ± 0.52 µg equivalent of the rutin/mL of mucilage aqueous solution containing 100 mg of MLM, which, in turn, reflects the richness of the mucilage with polyphenolic contents that undoubtedly influence its antioxidant potential.

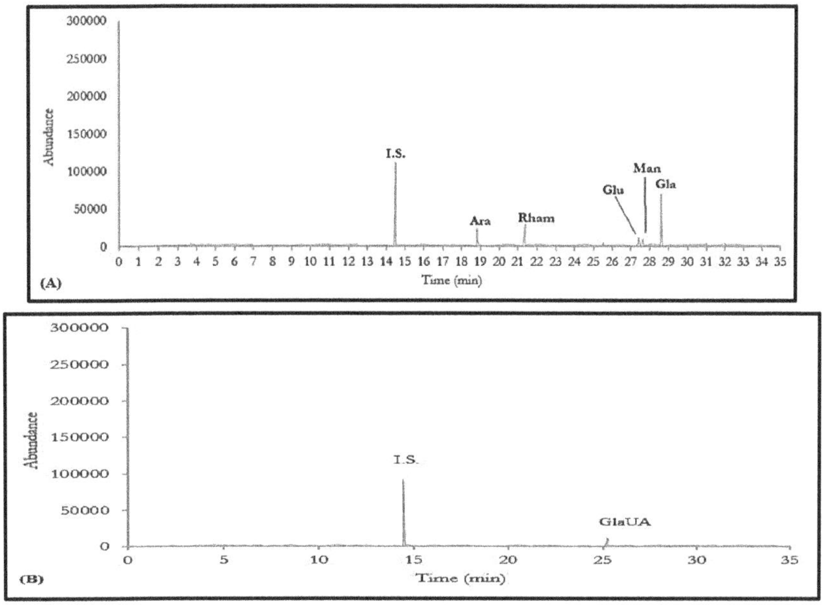

Figure 6. GC/MS chromatograms of neutral (**A**), and acidic (**B**) polysaccharides in *M. parviflora mucilage*, obtained from the leaves (MLM).

3.9. In Vitro Biological Evaluation of the Mucilage

3.9.1. In Vitro Antioxidant Activity Assessment Using the 1,1-Diphenyl-2-picrylhydrazyl (DPPH*) Radical Scavenging Capacity Assay

Oxidative stress in humans damages the macromolecules comprising nucleic acid, lipids and protein, and results in cellular injuries. The human body has its mechanisms for defending against oxidative stress. However, in the cases of severe oxidative stress, the defensive mechanism fails to protect the human body. Thorough research has shown that hydroxyl radical and superoxide anion radicals play an important role in aging and carcinogenesis. Therefore antioxidants should be used to protect the body against these fatal damages. Natural antioxidants are preferred over synthetic antioxidants, as synthetic antioxidants cause toxicity and carcinogenesis. Thus, pharmaceutical and food industries mainly target the discovery of natural antioxidants [39]. DPPH assay is widely used to screen natural antioxidants, due to its accuracy and feasibility, where DPPH is a stable free radical that produces a violet color in ethanol. Antioxidants reduce DPPH by donating protons to it. As a result, its purple color fades rapidly [63]. The antioxidant activity of MLM was evaluated using different concentrations (10–1000 μg/mL) employing ascorbic acid as a control (Table 3). Results displayed in Figure 7 showed that MLM revealed a considerable antioxidant activity, where the antioxidant potential of MLM increases with an increase in concentration displaying IC_{50} value 154.27 μg/mL, whereas that of ascorbic acid was 41.21 μg/mL. Antioxidant activity of MLM might be due to the hydrogen donating capacity, and it is highly correlated to TPC and TFC, as it was previously reported that phenolic compounds and flavonoids existing in plants are responsible for their pronounced antioxidant activity [64].

Table 3. Percentage inhibition of 1,1-diphenyl-2-picrylhydrazyl (DPPH*) radical by ascorbic acid and MLM.

Concentration (μg/mL)	Ascorbic Acid	MLM
10	12.20 ± 0.96	9.29 ± 1.38
20	18.69 ± 0.20	19.05 ± 1.09
40	40.34 ± 0.61	34.26 ± 1.84
80	71.21 ± 1.07	58.54 ± 1.46
100	80.27 ± 1.24	67.63 ± 2.72
200	87.35 ± 2.00	77.26 ± 0.94
400	91.60 ± 1.70	84.54 ± 0.56
800	94.87 ± 1.0	89.03 ± 0.39
1000	96.20 ± 1.08	91.07 ± 1.00

Results are expressed as mean ± standard deviation.

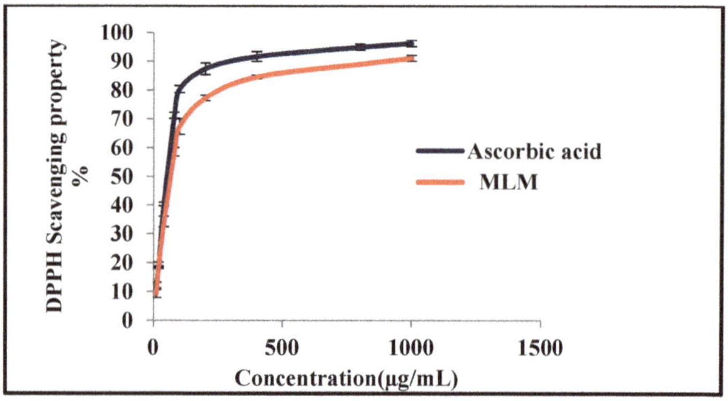

Figure 7. In vitro antioxidant activity assessment of MLM and compared to ascorbic acid using the 1,1-diphenyl-2-picrylhydrazyl (DPPH*) radical scavenging capacity assay.

3.9.2. In Vitro Determination of DNA Damage Protective Potential of the Mucilage

MLM was evaluated for its DNA damage protective ability where Fenton reagent was used as a DNA damaging reagent, as it can produce oxidative stress and destruct DNA. MLM protects DNA against oxidative damage induced by the Fenton reagent, as illustrated in Figure 8. The DNA band appears clearly in Lane 1 (DNA and distilled water), it completely disappears in Lane 2 (no DNA) and in Lane 3 (DNA with Fenton reagent that destroys DNA). Meanwhile, the DNA band began to reappear upon treatment with 50, 100 and 200 μg/mL of MLM in Lane 4–6, with a pronounced appearance in Lane 6, which contains the highest MLM concentration (200 μg/mL). MLM may inhibit the reaction of Fe^{2+} with hydrogen peroxide in the Fenton reagent, or it may donate the hydrogen atom or electron to inhibit the hydroxyl radicals [65].

3.9.3. In Vitro Evaluation of Sun Protection Factor of the Mucilage

When skin is continuously exposed to UV radiation, free radicals are produced that oxidize the biological molecules, trigger oxidative damage and cause cancer. Hence, photoprotective agents are widely used to protect against UV radiation, where herbal skin protective agents are preferred over synthetic ones, being safer, less expensive and highly welcomed by a large category of people all over the globe [66]. Sun protection factor (SPF) is an indicator used for quantitative measurement of the efficacy of sunscreen products. Flavonoids and phenolic compounds present in the plant are more responsible for antioxidant and photoprotective properties, where phenolic compounds are highly correlated with SPF [67]. The MLM displayed pronounced SPF, estimated by 5.56 ± 0.15 and 10.93 ± 0.15 at 200 and 400 μg/mL, respectively. This significant SPF is attributed to the presence of polysaccharides, where previous studies reported that polysaccharides are a photoprotective agent and the richness of the MLM with polyphenolic compounds, as evidenced by its high total phenolic and total flavonoid contents [68,69]. These results suggest that MLM could be perfectly used in sunscreen formulation to protect skin against UV rays. The absorbance of 200 and 400 μg/mL of MLM at different wavelengths is illustrated in Table 4.

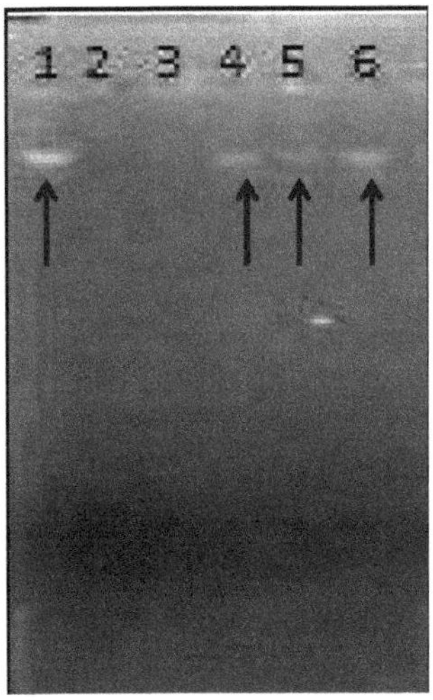

Figure 8. DNA damage protective effect of MLM. Lane 1: 4 μL DNA + 16 μL DW; **Lane 2**: 3 μL FR + 17 μL DW; **Lane 3**: 4 μL DNA + 3 μL FR + 13 μL DW; **Lane 4**: 4 μL DNA + 4 μL MLM (50 μg/mL) +3 μL FR + 9 μL DW; **Lane 5**: 4 μL DNA + 4 μL MLM (100 μg/mL) + 3 μL FR + 9 μL DW; **Lane 6**: 4 μL DNA + 4 μL MLM (200 μg/mL) + 3 μL FR+ 9 μL DW.

Table 4. Absorbance of *M. parviflora mucilage* leaves (MLM) at concentrations of 200 and 400 μg/mL, at different wavelengths.

Wavelength (λ nm)	EE × I (Normalize)	MLM (200 μg/mL)	MLM (400 μg/mL)
290	0.0150	0.595 ± 0.01	1.071 ± 0.03
295	0.0817	0.587 ± 0.04	1.142 ± 0.11
300	0.2874	0.506 ± 0.18	1.062 ± 0.05
305	0.3278	0.574 ± 0.02	1.099 ± 0.16
310	0.1864	0.572 ± 0.05	1.106 ± 0.04
315	0.0839	0.569 ± 0.11	1.098 ± 0.02
320	0.0180	0.588 ± 0.03	1.091 ± 0.10

Results are expressed as mean ± standard deviation.

4. Limitations and Future Research

Malva parviflora is a seasonal herb in Pakistan with an optimum yield that varies according to season, location, climate and soil. MLM contains a moisture content of 5.84%; therefore, there is a risk of microbial contamination. Thus, there is a need to adopt analytical techniques to isolate bioactive polysaccharides from this mucilage. Research is needed to incorporate MLM in various drug delivery systems and to study the biocompatibility of MLM. However, further in vivo studies are highly recommended to confirm their applications before being used in preclinical trials and the pharmaceutical industry, in order to demonstrate the health benefits of MLM. In addition, the isolation of polysaccharides from complex matrices with protein requires effective and selective methods for pretreatment and extraction. The structure, chemical fingerprint and biological function of the mucilage are important to understand the structure–function relationship. Modification of the structure can be done by combining advanced analytical tools with sophisticated tools.

Approval by food and drug regulatory authorities is mandatory to use MLM commercially. Regulatory authorities demand detailed toxicity and stability studies to approve mucilage and its products.

5. Conclusions

Herein, it was concluded that *M. parviflora* leaves contain a considerable amount of mucilage extracted in water, followed by the alcohol precipitation method. This mucilage consists of carbohydrates, proteins and amino acids, and it is free from toxic minerals. The extracted mucilage showed unique physicochemical, functional and micromeritic properties. The presence of the functional groups characteristic of polysaccharides was ascertained by the infrared (IR) and gas chromatography/mass spectrometry (GC/MS) analyses. GC revealed the presence of five neutral monosaccharides; namely, galactose, rhamnose, arabinose, glucose and mannose. Meanwhile, galacturonic acid is the only detected acidic monosaccharide. It showed noticeable antioxidant activity against the DPPH (1,1-diphenyl-2-picrylhydrazyl) radical. It also prevented oxidative damage to DNA caused by the Fenton reagent, as visualized in gel documentation system with considerable sun-protective properties. Hence, it can be used as a food supplement or as an ingredient in food and bio-industries; meanwhile, it can act as an attractive eco-friendly and sustainable biopolymer for the pharmaceutical industry. Further, it can be used as a natural antioxidant and protective therapy in skin diseases.

Supplementary Materials: The following are available online at https://www.mdpi.com/article/10.3390/polym13234251/s1, Figure S1: GC/MS chromatograms for standard neutral and acidic monosaccharides.

Author Contributions: Conceptualization, A.M., S.I. and S.H.K.; methodology A.M., S.I., S.S.E. and S.H.K.; software, F.S.Y.; resources, S.A.A., A.S. and S.S.E.; writing—original draft preparation, A.M., S.I. and S.H.K.; writing—review and editing, F.S.Y., S.S.E. and M.L.A.; supervision, F.S.Y. and M.L.A.; funding acquisition, S.A.A., A.S. and S.S.E. All authors have read and agreed to the published version of the manuscript.

Funding: This research was funded by the Deanship of Scientific Research (DSR) at King Abdulaziz University, Jeddah Saudi Arabia under grant number (RG-30-166-42).

Institutional Review Board Statement: Not applicable.

Informed Consent Statement: Not applicable.

Data Availability Statement: Data supporting reported results may be found with the authors.

Acknowledgments: The Deanship of Scientific Research (DSR) at King Abdulaziz University, Jeddah, Saudi Arabia has funded this project under grant number (RG-30-166-42). Therefore, all the authors acknowledge, with thanks, DSR for technical and financial support.

Conflicts of Interest: The authors declare no conflict of interest.

References

1. Choudhary, P.D.; Pawar, H.A. Recently investigated natural gums and mucilages as pharmaceutical excipients: An overview. *J. Pharm.* **2014**, *2014*, 204849. [CrossRef]
2. Tosif, M.M.; Najda, A.; Bains, A.; Kaushik, R.; Dhull, S.B.; Chawla, P.; Walasek-Janusz, M. A Comprehensive review on plant-derived mucilage: Characterization, functional properties, applications, and its utilization for nanocarrier fabrication. *Polymers* **2021**, *13*, 1066. [CrossRef]
3. Clifford, S.; Arndt, S.; Popp, M.; Jones, H. Mucilages and polysaccharides in *Ziziphus species* (Rhamnaceae): Localization, composition and physiological roles during drought-stress. *J. Exp. Bot.* **2002**, *53*, 131–138. [CrossRef]
4. Prajapati, V.D.; Jani, G.K.; Moradiya, N.G.; Randeria, N.P. Pharmaceutical applications of various natural gums, mucilages and their modified forms. *Carb. Polym.* **2013**, *92*, 1685–1699. [CrossRef] [PubMed]
5. Wadhwa, J.; Nair, A.; Kumria, R. Potential of plant mucilages in pharmaceuticals and therapy. *Curr. Drug Del.* **2013**, *10*, 198–207. [CrossRef]
6. Ameri, A.; Heydarirad, G.; Mahdavi Jafari, J.; Ghobadi, A.; Rezaeizadeh, H.; Choopani, R. Medicinal plants contain mucilage used in traditional Persian medicine (TPM). *Pharm. Biol.* **2015**, *53*, 615–623. [CrossRef] [PubMed]

7. Rebah, F.B.; Siddeeg, S. Cactus an eco-friendly material for wastewater treatment: A review. *J. Mater. Environ. Sci.* **2017**, *8*, 1770–1782.
8. Ngwuluka, N.C.; Ochekpe, N.A.; Aruoma, O.I. Naturapolyceutics: The science of utilizing natural polymers for drug delivery. *Polymers* **2014**, *6*, 1312–1332. [CrossRef]
9. Allegra, A.; Sortino, G.; Inglese, P.; Settanni, L.; Todaro, A.; Gallotta, A. The effectiveness of Opuntia ficus-indica mucilage edible coating on post-harvest maintenance of 'Dottato'fig (*Ficus carica* L.) fruit. *Food Pack. Shelf Life* **2017**, *12*, 135–141. [CrossRef]
10. Akbar, S.; Hanif, U.; Ali, J.; Ishtiaq, S. Pharmacognostic studies of stem, roots and leaves of *Malva parviflora* L. *Asian Pac. J. Trop. Biomed.* **2014**, *4*, 410–415. [CrossRef]
11. Dugani, A.; Dakhil, B.; Treesh, S. Protective effect of the methanolic extract of *Malva parviflora* L. leaves on acetic acid-induced ulcerative colitis in rats. *Saudi J. Gastroenterol.* **2016**, *22*, 226–233.
12. Azab, A. Malva: Food, medicine and chemistry. *Eur. Chem. Bull.* **2017**, *6*, 295–320. [CrossRef]
13. Farhan, H.; Rammal, H.; Hijazi, A.; Badran, B. Preliminary phytochemical screening and extraction of polyphenol from stems and leaves of a Lebanese plant *Malva parviflora* L. *Int. J. Curr. Pharm. Res.* **2012**, *4*, 55–59.
14. Fakhfakh, N.; Jdir, H.; Jridi, M.; Rateb, M.; Belbahri, L.; Ayadi, M.; Nasri, M.; Zouari, N. The mallow, *Malva aegyptiaca* L. (Malvaceae): Phytochemistry analysis and effects on wheat dough performance and bread quality. *LWT* **2017**, *75*, 656–662.
15. Mitich, L.W. Cheeseweed–The common mallows. *Weed Technol.* **1990**, *4*, 693–695. [CrossRef]
16. Pakravan, M.; Abedinzadeh, H.; Safaeepur, J. Comparative studies of mucilage cells in different organs in some species of *Malva*, Althaea and Alcea. *Pak. J. Biol. Sci. PJBS* **2007**, *10*, 2603–2605. [CrossRef] [PubMed]
17. Malviya, R. Extraction characterization and evaluation of selected mucilage as pharmaceutical excipient. *Pol. Med.* **2011**, *41*, 39–44.
18. Ma, F.; Zhang, Y.; Yao, Y.; Wen, Y.; Hu, W.; Zhang, J.; Liu, X.; Bell, A.E.; Tikkanen-Kaukanen, C. Chemical components and emulsification properties of mucilage from *Dioscorea opposita* Thunb. *Food Chem.* **2017**, *228*, 315–322. [CrossRef]
19. Raaman, N. *Phytochemical Techniques*; New India Publishing: New Dehli, India, 2006.
20. Bader, N.R. Sample preparation for flame atomic absorption spectroscopy: An overview. *Rasayan J. Chem.* **2011**, *4*, 49–55.
21. Sadasivam, S.; Manickam, A. *Biochemical Methods*; New Age International Publishers: New Delhi, India, 1996.
22. Lowry, O.H.; Rosebrough, N.J.; Farr, A.L.; Randall, R.J. Protein measurement with the Folin phenol reagent. *J. Biol. Chem.* **1951**, *193*, 265–275. [CrossRef]
23. Archana, G.; Sabina, K.; Babuskin, S.; Radhakrishnan, K.; Fayidh, M.A.; Babu, P.A.S.; Sivarajan, M.; Sukumar, M. Preparation and characterization of mucilage polysaccharide for biomedical applications. *Carb. Polym.* **2013**, *98*, 89–94. [CrossRef]
24. Kalegowda, P.; Chauhan, A.S.; Urs, S.M.N. Opuntia dillenii (Ker-Gawl) Haw cladode mucilage: Physico-chemical, rheological and functional behavior. *Carb. Polym.* **2017**, *157*, 1057–1064. [CrossRef] [PubMed]
25. Aslam, I.; Afridi, M.S.K. Pharmacognostic characterization of Beaumontia grandiflora (Roxb.) Wall. leaf for taxonomic identification for quality control of a drug. *J. Appl. Res. Med. Aromat. Plants* **2018**, *8*, 53–59. [CrossRef]
26. Schuck, P.; Jeantet, R.; Dolivet, A. *Analytical Methods for Food and Dairy Powders*; John Wiley & Sons: Hoboken, NJ, USA, 2012.
27. Killedar, S.; More, H.; Nadaf, S.; Pishawikar, S. Optimization of method for determination of swelling factor of ispaghula seeds. *J. Drug Metab. Toxicol.* **2016**, *7*, 3. [CrossRef]
28. Segura-Campos, M.; Acosta-Chi, Z.; Rosado-Rubio, G.; Chel-Guerrero, L.; Betancur-Ancona, D. Whole and crushed nutlets of chia (*Salvia hispanica*) from Mexico as a source of functional gums. *Food Sci. Technol.* **2014**, *34*, 701–709. [CrossRef]
29. Njintang, N.Y.; Boudjeko, T.; Tatsadjieu, L.N.; Nguema-Ona, E.; Scher, J.; Mbofung, C.M. Compositional, spectroscopic and rheological analyses of mucilage isolated from taro (*Colocasia esculenta* L. Schott) corms. *J. Food Sci. Technol.* **2014**, *51*, 900–907. [CrossRef] [PubMed]
30. Gemede, H.F.; Haki, G.D.; Beyene, F.; Rakshit, S.K.; Woldegiorgis, A.Z. Indigenous Ethiopian okra (Abelmoschus esculentus) mucilage: A novel ingredient with functional and antioxidant properties. *Food Sci. Nutr.* **2018**, *6*, 563–571. [CrossRef]
31. Arasi, M.A.G.; Rao, M.G. Physiochemical characterization of mucilage obtained from the fresh fruits of *Psidium guajava* L. *Int. J. Phytopharm.* **2015**, *5*, 30–36.
32. Sandeep, Y.; Sharma, P.K.; Goyal, N. Comparative study of mucilage extracted from seeds of Cassia fistula and gum Karaya. *Adv. Biol. Res.* **2015**, *9*, 177–181.
33. Singh, S.; Bothara, S.B. *Manilkara zapota* (Linn.) Seeds: A potential source of natural gum. *ISRN Pharm.* **2014**, *2014*. [CrossRef]
34. Punia, S.; Dhull, S.B. Chia seed (*Salvia hispanica* L.) mucilage (a heteropolysaccharide): Functional, thermal, rheological behaviour and its utilization. *Int. J. Biol. Macromol.* **2019**, *140*, 1084–1090. [CrossRef]
35. Gheribi, R.; Puchot, L.; Verge, P.; Jaoued-Grayaa, N.; Mezni, M.; Habibi, Y.; Khwaldia, K. Development of plasticized edible films from *Opuntia ficus-indica* mucilage: A comparative study of various polyol plasticizers. *Carb. Polym.* **2018**, *190*, 204–211. [CrossRef]
36. Xia, Y.-G.; Sun, H.-M.; Wang, T.-L.; Liang, J.; Yang, B.-Y.; Kuang, H.-X. A modified GC-MS analytical procedure for separation and detection of multiple classes of carbohydrates. *Molecules* **2018**, *23*, 1284. [CrossRef]
37. Ribarova, F.; Atanassova, M. Total phenolics and flavonoids in Bulgarian fruits and vegetables. *J. Univ. Chem. Technol. Metal.* **2005**, *40*, 255–260.
38. Saeed, N.; Khan, M.R.; Shabbir, M. Antioxidant activity, total phenolic and total flavonoid contents of whole plant extracts *Torilis leptophylla* L. *BMC Complement. Altern. Med.* **2012**, *12*, 221. [CrossRef] [PubMed]

39. Petera, B.; Delattre, C.; Pierre, G.; Wadouachi, A.; Elboutachfaiti, R.; Engel, E.; Poughon, L.; Michaud, P.; Fenoradosoa, T.A. Characterization of arabinogalactan-rich mucilage from *Cereus triangularis* cladodes. *Carb. Polym.* **2015**, *127*, 372–380. [CrossRef]
40. Aboulwafa, M.M.; Youssef, F.S.; Gad, H.A.; Sarker, S.D.; Nahar, L.; Al-Azizi, M.M.; Ashour, M.L. Authentication and discrimination of green tea samples using UV–vis, FTIR and HPLC techniques coupled with chemometrics analysis. *J. Pharm. Biomed. Anal.* **2019**, *164*, 653–658. [CrossRef] [PubMed]
41. Mamadalieva, N.Z.; Youssef, F.S.; Ashour, M.L.; Sasmakov, S.A.; Tiezzi, A.; Azimova, S.S. Chemical composition, antimicrobial and antioxidant activities of the essential oils of three Uzbek Lamiaceae species. *Nat. Prod. Res.* **2019**, *33*, 2394–2397. [CrossRef]
42. Kaur, R.; Arora, S.; Singh, B. Antioxidant activity of the phenol rich fractions of leaves of *Chukrasia tabularis* A. *Juss. Bioresource Technol.* **2008**, *99*, 7692–7698. [CrossRef]
43. Suva, M.A. Evaluation of sun protection factor of *Zingiber officinale* Roscoe extract by ultraviolet spectroscopy method. *J. Pharm. Sci. Technol.* **2014**, *3*, 95–97.
44. Arasi, M.; Rao, M.G.; Bagyalakshmi, J. The Comparison and Analysis of Two Extraction Methods for Polysaccharides in Psidium guajava L. Fruits. *Indian J. Pharm. Educ. Res.* **2016**, *50*, 218–224. [CrossRef]
45. Raheem, Z.H.; Jebor, A.A.A.; Mohammed, S.K. Evaluation of some heavy metal contamination in Malva parviflora L. plant and soil obtained from gardens of College of Agriculture-University of Baghdad. *Pak. J. Nut.* **2014**, *13*, 310. [CrossRef]
46. Farooq, U.; Malviya, R.; Sharma, P.K. Extraction and characterization of okra mucilage as pharmaceutical excipient. *Acad. J. Plant Sci.* **2013**, *6*, 168–172.
47. Rahim, H.; Sadiq, A.; Khan, S.; Chishti, K.A.; Amin, F.; Khan, M.A.; Abbas, S. Isolation and preliminary evaluation of Mulva Neglecta mucilage: A novel tablet binder. *Braz. J. Pharm. Sci.* **2016**, *52*, 201–210. [CrossRef]
48. Choy, S.; Prasad, K.; Wu, T.; Ramanan, R. A review on common vegetables and legumes as promising plant-based natural coagulants in water clarification. *Int. J. Environ. Sci. Technol.* **2015**, *12*, 367–390. [CrossRef]
49. Suvakanta, D.; Narsimha, M.P.; Pulak, D.; Joshabir, C.; Biswajit, D. Optimization and characterization of purified polysaccharide from *Musa sapientum* L. as a pharmaceutical excipient. *Food Chem.* **2014**, *149*, 76–83.
50. Boruah, A.K.; Nath, L.K. Extraction, purification and physicochemical evaluation of mucilage of *Chrysophyllum lanceolatum* (blume) dc fruits: An investigation for bioadhesive property. *Int. J. Pharm. Pharm. Sci.* **2016**, *8*, 282–288.
51. Liu, K. Effects of sample size, dry ashing temperature and duration on determination of ash content in algae and other biomass. *Algal Res.* **2019**, *40*, 101486. [CrossRef]
52. Samavati, V.; Lorestani, M.; Joolazadeh, S. Identification and characterization of hydrocolloid from Cordia myxa leaf. *Int. J. Biol. Macromol.* **2014**, *65*, 215–221. [CrossRef]
53. Thanatcha, R.; Pranee, A. Extraction and characterization of mucilage in *Ziziphus mauritiana* Lam. *Int. Food Res. J.* **2011**, *18*, 201–212.
54. Inglett, G.E.; Chen, D.; Lee, S. Rheological properties of barley and flaxseed composites. *Food Nutr. Sci.* **2013**, *4*, 26635. [CrossRef]
55. Trigui, I.; Yaich, H.; Sila, A.; Cheikh-Rouhou, S.; Bougatef, A.; Blecker, C.; Attia, H.; Ayadi, M. Physicochemical properties of water-soluble polysaccharides from black cumin seeds. *Int. J. Biol. Macromol.* **2018**, *117*, 937–946. [CrossRef] [PubMed]
56. Cerqueira, M.A.; Souza, B.W.; Simões, J.; Teixeira, J.A.; Domingues, M.R.M.; Coimbra, M.A.; Vicente, A.A. Structural and thermal characterization of galactomannans from non-conventional sources. *Carb. Polym.* **2011**, *83*, 179–185. [CrossRef]
57. Singh, A.V. A DSC study of some biomaterials relevant to pharmaceutical industry. *J. Ther. Anal. Calorim.* **2013**, *112*, 791–793. [CrossRef]
58. Vinod, V.; Sashidhar, R.; Suresh, K.; Rao, B.R.; Saradhi, U.V.; Rao, T.P. Morphological, physico-chemical and structural characterization of gum kondagogu (*Cochlospermum gossypium*): A tree gum from India. *Food Hydrocoll.* **2008**, *22*, 899–915. [CrossRef]
59. Behbahani, B.A.; Yazdi, F.T.; Shahidi, F.; Hesarinejad, M.A.; Mortazavi, S.A.; Mohebbi, M. Plantago major seed mucilage: Optimization of extraction and some physicochemical and rheological aspects. *Carb. Polym.* **2017**, *155*, 68–77. [CrossRef]
60. Derenne, A.; Derfoufi, K.-M.; Cowper, B.; Delporte, C.; Goormaghtigh, E. FTIR spectroscopy as an analytical tool to compare glycosylation in therapeutic monoclonal antibodies. *Anal. Chim. Acta* **2020**, *1112*, 62–71. [CrossRef]
61. Pereira, G.A.; Silva, E.K.; Araujo, N.M.P.; Arruda, H.S.; Meireles, M.A.A.; Pastore, G.M. Obtaining a novel mucilage from mutamba seeds exploring different high-intensity ultrasound process conditions. *Ultrasonics Sonochem.* **2019**, *55*, 332–340. [CrossRef]
62. Coates, J. Interpretation of Infrared Spectra a Practical Approach. In *Encyclopedia of Analytical Chemistry: Applications, Theory and Instrumentation*; Wiley: Hoboken, NJ, USA, 2006.
63. Bouaziz, F.; Koubaa, M.; Barba, F.J.; Roohinejad, S.; Chaabouni, S.E. Antioxidant properties of water-soluble gum from flaxseed hulls. *Antioxidants* **2016**, *5*, 26. [CrossRef] [PubMed]
64. Aslam, I.; Iqbal, J.; Peerzada, S.; Afridi, M.S.; Ishtiaq, S. Microscopic investigations and pharmacognostic techniques for the standardization of *Caralluma edulis* (Edgew.) Benth. ex Hook. f. *Microscopy Res. Tech.* **2019**, *82*, 1891–1902.
65. Golla, U.; Bhimathati, S.S.R. Evaluation of antioxidant and DNA damage protection activity of the hydroalcoholic extract of *Desmostachya bipinnata* L. *Stapf. Sci. World J.* **2014**, *2014*, 215084. [CrossRef]
66. Napagoda, M.T.; Malkanthi, B.M.A.S.; Abayawardana, S.A.K.; Qader, M.M.; Jayasinghe, L. Photoprotective potential in some medicinal plants used to treat skin diseases in Sri Lanka. *BMC Complement. Altern. Med.* **2016**, *16*, 479. [CrossRef] [PubMed]

67. Ebrahimzadeh, M.A.; Enayatifard, R.; Khalili, M.; Ghaffarloo, M.; Saeedi, M.; Charati, J.Y. Correlation between sun protection factor and antioxidant activity, phenol and flavonoid contents of some medicinal plants. *Iran. J. Pharm. Res. IJPR* **2014**, *13*, 1041–1047. [PubMed]
68. Li, H.; Li, Z.; Peng, L.; Jiang, N.; Liu, Q.; Zhang, E.; Liang, B.; Li, R.; Zhu, H. *Lycium barbarum* polysaccharide protects human keratinocytes against UVB-induced photo-damage. *Free Rad. Res.* **2017**, *51*, 200–210. [CrossRef] [PubMed]
69. Wei, X.; Liu, Y.; Xiao, J.; Wang, Y. Protective effects of tea polysaccharides and polyphenols on skin. *J. Agric. Food Chem.* **2009**, *57*, 7757–7762. [CrossRef]

polymers

Article

Physicochemical Properties and Digestion Resistance of Acetylated Starch Obtained from Annealed Starch

Ewa Zdybel [1], Aleksandra Wilczak [2], Małgorzata Kapelko-Żeberska [1,*], Ewa Tomaszewska-Ciosk [1], Artur Gryszkin [1], Anna Gawrońska [3] and Tomasz Zięba [1,3]

1 Department of Food Storage and Technology, Faculty of Food Science, Wroclaw University of Environmental and Life Sciences, Chełmońskiego 37, 51-630 Wrocław, Poland; ewa.zdybel@upwr.edu.pl (E.Z.); ewa.tomaszewska-ciosk@upwr.edu.pl (E.T.-C.); artur.gryszkin@upwr.edu.pl (A.G.); tomasz.zieba@upwr.edu.pl (T.Z.)

2 Department of Physico-Chemistry of Microorganisms, Faculty of Biological Sciences, University of Wroclaw, Przybyszewskiego 63-77, 51-148 Wrocław, Poland; aleksandra.wilczak@uwr.edu.pl

3 Institute of Sport, Tourism and Nutrition, Faculty of Biological Sciences, University of Zielona Góra, Licealna 9, 65-417 Zielona Góra, Poland; agawronska@uz.zgora.pl

* Correspondence: malgorzata.kapelko@upwr.edu.pl; Tel.: +48-71-320-7765

Abstract: One of the examples of physical starch modifications is the retention of a starch suspension in water having a temperature slightly lower than the pasting temperature (annealing). The aim of this study was to investigate the effect of the annealing process performed at various temperatures as the first stage of starch modification. The annealed starch preparations were then esterified using acetic acid anhydride. Finally, the annealed and acetylated starch preparations were determined for their properties. The annealing of starch before acetylation triggered changes in the properties of the modified preparations. It contributed to a higher degree of starch substitution with acetic acid residues and to the increased swelling power of starch. Both these properties were also affected by the annealing temperature. The highest resistance to amylolysis was found in the case of the starch preparation annealed at 53.5 °C and acetylated. The double modification involving annealing and acetylation processes increased the onset and end pasting temperatures compared to the acetylation alone. Similar observations were made for the consistency coefficient and yield point.

Keywords: annealing; acetylation; potato starch

Citation: Zdybel, E.; Wilczak, A.; Kapelko-Żeberska, M.; Tomaszewska-Ciosk, E.; Gryszkin, A.; Gawrońska, A.; Zięba, T. Physicochemical Properties and Digestion Resistance of Acetylated Starch Obtained from Annealed Starch. *Polymers* **2021**, *13*, 4141. https://doi.org/10.3390/polym13234141

Academic Editor: Arn Mignon

Received: 22 October 2021
Accepted: 24 November 2021
Published: 27 November 2021

Publisher's Note: MDPI stays neutral with regard to jurisdictional claims in published maps and institutional affiliations.

1. Introduction

Starch is a natural plant raw material applicable in many branches of the industry, which usually makes use of its modified preparations, i.e., starch subjected to various treatments to modify its properties. The methods of starch modification can be divided into physical, chemical, and enzymatic ones [1–5], but also their combinations are employed to achieve desired starch properties [6–8]. These modifications allow producing starch preparations with altered properties, including molecular weight, pasting temperature, solubility, viscosity, water binding capability, or resistance to acids and enzymes [9]. One of the examples of physical starch modifications is the retention of a starch suspension in water having a temperature slightly lower than the pasting temperature. This process has been defined as annealing [10].

Its goal is to increase the mobility of molecules; however, without making them turn into the gelatinized form. The annealing process results in the hydration of starch granules, which firstly affects their amorphous and then crystalline regions. Particle mobility increases by analogy, which intensifies the interactions between starch chains and leads to chain ordering. A homogeneous and stable structure of the entire granule is then developed. The swelling of the granule is limited and its form is preserved [10–13].

The effects of annealing on starch pasting characteristics determined with the DSC are well described in scientific literature [14] and include an increase in transition temperatures, a decrease in a pasting temperature range, and an increase in or no effect on process enthalpy. The annealing contributes to a diminished swelling power of starch granules [15]. In turn, studies have failed to demonstrate its explicit effect on the viscosity of pastes obtained from annealed starch of various botanical origins. The annealing has also been reported to increase the viscosity of pastes made of wheat starch and to decrease that of pastes made of lentil and oat starches [15].

Annealed starch has found multiple applications in the industry [16,17]. Due to its thermal stability and diminished tendency for retrogradation, it is used in food preserves and frozen foods [18]. In turn, Hormdok and Noomhorm [19] described the usability of hydrothermally-treated rice starch for pasta making. It is also worth adding that annealing can contribute to an increased content of resistant starch, especially when coupled with other modifications [20].

The scientific literature indicates the feasibility of performing the annealing process together with other modifications. Zhong et al. [21] investigated effects of rice starch annealing that was heated at a temperature of 45 °C for three days at a starch to water ratio of 1:9, and additionally microwaved. These modifications led to structural and physicochemical changes of starch as well as to increased initial pasting temperature and viscosity of the starch pastes produced. In turn, Iftikhar and Dutta [22] annealed rice starch that had earlier been retrograded; they kept starch in water at 50 °C for 72 h, at a starch to water ratio of 1:6, and by this means achieved new crystalline structures featuring higher thermal stability; whereas Chi et al. [23] modified maize and potato starches by heating dry starches and annealing them at 50 °C for 24 h, at a starch to water ratio of 1:4, and obtained modified preparations with increased resistance to enzymatic activity.

Acetylation is one of the methods of chemical modification that may be coupled with physical processes. It is usually performed using starch acetate (E1420) that is of great importance to the food industry [24] and can be used for a single modification as the so-called acetylated starch (E1420) or in combination with some other chemical modification methods. In turn, acetylated distarch phosphate (E1414), acetylated distarch adipate (E1422), and acetylated oxidized starch (E1451) are applied as food additives [25]. The acetylated starch is produced on the industrial scale by starch esterification in an aqueous suspension with acetic acid anhydride in an alkaline medium. Compared to native starch, it features a lower pasting temperature, pastes made of it are more viscous, and it forms stable gels resistant to retrogradation [26]. The acetylated starch is used as an additive to improve rheological properties of dough. Often, it is also an ingredient of pastas, mayonnaises, ketchups, cottage cheeses, low-fat products, and confectionery products. Oxidized acetylated starch is used to produce jelly candies owing to the clarity of pastes and gels and stability in acidic high-sugar solutions. Acetylated distarch phosphate has been applied in thermally-prepared food products, like soups, ketchups, sauces, cake fillings, or pasteurized cottage cheeses [27]. Acetylation of retrograded starch, precipitated during starch paste freezing and defrosting, is a patented method for resistant starch production [28].

Preparations of acetylated annealed potato starch may be deployed for a slow release of a therapeutic substance [29]. The properties of modified preparations are affected not only by the type but also by the order of performed modifications. Scientific literature lacks reports on the starch subjected to annealing followed by acetylation; hence, the aim of this study was to investigate the effect of the annealing process performed at various temperatures as the first stage of starch modification. The annealed starch preparations were then esterified using acetic acid anhydride. Finally, the annealed and acetylated starch preparations were determined for their properties.

2. Materials and Methods

The experimental material included Superior Standard potato starch (PEPEES Łomża S.A., Łomża, Poland) produced in 2020.

Differential scanning calorimetry (DSC) was employed to determine the pasting temperature of potato starch, which reached 62.79 °C.

2.1. Annealing

A suspension was prepared from native potato starch, which contained 10 g of starch per 100 g of the solution. The prepared suspension was kept for 48 h at temperatures of 51.0, 53.5, 56.0, 58.5, or 61 °C, under continuous stirring. Afterward, starch was rinsed three times with 5-L portions of distilled water. Starch precipitate was separated from the suspension using a stratos flow-centrifuge (Heraeus Sepatech, Germany), then dried at 30 °C for 24 h, ground, and sieved through a screen with mesh size of 400 μm.

2.2. Acetylation

Acetylation was applied to the annealed starch preparations and to native potato starch. Starch dry matter (200 g) and water (1000 g) were placed in a reactor and mixed to achieve a homogenous suspension, the pH value of which was adjusted to pH 8–9 using a 0.5 M NaOH solution. Afterward, 26 mL of acetic acid anhydride were instilled to the starch suspension under continuous stirring. At the same time, the 0.5 M NaOH solution was instilled into the solution to maintain the pH value of the reaction mixture at pH 8–9. Once the whole portion of the anhydride had been added, the mixture was stirred for 15 min and then acidified to pH 5.2–5.6 using a 2.8 M HCl solution. Afterward, the suspension was filtered through a Buchner funnel to enable starch precipitation. The resultant modified starch was rinsed with water until the excess of the reagent had been removed. Then, the preparations were dried at 30 °C for 24 h, ground, and sieved through a screen with mesh size of 400 μm.

2.3. Analytical Methods

2.3.1. Determination of the Acetylation Degree of Starch Preparations

Acetylated starch (10 g) was weighed and transferred to a 300 mL conical flask using 65 mL of distilled water. Starch and water were mixed, and then 2–3 drops of phenolphthalein and also 0.1 M NaOH solution were instilled to the mixture until it developed a light pink color which sustained for 1 min. Next, 25 mL of a 0.5 M NaOH solution were added to the mixture, which was then shaken for 35 min. Afterward, the mixture was titrated with a standard HCl solution [30].

The percentage of acetylation A was computed acc. to the following formula:

$$A = \frac{(P_0 - P_W) \cdot N_k \cdot 0.043 \cdot 100}{M_S} [\%] \tag{1}$$

where:

P_0—volume of a standard HCl solution used to titrate 25 mL of a 0.5 M NaOH solution [mL].
P_W—volume of a standard HCl solution used to titrate the sample [mL].
N_k—acid titer.
M_S—grams of starch dry matter in the sample.

The degree of acetylation was computed acc. to the following formula:

$$DS = \frac{160 \cdot A}{4300 - (42 \cdot A)} \tag{2}$$

2.3.2. Determination of the Swelling Power of Starch Preparations in Water at a Temperature of 20 °C

The starch preparation (0.3 g) was placed in a 4 mL test tube, to which 2.5 mL of distilled water were added. The content of the flask was conditioned at a temperature of 20 °C for 10 min. Afterward, the samples were centrifuged using an MPW-55 type laboratory centrifuge (MPW Instruments, Warsaw, Poland) at the speed of 5000 rpm and a temperature of 20 °C for 10 min. After centrifugation, the supernatant was collected, and the precipitate left in the test tubes was weighed [29].

Working swelling power was calculated using the following formula:

$$W = \frac{M_W - M_P}{M_P}[g/g] \tag{3}$$

where:

M_W—weight of the precipitate after centrifugation
M_P—weight of the sample.

2.3.3. Determination of the Resistance of Starch Preparations to the Action of Amyloglucosidase

A 0.72 g/100 g aqueous suspension was prepared from 38 g of starch. It was then heated to the boiling point, cooled, and then the volume of evaporated water was completed. Afterward, 34 mL of an acetate buffer were added, and the sample was placed in a water bath at a temperature of 37 °C. To enable hydrolysis, 4 mL of amyloglucosidase were added; its concentration was adjusted so as to ensure the complete saccharification of gelatinized native potato starch after 120 min of the process. After 120 min, 10 μL of the sample were collected and transferred into a test tube with 1 mL of a reagent for glucose concentration measurement (BioSystem S.A., Barcelona, Spain) containing glucose oxidase and peroxidase. The content of the flask was mixed and incubated at a temperature of 20 °C for 15 min. Then, absorbance was measured at a wavelength of λ = 500 nm using a CECIL CE 2010 colorimeter (Cecil Instruments, England). The measurement results were compared to those achieved for a control sample prepared in an analogous way but without the starch preparation. The absorbance value obtained was compared with the standard curve plotted based on the read outs of absorbance of solutions with a known glucose concentration [31].

Resistance of the starch preparations was computed using the following formula:

$$R = 100 - \frac{X \cdot 100}{0.4}[g/100\,g] \tag{4}$$

where:

R—resistance of starch preparations to amyloglucosidase action (g/100 g).
X—content of glucose read out from the standard curve (mg).

2.3.4. Determination of the Characteristics of Phase Transitions of the Starch Preparations Using Differential Scanning Calorimetry (DSC)

Determination was performed using a DSC 822E differential scanning calorimeteter (Mettler Toledo, Germany) [32]. A 10 g starch sample weighed exact to ±0.02 mg was placed on the bottom of a 100 μL aluminum crucible. Then, distilled water was poured into the crucible in the ratio of 3:1 respective to sample weight. The crucible was covered with a lid and conditioned at a temperature of 20 °C for 30 min, then placed in a measuring chamber having a temperature of 20 °C and heated to 100 °C at the heating rate of 4 °C/min. The onset and final temperatures of phase transition (°C) and the heat of this transition (J/g) were determined from thermograms.

2.3.5. Determination of the Flow Curves of Pastes Made of Starch Preparations

Analyses were conducted using an RS 6000 oscillating-rotating viscosimeter Haake (Karlsruhe, Germany) for 5 g/100 g starch suspensions that were heated at 96 °C for 30 min under continuous stirring [33].

Flow curves were plotted for the prepared pastes, at a measurement temperature of 50 °C and a shear rate of 1–300 s^{-1}. A hot paste was placed in a system of coaxial cylinders (Z38ALtype) of the RS 100 rheometer, then cooled, and relaxed at the measurement temperature for 15 min. The flow curves plotted were described using the following equations:

$$\text{Ostwald de Waele:} \quad \tau = K \cdots \dot{y}^{n} \tag{5}$$

$$\text{Casson:} \quad \tau^{0.5} = \tau_{oc}^{\ 0.5} + (\eta_c \cdots \dot{y})^{0.5} \tag{6}$$

where:

τ—shear stress (Pa)
K—consistency coefficient (Pa·s^n)
$\dot{y}$—shear rate (s^{-1})
n—flow index
τ_{oc}—yield point (Pa)
η_c—Casson's plastic viscosity (Pa·(s))

2.4. Statistical Analysis

Results were statistically analyzed using the Statistica 13.0 PL software package. Based on statistical computations (from at least three parallel replications), values of the least significant differences and standard deviation were calculated, and equations of flow curves were determined. For statistical evaluation, the results were subjected to one-way analysis of variance at a significance level of 0.05. Values of the least significant difference (LSD) between the means were computed using the Duncan's test at a significance level of 0.05 (StatSoft, Inc., Tulsa, OK, USA, 2011).

3. Results and Discussion

Figure 1 presents the degree of substitution of starch annealed at various temperatures and then acetylated. Although the annealing process has a significant effect on the possibility of starch substitution with acetic acid residues, there is paucity of literature data regarding the impact of starch annealing on its susceptibility to esterification. The heating of a native starch suspension leads to an increase in the volume of starch granules caused by slow water penetration into their interior. Presumably, the higher substitution degree, compared to the native starch, is due to the loosening of a starch granule structure induced by the breaking of hydrogen bridges accompanied by the release and increasing number of available hydroxyl groups that may be substituted [34]. Among the double-modified preparations, the lowest substitution degree was found for the starch heated at the highest tested temperature (61°), presumably due to starch crystallinity increase during annealing, which reduces starch susceptibility to acetylation [35].

Figure 2 depicts the effect of acetylation on the swelling power of annealed starch preparations. As a result of starch suspension heating at various temperatures, single-modified starch preparations were obtained having the swelling power of ca. 1 g/g. Their esterification caused an increase in their swelling power, except for the preparation annealed at the temperature of 53.5 °C. According to literature data, esterification increases potato starch affinity to water [36]; however, its impact on the earlier annealed starch has not been fully elucidated. Unpredictable changes that occur during annealing, like re-arrangement of starch granule structure principally in the amorphous region but also the formation of new crystallites or amylose leaching, trigger changes in the properties of annealed starch, which—when subjected to any additional chemical modification—can elicit various effects on the properties of the double-modified starch [37]. These changes can be especially interesting in the case of potato starch, which can develop advanced crystalline structures due to a high number of long chains (B2-B3) [38,39]. Furthermore, as Tomaszewska-Ciosk et al. [40] reports, the dynamics of crystallite formation depends on starch granule size. A broad range of potato starch granule sizes (5–100 μm) may contribute to the uneven increase in the swelling power of its annealed preparations.

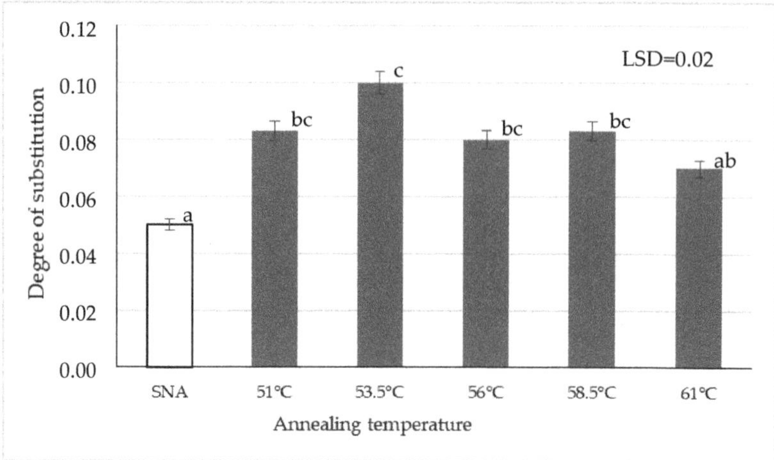

Figure 1. Degree of substitution of acetylated annealed starch preparations depending on the annealing temperature (Different letters mean homogeneous groups at $p < 0.05$).

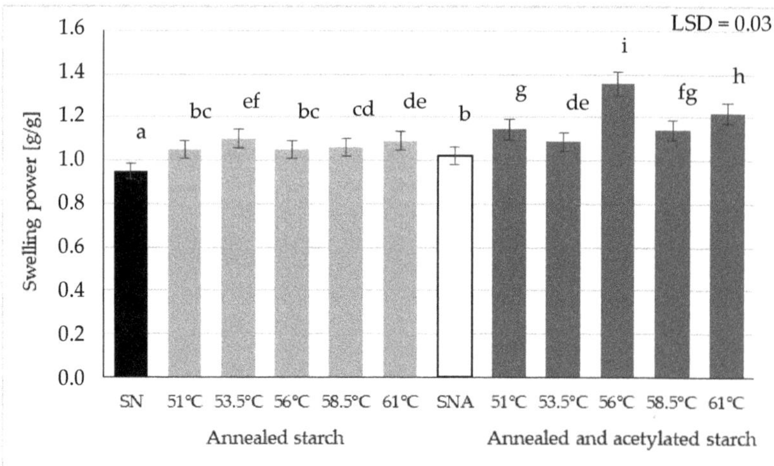

Figure 2. Swelling power of native starch (SN), annealed starch, native acetylated starch (SNA), and annealed and acetylated starch (Different letters mean homogeneous groups at $p < 0.05$).

The analysis of study results allows concluding that the increased water absorption capacity indicates debilitation of the association potential of polymer chains of starch and increased availability of hydroxyl groups for the formation of new bonds with water molecules. In addition, the micropores occurring naturally on the potato starch surface [41,42], presumably enlarge their sizes, thereby facilitating water penetration into the starch granule interior. However, structural changes evoked in starch granules by starch heating in the excess of water can also lead to both increased availability of hydrophilic groups and the formation of new crystallites, thus diminishing starch capability for water binding [34,36].

Some literature works report various effects of annealing on the swelling power of starches of various botanical origin. Pinto et al. [43] proved that, generally, annealing decreased the swelling power of pinhão starch analyzed in their study; however, this decrease was negligible compared to the non-modified raw material. A decreased swelling power of rice starch was also reported by Dias et al. [44] who annealed it at three different temperatures. In turn, the heating of four cultivars of white yam (a tropical tuber plant) at temperatures below the pasting temperature increased the water absorption capability of one cultivar and decreased it in the case of the three other cultivars studied [45].

Figure 3 presents the resistance of the analyzed starch preparations to the action of amyloglucosidase. The annealing of native potato starch at various temperatures resulted in the preparations being partly resistant to amyloglucosidase action. Literature works indicate that such hydrothermal modifications as heat moisture treatment (HMT) or annealing increase resistant starch content [29]. The aforementioned authors claim this increase to be due to the rearrangement of starch chains during heating in the excess of water. This type of heating leads to the strengthening of bonds between amylose and amylopectin, thereby increasing starch granule stability, which—in contrast—can inhibit enzymatic hydrolysis [23,46]. Furthermore, annealing can lead to the leaching of amylose chains from amorphous regions of starch granules being especially susceptible to hydrothermal modifications, which probably reinforces the crystalline structure of starch [28]. Ashogbon and Akintayo [37] claimed that the amorphous regions of semicrystalline starch granules represented regions most susceptible to the initial water absorption. Hydration of starch granules leads to their increased mobility in the amorphous regions, while the generated oscillating motion triggers changes in crystallization and interactions between amylose-amylose and amylose-amylopectin chains. With progress in heating/hydration, the initially weaker or imperfect crystallites successively disappear, whereas the other ones become more perfect upon fusion and recrystallization [37]. These are spontaneous, hardly-controllable processes. Many factors contribute to the ultimate effect. Starting from the temperature applied, through the raw material used, to water volume and heating duration.

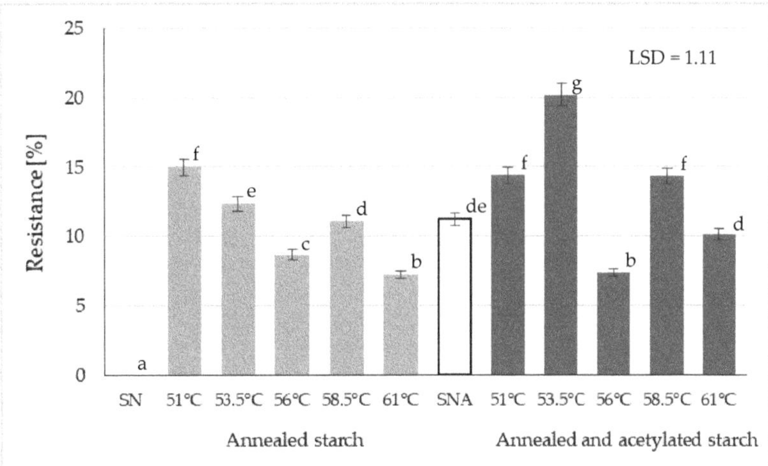

Figure 3. Resistance of native starch (SN), annealed starch, native acetylated starch (SNA), and annealed and acetylated starch (Different letters mean homogeneous groups at $p < 0.05$).

The resistance of the acetylated potato starch analyzed in the present study reached 11% (Figure 3). This result is consistent with our previous findings [28,47]. The annealing performed before acetylation caused no explicit increase in the resistance of double modified preparations compared to the starch subjected only to chemical modification. However, higher resistance could be noted in the case of the starch preparation annealed at 53.5 °C and acetylated, having a higher degree of substitution at carbon 2 and 3, which makes it difficult for the enzyme to access the hydrolyzed 1–4 glycosidic bond [47]. This shows that acetyl groups block the access of enzymes to the starch chain and that starch resistance increases along with the increasing number of acetyl groups.

Table 1 presents the thermal characteristics of annealed starch as well as acetylated and annealed starch preparations. As reported by other authors [48], starch acetylation decreases phase transition temperatures. This is, most likely, due to the weakening of interactions among starch chains as a result of substitution of hydroxyl groups with acetyl groups that impart amphiphilic character to starch [49], which is in turn associated with the disruption of the crystalline structure of starch during its esterification [26,49]. The acetylation decreased the onset and end pasting temperatures and the phase transition temperature also in the case of the analyzed starch preparations annealed at various temperatures. Even though the crystalline structure of starch had been strengthened by annealing, the successive acetylation of starch weakened its structure, like in the case of native starch.

Table 1. Thermal properties of starch preparations annealed at various temperatures and of annealed and acetylated starch preparations.

Preparation Type	Annealing Temperature	Onset Temperature (°C)	End Temperature (°C)	Gelatinization Enthalpy (J/g)
Annealed starch	51 °C	62.66	73.41	14.03
	53.5 °C	65.90	74.66	14.66
	56 °C	66.68	77.81	16.03
	58.5 °C	68.38	76.61	16.06
	61 °C	67.90	77.72	10.48
Acetylated starch	-	48.91	66.02	13.03
Acetylated annealed starch	51 °C	52.36	66.46	11.15
	53.5 °C	54.73	68.37	11.21
	56 °C	61.72	72.42	14.27
	58.5 °C	57.68	70.29	13.81
	61 °C	63.18	76.11	8.62
LSD		0.81	1.47	0.19

A Haake oscillating-rotary viscosimeter was used to measure the viscosity of pastes made of modified starch preparations (Table 2). Flow curves were plotted and rheological models were adjusted, which were further used to determine: the consistency coefficient denoting viscosity at the initial shearing phase, paste yield point describing the maximal stress of starch pastes at the null shear rate, and Casson's plastic viscosity indicating viscosity at the terminal shearing phase [28]. If various starch preparations are to be used in food technology, their rheological properties need to be determined in the first instance, including mainly changes in viscosity of pastes during cooling. This is especially important when starch preparations are applied as thickening agents. The knowledge of factors affecting rheological behavior of starch pastes allows optimizing and standardizing technological processes in the production of foods with starch additives [50,51].

Viscosity measurements demonstrated that the analyzed pastes were characterized by shear-thinned non-Newtonian flow, being typical of starch pastes. Viscosity at the initial shearing phase and yield point of single-modified annealed starch preparations depended on heating temperature. In the study by Wang et al. [52], the annealing of potato starch at a temperature of 50°C caused a viscosity increase, due to greater ordering of starch chains in the amorphous regions of starch. In addition, annealing promotes the swelling of starch granules, from which water is removed together with amylose chains, which may cause an increase in the viscosity of paste made of annealed starch [36].

Table 2. Rheological properties of starch preparations annealed at various temperatures and of annealed and acetylated starch preparations.

Preparation Type	Annealing Temperature	Consistency Coefficient [Pa · · · s^n]	Yield Point [Pa]	Casson's Plastic Viscosity [Pa · · · s]
Annealed starch	51 °C	9.77	25.06	0.42
	53.5 °C	14.00	34.69	0.40
	56 °C	21.78	50.17	0.36
	58.5 °C	20.35	47.20	0.36
	61 °C	9.43	22.64	0.19
Acetylated starch	-	4.78	12.41	0.28
Acetylated annealed starch	51 °C	6.66	17.24	0.30
	53.5 °C	6.45	16.71	0.28
	56 °C	10.67	26.20	0.28
	58.5 °C	8.5	21.96	0.37
	61 °C	13.53	30.78	0.20
LSD		1.09	3.27	0.06

Another step of modification, i.e., esterification of annealed preparations, contributed to a decrease in the values of coefficients characterizing paste viscosity at the initial shearing phase (except for the preparation annealed at the highest temperature tested), which is due to the diminished structural stability triggered by the introduction of acetyl groups to the starch chain [53,54]. The double modification resulted in lower values of the Casson's plastic viscosity, indicating greater susceptibility of the double-modified preparations to shear forces compared to the single-modified ones.

4. Conclusions

The annealing of starch before acetylation triggered changes in the properties of the modified preparations. It contributed to a higher degree of starch substitution with acetic acid residues and to the increased swelling power of starch. Both these properties were also affected by the annealing temperature. The highest resistance to amylolysis was found for the starch preparation annealed at 53.5 °C and acetylated. The double modification involving annealing and acetylation processes increased the onset and end pasting temperatures compared to the acetylation alone. Similar observations were made for the consistency coefficient and yield point. Preparations of this type may be suitable for the controlled release of bioactive or therapeutic substances.

Author Contributions: Conceptualization, T.Z. and E.Z.; methodology, M.K.-Ż.; software, E.T.-C.; validation, A.G. (Artur Gryszkin); formal analysis, A.W.; investigation, A.G. (Artur Gryszkin); resources, A.G. (Anna Gawrońska); data curation, E.T.-C.; writing—original draft preparation, E.Z.; writing—review and editing, M.K.-Ż.; visualization, T.Z.; supervision, A.W.; project administration, E.Z.; funding acquisition, M.K.-Ż. All authors have read and agreed to the published version of the manuscript.

Funding: The publication is co-financed under the Leading Research Groups support project from the subsidy increased for the period 2020–2025 in the amount of 2% of the subsidy referred to Art. 387 (3) of the Law of 20 July 2018 on Higher Education and Science, obtained in 2019.

Institutional Review Board Statement: Not applicable.

Informed Consent Statement: Not applicable.

Data Availability Statement: Data are contained within the article.

Conflicts of Interest: The authors declare no conflict of interest.

Sample Availability: Samples of the compounds are not available from the authors.

References

1. Ačkar, Đ.; Babić, J.; Jozinović, A.; Miličević, B.; Jokić, S.; Miličević, R.; Rajič, M.; Šubarić, D. Starch Modification by Organic Acids and Their Derivatives: A Review. *Molecules* **2015**, *20*, 19554–19570. [CrossRef]
2. Hoover, R.; Hughes, T.; Chung, H.J.; Liu, Q. Composition, molecular structure, properties, and modification of pulse starches: A review. *Food Res. Int.* **2010**, *43*, 399–413. [CrossRef]
3. van der Maarel, M.J.E.C.; Leemhuis, H. Starch modification with microbial alpha-glucanotransferase enzymes. *Carbohydr. Polym.* **2013**, *93*, 116–121. [CrossRef]
4. Castanha, N.; Matta, M.D.J.; Augusto, P.E.D. Potato starch modification using the ozone technology. *Food Hydrocoll.* **2017**, *66*, 343–356. [CrossRef]
5. Sweedman, M.C.; Tizzotti, M.J.; Schaefer, C.; Gilbert, R.G. Structure and physicochemical properties of octenyl succinic anhydride modified starches: A review. *Carbohydr. Polym.* **2013**, *92*, 905–920. [CrossRef]
6. Devi, R.; Sit, N. Effect of single and dual steps annealing in combination with hydroxypropylation on physicochemical, functional and rheological properties of barley starch. *Int. J. Biol. Macromol.* **2019**, *129*, 1006–1014. [CrossRef]
7. Hu, X.; Jia, X.; Zhi, C.; Jin, Z.; Miao, M. Improving properties of normal maize starch films using dual-modification: Combination treatment of debranching and hydroxypropylation. *Int. J. Biol. Macromol.* **2019**, *130*, 197–202. [CrossRef] [PubMed]
8. Li, H.; Gui, Y.; Li, J.; Zhu, Y.; Cui, B.; Guo, L. Modification of rice starch using a combination of autoclaving and triple enzyme treatment: Structural, physicochemical and digestibility properties. *Int. J. Biol. Macromol.* **2020**, *144*, 500–508. [CrossRef] [PubMed]
9. Tharanathan, R.N. Starch—Value Addition by Modification. *Crit. Rev. Food Sci.* **2005**, *45*, 371–384. [CrossRef]
10. Yao, T.; Sui, Z.; Janaswamy, S. Annealing. In *Physical Modifications of Starch*; Sui, Z., Kong, X., Eds.; Springer: Singapore, 2018; pp. 37–49. [CrossRef]
11. Rocha, T.S.; Felizardo, S.G.; Jane, J.-L.; Franco Celia, M.L. Effect of annealing on the semicrystalline structure of normal and waxy corn starches. *Food Hydrocoll.* **2012**, *29*, 93–99. [CrossRef]
12. Din, Z.U.; Xiong, H.; Fei, P. Physical and Chemical Modification of Starches: A review. *Crit. Rev. Food Sci. Nutr.* **2017**, *57*, 2691–2705. [CrossRef]
13. Piecyk, M.; Konarzewska, M.; Sitkiewicz, I. Effect of hydrothermal modification of annealing type on some selected properties of starch pea (*Pisum sativum*). *Żywność. Nauka. Technol. Jakość.* **2009**, *5*, 58–71. (In Polish)
14. Vamadevan, V.; Bertoft, E.; Soldatov, D.V.; Seetharaman, K. Impact on molecular organization of amylopectin in starch granules upon annealing. *Carbohydr. Polym.* **2013**, *98*, 1045–1055. [CrossRef]
15. Tester, R.F.; Debon, S.J. Annealing of starch—A review. *Int. J. Biol. Macromol.* **2000**, *27*, 1–12. [CrossRef]
16. Wang, H.; Xiao, N.; Wang, X.; Zhao, X.; Zhang, H. Effect of pregelatinized starch on the characteristics, microstructures, and quality attributes of glutinous rice flour and dumplings. *Food Chem.* **2019**, *283*, 248–256. [CrossRef]
17. Fu, Z.-Q.; Che, L.-M.; Li, D.; Wang, L.-J.; Adhikari, B. Effect of partially gelatinized corn starch on the rheological properties of wheat dough. *LWT Food Sci. Technol.* **2016**, *66*, 324–331. [CrossRef]
18. Adebowale, K.O.; Afolabi, T.A.; Olu-Owolabi, B.I. Hydrothermal treatments of Finger millet (*Eleusine coracana*) starch. *Food Hydrocoll.* **2005**, *19*, 974–983. [CrossRef]
19. Hormdok, R.; Noomhorm, A. Hydrothermal treatments of rice starch for improvement of rice noodle quality. *LWT Food Sci. Technol.* **2007**, *40*, 1723–1731. [CrossRef]
20. da Rosa Zavareze, E.; Dias, A.R.G. Impact of heat-moisture treatment and annealing in starches: A review. *Carbohydr. Polym.* **2011**, *83*, 317–328. [CrossRef]
21. Zhong, Y.; Xiang, X.; Zhao, J.; Wang, X.; Chen, R.; Xu, J.; Luo, S.; Wu, J.; Liu, C. Microwave pretreatment promotes the annealing modification of rice starch. *Food Chem.* **2020**, *304*, 125432. [CrossRef]
22. Iftikhar, S.A.; Dutta, H. Status of polymorphism, physicochemical properties and in vitro digestibility of dual retrogradation-annealing modified rice starches. *Int. J. Biol. Macromol.* **2019**, *132*, 330–339. [CrossRef]
23. Chi, C.; Li, X.; Lu, P.; Miao, S.; Zhang, Y.; Chen, L. Dry heating and annealing treatment synergistically modulate starch structure and digestibility. *Int. J. Biol. Macromol.* **2019**, *137*, 554–561. [CrossRef]
24. Xu, Y.; Miladinov, V.; Hanna, M.A. Synthesis and Characterization of Starch Acetates with High Substitution. *Cereal Chem.* **2004**, *81*, 735–740. [CrossRef]
25. EU Commission: Publications Office of the European Union. Commision Regulation (EU) No 1129/2011 of 11 November 2011. *Off. J. Eur. Union L2952* **2011**, *295*, 16–17.
26. Golachowski, A.; Zięba, T.; Kapelko-Żeberska, M.; Drożdż, W.; Gryszkin, A. Current research addressing starch acetylation. *Food Chem.* **2015**, *176*, 350–356. [CrossRef] [PubMed]
27. Leszczyński, W. The use of modified starches in the food industry, part I. *Przegląd Piek. Cukier.* **2006**, *54*, 54–57. (In Polish)
28. Zięba, T.; Juszczak, L.; Gryszkin, A. Properties of retrograded and acetylated starch preparations Part 2. Dynamics of saccharification with amyloglucosidase and rheological properties of resulting pastes and gels. *LWT Food Sci. Technol.* **2011**, *44*, 1321–1327. [CrossRef]
29. Zięba, T.; Wilczak, A.; Kobryń, J.; Musiał, W.; Kapelko-Żeberska, M.; Gryszkin, A.; Meisel, M. The Annealing of Acetylated Potato Starch with Various Substitution Degrees. *Molecules* **2021**, *26*, 2096. [CrossRef] [PubMed]
30. Zdybel, E.; Tomaszewska-Ciosk, E. Modification of starch with succinic acid residues. *Przemysł Chem.* **2015**, *94*, 1138–1141. (In Polish) [CrossRef]

31. Zięba, T.; Kapelko-Żeberska, M.; Gryszkin, A.; Wilczak, A.; Raszewski, B.; Spychaj, R. Effect of the botanical origin on properties of RS3/4 type resistant starch. *Polymers* **2019**, *11*, 81. [CrossRef] [PubMed]
32. Zdybel, E.; Zięba, T.; Tomaszewska-Ciosk, E.; Rymowicz, W. Effect of the Esterification of Starch with a Mixture of Carboxylic Acids from Yarrowia lipolitica Fermentation Broth on Its Selected Properties. *Polymers* **2020**, *12*, 1383. [CrossRef] [PubMed]
33. Zięba, T.; Kapelko, M.; Gryszkin, A. Selected properties of potato starch subjected to multiple physical and chemical modifications. *Pol. J. Food Nutr. Sci.* **2007**, *57*, 639–645.
34. Leszczyński, W. Resistant starch—Classification, structure, production. *Pol. J. Food Nutr. Sci.* **2004**, *54*, 37–50.
35. Chen, Z.; Schols, H.A.; Voragen, A.G.J. Differently sized granules from acetylated potato and sweet potato starches differ in the acetyl substitution pattern of their amylose populations. *Carbohydr. Polym.* **2004**, *56*, 219–226. [CrossRef]
36. Kapelko-Żeberska, M.; Zięba, T.; Spychaj, R.; Gryszkin, A. Selected Rheological Properties of RS3/4 Type Resistant Starch. *Pol. J. Food Nutr. Sci.* **2017**, *67*, 293–299. [CrossRef]
37. Ashogbon, A.O.; Akintayo, E.T. Recent trend in the physical and chemical modification of starches from different botanical sources: A review. *Starch* **2014**, *66*, 41–57. [CrossRef]
38. Leszczyński, W. Starch-raw material for industry, the structure and properties. *Adv. Agric. Sci. Probl. Issues* **2004**, *500*, 69–98. (In Polish)
39. Mathobo, V.M.; Silungwe, H.; Ramashia, S.E.; Anyasi, T.A. Effects of heat-moisture treatment on the thermal, functional properties and composition of cereal, legume and tuber starches-a review. *J. Food. Sci. Technol.* **2021**, *58*, 412–426. [CrossRef]
40. Tomaszewska-Ciosk, E.; Zdybel, E.; Drożdż, W. Model of retrogradation of extruded potato starch. *Starch* **2020**, *72*, 1900145. [CrossRef]
41. Juszczak, L.; Fortuna, T.; Krok, F. Non-contact atomic force microscopy of starch granules surface. Part I. Potato and tapioca starches. *Starch* **2003**, *55*, 1–7. [CrossRef]
42. Juszczak, L.; Fortuna, T.; Krok, F. Non-contact atomic force microscopy of starch granules surface. Part II. Selected cereal starches. *Starch* **2003**, *55*, 8–16. [CrossRef]
43. Pinto, V.Z.; Vanier, N.L.; Deon, V.G.; Moomand, K.; El Halal, S.L.M.; Zavareze, E.D.R.; Lim, L.T.; Dias, A.R.G. Effects of single and dual physical modifications on pinhão starch. *Food Chem.* **2015**, *187*, 98–105. [CrossRef]
44. Dias, A.R.G.; da Rosa Zavareze, E.; Spier, F.; Castro, L.A.S.; Gutkoski, L.C. Effects of annealing on the physicochemical properties and enzymatic susceptibility of rice starches with different amylose contents. *Food Chem.* **2010**, *123*, 711–719. [CrossRef]
45. Falade, K.O.; Ayetigbo, O.E. Effects of annealing, acid hydrolysis and citric acid modifications on physical and functional properties of starches from four yam (*Dioscorea* spp.) cultivars. *Food Hydrocoll.* **2015**, *43*, 529–539. [CrossRef]
46. Jayakody, L.; Hoover, R. Effect of annealing on the molecular structure and physicochemical properties of starches from different botanical origins–A review. *Carbohydr. Polym.* **2008**, *74*, 691–703. [CrossRef]
47. Zięba, T.; Kapelko, M.; Szumny, A. Effect of preparation method on the properties of potato starch acetates with an equal degree of substitution. *Carbohydr. Polym.* **2013**, *94*, 193–198. [CrossRef] [PubMed]
48. Wickramasinghe, H.A.M.; Yamamoto, K.; Yamauchi, H.; Noda, T. Effect of Low Level of Starch Acetylation on Physicochemical Properties of Potato Starch. *Food Sci. Biotechnol.* **2009**, *18*, 118–123.
49. Golachowski, A. Properties of acetylated starch obtained from SO_2-treated starch milk. *Electron. J. Pol. Agric. Univ.* **2003**, *6*, 1.
50. Adamczyk, G.; Krystyjan, M.; Dobosz, A.; Sikora, M. Thixotropic properties of starch. *Żywność Nauka Technol. Jakość* **2013**, *6*, 16–31. (In Polish)
51. Le Thanh-Blicharz, J.; Lubiewski, Z.; Voelkel, E.; Lewandowicz, G. Evaluation of rheological properties of commercial native starches. *Żywność Nauka Technol. Jakość* **2011**, *3*, 53–65. (In Polish)
52. Wang, S.; Wang, J.; Wang, S.; Wang, S. Annealing improves paste viscosity and stability of starch. *Food Hydrocoll.* **2017**, *62*, 203–211. [CrossRef]
53. Hong, J.; Zeng, X.A.; Brennan, C.S.; Brennan, M.; Han, Z. Recent Advances in Techniques for Starch Esters and the Applications: A Review. *Foods* **2016**, *5*, 50. [CrossRef] [PubMed]
54. Kapelko, M.; Zięba, T.; Michalski, A. Effect of the production method on the properties of RS3/RS4 type resistant starch. Part 2. Effect of a degree of substitution on the selected properties of acetylated retrograded starch. *Food Chem.* **2012**, *135*, 2035–2042. [CrossRef] [PubMed]

polymers

MDPI

Article

Antibacterial Films of Alginate-CoNi-Coated Cellulose Paper Stabilized Co NPs for Dyes and Nitrophenol Degradation

Yasir Anwar [1,2,*], Hani S. H. Mohammed Ali [1,2], Waseeq Ur Rehman [3], Hassan A. Hemeg [4] and Shahid Ali Khan [5,6,*]

1 Department of Biological Sciences, Faculty of Science, King Abdulaziz University, P.O. Box 80203, Jeddah 21589, Saudi Arabia; haniolfat@gmail.com
2 Princess Dr. Najla Bint Saud Al-Saud Center for Excellence Research in Biotechnology, King Abdulaziz University, P.O. Box 80203, Jeddah 21589, Saudi Arabia
3 Department of Chemistry, Government Post Graduate College Nowshera, Nowshera 24100, Pakistan; gwaseeq@gmail.com
4 Department of Medical Laboratory Technology, College of Applied Medical Sciences, Taibah University, Almadina Almunawra 30001, Saudi Arabia; hasanhemeg@hotmail.com
5 Center of Excellence for Advanced Materials Research (CEAMR), King Abdulaziz University, P.O. Box 80203, Jeddah 21589, Saudi Arabia
6 Department of Chemistry, University of Swabi, Swabi Anbar, Swabi 23561, Pakistan
* Correspondence: yasirpcsir2006@gmail.com (Y.A.); skhan@uoswabi.edu.pk (S.A.K.)

Citation: Anwar, Y.; Mohammed Ali, H.S.H.; Rehman, W.U.; Hemeg, H.A.; Khan, S.A. Antibacterial Films of Alginate-CoNi-Coated Cellulose Paper Stabilized Co NPs for Dyes and Nitrophenol Degradation. *Polymers* **2021**, *13*, 4122. https://doi.org/10.3390/polym13234122

Academic Editor: Arn Mignon

Received: 30 October 2021
Accepted: 22 November 2021
Published: 26 November 2021

Abstract: The development of a solid substrate for the support and stabilization of zero-valent metal nanoparticles (NPs) is the heart of the catalyst system. In the current embodiment, we have prepared solid support comprise of alginate-coated cellulose filter paper (Alg/FP) for the synthesis and stabilization of Co nanoparticles (NPs) named as Alg/FP@Co NPs. Furthermore, Alginate polymer was blended with 1 and 2 weight percent of CoNi NPs to make Alg-CoNi1/FP and Alg-CoNi2/FP, respectively. All these stabilizing matrixes were used as dip-catalyst for the degradation of azo dyes and reduction of 4-nitrophenol (4NP). The effect of initial dye concentration, amount of $NaBH_4$, and catalyst dosage was assessed for the degradation of Congo red (CR) dye by using Alg-CoNi2/FP@Co NPs. Results indicated that the highest k_{app} value (3.63×10^{-1} min^{-1}) was exhibited by Alg-CoNi2/FP@Co NPs and lowest by Alg/FP@Co NPs against the discoloration of CR dye. Furthermore, it was concluded that Alg-CoNi2/FP@Co NPs exhibited strong catalyst activity against CR, and methyl orange dye (MO) degradation as well as 4NP reduction. Antibacterial activity of the prepared composites was also investigated and the highest l activity was shown by Alg-CoNi2/FP@Co NPs, which inhibit 2.5 cm zone of bacteria compared to other catalysts.

Keywords: alginate; CoNi nanocomposite; cellulose paper; antibacterial potential; degradation

1. Introduction

Nanoscience received much interest in recent years due to their vast applications in sensing, drug delivery, antimicrobial activity, coating, bio-imaging, environmental remediation, and biomedical applications [1–6]. Besides their major applications in various technological sectors, they are used as catalysts in many industrial and chemical reactions like hydrogen production [7], oxygen evolution reaction, organic reactions [8], and nitroaromatics reduction [9–12]. The above diverse applications of metallic nanoparticles are because of their high surface to volume ratio, high surface energy, and small quantum size effect [13]. The quantum size effect gives an amazing reduction in the particle size of the nanoparticles beyond the threshold limit and leads to the change in the electronic environment of the particle. Owing to the change in the electronic environment, small particle size, high surface area, and high surface energy lead to an abrupt increase in particle activity. However, due to the high energy associated with the particle, there exist some drawbacks with the usage of these particles as the catalyst. For instance, aggregation and agglomeration of the particles are due to the physical entanglement and Wander Waal's forces [14]. The second most common problem in nanoparticles research is their separation after the reaction completion. Both the problems can be overcome by the use of a solid matrix that not only avoids particle aggregation, but also is easily separated from the reaction mixture. The separation of a catalyst by an easy method from the reaction mixture at any time is called a dip-catalyst. Moreover, the use of a solid support improves the stability of the nanoparticles. For example, using solid support not only avoids aggregation, but also prevents the leaching of nanoparticles, which otherwise cause's environmental issues. Many solid supports have been used for the stabilization of nanoparticles such as carbon [15], metal oxide [16], inorganic (e.g., silica (Li et al., 2013)), or organic materials [17]. Organic solid support is the most frequently used solid support for this purpose due to its inert nature to the reaction and ease of derivatization by introducing new functionality to the host material. However, among all these supporting materials, polymer-supported substrate for nanoparticles stabilization is one of the most important systems. For instance, Pd nanoparticles have been supported on polymethyl methacrylate spheres [18], and fabricated polystyrene microspheres supported Pt nanoparticles are examples of this [19]. Some approaches have been applied to use cellulose polymer as solid support for various metal nanoparticles. For example, some researchers have used cellulose cotton in the form of filter paper for metallic nanoparticle stabilization [20]. Cellulose is one of the most abundant natural organic polymers and is used in composite and packaging materials, and are also used in textile industries. The β 1,4 glycoside linked the β-D-anhydroglucopyranose monomers through one to four linkages. This linear cellulose molecule has three –OH groups in its monomer unit, which are responsible for the stabilization of metal nanoparticles. Furthermore, the cellulose polymer-supported NPs act as a dip-catalyst, and one can remove the catalyst from the reaction medium easily [21]. Although, cellulose microfibers have multiple –OH groups in their polymer network, but low process-ability, it was not considered for using directly as a stabilizing matrix for Co NPs. Furthermore, the alginate polymer swells easily and demonstrated a strong gelling characteristics to make hydrogels. Therefore, both cellulose microfibers and alginate polymer provide a good platform for the stabilization of NPs. Moreover, the use of alginate biopolymer to coat the surface of cellulose paper has markedly enhanced the active functionalities that further stabilize the metal nanoparticles. For instance, the presence of –OH and –COOH groups in the alginate biopolymer further increases the stabilization of the Co NPs [22–24]. It is reported that inorganic polymer hybrid material greatly enhanced the properties of the materials [25]. Therefore, CoNi nanocomposite was incorporated in the alginate polymer host material to achieve a hybrid composite.

Mostly the supported nanoparticles were used for the degradation of organic pollutants in an aqueous medium. Nitrophenols and azo dyes are among the most noticeable organic pollutants. The two common nitrophenols are 2- and 4-nitrophenols and various industries produce them in a large amount. If an industry produces one nitrophenol, the other will be formed in small quantities and vice versa. Both nitrophenols find large applications in pharmaceutical, agriculture, fabric, dye, cosmetic, and explosive industries. Nitrophenols are health hazards and many health issues are associated with them, for instance, high exposure to them can irritate eyes, nose, throat, skin, and possibly damage the eye. Furthermore, exposure to 4-nitrophenols can cause stomach disorders, enhance heartbeat, and may affect the central nervous system [26,27].

Furthermore, azo dyes contributed two-thirds of all the synthetic dyes and are used in large amounts in the pharmaceutical industry [28–30]. They are used in the pharmaceutical industry to color various drugs; however, carcinogenic and mutagenic activities of many azo dyes are reported in the literature. Azo dyes also find applications in the fabric and cosmetics industries. Moreover, the carcinogenic effect of most azo dyes depends on the structure of dye molecule and the way they are degraded [31,32].

In the current study, we have prepared alginate-based cellulose filter paper (FP) as solid support for t zero-valent Co NPs and used it in the degradation of CR and MO dyes as well as 4NP in the presence of borohydride in an aqueous medium.

2. Experimental Section

2.1. Reagents and Materials

Congo red, (CAS no. 573-58-0) methyl orange, (CAS no. 547-58-0), and $NaBH_4$ (CAS no. 16940-66-2) were purchased from DAEJUNG Company (Sasang-gu, Busan, Korea). Similarly, other chemicals such as, NaOH, $CoCl_2{\cdot}6H_2O$, and $NiCl_2{\cdot}6H_2O$ were acquired from Sigma Aldrich (Kawasaki, Kanagawa, Japan), and deionized water was used throughout the experiment. Sodium Alginate polymer (CAS no. 9005-38-3) was purchased from DAEJUNG Company (Sasang-gu, Busan, Korea), while Whatman filter paper was used as a source of cellulose paper and was obtained from Sigma Aldrich (Kawasaki, Kanagawa, Japan).

2.2. Synthesis of Catalyst

2.2.1. Synthesis of CoNi Nanocomposite

CoNi nanocomposite was synthesized via the sol-gel method using the previous procedure [20–22]. An equal amount of 0.5 mM solution of $CoCl_2{\cdot}6H_2O$, and $NiCl_2{\cdot}6H_2O$ was mixed in the stoichiometric ratio in a 1 L flask and mixed for 30 min. After that, a dilute solution of NaOH was added to the solution till pH 10, and heated for 7 h at 70 °C with uniform magnetic stirring. The product formed was washed with a 1:4 ethanol-water mixture, and heated in an oven for 8 h at 50 °C. The dried product was calcined and activated at 400 °C for 6 h.

2.2.2. Preparation of Alg/FP

The alginate polymer (2 g) was thoroughly dispersed in 50 mL distilled water through constant stirring to make a viscous solution. After that, 5 pieces (2×6 cm^2) of Whatman filter paper (FP) were put in the alginate solution for 6 h to coat the FP and then dried at room temperature.

2.2.3. Preparation of Alg-CoNi1/FP and Alg-CoNi2/FP

Both Alg-CoNi1/FP and Alg-CoNi2/FP were synthesized in three steps.

In the first step, 2 g alginate polymer was dissolved in 50 mL distilled water in two separate beakers to make their viscous solutions. In the second step, 1 and 2 weight% of CoNi composite was dispersed in each separate beaker and homogenized through continuous stirring, respectively. In the third step, the FP was placed in the above mixture for 6 h and then dried at room temperature.

2.2.4. Preparation of Alg-CoNi1/FP@Co and Alg-CoNi2/FP@Co NPs

The Alg-CoNi1/FP and Alg-CoNi2/FP substrate were placed in 0.5 mM Co (II) salt solution for 6 h and then rinsed with distilled water and dried. The dried Alg-CoNi1/FP and Alg-CoNi2/FP adsorbed Co (II) ions were dipped in 0.1 mM $NaBH_4$ fresh solution for 5 min, which converted Co ions to zero-valent Co NPs (Co^0). The as-synthesized Co NPs were directly used for the antibacterial studies and degradation of azo dyes or 4NP reduction as shown in Scheme 1.

The following equation indicated the conversion of Co (II) ions to their Co^0 NPs [33].

$$Co^{+2} + BH_4^- + 3H_2O + 2e^- \rightarrow Co^0 + H_3BO_3 + 7H^+$$

$$3Co^0 + 6H^+ \rightarrow 3Co^{+2} + 3H_2$$

Both Co^{+2} and Co^0 were supported on Alg/FP, Alg-CoNi1/FP, and Alg-CoNi2/FP solid substrate.

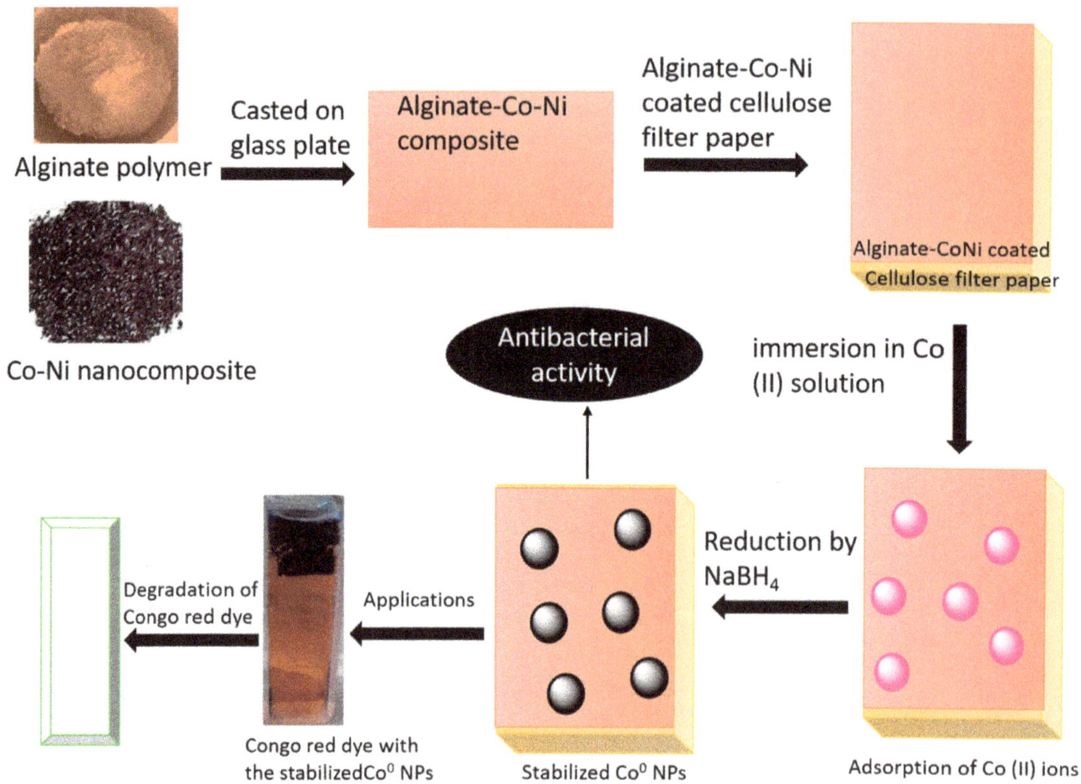

Scheme 1. Pictorial representation of the preparation of Alg-CoNi/FP@Co NPs and their applications against bacterial inhibition and degradation of dyes.

2.3. Antibacterial Activity

The antibacterial potential of Alg/FP, Alg/FP@Co, Alg-CoNi1/FP@Co, and Alg-CoNi2/FP@Co NPs was evaluated on a muller Hinton agar plates against *B. subtilis*. The antibacterial method was based on the Kirby–Bauer disk diffusion method with minor changes in the procedure [34]. The plates were prepared, sterilized, solidified, and then the culture of *B. subtilis* was spread thoroughly via a sterilized spreader on the whole plate. After that, each catalyst was cut in a specified dimension and then placed in the bacterial zone. The plates were placed at 37 °C for 24 h. After incubation, the zone of inhibition was measured. The experiments were performed in triplicate and the performance of the catalysts was determined by calculating the mean zone of inhibition around the nutrient agar disk.

2.4. Pollutants Degradation Experiment

All three catalysts were applied for the degradation of azo dyes and reduction of 4NP. The concentration of both dyes was 0.07 mM. In the batch experiment, 3 mL of each dye solution was put in a quartz cuvette with the addition of 0.5 mL of $NaBH_4$ solution and 25 mg of each catalyst. After that, the reaction was constantly monitored through UV-Vis spectrophotometer with a 1 min interval time. The decrease in absorbance at 495 and 464 nm for CR and MO dyes were recorded, respectively, until complete discoloration. The effect of initial dye concentration, $NaBH_4$ amount, and catalyst dosage was evaluated on the degradation of CR dye.

The degradation of dyes or nitrophenol in percent can be deduced from the following Equation (1).

$$\% \text{ reduction/degradation} = \frac{(C_0 - C_t)}{C_0} \times 100 \quad (1)$$

where C_0 and C_t represent the initial and final concentrations of the solution after passing time t.

While the rate constant k_{app} was deduced from the linear relationship of $\ln C_t/C_0$ vs. t as shown in Equation (2).

$$\ln \frac{C_t}{C_0} = -kt + C \quad (2)$$

2.5. Physiochemical Characterization

The crystalline nature of the catalyst was performed on X-ray diffraction technique (HighTech) with Cu Kα radiations source (λ = 0.154 nm) having 25 mA current and 40 kV voltage, while the scan range was recorded from 15–80 nm with step time 1 s and step size 0.05 degrees. FTIR was recorded on Autonuated total reflectance-Fourier transformed infrared spectroscopy machine of Thermo scientific Company from 400–4000 cm^{-1} range and FESEM on JEOL (JSM-7600F, Tokyo, Japan) and EDS (EDS oxford system, Oxford, UK). UV-Vis spectrophotometer of Thermo Scientific Evolution Company was used for the catalytic degradation experiment.

3. Results and Discussion

3.1. Characterizations

3.1.1. FESEM

The FESEM images of Alg/FP indicated the presence of alginate polymer on the surface of cellulose microfibers. Fibers are covered by a thin layer of the alginate polymer as shown in the Figure 1a. The Alg/FP@Co NPs indicated small particles on the surface of the alginate layer as well as cellulose microfibers. These small spherical particles indicated the presence of Co NPs (Figure 1c). The Alg-CoNi1/FP@Co NPs also showed a similar display of the alginate polymer covering the cellulose microfiber, where the CoNi are protruded from the alginate layer. Furthermore, Co NPs are stabilized on Alg-CoNi1/FP layer (Figure 1e). The same explanation is true for Alg-CoNi2/FP@Co as discussed for the Alg-CoNi1/FP@Co NPs (Figure 1g).

3.1.2. EDS

The EDS spectrum of all the catalysts is indicated in the inset of Figure 1. On the left side, the FESEM images of the catalyst are presented from which the right side elemental window are derived. For instance, Figure 1b indicated FESEM images and the EDS spectrum of Alg/FP, which indicated C and O as the main element along with Na, Ca, and Si. The C and O elements are 38.59 and 49.59 by mass%. Similarly, Alg/FP@Co NPs indicated C, O, and Co elements in 30.95, 56.74, and 4.61 by mass% (Figure 1d). The Alg-CoNi1/FP@Co NPs showing C, O, Ni, and Co elements in 31.61, 57.49, 0.13, and 2.77 by mass%, respectively (Figure 1f). Similarly, Alg-CoNi1/FP@Co NPs C, O, Ni, and Co elements in 24.18, 56.49, 0.70, and 6.35 by mass% respectively (Figure 1h). In all the elemental windows, Na has appeared in different ratios because the alginate polymer is used in the form of their sodium salt.

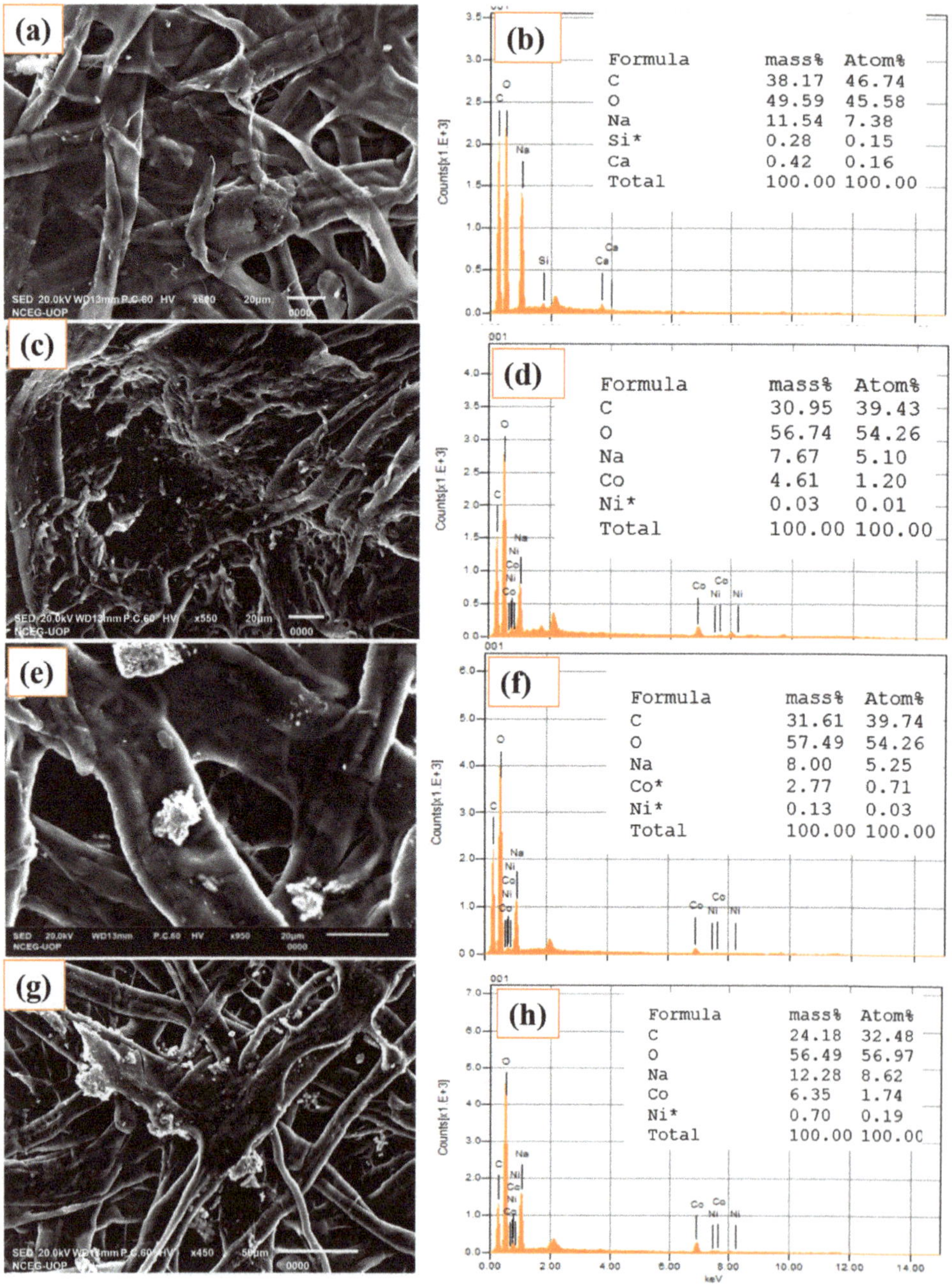

Figure 1. The left-hand side are the FESEM images that give the right-hand side elemental window with their mass and atomic percentage. The FESEM and EDS images of Alg/FP (**a**,**b**), Alg/FP@Co NPs (**c**,**d**), Alg-CoNi1/FP@Co (**e**,**f**), and Alg-CoNi2/FP@Co NPs (**g**,**h**), respectively. The star indicated in the EDS window are the Ni and Co. The EDS vertical axes indicate the scientific notation as 1 x 10^3.

3.1.3. FTIR

The FTIR spectra revealed many peaks below 1600 cm^{-1} due to the presence of various functionalities of the polymers backbone. Similarly, there are two peaks above 1600 cm^{-1}, which are displayed at 2897 and 3308 cm^{-1} due to the presence of C-H and O-H asymmetric stretching vibrations (Figure 2a). Furthermore, the broadness of peaks at 3308 cm^{-1} in all catalysts is due to the presence of multiple intra or inter-molecular H-binding in alginate and cellulose microfibers [21]. The absorbance at 1605 cm^{-1} suggests the adsorbed water on the polymer surface, while the acetate group exhibited peak at 1489 cm^{-1}. The acetate groups are present in both cellulose microfibers and alginate polymer host material. Moreover, the characteristic polysaccharides peaks appeared in the range of 1422–555 cm^{-1}. For instance, peaks at 1422, 1379, 1160, and 1096 cm^{-1} indicated the presence of $-CH_2$, $-CH_3$, and –OH and C-O bending vibrations [25]. Besides, the peak in the range of 1420–1430 cm^{-1} suggesting the crystalline nature of polysaccharide, while the amorphous amount appeared in the range of 842 cm^{-1} in the polymer backbone [35]. The absorbance peak at 555 cm^{-1} is the characteristic of the metal–oxygen bond [36], which is prominent in the Alg-CoNi2/FP@Co NPs. The fingerprint region also appeared in the range of 800–500 cm^{-1}, therefore, this peak also exhibited in another catalyst. The above discussion inferred that the synthesized catalysts have a polymer backbone with little difference in the absorbance peaks. The small difference in the absorbance peaks among the catalyst peaks and from the literature data is due to the interconnectivity of alginate polymer with CoNi catalyst and cellulose microfibers.

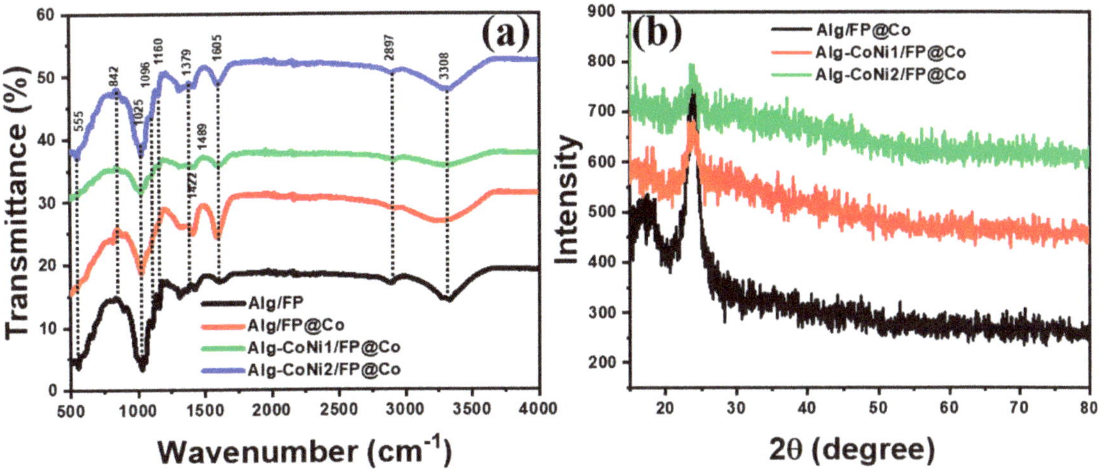

Figure 2. FTIR (**a**) and XRD (**b**) of Alg/FP, Alg-CoNi1/FP@Co, and Alg-CoNi2/FP@Co NPs.

3.1.4. XRD

XRD spectra indicated sharp crystalline peaks for cellulose microfibers at 2θ = 17.0° and 23.1°, which correspond to the crystal-I plane of cellulose microfibers at (110) and (200) planes, respectively (Figure 2b). The Co NPs did not appear in all the catalysts, suggesting their poor crystalline growth during the process at room temperature. Similar observations were also reported in the literature during the synthesis of Co and Cu NPs, respectively [37,38]. Therefore, we suggest that bigger particles of Co NPs formed due to aggregation.

3.2. *Antibacterial Characteristics*

Microbial contamination is a major issue in various technological fields such as the food industry, personal care products, water industry, medical devices, hospitals appliances, babies' toys, hospital appliances, surgical apparatus, food and beverages packaging, textiles industry, and many other daily life usages [39–41]. The synthesis and applications of antimicrobial materials are exponentially increasing and researchers are trying to find new materials that can inhibit or kill the microbes. Recently, the advent of nanoscale materials finds its numerous applications against bacterial and microbial killing or inhibition. Therefore, these materials are largely studied at both industrial and academic levels because they provided substantial importance to other materials. Once the material is proved as an antimicrobial agent its use and application became extended to other technological sectors. In this study, we have screened the Alg/FP, Alg/FP@Co, Alg-CoNi1/FP@Co, and Alg-CoNi2/FP@Co catalysts against the inhibition of *B. subtilis* Gram-positive bacterium. The highest zone of inhibition of 2.5 cm was achieved with Alg-CoNi2/FP@Co NPs and lowest with Alg/FP, which suggested the role of CoNi NPs in Alginate-cellulose filter paper network. Ting Tsai et al. studied the antibacterial activity of cellulose paper fabricated with Ag-coated Au NPs indicating strong antibacterial activity against *E. coli* JM109. The authors synthesized Ag-coated Au NPs in different sizes, where the particles with 15 nm showed an excellent antibacterial activity against *E. coli* [42]. Similarly, bacterial cellulose fabricated with Ag NPs showed good antibacterial potential against *E. coli*. Thus, it is suggested that cellulose materials can be modified with various inorganic filler to make it an efficient catalyst against bacterial inhibition [43]. The zone of inhibition *B. subtilis* by all catalysts is provided in Table 1.

Table 1. Zone of inhibition of *B. subtilis* by all four catalysts.

Catalyst	Zone of Inhibition (cm)
Alg/FP	0.3
Alg/FP@Co NPs	1.0
Alg-CoNi1/FP@Co NPs	1.7
Alg-CoNi2/FP@Co NPs	2.5

3.3. *Catalyst activity*

3.3.1. Discoloration of CR Dye

The as-synthesized Co NPs on the Alg/FP and Alg-CoNi1/FP@Co NPs and Alg-CoNi2/FP@Co NPs were used as solid matrix and were applied against the decolorization of CR dye. CR dye appeared at 495 nm in the UV-Vis. absorbance spectrum. This redshift is due to the presence of the diazo group, which upon treatment with $NaBH_4$ transformed to hydrazine products [44]. A literature survey discovered that $NaBH_4$ can decolorize CR dye, but the process is too slow and has no economic importance. The process is thermodynamically important, but not kinetically [44,45]. However, metal nanoparticles have proved to have an important effect on the degradation of CR dye. Alg/FP@Co, Alg-CoNi1/FP@Co, and Alg-CoNi2/FP@Co NPs were used against the degradation of CR dye in the presence of $NaBH_4$ as a reducing agent. As discussed earlier, the $NaBH_4$ has a negligible effect on the degradation of CR dye; however, it is thus required for the CR degradation along with the Co NPs to overcome the activation barrier of the reaction. Briefly, after the addition of the catalyst, there was a decline in the absorbance of CR dye at λ_{max} 495 nm, and some other peaks were arising at 344, 284, and 244 nm during the discoloration of CR dye. The appearance of new peaks predicted the formation of byproducts. These new peaks were due to the –COOH and $–NH_2$ groups, respectively [44]. As manifested in the inset of Figure 3a, the Alg/FP supported Co NPs degraded the CR dye in 20 min with k_{app} value 1.27×10^{-1} min^{-1} with a regression coefficient R^2 0.9382. The degradation percent of CR was 91.76% per 20 min (Table 2). Similarly, under the same experimental conditions, 91.65 and 90.41% of the CR dye solution was decolorized in 19 and 9 min by Alg-CoNi1/FP@Co and Alg-CoNi2/FP@Co NPs, respectively. The rate constant values of CR discoloration with Alg-CoNi1/FP@Co and Alg-CoNi2/FP@Co NPs were 1.46×10^{-1}, 3.63×10^{-1} min^{-1}, respectively. Based on the k_{app} values, the superior catalyst activity was displayed by Alg-CoNi2/FP@Co NPs compared to Alg/FP@Co NPs and Alg-CoNi1/FP@Co NPs. The UV-Vis. absorbance spectrum of CR dye catalyzed by Alg-CoNi1/FP@Co and Alg-CoNi2/FP@Co NPs manifested in Figure 3b,c, respectively, while the linear relationship based on $\ln C_t/C_0$ vs. time is depicted in Figure 3d. The slow rate of Alg/FP supported Co NPs against the degradation of CR dye was due to an induction period (t_0), as shown in Figure 3d. This period is characterized by the rearrangement of reactive sites of the catalyst such as faces and edges. Once these reactive sites are prepared for the chemical reactions, the rate of reaction is enhanced. A high t_0 value was observed for Alg/FP@Co NPs as compared to Alg/FP-Co-Ni1@Co NPs and Alg/FP-Co-Ni2@Co NPs, which further support the role of CoNi nanocatalyst in the Alg/FP polymer networks. It is further proposed that during the CR degradation, $NaBH_4$ and CR dye get adsorbed on the surface of Co NPs supported on Alg/FP or Alg-CoNi/FP. After the adsorption of $NaBH_4$ and CR dye, the Co NPs transferred the electrons provided by $NaBH_4$ in the form of H^{-1} ions to the dyes. Thus, Co NPs supported on the solid matrix provided a vast surface for the CR dye, $NaBH_4$, and electrons, where they play with each other and degraded the CR dye [46].

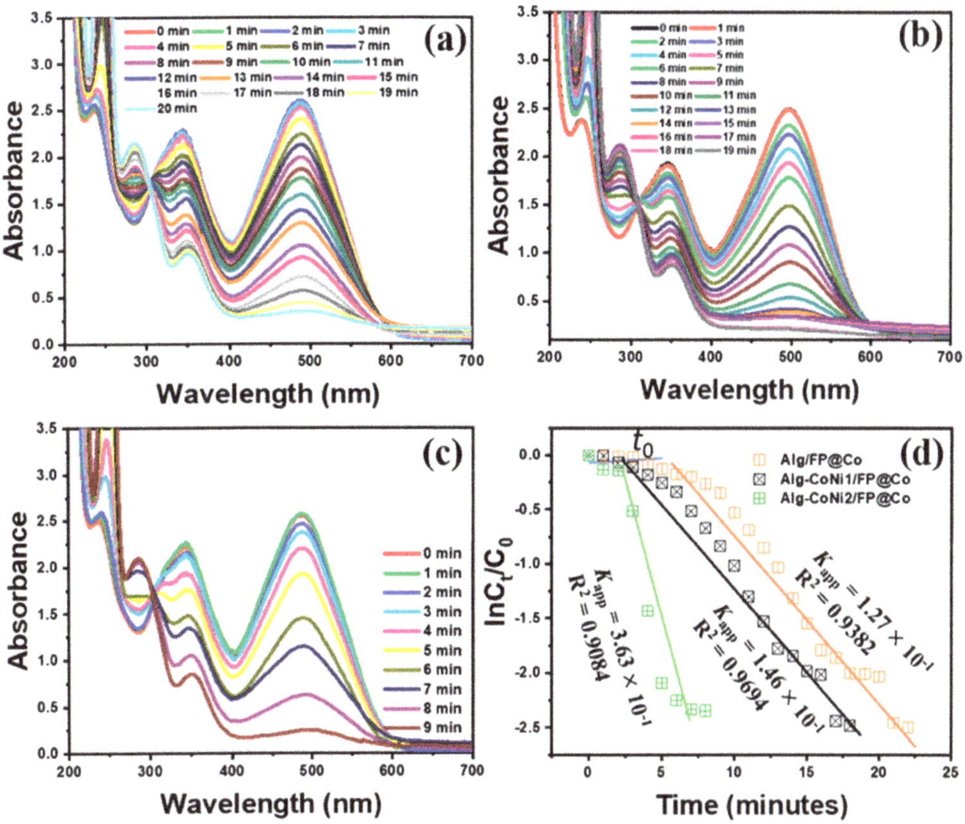

Figure 3. UV-Vis. absorbance spectra of CR dye discoloration in the presence of $NaBH_4$ using Alg/FP@Co (**a**), Alg-CoNi1/FP@Co (**b**), Alg-CoNi2/FP@Co NPs (**c**), and linear relationship of pseudo-1st order kinetics (**d**).

Table 2. Kinetics parameters based on pseudo-1st order and % degradation of CR, MO dyes, and 4NP.

Targeted Pollutants	Catalyst	k_{app} (min^{-1})	R^2	% Degradation
CR	Alg/FP@Co	1.27×10^{-1}	0.9382	91.76
	Alg-CoNi1/FP@Co	1.46×10^{-1}	0.9694	91.65
	Alg-CoNi2/FP@Co	3.63×10^{-1}	0.9084	90.41
MO	Alg/FP@Co	2.47×10^{-1}	0.9263	92.02
	Alg-CoNi1/FP@Co	3.10×10^{-1}	0.9371	91.55
	Alg-CoNi2/FP@Co	4.68×10^{-1}	0.9376	94.22
4NP	Alg/FP@Co	1.71×10^{-1}	0.9855	92.50
	Alg-CoNi1/FP@Co	2.56×10^{-1}	0.9636	91.06
	Alg-CoNi2/FP@Co	5.55×10^{-1}	0.9740	94.69

After concluding the high catalyst activity of Alg-CoNi/2FP@Co NPs, various factors such as effect of initial dye concentration, the effect of $NaBH_4$ and catalyst amount were studied on the degradation of CR dye by using Alg-CoNi/2FP@Co NPs in the presence of $NaBH_4$.

Effect of Concentration on CR Dye Degradation

Concentration has a major role in the degradation of pollutants because the reaction occurs on the surface of catalyst. It is well-known that increasing the amount of concentration will decrease the rate of reaction because high amount of pollutant molecules is available for the same amount of catalyst. Various concentrations of CR dye such as 0.03, 0.05, and 0.09 mM were studied by using 0.5 mL of $NaBH_4$ and 30 mg of the Alg-CoNi/2FP@Co NPs (Table 3).

Table 3. Effect of concentration, catalyst dosage, and reducing agent on the discoloration of CR dye using Alg-CoNi2/FP@Co NPs as a *dip*-catalyst.

Reaction Condition	Various Reaction Effect	Reaction Parameters	k_{app} (min^{-1})	R^2	% Degradation
0.5 mL $NaBH_4$ + 30 mg catalyst	Concentration	0.03 mM	3.63×10^{-1}	0.9568	86.8
		0.05 mM	2.57×10^{-1}	0.9139	91.65
		0.09 mM	3.63×10^{-1}	0.9084	90.41
3 mL of 0.09 mM CR dye solution + 30 mg catalyst	$NaBH_4$ (1 mM)	0.5 mL	3.63×10^{-1}	0.9084	90.42
		1 mL	3.68×10^{-1}	0.9076	89.91
		2 mL	6.98×10^{-1}	0.9508	94.85
3 mL of 0.09 mM CR dye solution + 0.5 mL of 1 mM $NaBH_4$ solution	Catalyst dosage	30 mg	3.68×10^{-1}	0.9084	90.42
		60 mg	4.28×10^{-1}	0.8897	90.42
		90 mg	5.97×10^{-1}	0.9074	93.82

Effect of $NaBH_4$ on the Discoloration of CR Dye

The effect of $NaBH_4$ was studied by changing the volume (0.5, 1, and 2 mL) of 1 mM $NaBH_4$ solution while keeping the same concentration and volume of CR dye (0.09 mM in 3.5 mL solution), and 30 mg of the Alg-CoNi/2FP@Co NPs. Increasing the volume of $NaBH_4$ solution from 0.5 to 2 mL, the rate of reaction also enhanced from 3.63×10^{-1}, 3.68×10^{-1}, and 6.98×10^{-1} min^{-1}, respectively. This indicated that the rate of reaction can be enhanced with increase in volume of $NaBH_4$, which suggests the prominent role of $NaBH_4$ in the degradation of CR dye (Figure 4a and Table 3). According to the Hinshelwood-Langmuir mechanism, both BH_4^- and reactant adsorbed on the surface of the NPs. Note that the adsorption of BH_4^- and reactant is a reversible process, which makes competition among both BH_4^- and reactant for the active sites of the NPs. Therefore, a high concentration of reactant slows down the reaction rate. As is obvious from Table 3, the slow rate was observed with high CR concentration. Similarly, a high volume of $NaBH_4$ increases the rate of reaction. For instance, a k_{app} 6.98 × 10^{-1} min^{-1} is observed when the amount of $NaBH_4$ is high. Both CR and BH_4^- accommodate on the surface of NPs in a reversible manner where the BH_4^- provided surface hydrogen to the CR and convert it to the hydrazine derivatives. These hydrazine derivatives further degraded the product. After the reaction completion, the product detached from the NPs surface and made it free for the next cycle.

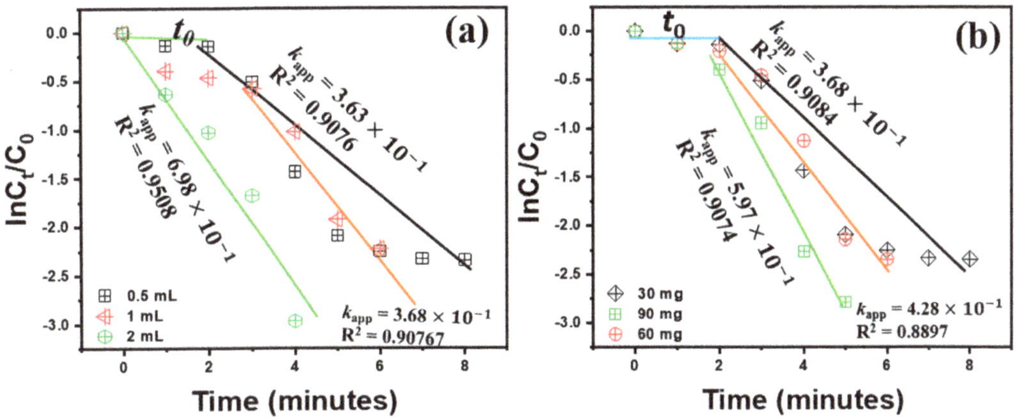

Figure 4. Pseudo first order kinetics plots of CR dye discoloration under the influence of different amount of $NaBH_4$ (30 mg catalyst, 0.09 mM CR dye) (**a**) and catalyst amount (0.05 mL of 1 mM $NaBH_4$ solution, 0.09 mM CR dye) (**b**).

Effect of Catalyst Dosage on the Discoloration of CR Dye

Optimizing the catalyst amount for the discoloration of reactant is highly desirable in the field of nanocatalysis. Increasing amount of the catalyst will increase the active sites of the catalyst for the molecules. Therefore, the reactant and BH_4^- accommodate easily on the surface of the catalyst. For instance, 30, 60, and 90 mg of the Alg-CoNi2/FP@Co NPs was used against CR dye degradation under the same experimental conditions. At 90 mg, the rate of reaction was k_{app} 5.97 × 10^{-1} min^{-1} compared to 3.68 × 10^{-1} min^{-1} with 30 mg of the catalyst. This suggested that at high amount of the catalyst, the rate of reaction is also high owing to the availability of more active sites for reaction (Table 3 and Figure 4b).

3.3.2. Discoloration of MO Dye

Similar experiments were also conducted for MO dye discoloration for evaluating all the catalysts activity as discussed for CR dye degradation. As clear from the absorbance spectrum of Figure 5a, the $NaBH_4$ has a negligible effect on the degradation of MO dye. Therefore, the catalyst was introduced along with $NaBH_4$ for the MO dye degradation. Initially, we used Alg/FP@Co NPs (Figure 5b), where the MO dye decolorized in 12 min. After that, the effect of CoNi composite was studied on Alg/FP. The catalyst Alg-CoNi1/FP@Co NPs (Figure 5c) has a significant effect where it took 9 min for MO discoloration. Interestingly, an increased amount of CoNi composite in Alg/FP increased the rate of reaction. For instance, Alg-CoNi2/FP@Co NPs decolorized MO dye in 7 min (Figure 5d). During the degradation of MO dye with all catalysts, it was observed that MO dye appeared at λ_{max} 464 nm, which can react with $NaBH_4$ to make their hydrazine product. The hydrazine product is oxidized after the addition of the respective catalyst and a new peak ascends at λ_{max} 251 nm, which is due to the formation of an amine functional group. The linear relationship lnC_t/C_0 vs. t (Figure 5e) indicates the superior catalyst activity of Alg-CoNi2/FP@Co NPs with k_{app} 4.68×10^{-1} min^{-1}. An induction period t_0 was observed in the degradation of MO dye. This time mainly encounters in catalysis reactions where the surface atoms organize themselves and expose their active sites such as faces, edges, and planes. Therefore, after this period the active sites become ready and the catalyst react very fast. Thus, induction period is very important for such a catalyst system. The percent degradation of MO dye with the respective catalyst indicated that approximately 7% of MO dye was degraded in 12 min by $NaBH_4$ alone, while in the same time Alg/FP@Co NPs degraded 92% dye in the presence of $NaBH_4$. Similarly, 91 and 94% efficiency was achieved with Alg-CoNi1/FP@Co and Alg-CoNi2/FP@Co NPs in 9 and 7 min, respectively (Figure 6f). This indicated that $NaBH_4$ alone cannot change the reaction, while the catalyst brought an actual change in the reactant nature. The present work for CR, MO, and 4NP was compared with the literature data as shown in Table 4.

Table 4. Comparative literature data for CR, MO, and 4NP degradation with the present data.

Catalyst	Targeted Pollutant	Concentration (mg/L)	Amount of Catalyst (g/L)	% Discoloration	References
CA-PS-ZnO	CR	50.00	—	95.0	[47]
Ba/Alg/CMC/TiO_2		30.00	1.20	95.0	[48]
Ba/Alg/CMC		30.00	1.20	58.0	[48]
Alg-CoNi2/FP@Co		48.72	8.44	90.4	Present work
cellulose/TiO_2 monolith	MO	20.00	0.19	—	[49]
PtNPs@KWP		20.00	—	99.0	[50]
Alg-CoNi2/FP@Co		22.89	8.44	94.2	Present work
CelA_C2–F-5%	4NP	0.10	10.0	85.7	[51]
Alg-CoNi2/FP@Co		21.00	8.44	94.6	Present work

3.3.3. Reduction of 4NP

4NP is considered a benchmark reaction for the evaluation of such a catalyst system. Therefore, all the catalyst was applied for the reduction of 4NP under a similar experimental procedure discussed above for CR and MO dyes degradation. 4NP is a toxic aromatic compound and strict restriction has been made on their use beyond the permissible limit, which was clariefied by the United State Enviornmental Protection Agency (U.S. EPA). This restriction is implemented on the use of nitrophenols because of their mutagenic and carcinogenic effect, as well as their adverse effect on lungs, kidneys, and CNS.

It was studied in detail that neither borohydride nor catalyst alone reduced nitrophenol because of the high activation barrier requirements. However, both catalyst and $NaBH_4$ together reduced the 4NP easily. As seen in the Figure 6a, Alg/FP@Co NPs can reduce 4NP to 4-aminophenol (4AmP) in 16 min, while Alg-CoNi1/FP@Co (Figure 6b) and Alg-CoNi2/FP@Co NPs (Figure 6c) took 10 and 6 min respectively, which further support the superior catalyst activity of Alg-CoNi2/FP@Co NPs. As depicted in all the absorbance spectra 4NP (318 nm) converted to 4-nitrophenolateanion having a λ_{max} value of 400 nm. Over time, the absorbance peak at 400 nm and their deep yellow color vanished with the rise of a new peak at 290 nm. This new peak is the indication of 4-AmP product [52,53]. Figure 6d exhibited a relationship of lnC_t/C_0 vs. time, which shows the highest rate constant value of 5.55×10^{-1} min^{-1} for Alg-CoNi2/FP@Co NPs.

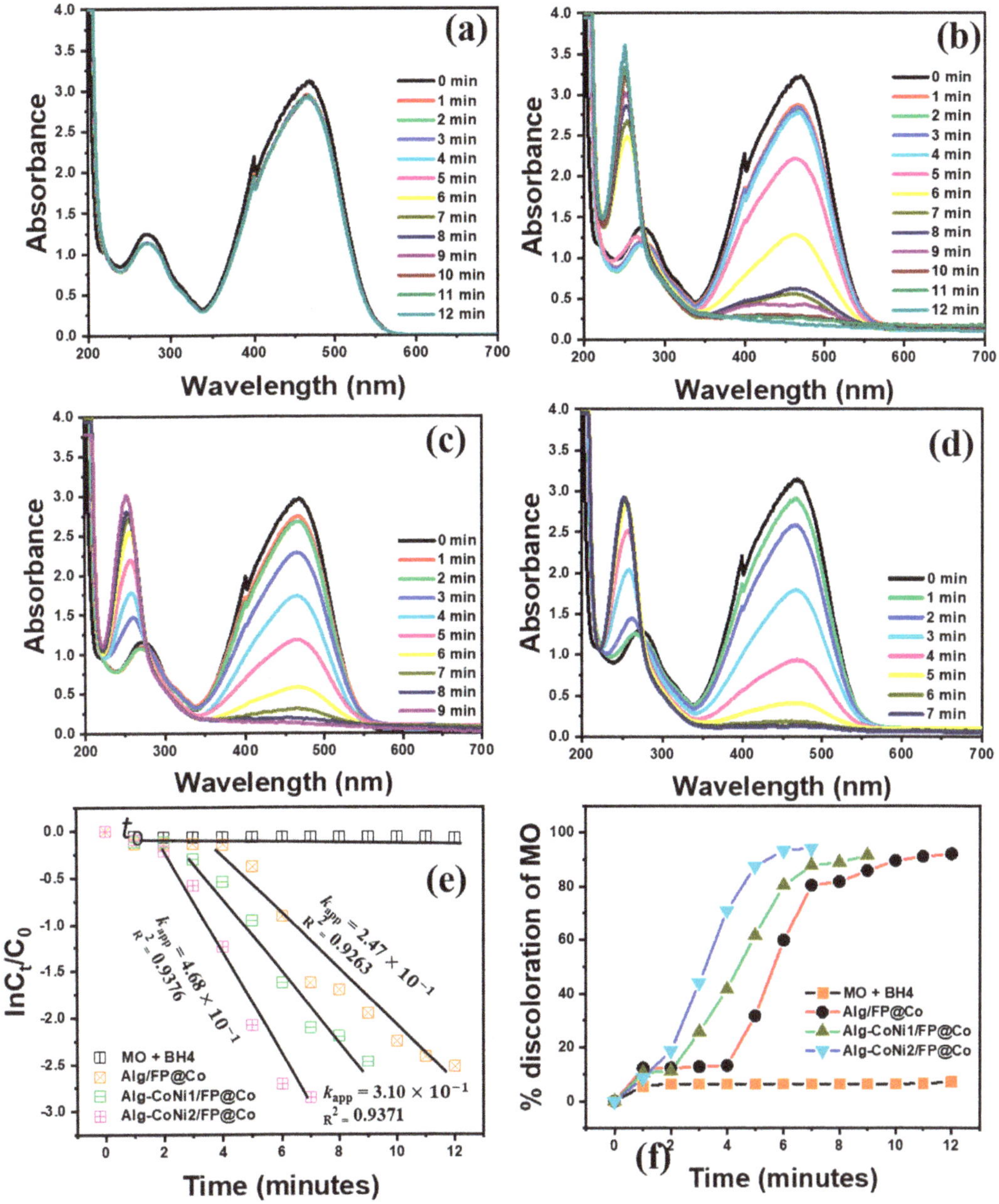

Figure 5. UV-Vis. absorbance spectra of MO dye discoloration in the presence of $NaBH_4$ (**a**), and Alg/FP@Co (**b**), Alg-CoNi1/FP@Co (**c**), and Alg-CoNi2/FP@Co NPs (**d**), linear relationship of pseudo-1st order kinetics (**e**), and percent degradation of the MO dye (**f**).

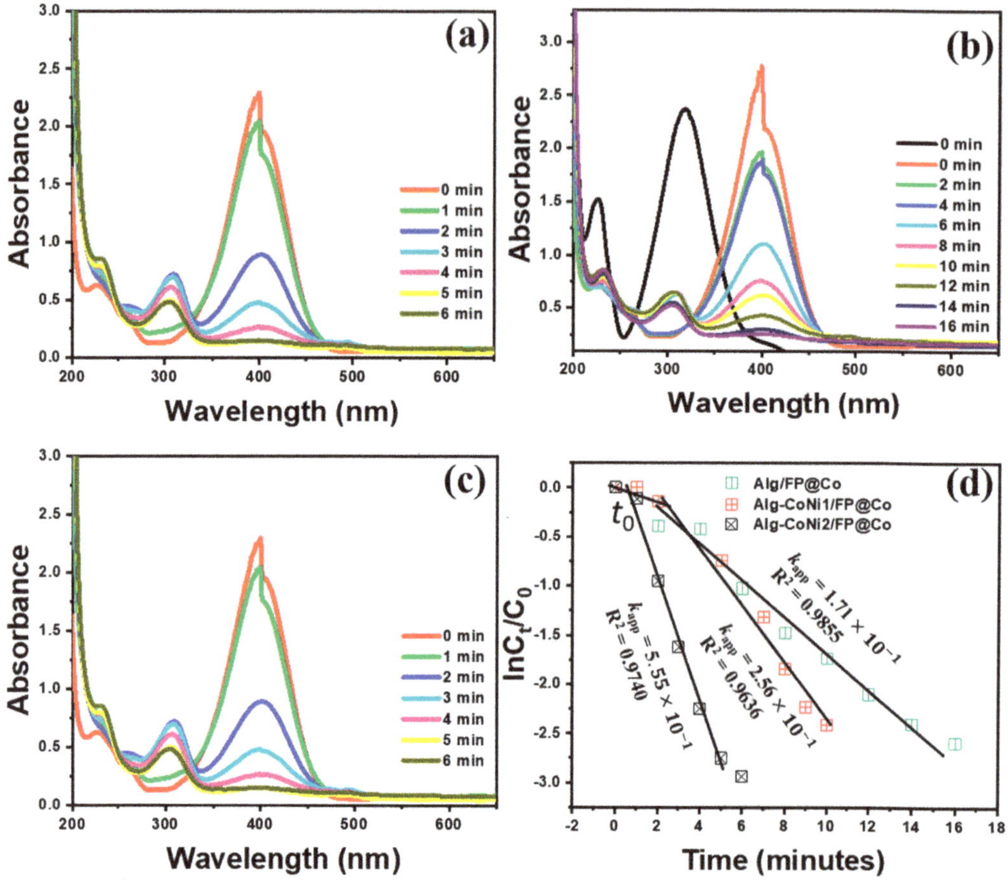

Figure 6. UV-Vis. absorbance spectra of 4NP reduction in the presence of $NaBH_4$ using Alg/FP@Co (**a**), Alg-CoNi1/FP@Co (**b**), and Alg-CoNi2/FP@Co NPs (**c**), and linear relationship of pseudo-1st order kinetics (**d**).

Based on the discoloration/reduction of azo dyes and nitrophenol it is inferred that CoNi composite played an important role in the reaction and facilitated the transfer of electrons for the degradation/reduction process.

4. Conclusions

In the current study, various solid supports comprised of Alg/FP, Alg-CoNi1/FP, and Alg/CoNi2/FP were designed for the stabilization of zero-valent Co NPs. The as-synthesized catalysts were used for the inhibition of *B. subtilis*, and the result revealed that Alg-CoNi2/FP@Co NPs exhibited the highest activity by inhibiting the bacterium zone to 2.5 cm. Furthermore, these catalysts were used as a dip-catalyst for the degradation of CR and MO dyes and reduction of 4NP. It was revealed that increasing the amount of CoNi nanocomposite increased the bacterial inhibition as well as the rate of dyes degradation and reduction of 4NP. These results suggest that CoNi nanocomposite has a major role in the polymer network for the chemical and biological studies. The CoNi nanocomposite probably facilitates the movement of electrons for the degradation of dyes. The rate constant values were deduced from the pseudo-first-order kinetics. The highest rate was displayed by Alg-CoNi2/FP@Co NPs for MO dye, which was 4.68×10^{-1} min^{-1}. Similarly, the k_{app} of Alg-CoNi2/FP@Co NPs against the degradation of CR dye is 3.63×10^{-1} min^{-1} and lowest shown by Alg/FP@Co NPs. Among these different catalysts, Alg-CoNi2/FP@Co NPs displayed the highest activity against both chemical and biological activities.

Author Contributions: Conceptualization, Y.A., W.U.R., and S.A.K.; methodology, W.U.R., and S.A.K.; software, S.A.K.; validation, Y.A., W.U.R., H.S.H.M.A., H.A.H., and S.A.K.; formal analysis, S.A.K.; investigation, S.A.K.; resources, Y.A., H.S.H.M.A., and S.A.K.; writing—original draft preparation, S.A.K.; writing—review and editing, Y.A., W.U.R., H.S.H.M.A., H.A.H., and S.A.K.; visualization, Y.A., W.U.R., H.S.H.M.A., H.A.H., and S.A.K.; supervision, S.A.K.; project administration, Y.A., H.S.H.M.A., and S.A.K.; funding acquisition, Y.A., H.S.H.M.A., and S.A.K.; All authors have read and agreed to the published version of the manuscript.

Funding: The Deputyship for Research & Innovation, Ministry of Education in Saudi Arabia for funding this research work through the project number IFPHI-004-130-2020 and King Abdulaziz University, DSR, Jeddah, Saudi Arabia.

Institutional Review Board Statement: No Human or animal experiment has carried out in this work.

Informed Consent Statement: No human studies has carried out in this project.

Data Availability Statement: N/A.

Acknowledgments: The authors extend their appreciation to the Deputyship for Research & Innovation, Ministry of Education in Saudi Arabia for funding this research work through the project number IFPHI-004-130-2020 and King Abdulaziz University, DSR, Jeddah, Saudi Arabia.

Conflicts of Interest: The authors declare no conflict of interest.

References

1. Sapsford, K.E.; Algar, W.R.; Berti, L.; Gemmill, K.B.; Casey, B.J.; Oh, E.; Stewart, M.H.; Medintz, I.L. Functionalizing nanoparticles with biological molecules: Developing chemistries that facilitate nanotechnology. *Chem. Rev.* **2013**, *113*, 1904–2074. [CrossRef] [PubMed]
2. Pezzato, C.; Maiti, S.; Chen, J.-Y.; Cazzolaro, A.; Gobbo, C.; Prins, L. Monolayer protected gold nanoparticles with metal-ion binding sites: Functional systems for chemosensing applications. *Chem. Commun.* **2015**, *51*, 9922–9931. [CrossRef] [PubMed]
3. Nie, Z.; Petukhova, A.; Kumacheva, E. Properties and emerging applications of self-assembled structures made from inorganic nanoparticles. *Nat. Nanotechnol.* **2010**, *5*, 15–25. [CrossRef]
4. Moritz, M.; Geszke-Moritz, M. The newest achievements in synthesis, immobilization and practical applications of antibacterial nanoparticles. *Chem. Eng. J.* **2013**, *228*, 596–613. [CrossRef]
5. Khan, S.A.; Ismail, M.; Anwar, Y.; Farooq, A.; Al Johny, B.O.; Akhtar, K.; Shah, Z.A.; Nadeem, M.; Raza, M.A.; Asiri, A.M. A highly efficient and multifunctional biomass supporting Ag, Ni, and Cu nanoparticles through wetness impregnation for environmental remediation. *Green Process. Synth.* **2019**, *8*, 309–319. [CrossRef]
6. Khan, S.A.; Khan, N.; Irum, U.; Farooq, A.; Asiri, A.M.; Bakhsh, E.M.; Khan, S.B. Cellulose acetate-Ce/Zr@ Cu0 catalyst for the degradation of organic pollutant. *Int. J. Biol. Macromol.* **2020**, *153*, 806–816. [CrossRef] [PubMed]
7. Figen, A.K. Improved catalytic performance of metal oxide catalysts fabricated with electrospinning in ammonia borane methanolysis for hydrogen production. *Int. J. Hydrog. Energy* **2019**, *44*, 28451–28462. [CrossRef]
8. Khan, S.A.; Khan, S.B.; Asiri, A.M.; Ahmad, I. Zirconia-based catalyst for the one-pot synthesis of coumarin through Pechmann reaction. *Nanoscale Res. Lett.* **2016**, *11*, 345. [CrossRef]
9. Sohni, S.; Khan, S.A.; Akhtar, K.; Khan, S.B.; Asiri, A.M.; Hashim, R.; Omar, A.M. Room temperature preparation of lignocellulosic biomass supported heterostructure (Cu+ Co@ OPF) as highly efficient multifunctional nanocatalyst using wetness co-impregnation. *Colloids Surf. A Physicochem. Eng. Asp.* **2018**, *549*, 184–195. [CrossRef]
10. Maslamani, N.; Khan, S.B.; Danish, E.Y.; Bakhsh, E.M.; Zakeeruddin, S.M.; Asiri, A.M. Carboxymethyl cellulose nanocomposite beads as super-efficient catalyst for the reduction of organic and inorganic pollutants. *Int. J. Biol. Macromol.* **2021**, *167*, 101–116. [CrossRef]
11. Khan, S.A.; Bakhsh, E.M.; Asiri, A.M.; Khan, S.B. Chitosan coated NiAl layered double hydroxide microsphere templated zero-valent metal NPs for environmental remediation. *J. Clean. Prod.* **2021**, *285*, 124830. [CrossRef]
12. Khan, S.A.; Baksh, E.M.; Akhtar, K.; Khan, S.B. A template of cellulose acetate polymer-ZnAl layered double hydroxide composite fabricated with Ni NPs: Applications in the hydrogenation of nitrophenols and dyes degradation. *Spectrochim. Acta Part A Mol. Biomol. Spectrosc.* **2020**, *241*, 118671. [CrossRef]
13. Akhtar, K.; Khan, S.A.; Khan, S.B.; Asiri, A.M. Nanomaterials and Environmental Remediation: A Fundamental Overview. *Nanomater. Environ. Appl. Fascin. Attrib.* **2018**, *2*, 1–36.

14. Henry, F.; Marchal, P.; Bouillard, J.; Vignes, A.; Dufaud, O.; Perrin, L. The effect of agglomeration on the emission of particles from nanopowders flow. In Proceedings of the 14th International Symposium on Loss Prevention and Safety Promotion in the Process Industry, Florence, Italy, 12–15 May 2013; pp. 811–816.
15. Subramanian, V.; Ordomsky, V.V.; Legras, B.; Cheng, K.; Cordier, C.; Chernavskii, P.A.; Khodakov, A.Y. Design of iron catalysts supported on carbon–silica composites with enhanced catalytic performance in high-temperature Fischer–Tropsch synthesis. *Catal. Sci. Technol.* **2016**, *6*, 4953–4961. [CrossRef]
16. Zhao, Y.; Yang, K.R.; Wang, Z.; Yan, X.; Cao, S.; Ye, Y.; Dong, Q.; Zhang, X.; Thorne, J.E.; Jin, L. Stable iridium dinuclear heterogeneous catalysts supported on metal-oxide substrate for solar water oxidation. *Proc. Natl. Acad. Sci. USA* **2018**, *115*, 2902–2907. [CrossRef] [PubMed]
17. Guo, X.; Hao, C.; Jin, G.; Zhu, H.Y.; Guo, X.Y. Copper nanoparticles on graphene support: An efficient photocatalyst for coupling of nitroaromatics in visible light. *Angew. Chem. Int. Ed.* **2014**, *53*, 1973–1977. [CrossRef] [PubMed]
18. Chun, Y.S.; Shin, J.Y.; Song, C.E.; Lee, S.-g. Palladium nanoparticles supported onto ionic carbon nanotubes as robust recyclable catalysts in an ionic liquid. *Chem. Commun.* **2008**, *8*, 942–944. [CrossRef] [PubMed]
19. Chen, C.W.; Chen, M.Q.; Serizawa, T.; Akashi, M. In-situ formation of silver nanoparticles on poly (N-isopropylacrylamide)-coated polystyrene microspheres. *Adv. Mater.* **1998**, *10*, 1122–1126. [CrossRef]
20. Kamal, T.; Khan, S.B.; Asiri, A.M. Synthesis of zero-valent Cu nanoparticles in the chitosan coating layer on cellulose microfibers: Evaluation of azo dyes catalytic reduction. *Cellulose* **2016**, *23*, 1911–1923. [CrossRef]
21. Khan, S.A.; Khan, S.B.; Farooq, A.; Asiri, A.M. A facile synthesis of CuAg nanoparticles on highly porous ZnO/carbon black-cellulose acetate sheets for nitroarene and azo dyes reduction/degradation. *Int. J. Biol. Macromol.* **2019**, *130*, 288–299. [CrossRef]
22. Saha, S.; Pal, A.; Kundu, S.; Basu, S.; Pal, T. Photochemical green synthesis of calcium-alginate-stabilized Ag and Au nanoparticles and their catalytic application to 4-nitrophenol reduction. *Langmuir* **2010**, *26*, 2885–2893. [CrossRef] [PubMed]
23. Mohammadi, P.; Heravi, M.; Daraie, M. Ag nanoparticles immobilized on new magnetic alginate halloysite as a recoverable catalyst for reduction of nitroaromatics in aqueous media. *Sci. Rep.* **2021**, *11*, 17124. [CrossRef]
24. Khan, S.B.; Ahmad, S.; Kamal, T.; Asiri, A.M.; Bakhsh, E.M. Metal nanoparticles decorated sodium alginate-carbon nitride composite beads as effective catalyst for the reduction of organic pollutants. *Int. J. Biol. Macromol.* **2020**, *164*, 1087–1098. [CrossRef] [PubMed]
25. Al-Ghamdi, Y.O.; Khan, S.A. Stabilization of zero-valent Au nanoparticles on carboxymethyl cellulose layer coated on chitosan-CBV 780 zeolite Y sheets: Assessment in the reduction of 4-nitrophenol and dyes. *Cellulose* **2020**, *27*, 8827–8841. [CrossRef]
26. Uberoi, V.; Bhattacharya, S.K. Toxicity and degradability of nitrophenols in anaerobic systems. *Water Environ. Res.* **1997**, *69*, 146–156. [CrossRef]
27. Brecken-Folse, J.A.; Mayer, F.L.; Pedigo, L.E.; Marking, L.L. Acute toxicity of 4-nitrophenol, 2,4-dinitrophenol, terbufos and trichlorfon to grass shrimp (*Palaemonetes* spp.) and sheepshead minnows (Cyprinodon variegatus) as affected by salinity and temperature. *Environ. Toxicol. Chem. Int. J.* **1994**, *13*, 67–77. [CrossRef]
28. Ahmed, M.S.; Kamal, T.; Khan, S.A.; Anwar, Y.; Saeed, M.T.; Asiri, A.M.; Khan, S.B. Assessment of Anti-bacterial Ni-Al/chitosan Composite Spheres for Adsorption Assisted Photo-Degradation of Organic Pollutants. *Curr. Nanosci.* **2016**, *12*, 569–575. [CrossRef]
29. Khan, S.A.; Khan, S.B.; Kamal, T.; Yasir, M.; Asiri, A.M. Antibacterial nanocomposites based on chitosan/Co-MCM as a selective and efficient adsorbent for organic dyes. *Int. J. Biol. Macromol.* **2016**, *91*, 744–751. [CrossRef]
30. Hussain, S.; Kamran, M.; Khan, S.A.; Shaheen, K.; Shah, Z.; Suo, H.; Khan, Q.; Shah, A.B.; Rehman, W.U.; Al-Ghamdi, Y.O. Adsorption, kinetics and thermodynamics studies of methyl orange dye sequestration through chitosan composites films. *Int. J. Biol. Macromol.* **2021**, *168*, 383–394. [CrossRef]
31. Gičević, A.; Hindija, L.; Karačić, A. Toxicity of azo dyes in pharmaceutical industry. In Proceedings of the International Conference on Medical and Biological Engineering, Banja Luka, Bosnia, 16–18 May 2019; pp. 581–587.
32. Brown, M.A.; De Vito, S.C. Predicting azo dye toxicity. *Crit. Rev. Environ. Sci. Technol.* **1993**, *23*, 249–324. [CrossRef]
33. Ahmad, Z.; Shah, S.A.; Khattak, I.; Ullah, H.; Khan, A.A.; Shah, R.A.; Khan, S.A.; Khan, S.B. Melia Azedarach impregnated Co and Ni zero-valent metal nanoparticles for organic pollutants degradation: Validation of experiments through statistical analysis. *J. Mater. Sci. Mater. Electron.* **2020**, *31*, 16938–16950. [CrossRef]
34. Hudzicki, J. *Kirby-Bauer Disk Diffusion Susceptibility Test Protocol*; American Society for Microbiology: Washington, DC, USA, 2009.
35. Hospodarova, V.; Singovszka, E.; Stevulova, N. Characterization of cellulosic fibers by FTIR spectroscopy for their further implementation to building materials. *Am. J. Anal. Chem.* **2018**, *9*, 303–310. [CrossRef]
36. Khan, S.A.; Khan, S.B.; Khan, L.U.; Farooq, A.; Akhtar, K.; Asiri, A.M. Fourier Transform Infrared Spectroscopy: Fundamentals and Application in Functional Groups and Nanomaterials Characterization. In *Handbook of Materials Characterization*; Springer: Berlin/Heidelberg, Germany, 2018; pp. 317–344.
37. Kamal, T.; Khan, S.B.; Haider, S.; Alghamdi, Y.G.; Asiri, A.M. Thin layer chitosan-coated cellulose filter paper as substrate for immobilization of catalytic cobalt nanoparticles. *Int. J. Biol. Macromol.* **2017**, *104*, 56–62. [CrossRef]
38. Zhu, J.; Hu, G.; Zhang, J. Preparation of Sn-Cu-graphene nanocomposites with superior reversible lithium ion storage. *Mater. Lett.* **2016**, *185*, 565–568. [CrossRef]
39. Fleming, K.; Randle, J. Toys-friend or foe? A study of infection risk in paediatric intensive care unit. *Paediatr. Nurs.* **2006**, *18*, 14–18. [CrossRef] [PubMed]

40. Mokoena, M.P.; Mutanda, T.; Olaniran, A.O. Perspectives on the probiotic potential of lactic acid bacteria from African traditional fermented foods and beverages. *Food Nutr. Res.* **2016**, *60*, 29630. [CrossRef]
41. Bello, B.A.; Khan, S.A.; Khan, J.A.; Syed, F.Q.; Mirza, M.B.; Shah, L.; Khan, S.B. Anticancer, antibacterial and pollutant degradation potential of silver nanoparticles from Hyphaene thebaica. *Biochem. Biophys. Res. Commun.* **2017**, *490*, 889–894. [CrossRef] [PubMed]
42. Tsai, T.-T.; Huang, T.-H.; Chang, C.-J.; Ho, N.Y.-J.; Tseng, Y.-T.; Chen, C.-F. Antibacterial cellulose paper made with silver-coated gold nanoparticles. *Sci. Rep.* **2017**, *7*, 3155. [CrossRef] [PubMed]
43. Pal, S.; Nisi, R.; Stoppa, M.; Licciulli, A. Silver-functionalized bacterial cellulose as antibacterial membrane for wound-healing applications. *ACS Omega* **2017**, *2*, 3632–3639. [CrossRef] [PubMed]
44. Ali, H.S.H.M.; Khan, S.A. Stabilization of Various Zero-Valent Metal Nanoparticles on a Superabsorbent Polymer for the Removal of Dyes, Nitrophenol, and Pathogenic Bacteria. *ACS Omega* **2020**, *5*, 7379–7391. [CrossRef]
45. Ali, H.S.M.; Anwar, Y.; Khan, S.A. Vigna radiata Impregnated Zero-Valent CuAg NPs: Applications in Nitrophenols Reduction, Dyes Discoloration and Antibacterial Activity. *J. Clust. Sci.* **2021**. [CrossRef]
46. Wunder, S.; Polzer, F.; Lu, Y.; Mei, Y.; Ballauff, M. Kinetic analysis of catalytic reduction of 4-nitrophenol by metallic nanoparticles immobilized in spherical polyelectrolyte brushes. *J. Phys. Chem. C* **2010**, *114*, 8814–8820. [CrossRef]
47. Rajeswari, A.; Christy, E.J.S.; Pius, A. New insight of hybrid membrane to degrade Congo red and Reactive yellow under sunlight. *J. Photochem. Photobiol. B Biol.* **2018**, *179*, 7–17. [CrossRef]
48. Thomas, M.; Naikoo, G.A.; Sheikh, M.U.D.; Bano, M.; Khan, F. Effective photocatalytic degradation of Congo red dye using alginate/carboxymethyl cellulose/TiO2 nanocomposite hydrogel under direct sunlight irradiation. *J. Photochem. Photobiol. A Chem.* **2016**, *327*, 33–43. [CrossRef]
49. Lucchini, M.A.; Lizundia, E.; Moser, S.; Niederberger, M.; Nystrom, G. Titania-cellulose hybrid monolith for in-flow purification of water under solar illumination. *ACS Appl. Mater. Interfaces* **2018**, *10*, 29599–29607. [CrossRef] [PubMed]
50. Islam, M.T.; Rosales, J.A.; Saenz-Arana, R.; Ghadimi, S.J.; Noveron, J.C. Rapid synthesis of ultrasmall platinum nanoparticles supported on macroporous cellulose fibers for catalysis. *Nanoscale Adv.* **2019**, *1*, 2953–2964. [CrossRef]
51. Melinte, V.; Chibac-Scutaru, A.-L.; Culica, M.E.; Coseri, S. Mineralization versus photoreduction of 4-nitrophenol under the influence of surface functionalized CeO_2 nanoparticles, hosted by versatile cellulose supports. *Appl. Surf. Sci.* **2021**, *565*, 150494. [CrossRef]
52. Lizundia, E.; Jimenez, M.; Altorfer, C.; Niederberger, M.; Caseri, W. Electroless plating of platinum nanoparticles onto mesoporous cellulose films for catalytically active free-standing materials. *Cellulose* **2019**, *26*, 5513–5527. [CrossRef]
53. Anwar, Y.; Ullah, I.; Ul-Islam, M.; Alghamdi, K.M.; Khalil, A.; Kamal, T. Adopting a green method for the synthesis of gold nanoparticles on cotton cloth for antimicrobial and environmental applications. *Arab. J. Chem.* **2021**, *14*, 103327. [CrossRef]

polymers

Article

Carboxymethyl Cellulose Hydrogel from Biomass Waste of Oil Palm Empty Fruit Bunch Using Calcium Chloride as Crosslinking Agent

Nur Fattima' Al-Zahara' Tuan Mohamood, Abdul Hakam Abdul Halim and Norhazlin Zainuddin *

Department of Chemistry, Faculty of Science, Universiti Putra Malaysia, Serdang 43400, Malaysia; fatimazahara.tm@gmail.com (N.F.A.-Z.T.M.); hakam5996@gmail.com (A.H.A.H.)
* Correspondence: norhazlin@upm.edu.my

Abstract: Carboxymethyl cellulose (CMC) is modified cellulose extracted from oil palm empty fruit bunch (OPEFB) biomass waste that has been prepared through etherification using sodium monochloroacetate (SMCA) in the presence of sodium hydroxide. In this research, CMC hydrogel was prepared using calcium chloride ($CaCl_2$) as the chemical crosslinker. Throughout the optimization process, four important parameters were studied, which were: (1) CMC concentration, (2) $CaCl_2$ concentration, (3) reaction time, and (4) reaction temperature. From the results, the best gel content obtained was 28.11% at 20% (w/v) of CMC with 1% (w/v) of $CaCl_2$ in 24 h reaction at room temperature. Meanwhile, the degree of swelling for CMC hydrogel was 47.34 g/g. All samples were characterized using FT-IR, XRD, TGA, and FESEM to study and compare modification on the OPEFB cellulose. The FT-IR spectrum of CMC hydrogel showed a shift of COO^- peaks at 1585 cm^{-1} and 1413 cm^{-1}, indicating the substitution of Ca^{2+} into the CMC molecular chains. The XRD diffractogram of CMC hydrogel showed no observation of sharp peaks, which signified an amorphous hydrogel phase. The CrI value also proved the decrement of the crystalline nature of CMC hydrogel. TGA–DTG thermograms showed that the T_{max} of CMC hydrogel at 293.33 °C is slightly better in thermal stability compared to CMC. Meanwhile, the FESEM micrograph of CMC hydrogel showed interconnected pores indicating the crosslinkages in CMC hydrogel. CMC hydrogel was successfully synthesized using $CaCl_2$ as a crosslinking agent, and its swelling ability can be used in various applications such as drug delivery systems, industrial effluent, food additives, heavy metal removal, and many more.

Keywords: oil palm biomass waste; anionic hydrogel; swelling; carboxymethyl cellulose; salt crosslinking agent

Citation: Tuan Mohamood, N.F.A.-Z.; Abdul Halim, A.H.; Zainuddin, N. Carboxymethyl Cellulose Hydrogel from Biomass Waste of Oil Palm Empty Fruit Bunch Using Calcium Chloride as Crosslinking Agent. *Polymers* **2021**, *13*, 4056. https://doi.org/10.3390/polym13234056

Academic Editor: Arn Mignon

Received: 31 August 2021
Accepted: 12 October 2021
Published: 23 November 2021

1. Introduction

Polysaccharides are polymeric and complex carbohydrates that are generated as repeating units of monosaccharides, the simplest carbohydrates, and are linked by glycosidic linkages. Polysaccharides make up the majority of biomass and are thought to account for more than 90% of the carbohydrate material in nature. This natural polymer is formed either linearly with a straight chain of monosaccharides or branched with arms contingent on the monosaccharide link and the location of the carbon to which it is attached [1]. Polysaccharides, particularly cellulose, have a linear to highly branched structure, with main walls composed of cellulose, hemicellulose, lignin, and pectin. Various chemical processes have been employed to change existing polysaccharide structures throughout the years to improve and tailor their features [2].

Cellulose, the most prevalent polymer on earth, gives plants strength and is a bio-renewable ecologically benign raw resource [3]. Cellulose is made up of two anhydroglucose units (AGU) that are joined together by β-1,4-glycosidic bonds [4]. Three hydroxyl groups of AGU play a significant part in leading the crystallinity of the polymer due to their capacity to create hydrogen bonds. Because of its high crystallinity and rigidity as a result of its long and linear chains with intermolecular hydrogen bonding, cellulose is insoluble in water and most organic solvents [5]. To overcome this critical issue and improve the water insolubility disadvantage that restricts cellulose's adaptability in numerous industries and applications, cellulose must undergo chemical modification such as esterification or etherification. The modification of cellulose is crucial in order to enhance its industrial demands, whereby alterations can result in a wide range of cellulose utilizations. The conversion of cellulose into carboxymethyl cellulose (CMC), hydroxyethyl cellulose (HEC), hydroxyethylmethyl cellulose (HEMC), cationic cellulose (CC), and a few more would undoubtedly boost the value of native cellulose because these cellulose derivatives are widely used in a range of industrial applications such as food additives, detergents, agriculture, pharmaceuticals, and many others. The alteration of cellulose into CMC will enhance its physicochemical properties such as biocompatibility, biodegradability, swelling power, and water solubility.

Malaysia, as one of the world's major oil palm producers, increased its planted area from 5.74 million hectares in 2016 to 5.81 million hectares in 2017 [6]. Biomass waste, particularly from the oil palm empty fruit bunch (OPEFB), has generated more than 18,000 tonnes and cellulose extracted from OPEFB has been determined to be 93% pure [7]. Since the purity of cellulose derived from OPEFB is high, it may be converted into a promising and profitable material with enormous opportunities in a variety of applications. OPEFB biomass waste has low economic value and most of the time poses a disposal challenge. Conventionally, OPEFB waste is often burnt, disposed of in landfills, or composted to organic fertilizer [8,9]. Maximizing waste energy recovery is beneficial for both environmental and economic reasons. The modification of OPEFB cellulose into CMC is crucial for improving the functional characteristics and performance of biomass waste cellulose.

Further modification of CMC into CMC hydrogel will raise the profitability of this polymeric material. The term "hydrogel" refers to a three-dimensional polymeric network of hydrophilic chains that can shrink, swell, and absorb a large amount of water [10]. It swells well in an aqueous solution but stays insoluble due to chemical or physical crosslinking between individual polymeric chains [11]. The rate of water absorption is determined by the presence of functional groups [12], availability of hydrophilic groups [13], state of water [14], and crystallinity of cellulose [15] in the polymeric hydrogel. The swelling property of the hydrogel has been widely used in a variety of applications, and advantages for their usage include the ability to encapsulate biomacromolecules such as proteins and DNA due to the hydrophobic interaction [16]. Hydrogel can be considered a smart material because of its capability to swell in different media, changing the structure due to certain external responses such as temperature, pH, ions, and substance concentration [17]. Interestingly, hydrogel can be produced by multiple routes of alteration techniques. For instance, CMC-HEC aerogel was successfully developed using a supercritical CO_2-assisted process. It was found that CMC-HEC aerogel can absorb water more than 500 times the original weight of the aerogel and is best used for agricultural applications [18], while CMC methacrylate hydrogel prepared from photopolymerization crosslinking efficiently showed better diffusivities of bovine serum albumin, which was studied as a controlled release device [19].

In this research, oil palm empty fruit bunch (OPEFB) cellulose is chosen due to its abundance availability from the oil palm milling process in Malaysia. Improper handling of this solid waste can contribute to environmental concerns, burdening industry operators with waste disposal difficulties and increasing operational costs. The main objectives of this research are to modify the OPEFB cellulose into CMC via carboxymethylation and convert it into CMC hydrogel via ionic crosslinking using $CaCl_2$ with thorough optimization. Since $CaCl_2$ has high solubility in an aqueous solution, the experiment was easy to handle. Up to date, there is no study reported on CMC hydrogel from OPEFB waste crosslinked with $CaCl_2$, and therefore, this study can contribute to the exploration of hydrogel from OPEFB.

2. Materials and Methods

2.1. Materials

Cellulose of OPEFB was supplied by Biorefinery Complex, Universiti Putra Malaysia, Serdang, Selangor, Malaysia. Sodium monochloroacetate (SMCA), methanol (100%), ethanol (95%), sodium hydroxide (NaOH, 99%) pellet, glacial acetic acid (99.8%), and calcium chloride ($CaCl_2$) were purchased from R&M Chemicals, Selangor, Malaysia, while isopropanol (IPA, 98%) was purchased from Sigma Aldrich, St. Louis, MO, USA. All chemicals used in this study are of analytical grade and used without further purification. Distilled water was used throughout the experiment.

2.2. Preparation of CMC

In a 250 mL beaker, 5.0 g of OPEFB cellulose was added and then followed by 10 mL of 30% (w/v) NaOH solution dropwise. Then, 100 mL of IPA was added to the mixture and mechanically stirred for an hour at room temperature. To complete the carboxymethylation reaction, 6.0 g of SMCA was added and the reaction continued for three hours at 45 °C. Then, the mixture was filtered, and the slurry of CMC was soaked in 300 mL of absolute methanol overnight. CMC was neutralized with glacial acetic acid then sieved and washed with 70% ethanol, followed by 99.7% ethanol. The washing process was repeated three times, and CMC was oven-dried for 24 h at 60 °C. CMC was kept in an air-tight container.

2.3. Preparation of CMC Hydrogel

CMC of 10–30% (w/v) was dissolved in 1–5% (w/v) of $CaCl_2$ solution. The mixture was stirred homogeneously until a paste-like solution of CMC–$CaCl_2$ was obtained and placed in a Petri dish. The CMC–$CaCl_2$ paste was left to crosslink for 12–96 h at room temperature up to 60 °C. The optimization of the preparation of CMC hydrogel was carried out by modifying and controlling various parameters using the one-variable-at-a-time (OVAT) approach to reach the optimum percentage of gel content. Figure 1 shows the diagrammatic scheme of OPEFB cellulose carboxymethylation reaction and hydrogel ionic crosslinking.

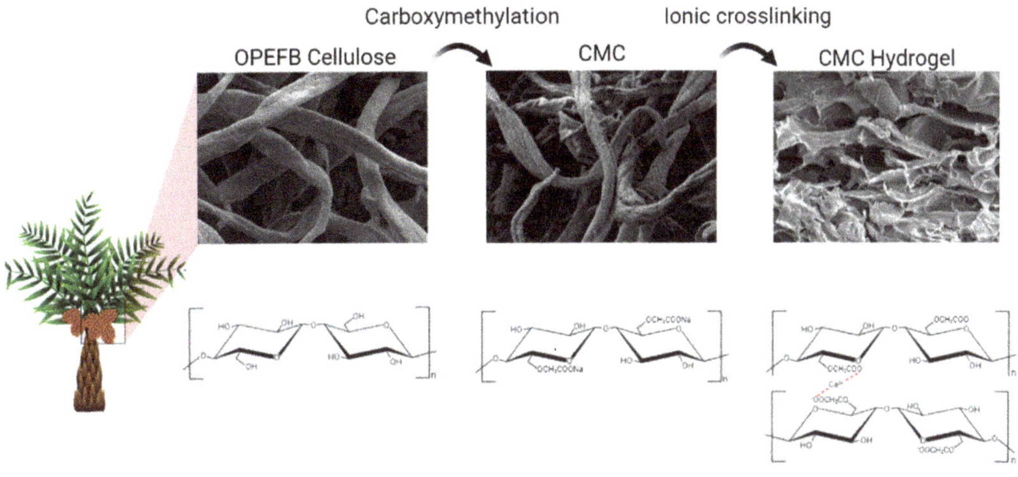

Figure 1. Diagrammatic scheme of conversion of OPEFB cellulose into CMC hydrogel.

2.4. Gel Content and Degree of Swelling of CMC Hydrogel

The gel content of CMC hydrogel was determined by measuring the insoluble part after the immersion of CMC hydrogel in distilled water for 72 h at room temperature. The distilled water was replaced every 24 h. The percentage of gel content was calculated using the following equation:

$$\text{Percentage of gel content} = (W_{da}/W_{db}) \times 100\% \quad (1)$$

where W_{da} is the weight of dried hydrogel after immersion and W_{db} is the weight of hydrogel before immersion.

The degree of swelling of CMC hydrogel was carried out by the immersion of the CMC hydrogel in distilled water for 72 h at room temperature. The hydrogel was weighed after it reached equilibrium and the degree of swelling was calculated using the following equation:

$$\text{Degree of swelling} = (W_s - W_d)/W_d \quad (2)$$

where W_s is the weight of swollen hydrogel and W_d is the weight of the dried hydrogel.

2.5. Characterization

FT-IR spectroscopy is a technique used to determine the presence of functional groups in the CMC and CMC hydrogel. In this study, FT-IR was obtained from a Spectrum 100 Perkin-Elmer FT-IR spectrophotometer with a wavenumber of 400–4000 cm^{-1}. The attenuated total reflection (ATR) sampling technique was used in conjunction with FT-IR spectroscopy and all samples were prepared in powder form.

The crystallinity of OPEFB cellulose, CMC, and CMC hydrogel was obtained from a Shimadzu XRD-6000 diffractometer with Cu Kα (λ = 1.5418 Å) radiation at room temperature. The machine was operated at 30 kV and 30 mA. The sample was placed in the aluminium sample holder and scattered intensity data were investigated in the scan range of 2–60° (2θ) with a continuous scanning rate of 2°/min. The crystallinity index (CrI) was calculated from the XRD diffractogram using the following equation [20] with the aid of OriginPro 2019 Graphing and Analysis software:

$$\text{Crystallinity Index} = (A_c/A_{c+a}) \times 100\% \quad (3)$$

where A_c is the total area of crystalline peak and A_{c+a} is the total area of crystalline and amorphous peaks.

TGA of OPEFB cellulose, CMC, and CMC hydrogel was studied using TGA/SDTA-851 Mettler Toledo. It was used to determine the decomposition stage of samples and was performed under nitrogen flow and atmosphere air from 50 °C to 800 °C at the rate of 5 °C/min. The amount of weight loss as a function of temperature was examined and data were presented as a thermogram of weight loss versus temperature.

FEI Nova NanoSEM 230 FESEM was used to study the surface morphology of OPEFB cellulose, CMC, and CMC hydrogel. Samples were sprinkled separately to the sample holder and were coated with gold before the analysis. A freeze-dried CMC hydrogel was used in this FESEM analysis.

In this study, the characterizations of the CMC hydrogel were performed using an optimized CMC hydrogel sample based on the optimization studies.

3. Results

3.1. Optimization of CMC Hydrogel

3.1.1. Effect of CMC Concentration on CMC Hydrogel

Figure 2 depicts a graph of the gel content and degree of swelling of CMC hydrogel at various CMC concentrations. The controlled variables for this parameter were the concentration of $CaCl_2$ at 1% (*w*/*v*) and 24 h reaction time at room temperature (27 °C). The gel content of CMC hydrogel steadily rose with CMC concentration starting at 10% (*w*/*v*) and reached a maximum of 20% (*w*/*v*) with 28.11%. It was found that at 25% (*w*/*v*) of CMC concentration and above, the percentage of gel content decreased. The increase in gel content up to 20% (*w*/*v*) of CMC may be due to higher crosslinkages between CMC chains. At high concentrations, CMC molecules are packed closer together, allowing Ca^{2+} ions to form linkages between the polymer chains. This finding is similar to the study by Fei et al. [16], who obtained a high gel fraction of hydrogel when the concentration of the CMC was from 5–30% only. However, as a higher concentration of CMC was used with a fixed amount of crosslinking agent throughout the experiment, the possibility for Ca^{2+} to create crosslinkages with CMC in the reaction was insufficient and limited, and thus gel content decreased. The schematic illustration of the crosslinking reaction between CMC and $CaCl_2$ is shown in Figure 3. In this study, 20% (w/v) was chosen as the optimal concentration for this reaction. The swelling of CMC hydrogel is inversely proportional to its gel content. The degree of swelling was the lowest at the greatest gel content, which was 46.78 g/g. The degree of swelling dropped from 10% to 20% (w/v) of CMC and then began to rise at 25% (w/v) and higher CMC concentrations. At a high gel content of CMC, a high number of crosslinkages occurred in the CMC hydrogel network. As a result, water diffusion into the hydrogel voids became more challenging, and thus the swelling decreased. On the other hand, as the gel content decreased at a high concentration of CMC, the increase in the swelling implies a weak hydrogel due to low gel content and more water trapped inside the pores of the CMC hydrogel.

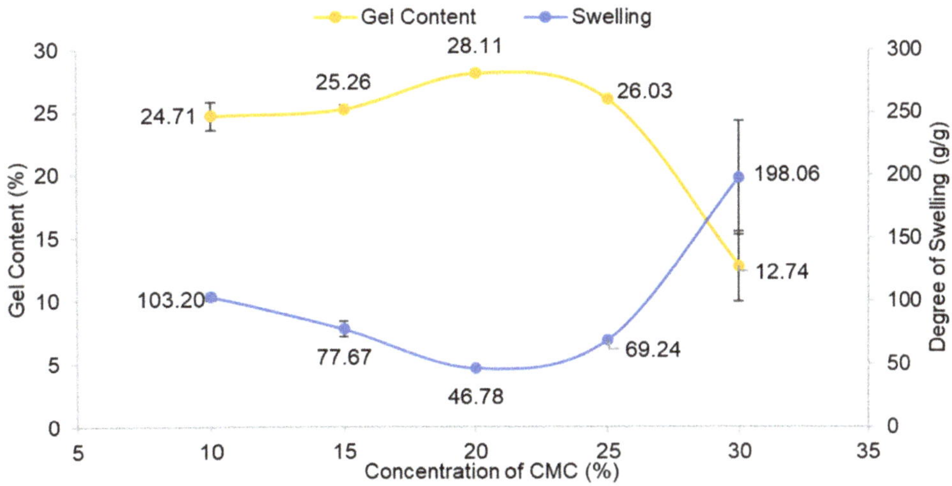

Figure 2. Effect of concentration of CMC on gel content and degree of swelling of CMC hydrogel. Error bar indicates the standard deviation of uncertainty.

3.1.2. Effect of $CaCl_2$ Concentration on CMC Hydrogel

The concentration of $CaCl_2$ varied from 1–5% (w/v), as illustrated in Figure 4. The controlled variables were 20% (w/v) CMC with 24 h reaction time and the experiment was carried out at room temperature. The highest percentage of gel content was 28.94% at 1% (w/v) $CaCl_2$ and it gradually decreased at higher concentrations of $CaCl_2$ solution. Sultana and Islam [21] published a similar discovery, stating that the ionic crosslinking of polymers increased as the electrostatic interaction between the ionic charges of polymer chains and multivalent cation Ca^{2+} increased. However, in this study, the gel content decreased as $CaCl_2$ concentration increased. Reversible reactions may occur at greater $CaCl_2$ concentrations, lowering the percentage of gel content. This finding is similar to the results of Che Nan et al. [22], which concluded that CMC chains lose their flexibility in high concentrations of $CaCl_2$ and thus decrease in hydrodynamic molecular size and make the polymer chains agglomerate. Furthermore, CMC dissolution in water would be inhibited since CMC could not be fully hydrated in water with a greater $CaCl_2$ concentration [23]. A similar trend in the swelling of CMC hydrogel where it is inversely proportional to the gel content was observed. As the gel content decreased, a lower degree of crosslinking occurred. Consequently, water molecules bound more easily to the surface of the hydrogel and diffused into the accessible voids, triggering hydrogel expansion owing to the water uptake of swelling response, as shown in Figure 5. According to Tolinski [24], the amount of interconnected polymer chains or the density of the gel component depends on the degree of crosslinking. The previous statement is supported by Maitra and Shukla [25], who stated that the swelling properties of hydrogel and transport of molecules are influenced by the degree of crosslinking in the polymeric chains of hydrogel.

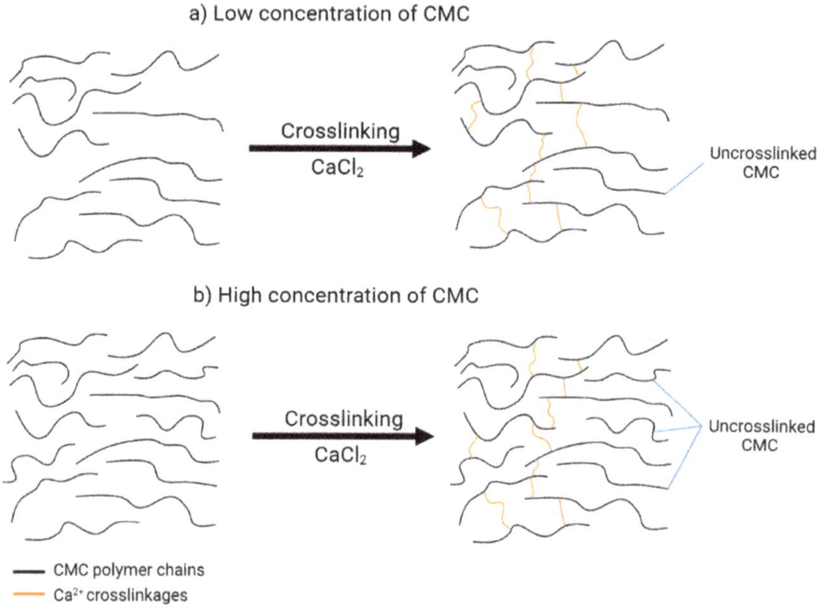

Figure 3. Schematic illustration of crosslinking reaction between CMC and $CaCl_2$ at (**a**) low concentration of CMC and (**b**) high concentration of CMC.

3.1.3. Effect of Different Reaction Times on CMC Hydrogel

The influence of reaction time was studied, as shown in Figure 6. The reaction time varied from 24 to 96 reaction hours, and the controlled variables for this parameter were 20% (w/v) of CMC and 1% (w/v) $CaCl_2$ performed at room temperature. The highest and best reaction time was at 24 h, which gave 27.27% of gel content. The trend displayed a gradual decrement of gel content when reaction time was prolonged. This might be due to the presence of NaCl (Na^+ from $-CH_2COONa$ combined with Cl^- ion) in the hydrogel, which reduced electrostatic attraction in the polymer chains when $CaCl_2$ was added at a longer reaction time. The chemical reaction is shown in Equation (4), as reported by Che Nan et al. [22]. Meanwhile, for the degree of swelling of CMC hydrogel, it showed a similar trend to the previous parameters; the effect of CMC and $CaCl_2$ concentration.

$$\text{Cell-OCH}_2\text{COO-Na}^+ + \text{CaCl}_2 \rightarrow \text{Cell-OCH}_2\text{COO-Ca}^{2+} - \text{-OOCCH}_2\text{O-Cell} + \text{NaCl} \quad (4)$$

3.1.4. Effect of Different Reaction Temperatures on CMC Hydrogel

Figure 7 displays the percentage of gel content and degree of swelling of CMC hydrogel at various reaction temperatures. The constant variables were 20% (w/v) of CMC, 1% (w/v) $CaCl_2$, and the reaction was completed at 24 h. This experiment was conducted at temperatures ranging from room temperature (27 °C) to 60 °C. The percentage of gel content of the CMC hydrogel dropped as the reaction temperature increased. It is found that room temperature gave the highest gel content of 27.61%. This might be due to the fact that the CMC hydrogel structure generated by $CaCl_2$ ionic crosslinking is more stable at lower reaction temperatures [26]. As the temperature rose, a reversible reaction may have triggered the molecules, causing the hydrogel to disintegrate slowly. As a result, a low gel content of the CMC hydrogel was obtained. Conversely, Figure 7 indicates a consistent pattern to the proceeding factor for the degree of swelling of CMC hydrogel, which is inversely proportioned to the percentage of gel content.

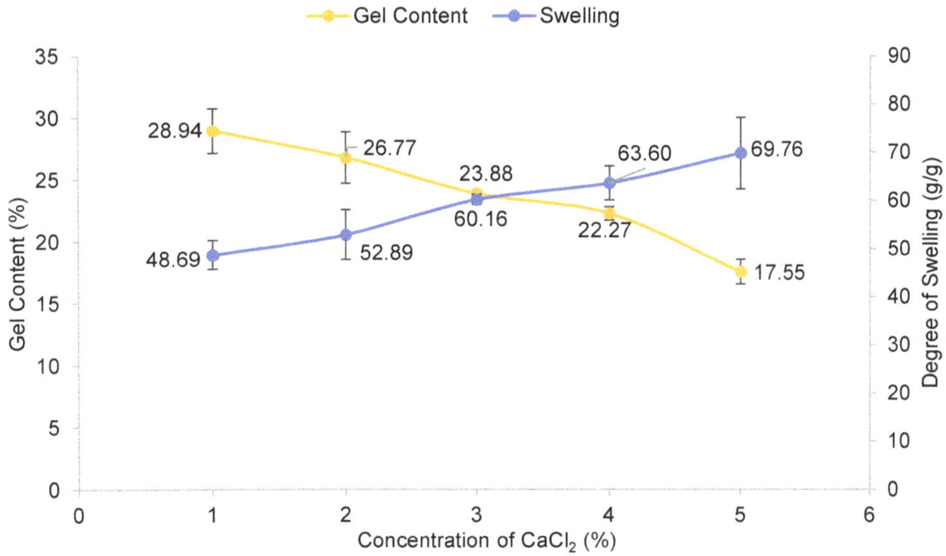

Figure 4. Effect of concentration of $CaCl_2$ on gel content and degree of swelling of CMC hydrogel. Error bar indicates the standard deviation of uncertainty.

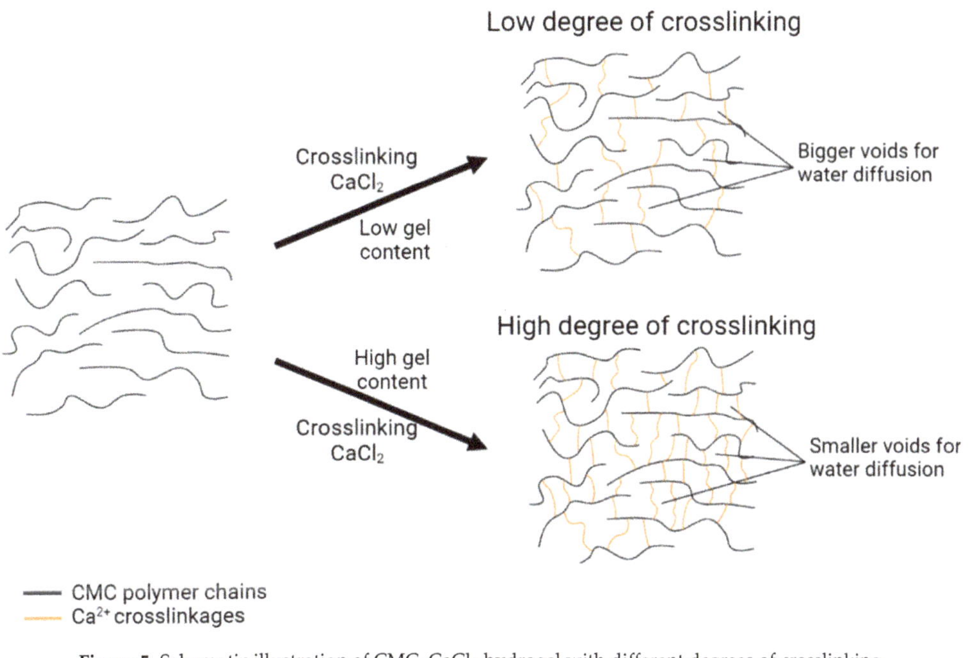

Figure 5. Schematic illustration of CMC–$CaCl_2$ hydrogel with different degrees of crosslinking.

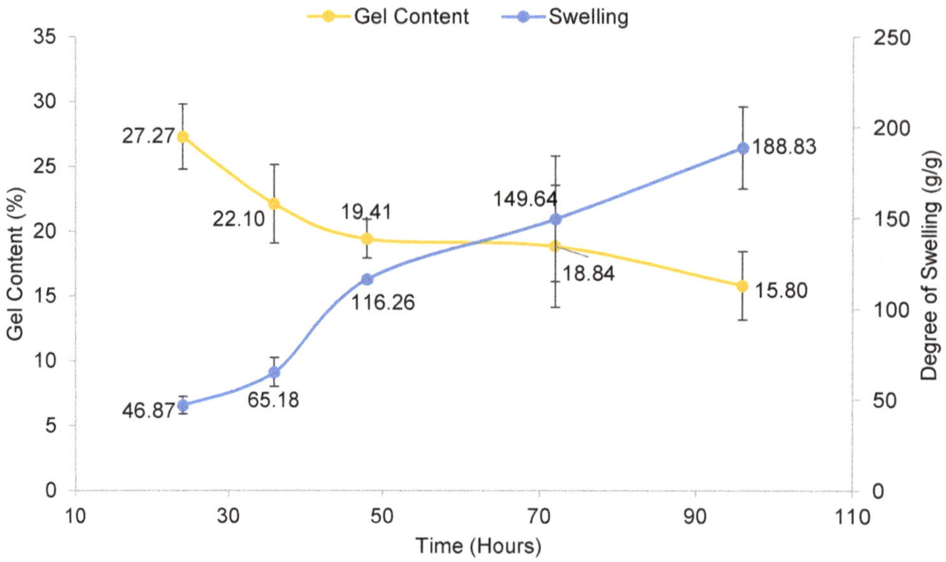

Figure 6. Effect of reaction time on gel content and degree of swelling of CMC hydrogel. Error bar indicates the standard deviation of uncertainty.

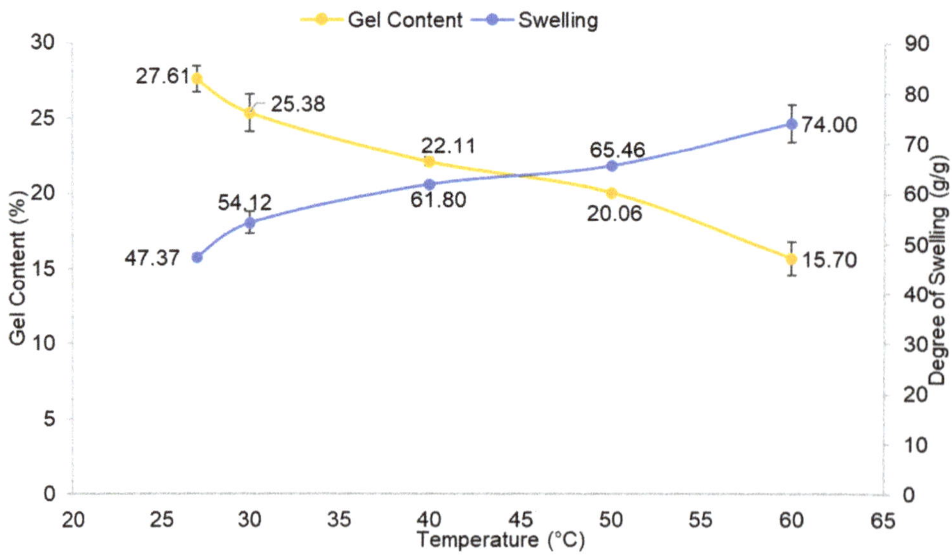

Figure 7. Effect of reaction temperature on gel content and degree of swelling of CMC hydrogel. Error bar indicates the standard deviation of uncertainty.

3.2. *Characterization*

3.2.1. Fourier Transform-Infrared Spectroscopy (FT-IR)

Figure 8 depicts the FT-IR spectrum of OPEFB cellulose with a strong absorption peak at 3334 cm^{-1}, which is assigned to the hydroxyl group of the polysaccharide chain, –OH stretching, while the appearance of a shoulder peak at 2893 cm^{-1} is assigned to the –CH stretching vibration. These two peaks are the indication of the main polysaccharide nature where it is composed of glucose units. A significant stretching C–O vibration of primary alcohols and ethers in the cellulose backbone of the cellulose chain was observed at 1055 cm^{-1}. According to Ngadi and Lani [27], the appearance of a peak at 1055 cm^{-1} might be attributed to the C–O–C pyranose ring stretching vibration and -CH stretching vibration at 3358 cm^{-1} and 2916 cm^{-1}, respectively. Strong absorption peaks at 1589 cm^{-1} and 1413 cm^{-1} indicated the appearance of the asymmetrical and symmetrical stretching of the –COO group from $-CH_2COONa$ of SMCA with a strong alkali solution of NaOH. By comparing to the OPEFB cellulose, the existence of the additional peak at 1589 cm^{-1} in CMC has proved that CMC was successfully synthesized from the OPEFB cellulose via carboxymethylation reaction. Similar research by Tulain et al. [28] also reported that in the range of 1500–1700 cm^{-1} of the absorbance band, carboxyl groups and their salts can be observed, which allied with Na–CMC. The peak observed from the CMC spectrum at 1325 cm^{-1} is assigned to –OH bending vibration. The FT-IR spectrum of CMC hydrogel revealed the same –OH and –CH absorption bands at 3348 cm^{-1} and 2895 cm^{-1}, respectively. However, these two bands are less intense compared to the OPEFB cellulose and CMC. The absorption peaks at 1585 cm^{-1} and 1423 cm^{-1} are attributed to the asymmetrical and symmetrical stretching of the –COO group. There is no significant difference in the FT-IR spectra of CMC and CMC hydrogel, where no new peak can be observed in CMC hydrogel. This could be due to the fact that crosslinking only involved the substitution of Na^+ with Ca^{2+} from $CaCl_2$ as the crosslinking agent. There was only slight shifting in the wavenumber after Ca^{2+} ionic crosslinking on the CMC chain from 1589 cm^{-1} to a lower wavenumber (1585 cm^{-1}) and 1413 cm^{-1} to a higher wavenumber (1423 cm^{-1}) due to the bidentate of the carboxylate group to Ca^{2+} [29]. In addition, both peaks showed less intensity after the ionic crosslinking reaction compared to the CMC, which reflects the decrement of the crystalline nature of the hydrogel [30]. The decrease in crystallinity of CMC hydrogel was proven by XRD analysis, which will be discussed in the next section.

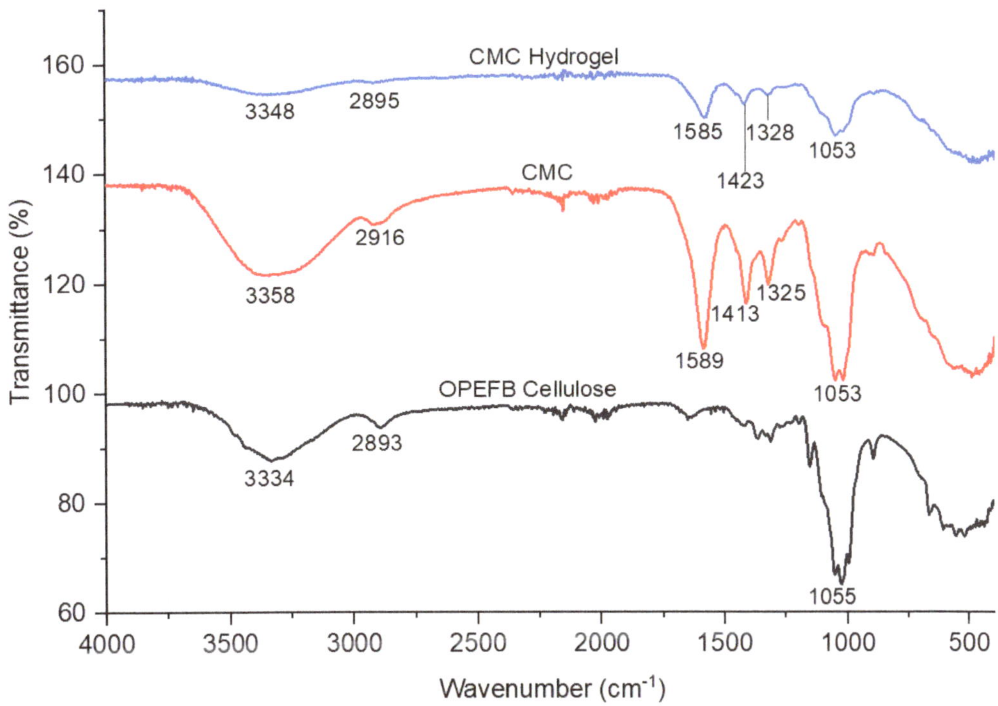

Figure 8. FT-IR spectra of OPEFB cellulose, CMC, and CMC hydrogel.

3.2.2. X-ray Diffraction (XRD)

The XRD diffraction of OPEFB cellulose, CMC, and CMC hydrogel is plotted in Figure 9, while the crystallinity index (CrI) is tabulated in Table 1. CrI was used to study the change in structure and degree of crystallinity in OPEFB cellulose, CMC, and CMC hydrogel.

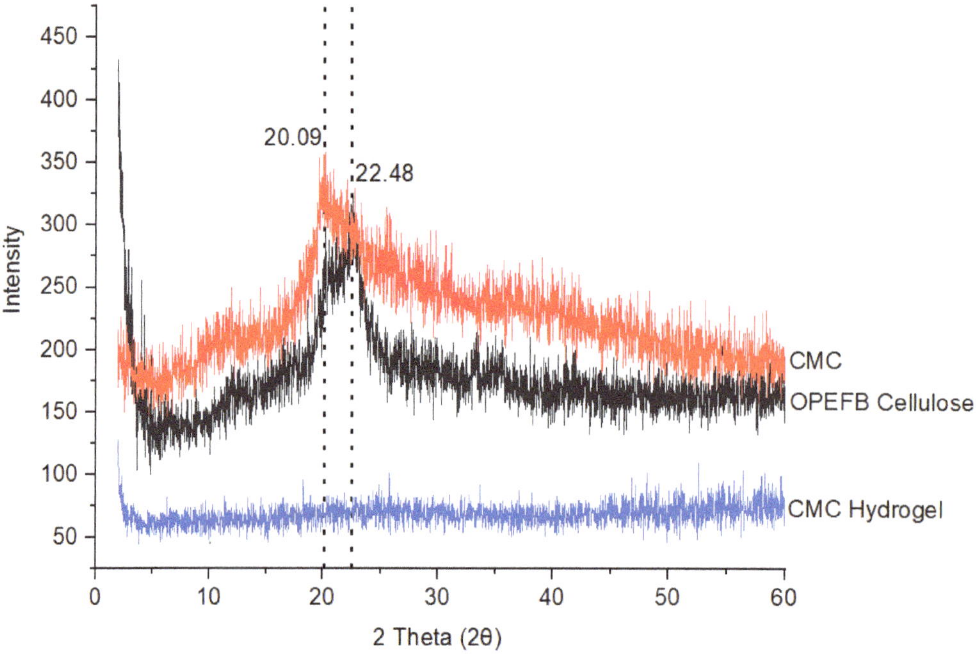

Figure 9. XRD diffractogram of OPEFB cellulose, CMC, and CMC hydrogel.

Table 1. Crystallinity index of OPEFB cellulose, CMC, and CMC hydrogel.

Sample	Crystallinity Index (%)
OPEFB cellulose	48.70
CMC	32.77
CMC hydrogel	4.24

The XRD pattern for OPEFB cellulose is shown at $2\theta = 22°$ and CMC at $2\theta = 20°$, which corresponds to the amorphous region and small crystallites in the cellulose granules. Baharin et al. [31] reported that the diffraction peak $2\theta = 22.6°$ of OPEFB cellulose is assigned to the crystalline phase, while Uyanga and Daoud [32] reported that commercial CMC with diffraction peak $2\theta = 20.03°$ indicates the amorphous phase. The crystallinity index (CrI) of OPEFB cellulose was found to be the highest among all samples at 48.70%, which indicates that there is a more crystalline phase in OPEFB cellulose compared to CMC and CMC hydrogel. The diffraction pattern of CMC showed that CMC has lower crystallinity compared to OPEFB cellulose as there was a less intense peak observed with a broad pattern at $2\theta = 20°$. The loss of crystalline nature is probably due to the replacement of hydroxyl groups in the samples with the inclusion of strong alkaline, NaOH, and water in the crystallites during carboxymethylation [33]. This finding is in good agreement with TGA data, which show that CMC has a lower maximum decomposition temperature compared to OPEFB cellulose due to the loss of the crystalline phase. The crystallinity reduction of the CMC has also caused this modified cellulose to experience excellent solubility in water [34]. Table 1 shows the CrI for OPEFB cellulose, CMC, and CMC hydrogel. The CrI for OPEFB cellulose and CMC was 48.70% and 32.77%, respectively. For the XRD pattern of CMC hydrogel, there was no appearance of a sharp peak, thus it can be concluded that CMC hydrogel was in the amorphous phase. The substitution of Ca^{2+} into the CMC formed a highly crosslinked CMC and thus disturbed the well-ordered cellulose structure and reduced the mobility of the cellulose chains. The CrI of the CMC hydrogel also showed a significant reduction at only 4.24%. This indicates that CMC hydrogel fully transformed into amorphous nature after some alterations of carboxymethylation and ionic crosslinking.

3.2.3. Thermogravimetry Analysis (TGA)

TGA measures the amount and rate of change in a material's weight in a monitored atmosphere as a function of temperature over time. It is mainly used to determine the material composition and predict thermal stability. In this study, all samples were analyzed at a temperature range of 50–800 °C for both TGA and DTG profiles, as shown in Figures 10 and 11, respectively. The thermogram profiles showed that OPEFB cellulose, CMC, and CMC hydrogel decomposed at two stages. The first decomposition stage took place in the range of 50–180 °C, which indicated the removal of water loosely bound in the sample, and the second decomposition stage took place at around 200–550 °C, which was correlated to the degradation of the cellulose backbone.

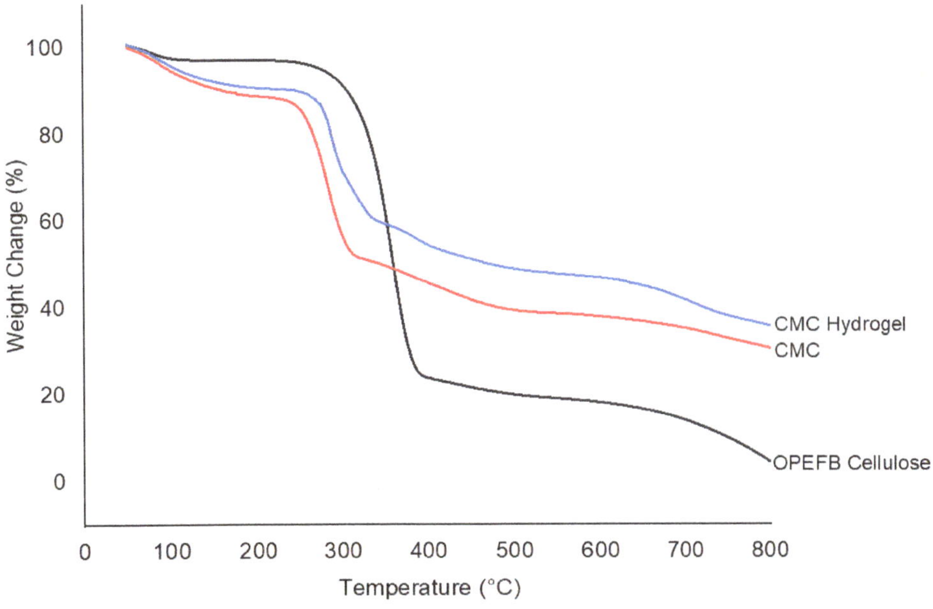

Figure 10. TGA thermograms of OPEFB cellulose, CMC, and CMC hydrogel.

The first decomposition stage for OPEFB cellulose, CMC, and CMC hydrogel was due to the loss of bound water, as well as the release of any volatile compounds from the surface and moisture entrapped inside the sample. The release of initial moisture is common in natural fiber. As depicted in TGA–DTG thermograms, small weight losses of OPEFB cellulose, CMC, and CMC hydrogel were observed at a temperature of 78.33 °C, 90.83 °C, and 96.66 °C, respectively. According to Mohtar et al. [35], any degradation or decomposition below 100 °C is not considered as a significant thermal event in the TGA study. The second decomposition stage was the most crucial stage for all three samples. The decomposition temperature of OPEFB cellulose started at 206.54 °C to 526.68 °C with a weight loss of 77.79% from its original weight, while CMC decomposed at the temperature range of 180.31–527.32 °C with a weight loss of more than half of the original weight after the modification of cellulose into CMC, which decomposed about 50.39%. For CMC hydrogel, after the ionic crosslinking reaction with $CaCl_2$, the decomposition stage occurred at a slightly broad temperature at the range of 180.13–530.42 °C with 51.73% weight loss. The weight loss of OPEFB cellulose was the highest compared to CMC and CMC hydrogel because lignocellulosic materials are chemically active, which decompose thermo-chemically in the range of 200 °C to 500 °C, hence it is easier for the cellulosic materials to decompose at this temperature range. This result is in good agreement with the previous study by Khalid et al. [36]. For CMC, the decomposition at this second stage might be due to the degradation of the cellulose backbone and the removal of CO_2 caused by decarboxylation [37]. Meanwhile, for CMC hydrogel, besides the elimination of CO_2 molecules in the polymeric backbone of hydrogel, it was also most probably the degradation of the crosslinked polymer network after the Ca^{2+} ionic crosslinking reaction together with the carbonization process at a temperature higher than 500 °C [38].

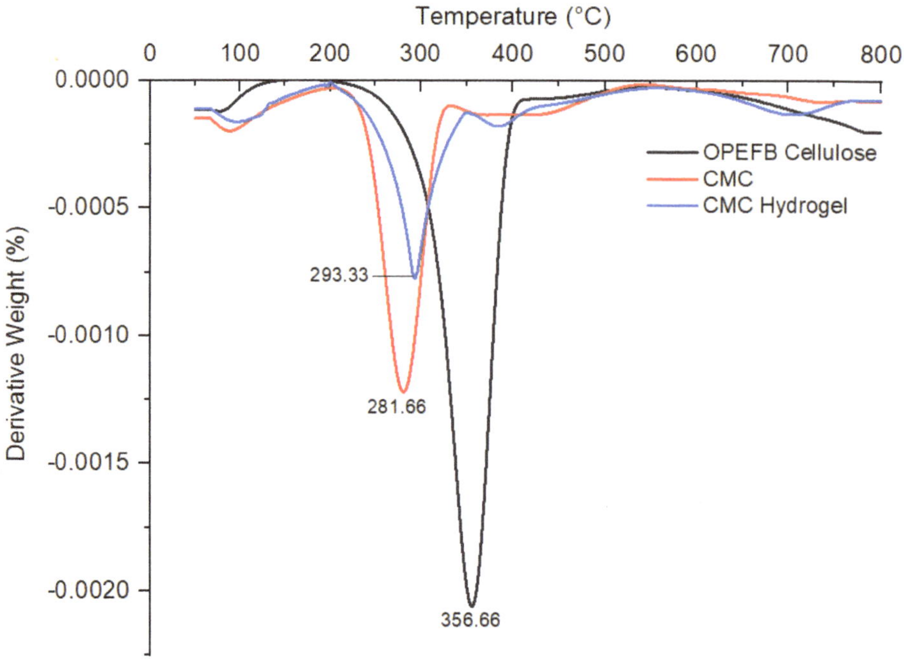

Figure 11. DTG thermograms of OPEFB cellulose, CMC, and CMC hydrogel.

According to Tudorachi et al. [39], the maximum degradation temperature, T_{max}, is the temperature where the degradation rate is maximum. The DTG thermogram of OPEFB cellulose showed the highest T_{max} at 356.66 °C compared to CMC at 281.66 °C and CMC hydrogel at 293.66 °C. This was due to the high crystalline nature and well-ordered structure of the OPEFB cellulose backbone, which made it more thermally stable than modified cellulose. This statement is supported by the CrI percentage in the XRD analysis. The T_{max} of CMC was the lowest because the modification affected the molecular structure and bonding energy, which caused the CMC to become thermally unstable. El-Sakhawy et al. [40], in their thermal study of CMC and CMC acetate, also found that CMC shifted to a lower temperature due to the reaction with NaOH during the carboxymethylation process, which increases the amorphous structure of CMC. The DTG thermogram shows that the T_{max} of CMC hydrogel was slightly higher than CMC. This might be due to the inclusion of Ca^{2+} ions into the CMC structure through ionic crosslinking, which made the CMC hydrogel backbone more stable. Seki et al. [41], in their study on different crosslinking reagents of CMC/HEC hydrogel, stated that the crosslinked CMC hydrogel has better T_{max} than CMC, which delayed thermal decomposition. From the TGA and DTG thermograms, it can be summarized that CMC hydrogel has a slightly better thermal stability in comparison with CMC.

3.2.4. Field Emission Scanning Electron Microscopy (FESEM)

FESEM is a type of electron microscope used to observe extremely fine morphology features on the surface of a whole or fractioned subject. A high-resolution sample image can be monitored and focused in high magnification using FESEM. Figure 12a–f show the images of OPEFB cellulose, CMC, and CMC hydrogel.

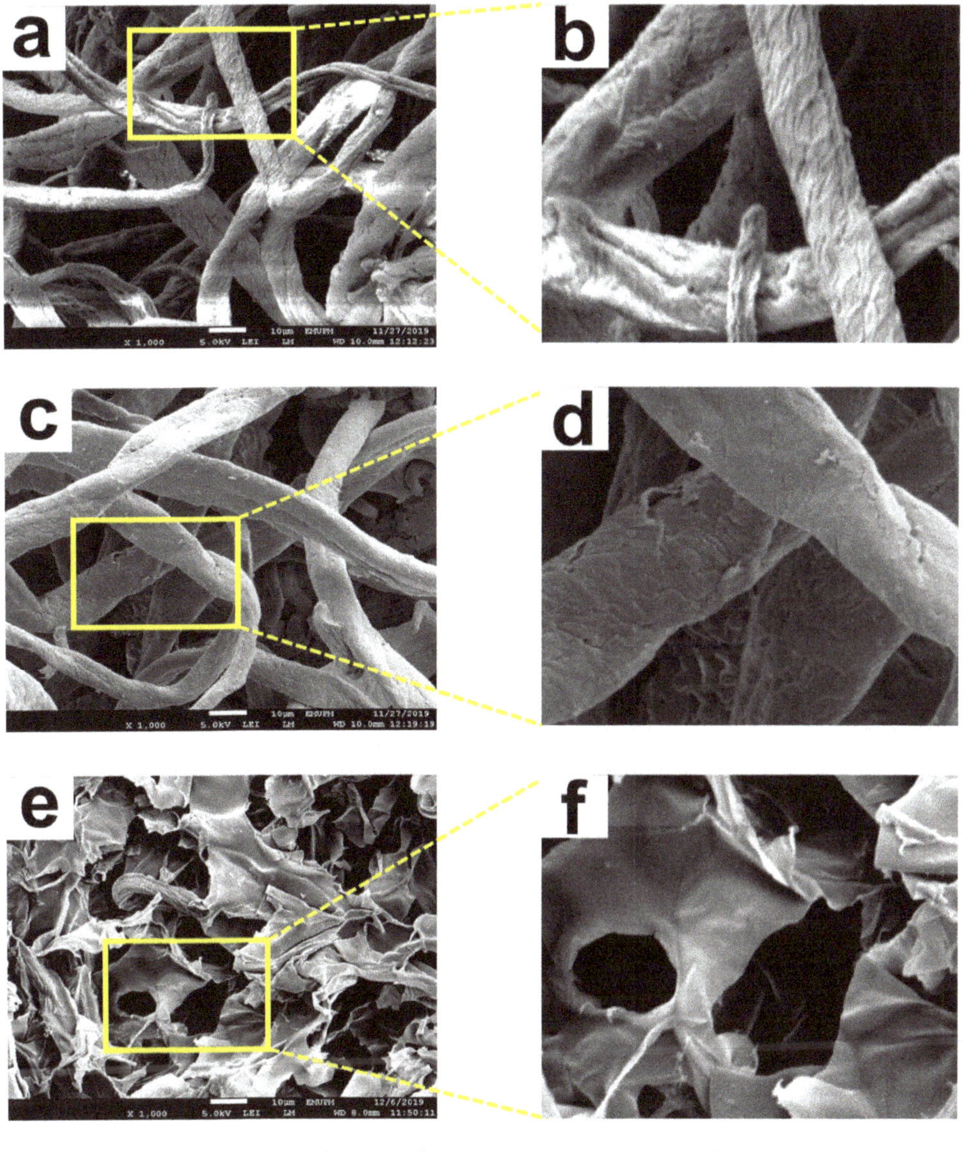

Figure 12. FESEM micrographs of OPEFB cellulose (**a**,**b**), CMC (**c**,**d**), and CMC hydrogel (**e**,**f**) at 1000×.

Figure 12a,b show the surface morphology of OPEFB cellulose, which reveals the rough surface of cellulose in fiber form. It is found that OPEFB cellulose fibers were twisted and ruptured. The rough string-like form of OPEFB cellulose fibers may be due to the use of strong chemicals and a high temperature during the cellulose extraction process [42]. Sreekala et al. [43] also stated that mercerization treatment during cellulose extraction to remove natural and artificial impurities such as hemicellulose, lignin, wax, and pectin produced a rough surface topology. Figure 12c,d show the FESEM micrographs of CMC. The CMC micrographs reveal that the surface of the modified cellulose had become less rough and wrinkled. In addition, the string-like OPEFB cellulose agglomerated and attached. The results obtained are consistent with a study by Muhamad Parid et al. [42], which stated that the decrease in roughness is due to the modification by the etherifying agent that reduced the crystalline nature of cellulose. Figure 12e,f show morphological images of freeze-dried CMC hydrogel. The CMC hydrogel has irregular sizes of macropores. According to Rahman et al. [44], the difference in pore size in hydrogel could be due to different crosslinking densities, which can lead to substantially lower or higher water uptake before freeze-drying. The majority of the CMC interfiber macropores were connected, which was desirable for water entrapment because diffusion and the presence of the pores enhanced the swelling ratio of the hydrogel.

4. Conclusions

In this study, macroporous CMC hydrogel was successfully synthesized and optimized from OPEFB cellulose through the OVAT method. The optimum gel content was 27.61% with 47.37 g/g degree of swelling. The best optimization conditions to produce CMC hydrogel with an optimum percentage of gel content were at 20% (w/v) of CMC, 1% (w/v) of $CaCl_2$, 24 h reaction time, and ambient temperature. The modification of OPEFB cellulose via carboxymethylation and ionic crosslinking can convert the abundantly available biomass waste from the oil palm industry and can be further utilized as a biodegradable sorbent material and drug carrier in various applications.

Author Contributions: Conceptualization, N.Z.; methodology and design, N.Z. and A.H.A.H.; software, N.F.A.-Z.T.M.; formal analysis and investigation, A.H.A.H. and N.F.A.-Z.T.M.; writing—original draft preparation, A.H.A.H. and N.F.A.-Z.T.M.; writing—review and editing, N.Z. and N.F.A.-Z.T.M.; supervision, N.Z. All authors have read and agreed to the published version of the manuscript.

Funding: This research was financially supported by the Ministry of Higher Education (MOHE), Malaysia (Trans-disciplinary grant scheme TRGS/2/2014/STG/UPM: VOT number: 5535401).

Institutional Review Board Statement: Not applicable.

Informed Consent Statement: Not applicable.

Data Availability Statement: Data sharing not applicable.

Acknowledgments: The authors are grateful for the outstanding support from all staff of the Department of Chemistry, Faculty of Science, UPM and Laboratory of Biopolymer and Derivatives, Institute of Tropical Forestry and Forest Product, UPM.

Conflicts of Interest: The authors declare that there is no conflict of interest regarding the publication of this paper.

References

1. Yadav, H.; Karthikeyan, C. Natural polysaccharides: Structural features and properties. In *Polysaccharide Carriers for Drug Delivery*; Woodhead Publishing: Sawston, CA, USA, 2019; pp. 1–17. [CrossRef]
2. Cheng, H.N.; Gu, Q.-M. Enzyme-Catalyzed Modifications of Polysaccharides and Poly(ethylene glycol). *Polymers* **2012**, *4*, 1311–1330. [CrossRef]
3. Thakur, V.K.; Thakur, M.K. Processing and characterization of natural cellulose fibers/thermoset polymer composites. *Carbohydr. Polym.* **2014**, *109*, 102–117. [CrossRef]

4. Klemm, D.; Heublein, B.; Fink, H.; Bohn, A. Fascinating Biopolymer and Sustainable Raw Material Angewandte. *Polym. Sci. Cellul.* **2005**, *44*, 3358–3393.
5. Lindman, B.; Karlström, G.; Stigsson, L. On the Mechanism of Dissolution of Cellulose. *J. Mol. Liq.* **2010**, *156*, 76–81. [CrossRef]
6. Kushairi, A.; Loh, S.K.; Azman, I.; Hishamuddin, E.; Ong-Abdullah, M.; Izuddin, Z.B.M.N.; Razmah, G.; Sundram, S.; Parveez, G.K.A. Oil Palm Economic Performance in Malaysia and R&D Progress in 2017. *J. Oil Palm Res.* **2018**, *30*, 163–195.
7. Megashah, L.N.; Ariffin, H.; Zakaria, M.R.; Hassan, M.A. Multi-step pretreatment as an eco-efficient pretreatment method for the production of cellulose nanofiber from oil palm empty fruit bunch. *Asia Pac. J. Mol. Biol. Biotechnol.* **2018**, *26*, 1–8. [CrossRef]
8. Hassan, M.A.; Abd-Aziz, S. Waste and Environmental Management in the Malaysian Palm Oil Industry. In *Palm Oil 2012*; AOCS Press: Urbana, IL, USA, 2012; pp. 693–711. [CrossRef]
9. Shuit, S.H.; Tan, K.T.; Lee, K.T.; Kamaruddin, A.H. Oil palm biomass as a sustainable energy source: A Malaysian case study. *Energy* **2009**, *34*, 1225–1235. [CrossRef]
10. Chen, G.; Tang, W.; Wang, X.; Zhao, X.; Chen, C.; Zhu, Z. Applications of Hydrogels with Special Physical Properties in Biomedicine. *Polymers* **2019**, *11*, 1420. [CrossRef] [PubMed]
11. Buhus, G.; Popa, M.; Desbrieres, J. Hydrogels Based on Carboxymethylcellulose and Gelatin for Inclusion and Release of Chloramphenicol. *J. Bioact. Compat. Polym.* **2009**, *24*, 525–545. [CrossRef]
12. Ambrosio, L.; Demitri, C.; Sannino, A. Superabsorbent Cellulose-Based Hydrogels for Biomedical Applications. In *Biomedical Hydrogels*; Woodhead Publishing Limited: Chicago, IL, USA, 2011; pp. 25–50. [CrossRef]
13. Bhattarai, N.; Gunn, J.; Zhang, M. Chitosan-based hydrogels for controlled, localized drug delivery. *Adv. Drug Deliv. Rev.* **2010**, *62*, 83–99. [CrossRef] [PubMed]
14. Fei, B.; Wach, R.; Mitomo, H.; Yoshii, F.; Kume, T. Hydrogel of biodegradable cellulose derivatives. I. Radiation-induced crosslinking of CMC. *J. Appl. Polym. Sci.* **2000**, *78*, 278–283. [CrossRef]
15. Sharma, S.; Sathasivam, T.; Rawat, P.; Pushpamalar, J. Lycopene-loaded nanostructured lipid carrier from carboxymethyl oil palm empty fruit bunch cellulose for topical administration. *Carbohydr. Polym. Technol. Appl.* **2021**, *2*, 100049. [CrossRef]
16. Young, S.; Wong, M.; Tabata, Y.; Mikos, A.G. Gelatin as a delivery vehicle for the controlled release of bioactive molecules. *J. Control. Release* **2005**, *109*, 256–274. [CrossRef]
17. Dong, M.; Chen, Y. The stimuli-responsive properties of hydrogels based on natural polymers. In *Hydrogels Based on Natural Polymers*; Elsevier: Amsterdam, The Netherlands, 2019; pp. 173–222. [CrossRef]
18. Baldino, L.; Zuppolini, S.; Cardea, S.; Diodato, L.; Borriello, A.; Reverchon, E.; Nicolais, L. Production of biodegradable superabsorbent aerogels using a supercritical CO2 assisted drying. *J. Supercrit. Fluids* **2019**, *156*, 104681. [CrossRef]
19. Reeves, R.; Ribeiro, A.; Lombardo, L.; Boyer, R.; Leach, J.B. Synthesis and Characterization of Carboxymethylcellulose-Methacrylate Hydrogel Cell Scaffolds. *Polymers* **2010**, *2*, 252–264. [CrossRef] [PubMed]
20. Rambo, M.K.D.; Ferreira, M.M.C. Determination of Cellulose Crystallinity of Banana Residues Using Near Infrared Spectroscopy and Multivariate Analysis. *J. Braz. Chem. Soc.* **2015**, *26*, 1491–1499. [CrossRef]
21. Sultana, S.; Islam, M.R. Effect of Mono- And Divalent Salts on The Properties of Carboxymethyl Cellulose Hydrogel Under Irradiation Technique. *Int. J. Chem. Sci.* **2012**, *10*, 627–634.
22. Che Nan, N.F.; Zainuddin, N.; Ahmad, M. Preparation and Swelling Study of CMC Hydrogel as Potential Superabsorbent. *Pertanika J. Sci. Technol.* **2019**, *27*, 489–498.
23. Khaled, B.; Abdelbaki, B. Rheological and Electrokinetic Properties of Carboxymethylcellulose-Water Dispersions in the Presence of Salts. *Int. J. Phys. Sci.* **2012**, *7*, 1790–1798.
24. Tolinski, M. Crosslinking. In *Additives for Polyolefins*; William Andrew Pub.: Oxford, UK, 2009; pp. 215–220.
25. Maitra, J.; Shukla, V.K. Cross-linking in Hydrogels—A Review. *Am. J. Polym. Sci.* **2014**, *4*, 25–31.
26. Nie, H.; Liu, M.; Zhan, F.; Guo, M. Factors on the preparation of carboxymethylcellulose hydrogel and its degradation behavior in soil. *Carbohydr. Polym.* **2004**, *58*, 185–189. [CrossRef]
27. Ngadi, N.; Lani, N.S. Extraction and Characterization of Cellulose from Empty Fruit Bunch (EFB) Fiber. *J. Teknol.* **2014**, *68*. [CrossRef]
28. Tulain, U.R.; Ahmad, M.; Rashid, A.; Iqbal, F.M. development and characterization of smart drug delivery System. *Acta Pol. Pharm.-Drug Res.* **2016**, *73*, 1009–1022.
29. Sharratt, W.N.; Lopez, C.G.; Sarkis, M.; Tyagi, G.; O'Connell, R.; Rogers, S.E.; Cabral, J.T. Ionotropic Gelation Fronts in Sodium Carboxymethyl Cellulose for Hydrogel Particle Formation. *Gels* **2021**, *7*, 44. [CrossRef]
30. Navarra, M.A.; Bosco, C.D.; Moreno, J.S.; Vitucci, F.M.; Paolone, A.; Panero, S. Synthesis and Characterization of Cellulose-Based Hydrogels to Be Used as Gel Electrolytes. *Membranes* **2015**, *5*, 810–823. [CrossRef] [PubMed]
31. Zailan, D.; Hashim, S.N. Factors Affecting Cellulose Dissolution of Oil Palm Empty Fruit Bunch and Kenaf Pulp in NaOH/Urea Solvent. *Sains Malays.* **2018**, *47*, 377–386.
32. Uyanga, K.A.; Daoud, W.A. Carboxymethyl cellulose-chitosan composite hydrogel: Modelling and experimental study of the effect of composition on microstructure and swelling response. *Int. J. Biol. Macromol.* **2021**, *181*, 1010–1022. [CrossRef] [PubMed]
33. Casaburi, A.; Rojo, M.; Cerrutti, P.; Vázquez, A.; Foresti, M.L. Carboxymethyl cellulose with tailored degree of substitution obtained from bacterial cellulose. *Food Hydrocoll.* **2018**, *75*, 147–156. [CrossRef]
34. He, X.; Wu, S.; Fu, D.; Ni, J. Preparation of sodium carboxymethyl cellulose from paper sludge. *J. Chem. Technol. Biotechnol.* **2008**, *84*, 427–434. [CrossRef]

35. Mohtar, S.S.; Busu, T.N.Z.T.M.; Noor, A.M.M.; Shaari, N.; Mat, H. An ionic liquid treatment and fractionation of cellulose, hemicellulose and lignin from oil palm empty fruit bunch. *Carbohydr. Polym.* **2017**, *166*, 291–299. [CrossRef]
36. Khalid, M.; Ratnam, C.T.; Luqman, C.A.; Salmiaton, A.; Choong, T.S.Y.; Jalaludin, H. Thermal and Dynamic Mechanical Behavior of Cellulose- and Oil Palm Empty Fruit Bunch (OPEFB)-Filled Polypropylene Biocomposites. *Polym. Technol. Eng.* **2009**, *48*, 1244–1251. [CrossRef]
37. Akar, E.; Altınışık, A.; Seki, Y. Preparation of pH- and ionic-strength responsive biodegradable fumaric acid crosslinked carboxymethyl cellulose. *Carbohydr. Polym.* **2012**, *90*, 1634–1641. [CrossRef] [PubMed]
38. Salleh, K.M.; Zakaria, S.; Sajab, M.S.; Gan, S.; Kaco, H. Superabsorbent hydrogel from oil palm empty fruit bunch cellulose and sodium carboxymethylcellulose. *Int. J. Biol. Macromol.* **2019**, *131*, 50–59. [CrossRef] [PubMed]
39. Tudorachi, N.; Lipsa, R.; Mustata, F.R. Thermal Degradation of Carboxymethyl Starch-g-Poly(Lactic Acid) Copolymer by TG−FTIR−MS Analysis. *Ind. Eng. Chem. Res.* **2012**, *51*, 15537–15545. [CrossRef]
40. El-Sakhawy, M.; Tohamy, H.-A.S.; Salama, A.; Kamel, S. Thermal properties of carboxymethyl cellulose acetate butyrate. *Cellul. Chem. Technol.* **2019**, *53*, 667–675. [CrossRef]
41. Seki, Y.; Altinisik, A.; Demircioğlu, B.; Tetik, C. Carboxymethylcellulose (CMC)–hydroxyethylcellulose (HEC) based hydrogels: Synthesis and characterization. *Cellulose* **2014**, *21*, 1689–1698. [CrossRef]
42. Parid, D.M.; Abd Rahman, N.A.; Baharuddin, A.S.; Mohammed, M.A.; Johari, A.M.; Razak, S.Z. Synthesis and Characterization of Carboxymethyl Cellulose from Oil Palm Empty Fruit Bunch Stalk Fibres. *BioResources* **2018**, *13*, 535–554.
43. Sreekala, M.S.; Kumaran, M.G.; Thomas, S. Oil Palm Fibers: Morphology, Chemical Composition, Surface Modification, and Mechanical Properties. *J. Appl. Polym. Sci.* **1997**, *66*, 821–835. [CrossRef]
44. Rahman, M.S.; Islam, M.M.; Islam, M.S.; Zaman, A.; Ahmed, T.; Biswas, S.; Sharmeen, S.; Rashid, T.U.; Rahman, M.M. Morphological Characterization of Hydrogels. In *Cellulose-Based Superabsorbent Hydrogels*; Springer International Publishing AG: New York, NY, USA, 2019; pp. 819–863.

polymers

MDPI

Article

Extrusion and Injection Molding of Poly(3-Hydroxybutyrate-co-3-Hydroxyhexanoate) (PHBHHx): Influence of Processing Conditions on Mechanical Properties and Microstructure

Chris Vanheusden [1,2], Pieter Samyn [3], Bart Goderis [4], Mouna Hamid [4], Naveen Reddy [1,2], Anitha Ethirajan [2,5], Roos Peeters [1,2] and Mieke Buntinx [1,2,*]

1 Materials and Packaging Research & Services, Institute for Materials Research (IMO-IMOMEC), Hasselt University, Wetenschapspark 27, 3590 Diepenbeek, Belgium; chris.vanheusden@uhasselt.be (C.V.); naveen.reddy@uhasselt.be (N.R.); roos.peeters@uhasselt.be (R.P.)
2 Division IMOMEC, IMEC vzw, Wetenschapspark 1, 3590 Diepenbeek, Belgium; anitha.ethirajan@uhasselt.be
3 Applied and Circular Chemistry, Institute for Materials Research (IMO-IMOMEC), Hasselt University, Agoralaan Gebouw D, 3590 Diepenbeek, Belgium; pieter.samyn@outlook.be
4 Polymer Chemistry and Materials, KU Leuven, Celestijnenlaan 200F, 3001 Leuven, Belgium; bart.goderis@kuleuven.be (B.G.); mouna.hamid@kuleuven.be (M.H.)
5 Nanobiophysics and Soft Matter Interfaces Group, Institute for Materials Research (IMO-IMOMEC), Hasselt University, Wetenschapspark 1, 3590 Diepenbeek, Belgium
* Correspondence: mieke.buntinx@uhasselt.be; Tel.: +32-11-29-21-55

Citation: Vanheusden, C.; Samyn, P.; Goderis, B.; Hamid, M.; Reddy, N.; Ethirajan, A.; Peeters, R.; Buntinx, M. Extrusion and Injection Molding of Poly(3-Hydroxybutyrate-co-3-Hydroxyhexanoate) (PHBHHx): Influence of Processing Conditions on Mechanical Properties and Microstructure. *Polymers* **2021**, *13*, 4012. https://doi.org/10.3390/polym13224012

Academic Editor: Dan Rosu

Received: 22 October 2021
Accepted: 15 November 2021
Published: 20 November 2021

Abstract: Biobased and biodegradable polyhydroxyalkanoates (PHAs) have great potential as sustainable packaging materials. However, improvements in their processing and mechanical properties are necessary. In this work, the influence of melt processing conditions on the mechanical properties and microstructure of poly(3-hydroxybutyrate-co-3-hydroxyhexanoate) (PHBHHx) is examined using a full factorial design of experiments (DoE) approach. We have found that strict control over processing temperature, mold temperature, screw speed, and cooling time leads to highly increased elongation at break values, mainly under influence of higher mold temperatures at 80 °C. Increased elongation of the moldings is attributed to relaxation and decreased orientation of the polymer chains together with a homogeneous microstructure at slower cooling rates. Based on the statistically substantiated models to determine the optimal processing conditions and their effects on microstructure variation and mechanical properties of PHBHHx samples, we conclude that optimizing the processing of this biopolymer can improve the applicability of the material and extend its scope in the realm of flexible packaging applications.

Keywords: polyhydroxyalkanoates; poly(3-hydroxybutyrate-co-3-hydroxyhexanoate; melt processing; extrusion; injection molding; mechanical properties; elongation at break; crystallization; DoE

1. Introduction

Fossil resource depletion and increased environmental awareness are driving the industry, scientific community, and general population to engage in developing and adopting more sustainable alternatives to conventional oil-based polymers [1]. Tough bioplastics currently represent only about 1% of the plastics produced annually, the market is forecasted to continuously grow to 2.87 million tons in 2025 [2]. Innovative biopolymers such as polyhydroxyalkanoates (PHAs) are one of the main drivers of growth in the field of biobased and biodegradable plastics. They are synthesized via specific bacteria and algae [3,4] from various substrates like glucose, vegetable oil, and glycerin under nutrient limitations as stress conditions [5]. Recently, more sustainable synthesis routes are using waste stream substrates from sugar, coffee, fruit, and milk production [6–8]. Although PHAs can be applied in a wide range of applications like packaging (e.g., films and cutlery), biomedical industry (e.g., drug carriers and tissue engineering) [9], and membrane technology (e.g., filtration) [10], challenges like high production cost, difficulties in processing and lack of clear mechanical property improvement limit their competition with conventional plastics [11].

Poly(3-hydroxybutyrate) (PHB), poly(3-hydroxybutyrate-co-3-hydroxyvalerate) (PHBV), poly(3-hydroxybutyrate-co-3-hydroxyhexanoate) (PHBHHx), and poly(3-hydroxybutyrate-co-4-hydroxybutyrate) (P3HB4HB) are PHA family members that are mostly investigated [12]. These thermoplastic polymers exhibit a wide variety of mechanical properties from hard crystalline to elastic, with medium gas permeability, depending on the co-monomer structure and content [13,14]. Co-polyesters containing PHB units with 3-hydroxyvalerate (3HV), 3-hydroxyhexanoate (3HHx), and 4-hydroxybutyrate (4HB) show reduced brittleness and better thermal stability above the melting point compared to the PHB homopolymer. PHBV and PHBHHx display slower crystallization rates than the homopolymer PHB because the co-monomer units are excluded from the PHB lattice during crystallization from the melt, which can be a challenge for efficient processing of PHBHHx, especially in methods with high cooling rates like injection molding [15,16]. Despite the rise in attempts to further improve both thermal and mechanical properties of PHAs with or without processing aids [17–20], current research regarding PHA processing and compound fabrication mostly relates to batch processing methodologies without the involvement of high temperatures and/or high shear rates, like solution casting [21–23], compression molding [24,25], or spinning techniques [26–28]. While these techniques are practical for use in a lab-scale context, they often have limited applicability on larger scales.

PHBHHx is more ductile compared to PHB and PHBV, with lower crystallinity, lower melting temperature, decreased Young's modulus and increased elongation at break (which is highly dependent on the 3HHx content) [12,29]. However, in practice, the mechanical properties such as tensile strength, elongation at break, and Young's modulus, together with the crystalline structures can strongly be affected by the applied melt processing method and conditions [30–33]. No studies are available to our knowledge that clearly relate the practical processing parameters in continuous melt processing like extrusion and/or injection molding of PHBHHx, with improvements in the mentioned mechanical parameters and how this, in turn, relates to the induced microstructure (orientation and crystallinity). Therefore, the current systematic and detailed study was undertaken, targeting enhanced mechanical properties of PHBHHx copolymers through optimization of the processing conditions. The obtained insights will allow for maximizing the mechanical properties and technical performance.

This paper has a twofold aim, i.e., (i) the optimization of PHBHHx melt processing targeting improved mechanical properties, including high tensile strength, high elongation at break, and—depending on the application—a low or high Young's modulus, and (ii) the correlation of these properties with the processing-induced microstructure. A systematic design of experiments (DoE) approach is used for predictive modeling and simultaneous optimization of four extrusion and injection molding parameters: processing temperature, mold temperature, screw speed, and cooling time. The model predictions are experimentally validated in test runs, targeting a high tensile strength and high elongation at break.

2. Materials and Methods

2.1. Materials

PHBHHx granulate (pellets) (trade name Aonilex X151A) containing 10.5 mol% 3HHx was kindly provided by Kaneka Corporation (Westerlo-Oevel, Belgium) [34]. The PHBHHx pellets were dried for at least three days at 65 °C in a circulating hot air oven prior to melt-processing to remove moisture.

2.2. General Sample Preparation

A lab-scale co-rotating twin-screw extruder (Process 11, Thermo Scientific, Karlsruhe, Germany) and piston injection molding apparatus (Haake MiniJet II, Thermo Scientific,

Karlsruhe, Germany) were used for the production of a dumbbell-shaped specimen. The dried granulate was fed through the feed hopper at a constant speed of 4 rpm into the twin-screw extruder, equipped with standard screws with a diameter of 11.0 mm and an L/D ratio of 40.0. The screw assembly includes feed screw elements (1.0 L/D), three mixing zones with kneading elements (1/4 L/D), and a discharge element (1 1/2 L/D) at the end (Figure 1A). A total granulate sample weight of about 10–15 g was used in order to make three replicates per processing condition. The melted polymer was immediately transferred from the extruder die-end into a heated injection cylinder (Figure 1B), which was mounted on top of a heated dumbbell-shaped mold in which the polymer was ultimately injected (Figure 1C). The injection unit pushes the polymer melt in the mold with a piston, resulting in samples of approximately 1.5 g. Afterward, the polymer samples were manually ejected from the mold and conditioned at 23 °C and 50% RH (relative humidity) for at least three days prior to characterization.

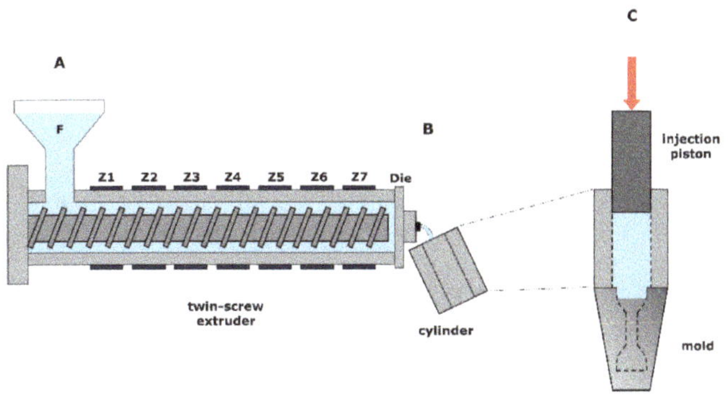

Figure 1. Three-step PHA melt processing with twin-screw extrusion through nine temperature zones (**A**), melt transfer into heated injection cylinder (**B**) and piston melt injection into a heated dumbbell-shaped mold (**C**).

2.3. Full Factorial Design (FFD) Methodology

A 2^k full factorial design (FFD) [35] with four factors at two levels and one center-point was composed to systematically investigate the main extrusion and injection molding parameters that could influence the mechanical properties of PHBHHx. The four independent factors include: (i) the mold temperature **M**, controlled by the injection unit, (ii) the (extrusion) screw speed **S**, the speed of the two co-rotating screws, (iii) the processing temperature **P**, which relates to the extrusion die and injection cylinder temperature, and (iv) the cooling time **C**, which relates to the holding time in the mold after injection and post pressure. The injection molding parameters, such as injection pressure, injection time, post pressure, and post pressure time were held constant for comparison and to obtain consistent samples (Table 1). These parameter settings resulted from an extensive pre-screening, aiming at defining all possible combinations of upper (+) and lower limit settings (−) that led to reproducible and testable samples for analysis. Extruding at temperatures above 160 °C combined with low mold temperatures, for example, resulted in clear deterioration of the mechanical properties, difficult demolding, and defective samples. All process parameters and settings are listed in Table 1.

Table 1. Extrusion and injection molding parameter settings in the FFD approach.

Parameter	Value
Feeding rate (rpm)	4
Screw speed (rpm)	Variable
Processing/Cylinder temperature (°C)	Variable
Mold temperature (°C)	Variable
Cooling time (s)	Variable
Injection pressure (bar)	200
Injection time (s)	15
Post pressure (bar)	150
Post pressure time (s)	10

The final FFD includes 17 experimental runs with low (−), middle (0), and high (+) levels of each independent factor variable (Table 2). The experimental runs were executed in a random order to minimize bias. The settings of the nine temperature zones during extrusion are shown in Table 3. The extruder barrel, die, screws, and injection molding cylinder were cleaned and purged with PHBHHx between individual test runs. Each experimental run includes three repetitions (n = 3).

Table 2. FFD test matrix with melt processing factor symbols (**P**, **M**, **S**, **C**), low (−), middle (0), and high (+) levels.

Independent Variable	Symbol	Low (−)	Middle (0)	High (+)
Processing temperature (°C)	**P**	145	152.5	160
Mold temperature (°C)	**M**	40	60	80
Screw speed (rpm)	**S**	50	100	150
Cooling time (s)	**C**	60	180	300

Table 3. Extrusion temperature profiles controlling nine temperature zones: seven extruder barrel zones (Z), one die temperature (D), and one feeder temperature (F).

F [1]	Z1	Z2	Z3	Z4	Z5	Z6	Z7	D
RT	140	140	140	140	145	145	145	145
RT	140	140	145	145	150	150	153	153
RT	140	140	150	150	155	155	160	160

[1] Feeder temperature is approximately 23 °C, no conditioning system was used for temperature and RH control.

Two specific combinations of settings of **P**, **M**, **S**, and **C** derived from modeling (A and B) were applied to produce samples for the validation testing (n = 5) using the confidence interval confirmation (CIC) approach [36,37].

To investigate the impact of the mold temperature in further detail, additional samples (n = 10) were processed at varying mold temperatures (**M** = 40, 50, 60, 70, 80 °C) with **P** = 145 °C, **S** = 50 rpm, and **C** = 60 s).

2.4. Predictive Modeling

The FFD was analyzed using JMP® Pro Version 15 statistical software (Marlow, UK). The experimental data were used to develop predictive models that correlate melt processing parameters and mechanical properties. These empirical models contain selected first-order terms (main factors) and second-order terms (interactions) as shown in Equation (1):

$$\hat{y} = \alpha_0 + \alpha_1 x_1 + \alpha_2 x_2 + \ldots + \alpha_{12} x_1 x_2 + \alpha_{13} x_1 x_3 + \ldots \quad (1)$$

where $\hat{y}$ is the model response (in this work a specific mechanical property) and α_1, α_2, $\alpha_{12}, \ldots$ the regression coefficients. The included main factors ($x_1, x_2, \ldots$) and interactions ($x_1 x_2, \ldots$) are selected if a backward stepwise regression analysis (standard least squares) indicates significance ($p < 0.05$). The actual experimental models are constructed by placing the midpoint of high (+) and low (−) factor values in the numerator of the coefficients and the difference between low/high and midpoint (0) values in the denominator, following JMP® Pro software.

2.5. Tensile Testing

Tensile tests on injection molded samples were executed following ISO 527-2 at 23 °C and 50% RH with a 10-M systems tensile tester (MTS, Adamel Lhomargy, Roissy-en-Brie, France). Sample geometry was selected according to specimen type 5A (ISO 527-2) with a gauge length of 20 ± 0.5 mm, a width of 4.0 ± 0.1 mm, and a thickness of 2.0 ± 0.2 mm as shown in Figure 2. Tensile testing was executed with manual grips, a 2 kN load cell, a constant crosshead speed of 1 or 20 mm/min, and a grip distance of 50 mm.

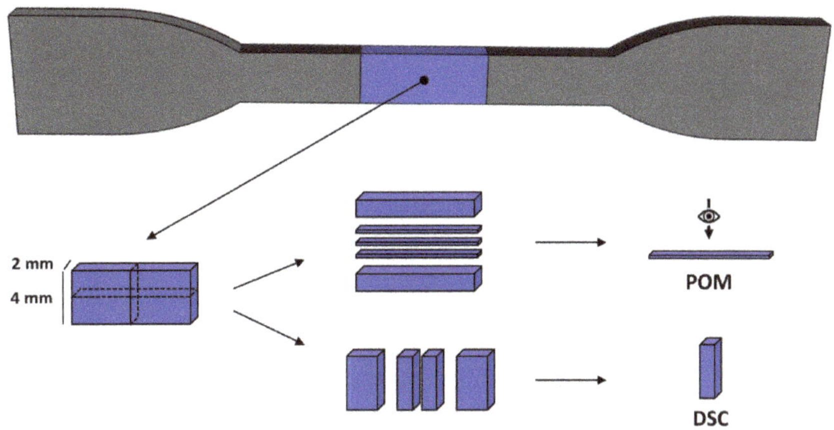

Figure 2. Dimensions of the dumbbell-shaped injection molded samples and cross-sections for POM and DSC analysis.

Tensile strength (TS) was calculated as the yield strength. The elongation at break (ε) used throughout this work is the nominal strain at break, which is calculated based on the grip distance and not the gauge length. The Young's modulus (E) was determined at a crosshead speed of 20 mm/min. Tensile strength, elongation at break, and Young's modulus were calculated with the Equations (2)–(4) respectively:

$$TS = \frac{F}{A} \tag{2}$$

$$\varepsilon = \frac{L - L_0}{L_0} \times 100 \tag{3}$$

$$E = \frac{FL_0}{A\Delta L} \times 100 \tag{4}$$

where L is the final specimen length, L_0 is the initial specimen length, F is the applied force, A is the initial cross-section area, and ΔL is the change in specimen length.

2.6. Polarized Optical Microscopy (POM)

POM was performed on injection molded samples to investigate the semicrystalline morphology and induced microstructure. A BH-2 type microscope (Olympus, Tokyo, Japan) equipped with a ProgRes® CF color camera and ProgRes® Capture Pro software was used. Specimens of the narrow part of the tensile bar center were retrieved and embedded in paraffin wax for increased cutting surface and adequate clamping in the microtome apparatus. The paraffine embedded specimens were cut into 40 µm thin sections along the polymer flow direction with an RM2125 RTS manual rotary microtome (Leica,

Wetzlar, Germany). The thin film samples were measured while the injection direction was at an angle of approximately 45° to the crossed polarizer and analyzer. The direction of measurement and microtome cutting of the injection molded samples are shown in Figure 2.

2.7. Differential Scanning Calorimetry (DSC)

PHBHHx's melting and crystallization behavior before and after processing was evaluated using DSC measurements under inert atmosphere (50 mL/min nitrogen) using a Q200 instrument (TA Instruments, New Castle, DE, USA). Samples of about 6 mg in sealed aluminum pans were heated from −30 °C to 150 °C, before being kept isothermal for 2 min. Then, the samples were cooled to −30 °C and kept constant for 2 min before heating up to 150 °C. The heating/cooling rate was 20 °C/min. The degree of crystallinity (X_c) was calculated using the following equation [38]:

$$X_c = \frac{\Delta H_m}{\Delta H_{m0}} \times 100 \quad (5)$$

where ΔH_m is the melting enthalpy of the formed crystals in the polymer and ΔH_{m0} is the melting enthalpy of the 100% crystalline polymer (115 J/g [16]). All measurements were performed in duplicate.

2.8. X-ray Diffraction (XRD)

The process-induced molecular orientation was assessed with XRD. Experiments were performed on a Xenocs Xeuss Mo Small Wide Angle Scattering instrument (Xenocs, Sassenage, France), comprising a GeniX 3D Molybdenum ultralow divergence X-ray beam delivery system (with wavelength λ = 0.71 Å) at a power of 50 kV−1 mA, a collimating assembly based on scatterless slits, a sample stage, a He flushed flight tube and a Mar345 image plate detector (MARresearch, Norderstedt, Germany). Injection molded samples were mounted perpendicular to the X-ray beam for the collection of 2D diffraction patterns at room temperature with an irradiation time of 30 min. 2D SAXS and WAXD data were simultaneously collected and depicted using the FIT2D software [39]. Angular calibration, azimuthal averaging and φ depended readouts were obtained using a β version of the ConeX software [40].

To quantify the degree of molecular orientation with respect to the injection direction (given by φ = 0), one can derive an orientation function. Orientation functions are best calculated from the φ dependent intensity of WAXD reflections. It is not recommended to extract quantitative orientation functions from SAXS data because in that case, crystal orientation effects cannot easily be separated from crystal size effects. Given the orthorhombic symmetry of the PHB unit cell, the normals to the 110 and 020 crystallographic planes occur at 90° with respect to the molecular chain axis. By consequence and assuming rotational symmetry around the injection axis, the molecular orientation function of chains within crystalline material can be calculated from the azimuthal evolution of the intensity of these reflections. As the foot of the 020 reflection at φ = 90° falls within the central blind area of the detector, it was decided to extract the molecular orientation function, *f*, from the 110 reflection according to Equation (6):

$$f = 1 - 3\langle cos^2\varphi_{110}\rangle \quad (6)$$

with

$$cos^2\varphi_{110} = \frac{\int_0^{90°} I(\varphi)_{110} sin\varphi cos^2\varphi d\varphi}{\int_0^{90°} I(\varphi)_{110} sin\varphi d\varphi} \quad (7)$$

In Equation (7), $I\varphi_{110}$ represents the intensity of the 110 reflection at a given azimuthal angle φ and integrated over 2θ. The peak integrals were obtained after separating them from the patterns using straight sectors.

2.9. Thermogravimetric Analysis (TGA)

TGA was performed to analyze the thermal processing window of the PHBHHx granulate using a TGA 55 (TA Instruments, New Castle, DE, USA). Samples of about 10 mg were weighted in high-temperature platinum pans and heated from 30 °C to 740 °C at a heating rate of 20 °C/min and a nitrogen gas flow of 80 mL/min.

2.10. Gel Permeation Chromatography (GPC)

GPC measurements were performed to determine the molar mass of PHBHHx before and after extrusion. $CHCl_3$-SEC (Size Exclusion Chromatography) was performed on an HLC-8320 GPC (Tosoh EcoSEC, Griesheim, Germany) comprising an autosampler and a PSS guard column SDV (50 × 8 mm^2), followed by two PSS SDV analytical linear M (5 µm, 300 × 8 mm^2) columns at 35 °C and a differential refractive index detector using $CHCl_3$ as the eluent with a flow rate of 1 mL/min. Toluene was used as a flow marker. The system was calibrated using linear, narrow polystyrene (PS) standards ranging from 3.70×10^2–7.775×10^5 g/mol PS (K = 4.9×10^{-5} dL/g and α = 0.794).

3. Results and Discussion

3.1. Full Factorial Design Analysis

The selection of the processing temperature **P**, mold temperature **M**, screw speed **S**, and cooling time **C** in the FFD is based on other PHA studies [33,41–43], preliminary DSC results, and practical considerations from an extensive pre-screening study. This selection of the processing parameters is important as mold and processing temperature are known to influence the mechanical properties of melt-processed bioplastics [44–46].

Table 4 shows a detailed summary of all executed test runs of the FFD matrix with factors and mechanical properties as a response. The average value of three sample measurements is used for calculations (n = 3). Large differences in elongation at break values are apparent between individual test runs. The maximum difference in elongation at break amounts to 263% elongation between individual runs nr. 1 and nr. 5. The Young's modulus varies between 872 MPa (run 15) and 1084 MPa (run 2) and the tensile strength ranges between 20.6 MPa (run 15) and 22.4 MPa (run 2).

Figure 3 shows the main effects plots for tensile strength, elongation at break, and Young's modulus, i.e., the mean response value at each factor level (+, 0, −) [36]. Main effects plots are a relative measure of factor significance on the system response. If the line from low to high factor levels is steep, factors have a high effect on the response. On the contrary, if the line is nearly horizontal, factors have a limited effect on the response. The sign of main effects relates to the direction of the line on the graph, showing a decrease or increase of the average response value. The center points in the main effects plot have to be interpreted with care because they are based on only one center point run (run 17). The main effects are calculated following Equation (8) [36]:

$$E_{f,X} = \overline{F}_{(+1)} - \overline{F}_{(-1)} \tag{8}$$

where $E_{f,X}$ is the main effect of parameter X on the studied response, $\overline{F}_{(+1)}$ is the average response at high-level factor setting, and $\overline{F}_{(-1)}$ is the average response at a low-level factor setting.

Table 4. FFD with factors, levels, and response data (mechanical properties) for tensile strength, elongation at break, and Young's modulus (n = 3, ±1 Standard Deviation (SD)).

Nr	P	M	S	C	TS (MPa)	ε (%)	E (MPa)
1	−	−	−	−	21.6 ± 0.4	103 ± 15	1014 ± 25
2	−	−	−	+	22.4 ± 0.1	113 ± 16	1084 ± 8
3	−	−	+	−	21.4 ± 0.6	194 ± 46	1026 ± 14
4	−	−	+	+	21.5 ± 0.5	143 ± 19	1043 ± 15
5	−	+	−	−	21.5 ± 0.5	366 ± 9	965 ± 26
6	−	+	−	+	21.5 ± 0.2	360 ± 22	906 ± 14
7	−	+	+	−	21.2 ± 0.6	346 ± 27	951 ± 31
8	−	+	+	+	21.3 ± 0.0	321 ± 66	947 ± 17
9	+	−	−	−	20.8 ± 0.7	143 ± 39	1014 ± 30
10	+	−	−	+	21.3 ± 0.2	183 ± 7	1021 ± 8
11	+	−	+	−	21.0 ± 0.3	141 ± 24	1011 ± 18
12	+	−	+	+	21.4 ± 0.4	189 ± 63	1031 ± 13
13	+	+	−	−	20.9 ± 0.4	285 ± 59	918 ± 38
14	+	+	−	+	20.8 ± 0.8	272 ± 59	919 ± 13
15	+	+	+	−	20.6 ± 0.1	297 ± 82	872 ± 27
16	+	+	+	+	20.9 ± 0.3	358 ± 23	889 ± 54
17	0	0	0	0	20.9 ± 0.3	318 ± 78	971 ± 5

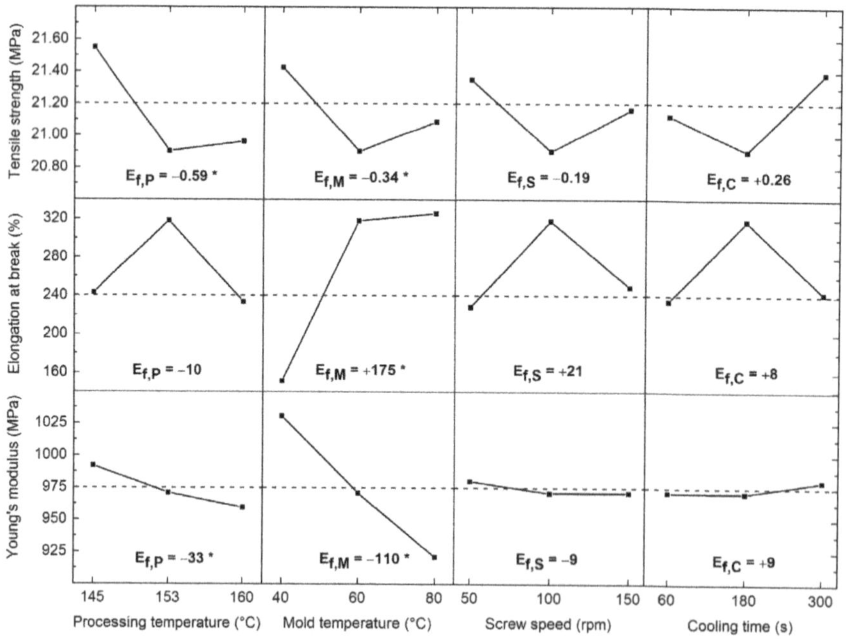

Figure 3. Main effects plot for tensile strength (TS), elongation at break (ε), and Young's modulus (E) showing average values across the FFD matrix for the parameters **M**, **P**, **S**, and **C**. Significant main effects $E_{f,X}$ are indicated with an asterisk.

From Figure 3, it can be seen that processing temperature **P** has the highest influence on tensile strength. Lower processing temperatures result in higher average tensile strength values (negative main effect). The main effect of processing temperature ($E_{f,P}$) on tensile strength equals −0.59 MPa. The effect of mold temperature **M** on tensile strength is lower and equals −0.34 MPa. The influence of screw speed **S** and cooling time **C** on tensile strength is more limited ($E_{f,S}$ = −0.19 MPa and $E_{f,C}$ = +0.26 MPa). It is also apparent from Figure 3 that mold temperature **M** has the highest impact on elongation at break.

High mold temperatures result in extensively increased elongation at break values. The main effect of mold temperature on elongation at break ($E_{f,M}$) equals +175%. On the contrary, the effect of processing temperature **P**, screw speed **S**, and cooling time **C** is limited ($E_{f,P} = -10\%$, $E_{f,S} = +21\%$ and $E_{f,C} = +8\%$). Further, an increased mold temperature **M** results in lower average Young's modulus values (negative main effect). The main effect of mold temperature on Young's modulus ($E_{f,M}$) is −110 MPa. The effect of processing temperature **P** ($E_{f,P} = -33$ MPa) is lower, while the effect of screw speed **S** ($E_{f,S} = -9$ MPa) and cooling time **C** ($E_{f,C} = +9$ MPa) are limited.

3.2. Effect of Processing Parameters on TS, ε, and E

Predictive models to estimate TS, ε, and E, are developed after statistical analysis in JMP® Pro by selecting significant parameters and interactions ($p < 0.05$). Tables 5–7 show the terms, parameter estimates, standard error, and t-ratio of the TS, ε, and E modeling respectively if the *p*-value is below 0.05 in the backward stepwise regression method.

Table 5. Significant terms, estimates, and regression details of the tensile strength (TS) model.

Model Term *	Parameter Estimate	Standard Error	t-Ratio	*p*-Value
Intercept (MPa)	21.24	0.07	311.99	<0.0001
P (°C)	−0.29	0.07	−4.19	0.0009
M (°C)	−0.17	0.07	−2.41	0.0306

* Model terms S, C, P × M, P × S, M × S, P × C, M × C and S × C show a *p*-value > 0.05. Therefore, they are excluded from the model development and not shown in the table.

Table 6. Significant terms, estimates, and regression details of the elongation at break (ε) model.

Model Term *	Parameter Estimate	Standard Error	t-Ratio	*p*-Value
Intercept (%)	243.06	9.68	25.11	<0.0001
M (°C)	87.25	9.98	8.74	<0.0001

* Model terms P, S, C, P × M, P × S, M × S, P × C, M × C and S × C show a *p*-value > 0.05. Therefore, they are excluded from the model development and not shown in the table.

Table 7. Significant terms, estimates, and regression details of Young's modulus (E) model.

Model Term *	Parameter Estimate	Standard Error	t-Ratio	*p*-Value
Intercept (MPa)	975.41	5.42	179.96	<0.0001
P (°C)	−16.31	5.59	−2.92	0.0112
M (°C)	−54.81	5.59	−9.81	<0.0001

* Model terms S, C, P × M, P × S, M × S, P × C, M × C and S × C show a *p*-value > 0.05. Therefore, they were excluded from the model development and not shown in the table.

Table 5 shows that the main factors processing temperature **P** and mold temperature **M** have a significant effect on tensile strength. No main factor interactions are present. Optimal settings for maximum tensile strength (MPa) values are calculated with the following predictive Equation (9):

$$TS = 21.24 - 0.29\left(\frac{P - 152.5}{7.5}\right) - 0.17\left(\frac{M - 60}{20}\right) \quad (9)$$

The intended goal is to increase the strength of the material. Therefore, maximized tensile strength is predicted at processing temperature **P** = 145 °C and mold temperature **M** = 40 °C and amounts to 21.70 MPa, within a 95% confidence interval (CI) of 21.44 MPa and 21.96 MPa.

Our findings of higher TS of PHBHHx samples at lower processing temperature **P** are not in total agreement with the results on PHBV processing of Vandi et al. [47]. In contrast to our results, they observed a lower TS of PHBV using a combination of low processing temperature **P** and low screw speed **S** and attributed this to an incomplete melt consolidation of the material. However, a combination of high processing temperature **P** > 190 °C and high screw speed **S** > 150 rpm also resulted in decreased tensile strength, similar to our results. This was attributed to a lower melt consolidation pressure at the extruder die due to higher melt flows when processing at higher temperatures. A decrease in PHBV mechanical performance at higher process temperatures was also reported elsewhere [33].

An explanation for a decrease in tensile strength at elevated process temperatures could be found in the fact that PHBHHx twin-screw extrusion and injection molding at high temperatures can lead to a reduction in molecular weight due to thermal degradation [18]. Random chain scission has been reported as the degradation mechanism causing a rapid decrease in molecular weight of PHAs during thermal treatment [46]. The theoretical processing window of the PHBHHx granulate is determined as the temperature region between the melting temperature of 130.5 °C (DSC) and the onset degradation temperature of 279.5 °C (TGA) and should be indicative for suitable melt processing of the polymer. Processing at temperatures of **P** = 145–160 °C seems suitable with a granulate peak degradation temperature of 295.3 °C (determined with TGA, and similar to the literature [48]). Although the processing temperature **P** was chosen well in the lower part of this theoretical range, the practical processing window is smaller because thermo-mechanical and thermo-oxidative degradation of the polymer might occur due to high shear (high rotational screw speed) at elevated temperatures [49].

Thermal degradation was quantified in separate GPC (SEC) experiments as the loss in molecular weight after processing at different temperatures. The weight average molecular weight (M_w) and number average molecular weight (M_n) of the unprocessed pellet and samples processed at different temperatures are shown in Table 8. The M_w after processing at 145 °C, 160 °C, 170 °C, and 180 °C decreases respectively with 13%, 23%, 19%, and 48%; the M_n does not decrease at **P** = 145 °C but decreases with 3, 8 and 37% at **P** = 160 °C, 170 °C, and 180 °C, respectively. This reduction in molecular weight might explain the decrease in TS when the temperature is increased (Figure 3). Processing at 180 °C seems not suitable due to a severe decrease in the polymer molecular weight. A decrease of molecular weight of PHB and PHBV (in the range of 4–53%) at elevated temperatures was reported earlier [50] and can decrease the tensile strength [51]. Decreased tensile strength of PLA by reduction of molecular weight was partially attributed to a decrease in chain length and number of chain entanglements [52]. Molecular weight decrease after processing at elevated temperatures was also reported for conventional PS and PMMA [53].

Table 8. Molecular weight details (M_w and M_n) of PHBHHx pellets and samples produced at **P** = 145 °C, 160 °C, 170 °C, and 180 °C determined with GPC (SEC).

Sample	M_w (g/mol)	M_n (g/mol)
Neat pellet	3.1×10^5	1.0×10^5
145 °C	2.7×10^5	1.1×10^5
160 °C	2.4×10^5	9.7×10^4
170 °C	2.5×10^5	9.2×10^4
180 °C	1.6×10^5	6.3×10^4

Concerning the elongation at break modeling shown in Table 6, the statistical analysis indicates that only the mold temperature **M** has a significant effect on this property. No main factor interactions are present. Optimal settings for maximum elongation (%) values are calculated with the following predictive Equation (10):

$$\varepsilon = 243.06 + 87.25\left(\frac{M - 60}{20}\right) \tag{10}$$

The intended goal is to increase the elongation of the material. Therefore, maximized elongation is predicted at mold temperature **M** = 80 °C and amounts to 330%, within a 95% CI of 301% and 360%.

Young's modulus is significantly affected by the main factors mold temperature **M** and processing temperature **P** (Table 7). No main factor interactions are apparent. Young's modulus values (MPa) are calculated with the following predictive Equation (11):

$$E = 975.41 - 16.31\left(\frac{\text{P} - 152.5}{7.5}\right) - 54.81\left(\frac{\text{M} - 60}{20}\right) \quad (11)$$

The intended Young's modulus value of the material is dependent on the application. Therefore, maximized Young's modulus is predicted at 145 °C processing temperature **P** and 40 °C mold temperature **M**, amounting to 1047 MPa (95% CI: 1026–1067 MPa). On the contrary, minimized Young's modulus is predicted at 160 °C processing temperature **P** and 80 °C mold temperature **M**, amounting to 904 MPa (95% CI: 884–925 MPa).

3.3. Predictive Model Validation

The mathematical models described in the previous section highlight the selection of significant processing parameters and their effect on the separate mechanical properties. By combining these models, optimal PHBHHx processing conditions to maximize the elongation at break (while maintaining high tensile strength) are selected as: processing temperature **P** = 145 °C, mold temperature **M** = 80 °C, screw speed **S** = 50 rpm, and cooling time **C** = 60 s (settings A). An optimal mold temperature **M** of 80 °C is selected because the effect of a low mold temperature **M** = 40 °C on tensile strength is limited. Since long(er) cooling times of samples in the mold (300 s) lead to increased injection molding cycle times and lower productivity, an optimal cooling time **C** of 60 s is selected because of the influence of this parameter on mechanical properties—especially elongation at break—is limited. According to the model and for comparison, processing temperature **P** = 160 °C, mold temperature **M** = 40 °C, screw speed **S** = 50 rpm, and cooling time **C** = 60 s, are considered as non-optimal processing conditions resulting in lower elongation at break and tensile strength (settings B), while minimizing the injection molding cycle time.

The goal of the FFD and model development is to predict optimal processing conditions for maximized mechanical properties (i.e., the intended use), not to predict the total design space in detail. Therefore, only a validation of the models for both predicted settings A and B is performed. The validation of the predictive models is executed following the confidence interval confirmation approach (CICon). This approach includes the calculation of confidence intervals using the confirmation runs themselves, rather than the confidence intervals of the experimental runs [36,37]. If the predicted values of the mechanical properties fall within this confidence interval (CI), the model is considered appropriate. The predicted mechanical properties are validated at a tensile test speed of 20 mm/min for both melt processing settings and are shown in Table 9. A 95% CI is used to assess the prediction quality of the model.

Table 9. Validation of TS, ε and E of PHBHHx samples produced using 2 different processing conditions: (A) **P** = 145 °C, **M** = 80 °C, **S** = 50 rpm, **C** = 60 s; (B) **P** = 160 °C, **M** = 40 °C, **S** = 50 rpm, **C** = 60 s (n = 5, α = 0.05).

	Tensile Speed (mm/min)	Tensile Strength (MPa)		Elongation at Break (%)		Young's Modulus (MPa)	
		Predicted	CI Observed	Predicted	CI Observed	Predicted	CI Observed
A	20	21.4	21.4–22.0	330	320–342	937	883–1007
	1	/	20.5–20.7	/	135–276	/	972–995
B	20	21.1	21.3–22.7	156	153–325	1014	940–1205
	1	/	20.0–20.1	/	19–93	/	1011–1027

The predictive models show good estimations for elongation at break and Young's modulus. For tensile strength, however, they show a slight underestimation for both test settings. The underestimations can be attributed to very small deviations of tensile strength in function of processing conditions, resulting in relatively small confidence intervals.

Higher prediction accuracy for tensile strength can be achieved by adding repetitions, replications, or extra points to the design.

In addition, samples from both validation runs are also compared at a tensile test speed of 1 mm/min. As the model is built on tensile properties measured at 20 mm/min, no prediction values for 1 mm/min are available in Table 9. Tensile testing at a lower speed (1 mm/min) shows similar results for Young's modulus at a test speed of 20 mm/min. However, testing the PHBHHx samples at low speed shows lower values of tensile strength and elongation at break compared to higher speeds.

Referring to the first objective, this study shows that PHBHHx can be melt-processed in products with TS ranging between 20–22 MPa, ε ranging between 19–342%, and E ranging between 883–1205 MPa depending on the processing conditions and tensile test method. The tensile strength of PHBHHx is relatively low compared to semi-crystalline PLA (50–70 MPa) [54], isotactic PP (29–39 MPa), PHB (40 MPa) [12], PHBV (30–38 MPa) [12] and LDPE (15–79 MPa) [12]. The elongation at break of PHBHHx is high compared to semi-crystalline PLA (4%) [54], PHBV (0.8–58%) [55–57] and PHB (3–8%) [12], and closer to ε values found for isotactic PP (500–900%) [13] and LDPE (150–600%) [12]. The Young's modulus of PHBHHx is higher compared to LDPE (50–100 MPa) and in line with values found for isotactic PP (1000–1700 MPa) [12], but lower as compared to semi-crystalline PLA (3000 MPa) [54], PHBV (700–2900 MPa) [12] and PHB (3500–4000) [12]. In conclusion, the flexibility of PHBHHx can be adapted within the range of flexible packaging materials by an appropriate selection of processing conditions.

3.4. *Influence of the Mold Temperature on Mechanical Properties*

The FFD and developed prediction models revealed that the mold temperature **M** has a significant impact on the mechanical properties of the melt-processed PHBHHx (Table 4). Figure 4 graphically shows that the elongation at break of injection molded samples significantly increases with mold temperature **M** up to 80 °C and deteriorates at low mold temperature **M** = 40 °C.

In order to investigate the underlying causes and to fully characterize mold temperature dependence, additional PHBHHx samples were produced at mold temperatures **M** = 40 °C, 50 °C, 60 °C, 70 °C and 80 °C (**P** = 145 °C, **S** = 50 rpm, **C** = 60 s). The mechanical properties were characterized by tensile testing at 20 mm/min. Figure 5 provides the experimental data for TS, ε, and E in relation to the mold temperature **M**. A clear trend of increasing elongation at break with mold temperature is observed, with stagnation at 60–70 °C. On the contrary, both the tensile strength and Young's modulus decrease with increasing mold temperature. Similarly, it is apparent from the FFD matrix (Table 4) that ε values increase on average from 103% when molding at 40 °C, to 366% when molding at 80 °C. In addition, there is less variation in ε for samples molded at 80 °C, which can be an indication of consistent microstructure development. The combination of increased elongation and decreased Young's modulus at higher mold temperatures shows that PHBHHx is rendered more ductile and less stiff when processed at optimal conditions (A) applying a higher mold temperature.

3.5. *Influence of the Mold Temperature on Process-Induced Microstructure*

The thermal and mechanical behavior of PHBHHx is strongly dependent on the crystal structure [29,32]. In order to investigate the observed impact of the mold temperature **M**, on TS, ε, and E properties, DSC, POM, and XRD measurements are performed to characterize the developed crystallinity and microstructure during processing.

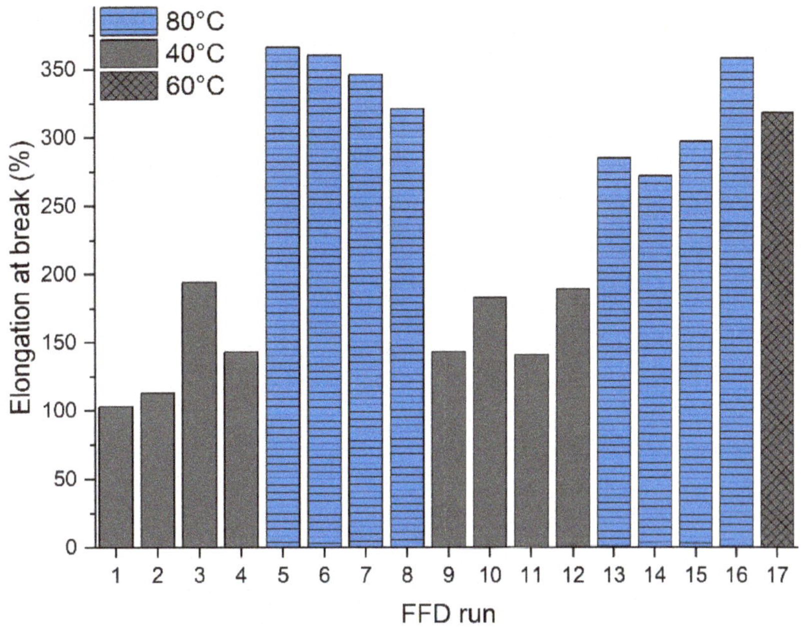

Figure 4. Individual value plot of the 17 run FFD for elongation at break (ε), grouped according to mold temperature **M** = 40 °C (grey), **M** = 80 °C (blue), and **M** = 60 °C center point run (grey, cross) showing increased elongation at break for high mold temperatures (only mean values are represented for clarity, statistical error is included in Table 4).

The DSC first heating curves of samples molded at temperatures of **M** = 40–80 °C are shown in Figure 6. The glass transition temperature (T_g) of all samples equals −1 to 1 °C (similar to the literature [48,58]). As no shift is observed, T_g is not shown in Figure 6. Three endothermal peaks are apparent in this first heating curve. The small peak I at ±84–89 °C can be attributed to the melting of secondary crystallites formed during secondary crystallization/molding. Peak II at ±107–119 °C is due to the melting of primary lamellae [16,59]. Peak III is due to the melting of lamellae formed through thickening and reorganization during DSC heating [16,32]. It is apparent from Figure 6 that the melting peak temperature of the primary lamellae (peak II) shifts to higher temperatures and the peak area increases when processed at elevated mold temperatures (less recrystallization during DSC). Allowing time to crystallize at higher temperatures in the mold leads to increased crystal sizes with higher melting points. In addition, the area of peak III decreases with increasing mold temperature, indicating a less exhaustive recrystallization process during heating. All peak temperatures are shown in Table 10.

The calculation of the processing-induced crystalline content is based on the sum of enthalpies (ΔH_m) of the three endothermic peaks. These peaks hide the exothermic heat due to recrystallization, which is expected to be as large as the endothermic melting enthalpy of the recrystallized crystals. As a result, the full integral actually corresponds to the melting enthalpy of the processing induced primary and secondary crystals. The crystallinity remains constant with increasing mold temperature **M**, around ±38–39% (Table 10). These results differ from a study on PLA, where almost no crystallization was observed after molding at room temperature or up to 50 °C [60,61]. Crystal formation and enlargement in PLA was achieved by post-annealing at temperatures of 80–120 °C, with a maximal increase of elongation at break when annealing occurred at 80 °C for 0.5 and 2 h [61]. As the mold temperature **M** has no effect on the crystallinity of the injection molded PHBHHx, other differences in microstructure, orientation effects, in particular, are investigated hereafter to account for the observed variations in mechanical properties.

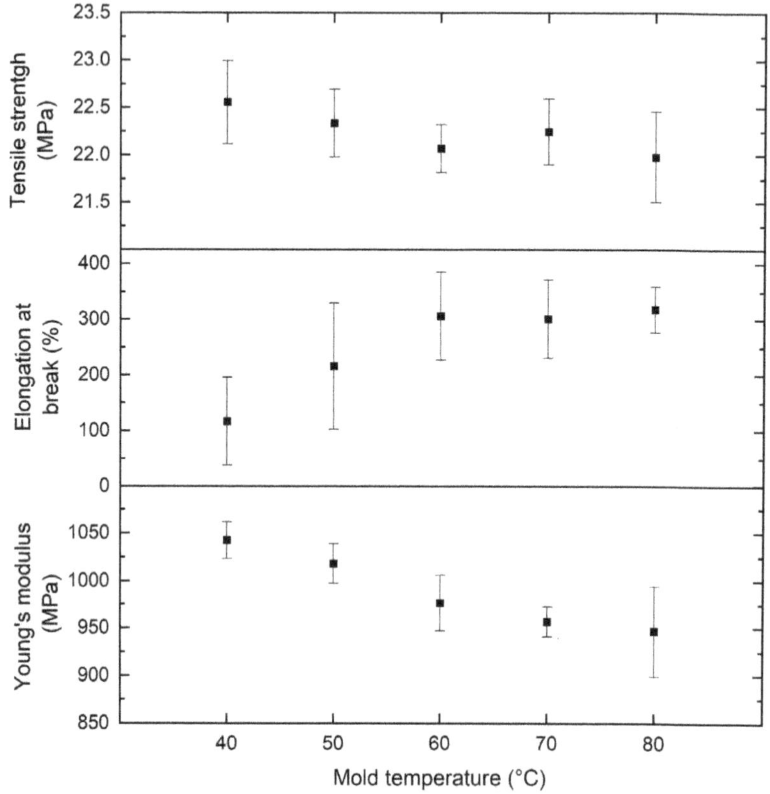

Figure 5. Correlation of mold temperature **M** and mean mechanical properties: tensile strength, elongation at break, and Young's modulus (n = 10, ± 1 SD).

Several studies have shown that melt processing parameters like mold temperature **M** and processing temperature **P** can induce order of magnitude variations in mechanical properties like elongation at break, due to microlayer development in the moldings [62–64]. A conventional three-layer process-induced morphology has been reported for a range of materials (PP, polyamide, poly(butylene terephthalate) …), with different semi-crystalline characteristics between the skin, shear, and core layers [65–67]. By increasing the mold temperature, the shear and skin layers reduce in thickness, providing more volume to the spherulitic core layer [68–70]. Morphological analysis on PHBHHx samples produced at mold temperatures **M** from 40 °C to 80 °C was performed in order to investigate the process-induced microstructure.

POM micrographs representing outer (I), middle (II), and inner (III) regions of a sample produced at **M** = 40 °C are shown in Figure 7. It can be seen that a distinctive skin-core layer structure is apparent, with high birefringence in the skin region and a gradual brightness decrease towards the core. The skin birefringence is an indication of molecular chain orientation [71]. The birefringence is absent towards the core because the polymer chains are randomly oriented, becoming optically isotropic [71]. The skin and shear layer are defined as regions A and B in Figure 7 based on birefringence intensity. The thickness of the skin (A) and shear layer (B) are respectively ±35 μm (±1.75% of sample thickness) and ±65 μm (±3.25% of sample thickness) on both sides of the sample. Hence, the total skin-like region is approximately 10% of the sample thickness. No spherulites are visible in the core region (no Maltese crosses) because it might be that the thin films (40 μm) are thicker than the spherulite size.

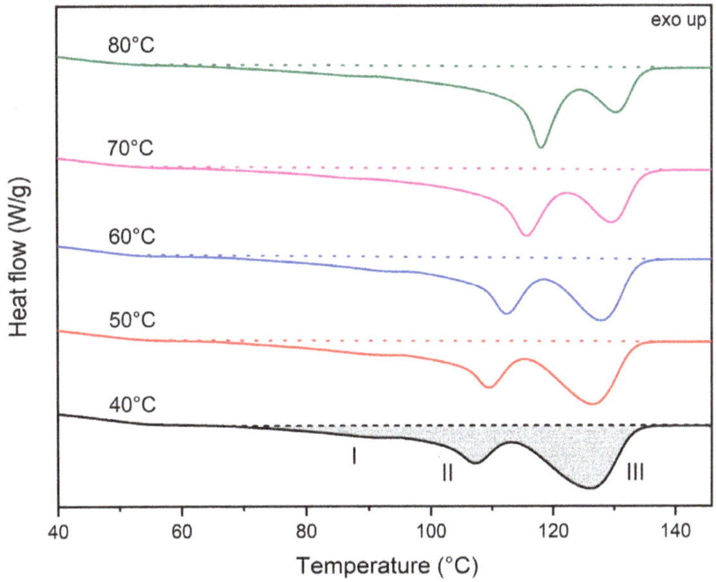

Figure 6. DSC first heating curves of PHBHHx samples with variable mold temperatures **M** = 40 to 80 °C with three endothermal peaks (I–III) for crystallinity calculations.

Table 10. Melting peak enthalpy ($\Delta H_m = \Delta H_{m,I} + \Delta H_{m,II} + \Delta H_{m,III}$), endothermal peak temperatures (T_I, T_{II}, and T_{III}) and percentage crystallinity (X_c) determined with DSC at 20 °C/min for mold temperatures **M** ranging from 40 °C to 80 °C (n = 2, ±1 SD).

Mold Temperature (°C)	m (J/g)	T_I (°C)	T_{II} (°C)	T_{III} (°C)	X_c (%)
40	43.8 ± 1.2	86.4 ± 0.1	107.4 ± 0.5	125.9 ± 0.0	38.1 ± 1.0
50	44.1 ± 2.5	87.7 ± 1.6	109.9 ± 0.7	126.6 ± 0.3	38.4 ± 2.1
60	43.8 ± 2.4	88.8 ± 0.1	112.4 ± 0.0	127.6 ± 0.0	38.1 ± 2.1
70	45.3 ± 0.4	84.2 ± 0.0	115.6 ± 0.1	129.5 ± 0.5	39.4 ± 0.3
80	44.2 ± 0.7	84.9 ± 1.0	118.5 ± 0.6	130.9 ± 0.6	38.4 ± 0.6

POM of samples produced at mold temperature **M** = 50 °C (Figure 8), shows a decrease in the thickness of the skin layer to ±10 μm (±0.50% of sample thickness) on both sides of the sample, without a distinguishable transition layer towards the core. The skin-like region is reduced from 10% at **M** = 40 °C to approximately 1% at **M** = 50 °C, with respect to the total sample thickness. The other POM micrographs in Figure 8 show that skin formation is absent at higher mold temperatures of **M** = 60, 70, and 80 °C. The oriented polymer chains are relaxed at elevated mold temperatures prior to solidification, which reduces orientational birefringence [71]. Hence, these moldings are characterized by a homogeneous and less oriented microstructure throughout the whole sample. At **M** = 80 °C, spherulites are visible because their size is increased compared to samples at **M** = 40 °C.

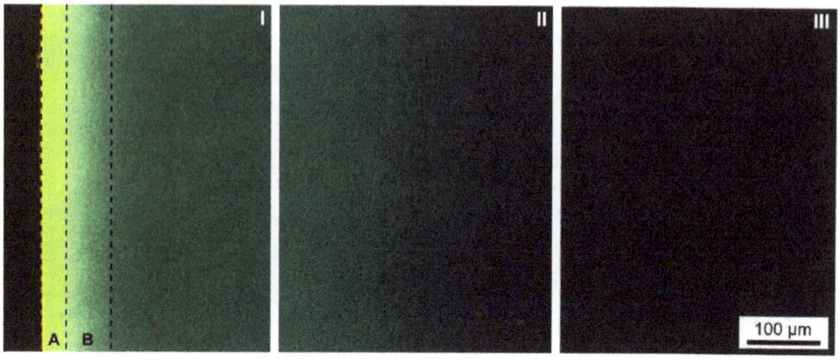

Figure 7. Optical analysis. POM micrographs on 40 µm thin sections with mold temperature **M** = 40 °C showing birefringence gradient existing in the different regions from the skin to the core (**I–III**). (**A**) and (**B**) represent the skin and shear-like layers, respectively.

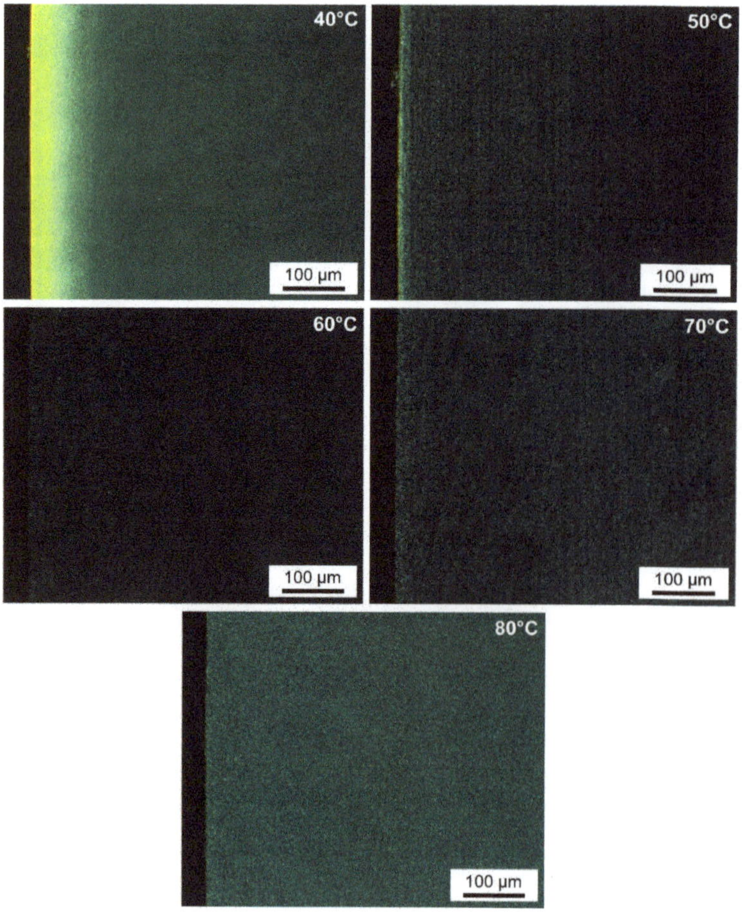

Figure 8. POM micrographs with 20× zoom on 40 µm thin sections for mold temperatures **M** ranging from 40–80 °C.

XRD data were obtained in transmission through the central part of the tensile test bar. As a result, these patterns are superpositions of the scattering coming from the bulk as well as the surface of the samples. The 2D scattering patterns in Figure 9 (left side) of **M** = 40 °C and 80 °C show typical PHB reflections with Miller indices 020, 110, 031, and 040 [72]. The individual scattering angle (2θ) scans along φ directions of 0°, 45°, 90°, 135°, and 180° (Figure 9, right side) are compared to the azimuthally averaged patterns (full lines) obtained by averaging the intensities over al φ directions. The injection molding flow direction is parallel to the meridional direction (φ = 0°).

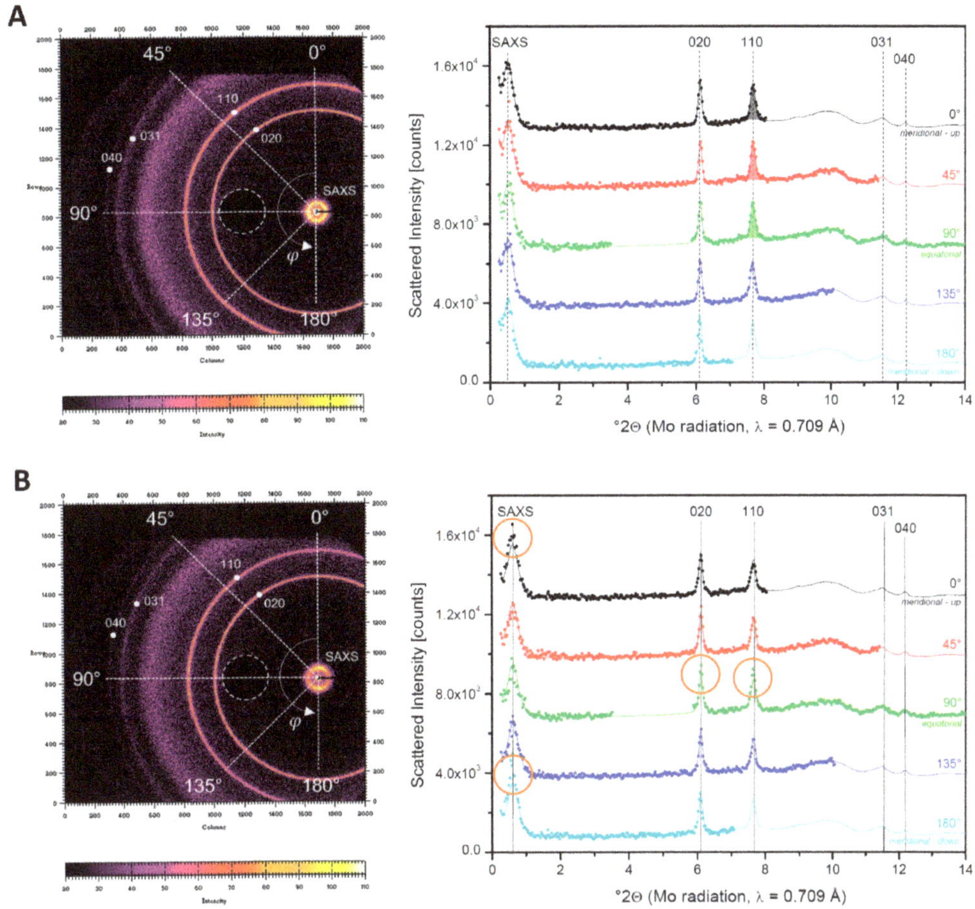

Figure 9. 2D scattering patterns of samples **M** = 80 °C (**A**) and **M** = 40 °C (**B**) with an indication of the PHB typical reflections with Miller indices (**left**) and corresponding individual scans along different φ directions (**right**). The white circled regions in the 2D scattering patterns are detector blind regions.

For **M** = 80 °C (Figure 9A), the scans over the different φ directions yield intensities equal to those of the azimuthal average. This means that no orientation is apparent.

For **M** = 40 °C (Figure 9B), the SAXS peaks are shifted to larger scattering angles compared to **M** = 80 °C, implying that the crystal-to-crystal separation for **M** = 40 °C is shorter. As the (DSC based) crystallinity for **M** = 40 °C is equal to that of **M** = 80 °C, one can conclude that the crystals for **M** = 40 °C, are thinner. No efforts were made to further quantify the size differences. The reflections in the WAXD part have not shifted drastically and peak intensities are comparable, indicative of a similar crystallinity as in samples with **M** = 80 °C and confirming the presented DSC data. The circled intensities of the angular scans (Figure 9B, right) are larger than the azimuthal average and point at (partial) molecular orientation along the injection axis (φ = 0°). It can be seen that the 020 and 110 reflections are stronger along the equator, meaning that a fraction of the chains (logically residing in the skin layer) are oriented parallel to the injection molding direction. However, the crystalline lamellar stacks are oriented perpendicular to this direction because the SAXS intensity is stronger along the meridional. This scattering behavior is typical of shish kebab structures as, e.g., observed earlier for polyethylene, which also has an orthorhombic crystal structure [73]. The orthogonal relation between the WAXD and SAXS reflections furthermore indicates that the chains within the crystals are oriented perpendicular to the lamellar surface direction.

It can qualitatively be derived from the scattering pattern that the orientation for **M** = 40 °C is not very large. This corresponds well with the POM finding that orientation is limited to the skin. The scattering from the core is isotropic, comparable to **M** = 80 °C. To quantify the degree of molecular orientation with respect to the injection direction (given by φ = 0), orientation functions f were derived from the φ dependent intensity of the 110 reflection. Intensities are obtained by integrating the 110 peaks after separating them from the patterns using straight sectors. To illustrate the integration procedure, the areas of the 110 reflections for φ equal to 0, 45, and 90 °C, are color shaded in Figure 9A. The values of the orientation function f in relation to mold temperature **M**, tensile properties and crystallinity are shown in Table 11. Zero values for f are obtained when the crystals and the chains they contain are oriented randomly and values of 1 correspond to perfect alignment with respect to the injection melt flow direction [74]. The molecular orientation is significant for **M** = 40 °C and 50 °C, but very small for the other mold temperatures **M** = 60–80 °C. The as-obtained molecular orientation parameter f only relates to the crystalline parts of the material. The overall molecular orientation is presumably less than given by f because amorphous matter more readily randomizes. This difference may be more important for **M** = 50 °C compared to **M** = 40 °C, as the POM experiments clearly indicate that the overall orientation in the skin is less developed for **M** = 50 °C. The increased molecular orientation at low mold temperature occurs due to polymer chain molecules being aligned, sheared, and stretched in the direction of flow during the injection. The frozen orientation is more retained at faster cooling rates of the polymer melt [75], i.e., at lower mold temperatures. The orientation effect is highly pronounced in polymer regions near the mold wall, where molecules are frozen in their stretched state, forming a skin-like layer. Taken together, the XRD and POM results suggest that processing at low mold temperature (**M** = 40 °C) induces skin formation and increased molecular orientation in the sample, while the latter is minimized at higher mold temperatures (**M** = 80 °C), i.e., at slow cooling rates.

Table 11. Overview on the influence of mold temperature on the orientation function f, tensile properties (TS, ε and E), and crystallinity content (X_c).

Mold Temperature (°C)	f *	TS (MPa)	ε (%)	E (MPa)	X_c (%)
40	0.14	22.6 ± 0.4	117 ± 79	1043 ± 19	38.1 ± 1.0
50	0.13	22.3 ± 0.4	216 ± 113	1018 ± 21	38.4 ± 2.1
60	0.07	22.1 ± 0.3	306 ± 79	977 ± 30	38.1 ± 2.1
70	0.04	22.2 ± 0.3	301 ± 70	957 ± 16	39.4 ± 0.3
80	0.02	22.0 ± 0.5	318 ± 41	947 ± 48	38.4 ± 0.6

* Error of the orientation function f is estimated to be ± 0.02.

The results of the microstructural analysis from DSC, POM, and XRD measurements indicate that the mold temperature changes the crystallization and orientation conditions of the melt during processing. Molding at low mold temperature (40 °C) produces small crystals with high molecular orientation in a skin-like structure comprising approximately 10% of the sample thickness. Due to the stretching and shearing of the polymer chains during injection molding, the inherent maximum stretch and strength of the chains are nearly reached compared to their initial relaxed state [75]. This orientation results in slightly increased tensile strength and Young's modulus but highly reduced elongation at break. On the contrary, the selected optimal processing condition (A) with higher mold temperature (80 °C) gives rise to slower cooling and a suitable time interval for the polymer chains to relax before solidification by crystallization, compared to molding at a lower temperature. Processing of PHBHHx at higher mold temperature reduces process-induced skin formation and molecular orientation. The produced samples have increased crystal size (derived from DSC and qualitatively from SAXS) and a homogeneous microstructure throughout the sample thickness (derived from POM). This microstructure results in increased elongation at break values because the randomly curled and oriented polymer chains can be stretched to a further extent [76]. In addition, increased elongation values at **M** = 80 °C are also likely to be related to lower residual stress of the polymer chains induced by slower cooling rates of the melt during molding [71,77].

In this study, a higher mold temperature was found to cause improved elongation at break due to a reduction of molecular orientation and skin formation. The increased elongation at break was not attributed to changes in the overall crystallinity. This finding is contrary to previous studies which have suggested that higher mold temperatures result in lower elongation at break, mainly due to the formation of larger spherulites and increased crystallinity. Fast cooling (quenching) of the PHB melt is related to the formation of smaller crystallites, while slow cooling results in larger spherulites [31]. These large spherulites are more brittle, giving rise to poor mechanical properties like short elongation and low impact strength. Increased crystallinity at elevated mold temperatures was reported for PLA [45] and isotactic PP [78,79], leading to decreased elongation at break and impact properties. This negative effect on mechanical properties was attributed to a volumetric increase of more brittle/rigid crystalline regions and a decrease of loosely arranged amorphous regions, causing brittle instead of ductile failure.

Several other attempts to improve the mechanical properties of PHAs were made previously by incorporating for example fillers and plasticizers. The addition of plasticizers in PHB can improve the elongation at break with values up to 45% but at the expense of tensile strength and Young's modulus [17]. Some talc-filled PHBHHx compounds remain very brittle with elongation values under ±5% [18], while the addition of *L*-phenylalanine [19] and ultrafine talc [20] nucleating agents have almost no effect on elongation at break. The limited increase of mechanical properties, i.e., elongation at break for PHAs in these studies highlights the importance of process optimization, as discovered in this work. As shown here, elongation at break values can be increased extensively with average values up to ±175% by optimizing process conditions. However, increasing the crystallization rate and nucleation by incorporation of fillers and the combination of efficient and optimal processing conditions can possibly improve and modify the mechanical properties of PHBHHx to a further extent. In addition, PHBHHx post-annealing can be a valuable approach to promote crystallization and improve mechanical properties, as previously shown for PLA [80].

4. Conclusions

In this study, the influence of melt processing parameters (processing temperature **P**, mold temperature **M**, screw speed **S**, and cooling time **C**) on the mechanical properties and microstructure of the injection molded PHBHHx was systematically investigated by a full factorial design of experiments approach. Increased tensile strength and elongation at break values are found when PHBHHx is processed at relatively low extrusion and injection molding temperature profiles of 140 to 145 °C (**P**), relatively high mold temperatures of 80 °C (**M**), low screw speeds of 50 rpm (**S**) and short cooling times in the mold of 60 s (**C**). Increased cooling in the mold of 300 s gives rise to slightly increased tensile strength but is not practical for use in industry. The statistically developed empirical models predict an optimal elongation at a break value of 330% at optimized processing conditions (A) with high mold temperatures compared to 156% at non-optimal processing conditions (B) with low mold temperatures.

This extensive increase in elongation at break of PHBHHx moldings at optimal processing conditions is mainly attributed to high mold temperatures, retarding the cooling in the mold. This results in a suitable time interval to allow chain relaxation prior to crystallization-induced solidification. An increase in the mold temperature does not change the sample crystallinity content under the optimized processing conditions. In contrast, molding at a lower temperature (40 °C) induces partial polymer orientation and skin-core formation, resulting in lower elongation values.

This study highlights the importance of optimal melt processing and the influence on both mechanical properties and developed microstructure. Further research includes the combination of process optimization, incorporation of fillers, and post-annealing to further improve the mechanical properties of these PHA materials. This knowledge can contribute to the development of innovative PHA materials for packaging and other applications, taking into account the importance of the relation between melt process parameters and the possibility to enhance the mechanical properties of PHBHHx by process optimization. Our results could be used as an initial guideline for appropriate processing at larger scales. However, because requirements regarding product dimensions, pressure, cycle time, and throughput are different at an industrial scale, a similar optimization strategy to fabricate specific products might be necessary. Processing of PHBHHx on industrial scale equipment can include sheet and film extrusion, as well as thin-walled injection molding for packaging applications, using appropriate mold and chill roll temperatures.

Author Contributions: Conceptualization, C.V., P.S. and M.B.; methodology, C.V., P.S., B.G. and M.B.; validation, C.V., P.S., B.G. and M.B.; formal analysis, C.V. and M.H.; resources, R.P.; writing—original draft preparation, C.V. and M.B.; writing—review and editing, C.V., M.B., B.G., P.S., A.E., N.R. and R.P.; visualization, C.V.; supervision, M.B.; funding acquisition, C.V., P.S., A.E. and M.B. All authors have read and agreed to the published version of the manuscript.

Funding: This research was funded by the Special Research Fund (BOF) of Hasselt University, grant number BOF20DOC06.

Institutional Review Board Statement: Not applicable.

Informed Consent Statement: Not applicable.

Data Availability Statement: Data is contained within the article.

Acknowledgments: The authors acknowledge Wouter Marchal (Analytical & Circular Chemistry, IMO-IMOMEC, UHasselt) for access to DSC and TGA experiments, Katrien Wauterixckx and Niels Hellings (BIOMED, UHasselt) for using the microtome and Greg Quintens for GPC experiments (Polymer Reaction Design group, UHasselt).

Conflicts of Interest: The authors declare no conflict of interest.

References

1. Filho, W.L.; Salvia, A.L.; Bonoli, A.; Saari, U.A.; Voronova, V.; Klõga, M.; Kumbhar, S.S.; Olszewski, K.; De Quevedo, D.M.; Barbir, J. An Assessment of Attitudes towards Plastics and Bioplastics in Europe. *Sci. Total Environ.* **2021**, *755*, 142732. [CrossRef]
2. European Bioplastics. *Bioplastics Market Development Update 2020*; European Bioplastics: Berlin, Germany, 2020; p. 1.
3. Możejko-Ciesielska, J.; Marciniak, P.; Szacherska, K. Polyhydroxyalkanoates Synthesized by Aeromonas Species: Trends and Challenges. *Polymers* **2019**, *11*, 1328. [CrossRef]
4. García, G.; Sosa-Hernández, J.E.; Rodas-Zuluaga, L.I.; Castillo-Zacarías, C.; Iqbal, H.; Parra-Saldívar, R. Accumulation of PHA in the Microalgae Scenedesmus Sp. under Nutrient-Deficient Conditions. *Polymers* **2020**, *13*, 131. [CrossRef]
5. Adeleye, A.T.; Odoh, C.K.; Enudi, O.C.; Banjoko, O.O.; Osiboye, O.O.; Toluwalope Odediran, E.; Louis, H. Sustainable Synthesis and Applications of Polyhydroxyalkanoates (PHAs) from Biomass. *Process Biochem.* **2020**, *96*, 174–193. [CrossRef]
6. Dutt Tripathi, A.; Paul, V.; Agarwal, A.; Sharma, R.; Hashempour-Baltork, F.; Rashidi, L.; Khosravi Darani, K. Production of Polyhydroxyalkanoates Using Dairy Processing Waste–A Review. *Bioresour. Technol.* **2021**, *326*, 124735. [CrossRef]
7. Ganesh Saratale, R.; Cho, S.-K.; Dattatraya Saratale, G.; Kadam, A.A.; Ghodake, G.S.; Kumar, M.; Naresh Bharagava, R.; Kumar, G.; Su Kim, D.; Mulla, S.I.; et al. A Comprehensive Overview and Recent Advances on Polyhydroxyalkanoates (PHA) Production Using Various Organic Waste Streams. *Bioresour. Technol.* **2021**, *325*, 124685. [CrossRef] [PubMed]

8. Meléndez-Rodríguez, B.; Torres-Giner, S.; Reis, M.A.M.; Silva, F.; Matos, M.; Cabedo, L.; Lagarón, J.M. Blends of Poly(3-Hydroxybutyrate-Co-3-Hydroxyvalerate) with Fruit Pulp Biowaste Derived Poly(3-Hydroxybutyrate-Co-3-Hydroxyvalerate-Co-3-Hydroxyhexanoate) for Organic Recycling Food Packaging. *Polymers* **2021**, *13*, 1155. [CrossRef] [PubMed]
9. Luo, Z.; Wu, Y.; Li, Z.; Loh, X.J. Recent Progress in Polyhydroxyalkanoates-Based Copolymers for Biomedical Applications. *Biotechnol. J.* **2019**, *14*, 1900283. [CrossRef]
10. Tomietto, P.; Loulergue, P.; Paugam, L.; Audic, J.-L. Biobased Polyhydroxyalkanoate (PHA) Membranes: Structure/Performances Relationship. *Sep. Purif. Technol.* **2020**, *252*, 117419. [CrossRef]
11. Li, Z.; Yang, J.; Loh, X.J. Polyhydroxyalkanoates: Opening Doors for a Sustainable Future. *NPG Asia Mater.* **2016**, *8*, e265. [CrossRef]
12. Ragaert, P.; Buntinx, M.; Maes, C.; Vanheusden, C.; Peeters, R.; Wang, S.; D'hooge, D.R.; Cardon, L. Polyhydroxyalkanoates for Food Packaging Applications. In *Reference Module in Food Science*; Elsevier: Amsterdam, The Netherlands, 2019; ISBN 978-0-08-100596-5.
13. Chen, G.-Q. A Microbial Polyhydroxyalkanoates (PHA) Based Bio- and Materials Industry. *Chem. Soc. Rev.* **2009**, *38*, 2434. [CrossRef]
14. Jost, V. Packaging Related Properties of Commercially Available Biopolymers–An Overview of the Status Quo. *Express Polym. Lett.* **2018**, *12*, 429–435. [CrossRef]
15. Bugnicourt, E.; Cinelli, P.; Lazzeri, A.; Alvarez, V.A. Polyhydroxyalkanoate (PHA): Review of Synthesis, Characteristics, Processing and Potential Applications in Packaging. *Express Polym. Lett.* **2014**. [CrossRef]
16. Vandewijngaarden, J.; Murariu, M.; Dubois, P.; Carleer, R.; Yperman, J.; Adriaensens, P.; Schreurs, S.; Lepot, N.; Peeters, R.; Buntinx, M. Gas Permeability Properties of Poly(3-Hydroxybutyrate-Co-3-Hydroxyhexanoate). *J. Polym. Environ.* **2014**, *22*, 501–507. [CrossRef]
17. Panaitescu, D.M.; Nicolae, C.A.; Frone, A.N.; Chiulan, I.; Stanescu, P.O.; Draghici, C.; Iorga, M.; Mihailescu, M. Plasticized Poly(3-Hydroxybutyrate) with Improved Melt Processing and Balanced Properties. *J. Appl. Polym. Sci.* **2017**, *134*. [CrossRef]
18. Post, W.; Kuijpers, L.J.; Zijlstra, M.; van der Zee, M.; Molenveld, K. Effect of Mineral Fillers on the Mechanical Properties of Commercially Available Biodegradable Polymers. *Polymers* **2021**, *13*, 394. [CrossRef]
19. Luo, R.; Xu, K.; Chen, G. Effects of L-Phenylalanine as a Nucleation Agent on the Nonisothermal Crystallization, Melting Behavior, and Mechanical Properties of Poly(3-Hydroxybutyrate-Co-3-Hydroxyhexanoate). *J. Appl. Polym. Sci.* **2008**, *110*, 2950–2956. [CrossRef]
20. Vandewijngaarden, J.; Murariu, M.; Dubois, P.; Carleer, R.; Yperman, J.; D'Haen, J.; Peeters, R.; Buntinx, M. Effect of Ultrafine Talc on Crystallization and End-Use Properties of Poly(3-Hydroxybutyrate- *Co* -3-Hydroxyhexanoate). *J. Appl. Polym. Sci.* **2016**, *133*. [CrossRef]
21. Pan, P.; Liang, Z.; Nakamura, N.; Miyagawa, T.; Inoue, Y. Uracil as Nucleating Agent for Bacterial Poly[(3-Hydroxybutyrate)-*Co*-(3-Hydroxyhexanoate)] Copolymers. *Macromol. Biosci.* **2009**, *9*, 585–595. [CrossRef] [PubMed]
22. Zhou, J.; Ma, X.; Li, J.; Zhu, L. Preparation and Characterization of a Bionanocomposite from Poly(3-Hydroxybutyrate-Co-3-Hydroxyhexanoate) and Cellulose Nanocrystals. *Cellulose* **2019**, *26*, 979–990. [CrossRef]
23. Díez-Pascual, A.M. Effect of Graphene Oxide on the Properties of Poly(3-Hydroxybutyrate-Co-3-Hydroxyhexanoate). *Polymers* **2021**, *13*, 2233. [CrossRef]
24. Abbas, M.; Buntinx, M.; Deferme, W.; Reddy, N.; Peeters, R. Oxygen Gas and UV Barrier Properties of Nano-ZnO-Coated PET and PHBHHx Materials Fabricated by Ultrasonic Spray-Coating Technique. *Nanomaterials* **2021**, *14*, 449. [CrossRef]
25. Robledo-Ortíz, J.R.; Martín del Campo, A.S.; Blackaller, J.A.; González-López, M.E.; Pérez Fonseca, A.A. Valorization of Sugarcane Straw for the Development of Sustainable Biopolymer-Based Composites. *Polymers* **2021**, *13*, 3335. [CrossRef] [PubMed]
26. Upson, S.J.; O'Haire, T.; Russell, S.J.; Dalgarno, K.; Ferreira, A.M. Centrifugally Spun PHBV Micro and Nanofibres. *Mater. Sci. Eng. C* **2017**, *76*, 190–195. [CrossRef]
27. Cherpinski, A.; Torres-Giner, S.; Cabedo, L.; Méndez, J.A.; Lagaron, J.M. Multilayer Structures Based on Annealed Electrospun Biopolymer Coatings of Interest in Water and Aroma Barrier Fiber-based Food Packaging Applications. *J. Appl. Polym. Sci.* **2018**, *135*, 45501. [CrossRef]
28. Râpă, M.; Stefan, M.; Popa, P.A.; Toloman, D.; Leostean, C.; Borodi, G.; Vodnar, D.C.; Wrona, M.; Salafranca, J.; Nerín, C.; et al. Electrospun Nanosystems Based on PHBV and ZnO for Ecological Food Packaging. *Polymers* **2021**, *13*, 2123. [CrossRef]
29. Cai, H.; Qiu, Z. Effect of Comonomer Content on the Crystallization Kinetics and Morphology of Biodegradable Poly(3-Hydroxybutyrate-Co-3-Hydroxyhexanoate). *Phys. Chem. Chem. Phys.* **2009**, *11*, 9569. [CrossRef]
30. El-Hadi, A.; Schnabel, R.; Straube, E.; Müller, G.; Riemschneider, M. Effect of Melt Processing on Crystallization Behavior and Rheology of Poly(3-Hydroxybutyrate) (PHB) and Its Blends. *Macromol. Mater. Eng.* **2002**, *287*, 363–372. [CrossRef]
31. El-Hadi, A.; Schnabel, R.; Straube, E.; Müller, G.; Henning, S. Correlation between Degree of Crystallinity, Morphology, Glass Temperature, Mechanical Properties and Biodegradation of Poly (3-Hydroxyalkanoate) PHAs and Their Blends. *Polym. Test.* **2002**, *21*, 665–674. [CrossRef]
32. Hu, Y.; Zhang, J.; Sato, H.; Noda, I.; Ozaki, Y. Multiple Melting Behavior of Poly(3-Hydroxybutyrate-Co-3-Hydroxyhexanoate) Investigated by Differential Scanning Calorimetry and Infrared Spectroscopy. *Polymer* **2007**, *48*, 4777–4785. [CrossRef]

33. Bossu, J.; Le Moigne, N.; Dieudonné-George, P.; Dumazert, L.; Guillard, V.; Angellier-Coussy, H. Impact of the Processing Temperature on the Crystallization Behavior and Mechanical Properties of Poly[R-3-Hydroxybutyrate-Co-(R-3-Hydroxyvalerate)]. *Polymer* **2021**, *229*, 123987. [CrossRef]
34. Vandewijngaarden, J.; Wauters, R.; Murariu, M.; Dubois, P.; Carleer, R.; Yperman, J.; D'Haen, J.; Ruttens, B.; Schreurs, S.; Lepot, N.; et al. Poly(3-Hydroxybutyrate-Co-3-Hydroxyhexanoate)/Organomodified Montmorillonite Nanocomposites for Potential Food Packaging Applications. *J. Polym. Environ.* **2016**, *24*, 104–118. [CrossRef]
35. Montgomery, D.C. *Design and Analysis of Experiments*; John Wiley & Sons: Hoboken, NJ, USA, 2017; ISBN 978-1-119-32093-7.
36. Antony, J. *Design of Experiments for Engineers and Scientists*, 2nd ed.; Elsevier Insights; Elsevier: London, UK, 2014; ISBN 978-0-08-099417-8.
37. Jensen, W.A. Confirmation Runs in Design of Experiments. *J. Qual. Technol.* **2016**, *48*, 162–177. [CrossRef]
38. Rastogi, V.; Samyn, P. Synthesis of Polyhydroxybutyrate Particles with Micro-to-Nanosized Structures and Application as Protective Coating for Packaging Papers. *Nanomaterials* **2016**, *7*, 5. [CrossRef] [PubMed]
39. Hammersly, A.P. *FIT2D V9.129 Reference Manual V3.1*; ESRF Internal Report; European Synchotron Radiation Facility: Grenoble, France, 1998.
40. Gommes, C.J.; Goderis, B. CONEX, a Program for Angular Calibration and Averaging of Two-Dimensional Powder Scattering Patterns. *J. Appl. Crystallogr.* **2010**, *43*, 352–355. [CrossRef]
41. Vandi, L.-J.; Chan, C.M.; Werker, A.; Richardson, D.; Laycock, B.; Pratt, S. Experimental Data for Extrusion Processing and Tensile Properties of Poly(Hydroxybutyrate-Co-Hydroxyvalerate) (PHBV) Polymer and Wood Fibre Reinforced PHBV Biocomposites. *Data Brief* **2019**, *22*, 687–692. [CrossRef]
42. de Almeida Neto, G.R.; Barcelos, M.V.; Ribeiro, M.E.A.; Folly, M.M.; Rodríguez, R.J.S. Formulation and Characterization of a Novel PHBV Nanocomposite for Bone Defect Filling and Infection Treatment. *Mater. Sci. Eng. C* **2019**, *104*, 110004. [CrossRef]
43. Mazur, K.; Kuciel, S. Mechanical and Hydrothermal Aging Behaviour of Polyhydroxybutyrate-Co-Valerate (PHBV) Composites Reinforced by Natural Fibres. *Molecules* **2019**, *24*, 3538. [CrossRef] [PubMed]
44. Thellen, C.; Coyne, M.; Froio, D.; Auerbach, M.; Wirsen, C.; Ratto, J.A. A Processing, Characterization and Marine Biodegradation Study of Melt-Extruded Polyhydroxyalkanoate (PHA) Films. *J. Polym. Environ.* **2008**, *16*, 1–11. [CrossRef]
45. Vadori, R.; Mohanty, A.K.; Misra, M. The Effect of Mold Temperature on the Performance of Injection Molded Poly(Lactic Acid)-Based Bioplastic. *Macromol. Mater. Eng.* **2013**, *298*, 981–990. [CrossRef]
46. Montano-Herrera, L.; Pratt, S.; Arcos-Hernandez, M.V.; Halley, P.J.; Lant, P.A.; Werker, A.; Laycock, B. In-Line Monitoring of Thermal Degradation of PHA during Melt-Processing by Near-Infrared Spectroscopy. *N Biotechnol.* **2014**, *31*, 357–363. [CrossRef] [PubMed]
47. Vandi, L.-J.; Chan, C.M.; Werker, A.; Richardson, D.; Laycock, B.; Pratt, S. Extrusion of Wood Fibre Reinforced Poly(Hydroxybu tyrate-Co-Hydroxyvalerate) (PHBV) Biocomposites: Statistical Analysis of the Effect of Processing Conditions on Mechanical Performance. *Polym. Degrad. Stab.* **2019**, *159*, 1–14. [CrossRef]
48. Dario Puppi; Andrea Morelli; Federica Chiellini Additive Manufacturing of Poly(3-Hydroxybutyrate-Co-3-Hydroxyhexanoate)/Poly(ε-Caprolactone) Blend Scaffolds for Tissue Engineering. *Bioengineering* **2017**, *4*, 49. [CrossRef]
49. Mysiukiewicz, O.; Barczewski, M.; Skórczewska, K.; Matykiewicz, D. Correlation between Processing Parameters and Degradation of Different Polylactide Grades during Twin-Screw Extrusion. *Polymers* **2020**, *12*, 1333. [CrossRef]
50. Gogolewski, S.; Jovanovic, M.; Perren, S.M.; Dillon, J.G.; Hughes, M.K. The Effect of Melt-Processing on the Degradation of Selected Polyhydroxyacids: Polylactides, Polyhydroxybutyrate, and Polyhydroxybutyrate-Co-Valerates. *Polym. Degrad. Stab.* **1993**, *40*, 313–322. [CrossRef]
51. Hoffmann, A.; Kreuzberger, S.; Hinrichsen, G. Influence of Thermal Degradation on Tensile Strength and Young's Modulus of Poly(Hydroxybutyrate). *Polym. Bull* **1994**, *33*, 355–359. [CrossRef]
52. Taubner, V.; Shishoo, R. Influence of Processing Parameters on the Degradation of Poly(L-Lactide) during Extrusion. *J. Appl. Polym. Sci.* **2001**, *79*, 2128–2135. [CrossRef]
53. Capone, C.; Di Landro, L.; Inzoli, F.; Penco, M.; Sartore, L. Thermal and Mechanical Degradation during Polymer Extrusion Processing. *Polym. Eng. Sci.* **2007**, *47*, 1813–1819. [CrossRef]
54. Farah, S.; Anderson, D.G.; Langer, R. Physical and Mechanical Properties of PLA, and Their Functions in Widespread Applications — A Comprehensive Review. *Adv. Drug Deliv. Rev.* **2016**, *107*, 367–392. [CrossRef]
55. Arcos-Hernández, M.V.; Laycock, B.; Donose, B.C.; Pratt, S.; Halley, P.; Al-Luaibi, S.; Werker, A.; Lant, P.A. Physicochemical and Mechanical Properties of Mixed Culture Polyhydroxyalkanoate (PHBV). *Eur. Polym. J.* **2013**, *49*, 904–913. [CrossRef]
56. Jost, V.; Langowski, H.-C. Effect of Different Plasticisers on the Mechanical and Barrier Properties of Extruded Cast PHBV Films. *Eur. Polym. J.* **2015**, *68*, 302–312. [CrossRef]
57. Nuchanong, P.; Seadan, M.; Khankrua, R.; Suttiruengwong, S. Thermal Stability Enhancement of Poly(Hydroxybutyrate-Co-Hydroxyvalerate) through *in Situ* Reaction. *Des. Monomers Polym.* **2021**, *24*, 113–124. [CrossRef]
58. Chen, G.-Q.; Wu, Q. The Application of Polyhydroxyalkanoates as Tissue Engineering Materials. *Biomaterials* **2005**, *26*, 6565–6578. [CrossRef] [PubMed]
59. Ding, C.; Cheng, B.; Wu, Q. DSC Analysis of Isothermally Melt-Crystallized Bacterial Poly(3-Hydroxybutyrate-Co-3-Hydroxyhexanoate) Films. *J. Therm. Anal. Calorim* **2011**, *103*, 1001–1006. [CrossRef]

60. Harris, A.M.; Lee, E.C. Improving Mechanical Performance of Injection Molded PLA by Controlling Crystallinity. *J. Appl. Polym. Sci.* **2008**, *107*, 2246–2255. [CrossRef]
61. Li, G.; Yang, B.; Han, W.; Li, H.; Lin, J.; Kang, Z. Tailoring the Thermal and Mechanical Properties of Injection-molded Poly (Lactic Acid) Parts through Annealing. *J. Appl. Polym. Sci.* **2020**, *138*, 49648. [CrossRef]
62. Fujiyama, M.; Azuma, K. Skin/Core Morphology and Tensile Impact Strength of Injection-Molded Polypropylene. *J. Appl. Polym. Sci.* **1979**, *23*, 2807–2811. [CrossRef]
63. Kantz, M.R.; Newman, H.D.; Stigale, F.H. The Skin-Core Morphology and Structure–Property Relationships in Injection-Molded Polypropylene. *J. Appl. Polym. Sci.* **1972**, *16*, 1249–1260. [CrossRef]
64. Katti, S.S.; Schultz, M. The Microstructure of Injection-Molded Semicrystalline Polymers: A Review. *Polym. Eng. Sci.* **1982**, *22*, 17. [CrossRef]
65. Pan, Y.; Guo, X.; Zheng, G.; Liu, C.; Chen, Q.; Shen, C.; Liu, X. Shear-Induced Skin-Core Structure of Molten Isotactic Polypropylene and the Formation of β-Crystal. *Macromol. Mater. Eng.* **2018**, *303*, 1800083. [CrossRef]
66. Rhoades, A.M.; Williams, J.L.; Wonderling, N.; Androsch, R.; Guo, J. Skin/Core Crystallinity of Injection-Molded Poly(Butylene Terephthalate) as Revealed by Microfocus X-Ray Diffraction and Fast Scanning Chip Calorimetry. *J. Therm. Anal. Calorim* **2017**, *127*, 939–946. [CrossRef]
67. Spoerer, Y.; Androsch, R.; Jehnichen, D.; Kuehnert, I. Process Induced Skin-Core Morphology in Injection Molded Polyamide 66. *Polymers* **2020**, *12*, 894. [CrossRef]
68. Liparoti, S.; Speranza, V.; Sorrentino, A.; Titomanlio, G. Mechanical Properties Distribution within Polypropylene Injection Molded Samples: Effect of Mold Temperature under Uneven Thermal Conditions. *Polymers* **2017**, *9*, 585. [CrossRef]
69. Liparoti, S.; Speranza, V.; Titomanlio, G.; Pantani, R. Effect of Rapid Mold Heating on the Structure and Performance of Injection-Molded Polypropylene. *Polymers* **2020**, *12*, 341. [CrossRef]
70. Liparoti, S.; Sorrentino, A.; Speranza, V. Morphology-Mechanical Performance Relationship at the Micrometrical Level within Molded Polypropylene Obtained with Non-Symmetric Mold Temperature Conditioning. *Polymers* **2021**, *13*, 462. [CrossRef]
71. Kobayashi, S.; Müllen, K. (Eds.) *Encyclopedia of Polymeric Nanomaterials*; Springer: Berlin/Heidelberg, Germany, 2015; ISBN 978-3-642-29647-5.
72. Sato, H.; Murakami, R.; Zhang, J.; Ozaki, Y.; Mori, K.; Takahashi, I.; Terauchi, H.; Noda, I. X-ray Diffraction and Infrared Spectroscopy Studies on Crystal and Lamellar Structure and Cho Hydrogen Bonding of Biodegradable Poly(Hydroxyalkanoate). *Macromol. Res.* **2006**, *14*, 408–415. [CrossRef]
73. Somani, R.H.; Yang, L.; Zhu, L.; Hsiao, B.S. Flow-Induced Shish-Kebab Precursor Structures in Entangled Polymer Melts. *Polymer* **2005**, *46*, 8587–8623. [CrossRef]
74. de Geus, M.; van der Meulen, I.; Goderis, B.; van Hecke, K.; Dorschu, M.; van der Werff, H.; Koning, C.E.; Heise, A. Performance Polymers from Renewable Monomers: High Molecular Weight Poly(Pentadecalactone) for Fiber Applications. *Polym. Chem.* **2010**, *1*, 525–533. [CrossRef]
75. Rosato, D.V.; Rosato, M.G. (Eds.) *Injection Molding Handbook*, 3rd ed.; Kluwer Academic Publishers: Boston, MA, USA, 2000; ISBN 978-0-7923-8619-3.
76. Osswald, T.A.; Menges, G. *Materials Science of Polymers for Engineers*, 3rd ed.; Hanser Publications: Munich, Germany, 2012; ISBN 978-1-56990-514-2.
77. Dar, U.A.; Xu, Y.J.; Zakir, S.M.; Saeed, M.-U. The Effect of Injection Molding Process Parameters on Mechanical and Fracture Behavior of Polycarbonate Polymer. *J. Appl. Polym. Sci.* **2017**, *134*. [CrossRef]
78. Rizvi, S.J.A. Effect of Injection Molding Parameters on Crystallinity and Mechanical Properties of Isotactic Polypropylene. *Int. J. Plast Technol.* **2017**, *21*, 404–426. [CrossRef]
79. Liu, Y.; Zhu, T.; Bi, J.; Hua, W.; Yu, T.; Jin, Y.; Zhao, D. Investigation on Microstructures and Mechanical Properties of Isotactic Polypropylene Parts Fabricated by Different Process Conditions with Different Aging Periods. *Polymers* **2020**, *12*, 2828. [CrossRef] [PubMed]
80. Simmons, H.; Tiwary, P.; Colwell, J.E.; Kontopoulou, M. Improvements in the Crystallinity and Mechanical Properties of PLA by Nucleation and Annealing. *Polym. Degrad. Stab.* **2019**, *166*, 248–257. [CrossRef]

polymers

MDPI

Article

Lignin Biopolymer for the Synthesis of Iron Nanoparticles and the Composite Applied for the Removal of Methylene Blue

Fang-Yi Peng, Pei-Wen Wang, Weisheng Liao and Ing-Song Yu *

Department of Materials Science and Engineering, National Dong Hwa University, Hualien 97401, Taiwan; 410522001@gms.ndhu.edu.tw (F.-Y.P.); 410722059@gms.ndhu.edu.tw (P.-W.W.); liao1427@alumni.uidaho.edu (W.L.)
* Correspondence: isyu@gms.ndhu.edu.tw; Tel.: +886-3-8903219

Abstract: In the current study, lignin, an abundant natural polymer, was dissolved in ethylene glycol and acidic H_2O to form nanoscale lignin. Then, zero-valent iron (ZVI) nanoparticles were synthesized in nanoscale lignin, producing a nZVI/n-lignin composite, via the borohydride reduction method. The use of nZVI/n-lignin for environmental remediation was tested by the removal of methylene blue in aqueous solutions at room temperature. The nZVI/n-lignin composite achieved a higher methylene blue removal ratio than that achieved by traditional nZVIs. Moreover, its excellent dispersibility in water and stability against oxidation in the air were observed. The functions of the nanoscale lignin in the composite material are (1) prevention of further growth and aggregation of the nZVI nanoparticles, (2) protection of nZVI from serious oxidation by H_2O/O_2, and (3) allowing better dispersibility of nZVI in aqueous solutions. These three functions are important for the field applications of nZVI/n-lignin, namely, to travel long distances before making contact with environmental pollutants. The present method for producing nZVI/n-lignin is straightforward, and the combination of nZVI and lignin is an efficient and environmentally friendly material for environmental applications.

Keywords: lignin; zero valent iron; nanoparticles; ethylene glycol; methylene blue

Citation: Peng, F.-Y.; Wang, P.-W.; Liao, W.; Yu, I.-S. Lignin Biopolymer for the Synthesis of Iron Nanoparticles and the Composite Applied for the Removal of Methylene Blue. *Polymers* **2021**, *13*, 3847. https://doi.org/10.3390/polym13213847

Academic Editor: Arn Mignon

Received: 12 October 2021
Accepted: 4 November 2021
Published: 7 November 2021

1. Introduction

The utilization of nanoscale zero-valent iron (nZVI) for in situ remediation of contaminated water and soil has been an active research area due to its low cost, high reactivity, good mobility, and environmental compatibility [1–3]. The synthesis of nZVI can be performed by top-down or bottom-up methods, such as the ball milling or lithography grinding of bulk iron materials, the hydrogen reduction of goethite/hematite at elevated temperatures, carbothermal reduction, electrolysis, pulsed plasma in liquids, and the reduction of ferric/ferrous salts in H_2O by polyphenolic plant extracts or other reducing agents [4]. Among these, the sodium borohydride reduction of iron salts in H_2O is the most widely studied method for the synthesis of reactive nZVI in academic research due to its simplicity [5]. The particle size of nZVI synthesized via the borohydride method usually ranges from ten to hundreds of nanometers, and it tends to gather into linear or fractal patterns because of its magnetic interaction, high surface energy, and van der Waals forces, which decreases its reactivity and mobility [6]. In addition, nZVI possesses a core–shell structure in which the surface is a thin layer of defective iron oxide that forms spontaneously during synthesis and continuously develops [5]. Therefore, using an inert environment during synthesis to prevent substantial oxide layer formation and modification of the nZVI surface to prevent particle aggregation is of great importance for the environmental applications of nZVI.

Ethylene glycol (EG) is an important raw material mainly used for the manufacture of polyester fibers and antifreeze, which can be mixed with H_2O at any ratio. Synthesis of nZVI using the borohydride method in H_2O with a small amount of EG can eliminate the need for an inert nitrogen or argon atmosphere and make the synthetic process simpler and less costly [7]. EG-functionalized nZVI provides better stability against oxidation and higher dispersibility in H_2O than those of bare nZVI. The particle sizes of EG-functionalized nZVI are normally in the range of 20 to 100 nm, which are slightly smaller than those of bare nZVI synthesized by the borohydride method. Moreover, a novel flowerlike Fe-EG nanostructure has also been synthesized by dissolving $FeCl_3{\cdot}6H_2O$ and $CH_3COONa{\cdot}3H_2O$ in EG, following a thermal treatment method [8]. This polymeric ferrous glycolate can play the role of a Fenton catalyst in the wastewater treatment aspect of environmental remediation.

To prevent the aggregation of nZVI, various modification methods have been proposed, such as the deposition of other metals on its surface (e.g., Pd or Ni), coating nZVI with stabilizers (e.g., starch, guar gum, carboxymethyl cellulose, or polyacrylic acid), emulsifying nZVI, and producing composite materials of nZVI in a matrix (e.g., silica, $CaCO_3$, montmorillonite, granular activated carbon, or carbon nanotube) [9]. In addition, supporting the nanoparticles with sustainable biopolymers could increase their stability and biocompatibility, as well as allowing them to attach different ligands to their surfaces [10]. Lignin-modified nZVI with a high specific surface area was first synthesized via the borohydride method in H_2O for the removal of arsenic from groundwater [11]. Recently, bentonite-supported organosolv lignin-stabilized nZVI was fabricated for the removal of hexavalent chromium from wastewater [12].

Lignin is a heterogenous phenylpropanoid macromolecule with a three-dimensionally branched structure that includes random crosslinks of monomeric units [13]. It is also the second most abundant biopolymer, formed by the free radical polymerization of three monolignols, coniferyl alcohol (G), sinapyl alcohol (S), and *p*-coumaryl alcohol (H) in the plant cell wall, giving it a structure containing many aliphatic and aromatic hydroxyls [14]. The β-O-4 ether linkage generally accounts for up to 50% of all the linkages in a lignin network. The paper industry generates large amounts of lignin (or Kraft lignin) as a major by-product, and it is mainly used as fuel [15]. Due to its non-toxicity, biodegradability, renewability, and its 3D network containing hydroxyl groups, lignin has also been used as a stabilizing and/or reducing agent in the synthesis of Pt [16], Pd [17,18], Ru, Re [19], Au [20], and Ag [21,22] nanoparticles for cross-coupling reactions, oxidation/reduction reactions, detection of Pb^{2+}, and as antibacterial materials.

There is global awareness of limited resources, and so the development of sustainable materials for reuse and recycling is urgent [23]. In this study, Kraft lignin was used as a starting material alongside various other types of lignin due to its easy availability and high abundance as a by-product of the paper industry. The particle size of Kraft lignin is usually in the micro-sized range [24], which still presents challenges for polymer blending or use as a solid support for metal nanoparticles. Since Kraft lignin is only soluble in H_2O at pH > 10, while it is soluble in EG at up to 70 wt% [25], EG was selected to dissolve the Kraft lignin, which could subsequently be converted to lignin nanoparticles simply by using the antisolvent or acid precipitation method [26,27]. Therefore, nanoscale lignin was synthesized by dissolving commercial micro-sized Kraft lignin in EG, which was then added to acidic H_2O containing iron salts via the antisolvent and acid precipitation methods. Afterward, the iron ions trapped in the nanoscale lignin were reduced to nZVI by the borohydride method. By using the nanoscale lignin to stabilize the nZVI, the resulting composite, called nZVI/n-lignin, may possess a higher colloidal stability and better mobility than that of the bare nZVI and the EG-functionalized nZVI. Characterizations of nZVI/n-lignin were performed by transmission electron microscopy, X-ray photoelectron spectroscopy, and an X-ray diffractometer. The dispersibility in water and stability against oxidation of nZVI/n-lignin were compared with those of bare and EG-functionalized nZVIs. Finally, the effectiveness of nZVI/n-lignin for the removal of methylene blue in H_2O was investigated, and the process was monitored via the characteristic UV–Vis absorption peaks of the cationic dye. This nZVI/n-lignin composite material could be employed for application in environmental remediation in the future.

2. Materials and Methods

2.1. Chemicals

Iron (III) chloride hexahydrate ($FeCl_3 \cdot 6H_2O$, ≧99%), Kraft lignin, ethylene glycol (EG), sodium borohydride ($NaBH_4$, ≧98%), and ethanol were purchased from Sigma-Aldrich (Milwaukee, WI, USA). Methylene blue solution (0.1%) was supplied by Choneye Pure Chemicals (Tamil Nadu, India).

2.2. Preparation of Nanoscale Lignin-Stabilized nZVI

The experimental procedure is shown in Figure 1. An amount of 40 mg of Kraft lignin was first dissolved in 2 mL EG to form nano-sized lignin, which could be observed by transmission electron microscopy. The dissolution process happened quickly, producing a brownish EG solution. After the dissolution, the EG solution containing the dissolved Kraft lignin was added to a 30 mL aqueous solution alongside 0.006 M $FeCl_3$. After the solution had been mixed for 10 min, 0.133 g $NaBH_4$ was added to it. Tiny black particles displaying magnetic behavior were formed quickly. After a short period, the DI-H_2O-containing EG and nZVI/n-lignin were separated by a magnet. The nZVI/n-lignin composite was washed with ethanol three times. The washed nZVI/n-lignin was dried in an oven at a temperature of 40 °C for 20 min. Using the same method, the bare nZVI and EG-functionalized nZVI were also synthesized for comparison with nZVI/n-lignin composite. These nZVIs were then stored under atmospheric conditions for the subsequent characterizations and the degradation experiment in methylene blue solution. Bare nZVI, EG-functionalized nZVI, and nZVI/n-lignin were characterized by transmission electron microscopy (TEM, JEOL JEM-2010, Tokyo, Japan), X-ray diffractometer (XRD, Rigaku D/MAX-3C OD-2988N, Neu-Isenburg, Germany), and X-ray photoelectron spectroscopy (XPS, Thermo Scientific K-Alpha, Waltham, MA, USA). The XPS depth profile analyses were also conducted by argon ion etching using an energy level of 3000 eV for 1 min.

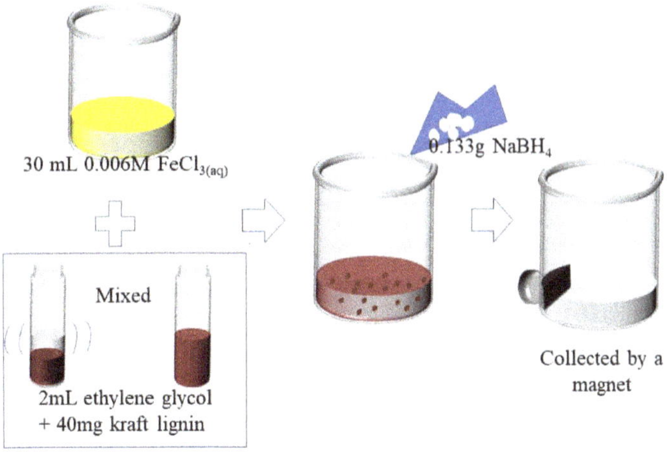

Figure 1. Experimental procedure for the synthesis of nZVI/n-lignin composite material.

2.3. Dispersibility and Stability against Oxidation of nZVI/n-Lignin

To verify the applicability of nZVI/n-lignin, the synthesized bare nZVI, EG-functionalized nZVI and nZVI/n-lignin were tested in the aqueous solutions for 7 days. The dispersibility of the two nZVIs and nZVI/n-lignin in H_2O was observed. Moreover, their stability against oxidation in air and water was verified. After the synthesis of materials, the three dried materials were stored in atmospheric conditions for 7 days and placed in water for 7 days. Then, the powders were measured by XRD to check the characteristic peaks of zero-valent iron and iron oxides.

2.4. Removal of Methylene Blue over nZVI/n-Lignin in Aqueous Solutions

For the removal ratio of methylene blue, 0.5 mL methylene blue solution was added to DI-H_2O to give a total volume of 50 mL, with a concentration of 3 ppm. Forty milligrams each of the bare nZVI, EG-functionalized nZVI, and nZVI/n-lignin powders were added to the solution to start the degradation and adsorption of methylene blue dye. The absorption of the filtered solution after a certain duration was recorded by an UV–Vis spectrometer (Jasco, V-650, Hachioji, Tokyo, Japan) to monitor the degradation and adsorption process of methylene blue. The absorption intensity of the methylene blue solution at the wavelength of 664 nm decreased as the reaction time increased.

3. Results and Discussion

3.1. Characterization of Bare nZVI, EG-Functionalized nZVI, and nZVI/n-Lignin

EG was selected to dissolve Kraft lignin, which could be subsequently converted to lignin nanoparticles simply by using the antisolvent precipitation or acid precipitation method. After the addition of the lignin dissolved in EG to the acidic H_2O containing $FeCl_3$ (pH ~ 2), nanoscale lignin was formed, suspended in the H_2O. Once $NaBH_4$ was added, tiny black particles (nZVI/n-lignin) were formed instantly. After the synthesis for 30 min, the solution with nZVI/n-lignin was still turbid, as shown in Figure 2a (right). For comparison, bare nZVI synthesized in H_2O (Figure 2a, left) and EG-functionalized nZVI synthesized in H_2O containing a small amount of EG (Figure 2a, middle) were also prepared. Iron particles in both the solution with bare nZVI and the solution with EG-functionalized nZVI aggregated and formed bigger clusters in H_2O. To collect the powders of the nZVIs and nZVI/n-lignin, the three solutions were then washed three times with ethanol and put into an oven at a temperature of 40 °C for 20 min to evaporate the ethanol. The nZVI and nZVI/n-lignin powders are shown in Figure 2b. The bare nZVI (left) and EG-functionalized nZVI (middle) both show an obviously fractal pattern and are granular, but nZVI/n-lignin produced a very fine powder. Afterward, these three samples were placed for several weeks under atmospheric conditions. As Figure 2c shows, the powders of the bare nZVI (left) and EG-functionalized nZVI (middle) became yellowish, indicating their oxidation to form iron oxide. On the other hand, nZVI/n-lignin (right) maintained its black color and magnetic property, without obvious aggregation. These observations initially indicate that nanoscale lignin can prevent nZVI aggregation and protect nZVI from serious oxidation in the air.

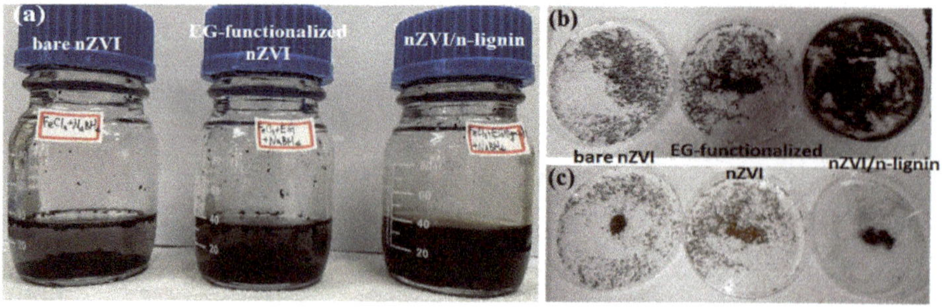

Figure 2. (**a**) After the synthesis for 30 min, the solution of the bare nZVI (left), EG-functionalized nZVI (middle), and nZVI/n-lignin (right). (**b**) After drying in the oven, the powders of the bare nZVI (left), EG-functionalized nZVI (middle), and nZVI/n-lignin (right). (**c**) After several weeks under atmospheric conditions, the powders of the bare nZVI (left), EG-functionalized nZVI (middle), and nZVI/n-lignin (right).

For comparison with the nZVI/n-lignin composite, nanoscale lignin was first synthesized by the same procedure as was used for nZVI/n-lignin, described in Section 2.2, without the reduction step using $NaBH_4$. TEM images displaying lignin clusters with a fractal shape are shown in Figure 3a. In Figure 3b, lignin with an irregular shape and a size of 100 nm is shown at higher magnification, which can prove the presence of nanoscale lignin in the mixed solutions of EG and H_2O. Figure 3c,d are the TEM images of bare nZVI and EG-functionalized nZVI, respectively. Both nZVI materials are arranged in a chain morphology pattern and form large clusters, mainly due to the magnetic interactions of individual particles. For the TEM observation of the nZVI/n-lignin composite, the small ion nanoparticles gather into fractal nanoscale lignin as shown in Figure 3e. In the TEM image at a higher magnification, Figure 3f, the diameter of the nanoparticles is mostly less than 100 nm. The nZVI/n-lignin composite, in which nZVI particles are confined in the structure of nanoscale lignin, is a magnetic substance and has a smaller size than that of the bare and EG-functionalized nZVIs. In addition, the long-chain patterns of the bare and EG-functionalized nZVIs were not observed in the case of the nZVI/n-lignin composite, which can also explain its dispersibility in H_2O.

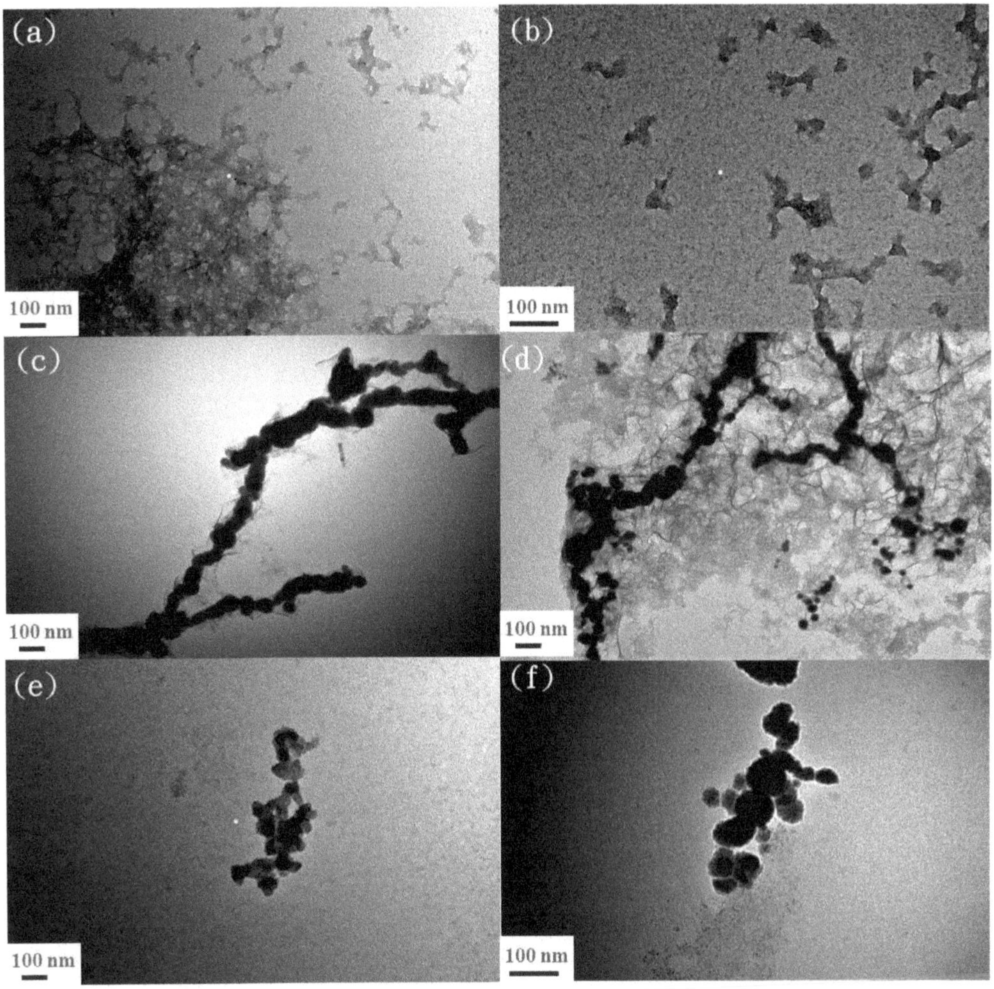

Figure 3. TEM images: (**a**) nanoscale lignin at a magnification of 100,000×, (**b**) nanoscale lignin at a magnification of 200,000× (**c**) bare nZVI at a magnification of 100,000×, (**d**) EG-functionalized nZVI at a magnification of 100,000×, (**e**) nZVI/n-lignin at a magnification of 100,000×, and (**f**) nZVI/n-lignin at a magnification of 200,000×.

To investigate the composition and chemical states of nZVI/n-lignin, XPS measurements of nZVI/n-lignin are shown in Figure 4. The XPS data were analyzed by the software Thermo Avantage. Two strong peaks were present for C 1s and O 1s, and a minor peak was observed for Fe 2p. Raw Kraft lignin showed strong peaks of binding energy around 285 eV for C 1s and 531 eV for O 1s due to the lignin structure containing the C–C, C–H, C-OH, C-O-C, O–O, Ph-OH, and Ph-C structures [24]. Obviously, the XPS spectra of nZVI/n-lignin for C 1s peaking at 285.53 eV (Figure 4a) and O 1s peaking at 531.03 eV (Figure 4b) are a result of the lignin structure, which contains three basic monolignol structures, coniferyl alcohol (G), sinapyl alcohol (S), and *p*-coumaryl alcohol (H). The photoelectron peaks for Fe 2p of nZVI/n-lignin are less intensive than those for C 1s and O 1s, as shown in Figure 4c. For the XPS spectrum of the bare nZVI, the peaks of binding energy at 710.56, 719.26, and 723.91 eV are for iron oxide and iron hydroxide [28], while the peak at 706.01 eV is for zero-valent iron [29]. As Figure 4c shows, only peaks for iron oxide and hydroxide were observed on the surface of nZVI/n-lignin, and there was no significant signal for zero-valent iron. To confirm the existence of zero-valence iron, argon ion etching on nZVI/n-lignin was performed during XPS analysis. A strong peak at 706.01 eV for zero-valent iron appeared, and the signal intensity for iron oxide and iron hydroxide at 710.56, 719.26, and 723.91 eV also increased in a similar manner to the XPS spectrum of Fe 2p, as shown in Figure 4d. Furthermore, the peak intensity for C 1s and O 1s decreased dramatically after ion etching, as shown in the insets of Figure 4a,b. Therefore, the XPS result indicates that the surface of nZVI is a thin layer of iron oxide/hydroxide and that the whole nZVI particle is surrounded by a lignin structure for providing steric stabilization. Desalegn et al. synthesized nZVI using mango peel extract [30]. XPS depth-profiling analysis by ion etching also shows increased peak intensity for nZVI around 706 eV.

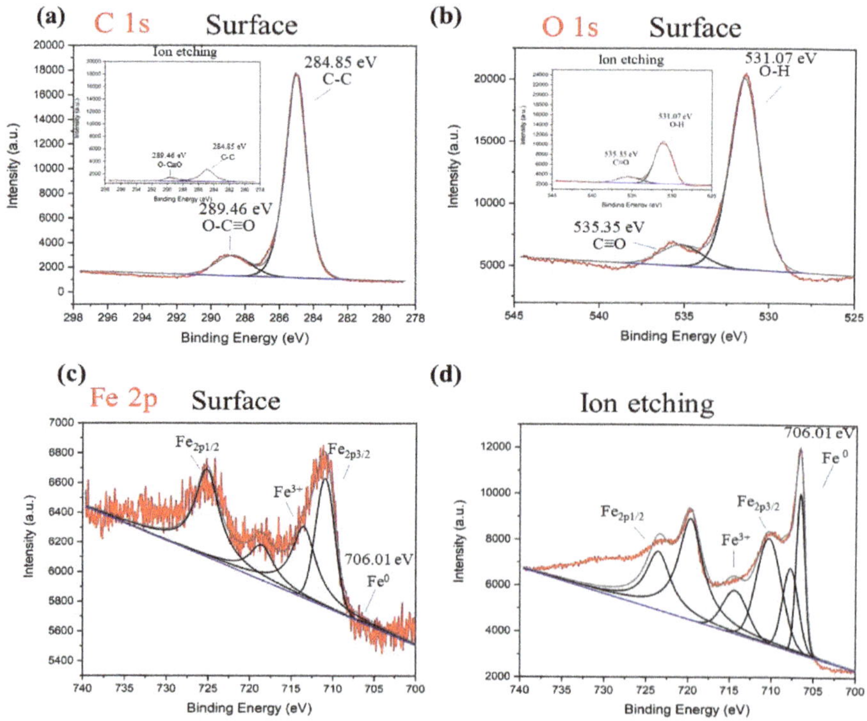

Figure 4. XPS analysis of nZVI/n-lignin: (**a**) C 1s scan, (**b**) O 1s scan, (**c**) Fe 2p scan, and (**d**) Fe 2p scan after argon ion etching. The insets in (**a**,**b**) are C 1s and O 1s of nZVI/n-lignin after argon ion etching.

To identify the presence of nZVI in the lignin, XRD measurements of the bare nZVI, EG-functionalized nZVI, and nZVI/n-lignin were also conducted. A strong peak of lignin was observed at the 2θ scan of 29.4° (not shown). As Figure 5a shows, three peaks at the 2θ scan of 44.9°, 65.0° and 82.3° are clear for the XRD result of the bare nZVI, which indicate the iron crystal orientations of (110), (200), and (211) planes, respectively [31,32]. In Figure 5b, the noise of the XRD spectra increases due to the smaller size of the EG-functionalized nZVI. Moreover, the XRD analysis of nZVI/n-lignin is shown in Figure 5c. A broad peak in the noisy XRD spectrum is observed near 44.9°, indicating the presence of nano-sized α-Fe^0 due to the formation of the composite with three-dimensional-branched lignin in the EG solution.

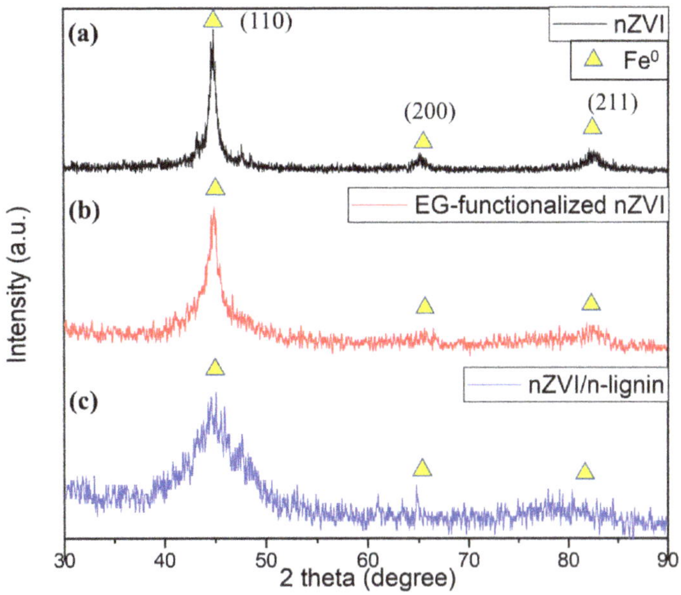

Figure 5. XRD analysis: (**a**) bare nZVI, (**b**) EG-functionalized nZVI, and (**c**) nZVI/n-lignin.

3.2. Dispersibility and Stability against Oxidation of nZVI/n-Lignin

The dispersibility of the bare nZVI, EG-functionalized nZVI and nZVI/n-lignin in aqueous solutions over different durations was tested. After the synthesis and washing processes, the three samples were put into water and dispersed by an ultrasonicator. After sonication for a short time, optical images of the solutions were taken, as shown in Figure 6a. The results for the solutions after standing for 60 min are shown in Figure 6b. The bare and EG-functionalized nZVIs formed large clusters that reduced their dispersibility in H_2O. However, the solution with nZVI/n-lignin maintained its black color, which indicates that nZVI/n-lignin was still well dispersed and suspended in the H_2O. The results for the solutions after standing for one day are shown in Figure 6c. The solutions became yellowish, indicating the oxidation of nZVIs. When nZVIs are put into water, they start to be oxidized by H_2O/O_2 to form iron oxide/hydroxide according to the following equations [31,32]: (1) $Fe^0 + 2H_2O \rightarrow Fe^{2+} + H_2 + 2OH^-$; (2) $2Fe^0 + O_2 + 2H_2O \rightarrow 2Fe^{2+} + 4OH^-$; (3) $6Fe^{2+} + O_2 + 6H_2O \rightarrow 2Fe_3O_{4(s)} + 12H^+$; (4) $Fe^{2+} + 2OH^- \rightarrow Fe(OH)_{2(s)}$; (5) $6Fe(OH)_{2(s)} + O_2 \rightarrow 2Fe_3O_4 + 6H_2O$. According to our observations, obvious precipitation of the solution with nZVI/n-lignin only happened after it had been standing for several days. The result shows that nZVI/n-lignin possesses better dispersibility in H_2O than that of the bare nZVI and EG-functionalized nZVI, probably because ion nanoparticles imbed in nano-sized lignin to prevent the formation of a chain-like pattern.

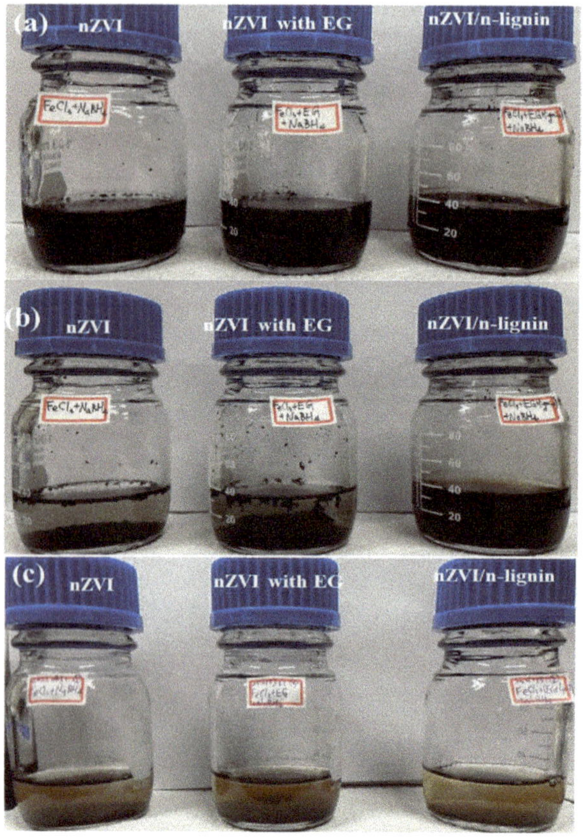

Figure 6. Dispersibility in H_2O of bare nZVI, EG-functionalized nZVI, and nZVI/n-lignin: (**a**) after the sonication, (**b**) standing for 60 min, and (**c**) standing for 1 day.

After the synthesis, the three dried materials were stored in atmospheric conditions for 7 days. Then, the powders were analyzed by XRD. The XRD measurements of the three aged materials in atmospheric conditions are shown in Figure 7a. The peak intensity of the nZVIs decreased dramatically, especially for the bare and EG-functionalized nZVIs, because of the oxidation of nZVI in the air. The XRD of the nZVI/n-lignin stored in atmospheric conditions for 7 days did not change, which indicates the better stability against oxidation of nZVI/n-lignin in the air. In a separate experiment, the three materials after synthesis were placed in H_2O for 7 days. The XRD analysis of the three aged materials placed in H_2O for 7 days is shown in Figure 7b. All the peaks from the nZVIs disappeared and were replaced by peaks at 35.52°, 43.17°, 47.27°, 53.56°, 57.10°, and 62.7°, indicating the formation of Fe_3O_4. These characteristic peaks in the X-ray diffraction indicate the crystal orientations of (311), (400), (331), (422), (511), and (440), respectively. The XRD results show the better stability against oxidation of nZVI/n-lignin and the oxidation reaction of nZVIs to iron oxide.

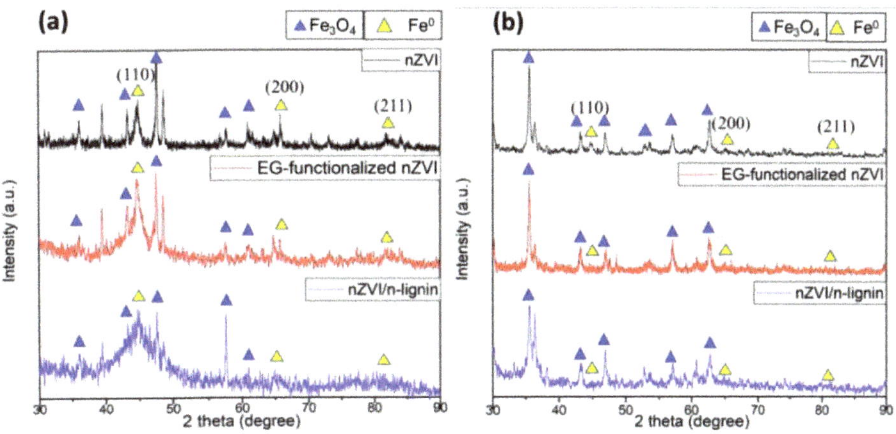

Figure 7. XRD analysis of aged bare nZVI, EG-functionalized nZVI, and nZVI/n-lignin: (**a**) exposed to air for 7 days and (**b**) put in water for 7 days.

3.3. Removal of Methylene Blue in Aqueous Solutions

Besides the synthesis, the dispersibility in water, and the antioxidation in the air of nZVI/n-lignin, the removal of methylene blue in H_2O by nZVI/n-lignin was also investigated. In the presence of nZVIs, the removal mechanisms of methylene blue (MB) include reduction into colorless leuco-MB, precipitation as Fe(II)-MB, absorption as ZVI-MB, and degradation using hydroxyl radicals [33]. Moreover, the methylene blue dye in the solution can be adsorbed by biopolymers such as lignin and chitosan [10,34]. The absorption mechanism for dyes and lignin may involve electrostatic interactions between the MB molecule with positively charged nitrogen and the dissociated functional groups of lignin [35]. Therefore, the inset of Figure 8a shows an image of the methylene blue aqueous solutions. After 40 mg of nZVI/n-lignin was added to the methylene blue solution and stirred vigorously with a magnetic bar, the color of the solution quickly faded within a short time until a yellow color appeared, indicating the removal of the methylene blue as shown in Figure 8a. The decolorization of methylene blue from blue to yellow is consistent with reported studies [36]. The removal ratio of methylene blue can be monitored by a UV–Vis spectrometer. The UV–Vis spectrum of the methylene blue in H_2O had a strong absorption peak at the wavelength of 664 nm and a shoulder peak at around 615 nm from the dimer of methylene blue [33]. The UV–Vis spectra between 400 and 800 nm of the methylene blue degraded and adsorbed by the nZVI/n-lignin composite in H_2O are shown in Figure 8b. The UV–Vis spectra of methylene blue gradually decreased to no absorption, indicating that it had been fully removed by nZVI/n-lignin. The absorption of the solutions at 664 nm as a function of time is shown in Figure 8c. Two regions are observed. In the first region (0~10 min), the removal rate of methylene blue, including degradation and adsorption, is very fast. After 30 min, the removal rate becomes slower, and methylene blue is removed completely after 90 min. The reason for the slower decreasing rate and fluctuation in the second region may come from the degradation product and iron oxidation product diffusing throughout the lignin structure, interfering in the contact between the methylene blue and nZVI within the lignin structure.

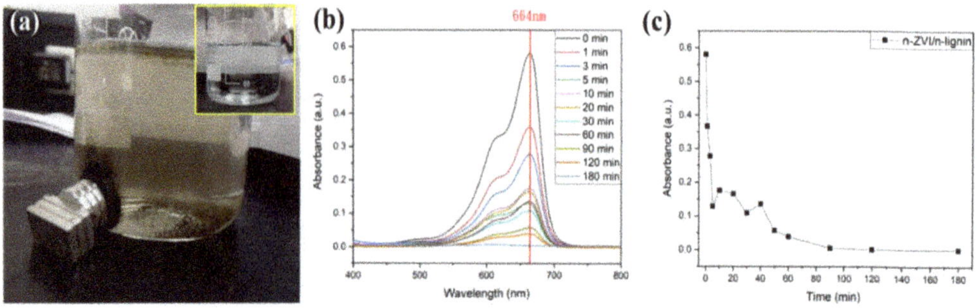

Figure 8. (**a**) The image of the degradation of methylene blue in aqueous solutions by nZVI/n-lignin. The inset is the image before the addition of nZVI/n-lignin. (**b**) UV–Vis spectra of high ratio test and (**c**) the absorption as a function of time at the wavelength of 664 nm.

In practical application, once the nZVI is added to ground water in the field for pollutant remediation, the oxidation reaction between nZVI and H_2O/O_2 starts, which consumes the nZVI before it degrades the pollutants. The nZVI's stability against H_2O/O_2 oxidation is of great importance. In our study, the applicability of bare nZVI, EG-functionalized nZVI, and nZVI/n-lignin was compared according to the removal ratio of methylene blue in aqueous solutions. After the addition of 40 mg each of bare nZVI, EG-functionalized nZVI, and nZVI/n-lignin to the methylene blue solution, their absorption rates were measured by UV–Vis spectra, as shown in Figure 9a. The two nZVIs and nZVI/n-lignin could remove methylene as the duration increased. Of the three, the reaction of the 40 mg nZVI/n-lignin composite could persevere for the longest duration, although just a small amount of nZVI is in nanoscale lignin. On the other hand, from the removal ratio of methylene blue shown in Figure 9b, nZVI/n-lignin performed best in the removal of methylene blue in aqueous solutions. This is because of the better dispersibility in water and stability against oxidation of the nZVI/n-lignin composite.

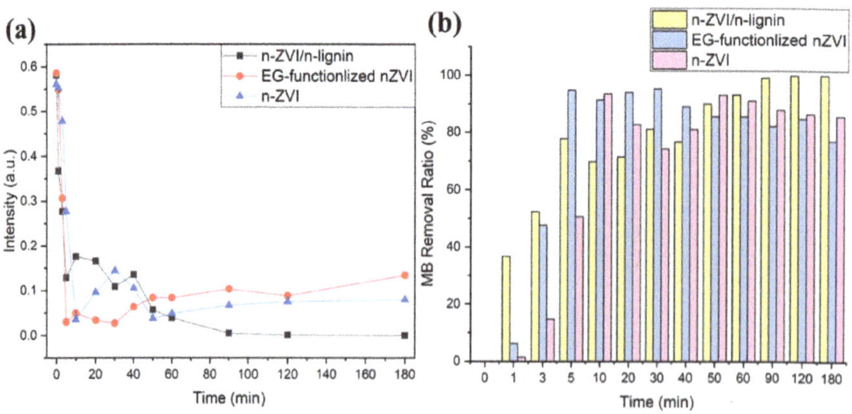

Figure 9. (**a**) UV–Vis absorption intensity at the wavelength of 664 nm for bare nZVI, EG-functionalized nZVI, and nZVI/n-lignin. (**b**) The removal ratio of methylene blue for bare nZVI, EG-functionalized nZVI, and nZVI/n-lignin.

4. Conclusions

In our study, sustainable lignin biopolymer was employed for the synthesis of iron nanoparticles for the removal of environmental pollutants. nZVI is the most frequently studied material for the water treatment aspect of environmental engineering. In real field applications, the environmental pollutants in a contaminated area are far away from the nZVI injection sites. Consequently, the prevention of nZVI aggregation and the avoidance of strong oxidation by H_2O/O_2 are of great importance. The agglomeration of the nZVI results in reduced degradation activity and poor mobility. Strongly oxidized behavior results in the consumption of the nZVI before it makes contact with the pollutants away from the injection sites. In this study, the nanoscale lignin-protected nZVI (nZVI/n-lignin) was first proposed to solve the above-mentioned issues. According to the analysis of the TEM and XPS measurements, the structure of nZVI/n-lignin is nanoscale lignin surrounding nZVI particles, which consist of zero-valent iron as the core with a thin layer of defective iron oxide/hydroxide on the surface. The traditional chain-like pattern and large clusters of bare nZVI can be avoided in the case of nZVI/n-lignin probably because of the steric stabilization alleviating magnetic interactions. This feature, combined with the use of nano-sized lignin rather than the use of micro-sized lignin, makes nZVI/n-lignin possess much better dispersibility and mobility in H_2O than those of bare nZVI. The structure of nano-lignin could provide more heterogeneous nucleation sites for the growth of ion nanoparticles. From their optical images, TEM, and XRD analysis, it was observed that nZVI/n-lignin possesses better stability against oxidation by H_2O/O_2 than those of bare and EG-functionalized nZVIs. The as-synthesized nZVI/n-lignin maintains its degradation ability even after it has been stored for several weeks in atmospheric conditions or several hours in H_2O, while bare nZVI is quickly oxidized to form iron oxide/hydroxide in a much shorter time. Compared with nZVIs, the nZVI/n-lignin composite may have better mobility in ground water while maintaining its degradation ability toward pollutants far away from the injection sites. In summary, the presented synthetic method for nZVI/n-lignin involving a dissolution and precipitation approach with ethylene glycol/acidic H_2O is easy to carry out, and it is environmentally friendly. It could be used as a universal approach for synthesizing nanoscale lignin-stabilized zero-valent metal and metal oxide nanoparticles for various applications.

Author Contributions: Conceptualization, W.L. and I.-S.Y.; methodology, W.L.; resources, I.-S.Y.; data curation, F.-Y.P. and P.-W.W.; writing—original draft preparation, W.L. and I.-S.Y.; writing—review and editing, I.-S.Y. All authors have read and agreed to the published version of the manuscript.

Funding: This research was funded by Ministry of Science and Technology, Taiwan, grant number MOST 109-2221-E-259-004-MY3.

Institutional Review Board Statement: Not applicable.

Informed Consent Statement: Not applicable.

Data Availability Statement: The data presented in this study are available on request from the corresponding author.

Acknowledgments: All authors acknowledge the financial support of Ministry of Science and Technology, Taiwan.

Conflicts of Interest: The authors declare no conflict of interest.

References

1. Crane, R.A.; Scott, T.B. Nanoscale zero-valent iron: Future prospects for an emerging water treatment technology. *J. Hazard. Mater.* **2012**, *211–212*, 112–125. [CrossRef] [PubMed]
2. Stefaniuk, M.; Oleszczuk, P.; Ok, Y.S. Review on nanozerovalent iron (nZVI): From synthesis to environmental applications. *Chem. Eng. J.* **2016**, *287*, 618–632. [CrossRef]
3. Pasinszki, T.; Krebsz, M. Synthesis and application of zero-valent iron nanoparticles in water treatment, environmental remediation, catalysis and their biological effects. *Nanomaterials* **2020**, *10*, 917. [CrossRef] [PubMed]
4. Zhao, X.; Liu, W.; Cai, Z.; Han, B.; Qian, T.; Zhao, D. An overview of preparation and applications of stabilized zero-valent iron nanoparticles for soil and groundwater remediation. *Water Res.* **2016**, *100*, 245–266. [CrossRef]
5. Yan, W.; Lien, H.L.; Koel, B.E.; Zhang, W. Iron nanoparticles for environmental clean-up: Recent developments and future outlook. *Environ. Sci. Process. Impacts* **2013**, *15*, 63–77. [CrossRef] [PubMed]

6. Zou, Y.; Wang, X.; Khan, A.; Wang, P.; Liu, Y.; Alsaedi, A.; Hayat, T.; Wang, X. Environmental remediation and application of nanoscale zero-valent iron and its composites for the removal of heavy metal ions: A review. *Environ. Sci. Technol.* **2016**, *50*, 7290–7304. [CrossRef] [PubMed]
7. Ruiz-Torres, C.A.; Araujo-Martinez, R.F.; Martinez-Castanon, G.A.; Morales-Sanchez, J.E.; Guajardo-Pacheco, J.M.; Gonzalez-Hernandez, J.; Lee, T.-J.; Shin, H.-S.; Hwang, Y.; Ruiz, F. Preparation of air stable nanoscale zero valent iron functionalized by ethylene glycol without inert condition. *Chem. Eng. J.* **2018**, *336*, 112–122. [CrossRef]
8. Fan, T.; Li, Y.; Zhang, H. Surfactant-free solvothermal synthesis of 3D flowerlike iron alkoxide (Fe-EG) micro/nanostructures: Structure, formation mechanism, and fenton oxidation of azo dyes. *Ind. Eng. Chem. Res.* **2017**, *56*, 11684–11696. [CrossRef]
9. Xiang, S.; Cheng, W.; Chi, F.; Nie, X.; Hayat, T.; Alharbi, N.S. Photocatalytic removal of U(VI) from wastewater via synergistic carbon-supported zero-valent iron nanoparticles and *S. Putrefaciens*. *ACS Appl. Nano Mater.* **2020**, *3*, 1131–1138. [CrossRef]
10. Gozdecka, A.; Wiacek, A.E. Effect of UV radiation and chitosan coating on the adsorption-photocatalytic activity of TiO_2 particles. *Mater. Sci. Eng. C* **2018**, *93*, 582–594. [CrossRef]
11. Jungcharoen, P.; O'Carroll, D.; Anotai, J.; Phenrat, T. Synthesis of Lignin-modified nanoscale zerovalent iron applied to arsenic removal. In Proceedings of the Full Paper Proceeding of International Conference on Engineering and Technology, Computer, Basics and Applied Sciences, ECBA-2017, Marrakesh, Morocco, 7–8 January 2017; Volume 3, pp. 1–6.
12. Wang, Z.; Chen, G.; Wang, X.; Li, S.; Liu, Y.; Yang, G. Removal of hexavalent chromium by bentonite supported organosolv lignin-stabilized zero-valent iron nanoparticles from wastewater. *J. Clean. Prod.* **2020**, *267*, 122009. [CrossRef]
13. Ganewatta, M.S.; Lokupitiya, H.N.; Tang, C. Lignin biopolymers in the age of controlled polymerization. *Polymers* **2019**, *11*, 1176. [CrossRef]
14. Sun, Z.; Fridrich, B.; de Santi, A.; Elagovan, S.; Barta, K. Bright side of lignin depolymerization: Toward new platform chemicals. *Chem. Rev.* **2018**, *118*, 614–678. [CrossRef] [PubMed]
15. Azadi, P.; Inderwildi, O.R.; Farnood, R.; King, D.A. Liquid fuels, hydrogen and chemicals from lignin: A critical review. *Renew. Sustain. Energy Rev.* **2013**, *21*, 506–523. [CrossRef]
16. Coccia, F.; Tonucci, L.; Bosco, D.; Bressan, M.; d'Alessandro, N. One-pot synthesis of lignin-stabilized platinum and palladium nanoparticles and their catalytic behavior in oxidation and reduction reactions. *Green Chem.* **2012**, *14*, 1073–1078. [CrossRef]
17. Coccia, F.; Tonucci, L.; d'Alessandro, N.; D'ambrosio, P.; Bressan, M. Palladium nanoparticles, stabilized by lignin, as catalyst for cross-coupling reactions in water. *Inorg. Chim. Acta* **2013**, *399*, 12–18. [CrossRef]
18. Marulasiddeshwara, M.B.; Raghavendra, P.K. Hydrogenation of carbonyl compounds to alcohols catalyzed by lignin supported palladium nanoparticles. *Mater. Today Proc.* **2019**, *9*, 295–305. [CrossRef]
19. Rak, M.J.; Friscic, T.; Moores, A. Mechanochemical synthesis of Au, Pd, Ru, and Re nanoparticles with lignin as a bio-based reducing agent and stabilizing matrix. *Faraday Discuss.* **2014**, *170*, 155–167. [CrossRef] [PubMed]
20. Han, G.; Wang, X.; Hamel, J.; Zhu, H.; Sun, R. Lignin-AuNPs liquid marble for remotely-controllable detection of Pb^{2+}. *Sci. Rep.* **2016**, *6*, 38164. [CrossRef]
21. Saratale, R.G.; Saratale, G.D.; Ghodake, G.; Cho, S.-K.; Kadam, A.; Kumar, G.; Jeon, B.-H.; Pant, D.; Bhatnagar, A.; Shin, H.S. Wheat straw extracted lignin in silver nanoparticles synthesis: Expanding its prophecy towards antineoplastic potency and hydrogen peroxide sensing ability. *Int. J. Biol. Macromol.* **2019**, *128*, 391–400. [CrossRef]
22. Kwon, G.-J.; Han, S.-Y.; Park, C.-W.; Park, J.-S.; Lee, E.-A.; Kim, N.H.; Alle, M.; Bandi, R.; Lee, S.H. Adsorption characteristic of Ag nanoparticles on cellulos nanofibers with different chemical compositions. *Polymers* **2020**, *12*, 164. [CrossRef] [PubMed]
23. Shirvanimoghaddam, K.; Czech, B.; Wiacek, A.E.; Bundyra, W.C.; Naebe, M. Sustainable carbon microtube derived from cotton waste for environmental applications. *Chem. Eng. J.* **2019**, *361*, 1605–1616. [CrossRef]
24. Myint, A.A.; Lee, H.W.; Seo, B.; Son, W.-S.; Yoon, J.; Yoon, T.J.; Park, H.J.; Yu, J.; Yoon, J.; Lee, Y.W. One pot synthesis of environmentally friendly lignin nanoparticles with compressed liquid carbon dioxide as an antisolvent. *Green Chem.* **2016**, *18*, 2129. [CrossRef]
25. Mu, L.; Shi, Y.; Wang, H.; Zhu, J. Lignin in ethylene glycol and poly(ethylene glycol): Fortified lubricants with internal hydrogen bonding. *ACS Sustain. Chem. Eng.* **2016**, *4*, 1840–1849. [CrossRef]
26. Yang, M.; Zhao, W.; Singh, S.; Simmons, B.; Cheng, G. On the solution structure of kraft lignin in ethylene glycol and its implication for nanoparticle preparation. *Nanoscale Adv.* **2019**, *1*, 299–304. [CrossRef]
27. Yang, W.; Fortunati, E.; Gao, D.; Balestra, G.M.; Giovanale, G.; He, X.; Torre, L.; Kenny, J.M.; Puglia, D. Valorization of Acid isolated high yield lignin nanoparticles as innovative antioxidant/antimicrobial organic materials. *ACS Sustain. Chem. Eng.* **2018**, *6*, 3502–3514. [CrossRef]
28. Yamashita, T.; Hayes, P. Analysis of XPS spectra of Fe^{2+} and Fe^{3+} ions in oxide materials. *Appl. Surf. Sci.* **2008**, *254*, 2441–2449. [CrossRef]
29. Sun, Y.-P.; Li, X.-Q.; Cao, J.; Zhang, W.-X.; Wang, H.P. Characterization of Zero-valent Iron Nanoparticles. *Adv. Colloid Interface Sci.* **2006**, *120*, 47–56. [CrossRef]
30. Desalegn, B.; Megharaj, M.; Chen, Z.; Naidu, R. Green synthesis of zero valent iron nanoparticle using mange peel extract and surface characterization using XPS and GC-MS. *Heliyon* **2019**, *5*, e01750. [CrossRef]
31. Liu, A.; Liu, J.; Han, J.; Zhang, W.-X. Evolution of nanoscale zero-valent iron (nZVI) in water: Microscopic and spectroscopic evidence on the formation of nano- and micro-structure iron oxides. *J. Hazard. Mater.* **2017**, *322*, 129–135. [CrossRef]

32. Pullin, H.; Springell, R.; Parry, S.; Scott, T. The effect of aqueous corrosion on the structure and reactivity of zero-valent iron nanoparticles. *Chem. Eng. J.* **2017**, *308*, 568–577. [CrossRef]
33. Hamdy, A.; Mostafa, M.K.; Nasr, M. Zero-valent iron nanoparticles for methylene blue removal from aqueous solutions and textile wastewater treatment, with cost estimation. *Water Sci. Technol.* **2018**, *78*, 367. [CrossRef]
34. Chen, F.; Hu, X.; Tu, X.; Chen, L.; Liu, X.; Tan, L.; Mao, Y.; Shi, J.; Teng, X.; He, S.; et al. High-yield production of lignin-drived functional carbon nanosheet for dye adsorption. *Polymers* **2020**, *12*, 797. [CrossRef] [PubMed]
35. Goliszek, M.; Wiacek, A.E.; Wawrzkiewicz, M.; Sevastyanova, O. The impact of lignin addition on the properties of hybrid microspheres based on trimethoxyvinylsiane and divinylbenzene. *Eur. Polym. J.* **2019**, *361*, 1605–1616. [CrossRef]
36. Mao, Z.; Wu, Q.; Wang, M.; Yang, Y.; Long, J.; Chen, X. Tunable synthesis of SiO_2-encapsulated zero-valent iron nanoparticles for degradation of organic dyes. *Nanoscale Res. Lett.* **2014**, *9*, 501. [CrossRef] [PubMed]

polymers

MDPI

Article

Polymeric Microparticles of Calcium Pectinate Containing Urea for Slow Release in Ruminant Diet

Myrla Melo [1], André da Silva [1,*], Edson Silva Filho [2], Ronaldo Oliveira [3], Jarbas Silva Junior [3], Juliana Paula Oliveira [1], Antônio Vaz [1], José Moura [1], José Pereira Filho [1] and Leilson Bezerra [1,*]

1 Department of Animal Science, Federal University of Campina Grande, Avenida Universitária, s/n-Jatobá, Patos 58708-110, Brazil; myrla_kristy@hotmail.com (M.M.); juli.foo@gmail.com (J.P.O.); antonio.fernando@professor.ufcg.edu.br (A.V.); jose.fabio@ufcg.edu.br (J.M.); jmpfpiaui@gmail.com (J.P.F.)

2 Chemistry Department, Federal University of Piaui, Ininga, s/n, Teresina 64049-550, Brazil; edsonfilho@ufpi.edu.br

3 Department of Animal Science, Federal University of Bahia, Av. Adhemar de Barros, 500, Ondina, Salvador City 40170-110, Brazil; ronaldooliveira@ufba.br (R.O.); miguelreges@gmail.com (J.S.J.)

* Correspondence: andre.leandro@ufcg.edu.br (A.d.S.); leilson@ufpi.edu.br (L.B.)

Abstract: In ruminant feeding, mechanisms for controlling the rate of ammonia release in the rumen are important for increasing the efficiency of transforming dietary nitrogen into microbial protein. Three microencapsulated formulations, with increased urea concentrations of 10 (MPec1), 20 (MPec2) and 30% (MPec3) from the w/w, based on the mass of citrus pectin solution, employ the external ionic gelation/extrusion technique. The properties of microencapsulated urea were examined as a completely randomized design with 5 treatments each with 10 replicates for evaluation, and the ratios of dietary to free urea were compared using 5 fistulated male Santa Ines sheep in a Latin 5 × 5 square design. The degradation kinetics showed that the rate of controlled release from the microencapsulated systems was significantly reduced compared with that of free urea ($p < 0.05$). The population density of ruminal protozoa increased when sheep received the microencapsulated urea ($p < 0.05$). The disappearance of dry matter and crude protein reached a degradation plateau during the first minutes for the MPec1 and MPec2 systems and was slower for MPec3. The MPec1 and MPec2 systems presented higher ($p < 0.05$) blood serum concentrations of albumin, urea nitrogen (BUN), creatinine and total cholesterol and did not affect ($p > 0.05$) the other blood metabolites. The MPec2 systems are recommended because they consist of microspheres with more ($p < 0.05$) controlled core release, delaying the peak of urea released in the rumen and BUN without affecting ($p < 0.05$) ruminal pH and temperature. Microencapsulation with calcium pectinate provided better utilization of urea, reducing the risk of ruminant intoxication.

Keywords: biochemistry; slow release; pH and rumen temperature; protozoa

Citation: Melo, M.; da Silva, A.; Silva Filho, E.; Oliveira, R.; Silva Junior, J.; Oliveira, J.P.; Vaz, A.; Moura, J.; Pereira Filho, J.; Bezerra, L. Polymeric Microparticles of Calcium Pectinate Containing Urea for Slow Release in Ruminant Diet. *Polymers* **2021**, *13*, 3776. https://doi.org/10.3390/polym13213776

Academic Editor: Arn Mignon

Received: 31 August 2021
Accepted: 27 October 2021
Published: 31 October 2021

Publisher's Note: MDPI stays neutral with regard to jurisdictional claims in published maps and institutional affiliations.

1. Introduction

The use of urea ($(NH_2)_2CO$) as a source of nonprotein nitrogen (NPN) has been applied on a large scale in ruminant diets, mainly to reduce production costs. $(NH_2)_2CO$ is hydrolyzed by ureolytic bacteria in the rumen into ammonia nitrogen (NH_3-N) and incorporated by ruminal microorganisms for microbial protein production, which is very important to ruminants [1]. The NH_3-N peak in the rumen when urea is supplied usually occurs 1 to 2 h after feeding, while for true protein sources, this peak occurs approximately 3 to 5 h after feeding [1,2]. However, there is concern about its use at high levels in animal diets due to the low utilization of NH_3-N by ruminal microorganisms and, especially, the possibility of the intoxication of the ruminant herd [3].

The low palatability of urea and its intoxication capacity when converted to ammonia in the rumen by ruminal microorganisms limits its inclusion in concentrate mixtures to 2% or 40 g/100 kg body weight, and the supply to the animals must be conducted in covered troughs to avoid volatilization [4,5].

The speed of ammonia release in the rumen is the determining factor in the transformation of dietary nitrogen into microbial protein [6,7], and the adaptation of the ruminal environment of animals to urea is essential. Thus, the urea microencapsulation technique can gradually release this ingredient in the rumen environment, reducing the risk of animal poisoning and improving the synchronism of nutrients in the rumen without compromising productive performance [8]. Several materials are being tested as encapsulating systems with very promising results [6,9,10].

One of these materials is citrus pectin, a heteropolysaccharide contained in the middle and primary lamellae and cell walls of land plants [11,12], which has attracted significant interest in the food industry [13]. Chemically, pectins are biopolymers composed of homogalacturonan, xylogalacturonan and rhamnogalacturonans, with the latter being divided into two types: rhamnogalacturonans and rhamnogalacturonans [11,14]. The chemical composition of citrus pectin allows its ability to form gels due to the differences in the size of the polygalacturonic acid chain and the degree of esterification of its carboxylic groups, which can be used in the production of several materials as it is a natural product and easy to acquire [12].

Shtriker et al. [15] observed that intake from diets containing soluble fibers from citrus pectin resulted in gut microbiota comprising a healthier flora and led to positive effects on weight, glycemic control and liver β oxidation via adenosine monophosphate-activated protein kinase in mice. Ben-Ghedalia et al. [16] investigated a pectin-rich diet in sheep and its effect on rumen parameters and nitrogen (N) metabolism. It was observed that the dried citrus pectin pulp can provide favorable conditions for cellulolysis in the rumen with a positive effect on the supply of N to the intestine. Thus, if used as an encapsulating agent, citrus pectin might allow the protection of urea, promoting its slower release, improving its use and reducing environmental contamination by N excretion from the urea cycle.

The rapid degradation of urea in the rumen by microbial urease, with risk of intoxication due to the formation of ammonia, as well as the need for the adaptation of animals fed diets containing urea, have motivated studies to develop products that allow the gradual release of urea in the ruminal environment to increase the production of microbial protein and reduce intoxication cases [17,18]. Therefore, the technology of the microencapsulation of feed has become an interesting option because it allows a greater synchronism of the degradation rate of this compound in the rumen [6,9,10], keeping the NH_3-N concentration constant over a long period and, consequently, allowing a greater use of the feed by the animal [19]. Thus, the aim was to test the hypothesis that calcium pectinate can be used as a microencapsulation agent for efficient and slow release of urea and to increase the addition of urea in the sheep diet without harming the microflora and the ruminal environment, as well as reducing the risk of intoxication to the sheep.

2. Materials and Methods

2.1. Obtaining and Characterization of Microencapsulated Systems

The microencapsulated formulations experiment was examined as a completely randomized design with 5 treatments each (free urea, microencapsulated urea into calcium pectinate at different levels (MPec1, MPec2 and MPec3) and calcium pectinate encapsulating matrix free as a control) with 10 replicates.

Three microencapsulated formulations from calcium pectinate containing urea (U) were developed using citrus pectin as an encapsulating agent with a fixed concentration of 5% (*w*/*v*), defined from a previous test, and increased urea concentrations of 10 (MPec1), 20 (MPec2) and 30% (MPec3) from the *w*/*w*, based on the weight of citrus pectin solution, employing the external ionic gelation/extrusion technique.

To prepare the respective solutions, the corresponding masses of urea and pectin were duly weighed for each formulation and dissolved in distilled water. Then, for each system, the urea solution was slowly added to the pectin solution and stirred with a glass rod until completely homogenized. Each solution of the core/encapsulant mixture was subsequently extruded with the aid of a plastic syringe in a previously prepared 3% (*w*/*v*) calcium chloride crosslinking bath to form calcium pectinate microparticles. The extrusion was carried out from a fixed height of 10 cm, and the microspheres remained in contact with the crosslinking solution for 30 min under constant magnetic stirring centrifuged at 400× *g*. Finally, the microspheres were separated with the aid of a sieve, washed with distilled water, transferred to a plastic tray and dried in an oven at 45 °C for 24 h.

Subsequently, micrographs of calcium pectinate microparticles with and without urea were obtained by optical microscopy, in a Medilux® microscope (Barneveld, The Netherlands) and by stereomicroscopy. For scanning under an optical microscope, the samples were fixed on a cover slip with adjusted lighting and 40× magnification.

The microencapsulation yield was based on the masses of urea and pectin solution before and after ionic extrusion/gelation, calculated using the following equation:

$$MY = (MF/MI) \times 100 \quad (1)$$

where *MY* = microencapsulation yield; *MF* = final mass of the microencapsulated product after extrusion/crosslinking; and *MI* = initial mass of urea and pectin solution.

The microencapsulation efficiency evaluated the retention capacity of the calcium pectinate matrix and was determined based on the urea content inserted and the content retained after the process. The microencapsulation efficiency was calculated using the following equation:

$$ME = (Uactual/Utheoretical) \times 100 \quad (2)$$

where *ME* = microencapsulation efficiency; *Uactual*: actual retained urea content; *Utheoretical*: Urea content inserted.

Urea was quantified according to the AOAC Kjeldahl method [20]. The data obtained were analyzed to quantify the total nitrogen using the following equation:

$$\%N = V \times M \times F \times 0.014 \times 100/\mathrm{m} \quad (3)$$

where M = molarity of hydrochloric acid, 0.02 N; F = hydrochloric acid correction factor = 1.00; 0.014, milliequivalent weight of nitrogen (g); V = volume of hydrochloric acid used in the titration, in mL; m = sample weight (g).

Thermogravimetry (TG) and differential scanning calorimetry (DSC) curves for urea, calcium pectinate and microencapsulated systems were obtained simultaneously in a thermal analyzer (SDT Q600, V20.9 Build 2, Columbus, OH, USA), under an inert atmosphere, flow of 100 mL/min, heating rate of 10 °C/min, from 30 to 600 °C, using a platinum crucible containing approximately 8.0 mg of sample. T_{onset} was considered to evaluate the thermal stability of the materials studied from the TG curves. The temperature peaks were considered to extract the events from DSC curves.

2.2. Ethical Considerations, Animals, Diets and General Procedures

The experimental trial was developed in strict accordance with the recommendations contained in the Guide of the National Council for the Control of Experiences in Animals, Brazil, and the protocol was approved by Permit Number 116/2018 [21].

Five rumen-fistulated sheep (initial average body weight (BW) 30.4 ± 6 kg) and age of 28 ± 2 months housed in individual pens (1.0 × 1.5 m) fitted with feeders and waterers were assigned in a 5 × 5 Latin square design, with five consecutive 16-day periods divided into 14-day adaptation and 2-day (48 h) sampling periods. All animals were previously treated for internal and external parasites with ivermectin (Ivomec gold, Merial®, Salvador, Bahia, Brazil) and vaccinated against clostridiosis (Sintoxan, Merial®, Sao Paulo, Brazil).

The formulated diets were prepared according to National Research Council (NRC) [22] guidelines for sheep maintenance and growth with a weight gain of 40 g/d and an average weight of 30 kg. The sheep received water ad libitum and were fed twice daily (08:00 and 16:00 h) with total mixed ration (TMR). The feed refusals were collected and weighed daily, and the amount of feed offered was adjusted to allow a 100 g/kg refusal. The chemical composition and proportion of ingredients and experimental diets are presented in Table 1.

The diet provided presented ≅ 10.5% crude protein (CP) and ≅35% non-fibrous carboydrates (FC), meeting synchronism of energy and protein [22] and an amount of fiber (≅50) meeting the minimum amount for rumination of the animals [23].

Table 1. Chemical composition and proportion of ingredients and experimental diets containing levels microencapsulated urea into calcium pectinate matrix of formulations (MPec1, MPec2 and MPec3).

Item	**Ingredients**							
Chemical Composition (g/kg DM)	**Tifton-85 Hay**	**Soybean Meal**	**Ground Corn**	**MPec1**	**MPec2**	**MPec3**	**MPec [3]**	**Urea**
Dry matter (g/kg as fed)	872	907	889	990	990	990	990	-
Ash	60.3	70.1	15.1	-	-	-	-	-
Crude protein	54.1	489	93.9	740	796	875	77.3	2800
Ether extract	10.8	54.9	30.1	24.3	24.3	24.3	24.3	-
Neutral detergent fiber	728	198	158	219	219	219	219	-
Acid detergent fiber	371	95.2	43.0	-	-	-	-	-
Nonfibrous carbohydrate	138	487	751	59.5	59.5	59.5	59.5	-

	Experimental diets				
Ingredient proportion (g/kg DM)	Urea	MPec1	MPec2	MPec3	MPec
Tifton-85 hay	600	600	600	600	600
Soybean meal	60.0	80.0	80.0	80.0	90.0
Ground corn	325	305	305	305	295
Encapsulation material [1]	5.00	5.00	5.00	5.00	5.00
Mineral mixture [2]	10.0	10.0	10.0	10.0	10.0
Chemical composition (g/kg DM)					
Dry matter (g/kg as fed)	881	882	882	882	882
Ash	55.2	56.3	56.3	56.3	5.68
Crude protein	106	104	104	105	105
Ether extract	19.6	2.02	2.02	2.02	2.04
Neutral detergent fiber	500	501	501	501	502
Acid detergent fiber	243	244	244	244	244
Nonfibrous carbohydrate	356	354	354	354	351

[1] Urea free, pectinate or microencapsulated urea at different levels. [2] Presenting guaranteed the following levels of active elements: 120 g calcium, 87 g phosphorus, 147 g sodium, 18 g sulfur, 590 mg copper, 40 mg cobalt, 20 mg chromium, 1.8 g iron, 80 mg iodine, 1.3 g manganese, 15 mg selenium, 3.8 g zinc, 300 mg molybdenum and a maximum of 870 mg fluoride. Solubility of phosphorus citric acid: 2 to 95%. [3] Encapsulating matrix based on citrus pectin produced on the basis of ionic gelation/extrusion technique.

Samples of the ingredients and refusals were pre-dried in a forced-air ventilation oven at 55 °C for 72 h. Then, samples of the ingredients and refusals were ground in a Wiley knife mill with a sieve size of 1 mm. The samples were stored in plastic jars with lids, labeled and subjected to analyses (triplicate) to determine the dry matter (DM; method 967.03), ash (method 942.05), crude protein (CP; method 981.10) and ether extract (EE; method 920.29) contents [20]. Analyses for the determination of neutral detergent fiber (NDF) and acid detergent fiber (ADF) were performed according to Van Soest et al. [23], with changes proposed by Senger et al. [24] to include the use of an autoclave. The autoclave temperature was maintained at 110 °C for a period of 40 min. The nonfiber carbohydrates (NFCs) were determined by the following equation calculated by Hall [25] (2000):

$$NFC = 100 - [(CP - CP\ from\ urea + urea) + NDF + EE + Ash. \quad (4)$$

2.3. Degradation Kinetics and Ruminal and Blood Serum Parameters

The in situ technique was used for ruminal degradation from the nonwoven fabric bags ((TNT—100 g/m^2 (polypropylene)) with dimensions of 4.5 × 4.5 cm, sealed by a sealing machine. Each bag contained approximately 1.0 g of sample (microencapsulated urea into calcium pectinate at different levels (MPec1, MPec2 and MPec3) and free urea and calcium pectinate encapsulating matrix free as a control). Each bag and sample were oven-dried at 55 °C for 24 h and weighed to determine the dry matter content.

During the incubation period, the samples in the bags corresponded to three treatments with the MPec produced, free urea and calcium pectinate, placed in duplicate for each sheep (5) at the respective time intervals (0.25, 0.5, 1, 3, 6, 12, 24 and 48 h), as well as a blank, with two bags per time interval (8) for each animal, totaling 16 bags (identified with graphite). The samples were placed at the final time (48 h), and gradually, the remaining samples were placed, so that the contents spent their time in the rumen and at the end they were all removed at once. The bags for time (0 h) zero (used to determine the readily soluble fraction) were introduced into water for 20 s and then removed, receiving the same treatment as the others. In adaptation period, four bags were inserted during 48 h as control (blank) and were used to determine the ruminal content that adheres to the bags for subsequent correction of the dry matter Afterward, they were washed in running water, placed in an oven with forced ventilation at 55 °C for 72 h and then weighed on an analytical balance. The degradation profiles were calculated by the nonlinear least-squares approach using the exponential model described by Ørskov and McDonald [26].

The in situ ruminal degradation kinetics of dry matter (DM) and nitrogen (N) were calculated using the exponential model of Ørskov and McDonald [26]:

$$PD = a + b\,[1 - e^{(-c \times t)}] \quad (5)$$

where PD is the potential degradation at time t; a is the readily water-soluble fraction; b is the water-insoluble but potentially degradable fraction; c is the fraction degradation rate b; and t is the incubation time in minutes. After processing and incubation of the samples, the nitrogen profile was analyzed according to the AOAC Kjeldahl method [20].

The effective DM and N degradability (ED) in the rumen was calculated using the following equation:

$$ED = a + [(b \times c)/c + k)] \quad (6)$$

where a is the soluble fraction, b is the potentially degradable insoluble fraction, c is the degradation rate constant and k is the passage rate of the digesta from the rumen. The ED was calculated using a single rumen-reticulum particle passage rate of 8%/h. The undegradable fraction (U) was calculated from a and b fractions:

$$U = 100 - (a + b) \quad (7)$$

For the ruminal parameters, all sampling of the ruminal content was performed during the collection of bags. The degradability test was performed at each time of incubation of the microencapsulated samples: 0, 0.25, 0.5, 1, 3, 6, 12, 24 and 48 h. The collection of ruminal fluid occurred manually at different locations in the rumen through the fistula, obtaining a sample of approximately 200 mL. After each collection, the content went through the filtration process, which was separated into containers (20 mL) with a specific amount for each parameter evaluated. The pH and temperature were immediately measured with the aid of a pH meter and a portable digital thermometer that was previously calibrated.

Each sample collector contained 2 mL of filtered inoculum and 4 mL of M.S.F solution (35% formaldehyde, methyl green and sodium chloride). To carry out protozoan counting, the samples were initially homogenized with the aid of a magnetic stirrer, after which the reading was carried out in a Neubauer chamber in which 10 μL was pipetted into each counting area of the chamber and a cover slip was placed on top to improve the visualization of protozoa. The reading was carried out under an optical microscope (Lumen) at 40× in the C field. In the center of these chambers, there were several perpendicular lines with markings in quadrants, so there were 4 readings in each quadrant. The final count results were calculated by equation:

$$N \times 3 \times 10{,}000 = NP/1\ \text{mL} \quad (8)$$

where N: mean of the readings of quadrant C (uppercase) in mL; 3: inoculum dilution; 10,000: constant; NP: protozoan population count.

2.4. Blood Metabolites

The blood collection procedure of the five animals was performed by jugular venipuncture from each animal in tubes using a vacuum system (Becton, Dickson and Co., São Paulo, SP, Brazil) at the same incubation times of the degradability test. Blood collection occurred simultaneously with the ruminal incubation times of the microencapsulated systems. One #16 catheter (Medical supply®, Sao Paulo, Brazil) was inserted in each animal to facilitate collection and meet animal welfare requirements. Samples were temporarily kept at room temperature until clot retraction and then centrifuged (Centrifuge 90–1 model, Coleman®, São Paulo, Brazil) at 2500× g for five min to generate blood serum. Finally, the serum was stored at −20 °C in Eppendorf Tubes® (Sigma-Aldrich, São Paulo, Brazil) until analysis. The serum metabolites were measured using commercial kit tests to measure total protein, albumin, blood urea nitrogen (BUN), creatinine, aspartate aminotransferase (AST), cholesterol and triglycerides from the means obtained in an automated Cobas C111 biochemical apparatus (Roche, Ludwigsburg, Germany) in enzymatic or colorimetric kinetic assays. The electrolytic analysis was performed in an automated analyzer (Max Ion-Medmax, Shenzhen, China) by the direct ion selectivity method for calcium, chlorine, sodium and potassium.

2.5. Statistical Analysis

The microencapsulated systems were examined as a completely randomized design with 5 treatments (MPec1, MPec2 and MPec3, free urea and calcium pectinate) and 10 replicates. For evaluation of ruminal parameters and blood parameters, a 5 × 5 Latin square design (5 treatments and 5 periods) was used, and the results were analyzed and measurements were repeated over time (0, 0.25, 0.5, 1, 3, 6, 12, 24 and 48 h relative to the incubation time) using the MIXED procedure of SAS [27] following a model including experimental treatment, incubation time, animal and interactions according to following Equation:

$$Yijk = m + Ti + aj + Pk + eijk, \tag{9}$$

where $Yijk$ is the dependent variable measured in animal j, which was subjected to the i treatment in period k; μ is the general mean; Ti is the fixed effect of treatment I; aj is the random effect of animal j; Pk is the random effect of period k; and $eijk$ is the unobserved random error assuming normal distribution. All means were compared using the Tukey test, with a critical level of 5% probability being adopted for type I error.

3. Results

3.1. Characterization of Microencapsulated Systems

The microencapsulated systems and calcium pectinate (empty pectin microparticles) were evaluated for their microstructure under stereoscopic and optical microscopy. The micrographs (Figure 1) demonstrated that compared to calcium pectinate microparticles, all systems were whitish in color due to the presence of urea. The image of the three systems showed no porosity or cracks, although MPec1 and MPec2 have a more regular particle shape. The MPec3 microencapsulated system showed a rough surface, which suggests the incidence of surface urea.

The addition of urea from the microencapsulated systems to calcium pectinate increased ($p \leq 0.05$) the microencapsulated yield to 92.2, 93.3 and 97.1% for MPec1, MPec2 and MPec3, respectively (Table 2). There was a reduction ($p \leq 0.05$) in the encapsulation efficiency with the increasing addition of urea, with values of 262, 218 and 264% and actual urea retention of 25.2, 28.4 and 31.1% for MPec1, MPec2 and MPec3, respectively.

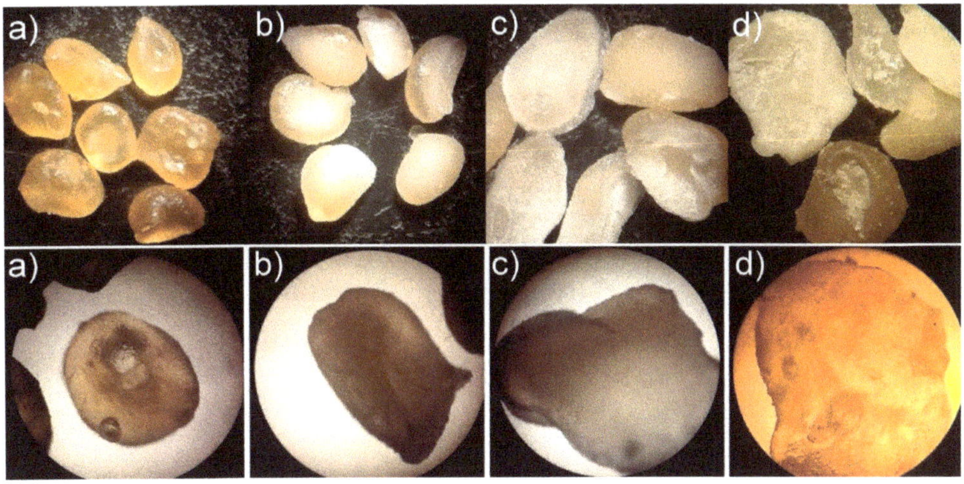

Figure 1. Micrographs obtained by stereomicroscope (Top) and optical microscope (Bottom) of (**a**) calcium pectinate and microencapsulated systems MPec1 (**b**), MPec2 (**c**) and MPec3 (**d**).

Table 2. Mean and standard deviation (±SD) of the total nitrogen, crude protein, theoretical and actual urea contents and microencapsulation efficiency of microencapsulated urea into calcium pectinate matrix of formulations (MPec1, MPec2 and MPec3).

Variables (%)	Microencapsulated Urea			*p*-Value [1]
	MPec1	Mpec2	Mpec3	
Microencapsulation yield	92.2 [a] ± 0.02	93.3 [b] ± 0.03	97.1 [c] ± 0.04	<0.01
Microencapsulation efficiency	262 [a] ± 1.17	218 [b] ± 1.12	264 [a] ± 1.16	<0.01
Nitrogen (N) total	11.2 [a] ± 0.08	12.8 [b] ± 0.09	14.0 [c] ± 0.10	<0.01
Crude protein	70.1 [c] ± 0.88	80.2 [b] ± 1.06	87.5 [a] ± 0.79	<0.01
Theoretical urea content	10.0 [c] ± 0.0	20.0 [b] ± 0.0	30.0 [a] ± 0.0	<0.01
Actual urea content	25.2 [b] ± 0.48	28.4 [a] ± 0.32	31.1 [a] ± 0.30	<0.01

[1] Means followed by the different letters differ (a; b and c) differ from Tukey's test when $p \leq 0.05$.

The TG curves (Figure 2a) demonstrated in general that urea, calcium pectinate and the microencapsulated systems presented two stages of mass loss, the first one due to moisture loss in the case of pectin and microparticles. The DSC curves (Figure 2b) show an endothermic event below 100 °C for calcium pectinate and the microencapsulated systems, corresponding to moisture loss, as also seen in the TG curves. Urea showed endothermic events at 130 and 212 °C, corresponding to its melting and thermal degradation, respectively. Pectin showed two endothermic events at 184 and 213 °C and one exothermic event at 240 °C, corresponding to its thermal degradation. A higher melting temperature (162 °C) was observed for the urea in MPec1 compared to free urea (130 °C), while the urea in MPec2 and MPec3 did not show melting events, suggesting simultaneous melting and thermal degradation.

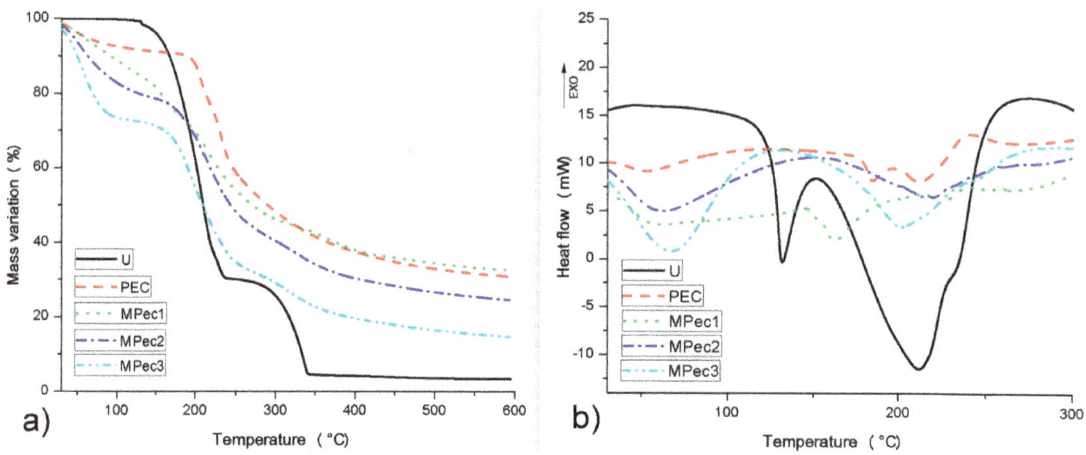

Figure 2. (**a**) Thermogravimetric (TG) and (**b**) differential scanning calorimetry (DSC) curves of the bioactive films obtained from the incorporation of Cashew Nut Shell Liquid into a matrix of sodium alginate.

3.2. Degradation Kinetics and Ruminal and Blood Serum Parameters

There was a greater disappearance ($p \leq 0.05$) of the initial soluble fraction (a) of DM in the MPec3 treatment compared to other microencapsulation systems and calcium pectinate and a smaller disappearance of the crude protein fraction (a) in the MPec3 treatment (Table 3).

Table 3. Rumen degradation profiles in sheep fed with microencapsulated urea intro calcium pectinate matrix of formulations (MPec1, MPec2 and MPec3) and encapsulating matrix (Mpec).

Variables	Microencapsulated Urea				p-Value
	MPec1	MPec2	MPec3	MPec	
Dry matter (DM)					
a [1] (%)	14.5 [b] ± 3.42	16.5 [b] ± 1.49	21.1 [c] ± 6.58	11.9 [a] ± 3.61	<0.01
b [2] (%)	66.2 [ab] ± 12.00	68.2 [ab] ± 1.55	63.5 [b] ± 12.26	70.1 [a] ± 6.14	<0.01
U [3] (%)	19.2 [a] ± 5.88	15.3 [b] ± 3.04	15.4 [b] ± 5.90	18.0 [a] ± 8.02	<0.01
c [4] ($\%h^{-1}$)	4.48 [a] ± 2.37	0.35 [c] ± 0.25	0.66 [c] ± 0.25	2.67 [b] ± 0.89	<0.01
ED [5] (%)	36.4 ± 11.25	19.2 ± 3.74	25.9 ± 5.86	29.1 ± 8.02	0.34
Crude Protein (CP)					
a [1] (%)	63.1 [b] ± 18.02	56.3 [b] ± 7.92	12.7 [c] ± 4.10	97.3 [a] ± 0.05	<0.01
b [2] (%)	33.6 [b] ± 10.95	40.6 [b] ± 8.92	79.5 [a] ± 10.58	2.10 [c] ± 0.33	<0.01
U [3] (%)	3.33 [ab] ± 1.04	3.15 [ab] ± 1.12	7.81 [a] ± 6.11	0.56 [b] ± 0.27	<0.01
c [4] ($\%h^{-1}$)	7.17 [b] ± 3.93	3.57 [c] ± 0.69	9.08 [a] ± 1.80	9.30 [a] ± 4.87	<0.01
ED [5] (%)	80.1 [b] ± 3.66	68.5 [c] ± 6.63	54.7 [d] ± 0.74	98.2 [a] ± 0.58	<0.01

Means followed by the different letters differ (a; b and c) from Tukey's test when $p \leq 0.05$; [1] soluble/rapidly degradable fraction; [2] slowly degradable fraction; [3] undegradable fraction; [4] degradation rate (c) of fraction "b" expressed in %/h; [5] effective degradation considering a pass rate of 8%/h). MPec is encapsulating matrix free based on citrus pectin produced on the basis of ionic gelation/extrusion technique.

Regarding the potentially degradable insoluble fraction (b), the behavior was the opposite; the greatest disappearance ($p \leq 0.05$) of DM occurred in pectinate calcium encapsulating matrix free, which differed only from the microencapsulated system with 30% urea (MPec3), while the fraction (b) of crude protein disappearance was higher ($p \leq 0.05$) in MPec3 compared to the other microencapsulated systems, which formed similar (MPec1 and MPec2) and had a greater disappearance of fractions (b) in relation to pectinate. Regarding the insoluble fraction (U), there was a greater disappearance of DM ($p \leq 0.05$) in pectinate and MPec1, which was superior to the MPec2 and MPec3 treatments. In contrast, the disappearance of fraction (U) of the major crude protein in MPec3 differed from pectinate and the other two formulations.

The DM effective degradation (ED) did not differ between treatments. However, the highest ED of CP was observed for calcium pectinate encapsulating matrix free, followed by the respective additions of microencapsulated urea (MPec1, MPec2 and MPec3).

Lower DM and CP disappearances (potential degradation—PD) were observed in calcium pectinate encapsulating matrix free, and there was no difference between the PD of the microencapsulated systems (Figure 3).

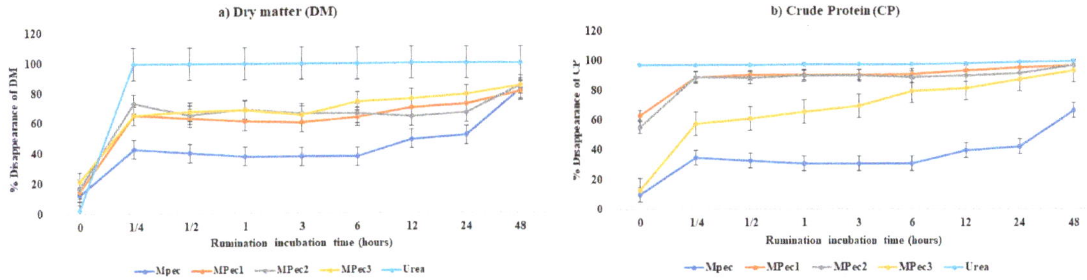

Figure 3. Mean disappearance of (**a**) dry matter (DM) and (**b**) crude protein (CP) in sheep from the urea and encapsulating matrix (MPec) free and microencapsulated urea in calcium pectinate matrix (MPec1, MPec2 and MPec3) incubated in sheep rumen at different times (Significant when $p < 0.05$).

Disappearance reached the degradation plateau at 0.25 h and then remained constant until 48 h of incubation, when all treatments reached a disappearance peak above 85%. There was a greater DM disappearance in free urea, as expected, and within 0.25 h, >99% of the urea had already disappeared confirming of the rapid disappearance after the first minutes of ruminal incubation.

There was no interaction ($p > 0.05$) between the treatments and the ruminal incubation for rumen pH (Figure 4a). The use of microspheres containing urea in their cores did not affect ruminal pH, with mean values varying between 7.02 and 7.26 between time periods from 0.5 and 1 h.

There was an effect of the treatments on the mean temperature values ($p \leq 0.05$) of the ruminal fluid (Figure 4b). At the incubation times of 0, 0.5 and 1 h, MPec1 promoted significant changes ($p \leq 0.05$) at the other times evaluated compared to the other treatments, while MPec3 stood out at a higher temperature (37 °C) compared to the other microencapsulated systems ($p \leq 0.05$), urea and encapsulating matrix free, with a significant effect at 0.25, 0.5 and 1 h. Calcium pectinate, free urea and MPec2 did not differ in temperature as a function of collection time. Likewise, it was evident that free urea did not differ significantly from the three microencapsulated systems or from calcium pectinate.

Regardless of the collection time, the inclusion of MPec1 followed by MPec2 presented the higher ($p \leq 0.05$) counts of protozoa at time periods between 0 and 48 h, and the higher counts (peak) occurred at 6 h. These two systems present practically the same protozoa count at 48 h. There was no difference ($p > 0.05$) in the protozoan counts between the calcium pectinate encapsulating matrix free, MPec3 and urea treatments (Figure 4c). For the inclusion of MPec3 in the rumen of sheep, the number of protozoa in the rumen environment was reduced compared to the other two systems.

The MPec1 and MPec2 microencapsulated systems presented higher blood serum concentrations of albumin, nitrogen (BUN), creatinine and total cholesterol (Table 4) in sheep compared to MPec3 and to free urea, which did not differ from others treatments ($p \leq 0.05$).

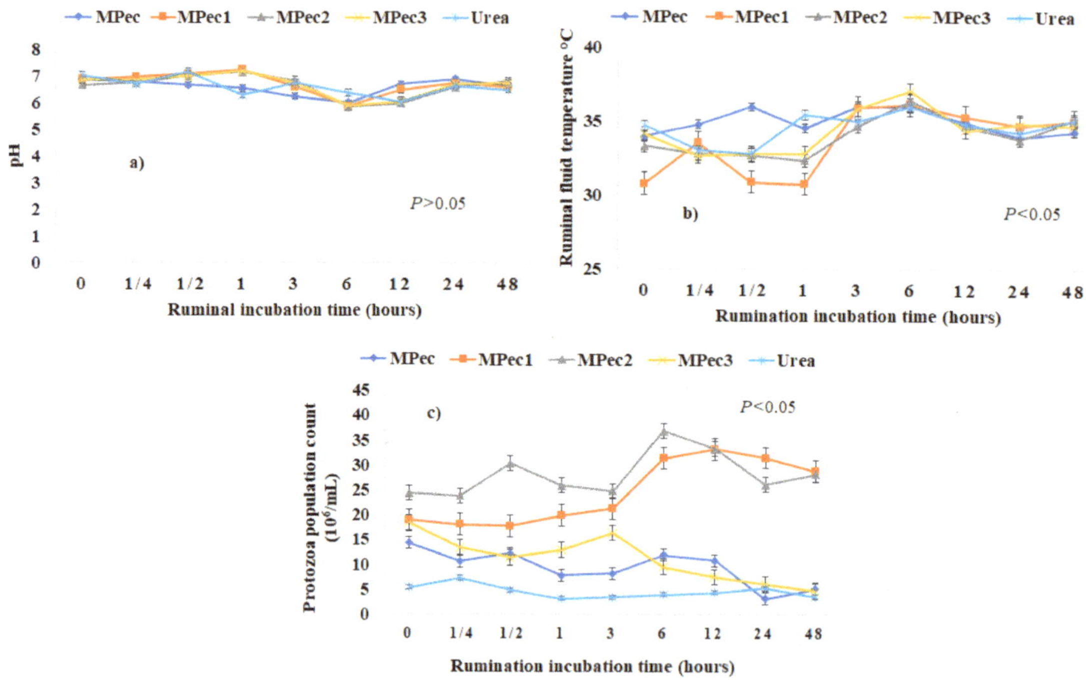

Figure 4. Mean variation of ruminal (**a**) pH; (**b**) ruminal temperature; and (**c**) total count of protozoa from incubation of urea and encapsulating matrix (MPec) free and microencapsulated urea in calcium pectinate matrix (MPec1, MPec2 and MPec3) in sheep rumen at different times.

Table 4. Mean and standard deviation (SD) of blood serum metabolites in sheep fed with microencapsulated urea into calcium pectinate matrix of formulations (MPec10, MPec2 and MPec3) and urea and calcium pectinate encapsulating matrix (MPec) free.

Metabolites	Microencapsulated Urea			Free		p-Value
	MPec1	MPec2	MPec3	Urea	MPec	
Albumin (g/dL)	$2.78^{a} \pm 0.11$	$2.69^{a} \pm 0.20$	$2.21^{b} \pm 0.35$	$2.36^{b} \pm 0.24$	$2.23^{b} \pm 0.32$	0.03
Total protein (g/dL)	$6.53^{a} \pm 0.27$	$6.35^{a} \pm 0.45$	$5.23^{b} \pm 0.78$	$5.54^{b} \pm 0.61$	$5.32^{b} \pm 0.81$	<0.01
BUN [3] (mg/dL)	$52.7^{b} \pm 6.49$	$53.7^{b} \pm 8.17$	$58.0^{a} \pm 9.13$	$52.4^{b} \pm 11.26$	$37.8^{c} \pm 6.28$	<0.01
Creatinine (mg/dL)	$0.64^{a} \pm 0.09$	$0.64^{a} \pm 0.14$	$0.53^{b} \pm 0.14$	$0.50^{b} \pm 0.10$	$0.47^{b} \pm 0.10$	0.028
Cholesterol (mg/dL)	$29.0^{a} \pm 5.87$	$29.4^{a} \pm 8.81$	$22.8^{b} \pm 6.09$	$24.5^{b} \pm 5.11$	$22.6^{b} \pm 3.62$	<0.01
Triglycerides (mg/dL)	14.9 ± 6.40	16.1 ± 3.78	14.2 ± 4.63	13.4 ± 4.85	15.1 ± 5.43	0.64
Calcium (mmol/L)	2.68 ± 0.39	2.72 ± 0.46	2.93 ± 0.22	2.76 ± 0.31	2.64 ± 0.35	0.53
Chlorine (mmol/L)	110 ± 6.19	110 ± 5.12	112 ± 3.25	115 ± 17.26	109 ± 16.92	0.55
Potassium (mmol/L)	5.58 ± 2.57	8.64 ± 6.41	10.3 ± 3.15	4.97 ± 2.45	4.67 ± 0.99	0.48
Sodium (mmol/L)	141 ± 9.72	136 ± 9.38	132 ± 3.18	144 ± 21.68	136 ± 21.50	0.23
AST [4] (U/L)	$59.2^{ab} \pm 13.31$	$53.7^{ab} \pm 10.40$	$41.6^{b} \pm 10.66$	$65.9^{a} \pm 11.52$	$57.1^{ab} \pm 10.69$	<0.01

Means followed by the different letters differ (a; b and c) from Tukey's test when $p \leq 0.05$; [3] Blood urea nitrogen; [4] Aspartate aminotransferase.

The total protein presented a greater serum concentration in the MPec1 and MPec2 microencapsulated systems (Table 4), with total protein values of 6.73 and 6.67 mg/dL, respectively, at 0.25 h (Figure 5a). There was a decrease ($p \leq 0.05$) in albumin values (Figure 5b) for the inclusion of MPec3 treatment at time periods 0 and 0.25 h (1.95 and 1.96 g/dL, respectively), as well as in the use of free urea at the time period of 0.5 h and for calcium pectinate encapsulating matrix free at the time period of 1 h, presenting the lowest blood concentration observed (1.91 g/dL).

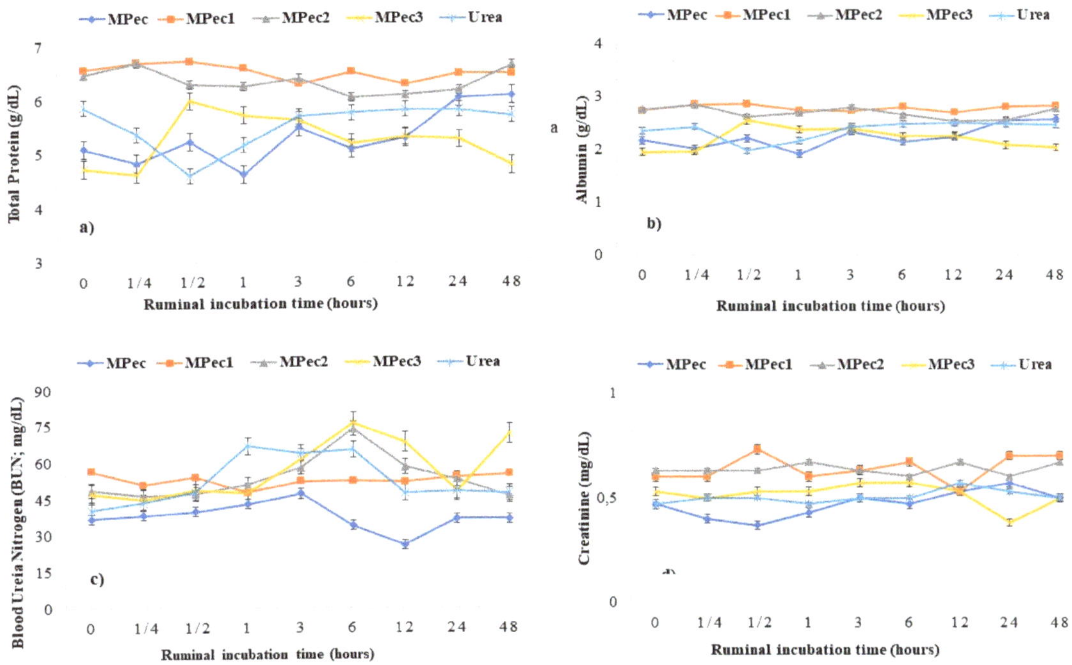

Figure 5. Mean and standard deviation (SD) of serum concentrations of (**a**) total protein; (**b**) albumin; (**c**) BUN; and (**d**) creatinine in blood sheep from incubation of urea and encapsulating matrix (MPec) free and microencapsulated urea in calcium pectinate matrix (MPec1, MPec2 and MPec3) at different times.

The peak of BUN occurred at six hours after incubation for the MPec2 and MPec3 treatments ($p \leq 0.05$), while for the other treatments, this peak occurred at three hours after incubation (Figure 5c). The lowest value ($p \leq 0.05$) was observed for serum creatinine concentration in sheep (Figure 5d) for the use of free calcium pectinate encapsulating matrix free at 0.25 h, with 0.40 mg/dL, and the highest concentration (0.73 mg/dL) was at 0.5 h in the MPec1 system.

The concentrations of triglycerides and calcium, chlorine, potassium and sodium electrolytes were not affected ($p > 0.05$) by the microencapsulated systems (MPec1, MPec2 and MPec3) or by encapsulating matrix free and urea. The microencapsulated system MPec3 (43.6 U/L) had a lower concentration ($p \leq 0.05$) of AST enzymes than the system with free urea (65.9 U/L), but there was no difference between the other treatments. There was no significant effect ($p > 0.05$) for the enzyme AST in relation to the incubation time.

4. Discussion

All microencapsulated systems showed a high microencapsulation yield, indicating that external ionic gelation is an adequate technique for urea microencapsulation, and citrus pectin was shown to be a viable encapsulation matrix. Noh et al. [28], in their study of microencapsulating multiple hydrophobic and hydrophilic active agents, described the potential use of pectin in microcapsule formulations as protection of active agents by gelation by electrostatic crosslinking.

Regarding the values of microencapsulation efficiency over 100%, the actual urea increase is related to the microencapsulation technique used, since in the microsphere drying process, the water present is evaporated and the core content is concentrated. It was observed that the microencapsulation efficiency decreased as the urea content increased, indicating an advantage for the lower levels inserted. This is because each encapsulating material has a retention limit, as well as the influence of the aqueous medium for preparing the microparticles, in which there may already be an early release of urea given its high solubility in water. Nevertheless, all three systems showed good results. When evaluating the microencapsulation efficiency of urea as a nucleus, Medeiros et al. [6] and Carvalho Neto et al. [10] obtained values above 98%.

It was observed from the micrographs that the higher the urea content inserted, the more irregular, thinner and larger the particle became, assuming a flattened shape. Morphological evaluations are important factors to study the protection of the encapsulated content, its uniformity in size and its release conditions. This is because the effectiveness of the use of microspheres directly depends on the properties of the microencapsulating agent, which should not allow the release of the nucleus before the desired time [6,7].

T_{onset} is the most commonly used parameter to estimate a material's thermal stability and is considered the point where thermal degradation begins. According to the data extracted from the TG curves (Figure 2a), free urea presented T_{onset} at 164 °C, a value approximately that (170.5 °C) found by Carvalho et al. [9]. The urea in microencapsulated systems showed the beginning of thermal degradation at a higher temperature when compared to free urea, considering the T_{onset}, whose values were 180, 181 and 169 °C for MPec1, MPec2 and MPec3, respectively. The DSC curves confirmed a better thermal stability for urea after microencapsulation. This showed the effectiveness of urea protection by citrus pectin in the form of calcium pectinate encapsulating matrix free, especially to MPec1 and MPec2, which presented more effective protections against thermal degradation, probably due to the influence of the particle microstructure (more regular and thicker), according to the visual analysis itself and as already indicated in the micrographs. Urea in all microencapsulated systems had better thermal stability compared to free urea, with emphasis on MPec1 and MPec2, whose core protections can be reproduced in the rumen, enabling gradual release and, therefore, better use and less risk of intoxication.

Urea microencapsulation from the citrus pectin promoted a gradual disappearance of the DM (Figure 3) after the first 25 min, regardless of the system, lower than urea and greater than pectinate. From 0.25 h, there was a disappearance of the stability of material for both DMs. This behavior indicates that it was released more slowly in the rumen, because, in the first 25 min, the free urea was almost totally released, whereas in the microencapsulated system, the release above 80% only occurred at 48 h.

Despite the fact that protozoa make up a large portion of the rumen biomass, their role in ruminal fermentation and their contribution to the metabolism and nutrition of the host is still an area of substantial controversy [29]. The higher concentration of ammonia promoted by free urea and MPec3 is probably the most consistent factor to explain protozoal elimination and seems to be due to decreased bacterial protein breakdown and feed protein degradability in the absence of protozoan rumen [29,30]. Moreover, it has been demonstrated that although bacterial predation by rumen protozoa is dependent on the protozoal size, a lower protozoa population is beneficial due to an increase in bacteria population and can result in lower methane emission, because one of the many important symbiotic associations formed in the rumen is that of the relationship between methanogenic archaea and ciliated protozoa [31].

Regarding free urea, in the early time periods (0.25 and 0.5 h), almost all the incubated content disappeared, with values of approximately 99.9 and 99.6%, respectively, which differs from microencapsulated systems in which up to 48 h showed degradability of 80.8, 84.7 and 84.7% for MPec1, MPec2 and MPec3, respectively, indicating that all treatments showed a gradual release. Furthermore, when we analyzed BUN, it was observed that MPec2 and MPec3 were delayed, reaching the highest blood concentrations only after 6 h of incubation, whereas MPec1 and free urea peaked 3 h after incubation. These results confirm that citrus pectin was a suitable wall material to protect urea through the acquisition of calcium pectinate microparticles, since the protection promoted a delay in ruminal release, as well as in serum levels of BUN. According to Patra and Aschenbach [5], the forage takes about 4 to 6 h to start its degradation, while urea, after 4 h of ruminal incubation, practically disappears. This statement corroborates our findings, in which the urea disappeared practically in the first 15 min, probably due to being inserted in the bag without feed mixture. However, when more pectinate was added in the formulation (30%), a slower release occurred.

In general, the MPec1 and MPec2 systems showed a more gradual degradation rate after 3 h, suggesting a more effective protection for urea. This may have occurred due to the higher core content of the MPec3 system, since it increases the possibility of urea being closer to the particle surface and therefore being more easily released/degraded. DM disappearance increases with the concentration of urea used in the system because the increase in urea content increases the amount of NPN that is highly soluble in water.

The lower values of the indigestible fraction (U) of crude protein can be explained because the microencapsulated systems had a higher value of the soluble fraction. The MPec1 and MPec2 systems remained similar, while MPec3 presented a higher value, expressing its low degradation potential compared to the other studied systems. Regarding the degradation rate, the MPec2 system presented the lowest values, with 0.35 and 3.57% for DM and crude protein, respectively. Faster passage rates favor greater efficiency in the growth of ruminal microorganisms. The MPec3 presented a lower soluble CP and ED of CP than MPec1 and MPec2, implying that MPec3 can release urea slowly but has more bypass urea to the lower gut. While MPec1 has more soluble CP and a higher ED of CP, this implies that MPec1 releases urea faster (a = 63.1%) than MPec3 (a = 12.7%), and the majority of urea in MPec1 is degraded in the rumen (ED = 80%).

The use of microspheres containing urea in their core only interfered with the ruminal pH in a gradual way, with mean values varying between 7.02 and 7.26 between times from 0.5 and 1 h, with no significant changes between the evaluated treatments. It is important to emphasize that regardless of the system, the lowest pH values were observed at 3 h and 6 h. Van Soest [3] states that for an environment favorable to bacterial proliferation, it is necessary that the ruminal pH ranges between 6.0 and 7.0 and that a pH ranging between 5.5 and 6.0 significantly reduces the activity of the ruminal microbiota in the rumen. The MPec1 and MPec2 systems presented the best ruminal pH ranges.

From the characterizations and visual analyses, MPec3 showed a lower capacity to protect the nucleus, which favors the rapid release of nitrogen, consequently implicating in microbial growth [32]. It was observed that the number of protozoa per milliliter significantly varied between the studied treatments and incubation times, noting that the use of more citrus pectin in microencapsulated systems favored the increase in the population density of protozoa, because it probably enabled a more favorable ruminal environment, since ammonia in larger quantities slows the growth of microorganisms [4,33]. These results indicate that the gradual release of urea promotes better efficiency in the use of nitrogen and the maintenance of a stable environment, since the use of free urea exhibited inferior and differentiated behavior for the conditions of ruminal adaptation. Ammonia is produced from dietary protein, or urea is used by the ruminal microorganisms for their growth, which is subsequently available to the host as a microbial protein [5].

The MPec1 and MPec2 systems had the highest serum concentrations of total protein, albumin and creatinine, probably due to the greater amount of encapsulating matter, thus allowing for a faster rate of ammonia release compared MPec3 but in a smaller amount. This provided a more favorable ruminal environment for the growth of microorganisms and allowed a greater count of the protozoan population [5]. Urea is synthesized in the liver in amounts proportional to the concentration of ammonia produced in the rumen, and its concentration is directly related to the protein levels in the feed and the energy–protein ratio in the diet [2,34]. Calomeni et al. [2], quantifying the effects of feeding polymer-coated slow-release urea on blood parameters, noted similar values for albumin analysis (2.37 and 2.34 mg/dL) and higher creatinine values (1.26 and 1.27 mg/dL) for slow-release urea and conventional urea, respectively.

The addition, the MPec3 system increased the concentration of BUN, probably due to the higher concentration of urea in the microencapsulated system, and, consequently, a greater production of ruminal ammonia and the greater difficulty microorganisms have in using ammonia [4,5,33]. The consequence was a reduction in the total microorganism count. The serum urea concentration in the MPec3 treatment was at the maximum threshold (58 mg/dL) of the stipulated physiological standards (23 to 58 mg/dL) for the sheep breed [2,34,35]. Ziguer et al. [36], for instance, found average values of 62.45 and 63.82 mg/dL. Other research using the microencapsulation of urea from other materials has shown the range of serum urea in sheep to be greater compared to the literature [6,9,37]. Perhaps this is due to the longer continuing release of urea into the rumen [7,9,37]. However, it is still too early to state from only published articles that this is the range for the use of slow-release urea. It is noteworthy that no animal showed signs of toxicity, but the excess of free ammonia in the rumen is undesirable, as it can lead to energy losses by the animal to release this ammonia through the urine [9,34]. According to Kozloski [38], blood urea has a positive correlation with the concentration of ammonia in the rumen and with the use of amino acids (alanine, glutamine and glycine) in the liver. Calomeni et al. [2] observed mean values of 45.5 mg/dL with the use of microencapsulated slow-release urea in the sheep diet.

5. Conclusions

The calcium pectinate microparticles containing urea studied in sheep diet for slow release improved effective and potential degradability and protozoa population count, without affecting blood metabolite concentrations. The MPec2 formulation presented microspheres with a more controlled urea release. Thus, the use of citrus pectin as a urea wall material to obtain calcium pectinate microencapsulated systems is a promising alternative, as it can provide a better use of urea, reduce the risk of animal poisoning, as it did not change the liver metabolic enzymes, and favor the safe administration of larger amounts of this ingredient to ruminants.

Author Contributions: Conceptualization, A.d.S. and L.B.; formal analysis, M.M., E.S.F., J.S.J., J.P.O., A.V., J.M. and J.P.F.; methodology, M.M. and J.P.O.; resources, R.O., J.S.J. and L.B.; investigation; M.M., A.d.S. and L.B.; supervision, A.d.S. and L.B.; writing—original draft, M.M., A.d.S. and R.O.; writing—review and editing, E.S.F., A.V. and J.M. All authors have read and agreed to the published version of the manuscript.

Funding: This research was funded by the National Council for Scientific and Technological Development (Brazil), and the Coordination for the Improvement of Greater Education Personnel (CAPES).

Institutional Review Board Statement: The study was conducted according to the guidelines of the Declaration of Helsinki, and approved by the Ethics Committee of Federal University of Campina Grande (protocol code nº 116-2018 and date of approval 30 November 2018).

Informed Consent Statement: Not applicable.

Data Availability Statement: The data that support the findings of this study are available from the corresponding author upon reasonable request.

Acknowledgments: The authors would like to thank the research supported granted by the CNPq-Brazil (National Council for Scientific and Technological Development of Brazil), by the CAPES-Brazil (Coordination for the Improvement of Greater Education Personnel of Brazil) and by the FAPESB (Bahia State Research Foundation).

Conflicts of Interest: The authors declare no conflict of interest.

References

1. Geron, L.J.V.; Garcia, J.; de Aguiar, S.C.; da Costa, F.G.; da Silva, A.P.; Sousa Neto, E.L.; de Carvalho, J.T.H.; Roberto, L.S.; Coelho, K.S.M.; Santos, I.S. Effect of slow release urea in sheep feed on nitrogen balance. *Semina Ciencias Agr.* **2018**, *39*, 683–696. [CrossRef]
2. Calomeni, G.D.; Gardinal, R.; Ventureli, B.C.; de Freitas, J.E., Jr.; Vendramini, T.H.A.; Takiya, C.S.; de Souza, H.N.; Rennó, F.P. Effects of polymer-coated slow-realese urea on performance, ruminal fermentation, and blood, metabolites in dairy cows. *Rev. Bras. Zootec.* **2015**, *44*, 327–334. [CrossRef]
3. Van Soest, P.J. *Nutritional Ecology of the Ruminant*; Cornell University: Ithaca, NY, USA, 1994.

4. Patra, A.K. *Rumen Microbiology: From Evolution to Revolution*; Puniya, A.K., Singh, R., Kamra, D.N., Eds.; Springer: Basingstoke, Reino Unido, 2015; pp. 329–341.
5. Patra, A.K.; Aschenbach, J.R. Ureases in the gastrointestinal tracts of ruminant and monogastric animals and their implication in urea-N/ammonia metabolism: A review. *J. Adv. Res.* **2018**, *13*, 39–50. [CrossRef]
6. De Medeiros, T.T.B.; Silva, A.M.D.A.; Da Silva, A.L.; Bezerra, L.R.; Agostini, D.L.D.S.; De Oliveira, D.L.V.; Mazzetto, S.E.; Kotzebue, L.R.V.; Oliveira, J.R.; Souto, G.S.B.; et al. Carnauba wax as a wall material for urea microencapsulation. *J. Sci. Food Agric.* **2019**, *99*, 1078–1087. [CrossRef]
7. Campos, A.C.; Silva, A.L.; Silva, A.M.; Filho, J.M.A.; Costa, T.; Filho, J.M.P.; Oliveira, J.P.; Bezerra, L.R. Dietary replacement of soybean meal with lipid matrix-encapsulated urea does not modify milk production and composition in dairy goats. *Anim. Feed. Sci. Technol.* **2020**, *274*, 114763. [CrossRef]
8. Souza, V.L.; Almeida, R.; Silva, D.F.F.; Piekarski, P.R.B.; Jesus, C.P.; Pereira, M.N. Effects of partial replacement of soybean meal by protected urea on milk yield and composition. *Arq. Bras. Med. Vet. Zoot.* **2010**, *62*, 1415–1422. [CrossRef]
9. Carvalho, A.D.B.; da Silva, A.; Netto, A.; de Medeiros, T.; Filho, J.A.; Agostini, D.D.S.; de Oliveira, D.; Mazzetto, S.; Kotzebue, L.; Oliveira, J.; et al. Effect of slow-release urea microencapsulated in beeswax and its inclusion in ruminant diets. *Small Rumin. Res.* **2019**, *179*, 56–63. [CrossRef]
10. de Carvalho Neto, J.P.; Bezerra, L.R.; da Silva, A.L.; de Moura, J.F.P.; Pereira Filho, J.M.; da Silva Filho, E.C.; Guedes, A.F.; Araújo, M.J.; Edvan, R.L.; Oliveira, R.L. Methionine microencapsulated with a carnauba (*Copernicia prunifera*) wax matrix for protection from degradation in the rumen. *Livest. Sci.* **2019**, *228*, 53–60. [CrossRef]
11. Mohnen, D. Pectin structure and biosynthesis. *Curr. Opin. Plant. Biol.* **2008**, *11*, 266–277. [CrossRef]
12. Zema, D.A.; Calabrò, P.S.; Folino, A.; Tamburino, V.; Zappia, G.; Zimbone, S.M. Valorisation of citrus processing waste: A review. *Waste Manag.* **2018**, *80*, 252–273. [CrossRef] [PubMed]
13. Huang, J.; Liao, J.; Qi, J.; Jiang, W.; Yang, X. Structural and physicochemical properties of pectin-rich dietary fiber prepared from citrus peel. *Food Hydrocoll.* **2021**, *110*, 106140. [CrossRef]
14. Morris, G.A.; Ralet, M.C.; Bonnin, E.; Thibault, J.F.; Harding, S.E. Physical characterisation of the rhamnogalacturonan and homogalacturonan fractions of sugar beet (*Beta vulgaris*) pectin. *Carbohydr. Polym.* **2010**, *82*, 1161–1167. [CrossRef]
15. Shtriker , M.G.; Hahn, M.; Taieb, E.; Nyska, A.; Moallem, U.; Tirosh, O.; Madar, Z. Fenugreek galactomannan and citrus pectin improve several parameters associated with glucose metabolism and modulate gut microbiota in mice. *Nutrition* **2018**, *46*, 134–142. [CrossRef]
16. Ben-Ghedalia, D.; Yosef, E.; Miron, J.; Est, Y. The effects of starch- and pectin-rich diets on quantitative aspects of digestion in sheep. *Anim. Feed Sci. Technol.* **1989**, *24*, 289–298. [CrossRef]
17. Alves, E.M.; Magalhães, D.R.; Freitas, M.A.; Santos, E.D.J.D.; Pereira, M.L.A.; Pedreira, M.D.S. Nitrogen metabolism and microbial synthesis in sheep fed diets containing slow release urea to replace the conventional urea. *Acta Sci. Anim. Sci.* **2014**, *36*, 55–62. [CrossRef]
18. Alves, F.J.L.; Ferreira, M.A.; Urbano, S.A.; Andrade, R.P.X.; Silva, A.E.M.; Siqueira, M.C.B.; Oliveira, J.P.F.; Silva, J.L. Performance of lambs fed alternative protein sources to soybean meal. *Rev. Bras. Zootec.* **2016**, *45*, 145–150. [CrossRef]
19. Nedovica, V.; Kalusevica, A.; Manojlovicb, V.; Levica, S.; Bugarskib, B. An overview of encapsulation technologies for food applications. *Proc. Food Sci.* **2010**, *1*, 1806–1815. [CrossRef]
20. AOAC. *Official Methods of Analysis of AOAC*, 12th ed.; AOAC International: Gaithersburg, MD, USA, 1990.
21. Brazil. Resolução Normativa No. 39, de 20 de Junho de 2018. Ministério da Ciência, Tecnologia, Inovações e Comunicações. Dispõe Sobre Restrições ao Uso de Animais em Procedimentos Classificados com Grau de Invasividade 3 e 4, em Complemento à Diretriz Brasileira Para o Cuidado e a Utilização de Animais em Atividades de Ensino ou de Pesquisa Científica-DBCA. Available online: http://www.mctic.gov.br/mctic/opencms/institucional/concea/arquivos/ (accessed on August 2021).
22. NRC. *Nutrient Requirements of Small Ruminants: Sheep, Goats, Cervids and New World Camelids*, 3rd ed.; The National Academies Press: Washington, DC, USA, 2007.
23. Van Soest, P.J.; Robertson, J.B.; Lewis, B.A. Methods for dietary fiber, neutral detergent fiber, and nonstarch polyssacharides in relation to animal nutrition. *J. Dairy Sci.* **1991**, *74*, 3583–3597. [CrossRef]
24. Senger, C.C.D.; Kozloski, G.V.; Snachez, L.M.B.; Mesquita, F.R.; Alves, T.P.; Castagnino, D.S. Evalution of autoclave procedures for fibre analysis in forage and concetrate feedstuffs. *Anim. Feed Sci. Technol.* **2008**, *146*, 169–174. [CrossRef]
25. Hall, M.B. *Neutral Detergent-Soluble Carbohydrates. Nutritional Relevance and Analysis: A Laborary Manual*; University of Florida: Gainesville, FL, USA, 2000.
26. Ørskov, E.R.; McDonald, T. The estimation of protein degradability in the rumen from incubation measurements weighted according to rate passage. *J. Agric. Sci.* **1979**, *92*, 499–503. [CrossRef]
27. SAS. *SAS Systems for Windows*; Version 9.1; SAS Institute Inc.: Cary, NC, USA, 2003.
28. Noh, J.; Kim, J.; Kim, J.S.; Chung, Y.S.; Chang, S.T.; Park, J. Microencapsulation by pectin for multi-components carriers bearing both hydrophobic and hydrophilic active agents. *Carbohydr. Polym.* **2018**, *182*, 172–179. [CrossRef] [PubMed]
29. Williams, A.G.; Coleman, G.S. *The Rumen Protozoa*; Springer: New York, NY, USA, 1992.
30. Newbold, C.J.; De La Fuente, G.; Belanche, A.; Ramos-Morales, E.; McEwan, N. The Role of Ciliate Protozoa in the Rumen. *Front. Microbiol.* **2015**, *6*, 1313. [CrossRef] [PubMed]

31. Ng, F.; Kittelmann, S.; Patchett, M.L.; Attwood, G.T.; Janssen, P.H.; Rakonjac, J.; Gagic, D. An adhesin from hydrogenutilizing rumen methanogen *Methanobrevibacter ruminantium* M1 binds a broad range of hydrogen-producing microorganisms. *Environ. Microbiol.* **2016**, *18*, 3010–3021. [CrossRef] [PubMed]
32. Alcaide, E.M.; Martín-García, A.I.; Aguilera, J.F. A comparative study of nutrient digestibility, kinetics of degradation and passage and rumen fermentation pattern in goats and sheep offered good quality diets. *Livest. Prod. Sci.* **2000**, *64*, 215–223. [CrossRef]
33. Patra, A.K.; Saxena, J. The effect and mode of action of saponins on the microbial populations and fermentation in the rumen and ruminant production. *Nutr. Res. Rev.* **2009**, *22*, 204–219. [CrossRef] [PubMed]
34. Bezerra, L.R.; Oliveira, W.D.C.; Silva, T.P.D.; Torreao, J.N.C.; Marques, C.A.T.; Araujo, M.J.; Oliveira, R.L. Comparative hematological analysis of Morada Nova and Santa Inês ewes in all reproductive stages. *Pesq. Vet. Bras.* **2017**, *37*, 408–414. [CrossRef]
35. Kaneko, J.J.; Harvey, J.W.; Bruss, M.L. *Clinical Biochemistry of Domestic Animals*, 6th ed.; Academic: San Diego, CA, USA, 2008.
36. Ziguer, E.A.; Roll, V.F.B.; Bermudes, R.F.; Montagner, P.; Pfeiter, L.F.M.; Pino, F.A.B.D.; Corrêa, M.N.; Dionello, N.J.L. Performance and metabolic pattern of feedlot lambs using soyabean hulls associated to different non-protein nitrogen sources. *Rev. Bras. Zootec.* **2012**, *41*, 449–456. [CrossRef]
37. Netto, A.J.; Silva, A.M.D.A.; Bezerra, L.R.; Carvalho, A.D.B.; Agostini, D.L.D.S.; de Oliveira, D.L.V.; Mazzetto, S.E.; Kotzebue, L.R.V.; Oliveira, J.R.; Oliveira, R.L.; et al. Lipid microspheres containing urea for slow release of nonprotein N in ruminant diets. *Anim. Prod. Sci.* **2021**, acepted.
38. Kozloski, G.V. *Bioquímica dos Ruminantes*, 3rd ed.; Editora UFSM: Santa Maria, Brazil, 2011.

polymers

MDPI

Article

Transformation of Oil Palm Waste-Derived Cellulose into Solid Polymer Electrolytes: Investigating the Crucial Role of Plasticizers

Cheyma Naceur Abouloula [1,*], Muhammad. Rizwan [2], Vidhya Selvanathan [3], Rosiyah Yahya [4], Khaled Althubeiti [5], Hend I. Alkhammash [6], Md. Akhtaruzzaman [3,*] and A. Oueriagli [1]

1 MEE Lab, Faculty of Science Semlalia, University of Cadi Ayyad, Marrakesh 40090, Morocco; oueriagli@uca.ac.ma
2 Department of Chemistry, The University of Lahore, Lahore 54000, Pakistan; rizi_chem1981@hotmail.com
3 Solar Energy Research Institute (SERI), Universiti Kebangsaan Malaysia (UKM), Bangi 43600, Malaysia; vidhya@ukm.edu.my
4 Department of Chemistry, Faculty of Science, University of Malaya, Kuala Lumpur 50603, Malaysia; rosiyah@um.edu.my
5 Department of Chemistry, College of Science, Taif University, Taif 21944, Saudi Arabia; k.althubeiti@tu.edu.sa
6 Department of Electrical Engineering, College of Engineering, Taif University, Taif 21944, Saudi Arabia; Khamash.h@tu.edu.sa
* Correspondence: cheyma.naceur@ced.uca.ac.ma (C.N.A.); akhtar@ukm.edu.my (M.A.)

Abstract: This study explores the possibility of transforming lignocellulose-rich agricultural waste materials into value-added products. Cellulose was extracted from an empty fruit bunch of oil palm and further modified into carboxymethyl cellulose (CMC), a water-soluble cellulose derivative. The CMC was then employed as the polymeric content in fabrication of solid polymer electrolyte (SPE) films incorporated with lithium iodide. To enhance the ionic conductivity of the solid polymer electrolytes, the compositions were optimized with different amounts of glycerol as a plasticizing agent. The chemical and physical effects of plasticizer content on the film composition were studied by Fourier transform infrared (FTIR) and X-ray diffraction (XRD) analysis. FTIR and XRD analysis confirmed the interaction plasticizer with the polymer matrix and the amorphous nature of fabricated SPEs. The highest ionic conductivity of 6.26×10^{-2} S/cm was obtained with the addition of 25 wt % of glycerol. By fabricating solid polymer electrolytes from oil palm waste-derived cellulose, the sustainability of the materials can be retained while reducing the dependence on fossil fuel-derived materials in electrochemical devices.

Keywords: carboxymethyl cellulose; glycerol; polymer electrolyte; ionic conductivity

Citation: Abouloula, C.N.; Rizwan, M.; Selvanathan, V.; Yahya, R.; Althubeiti, K.; Alkhammash, H.I.; Akhtaruzzaman, M.; Oueriagli, A. Transformation of Oil Palm Waste-Derived Cellulose into Solid Polymer Electrolytes: Investigating the Crucial Role of Plasticizers. *Polymers* **2021**, *13*, 3685. https://doi.org/10.3390/polym13213685

Academic Editor: Arn Mignon

Received: 8 September 2021
Accepted: 19 October 2021
Published: 26 October 2021

1. Introduction

Malaysia is one of the world's leading producers of palm oil, giving rise to the production of enormous amounts of agricultural waste from the oil palm industry. For every 100 tonnes of fresh oil palm fruit bunches processed, about 22 tonnes of crude palm oil and 26 tonnes of empty fruit bunches are produced. These oil palm wastes are rich in cellulose, a natural polymer with excellent mechanical properties which can be exploited for various material developments. One of such novel applications is the employment of cellulose as the polymeric material for fabrication of solid polymer electrolyte (SPE) films. SPEs are commonly used in various electrochemical devices such as in batteries [1], electrochromic devices [2,3], supercapacitors [4,5] and solar cells [6]. By far, most commercially available SPE are made up of petroleum-derived synthetic polymers. By substituting the non-renewable material with oil palm-derived cellulose, the ecological footprint of the product can be reduced while simultaneously converting the biomass into a useful component.

Since cellulose in its native state is practically insoluble in most solvents, the polymer is often chemically modified to soluble derivatives such as carboxymethyl cellulose (CMC).

Upon functionalization of the cellulose, it can be easily converted into conductive films by adding inorganic salts. Previously, cellulose derivatives-based SPEs have been exploited for fabrication of SPEs. For instance, Selvakumar and co-workers have used cellulose acetate (CA) as a biodegradable polymer electrolyte for supercapacitors. The electrochemical properties and degradation tests of CA-based SPEs were investigated, then used for the fabrication of supercapacitors that showed good capacitive nature and stability during cycling [7].

The application of biopolymerbased SPE is restricted by its low ionic conductivity at ambient temperature. Hence, most of the studies are dedicated to enhancing the ionic conductivity of the films by various strategies including blending with other polymers, insertion of fillers, and plasticizers. Amongst these methods, the use of plasticizer to produce higher ionic conductivity has shown appreciable results in recent research. To enhance the ionic conductivity of SPEs, it was found that the choice of a good plasticizer with a low molecular weight and high dielectric constant or polar solvent, such as propylene carbonate, ethylene carbonate, is highly effective [8,9]. The addition of a plasticizer leads to an increase in the amorphousness of polymer electrolyte and dissociate ion pairs into free cations and anions, thereby leading to an overall enhancement in conductivity. A high plasticizer/salt ratio modifies the morphological, thermal, and electrochemical performance of the matrix, which also leads to the enhancement of conductivity. Chinnam et al. have prepared SPEs by mixing methyl cellulose, $LiClO_4$, and polyethylene glycol (PEG), to be used as a binder for Li-ion batteries and replace the commonly used materials, i.e., polyvinylidene fluoride–hexafluoropropylene (PVDF–HFP) or polyvinylidene fluoride (PVDF). The fabricated SPE showed good electrochemical stability and interfacial resistance with an optimum ambient conductivity of 1.6×10^{-5} S/cm [10].

Although previous literature has established the prospects of cellulose as polymer host in fabrication of SPE films, most of these studies have employed commercially available cellulose powder. Like any other natural polymer, the physical and chemical properties of cellulose films depend on the source from which they are derived. Thus, to fully exploit agrowaste-derived cellulose as a precursor material for SPE application, it is essential to ensure that the films are fabricated from the waste material itself. There are numerous studies on fabrication of cellulose films from different agricultural by-products including sugarcane bagasse [11], pineapple peel [12], *Ficus natalensis* bark cloth [13], rice husks [14] and corn husks [15]. However, research focussing on agrowaste-derived cellulose for SPE fabrication have been scarce. More importantly, the electrical performance of such waste-based SPE are poor, disabling their application into electrochemical devices. For instance, carboxymethyl cellulose film fabricated from coconut husk recorded ionic conductivity of 4.82×10^{-4} S/cm [16]. Similarly, biopolymer electrolytes comprised of kenaf fibre-derived CMC and ammonium acetate achieved the highest conductivity at ambient temperature of 5.77×10^{-4} S/cm [17].

Therefore, this study is dedicated to explore the potential of elevating the electrical conductivity of oil palm waste fibre-derived cellulose films. To achieve this, two main strategies were adopted. Firstly, the extracted cellulose chemically transformed into carboxymethyl cellulose. Our previous study has verified that the CMC derivative is essential to impart solubility and diminish high crystallinity of the native oil palm-based cellulose. However, the ionic conductivity of lithium iodide-incorporated CMC films can only achieve 5.58×10^{-3} S/cm. In this work, glycerol was infused as a plasticizer to improve the film's conductivity. Different amounts of glycerol were added to the polymer electrolyte matrix to identify the best composition ratio with good physico-chemical properties and high ionic conductivity for potential use in electrochemical devices.

2. Materials and Methods

2.1. Materials

Raw fibres of EFB oil palm were collected from Malaysian Agricultural Research and Development Institute (MARDI) (Selangor, Malaysia). NaOH, glacial acetic acid, monochloroacetic acid, and isopropanol were supplied by Merck (Darmstadt, Germany). Sodium chlorite, ethanol, methanol, glycerol, and lithium iodide were purchased from Sigma-Aldrich Co. (Saint Louis, MO, USA). The used chemicals were of analytical grade and were used without further purification.

2.2. Extraction Cellulose from Oil Palm Waste

The extraction of cellulose from raw EFB oil palm fibres is comprised of two steps, as reported in our previous work [18]. Firstly, the fibres were treated in an aqueous solution of 4 wt % of NaOH at 80 °C for 3 h. Then, the solid was filtered and washed several times with distilled water. The process was repeated thrice. This was followed by bleaching in equal parts of acetate buffer, aqueous chlorite (1.7% w/v), and distilled water, at 120 °C for 4 h. The bleaching treatment was repeated twice.

2.3. Carboxymethylation of Cellulose

The extracted cellulose powder was dispersed in isopropanol with subsequent addition of 30% w/v solution of NaOH and kept under stirring for 30 min at ambient temperature. Monochloroacetic acid was added to the suspension followed by stirring for 90 min. the mixture was then allowed to retain for 90 min at 55 °C. The mixture was carefully separated, suspended in methanol and the pH was adjusted with glacial acetic acid before being washed multiple times in ethanol and methanol. The solid was left to dry overnight at 50 °C.

2.4. Fabrication of CMC-Based Solid Polymer Electrolyte

The plasticized films were fabricated by preparing an aqueous solution containing dissolved CMC, LiI and glycerol. For every film, 0.2 g of CMC was mixed with 0.13 g of LiI, as per the optimized composition of LiI from our previously reported work [11]. Different amounts of glycerol, ranging from 5 to 30% (as outlined in Table 1) were added into the aqueous CMC-LiI solution to form a homogeneously distributed solution. The solutions were cast onto moulding dish followed by drying at room temperature. Figure 1 depicts the formation of plasticized CMC-based SPEs from oil palm-derived cellulose.

Figure 1. Transformation of oil palm waste into solid polymer electrolyte films.

Table 1. Composition of plasticized SPEs with various masses of glycerol.

Designations	CMC (g)	LiI (g)	Glycerol	
			(g)	w/w (%)
B1	0.2	0.13	0.02	5
B2			0.03	10
B3			0.05	15
B4			0.07	20
B5			0.08	25
B6			0.10	30

2.5. Characterization

Changes in structural and chemical properties of the extrcated cellulose and its corresponding films were analysed using a PerkinElmer Spotlight 400 FTIR spectrometer (UK). The FTIR measurement was performed with the resolution of 2 cm^{-1} and 32 scans by scanning the samples from 650 to 4000 cm^{-1}. X-ray diffractograms of both powder and film samples were characterized utilizing an Empyrean diffractometer (PANalytical, Al-melo, Netherlands). The X-ray diffraction data was scanned at 2θ angle between 10° and 60° with a step size of 0.026°. The equation below was used to evaluate the crystallinity index (*CrI*) using an empirical method:

$$CrI\ (\%) = \frac{I_{002} - I_{am}}{I_{002}} \times 100 \tag{1}$$

where I_{002} indicates the highest intensity of the (002) lattice diffraction peak and I_{am} is attributed to the amorphous section of the polymer. In this study, the diffraction peak for the highest intensity peak and the amorphous portion was measured at 2θ = 22.5° and 18.0° respectively. The ionic conductivity of the polymer electrolyte films were evaluated using the electrochemical impedance spectroscopy (EIS) technique according the the equation below. The bulk resistance of the samples were measured using the 3532-50 LCR Hi-Tester, Hioki (Nagano, Japan), with a frequency scan from 50 Hz to 5 MHz and within 30–80 °C.

$$\sigma = t/(R_b \times A) \tag{2}$$

where t is the thickness of the sample measured, R_b is the bulk resistance of the sample, and A is the contact surface area. All the EIS measurements were repeated three times for each sample, and the average value was calculated.

3. Results and Discussion

3.1. Synthesis of Carboxymethyl Cellulose from Oil Palm Waste

CMC was synthesized by reacting cellulose and monochloroacetic acid in an alkali environment. The yield % of the obtained CMC was calculated based on the theoretical amount of reacted cellulose with the carboxymethyl group following the equation:

$$Yield\ \%\ of\ CMC = \frac{Weight\ of\ CMC\ obtained\ (g)}{Theoretical\ weight\ of\ CMC\ (g)} \times 100 \tag{3}$$

The yield % of CMC calculated following the above equation was found to be 84.84%. The chemical identity of the cellulose extracted from oil palm waste and the CMC derivative was confirmed via NMR, XRD, and DSC analysis in our previous publication [18].

The purity of cellulose extracted from oil palm fibres was verified by FTIR analysis of the sample, as shown in Figure 2a. The bands between 1000 and 1500 cm^{-1} correspond to characteristic absorption of cellulose. The bands at 1429 cm^{-1}, 1163 cm^{-1}, 1032 cm^{-1}, and 892 cm^{-1} are attributed to C-H bending and C-O-C stretching of β-linked glucose subunits.

Upon carboxymethylation, a significant absorption band appeared at 1588 cm^{-1}, which is assigned to the carboxyl group, thus verifying the transformation of cellulose into CMC (Figure 2b).

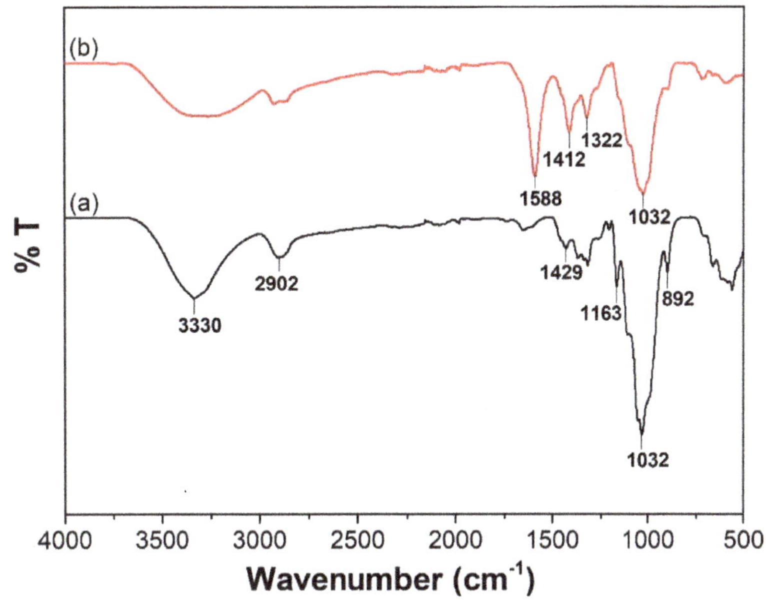

Figure 2. FTIR spectra of (**a**) cellulose extracted from oil palm waste and (**b**) carboxymethyl cellulose.

3.2. Chemical Interactions within the Polymer Film

The FTIR spectra of plasticized SPEs in Figure 3 show appreciable changes in absorbance with changing proportion of glycerol in the composition of SPEs. The scheme in Figure 3 outlines the possible site of interactions between the polymer chains and the plasticizer molecules and the ionic species present with the electrolyte matrix. Evidently, significant changes in the FTIR spectra of the polymer films were observed for absorption bands corresponding to hydroxyl, carboxyl, and ether groups of the CMC backbone, indicating possible complexation at these sites.

The addition of glycerol affected the O-H stretching vibration of SPEs, which shifted from 3330 cm^{-1} in CMC to 3368 cm^{-1} in the films. This is expected primarily due to the hydrogen bonding interaction between the hydroxyls in glycerol and in the polysaccharide chains. The presence of glycerol also affected the carboxyl group absorption at 1588 cm^{-1}, causing an upshift to 1612 cm^{-1}. The stretching vibrations of C-O in the glucose subunits experienced an upshift from 1032 cm^{-1} to 1048 cm^{-1}. It is evident that the main chemical interactions between the polymer and the plasticizer occur within an oxygen-associated functional group such as (-COO, -OH,-C-O), which serve as chemically active sites, as explained by Liu et al. [19]. The trend in the FTIR spectra upon plasticization tallies with previously reported cellulose-based films [20,21].

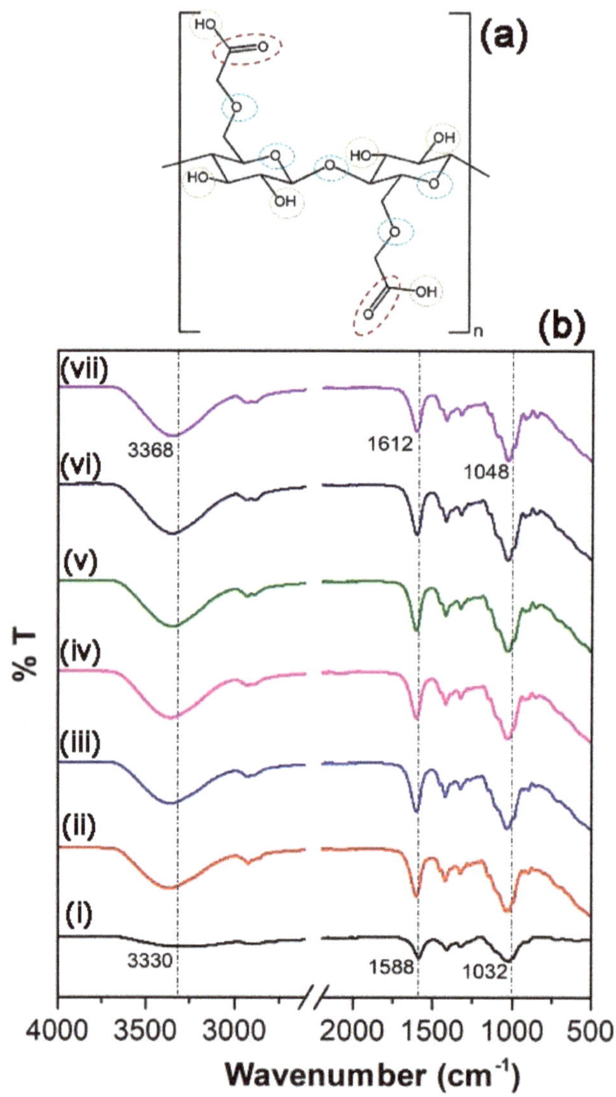

Figure 3. (**a**) Chemical scheme of possible complexation site in the polymer chain and (**b**) FTIR spectra of (i) CMC powder and CMC-LiI films with (ii) 5%, (iii) 10%, (iv) 15%, (v) 20%, (vi) 25%, and (vii) 30% glycerol content.

3.3. *Crystallinity Analysis*

The addition of glycerol is complemented with modification in the crystallinity of the polymer chains within the SPEs, as shown in Figure 4. The increase of glycerol in fabricated SPEs leads to a wider peak and a decrease of its intensity to reach a maximum with the addition of 25% of glycerol content. These changes indicate that the addition of glycerol helps to increase the amorphousness of the plasticized SPEs membrane [22]. The decline in crystallinity is further confirmed by degree of crystallinity calculation, as tabulated in Table 2, which shows a downward trend up to 25% glycerol. Beyond 25% of glycerol, the crystallinity of CMC-LiI-Gly SPEs increases, which can be explained as the impact of plasticizer overloading, leading to phase separation between the different components.

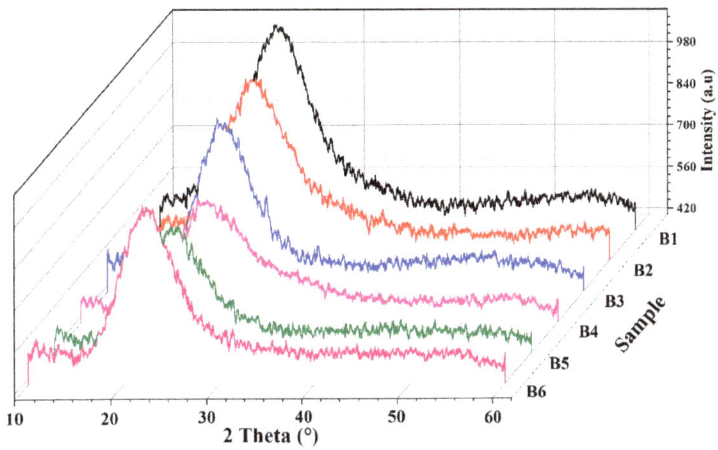

Figure 4. X–ray diffractograms of CMC–LiI–Gly-based SPEs.

Table 2. Degree of crystallinity of plasticized SPEs with various mass of glycerol.

Sample	% Glycerol (*w*/*w*)	Degree of Crystallinity (%)
CMC powder	-	36.9
B1	5	25.0
B2	10	22.9
B3	15	20.1
B4	20	18.3
B5	25	17.6
B6	30	23.8

3.4. Conductivity Analysis

To obtain better results and enhance ionic conductivity of CMC-LiI SPEs, glycerol was added as a plasticizer owing to its high dielectric constant ε_r = 41.14 [23]. The addition of plasticizer to the SPEs does not supply more ions, instead, it facilitates more dissociation of salt and generates separate conducting pathways for the migration of free ions.

The ionic conductivity of CMC-LiI SPEs was improved by adding glycerol as a liquid plasticizer, and the highest value (6.26×10^{-2} S/cm) was achieved with the addition of 25% glycerol (Figure 5). This increase is due to the high dielectric constant of glycerol and its low molecular weight, which modify the physical properties of SPEs [24]. Beyond 25% of glycerol content, the ionic conductivity dramatically decreased. This decrease may be due to the bulk quantity of the plasticizer, which promotes the polymer/plasticizer phase separation, an observation that is in agreement with the increased crystallinity of the film [25].

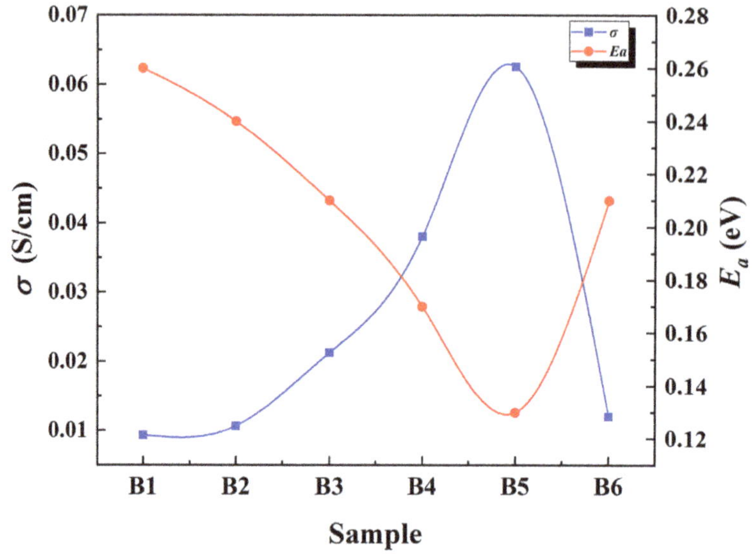

Figure 5. The trend in the ionic conductivity with different amounts of glycerol.

Interestingly, the ionic conductivity values recorded in this work are better than previously reported commercial cellulose-based films and are on par with synthetic polymer-based SPEs, as shown in Table 3. This, in fact, consolidates that, with appropriate chemical modifications, cellulose derivatives of agricultural waste origin can be tailored to form functional materials with desirable physico-chemical properties.

Table 3. Recent literature on solid polymer electrolytes.

Polymeric Content	Salt	Additives	Ionic Conductivity ($S\ cm^{-1}$)	Reference
Carboxymethyl cellulose	Lithium tetrafluoroborate	Glycerol	0.0037	[24]
Carboxylated cellulose	Lithium hexafluorophosphate	Ethylene carbonate + dimethyl carbonate + ethyl methyl carbonate	0.0018	[26]
Carboxymethyl cellulose	-	Polyaniline	0.018	[27]
Poly(ethylene oxide)	Sodium iodide	1-methyl-3-propylimidazolium iodide	0.0094	[28]
Poly(vinylidene fluoride-hexafluoro propylene	-	1-ethyl-3-methylimidazolium tetrafluoroborate + graphene oxide	0.025	[29]
Carboxymethyl cellulose	Lithium iodide	Glycerol	0.063	This study

The SPE film fabricated in this work attains an excellent ionic conductivity value, thus enabling the application of the films in various electrochemical devices. In recent times, biobased solid polymer electrolytes with the conductivity of magnitude within 10^{-3} and 10^{-2} S/cm have been proven to show good performance upon incorporation into full devices. For instance, incorporation of cellulose acetate and ammonium nitrate in different ratios were designed to prepare eco-friendly biopolymer electrolytes and the highest ionic conductivity of 1.02×10^{-3} S/cm was obtained [30]. The film with the highest ionic conductivity was used in the development of a primary proton battery and proton exchange membrane fuel cell. Monisha and co-workers found that the performances of the fabricated biopolymer electrolyte were optimum and could replace the standard Nafion 117 membrane. In another work, cellulose triacetate, poly(polyethylene glycol methacrylate) and ionic liquid were combined to make a solid polymer electrolyte and showed a high ionic conductivity of 5.24×10^{-3} S/cm [31]. The fabricated solid polymer electrolytes were tested on a Li cell and were found to be highly stable up to 5 V and safer with high performance for lithium batteries. Promising solid polymer electrolytes have been also prepared by blending methylcellulose and chitosan doped with ammonium thiocyanate for electrochemical double-layer capacitor (EDLC) application [32]. The electrolyte with the highest DC conductivity of 2.81×10^{-3} S/cm was used as an electrode separator in EDLC and exhibited good performance. Hence, the oil palm-derived CMC films incorporated with glycerol and lithium iodide have prospects to be used as a biodegradable SPE in place of conventional synthetic polymer-based films.

However, to fully realize the potential of the films, it is vital to thoroughly investigate the thermal and chemical stability of the films in the future. In addition to that, examining physical parameters such as tensile strength, flexibility, adhesiveness, and hydrophilicity will be useful to improve the integration of the films as SPE in electrochemical devices.

3.5. Temperature-Dependent Conductivity Analysis

Practically, the fabricated SPEs are expected to be exposed to high temperature. Thus, it is vital to evaluate the temperature-dependent ionic conductivity values of plasticized SPEs. Figure 6 shows the ionic conductivity of the SPEs with the increase of temperature from 30 °C to 80 °C. The ionic conductivity increases linearly with temperature. This linear behaviour shows that there is no phase transition in the polymer matrix within that temperature range [24]. Since the conductivity behaviour at elevated temperature obeys the Arrhenius equation, the activation energy, E_a, was calculated using the following equation:

$$\sigma = \sigma_0 \exp\left[-\frac{E_a}{kT}\right] \quad (4)$$

where σ_0 is the pre-exponential factor, E_a is the activation energy of ionic conduction, k is the Boltzmann constant and T is the temperature in Kelvin.

Based on the calculation, the E_a was found to be in the range of 0.18–0.22 eV, which reflects the amorphous nature of the SPEs. The increase in ionic conductivity behaviour of SPEs with the rise of temperature is due to the dislocation and migration of ions across the SPEs. Furthermore, the heat provides more energy for the segmental motion of polymer chains, and thus facilitates ions transportation. The increase in conductivity can be translated by the increase of ions motion and chain flexibility of polymers due to the rise of temperature.

The Anderson–Stuart model defined that an ion within a matrix occur when the ion is equipped with enough energy to detach itself from a site and suffiecient kinetic energy to support its motion [33,34]. Thus, the activation energy is the main factor in influencing the overall ionic conductivity of a sample at any given temperature. With the increase of glycerol, E_a decreases up to a minimum with the addition of 25 wt %, followed by an increase of its energy. Figure 5 clearly shows that the ionic conductivity and E_a have an inverse relationship, where the conductivity is highest with lowest value of E_a at 65% salt and 25 wt % plasticizer. This result can be attributed to the presence of substantial amount of amorphous portions within the polymer film upon inclusion of salt and plasticizer, as depicted by XRD analysis, aiding in the mobility of ions via the transitional motion of the polymer chains contained within amorphous regions [35]. Selvasekarapandian and co-workers reported that the density of ions in the electrolyte increases with the increase in salt or glycerol content. Thus, the energy barrier for the proton transport decreases, leading to a decrease in the E_a value. These results are in agreement with the values obtained in this work.

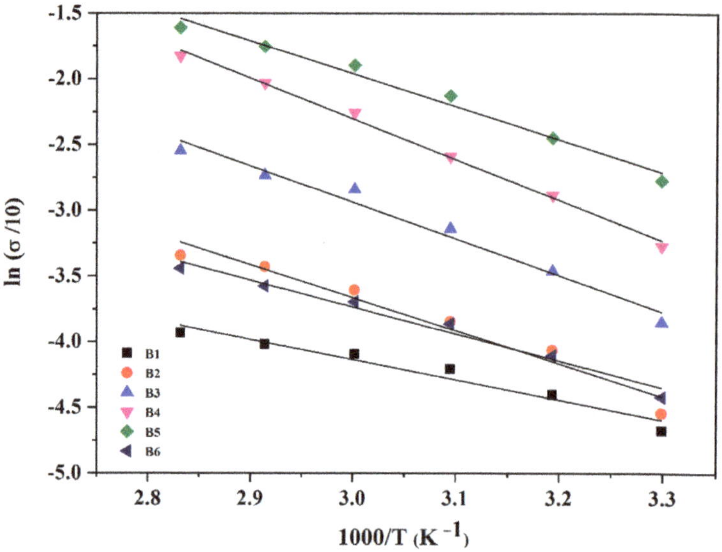

Figure 6. The trend in the ionic conductivity with different amounts of glycerol.

4. Conclusions

The CMC-LiI-based SPEs plasticized with glycerol were successfully fabricated using a solution-cast method. The interactions between polymer, salt, and plasticizer were investigated by FTIR spectroscopy. The XRD showed that the amorphousness of the SPEs facilitates the ions movement, which reflected the increase of ionic conductivity. The optimum ionic conductivity was found to be 6.26×10^{-2} S/cm with a composition of 25 wt % of glycerol. This translates the appreciable effect of the addition of glycerol as a plasticizer. Thus, the plasticized SPEs can be used in electrochemical devices.

Author Contributions: Conceptualization, A.O.; methodology, C.N.A.; investigation, M.R.; resources, H.I.A.; writing—original draft preparation, C.N.A.; writing—review and editing, V.S., K.A., H.I.A.; supervision, R.Y.; project administration, M.A.; funding acquisition, K.A. All authors have read and agreed to the published version of the manuscript.

Funding: V. Selvanathan extends their appreciation to Universiti Kebangsaan Malaysia for the Modal Insan postdoctoral fellowship (MI 2019-014). The authors also extend their appreciation to The University Researchers Supporting Project Number (TURSP-2020/241), Taif University, Taif, Saudi Arabia.

Institutional Review Board Statement: Not applicable.

Informed Consent Statement: Not applicable.

Data Availability Statement: Not applicable.

Acknowledgments: V.S. extends their appreciation to Universiti Kebangsaan Malaysia for the Modal Insan postdoctoral fellowship (RFA 1). The authors also extend their appreciation to The University Researchers Supporting Project Number (TURSP-2020/241), Taif University, Taif, Saudi Arabia.

Conflicts of Interest: The authors declare no conflict of interest.

References

1. Xiao, S.; Yang, Y.; Li, M.; Wang, F.; Chang, Z.; Wu, Y.; Liu, X. A composite membrane based on a biocompatible cellulose as a host of gel polymer electrolyte for lithium ion batteries. *J. Power Sources* **2014**, *270*, 53–58. [CrossRef]
2. Wang, W.; Guan, S.; Li, M.; Zheng, J.; Xu, C. A novel hybrid quasi-solid polymer electrolyte based on porous PVB and modified PEG for electrochromic application. *Org. Electron.* **2018**, *56*, 268–275. [CrossRef]
3. Ledwon, P.; Andrade, J.R.; Lapkowski, M.; Pawlicka, A. Hydroxypropyl cellulose-based gel electrolyte for electrochromic devices. *Electrochim. Acta* **2015**, *159*, 227–233. [CrossRef]
4. Yang, H.; Liu, Y.; Kong, L.; Kang, L.; Ran, F. Biopolymer-based carboxylated chitosan hydrogel film crosslinked by HCl as gel polymer electrolyte for all-solid-sate supercapacitors. *J. Power Sources* **2019**, *426*, 47–54. [CrossRef]
5. Peng, S.; Jiang, X.; Xiang, X.; Chen, K.; Chen, G.; Jiang, X.; Hou, L. High-performance and flexible solid-state supercapacitors based on high toughness and thermoplastic poly(vinyl alcohol)/NaCl/glycerol supramolecular gel polymer electrolyte. *Electrochim. Acta* **2019**, *324*, 134874. [CrossRef]
6. Dissanayake, M.; Jaseetharan, T.; Senadeera, G.; Mellander, B.-E.; Albinsson, I.; Furlani, M.; Kumari, J. Solid-state solar cells co-sensitized with PbS/CdS quantum dots and N719 dye and based on solid polymer electrolyte with binary cations and nanofillers. *J. Photochem. Photobiol. A Chem.* **2021**, *405*, 112915. [CrossRef]
7. Selvakumar, M.; Bhat, D.K. LiClO4doped cellulose acetate as biodegradable polymer electrolyte for supercapacitors. *J. Appl. Polym. Sci.* **2008**, *110*, 594–602. [CrossRef]
8. Gupta, S.; Varshney, P.K. Effect of plasticizer on the conductivity of carboxymethyl cellulose-based solid polymer electrolyte. *Polym. Bull.* **2019**, *76*, 6169–6178. [CrossRef]
9. Ma, X.; Yu, J.; He, K.; Wang, N. The Effects of Different Plasticizers on the Properties of Thermoplastic Starch as Solid Polymer Electrolytes. *Macromol. Mater. Eng.* **2007**, *292*, 503–510. [CrossRef]
10. Chinnam, P.R.; Zhang, H.; Wunder, S.L. Blends of Pegylated Polyoctahedralsilsesquioxanes (POSS-PEG) and Methyl Cellulose as Solid Polymer Electrolytes for Lithium Batteries. *Electrochim. Acta* **2015**, *170*, 191–201. [CrossRef]
11. Vanitjinda, G.; Nimchua, T.; Sukyai, P. Effect of xylanase-assisted pretreatment on the properties of cellulose and regenerated cellulose films from sugarcane bagasse. *Int. J. Biol. Macromol.* **2019**, *122*, 503–516. [CrossRef] [PubMed]
12. Dai, H.; Ou, S.; Huang, Y.; Huang, H. Utilization of pineapple peel for production of nanocellulose and film application. *Cellulose* **2018**, *25*, 1743–1756. [CrossRef]
13. Farooq, A.; Jiang, S.; Farooq, A.; Naeem, M.A.; Ahmad, A.; Liu, L. Structure and properties of high quality natural cellulose nano fibrils from a novel material Ficus natalensis barkcloth. *J. Ind. Text.* **2019**, *51*, 668–680. [CrossRef]
14. Hayatun, A.; Jannah, M.; Ahmad, A.; Taba, P. Synthetic Bioplastic Film from Rice Husk Cellulose. *J. Phys. Conf. Ser.* **2020**, *1463*, 012009. [CrossRef]
15. Bernhardt, D.C.; Pérez, C.D.; Fissore, E.N.; De'Nobili, M.D.; Rojas, A.M. Pectin-based composite film: Effect of corn husk fiber concentration on their properties. *Carbohydr. Polym.* **2017**, *164*, 13–22. [CrossRef]
16. Chua, K.Y.; Azzahari, A.D.; Abouloula, C.N.; Sonsudin, F.; Shahabudin, N.; Yahya, R. Cellulose-based polymer electrolyte derived from waste coconut husk: Residual lignin as a natural plasticizer. *J. Polym. Res.* **2020**, *27*, 115. [CrossRef]
17. Rani, M.S.A.; Rudhziah, S.; Ahmad, A.; Mohamed, N.S. Biopolymer Electrolyte Based on Derivatives of Cellulose from Kenaf Bast Fiber. *Polymers* **2014**, *6*, 2371–2385. [CrossRef]
18. Abouloula, C.N.; Rizwan, M.; Selvanathan, V.; Abdullah, C.I.; Hassan, A.; Yahya, R.; Oueriagli, A. A novel application for oil palm empty fruit bunch: Extraction and modification of cellulose for solid polymer electrolyte. *Ionics* **2018**, *24*, 3827–3836. [CrossRef]
19. Liu, H.; Adhikari, R.; Guo, Q.; Adhikari, B. Preparation and characterization of glycerol plasticized (high-amylose) starch–chitosan films. *J. Food Eng.* **2013**, *116*, 588–597. [CrossRef]
20. Rudhziah, S.; Rani, M.; Ahmad, A.; Mohamed, N.; Kaddami, H. Potential of blend of kappa-carrageenan and cellulose derivatives for green polymer electrolyte application. *Ind. Crop. Prod.* **2015**, *72*, 133–141. [CrossRef]
21. Colò, F.; Bella, F.; Nair, J.R.; Destro, M.; Gerbaldi, C. Cellulose-based novel hybrid polymer electrolytes for green and efficient Na-ion batteries. *Electrochim. Acta* **2015**, *174*, 185–190. [CrossRef]
22. Yusof, Y.M.; Majid, N.A.; Kasmani, R.M.; Illias, H.A.; Kadir, M.F.Z. The Effect of Plasticization on Conductivity and Other Properties of Starch/Chitosan Blend Biopolymer Electrolyte Incorporated with Ammonium Iodide. *Mol. Cryst. Liq. Cryst.* **2014**, *603*, 73–88. [CrossRef]
23. Akerlof, G. Dielectric constants of some organic solvent-water mixtures at various temperatures. *J. Am. Chem. Soc.* **1932**, *54*, 4125–4139. [CrossRef]
24. Gupta, S.; Varshney, P.K. Effect of plasticizer concentration on structural and electrical properties of hydroxyethyl cellulose (HEC)-based polymer electrolyte. *Ionics* **2017**, *23*, 1613–1617. [CrossRef]
25. Machado, G.; Ferreira, H.; Pawlicka, A. Influence of plasticizer contents on the properties of HEC-based solid polymeric electrolytes. *Electrochim. Acta* **2005**, *50*, 3827–3831. [CrossRef]
26. Gou, J.; Liu, W.; Tang, A. A renewable gel polymer electrolyte based on the different sized carboxylated cellulose with satisfactory comprehensive performance for rechargeable lithium ion battery. *Polymer* **2020**, *208*, 122943. [CrossRef]
27. Khong, C.-H.; Lee, M.L.-Y.; Ahmad, I.; Phang, S.-W. Development of grafted rubber/polyaniline/carboxymethyl cellulose film as green conductive polymer film. *Polym. Bull.* **2021**, 1–18. [CrossRef]

28. Syairah, A.; Khanmirzaei, M.H.; Saidi, N.M.; Farhana, N.K.; Ramesh, S.; Ramesh, K. Effect of different imidazolium-based ionic liquids on gel polymer electrolytes for dye-sensitized solar cells. *Ionics* **2019**, *25*, 2427–2435. [CrossRef]
29. Yang, X.; Zhang, F.; Zhang, L.; Zhang, T.; Huang, Y.; Chen, Y. A High-Performance Graphene Oxide-Doped Ion Gel as Gel Polymer Electrolyte for All-Solid-State Supercapacitor Applications. *Adv. Funct. Mater.* **2013**, *23*, 3353–3360. [CrossRef]
30. Monisha, S.; Mathavan, T.; Selvasekarapandian, S.; Benial, A.M.F.; Aristatil, G.; Mani, N.; Premalatha, M.; Pandi, D.V. Investigation of bio polymer electrolyte based on cellulose acetate-ammonium nitrate for potential use in electrochemical devices. *Carbohydr. Polym.* **2017**, *157*, 38–47. [CrossRef]
31. Kale, S.B.; Nirmale, T.C.; Khupse, N.D.; Kale, B.B.; Kulkarni, M.V.; Pavitran, S.; Gosavi, S.W. Cellulose-Derived Flame-Retardant Solid Polymer Electrolyte for Lithium-Ion Batteries. *ACS Sustain. Chem. Eng.* **2021**, *9*, 1559–1567. [CrossRef]
32. Aziz, S.B.; Hamsan, M.H.; Abdullah, R.M.; Kadir, M.F.Z. A Promising Polymer Blend Electrolytes Based on Chitosan: Methyl Cellulose for EDLC Application with High Specific Capacitance and Energy Density. *Molecules* **2019**, *24*, 2503. [CrossRef] [PubMed]
33. Selvanathan, V.; Ruslan, M.H.; Aminuzzaman, M.; Muhammad, G.; Amin, N.; Sopian, K.; Akhtaruzzaman, M. Resorcinol-formaldehyde (RF) as a novel plasticizer for starch-based solid biopolymer electrolyte. *Polymers* **2020**, *12*, 2170. [CrossRef] [PubMed]
34. Abouloula, C.N.; Rizwan, M.; Selvanathan, V.; Hassan, A.; Yahya, R.; Oueriagli, A. Oil palm waste based phthaloyl cellulose: A product of photosynthesis as an electrolyte of photovoltaics. *Cellulose* **2019**, *26*, 1605–1617. [CrossRef]
35. Shukur, M.F.; Ithnin, R.; Kadir, M.F.Z. Ionic conductivity and dielectric properties of potato starch-magnesium acetate biopolymer electrolytes: The effect of glycerol and 1-butyl-3-methylimidazolium chloride. *Ionics* **2016**, *22*, 1113–1123. [CrossRef]

polymers

Article

Laccase-Mediator System Using a Natural Mediator as a Whitening Agent for the Decolorization of Melanin

Saerom Park †, Dahun Jung †, Hyejin Do, Jonghyeon Yun, Dongjun Lee, Soeun Hwang and Sang Hyun Lee *

Department of Biological Engineering, Konkuk University, Seoul 05029, Korea; angel4y@naver.com (S.P.); 1wjdekgns@naver.com (D.J.); dkssud12370@naver.com (H.D.); dbsjh6158@gmail.com (J.Y.); dlehdwns1213@gmail.com (D.L.); thdmsdlrh@naver.com (S.H.)

* Correspondence: sanghlee@konkuk.ac.kr; Tel.: +82-2-2049-6269

† These authors contributed equally.

Abstract: In this study, a laccase-mediator system (LMS) using a natural mediator was developed as a whitening agent for melanin decolorization. Seven natural mediators were used to replace synthetic mediators and successfully overcome the low redox potential of laccase and limited access of melanin to the active site of laccase. The melanin decolorization activity of laccases from *Trametes versicolor* (lacT) and *Myceliophthora thermophila* (lacM) was significantly enhanced using natural mediators including acetosyringone, syringaldehyde, and acetovanillone, which showed low cytotoxicity. The methoxy and ketone groups of natural mediators play an important role in melanin decolorization. The specificity constants of lacT and lacM for melanin decolorization were enhanced by 247 and 334, respectively, when acetosyringone was used as a mediator. LMS using lacM and acetosyringone could also decolorize the melanin present in the cellulose hydrogel film, which mimics the skin condition. Furthermore, LMS could decolorize not only synthetic eumelanin analogs prepared by the oxidation of tyrosine but also natural melanin produced by melanoma cells.

Keywords: laccase; melanin; decolorization; natural mediators

Citation: Park, S.; Jung, D.; Do, H.; Yun, J.; Lee, D.; Hwang, S.; Lee, S.H. Laccase-Mediator System Using a Natural Mediator as a Whitening Agent for the Decolorization of Melanin. *Polymers* **2021**, *13*, 3671. https://doi.org/10.3390/polym13213671

Academic Editor: Cédric Delattre

Received: 6 October 2021
Accepted: 22 October 2021
Published: 25 October 2021

Publisher's Note: MDPI stays neutral with regard to jurisdictional claims in published maps and institutional affiliations.

1. Introduction

Laccases (EC 1.10.3.2, benzenediol: dioxygen oxidoreductases) are multicopper proteins that catalyze the oxidation of various phenolic and non-phenolic compounds via a radical-catalyzed reaction mechanism by the reduction of molecular oxygen [1,2]. Laccases have been used as biocatalysts for biodegradation processes, such as the bioremediation of dyes [3,4], pharmaceuticals [5,6], and herbicides [7], and delignification [8–10]. Laccases have also been used to catalyze the polymerization of dye precursors and organic compounds [11]. In particular, their attractive properties, such as low substrate specificity, the use of oxygen as the final electron acceptor, generation of water as a by-product, and no demand (or no production) of peroxides, make them interesting in biotechnological and environmental fields [1,11,12].

Four copper ions at the active site are involved in the catalytic activity of laccase. "Blue" copper (T1 site) oxidizes the substrate, and the trinuclear copper cluster (T2/T3) receives the electrons from the T1 site to reduce the molecular oxygen [1,12,13]. In particular, the redox potential of the T1 site Cu is considered as a major factor in determining the catalytic ability of laccases [14]. Laccases possess a relatively low redox potential (0.4–0.8 V) compared to ligninolytic peroxidases (over 1 V) such as manganese peroxidase and lignin peroxidase. Laccases cannot directly oxidize non-phenolic substrates with redox potential above 1.3 V [13,14]. Therefore, to overcome the limitations of laccase, laccase-mediator systems (LMS) using small molecular compounds, such as 2,2′-azinobis(3-ethylbenzthiazoline-6-sulphonate) (ABTS), 1-hydroxybenzotriazole (HOBt), violuric acid (VLA), *N*-hydroxyphthalimide (HPI), *N*-hydroxyacetanilide (NHA), and TEMPO, which act as redox mediators, have been suggested [15–17].

These mediators permit the oxidation of bulky compounds via different oxidation routes. The laccase-ABTS system oxidizes substrates by generating a cationic ABTS radical via an electron transfer (ET) mechanism. LMSs with HOBt, VLA, HPI, or NHA produce nitroxyl radicals via the hydrogen atom transfer (HAT) mechanism [1,12,17]. Furthermore, mediators such as TEMPO and its analogs react via ionic pathways to generate oxoammonium ions [1,12,18]. The use of these mediators can oxidize a wide range of compounds in various applications, such as dye degradation [3,4], drug degradation [5,6], and lignin degradation [8–10]. Nevertheless, the applications of synthetic mediators in industrial fields have been limited due to their potential toxicity, high cost, and enzyme inactivation effect. Recently, lignin-derived phenolic molecules as natural mediators (e.g., syringaldehyde, acetosyringone, vanillin, acetovanillone, methylvanillate, ferulic acid, sinapic acid, *p*-coumaric acid, etc.) have been studied to replace synthetic mediators [1,12]. The advantages of natural mediators are low cost and low toxicity because they are obtained from natural and renewable sources [19].

Melanin is a group of natural pigments produced by melanogenesis through the oxidative polymerization of tyrosine by melanocytes. Natural melanin can be classified into five categories of eumelanin, pheomelanin, allomelanin, pyomelanin, and neuromelanin [20]. Recently, various medical and electrochemical applications using melanin or melanin precursors have been studied [20,21]. The human skin color is mostly determined by the presence of melanin. In the cosmetic industry, the direct depigmentation of melanin using enzymes has been proposed for the development of skin-whitening agents. Several peroxidases have been studied to decolorize melanin. Woo et al. showed that synthetic melanin can be directly decolorized by lignin peroxidase from *P. chrysosporium* [22]. The Keneko and Mohorčič groups also reported the enzymatic decolorization of melanin by manganese peroxidase isolated from fungi (*Sporotrichum pruinosum* and *Phlebia radiata*) [23,24]. Kim et al. reported that crude enzyme mixtures containing manganese peroxidase, lignin peroxidase, and laccase showed melanin depigmentation activity [25]. When peroxidases decolorize melanin, they require hydrogen peroxide (H_2O_2) as a cofactor that can irritate the skin. Thus, to reduce the usage of H_2O_2, glucose oxidase or laccase was introduced into the enzyme combination system [26,27]. Laccases can decolorize melanin without the use of hydrogen peroxide. Khammuang and Sarnthima reported that laccase from *Lentinus polychrous* Lév showed melanin decolorization activity using ABTS, vanillin, and vanillic acid as mediators [28].

In this study, LMS using a natural mediator was developed as a whitening agent for melanin decolorization. Various lignin-derived phenolic molecules have been tested as mediators of melanin decolorization to replace synthetic mediators. The effect of mediator concentration and pH on melanin decolorization in LMS using commercially available laccases from *Trametes versicolor* and *Myceliophthora thermophila* was investigated, and the cytotoxicity of natural mediators was also investigated. The enhancing effect of natural mediators was quantitatively analyzed by a kinetic study of the melanin decolorization reaction using LMS. Furthermore, the decolorization of melanin in the cellulose hydrogel film, which mimics the skin condition, and the decolorization of natural melanin produced by melanoma cells was also studied.

2. Materials and Methods

2.1. Materials

Laccases from *Trametes versicolor* (lacT), laccase from *Myceliophthora thermophila* (lacM), synthetic melanin, acetosyringone, syringaldehyde, vanillin, *p*-coumaric acid, acetovanillone, vanillic acid, vanillyl alcohol, 1-hydroxybenzotriazole hydrate (HOBt), sodium phosphate dibasic, penicillin, streptomycin, phosphate-buffered saline, neutral red (NR), microcrystalline cellulose (MCC), and 1-ethyl-3-methylimidazolium acetate ([Emim][Ac]) were purchased from Sigma-Aldrich (St Louis, MO, USA). Citric acid was obtained from Junsei (Tokyo, Japan). Trypsin-EDTA, fetal bovine serum, and DMEM were obtained from

Thermo Fisher Scientific (Waltham, MA, USA). All chemicals used in this study were of analytical grade and were used without further purification.

2.2. Melanin Decolorization by LMS

The saturated melanin solution (1.4 mg/mL) was prepared by the dissolution of 3 mg synthetic melanin in 1.3 mL of 10 mM NaOH. The solution was centrifuged at 8500 rpm for 5 min to remove the undissolved melanin, and the supernatant was diluted with 0.1 M citric acid phosphate buffer (pH 3, 4, 5, 5.5, 6, or 7) and used as a substrate solution for LMS. The concentration of melanin in the substrate solution was 63 μg/mL and spectrophotometrically confirmed at 475 nm. The 0.8 mL of melanin substrate solution was mixed with 0.1 mL mediator solution (0–1 mM) in a 1.5 mL Eppendorf tube. The melanin decolorization reaction was initiated by adding 0.1 mL of laccase solution (15.8 μg (0.6 U) lacT or 19.2 μg (1.8 U) lacM) to a mixture of melanin and the mediator at 25 °C in a shaking water bath at 120 rpm. After the reaction, the reaction mixture was centrifuged, and the absorbance of the supernatant was measured at 475 nm. The decolorization yield (%) was calculated using the following equation:

$$\text{Decolorization (\%)} = (A_0 - A_t)/A_0 \times 100, \quad (1)$$

where A_0 is the absorbance of the melanin solution before the decolorization reaction and A_t is the absorbance of the melanin solution after the decolorization reaction.

The protein content of the laccase solution was determined by the BCA method. One unit (U) corresponds to the amount of laccase that converts 1 μmol of ABTS per minute at pH 5.5 and 25 °C.

2.3. Kinetic Study of Melanin Decolorization by LMS

To determine the kinetic constants for the melanin decolorization reaction by LMS, the initial rates of laccase (0.6 U lacT or 1.8 U lacM) with or without 0.1 mM acetosyringone were measured using various melanin concentrations (0–420 μg/mL). The melanin solution was prepared by diluting completely dissolved melanin (1 mg/mL) in 10 mM NaOH. The kinetic constants were obtained from the Michaelis–Menten equation using the non-linear regression function of SigmaPlot 12 (Systat Software, San Jose, CA, USA).

2.4. Cytotoxicity of Natural Mediators

The B16F10 melanoma cell line (Korea Cell Line Bank, Seoul, Korea) was used to determine the cytotoxicity of natural mediators for LMS. A neutral red (NR) assay was performed to measure the cytotoxicity of the mediators [29]. NR measures the viability of live cell lysosomes. Melanoma cells with a concentration of 3×10^4 cells were dispensed into each well of a 96-well plate. After 24 h of cultivation, the cells were treated with natural mediators (1, 2, 5, 10, 22, and 46 mM). After additional cultivation for 2 days, the cells were treated with 50 μg/mL NR solution dissolved in DMEM and incubated for 3 h. After removing the supernatant through suction, an NR desorb solution (1% glacial acetic acid, 49% ethanol, and 50% distilled water) was used for color extraction. After the extraction process, the change in absorbance was measured at 540 nm.

2.5. Preparation and Decolorization of the Melanin/Cellulose Hydrogel Film

To prepare the melanin/cellulose film, 0.5% (*w*/*v*) synthetic melanin was dissolved in [Emim][Ac] under ultrasound irradiation for 10 min. The melanin solution was centrifuged at 8500 rpm for 20 min to remove insoluble melanin, and then 7 wt % of cellulose was dissolved in the supernatant at 100 °C for 2 h with stirring. The mixture solution was cast on a glass slide to a thickness of 0.3 mm using an applicator/1117 (Mitutoyo Corp., Kawasaki, Japan), and dissolved melanin and cellulose were regenerated with distilled water. The prepared film was washed with 0.1 M citric acid phosphate buffer (pH 5.5) until no absorbance of [Emim][Ac] was measured at 211 nm. The melanin/cellulose hydrogel film was stored in 0.1 M citric acid phosphate buffer (pH 5.5) until further use.

To measure the decolorization activity of LMS for the melanin/cellulose film, the prepared hydrogel film was cut into a 1 × 2 cm sheet. The hydrogel film was immersed in 4 mL of 0.1 M citric acid phosphate buffer (pH 5.5); subsequently, 0.5 mL of 1 mM acetosyringone and 0.5 mL of lacM solution (2.5 U) were added to the buffer. The decolorization reaction was carried out in a water bath with shaking at 120 rpm and 25 °C for 3 h. After the reaction, the film was washed with distilled water and attached to the inner side of the cuvette to measure the change in the spectra in the range of 400–800 nm using a UV/Vis spectrophotometer. Control reactions without lacM or mediators were also conducted under the same conditions. The release of melanin from the film or color change of the melanin/cellulose film was not detected under the reaction conditions. Furthermore, the change in color parameters (L*, a*, and b* values) of the melanin/cellulose film after the decolorization reaction by LMS was also recorded using a colorimeter (KONICA MINOLTA, Tokyo, Japan). The ΔL (metric lightness difference), ΔE (total color difference), YI (yellowness index), and WI (whiteness index) values were obtained using the following equations [30–32]:

$$\Delta L = L_{after} - L_{before}, \tag{2}$$

$$\Delta E = [(\Delta L)^2 + (a_{after} - a_{before})^2 + (b_{after} - b_{before})^2]^{0.5}, \tag{3}$$

$$YI = (142.86 \times b^*)/L^*, \tag{4}$$

$$WI = 100 - [(100 - L^*)^2 + a^{*2} + b^{*2}]^{0.5}, \tag{5}$$

where L_{after}, a_{after}, b_{after}, L_{before}, a_{before}, and b_{before} are the mean color values after and before the decolorization reaction, respectively.

2.6. Preparation of Natural Melanin

Natural melanin was obtained from B16F10 melanoma cells. The cells were treated with alpha-melanocyte-stimulating hormone to produce melanin. After 4 days of incubation, the cells were captured using trypsin-EDTA and sonicated for 10 min. The supernatant was obtained by centrifugation at 8000 rpm for 10 min and then adjusted to pH 1.5 using 6 M HCl. The solution was boiled at 100 °C for 4 h to hydrolyze the residual protein fractions. The solution containing natural melanin was washed with acetone, followed by chloroform and ethanol, and then washed with deionized water to eliminate residues, such as cells, media components, and protein fractions [33,34]. All washing processes were performed more than twice. Finally, natural melanin was obtained by freeze-drying and used as a substrate for LMS.

3. Results and Discussion

3.1. The Effect of Mediators on the Melanin Decolorization by LMS

The effect of various mediators on the melanin decolorization reaction by LMS was investigated using two laccases from *T. versicolor* (lacT) and *M. thermophila* (lacM) (Figure 1).

When lacT was used without a mediator for melanin decolorization, the decolorization yield was only 1% after 5 h of reaction. When HOBt was used as a mediator for lacT, the decolorization yield was slightly enhanced to 2% after 5 h of reaction. The use of various synthetic mediators, such as HOBt, ABTS, VLA, and TEMPO, in the laccase-catalyzed oxidation of phenolic or non-phenolic compounds significantly enhanced the reaction rates [10,15]. When the access of target compounds into the active site of laccase is limited by their steric hindrance, mediator radicals formed by laccase can efficiently oxidize the target compounds by the electron transfer or hydrogen atom transfer mechanism [12]. HOBt is one of the most commonly used synthetic mediators in LMS due to its high redox potential (1.1 V) [6]. However, HOBt is not a good cosmetic ingredient because of its potential cell toxicity and ability to inactivate laccase. Thus, we selected seven natural mediators, acetosyringone, syringaldehyde, *p*-coumaric acid, vanillin, vanillic acid, vanillyl alcohol, and acetovanillone, for the melanin decolorization reaction by LMS. Interestingly, all of the natural mediators act as more efficient mediators than HOBt for melanin decolorization by lacT. When acetosyringone, syringaldehyde, and *p*-coumaric acid were used, the decolorization yields were 28%, 22%, and 18%, respectively, after 5 h of reaction. These results clearly demonstrate the usefulness of natural mediators for the melanin decolorization reaction by LMS. The mediators in the LMS are oxidized to mediator radicals by laccase, and the mediator radicals induce the oxidation and decolorization of melanin. When lacT was used without a mediator for melanin decolorization during a sufficient reaction time, which could reach the equilibrium state, the decolorization yield was 7% after 24 h of reaction. The natural mediators, except vanillic acid, act as more efficient mediators than HOBt for melanin decolorization by lacT after 24 h of reaction. The decolorization yield after 24 h reaction using vanillic acid as a mediator was lower than that after 5 h of reaction. This may be caused by the low stability of the oxidized radical form of vanillic acid. When acetosyringone, syringaldehyde, and acetovanillone were used, the decolorization yields were 34%, 30%, and 31%, respectively, after a 24 h reaction. *p*-Coumaric acid was more efficient in enhancing the initial reaction rate than acetovanillone, while acetovanillone induced a higher decolorization yield at the equilibrium state than *p*-coumaric acid.

The effect of the mediator on the decolorization reaction by LMS using lacM was also very similar to that obtained by LMS using lacT. When lacM was used without a mediator for melanin decolorization, the decolorization yield was only 2% after 5 h of reaction. HOBt as a mediator for lacM did not enhance the decolorization yield during the 5 h reaction. All of the natural mediators, except *p*-coumaric acid and vanillin, acted as efficient mediators of melanin decolorization by lacM. When acetosyringone and syringaldehyde were used, the decolorization yields were 25% and 22%, respectively, after 5 h of reaction. *p*-Coumaric acid and vanillin were used as efficient mediators for lacT, but they could not efficiently enhance the decolorization rate in LMS using lacM. This may be caused by the lower substrate specificity of lacM for *p*-coumaric acid and vanillin. The decolorization yields after 24 h of reaction of lacM with *p*-coumaric acid and vanillin were similar to those obtained by lacT. This indicates that the oxidized forms of *p*-coumaric acid and vanillin can efficiently decolorize melanin, although their oxidation rate by lacM was much lower than that by lacT. When lacM without a mediator was used for melanin decolorization during a sufficient reaction time, the decolorization yield was 5% after 24 h of reaction. The natural mediators, except vanillic acid, also act as more efficient mediators than HOBt for melanin decolorization by lacM after 24 h of reaction. When acetosyringone, syringaldehyde, and acetovanillone were used as mediators for lacM, the decolorization yields were 34%, 28%, and 31%, respectively, after 24 h of reaction. When vanillic acid was used as a mediator for both lacT and lacM, it showed the lowest decolorization yield. This may be caused by the low stability of the oxidized radical form of vanillic acid. Khammuang and Sarnthima reported that vanillin and vanillic acid could be used as mediators for melanin decolorization using laccase from *Lentinus polychrous* [28]. However, they showed much lower decolorization activity for melanin than acetosyringone when they were used as mediators for lacT and lacM.

These results clearly indicate that natural mediators are more efficient for melanin decolorization by LMS than HOBt. HOBt has been considered as an efficient synthetic mediator for laccase because of its high redox potential and the catalytic role of the N-OH group of HOBt [5]. The efficiency of mediators to oxidize target substrates is highly dependent on the ability to form stable radicals as well as the steric hindrance caused by bulky alkyl substituents rather than the redox potential of the mediators [19,35]. The low stability of the oxidized intermediate of HOBt has been determined through cyclic voltammetry [6]. Therefore, the low decolorization yield by LMS using HOBt may be caused by the low stability of HOBt under the reaction conditions of laccase. Although the redox potential of syringaldehyde was lower than that of HOBt, syringaldehyde showed relatively higher stability than HOBt [6].

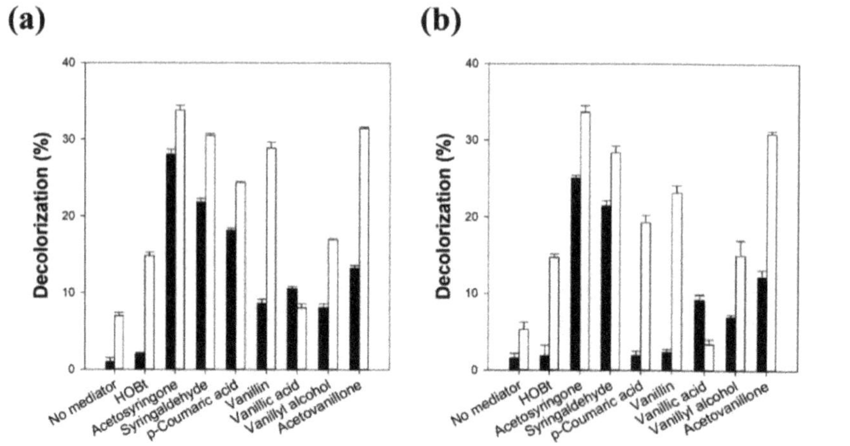

Figure 1. The effect of mediators on the melanin decolorization by LMS using laccase from *T. versicolor* (**a**) and *M. thermophila* (**b**). Black and white bars represent 5 h and 24 h reactions, respectively. Reaction conditions: 50 μg/mL melanin, 0.1 mM mediator, and 0.1 M citric acid phosphate buffer (pH 5.5) at 25 °C.

As shown in Figure 2, the natural mediators used in this work have various substituents (e.g., hydroxyl, methoxy, carboxyl, ketone, or aldehyde) at different positions on the benzene ring [12,19]. Mediators with two methoxy groups (acetosyringone and syringaldehyde) showed higher decolorization rates than those with one methoxy group. The decolorization rate obtained by *p*-coumaric acid with no methoxy group was dependent on the type of laccase. The *p*-coumaric acid with lacT showed a higher decolorization rate than those with one methoxy group, while *p*-coumaric acid with lacM showed the lowest decolorization rate in the 5 h reaction. Fillat et al. also showed similar results for the decolorization of flexographic inks by fungal laccases with natural mediators [36]. The phenolic natural mediators (acetosyringone, methyl syringate, and syringaldehyde) with two methoxy substituents in the ring were oxidized by laccase faster than *p*-coumaric acid with no methoxy group. This indicates that methoxy groups play a more important role as electron donors than the double bond of the lateral chain of *p*-coumaric acid. When the mediators with one methoxy group were compared, the decolorization yield increased in the following order: acetovanillone > vanillin > vanillyl alcohol > vanillic acid. Acetovanillone, which has a ketone group, showed a higher decolorization rate and yield than the mediators with aldehyde, hydroxyl, and carboxyl groups. Acetosyringone with a ketone group also showed a higher decolorization rate and yield than syringaldehyde with an aldehyde group.

In the following experiments, acetosyringone, syringaldehyde, and acetovanillone, which showed high melanin decolorization ability, were selected as mediators for LMS to decolorize melanin. The effect of the mediator on the decolorization reaction by LMS was investigated over time (Figure S1). The decolorization reaction using lacT with acetosyringone, syringaldehyde, and acetovanillone resulted in 21%, 18%, and 1% decolorization yields after 1 h of reaction, respectively. The decolorization reaction using lacM with acetosyringone and syringaldehyde resulted in 19% and 18% decolorization yields after 1 h of reaction, respectively. Both laccases showed similar reaction profiles when the same mediator was used. Acetosyringone and syringaldehyde significantly enhanced the decolorization rate during the initial reaction. These results show that acetosyringone and syringaldehyde containing dimethoxy groups were more efficient in enhancing the initial rate of decolorization by LMS than acetovanillone containing one methoxy group. Fillat et al. also reported that the methoxy groups in the ring structures of mediators act as accelerators for the oxidation of substrates [36]. On the other hand, the decolorization yield after 24 h of reaction by acetovanillone was similar to that by syringaldehyde, although acetovanillone moderately enhanced the reaction rate.

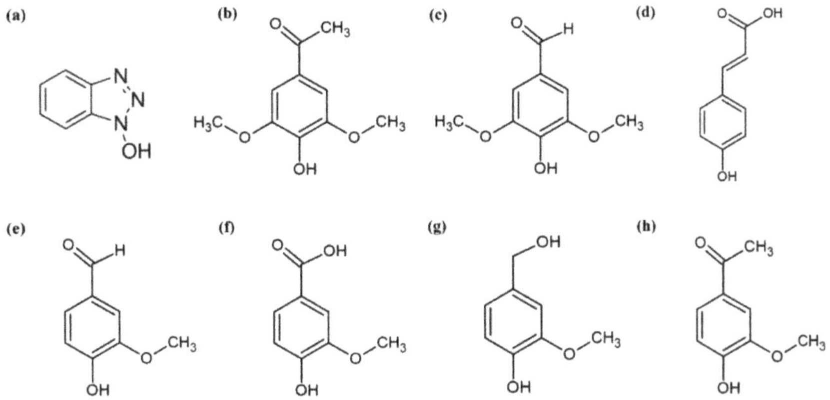

Figure 2. The structure of mediators used for LMS. (**a**) HOBt, (**b**) acetosyringone, (**c**) syringaldehyde, (**d**) *p*-coumaric acid, (**e**) vanillin, (**f**) vanillic acid, (**g**) vanillyl alcohol, (**h**) acetovanillone.

3.2. Effect of Mediator Concentration on Melanin Decolorization by LMS

The effect of the mediator concentration on the decolorization yield by LMS was investigated (Figure 3). The LMS using lacT and 100 μM acetosyringone showed the highest decolorization yield, which was 4.4-fold higher than that without the mediator. The LMS using lacM and 100 μM acetosyringone showed the highest decolorization yield, which was 6.7-fold higher than that without the mediator. When lacT was used for LMS, the decolorization yields using acetosyringone, syringaldehyde, and acetovanillone increased with increasing concentration and then reached a maximum at concentrations of 100, 50, and 200 μM, respectively (Figure 3a). Acetosyringone and syringaldehyde more efficiently enhanced the decolorization yield at a lower concentration than acetovanillone. When lacM was used for LMS, the effect of mediator concentration was very similar to the result obtained with lacT. The decolorization yields using acetosyringone, syringaldehyde, and acetovanillone reached a maximum at concentrations of 100, 50, and 200 μM, respectively (Figure 3b). A mediator concentration of over 200 μM significantly decreased the decolorization yield (data not shown). Therefore, a mediator of 100 μM was chosen as the optimal concentration for the following experiments. Lloret et al. reported that the optimal mediator concentration should be used because laccase can be inactivated by a high concentration of mediator [6], whereas Khammuang and Sarnthima reported that the melanin decolorization activity of LMS using vanillin and vanillic acid was not significantly influenced by high concentrations of up to 10 mM [28]. The optimal mediator concentration may be dependent on the mediator type and the target compound of the laccase-catalyzed reaction [5,6]. When the LMS catalyzed the degradation of isoproturon, the degradation yield increased with increasing concentration of acetosyringone, while the concentrations of vanillin and syringaldehyde were not related to the degradation yield [7].

3.3. Cytotoxicity of Natural Mediators

To use natural mediators as skin-whitening cosmetic ingredients, the cytotoxicity of mediators (acetosyringone, syringaldehyde, and acetovanillone) was investigated using the B16F10 melanoma cell line. The natural mediators were treated on the cultured cells and the cell viability was measured by the NR assay. When mediators of over 22 mM were added to the cultured cells, they considerably reduced cell viability (Figure 4). When the concentration of the mediator was higher than 5 mM, cell viability increased in the following order: acetosyringone > syringaldehyde > acetovanillone. Furthermore, mediators of less than 1 mM showed no inhibitory effect on B16F10 melanoma cells. These results indicate that the optimal mediator concentration of 0.1 mM used in this work showed negligible cytotoxicity.

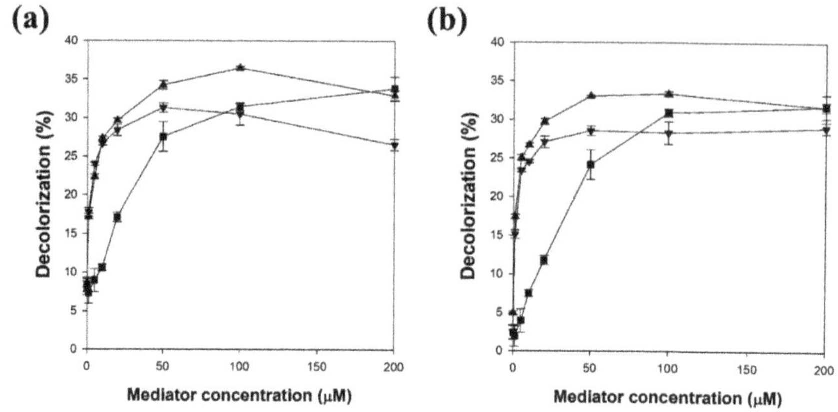

Figure 3. The effect of mediator concentration on the melanin decolorization by LMS using laccase from *T. versicolor* (**a**) and *M. thermophila* (**b**). ▲: acetosyringone, ▼: syringaldehyde, ■: acetovanillone. Reaction conditions: 50 μg/mL melanin and 0.1 M citric acid phosphate buffer (pH 5.5) at 25 °C for 24 h.

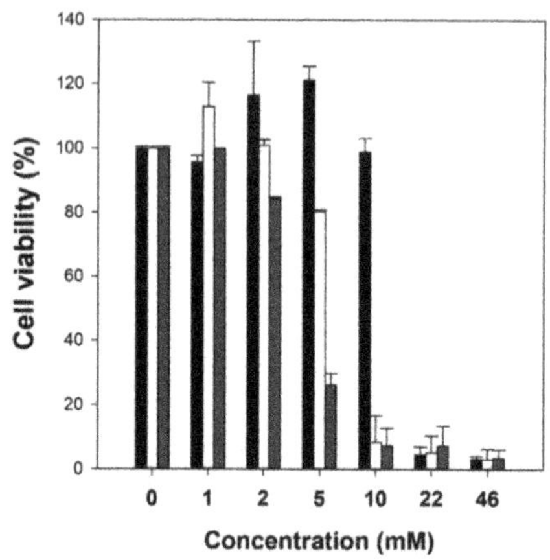

Figure 4. Cytotoxicity of natural mediators used for LMS. Black bars: acetosyringone, white bars: syringaldehyde, gray bars: acetovanillone.

3.4. *The Effect of pH on Melanin Decolorization by LMS*

Figure 5 shows the effect of pH on the LMS decolorization yield. The lacT without a mediator showed the highest decolorization yield at pH 4, and the decolorization yield decreased with increasing pH (Figure 5a). This profile is similar to the effect of pH on laccase-catalyzed ABTS oxidation [37]. In general, lacT has optimal activity at acidic conditions of less than pH 5, and its activity decreases with an increase in pH. At higher pH, the hydroxide anions combine with the T2/T3 coppers of lacT and disturb the electron transfer and cause a decrease in catalytic activity. However, the effect of pH on LMS is more complicated and can be influenced by the activity and stability of laccase and oxidized mediators [1,7]. The decolorization yield of lacT with acetosyringone increased with increasing pH up to 7 (41%). The optimal decolorization yields of lacT with syringaldehyde and acetovanillone were 32% and 42%, respectively, at pH 6. Although lacT showed lower catalytic activity at higher pH, melanin decolorization increased with increasing pH. These results could be explained by the high activity and stability of the oxidized forms of natural mediators at high pH.

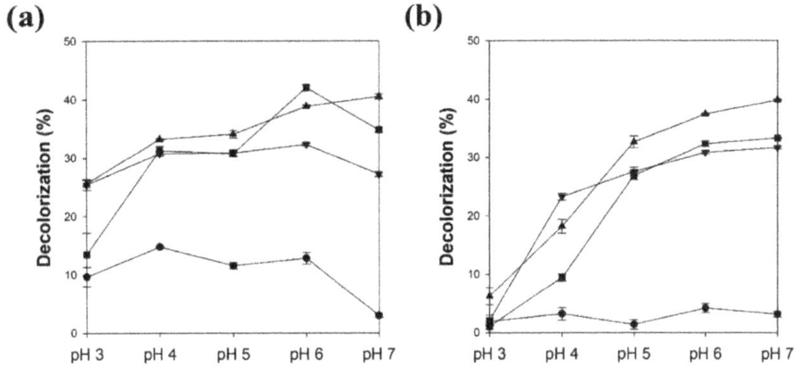

Figure 5. The effect of pH on melanin decolorization by LMS using laccase from *Trametes versicolor* (**a**) and *M. thermophila* (**b**). •: no mediator, ▲: acetosyringone, ▼: syringaldehyde, ■: acetovanillone. Reaction conditions: 50 μg/mL melanin and 0.1 mM mediator at 25 °C for 24 h.

The lacM without a mediator showed the highest decolorization yield at pH 6, and lacM showed a lower decolorization yield in all pH ranges compared to lacT. The optimal pH of lacM was approximately 6 for the oxidation of ABTS [38]. Therefore, the optimal pH of lacM without a mediator was similar to that of lacM for ABTS oxidation. When the natural mediators were used with lacM, the decolorization yields were highly dependent on the reaction pH (Figure 5b). The decolorization yield of lacM with natural mediators increased with increasing pH up to pH 7. The maximum decolorization yields of lacM with acetosyringone, syringaldehyde, and acetovanillone were 40%, 32%, and 33%, respectively, at pH 7. The high melanin decolorization activity at neutral pH makes the use of LMS with natural mediators better for skin, because the optimal pH range is similar to that of normal skin (around 5.5).

3.5. Kinetic Study of Melanin Decolorization by LMS

A kinetic study of melanin decolorization by LMS with acetosyringone was investigated quantitatively to understand the enhancing effect of natural mediators (Table 1). The K_m value of lacT without a mediator was 10.6-fold higher than that of lacT with acetosyringone. This means that the affinity for melanin was highly enhanced by the use of a mediator. The k_{cat} value of lacT with the mediator was 22.6-fold higher than that of lacT without a mediator. This indicates that the decolorization rate was significantly increased by the mediator. The specificity constant (k_{cat}/K_m) of lacT was enhanced 247 times by using acetosyringone as a mediator. These results clearly show that the limited access of melanin to the active site of laccase was overcome by acetosyringone. The K_m value of lacM without a mediator was 2.4-fold higher than that of lacM with acetosyringone. The affinity for melanin can be enhanced by using a mediator. However, the increasing effect for melanin affinity by lacM with acetosyringone was lower than that by lacT with acetosyringone. The k_{cat} value of lacM was approximately 26% of that of lacT. The lacM without a mediator showed very low activity for melanin decolorization. However, the k_{cat} value of lacM with the mediator was 161-fold higher than that of lacM without a mediator. The decolorization rate by lacM was significantly increased by the mediator. Therefore, the specificity constant of lacM was 334 times enhanced using acetosyringone as a mediator. These results clearly demonstrate the usefulness of acetosyringone as a mediator of laccases for melanin decolorization.

Table 1. Kinetic constants of laccase from *T. versicolor* and *M. thermophila* for the melanin decolorization reaction using acetosyringone as a mediator.

		K_m (μg/mL)	k_{cat} (/h)	k_{cat}/K_m ($\times 10^{-3}$ mL/μg/h)
Laccase from	w/o mediator	284.6	0.44	1.5
T. versicolor	w/mediator	26.8	9.93	371.0
Laccase from	w/o mediator	319.7	0.11	0.4
M. thermophila	w/mediator	132.8	17.75	133.7

3.6. Decolorization of the Melanin/Cellulose Film by LMS

A melanin/cellulose composite hydrogel film was prepared to mimic melanin in the skin. The melanin/cellulose hydrogel film could be prepared by the co-dissolution of melanin and cellulose in [Emim][Ac], which is followed by regeneration with water. The obtained film exhibited a transparent dark brown color. The decolorization of the melanin/cellulose hydrogel film was performed by lacM with acetosyringone in 0.1 M citric acid phosphate buffer (pH 5.5). The color of the melanin/cellulose film changed from dark brown to pale brown after the LMS-catalyzed reaction (Figure 6a). When the absorbance of the melanin/cellulose film was measured after the decolorization reaction in the range of 400–700 nm, the absorbance of the melanin/cellulose film decreased significantly over the entire wavelength range (Figure 6b). The decolorization of melanin was also confirmed through the measurement of color values (L*, a*, and b*) of the melanin/cellulose film using a colorimeter. The ΔL, ΔE, YI, and WI values were calculated from the color parameters (Table 2). After the melanin decolorization reaction by LMS, the L* (lightness) value of the melanin/cellulose film was considerably increased, while the a* (redness) and b* (yellowness) values decreased slightly. The ΔE value representing the color difference between the samples was 31.1. An ΔE value greater than 12 indicated that the colors of the film before and after the reaction are quite different from each other [31]. The yellowness (YI) of the film decreased from 209 to 92 after the decolorization reaction, while the whiteness (WI) increased from 16 to 43. Łopusiewicz et al. reported that the YI of poly(lactic acid)/melanin film increased with increasing melanin content, whereas the WI of the film decreased [32]. Therefore, the changes in the color properties quantitatively explain the decolorization of melanin in the melanin/cellulose hydrogel film. These results show that LMS can efficiently decolorize melanin in a cellulose hydrogel environment.

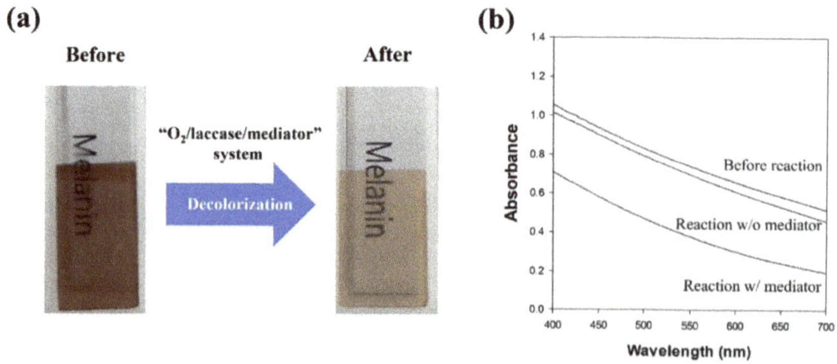

Figure 6. (**a**) Photo images and (**b**) spectrum of the melanin/cellulose films before and after the decolorization reaction by LMS using laccase from *M. thermophila* and acetosyringone.

Table 2. Color parameters and indices of the melanin/cellulose hydrogel films before and after the decolorization reaction.

Decolorization of Melanin/Cellulose Film	L*	a*	b*	ΔL	ΔE	YI	WI
Before	28.3	13.3	41.4	-	-	208.9	16.2
After	59.2	12.2	38.0	30.9	31.1	91.6	43.0

3.7. Decolorization of Natural Melanin Produced by Melanoma Cells

Natural melanin in human skin is divided into eumelanin (black to brown) and pheomelanin (yellow to red), while the synthetic melanin prepared by the oxidation of tyrosine with hydrogen peroxide is an analog of eumelanin. Therefore, the decolorization of natural melanin produced by melanoma cells was also investigated in this work. The decolorization of synthetic melanin and natural melanin by LMS with lacM and acetosyringone was compared by measuring the absorbance in the range of 400–700 nm (Figure 7). The absorbance of synthetic melanin rapidly decreased with increasing reaction time at all wavelengths. However, the absorbance of decolorized natural melanin showed a different pattern compared to that of decolorized synthetic melanin. The absorbance of decolorized natural melanin decreased at wavelengths greater than 450 nm, while the absorbance increased at wavelengths less than 450 nm. More studies are required to understand the increase in absorbance of decolorized natural melanin.

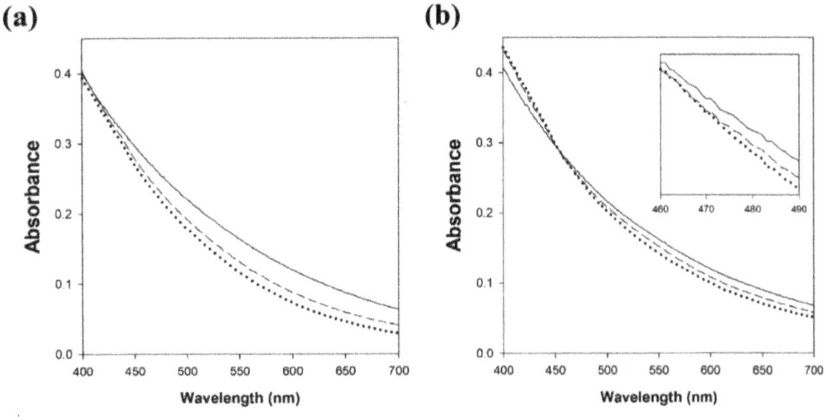

Figure 7. Changes in spectra of the synthetic melanin (**a**) and natural melanin (**b**) after the decolorization reaction by LMS using laccase from *M. thermophila* and acetosyringone. Solid line, dashed line, and dotted line represent no reaction, 15 min reaction, and 60 min reaction, respectively.

4. Conclusions

In this study, melanin decolorization was achieved by using the "O_2/laccase/mediator" system, since laccase showed low catalytic activity for the direct oxidation of melanin due to its low redox potential and limited access of melanin into the active site of laccase. Seven kinds of natural mediators were successfully used to replace synthetic mediator (HOBt) for melanin decolorization by LMS using lacT and lacM. Among the tested natural mediators, acetosyringone and syringaldehyde, containing two methoxy groups, showed high decolorization rates and yields. Acetovanillone containing one methoxy group and one ketone group also showed a high decolorization yield in the equilibrium state. The LMS with natural mediators showed high decolorization activity at the pH of normal skin, and the cytotoxicity of natural mediators was very low. A kinetic study of LMS using acetosyringone for melanin decolorization showed that acetosyringone efficiently overcame the limitations of lacT and lacM by increasing the affinity for melanin and decolorization activity. LMS with acetosyringone decolorized the melanin present in the cellulose hydrogel film, which mimics skin. Furthermore, LMS with acetosyringone could decolorize not only synthetic eumelanin analogs prepared by the oxidation of tyrosine but also natural melanin produced by melanoma cells. Thus, LMS using natural mediators can be used as an effective skin-whitening agent in the cosmetics industry.

Supplementary Materials: The following are available online at https://www.mdpi.com/article/10.3390/polym13213671/s1, Figure S1: The melanin decolorization by LMS using laccase from *T. versicolor* and *M. thermophila*, Figure S2: The initial rate for the melanin decolorization reaction by laccase from *T. versicolor* and *M. thermophila*.

Author Contributions: Conceptualization, S.P., D.J. and S.H.L.; Data curation, S.H.L.; Formal analysis, S.P.; Funding acquisition, S.H.L.; Investigation, S.P., D.J. and H.D.; Methodology, S.P., D.J. and H.D.; Project administration, S.H.L.; Resources, S.H.L.; Software, J.Y., D.L. and S.H.; Supervision, S.H.L.; Validation, J.Y., D.L., S.H. and S.H.L.; Visualization, J.Y., D.L. and S.H.; Writing—original draft, S.P. and D.J.; Writing—review and editing, S.H.L. All authors have read and agreed to the published version of the manuscript.

Funding: This work was supported by Konkuk University in 2018.

Institutional Review Board Statement: Not applicable.

Informed Consent Statement: Not applicable.

Data Availability Statement: Not applicable.

Conflicts of Interest: The authors declare no conflict of interest.

References

1. Strong, P.J.; Claus, H. Laccase: A review of its past and its future in bioremediation. *Crit. Rev. Environ. Sci. Technol.* **2011**, *41*, 373–434. [CrossRef]
2. Mayer, A.M.; Staples, R.C. Laccase: New functions for an old enzyme. *Phytochemistry* **2002**, *60*, 551–565. [CrossRef]
3. Camarero, S.; Ibarra, D.; Martínez, M.J.; Martínez, A.T. Lignin-Derived Compounds as Efficient Laccase Mediators for Decolorization of Different Types of Recalcitrant Dyes. *Appl. Environ. Microbiol.* **2005**, *71*, 1775–1784. [CrossRef] [PubMed]
4. Campos, R.; Cavaco-Paulo, A.; Kandelbauer, A.; Robra, K.H.; Gübitz, G.M. Indigo degradation with purified laccases from *Trametes hirsuta* and *Sclerotium rolfsii*. *J. Biotechnol.* **2001**, *89*, 131–139. [CrossRef]
5. Ostadhadi-Dehkordi, S.; Tabatabaei-Sameni, M.; Forootanfar, H.; Kolahdouz, S.; Ghazi-Khansari, M.; Faramarzi, M.A. Degradation of some benzodiazepines by a laccase-mediated system in aqueous solution. *Bioresour. Technol.* **2012**, *125*, 344–347. [CrossRef]
6. Lloret, L.; Eibes, G.; Lú-Chau, T.A.; Moreira, M.T.; Feijoo, G.; Lema, J.M. Laccase-catalyzed degradation of anti-inflammatories and estrogens. *Biochem. Eng. J.* **2010**, *51*, 124–131. [CrossRef]
7. Zeng, S.; Qin, X.; Xia, L. Degradation of the herbicide isoproturon by laccase-mediator systems. *Biochem. Eng. J.* **2017**, *119*, 92–100. [CrossRef]
8. Ibarra, D.; Romero, J.; Martínez, M.J.; Martínez, A.T.; Camarero, S. Exploring the enzymatic parameters for optimal delignification of eucalypt pulp by laccase-mediator. *Enzyme Microb. Technol.* **2006**, *39*, 1319–1327.
9. Fillat, A.; Colom, J.F.; Vidal, T. A new approach to the biobleaching of flax pulp with laccase using natural mediators. *Bioresour. Technol.* **2010**, *101*, 4104–4110. [CrossRef] [PubMed]
10. Camarero, S.; Ibarra, D.; Martínez, A.T.; Romero, J.; Gutiérrez, A.; del Río, J.C. Paper pulp delignification using laccase and natural mediators. *Enzyme Microb. Technol.* **2007**, *40*, 1264–1271. [CrossRef]
11. Zerva, A.; Simić, S.; Topakas, E.; Nikodinovic-Runic, J. Applications of Microbial Laccases: Patent Review of the Past Decade (2009–2019). *Catalysts* **2019**, *9*, 1023. [CrossRef]
12. Cañas, A.I.; Camarero, S. Laccases and their natural mediators: Biotechnological tools for sustainable eco-friendly processes. *Biotechnol. Adv.* **2010**, *28*, 694–705. [CrossRef] [PubMed]
13. Shleev, S.; Tkac, J.; Christenson, A.; Ruzgas, T.; Yaropolov, A.I.; Whittaker, J.W.; Gorton, L. Direct electron transfer between copper-containing proteins and electrodes. *Biosens. Bioelectron.* **2005**, *20*, 2517–2554. [CrossRef] [PubMed]
14. Christopher, L.P.; Yao, B.; Ji, Y. Lignin biodegradation with laccase-mediator systems. *Front. Energy Res.* **2014**, *2*, 1–13. [CrossRef]
15. Bourbonnais, R.; Paice, M.G. Oxidation of non-phenolic substrates: An expanded role for lactase in lignin biodegradation. *FEBS Lett.* **1990**, *267*, 99–102. [CrossRef]
16. Srebotnik, E.; Hammel, K.E. Degradation of nonphenolic lignin by the laccase/1-hydroxybenzotriazole system. *J. Biotechnol.* **2000**, *81*, 179–188. [CrossRef]
17. Baiocco, P.; Barreca, A.M.; Fabbrini, M.; Galli, C.; Gentili, P. Promoting laccase activity towards non-phenolic substrates: A mechanistic investigation with some laccase–mediator systems. *Org. Biomol. Chem.* **2003**, *1*, 191–197. [CrossRef]

18. Fabbrini, M.; Galli, C.; Gentili, P. Comparing the catalytic efficiency of some mediators of laccase. *J. Mol. Catal. B-Enzym.* **2002**, *16*, 231–240. [CrossRef]
19. González, M.D.; Vidal, T.; Tzanov, T. Electrochemical study of phenolic compounds as enhancers in laccase-catalyzed oxidative reactions. *Electroanalysis* **2009**, *21*, 2249–2257. [CrossRef]
20. Yang, L.; Guo, X.; Jin, Z.; Guo, W.; Duan, G.; Liu, X.; Li, Y. Emergence of melanin-inspired supercapacitors. *Nano Today* **2021**, *37*, 101075. [CrossRef]
21. Caldas, M.; Santos, A.C.; Veiga, F.; Rebelo, R.; Reis, R.L.; Correlo, V.M. Melanin nanoparticles as a promising tool for bi-omedical applications—A review. *Acta Biomater.* **2020**, *105*, 26–43. [CrossRef]
22. Woo, S.H.; Cho, J.S.; Lee, B.S.; Kim, E.K. Decolorization of melanin by lignin peroxidase from *Phanerochaete chrysosporium*. *Biotechnol. Bioprocess Eng.* **2004**, *9*, 256–260. [CrossRef]
23. Kaneko, S.; Cheng, M.; Murai, H.; Takenaka, S.; Murakami, S.; Aoki, K. Purification and Characterization of an Extracellular Laccase from *Phlebia radiata* Strain BP-11-2 That Decolorizes Fungal Melanin. *Biosci. Biotechnol. Biochem.* **2009**, *73*, 939–942. [CrossRef]
24. Mohorčič, M.; Friedrich, J.; Renimel, I.; André, P.; Mandin, D.; Chaumont, J.-P. Production of melanin bleaching enzyme of fungal origin and its application in cosmetics. *Biotechnol. Bioprocess Eng.* **2007**, *12*, 200–206. [CrossRef]
25. Kim, B.S.; Blaghen, M.; Hong, H.; Lee, K. Purification and characterization of a melanin biodegradation enzyme from *Geotrichum* sp. *Int. J. Cosmetic Sci.* **2016**, *38*, 622–626. [CrossRef] [PubMed]
26. Sung, H.J.; Khan, M.F.; Kim, Y.H. Recombinant lignin peroxidase-catalyzed decolorization of melanin using in-situ generated H_2O_2 for application in whitening cosmetics. *Int. J. Biol. Macromol.* **2019**, *136*, 20–26. [CrossRef]
27. Shin, S.K.; Hyeon, J.E.; Joo, Y.; Jeong, D.W.; You, S.K.; Han, S.O. Effective melanin degradation by a synergistic laccase-peroxidase enzyme complex for skin whitening and other practical applications. *Int. J. Biol. Macromol.* **2019**, *129*, 181–186. [CrossRef]
28. Khammuang, S.; Sarnthima, R. Decolorization of synthetic melanins by crude laccases of *Lentinus polychrous* Lév. *Folia Microbiol.* **2013**, *58*, 1–7. [CrossRef] [PubMed]
29. Repetto, G.; del Peso, A.; Zurita, J. Neutral red uptake assay for the estimation of cell viability/cytotoxicity. *Nat. Protoc.* **2008**, *3*, 1125–1131. [CrossRef] [PubMed]
30. Castellar, M.R.; Obón, J.M.; Fernández-López, J.A. The isolation and properties of a concentrated red-purple betacyanin food colourant from *Opuntia stricta* fruits. *J. Sci. Food Agric.* **2006**, *86*, 122–128. [CrossRef]
31. Cserhalmi, Z.; Sass-Kiss, Á.; Tóth-Markus, M.; Lechner, N. Study of pulsed electric field treated citrus juices. *Innov. Food Sci. Emerg. Technol.* **2006**, *7*, 49–54. [CrossRef]
32. Łopusiewicz, L.; Jędra, F.; Mizielińska, M. New poly(lactic acid) active packaging composite films incorporated with fungal melanin. *Polymers* **2018**, *10*, 386. [CrossRef] [PubMed]
33. Pralea, I.E.; Moldovan, R.C.; Petrache, A.M.; Ilieș, M.; Hegheș, S.C.; Ielciu, I.; Nicoară, R.; Moldovan, M.; Ene, M.; Radu, M.; et al. From extraction to advanced analytical methods: The challenges of melanin analysis. *Int. J. Mol. Sci.* **2019**, *20*, 3943. [CrossRef] [PubMed]
34. Tran-Ly, A.N.; Reyes, C.; Schwarze, F.W.M.R.; Ribera, J. Microbial production of melanin and its various applications. *World J. Microbiol. Biotechnol.* **2020**, *36*, 1–9. [CrossRef] [PubMed]
35. Medina, F.; Aguila, S.; Baratto, M.C.; Martorana, A.; Basosi, R.; Alderete, J.B.; Vazquez-Duhalt, R. Prediction model based on decision tree analysis for laccase mediators. *Enzyme Microb. Technol.* **2013**, *52*, 68–76. [CrossRef]
36. Fillat, U.; Prieto, A.; Camarero, S.; Martínez, A.T.; Martínez, M.J. Biodeinking of flexographic inks by fungal laccases using synthetic and natural mediators. *Biochem. Eng. J.* **2012**, *67*, 97–103. [CrossRef]
37. Hong, J.; Jung, D.; Park, S.; Oh, Y.; Oh, K.K.; Lee, S.H. Immobilization of laccase via cross-linked enzyme aggregates prepared using genipin as a natural cross-linker. *Int. J. Biol. Macromol.* **2021**, *169*, 541–550. [CrossRef]
38. Berka, R.M.; Schneider, P.; Golightly, E.J.; Brown, S.H.; Madden, M.; Brown, K.M.; Halkier, T.; Mondorf, K.; Xu, F. Characterization of the gene encoding an extracellular laccase of *Myceliophthora thermophila* and analysis of the recombinant enzyme expressed in *Aspergillus oryzae*. *Appl. Environ. Microbiol.* **1997**, *63*, 3151–3157. [CrossRef]

polymers

MDPI

Article

Ruminant-Waste Protein Hydrolysates and Their Derivatives as a Bio-Flocculant for Oil Sands Tailing Management

Jesse Yuzik [1], Vinay Khatri [1], Michael Chae [1], Paolo Mussone [2] and David C. Bressler [1,*]

[1] Department of Agricultural, Food and Nutritional Science, Faculty of Agricultural, Life and Environmental Sciences, University of Alberta, Edmonton, AB T6G 2P5, Canada; yuzik@ualberta.ca (J.Y.); vkhatri@ualberta.ca (V.K.); mchae@ualberta.ca (M.C.)

[2] Applied BioNanotechnology Industrial Research Chair, Industry Solutions, Northern Alberta Institute of Technology, 10210 Princess Elizabeth Ave., NW, Edmonton, AB T5G 0Y2, Canada; PMUSSONE@nait.ca

* Correspondence: dbressle@ualberta.ca

Abstract: Reclamation of tailings ponds is a critical issue for the oil industry. After years of consolidation, the slurry in tailings ponds, also known as fluid fine tailings, is mainly comprised of residual bitumen, water, and fine clay particles. To reclaim the lands that these ponds occupy, separation of the solid particles from the liquid phase is necessary to facilitate water removal and recycling. Traditionally, synthetic polymers have been used as flocculants to facilitate this process, but they can have negative environmental consequences. The use of biological polymers may provide a more environmentally friendly approach to flocculation, and eventual soil remediation, due to their natural biodegradability. Peptides derived from specified risk materials (SRM), a proteinaceous waste stream derived from the rendering industry, were investigated to assess their viability for this application. While these peptides could achieve >50% settling within 3 h in bench-scale settling tests using kaolinite tailings, crosslinking peptides with glutaraldehyde greatly improved their flocculation performance, leading to a >50% settling in only 10 min. Settling experiments using materials obtained through different reactant ratios during crosslinking identified a local optimum molar reactant ratio of 1:32 (peptide amino groups to glutaraldehyde aldehyde groups), resulting in 81.6% settling after 48 h. Taken together, these data highlight the novelty of crosslinking waste-derived peptides with glutaraldehyde to generate a value-added bioflocculant with potential for tailings ponds consolidation.

Keywords: flocculant; crosslinking; peptides; glutaraldehyde; specified risk materials

Citation: Yuzik, J.; Khatri, V.; Chae, M.; Mussone, P.; Bressler, D.C. Ruminant-Waste Protein Hydrolysates and Their Derivatives as a Bio-Flocculant for Oil Sands Tailing Management. *Polymers* **2021**, *13*, 3533. https://doi.org/10.3390/polym13203533

Academic Editor: Arn Mignon

Received: 31 August 2021
Accepted: 8 October 2021
Published: 14 October 2021

1. Introduction

In 2019, the Alberta oil sands generated 2.95 million barrels per day, making them the source of 63% of Canada's oil production [1]. Isolation of a hydrocarbon-rich bitumen stream from oil sands is achieved through the Clarke hot water extraction method, in which hot water is injected into bitumen-rich layers of the ground to facilitate separation of bitumen and sediments [2]. The remaining byproduct, which consists of water with fine clays and some residual bitumen, is called tailings [2]. These tailings are stored indefinitely in tailings ponds, which are large, man-made ponds supported by dykes. With the inventory of these ponds growing by the day, it is essential to find economic and environmentally friendly ways to manage the tailings and reclaim the lands that they occupy.

Coagulation and flocculation are two basic strategies for consolidation and reclamation of tailings ponds [3]. Coagulants are multivalent cations (Ca^{2+}, Al^{3+}, and Fe^{3+}) that reduce repulsion between solid particles by electrostatic, steric, and hydration forces, facilitating improved flocculation [4]. Flocculants are typically high molecular weight polymers that can bridge gaps between particles in a colloid to allow them to interact with each other and form larger particles (i.e., flocs), which can then settle out of solution [4,5]. Charged polymers are often used for the flocculation of the fine, negatively charged particles contained in tailings, forming strong ionic interactions that promote flocculation [4,6].

Currently, gypsum, a coagulant, and hydrolyzed polyacrylamide (HPAM), a flocculant, are widely used for the reclamation of tailings ponds. However, the synthetic polymer HPAM accumulates in the environment as it is not biodegradable and can also contain unreacted acrylamide, a known neurotoxin and potential carcinogen [7,8]. The development of bioflocculants can alleviate these concerns due to their improved biodegradability and sustainability. For example, chitosan is one of the most popular and highly studied bioflocculants due to its cationic nature [9,10]. Peptides, the building blocks of proteins, also have bioflocculant potential as charged groups inherent on the peptide backbone can interact with charged clay surfaces. However, due to their low molecular weight, peptides are often subjected to crosslinking, creating larger molecular weight polymers that are necessary to improve floc strength and formation rates, and thus induce rapid flocculation [11–13]. Glutaraldehyde is a well-known chemical crosslinker for peptides, with its two aldehyde groups known to react readily with amino groups of peptides at ambient conditions [14,15].

Specified risk materials (SRM) are protein-rich waste products from the beef rendering industry that can be transformed into high value products, including flocculants for wastewater treatment. SRM are certain tissues of livestock (i.e., brain, eyes, and spinal column) that have the potential to contain prion diseases, such as bovine spongiform encephalopathy (BSE), more commonly known as mad cow disease [16]. In 2007, the Canadian Food Inspection Agency banned the use of these materials in animal feeds, pet foods, and fertilizers, and mandated their disposal via land-filling or incineration [17]. However, this waste material can be rendered safe to use in other applications after it undergoes a thermal or alkaline hydrolysis, generating peptides in the process. Adding value to SRM-derived peptides would convert a costly and potentially dangerous waste with negative environmental impacts into a product that could ultimately provide an alternative revenue source for the rendering industry. Recent valorization approaches that have been examined by our laboratory include the conversion of SRM-derived peptides to plywood adhesives, torrefied wood pellet binders, tackifiers, and flocculants [12,18–23].

In this study, an SRM peptide–based flocculating agent was developed for tailings ponds treatment and recovery. Application of glutaraldehyde crosslinking to increase the molecular weight of these peptides was investigated and the resulting materials were subjected to bench-scale settling tests using synthetic kaolinite tailings. Finally, the use of different ratios of reagents during the crosslinking of peptides with glutaraldehyde was also assessed. Overall, this work provides insights into the development of a novel, biodegradable flocculant utilizing animal waste protein.

2. Materials

SRM were obtained from a large multinational rendering corporation in western Canada. The kaolinite clay used for bench-scale settling tests (i.e., synthetic kaolinite tailings) was Polygloss 90, which had a median particle size of 0.4 microns, and was obtained from KaMin LLC (Wrens, GA, USA). Calcium sulphate dihydrate (gypsum) (98+%) was obtained from Fisher Scientific (Fair Lawn, NJ, USA). Whatman no. 4 (pore size 20–25 mm) (GE Healthcare, Chicago, IL, USA) filter paper was obtained from Fisher Scientific and was used for all the vacuum filtration processes. A Buchi B290 Spray Dryer (BUCHI Corporation New Castle, DE, USA) was used to recover hydrolyzed peptides. The glutaraldehyde solution (50% Certified), methanol (HPLC grade, 99.9%), and concentrated HCl (Certified ACS Plus, 36.5 to 38.0%) were purchased from Fisher Scientific (Fair Lawn, NJ, USA).

2.1. Hydrolysis of SRM

SRM were hydrolyzed as per the protocol described by Mekonnen et al., 2015 [23]. In brief, hydrolysis took place by addition of distilled water to 1 kg of SRM in a 1:1 (wt/wt) ratio, followed by heating at 180 °C at ≥174 psi for 40 min in a 5.5 L reactor (Parr 4582, Parr Instrument Company, Moline, IL, USA) with constant stirring at 200 rpm, as per Canadian Food Inspection Agency regulations. After hydrolysis, the material is considered nonhazardous and was further processed to remove lipids and other insoluble material by centrifugation (Avanti J-26 XP high-performance centrifuge, Beckman Coulter Canada LP, Mississauga, ON, Canada) at 7000× *g* for 30 min, followed by a second round of centrifugation at 7000× *g* for 10 min. After each round of centrifugation, the liquid phase was decanted, and the solids were removed. Residual solids were further removed by vacuum filtration (Whatman No. 4 filter paper, Oakville, ON, Canada) of the liquid phase. After processing, the peptide mixture was spray-dried to remove water, with drying conditions consisting of an inlet temperature of 170 °C and an outlet temperature of 80 °C. The final product was a light brown powder.

2.2. Preparation of Synthetic Process Water

Synthetic process water (SPW) was used in our bench-scale settling tests to replicate the ion concentrations of a model oil sands tailings pond. It was prepared by adding the compounds to distilled water in the following concentrations: 0.055 g/L $CaCl_2{\cdot}2H_2O$; 0.028 g/L KCl; 0.545 g/L NaCl; 0.443 g/L Na_2SO_4; 0.084 g/L $MgCl_2{\cdot}6H_2O$; 0.895 g/L $NaHCO_3$. Before the kaolinite slurry was prepared, the pH of the SPW was adjusted to 8.00 with concentrated HCl using a pH meter (Fisherbrand™ accumet™ AB15 Basic, Thermo Fisher Scientific, Ottawa, ON, Canada).

2.3. Preparation of Synthetic Kaolinite Tailings

To prepare the 4% (wt/wt) synthetic kaolinite tailings (i.e., kaolinite clay slurries), 10 g of kaolinite clay was added to 240 g of synthetic process water (SPW). When necessary, the SPW contained 300 ppm of gypsum ($2H_2O{\cdot}CaSO_4$) as a coagulant. The slurries were then mixed on a jar tester (PB-900 jar tester from Phipps and Bird, Richmond, VA, USA) at 300 rpm for 30 min. Following this, the slurry was left to sit for 90 min to allow for full hydration of the clay particles. The mixture was then mixed again for 5 min immediately before use.

2.4. Bench-Scale Flocculation Experiments

Bench-scale flocculation experiments were carried out in 250 mL graduated cylinders. A quantity of 250 g of synthetic kaolinite slurry, corresponding to 247 mL, was added to the graduated cylinders. Various treatments were added to the cylinders and the contents of the cylinders were then homogenized by plunging 20 times using a custom plunger, made by welding a metal rod to a washer. Flocculants were added on a wt/wt basis. Settling was recorded by measuring the height of the mudline over time. An example is shown in Figure 1.

2.5. Turbidity Measurements

The turbidity of the release water was measured using a turbidimeter (Orion AQ4500 Thermo Fisher Scientific, Grand Island, NY, USA) after 48 h of settling by withdrawing the supernatant with a pipette approximately halfway between the surface of the solution and the mudline. Samples were measured using the IR ratio mode to eliminate any interference from the color of the sample. The principle behind this measurement is that there is one light source and two detectors to measure the scattering of light. The light source sends a beam of light into the sample and is detected by detector 1, which is situated directly across from the light source, to record a reference measurement. At the same time, a second detector records the light that was scattered at a 90° angle to the light source, which is qualified as the active signal. An algorithm then determines the turbidity based on the ratio of these measurements and compensates for the color absorption in the reference beam. The range of the measurement method described by the manufacturer is from 0 to

4000 NTU (Nephelometric Turbidity Unit). The wavelength used for this measurement was 860 nm.

2.6. Crosslinking Reaction of Peptides with Glutaraldehyde

To increase the molecular weight of the peptides, they were reacted with the crosslinking agent glutaraldehyde. The peptides were dispersed in methanol, followed by a dropwise addition of the glutaraldehyde solution. This reaction was carried out at room temperature for 2 h with stirring. The product was recovered by vacuum filtration, followed by washing with methanol. Residual solvent was evaporated from the product in a fume hood at room temperature for 48 h. After drying, the resulting solid was ground to a powder with a mortar and pestle. The resulting product could be characterized as a fine brown colored powder.

A reaction scheme for the glutaraldehyde crosslinking reaction is proposed in Figure 2. The glutaraldehyde can readily react with the amino group of the peptide to form the peptide–glutaraldehyde intermediate molecule. This reaction can occur at the other hydroxyl group on the peptide–glutaraldehyde intermediate resulting in the formation of a crosslinked peptide.

Figure 1. Example of a bench-scale flocculation experiment.

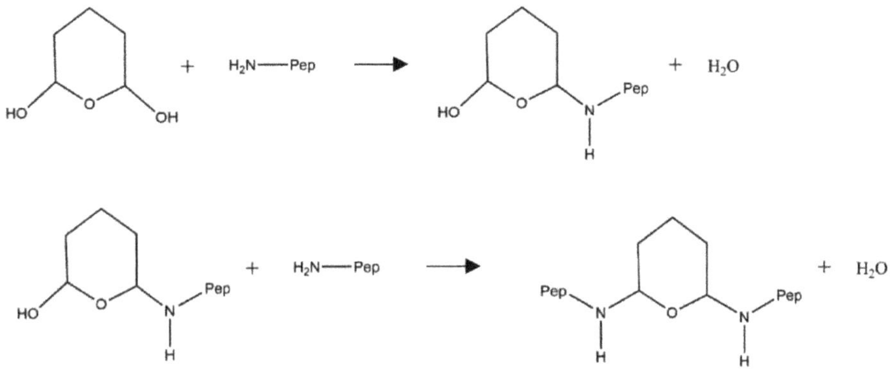

Figure 2. Crosslinking reaction scheme of glutaraldehyde with peptides. Glutaraldehyde can react with the amino groups of two peptide molecules to form a larger crosslinked structure. Adapted from Migneault et al. [24].

2.7. Varying the Crosslinking Ratio

The ratio of the reactants used during the crosslinking reaction was varied to determine the effect that altering the crosslinker amount would have on the products. An estimation of the amount of the amino and carboxylic acid groups of the SRM peptides was determined in a previous study to be 0.6 mmol/g for the amino groups and 1.6 mmol/g for the carboxylic acid groups [22]. The reactant ratio was varied by keeping the peptide amount constant and varying the amount of glutaraldehyde added based on the ratio of the aldehyde groups to the amino groups of the peptides. The reactant ratio was based on the reaction taking place solely with the amino groups, which was also shown in previous studies [14,15,24,25]. Therefore, the reactants were varied as shown in Table 1.

Table 1. Variation of the molar ratios of the reactants in the glutaraldehyde–peptide crosslinking reactions. The ratios were based on the estimated amino groups of the peptides to the aldehyde functional groups of glutaraldehyde. Reactions were performed in triplicate.

Reactant Ratio	Peptides (g)	Glutaraldehyde (50% Solution) (mL)	Methanol (mL)
1:2	4.00	0.426	100
1:4	4.00	0.852	100
1:8	4.00	1.70	100
1:16	4.00	3.41	100
1:32	4.00	6.82	100
1:64	4.00	13.6	100

2.8. Thermal Gravimetric Analysis

Thermal gravimetric analysis (TGA) was used to determine the thermal stability of the peptide samples using a TGA Q50 from TA instruments (New Castle, DE, USA). The temperature was ramped to 100 °C with an increment of 10 °C/min, followed by an isothermal stage for 10 min, and finally the temperature was increased to 500 °C with an increment of 5 °C/min.

2.9. Size Exclusion High Performance Liquid Chromatography

Size exclusion chromatography–high performance liquid chromatography (SEC-HPLC) was done using a 1200 Series LC System from Agilent Technologies (Mississauga, ON, USA) with a diode array detector to measure in the UV range at 210 nm, based on a method from previous research [11]. Superdex Peptide 10/300 GL and Superdex 200 Increase 10/300 GL columns (GE Healthcare Biosciences AB, Uppsala, Sweden) were used in series to enhance separation of the molecules. The Superdex Peptide column has a separation range from 100–7000 Da and the Superdex 200 column has a range of

10–600 kDa. The mobile phase used was a 0.15 M Na_2HPO_4 solution adjusted to pH 9 with NaOH. Sample injection was done at 0.5 mL/min with an injection volume of 20 μL. A series of standards of blue dextran (2000 kDa), alcohol dehydrogenase (150 kDa), albumin (66 kDa), carbonic anhydrase (29 kDa), cytochrome C (12.4 kDa), aprotinin (6.5 kDa), and Vitamin B-12 (1.36 kDa) were used to standardize the retention times of the products and were obtained from Sigma-Aldrich (St. Louis, MO, USA).

2.10. Sodium Dodecyl Sulfate-Polyacrylamide Gel Electrophoresis

The method used for gel electrophoresis was a tricine-SDS PAGE approach based on a protocol by Schägger and was run on a mini-protean II electrophoresis unit (Bio-Rad Laboratories, Richmond, CA, USA) [26]. Modifications were made to the protocol to obtain better results with the crosslinked peptides. To improve the solubility of the crosslinked peptides and loading into the gel, the samples were incubated in the 1X sample buffer solution and mixed for 2 h on a nutating mixer prior to gel loading. A quantity of 20 μL of the samples were loaded into the gel. The concentrations of the peptide sample solutions were optimized to allow proper visualization. The 0.5% and 0.1% (wt/wt) solutions were prepared for the peptides and a 10X more dilute solution was required for the crosslinked peptides at 0.05% and 0.01% (wt/wt). To obtain the lower concentration samples, dilution with the 1X sample buffer solution was performed. Once loaded, the samples were run at 100 V until they left the loading gel and were run at 200 V until the separation was complete. The gel was immediately transferred to a container to be visualized by a silver staining kit. The silver staining kit was a Pierce Silver Staining Kit obtained from Thermo Scientific and the method used was defined by the manufacturer [27]. After staining, the gels were imaged on an AlphaImager HP gel visualizing system from Alpha Innotech (San Leandro, CA, USA).

2.11. Fourier-Transform Infrared Spectroscopy

Assessment of the functional groups present within the peptides and the glutaraldehyde-crosslinked peptides was achieved using Fourier-Transform infrared spectroscopy (FTIR). Analysis was performed using an Attenuated Total Reflectance (ATR) model ALPHA from Bruker Optik GmbH (Rudolf-Plank-StraBe 27, 76275 Ettlingen, Germany). ATR was used as it provides a fast and convenient way to analyze solid samples.

2.12. Statistical Analyses

Statistical analyses were performed using Prism 9.1.2. (San Diego, CA, USA). Data were analyzed using one-way analysis of variance (ANOVA), with a Tukey HSD test applied post hoc at a 95% confidence level. To facilitate statistical analyses, all experiments were performed in triplicate unless otherwise indicated.

3. Results and Discussion

3.1. Glutaraldehyde Crosslinking Reaction with Peptides

Previous work had established that SRM-derived peptides could improve flocculation of a synthetic kaolinite slurry, albeit at relatively slow rates [12]. To improve flocculation, a glutaraldehyde crosslinking reaction was investigated, with the goal to increase the molecular weight of the peptides since larger molecules have been shown to have better flocculation characteristics [5,13,28]. The crosslinking chemical glutaraldehyde was investigated due to its relatively low cost and high reactivity. When attempting this reaction in an aqueous environment, the recovered products were tightly bound to the filter paper, resulting in a low recovery and contamination with filter paper. Therefore, different solvents were then investigated with methanol eventually chosen due to its high miscibility in water and its polar protic nature. Methanol has a pKa of 15.5, which is similar to the pKa of water at 15.7. In addition, the peptides were less soluble in methanol compared to water, due to its lower polarity, which allowed for better product recovery. The SRM-derived peptides were reacted with glutaraldehyde at room temperature for two hours in methanol at a molar amine–aldehyde ratio of 1:8 and others as described in Table 1.

3.2. Characterization of the Glutaraldehyde-Crosslinked Peptides

The products generated from crosslinking of peptides with glutaraldehyde were characterized in order to identify any changes to the structure of the peptides that occurred after the reaction. Theoretically, the crosslinking reaction should have occurred at the amino groups of the peptides to create more thermally stable, higher molecular weight products (Figure 2). Several analytical techniques were used to characterize the peptide–glutaraldehyde-crosslinked products including TGA, SDS-PAGE, and SEC-HPLC. It should also be noted that ATR-FTIR was also performed (Supplemental Figure S1), but since crosslinking of peptides with glutaraldehyde does not result in the formation of new types of chemical bonds, these data cannot be used to assess the occurrence of crosslinking.

3.3. TGA of Glutaraldehyde-Crosslinked Peptides

The first analysis that was used to characterize the glutaraldehyde-crosslinked peptides was TGA (Figure 3). The onset of degradation can be determined by finding the intercept of two extrapolated lines, one from the flat base before decomposition, and the other from the sharpest slope, as described by ASTM standard method E2550–17. For the unmodified peptides, the onset of degradation temperature was 249 °C, while for the glutaraldehyde-crosslinked peptides it was 263 °C. This shift signifies that the glutaraldehyde-crosslinked peptides were more thermally stable and required a higher temperature to break the additional bonds formed during the reaction with glutaraldehyde. Another interesting note is that at 500 °C the unmodified peptides had a lower weight % remaining than the crosslinked peptides, indicating that the material had increased thermal stability at higher temperatures, again likely due to the greater number of chemical bonds in the crosslinked product. The increased bonding suggests that higher molecular weight compounds were formed during the crosslinking of peptides with glutaraldehyde.

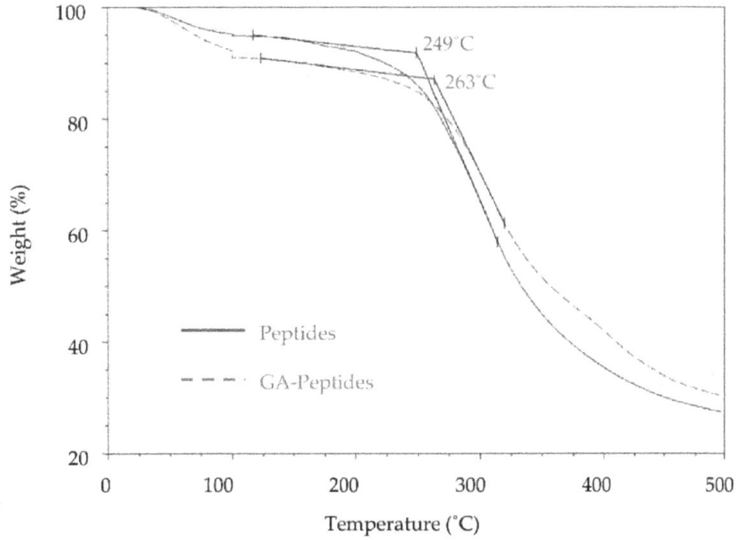

Figure 3. TGA data acquired using peptides (Peptides) or glutaraldehyde-crosslinked peptides (GA-Peptides; 1:8 ratio). The onset of the thermal degradation was determined using the ASTM standard method E2550-17.

3.4. SEC-HPLC Analysis of Glutaraldehyde-Crosslinked Peptides

To confirm the changes in the molecular weight of the peptides after the glutaraldehyde crosslinking reaction, SEC-HPLC was attempted. However, the material recovered after the glutaraldehyde crosslinking reaction was much less soluble. While this physical change further supports the occurrence of crosslinking, it is also problematic because SEC-HPLC requires that the analyte is completely soluble in the mobile phase. Nevertheless, the soluble material present after crosslinking was analyzed (Figure 4). From these data, the soluble portion of the glutaraldehyde-crosslinked peptides are observed to be distributed in a higher molecular weight region than the unmodified peptides. This further suggests that the crosslinking reaction of peptides with glutaraldehyde was successful.

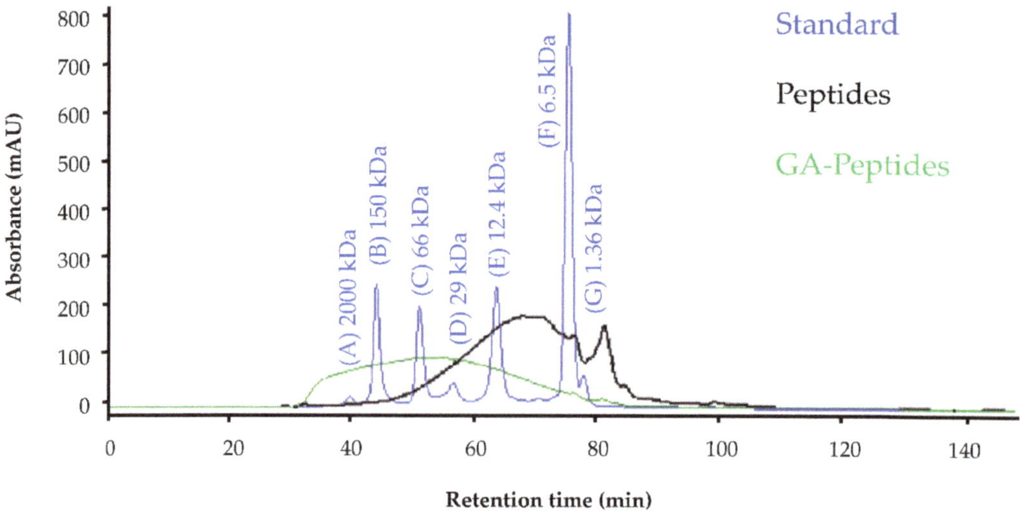

Figure 4. Results from the SEC-HPLC analysis of the peptides (Peptides) and the soluble portion of the glutaraldehyde-crosslinked peptides (GA-Peptides; 1:8 ratio). A 0.15 M Na_2HPO_4 solution was used as the mobile phase using HPLC grade reagents. Standards used to assess the retention times of the products were (A) blue dextran (2000 kDa), (B) alcohol dehydrogenase (150 kDa), (C) albumin (66 kDa), (D) carbonic anhydrase (29 kDa), (E) cytochrome c (12.4kDa), (F) aprotinin (6.5 kDa), and (G) Vitamin B-12 (1.36 kDa).

3.5. SDS-PAGE Analysis of Glutaraldehyde-Crosslinked Peptides

To further confirm the crosslinking of the SRM-derived peptides with glutaraldehyde, a commonly used molecular biology technique, sodium dodecyl sulfate–polyacrylamide gel electrophoresis (SDS-PAGE), was also performed. One of the advantages of SDS-PAGE over SEC-HPLC is that it uses a surfactant, sodium dodecyl sulphate (SDS), which is normally added to help denature the protein structure. It was found that the SDS was also able to improve the solubility of the crosslinked peptides in solution.

A silver staining procedure was used to stain the gel after SDS-PAGE was performed to visualize the samples. As shown in the gel in Figure 5, two loading amounts of both the peptides and glutaraldehyde-crosslinked peptides were analyzed to obtain a better visualization of the products in different molecular weight ranges. As shown in the lane containing 0.01% glutaraldehyde-crosslinked peptide (GA-Peptides; 0.01%), most of the material observed was in the $\geq$75 kDa range, which was much higher than the range observed in the 0.1% peptide treatment lane (Peptides: 0.1% Pep). This further confirmed that the glutaraldehyde-crosslinked peptides had a much higher molecular weight than the unmodified peptides, providing additional evidence that the crosslinking reaction with glutaraldehyde was successful.

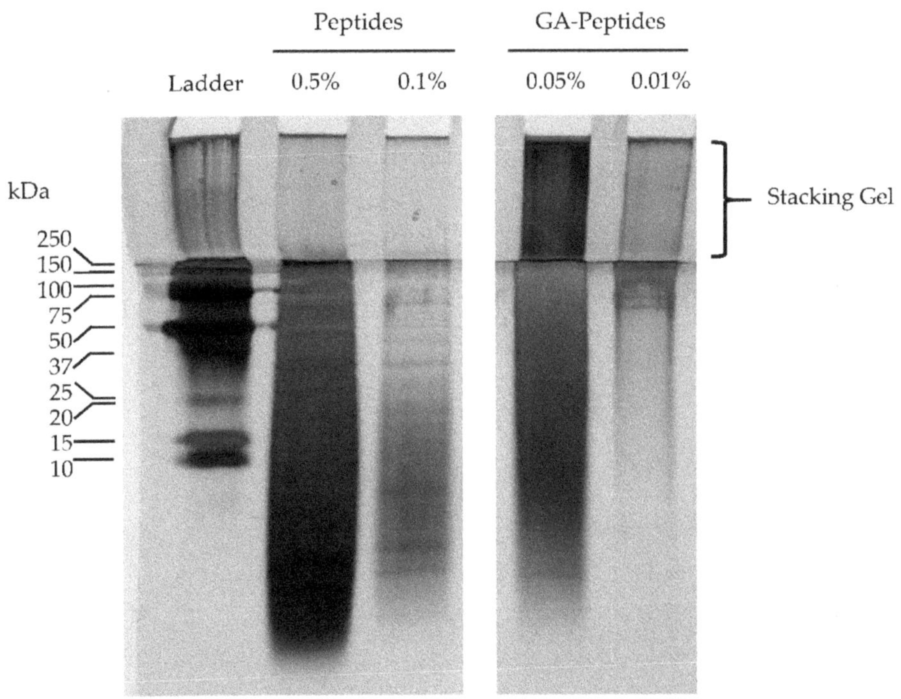

Figure 5. SDS-PAGE of unmodified peptides and glutaraldehyde-crosslinked peptides (1:8 ratio) after silver staining. The samples include unmodified peptides (Peptides) and the glutaraldehyde-crosslinked peptides (GA-Peptides). The 0.5% and 0.05% solutions were prepared by adding 5 mg and 0.5 mg, respectively, of sample to 1 mL of 1X sample buffer. The 0.1% and 0.01% solutions were prepared through 1/5 dilutions of the 0.5% and 0.05% solutions, respectively. Samples were dissolved in sample buffer for 2 h prior to loading of 20 μL in each lane. It should be noted that all lanes shown were from the same gel, but the middles lanes were removed as they contained samples that were not relevant to this study.

Also of note, there is a visible amount of material in the crosslinked peptide samples found in the stacking gel, which likely has a molecular weight that is ≥250 kDa. This material may be part of the insoluble material that was not able to dissolve in the HPLC mobile phase and was therefore not analyzed. This would explain why there was a difference in the range of the molecular weights from the SDS-PAGE and the SEC-HPLC, and thus demonstrates the advantage to using SDS-PAGE when analyzing the glutaraldehyde-crosslinked peptides.

3.6. Flocculation of a Kaolinite Clay Slurry with Glutaraldehyde-Crosslinked Peptides

The crosslinked products that were characterized above were then tested in flocculation experiments. The main objective of this experiment was to determine whether application of the glutaraldehyde-crosslinked peptides to a model kaolinite clay slurry would result in a faster settling rate than the unmodified peptides. As shown in Figure 6 and Supplemental Table S1, after only five minutes, the glutaraldehyde-crosslinked peptides led to a significantly higher amount of settling, with and without gypsum, than the unmodified peptides and the other treatments. In fact, it took 120 min of settling before the unmodified peptides reached a similar settling volume to that achieved after 5 min using the glutaraldehyde-crosslinked peptides. These data clearly demonstrate that crosslinking SRM-derived peptides with glutaraldehyde dramatically improved their settling performance. Even after 48 h, the settling achieved using the glutaraldehyde-crosslinked peptides (either with or without gypsum) was significantly higher than that achieved using the unmodified peptides. Crosslinking of peptides with glutaraldehyde leads to an increase in molecular weight, which results in improved bridging of suspended solid particles. Improved flocculation resulting from an increase in molecular weight of the flocculant has been observed in many other systems described in the literature [12,13,29].

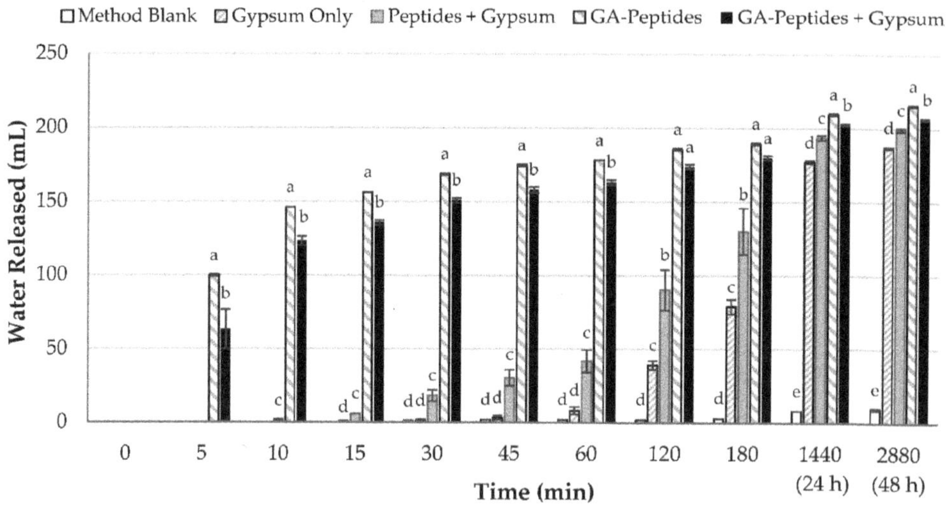

Figure 6. Flocculation using glutaraldehyde-crosslinked peptides (1:8 ratio). A 4% (wt/wt) kaolinite slurry was used. Gypsum was added as a coagulant when necessary, at a concentration of 300 ppm. Flocculants were added at a 3% dosage (wt/wt), and the difference between the mudline and the initial height of the slurry was recorded as the water released. As controls, the kaolinite slurry was subjected to settling tests after no flocculant addition (Method Blank) or with the addition of gypsum alone (Gypsum Only). Unmodified peptides were also used as a flocculant with gypsum (Peptides + Gypsum). For the peptides that were crosslinked with glutaraldehyde, settling experiments were performed in both the absence (GA- Peptides) and presence (GA-Peptides + Gypsum) of gypsum. Experiments were performed in triplicate, except for GA-Peptides, which was performed in duplicate. Within each time point, columns with different letters (a-e) above them are significantly different at a 95% confidence level.

Interestingly, the settling data showed that the peptides crosslinked with glutaraldehyde did not require gypsum as a coagulant to settle the slurry, as is the case with the incumbent industrial flocculant, HPAM. In fact, the peptides crosslinked with glutaraldehyde had a faster settling rate when gypsum was excluded from the settling cylinder (Figure 6). However, when comparing the turbidity data after 48 h of settling, the treatment without gypsum had a much higher turbidity than the treatment with gypsum (Figure 7). This indicates that there is a trade-off between a slightly faster settling rate and a lower final turbidity, when gypsum is absent or present. This trade-off has also been shown in other flocculation studies. Wang et al. (2010) looked at two versions of polyacrylamide (PAM), one cationic and one anionic, for tailings settling [30]. They found that the PAM with the higher initial settling rate (43 m/h) had a higher turbidity (470 NTU) compared to the other PAM, which had slower initial settling rate (20 m/h), but a lower turbidity (68 NTU). When the turbidity of the release water is higher it indicates that there are some particles remaining in solution and, thus, when gypsum is present, fewer particles are remaining in the solution. Ultimately, the use of gypsum will depend on the application, and whether the settling rate or turbidity is the more crucial factor.

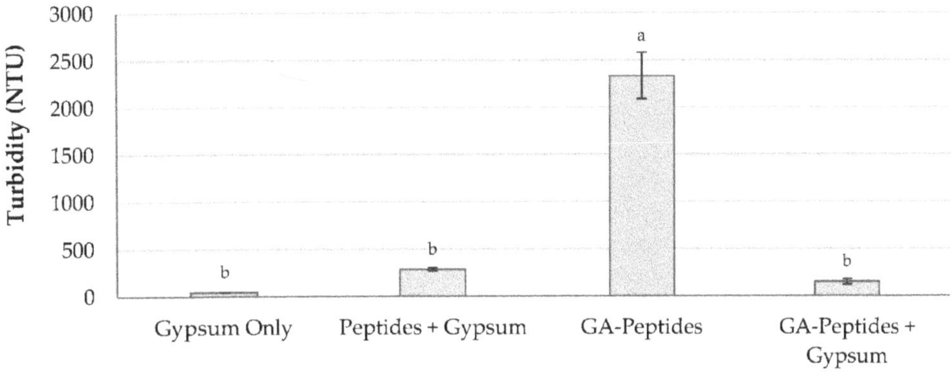

Figure 7. Turbidity after flocculation with glutaraldehyde-crosslinked peptides (1:8 ratio). The turbidity averages of the different treatments after 48 h of settling in a 4% (wt/wt) kaolinite clay system. Gypsum was added as a coagulant at a concentration of 300 ppm. Experiments were performed in triplicate, except for GA-Peptides, which was performed in duplicate. Columns with different letters (a-b) above them have significantly different turbidity at a 95% confidence level.

3.7. *Varying Reactant Ratio during Peptide Crosslinking*

Next, the ratio of the two functional groups present in the reactants, glutaraldehyde and peptides, was differed and the effect on settling performance was studied. In these experiments, the amount of the peptide amino groups was kept consistent and the amount of glutaraldehyde was varied. It should be noted that due to the large number of measurements required to assess flocculation performance in graduated cylinders, particularly within the first 60 min of the experiment, only two formulations (and associated controls, all performed in triplicate) could be tested in a single experiment. Thus, the 1:2 and 1:4 formulations were tested first, followed by the 1:18 and 1:16 formulations, and finally the 1:32 and 1:64 formulations.

As shown in Figure 8A and Supplemental Table S2, the product generated from a 1:4 ratio outperformed that produced using a 1:2 ratio, both in terms of the settling rate and the final water release after 48 h. Conversely, the 1:16 ratio flocculant resulted in a higher amount of settling compared to the 1:8 ratio material for the first 10 min of settling (Figure 8B and Supplemental Table S3). However, after this point there was no longer a statistically significant difference between the groups until 48 h of settling, where the product from the 1:8 ratio induced slightly more settling than that from the 1:16 ratio. Finally, the material generated from a 1:32 ratio outperformed that from the 1:64 ratio with regards to settling rate and final water release (Figure 8C and Supplemental Table S4).

Although the results from the three panels cannot be directly compared, there was a general trend suggesting that increasing the reactants ratio higher than 1:2 results in a product with better flocculation properties; conversely, decreasing the ratio from 1:32 to 1:64 led to a product with reduced flocculation performance. Taken together, these data suggest that the optimal reactant ratio was between 1:2 and 1:64. Furthermore, looking at the data obtained after 5 min of settling, the general trend revealed that as the reactant ratio was increased from 1:2 to 1:32, there was substantial improvement in the settling performance of the crosslinked peptide before falling off again when the reactant ratio was decreased to 1:64. This suggests that a local optimum in settling performance was reached with the 1:32 reactants ratio. It should be noted that other variables during crosslinking and settling experiments (i.e., pH, temperature, etc.) could impact flocculation performance and floc morphology. These variables will be studied in future experiments.

The turbidity of the released water was also measured after 48 h of settling with the flocculants generated using different molar ratios of reactants. As shown in Figure 9A, there was no statistical difference in turbidity observed between the flocculants generated at a 1:2 and 1:4 reactants ratio. A large improvement in turbidity was attained when the ratio was reduced to 1:8 and 1:16, with the latter ratio generating a product with a slightly lower turbidity. Further decreases in the reactant ratio to 1:32 and 1:64 led to successive increases in turbidity, with the 1:64 ratio resulting in the highest turbidity of all ratios tested. Taken together, these data demonstrate that the flocculant derived using a reactants ratio of 1:16 displayed the best water clarity, performing slightly better than that generated from a reactants ratio of 1:8.

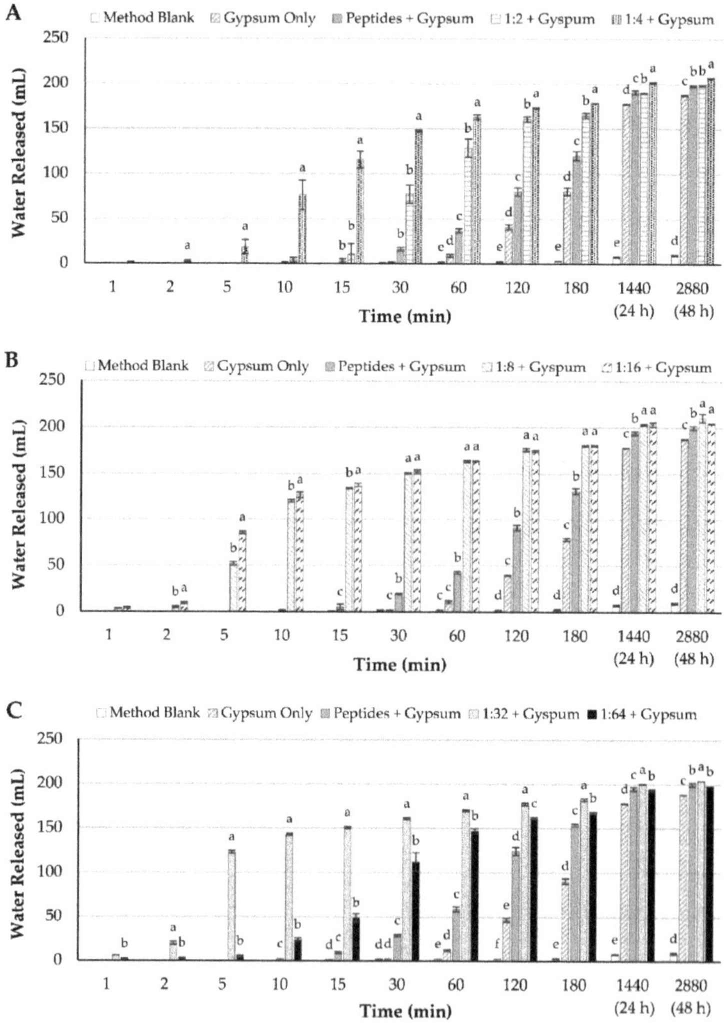

Figure 8. Flocculation performance of glutaraldehyde-crosslinked peptides generated using different ratios of reactant functional groups (amino groups in peptides–aldehyde groups in glutaraldehyde): 1:2 and 1:4 (**A**); 1:8 and 1:16 (**B**); and 1:32 and 1:64 (**C**). Flocculation of a 4% (wt/wt) kaolinite slurry was observed over a 48-h time period. For the method blank, no additional materials were added. Where indicated, gypsum was added as a coagulant at a concentration of 300 ppm. The various flocculants were added at 3% (wt/wt). Unmodified peptides (Peptides) were used as a baseline to assess the impacted of crosslinking. The difference between the mudline and the initial height of the slurry was recorded as the water released. Experiments were performed in triplicate. Within each time point, columns with different letters (a–e) above them are significantly different at a 95% confidence level.

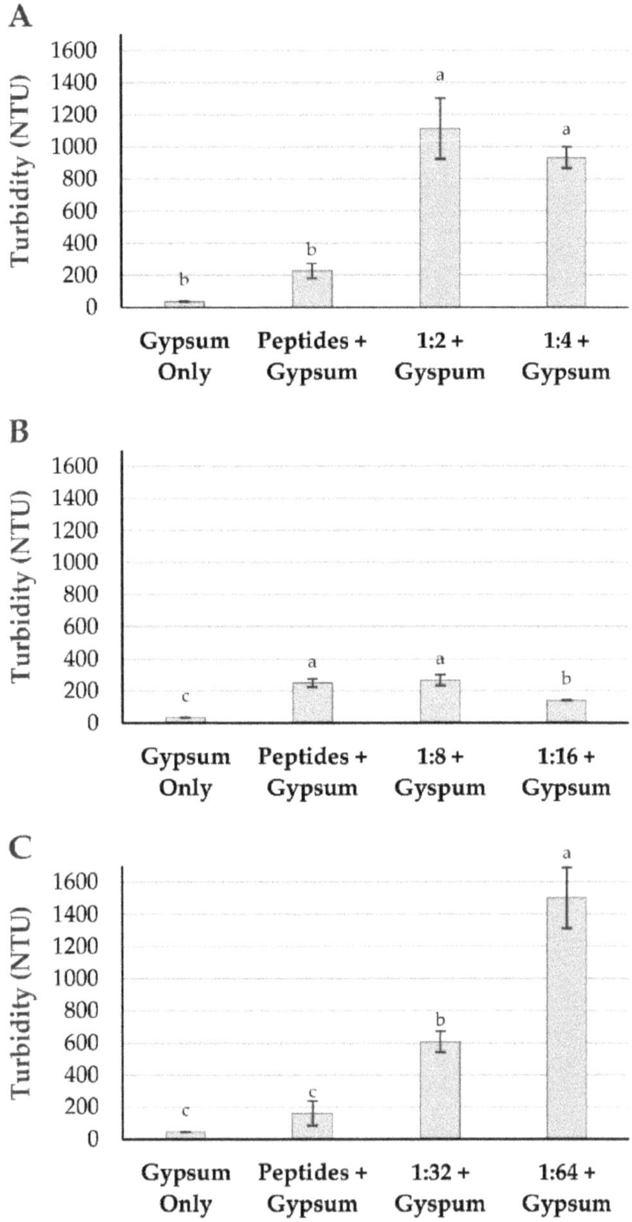

Figure 9. Turbidity of release water after flocculation using unmodified peptides (Peptides) or glutaraldehyde-crosslinked peptides generated using different ratios of reactant functional groups (amino groups in peptides–aldehyde groups in glutaraldehyde): 1:2 and 1:4 (**A**); 1:8 and 1:16 (**B**); and 1:32 and 1:64 (**C**). In flocculation experiments, gypsum was added as a coagulant at a concentration of 300 ppm. Turbidity data are reported as averages of triplicates after 48 h of settling in a 4% (wt/wt) kaolinite clay system. Within each panel, columns with different letters (a–c) above them have significantly different turbidity at a 95% confidence level.

4. Conclusions

This research provides a proof-of-concept that waste materials from the beef rendering industry could be valorized through conversion into an effective bioflocculant for tailings ponds applications. Peptides were successfully crosslinked with glutaraldehyde leading to molecules with substantially higher molecular weight, as determined through liquid chromatography, thermogravimetric analysis, and gel electrophoresis. The settling rates achieved using the crosslinked peptides were significantly enhanced relative to what was observed using the unmodified peptides, particularly within the first hour of the experiment. Furthermore, the release water generated using the crosslinked peptides was significantly less turbid than that obtained using unmodified peptides. Altering the ratio of amino groups in the peptides and aldehyde groups in glutaraldehyde during crosslinking had a significant impact on the settling performance of the resulting flocculant, and also impacted the turbidity of the release water. Thus, selection of the appropriate amino to aldehyde ratio depends on whether the settling rate, final water release, or final turbidity is of greatest importance for a particular application. Taken together, these data demonstrate the great potential of using glutaraldehyde-crosslinked peptides as a bioflocculant for tailings management.

Supplementary Materials: The following are available online at https://www.mdpi.com/article/10.3390/polym13203533/s1.

Author Contributions: Conceptualization, J.Y., V.K., M.C., P.M. and D.C.B.; methodology, J.Y., V.K., M.C. and D.C.B.; validation, J.Y., M.C., P.M. and D.C.B.; formal analysis, J.Y.; investigation, J.Y.; resources, M.C., P.M. and D.C.B.; data curation, J.Y.; writing—original draft preparation, J.Y., V.K. and M.C.; writing—review and editing, J.Y., V.K., M.C., P.M. and D.C.B.; visualization, J.Y.; supervision, M.C. and D.C.B.; project administration, M.C. and D.C.B.; funding acquisition, M.C., P.M. and D.C.B. All authors have read and agreed to the published version of the manuscript.

Funding: This research was funded by the Alberta Prion Research Institute (APRI) of Alberta Innovates (No. 201700003), Alberta Agriculture and Forestry and Results Driven Agriculture Research (RDAR) (No. 2017A004R), and the Natural Sciences and Engineering Research Council of Canada (NSERC) (No. RGPIN-2019-04184).

Institutional Review Board Statement: Not applicable.

Informed Consent Statement: Not applicable.

Data Availability Statement: Not applicable.

Conflicts of Interest: The authors declare no competing financial interest.

References

1. Natural Resources Canada Crude Oil Facts. Available online: https://www.nrcan.gc.ca/science-data/data-analysis/energy-data-analysis/energy-facts/crude-oil-facts/20064 (accessed on 4 June 2021).
2. Allen, E.W. Process water treatment in Canada's oil sands industry: I. Target pollutants and treatment objectives. *J. Environ. Eng. Sci.* **2008**, *7*, 123–138. [CrossRef]
3. Masliyah, J.H.; Czarnecki, J.A.; Xu, Z.; Dabros, M. *Handbook on Theory and Practice of Bitumen Recovery from Athabasca Oil Sands*; Kingsley Publishing Services: Alberta, AB, Canada, 2011; ISBN 9781926832166.
4. O'Gorman, J.V.; Kitchener, J.A. The flocculation and de-watering of kimberlite clay slimes. *Int. J. Miner. Process.* **1974**, *1*, 33–49. [CrossRef]
5. Fleer, G.J.; Lyklema, J. Polymer adsorption and its effect on the stability of hydrophobic colloids. III. Kinetics of the flocculation of silver iodide sols. *J. Colloid Interface Sci.* **1976**, *55*, 228–238. [CrossRef]
6. Feng, L.; Stuart, M.C.; Adachi, Y. Dynamics of polyelectrolyte adsorption and colloidal flocculation upon mixing studied using mono-dispersed polystyrene latex particles. *Adv. Colloid Interface Sci.* **2015**, *226*, 101–114. [CrossRef]
7. Guezennec, A.G.; Michel, C.; Bru, K.; Touze, S.; Desroche, N.; Mnif, I.; Motelica-Heino, M. Transfer and degradation of polyacrylamide-based flocculants in hydrosystems: A review. *Environ. Sci. Pollut. Res.* **2015**, *22*, 6390–6406. [CrossRef] [PubMed]
8. Xiong, B.; Loss, R.D.; Shields, D.; Pawlik, T.; Hochreiter, R.; Zydney, A.L.; Kumar, M. Polyacrylamide degradation and its implications in environmental systems. *npj Clean Water* **2018**, *1*, 17. [CrossRef]
9. Divakaran, R.; Sivasankara Pillai, V.N. Flocculation of kaolinite suspensions in water by chitosan. *Water Res.* **2001**, *35*, 3904–3908. [CrossRef]

10. Renault, F.; Sancey, B.; Charles, J.; Morin-Crini, N.; Badot, P.M.; Winterton, P.; Crini, G. Chitosan flocculation of cardboard-mill secondary biological wastewater. *Chem. Eng. J.* **2009**, *155*, 775–783. [CrossRef]
11. Mekonnen, T.H.; Mussone, P.G.; Stashko, N.; Choi, P.Y.; Bressler, D.C. Recovery and characterization of proteinacious material recovered from thermal and alkaline hydrolyzed specified risk materials. *Process Biochem.* **2013**, *48*, 885–892. [CrossRef]
12. Zhu, Y.; Gong, Y.; Kaminsky, H.; Chae, M.; Mussone, P.; Bressler, D.C. Using Specified Risk Materials-Based Peptides for Oil Sands Fluid Fine Tailings Management. *Materials* **2021**, *14*, 1582. [CrossRef]
13. Jankovics, L. Effect of agitation and molecular weight on polymer adsorption and deflocculation. *J. Appl. Polym. Sci.* **1965**, *9*, 545–552. [CrossRef]
14. Korn, A.H.; Feairheller, S.H.; Filachoine, E.M. Glutaraldehyde: Nature of the reagent. *J. Mol. Biol.* **1972**, *65*, 525–529. [CrossRef]
15. Okuda, K.; Urabe, I.; Yamada, Y.; Okada, H. Reaction of glutaraldehyde with amino and thiol compounds. *J. Ferment. Bioeng.* **1991**, *71*, 100–105. [CrossRef]
16. Canadian Food Inspection Agency Meat Hygiene Manual of Procedures. Available online: https://inspection.canada.ca/food-safety-for-industry/archived-food-guidance/meat-and-poultry-products/manual-of-procedures/chapter-5/eng/1395150894222/1395150895519?chap=4 (accessed on 29 August 2021).
17. Canadian Food Inspection Agency Evaluation of the Canadian Food Inspection Agency's Enhanced Feed Ban Program. Available online: https://inspection.canada.ca/about-cfia/transparency/corporate-management-reporting/audits-reviews-and-evaluations/enhanced-feed-ban-program/report/eng/1373917327288/1373917408136 (accessed on 30 August 2021).
18. Shui, T.; Adhikari, B.B.; Chae, M.; Bressler, D.C. Evaluation of thermally hydrolyzed specified risk materials cross-linked with glutaraldehyde for tackifier applications. *Prog. Org. Coatings* **2020**, *140*, 105535. [CrossRef]
19. Shui, T.; Khatri, V.; Chae, M.; Sokhansanj, S.; Choi, P.; Bressler, D.C. Development of a torrefied wood pellet binder from the cross-linking between specified risk materials-derived peptides and epoxidized poly (vinyl alcohol). *Renew. Energy* **2020**, *162*, 71–80. [CrossRef]
20. Shui, T.; Chae, M.; Bressler, D.C. Cross-Linking of Thermally Hydrolyzed Specified Risk Materials with Epoxidized Poly (Vinyl Alcohol) for Tackifier Applications. *Coatings* **2020**, *10*, 630. [CrossRef]
21. Adhikari, B.B.; Kislitsin, V.; Appadu, P.; Chae, M.; Choi, P.; Bressler, D.C. Development of hydrolysed protein-based plywood adhesive from slaughterhouse waste: Effect of chemical modification of hydrolysed protein on moisture resistance of formulated adhesives. *RSC Adv.* **2018**, *8*. [CrossRef]
22. Adhikari, B.B.; Appadu, P.; Kislitsin, V.; Chae, M.; Choi, P.; Bressler, D.C. Enhancing the Adhesive Strength of a Plywood Adhesive Developed from Hydrolyzed Specified Risk Materials. *Polymers* **2016**, *8*, 285. [CrossRef]
23. Mekonnen, T.H.; Mussone, P.G.; Choi, P.; Bressler, D.C. Development of proteinaceous plywood adhesive and optimization of its lap shear strength. *Macromol. Mater. Eng.* **2015**, *300*, 198–209. [CrossRef]
24. Migneault, I.; Dartiguenave, C.; Bertrand, M.J.; Waldron, K.C. Glutaraldehyde: Behavior in aqueous solution, reaction with crosslinking. *Biotechniques* **2004**, *37*, 790–802. [CrossRef]
25. Broun, G.B. Chemically aggregated enzymes. In *Immobilized Enzymes*; Academic Press: Cambridge, MA, USA, 1976; Volume 44, pp. 263–280, ISBN 0076-6879.
26. Schägger, H. Tricine–SDS-PAGE. *Nat. Protoc.* **2006**, *1*, 16. [CrossRef] [PubMed]
27. Thermo Fisher Scientific Inc. *PierceTM Silver Stain Kit: Instructions*; Thermo Fisher Scientific Inc.: Waltham, MA, USA, 2016.
28. Nasser, M.S.; James, A.E. The effect of polyacrylamide charge density and molecular weight on the flocculation and sedimentation behaviour of kaolinite suspensions. *Sep. Purif. Technol.* **2006**, *52*, 241–252. [CrossRef]
29. Vajihinejad, V.; Gumfekar, S.P.; Bazoubandi, B.; Rostami Najafabadi, Z.; Soares, J.B.P. Water Soluble Polymer Flocculants: Synthesis, Characterization, and Performance Assessment. *Macromol. Mater. Eng.* **2019**, *304*, 1800526. [CrossRef]
30. Wang, X.W.; Feng, X.; Xu, Z.; Masliyah, J.H. Polymer aids for settling and filtration of oil sands tailings. *Can. J. Chem. Eng.* **2010**, *88*, 403–410. [CrossRef]

polymers

MDPI

Article

Comparative Study on Different Modified Preparation Methods of Cellulose Nanocrystalline

Xinhui Wang [1], Na Wang [1], Baoming Xu [1], Yili Wang [1], Jinyan Lang [1], Junliang Lu [1,2], Guorong Chen [3] and Heng Zhang [1,2,3,*]

1 College of Marine Science and Biological Engineering, Qingdao University of Science & Technology, Qingdao 266042, China; wxhameq@163.com (X.W.); wlalala21@163.com (N.W.); 14763738886@163.com (B.X.); 15689122781@163.com (Y.W.); ljy17806248212@163.com (J.L.); juling_lu@163.com (J.L.)
2 Guangdong Provincial Key Lab of Green Chemical Product Technology, Guangzhou 510640, China
3 Fujian Provincial Key Laboratory of Fire Retardant Materials, Xiamen 361005, China; grchen@xmu.edu.cn
* Correspondence: hgzhang@qust.edu.cn; Tel.: +86-1595-480-3800

Abstract: Different modification process routes are used to improve the modified cellulose nanocrystalline (MCNC) with higher fatty acid by esterification reaction and graft polymerization to obtain certain hydrophobic properties. Two preparation methods, product structure and surface activity, are compared and explored. Experimental results show that the modified product is still at the nanometer level and basically retains the crystal structure of the raw cellulose nanocrystalline (CNC). The energy consumption of the two preparation methods is low; however, the esterification method with co-reactant requires short reaction time, and the degree of substitution of the product is high. The modified product prepared by grafting polymerization method has a high HLB value and amphiphilicity, which can effectively reduce the surface tension of water. Therefore, it can be used as a green and environmentally friendly surface-active substance.

Keywords: CNC; esterification reaction; graft copolymerization; hydrophobic modification

Citation: Wang, X.; Wang, N.; Xu, B.; Wang, Y.; Lang, J.; Lu, J.; Chen, G.; Zhang, H. Comparative Study on Different Modified Preparation Methods of Cellulose Nanocrystalline. *Polymers* **2021**, *13*, 3417. https://doi.org/10.3390/polym13193417

Academic Editor: Arn Mignon

Received: 12 September 2021
Accepted: 1 October 2021
Published: 5 October 2021

1. Preface

Nanocellulose is a degradation product of cellulose, and its basic structural unit is anhydroglucose. Nanocellulose carries a large number of hydroxyl groups. It not only inherits some of the cellulose characteristics, such as hydrophilicity, renewability, and degradability, but also has the characteristics of nanomaterials, such as high specific surface area, hyperfine structure, and high transparency. Therefore, nanocellulose is a new type of green environmental protection material with great development prospects [1–3].

The hydroxyl group carried by nanocellulose provides excellent modification sites. Therefore, the physical or chemical modification of nanocellulose can provide new functions and prepare new nanocellulose functional materials. Most modification techniques of cellulose are also applicable to nanocellulose given that the basic structural units of nanocellulose and cellulose are the same. Among them, physical adsorption is a commonly used polymer modification scheme. This modification method is convenient to operate, and the entire production process is simple and can effectively retain the integrity of nanocellulose. Qing et al. [4] used cetyltrimethylammonium bromide to adsorb on nanocellulose for hydrophobic drug delivery. The product has been tested, and the result proves that it has stable dispersibility and stability in organic solvents, and the raw material nanocellulose has improved in hydrophobicity. Shimizu et al. [5] made 2, 2, 6, 6-tetramethylpiperidine-1-oxy oxidized cellulose nanofibers and quaternary alkylammonium adsorbate by casting method to produce hydrophobically modified nanocellulose film. The results of hydrophobic tests show that the contact angle between the product and water has increased from the original 50° to 100°. This method improves the hydrophobicity of the product and can simply and effectively achieve the hydrophobic modification of the product. However, nanocellulose and the introduced functional groups are combined in the form of van der Waals forces or hydrogen bonds, which is instability. Because the modified products of physical modification have weak shear resistance, and the adsorbed substances are easy to dissociate under a certain external force [6].

The stability of chemically modified nanocellulose is better because the strength of chemical bonds is higher than van der Waals forces and hydrogen bonds. The commonly used chemical modification methods include oxidation, silanization, cationization, esterification, and graft copolymerization. Among them, TEMPO (2,2,6, 6-tetramethylpiperidine-1-oxygen radical) oxidation modification can oxidize the hydroxyl methyl on the surface of nanocellulose to carboxyl group and improve its water solubility. In modified polysaccharide, only the hydroxymethyl on the surface of the polysaccharide is oxidized without affecting the hydroxyl groups on the surface [7]. Araki et al. [8] used TEMPO-modified hydrochloric acid hydrolyzed nanocellulose to prepare homogeneous aqueous suspension. Silanization is widely used in surface chemical modification of cellulose whiskers, and the dispersion of the modified nanoparticles in organic solvents is improved; however, the morphology of the modified nanoparticles slightly changes, and a swelling phenomenon occurs [9]. Grunert et al. [10] modified cellulose nanocrystalline (MCNC) with trimethylsilane and compounded with cellulose acetate butyrate. The modified nanocellulose can evidently enhance the energy storage modulus of cellulose acetate butyrate. The hydroxyl groups of nanocellulose can be cationically modified by derivatization reaction. Zaman et al. [11] modified the nanocellulose hydrolyzed by sulfuric acid with glycidyl trimethylammonium chloride and adjusted the surface charge density of the modified product by changing the water content in the reaction system to be better dispersed in the water system.

Esterification and graft copolymerization are common methods for chemical modification of nanocellulose. Esterification is a chemical reaction between nanocellulose and organic acids or inorganic acids. The reaction is essentially the reaction of hydroxyl and carboxyl groups to generate ester group, sometimes with acyl halide and anhydride as the reactants; however, the final result is the generation of ester group. Lakovaara et al. [12] obtained high strength and sustainable films with increased hydrophobicity and good mechanical properties by the esterification of cellulose nanofiber and all-cellulose composite films with n-octylsuccinic anhydride. In addition, the experimental method is relatively mild. Beaumont et al. [13] presented a general concept of wet surface esterification of cellulose using acyl imidazoles, which means acetyl groups are introduced directly onto never-dried, water-swollen cellulose fibres. This method can obtain cellulose fibers with higher hydrophobicity in simple and gentle process. Shang et al. [14] prepared super-hydrophobicity/super-oleophilicity with good adsorption capacity through the freeze-drying process and subsequent esterification. The reaction process of it was facile and environmental-friendly approach. Spinella et al. [15] prepared functionalized modified product by direct melt blending of nanocellulose and lactic acid through Fischer esterification reaction, which improved the dispersion of original nanocellulose in polymer matrix, increased the contact angle with water, and expanded the application range of this material. Yuan et al. [16] acetylated nanocellulose by using alkyl succinic anhydride. After modification, the hydrophobicity of nanoparticles was significantly improved and was easily dispersed into polar solvents with different dielectric constants. Menezes et al. [17] reacted the nanocellulose with organic fatty acid chlorides of different hydrocarbon chain lengths to prepare esterification-modified products with high substitution degree and high graft density. Berlioz et al. [18] reacted BNC with hexadecyl chloride vapor to obtain the same esterified product of nanocellulose with high substitution degree. This reaction is a highly efficient solvent-free esterification reaction, and the crystal structure of the modified nanocellulose is insignificantly changed. Eyholzer et al. [19] used sulfuric acid as the catalyst to transfer the reaction of ester group of nanocellulose in hexane–alcohol medium, and hexyl monomer replaced carboxymethyl through carboxymethylation.

Graft copolymerization can directly react the functional groups of the graft with the active hydroxyl group of the nanocellulose to form the modified products, or it can initially react with the active hydroxyl group of the nanocellulose by the small molecular monomer, and then perform chemical modification by the polymerization of the small molecular monomer grafted on the nanocellulose, namely, chain growth. Goffin et al. [20] grafted PLA on the surface of cellulose nanocrystalline through ring-opening polymerization. The length of grafting chain in the modified products is short due to the crystallization of PLA on the surface of nanocellulose. Westlund et al. [21] prepared graft products with special functions by modifying nanocellulose with tris-[2-(dimethylamino) ethyl] amine as ligand and 11-(4′-cyanophenyl-4″-phenoxy) undecyl acrylonitrile and methyl acrylate as monomers. Lee et al. [22] grafted acrylamide onto the surface of nanocellulose by ultraviolet light initiation by using benzophenone as initiator. The grafting ratio of modified products increased with the increase in initiator concentration. The introduction of acrylamide chain improved the dispersion ability of nanocellulose. Morandi et al. [23] grafted polyphenylene onto the surface of nanocellulose by using ethyl bromoisobutyrate as initiator. The modified products showed chiral nematic structure in thermotropic and lyotropic liquid crystal and had strong adsorption for 1,2, 4-trichlorobenzene.

Among nanocellulose modification methods, esterification and graft copolymerization have more mature reaction routes and do not require harsh reaction conditions; thus, they have the most industrial application prospects. In our previous studies, we investigated the graft copolymerization method of nanocellulose, grafting high fatty acids with $FeSO_4/H_2O_2$ system or $2(NH_4)_2SO_4{\cdot}Ce(SO_4)_2{\cdot}4H_2O$ as initiator. Under the same reaction conditions, the initiation efficiency of $FeSO_4/H_2O_2$ was better than that of tetravalent cerium ion in ammonium cerium sulfate, and the degree of substitution (DS) of the obtained product was relatively high [24–26]. However, only a few studies compared the esterification method with that using co-reactant. The esterification reaction modified cellulose nanocrystalline has the characteristics of facile and eco-friendly. Therefore, low reaction temperature and green reagent were used in this experiment. Nanocellulose and higher fatty acids were used as raw materials to graft alkanes onto nanocellulose by introducing co-reactants and initiators, respectively. The performance of reaction products was evaluated by comparing the degree of difficulty of the two reaction types. The differences in surface activity of the products were analyzed on the basis of the reaction mechanism, thereby providing a feasible process route for the modification of nanocellulose.

2. Preparation and Characterization of Cellulose Nanocrystalline (CNC) and Modified Products

2.1. Chemicals and Materials

Trifluoroacetic anhydride was purchased from Shanghai Aladdin Biochemical Technology Co., Ltd. Ferrous sulfate was obtained from China Shanghai Pierce Chemical Reagent Co., Ltd. Hydrogen peroxide at 30% (*w*/*w*) was purchased from Tianjin Dingshengxin Chemical Co., Ltd. Chloroform, sodium hydroxide, and absolute ethanol were purchased from Sinopharm Chemical Reagent Co., Ltd. Lauric acid was obtained from Tianjin BASF Chemical Co., Ltd. Palmitic acid was obtained from Tianjin Bodi Chemical Co., Ltd. Stearic acid was purchased from Tianjin Beichen Fangzheng Reagent Factory. Concentrated sulfuric acid, hydrochloric acid, was purchased from Yantai Sanhe Chemical Reagent Co., Ltd. These chemicals are all analytical pure. Pharmaceutical grade microcrystalline cellulose was purchased from Chengdu Kelong Chemical Reagent Factory. The MD3500 dialysis bag was purchased from an American company, Viskase.

2.2. Preparation of CNC

Concentrated sulfuric acid aqueous solution at 60% (*w*/*w*) was prepared and naturally cooled. It was directly added to a flask containing microcrystalline cellulose (MCC) when its temperature dropped to approximately 45 °C. The magnetic heater was heated to 45 °C, and the speed was adjusted to 500 rpm. Timing after the mixture is evenly stirred, the whole reaction is protected by nitrogen. After 2.5 h, heating was stopped, the mixed solution was transferred to a beaker, and a large amount of deionized water was added to terminate the acid-catalyzed hydrolysis reaction. Finally, the milky white product mixed with microcrystalline cellulose and CNC obtained after adding deionized water was transferred to a 50 mL centrifuge tube. The turbid liquid was centrifuged at the speed of 5000 rpm to separate the upper layer of acid-containing waste liquid, and deionized water was added to continuously dilute the concentration of the sulfuric acid aqueous solution. This process is repeated until the supernatant liquid became turbid, and then the supernatant liquid containing the desired CNC was collected.

The collected upper layer of CNC colloid was concentrated by vacuum distillation, and then transferred to the dialysis bag (molecular weight 3500) under 30 °C deionized water environment with magnetic stirring; dialysis was repeated until the pH became neutral. Finally, a transparent and light blue glowing CNC colloid was obtained.

2.3. *Preparation of MCNC*

2.3.1. Preparation of MCNC by Esterification Reaction

Trifluoroacetic anhydride was used as the co-reactant to activate three higher fatty acids in situ to produce a mixed acid anhydride intermediate. Then, the dried CNC powder was added to a four-necked flask containing mixed acid anhydride intermediate to generate MCNC at different temperatures and reaction times. The system should be anhydrous and ethanol-free because trifluoroacetic anhydride is chemically active and can react with water and absolute ethanol. Through preliminary experiments and exploration, chloroform was selected as the liquid phase system of the reaction. The entire reaction process was under nitrogen protection. When the reaction was over, the mixture was cooled to room temperature, and the substances that did not participate in the reaction were removed by alternate washing with absolute ethanol and water, and the resulting product was separated. Then, ethanol was used as the solvent to extract the product using a Soxhlet extractor. Finally, the product was vacuum dried at 40 °C. The reaction mechanism is shown in Equation (1).

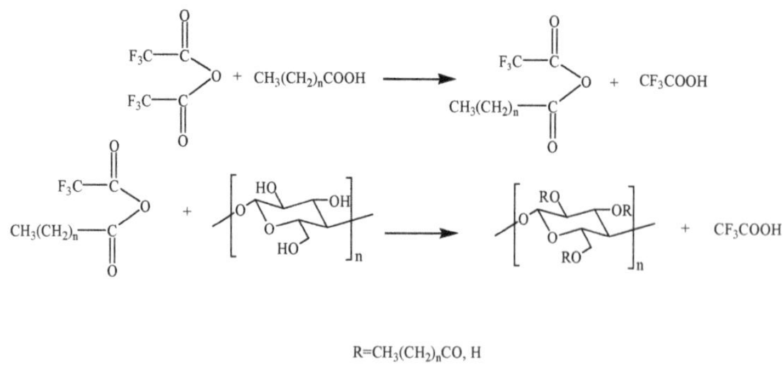

(1)

2.3.2. Preparation of MCNC by Graft Copolymerization

An appropriate amount of self-made CNC colloidal solution was obtained and mixed with an ethanol solution mixed with dissolved fatty acids. The reaction was carried out in a flask, stirring and mixing continuously for 0.5 h in the water bath environment at 50 °C to obtain even mixture. Then, a quantitative initiator was added, and the reaction temperature and time were changed to carry out the chemical modification reaction of CNC. The entire reaction process was under nitrogen protection. After the reaction, a mixed suspension of MCNC was obtained, and the suspension was separated and washed alternately with alcohol and water. Finally, the product was vacuum dried at a temperature of 40 °C.

In the process of MCNC preparation with $FeSO_4/H_2O_2$ as the initiator, Fe^{2+} and H_2O_2 undergo a redox reaction, thereby triggering H_2O_2 to generate active free radicals. The free radicals cause fatty acids and CNC to form two active free radicals. Finally, these active free radicals react with each other to complete the chemical modification of CNC [27–29]. The reaction mechanism of this process is shown in Equation (2).

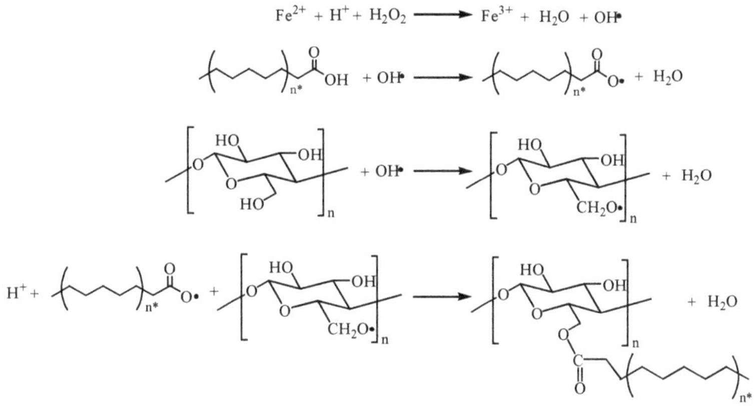

(2)

2.4. Test Method

2.4.1. Determination and Calculation of DS of MCNC

The synthesized MCNC was repeatedly washed with ethanol and deionized water, filtered, and dried to obtain the product, and the DS was determined by the saponification method [30,31].

2.4.2. X-ray Diffraction

CNC and MCNC powder samples were obtained for X-ray diffraction measurement. At room temperature, the measured substance was tested with a DX-2700 XRD tester (Bruker Corporation, Ettlingen, Germany) at a scanning rate of 0.02°/min. The diffraction setting parameters are as follows: Cu target, 40 kV, and 30 mA; the diffraction angle 2θ range is set to 10°–60°.

The crystallinity index CrI value can be calculated by Segal equation (Equation (3)) [32], as follows:

$$\text{CrI} = \frac{\text{I}_{002} - \text{I}_{\text{am}}}{\text{I}_{002}} \times 100\% \tag{3}$$

(CrI: relative crystallinity index; I_{002}: maximum intensity of 002 lattice diffraction; I_{am}: diffraction intensity of the same unit at 2θ = 18°)

2.4.3. Determination of Particle Size

CNC colloid and MCNC suspension are diluted and ultrasonically dispersed for testing. Malvern Zetasizer Nano-ZS90-type Malvern laser particle size analyzer (Malvern (China) instrument Co. Ltd., Shanghai, China) is used. The particle refractive index is set to 1.470, water is used as the dispersant, the temperature of the test system is 25 °C, the count rate is 349.8 kcps, and the time is set to 60 s.

2.4.4. Calculation of the HLB Value of MCNC

Equation (4) is used to calculate the HLB value of the MCNC.

$$\begin{aligned} HLB &= \frac{\text{Hydrophilic group quality}}{\text{Surfactant quality}} \times \frac{100}{5} \\ &= \frac{\text{The quality of the hydrophilic group}}{\text{Lipophilic base quality} + \text{Hydrophilic group quality}} \times \frac{100}{5} \\ &= \frac{\text{A} - \text{DS}}{(\text{B}-1) \times DS + A} \times \frac{100}{5} \end{aligned} \tag{4}$$

(A: relative molecular mass of glucoside unit; B: relative molecular mass of fatty acid groups ($C_{12}H_{23}O$, $C_{16}H_{33}O$, $C_{18}H_{37}O$) connected to CNC in the form of ester bonds; DS: degree of substitution of MCNC).

2.4.5. Determination of Surface Tension

The CNC and MCNC test liquids with different contents are configured, and the surface tension of the QBZY-2 automatic surface tension meter (Shanghai Fangrui Instrument Co., Ltd., Shanghai, China) was used to determine the surface tension.

3. Results and Discussion

3.1. Factors Affecting the DS of MCNC

The saponification method was used to determine the DS to explore the influence of different reaction conditions on the DS of the modified product. The co-reactant or initiator, three higher fatty acids, and CNC are combined, and the corresponding temperature change interval and time change interval are selected to prepare MCNC under different reaction conditions.

3.1.1. Effect of Co-Reactant Content on the DS of MCNC

Trifluoroacetic anhydride acts as a co-reactant to react with three types of higher fatty acids to generate active intermediates, namely, mixed acid anhydrides. MCNC was prepared by the reaction of mixed anhydride and CNC. The amount of fatty acid was the same as that of the co-reactant according to the esterification reaction mechanism. When lauric acid was used as the grafting substance, the temperature should be controlled at 50 °C, and the reaction time should be 5 h. When palmitic acid and stearic acid were used as grafting substances, the reaction temperature was controlled at 55 °C and the reaction time was 6 h. The test results are as follows.

Figure 1 shows that the DS of MCNC is related to the feed ratio of the substances participating in the reaction and the hydrocarbon chain length of the hydrophobic groups of MCNC. As the adding ratio of anhydroglucose unit (AGU), co-reactant and fatty acid increases, the DS of MCNC gradually increases. When the feeding ratio reaches 1:3:3, the curve tends to be flat. On the basis of the esterification reaction mechanism, the increase in the feed ratio increases the concentration of the mixed acid anhydride in the reaction system, thereby increasing the probability of reacting with the CNC hydroxyl group. Thus, the degree of product substitution increases. However, when the feeding ratio is excessive, although the concentration of the mixed acid anhydride still increases on a macro scale, the contact concentration of the mixed acid anhydride with the hydroxyl group gradually becomes saturated; thus, the DS does not change evidently. The DS of MCNC-C12 is higher than that of MCNC-C16 and MCNC-C18; it is determined by the steric hindrance of higher saturated fatty acids. With the continuous increase in fatty acid hydrocarbon chain length, its steric hindrance effect gradually becomes evident, inhibiting the increase in substitution degree. Therefore, the DS of the three MCNC from high to low is MCNC-C12, MCNC-C16, and MCNC-C18, under the same conditions.

3.1.2. Effect of Initiator and Content on the DS of MCNC

$FeSO_4/H_2O_2$ was selected as the initiator, the content of the added initiator was changed at a specific reaction temperature and time, and lauric acid, palmitic acid, and stearic acid were used to react with the CNC. CNC reacted with lauric acid in the 50 °C water bath environment for 8 h, CNC and palmitic acid reacted in the 55 °C water bath environment for 20 h, and CNC reacted with stearic acid in the 60 °C water bath environment for 24 h. Different initiator types were added, and the addition amount was changed to prepare MCNC with different DS under these reaction conditions. The DS is measured, and the results are shown in the figure.

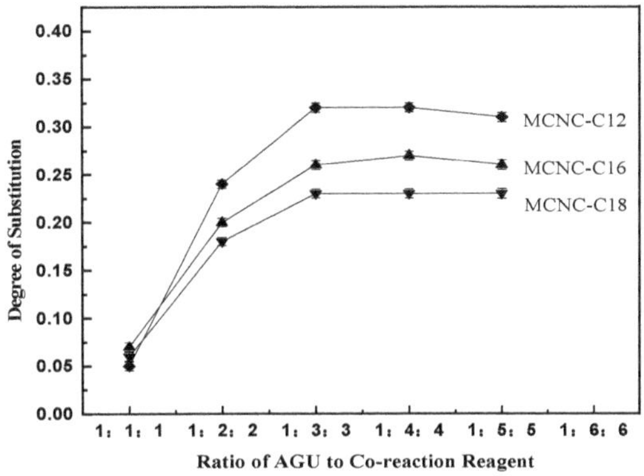

Figure 1. Effect of the ratio of co-reaction reagent on substitution degree of MCNC. (MCNC-C12: MCNC grafted with lauric acid; MCNC-C16: MCNC grafted with palmitic acid; MCNC-C18: MCNC grafted with stearic acid.)

Figure 2 shows that under the $FeSO_4/H_2O_2$ redox initiation system, as the amount of initiator increases, the DS of lauric acid, palmitic acid, and stearic acid esterified CNC initially increases, and then gradually tends to be flat. When the amount of initiator is 0.8%, lauric acid as the graft can obtain the best DS of 0.24 under this preparation method. In addition, the amount of initiator continues to increase, and the DS tends to be stable. The substitution effect of palmitic acid and stearic acid is the same as that of lauric acid. The best DS is obtained when the dosage of the initiator is 1.2%. The substitution degrees of palmitic acid and stearic acid are 0.16 and 0.12, respectively.

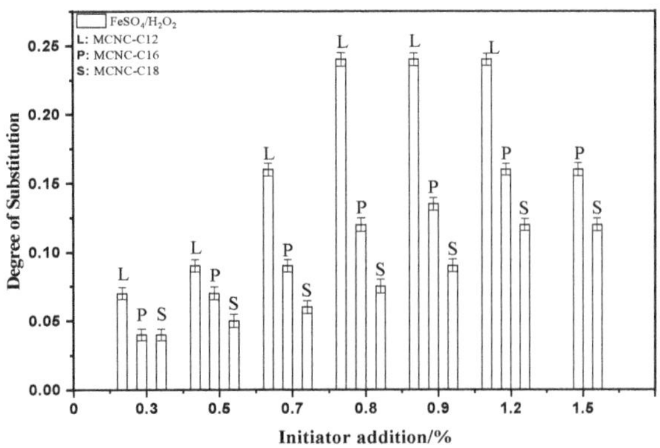

Figure 2. Effect of initiator type and dosage on substitution degree of MCNC. (MCNC-C12: MCNC grafted with lauric acid; MCNC-C16: MCNC grafted with palmitic acid; MCNC-C18: MCNC grafted with stearic acid.)

The main reason for this phenomenon is the steric hindrance of the system on the glucoside unit. When the amount of initiator is not optimal, the initiation efficiency of the reaction system gradually increases as the content of initiator increases, the concentration of initiated active free radicals increases, and the degree of reaction substitution increases. In addition, the $FeSO_4/H_2O_2$ system generates free radicals on the primary hydroxyl group at the C_6 position of the glucoside unit. When the amount of initiator is higher than the optimal amount, the DS of the product tends to be stable. At this time, although increasing the amount of initiator can still increase the concentration of free radicals, the space steric hindrance effect of higher fatty acids greatly affects the esterification efficiency between free radicals, and the steric hindrance effect increases with the increase in the length of the fatty acid hydrocarbon chain, thereby limiting the continued increase in the DS of the product.

Therefore, the steric hindrance effect of higher fatty acid becomes an important factor affecting the DS when the free radical content reaches a certain concentration. Under the $FeSO_4/H_2O_2$ system, the DS of lauric acid, palmitic acid, and stearic acid MCNC also decreases.

3.1.3. Effect of Reaction Time on Substitution Degree of MCNC

Lauric acid was used as raw material, and trifluoroacetic anhydride was used as co-reaction agent; they are mixed in a molar ratio of 1:1. The mixture was prepared at 4 h, 5 h, 6 h, 7 h, and 8 h, by changing the reaction time in a water bath at 50 °C with the molar ratio of 3:1 to CNC. The preparation method of palmitic acid and stearic acid as raw materials was similar to that of lauric acid, but the reaction temperature was raised to 55 °C. After the reaction, an appropriate amount of stearic acid MCNC was obtained, and the DS was tested by saponification method.

With 0.8% $FeSO_4/H_2O_2$ as the initiator, lauric acid was reacted with CNC in a water bath at 50 °C to prepare MCNC-C12. The reaction time was set at 6 h, 7 h, 8 h, 9 h, and 10 h, according to the same experimental method to prepare MCNC-C16 and MCNC-C18. The reaction temperature was set to 55 °C, and 1.2% $FeSO_4/H_2O_2$ was selected as the initiator. The reaction times of 16 h, 18 h, 20 h, 22 h, and 24 h were set for the graft copolymerization of CNC with palmitic acid as the raw material. The reaction temperature was set to 55 °C, and 1.2% $FeSO_4/H_2O_2$ was selected as the initiator. The reaction times of 16 h, 18 h, 20 h, 22 h, 24 h, and 25 h were set for the graft copolymerization of CNC with stearic acid as the raw material. After the reaction, an appropriate amount of stearic acid MCNC was obtained, and the DS was tested by saponification method. The test results are as follows.

Figure 3 shows that under the same conditions, the DS of lauric acid was higher than that of palmitic acid, whereas that of stearic acid was the lowest. The esterification produced the mixture of anhydrides with higher activity than fatty acids in the first place; thus, its substitution degree is higher than the graft copolymerization under the same conditions. Moreover, the time interval for the preparation of three fatty acid esterified CNC by this method is more concentrated. However, in the graft copolymerization, the reaction time was longer because the steric hindrance caused by the over-long hydrocarbon chains of palmitic acid and stearic acid was much higher than that of lauric acid. In addition, their reaction activity was lower than that of anhydride. In general, the reaction time of the modified products of lauric acid was lower than that of the modified products of the two other acids. In summary, in the system with trifluoroacetic anhydride as co-reactive agent, the optimal reaction time with lauric acid as the raw material is 5 h, palmitic acid as the raw material is 6 h, and stearic acid as the raw material is 6 h. However, in the initiator system, the optimal reaction time of the three fatty acids was 8 h, 20 h, and 24 h.

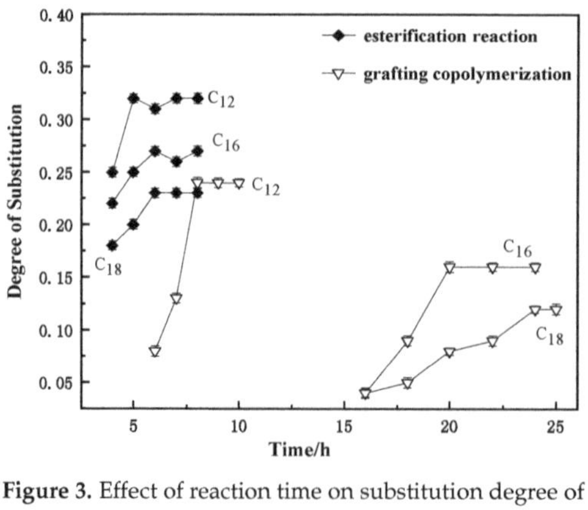

Figure 3. Effect of reaction time on substitution degree of MCNC.

3.1.4. Effect of Temperature on the DS of MCNC

Trifluoroacetic anhydride was considered the co-reactant, and CNC was mixed at a molar ratio of 3:1. The amounts of fatty acids involved in the reaction were the same as that of the co-reactant. When lauric acid was used, the reaction time was controlled at 5 h, and the reaction temperatures were set at 40 °C, 45 °C, 50 °C, 55 °C, and 60 °C. When soft fatty acid was used, the reaction time was controlled at 6 h, and the reaction temperatures were set at 40 °C, 45 °C, 50 °C, 55 °C, and 60 °C. When stearic acid was used, the reaction time was controlled at 6 h, and the reaction temperatures were set at 40 °C, 45 °C, 50 °C, 55 °C, and 60 °C; when stearic acid was used, the reaction time was controlled at 6 h, and the reaction temperatures were set at 40 °C, 45 °C, 50 °C, 55 °C, and 60 °C.

At 8 h reaction time, CNC was graft-modified with lauric acid by adding 0.8% $FeSO_4/H_2O_2$ at 35 °C, 40 °C, 45 °C, 50 °C, 55 °C, 60 °C, and 65 °C. At 20 h reaction time, CNC was graft-modified with soft stearic acid by adding 1.2% $FeSO_4/H_2O_2$ at 35 °C, 40 °C, 45 °C, 50 °C, 55 °C, 60 °C, and 65 °C. At 24 h reaction time, CNC was graft modified with stearic acid by adding 1.2% $FeSO_4/H_2O_2$ at 35 °C, 40 °C, 45 °C, 50 °C, 55 °C, 60 °C, and 65 °C. Appropriate amounts of the prepared fatty acid modification products were taken and tested for their DS by saponification method.

Figure 4 reflects the effect of reaction temperature on the degree of product substitution when grafting CNC by both preparation methods. The trend of the curves in the figure is the same as that of the effect of reaction time on the DS of the modified products. Lauric acid has a higher DS than soft and stearic acid under the same conditions due to the steric hindrance effect of the polymer. Relatively high DS were obtained at 50 °C to 60 °C for both preparations, but the reaction temperature was relatively higher for stearic acid because as the hydrocarbon chains of the fatty acids involved in the reaction grow, their steric hindrance becomes progressively greater. When the molecular weight of the fatty acid reaches 18 carbons, the steric hindrance becomes an important factor in the DS of the modified reaction, and the reaction system requires a higher temperature to better activate the reactants.

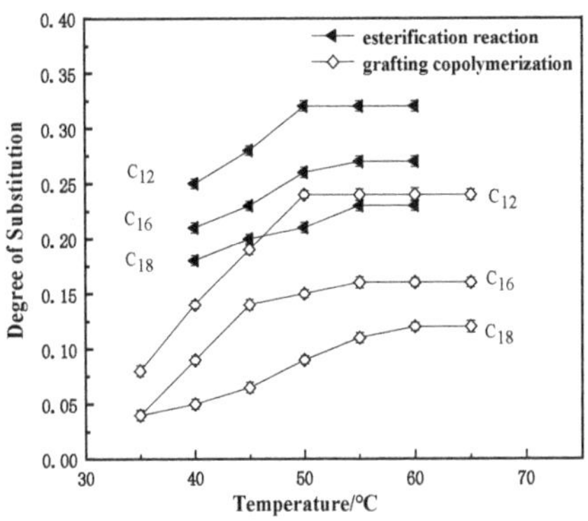

Figure 4. Effect of reaction temperature on substitution degree of MCNC.

In summary, when using trifluoroacetic anhydride as the co-reactant, the optimum reaction conditions for MCNC-C12 were 50 °C, 5 h, and 1:3 feeding ratio, and the degree of product substitution was 0.32. The optimum reaction conditions for MCNC-C16 were 55 °C, 6 h, and 1:3 feeding ratios; the DS was 0.27. The optimum reaction conditions for MCNC-C18 were 55 °C, 6 h, and 1:3 feeding ratio, and the degree of product substitution was 0.23. In the $FeSO_4/H_2O_2$ redox system, the highest DS of 0.24 was achieved by adding 0.8% initiator at a reaction temperature of 50 °C for 8 h. The highest DS of 0.24 was achieved by adding 1.2% initiator at a reaction temperature of 55 °C for MCNC-C16 for 20 h. In the $FeSO_4/H_2O_2$ redox system, the highest DS of 0.24 was achieved by adding 0.8% initiator at a reaction temperature of 50 °C for MCNC-C12. The maximum DS was 0.16 for 20 h at

55 °C with 1.2% initiator and 0.12 for 24 h at 60 °C with 1.20% initiator for MCNC-C18. The differences in the DS between the two methods were due to the different principles of modifying CNC with co-reactants and initiators. The nature of the modified product by introducing a co-reactant is the reaction between the mixed anhydride and the hydroxyl groups of the CNC, which are at the C_6, C_2, and C_3 positions. On the contrary, the use of an initiator for the preparation of the modified product involves the initiation of the hydroxyl group at the C_6 position, and the number of reaction sites is different between the two methods; hence a difference is found in the DS.

3.2. X-ray Diffraction Analysis of MCNC

X-ray diffraction analysis of CNC and MCNC prepared by different methods. The test results are shown in the following Figure 5:

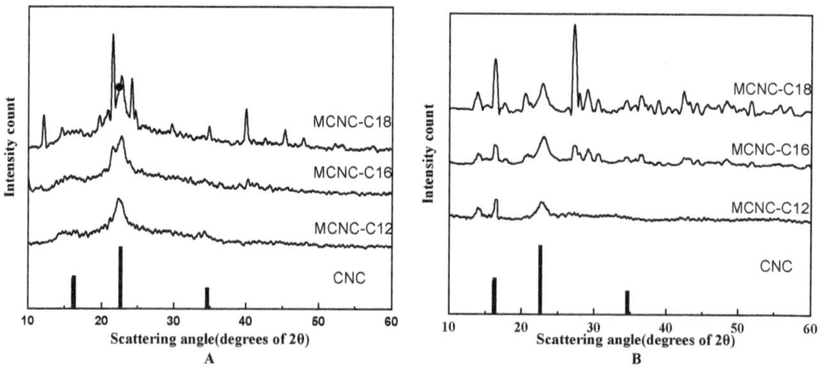

Figure 5. X-ray diffraction pattern of fatty acid MCNC. ((**A**): diffraction curve of MCNC crystal prepared by esterification method, and (**B**): diffraction curve of MCNC crystal prepared by graft copolymerization method).

Regardless of whether the esterification substance with CNC is lauric acid, palmitic acid, or stearic acid, the essence of the grafted CNC is the same. The difference in steric hindrance caused by the chain length causes different crystallinity. Therefore, the XRD diffraction peak value of various products prepared under the same system is different, but the peak position is approximately the same. The results show that the diffraction peak of the MCNC is weakened at the intensity of I_{002}, and the crystallinity of the modified products decreases. However, the products of the grafted MCNC with three types of advanced fatty acids still have diffraction peaks at the intensity of I_{002}, I_{am}, and I_{004}, indicating that the products still maintain the basic crystal structure of CNC. The new diffraction peak generated by MCNC and the decrease in the area of the diffraction peak at the intensity of I_{002} also prove the chemical modification of CNC.

3.3. MCNC Particle Size Test

MCNC was prepared under the optimal reaction conditions to detect the particle size of CNC and three types of MCNC, using trifluoroacetic anhydride as co-reactant and $FeSO_4/H_2O_2$ as initiator. The test results are as follows:

Figure 6 shows that the average particle size of CNC modified by fatty acids exceeds the average particle size of the original CNC (168 nm), which is in the range of 240–260 nm. However, the specific surface area is lower than the value of unmodified CNC (19,260 m^2/kg), which is in the range of 17,400–17,500. In general, regardless of the advanced fatty acid and method selected to prepare MCNC, the difference between the particle size, particle size distribution coefficient, and specific surface area is very small, and can even be approximately considered the same. However, because the long chain of carbon and hydrogen replaced the original hydrogen atoms after esterification of fatty acids and CNC, the molecular weight increased to varying degrees, and the spatial scale of molecular structure increased. Thus, the particle size of the modified products still increased slightly; it was higher than that of the unmodified CNC. Nevertheless, the esterified cellulose particles are still nanometer in size. The figure shows that the grafting of long-chain fatty acids did not significantly increase the original nanometer particle size, but only increased the particle size in a small range. In addition, modified products with wider particle size distribution are greater than the original CNC size distribution coefficient. This finding shows that during the esterification reaction of CNC, a certain degree of flocculation has occurred. The reduction of the specific surface area of the MCNC causes the original CNC to be grafted with long hydrocarbon chains, weakening their reactivity to a certain extent. Therefore, CNC is still at the nanoscale level and has large specific surface area after esterification and grafting of carbon and hydrogen long chains. This condition lays a good foundation for their subsequent functional applications.

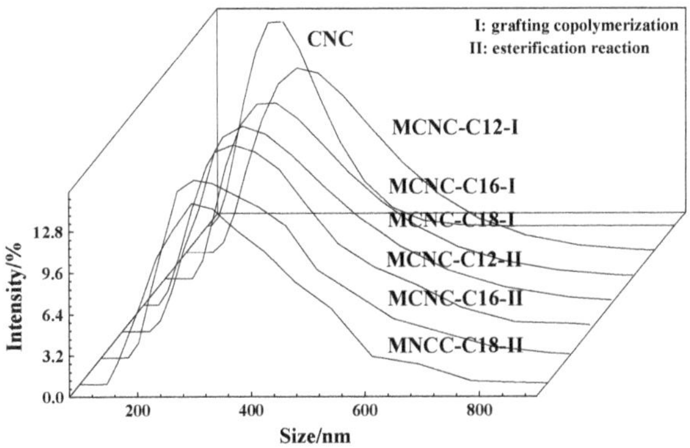

Figure 6. Particle size distribution of MCNC.

3.4. Calculation of HLB Value of MCNC

In accordance with the HLB value equation (Equation (4)), the HLB values of MCNC under various optimal reaction conditions were calculated and shown in Table 1.

Table 1. HLB value of fatty acid graft MCNC.

MCNC	HLB Value	
	Esterification	Grafting Copolymerization
MCNC-C12	14.68	15.73
MCNC-C16	14.29	16.17
MCNC-C18	14.49	16.69

By comparison, the HLB values of MCNC prepared by grafting copolymerization are slightly higher than that of MCNC prepared by esterification. The HLB values of MCNC are affected by the DS and the length of the grafted hydrocarbon chain. With the increase in DS, the HLB values show a decreasing trend. The increase in DS increases the number of hydrophobic groups, and the modified products become more lipophilic. Thus, the MCNC gradually tend to be hydrophobic, and the HLB values decreases. However, due to the different molecular weights of the grafted fatty acids, the calculation results of HLB value would be affected to some extent. In addition, the phenomenon indicates that the HLB value of low-DS MCNC is higher than that of low-DS MCNC. In general, all modified products are hydrophilic and lipophilic. The raw CNC contains a large number of unsubstituted hydroxyl groups due to the low DS of MCNC. Thus, modified products tend to be hydrophilic. Therefore, the amphiphilicity of MCNC is determined by the DS and the length of the grafted hydrocarbon chain.

3.5. *Influence of MCNC Concentration on Surface Tension of Water*

CNC and MCNC were configured into different concentrations of suspension to be tested. During the test, the suspension was pretreated with ultrasound, and the surface tension of suspension was measured at 25 °C. The following results were obtained.

Figure 7 shows that CNC and their modified products can effectively reduce the surface tension of water, and the modified products have a more evident reduction effect on the surface tension. With the gradual increase in the concentration of various substances in water, the decreasing trend of surface tension changes from rapid to smooth. When a certain concentration is reached, continuously increasing its concentration does not cause a significant change in the surface tension of the water. The concentration value at this point is called the critical micelle concentration (CMC). In Figure 7, the CMC of the CNC and the MCNC appear near 12.5 mg/L, and the surface tension reduction intensity increases as the length of the hydrocarbon chain grafted on the CNC increases. Thus, MCNC-C18 has the best effect on reducing the surface tension of water, followed by MCNC-C16 and MCNC-C12. The figure shows that the carbon chain length slightly affects the critical micelle concentration because the main molecular chain skeleton of CNC is a macromolecular structure. At the same concentration of the same modified products, the esterification method has better surface tension reduction effect than the grafting copolymerization method. Its essence is due to the higher DS of the products prepared by the esterification, and the change in the carbon chain length slightly affects the overall structure. In addition, the structure analysis indicates that the MCNC cannot ionize in water; thus, MCNC belongs to the non-ionic surfactant.

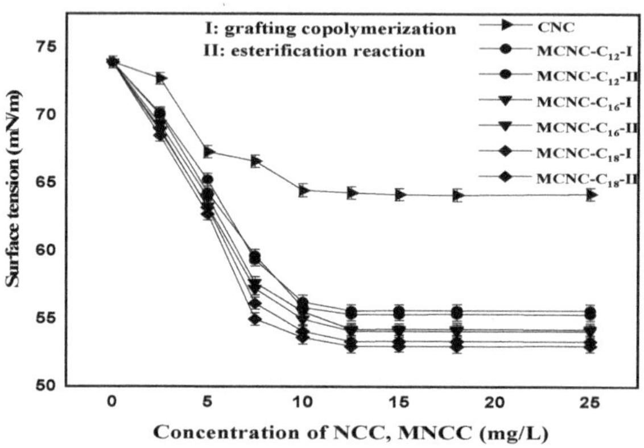

Figure 7. Effect of MCNC concentration on surface tension of MCNC suspension.

4. Conclusions

The MCNC prepared by esterification and grafting copolymerization is compared. Both methods can complete the reaction with low energy consumption, but the preparation time of the esterification method is shorter and the DS of MCNC is higher. The X-ray diffraction and particle size analysis of the modified products show that the products prepared by the two methods maintain the crystal structure of the raw CNC, and they are at the nanometer level. The measurement results of HLB value and surface tension prove that the modified products have hydrophilic and lipophilic properties, and the effect of reducing the surface tension of water is better than that of raw CNC. Because of good biodegradability, the modified product can be used as a new type of green surface-active substance, which are endowed with emulsion, dispersion, wetting, washing, sterilization, water resistant, antistatic, solubilization and stable features. Further, it will be widely concerned in food, chemical, agriculture, construction and other fields, to achieve high value utilization of biomass resources.

Author Contributions: Methodology, J.L. (Jinyan Lang); software, J.L. (Junliang Lu); validation, N.W. and X.W.; data curation, Y.W.; writing—original draft preparation, J.L. (Jinyan Lang); writing—review and editing, N.W. and X.W.; visualization, B.X. and G.C.; supervision, H.Z.; project administration, H.Z.; funding acquisition, H.Z. All authors have read and agreed to the published version of the manuscript.

Funding: This work was supported by the Shandong Provincial Key Research and Development Program (SPKR&DP) (Grant No. 2019GGX102029), the Research Fund Program of Guangdong Provincial Key Lab of Green Chemical Product Technology (Grant No. GC202112) and open fund of the Key Laboratory of Fire-Retardant Materials of Fujian Province (Grant No. FH202101).

Institutional Review Board Statement: Not applicable.

Informed Consent Statement: Not applicable.

Data Availability Statement: The data presented in this study are available on request from the corresponding author.

Acknowledgments: This work was supported by the Shandong Provincial Key Research and Development Program (SPKR&DP) (Grant No. 2019GGX102029), the Research Fund Program of Guangdong Provincial Key Lab of Green Chemical Product Technology (Grant No. GC202112) and open fund of the Key Laboratory of Fire-Retardant Materials of Fujian Province (Grant No. FH202101). Thanks to Shuyin Tan, Mengdi Ru, Xun Zhu, Jufang Zhang and Jingxue Yang, for their participation in the translation of this manuscript.

Conflicts of Interest: The authors declare no conflict of interest.

References

1. Yan, C.F.; Yu, H.Y.; Yao, J.M. One-step extraction and functionalization of cellulose nanospheres from lyocell fibers with cellulose II crystal structure. *Cellulose* **2015**, *22*, 3773–3788. [CrossRef]
2. Paakko, M.; Ankerfo, R.; Kosonen, H.; Nykänen, A.; Ahola, S.; Österberg, M.; Ruokolainen, J.; Laine, J.; Larsson, P.T.; Ikkala, O.; et al. Enzymatic hydrolysis combined with mechanical shearing and high-pressure homogenization for nanoscale cellulose fibrils and strong gels. *Biomacromolecules* **2007**, *8*, 1934–1941. [CrossRef] [PubMed]
3. Klemm, D.; Kramer, F.; Moritz, S.; Lindström, T.; Ankerfors, M.; Gray, D.; Dorris, A. Nanocrystalline celluloses: A new family of nature-based materials. *Angew. Chem. Int. Ed. Engl.* **2011**, *50*, 5438–5466. [CrossRef]
4. Qing, W.; Wang, Y.; Wang, Y.; Zhao, D.; Liu, X.; Zhu, J. The Modified Nanocrystalline Cellulose for Hydrophobic Drug Delivery. *Appl. Surf. Sci.* **2016**, *366*, 404–409. [CrossRef]
5. Shimizu, M.; Saito, T.; Fukuzumi, H.; Isogai, A. Hydrophobic, ductile, and transparent nanocellulose films with quaternary alkylammonium carboxylates on nanofibril surfaces. *Biomacromolecules* **2014**, *15*, 4320–4325.
6. Lim, C.; Mei, C.T.; Li, M.C.; Mei, C.; Xu, X.; Lee, S.; Wu, Q. Cationic surface modification of cellulose nanocrystals: Toward tailoring dispersion and interface in carboxymethyl cellulose films. *Carbohydr Polym* **2016**, *107*, 200–210.
7. Nooy, A.; Besemer, A.C.; Bekkum, H.V. Highly selective tempo mediated oxidation of primary alcohol groups in polysaccharides. *Recl. Trav. Chim. Pays-Bas* **1994**, *113*, 165–166. [CrossRef]
8. Araki, J.; Wada, M.; Kuga, S. Stabilization of a cellulose microcrystal suspension by poly(ethylene glycol) grafting. *Langmuir* **2001**, *17*, 21–27. [CrossRef]
9. Susheel, K.; Sami, B.; Celli, A.; Kango, S. Nanofibrillated cellulose: Surface modification and potential applications. *Colloid Polym. Sci.* **2014**, *292*, 5–31.
10. Grunert, M.; Winter, W.T. Nanocomposites of cellulose acetate butyrate reinforced with cellulose nanocrystals. *J. Polym. Environ.* **2002**, *10*, 27–30. [CrossRef]
11. Zaman, M.; Xiao, H.; Chibante, F.; Ni, Y. Synthesis and characterization of cationically modified nanocrystalline cellulose. *Carbohydr. Polym.* **2012**, *89*, 163–170. [CrossRef] [PubMed]
12. Lakovaara, M.; Sirvi, J.A.; Ismail, M.Y.; Liimatainen, H.; Sliz, R. Hydrophobic modification of nanocellulose and all-cellulose composite films using deep eutectic solvent as a reaction medium. *Cellulose* **2021**, *28*, 5433–5447.
13. Beaumont, M.; Winklehner, S.; Veigel, S.; Mundigler, N.; Gindl-Altmutter, W.; Potthast, A.; Rosenau, T. Wet esterification of never-dried cellulose: A simple process to surface-acetylated cellulose nanofibers. *Green Chem.* **2020**, *22*, 5605–5609. [CrossRef]
14. Shang, Q.; Chen, J.; Hu, Y.; Yang, X.; Zhou, Y. Facile Fabrication of Superhydrophobic Cross-Linked Nanocellulose Aerogels for Oil–Water Separation. *Polymers* **2021**, *13*, 625. [CrossRef]
15. Spinella, S.; Giada, L.R.; Liu, B.; John, D.; Youssef, H.; Jane-Marie, R.; Philippe, D.; Richard, A.G. Modification of cellulose nanocrystals with lactic acid for direct melt blending with pla. *AIP Conf. Proc.* **2015**, *1664*, 070019.
16. Yuan, H.; Nishiyama, Y.; Wada, M.; Kuga, S. Surface acylation of cellulose whiskers by drying aqueous emulsion. *Biomacromolecules* **2006**, *7*, 696–700. [CrossRef]

17. Curvelo, A.; Menezes, A.; Dufresne, A.; Siqueira, G. Extrusion and characterization of functionalized cellulose whiskers reinforced polyethylene nanocomposites. *Polymer* **2009**, *50*, 4552–4563.
18. Berlioz, S.; Molina-Boisseau, S.; Nishiyama, Y.; Heux, L. Gas-phase surface esterification of cellulose microfibrils and whiskers. *Biomacromolecules* **2009**, *10*, 2144–2151. [CrossRef] [PubMed]
19. Eyholzer, C.; Tingaut, P.; Zimmermann, T.; Oksman, K. Dispersion and reinforcing potential of carboxymethylated nanofibrillated cellulose powders modified with 1-hexanol in extruded poly (lactic acid)(PLA) composites. *J. Polym. Environ.* **2012**, *20*, 1052–1062. [CrossRef]
20. Goffin, A.; Raquez, J.; Duquesne, E.; Siqueira, G.; Habibi, Y.; Dufresne, A.; Dubois, P. From interfacial ring-opening polymerization to melt processing of cellulose nanowhisker-filled polylactide-based nanocomposites. *Biomacromolecules* **2011**, *12*, 2456–2465. [CrossRef]
21. Westlund, R.; Carlmark, A.; Hult, A.; Malmstrm, E.; Saez, I.M. Grafting liquid crystalline polymers from cellulose substrates using atom transfer radical polymerization. *Soft Matter* **2007**, *3*, 866–871. [CrossRef] [PubMed]
22. Lee, K.M.; Koerner, H.; Vaia, R.A.; Bunning, T.J.; White, T.J. Light-activated shape memory of glassy, azobenzene liquid crystalline polymer networks. *Soft Matter* **2011**, *7*, 4318–4324. [CrossRef]
23. Morandi, G.; Heath, L.; Thielemans, W. Cellulose nanocrystals grafted with polystyrene chains through surface-initiated atom transfer radical polymerization (SI-ATRP). *Langmuir* **2009**, *25*, 8280–8286. [CrossRef] [PubMed]
24. Zhang, H.; Yang, H.Y.; Lu, J.; Lang, J.; Gao, H. Study on Stability and Stability Mechanism of Styrene-Acrylic Emulsion Prepared Using Nanocellulose Modified with Long-chain Fatty Acids. *Polymers* **2019**, *11*, 1131. [CrossRef] [PubMed]
25. Lu, J.L.; Lang, J.Y.; Lan, P.; Yang, H.; Zhang, H. Evalution of surface activity of hydrophobic modified nanocrystalline cellulose. *Cellulose* **2020**, *27*, 9299–9309. [CrossRef]
26. Lu, J.L.; Lang, J.Y.; Wang, N.; Wang, X.; Lan, P.; Zhang, H. Preparation and thermostability of hydrophobic modified nanocrystalline cellulose. *Nord. Pulp. Paper Res. J.* **2021**, *36*, 157–165. [CrossRef]
27. Thakur, V.K.; Thakur, M.K.; Gupta, R.K. Rapid synthesis of graft copolymers from natural cellulose fibers. *Carbohydr. Polym.* **2013**, *98*, 820–828. [CrossRef] [PubMed]
28. Rahman, M.L.; Rohani, N.; Mustapa, N.; Yusoff, M.M. Synthesis of polyamidoxime chelating ligand from polymer-grafted corn-cob cellulose for metal extraction. *J. Appl. Polym. Sci.* **2014**, *131*, 40833. [CrossRef]
29. Lee, H.J.; Ramaraj, B.; Yoon, K.R. Esterification on solid support by surface-initiated ring-opening polymerization of ε-caprolactone from benzylic hydroxyl-functionalized Wang resin bead. *J. Appl. Polym. Sci.* **2010**, *111*, 839–844. [CrossRef]
30. Huang, F.Y.; Feng, Y. Thermal Properties and Thermal Degradation of Cellulose Tri-Stearate (CTs). *Polymers* **2012**, *4*, 1012–1024. [CrossRef]
31. Freire, C.; Silvestre, A.; Neto, C.P.; Rocha, R.M. An efficient method for determination of the degree of substitution of cellulose esters of long chain aliphatic acids. *Cellulose* **2015**, *12*, 449–458. [CrossRef]
32. Segal, L.; Creely, J.J.; Martin, A.E.; Conrad, C.M. An Empirical Method for Estimating the Degree of Crystallinity of Native Cellulose Using the X-Ray Diffractometer. *Text. Res. J.* **1959**, *29*, 786–794. [CrossRef]

polymers

MDPI

Review

Surface Modified Nanocellulose and Its Reinforcement in Natural Rubber Matrix Nanocomposites: A Review

Nik Muhammad Faris Hakimi [1], Seng Hua Lee [1,*], Wei Chen Lum [2,*], Siti Fatahiyah Mohamad [3], Syeed SaifulAzry Osman Al Edrus [1,*], Byung-Dae Park [4] and Anis Azmi [2]

1 Institute of Tropical Forestry and Forest Product, Universiti Putra Malaysia, Serdang 43400, Malaysia; nikfaris16.95@gmail.com
2 Institute for Infrastructure Engineering and Sustainable Management (IIESM), Universiti Teknologi MARA, Shah Alam 40450, Malaysia; anisazmi8855@gmail.com
3 Radiation Processing and Technology Division, Malaysia Nuclear Agency, Kajang 43000, Malaysia; fatahiyah@nuclearmalaysia.gov.my
4 Department of Wood and Paper Science, Kyungpook National University, Daegu 41566, Korea; byungdae@knu.ac.kr
* Correspondence: lee_seng@upm.edu.my (S.H.L.); lumweichen@outlook.com (W.C.L.); saifulazry@upm.edu.my (S.S.O.A.E.)

Abstract: Natural rubber is of significant economic importance owing to its excellent resilience, elasticity, abrasion and impact resistance. Despite that, natural rubber has been identified with some drawbacks such as low modulus and strength and therefore opens up the opportunity for adding a reinforcing agent. Apart from the conventional fillers such as silica, carbon black and lignocellulosic fibers, nanocellulose is also one of the ideal candidates. Nanocellulose is a promising filler with many excellent properties such as renewability, biocompatibility, non-toxicity, reactive surface, low density, high specific surface area, high tensile and elastic modulus. However, it has some limitations in hydrophobicity, solubility and compatibility and therefore it is very difficult to achieve good dispersion and interfacial properties with the natural rubber matrix. Surface modification is often carried out to enhance the interfacial compatibilities between nanocellulose and natural rubber and to alleviate difficulties in dispersing them in polar solvents or polymers. This paper aims to highlight the different surface modification methods employed by several researchers in modifying nanocellulose and its reinforcement effects in the natural rubber matrix. The mechanism of the different surface medication methods has been discussed. The review also lists out the conventional filler that had been used as reinforcing agent for natural rubber. The challenges and future prospective has also been concluded in the last part of this review.

Keywords: nanomaterials; surface modification; latex; lignocellulosic fibers; conventional fillers

Citation: Hakimi, N.M.F.; Lee, S.H.; Lum, W.C.; Mohamad, S.F.; Osman Al Edrus, S.S.; Park, B.-D.; Azmi, A. Surface Modified Nanocellulose and Its Reinforcement in Natural Rubber Matrix Nanocomposites: A Review. *Polymers* **2021**, *13*, 3241. https://doi.org/10.3390/polym13193241

Academic Editor: Arn Mignon

Received: 12 September 2021
Accepted: 22 September 2021
Published: 24 September 2021

Publisher's Note: MDPI stays neutral with regard to jurisdictional claims in published maps and institutional affiliations.

1. Introduction

Natural rubber is a distinctive biopolymer of significant economic importance thanks to its high molecular weight and many others minor components that are existing in the latex. Figure 1 displays the latex (white colored colloidal suspension of rubber particles) being collected from a Para rubber tree (*Hevea brasiliensis*). The rubber tree is still the main source of natural latex as it produces a large amount of high molecular weight latex. Natural rubber could be produced from the latex found in more than 2500 plant species, guayule (*Parthenium argentatum* Gray) and Russian dandelion (*Taraxacum koksaghyz*) being the most promising alternatives to rubber trees [1]. Up until now, natural rubber is still irreplaceable by any other synthetic materials thanks to its resilience, elasticity, abrasion and impact resistance, efficient heat dispersion and malleability at cold temperature [2,3].

Figure 1. Latex being collected from a tapped rubber tree (PRA (2007) Récolte du latex sur un hévéa au Cameroun. Freely redistribute under Creative Commons Attribution-Sharealike 3.0 Unported License. Figure available from https://commons.wikimedia.org/wiki/File:Latex_-_Hevea_-_Cameroun.JPG [4]).

Natural rubber is an important commodity, particularly to those of the people in the Southeast Asian region. In 2020, the total natural rubber production amounted to 13.008 million metric tons. Asia is the leading natural rubber producing continent worldwide where Thailand, Indonesia, Vietnam, China, India and Malaysia are among the main producers of natural rubber (Figure 2). Around 81% of natural rubber is produced by these six countries with Thailand being the leading producer followed by Indonesia [5]. With very high cost performance, natural rubber becomes a very versatile material that is being used in the production of more than 40,000 products. The most common products include surgical gloves and tires. Automobiles is the leading industry that drives the global market of natural rubber. According to the report by Mordor Intelligence [6], in the year of 2019, automobile tires made up to 46% of the total industrial applications of natural rubber worldwide, followed by footwear (17%), tubes (15%), latex (10%) and others (12%). The demand of natural rubber keeps increasing over years, corresponding to the increased applications of natural rubber. Expert Market Research [7] forecasts the demand for natural rubber is expected to grow at a compound annual growth rate (CAGR) rate of 4.8% in the period of 2021–2026. By 2026, a whopping volume of 20.1 million metric tons is anticipated in order to cater to the demand worldwide.

Natural rubber is one of the most significant elastomers due to its versatility and application volume. Unfortunately, natural rubber has been identified with some drawbacks such as low modulus and strength. To account for that, crosslinking and adding reinforcing fillers of various sources and aggregate size/aspect ratio, such as cellulose nanocrystals, can improve and modify the mechanical characteristics of natural rubber. Carbon nanotubes, ceramics, and natural fibers are some examples of nanoparticles that can provide polymeric matrices with particular characteristics. When nanofiller is added to a matrix, it can change its mechanical characteristics and its crystallinity and permeability [8]. In the past few decades, natural rubber composites reinforced with various fillers such as lignocellulosic fibers and nanocellulose in replacing non-renewable carbon black has been the main focus of researchers worldwide [9]. In recent years, the usage of nano sized cellulose in reinforcing natural rubber has gained the most attention of researchers. However, the surface characteristic of the nanocellulose is one of the major challenges as it greatly affects the dispersion and interfacial properties of the resultant natural rubber nanocomposite. On that account, surface modification is often employed to modify the nanocellulose in order to achieve better dispersion and interfacial properties with the natural rubber matrix.

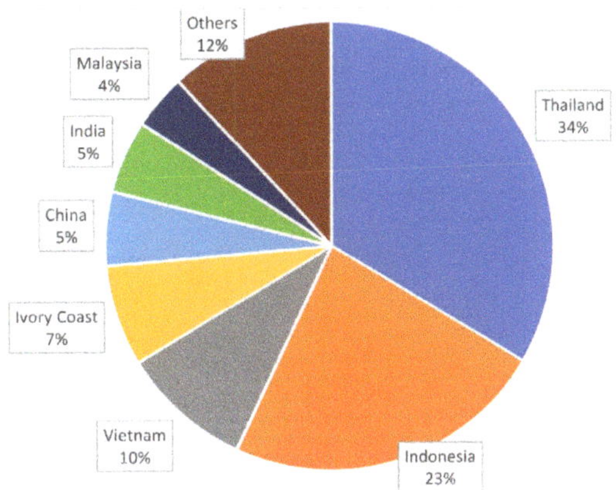

Figure 2. Natural rubber producing countries worldwide in 2020 (Statista 2021).

Some reviews have been conducted by several researchers on the topic. The most recent review by Low et al. [10] discussed the recent developments in nanocellulose reinforced rubber matrix composites. Property enhancement of rubber composites (natural rubber and synthetic rubber) as a result of nanocellulose reinforcement has been highlighted. Kargarzadeh et al. [11] reviewed the recent developments in nanocellulose reinforced polymer nanocomposites where the topic was not confined only to nanocellulose reinforced rubber. Nanocellulose reinforced thermoset composites such as epoxy, unsaturated polyester, polyethylene terephthalate (PET), phenol-, melamine-, and urea-formaldehyde resins have also been highlighted in the review. On the other hand, Zhou et al. [9] compiled a review on the rubber composites reinforced with various wood flour and lignocellulosic fibers including oil palm, hemp, husk, bamboo, bagasse, jute etc. However, a limited number of studies on nanocellulose reinforced rubber nanocomposites were reported at the time of writing. A book chapter by Nunes [12] also focused on the rubber nanocomposites with nanocellulose where the structure and properties of different rubber nanocomposites such as natural rubber, epoxidized natural rubber, polybutadiene rubber, ethylene propylene diene methylene rubber and so on were discussed. However, there is still a need for a compilation emphasizing on the surface modification of the nanocellulose and its resultant natural rubber composites. Therefore, this review focuses on summarizing the recent surface modification methods employed on the nanocellulose and its applications in the production of natural rubber nanocomposites. The conventional fillers used in the natural rubber based nanocomposite are also discussed. Some challenges and future perspectives are also highlighted at the last part of this review.

2. Conventional Fillers for Natural Rubber Nanocomposite

Natural rubber is traditionally incorporated with carbon black and silica during the vulcanization process to form rubber composite with enhanced properties. Despite their capacities as rubber composite reinforcements, carbon black and silica are not bio-degradable and require significant energy to manufacture. Carbon black being a petrochemical based product is also non-renewable. Therefore, in the route to search for a renewable and environmentally friendly filler, lignocellulosic fiber has gained attention and traction in the rubber composite industry due to its natural properties. However, the fundamental problem of the lignocellulosic fiber reinforced rubber composite is the poor compatibility of the hydrophilic wood flour or other natural fibers with the hydrophobic rubber matrix, which results in the composite's mediocre mechanical qualities. In order to improve the compatibility between natural rubber matrix and the filler used, various modification approaches of the fillers used are often conducted. In recent years, the advancement of nanotechnology has propelled surface modified nanocellulose as a viable filler for rubber composite. The role of nanocellulose as a viable filler for rubber composite will be discussed in detailed in the next section. In this section, the main findings of natural rubber composite reinforced with conventional filler (i.e., carbon black, silica and lignocellulosic fibers) from available literatures are discussed and summarized in Tables 1–3.

Table 1. The summary of literature using carbon black as the main filler for natural rubber composite.

Filler/Source	Treatments	Variables	Properties Tested and Findings	References
Carbon black (N220), silica	-	Carbon black to silica ratio (30:15, 40:15, 50:15)	**Tensile and Tear Strength** Generally, tensile and tear strength increased as the carbon black to silica ratio increased and peaked at 40 phr carbon black content before decreased slightly. **Abrasion Resistance and Modulus** The composite with the highest carbon black content (50 phr) exhibited the highest abrasion resistant index (ARI) and modulus. The ARI and modulus increased proportionally to the carbon black content.	[13]
Carbon black	-	Type of carbon black (HAF and ECF)	**Curing Characteristic** The cure rate index (CRI) for rubber composite added with both type of carbon black showed significant lower value than rubber composite without reinforcement. **Tensile Strength** Both types of carbon black filler (HAF and ECF) decreased the tensile strength of natural rubber composite.	[14]
Carbon black	N-tert-butyl-2-benzothiazole sulfenamide (TBBS)	TTBS concentration (1.0, 1.4, 1.8, 2.2, 2.6 phr)	**Curing Characteristic** The cure rate index (CRI) was similar between carbon black (40 phr) and silica (3.2 phr) filled rubber composite. The CRI decreased with increasing TTBS concentration. **Tensile Strength and Modulus** Tensile strength and modulus of carbon black filled rubber composite (40 phr) showed higher tensile strength than silica reinforced rubber composite (3.2 phr) for all TTBS concentration. The tensile strength for carbon black filled rubber composite peaked at 2.0 phr TTBS concentration then decreased with higher concentration, while the modulus value continued to increase with increasing concentration.	[15]

2.1. Carbon Black

Table 1 summarizes the main findings of natural rubber composite reinforced with conventional carbon black filler from available literatures. According to Table 1, most researchers studied the effects of types of carbon black, the concentration of carbon black and also the treatment employed to the carbon black used to enhance the curing characteristic and mechanical properties of rubber hybrid composites.

Generally, the application of carbon black filler increases the curing time of the natural rubber composite as shown by the lower cure rate index (CRI). It was also found that the tensile strength of natural rubber composite peaked at 40 phr carbon black content and then decreased with higher carbon black content. However, the modulus increased proportionally with the filler concentration [14,15].

Sivaselvi et al. [13] studied the effect of carbon black ratio in hybrid filler on the mechanical properties of rubber composite. The authors concluded that the rubber composite reinforced with the highest carbon black content exhibited the best abrasion resistance index (67%) and tensile and tear strength. This can be explained by the good bonding of the rubber matrix provided by the carbon black particles. The small particle size of carbon black blended well with natural rubber and did not hinder the vulcanization process therefore impart good mechanical properties. Thus, the high carbon black incorporated samples had improved mechanical properties. However, the authors noted that to achieve the optimum mechanical properties in every aspect, a perfect ratio of concentration of fillers is important. Salaeh and Nakason [14] conducted an experiment to study the effects of the types of carbon black filler, high-abrasion furnace (HAF) and extra conductive furnace (ECF) on the curing characteristic and tensile strength of natural rubber composite. They discovered that because ECF has a finer structure than HAF, it absorbs more curing agent and accelerator molecules during vulcanization, delaying the crosslinking reaction. They also concluded that a high concentration level of carbon black content of 50 phr reduced the tensile strength of rubber composite. This could be because a high carbon black content reduces the volume fraction of the rubber component that forms the composite's continuous matrix [16].

2.2. Silica

Table 2 summarizes the main findings of natural rubber composite reinforced with conventional silica-based filler from available literatures. According to Table 2, most researchers studied the effects of type of silica, loadings of silica and also treatment of the silica used on the curing characteristic and mechanical properties of rubber hybrid composites.

Table 2. The summary of literatures using silica as main fillers for rubber composite.

Filler/Source	Treatments	Variables	Properties Tested and Findings	References
Silica fume	-	Silica loading (10, 20, 30, 40, 50 phr)	**Curing Characteristic** The cure time of all the reinforced composites increased with the increase in silica fume loading **Tensile Strength and Modulus** Composite added with 20 phr silica showed improved tensile strength. The tensile strength dropped slightly with further increase of silica fume loading. However, the composite's modulus increased proportionally with silica fume loading. **Tear Strength** Tear strength increased up to 20 phr of silica fume loading and then decreased with further increases in loading.	[17]

Table 2. *Cont.*

Filler/Source	Treatments	Variables	Properties Tested and Findings	References
Silica, Silica-graphene oxide (SiO_2@GO)	-	Silica loading of 3 phr	**Tensile strength, Modulus and Elongation at Break** Composite added with silica exhibited insignificant change in tensile strength due to the low content of silica used. However, composite consisted of hybrid silica-graphene oxide filler, showed an increase in tensile strength, elongation at break, and modulus.	[18]
Silica from sugarcane bagasse ash	Drying technique	Freeze drying (FD) and heat drying (HD)	**Curing Characteristic** Scorch time of the rubber composites increased with an increase in silica content. **Tensile Strength** Tensile strength of the tested composites increased with increase in silica content but tended to reduce slightly at higher silica content. **Modulus, Hardness and Elongation at Break** The modulus and hardness of rubber composite increased while elongation at break decreased with an increase in silica content	[19]
Silica, styrene	-	Particle size	**Tensile Strength, Modulus and Hardness** All the mechanical properties of reinforced natural rubber increased compared to rubber compound without reinforcement.	[20]
Silica -precipitated silica (PS), autonomous monodisperse silica (AS)	Silane treatment	Silica dimension and polydispersity	**Tensile Strength and Abrasive Resistance** PS reinforced rubber showed better tensile strength and abrasive resistance than AS reinforced rubber.	[21]
Silica from rice husk	Alkali treatment	Silica loading of 60 phr	**Curing Characteristic** The result shows that the curing time of the rubber composites decrease with increasing silica loading. **Tensile Strength** The tensile strength was determined at the break point of the specimen. Results show the addition of silica in natural rubber matrix resulted in the improvement in the tensile properties **Tear Strength and Hardness** The addition of silica as filler increases the tear strength as well as the hardness of the rubber composite as the silica loading increases.	[22]

Table 2. *Cont.*

Filler/Source	Treatments	Variables	Properties Tested and Findings	References
Precipitated silica (PSi) and fly ash silica (FASi)	-	Silica loading of 0–75 phr	**Curing Characteristic** The cure time and minimum and maximum torques of the rubber compounds were unaffected by silica loadings from 0 to 30 phr, and above these concentrations, the values progressively increased with increasing PSi loading but the FASi remained the same. 0 to 30 phr FASi could be recommended to natural rubber but not styrene–butadiene rubber **Tensile Strength and Elongation at Break** The tensile strength and elongation at break of rubber composite filled with untreated silica decreased with increasing silica content.	[23]

Table 2 shows that the curing time of rubber composites is affected by the type of silica filler and the amount of silica applied. According to certain researches, the higher the silica level, the longer the curing period. Other researches, on the other hand, found the opposite. The mechanical properties of silica reinforced rubber composite are more uniform among studies and show a similar trend to carbon black reinforced rubber composite. The addition of silica increases the mechanical properties of rubber composite compared to the rubber composite without reinforcement. However, the strength of the material is not positively proportionate to the silica content added. The strength peaks at certain silica content before it decreases as more silica is added.

The curing characteristic of rubber composite is influenced by the filler content. Boonmee and Jarukumjor [19] noted that the scorch time of natural composites was increased with an increase in silica nanoparticle content. Dileep et al. [17] also reported similar findings, which stated that at higher loading of silica fume the curing time increased. This might be cause by the disturbance of vulcanization process by silica particles surface. Sombatsompop et al. [23] found out that the cure time of rubber composite was not affected by the silica concentration below 15 phr but decreased abruptly at further concentration. At lower silica concentration, the silica particles were separated from one another, forming a dispersed gel through the rubber matrix without hindering the polymerization process. This in agreement with results obtain by Ahmed et al. [22] which observed that curing time decreased with an increase in silica loading.

Thuong et al. [20] suggested that silica filler has a more prominent effect than styrene filler in their study. The rubber composite incorporated with both silica and styrene exhibit outstanding tensile strength than the samples reinforced with styrene alone. Although the inclusion of silica to rubber composites has the potential to increase mechanical properties, the amount employed is crucial. The tear strength of silica filled natural rubber composite significantly improved compared to natural rubber composite without added silica according to Dileep et al. [17]. Tear strength of all silica filled natural rubber composite increases up to 20 phr of silica loading and then decreases slightly. This phenomenon is explained by agglomeration of the silica particles at higher loading which results in poor dispersion and consequently lower mechanical properties. This trend is in agreement with the results obtained by Sadequl et al. [24]. In another research by Charoenchai et al. [18], the mechanical properties of neat natural rubber did not show significant change when compared to reinforced rubber composites due to low content of silica used in the system. Sombatsompop et al. [23] concluded that the tensile strength stabilized at 30 phr of silica content and further increment of silica content only increased the crosslink density of the vulcanizates without providing any mechanical enhancement. The decrease in mechanical properties was due to low interaction between filler and the rubber component.

2.3. Lignocellulosic Fibre

Table 3 summarizes the main findings of rubber composite reinforced with lignocellulosic fiber filler from available literatures.

Table 3. The summary of literatures using lignocellulosic fiber as fillers for rubber composite.

Lignocellulosic Fiber/Filler	Treatments	Variables	Properties Tested and Findings	References
Torrefied almond shells (TAS) and torrefied rice hulls (TRH)	torrefaction	Carbon black to torrefied filler loadings ratio (40:20, 30:30, 20:40)	**Curing Characteristic** The curing time increased with higher loadings of both torrefied fillers. **Tensile Strength and Modulus** Generally, carbon black filled composite showed better mechanical properties than torrefied filler reinfored composite. The tensile strength of TAS filled natural rubber composite decreased with decreasing carbon black to torrefied filler ratio, while TRH filled natural rubber composite and reached the lowest tensile strength at 30:30 ratio before increasing significantly at 40:20 ratio. The modulus of TAS filled natural rubber composite showed similar trend with its tensile strength. On the other hand, the modulus of TRH rubber composite showed the lowest modulus at 40:20 ratio before increasing gradually at 30:30 and 20:40 ratio.	[25]
Horsetail (*Equisetum Arvense*)		Horsetail filler loading (10, 20, 30, 40 50 phr)	**Elongation at Break** The elongation at break increases with increasing horsetail loading from 10 phr to 50 phr. **Tensile Strength** Tensile strength of rubber composite reinforced with horsetail show higher value than the pure natural rubber sample. Initially, tensile strength increases with the addition of horsetail filler then drop slightly as the loading increases.	[26]
Cereal straw	Silane treatment	Silanes coupling agents (PTES, VTES, TESPTS), filler loading (10, 20, 30 phr)	**Elongation at Break** The elongation at break decreased with increasing cereal straw filler from 10 to 30 phr) **Tensile Strength** Rubber composites with silanes modified filler show improved tensile strength compared to composite with unmodified filler and the control specimens. The tensile strength increased and peaked at 10 phr loading and decreased with higher filler content.	[27]

Table 3. *Cont.*

Lignocellulosic Fiber/Filler	Treatments	Variables	Properties Tested and Findings	References
Hemp fibre	Silane (Si69) and permanganate ($KMnO_4$) treatment	Filler loading (5, 10, 15 phr)	**Curing Characteristic** Generally, the curing time of rubber composites increases with increasing hemp loading. The curing time for rubber composites filled with both silane or permanganate treated fibre is longer than the untreated counterparts. **Tensile Strength and Modulus** This finding shows increased tensile strength of silane treated fibre rubber composites compared with untreated and permanganate treated hemp fiber filled rubber composites. Tensile strength increased with filler loading and peaked at 10 phr before showing a decrease trend at 15 phr. The modulus however demonstrated an continual increase trend with higher filler loading.	[28]
Coconut shell powder	Alkali treatment	Filler loading (10, 20, 30, 40 phr)	**Curing Characteristic** Curing time was found to decrease consistently with increasing filler loading **Tensile Strength and Modulus** Tensile strength of the natural rubber composite was highest at 10 phr loading and decreased with increasing filler loading of 20 to 40 phr. Generally, the modulus showed similar trend with tensile strength.	[29]
Rice husk	Electron beam irradiation	Irradiation dosage	**Tensile Strength and Modulus** Max stress increases with irradiation dosage until about 20 kGy and decreases with further increase of radiation. However, the modulus seems to maximize at about 30 kGy of radiation.	[30]
Wood flour	Corona treatment in air and in ammonia	Filler loading (10, 20, 30, 40, 50, 60, 70 phr)	**Curing Characteristics** The higher the filler loading the longer the curing time. Nonetheless, the addition of treated wood flour treated with corona in air and ammonia did not significantly affect the vulcanization process of the rubber matrix. **Tensile Strength and Modulus** The tensile strength increased with filler laoding at 10 phr and decreased consistently with filler loading of 20, 30, 40 and 50 phr. However, the tensile strength increased again at 70 phr filler loading	[31]

Table 3. *Cont.*

Lignocellulosic Fiber/Filler	Treatments	Variables	Properties Tested and Findings	References
Oil palm wood flour		Filler loading	**Tensile Strength and Modulus** The results show that increasing the concentration of oil palm wood flour increases the tensile modulus. However, the tensile strength and the elongation at break show a reverse trend.	[32]

According to Table 3, the curing characteristic and mechanical properties vary widely and are influenced by various variables or a combination of variables such as the type of lignocellulosic fiber filler, the treatment applied on the filler as well as the loadings of the filler used in natural rubber composites. Based on Table 3, it can be summarized that the curing time of natural rubber composite increased with increasing lignocellulosic fiber filler loading. Generally, natural rubber composite reinforced with 10 phr filler loading exhibited the strongest tensile properties. Further increment in filler loading showed a decline in tensile strength but increased the modulus of the rubber composite.

Torres et al. [25] and Moonart and Utara [28] concluded that the curing time increased consistently with increasing filler loading. The increase in curing time suggested that a retardation effect occurred during the curing process. They speculate that the accelerators involved in speeding up the curing process were trap in the porous structure of lignocellulosic filler, thus affected the curing characteristic of rubber composite. Sareena et al. [29] however reported a reverse trend. Vladkova et al. [31] reported that the addition of corona treated lignocellulosic filler did not significantly influence the curing time. However, increasing the filler loading did increase the curing time of natural rubber composite. The mechanisms of this phenomena are similar to those of carbon black and silica filler.

In 1998, Ismail and Nurdin [32] conducted research to determine the tensile properties of rubber composite reinforced with oil palm wood flour. They discovered that when the concentration of oil palm wood flour increased, the tensile modulus increased, however the tensile strength and elongation at break decreased. Although scanning electron microscopy has demonstrated an improvement in the surface interaction between oil palm wood flour and rubber components, systematic research on the influence of wood flour in rubber composites is lacking. According to Vladkova et al. [31], corona treatment under ideal operating circumstances can improve the efficiency of wood flour as a filler in natural rubber composites. Corona treatment in air, according to the authors, is a more effective method for increasing the wood flour reinforcing effect in nonpolar rubbers. Corona treatment in ammonia, on the other hand, causes a more complex change in the chemical composition of the wood surface due to the suppression of surface oxidation and the build-up of nitrogen-containing groups.

Sareena et al. [29] studied the effect of coconut shell powder loadings on the mechanical properties of rubber composite. They noted that samples with filler loading at 10 phr exhibited highest tensile strength and showed highest modulus value as this loading concentration provides a large interfacial area of contact, resulting in better interfacial adhesion. Any higher filler loading will cause weak interaction and bonding between the filler particles and the natural rubber component which was responsible for the decline of tensile strength. This is in agreement with Miedzianowska et al. [27], which observed in their study that initial increase of cereal straw filler (10 phr) also increased the tensile strength but further increment (20 and 30 phr) reduced the tensile strength slightly. However, it is worth noting that the cereal filler used was treated with silanes coupling agents i.e., propyltriethoxysilane (PTES), vinyltriethoxysilane (VTES) and 3,3′-Tetrathiobis (propyl-triethoxysilane) (TESPTS). The TESPTS and VTES modified rubber composite exhibited the most significant changes in mechanical properties. Improvement in the mechanical properties of composites may result from increased adhesion of fibers to the polymer, and thus stronger interfacial interactions influencing the increase in cross-linking density. Torres et al. [25] studied the effects of carbon black to torrefied fillers ratio on the tensile strength and modulus of natural rubber composite. The authors observed that the introduction of higher torrefied fillers loading reduced the tensile strength and modulus of the hybrid natural rubber composite. They suggested that the residue of hydroxyl or carbonyl group after torrefaction treatment might reduce the interaction between rubber matrix and filler (almond shells or rice hulls). Additionally, the torrefied filler might be unable to support stresses transferred from the rubber matrix. This is in line with the findings by Masłowski et al. [26]. The authors also noted that for rubber composite filled with 10 phr of horsetail, the tensile strength increased by a significant 27%. Nevertheless, further addition of horsetail content triggered a slight decrease in the tensile strength. It is worth noting that, with 50 phr horsetail loading, the tensile strength was the lowest but still slightly higher than natural rubber composite without the addition of horsetail filler. Figure 3 summarizes the effects of different types of filler and filler loadings on the properties of natural rubber composites.

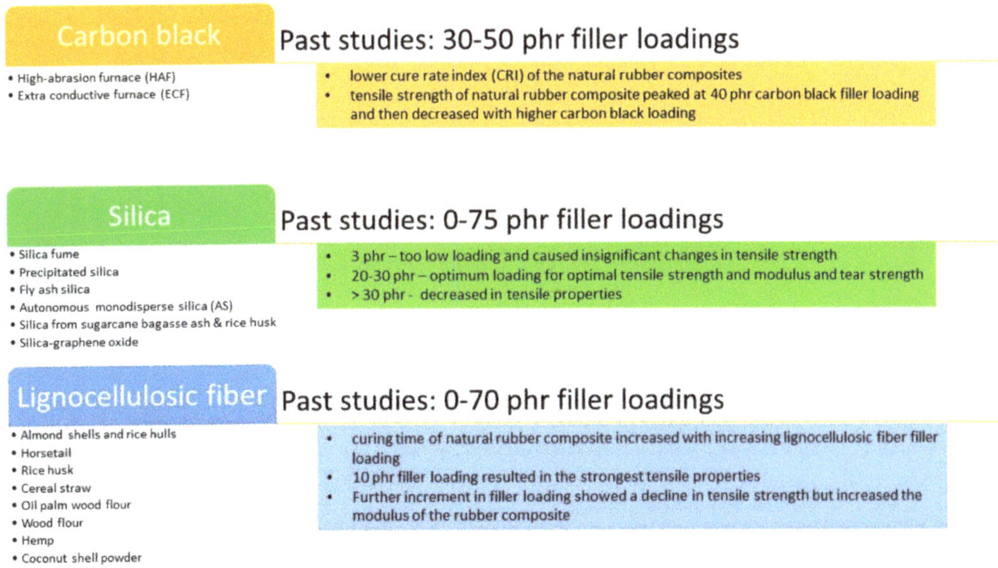

Figure 3. Summary of the effects of different types of filler and filler loadings on the properties of natural rubber composites.

3. Types of Nanocellulose Fillers from Natural Fiber

As previously stated, carbon black is regarded as the most commercially suitable filler for natural rubber. However, in recent years, nanocellulose has demonstrated excellent potential in replacing carbon black. In fact, nanocellulose (NC), cellulose in the form of nanostructures, has been proven as one of the most promising sustainable materials of the future in recent decades. Kulshrestha et al. [33] demonstrated that the reinforcing ability of a hybrid filler system including 2 phr CNFs and 50 phr carbon black was practically equivalent to that of 65 phr carbon black in natural rubber compounds in terms of mechanical behavior. As a result, 2 phr CNFs might potentially replace roughly 15 phr carbon black in natural rubber formulations. Furthermore, when subjected to repeated loading cycles, natural rubber nanocomposite reinforced with CNC maintains stiffness better than neat natural rubber [34]. In their study, the stress softening effect of natural rubber nanocomposite reinforced with thiol-modified CNC was 4–6 times better than natural rubber reinforced with carbon black as reported by Harwood et al. [35]. There are many types of nanocellulose which can be generally classified into microfibrillated cellulose (MFC), cellulose nanocrystals (CNCs) and bacterial nanocellulose (BNC).

3.1. Microfibrillated Cellulose (MFC)

Generally, MFC is defined as cellulose nanomaterials consisting of fibrils and also known as cellulose nanofibrils (CNFs) and nanofibrillated cellulose (NFC) [36–39]. It has the same properties but different in size, with diameters of 5–30 nm and up to several microns long [40]. Muqeet et al. [41] stated the CNFs can be easily modified through substitution reaction because of their abundant hydroxyl groups. Besides that, cellulose-based nanomaterials have also been studied extensively as adsorbents, owing to their high surface area to volume ratio and the flexibility of functionalization in various fashions.

3.2. Cellulose Nanocrystals (CNCs)

CNCs, sometimes knows as nanocrystalline cellulose (NCC) and cellulose nanowhiskers (CNW), are widely used and preferable for industrial needs, with their characteristic of high mechanical strength plus flexibility of surface chemistry [42,43]. Under magnification, CNC represents a rod-shaped like crystalline form [44]. The main contribution in CNC properties is their cholesteric (Ch) liquid crystalline structures in liquid form. This ability helps the industry in terms of humidity sensors, optical encryptors, structural pigments, light shutters, and templates to synthesize inorganic materials. The CNC are becoming the industrial favorite component used due to their renewable characteristic and significant application as a reinforcing agent. Besides that, in the form of the solidified film, the CNC appears as a structural color that can be altered by additives (nanoparticles, organic dyes, and surfactants) [45]. Compared to CNFs, CNC has a stiffer structure due to higher crystallinity resulting from the removal of the amorphous region during the acid hydrolysis [46]. CNC has 3 to 5 nm of width and 100 nm of length up to several micrometers [46]. The CNC particles are highly crystalline materials varying from 54 to 88% in crystallinity and are 100% cellulose with a higher thermal stability value (~260 °C), larger aspect ratio (10 to 70), and lower density value (1.5–1.6 g/cm^3) [46].

3.3. Bacterial Nanocellulose (BNC)

BNC is the extracellular product that produces by bacteria with high purity [47,48]. BNC has a diameter size of 20–100 nm and ribbon-like shape [40]. Gluconoacetobacter is an example of a bacteria type that BNC is extracted from, and it was grown in liquid culture media [46]. Carbon is used as the energy source for the bacteria culture and the nitrogen for the growing culture. BNC has almost similar properties to CNC and CNF, but more often, it is lightweight, non-toxic, and controls microfibril formation depending on the bacterial culture parameters [46]. The application of BNC is famous in biomedical applications such as wound healing and regenerative medicine.

4. Surface Modification of Nanocellulose for Natural Rubber Nanocomposites

Chemical modification is often carried out to enhance the interfacial compatibilities between nanocellulose and natural rubber and to alleviate difficulties in dispersing them in polar solvents or polymers [49]. Hydroxyl groups are the sole functional group that exists in nanocellulose and therefore the surface modification of nanocellulose is highly dependent of the reactivity of these functional groups. In the hydrogen bonding-induced aggregation of materials, the nanoscale structure is critical. Under certain conditions, cellulose can be chemically changed in the presence of active sites in chains. The cellobiose ring is made up of three hydroxyl groups: secondary (C2 and C3) and primary (C6) alcohol groups, which allow it to be substituted for other functional groups and lengthy chains, and it may also be oxidized [50].

Figure 4 illustrates several surface modification methods of the nanocellulose which including esterification/acetylation, silylation, TEMPO-mediated oxidation, sulfonation, phosphorylation, amidation, carbamation, grafting-onto, grafting-from and non-covalent cross-linking. A comprehensive discussion on these methods could be found in the review compiled by Ghasemlou et al. [51].

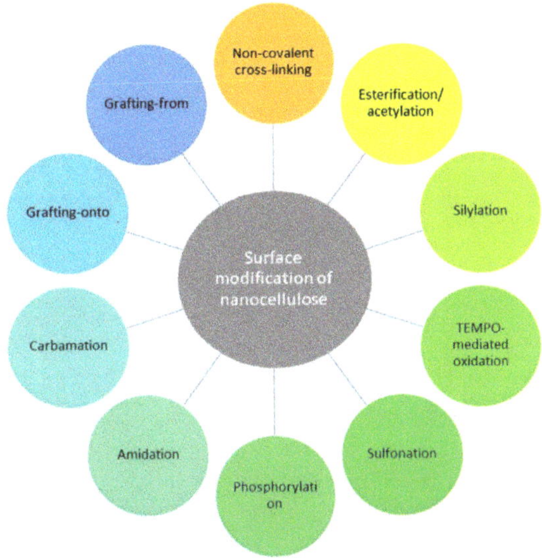

Figure 4. Surface modification methods of nanocellulose (summarized from Ghasemlou et al. [51]).

Based on the available literatures, in the production of nanocellulose reinforced natural rubber nanocomposites, esterification, silylation and TEMPO-mediated oxidation are the most widely used surface modification methods to modify the nanocellulose. Table 4 summarizes the surface modification done on the nanocellulose and their reinforcement effect on the natural rubber nanocomposite.

Table 4. Recent reports on surface modified nanocellulose (NC) reinforced natural rubber nanocomposites.

Nanocellulose	Surface Modification	Nanocomposite	Findings	References
Nanocrystalline cellulose (NCC) from commercial microcrystalline cellulose (MCC)	Silylation—3-aminopropyl-triethoxysilane	Natural rubber (NR)/NCC/silica nanocomposite at different ratios of NCC and silica (0:30, 5:25, 10:20, 15:15, 20:10 and 25:5)	- Enhanced dispersion and interfacial strength for modified NCC - Nanocomposites reinforced with modified NCC exhibited higher tensile strength, modulus, elongation at break, tear strength and hardness compared to that of unmodified NCC	[52]
Cellulose nanocrystals (CNC) from kenaf bast fiber	Silylation—0.5 wt% silane	CNC reinforced unsaturated polyester resins (UPR) toughened with liquid natural rubber (LNR) at 2, 4 and 6 wt% CNC content	- Compatibility between UPR and LNR has been improved after reinforced with modified CNC - Lower tensile strength and modulus but higher impact energy was observed for modified CNC compared to that of unmodified CNC nanocomposite - Viscoelastic behavior and thermal resistance of the modified CNC nanocomposite was slightly lower than unmodified CNC	[53]
Crystalline nanocellulose (CNC) from ramie fiber	Silylation—Different organosilanes: (3-Aminopropyl)triethoxysilane (APTES), 3-aminopropyl-triethoxysilane, bis[3-(triethoxysilyl)propyl] tetrasulfide (TESPT), (3-mercaptopropyl) trimethoxysilane (MPTMS)	Natural rubber (NR)/CNC nanocomposite at 2.5 and 5 wt% CNC content	- Crystallinity index of unmodified CNC, APTES-MCNC, TESPT-MCNC, and MPTMS-MCNC was 74%, 66%, 70% and 51%, respectively. - Tensile properties of the natural rubber reinforced with modified CNC was significantly higher than that of unmodified CNC	[54]
Cellulose nanofibrils (CNF) from Napier grass stem	Silylation—Bis-(triethoxysilyl-propyl) tetrasulfide (TESPT)	Natural rubber (NR)/CNF nanocomposite at 0.5, 1, 3, 5 and 10 wt% filler loading	- Lower degree of crystallinity of CNF after modified with TESPT (74.3% in unmodified CNF and 69.5% in modified CNF) - TESPT-modified CNF has lower onset temperature and maximum decomposition temperature compared to unmodified CNF - Better rubber-filler interaction (indicates by bound rubber content) was observed in modified CNF filled nanocomposite - Greater reinforcing effect was shown by modified CNF as higher modulus and hardness and tensile strength of the resultant nanocomposites was recorded - Performance of nanocomposite improved with increasing filler loading up to 5 wt% and level-off beyond this loading level	[55]
Nanocrystalline cellulose (NCC) and nanofiber cellulose (NFC) from lower part of empty fruit bunches	TEMPO-mediated oxidation—2,2,6,6-tetramethyl-1-piperidinyloxyl	Polypropylene (PP)/cyclic natural rubber (CNR)/NCC or NFC nanocomposite with 1, 2 and 3 wt% NCC content	- Addition of NCC decreases the tensile strength and modulus of the PP/CNR nanocomposites by 13 and 56%, respectively and 56% higher in elongation at break was recorded - 16 and 25% increment in tensile strength and modulus was recorded when NFC was added and 5% decrement in elongation at break was observed - Nanocomposite added with NCC has better thermal stability than NFC	[56]

Table 4. *Cont.*

Nanocellulose	Surface Modification	Nanocomposite	Findings	References
Commercial cellulose nanofibers (CNF)	TEMPO-mediated oxidation—2,2,6,6-tegramethylpyperidine-1-oxyl	TEMPO-CNF/ nitrile-butadiene rubber (NBR) and sheets and carboxy group-containing nitrile-butadiene rubber (XNBR) composite	- Tensile strength, storage modulus at 23 °C, work of fracture, and elongation at break of the TEMPO-CNF/XNBR nanocomposites are higher than that of control XNBR nanocomposites	[57]
Nanofibrillated cellulose (NFC) from bleached softwood bisulfite pulp	TEMPO-mediated oxidation—2,2,6,6-tetramethylpiperidine-1-oxyl	NFC reinforced latex nanocomposite at 1, 2, 3, 4 and 5 wt% NFC content	- The ultimate strength and elastic modulus of the nanocomposites improved significantly with addition of up to 3 wt% NFC content while 5 wt% resulted in the highest values	[58]
Cellulose Nanocrystals (CNC) from cotton cellulose	Esterification of thiols—mercaptoundecanoic acid mixed with acetic anhydride, glacial acetic acid, and concentrated sulfuric acid	Natural rubber/CNC nanocomposite at 2, 5 and 10 wt% CNC content	- Smooth and homogenous surface structure similar to that of the neat natural rubber was observed after incorporation of modified CNC - Higher crosslink density and significantly increased tensile strength and strain-to-failure observed in modified CNC nanocomposite compared to unmodified CNC - The improvement was significant at 5 wt% CNC content but level-off when 10 wt% CNC content was used	[34]
Cellulose nanofibers (CNF) from never-dried bleached softwood kraft pulp	Esterification—unsaturated fatty acids (oleic acid and stearic acid)	sulfur-vulcanized natural rubber reinforced with CNF at 1, 3 and 5 wt%	- Good dispersion of modified CNF was observed as the resultant natural rubber was transparent at CNF content of 5 wt% - 2 to 3 fold increment in Young's modulus when 1 and 3 wt% modified CNF was added to the natural rubber compared to unmodified CNF. At higher CNF content (5 wt%), the increment is more significant - Oleoyl CNF displayed higher interaction with natural rubber compared to stearoyl CNF	[59]
Nanocrystalline cellulose (NCC) from softwood pulp	Non-covalent surface modification—cationic surfactant, cetyltrimethyl ammonium bromide (CTMAB)	Natural rubber (NR) composite reinforced with NCC at 5, 10, 15 and 20 wt%	- Lower crystallinity was observed in modified NCC (69.4%) compared to unmodified NCC (72.2%) - Better dispersion in NR was observed for modified NCC - Modified NCC accelerated the vulcanization process of NR - At lower NCC content (≤10 wt%), NR composites reinforced with modified NCC displayed superior tensile strength, tear strength and abrasion resistance than their unmodified NCC counterparts	[60]
Bacterial cellulose (BC) from modified Hestrin Shran liquid culture medium	Admicelar polymerization of styrene at the surface of the BC nanofibers	Natural-rubber based nanocomposites reinforced with bacterial cellulose (BC) and bacterial cellulose coated with polystyrene (BCPS) at 1, 2.5, 5 and 10 wt%	- Nanocomposites with low BC and BCPS fibers content (1 and 2.5 wt%) did not show significant improvement compared to natural rubber - At higher fiber content (5 and 10 wt%), the tensile strength and Young's modulus of the nanocomposites increased significantly - No significant difference was found between BC and BCPS	[61]

4.1. Esterification

Esterification is the easiest reaction in order to remove the hydroxyl groups from nanocellulose. Esterification is a chemical reaction between acid (carboxylic acid) and alcohol (or other -OH) to form ester and water [62]. In a study by Kanoth et al. [34], cross-linkable mercapto group has been added covalently on the surface of cellulose nanocrystals (CNC) via esterification with 11-mercaptoundecanoic acid. The study intended to explore the synergistic effect of CNC as both reinforcing filler and crosslinker agent. The surface-grafted thiol (-SH) groups from 11-mercaptoundecanoic acid on CNC lead to thiol-ene coupling reaction with alkene groups in the molecules of natural rubber matrix to form covalent crosslinks (Figure 5). As a result, an effective chemical bonding between the CNC and natural rubber matrix interface is formed. Therefore, natural rubber nanocomposite reinforced with modified CNC has good dispersion and higher crosslink density than unmodified CNC. A uniaxial tensile test was conducted to evaluate the mechanical strength of the natural rubber/CNC nanocomposites. At similar nanocellulose content (10 wt%), natural rubber nanocomposite reinforced with modified CNC has significantly better tensile strength (10.2 MPa vs. 4.2 MPa), strain-to-failure (750% vs. 1210%), modulus (1.75 MPa vs. 1.86 MPa) and work-of-fracture (1.56 MJ m^{-3} vs. 4.60 MJ m^{-3}) than that of natural rubber nanocomposite reinforced with unmodified CNC. In addition, natural rubber nanocomposite reinforced with modified CNC also has better preservation of stiffness than unmodified CNC nanocomposites when subjected to repeated loading cycles. The authors reported that the stress softening effect of natural rubber nanocomposite reinforced with thiol-modified CNC in their study was 4–6 times better than the natural rubber reinforced with carbon black as reported by Harwood et al. [35].

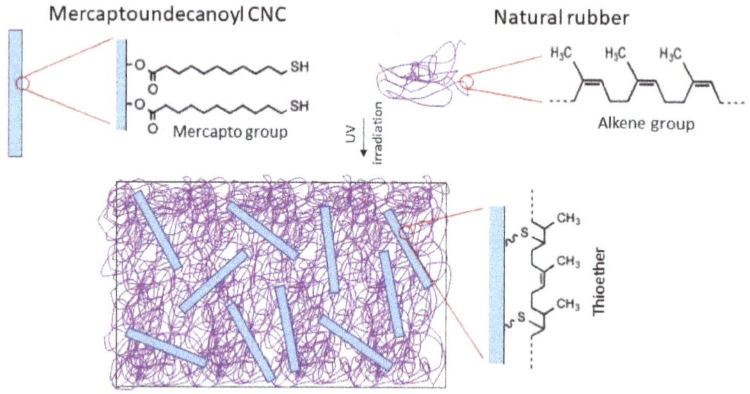

Figure 5. Illustration of the thiol-ene coupling reaction between natural rubber and modified cellulose nanocrystals in nanocomposite (adapted with permission from Kanoth et al. [34]. Copyright 2015 American Chemical Society, Washington, United States).

Apart from that, esterification of nanocellulose could also be carried out using unsaturated fatty acids such as oleic acid and stearic acid. Natural rubber reinforced with oleic acid and stearic acid modified cellulose nanofibers (CNF) at CNF content of 5 wt% portrayed a transparent feature, indicates well dispersion of modified CNF in natural rubber [59]. Esterification has enhanced the reinforcing efficiency of the CHF as natural rubber reinforced with modified CNF displayed better Young's modulus than unmodified CNF at any level of CNF content (1, 3 and 5 wt%). It is noteworthy that oleoyl CNF with double bonds resulted in better mechanical behavior of the natural rubber composite compared to that of stearoyl CNF. Lower swelling of 241.1% for the oleoyl CNF/natural rubber nanocomposite was also observed compared to 272% in stearoyl CNF/natural nanocomposite. It can be concluded that oleoyl CNF with double bonds achieves higher level of interaction with natural rubber.

4.2. Silylation

Silyation is one of the frequently used modification methods to surface modified nanocellulose where silyl groups are introduced to the surface of nanocellulose. Figure 6 displays the grating mechanism of different amino silanes on the hydroxyl groups of cellulose of cellulose nanofiber (CNF) films. Grafting of amino silanes happened on the accessible amorphous regions of CNF. Therefore, fibers thickening and formation of three dimension al silane networks at the surface of CNF films could be observed as result of self-condensation reaction [63].

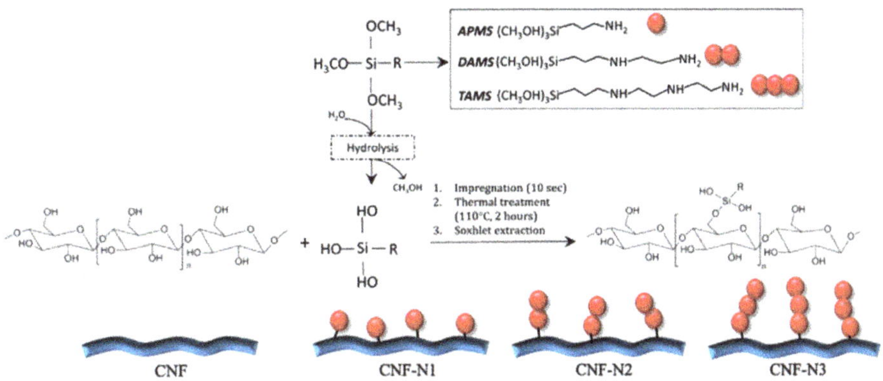

Figure 6. Schematic illustration of the mechanism for aqueous-based silylation reaction with amino silanes on the surface of cellulose nanofibers films (Saini et al. [63]; with permission from Elsevier Science Ltd., Amsterdam, Netherlands).

Xu et al. [52] modified nanocrystalline cellulose (NCC) with 3-aminopropyl-triethoxysilane and the modified NCC were reinforced into natural rubber matrix with different NCC:silica ratio. Addition of both unmodified NCC and modified NCC accelerated the vulcanization rate of natural rubber. The Payne effect, a particular feature of the stress-strain behavior of rubber, decreased along with increasing content of unmodified and modified NCC. Meanwhile, the mechanical properties of the natural rubber nanocomposites were significantly improved by the addition of modified NCC compared to that of unmodified NCC as a result of a more uniform dispersion of modified NCC in the natural rubber matrix. By the means of scanning electron microscopy (SEM) micrographs, it was revealed that the unmodified NCC aggregated more heavily in the natural rubber matrix. On the contrary, more uniform dispersion and lesser aggregations were found in the case of modified NCC, resulted in better bonding between natural rubber and nanocellulose. Consequently, nanocomposites reinforced with modified NCC portrayed better performance on the macroscale.

The crystallinity of the nanocellulose was reduced by chemical modification and subsequently weakened its reinforcement potential in polymer composites [64]. However, Singh et al. [54] proved that the silylation treatment did not affect the crystallinity of the CNC adversely as the modification only occurred mostly at the surface of the CNC. In a study by Somseemee et al. [55], the crystallinity of the CNF modified by Bis-(triethoxysilyl-propyl) tetrasulfide (TESPT) showed a slight decrease from 74.3% to 69.5%. Similar to the previous studies, better dispersion is observed for modified CNC in the natural rubber matrix. Unmodified CNC exhibited agglomerated nature owing to the internal hydrogen binding between the spherical CNC particles. In addition, non-polar natural rubber also restricted the well-dispersion of polar CNC in the natural rubber matrix. This problem was overcome by the surface modification of CNC with (3-Aminopropyl)triethoxysilane (APTES), where the polar nature of the CNC was reduced after modification [54]. Moreover, aliphatic chains from APTES could coat over the surface of CNC and improve its compatibility with natural rubber. The chemical reaction between the hydroxyl group of nanocellulose and hydrolyzable alkoxy group of TESPT has contributed to the improvement in compatibility of nanocellulose with natural rubber (Figure 7).

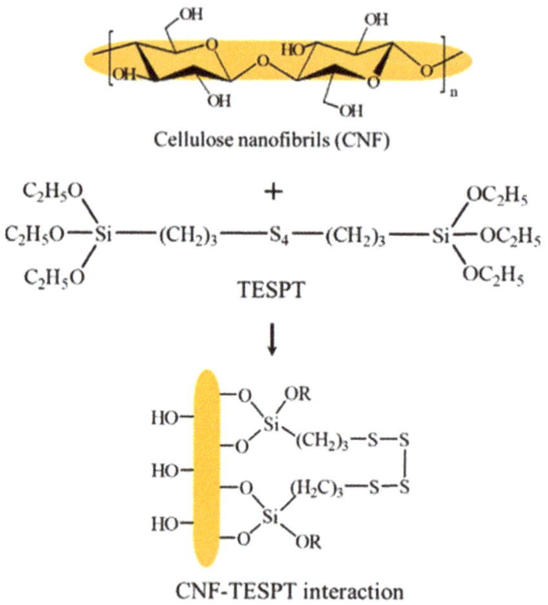

Figure 7. Proposed mechanism of interaction between CNF and TESPT (Somseemee et al. [55]; with permission from Elsevier Science Ltd., Amsterdam, Netherlands).

Nevertheless, some contrary findings have also been reported. Kargarzadeh et al. [53] modified cellulose nanocrystals (CNCs) using 0.5 wt% silane. The modified CNC was then used to prepared unsaturated polyester resin (UPR) nanocomposites toughened with liquid natural rubber (LNR). The study revealed that the reinforcement with both unmodified and modified CNC improved the tensile properties, impact energy, viscoelastic behavior and thermal resistance of the nanocomposite. However, when comparing between unmodified and modified CNC, nanocomposites reinforced with silane modified CNC displayed lower tensile strength and modulus compared to that of its unmodified CNC counterparts. The authors attributed the decrement in tensile properties to the weaker intramolecular interactions between the hydroxyl groups of the modified CNC with the LNR. Therefore, a CNC network failed to form and the stress transferring between rubber and UPR was lessened which subsequently led to inferior tensile strength. In addition, the fact that the crystallinity of CNC was reduced by the silane treatment and its subsequent lesser stiffening effects has also contributed to the decrement in tensile strength of the nanocomposites [65]. Additionally, the lesser crystalline and rigid characteristics of silane modified CNC also resulted in lower glass transition temperature (T_g) compared to that of unmodified CNC. Nevertheless, higher impact properties were observed from the nanocomposites reinforced with modified CNC, mainly due to their lower propensity toward self-aggregation. Correspondingly, nanocomposites with a more flexible filler-matrix interface and reduced stress concentrations could be fabricated. Based on the findings, it was concluded that the use of CNC modified with silane is less preferential as the treatment weakens the interaction between CNC and LNR.

4.3. TEMPO-Mediated Oxidation

According to Pierre et al. [66], TEMPO-mediated oxidation is deemed as an effective treatment to enhance the stability of nanocellulose suspensions via bestowing negative charges onto the surface of nanocellulose. As shown in Figure 8, during the TEMPO-mediated oxidation, the C6 hydroxy groups of nanocellulose transform into carboxylate groups selectively, in the presence of sodium hypochlorite (NaClO) and sodium bromide (NaBr) [51]. These carboxylate groups could then assist in obtaining nanocellulose with different functionalities by acting as a platform to gather metals ions by ion exchange [67].

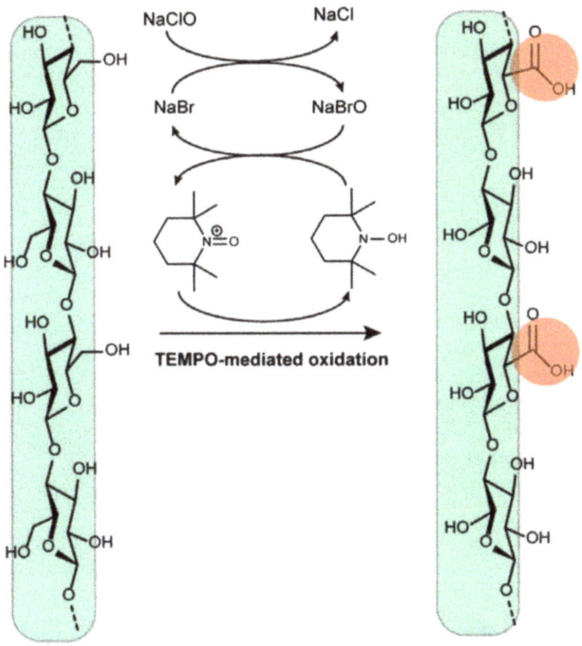

Figure 8. Schematic illustration of the TEMPO-mediated oxidation where it selectively transforms C6-hydroxy groups of nanocellulose to carboxylate groups (Ghasemlou et al. 2021; with permission from Elsevier Science Ltd., Amsterdam, Netherlands).

Mahendra et al. [56] TEMPO-oxidized nanocrystalline cellulose (NCC) and nanofiber cellulose (NFC) from lower part of empty fruit bunches and reinforced them into polypropylene (PP)/cyclic natural rubber (CNR) (80/20) nanocomposites. NCC produced in this study has larger diameter and lower aspect ratio (L/D) compared to that of the NFC. Consequently, owing to its larger dimensions, NCC has higher degradation temperature than NFC. Generally, PP/CNR nanocomposites exhibited inferior tensile strength when modified NCC and NFC was added and the decrement in tensile strength increased with increasing nanocellulose content from 1 to 3 wt%. The reduction in tensile strength is a common observation in natural rubber composites reinforced with lignocellulosic fibers. The fibers are more rigid and therefore will experience fractures earlier than the natural rubber matrix. Similar observation was found for Young's modulus in the PP/NCR nanocomposite reinforced with NCC. However, Young's modulus of the PP/NCR nanocomposite reinforced with NCC reinforced with NFC was higher and increased along with increasing nanocellulose content. As NFC has higher aspect ratio than NCC, the interfacial contact surface of NFC and CNR are better and hence enhanced reinforcement effect was obtained. SEM micrographs also proved that NFC resulted in a better dispersion in CNR matrix compared to that of NCC. On the contrary, NCC reinforced PP/CNR nanocomposite possesses higher thermal stability than that of its NFC counterpart.

4.4. Other Modifications

In a study by Jiang et al. [60], nanocrystalline cellulose (NCC) modified by cetyltrimethyl ammonium bromide (CTMAB) has been reinforced into natural rubber at an NCC content of 0, 5, 10, 15 and 20 wt%. Modification with CTMAB has increased the size of the NCC as CTMAB was absorbed on the surface of NCC. Additionally, the charge on the surface of NCC became positive after modification. In comparison to unmodified NCC, CTMAB modified NCC exhibited slightly lower crystallinity index. However, the crystallinity of NCC is still considered high after modification. The authors also reported that the modified NCC may have assisted in accelerating the vulcanization of natural rubber and shorter curing time was observed in modified NCC nanocomposites. One of the prominent observations is the improvement of dispersion of NCC in the natural rubber as the surfactant on the NCC surface prevented the agglomeration of NCC. Nevertheless, it only could be observed when low NCC content (5 and 10 wt%) was used. Modification has improved the interfacial interaction between NCC and natural rubber as proved by the decrement of tan δ in glass transition region. Similar findings were also recorded for mechanical properties of the nanocomposites. At 5 and 10 wt% NCC content, natural rubber nanocomposites reinforced with modified NCC displayed better tensile strength, tear strength and abrasion resistance compared to that of the unmodified NCC. The improvement in mechanical properties is mainly attributed to improved dispersion and enhanced interfacial interaction as a result of surface medication of NCC. When the NCC content was 15% and above, the dispersion reduced and resulted in inferior mechanical properties. Moreover, addition of modified NCC also resulted in better wet-skid resistance and aging resistance of the resultant nanocomposites. Overall, addition of 10% CTMAB modified NCC led to the optimum properties.

Apart from the aforementioned methods, the surface of nanocellulose could also be decorated with polystyrene. A study by Trovatti et al. [61] demonstrated admicelar polymerization of polystyrene at the surface of the bacterial cellulose (BC) nanofibers. The purpose of the treatment was to ensure a better dispersion into the natural rubber matrix by reducing its hydrophilicity and polarity. The polystyrene coated BC nanofibers remained its original morphology but with more hydrophobic nature and decreased surface energy as a result of partially masking of hydroxyl groups by the polystyrene sleeve. Tensile strength and Young's modulus of the nanocomposites reinforced with unmodified BC nanofibers and polystyrene coated BC nanofibers exhibited significant improvement. However, no significant difference in terms of mechanical performance between the two. It should be noted that, at lower BC nanofibers content (1 and 2.5 wt%), the resultant nanocomposite did not differ much with the neat natural rubber. The advantages of the incorporation of BC nanofibers could only be shown when higher content (5 and 10 wt%) of BC nanofibers were added to the natural rubber matrix.

5. Conclusions, Challenges and Future Perspectives

As the demand from automotive industry keep increasing over the years, the application and supply of natural rubber is undoubtedly pivotal. However, environmental issues and health hazards that triggered by the rubber industry is a problem that has come to the foreground in recent years. More and more stringent environmental regulations have been imposed to prevent the rubber industry to produce some types of rubber. For instance, butyl rubber has been classified as a major source of Hazardous Air Pollutants (HAP) emission by The Environmental Protection Agency (EPA) during its production process. Prolonged and acute exposure to the chemicals used in the production of industrial rubber could be detrimental and lethal to humans' health. Moreover, the majority of the raw materials used are petroleum-based and non-renewable and therefore lead to high environmental impact [68].

Therefore, the development of green-based natural rubber materials has become more and more important in ensuring environmental safety. The demand for eco-friendly rubber is increasing worldwide. At this moment, the advantages of nanocellulose derived from renewable resources as reinforcing filler to natural rubber has become irreplaceable. However, the dispersion of nanocellulose into the natural rubber matrix is one of the main challenges during the production of nanocellulose reinforced nanocomposites [69]. Another problem in the production of these nanocomposites is the poor compatibility of hydrophobic natural rubber and hydrophilic nanocellulose [70]. Various surface modification methods of nanocellulose have been reported in this review. Encouraging breakthrough and improved nanocellulose dispersion into the natural rubber matrix via esterification, silylation and TEMPO-mediated oxidization has been reported. Better interfacial compatibility was achieved and subsequently nanocomposites with improved mechanical, curing, dynamic mechanical and thermal properties were produced.

Even so, the commercial suitability and the ability of large-scale processing of these nanocomposites is still questionable [71]. Apart from that, the cumbersome process also makes the production of nanocellulose intensively laborious. Consequently, manufacturers are always facing dilemma in judging and weighing between the advantages bestowed by nanocellulose and the disadvantages of arduous processing of nanocellulose reinforced natural rubber nanocomposites. Volatile global rubber price is also an issue that must be paid attention to as the price of rubber is anticipated to slowly rise up to the year 2030.

Despite the aforementioned challenges, nanocellulose based natural rubber composites are still considered as very promising future materials to be used in many different sectors. The evidence from several past studies in this review suggests that the natural rubber nanocomposites reinforced with surface modified nanocellulose could be potentially used in elastic packaging in food and medical applications owing to their excellent mechanical properties and thermal stability. Meanwhile, thanks to the environmentally friendly nature and biodegradability of nanocellulose, plenty of green composites with superior thermal, mechanical and barrier properties could be produced. Nanocellulose could serve as an ideal filler in the automobile industry in producing green tires [72].

Author Contributions: Conceptualization, S.H.L., S.F.M. and S.S.O.A.E.; data curation, N.M.F.H., W.C.L. and S.H.L.; writing—original draft preparation, N.M.F.H., W.C.L. and S.H.L.; writing—review and editing, S.F.M., W.C.L., S.H.L. and A.A.; supervision, B.-D.P., S.F.M. and S.S.O.A.E.; funding acquisition, S.H.L. and B.-D.P. All authors have read and agreed to the published version of the manuscript.

Funding: This research was funded by the Fundamental Research Grant Scheme (FRGS 2018-1), Reference code: FRGS/1/2018/WAB07/UPM//1 and Higher Institution Centre of Excellence (HICoE) (UPM/INTROP/100-13/9/3/6369113/H), provided by Ministry of Higher Education, Malaysia.

Institutional Review Board Statement: Not applicable.

Informed Consent Statement: Not applicable.

Data Availability Statement: Not applicable.

Acknowledgments: The authors would like to acknowledge the Library of Universiti Putra Malaysia for providing access to the literatures.

Conflicts of Interest: The authors declare no conflict of interest.

References

1. Van Beilen, J.B.; Poirier, Y. Establishment of new crops for the production of natural rubber. *Trends Biotechnol.* **2007**, *25*, 22–29. [CrossRef]
2. Cornish, K. Similarities and differences in rubber biochemistry among plant species. *Phytochemistry* **2001**, *57*, 1123–1134. [CrossRef]
3. Cataldo, F. Guayule rubber: A new possible world scenario for the production of natural rubber. *Prog. Rubber Plast. Technol.* **2000**, *16*, 31–59.
4. Wikipedia. Natural Rubber. Available online: https://en.wikipedia.org/wiki/Natural_rubber (accessed on 2 September 2021).
5. Statista. Leading Natural Rubber Producing Countries Worldwide in 2019 and 2020. Available online: https://www.statista.com/statistics/275397/caoutchouc-production-in-leading-countries/ (accessed on 29 August 2021).
6. Mordor Intelligence. Natural Rubber Market—Growth, Trends, COVID-19 Impact, and Forecasts (2021–2026). Available online: https://www.mordorintelligence.com/industry-reports/natural-rubber-market (accessed on 23 August 2021).

7. Expert Market Research. Global Natural Rubber Market Outlook. Available online: https://www.expertmarketresearch.com/reports/natural-rubber-market (accessed on 11 September 2021).
8. Mariano, M.; El Kissi, N.; Dufresne, A. Cellulose nanocrystal reinforced oxidized natural rubber nanocomposites. *Carbohydr. Polym.* **2016**, *137*, 174–183. [CrossRef] [PubMed]
9. Zhou, Y.; Fan, M.; Chen, L.; Zhuang, J. Lignocellulosic fibre mediated rubber composites: An overview. *Compos. Part B Eng.* **2015**, *76*, 180–191. [CrossRef]
10. Low, D.Y.S.; Supramaniam, J.; Soottitantawat, A.; Charinpanitkul, T.; Tanthapanichakoon, W.; Tan, K.W.; Tang, S.Y. Recent developments in nanocellulose-reinforced rubber matrix composites: A review. *Polymers* **2021**, *13*, 550. [CrossRef] [PubMed]
11. Kargarzadeh, H.; Mariano, M.; Huang, J.; Lin, N.; Ahmad, I.; Dufresne, A.; Thomas, S. Recent developments on nanocellulose reinforced polymer nanocomposites: A review. *Polymer* **2017**, *132*, 368–393. [CrossRef]
12. Nunes, R.C. Rubber Nanocomposites with Nanocellulose. In *Progress in Rubber Nanocomposites*; Thomas, S., Maria, H.J., Eds.; Woodhead Publishing: Sawston, UK, 2017; pp. 463–494.
13. Sivaselvi, K.; Varma, V.S.; Harikumar, A.; Jayaprakash, A.; Sankar, S.; Krishna, C.Y.; Gopal, K. Improving the mechanical properties of natural rubber composite with carbon black (N220) as filler. *Mater. Today Proc.* **2021**, *42*, 921–925. [CrossRef]
14. Salaeh, S.; Nakason, C. Influence of modified natural rubber and structure of carbon black on properties of natural rubber compounds. *Polym. Compos.* **2012**, *33*, 489–500. [CrossRef]
15. Choi, S.S.; Nah, C.; Jo, B.W. Properties of natural rubber composites reinforced with silica or carbon black: Influence of cure accelerator content and filler dispersion. *Polym. Int.* **2003**, *52*, 1382–1389. [CrossRef]
16. Abdul Wahab, M.K.; Ismail, H.; Othman, N. Effects of dynamic vulcanization on the physical, mechanical, and morphological properties of high-density polyethylene/(natural rubber)/(thermoplastic tapioca starch) blends. *J. Vinyl Addit. Technol.* **2012**, *18*, 192–197. [CrossRef]
17. Dileep, P.; Varghese, G.A.; Sivakumar, S.; Narayanankutty, S.K. An innovative approach to utilize waste silica fume from zirconia industry to prepare high performance natural rubber composites for multi-functional applications. *Polym. Test.* **2020**, *81*, 106172. [CrossRef]
18. Charoenchai, M.; Tangbunsuk, S.; Keawwattana, W. Silica-graphene oxide nanohybrids as reinforcing filler for natural rubber. *J. Polym. Res.* **2020**, *27*, 1–16. [CrossRef]
19. Boonmee, A.; Jarukumjorn, K. Preparation and characterization of silica nanoparticles from sugarcane bagasse ash for using as a filler in natural rubber composites. *Polym. Bull.* **2020**, *77*, 3457–3472. [CrossRef]
20. Thuong, N.T.; Dung, T.A.; Yusof, N.H.; Kawahara, S. Controlling the size of silica nanoparticles in filler nanomatrix structure of natural rubber. *Polymer* **2020**, *195*, 122444. [CrossRef]
21. Xia, L.; Song, J.; Wang, H.; Kan, Z. Silica nanoparticles reinforced natural rubber latex composites: The effects of silica dimension and polydispersity on performance. *J. Appl. Polym. Sci.* **2019**, *136*, 47449. [CrossRef]
22. Ahmed, K.; Nizami, S.S.; Riza, N.Z. Reinforcement of natural rubber hybrid composites based on marble sludge/Silica and marble sludge/rice husk derived silica. *J. Adv. Res.* **2014**, *5*, 165–173. [CrossRef]
23. Sombatsompop, N.; Thongsang, S.; Markpin, T.; Wimolmala, E. Fly ash particles and precipitated silica as fillers in rubbers. *I. Untreated fillers in natural rubber and styrene–butadiene rubber compounds. J. Appl. Polym. Sci.* **2004**, *93*, 2119–2130.
24. Sadequl, A.M.; Poh, B.T.; Ishiaku, U.S. Effect of filler loading on the mechanical properties of epoxidized natural rubber (ENR 25) compared with natural rubber (SMR L). *Int. J. Polym. Mater.* **1999**, *43*, 261–278. [CrossRef]
25. Torres, L.F.; McCaffrey, Z.; Washington, W.; Williams, T.G.; Wood, D.F.; Orts, W.J.; McMahan, C.M. Torrefied agro-industrial residue as filler in natural rubber compounds. *J. Appl. Polym. Sci.* **2021**, *138*, 50684. [CrossRef]
26. Masłowski, M.; Miedzianowska, J.; Czylkowska, A.; Strzelec, K. Horsetail (*Equisetum arvense*) as a functional filler for natural rubber biocomposites. *Materials* **2020**, *13*, 2526. [CrossRef] [PubMed]
27. Miedzianowska, J.; Masłowski, M.; Rybiński, P.; Strzelec, K. Properties of chemically modified (selected silanes) lignocellulosic filler and its application in natural rubber biocomposites. *Materials* **2020**, *13*, 4163. [CrossRef]
28. Moonart, U.; Utara, S. Effect of surface treatments and filler loading on the properties of hemp fiber/natural rubber composites. *Cellulose* **2019**, *26*, 7271–7295. [CrossRef]
29. Sareena, C.; Ramesan, M.T.; Purushothaman, E. Utilization of coconut shell powder as a novel filler in natural rubber. *J. Reinf. Plast. Compos.* **2012**, *31*, 533–547. [CrossRef]
30. Chong, E.L.; Ahmad, I.; Dahlan, H.M.; Abdullah, I. Reinforcement of natural rubber/high density polyethylene blends with electron beam irradiated liquid natural rubber-coated rice husk. *Radiat. Phys. Chem.* **2010**, *79*, 906–911. [CrossRef]
31. Vladkova, T.G.; Dineff, P.D.; Gospodinova, D.N.; Avramova, I. Wood flour: New filler for the rubber processing industry. IV. Cure characteristics and mechanical properties of natural rubber compounds filled by non-modified or corona treated wood flour. *J. Appl. Polym. Sci.* **2006**, *101*, 651–658. [CrossRef]
32. Ismail, H.; Nurdin, H.I. Tensile properties and scanning electron microscopy examination of the fracture surface of oil palm wood flour/natural rubber composites. *Iran. Polym. J.* **1998**, *7*, 53–58.
33. Kulshrestha, U.; Gupta, T.; Kumawat, P.; Jaiswal, H.; Ghosh, S.B.; Sharma, N.N. Cellulose nanofibre enabled natural rubber composites: Microstructure, curing behaviour and dynamic mechanical properties. *Polym. Test.* **2020**, *90*, 106676. [CrossRef]

34. Kanoth, P.B.; Claudino, M.; Johansson, M.; Berglund, L.A.; Zhou, Q. Biocomposites from natural rubber: Synergistic effects of functionalized cellulose nanocrystals as both reinforcing and cross-linking agents via free-radical thiol–ene chemistry. *ACS Appl. Mater. Interfaces* **2015**, *7*, 16303–16310. [CrossRef]
35. Harwood, J.A.C.; Mullins, L.; Payne, A.R. Stress softening in natural rubber vulcanizates. Part ii. Stress softening effects in pure gum and filler loaded rubbers. *J. Appl. Polym. Sci.* **1965**, *9*, 3011–3021. [CrossRef]
36. Zinge, C.; Kandasubramanian, B. Nanocellulose based biodegradable polymers. *Eur. Polym. J.* **2020**, *133*, 109758. [CrossRef]
37. Zhang, K.; Barhoum, A.; Xiaoqing, C.; Li, H.; Samyn, P. Cellulose Nanofibers: Fabrication and Surface Functionalization Techniques. In *Handbook of Nanofibers*; Barhoum, A., Bechelany, M., Makhlouf, A.S.H., Eds.; Springer Nature Switzerland AG: Cham, Switzerland, 2019; pp. 409–449.
38. Gopakumar, D.A.; Thomas, S.; Grohens, Y. Nanocelluloses as Innovative Polymers for Membrane Applications. In *Multifunctional Polymeric Nanocomposites Based on Cellulosic Reinforcements*; Puglia, D., Fortunati, E., Kenny, J.M., Eds.; William Andrew Publishing: Norwich, NY, USA, 2016; pp. 253–275.
39. Supramaniam, J.; Wong, S.K.; Leo, B.F.; Tan, L.T.H.; Goh, B.H.; Tang, S.Y. Unravelling the swelling behaviour and antibacterial activity of palm cellulose nanofiber-based metallic nanocomposites. *IOP Conf. Ser. Mater. Sci. Eng.* **2020**, *778*, 012027. [CrossRef]
40. Mokhena, T.C.; John, M.J. Cellulose nanomaterials: New generation materials for solving global issues. *Cellulose* **2019**, *27*, 1149–1194. [CrossRef]
41. Muqeet, M.; Mahar, R.B.; Gadhi, T.A.; Halima, N.B. Insight into cellulose-based-nanomaterials—A pursuit of environmental remedies. *Int. J. Biol. Macromol.* **2020**, *163*, 1480–1486. [CrossRef] [PubMed]
42. George, J.; Sabapathi, S.N. Cellulose nanocrystals: Synthesis, functional properties, and applications. *Nanotechnol. Sci. Appl.* **2015**, *8*, 45–54. [CrossRef] [PubMed]
43. Abitbol, T.; Rivkin, A.; Cao, Y.; Nevo, Y.; Abraham, E.; Ben-Shalom, T.; Lapidot, S.; Shoseyov, O. Nanocellulose, a tiny fiber with huge applications. *Curr. Opin. Biotechnol.* **2016**, *39*, 76–88. [CrossRef]
44. Vollick, B.; Kuo, P.Y.; Alizadehgiashi, M.; Yan, N.; Kumacheva, E. From structure to properties of composite films derived from cellulose nanocrystals. *ACS Omega* **2017**, *2*, 5928–5934. [CrossRef] [PubMed]
45. Leng, J.; Li, G.; Ji, X.; Yuan, Z.; Fu, Y.; Li, H.; Qin, M.; Moehwald, H. Flexible latex photonic films with tunable structural colors templated by cellulose nanocrystals. *J. Mater. Chem. C* **2018**, *6*, 2396–2406. [CrossRef]
46. Farooq, A.; Patoary, M.K.; Zhang, M.; Mussana, H.; Li, M.; Naeem, M.A.; Mushtaq, M.; Farooq, A.; Liu, L. Cellulose from sources to nanocellulose and an overview of synthesis and properties of nanocellulose/zinc oxide nanocomposite materials. *Int. J. Biol. Macromol.* **2020**, *154*, 1050–1073. [CrossRef]
47. Stanisławska, A. Bacterial Nanocellulose as a Microbiological Derived Nanomaterial. *Adv. Mater. Sci.* **2016**, *16*, 45–57. [CrossRef]
48. Klemm, D.; Cranston, E.D.; Fischer, D.; Gama, M.; Kedzior, S.A.; Kralisch, D.; Kramer, F.; Kondo, T.; Lindström, T.; Nietzsche, S.; et al. Nanocellulose as a natural source for groundbreaking applications in materials science: Today's state. *Mater. Today* **2018**, *21*, 720–748. [CrossRef]
49. Gelir, A.; Yargi, O.; Yuksel, S.A. Elucidation of the pore size and temperature dependence of the oxygen diffusion into porous silicon. *Thin Solid Films* **2017**, *636*, 602–607. [CrossRef]
50. Kargarzadeh, H.; Mariano, M.; Gopakumar, D.; Ahmad, I.; Thomas, S.; Dufresne, A.; Huang, J.; Lin, N. Advances in cellulose nanomaterials. *Cellulose* **2018**, *25*, 2151–2189. [CrossRef]
51. Ghasemlou, M.; Daver, F.; Ivanova, E.P.; Habibi, Y.; Adhikari, B. Surface modifications of nanocellulose: From synthesis to high-performance nanocomposites. *Prog. Polym. Sci.* **2021**, *119*, 101418. [CrossRef]
52. Xu, S.H.; Gu, J.; Luo, Y.F.; Jia, D.M. Effects of partial replacement of silica with surface modified nanocrystalline cellulose on properties of natural rubber nanocomposites. *Express Polym. Lett.* **2012**, *6*, 14–25. [CrossRef]
53. Kargarzadeh, H.; Sheltami, R.M.; Ahmad, I.; Abdullah, I.; Dufresne, A. Cellulose nanocrystal reinforced liquid natural rubber toughened unsaturated polyester: Effects of filler content and surface treatment on its morphological, thermal, mechanical, and viscoelastic properties. *Polymer* **2015**, *71*, 51–59. [CrossRef]
54. Singh, S.; Dhakar, G.L.; Kapgate, B.P.; Maji, P.K.; Verma, C.; Chhajed, M.; Rajkumar, K.; Das, C. Synthesis and chemical modification of crystalline nanocellulose to reinforce natural rubber composites. *Polym. Adv. Technol.* **2020**, *31*, 3059–3069. [CrossRef]
55. Somseemee, O.; Sae-Oui, P.; Siriwong, C. Reinforcement of surface-modified cellulose nanofibrils extracted from Napier grass stem in natural rubber composites. *Ind. Crops Prod.* **2021**, *171*, 113881. [CrossRef]
56. Mahendra, I.P.; Wirjosentono, B.; Tamrin, T.; Ismail, H.; Mendez, J.A.; Causin, V. The effect of nanocrystalline cellulose and TEMPO-oxidized nanocellulose on the compatibility of polypropylene/cyclic natural rubber blends. *J. Thermoplast. Compos. Mater.* **2020**. published online. [CrossRef]
57. Noguchi, T.; Niihara, K.I.; Kurashima, A.; Iwamoto, R.; Miura, T.; Koyama, A.; Endo, M.; Marubayashi, H.; Kumagai, A.; Jinnai, H.; et al. Cellulose nanofiber-reinforced rubber composites prepared by TEMPO-functionalization and elastic kneading. *Compos. Sci. Technol.* **2021**, *210*, 108815. [CrossRef]
58. Nechyporchuk, O.; Pignon, F.; Do Rego, A.M.; Belgacem, M.N. Influence of ionic interactions between nanofibrillated cellulose and latex on the ensuing composite properties. *Compos. Part B Eng.* **2016**, *85*, 188–195. [CrossRef]
59. Kato, H.; Nakatsubo, F.; Abe, K.; Yano, H. Crosslinking via sulfur vulcanization of natural rubber and cellulose nanofibers incorporating unsaturated fatty acids. *RSC Adv.* **2015**, *5*, 29814–29819. [CrossRef]

60. Jiang, W.; Shen, P.; Yi, J.; Li, L.; Wu, C.; Gu, J. Surface modification of nanocrystalline cellulose and its application in natural rubber composites. *J. Appl. Polym. Sci.* **2020**, *137*, 49163. [CrossRef]
61. Trovatti, E.; Carvalho, A.J.F.; Ribeiro, S.J.L.; Gandini, A. Simple green approach to reinforce natural rubber with bacterial cellulose nanofibers. *Biomacromolecules* **2013**, *14*, 2667–2674. [CrossRef] [PubMed]
62. Lee, S.H.; Md Tahir, P.; Lum, W.C.; Tan, L.P.; Bawon, P.; Park, B.-D.; Osman Al Edrus, S.S.; Abdullah, U.H. A review on citric acid as green modifying agent and binder for wood. *Polymers* **2020**, *12*, 1692. [CrossRef]
63. Saini, S.; Belgacem, M.N.; Bras, J. Effect of variable aminoalkyl chains on chemical grafting of cellulose nanofiber and their antimicrobial activity. *Mater. Sci. Eng. C* **2017**, *75*, 760–768. [CrossRef] [PubMed]
64. Khanjanzadeh, H.; Behrooz, R.; Bahramifar, N.; Wolfgang, G.; Bacher, M.; Edler, M.; Griesser, T. Surface chemical functionalization of cellulose nanocrystals by 3-aminopropyltriethoxysilane. *Int. J. Biol. Macromol.* **2018**, *106*, 1288–1296. [CrossRef] [PubMed]
65. Kargarzadeh, H.; Sheltami, R.M.; Ahmad, I.; Abdullah, I.; Dufresne, A. Cellulose nanocrystal: A promising toughening agent for unsaturated polyester nanocomposite. *Polymer* **2015**, *56*, 346–357. [CrossRef]
66. Pierre, G.; Punta, C.; Delattre, C.; Melone, L.; Dubessay, P.; Fiorati, A.; Pastori, N.; Galante, Y.M.; Michaud, P. TEMPO-mediated oxidation of polysaccharides: An ongoing story. *Carbohydr. Polym.* **2017**, *165*, 71–85. [CrossRef] [PubMed]
67. Isogai, A.; Zhou, Y. Diverse nanocelluloses prepared from TEMPO-oxidized wood cellulose fibers: Nanonetworks, nanofibers, and nanocrystals. *Curr. Opin. Solid State Mater. Sci.* **2019**, *23*, 101–106. [CrossRef]
68. Markets and Markets. Industrial Rubber Market by Application (Automotive, Building & Construction, Industrial Manufacturing, Polymer Modification, Wire & Cable, Electrical & Electronics, Bitumen Modification), Type, Product, and Region—Global Forecast to 2022. Available online: https://www.marketsandmarkets.com/Market-Reports/industrial-rubber-market-42187401.html (accessed on 23 August 2021).
69. Oksman, K. Nanocelluloses and their use in composite materials. *Express Polym. Lett.* **2012**, *6*, 687. [CrossRef]
70. Siro, I.; Plackett, D. Microfibrillated cellulose and new nanocomposite materials: A review. *Cellulose* **2010**, *17*, 459–494. [CrossRef]
71. Roy, K.; Pongwisuthiruchte, A.; Debnath, S.C.; Potiyaraj, P. Application of cellulose as green filler for the development of sustainable rubber technology. *Curr. Res. Green Sustain. Chem.* **2021**, *4*, 100140. [CrossRef]
72. Dominic, M.; Joseph, R.; Begum, P.S.; Kanoth, B.P.; Chandra, J.; Thomas, S. Green tire technology: Effect of rice husk derived nanocellulose (RHNC) in replacing carbon black (CB) in natural rubber (NR) compounding. *Carbohydr. Polym.* **2020**, *230*, 115620. [CrossRef] [PubMed]

polymers

Article

Phenol–Hyaluronic Acid Conjugates: Correlation of Oxidative Crosslinking Pathway and Adhesiveness

Jungwoo Kim [1,2], Sumin Kim [1,2], Donghee Son [3,4,5,*] and Mikyung Shin [1,2,5,*]

1 Department of Biomedical Engineering, Sungkyunkwan University (SKKU), Suwon 16419, Korea; wjddn1998@naver.com (J.K.); ally0618@naver.com (S.K.)
2 Department of Intelligent Precision Healthcare Convergence, Sungkyunkwan University (SKKU), Suwon 16419, Korea
3 Department of Electrical and Computer Engineering, Sungkyunkwan University (SKKU), Suwon 16419, Korea
4 Department of Superintelligence Engineering, Sungkyunkwan University (SKKU), Suwon 16419, Korea
5 Center for Neuroscience Imaging Research, Institute for Basic Science (IBS), Suwon 16419, Korea
* Correspondence: daniel3600@g.skku.edu (D.S.); mikyungshin@g.skku.edu (M.S.)

Abstract: Hyaluronic acid (HA) is a natural polysaccharide with great biocompatibility for a variety of biomedical applications, such as tissue scaffolds, dermal fillers, and drug-delivery carriers. Despite the medical impact of HA, its poor adhesiveness and short-term in vivo stability limit its therapeutic efficacy. To overcome these shortcomings, a versatile modification strategy for the HA backbone has been developed. This strategy involves tethering phenol moieties on HA to provide both robust adhesiveness and intermolecular cohesion and can be used for oxidative crosslinking of the polymeric chain. However, a lack of knowledge still exists regarding the interchangeable phenolic adhesion and cohesion depending on the type of oxidizing agent used. Here, we reveal the correlation between phenolic adhesion and cohesion upon gelation of two different HA–phenol conjugates, HA–tyramine and HA–catechol, depending on the oxidant. For covalent/non-covalent crosslinking of HA, oxidizing agents, horseradish peroxidase/hydrogen peroxide, chemical oxidants (e.g., base, sodium periodate), and metal ions, were utilized. As a result, HA–catechol showed stronger adhesion properties, whereas HA–tyramine showed higher cohesion properties. In addition, covalent bonds allowed better adhesion compared to that of non-covalent bonds. Our findings are promising for designing adhesive and mechanically robust biomaterials based on phenol chemistry.

Keywords: hyaluronic acid; phenol; adhesive hydrogels

Citation: Kim, J.; Kim, S.; Son, D.; Shin, M. Phenol–Hyaluronic Acid Conjugates: Correlation of Oxidative Crosslinking Pathway and Adhesiveness. *Polymers* **2021**, *13*, 3130. https://doi.org/10.3390/polym13183130

Academic Editor: Arn Mignon

Received: 16 August 2021
Accepted: 14 September 2021
Published: 16 September 2021

1. Introduction

Hyaluronic acid (HA) is a natural polysaccharide constructed from two alternating units of N-acetyl-D-glucosamine and D-glucuronic acid [1,2]. It is an important component of the extracellular matrix and plays a role in wound healing and in controlling the release of growth factors [3–5]. Previous research has further shown that HA is very versatile in its use in medical treatment and tissue engineering because of its high biocompatibility, biodegradability, viscoelasticity, and non-toxic characteristics [5,6]. These properties make HA an ideal biomaterial for injectable hydrogels, wound patches, 3D bioprinting, tissue scaffolds, and drug delivery [7–11].

However, HA is currently limited in its use, owing to its relatively weak mechanical properties that prevent HA gelation into a hydrogel [8,12–14]. Furthermore, HA is repeatedly enzymatically degraded in a physiological environment because it is vulnerable to hyaluronidase in vivo [13,15]. This is an essential hurdle that must be overcome to further utilize HA in tissues, such as photo-crosslinking, Schiff base crosslinking, and click chemistry crosslinking, to extend the duration in vivo and obtain better mechanical properties [16–18]. However, the functionalized molecule has no chemical moieties with adhesive properties. These methods do not have the adhesive property needed for certain medical applications such as implant materials for surgical recovery [11,19]. Adhesive properties enable the research to be conducted in the past few years to develop an adhesive HA-derived hydrogel for medical treatment [20].

To achieve good adhesion and mechanical properties simultaneously, polyphenol modification has been introduced. When polyphenol is conjugated with HA, the adhesive strength increases, and polyphenol can be crosslinked. Representative materials of polyphenols with these properties include catechol, tyramine, and gallol, and they can adhere to various substrates through several interactions such as π-π stacking, hydrogen bonding, electrostatic interaction, and catechol metal correlation [21,22]. In addition, these can be oxidized under basic condition (e.g., NaOH) and treatment of $NaIO_4$ or horseradish peroxidase (HRP) to crosslink via covalent bonds, or form coordination complex with metal ions [23]. However, the problem is that the catechol moieties are simultaneously involved in both crosslinking HA backbones and showing their adhesive properties. Crosslinking of phenol molecules causes change of chemical structure of phenol, so it has a possibility of losing adhesion. However, comparative analysis of cohesion and adhesion ability, according to pathway or degree of polyphenol conjugates (HA–Ca and HA–Ty), then made hydrogels using several oxidants (Figure 1). To investigate different crosslinking pathway, biological oxidant (horse radish peroxidase/hydrogen peroxide), chemical oxidant ($NaIO_4$, Ammonium persulfate (APS), NaOH), and metal ions ($FeCl_3$) were used. Depending on the oxidative pathway of phenol or catechol for gelation, the cohesive and adhesive strength of the HA hydrogels can be balanced because the physical amount and chemical status of these moieties involved in the crosslinking of the polymeric chains would be different. Therefore, in this study, we focused on the comparison of the cohesive and adhesive properties of crosslinked hydrogels.

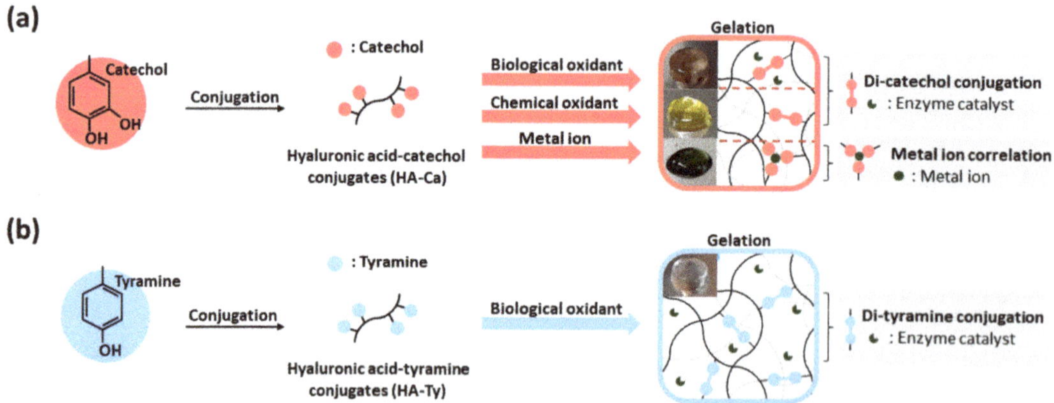

Figure 1. Schematic description of hyaluronic acid–polyphenol hydrogels. (**a**) Synthesis of HA–Ca conjugates (left) and their gelation by three types of oxidants (right). Biological oxidant and chemical oxidant induce crosslinking of catechol via di-catechol conjugation, and metal ion can be coordinated with catechol. (**b**) Synthesis of HA–Ty conjugates (left) and the gelation crosslinked by generating di-tyramine under a biological oxidant (right).

2. Materials and Methods

2.1. Materials

Sodium hyaluronate (Molecular weight = 200 kDa) was purchased from Lifecore Biomedical (Chaska, MN, USA). Dopamine hydrochloride with catechol and amine groups, tyramine hydrochloride with phenol and amine groups, N-hydroxysuccinimide (NHS), 2-(N-morpholino) ethanesulfonic acid (MES) solution (1 M), hydrochloric acid (HCl), sodium hydroxide (NaOH), horseradish peroxidase (HRP), hydrogen peroxide solution (H_2O_2), sodium periodate ($NaIO_4$), ammonium persulfate (APS), and iron (III) chloride ($FeCl_3$)

were purchased from Sigma-Aldrich (St. Louis, MO, USA). 1-(3-Dimethylaminopropyl)-3-ethylcarbodiimide hydrochloride (EDC-HCl) was purchased from Tokyo Chemical Industry (Tokyo, Japan). Phosphate-buffered saline (PBS, 10X, pH 7.2) was purchased from Welgene (Gyeongsan, Korea). Sodium chloride (NaCl) was purchased from Daejung (Siheung, Korea). Anhydrous ethyl alcohol was purchased from Samchun Pure Chemical (Pyeongtaek, Korea). SpectraPor 1 Dialysis Membrane (Standard RC tubing, molecular weight cut-off (MWCO) = 6–8 kDa) was purchased from Spectrum (Rancho Dominguez, CA, USA).

2.2. Synthesis and Characterization of Hyaluronic Acid-Catechol and Hyaluronic Acid-Tyramine Conjugates

For the synthesis of HA–Ca, dopamine hydrochloride was conjugated with a sodium HA backbone by EDC/NHS coupling. HA (500 mg) was dissolved in 55 mL of MES buffer (0.1 M, pH 4.6). The solution was stirred for 15 min under a nitrogen atmosphere to remove dissolved oxygen capable of triggering the oxidation of catechol. After HA was fully dissolved, 190 mg/mL EDC and 115 mg/mL NHS were separately dissolved in 1 mL of MES buffer (0.1 M, pH 4.6) and then added into the reaction solution using syringes. After 10 min, 190 mg of dopamine hydrochloride dissolved in 2 mL MES buffer (0.1 M, pH 4.6) was injected into the reaction solution using a syringe. The solution was stirred for 12 h at room temperature to facilitate the reaction, with a final pH of 5.5 for prevention of further oxidation of catechol groups. To remove any unreacted free molecules, dialysis was performed using a 6–8 kDa MWCO membrane in 5 L of 100 mM NaCl solution (dissolved in acidified deionized distilled water, pH 5) for 2 days and then dialyzed in deionized distilled water for 4 h. After dialysis, the solution was lyophilized for 6 days at −80 °C, 5 mTorr. In addition, for the synthesis of HA–Ty, dopamine hydrochloride was conjugated with a sodium HA backbone by EDC/NHS coupling, following a previous report [24] and some adjustments were made. HA (500 mg) was dissolved in 50 mL of MES buffer (0.1 M, pH 5.5). After HA was fully dissolved, 20.2 mg of tyramine hydrochloride was added and stirred for 10 min. Subsequently, 190 mg/mL EDC and 237 mg/mL NHS were added together. The pH was adjusted to 4.7 with 0.1 M NaOH for optimal amide coupling reaction. After overnight reaction with constant stirring at room temperature, dialysis was performed using a 6–8 kDa MWCO membrane in 5 L of 100 mM NaCl solution (dissolved in acidified deionized water (DW), pH 5) for 2 days, dialyzed with 25% ethanol for 2 days, and then dialyzed with DW for 1 day. After dialysis, the solution was lyophilized for 6 days at −80 °C, 5 mTorr. The degree of catechol or tyramine substitution (DOS%) was analyzed by both ^{1}H NMR spectroscopy (300 MHz; Varian, Palo Alto, CA, USA) and UV-vis spectroscopy (Agilent 8453; Agilent Technologies, Santa Clara, CA, USA). For obtaining ^{1}H NMR spectra, each polymer was dissolved in deuterium oxide (D_2O) at a concentration of 10 mg/mL. Additionally, for UV-vis spectra, the polymer solutions dissolved in DW were prepared at a concentration of 5 mg/mL. The absorbance at the wavelength of 280 nm (A_{280} for catechol) or 275 nm (A_{275} for tyramine) was detected. The calibration curves were established using dopamine (the concentration ranging from 15.6 µg/mL to 62.5 µg/mL) or tyramine (the concentration ranging from 3.9 µg/mL to 62.5 µg/mL). Additionally, the sample purity was confirmed by diffusion ordered spectroscopy (DOSY) (Bruker, German).

2.3. Preparation of Hydrogels

2.3.1. HRP-Induced HA–Ca and HA–Ty Hydrogels

HRP was used as a biological oxidant for H_2O_2. To investigate the change in hydrogel properties based on the concentration of the oxidants, stock solutions of H_2O_2 (0.5 mg/mL, 1 mg/mL, and 1.5 mg/mL in pH 6 PBS for obtaining catechol (Ca): H_2O_2 molar ratios of 1:0.5, 1:1.0, 1:1.5, respectively) and HRP (2 unit/mL, 6 unit/mL, 18 unit/mL to make 0.1 unit/mL, 0.3 unit/mL, and 0.9 unit/mL hydrogel solution) were prepared. All hydrogels were 2 wt%.

HRP-induced HA–Ca gel was prepared as follows. HA–Ca (4 mg) was fully dissolved in 176 µL of DW, 10 µL of HRP, and 10 µL of H_2O_2 stock solution. After 12 h, 2 wt%

HRP-induced HA–Ca gels were fabricated. HRP-induced HA–Ty gel was prepared using the same protocol, and HA–Ca was substituted with HA–Ty.

2.3.2. Detection of the Free Dopamine Not Involved in HRP-Induced HA–CA Hydrogels

HRP-induced HA–Ca hydrogel (HRP 0.9 unit/mL and the molar ratio of Ca to H_2O_2, (1) was prepared as previous methods. The hydrogel was placed in transwell insert (24 well, 8 μm pore, Corning) and exposed to 1 mL of DW. After 12 h, the absorption spectra of the released sample solutions from the hydrogel were analyzed by UV-vis spectroscopy.

2.3.3. Chemical Oxidant-Induced HA–Ca Hydrogels

For triggering oxidative crosslinking of catechols, NaOH was used as a basic additive, and $NaIO_4$ and APS were utilized as the chemical oxidants. To investigate the change in gel properties based on the concentration of the oxidants, stock solutions of $NaIO_4$ (0.32 mg/mL, 1.6 mg/mL, 3.2 mg/mL, 9.6 mg/mL for obtaining catechol: $NaIO_4$ molar ratios of 10:1. 2:1. 1:1 and 1:3, respectively) and APS (170 mg/mL, 340 mg/mL, and 680 mg/mL to obtain molar ratios of Ca:APS molar ratios of 1:50, 1:100, and 1:200, respectively) were prepared. All hydrogels were 2 wt%.

The NaOH-induced HA–Ca hydrogel was prepared using the following steps. HA–Ca (4 mg) was fully dissolved in 190 μL of DW, and the pH was adjusted by adding 6 μL of NaOH solution. After 24 h, hydrogels were prepared. The $NaIO_4$-induced HA–Ca hydrogel was prepared using the following steps. HA–Ca (4 mg) was fully dissolved in 176 μL of DW, and 20 μL of $NaIO_4$ stock solution was added to fabricate $NaIO_4$-induced HA–Ca gels. After 3 h, the $NaIO_4$-induced HA–Ca hydrogels were prepared. The APS-induced HA–Ca hydrogel was also prepared using the same protocol by substituting $NaIO_4$ with APS. The APS-induced HA–Ty hydrogel was also prepared using the same protocol by substituting HA–Ca with HA–Ty.

2.3.4. $FeCl_3$-Induced HA–Ca Hydrogels

To investigate the change in hydrogel properties based on the concentration of the oxidants, stock solutions of $FeCl_3$ (2.0 mg/mL, 4.0 mg/mL, and 8.0 mg/mL for obtaining Ca:Fe^{3+} molar ratios of 2:1, 1:1, 1:2, respectively) were prepared. All hydrogels were 2 wt%. For gelation, HA–Ca (4 mg) was fully dissolved in 170 μL of DW. Subsequently, 20 μL of the specified concentration of $FeCl_3$ solution was added and 6 μL of NaOH solution was added to adjust the pH. The gelation occurred after 18 h.

2.4. Morphological Analysis and Chemical Element Mapping of HA–Ca or HA–Ty Hydrogels

To analyze cross-sectional morphology of the lyophilized HA–Ca or HA–Ty hydrogels, scanning electron microscopy (SEM; JSM7600F, Japan) equipped with an energy-dispersive X-ray spectroscopy (EDS) instrument was used.

2.5. Rheological Characterization

The rheological properties of the hydrogels were determined using a Discovery Hybrid Rheometer 2 (TA Instrument, New Castle, DE, USA) with a 20 mm parallel plate geometry and a gap size of 300 μm. The storage modulus (G') and loss modulus (G'') of the hydrogels as a function of the frequency (0.1–10 Hz) were performed at a strain of 1% at 25 °C. To test shear viscosity as a function of strain (From 0.01 to 100%), HRP-induced HA–Ty hydrogels, HRP-induced HA–Ca hydrogels, and $FeCl_3$-induced HA–Ca hydrogels were performed at 25 °C.

2.6. Compression Test

To compare the morphologies of the HRP-induced HA–Ca hydrogel and HA–Ty hydrogel after compression, a compression test was performed. The method for preparing the hydrogel is described in Section 2.3.1. In addition, HRP-induced HA–Ca hydrogel (2 wt%) was prepared with the concentration of 0.9 unit/mL HRP and Ca:H_2O_2 molar ratio of 1:1 as a final concentration. HA–Ty hydrogel (200 mg, 2 wt%) was also prepared with the same concentration of HRP and H_2O_2. After the preparation of the hydrogels, a weight of 700 g was placed on the gels to compress them for 10 min. After the removal of the weight from the gels, the shapes before and after compression were compared.

2.7. Swelling Behavior

To examine the swelling kinetics of the HA–Ca and HA–Ty hydrogels crosslinked by HRP/H_2O_2 catalyzed reaction (HRP 0.9 unit/mL and the molar ratio of Ca: H_2O_2, 1:1), each hydrogel was swollen in DW. At a pre-determined time interval (0, 0.5, 1, 2, 4, 8, 16, and 24 h), we measured the weight of each hydrogel after removal of superficial moisture. The swelling ratio (%) was calculated as the ratio of swollen weight of hydrogel to their initial weight. All experiments were triplicate.

2.8. Chemical Analysis of HA–Ca Crosslinking Depending on Oxidative Pathway

For investigating HA–Ca crosslinking chemistry, UV-vis spectroscopy data were collected. The method for preparing the hydrogel is described in Section 2.3.2 of this report, except for the incubation time. To obtain UV-vis spectrum data, solutions were prepared under the condition that color changes but does not form a hydrogel or color changes but before gelation time. Therefore, the NaOH-induced HA–Ca hydrogel (pH 12) was prepared with an incubation time of 4 h. $NaIO_4$-induced HA–Ca hydrogel (Ca:$NaIO_4$ molar ratio of 1:3) was prepared with an incubation time of 12 h. APS-induced HA–Ca hydrogel (Ca:APS molar ratio of 1:200) was prepared with an incubation time of 30 min. HRP-induced HA–Ca hydrogel (0.3 unit/mL of HRP, Ca:H_2O_2 molar ratio of 1:1) was prepared with an incubation time of 11 h. UV-vis spectra were recorded using an Agilent 8453 UV-vis spectrometer (Agilent Technologies, Santa Clara, CA, USA).

2.9. Adhesion Strength Characterization of the Hydrogels

Tensile adhesion of the hydrogels was determined using a universal testing machine (34SC-1, Instron, IL, USA). The substrate was prepared using 30 mm × 10 mm × 0.1 mm and PET film. The samples were placed between two substrates and pressed with a weight of 1 kg for 15 min. The overlapped area was 10 mm × 10 mm, and the crosshead speed was 20 mm/min. Each sample test was repeated five times.

2.10. Degradation Test

To investigate degradation profile of the HA–Ca hydrogels, we prepared the hydrogels crosslinked using three different oxidation methods, such as gelation in basic condition (pH 10) and under treatment of $NaIO_4$ or APS. For gelation triggered by either $NaIO_4$ or APS, the molar ratio of catechol to each oxidant as 1 to 1 for $NaIO_4$ and 1 to 100 for APS was utilized. After 0 (initial hydrogels) or 24 h of swelling in DW, these hydrogels were lyophilized over 12 h at −80 °C, and then the weight of dried samples was measured. Finally, the degradation (%) was calculated by the weight changes after soaking in DW compared to initial sample weight.

2.11. Statistical Analysis

All statistically analyzed data were determined using Student's unpaired *t*-test. Statistically significant differences were considered when the *p*-value was less than 0.05.

3. Results and Discussion

3.1. Preparation of HA–Ca and HA–Ty Polymers and Cohesion Properties of HRP-Induced Each Hydrogel

To synthesize the desired hydrogels, modified HA was initially prepared and characterized HA was individually conjugated with catechol and tyramine to obtain HA–Ca and HA–Ty, respectively. The two modified HAs, HA–Ca and HA–Ty, were synthesized via the EDC/NHS coupling reaction (Supplementary Figure S1a,b). For HA–Ca, the DOS%

of catechol was 3.7%, which was calculated by integral values of protons in aromatic rings of catechol compared to protons of HA backbone in ^{1}H NMR spectra. Additionally, the DOS% analyzed by UV-vis spectra was 4.0%, like that of ^{1}H NMR result. For HA–Ty, the DOS% of tyramine was 4.6% from ^{1}H NMR spectroscopy and 4.5% from UV-vis spectroscopy (Figure 2a,b). When the catechol is tethered on polysaccharide, a few free catechol derivatives can be intercalated among the polymeric chains due to their intrinsic adhesiveness [24]. As shown in the results of DOSY (Supplementary Figure S1c,d), all proton signals in HA–Ty showed similar diffusion velocity (~10^{-12} m^2/s), which indicates high purity of the polymer without free tyramine molecules (Supplementary Figure S1c). In contrast, a part of proton signals in HA–Ca exhibited fast diffusion behavior (~10^{-10} m^2/s) (red dashed box, Supplementary Figure S1d), referring a certain degree of free dopamine entrapped in the polymer. For quantitative analysis of the free dopamine, we examined the dopamine amount not involved in gelation (e.g., HRP-triggered HA–Ca hydrogels) using UV-vis spectroscopy. As a result, 0.75% of free dopamine in total ~4 DOS% was present in the gels (Supplementary Figure S2), which might be low not to significantly affect cohesive and adhesive strength of the hydrogels. Considering similar DOS% in both HA–Ca and HA–Ty, the polymers were utilized for further gelation to compare the cohesion and adhesion properties by each polyphenol (Supplementary Figure S2).

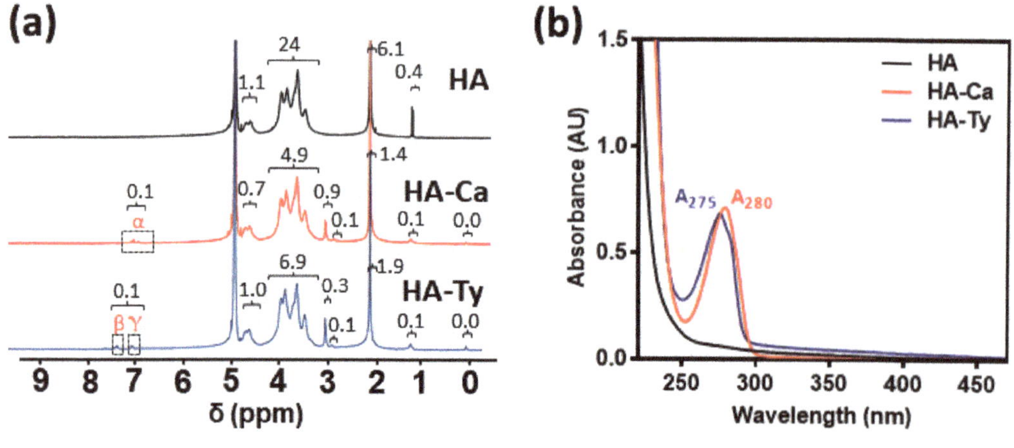

Figure 2. Characterization of HA–Ca and HA–Ty conjugates to evaluate degree of polyphenol substitution (%) on the HA backbone. (**a**) ^{1}H NMR spectrum of HA (black), HA–Ca (yellow), and HA–Ty (blue). The 'α' protons indicate the protons adjacent to two hydroxyl groups in catechol moieties. The 'β and γ' protons indicate the protons adjacent to hydroxyl group in tyramine moieties. (**b**) UV–vis spectra of HA (black), HA–Ca (red), and HA–Ty (blue) solutions. The absorbance at the wavelength of 280 nm (A_{280}) means the presence of catechol, and the absorbance at 275 nm (A_{275}) indicates tyramine.

Each HA–Ca or HA–Ty hydrogel was obtained upon addition of HRP and H_2O_2 into the modified HA solutions, causing di–catechol and di–tyramine crosslinking in each solution (Figure 3a) [25,26]. During such crosslinking reaction, di–catechol and di–tyramine bonds form the HA polymeric network (e.g., hydrogels) with different color appearances. Before gelation, both HA–Ca and HA–Ty solutions were initially transparent; however, after gelation, the HA–Ca hydrogel had an observable reddish-brown hue, whereas HA–Ty hydrogels remained transparent (Figure 3b). This might result from the formation of di–catechol (Supplementary Figure S3) [27]. In addition, for morphological analysis of each hydrogel, the cross-sectional images and chemical element mapping of the dried hydrogels were observed by SEM and EDS, respectively (Supplementary Figure S4). Both hydrogels exhibited typical microporous structures with ~40 μm of pores, and all elements (e.g., carbon, nitrogen, and oxygen of each polymer) were distributed in the overall area.

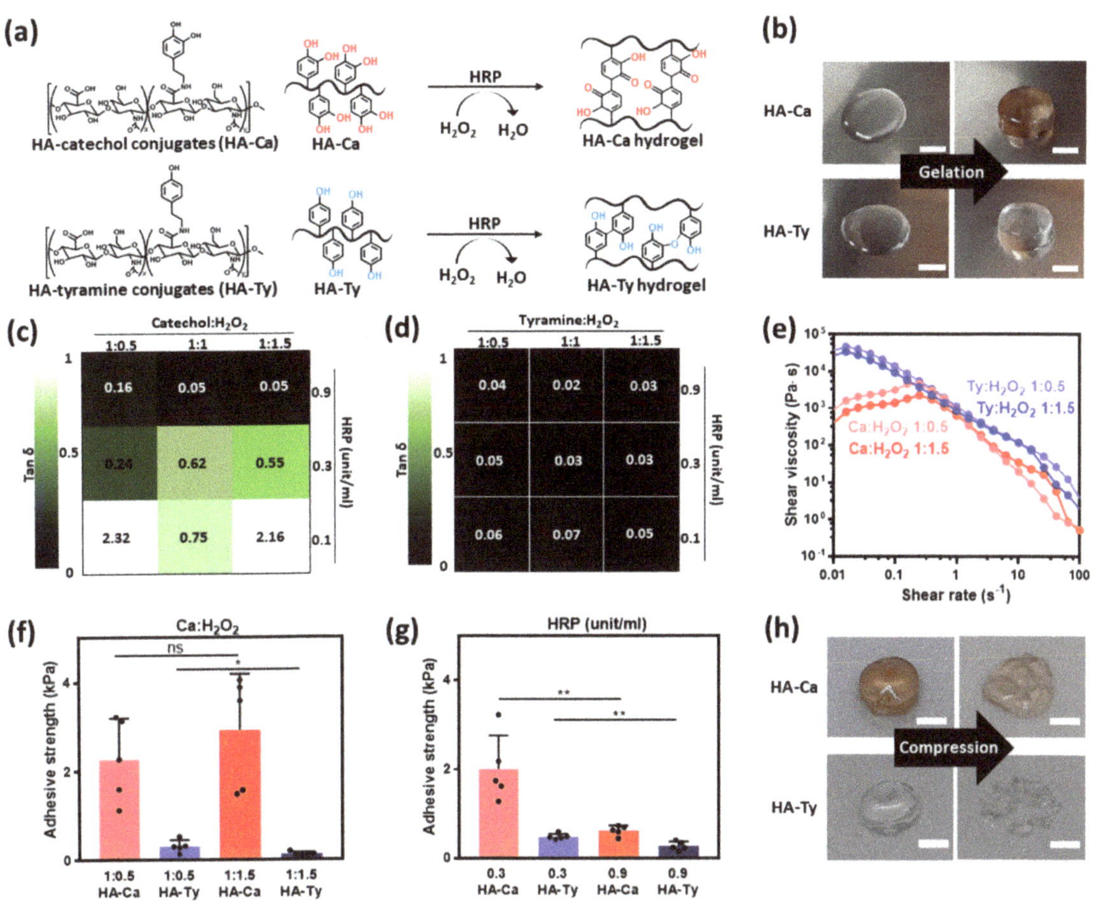

Figure 3. Comparative observation of HA–Ca and HA–Ty hydrogel which are crosslinked by biological oxidants. (**a**) HA–Ca (top) and HA–Ty (bottom) are crosslinked by horseradish peroxidase (HRP) and hydrogen peroxide (H_2O_2) to form HA–Ca and HA–Ty hydrogel, respectively. (**b**) Gelation of HA–Ca solution (top) and HA–Ty solution (bottom) before (left) and after gelation (right) (scale bar = 5 mm). Rheological properties of (**c**) HA–Ca and (**d**) HA–Ty hydrogels with different HRP and H_2O_2 concentrations. (**e**) Shear viscosity of HA–Ca (reddish) and HA–Ty (bluish) as a function of shear rate in different ratios of H_2O_2 showing a shear-thinning property. Adhesion strength of HA–Ca and HA–Ty hydrogels in (**f**) different H_2O_2 molar ratio at 0.3 unit/mL of HRP concentration and (**g**) different concentrations of HRP at H_2O_2 molar ratio of 1:1 (n = 5, mean ± SD) (* $p < 0.05$, ** $p < 0.01$, ns = not significant). (**h**) Images of HA–Ca hydrogel (top) and HA–Ty hydrogel (bottom) before (left) and after (right) compression (scale bar = 5 mm).

Furthermore, the resulting HA–Ca hydrogels and HA–Ty hydrogels showed comparatively different storage moduli and tan δ values. HA–Ca hydrogels with different HRP concentrations (from 0.1 to 0.9 unit/mL) and different molar ratios of catechol to H_2O_2 (from 1:0.5 to 1:1.5) were compared to HA–Ty hydrogels prepared in the same condition. At the beginning of the study, we described a hydrogel with a tan $\delta \leq 0.05$ as being stiff. As regards the HA–Ca hydrogel that contained less than 0.1 unit/mL concentration of HRP, a negligible difference existed in the storage modulus and tan δ owing to its inability to form a stable hydrogel structure. By increasing the concentration, hydrogels with a concentration of over 0.3 unit/mL resulted in an HA–Ca hydrogel with biological activity. HA–Ca formed hydrogels when the concentration of HRP was 0.3 unit/mL. However, when the molar ratio of H_2O_2 is increased, a softer hydrogel is formed. This implies that when the concentration of H_2O_2 exceeds an optimum ratio, excess H_2O_2 will interfere with hydrogel formation rather than supporting it.

When the HRP concentration is over 0.9 unit/mL, it provides conditions that are adequate for HA–Ca to form stiff hydrogels. A significant observation based on these measurements is that the minimum ratio of H_2O_2 (1:1 molar ratio and HRP 0.9 unit/mL) must be obtained to create a stiff hydrogel in HA–Ca (tan $\delta = 0.05$ and $G' = 290$ Pa at 1 Hz). Figure 3c shows that when the concentration of H_2O_2 was 1:0.5, even though HRP content was 0.9 unit/mL, it still failed to show stiff hydrogel formation (tan $\delta = 0.16$ and $G' = 93$ Pa at 1 Hz). Stiff hydrogels were present in HA–Ca only when the HRP concentration was 0.9 unit/mL and the H_2O_2 concentration was either 1:1 and 1:1.5. This reaction occurs for two reasons: (i) the larger quantity of H_2O_2 and HRP resulted in a higher degree of crosslinking in the hydrogel, and (ii) no H_2O_2 remained in the hydrogel, preventing it from interfering in the hydrogel formation (Figure 3d, Supplementary Figure S5). The HA–Ca and HA–Ty hydrogels crosslinked by HRP reaction had different cohesion properties, which correspond to different crosslinking density. For details, the HA–Ty hydrogels possess much higher crosslinking ratio than that of HA–Ca, affecting their swelling kinetics [28]. To demonstrate this, we checked the swelling kinetics of each hydrogel as a function of time. As shown in Supplementary Figure S6, the HA–Ca hydrogels showed higher swelling ratio (398% after 8-h incubation) than that of HA–Ty (1230% after 8-h incubation).

To compare the shear thinning property, shear viscosity was observed based on the shear rate. Shear thinning is crucial for 3D bioprinting and injection using syringes. Both HA–Ca and HA–Ty hydrogels exhibited shear viscosities similar to those of the HA–Ty hydrogel when the shear rate is over 0.25 s^{-1}. Under 0.25 s^{-1}, HA–Ca showed shear thickening, whereas HA–Ty showed shear-thinning properties (Figure 3e).

3.2. *Adhesion Properties of HRP-Induced HA–Ca and HA–Ty Hydrogels*

To verify the correlation between cohesion and adhesion strength, a versatile molar ratio of H_2O_2 was used to fabricate hydrogels, and the concentration of HRP was fixed at 0.3 unit/mL. We hypothesized that the stronger cohesion is induced by the greater number of phenol moieties used for crosslinking, which would decrease the adhesion properties. In previous studies, HA–Ca and HA–Ty made with a 1:0.5 H_2O_2 molar ratio showed a higher storage modulus ($G' = 47$ Pa at HA–Ca hydrogel, $G' = 3{,}172$ Pa at HA–Ty hydrogel) than 1:1.5 ($G' = 21$ Pa at HA–Ca hydrogel, $G' = 1446$ Pa at HA–Ty hydrogel) (Tables 1 and 2). However, the result did not show significant differences because the difference in concentration of H_2O_2 was not significant. This shows that the structural differences between crosslinked catechol and tyramine cause differences in the adhesion strength of the hydrogel (Figure 3f).

Table 1. Storage modulus (G′) of HA–Ca hydrogels at different HRP concentrations and the stoichiometric ratio of H_2O_2 to 0.5, 1, or 1.5.

G′ (Pa)	H_2O_2 1:0.5	H_2O_2 1:1	H_2O_2 1:1.5
HRP 0.3 unit/mL	93	290	205
HRP 0.9 unit/mL	47	18	21

Table 2. Storage modulus (G′) of HA–Ty hydrogels at different HRP concentrations and the stoichiometric ratio of H_2O_2 to 0.5, 1, or 1.5.

G′ (Pa)	H_2O_2 1:0.5	H_2O_2 1:1	H_2O_2 1:1.5
HRP 0.3 unit/mL	5146	4736	4305
HRP 0.9 unit/mL	3172	1551	1446

To obtain the adhesion strength, a lap-shear test was performed. When the concentration of H_2O_2 was fixed at a 1:1 molar ratio while changing the concentration of HRP, the results supported our initial hypothesis, where an increase in the degree of crosslinking in the hydrogel will result in a correlative decrease in the adhesion strength. The HA–Ca hydrogel resulted in a stronger adhesion strength at 0.3 unit/mL of HRP concentration of 0.9 unit/mL. Moreover, 0.3 unit/mL HRP-induced HA–Ca gel showed low G′ (18 Pa) compared to 0.9 unit/mL HRP-induced HA–Ca hydrogel (G′ = 290 Pa). This indicates that the cohesion becomes strong owing to more crosslinking, and the adhesion becomes weak (Figure 3g, Table 1). In addition, because of their crosslinking structure, HA–Ca and HA–Ty had different deformation shapes. When a weight of 700 g was applied to the gel for 10 min, HA–Ca showed plastic deformation, but HA–Ty was ruptured. This unexpected result might be attributed to the different structures of HA–Ca and HA–Ty being ruptured. This result can also be attributed to the different structures of HA–Ca and HA–Ty after crosslinking. This is because the HA–Ca hydrogel has one carbonyl group and two hydroxyl groups, which can exhibit adhesive properties via hydrogen bonding. Thus, HA–Ca interacts with each other through non-covalent bonding. They can lump together, even after compression (Figure 3h). These adhesion properties were supported by previous adhesion tests (Figure 3f,g).

3.3. Cohesion Properties of Chemical Oxidant-Induced HA–Ca Hydrogels

Previous studies have suggested several chemical oxidants, such as $NaIO_4$, and APS, which can oxidize and crosslink catechol moieties. In addition, the basic condition (e.g., NaOH) can induce di–catechol crosslinking. To investigate the effect of the oxidants, the rheological properties of each oxidizing agent were investigated. These oxidants are known to convert catechol into di–catechol (Figure 4a) [29,30]. However, the colors of the prepared hydrogels were different. The differences were verified using UV-vis spectroscopy and photographic images. HA–Ca hydrogels prepared in NaOH appeared a deep brown color with an absorption peak at 325 nm. This peak is observed when semi-quinone is generated, and it can initiate di–catechol crosslinking [31]. The hydrogel induced by APS was bright yellow and had a narrow absorption peak at 400 nm. This peak indicates the quinone form generated by the oxidation of catechol which initiates di–catechol, similar to semiquinone [32]. The hydrogel with $NaIO_4$ showed a slightly yellowish hue with a broad absorption peak at 425 nm. This peak indicates the formation of di–catechol (Figure 4a,b) [33].

Variable crosslinking density can be achieved by tuning the number of oxidants. However, in the case of NaOH, the data are expressed as pH instead of the amount of NaOH. When the pH was increased from 5 to 9, HA–Ca formed a hydrogel (storage modulus > loss modulus), which indicated that the cohesive ability increased with an increase in pH. However, as the pH exceeded a certain level, the hydrogel did not form ($G' < G''$ at pH 12). This weakening of the cohesive ability is likely caused by heterogeneous gelation induced by an excessively high pH (Figure 4c).

Compared to the hydrogel with NaOH, the hydrogel with APS was more elastic. The APS-added HA–Ca showed elastic hydrogel (Ca:APS was 1:100 and 1:200) when the molar ratio was above 1:50. This suggests that the cation–π interaction helps to form a stiff hydrogel. (Figure 4c,d) [34]. In addition, APS induced a transparent HA–Ty hydrogel. Tyramine is known to be crosslinked in the presence of an enzyme catalyst such as HRP, and hydrogel formation of a polymer-tyramine conjugate induced by APS has not been reported. This unexpected result also suggests that the π–cation interaction may promote the cohesive properties of hydrogels, unlike other chemical oxidants (Supplementary Figure S7). The $NaIO_4$ added HA–Ca hydrogel acts similar to NaOH. Furthermore, as the molar ratio of Ca:$NaIO_4$ was increased from 10:1 to 1:1, the cohesive ability of the solution increased, eventually leading to the formation of the hydrogel at a 1:1 molar ratio ($G' > G''$). However, as the molar ratio of $NaIO_4$ is further increased, the cohesive ability decreases, which is similar to NaOH, where the weakening of its cohesive ability is caused by the aforementioned heterogeneous gelation (Figure 4e).

Meanwhile, such chemical oxidants (e.g., NaOH) can degrade HA backbone [35,36]. According to the results, to evaluate the recovered dry weight of HA–Ca after swelling of

24 h (Supplementary Figure S8), the degree of degradation was ~ 20% for NaOH-induced HA–Ca, ~ 11% for APS-induced one, and less than ~2% for the $NaIO_4$-induced one. That is, a certain degree of HA backbone can be degraded by those chemical oxidants, yet it was approximately less than only 20% of total hydrogel weights.

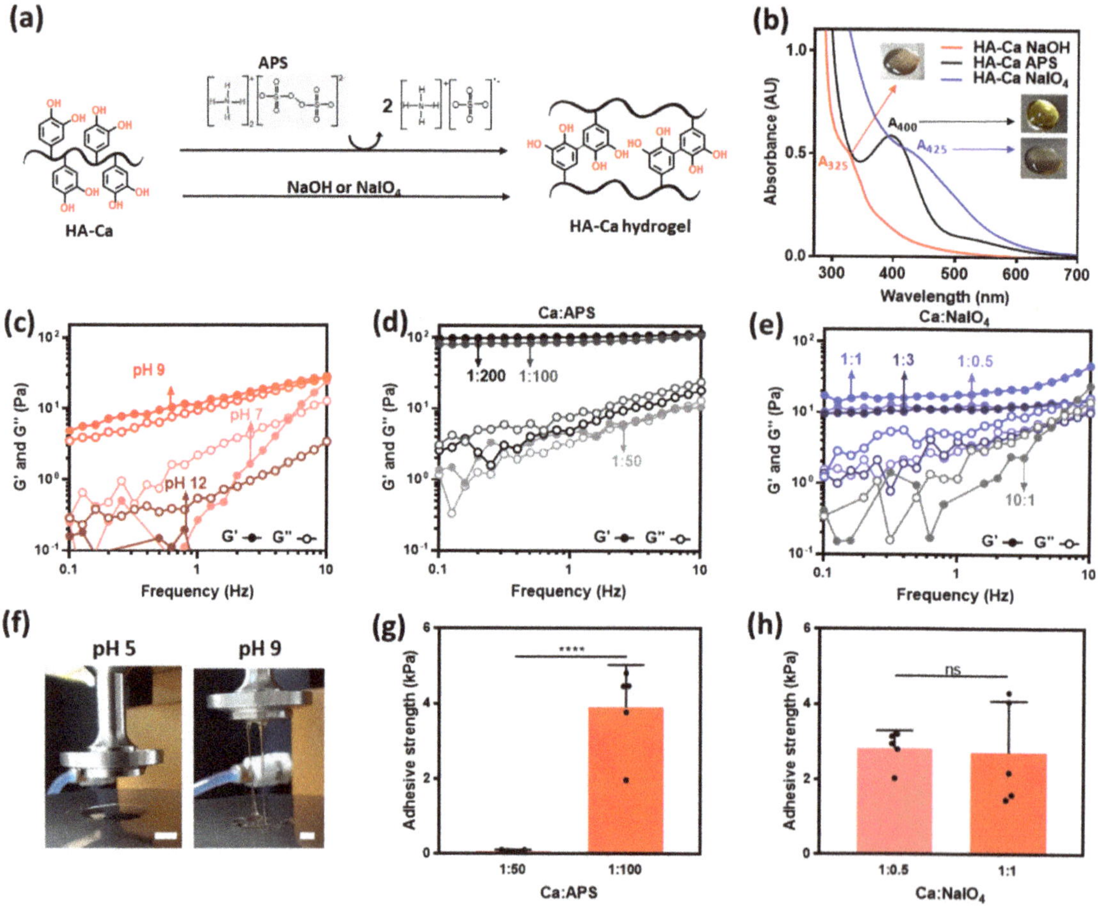

Figure 4. Adhesive and cohesive properties of chemical oxidants-induced hydrogels. (**a**) Catechol forms di-catechol covalent bonds owing to APS, $NaIO_4$, NaOH. (**b**) UV-vis spectra of chemical crosslinking-induced HA–Ca solutions by three different oxidants. Frequency sweep-storage (G′) and loss (G″) moduli of HA–Ca hydrogels at 1% strain, oxidized with (**c**) NaOH, (**d**) $NaIO_4$, and (**e**) APS with different molar ratios of catechol and the oxidant. (**f**) Images of NaOH-induced hydrogel with different pH (scale bar = 5 mm). Adhesion strength of various oxidants-induced HA–Ca hydrogels on PET substrate in (**g**) different Ca:APS molar ratio and (**h**) different Ca:$NaIO_4$ molar ratio (n = 5, mean ± SD) (**** $p < 0.0001$, ns = not significant).

3.4. Adhesion Properties of Chemical Oxidant-Induced HA–Ca Hydrogels

Because NaOH-induced hydrogels were formed at a pH of 7, the comparison of different pH conditions of the hydrogel was impossible. Therefore, adhesion strength was tested only for APS-induced hydrogels and $NaIO_4$-induced hydrogels, and the stickiness according to the pH change is shown in Figure 4f. Stickiness increasingly appeared at pH 7 compared to pH 5. In the case of APS-induced hydrogels, two APS molar ratios (1:50 and 1:100) were tested. A molar ratio of 1:100 showed higher adhesion strength (3.9 ± 0.5 kPa) than that of 1:50 (0.08 ± 0.0 kPa). The reason for this result is that gel formation is difficult at a ratio of 1:50, which means the phase is almost same as liquid. This means that adhesive force did not appear if no minimal cohesive force existed (Figure 4g).

In the case of the $NaIO_4$-induced HA–Ca hydrogel, the 1:0.5, and 1:1 molar ratio of Ca:$NaIO_4$ did not show significant differences in adhesion strength. This is because the difference in G′ (12 Pa at 1:1 molar ratio, 18 Pa at 1:0.5, molar ratio) is not large; thus, the degree of crosslinking was not significantly different (Figure 4h).

3.5. Cohesion Properties of Fe^{3+}-Induced HA–Ca Hydrogels

Catechol interacts with ferric (Fe^{3+}) ions via coordination bonds to form mono-, bis-, and tri-complexes. Based on the pH, this catechol-Fe^{3+} ion complexation can form mono-, bis-, and tris complexes (Figure 5a) [37]. In addition, bis and tri complexes increase cohesion strength because they can grab onto other polymers. At pH 5 and 7, the hydrogels have an observable dark green hue and a brown hue at pH 10. This brown color indicates the formation of a tri complex (Figure 5b) [38].

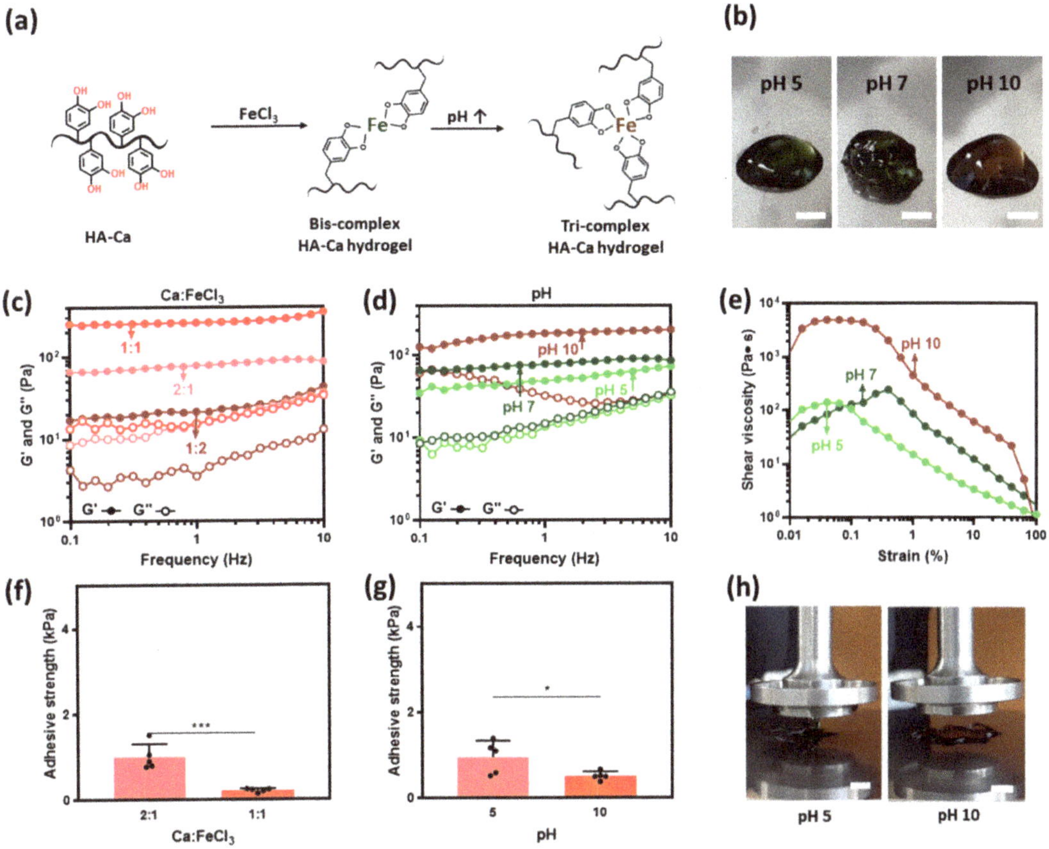

Figure 5. Adhesive and cohesive properties of $FeCl_3$-induced hydrogels. (**a**) Catechol forms a non-covalent coordination bond owing to Fe^{3+} ion. (**b**) Photos of $FeCl_3$-induced HA–Ca hydrogel prepared at different pH (5, 7, and 10) (scale bar = 5 mm). (**c**) in different pH at Ca:Fe^{3+} molar ratio of 2:1 and (**d**) in different Fe^{3+} molar ratios at pH 7. (**e**) Shear-thinning properties of the hydrogel in different pH at Ca:Fe^{3+} molar ratio of 2:1. Adhesion strength of the hydrogels in (**f**) different catechol:Fe^{3+} molar ratios at pH 7, (**g**) different pH at Ca:Fe^{3+} molar ratio of 2:1 (n = 5, mean ± SD) (* $p < 0.05$, *** $p < 0.001$), and (**h**) their images at pH 5 (left) and pH 10 (right) (scale bar = 5 mm).

To tune the crosslinking density, the HA–Ca solution was crosslinked with a variable amount of Fe^{3+} ions. The molar ratio of catechol and Fe^{3+} ions was modulated (2:1, 1:1, and 1:2) while fixing the pH to 7. Among the three conditions, a 1:1 concentration is where the cohesion strength was the highest. This result might be contrary to previous reports [22] in which a ratio of 2:1 can completely form bis-complexes. A possible explanation is that the distance between catechol is too far for crosslinking between them to properly form, thus a higher concentration of mono complex instead of bis-complex will form. In the case of 1:2, the storage modulus was decreased because catechol makes more mono complexes, owing to excess Fe^{3+} ions (Figure 5c).

In addition, the cohesive ability was measured based on the change in pH when the amount of Fe^{3+} ions was at a constant 2:1 molar ratio. Hydrogel formation was observed throughout the entire pH range (pH 5–10), with the largest storage modulus (185 Pa) observed at pH 10. This is because the tri-complex that is formed exhibits a much denser crosslinking (Figure 5d) [39].

To verify the shear thinning property, the shear viscosity based on the shear rate was evaluated. From a shear rate between 0 s^{-1} and 0.6 s^{-1}, shear thickening properties are observed throughout the pH range (pH 5 to 10). However, over 0.6 s^{-1}, shear-thinning properties were observed at all pH ranges (Figure 5e).

3.6. Adhesion Properties of Fe^{3+} Induced HA–Ca Hydrogels

Fe^{3+} ions improve the cohesion strength by coordinating with the hydroxyl groups of catechol. Because hydrogen bonding is the strongest interaction among molecular-molecular interactions, it was expected that the adhesion strength would drop rapidly as the hydroxyl group forms coordination bonds with Fe^{3+} ions. Therefore, it was hypothesized that the adhesive property of the hydrogel would be lowered based on the degree of crosslinking with Fe^{3+} ions. Previous data showed that the degree of crosslinking can vary based on the pH and the amount of Fe^{3+} ions. To determine the cohesive property variation based on each variable, a rheological test was performed in the two groups. The first group was used to investigate the effect of pH while fixing the molar ratio of Fe^{3+} to 2:1 (=Ca: Fe^{3+}). The second group was used to investigate the effect of Fe^{3+} ions while maintaining the pH at 7. As expected, it was observed that the adhesion strength of the more crosslinked hydrogel (1:1 catechol: Fe^{3+} molar ratio) was lower than that of the less crosslinked hydrogel (2:1 molar ratio) (Figure 5f). Similar results were obtained when the pH was changed while keeping the molar ratio of $FeCl_3$ constant. When pH 5 and pH 10 were compared, the storage modulus was larger at pH 10; however, the adhesion strength was smaller (1.0 kPa at pH 5 and 0.5 kPa at pH 10) (Figure 5g) and can be visibly seen in Figure 5h. The reason is that when the same amount of catechol and Fe^{3+} ions exist, the tri-complex formed can grab more catechol by forming a tri-complex. These results indicate that adhesion decreases with the degree of cohesion.

In summary, we reported the correlation between cohesion and adhesion strength based on the crosslinking pathways and crosslinking density. When the crosslinking mechanisms are compared, non-covalently bonded hydrogels obtained via metal–catechol coordination showed similar storage modulus; however, adhesion strength was lower than that of the covalently bonded hydrogel.

Depending on the type of oxidant, the HA–Ty hydrogel crosslinked with HRP used by biological oxidants showed the highest as well as the lowest cohesion strength. In the case of HA–Ca, the radical scavenging ability was stronger than that of HA–Ca, allowing it to form less crosslinking with the biological oxidant, HRP. However, the adhesion was higher than that of the HA–Ty hydrogel. In comparison with HA–Ca, the $NaIO_4$ hydrogel showed high adhesion strength and low cohesion strength. In contrast, the Fe^{3+} ion hydrogel had a fine cohesion strength; however, the adhesion strength was weak. Among the hydrogels tested, the APS hydrogel exhibited the best adhesion strength and showed a good storage modulus, similar to that of the Fe^{3+} hydrogel. In situations where both good cohesion and good adhesion are required, the APS-induced hydrogel is an option (Figure 6).

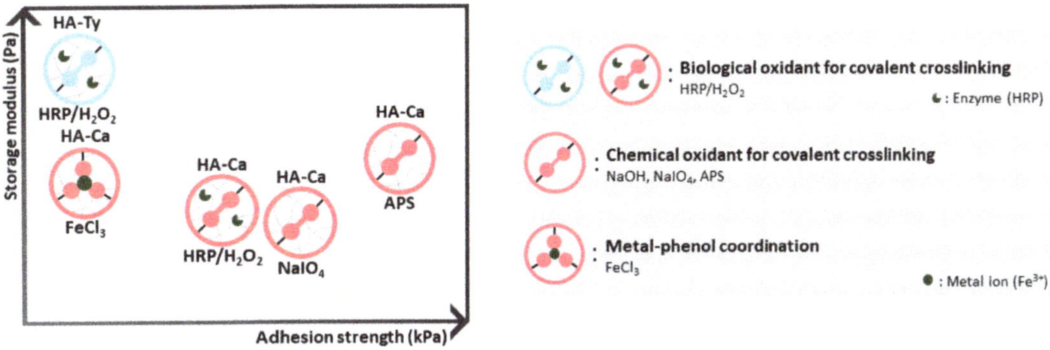

Figure 6. Correlation of adhesion and cohesion strength varying by crosslinking mechanisms. Hydrogels are indicated in the graph (left) based on the level of adhesion strength and storage modulus.

4. Conclusions

In conclusion, the correlation between cohesive and adhesive strength of phenol–HA hydrogels depending on their crosslinking pathway was investigated. Regarding mechanical properties of the hydrogels crosslinked by enzymatic reaction, HA–Ty formed a stiff hydrogel with high storage modulus of ~10^3 Pa when compared to that HA–Ca. However, their adhesiveness was 21 times lower than that of HA–Ca. When the HA–Ca was covalently crosslinked by APS, the storage modulus increased up to ~10^2 Pa, which was still lower than that of HA–Ty. That is, HA–Ty hydrogels showed strong cohesion yet weak adhesion. Among different oxidation methods (e.g., covalent or non-covalent bonds) to crosslink HA–Ca, the best option to achieve strong adhesion and cohesion was APS-triggered crosslinking. Although metal coordination network with catechol improved cohesion, their adhesive strength did not increase because most of catechols to show adhesiveness were strongly bound to metal ions. The di–catechol covalent bonds can enhance adhesion of the hydrogels due to prevention of cohesive failure. Our finding would be useful for choosing design rationale of the phenol-conjugated polymers with both robust adhesion and cohesion.

Supplementary Materials: The following are available online at https://www.mdpi.com/article/10.3390/polym13183130/s1, Supplementary Figure S1: Synthesis and characterization of hyaluronic acid (HA)-based conjugates. Supplementary Figure S2: UV-vis spectra of free dopamine (A_{280}) released from the HRP-induced HA–Ca hydrogels. Supplementary Figure S3: UV-vis spectra of HRP-induced HA–Ca hydrogel. Supplementary Figure S4: SEM images (1st photos) and EDS mapping (2nd image for carbon (C), 3rd image for nitrogen (N), and 4th image for oxygen (O)). Supplementary Figure S5: Rheological characterization of HRP-induced HA–Ca hydrogels in different molar ratios of Ca:H_2O_2. Supplementary Figure S6: Swelling ratio (%) of HRP/H_2O_2-induced HA–Ca (red) and HA–Ty (blue) hydrogels as a function of time. Supplementary Figure S7: Rheological characterization of APS-induced HA–Ty hydrogels in different molar ratios of Ty:H_2O_2. Supplementary Figure S8: Degradation of HA–Ca hydrogels after swelling of 24 h.

Author Contributions: Conceptualization, M.S.; methodology, M.S., J.K. and S.K.; software, J.K.; validation, J.K.; formal analysis, M.S. and J.K.; investigation, J.K.; resources, M.S. and J.K.; data curation, M.S.; writing—original draft preparation, J.K.; writing—review and editing, M.S.; visualization, J.K.; supervision, M.S. and D.S.; project administration, M.S. and D.S.; funding acquisition, M.S. All authors have read and agreed to the published version of the manuscript.

Funding: This study was supported by the National Research Foundation of Korea (NRF) through a grant funded by the Korean government (MSIT) (NRF-2020R1C1C1003903 to M.S. and NRF-2020R1C1C1005567 to D.S.).

Institutional Review Board Statement: Not applicable.

Informed Consent Statement: Not applicable.

Data Availability Statement: The data presented in this study are available in the article.

Conflicts of Interest: The funders had no role in the study design; collection, analyses, or data interpretation; in the writing of the manuscript; or in the decision to publish the results.

References

1. Smith, A.M.; Moxon, S.; Morris, G. Biopolymers as wound healing materials. In *Wound Healing Biomaterials*, 1st ed.; Ågren, M., Ed.; Woodhead Publishing: Sawston, UK, 2016; Volume 2, pp. 261–287.
2. Song, W.; Lee, B.H.; Tan, L.P.; Li, H. Cardiovascular engineering materials in translational medicine. In *Biomaterials in Translational Medicine*, 1st ed.; Yang, L., Bhaduri, S., Webster, T., Eds.; Woodhead Publishing: Sawston, UK, 2018; Volume 1, pp. 57–91.
3. Iio, K.; Furukawa, K.-I.; Tsuda, E.; Yamamoto, Y.; Maeda, S.; Naraoka, T.; Kimura, Y.; Ishibashi, Y. Hyaluronic acid induces the release of growth factors from platelet-rich plasma. *Asia-Pac. J. Sports Med. Arthrosc. Rehabil. Technol.* **2019**, *4*, 27–32. [CrossRef]
4. Voinchet, V.; Vasseur, P.; Kern, J. Efficacy and safety of hyaluronic acid in the management of acute wounds. *Am. J. Clin. Dermatol.* **2006**, *7*, 353–357. [CrossRef] [PubMed]
5. Khunmanee, S.; Jeong, Y.; Park, H. Crosslinking method of hyaluronic-based hydrogel for biomedical applications. *J. Tissue Eng.* **2017**, *8*, 2041731417726464. [CrossRef]
6. Fallacara, A.; Baldini, E.; Manfredini, S.; Vertuani, S. Hyaluronic acid in the third millennium. *Polymers* **2018**, *10*, 701. [CrossRef]
7. Shin, J.; Choi, S.; Kim, J.H.; Cho, J.H.; Jin, Y.; Kim, S.; Min, S.; Kim, S.K.; Choi, D.; Cho, S.W. Tissue Tapes—Phenolic Hyaluronic Acid Hydrogel Patches for Off-the-Shelf Therapy. *Adv. Funct. Mater.* **2019**, *29*, 1903863. [CrossRef]
8. Trombino, S.; Servidio, C.; Curcio, F.; Cassano, R. Strategies for hyaluronic acid-based hydrogel design in drug delivery. *Pharmaceutics* **2019**, *11*, 407. [CrossRef] [PubMed]
9. Petta, D.; Grijpma, D.W.; Alini, M.; Eglin, D.; D'Este, M. Three-dimensional printing of a tyramine hyaluronan derivative with double gelation mechanism for independent tuning of shear thinning and postprinting curing. *ACS Biomater. Sci. Eng.* **2018**, *4*, 3088–3098. [CrossRef] [PubMed]
10. Le Thi, P.; Son, J.Y.; Lee, Y.; Ryu, S.B.; Park, K.M.; Park, K.D. Enzymatically crosslinkable hyaluronic acid-gelatin hybrid hydrogels as potential bioinks for tissue regeneration. *Macromol. Res.* **2020**, *28*, 400–406. [CrossRef]
11. Zhou, L.; Dai, C.; Fan, L.; Jiang, Y.; Liu, C.; Zhou, Z.; Guan, P.; Tian, Y.; Xing, J.; Li, X.; et al. Injectable Self-Healing Natural Biopolymer-Based Hydrogel Adhesive with Thermoresponsive Reversible Adhesion for Minimally Invasive Surgery. *Adv. Funct. Mater.* **2021**, *31*, 2007457. [CrossRef]
12. Stern, R.; Asari, A.A.; Sugahara, K.N. Hyaluronan fragments: An information-rich system. *Eur. J. Cell Biol.* **2006**, *85*, 699–715. [CrossRef] [PubMed]
13. Buhren, B.A.; Schrumpf, H.; Hoff, N.-P.; Bölke, E.; Hilton, S.; Gerber, P.A. Hyaluronidase: From clinical applications to molecular and cellular mechanisms. *Eur. J. Med. Res.* **2016**, *21*, 1–7. [CrossRef] [PubMed]
14. Burdick, J.A.; Prestwich, G.D. Hyaluronic acid hydrogels for biomedical applications. *J. Adv. Mater.* **2011**, *23*, H41–H56. [CrossRef] [PubMed]
15. Paap, M.K.; Silkiss, R.Z. The interaction between hyaluronidase and hyaluronic acid gel fillers-a review of the literature and comparative analysis. *Plast. Aesthet. Res.* **2020**, *7*, 220514027. [CrossRef]
16. Wang, G.; Cao, X.; Dong, H.; Zeng, L.; Yu, C.; Chen, X. A hyaluronic acid based injectable hydrogel formed via photo-crosslinking reaction and thermal-induced diels-alder reaction for cartilage tissue engineering. *Polymers* **2018**, *10*, 949. [CrossRef]
17. Yoon, H.Y.; Koo, H.; Choi, K.Y.; Kwon, I.C.; Choi, K.; Park, J.H.; Kim, K. Photo-crosslinked hyaluronic acid nanoparticles with improved stability for in vivo tumor-targeted drug delivery. *Biomaterials* **2013**, *34*, 5273–5280. [CrossRef] [PubMed]
18. Hu, X.; Gao, Z.; Tan, H.; Wang, H.; Mao, X.; Pang, J. An injectable hyaluronic acid-based composite hydrogel by DA click chemistry with pH sensitive nanoparticle for biomedical application. *Front. Chem.* **2019**, *7*, 477. [CrossRef] [PubMed]
19. Zaokari, Y.; Persaud, A.; Ibrahim, A. Biomaterials for adhesion in orthopedic applications: A review. *Eng. Regen.* **2020**, *1*, 51–63. [CrossRef]
20. Lee, J.S.; Cho, J.H.; An, S.; Shin, J.; Choi, S.; Jeon, E.J.; Cho, S.-W. In situ self-cross-linkable, long-term stable hyaluronic acid filler by gallol autoxidation for tissue augmentation and wrinkle correction. *Chem. Mater.* **2019**, *31*, 9614–9624. [CrossRef]
21. Dai, Q.; Geng, H.; Yu, Q.; Hao, J.; Cui, J. Polyphenol-based particles for theranostics. *Theranostics* **2019**, *9*, 3170. [CrossRef]
22. Yang, J.; Stuart, M.A.C.; Kamperman, M. Jack of all trades: Versatile catechol crosslinking mechanisms. *Chem. Soc. Rev.* **2014**, *43*, 8271–8298. [CrossRef]
23. Loebel, C.; Szczesny, S.E.; Cosgrove, B.D.; Alini, M.; Zenobi-Wong, M.; Mauck, R.L.; Eglin, D. Cross-linking chemistry of tyramine-modified hyaluronan hydrogels alters mesenchymal stem cell early attachment and behavior. *Biomacromolecules* **2017**, *18*, 855–864. [CrossRef]
24. Gan, D.; Xu, T.; Xing, W.; Wang, M.; Fang, J.; Wang, K.; Ge, X.; Chan, C.W.; Ren, F.; Tan, H.; et al. Mussel-inspired dopamine oligomer intercalated tough and resilient gelatin methacryloyl (GelMA) hydrogels for cartilage regeneration. *J. Mater. Chem. B* **2019**, *7*, 1716–1725. [CrossRef] [PubMed]

25. Abu-Hakmeh, A.; Kung, A.; Mintz, B.R.; Kamal, S.; Cooper, J.A.; Lu, X.L.; Wan, L.Q. Sequential gelation of tyramine-substituted hyaluronic acid hydrogels enhances mechanical integrity and cell viability. *Med. Biol. Eng. Comput.* **2016**, *54*, 1893–1902. [CrossRef] [PubMed]
26. Thi, T.T.H.; Lee, Y.; Le Thi, P.; Park, K.D. Engineered horseradish peroxidase-catalyzed hydrogels with high tissue adhesiveness for biomedical applications. *J. Ind. Eng. Chem.* **2019**, *78*, 34–52.
27. Park, M.K.; Li, M.-X.; Yeo, I.; Jung, J.; Yoon, B.-I.; Joung, Y.K. Balanced adhesion and cohesion of chitosan matrices by conjugation and oxidation of catechol for high-performance surgical adhesives. *Carbohydr. Polym.* **2020**, *248*, 116760. [CrossRef] [PubMed]
28. Larrañeta, E.; Henry, M.; Irwin, N.J.; Trotter, J.; Perminova, A.A.; Donnelly, R.F. Synthesis and characterization of hyaluronic acid hydrogels crosslinked using a solvent-free process for potential biomedical applications. *Carbohydr. Polym.* **2018**, *181*, 1194–1205. [CrossRef]
29. Hong, S.; Yang, K.; Kang, B.; Lee, C.; Song, I.T.; Byun, E.; Park, K.I.; Cho, S.W.; Lee, H. Hyaluronic acid catechol: A biopolymer exhibiting a pH-dependent adhesive or cohesive property for human neural stem cell engineering. *Adv. Funct. Mater.* **2013**, *23*, 1774–1780. [CrossRef]
30. Ryu, J.H.; Hong, S.; Lee, H. Bio-inspired adhesive catechol-conjugated chitosan for biomedical applications: A mini review. *Acta Biomater.* **2015**, *27*, 101–115. [CrossRef]
31. Lee, F.; Chung, J.E.; Kurisawa, M. An injectable enzymatically crosslinked hyaluronic acid–tyramine hydrogel system with independent tuning of mechanical strength and gelation rate. *Soft Matter* **2008**, *4*, 880–887. [CrossRef]
32. Guo, Z.; Mi, S.; Sun, W. The multifaceted nature of catechol chemistry: Bioinspired pH-initiated hyaluronic acid hydrogels with tunable cohesive and adhesive properties. *J. Mater. Chem. B.* **2018**, *6*, 6234–6244. [CrossRef]
33. Lee, B.P.; Dalsin, J.L.; Messersmith, P.B. Synthesis and gelation of DOPA-modified poly (ethylene glycol) hydrogels. *Biomacromolecules* **2002**, *3*, 1038–1047. [CrossRef] [PubMed]
34. Ferretti, A.; Prampolini, G.; d'Ischia, M. Noncovalent interactions in catechol/ammonium-rich adhesive motifs: Reassessing the role of cation-π complexes? *Chem. Phys. Lett.* **2021**, *779*, 138815. [CrossRef]
35. Hong, B.M.; Park, S.A.; Park, W.H. Effect of photoinitiator on chain degradation of hyaluronic acid. *Biomater. Res.* **2019**, *23*, 1–8. [CrossRef] [PubMed]
36. De Souza, A.B.; Chaud, M.V.; Santana, M.H.A. Hyaluronic acid behavior in oral administration and perspectives for nanotechnology-based formulations: A review. *Carbohydr. Polym.* **2019**, *222*, 115001. [CrossRef]
37. Bijlsma, J.; de Bruijn, W.J.; Hageman, J.A.; Goos, P.; Velikov, K.P.; Vincken, J.-P. Revealing the main factors and two-way interactions contributing to food discolouration caused by iron-catechol complexation. *Sci. Rep.* **2020**, *10*, 1–11.
38. Lee, J.; Chang, K.; Kim, S.; Gite, V.; Chung, H.; Sohn, D. Phase controllable hyaluronic acid hydrogel with iron (III) ion—Catechol induced dual cross-linking by utilizing the gap of gelation kinetics. *Macromolecules* **2016**, *49*, 7450–7459. [CrossRef]
39. Holten-Andersen, N.; Harrington, M.J.; Birkedal, H.; Lee, B.P.; Messersmith, P.B.; Lee, K.Y.C.; Waite, J.H. pH-induced metal-ligand cross-links inspired by mussel yield self-healing polymer networks with near-covalent elastic moduli. *Proc. Natl. Acad. Sci. USA* **2011**, *108*, 2651–2655. [CrossRef]

polymers

MDPI

Article

Production of Thermophilic Chitinase by *Paenibacillus* sp. TKU052 by Bioprocessing of Chitinous Fishery Wastes and Its Application in *N*-acetyl-D-glucosamine Production

Chien Thang Doan [1,2], Thi Ngoc Tran [1,3] and San-Lang Wang [2,4,*]

1 Faculty of Natural Sciences and Technology, Tay Nguyen University, Buon Ma Thuot 630000, Vietnam; dcthang@ttn.edu.vn (C.T.D.); ttngoc@ttn.edu.vn (T.N.T.)
2 Department of Chemistry, Tamkang University, New Taipei City 25137, Taiwan
3 Doctoral Program in Applied Sciences, College of Science, Tamkang University, New Taipei City 25137, Taiwan
4 Life Science Development Center, Tamkang University, New Taipei City 25137, Taiwan
* Correspondence: sabulo@mail.tku.edu.tw; Tel.: +886-2-2621-5656; Fax: +886-2-2620-9924

Citation: Doan, C.T.; Tran, T.N.; Wang, S.-L. Production of Thermophilic Chitinase by *Paenibacillus* sp. TKU052 by Bioprocessing of Chitinous Fishery Wastes and Its Application in *N*-acetyl-D-glucosamine Production. *Polymers* **2021**, *13*, 3048. https://doi.org/10.3390/polym13183048

Academic Editor: Arn Mignon

Received: 18 August 2021
Accepted: 7 September 2021
Published: 9 September 2021

Abstract: The bioprocessing of chitinous fishery wastes (CFWs) to chitinases through fermentation approaches has gained importance owing to its great benefits in reducing the enzyme production cost, and utilizing chitin waste. In this work, our study of the chitinase production of *Paenibacillus* sp. TKU052 in the presence of different kinds of CFWs revealed a preference for demineralized crab shells powder (deCSP); furthermore, a 72 kDa chitinase was isolated from the 0.5% deCSP-containing medium. The *Paenibacillus* sp. TKU052 chitinase displayed maximum activity at 70 °C and pH 4–5, while Zn^{2+}, Fe^{3+}, Triton X-100, Tween 40, and SDS exerted a negative effect on its activity, whereas Mn^{2+} and 2-mercaptoethanol were found to potentially enhance the activity. Among various kinds of polysaccharide, *Paenibacillus* sp. TKU052 chitinase exhibited the best catalytic activity on colloidal chitin (CC) with K_m = 9.75 mg/mL and V_{max} = 2.43 µmol/min. The assessment of the hydrolysis of CC and *N*-acetyl chitooligosaccharides revealed that *Paenibacillus* sp. TKU052 chitinase possesses multiple catalytic functions, including exochitinase, endochitinase, and *N*-acetyl-β-D-glucosaminidase activities. Finally, the combination of *Paenibacillus* sp. TKU052 chitinase and *Streptomyces speibonae* TKU048 *N*-acetyl-β-D-glucosaminidase could efficiently convert CC to *N*-acetyl-D-glucosamine (GlcNAc) with a production yield of 94.35–98.60% in 12–24 h.

Keywords: chitinous fishery wastes; chitinase; crab shells; *Paenibacillus*; *N*-acetyl-D-glucosamine

1. Introduction

Chitin is a linear polysaccharide made of *N*-acetyl-D-glucosamine (GlcNAc) units linked through the β-1,4-glycosidic linkages [1,2]. In nature, it is the second most abundant polysaccharide after cellulose, and is also a structural constituent of crustacean exoskeletons, squid pens, fungal cell walls, etc. [3,4]. In the fishery industry, a massive amount of chitinous fishery wastes (CFWs) are generated from shrimp shells, crab shells, and squid pens, and may cause serious environmental pollution if they are not utilized effectively [5]. The chitin extraction process from CFWs commonly involves chemical deproteinization (using strong alkalis), demineralization (using strong acids), and bleaching (using oxidants) [6], thereby potentially releasing hazardous and protein-rich wastewater. On the other hand, CFWs mainly contain chitin and proteins, which can directly serve as nutrient components for microbial fermentation [7–10]. This "green technique" has shown great potential owing to the capability of microbes in utilizing CFWs as appropriate C/N sources for producing various metabolites, such as enzymes [11–14], antioxidants [15], enzymes inhibitors [16,17], antimicrobial agents [18], etc. Concurrently, the bioremediation and chitin waste utilization aspects may be resolved through this technique [19].

GlcNAc is of interest due to its potential use in the biomedical, pharmaceutical, food, biofuel, and chemical industries [20,21]. It is commonly produced through the enzymatic and chemical hydrolysis of chitin. Compared to chemical methods requiring harsh conditions (concentrated acid and high temperature), the enzymatic methods can be operated in mild conditions, and are thus highly preferred in terms of being environmentally friendly [22]. Chitinases (EC 3.2.2.14) belong to a group of hydrolytic enzymes that catalyze the degradation of chitin by breaking down the β-1,4-glycosidic linkages between GlcNAc units [23] and have thus received much attention in the production of GlcNAc. The two categories of chitinases based on the mechanism of chitin hydrolysis include endochitinases (EC 3.2.1.14) and exochitinases (EC 3.2.1.52) [2]. While endochitinases randomly break down the β-1,4-glycosidic linkages of chitin at internal sites to release *N*-acetyl chitooligosaccharides (*N*-acetyl COSs), exochitinases—further split into two subcategories: chitobiosidases (EC 3.2.1.29) and *N*-acetyl-β-D-glucosaminidases (EC 3.2.1.30)—act at the endpoint of *N*-acetyl COSs to release $(GlcNAc)_2$ (chitobiosidases) or GlcNAc (*N*-acetyl-β-D-glucosaminidases) [21]. A high yield of GlcNAc can be obtained by hydrolyzing chitin using a combination of different kinds of chitinases [22].

A wide spectrum of organisms exhibits the capacity to produce chitinase, including viruses, bacteria, insects, fungi, plants, and mammals. Among these, chitinases from microorganisms are of high industrial value, thereby attracting interest in the exploitation of their production processes [24]. For chitinase production by microbes, chitin is of considerable importance by serving as a potential carbon and nitrogen (C/N) source or an enzyme inducer [25]. Accordingly, the nutritional and enzyme-inducing functions of chitin could be excellently replaced by CFWs, which are deemed as cheap and abundantly available chitin sources. The use of CFWs for medium fermentation to produce microbial enzymes has several advantages, such as lower production costs, superior productivity, and simpler techniques [26]. Several previous studies have reported the production of microbial chitinases using different kinds of CFWs, such as shrimp shells [27,28], shrimp heads [21], crab shells [29], and squid pens [12,30,31]. *Paenibacillus,* a common chitinase-producing bacterial genus, however, has rarely been reported for chitinase production using CFWs, especially crab shells, as the unique source of C/N. As such, the evaluation of chitinase production by *Paenibacillus* using this kind of waste is of significant value.

It is of great importance to obtain novel chitinases with a low-cost production process and appropriate characteristics for industrial applications. Therefore, the objective of this study is to produce chitinase by *Paenibacillus* sp. TKU052 using CFWs, which are cheap materials, as the sole C/N source. The CFWs included in this study were shrimp head powder (SHP), demineralized shrimp shell powder (deSSP), demineralized crab shell powder (deCSP), squid pen powder (SPP), and shrimp shell powder (SSP). Furthermore, the obtained chitinase was examined for its valuable biochemical properties. The chitinase produced by *Paenibacillus* sp. TKU052 was a thermophilic enzyme with multi-functional activities, and thus its potential for saccharification of chitin was assessed. Accordingly, *Paenibacillus* sp. TKU052 chitinase was combined with *Streptomyces speibonae* TKU048 *N*-acetyl-β-D-glucosaminidase to effectively hydrolyze colloidal chitin (CC) to produce GlcNAc.

2. Materials and Methods

2.1. Materials

Paenibacillus sp. TKU052 was the same strain as was used in our previous work [9]. Shrimp heads were purchased from Fwu-Sow Industry (Taichung, Taiwan). Crab shells, squid pens, and shrimp shells were purchased from Shin-Ma Frozen Food Co. (I-Lan, Taiwan). The demineralization of shrimp shells and crab shells was described in our previous work [9]. Macro-Prep High S resin was purchased from Bio-Rad (Hercules, CA, USA). Chitin, 75% DDA (degree of deacetylation) chitosan, GlcNAc, 2-mercaptoethanol (2-ME), ethylenediaminetetraacetic acid (EDTA), D_2O, Congo Red, pectin, starch, xylan, carboxymethyl cellulose, gum arabic, β-1,3-glucan, dextran and 3,5-dinitrosalicylic acid

(DNS) were purchased from Sigma (St. Louis, MO, USA). Tween 20, Tween 40, Triton X-100, and SDS were purchased from Merck (Darmstadt, Germany). α-chitin powder and β-chitin powder were purchased from Charming and Beauty Co. (Taipei, Taiwan). Other chemicals used were of the highest possible quality.

2.2. Chitinase Assay

The activity of *Paenibacillus* sp. TKU052 chitinase was determined using CC (1%) as the substrate. Briefly, 200 μL of the reaction solution, consisting of 100 μL of CC (prepared in 100 mM sodium acetate buffer, pH = 5) and 100 μL of the enzyme, was incubated at 37 °C for 30 min. Then, 750 μL of DNS reagent was added to the reaction solution, and the mixture was heated at 100 °C for 10 min. The resulting solution was centrifuged at 13,000 rpm (10 min) to obtain a clear liquid. Then, 250 μL of the liquid was transferred to a 96-well plate and quantified by a microplate reader (Bio-Rad, Hercules, CA, USA) at 515 nm. GlcNAc was used as the reference to calculate the amount of reducing sugar in the sample solution. One chitinase unit is the amount of enzyme that catalyzes the degradation of CC to liberate 1 μM of reducing sugar at 37 °C in 1 min.

2.3. Chitinase Production and Purification

Paenibacillus sp. TKU052 was grown in a culture medium consisting of 0.05% $MgSO_4$, 0.1% K_2HPO_4, and 1% of each CFW type (deCSP, SHP, deSSP, SSP, and SPP) or chitin at the following cultural conditions: culture temperature of 37 °C and shaking speed of 150 rpm. The culture medium was tested for its chitinase activity every 24 h. Different amounts of deCSP, from 0.25% to 3%, were also used to assess the suitable C/N concentration for chitinase production at 37 °C and 150 rpm.

One liter of *Paenibacillus* sp. TKU052 culture supernatant obtained from 5-day old culture medium was used for purifying the chitinase produced. $(NH_4)_2SO_4$ was added to the supernatant for 60% saturation, and the mixture was kept at 4 °C overnight. The precipitate was obtained by centrifugation (4 °C, 13,000 rpm, 30 min), resuspended in 25 mM sodium phosphate buffer (pH = 5.8), dialyzed against a similar buffer, and loaded onto a Macro-Prep High S column. The elution was performed with the NaCl gradients of 0–0.1 M and 0.1–1 M. The activity fraction was concentrated by means of the freeze-drying method and loaded into a Hitachi Chromaster HPLC (High-Performance Liquid Chromatography, Hitachi, Tokyo, Japan) apparatus coupled with KW-802.5 column under the following conditions: solvent: 25 mM sodium phosphate buffer (pH = 5.8); flow rate: 0.6 mL/min; sample volume: 50 μL; column temperature: 20 °C; ultraviolet detector at the wavelength of 280 nm. The molecular weight of the *Paenibacillus* sp. TKU052 chitinase was determined according to the Laemmli method using a 10% resolving gel [9]. Zymogram of *Paenibacillus* sp. TKU052 chitinase was performed on 0.05% chitin acrylamide gel and using 0.01% Congo Red as the staining solution [30].

2.4. Effect of Temperature and pH

The optimum temperature for *Paenibacillus* sp. TKU052 chitinase was assayed in the range of 30–80 °C. The optimum pH for the enzyme was assayed in the range of pH = 3–10 using various buffers at the same concentration of 50 mM. The buffers included Na_2CO_3-$NaHCO_3$ buffer (pH = 9–10), sodium phosphate buffer (pH = 6–8), sodium acetate buffer (pH = 4–5), and glycine-HCl buffer (pH = 3). The thermal and pH stabilities of *Paenibacillus* sp. TKU052 chitinase was based on the residual activity after treating the enzyme at different temperatures or pH for 1 h. The residual activity was measured at pH = 5 and 37 °C following the method described above.

2.5. Effect of Various Ions and Chemicals

Paenibacillus sp. TKU052 chitinase was incubated with various chemicals (including Fe^{2+}, Ca^{2+}, Ba^{2+}, Mg^{2+}, Cu^{2+}, Fe^{3+}, Zn^{2+}, EDTA, 2-ME, Tween 20, Tween 40, Triton X-100, and SDS) at 20 °C for 30 min. The chemicals were tested at a final concentration of 5 mM, while the surfactants were used at 5%. The residual activity of *Paenibacillus* sp. TKU052 chitinase was measured as described above.

2.6. Substrate Specificity

This test was performed following the chitinase assay described above, but CC was alternately replaced by α-chitin powder, β-chitin powder, 75% DDA chitosan, 100% DDA chitosan, pectin, starch, xylan, carboxymethyl cellulose, gum arabic, β-1,3-glucan, and dextran. The K_m and V_{max} of *Paenibacillus* sp. TKU052 chitinase were estimated using Lineweaver–Burk plots with a final CC concentration range of 0.5–2.5 mg/mL.

2.7. The Pattern of Hydrolysis

CC and *N*-acetyl COSs with the degrees of polymerization (DP) = 2–6 were used as the enzyme substrates to explore the hydrolysis mechanism of *Paenibacillus* sp. TKU052 chitinase. At different time intervals, the reaction solution was analyzed by the HPLC process described in Section 2.9.

2.8. Production of GlcNAc

The reaction solution (20 mL), consisting of 1% CC, 100 mM sodium acetate buffer (pH 5), and 10 U/mL of *Paenibacillus* sp. TKU052 chitinase or 10 U/mL of *S. speibonae* TKU048 *N*-acetyl-β-D-glucosaminidase or 5 U/mL of *Paenibacillus* sp. TKU052 chitinase combined with 5 U/mL of *S. speibonae* TKU048 *N*-acetyl-β-D-glucosaminidase were incubated at 60 °C and a shaking speed of 150 rpm. At different time intervals, 0.2 mL of the solution was withdrawn to quantify the concentration of GlcNAc and *N*-acetyl COSs by the HPLC method (described in Section 2.9). The GlcNAc yield was calculated using the formula:

$$\text{GlcNAc yield} = \text{Amount of GlcNAc}/\text{Amount of substrate}\ (\%) \quad (1)$$

2.9. HPLC Analysis

GlcNAc and *N*-acetyl COSs were analyzed by HPLC using a Hitachi Chromaster HPLC apparatus coupled with a KS-802 column under the following conditions: solvent: water; flow rate: 0.6 mL/min; sample volume: 20 μL; column temperature: 80 °C; ultraviolet detector at the wavelength of 205 nm.

2.10. Proton Nuclear Magnetic Resonance (^{1}H-NMR) Analysis

^{1}H-NMR analysis was performed using a Bruker 600 Ultrashield NMR spectrophotometer (Bruker, New Taipei City, Taiwan). The solvent used was D_2O and the chemical shifts are shown in ppm (parts per million).

3. Results and Discussion

3.1. Chitinase Production

Paenibacillus sp. TKU052 was isolated from the soil of Tamkang University using SPP-containing medium and was found to be closely related to *P. tyrfis* MSt1 (99.3%) and *P. elgii* SD17 (99.3%) according to 16S rDNA partial base sequence analysis [9]. In the current work, *Paenibacillus* sp. TKU052 created a zone of hydrolysis on the CC plate after three days of incubation (Figure 1), indicating that this strain secretes extracellular chitinolytic enzymes to degrade the surrounding CC. The chitinase-producing ability of *P. elgii* and *P. tyrfis* species is rarely reported [32,33]. More so, none of these species have been evaluated for their chitinase production on a medium containing CFWs and the application of their enzymes in GlcNAc production. As such, this work may provide an insight into the conversion of CFWs to produce chitinase by *Paenibacillus* sp. TKU052 strain, as well as the biochemical properties and the application of the obtained chitinase in GlcNAc production.

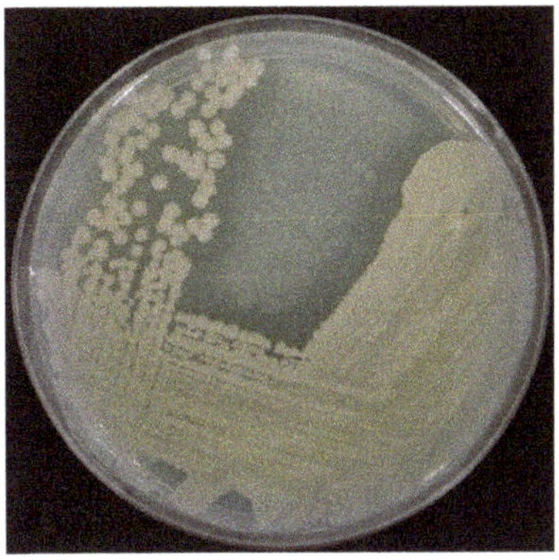

Figure 1. The chitinase-producing ability of *Paenibacillus* sp. TKU052 on CC agar medium.

The chitinase production of *Paenibacillus* sp TKU052 on a medium containing CFWs as the unique C/N source was explored. The chitinous materials used were deCSP, deSSP, SPP, SHP, SSP, and chitin. In addition, nutrient broth, a non-chitinous medium, was also used to test the enzyme productivity of *Paenibacillus* sp TKU052. As shown in Figure 2a, the maximal chitinase activity of deCSP-supplemented medium was 0.53 (day 5), 0.55 (day 6), and 0.55 U/mL (day 7), that of deSSP-supplemented medium was 0.32 U/mL (day 7); that of SPP-supplemented medium was 0.29 U/mL (day 4); that of SHP-supplemented medium was 0.30 U/mL (day 4); that of SSP-supplemented medium was 0.29 U/mL (day 5); that of chitin-supplemented medium was 0.36 U/mL (day 7); and finally, that of nutrient broth medium was 0.07 (day 4) and 0.06 U/mL (day 5). According to the result, *Paenibacillus* sp. TKU052 exhibited higher chitinase productivity toward chitin-containing media, suggesting that chitin may be a key factor for chitinase production by *Paenibacillus* sp. TKU052. Among those, the highest chitinase productivity of *Paenibacillus* sp. TKU052 was found in the deCSP-supplemented medium. Chitin, especially in colloidal form, is a widely used supplement to the chitinase-producing medium as an inducer or a nutrient source. Chemical treatment is a mandatory step to produce chitin from chitinous materials, thereby possibly leading to environmental pollution and increased production cost of the processes using chitin [34,35]. Therefore, CFWs are being assessed as an effective alternative for the cost-effective production of microbial chitinases. In addition, the use of chitinous wastes for the production of microbial chitinases by various bioprocesses is regarded as being a significant part of the utilization of these waste sources. In this study, deCSP exhibited good results and was chosen as the best C/N source for chitinase production by *Paenibacillus* sp. TKU052 and further analyses.

The effect of the deCSP amount on chitinase productivity of *Paenibacillus* sp. TKU052 was examined over a range of 0.25–3%, and the results are shown in Figure 2b. The maximal chitinase activity of 0.25% deCSP-supplemented medium was 0.43 U/mL (day 5); of 0.5% deCSP-supplemented medium was 0.59 U/mL (day 5); that of 1% deCSP-supplemented medium was 0.54 U/mL (day 5); that of 2% deCSP-supplemented medium was 0.47 U/mL (day 7); and that of 3% deCSP-supplemented medium was 0.26 U/mL (day 6). The results indicate that 0.5% deCSP is the most appropriate C/N amount for chitinase production by *Paenibacillus* sp. TKU052.

With its abundance, wasted crab shell is of interest for the preparation of various bioactive compounds, such as chitin, protease, prodigiosin, antioxidant, antidiabetic and anticancer agents [9,36]. In this study, this kind of CFW was also found to be the best C/N source for chitinase production by *Paenibacillus* sp. TKU052.

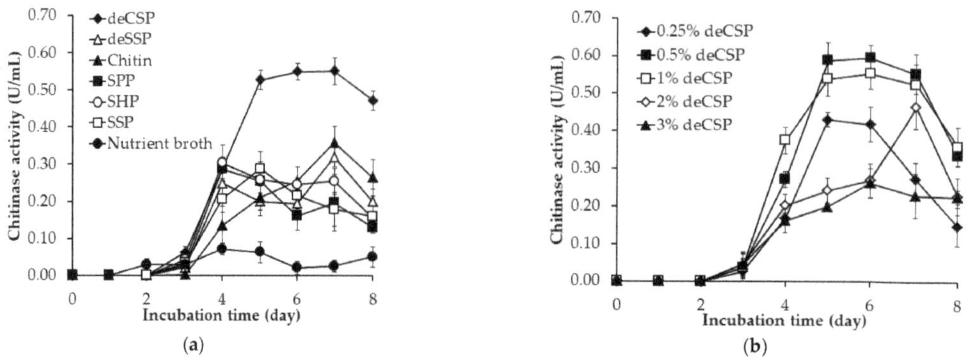

Figure 2. Screening of suitable C/N source for chitinase production by *Paenibacillus* sp. TKU052: (**a**) different kinds of CFWs and (**b**) different amounts of deCSP were examined. The error bar is the standard deviation of three replicates.

3.2. Chitinase Purification

The supernatant of a 5-day old culture of *Paenibacillus* sp. TKU052 grown using 0.5% deCSP-supplemented medium was obtained by centrifugation. Chitinase in the supernatant was purified as follows: 60% $(NH_4)_2SO_4$ precipitation, cation exchange chromatography using Macro-Prep High S resin, and size exclusion chromatography using HPLC coupling with KW-802.5 column. The crude enzyme was loaded onto a Macro-Prep High S column equilibrated by sodium phosphate buffer (25 mM, pH = 5.8). According to the enzyme elution profile, only one chitinase peak was eluted at 10–70 mM NaCl (fractions 129–166) (Figure 3a). The HPLC profile of the chitinase fraction obtained from cation exchange chromatography is shown in Figure 3b. The fraction containing the chitinase activity was eluted at a retention time of 13.3 min. The fractions that exhibited the chitinase activity were pooled and analyzed by the HPLC system using a KW-802.5 column. Accordingly, only one protein peak at a retention time of 13.3 min was observed, suggesting that the obtained chitinase was homogeneous (Figure 3c). The result of the purification of *Paenibacillus* sp. TKU052 chitinase is summarized in Table 1. Enzyme purity was estimated to be 19.47-fold greater than that of the cultural supernatant. The purified chitinase had a specific activity of 2.55 U/mg, with a yield of about 4.01%.

The homogeneity of the purified chitinase was also examined by SDS-PAGE and chitin zymography. A unique protein band (Figure 4a) and chitinolytic band (Figure 4b) at 72 kDa were observed for the purified chitinase indicating that the *Paenibacillus* sp. TKU052 chitinase has a molecule weight (MW) of 72 kDa. The MW of *Paenibacillus* sp. TKU052 chitinase is similar to that of chitinases/chitosanases from *P. barengoltzii* (74.6 kDa) [2], *P. ehimensis* MA2012 (>100, 100, 72, 65, 60, 50, 37, and 35 kDa) [37], *P. timonensis* LK-DZ15 (70 kDa) [38], *P. pasadennesis* CS0611 (69 kDa) [39], *P. elgii* HOA73 (68 kDa) [33], *Paenibacillus* sp. TKU042 (70 kDa) [35], but much higher than that from *Paenibacillus* sp. TKU047 (23 kDa) [30], *P. thermoaerophilus* TC22-2b (48 kDa) [40], *P. dendritiformis* (31 kDa) [41], *P. macerans* TKU029 (63 kDa) [42], and *P. pasadenensis* NCIM 5434 (35 kDa) [43]. Generally, the MW of chitinases/chitosanases from *Paenibacillus* species is in the range of 30–70 kDa [30].

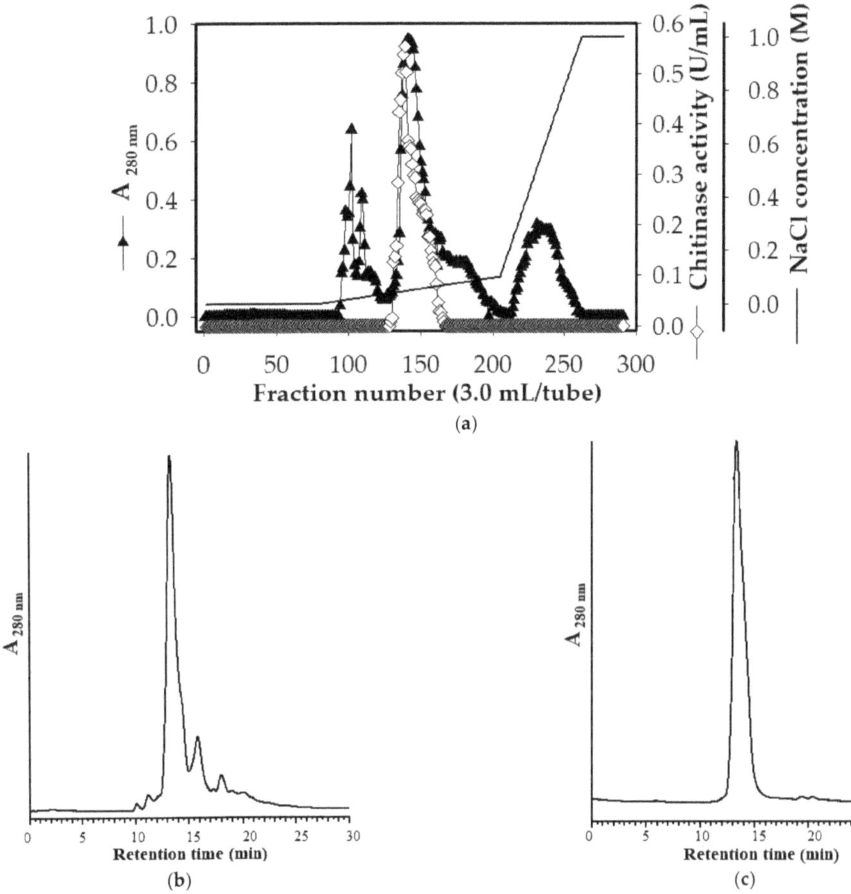

Figure 3. A typical cation exchange chromatography profile of the crude enzyme obtained from the supernatant of *Paenibacillus* sp. TKU052 culture medium supplemented with 0.5% deCSP (**a**); HPLC profiles of chitinase fraction obtained from the cation exchange chromatography step (**b**) and the HPLC step (**c**).

Table 1. A summary of the purification of *Paenibacillus* sp. TKU052 chitinase.

Step	Total Protein (mg)	Total Activity (U)	Specific Activity (U/mg)	Recovery (%)	Purification (Fold)
Cultural supernatant	2731.95	358.00	0.13	100.00	1.00
$(NH_4)_2SO_4$ precipitation	395.94	164.40	0.42	45.92	3.17
Macro-Prep High S	39.67	43.56	1.10	12.17	8.38
KW-802.5	5.63	14.36	2.55	4.01	19.47

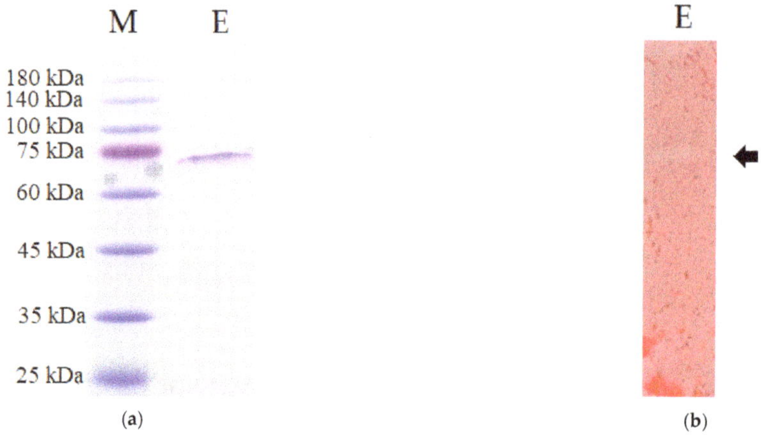

Figure 4. SDS-PAGE (**a**) and chitin zymography (**b**) profile of *Paenibacillus* sp. TKU052 chitinase. M, protein markers; E, the purified enzyme. The arrow indicates the enzyme location.

3.3. Effect of Temperature and pH

The temperature was applied in the range of 30–80 °C to explore the influence of temperature on *Paenibacillus* sp. TKU052 chitinase activity. The optimal temperature of *Paenibacillus* sp. TKU052 chitinase was found to be 70 °C and it was stable up to 60 °C (Figure 5a). Due to the outstanding functioning at temperatures $\geq$ 50 °C, *Paenibacillus* sp. TKU052 chitinase could be categorized as a thermophilic enzyme. By far, thermophilic chitinolytic enzymes have gained the most attention as they can tolerate and act at elevated temperatures. The reaction at a high temperature can provide the benefits of enhancing the solubility/dispersibility of compounds and preventing microbial contamination [44]. The thermophilic enzymes have been examined for various aspects, including the source of the enzyme, the hydrolysis mechanism of the enzyme at a high temperature, genetic engineering, and industrial applications. Accordingly, several *Paenibacillus* species strains have been found to be capable of producing thermophilic chitinolytic enzymes, with promising potential in the production of chitooligosaccharides and GlcNAc, such as *P. barengoltzii* chitinase [2], *P. thermoaerphilus* TC22-2b chitinase [40], *P. mucilaginosus* TKU032 chitosanase [45], and *Paenibacillus* sp. 1794 chitosanase [46]. In this study, *Paenibacillus* sp. TKU052 chitinase exhibited a comparable or even better thermal stability and thermal activity than most of *Paenibacillus* chitinases/chitosanases [1,33,40,46,47], indicating that this enzyme may be useful for applications in which a thermophilic chitinase is prioritized.

The influence of pH on the activity of *Paenibacillus* sp. TKU052 chitinase is shown in Figure 5b. The maximal activity of the enzyme was observed within the pH range of 4–5, and it was stable within the pH range of 4–6, suggesting that *Paenibacillus* sp. TKU052 chitinase is an acidic chitinase. Likewise, several chitinolytic enzymes from *Paenibacillus* species are functional in acidic conditions, such as *P. thermoaerophilus* TC22-2b. (pH = 4) [40], *P. mucilaginosus* TKU032 (pH = 6) [45], *Paenibacillus* sp. 1794 (pH = 4.8) [46], *P. illinoisensis* KJA-424 (pH = 5) [48], *P. timonensis* LK-DZ15 (pH = 4.5) [38], *P. pasadenensis* CS0611 (pH = 5) [39], *P. barengoltzii* CAU904 (pH = 3.5) [49], *Paenibacillus* sp. D1 (pH = 5) [50], and *Paenibacillus* sp. BISR-047 (pH = 5) [51]. However, some chitinolytic enzymes from *Paenibacillus* species have their optimum pH at neutral and alkaline conditions [30,33,41,42]. Due to its pH stability and acidic pH optimum, *Paenibacillus* sp. TKU052 chitinase could be useful for the applications set up in acidic conditions.

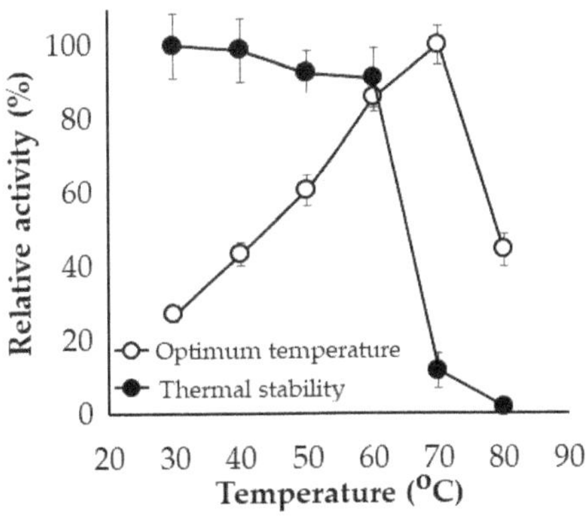

(a)

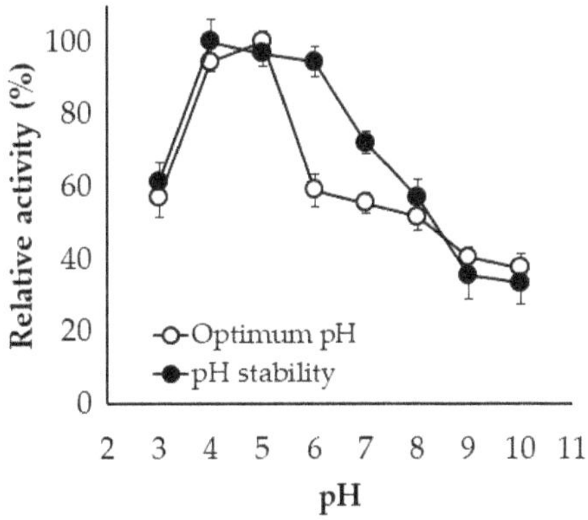

(b)

Figure 5. Effect of temperature (**a**) and pH (**b**) on the activity of *Paenibacillus* sp. TKU052 chitinase. The error bar is the standard deviation of three replicates.

3.4. Effect of Various Ions and Chemicals

The activity of *Paenibacillus* sp. TKU052 chitinase incubated with various chemicals was examined, and the results are shown in Table 2. Among the metal ions, Fe^{2+}, Ca^{2+}, Ba^{2+}, Mg^{2+}, Cu^{2+} did not significantly affect the activity of *Paenibacillus* sp. TKU052 chitinase, while Fe^{3+} and Zn^{2+} showed an inhibitory effect by reducing the enzyme activity to 26.72% and 59.51%, respectively, of its initial activity. The negative effect of heavy metal ions may be due to their ability to destroy the tertiary structure of the protein, thereby inactivating the enzyme [1]. In contrast, Mn^{2+} showed an enhancing effect on the activity of *Paenibacillus* sp. TKU052 chitinase (relative activity of 142.57%), similar to its positive effect on *P. chitinolyticus* UMBR 0002 chitinase [1] and *Paenibacillus* sp. TKU047 chitosanase [30]. Surfactants such as SDS, Triton X-100, and Tween 40 showed a negative effect on *Paenibacillus* sp. TKU052 chitinase (relative activity of 28.84%, 78.96%, and 65.71%, respectively), whereas Tween 20 did not have a significant effect. *Paenibacillus* sp. TKU052 chitinase was not significantly affected by EDTA, a metal ion chelator, indicating that the activity of the enzyme is not dependent on metal ions. Interestingly, in the presence of 2-ME, the activity of *Paenibacillus* sp. TKU052 chitinase significantly increased to 129.70% of its initial activity, demonstrating that cysteine residues are not involved in the formation of its catalytic center [52]. Earlier studies have reported strong inhibition of several chitinases by 2-ME, such as ChiA-Pt70 from *P. timonensis* LK-DZ15 [38] and CHI from *P. chitinolyticus* UMBR 0002 [1].

Table 2. Effect of various chemicals on the activity of *Paenibacillus* sp. TKU052 chitinase.

Chemical	Relative Activity (%)
None	100.00 ± 7.23
Fe^{2+}	102.31 ± 3.85
Fe^{3+}	26.72 ± 4.12
Ca^{2+}	102.13 ± 6.37
Ba^{2+}	107.57 ± 3.12
Mn^{2+}	142.57 ± 5.02
Mg^{2+}	102.65 ± 2.18
Cu^{2+}	109.93 ± 6.82
Zn^{2+}	59.51 ± 4.30
SDS	28.84 ± 3.30
Triton X-100	78.96 ± 7.48
Tween 40	65.71 ± 11.06
Tween 20	101.16 ±12.15
EDTA	108.69 ± 7.21
2-ME	129.70 ± 5.45

The data are presented as mean ± standard deviation.

3.5. Substrate Specificity

Paenibacillus sp. TKU052 chitinase exhibited the highest relative activity towards CC (100.00%), followed by β-chitin powder (16.20%), 75% DDA chitosan (8.64%), and α-chitin powder (6.65%) and no activity towards other polysaccharide substrates (Table 3). This result indicates that the activity of *Paenibacillus* sp. TKU052 chitinase is specific to the linkages of GlcNAc-GlcNAc. Likewise, a similar phenomenon was observed earlier in chitinases from *P. chitinolyticus* UMBR 0002 [1], *P. barengoltzii* CAU904 [49], and *P. elgii* HOA73 [33]. In addition, the physical form of chitin significantly affected the activity of *Paenibacillus* sp. TKU052 chitinase, wherein the colloidal form of chitin was the most suitable substrate, followed by β-chitin powder and α-chitin powder. Most *Paenibacillus* chitinases exhibit the highest chitinolytic activity on CC; however, PeChi68, a chitinase from *P. elgii* HOA73, was reported to prefer chitin powder over CC [33]. Finally, the kinetic parameters (K_m and V_{max}) of *Paenibacillus* sp. TKU052 chitinase toward CC was estimated to be 9.75 mg/mL and 2.43 μmol/min, respectively.

Table 3. The activity of *Paenibacillus* sp. TKU052 chitinase on different kinds of substrates.

Substrate	Relative Activity (%)
CC	100.00 ± 8.37
α-chitin powder	6.65 ± 0.48
β-chitin powder	16.20 ± 6.36
75% DDA chitosan	8.64 ± 4.46
100% DDA chitosan	N.A.
Pectin	N.A.
Starch	N.A.
Xylan	N.A.
Carboxymethyl cellulose	N.A.
Gum arabic	N.A.
β-1,3-glucan	N.A.
Dextran	N.A

N.A., no activity. The data are presented as mean ± standard deviation.

3.6. Hydrolysis Pattern

To explore the hydrolysis mechanism of *Paenibacillus* sp. TKU052 chitinase, CC and *N*-acetyl COSs (DP = 2–6) were used as the enzyme substrates. The results are shown in Figure 6. During the first stage (0–6 h), *Paenibacillus* sp. TKU052 chitinase hydrolyzed CC to yield $(GlcNAc)_2$ and GlcNAc, with $(GlcNAc)_2$ being the most produced product. This indicates that *Paenibacillus* sp. TKU052 chitinase is an exochitinase. From 24 h onwards, the $(GlcNAc)_2$ gradually decreased over time along with the accumulation of GlcNAc, indicating a clear outcome of the hydrolysis of $(GlcNAc)_2$ to form GlcNAc, which means *Paenibacillus* sp. TKU052 chitinase also possesses *N*-acetyl-β-D-glucosaminidase activity. However, the hydrolysis of $(GlcNAc)_2$ to form GlcNAc of *Paenibacillus* sp. TKU052 chitinase was slow and incomplete, despite the extension of the incubation time (up to 96 h), indicating a poor *N*-acetyl-β-D-glucosaminidase activity of *Paenibacillus* sp. TKU052 chitinase. In addition, a trace peak of $(GlcNAc)_3$ was detected after 0.1 h of the reaction time, suggesting that *Paenibacillus* sp. TKU052 chitinase may have some endochitinase activity. The decrease in the $(GlcNAc)_3$ peak also indicates that the generated $(GlcNAc)_3$ was further degraded by *Paenibacillus* sp. TKU052 chitinase. The hydrolysis mechanism of *Paenibacillus* sp. TKU052 chitinase on *N*-acetyl COSs is presented in Figure 6b. *Paenibacillus* sp. TKU052 chitinase degrades $(GlcNAc)_2$ slowly to form GlcNAc, indicating its poor *N*-acetyl-β-D-glucosaminidase activity, while it could rapidly degrade $(GlcNAc)_3$, $(GlcNAc)_4$, $(GlcNAc)_5$, and $(GlcNAc)_6$ to form $(GlcNAc)_2$ and GlcNAc, which confirms that it is an exochitinase. In addition, *Paenibacillus* sp. TKU052 chitinase rapidly degraded $(GlcNAc)_4$ to form $(GlcNAc)_2$, indicating that it could cleave $(GlcNAc)_4$ at the second glycosidic linkage. *Paenibacillus* sp. TKU052 chitinase could also rapidly degrade $(GlcNAc)_5$ to form $(GlcNAc)_3$ and $(GlcNAc)_2$, and could degrade $(GlcNAc)_6$ to form $(GlcNAc)_4$, $(GlcNAc)_3$, and $(GlcNAc)_2$, indicating that it cleaved these oligomers at the second and third glycosidic linkages. These results confirm that *Paenibacillus* sp. TKU052 chitinase may have both exochitinase and endochitinase properties. The hydrolysis mechanism of *Paenibacillus* sp. TKU052 chitinase is different from chitinases produced by *P. barengoltzii* [53], *S. speibonae* TKU048 [20], and *P. chitinolyticus* UMBR 0002 [1] but similar to those from *Chitinolyticbacter meiyuanensis* SYBC-H1 (*Cm*Chi1), which has been efficiently used to prepare GlcNAc from CC [5]. In conclusion, the result suggests that *Paenibacillus* sp. TKU052 chitinase may be a multi-functional chitinase, having exochitinase, endochitinase, and *N*-acetyl-β-D-glucosaminidase activities.

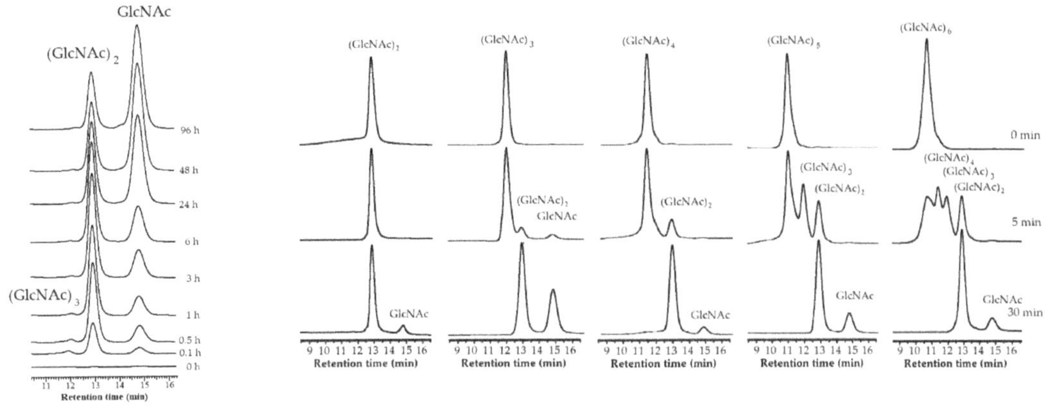

(a) (b)

Figure 6. Hydrolysis pattern of *Paenibacillus* sp. TKU052 chitinase toward colloidal chitin (CC) (**a**) and *N*-acetyl COSs (**b**).

3.7. GlcNAc Production

The synergy between chitinase with both exochitinase and endochitinase activities and *N*-acetyl-β-D-glucosaminidase may be required for the complete degradation of chitin to GlcNAc [20]. According to the previous experiment, *Paenibacillus* sp. TKU052 chitinase could rapidly degrade CC to form $(GlcNAc)_2$ and GlcNAc. As a result, this enzyme can effectively assist the *N*-acetyl-β-D-glucosaminidases to hydrolyze chitin and produce

GlcNAc. In this study, we used a combination of *Paenibacillus* sp. TKU052 chitinase and *S. speibonae* TKU048 *N*-acetyl-β-D-glucosaminidase to assess the possibility of producing GlcNAc from CC. The products of the hydrolysis process were analyzed by the HPLC method, and the results are presented in Figure 7. In the first hour, GlcNAc and $(GlcNAc)_2$ were generated at the concentrations of 3.44 and 2.58 mg/mL, respectively. In the next few hours, the concentration of $(GlcNAc)_2$ significantly decreased and reduced to zero after 12 h, whereas the concentration of GlcNAc gradually increased over time and reached its maximum value after 24 h (9.86 mg/mL). According to the estimates, the GlcNAc production process could achieve a yield of 98.60% within a reaction time of 24 h. In addition, 9.44 mg/mL of GlcNAc, indicating a production yield of 94.35%, without *N*-acetyl COSs could be observed after a reaction time of 12 h. In addition, we also explored the hydrolysis processes of CC using *Paenibacillus* sp. TKU052 chitinase and *S. speibonae* TKU048 *N*-acetyl-β-D-glucosaminidase separately. *Paenibacillus* sp. TKU052 chitinase could not completely degrade $(GlcNAc)_2$, and the reaction solution still contained both $(GlcNAc)_2$ and GlcNAc at 0.83 and 9.03 mg/mL concentrations, respectively, (after 240 h). In the case of *S. speibonae* TKU048 *N*-acetyl-β-D-glucosaminidase, 6.10 mg/mL of GlcNAc without *N*-acetyl COSs was obtained from the CC hydrolysis with a 60.99% yield after 96 h. This indicates that a novel combination of *Paenibacillus* sp. TKU052 chitinase and *S. speibonae* TKU048 *N*-acetyl-β-D-glucosaminidase could efficiently improve the GlcNAc production from CC. The efficiency of the present process was compared with those in earlier reports (Table 4). Among them, the combination of *Paenibacillus* sp. TKU052 chitinase and *S. speibonae* TKU048 *N*-acetyl-β-D-glucosaminidase reveals a comparable result with a short required reaction time (12–24 h) and high GlcNAc production yield (94.35–98.60%). In addition, the absence of *N*-acetyl COSs in the final reaction solution could also simplify the GlcNAc purification step. In this study, GlcNAc was obtained from the chitin hydrolysate solution by one-step purification using HPLC coupled with a KS-802 column. The HPLC result revealed only one peak at a retention time of 14.7 min (Figure 8a), indicating that the molecular weight of the product is similar to that of the GlcNAc standard. The obtained GlcNAc was analyzed by ^{1}H-NMR spectroscopy, and as shown in Figure 8b, the obtained GlcNAc had a similar ^{1}H-NMR profile as that of the profile of commercial GlcNAc from Sigma (St. Louis, MO, USA), confirming the purity of the product. This result, therefore, provides a promising enzymatic method for the production of GlcNAc from CC.

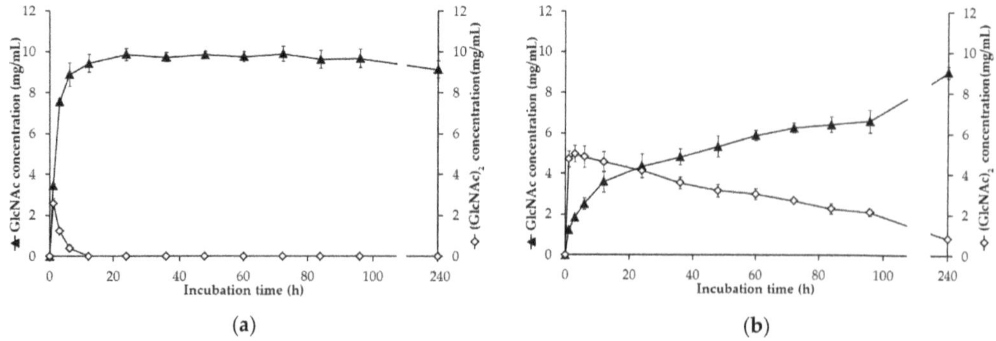

Figure 7. *Cont.*

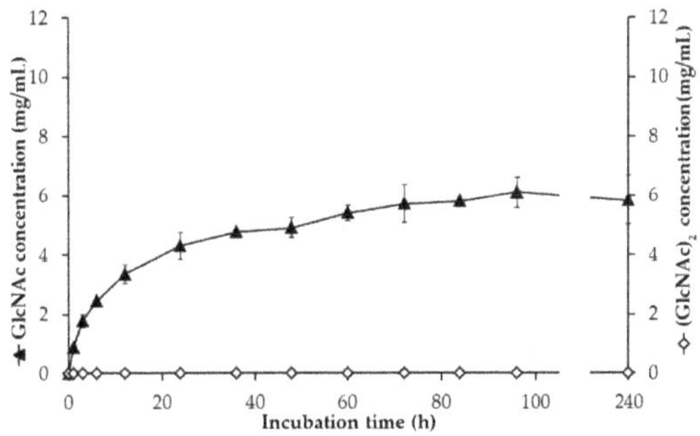

(c)

Figure 7. Time courses of GlcNAc and $(GlcNAc)_2$ production generated from the hydrolysis of colloidal chitin (CC) catalyzed by a mixture of *Paenibacillus* sp. TKU052 chitinase and *Streptomyces speibonae* TKU048 *N*-acetyl-β-D-glucosaminidase (**a**), and independently by *Paenibacillus* sp. TKU052 chitinase (**b**), and *S. speibonae* TKU048 *N*-acetyl-β-D-glucosaminidase (**c**). The error bar indicates the standard deviation of three replicates.

Table 4. Comparison of enzymatic GlcNAc production by different microbial chitinases.

Enzyme	Substrate	Time Consumed	Yield (%)	Ref.
Paenibacillus sp. TKU052 chitinase and *S. speibonae* TKU048 *N*-acetyl-β-D-glucosaminidase	CC	12–24 h	94.35–98.60	This study
S. speibonae TKU048 *N*-acetyl-β-D-glucosaminidase	β-chitin powder	96 h	73.64	[20]
Serratia marcescens chitinases (SmChiA, SmChiB, and SmChiC) and *Ostrinia furnacalis N*-acetyl-d-glucosaminidase (OfHex1)	*Asperillus niger* mycelia powder	24 h	93	[22]
Streptomyces coelicolor A3(2) chitinase C (ScChiC) and *N*-acetylhexosaminidase (ScHEX)	Crystalline chitin	8 h	90	[54]
Chitinolyticbacter meiyuanensis SYBC-H1 chitinase (CmChi1)	CC	24 h	98	[5]
C. meiyuanensis SYBC-H1 chitinase	Chitin powder	4 days	near 100	[55]
Myceliophthora thermophila C1 β-*N*-acetylglucosaminidase (M*th*NAG) and chitinase Chi1	Swollen chitin	24 h	37.8	[56]
Aspergillus terreus chitinase	Ground prawn shell Chitin flakes Colloidal prawn shell Swollen chitin	5 days	30 73 80 92	[57]
Aeromonas sp. PTCC1691 crude enzyme	CC	24 h	79	[58]
Aeromonas caviae CHZ306 enzyme cocktail	CC	6 h	90	[59]
Aeromonas sp. GJ-18 crude enzyme	Swollen chitin	5–9 days	83.0–94.9	[60]
P. barengoltzii β-*N*-acetylglucosaminidase (PbNag39) and chitinase (PbChi70)	Powdery chitin CC	24 h	75.3 97.0	[61]
P. barengoltzii chitinase (PbChi74) and *Rhizomucor miehei* β-*N*-acetylglucosaminidase (RmNAG)	CC	24 h	92.6	[2]
P. illinoisensis KJA-424 crude enzyme	Swollen chitin	24 h	62.2	[62]
Streptomyces alfalae β-*N*-acetylhexosaminidase (SaHEX) and a commercial chitinase (SgCtn)	CC	6 h	93.7	[63]
Streptomyces violascens β-*N*-acetylglucosaminidases (SvNag2557 and SvNag4755)	CC Ionic liquid pretreated chitin		80.2 73.8	[64]

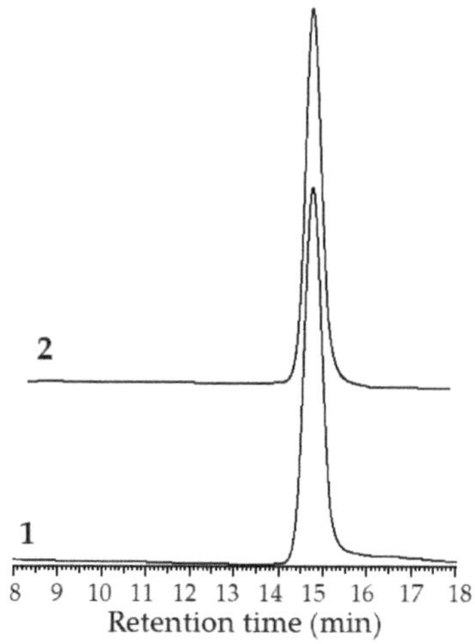

(a)

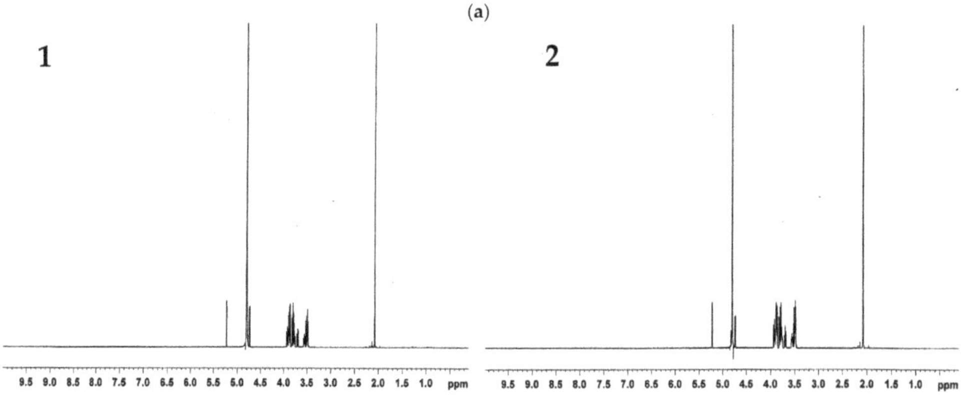

(b)

Figure 8. HPLC (**a**) and ^{1}H-NMR (**b**) profiles of the obtained and commercial GlcNAc (1, obtained GlcNAc; 2, commercial GlcNAc).

4. Conclusions

The exploitation of CFWs to obtain high-quality products through microbial fermentation is increasing gradually. In this study, deCSP was found to be the best potential C/N source for chitinase production by *Paenibacillus* sp. TKU052 among several CFWs, and a chitinase with an MW of 72 kDa was purified from the 0.5% deCSP-containing medium. Interestingly, the pure chitinase exhibited some valuable properties, such as high catalytic function at acidic pH and elevated-temperature conditions (pH = 4–5, and 70 °C), and can also carry out multi-functional (exochitinase, endochitinase, and *N*-acetyl-β-D-glucosaminidase) activities. These characteristics make *Paenibacillus* sp. TKU052 chitinase a promising candidate for the saccharification of chitin under mild conditions. Remarkably, the coupling of *Paenibacillus* sp. TKU052 chitinase and *S. speibonae* TKU048 *N*-acetyl-β-D-glucosaminidase could efficiently improve the GlcNAc production from CC with a yield of 94.35–98.60% in 12–24 h. As a result, the bioprocessing of deCSP by *Paenibacillus* sp. TKU052 may be potentially useful in producing chitinase as a tool for the bioconversion of chitin to GlcNAc.

Author Contributions: Conceptualization, C.T.D. and S.-L.W.; methodology, C.T.D., T.N.T. and S.-L.W.; software, C.T.D. and T.N.T.; validation, C.T.D., T.N.T. and S.-L.W.; formal analysis, C.T.D., T.N.T. and S.-L.W.; investigation, C.T.D., T.N.T. and S.-L.W.; resources, S.-L.W.; data curation, C.T.D., T.N.T. and S.-L.W.; writing—original draft preparation, C.T.D.; writing—review and editing, C.T.D., T.N.T. and S.-L.W.; visualization, C.T.D., T.N.T. and S.-L.W.; supervision, S.-L.W.; project administration, S.-L.W.; funding acquisition, S.-L.W. All authors have read and agreed to the published version of the manuscript.

Funding: This study was supported in part by a grant from the Ministry of Science and Technology, Taiwan (MOST 109-2313-B-032-002-; MOST 110-2320-B-032-001; MOST 110-2923-B-032-001-; MOST 110-2811-B-032-001).

Institutional Review Board Statement: Not applicable.

Informed Consent Statement: Not applicable.

Data Availability Statement: The data presented in this study are available on request from the corresponding author.

Conflicts of Interest: The authors declare no conflict of interest.

References

1. Liu, C.; Shen, N.; Wu, J.; Jiang, M.; Shi, S.; Wang, J.; Wei, Y.; Yang, L. Cloning, expression and characterization of a chitinase from *Paenibacillus chitinolyticus* strain UMBR 0002. *PeerJ* **2020**, *8*, e8964. [CrossRef]
2. Fu, X.; Yan, Q.; Yang, S.; Yang, X.; Guo, Y.; Jiang, Z. An acidic, thermostable exochitinase with β-*N*-acetylglucosaminidase activity from *Paenibacillus barengoltzii* converting chitin to *N*-acetyl glucosamine. *Biotechnol. Biofuels* **2014**, *7*, 174. [CrossRef] [PubMed]
3. Yadav, M.; Goswami, P.; Paritosh, K.; Kumar, M.; Pareek, N.; Vivekanand, V. Seafood waste: A source for preparation of commercially employable chitin/chitosan materials. *Bioresources* **2019**, *6*, 8–28. [CrossRef]
4. Santos, V.P.; Marques, N.S.S.; Maia, P.C.S.V.; Lima, M.A.B.d.; Franco, L.d.O.; Campos-Takaki, G.M.d. Seafood waste as attractive source of chitin and chitosan production and their applications. *Int. J. Mol. Sci.* **2020**, *21*, 4290. [CrossRef] [PubMed]
5. Zhang, A.; He, Y.; Wei, G.; Zhou, J.; Dong, W.; Chen, K.; Quyang, P. Molecular characterization of a novel chitinase *Cm*Chi1 from *Chitinolyticbacter meiyuanensis* SYBC-H1 and its use in *N*-acetyl-d-glucosamine production. *Biotechnol. Biofuels* **2018**, *11*, 179. [CrossRef] [PubMed]
6. Deng, J.J.; Zhang, M.S.; Li, Z.W.; Lu, D.L.; Mao, H.H.; Zhu, M.J.; Li, J.Z.; Luo, X.C. One-step processing of shrimp shell waste with a chitinase fused to a carbohydrate-binding module. *Green Chem.* **2020**, *22*, 6862–6873. [CrossRef]
7. Wang, C.H.; Doan, C.T.; Nguyen, V.B.; Nguyen, A.D.; Wang, S.L. Reclamation of fishery processing waste: A mini-review. *Molecules* **2019**, *24*, 2234. [CrossRef]
8. Doan, C.T.; Tran, T.N.; Nguyen, V.B.; Vo, T.P.K.; Nguyen, A.D.; Wang, S.L. Chitin extraction from shrimp waste by liquid fermentation using an alkaline protease-producing strain, *Brevibacillus parabrevis*. *Int. J. Biol. Macromol.* **2019**, *131*, 706–715. [CrossRef] [PubMed]
9. Doan, C.T.; Tran, T.N.; Nguyen, V.B.; Nguyen, A.D.; Wang, S.-L. Utilization of seafood processing by-products for production of proteases by *Paenibacillus* sp. TKU052 and their application in biopeptides' preparation. *Mar. Drugs* **2020**, *18*, 574. [CrossRef]
10. Nguyen, T.H.; Wang, S.L.; Nguyen, D.N.; Nguyen, A.D.; Nguyen, T.H.; Doan, M.D.; Ngo, V.A.; Doan, C.T.; Kuo, Y.H.; Nguyen, V.B. Bioprocessing of marine chitinous wastes for the production of bioactive prodigiosin. *Molecules* **2021**, *26*, 3138. [CrossRef]
11. Das, S.; Roy, D.; Sen, R. Utilization of chitinaceous wastes for the production of chitinase. *Adv. Food Nutr. Res.* **2016**, *78*, 27–46. [PubMed]
12. Tran, T.N.; Doan, C.T.; Nguyen, V.B.; Nguyen, A.D. The isolation of chitinase from *Streptomyces thermocarboxydus* and its application in the preparation of chitin oligomers. *Res. Chem. Intermed.* **2019**, *45*, 727–742. [CrossRef]
13. Wang, S.L.; Yu, H.T.; Tsai, M.H.; Doan, C.T.; Nguyen, V.B.; Do, V.C.; Nguyen, A.D. Conversion of squid pens to chitosanases and dye adsorbents via *Bacillus cereus*. *Res. Chem. Intermed.* **2018**, *44*, 4903–4911. [CrossRef]
14. Doan, C.T.; Tran, T.N.; Nguyen, M.T.; Nguyen, V.B.; Nguyen, A.D.; Wang, S.L. Anti-α-glucosidase activity by a protease from *Bacillus licheniformis*. *Molecules* **2019**, *24*, 691. [CrossRef] [PubMed]
15. Liang, T.W.; Tseng, S.C.; Wang, S.L. Production and characterization of antioxidant properties of exopolysaccharides from *Paenibacillus mucilaginosus* TKU032. *Mar. Drugs* **2016**, *14*, 40. [CrossRef] [PubMed]
16. Hsu, C.H.; Nguyen, A.D.; Chen, Y.W.; Wang, S.L. Tyrosinase inhibitors and insecticidal materials produced by *Burkholderia cepacia* using squid pen as the sole carbon and nitrogen source. *Res. Chem. Intermed.* **2014**, *40*, 2249–2258. [CrossRef]
17. Nguyen, V.B.; Wang, S.L. Reclamation of marine chitinous materials for the production of α-glucosidase inhibitors via microbial conversion. *Mar. Drugs* **2017**, *15*, 350. [CrossRef] [PubMed]
18. Liang, T.W.; Wu, C.C.; Cheng, W.T.; Chen, Y.C.; Wang, C.L.; Wang, I.L.; Wang, S.L. Exopolysaccharides and antimicrobial biosurfactants produced by *Paenibacillus macerans* TKU029. *Appl. Biochem. Biotechnol.* **2014**, *172*, 933–950. [CrossRef] [PubMed]

19. Shereena, E.K.; Nisha, M.K.; Gaayathiri Devi, E. Chitinase production by *Aspergillus terreus* from marine wastes and its efficacy in antifungal activity. *Int. J. Adv. Res.* **2020**, *8*, 1399–1404.
20. Chen, J.K.; Shen, C.R.; Liu, C.L. *N*-acetylglucosamine: Production and applications. *Mar. Drugs* **2010**, *8*, 2493–2516. [CrossRef]
21. Tran, T.N.; Doan, C.T.; Nguyen, M.T.; Nguyen, V.B.; Vo, T.P.K.; Nguyen, A.D.; Wang, S.L. An exochitinase with *N*-acetyl-β-glucosaminidase-like activity from shrimp head conversion by *Streptomyces speibonae* and its application in hydrolyzing β-chitin powder to produce *N*-acetyl-D-glucosamine. *Polymers* **2019**, *11*, 1600. [CrossRef]
22. Zhu, W.; Wang, D.; Liu, T.; Yang, Q. Production of *N*-Acetyl-d-glucosamine from mycelial waste by a combination of bacterial chitinases and an insect *N*-acetyl-D-glucosaminidase. *J. Agric. Food Chem.* **2016**, *64*, 6738–6744. [CrossRef]
23. Salas-Ovilla, R.; Gálvez-López, D.; Vázquez-Ovando, A.; Salvador-Figueroa, M.; Rosas-Quijano, R. Isolation and identification of marine strains of *Stenotrophomona maltophilia* with high chitinolytic activity. *PeerJ.* **2019**, *7*, e6102. [CrossRef]
24. Bhattacharya D, Nagpure A, Gupta RK: Bacterial chitinases: Properties and potential. *Crit. Rev. Biotechnol.* **2007**, *27*, 21–28. [CrossRef]
25. Dahiya, N.; Tewari, R.; Hoondal, G.S. Biotechnological aspects of chitinolytic enzymes: A review. *Appl. Microbiol. Biotechnol.* **2006**, *71*, 773–782. [CrossRef]
26. Ben Rebah, F.; Miled, N. Fish processing wastes for microbial enzyme production: A review. *3 Biotech* **2013**, *3*, 255–265. [CrossRef] [PubMed]
27. Ghorbel-Bellaaj, O.; Manni, L.; Jellouli, K.; Hmidet, N.; Nasri, M. Optimization of protease and chitinase production by *Bacillus cereus* SV1 on shrimp shell waste using statistical experimental design. Biochemical and molecular characterization of the chitinase. *Ann. Microbiol.* **2021**, *62*, 1255–1268. [CrossRef]
28. Wang, S.L.; Li, J.Y.; Liang, T.W.; Hsieh, J.L.; Tseng, W.N. Conversion of shrimp shell by using *Serratia* sp. TKU017 fermentation for the production of enzymes and antioxidants. *J. Microbiol. Biotechnol.* **2010**, *20*, 117–126. [CrossRef] [PubMed]
29. Cheba, B.A.; Zaghloul, T.I.; El-Mahdy, A.R. Demineralized crab and shrimp shell powder: Cost effective medium for *Bacillus* Sp. R2 growth and chitinase production. *Procedia Manuf.* **2018**, *22*, 413–419. [CrossRef]
30. Doan, C.T.; Tran, T.N.; Nguyen, V.B.; Tran, T.D.; Nguyen, A.D.; Wang, S.-L. Bioprocessing of squid pens waste into chitosanase by *Paenibacillus* sp. TKU047 and its application in low-molecular weight chitosan oligosaccharides production. *Polymers* **2020**, *12*, 1163. [CrossRef] [PubMed]
31. Wang, S.L.; Hsu, W.H.; Liang, T.W. Conversion of squid pen by *Pseudomonas aeruginosa* K187 fermentation for the production of *N*-acetyl chitooligosaccharides and biofertilizers. *Carbohydr. Res.* **2010**, *345*, 880–885. [CrossRef] [PubMed]
32. Tariq, A.L.; Sarath Chandran, R.; Reyaz, A.L. Extracellular chitinolytic enzyme producing *Paenibacillus elgii* TS33 isolated from shrimp shell waste. *Int. J. Pharma. Res. Health Sci.* **2017**, *5*, 2064–2069.
33. Kim, Y.H.; Park, S.K.; Hur, J.Y.; Kim, Y.C. Purification and characterization of a major extracellular chitinase from a biocontrol bacterium, *Paenibacillus elgii* HOA73. *Plant Pathol. J.* **2017**, *33*, 318–328. [CrossRef]
34. Lee, D.H.; Doan, C.T.; Tran, T.N.; Nguyen, V.B.; Nguyen, A.D.; Wang, C.L.; Wang, S.L. Proteases production and chitin preparation from the liquid fermentation of chitinous fishery by-products by *Paenibacillus elgii*. *Mar. Drugs* **2021**, *19*, 477. [CrossRef]
35. Doan, C.T.; Tran, T.N.; Nguyen, V.B.; Nguyen, A.D.; Wang, S.L. Conversion of squid pens to chitosanases and proteases via *Paenibacillus* sp. TKU042. *Mar. Drugs* **2018**, *16*, 83. [CrossRef]
36. Nguyen, V.B.; Nguyen, D.N.; Nguyen, A.D.; Ngo, V.A.; Ton, T.Q.; Doan, C.T.; Pham, T.P.; Tran, T.P.H.; Wang, S.L. Utilization of crab waste for cost-effective bioproduction of prodigiosin. *Mar. Drugs* **2020**, *18*, 523. [CrossRef]
37. Seo, D.J.; Lee, Y.S.; Kim, K.Y.; Jung, W.J. Antifungal activity of chitinase obtained from *Paenibacillus ehimensis* MA2012 against conidial of *Collectotrichum gloeosporioides* in vitro. *Microb. Pathog.* **2016**, *96*, 10–14. [CrossRef]
38. Yahiaoui, M.; Laribi-Habchi, H.; Bouacem, K.; Asmani, K.L.; Mechri, S.; Harir, M.; Bendif, H.; Fertas, R.A.E.; Jaouadi, B. Purification and biochemical characterization of a new organic solvent-tolerant chitinase from *Paenibacillus timonensis* strain LK-DZ15 isolated from the Djurdjura Mountains in Kabylia, Algeria. *Carbohydr. Res.* **2019**, *483*, 107747. [CrossRef] [PubMed]
39. Guo, X.; Xu, P.; Zong, M.; Lou, W. Purification and characterization of alkaline chitinase from *Paenibacillus pasadenensis* CS0611. *Chin. J. Catal.* **2017**, *38*, 665–672. [CrossRef]
40. Ueda, J.; Kurosawa, N. Characterization of an extracellular thermophilic chitinase from *Paenibacillus thermoaerophilus* strain TC22-2b isolated from compost. *World J. Microbiol. Biotechnol.* **2015**, *31*, 135–143. [CrossRef] [PubMed]
41. Sun, H.; Mao, X.; Guo, N.; Zhao, L.; Cao, R.; Liu, Q. Discovery and characterization of a novel chitosanase from *Paenibacillus dendritiformis* by phylogeny-based enzymatic product specificity prediction. *J. Agric. Food Chem.* **2018**, *66*, 4645–4651. [CrossRef]
42. Doan, C.T.; Tran, T.N.; Nguyen, V.B.; Nguyen, A.D.; Wang, S.L. Reclamation of marine chitinous materials for chitosanase production via microbial conversion by *Paenibacillus macerans*. *Mar. Drugs* **2018**, *16*, 429. [CrossRef]
43. Loni, P.P.; Patil, J.U.; Phugare, S.S.; Bajekal, S.S. Purification and characterization of alkaline chitinase from *Paenibacillus pasadenensis* NCIM 5434. *J. Basic Microbiol.* **2014**, *54*, 1080–1089. [CrossRef] [PubMed]
44. Mathew, G.M.; Madhavan, A.; Arun, K.B.; Sindhu, R.; Binod, P.; Singhania, R.R.; Sukumaran, R.K.; Pandey, A. Thermophilic chitinases: Structural, functional and engineering attributes for industrial applications. *Appl. Biochem. Biotechnol.* **2021**, *193*, 142–164. [CrossRef] [PubMed]

45. Doan, C.T.; Tran, T.N.; Nguyen, V.B.; Nguyen, A.D.; Wang, S.L. Production of a thermostable chitosanase from shrimp heads via *Paenibacillus mucilaginosus* TKU032 conversion and its application in the preparation of bioactive chitosan oligosaccharides. *Mar. Drugs* **2019**, *17*, 217. [CrossRef] [PubMed]
46. Zitouni, M.; Fortin, M.; Scheerle, R.K.; Letzel, T.; Matteau, D.; Rodrigue, S.; Brzezinski, R. Biochemical and molecular characterization of a thermostable chitosanase produced by the strain *Paenibacillus* sp. 1794 newly isolated from compost. *Appl. Microbiol. Biotechnol.* **2013**, *97*, 5801–5813. [CrossRef]
47. Zhang, W.; Ma, J.; Yan, Q.; Jiang, Z.; Yang, S. Biochemical characterization of a novel acidic chitinase with antifungal activity from *Paenibacillus xylanexedens* Z2-4. *Int. J. Biol. Macromol.* **2021**, *182*, 1528–1536. [CrossRef]
48. Jung, W.J.; Kuk, J.K.; Kim, K.Y.; Kim, T.H.; Park, R.D. Purification and characterization of chitinase from *Paenibacillus illinoisensis* KJA-424. *J. Microbiol. Biotechnol.* **2005**, *15*, 274–280.
49. Fu, X.; Yan, Q.; Wang, J.; Yang, S.; Jiang, Z. Purification and biochemical characterization of novel acidic chitinase from *Paenibacillus barengoltzii*. *Int. J. Biol. Macromol.* **2016**, *91*, 973–979. [CrossRef]
50. Singh, A.K.; Mehta, G.; Chhatpar, H.S. Optimization of medium constituents for improved chitinase production by *Paenibacillus* sp. D1 using statistical approach. *Lett. Appl. Microbiol.* **2009**, *49*, 708–714. [CrossRef]
51. Meena, S.; Gothwal, R.K.; Saxena, J.; Nehra, S.; Mohan, M.K.; Ghosh, P. Effect of metal ions and chemical compounds on chitinase produced by a newly isolated thermotolerant *Paenibacillus* sp. BISR-047 and its shelf-life. *Int. J. Curr. Microbiol. Appl. Sci.* **2015**, *4*, 872–881.
52. Alves, T.B.; de Oliveira Ornela, P.H.; de Oliveira, A.H.C.; Jorge, J.A.; Guimarães, L.H.S. Production and characterization of a thermostable antifungal chitinase secreted by the filamentous fungus *Aspergillus niveus* under submerged fermentation. *3 Biotech* **2018**, *8*, 369. [CrossRef]
53. Yang, S.; Fu, X.; Yan, Q.; Guo, Y.; Liu, Z.; Jiang, Z. Cloning, expression, purification and application of a novel chitinase from a thermophilic marine bacterium *Paenibacillus barengoltzii*. *Food Chem.* **2016**, *192*, 1041–1048. [CrossRef]
54. Nguyen-Thi, N.; Doucet, N. Combining chitinase C and *N*-acetylhexosaminidase from *Streptomyces coelicolor* A3(2) provides an efficient way to synthesize *N*-acetylglucosamine from crystalline chitin. *J. Biotechnol.* **2016**, *220*, 25–32. [CrossRef] [PubMed]
55. Zhang, A.; Gao, C.; Wang, J.; Chen, K.; Quyang, P. An efficient enzymatic production of *N*-acetyl-D-glucosamine from crude chitin powders. *Green Chem.* **2016**, *18*, 2147–2154. [CrossRef]
56. Krolicka, M.; Hinz, S.W.A.; Koetsier, M.J.; Eggink, G.; van den Broek, L.A.M.; Boeriu, C.G. β-N-Acetylglucosaminidase MthNAG from *Myceliophthora thermophila* C1, a thermostable enzyme for production of *N*-acetylglucosamine from chitin. *Appl. Microbiol. Biotechnol.* **2018**, *102*, 7441–7454. [CrossRef] [PubMed]
57. Das, S.; Dey, P.; Roy, D.; Maiti, M.K.; Sen, R. *N*-Acetyl-D-glucosamine production by a chitinase of marine fungal origin: A case study of potential industrial significance for valorization of waste chitins. *Appl. Biochem. Biotechnol.* **2019**, *187*, 407–423. [CrossRef]
58. Jamialahmadi, K.; Behravan, J.; Fathi Najafi, M.; Tabatabaei Yazdi, M.; Shahverdi, A.R.; Faramarzi, M.A. Enzymatic production of *N*-Acetyl-D-glucosamine from chitin using crude enzyme preparation of *Aeromonas* sp. PTCC1691. *Biotechnology* **2011**, *10*, 292–297. [CrossRef]
59. Cardozo, F.A.; Facchinatto, W.M.; Colnago, L.A.; Campana-Filho, S.P.; Pessoa, A. Bioproduction of *N*-acetyl-glucosamine from colloidal α-chitin using an enzyme cocktail produced by *Aeromonas caviae* CHZ306. *World J. Microbiol. Biotechnol.* **2019**, *35*, 114. [CrossRef]
60. Kuk, J.H.; Jung, W.J.; Jo, G.H.; Kim, Y.C.; Kim, K.Y.; Park, R.D. Production of *N*-acetyl-beta-D-glucosamine from chitin by *Aeromonas* sp. GJ-18 crude enzyme. *Appl. Microbiol. Biotechnol.* **2005**, *68*, 384–389. [CrossRef]
61. Liu, Y.; Jiang, Z.; Ma, J.; Ma, S.; Yan, Q.; Yang, S. Biochemical characterization and structural analysis of a β-*N*-acetylglucosaminidase from *Paenibacillus barengoltzii* for efficient production of *N*-acetyl-d-glucosamine. *J. Agric. Food Chem.* **2020**, *68*, 5648–5657. [CrossRef] [PubMed]
62. Jung, W.J.; Souleimanov, A.; Park, R.D.; Smith, D.L. Enzymatic production of *N*-acetyl chitooligosaccharides by crude enzyme derived from *Paenibacillus illioisensis* KJA-424. *Carbohydr. Polym.* **2007**, *67*, 256–259. [CrossRef]
63. Lv, C.; Gu, T.; Xu, K.; Gu, J.; Li, L.; Liu, X.; Zhang, A.; Gao, S.; Li, W.; Zhao, G. Biochemical characterization of a β-*N*-acetylhexosaminidase from *Streptomyces alfalfae* and its application in the production of *N*-acetyl-D-glucosamine. *J. Biosci. Bioeng.* **2019**, *128*, 135–141. [CrossRef] [PubMed]
64. Li, J.; Gao, K.; Secundo, F.; Mao, X. Biochemical characterization of two β-*N*-acetylglucosaminidases from *Streptomyces violascens* for efficient production of *N*-acetyl-d-glucosamine. *Food Chem.* **2021**, *364*, 130393. [CrossRef] [PubMed]

polymers

MDPI

Article

Hydrophobic Modification of Biopolymer Aerogels by Cold Plasma Coating

Baldur Schroeter [1,*], Isabella Jung [1], Katharina Bauer [2], Pavel Gurikov [3] and Irina Smirnova [1]

1 Institute for Thermal Separation Processes, Hamburg University of Technology, Eißendorfer Straße 38, 21073 Hamburg, Germany; Isabella.Jung@tuhh.de (I.J.); Irina.Smirnova@tuhh.de (I.S.)
2 Diener Electronic GmbH & Co. KG, Nagolder Straße 61, 72224 Ebhausen, Germany; Katharina-bauer@mein.gmx
3 Laboratory for Development and Modelling of Novel Nanoporous Materials, Hamburg University of Technology, Eißendorfer Straße 38, 21073 Hamburg, Germany; Pavel.Gurikov@tuhh.de
* Correspondence: baldur.schroeter@tuhh.de

Abstract: The aim of this work was to evaluate the potential of cold plasma polymerization as a simple, fast and versatile technique for deposition of protective hydrophobic and oleophobic polymer layers on hydrophilic biopolymer aerogels. Polymerization of different fluorinated monomers (octafluorocyclobutane C_4F_8 and perfluoro-acrylates PFAC-6 and PFAC-8) on aerogel monoliths derived from alginate, cellulose, whey protein isolate (WPI) and potato protein isolate (PPI) resulted in fast and significant surface hydrophobization after short process times of 5 min and led to superhydrophobic surfaces with static water contact angles up to 154° after application of poly-C_4F_8 coatings. Simultaneous introduction of hydro- and oleophobicity was possible by deposition of perfluoro-acrylates. While the porous structure of aerogels stayed intact during the process, polymerization inside the aerogels pores led to the generation of new porous moieties and resulted therefore in significant increase in the specific surface area. The magnitude of the effect depended on the individual process settings and on the overall porosity of the substrates. A maximization of specific surface area increase (+179 m^2/g) was obtained by applying a pulsed wave mode in the C_4F_8-coating of alginate aerogels.

Keywords: aerogels; biopolymers; cold plasma coating; hydrophobization; pore structure

Citation: Schroeter, B.; Jung, I.; Bauer, K.; Gurikov, P.; Smirnova, I. Hydrophobic Modification of Biopolymer Aerogels by Cold Plasma Coating. *Polymers* **2021**, *13*, 3000. https://doi.org/10.3390/polym13173000

Academic Editor: Arn Mignon

Received: 19 August 2021
Accepted: 2 September 2021
Published: 4 September 2021

1. Introduction

Aerogels are solid materials with high porosity, low density and high specific surface areas, which can be produced from a variety of starting materials such as silica, synthetic polymers and biopolymers. Due to sustainability aspects as well as low material costs, high availability, biocompatibility and non-toxicity of the educts, research on biopolymer-based aerogels has seen significantly increased importance in the last decade [1,2]. The typical steps of biopolymer–aerogel synthesis involve the formation of a hydrogel in an aqueous phase, followed by a solvent exchange and subsequent supercritical drying or, in special cases, freeze drying [3,4]. Most biopolymer-based aerogels are derived from polysaccharides, common examples being alginate and cellulose aerogels. They contain low initial biopolymer content of approximately 1.0–7.0 wt%, demonstrate high porosities (ε) up to 99%, low envelope densities (ρ_e) of approximately 0.07–0.5 g/cm^3 and specific surface areas in the typical range from 200–600 m^2/g [4]. Recent studies showed that aerogels with considerable surface areas of approximately 300–400 m^2/g and significantly higher biopolymer content of 10–20 wt% can be obtained by a controlled gelation of proteins such as whey protein isolate (WPI) and potato protein isolate (PPI) [5,6]. Since all biopolymer aerogels are characterized by a particularly high surface concentration and a variety of functionalities on the surface, such as OH-, carboxy-, amine- and amide groups, they show high potential in adsorption and separation processes and offer the possibility of targeted post-functionalization. Due to the combination of tailorable microstructures and high porosities, further application areas are as high performance thermal insulation materials and carrier matrices for targeted drug delivery [7,8]. Although, from a material standpoint, aerogels offer high variety and potential, the use of aerogels poses challenges in handling and storage: high polarity (in case of polysaccharide gels) and open-porous structure (for all aerogel types) makes biopolymer aerogels especially susceptible to moisture that can penetrate into the pores, resulting in changes of the internal surface and pore collapse. Furthermore, contact of biopolymer aerogels, e.g., based on alginate and cellulose, with polar liquids may lead to destruction of the 3D structure, and in the case of non-polar liquids, may lead to uptake of the liquid into the pores. Since surface sensitivity, storage and aging behavior are crucial properties, strategies for aerogel protection are of high relevance.

Hydrophobic post modifications of dry biopolymer aerogels may offer protection against polar substances and can be applied principally in two ways: application of an (closed) outer layer on the aerogel's surface or by modification of the inner porous surface, whereas the favored strategy depends on the final application. The former can be achieved via different strategies, e.g., chemical vapor deposition (CVD) of coating materials from the gas phase or spray coating of aerogel particles with polymeric solutions/dispersions [7]. The drawbacks of these methods include the limitations of specific material systems, the need for high temperatures in particular cases, the high consumption of educts (for CVD) and the possibility of pore collapse due to solvent entering the non-protected aerogels pores during the process (in spray coating process) [7,9–11].

A promising solvent-free alternative approach is the deposition of hydrophobic polymer layers via cold plasma coating processes, wherein dry aerogels are exposed to cold plasma generated by glow discharging of a gaseous hydrophobizing agents in moderate vacuum [7]. Plasma polymerization is a complex process, involving the simultaneous generation and reaction of short-lived species. Important process parameters that determine the fragmentation of monomers and crosslinking degree of the formed polymer films as well as the film formation kinetics are deposition time (t), monomer flow rate, system pressure and power input (p_{input}) [9,12]. The power input can be controlled principally via two ways: by adjusting the glow discharge power in a continuous wave mode (CW) or by modulating the power in a pulsed wave mode (PW), whereas the chemical structure and composition of PW plasma polymer coatings depends on the duty cycle (DC) [9,13], which reflects the ratio of pulse-on and pulse-off times:

$$DC\ (\%) = \frac{t_{on}}{t_{on} + t_{off}} \cdot 100 \quad (1)$$

where t_{on} = pulse on time, t_{off} = pulse off time.

The thickness of polymer layers in cold plasma polymerization depends heavily on the mass deposition rate of solid from the gas/plasma phase, which is therefore a crucial parameter for the formation of stable layers and depends on the individual monomer characteristics and operational parameters as well as on plasma chamber design [12]. In addition, a freshly deposited material is subject to fragmentation and ablation by the plasma, from which the deposition occurred [12].

Cold plasma coating processes provide intrinsic advantages, such as short processing times and efficiency in terms of educt consumption; no need for wet chemicals, solvents and purification steps (and therefore environmental pollution-free); controlled attachment of nano-meter scaled layers; high variety of usable coating materials with different functionalities; and the possibility of applying controlled and tunable wettability to different surfaces [9,14–17]. Deposition of thin polymer coatings by cold plasma technology is very flexible, allowing functionalization of practically any kind of substrate, including polymers, glass, ceramics and metals, whereas the adhesion of a plasma-polymerized film to a substrate is the most important factor for the success of the surface modification process [18]. In general, better adhesion is achieved by slower deposition rates, small film thicknesses and with polar substrates [18]. Although the above given aspects suggest that cold plasma coating is a promising tool for hydrophobic post modification of polar biopolymer aerogels, only a few examples of the combination of cold plasma hydrophobization with aerogels substrates are documented in the literature. Hydrophobization of the external surface of hydrophilic cellulose and alginate aerogel substrates obtained by freeze drying was achieved by the use of trimethylchlorosilane and tetrachloromethane monomers in short process times of approximately 3–20 min (CW mode, input power ~150–250 W) in several works, leading to protection of the surfaces against water and resulting in static water contact angles (θ) in the range of approximately 102–150° [19–22]. While the mentioned studies demonstrate the principal possibility of achieving hydrophobic to superhydrophobic surfaces on porous biopolymer substrates, it remains unclear if the modification is limited to the outer surface or if the inner porous structure is also affected. Additionally, freeze-drying was used as the drying technique in all cases, which is known to lead to significantly lower specific surface areas compared with supercritical drying [1,3,23,24]. Consequently, specific surface areas in the cited examples can be expected to be low. Furthermore, pore size is expected to be rather larger (a few micrometers) than in typical aerogels from supercritical drying (a few tens of nanometers) [23].

A promising non-toxic alternative to the use of chloromethanes and chlorosilanes is the application of fluorocarbon-based polymer coatings since they exhibit extremely low surface energies. Fluoropolymer layers satisfy the widely accepted regulatory assessment criteria to be considered as "polymers of low concern" [25]. The use of octafluorocyclobutane (C_4F_8) as a coating agent has been documented to result in smooth, relatively robust and hydrophobic surfaces due to the formation of highly cross-linked CF_2 and CF_3 groups, whereas the layer thickness, deposition rates and crosslinking degree can be controlled by the power input energy and input mode [26]. Superhydrophobic surfaces can be obtained by combination of C_4F_8-plasma coating and nano-texturing of the substrates [16]. Simultaneous introduction of (super)-hydrophobicity and oleophobicity is possible by using perfluoro-acrylates such as 1H,1H,2H,2H-perfluorooctyl acrylate (PFAC-6) and 1H,1H,2H,2H-Heptadecafluorodecylacrylate (PFAC-8) as precursors [9,13].

The aim of this work was to evaluate the potential of cold plasma coating processes as a simple and efficient one step-technique for post modification of biopolymer aerogels with high surface areas produced via supercritical drying. The main question to be answered was whether it is possible to achieve liquid repellency of hydrophilic and open porous aerogel surfaces using fluorocarbon-based coating materials against both polar and non-polar liquids. In addition, influences of the coating process on the inner pore structure were evaluated.

Our strategy towards plasma-modified biopolymer aerogels involved the following steps: (1) Synthesis of biopolymer aerogel monoliths with different porosities and surface energies via gelation, solvent exchange and supercritical drying. Alginate and cellulose-based aerogels were used as model systems for highly hydrophilic gels with low intrinsic biopolymer content $\leq$6 wt% and high porosity. PPI- and WPI-based aerogels were used as substrates with higher intrinsic biopolymer content $\geq$15 wt% and therefore lower porosities as well as lower surface energies and additional functionalities compared with the polysaccharides. (2) Cold plasma coating of the aerogel substrates using different fluorinated precursors (C_4F_8, PFAC-6, PFAC-8) and variation of the process parameters time, input power and input mode (CW and PW). (3) Characterization of the aerogels surface and textural properties prior to and after the coating process.

2. Materials and Experimentation

2.1. Preparation of Aerogel Substrates

Aerogels based on alginate (Hydagen 500, BASF, Ludwigshafen, Germany), microcrystalline cellulose type II (JRS Pharma GmbH & Co. KG, Rosenberg, Germany, Vivapure®, 101), whey protein isolate (WPI, Agropur, BiPro 9500, Longueuil, Quebec, Canada) and potato protein isolate (PPI, Avebe, Veendam, The Netherlands, Solanic 200) were prepared via dissolution in water, gelation, solvent exchange with ethanol and supercritical drying with CO_2, as detailed below. Different biopolymer contents were set for the different biopolymers (Table 1). An overall quantity of 500 g solution was produced from each biopolymer solution.

Table 1. Biopolymer content and gelation conditions for production of hydrogels from different biopolymers.

Material [–]	Biopolymer Content (wt%)	Gelation Method [-]	Post Treatment After Hydrogel Formation [-]
alginate	2.5	ionotropic in presence of Ca^{2+}	completion of gelation with aqueous $CaCl_2$ solution
cellulose	6.0	Thermal 60 °C, 30 min	washing with demineralized water
PPI	15	Thermal 85 °C, 30 min	none
WPI	20	Thermal 90 °C, 30 min	none

Dissolution of alginate, WPI and PPI was carried out by adding the educts to demineralized water followed by stirring at room temperature (500 rpm, overhead stirrer, 30 min). Dissolution of cellulose was carried out in aqueous NaOH solution at low temperature, following the procedure described in previous works [27]. The gelation of cellulose, WPI and PPI was induced thermally at different temperatures (Table 1) and gelation of alginate was carried out Ca^{2+}-induced via internal setting method. Prior to gelation, biopolymer solutions were divided and transferred (5 mL each) into round cups (SecurTainer™ III, Thermo Fisher Scientific, Waltham, MA, USA) with a diameter of 28 mm, where the gelation took place. For thermal gelation, cups filled with solution were sealed and placed in a preheated oven for 30 min. Cellulose solutions showed color change from white to yellow during the gelation process. Obtained cellulose hydrogels were therefore washed with demineralized water until complete white gels were obtained. Ionotropic gelation of alginate via internal setting method was carried out as follows: The alginate solution was mixed with 0.25 wt% $CaCO_3$ (g $CaCO_3$/g alginate solution) via overhead stirrer (500 rpm, 30 min). In the next step, 0.4 wt% glucono-δ-lacton (GdL, Merck KGaA, g GdL/g alginate solution) was added to the alginate—$CaCO_3$ solution, mixed for 2 min and quickly transferred into cups, before gelation occurred, in several minutes at room temperature. The so-formed weak gels were transferred to an aqueous $CaCl_2$ solution (10 g $CaCl_2 \cdot 2\,H_2O$, Carl Roth GmbH & Co. KG, in 1 L demineralized water) in order to complete the gelation. For solvent exchange, alginate hydrogels were immersed stepwise in mixtures of ethanol (EtOH, Carl Roth GmbH & Co. KG, Karlsruhe, Germany) and water of different concentrations (30, 50, 70 wt% EtOH/water and 99.9 wt% anhydrous EtOH) until a minimum final EtOH concentration of 97.0 wt% inside the substrates was achieved (determined by density measurements, Anton Paar, DMA 4500 M).

Solvent exchange of WPI, PPI and cellulose hydrogels was carried out via direct solvent exchange by immersing the substrates in 99.9 wt% anhydrous EtOH until a minimum final concentration of 97.0 wt% ethanol inside the hydrogel slabs was achieved. After the solvent exchange, alcogels were placed in a high pressure autoclave with an overall volume of 3.9 L for supercritical drying with CO_2. The supercritical drying was carried out at a temperature of 60 °C, pressure of 120 bar, under continuous flow of CO_2 (flow rate = 120–160 g/min) through the autoclave until complete extraction of EtOH was achieved after 6 h. The dry aerogels were collected after slow depressurization (1 bar/min) of the autoclave and stored in sealed vessels in a desiccator over calcined silica gel prior to plasma coating.

2.2. Coating of Aerogel Particles via Cold Plasma Process

Cold plasma coating of aerogel substrates was carried out using a Tetra 100 plasma equipment provided by Diener Electronic GmbH & Co.KG, Ebhausen, Germany, with a maximum specific power of 300 W. The substrates were placed on a perforated metal plate that was subsequently fixed in the process chamber (dimensions: length and width = 400 mm, height = 625 mm) at a height of 100 mm. The temperature in the process chamber was adjusted to 40 °C, and a vacuum was set. Subsequently, the system pressure was adjusted to 0.15 mbar via introduction of argon into the chamber. Three different monomers were used: 1H,1H,2H,2H-Perfluordecyl-acrylat (PFAC-8), 1H,1H,2H,2H-perfluorooctylacrylate (PFAC-6) and Octafluorocyclobutane (C_4F_8). The liquid monomers PFAC-6 and PFAC-8 were injected into the chamber by a dosing pump with a constant flow rate of 30 µL/min. For the gaseous C_4F_8, a mass flow of 6 sccm was set. Each process was initiated by starting the glow discharge at the accorded input power and mode (CW or PW). Three different process settings were applied for polymerization of all monomers (Table 2). In PW mode, t_{on} and t_{off} times were set to 15 µs each, resulting in a duty cycle of 50%.

Table 2. Process modes used for plasma polymerization of different monomers.

Mode [-]	p_{input} (W)	Duty Cycle (%)	Monomer [-]	Flow Rate [-]
PW	300	50	C_4F_8	6 sccm
CW	90	100	PFAC-6	30 µL/min
CW	30	100	PFAC-8	30 µL/min
PW	300	50	C_4F_8	6 sccm
CW	90	100	PFAC-6	30 µL/min
CW	30	100	PFAC-8	30 µL/min
PW	300	50	C_4F_8	6 sccm
CW	90	100	PFAC-6	30 µL/min
CW	30	100	PFAC-8	30 µL/min

Different deposition times were set for all experiments in the range of 5–50 min by removing samples at the accorded process times. For this purpose, the process was briefly stopped and restarted after sampling.

2.3. *Specific Surface Area, Mesopore Volume, Mesopore Diameter*

Characterization of aerogels microstructural properties was carried out by low-temperature N_2 adsorption-desorption analysis (Nova 3000e Surface Area Analyzer, Quantachrome Instruments, Boynton Beach, FL, USA). Parts of the monoliths were cut or broken into pieces, which included parts of the outer layer as well as the inner pore structure. An overall sample mass of 20–30 mg was used and all samples were degassed under vacuum at 60 °C for at least 6 h prior to each analysis. The Brunauer-Emmett-Teller (BET) method was used to determine the specific surface area S_V and BET constant C as a single determination (plasma-treated samples) and double determination (non-treated samples). The relative standard of 5.8% (S_V) and 15.1% (C) for the individual BET results were estimated from the average error of a 4-fold determination. The pore volume of the mesopores V_{meso} and mean pore diameter of mesopores d_{meso} were determined by the Barrett-Joyner-Halendia (BJH) method as single determinations under estimation of a relative measurment error of 5%.

2.4. *Density and Porosity of Aerogels*

The envelop density ρ_e was determined by weighting aerogels monoliths on a fine balance (3-fold determination) according to:

$$\rho_e = \frac{m}{V} \tag{2}$$

with m being the sample weight and V the sample volume. Skeletal density ρ_s was determined via helium pycnometry (Multivolume Micromeritics 1305, 4-fold determination) at room temperature. The overall porosity ε was calculated from envelope and skeletal densities as follows:

$$\varepsilon = \left(1 - \frac{\rho_e}{\rho_s}\right) \cdot 100 \tag{3}$$

2.5. Contact Angle Measurements

The static contact angle (STA) was determined using a drop shape analyzer (OCA 15EC) with the software SCA20 for OCA. All contact angles were taken with 1 μL (water) and 2 μL (*n*-hexadecane, Merck KGaA) of solution immediately after deposition of the droplet on the surface (detachment from syringe tip) and as average of contact angles at both sides. The standard error of 2.3% for the individual results was estimated from the average error of four double determinations. For the identification of sinking-in and deformation events, the drop was left on the surface for 60 s, and the respective effects were determined after this time.

2.6. Scanning Electron Microscopy and Energy Dispersive X-ray Spectroscopy

The surface properties and inner structure of the aerogels were characterized via scanning electron microscopy (SEM) and energy dispersive X-ray spectroscopy (EDS) (Zeiss Supra VP55, Jena, Germany). Samples were sputtered with a thin layer of gold (ca. 6 nm, Sputter Coater SCD 050, BAL-TEC) before analysis was started. The measurements were carried out under high vacuum at an accelerating voltage of 4–5 kV.

3. Results and Discussion

3.1. Properties of Non-Treated Aerogels

Aerogels were produced from four different biopolymers, whereas different porosities ε and envelope densities were obtained by variation of the solid content in the stock solutions (c_{biop}). Specific surface areas were in the range of approximately 300 m^2/g for polysaccharide gels and 200 m^2/g for the protein-based materials (Table 3).

Table 3. Properties of non-treated aerogels produced from different biopolymers.

Material [-]	c_{biop} (wt%)	S_V (m^2/g)	d_{meso} (nm)	V_{meso} (cc/g)	V_{macro} (cc/g)	ρ_s (g/cm^3)	ρ_e (g/cm^3)	ε (%)	C [-]
Alginate	2.5	301 ± 18	13.7 ± 0.7	1.82 ± 0.09	1.898 ± 0.020	0.0961 ± 1·10^{-4}	1.90 ± 2·10^{-2}	95.0 ± 1.0	148 ± 22
Cellulose	6.0	289 ± 17	17.4 ± 0.9	1.69 ± 0.09	1.613 ± 0.049	0.1754 ± 6·10^{-4}	1.61 ± 5·10^{-2}	89.0 ± 3.0	97 ± 15
PPI	15.0	200 ± 12	13.9 ± 0.7	1.23 ± 0.06	1.345 ± 0.012	0.2865 ± 3·10^{-4}	1.34 ± 1·10^{-2}	78.7 ± 0.8	46 ± 7
WPI	20.0	201 ± 12	7.1 ± 0.4	0.72 ± 0.04	1.357 ± 0.009	0.4956 ± 2·10^{-4}	1.36 ± 1·10^{-2}	63.5 ± 0.5	46 ± 7

Increasing the solids content from 2.5 to 20 wt% resulted as expected in a linear increase (R^2 0.939) in the envelope density of the resulting aerogels, from 0.10 to 0.50 g/cm^3, respectively. In contrast, the skeletal density decreased linearly with increasing solids content in the range of 2.5–15 wt% (R^2 0.977). The total porosity ε increased linearly (R^2 0.907) with decreasing c_{biop} (Figure S1). The contribution of micro-, meso- and macropores to the overall pore volume was determined as follows. The t-plot analysis of nitrogen desorption data showed that none of the aerogels obtained a significant amount of micropores. The macropore volume V_{macro} could therefore be calculated from ρ_e and V_{meso}:

$$V_{macro} = \frac{1}{\rho_e} - V_{meso} \quad (4)$$

The division of the total porosity into total macro- and mesopore volumes showed that both V_{meso} and V_{macro} were higher at lower c_{biop}, while the ratio between the pore volumes V_{macro}:V_{meso} changed as an exponential function of c_{biop}: aerogels with lower c_{biop} contained higher macropore volume up to 83% at c_{biop} = 2.5 wt% (Figure 1f). These findings were also supported by SEM analysis, which provided qualitative insights into the macroporous aerogel structures: a continuous fibrous macroporous and interconnected pore network could be seen in alginate aerogels, whereas average macropores size range (237 ± 118 nm) could roughly be estimated based on image analysis (Figure 1a and

Figure S2). In contrast, cellulose gels obtained a denser pore structure, which was interrupted by gaps in the size range of some micrometers (Figure 1b). Protein aerogels showed a generally denser network due to the higher biopolymer content: PPI gels consisted of a fibrous network with visible macropores in the size range of approximately 100 nm (Figure 1c), WPI gels formed a globular structures (Figure 1d). In the mesoporous range, unimodal pore size distributions were obtained, with significantly narrowest distribution and smallest diameter for the WPI aerogel (Figure 1e).

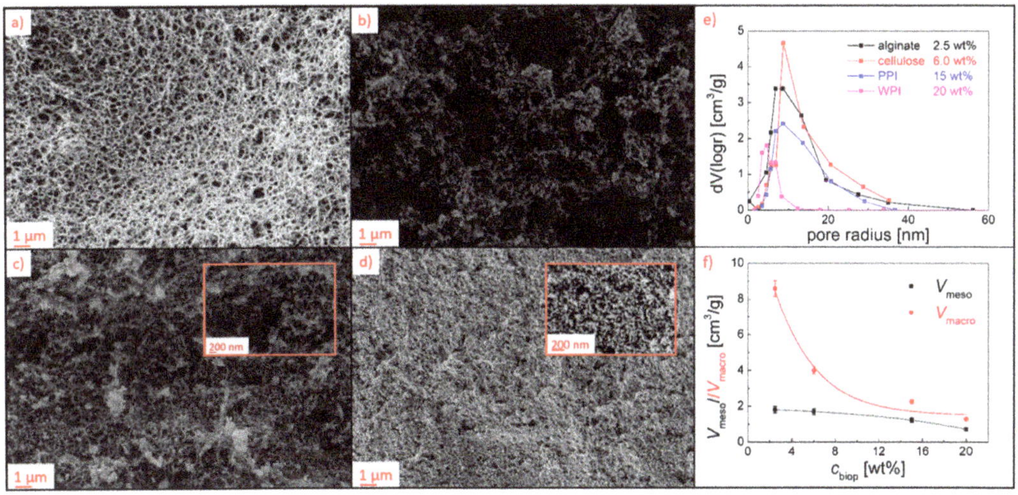

Figure 1. (**a–d**) SEM pictures of non-treated aerogels surfaces: (**a**) alginate, (**b**) cellulose, (**c**) PPI, (**d**) WPI, (**e**) pore size distributions, (**f**) macro- and mesopore volumes.

Gels derived from different biopolymers exhibited different surface polarities: This was reflected in water contact angle measurements of the pure aerogels, which led to immediate collapse of the gel when being exposed to a liquid water droplet in the case of the highly polar polysaccharide gels. Gelation of both WPI and PPI was influenced by hydrophobic interactions between hydrophobic amino acids side chains [5,28], which contributed to an overall lower surface polarity. Therefore, static water contact angles could be determined for non-treated protein gels ($83 \pm 2°$), which showed immediate discoloration after application of the droplet, while no collapse and significant deformation occurred. Insights about the surface polarity in the pores can be provided via the BET *C*-constant, which was exponentially related to the energy of monolayer adsorption. High values of *C* (~150) are generally associated with adsorption on high-energy surface sites (in the absence of micropores), whereas the overall value of the *c*-constant decreases in presence of increasing non-polar moieties, as shown for silica-casein hybrid aerogels by Lazar et al. [29,30] In this work, the order of *C*-constant values (alginate = 148 > cellulose = 97 > WPI and PPI = 46) corresponded well with the expected surface energy of the aerogels.

3.2. Liquid Repellency of Coated Aerogels

A systematic screening was conducted to investigate the effects of cold plasma process time t (t = 5–50 min), input mode and input power in three different process modes (PW, p_{input} = 300 W; CW, p_{input} = 90 W; CW, p_{input} = 30 W) on the surface wettability of all different combinations of aerogel substrates and coating materials (C_4F_8, PFAC-6, PFAC-8). Surface water wettability was tested via determination of the static contact angle θ of water for all samples. Four different cases could be distinguished after deposition of water droplets on the aerogel substrates: (1) Immediate sinking in of the droplet and collapse of the gel at the site of application in the case of polysaccharide gels (unsuccessful process setting). (2) Immediate discoloration of the gels surface in the case of protein gels (unsuccessful process setting). (3) Deformation of the substrate at the site of application during the measurement, resulting in formation of a dimple and quick change of the obtained contact angle over time, whereas no absorption of the droplet or destruction of the aerogel took place. (4) No deformation at the droplet application site and stable contact angle over a timeframe of ≥1 min. Two process times were defined accordingly: $t(1)$ as the minimum process time, after which a contact angle could be measured without substrate wetting or destruction, but under substrate deformation (case 3), and $t(2)$ as the minimum process time, which was needed to obtain contact angles without further deformation of the substrate over time (case 4). Water surface wettability was achieved for all combinations of substrates and coatings materials and resulted in values of θ in a broad range from 78–154°, whereas deformation of the substrate occurred only in the case of the polysaccharide aerogels (Figure 2, Table 4).

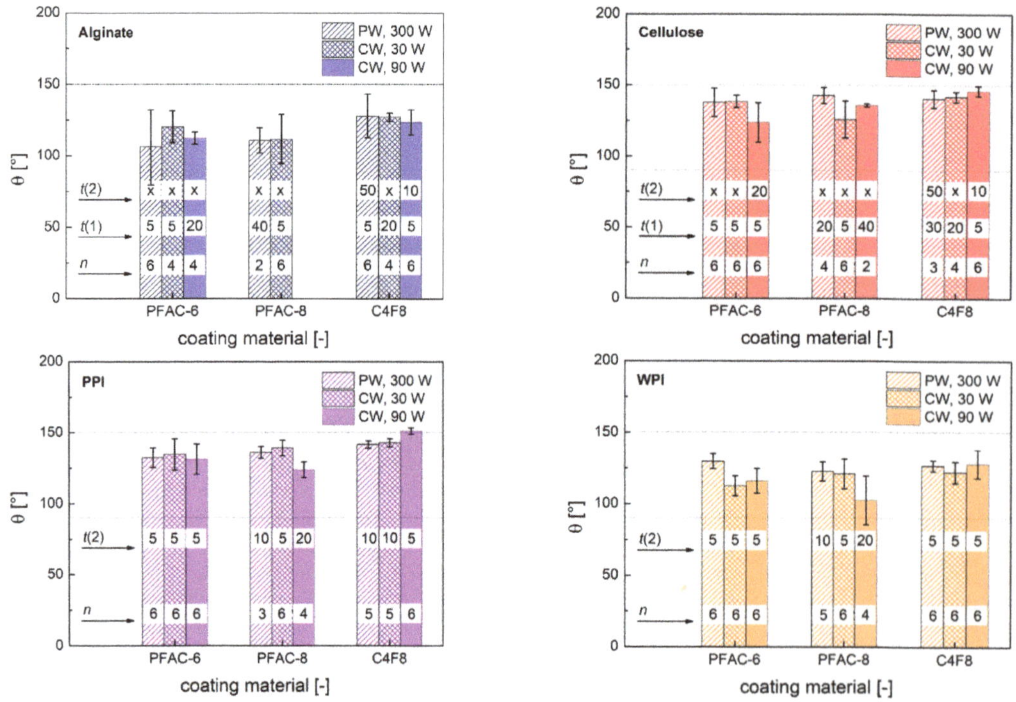

Figure 2. Static water contact angles θ depending on coating material and aerogel type. Dashed lines mark the regions of hydrophobicity (θ = 90–150°) to superhydrophobicity (θ ≥ 150°). All values represent the average of successful settings n (cases 3 and 4) at different process times; error bars represent the standard deviation of the averaged values. PW (pulsed wave mode) and CW (continuous wave mode) stand for the power input mode; accordingly, numbers represent the input energy. Numbers in columns stand for the minimum process times (min) needed to achieve the corresponding case; x denotes that the corresponding case did not occur under the given process conditions.

The latter can be explained by the different mechanical stabilities of the gels: Protein gels with comparably high solids content showed a rigid network and no plasticity. In contrast, alginate gels were most susceptible against deformation or pore collapse since they exhibited the highest intrinsic surface energy and lowest solids content. We conclude that a sufficient thickness of the polymer layer is needed in case of polysaccharide aerogels to reach the time $t(2)$. Significant surface hydrophobization of all combinations of substrate coating material was achieved after a short processing time $t(1)$ = 5 min. Coating with

C_4F_8 in CW mode and p_{input} = 90 W resulted in the highest contact angles in the case of cellulose, PPI and WPI (Table 4). In addition, process times of $t(2)$ = 10 min were sufficient in preventing alginate and cellulose aerogels deformation in the same mode.

Table 4. Range of static water contact angles on different aerogel substrates.

Aerogel [-]	Range of θ (°)	Raw Data θ Maximum [-]	Conditions Maximum θ [-]	Conditions for Non-Deformed Substrates [-]
alginate	78–139		t = 10 min, C_4F_8 PW, 300 W	C_4F_8, CW, 90 W, 10 min C_4F_8, PW, 300 W, 50 min
cellulose	102–151		t = 40 min, C_4F_8 CW, 300 W	C_4F_8, CW, 90 W, 10 min C_4F_8, PW, 300 W, 50 min PFAC-6, CW, 90 W, 20 min
PPI	116–154		t = 5 min, C_4F_8 CW, 90 W	all
WPI	103–142		t = 10 min, C_4F_8 CW, 90 W	all

Stable layer formation was also achieved in PW/p_{input} = 300 W mode after a significantly higher process time of $t(2)$ = 50 min. CW plasma polymerization generally led to higher monomer fragmentation and crosslinking degree of the resulting polymer compared with PW mode and was reported to favor the formation of low energy CF_3 groups in the case of C_4F_8 polymerization, while PW plasma film thickness rate was significantly lower [9,13]. Our results show accordingly faster formation of stable layers on polysaccharide aerogels in CW/p_{input} = 90 W mode of C_4F_8 polymerization. In contrast, deformation of polysaccharide aerogels was noticed at p_{input} = 30 W at all process times. The observed trend may be attributed to changes in monomer deposition rate since deposition rate was found to increase with rising average power under CW conditions [31]. In contrast, higher input power of 90 W did not lead to the formation of stable layers in the case of CW PFAC-8 polymerization and led to delayed $t(2)$. A reversed trend depending on deposition rate on input power was also reported by Coulson et al. for PW cold plasma polymerization of 1H,1H,2H,2H-heptadecafluorodecyl acrylate and was consistent with fragmentation and damage of the long perfluoro-alkyl chains [32]. Lower influence of process parameters was observed for PFAC-6 monomer with shorter perfluoro-alkyl chains, whereas CW/p_{input} = 90 W mode provided similar results compared with the other modes and resulted in stable layer formation on cellulose aerogels with $t(2)$ = 20 min. Summarized, preferable process conditions to achieve liquid water repellency were determined for each coating material. The conditions were found to depend on the process parameters as well as on the nature of monomers and substrates.

Oleophobic repellency was tested by determination of the static contact angle θ_{ol} using n-hexadecane as probe liquid. Protein aerogels are well known for their good oil uptake capability [33]. Immediate absorption of n-hexadecane droplets was therefore observed in case of non-treated gels and also of poly-C_4F_8 coated gels, since longer perfluorinated chain lengths are necessary to obtain oleophobic behavior [32]. PFAC-6 (PW, p_{input} = 300 W) and PFAC-8 (CW, p_{input} = 30 W) coatings resulted in repellency of n-hexadecane with a range of θ_{ol} = 71–134° on all tested substrates after short process times of 5 min (Table 4). In accordance with the results for water repellency, no general influence of the process time was determined, while the value of θ_{ol} depended mainly on the substrate/coating combination (Figure 3). Significantly higher values of θ_{ol} were obtained by application of poly-PFAC-6 on protein aerogels compared with poly-PFAC-8 on protein aerogels, while no differences were found in case of polysaccharide aerogels (Figure 3).

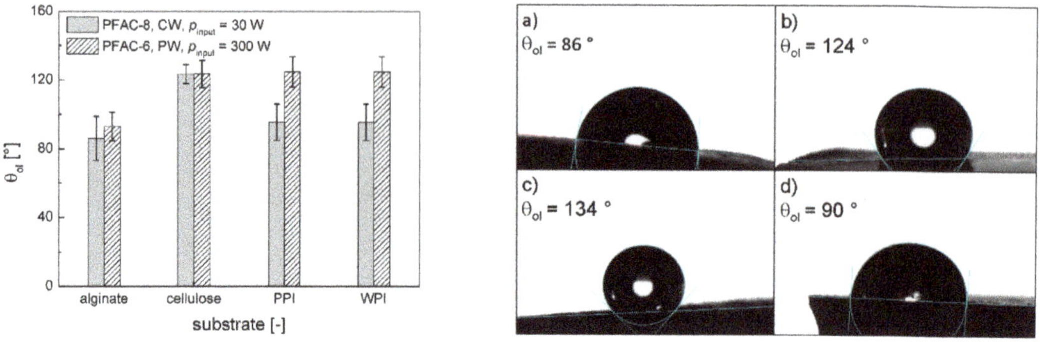

Figure 3. θ_{ol} for different substrate/coating combinations. Values in the overview (**left**) represent the average at different process times under a given process condition (as indicated in the legend); error bars represent the standard deviation (n = 6). Raw data (**right**) represent exemplary process settings: (**a**) alginate, t = 5 min, PFAC-8, CW, p_{input} = 30 W, (**b**) cellulose, t = 10 min, PFAC-6, PW, p_{input} = 300 W, (**c**) PPI, t = 5 min, PFAC-6, PW, p_{input} = 300 W, (**d**) PPI, t = 5 min, PFAC-8, CW, p_{input} = 30 W.

It is notable that no deformation of polysaccharide substrates took place after application of n-hexadecane droplets. Therefore, substrate deformation did not result from the droplets weight: influences such as the electrostatic attraction between droplet and substrate (polar–polar interactions) may play an important role in the deformation process and require a separate investigation.

3.3. Coating Properties

The surface microstructure and homogeneity of the coating layers were qualitatively analyzed via SEM. Film quality and structure depended on the individual combinations of process parameters, substrate/coating materials combination as well as the process time as exemplified in Figure 4. Additional examples are provided in the Supplementary Materials (Figures S3–S17).

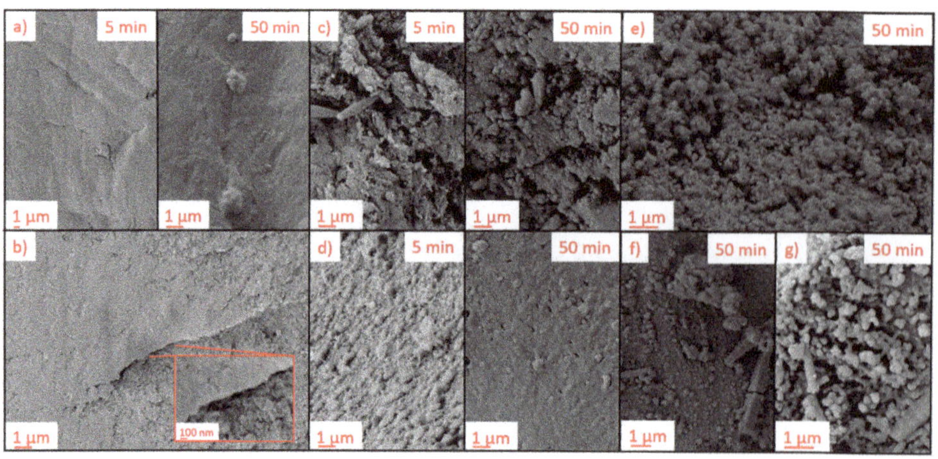

Figure 4. SEM pictures of external coating on different aerogels: (**a**) alginate, C_4F_8, CW, p_{input} = 90 W, (**b**) WPI, C_4F_8, CW, p_{input} = 90 W, (**c**) cellulose, C_4F_8, CW, p_{input} = 90 W, (**d**) PPI, C_4F_8, CW, p_{input} = 90 W, (**e**) alginate PFAC8, CW, p_{input} = 30 W, (**f**) PPI, PFAC-6, CW, p_{input} = 90 W, (**g**) PPI, PFAC-8, CW, p_{input} = 30 W.

Application of C_4F_8 in CW/p_{input} = 90 W mode on alginate aerogels resulted in the fast formation of homogeneous surface coating layers after short process times of 5 min, while cellulose and PPI aerogels were not completely covered (cf. Figure 4a,c,d). Increasing the process time to 50 min resulted in generally smooth coatings with visible aggregates on the top for alginate and PPI aerogels, while inhomogeneities were observed on the coated cellulose aerogel (Figure 4a,c,d). WPI aerogels exhibited non-covered regions with open pores, even after the maximum process time of 50 min (Figure 4d), which speaks to a generally insufficient adhesion between coating materials and WPI. The results indicate that initially homogeneous aerogel surfaces without micro-gaps in the pore structure and high polarity of the substrates were beneficial for the fast formation of uniform coating layers. Furthermore, we can conclude that complete film formation is not necessarily required to obtain water repellent surfaces, since contact angles in the range of approximately 120–140° were also determined (under substrate deformation) in cases of non-complete coverage of the pores (see examples Figures S10 and S13). In contrast with C_4F_8-based coatings, poly-PFAC-6 and PFAC-8 showed rougher surfaces with crystallization of the material in some cases (Figure 4e–g), whereas a generally higher degree of crystallization was found for poly-PFAC-8. This observation may be explained by the different side chain lengths of the PFAC-monomers, which resulted in different reorganization and crystallization properties, leading to enhanced crystallization and formation of ordered structures for side chain lengths $\geq$8 [34].

3.4. Textural Properties

Quantification of the influence of the cold plasma coating process on the microstructure was possible by comparing the specific surface area S_V of aerogels before and after post-modification: The decrease in the specific surface area could principally result from pore collapse or from the contribution of additional mass from non-porous coating material. Both effects were expected to be non-significant in the case of cold plasma polymerization due to the small amount of added material and absence of liquids and moisture during the process.

A negative effect of the coating process on the specific surface area was indeed not detected, showing that the porous structure remained intact during the process under all process conditions. In contrast, a significant increase in ΔS_V in specific surface areas of treated substrates up to ΔS_V = +179 m^2/g (corresponding to an increase up to +61%) was determined (Figure 5). The average change of S_V depended strongly on the aerogel nature (+ ΔS_v alginate > cellulose > PPI > WPI), whereas the individual combinations with coating materials, the process settings and the process time also played important roles. A generally equal influence of the process time was found for all individual settings: The maximum ΔS_V was reached after a process time of 10 min, followed by steady state conditions in the range of t = 10–40 min and a slight decrease when running the process for 50 min. Furthermore, the highest changes occurred using C_4F_8 monomer in PW/p_{input} = 300 W mode in all cases, which is therefore presented separately (circles in Figure 5).

The detected increase in the BET surface area was a clear indication of polymerization taking place inside of the pores, resulting in the formation of new, hydrophobic zones that could be detected with nitrogen adsorption experiments. This was also reflected in a lowering of the BET C constant compared with the pristine polysaccharide aerogels, which is indicative of the deposition of non-polar coating material in the pores (Figure 5). The maximum change in specific surface area ΔS_V was quantified for each individual setting by averaging the values of S_V in the steady state range from t = 10–40 min. The ΔS_V was related to the porous structures of the different substrates and increased exponentially (R^2 0.902–0.999) with the overall porosity and linearly (R^2 0.984–0.999) with the overall pore volume of the aerogels (Figures S18 and S19). An exponential increase in ΔS_V by increasing the mesopore volume and a linear increase by increasing the macropore volume show the significance as well as individual contributions of differently sized pores in the process (Figure 6). We surmise that an interconnected macroporous network, such as that observed for alginate aerogels (see Figure 1a), should contribute to a relatively unhindered transport of activated monomer into inner aerogels parts but should also provide comparably low surface area for the coating processes (transport-pores). In contrast, mesopores provided high specific surface area on which reactions with activated monomer could occur, but could only transport the activated monomer into the aerogel to a limited extent (reaction-pores).

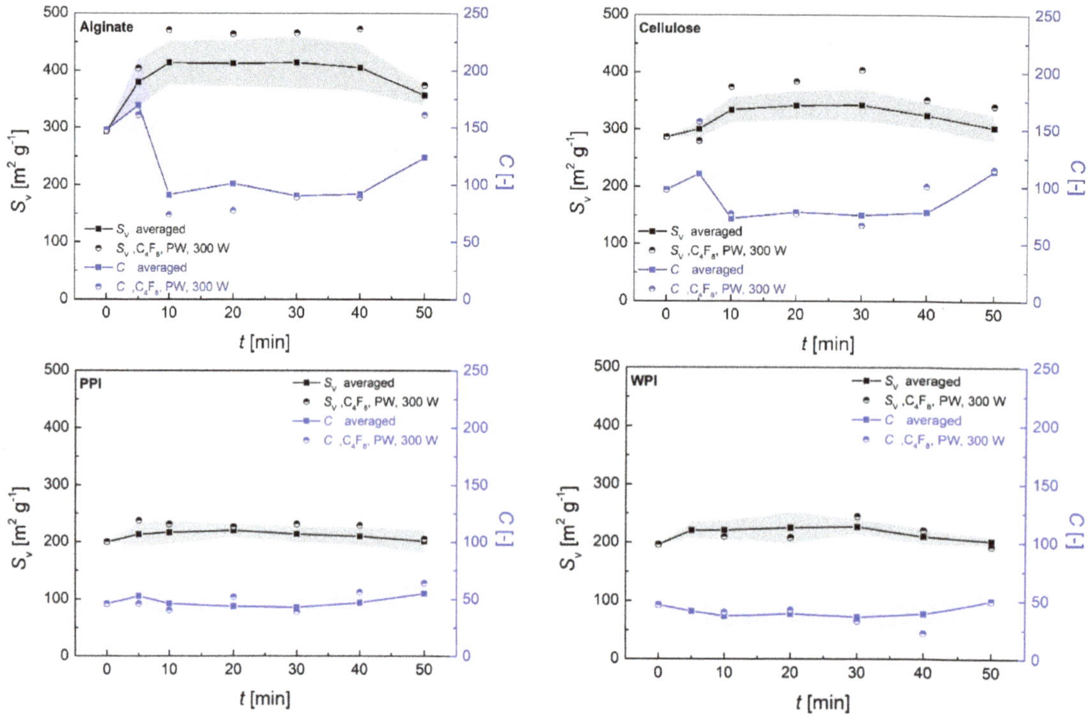

Figure 5. Changes of specific surface area and C-constant depending on process time and process conditions. Circles represent the values obtained with C_4F_8 monomer in PW mode with p_{input} = 90 W. Squares represent the averaged values from all coating materials and process conditions, excluding PFAC-8 coating in CW mode with p_{input} = 300 W. Error areas represent the standard deviation from averaged values; solid lines are drawn to guide the eyes.

Transport of coating material into the pores was verified via EDS spectroscopy of cut-open substrates, an example of which was carried out for C_4F_8-coating. The thin coating layer and the inner porous part were clearly distinguishable on the corresponding SEM images and the distance *d* (max. 112 μm) from the outer layer was determined accordingly (Figure 7 left). The amount of fluorine was calculated as relative abundance r_F in relation to oxygen and carbon. Significant amounts of fluorine were detected in the inner part of the aerogels, which is a proof of the penetration of monomer molecules into the substrates. The overall fluorine content in the pores was related to the distance *d* from the outer layer, while the initial value of r_F on the surface varied significantly with the aerogel nature (Figure 7, right). The latter can be expected to depend mainly on the adhesion between monomer and substrate, resulting in significantly higher values of r_F for the more polar polysaccharide aerogels. The exponential decay of r_F spoke to a significant mass transport limitation of monomer diffusion into the bulk of the aerogels. The value of r_F remained mostly constant in the inner part of the substrates.

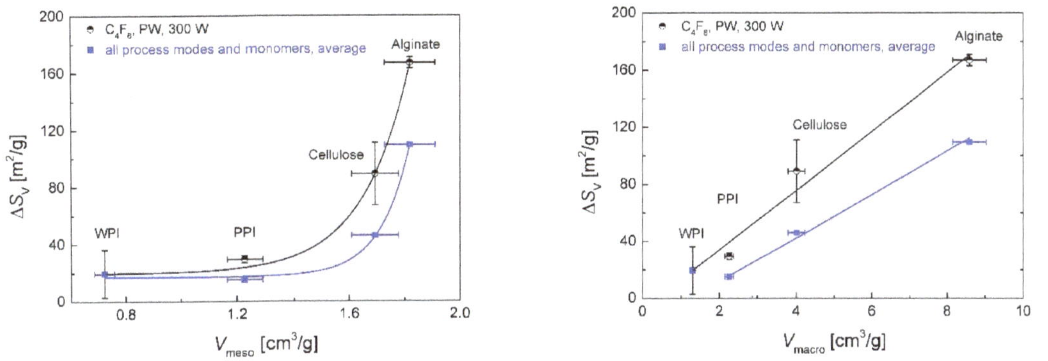

Figure 6. Increase in ΔS_V depending on the mesopore volume (**left**) and the macropore volume (**right**). Circles represent the values obtained with C_4F_8 monomer in PW mode with p_{input} = 300 W. Squares represent the averaged values from all coating materials and process conditions, excluding PFAC-8 coating in CW mode with p_{input} = 90 W. Straight lines represent exponential (**left**) and linear (**right**) fitting. Error areas represent the standard deviation of ΔS_V from averaged values at different process conditions; error bars represent the relative measurement errors (*x* error) and standard deviation of ΔS_V for the individual process condition (*y* error).

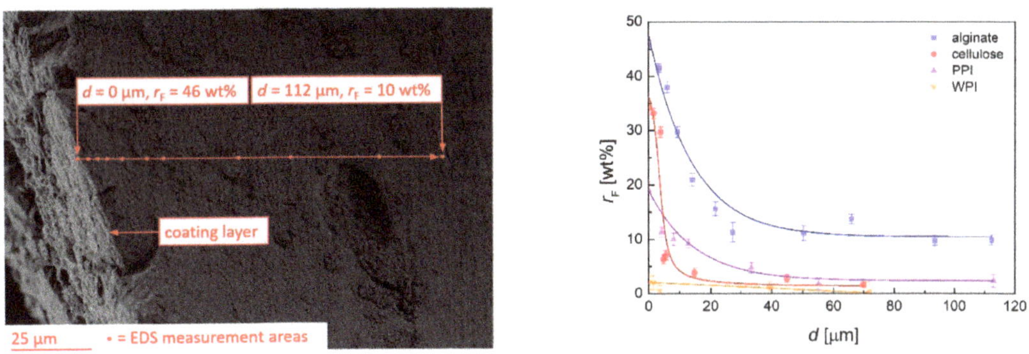

Figure 7. SEM picture of cut-open alginate aerogel substrate with external coating and inner porous part (**left**). Points mark the areas according to EDS measurements. Fluorine content depending on the distance from outer layer (**right**). Error bars represent the error of the individual measurements; solid lines illustrate the course of r_F for different aerogels. All measurements were performed for aerogel substrates treated under the following process condition: C_4F_8, CW, p_{input} = 90 W, t = 50 min.

The alginate aerogel shows a higher fluorine content in the inner part ($r_F = 11.2 \pm 1.6$ wt% at $d \geq 50$ µm) and a lower slope of r_F-decay as compared to the cellulose aerogel. Our results suggest that a high overall porosity and macroporous content are in case of good monomer adhesion decisive for the monomer transport into the material. In case of a lower adhesion of the monomer (the case of protein aerogels), transport in the pores is relatively unhindered on the one hand, but the overall fluorine content on the aerogels surface and in the pores is significantly lower when compared to the alginate aerogel on the other hand. Summarized we conclude that an interconnected macroporous network combined with high surface adhesivity are both necessary to achieve a good monomer transport into the materials on the one hand, while enabling sufficient monomer deposition onto the pore walls on the other hand.

Changes of the porous structure caused by the plasma deposition of the coating materials were determined via BJH desorption analysis. Pore size distributions of alginate aerogels showed that smaller pores with a range of pore radius $r_{pore} \approx$ 3–7 nm disappeared after a short process time of 5 min, while new mesopores with $r_{pore} \approx$ 15–25 nm were generated (Figure 8 top, Figure S20). The generation of new pores in the same size range was observed for cellulose aerogels, whereas the distributions generally broadened, and additional pore volume was also obtained in the range of low r_{pore} (Figure 8 bottom, Figure S21).

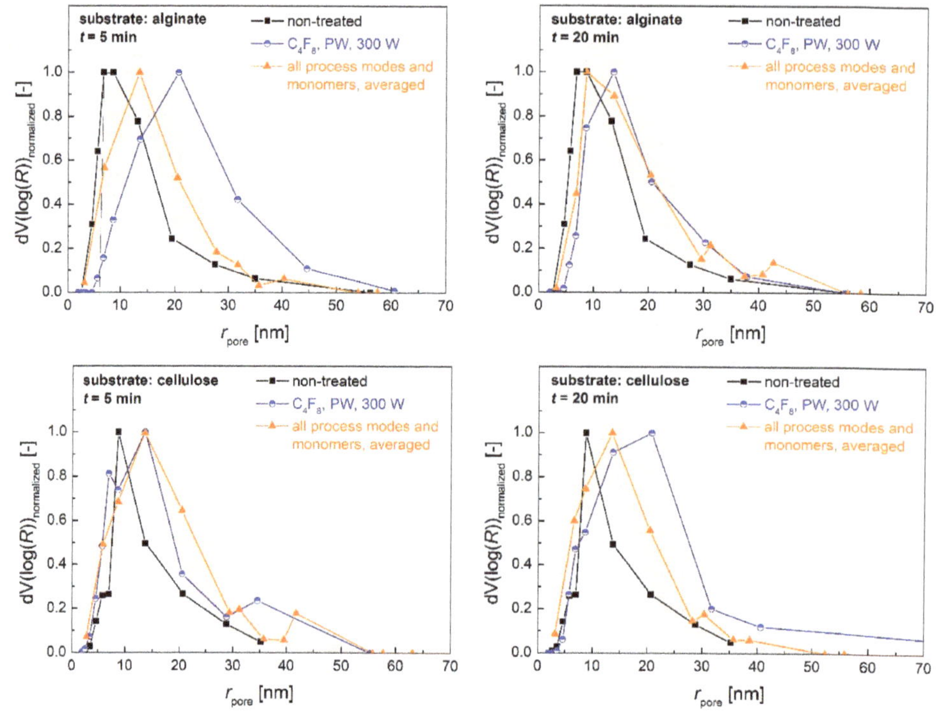

Figure 8. Comparison of BJH pore size distributions (normalized) from alginate (**top**) and cellulose (**bottom**) aerogels prior to and after the plasma coating process. Round data points represent the values obtained after coating with C_4F_8 in PW mode with p_{input} = 300 W; triangular data points represent the averaged values obtained after all other coating process conditions, excluding PFAC-8 coating in CW mode with p_{input} = 90 W. Error areas represent the standard deviation from averaged values.

It is of interest to try elucidating the mechanism of the coating given the diversity in nature and properties of the studied aerogels. Although a complete elucidation of all steps involved in the cold plasma coating of aerogels is beyond the scope of this study, we discuss here probable elementary processes that stack up the entire coating process and exclude unlikely ones. We begin by building upon the observed increase in the *specific* surface area. Because the nitrogen porosimetry can detect pores with $r_{pore} \lesssim 75$ nm, one may speculate that the increase in the specific surface area in the range of a processing time of 10–40 min is due to the deposition onto the macropores ($r_{pore} > 75$ nm), which now become detectable by N_2 porosimetry. To test out this hypothesis, we estimated the amount of the material required to convert all macropores into mesopores as follows. We consider here an extreme case where all macropores are converted to mesopores. The mesopores remain however non-penetrable to the monomer, i.e., their contribution to the total specific surface area does not change during the plasma coating.

When coating material with a mass m is deposited in macropores, the macropore volume of the coated aerogel is

$$V_{macro} = V_{macro,0} - m/\rho_{sk} \quad (5)$$

where $V_{macro,0}$ is the macropore volume of the pristine aerogel, and ρ_{sk} is the skeletal density of the coating (assumed to be equal 2.2 g/cm^3 as for Teflon). In the following, the subscript "0" stands for pristine aerogel. For our purposes it is sufficient to treat a macroporous aerogel network within the cylindrical pore model [35]. According to this model, there is one efficient cylindrical pore with a diameter d_{pore} and length h_{pore}. The macropore volume is then given by

$$V_{macro} = \pi/4 d_{pore}^2 h_{pore} \tag{6}$$

The specific surface area in the cylindrical pore model can be easily calculated as [27]

$$S_V = 4V_{macro}/d_{pore} \tag{7}$$

For example, pristine alginates have V_{macro} = 8.6 cm^3/g (see Figure 6) and the average macropore size is d_{pore} = 237 nm (see Section 3.1), which yields via Equation (7) a value $S_{V,0}$ = 145 m^2/g. This value corresponds to non-detectable specific surface area due to macropores only. If we then assume that the pore length h_{pore} does not change significantly during the plasma coating, from Equation (6) we readily have

$$d_{pore} = d_{pore,0}\sqrt{V_{macro}/V_{macro,0}} \tag{8}$$

Substituting Equations (5) and (8) in Equation (7) we finally obtain for the specific surface area of the coated aerogel:

$$S_V = \frac{S_{V,0}}{m_0}\frac{1}{m + m_0}\sqrt{1 - \frac{m}{\rho_{sk} V_{macro,0}}} \tag{9}$$

Because the *specific* surface area (m^2/g) is what is always measured, m_0 = 1 g in Equation (9). Analysis of Equation (9) shows that the specific surface area S_V is a decreasing function of the deposited mass m. Therefore, the specific surface area can only decrease in the coating process, i.e., $\Delta S_V = S_V - S_{V,0}$ is always negative. Results for changes of a specific surface contradict this picture since ΔS_V is positive (see Figures 5 and 6). Therefore, we are forced to search for alternative mechanisms.

One probable explanation for the observed increase in the specific surface area is that the polymer deposition takes place where the mass transport through the porous network becomes limited, e.g., where two pores with significantly different diameters d_1 and d_2 are coupled to each other (Figure 9). Deposition of a film with a thickness h would create and extra surface $\pi d^2/2$ at each coupling point (given that $d_1 < d_2$). The total surface area for N coupling points is

$$S_{V,0} m_0 + N\pi d^2/2 \tag{10}$$

whereas the total increase in mass is

$$m_0 + \rho_{sk} N \pi d^2 h/4 \tag{11}$$

The deposited polymer layer can in principle be mesoporous with its own specific surface area of $S_{V,coat}$. This is what would add an extra term $S_{V,coat}\rho_{sk} N\pi d^2 h/4$ in Equation (10).

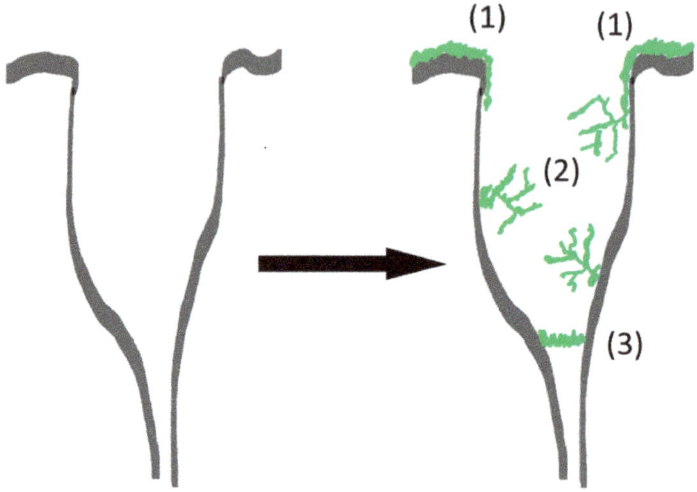

Figure 9. Schematic representation of two coupled pores. The large pore (macropore) is exposed to the gas phase with activated monomer (not shown). A layer at the outer surface (1), porous structures from the deposited polymer (2) and a thin layer at the coupling point of two pores (3) are shown. The layer (1) does not lead to an increase in the specific surface area, while the structures (2) and (3) do. Deposition of the layer (3) may be due to a limited mass transfer through the smaller pore.

The ratio of the expressions Equations (10) and (11) gives the specific surface area of the coated aerogel. Calculations show that for thin layers ($\sim$1 nm), the gain in surface area overweighs the gain in the coating mass. Therefore, this mechanism agrees qualitatively with our experimental observations. Further insights are provided via SEM images of the pore structures from pristine and coated aerogels, wherein coating material and alginate fibers are clearly distinguishable (Figure 10a,b). High resolution pictures indicate that the deposited polymer layer has an intrinsic porosity: New adsorption sites are provided by narrow crevices and mesopores formed by growing coating material (Figure 10c,d and Figures S23 and S24). Furthermore, change of alginate fiber thickness due to deposited coating material could principally cause additional changes of the specific surface area in case of mesopores, as schematically indicated in Figure 9. Since any deposition of coating material layers onto fibrils leads also to a drop of the specific surface area (no matter how thin the layer is), we come to the conclusion that the only ways to create additional surface are as mentioned above: the deposition of thin films in pore coupling points and by intrinsic porosity of the polymer layers.

Quantitative comparisons going beyond the basic calculations presented above are not possible at the moment since the parameters N, h and $S_{V,coat}$ are unknown. Elucidation of a quantitative model of the cold plasma coating is a part of our ongoing work. Generally, we believe that cold plasma coating is a very well suited method for the post-processing of an aerogel surface and worth further investigations—for instance, by extension to additional aerogel types and monomers.

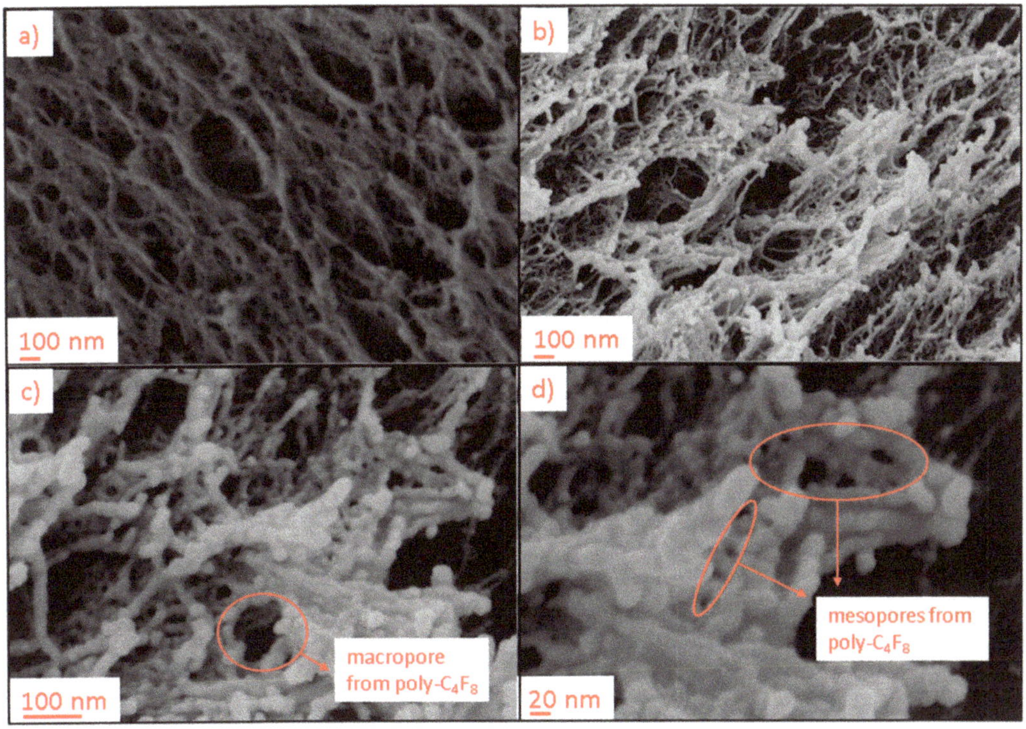

Figure 10. SEM pictures: (**a**) porous structure of non-treated alginate aerogel; (**b**) porous structure of treated alginate aerogel, coating material C_4F_8, CW, p_{input} = 90 W, t = 50 min, distance from outer layer = 134 µm; (**c**) and (**d**) high-resolution pictures of poly-C_4F_8 in alginate aerogel and new pores generated thereof.

4. Conclusions

It was shown first that cold plasma coating with fluorinated monomers is a suitable method for fast, simple and material-efficient hydrophobic and oleophobic post-modification of various open porous biopolymer aerogels with high specific surface areas (200–300 m^2/g). Significant effects were achieved after short process times of 5–10 min, and no further purification of the substrates was necessary. The wide applicability of the process was demonstrated by coating four different aerogels with different polarities and total porosities (64–95%) as well as macro-to-mesoporous fractions. It was possible to coat all substrates with layers of three different fluorinated coatings (C_4F_8, PFAC-6, PFAC-8), resulting in hydrophobic to superhydrophobic liquid water-repellent surfaces, which showed static water contact angles up to 154°. While application of water droplets on non-treated alginate and cellulose gels resulted in immediate pore collapse at the site of droplet application, protection of the gels was generally achieved after process times of 5 min. The individual results depended on the specific monomer/substrate pair and the process parameters (input power, input mode and process time). The polymerization of perfluorinated acrylates resulted in oil repellent surfaces, as verified by contact angle measurements with n-hexadecane. In many cases, coated polysaccharide aerogels showed a slight deformation at the site where water droplet was applied. Prevention of this effect was achieved via C_4F_8 polymerization in CW mode and input power of 90 W after process times $\geq$10 min.

Furthermore, it was demonstrated that new adsorption sites in the pores were formed during the process, resulting in a significant increase of up to 61% in the specific surface area. Maximization of ΔS_V was achieved in the PW mode, using C_4F_8 as monomer. A steady state was reached after t = 10–40 min, resulting in constant values of ΔS_V under the given process conditions. Relations of the ΔS_V with the overall aerogel porosity as well as individual macro- and mesopore volumes were identified. It was shown that high porosities $\geq 90\%$ are necessary for achieving high effects of $\Delta S_V \geq 100\ m^2/g$. While changes of the BET C constant indicated the generation of moieties with lower surface energies in polysaccharide gels, the penetration of coating material into the pores was proven via EDS analysis of cut-open substrates. The results indicate that two main effects determine the intrusion depth and final fluorine concentration in the pores: (1) the adhesivity of the substrate material to the activated monomers and monomer fragments; (2) the composition of the porous structure, whereas an interconnected, highly macroporous network is beneficial for a good transport of monomer into the structure. The dynamic changes of the mesoporous pore structure during the process, as determined via BJH desorption analysis, led in the case of alginate aerogels to a decrease in pore volume from small mesopores in the range of $r_{pore} \leq 10$ nm and an increase in the range of r_{pore} approximately 15–60 nm. Estimations based on geometric and mass balance aspects showed that the observed increase in the specific surface area may result from two processes: (i) polymer deposition at coupling points of macro- and mesopores and (ii) due to intrinsic porosity of the deposited polymer layer.

Summarized, it was shown that surface plasma coating of biopolymer aerogels with fluorinated monomers leads to modification of the external surfaces as well as to modification of the pore structure of aerogels. An increase in the specific surface area via coating processes has not yet been reported and may be a significant step towards further improvement of biopolymer aerogel properties. While the results of this study provide first considerations about the key process parameters and material properties for pore modification, additional systematic studies are necessary in order to quantify the influences of power input and input mode from the process side and substrate adhesivity and pore structure from the materials side.

Supplementary Materials: The following are available online at https://www.mdpi.com/article/10.3390/polym13173000/s1. Additional data (Figures S1–S24) are reported in the electronic supplementary data.

Author Contributions: B.S. designed and performed the experiments, wrote the paper and analyzed the data; I.J. designed and performed the experiments and revised the paper; K.B. designed and performed the experiments; P.G. wrote and revised the paper; I.S. wrote and revised the paper. All authors have read and agreed to the published version of the manuscript.

Funding: The authors wish to acknowledge support for this research received from the German Research Foundation (DFG) under project SM 82/18-1 and from Hamburg University of Technology under the I3 junior project "Strategy development for coating of open-porous materials with low density using ionized plasma".

Informed Consent Statement: Not applicable.

Conflicts of Interest: There are no conflict to declare. The authors declare no competing financial interests.

References

1. Zhao, S.; Malfait, W.J.; Guerrero-Alburquerque, N.; Koebel, M.M.; Nyström, G. Biopolymer Aerogels and Foams: Chemistry, Properties, and Applications. *Angew. Chem.-Int. Ed.* **2018**, *57*, 7580–7608. [CrossRef] [PubMed]
2. Smirnova, I.; Gurikov, P. Aerogel production: Current status, research directions, and future opportunities. *J. Supercrit. Fluids* **2017**, *134*, 228–233. [CrossRef]
3. Ganesan, K.; Budtova, T.; Ratke, L.; Gurikov, P.; Baudron, V.; Preibisch, I.; Niemeyer, P.; Smirnova, I.; Milow, B. Review on the production of polysaccharide aerogel particles. *Materials* **2018**, *11*, 2144. [CrossRef]
4. García-González, C.A.; Alnaief, M.; Smirnova, I. Polysaccharide-based aerogels—Promising biodegradable carriers for drug delivery systems. *Carbohydr. Polym.* **2011**, *86*, 1425–1438. [CrossRef]
5. Andlinger, D.J.; Bornkeßel, A.C.; Jung, I.; Schroeter, B.; Smirnova, I.; Kulozik, U. Microstructures of potato protein hydrogels and aerogels produced by thermal crosslinking and supercritical drying. *Food Hydrocoll.* **2021**, *112*, 106305. [CrossRef]

6. Kleemann, C.; Selmer, I.; Smirnova, I.; Kulozik, U. Tailor made protein based aerogel particles from egg white protein, whey protein isolate and sodium caseinate: Influence of the preceding hydrogel characteristics. *Food Hydrocoll.* **2018**, *83*, 365–374. [CrossRef]
7. Smirnova, I.; Gurikov, P. Aerogels in chemical engineering: Strategies toward tailor-made aerogels. *Annu. Rev. Chem. Biomol. Eng.* **2017**, *8*, 307–334. [CrossRef]
8. Ebert, H.P.; Braxmeier, S.; Reichenauer, G.; Hemberger, F.; Lied, F.; Weinrich, D.; Fricke, M. Intercomparison of Thermal Conductivity Measurements on a Nanoporous Organic Aerogel. *Int. J. Thermophys.* **2021**, *42*, 1–18. [CrossRef]
9. Iqbal, M.; Dinh, D.K.; Abbas, Q.; Imran, M.; Sattar, H.; Ul Ahmad, A. Controlled Surface Wettability by Plasma Polymer Surface Modification. *Surfaces* **2019**, *2*, 26. [CrossRef]
10. Alnaief, M.; Antonyuk, S.; Hentzschel, C.M.; Leopold, C.S.; Heinrich, S.; Smirnova, I. A novel process for coating of silica aerogel microspheres for controlled drug release applications. *Microporous Mesoporous Mater.* **2012**, *160*, 167–173. [CrossRef]
11. Antonyuk, S.; Heinrich, S.; Gurikov, P.; Raman, S.; Smirnova, I. Influence of coating and wetting on the mechanical behaviour of highly porous cylindrical aerogel particles. *Powder Technol.* **2015**, *285*, 34–43. [CrossRef]
12. Yasuda, H. Plasma polymerization and plasma modification of polymer surfaces. In *New Methods of Polymer Synthesis*; Springer: Berlin, Germany, 1995; pp. 161–196. [CrossRef]
13. Muzammil, I.; Li, Y.P.; Li, X.Y.; Lei, M.K. Duty cycle dependent chemical structure and wettability of RF pulsed plasma copolymers of acrylic acid and octafluorocyclobutane. *Appl. Surf. Sci.* **2018**, *436*, 411–418. [CrossRef]
14. Dimitrakellis, P.; Gogolides, E. Hydrophobic and superhydrophobic surfaces fabricated using atmospheric pressure cold plasma technology: A review. *Adv. Colloid Interface Sci.* **2018**, *254*, 1–21. [CrossRef]
15. Yoshinari, M.; Matsuzaka, K.; Inoue, T. Surface modification by cold-plasma technique for dental implants-Bio-functionalization with binding pharmaceuticals. *Jpn. Dent. Sci. Rev.* **2011**, *47*, 89–101. [CrossRef]
16. Di Mundo, R.; De Benedictis, V.; Palumbo, F.; d'Agostino, R. Fluorocarbon plasmas for nanotexturing of polymers: A route to water-repellent antireflective surfaces. *Appl. Surf. Sci.* **2009**, *255*, 5461–5465. [CrossRef]
17. Sahin, H.T.; Manolache, S.; Young, R.A.; Denes, F. Surface fluorination of paper in CF4-RF plasma environments. *Cellulose* **2002**, *9*, 171–181. [CrossRef]
18. Chan, C.M.; Ko, T.M.; Hiraoka, H. Polymer surface modification by plasmas and photons. *Surf. Sci. Rep.* **1996**, *24*, 1–54. [CrossRef]
19. Lin, R.; Li, A.; Zheng, T.; Lu, L.; Cao, Y. Hydrophobic and flexible cellulose aerogel as an efficient, green and reusable oil sorbent. *RSC Adv.* **2015**, *5*, 82027–82033. [CrossRef]
20. Shi, J.; Lu, L.; Guo, W.; Sun, Y.; Cao, Y. An environment-friendly thermal insulation material from cellulose and plasma modification. *J. Appl. Polym. Sci.* **2013**, *130*, 3652–3658. [CrossRef]
21. Cheng, Y.; Lu, L.; Zhang, W.; Shi, J.; Cao, Y. Reinforced low density alginate-based aerogels: Preparation, hydrophobic modification and characterization. *Carbohydr. Polym.* **2012**, *88*, 1093–1099. [CrossRef]
22. Zheng, T.; Li, A.; Li, Z.; Hu, W.; Shao, L.; Lu, L.; Cao, Y.; Chen, Y. Mechanical reinforcement of a cellulose aerogel with nanocrystalline cellulose as reinforcer. *RSC Adv.* **2017**, *7*, 34461–34465. [CrossRef]
23. Baudron, V.; Gurikov, P.; Smirnova, I.; Whitehouse, S. Porous starch materials via supercritical-and freeze-drying. *Gels* **2019**, *5*, 12. [CrossRef] [PubMed]
24. Rodríguez-Dorado, R.; López-Iglesias, C.; García-González, C.A.; Auriemma, G.; Aquino, R.P.; Del Gaudio, P. Design of Aerogels, Cryogels and Xerogels of Alginate: Effect of Molecular Weight, Gelation Conditions and Drying Method on Particles' Micromeritics. *Molecules* **2019**, *24*, 1049. [CrossRef] [PubMed]
25. Henry, H.J.; Carlin, P.C.; Hammerschmidt, J.A.; Buck, R.C.; Buxton, L.W.; Fiedler, H.; Seed, J.; Hernandez, O. A Critical Review of the Application of Polymer of Low Concern and Regulatory Criteria to Fluoropolymers. Integrated Environmental Assessment and Management. *Integr. Environ. Assess. Manag.* **2018**, *14*, 316–334. [CrossRef]
26. Martin, I.T.; Malkov, G.S.; Butoi, C.I.; Fisher, E.R. Comparison of pulsed and downstream deposition of fluorocarbon materials from C_3F_8 and c-C_4F_8 plasmas. *J. Vac. Sci. Technol. A* **2004**, *22*, 227–235. [CrossRef]
27. Schroeter, B.; Yonkova, V.P.; Niemeyer, N.A.M.; Jung, I.; Preibisch, I.; Gurikov, P.; Smirnova, I. Cellulose aerogel particles: Control of particle and textural properties in jet cutting process. *Cellulose* **2021**, *28*, 223–239. [CrossRef]
28. Bowland, E.L.; Allen Foegeding, E.; Hamann, D.D. Rheological analysis of anion-induced matrix transformations in thermally induced whey protein isolate gels. *Top. Catal.* **1995**, *9*, 57–64. [CrossRef]
29. Thommes, M.; Kaneko, K.; Neimark, A.V.; Olivier, J.P.; Rodriguez-Reinoso, F.; Rouquerol, J.; Sing, K.S.W. Physisorption of gases, with special reference to the evaluation of surface area and pore size distribution (IUPAC Technical Report). *Pure Appl. Chem.* **2015**, *87*, 1051–1069. [CrossRef]
30. Lázár, I.; Forgács, A.; Horváth, A.; Király, G.; Nagy, G.; Len, A.; Dudás, Z.; Papp, V.; Balogh, Z.; Moldován, K.; et al. Mechanism of hydration of biocompatible silica-casein aerogels probed by NMR and SANS reveal backbone rigidity. *Appl. Surf. Sci.* **2020**, *531*, 147232. [CrossRef]
31. Yasuda, H. Operational Parameters of Plasma Polymerization. In *Plasma Polymerization*; Academic Press: New York, NY, USA, 1985; pp. 277–333. [CrossRef]
32. Coulson, S.R.; Woodward, I.S.; Badyal, J.P.S.; Brewer, S.A.; Willis, C. Ultralow surface energy plasma polymer films. *Chem. Mater.* **2000**, *12*, 2031–2038. [CrossRef]

33. Manzocco, L.; Mikkonen, K.S.; García-González, C.A. Aerogels as porous structures for food applications: Smart ingredients and novel packaging materials. *Food Struct.* **2021**, *28*, 100188. [CrossRef]
34. Honda, K.; Morita, M.; Takahara, A. Surface molecular aggregation structure and surface properties of poly (fluoroalkyl acrylate) thin films. *Kobunshi Ronbunshu* **2007**, *64*, 181–190. [CrossRef]
35. Rege, A.; Schestakow, M.; Karadagli, I.; Ratke, L.; Itskov, M. Micro-mechanical modelling of cellulose aerogels from molten salt hydrates. *Soft Matter* **2016**, *12*, 7079–7088. [CrossRef] [PubMed]

polymers

MDPI

Article

Chitosan-Urea Nanocomposite for Improved Fertilizer Applications: The Effect on the Soil Enzymatic Activities and Microflora Dynamics in N Cycle of Potatoes (*Solanum tuberosum* L.)

Rohini Kondal [1], Anu Kalia [2,*], Ondrej Krejcar [3,4], Kamil Kuca [5,*], Sat Pal Sharma [6], Karanvir Luthra [1], Gurmeet Singh Dheri [7], Yogesh Vikal [8], Monica Sachdeva Taggar [9], Kamel A. Abd-Elsalam [10] and Carmen L. Gomes [11]

1 Department of Microbiology, Punjab Agricultural University, Ludhiana 141004, Punjab, India; rohini.kondal13@gmail.com (R.K.); luthrakaran11@gmail.com (K.L.)
2 Electron Microscopy and Nanoscience Laboratory, Department of Soil Science, Punjab Agricultural University, Ludhiana 141004, Punjab, India
3 Center for Basic and Applied Science, Faculty of Informatics and Management, University of Hradec Kralove, 50003 Hradec Kralove, Czech Republic; ondrej.krejcar@uhk.cz
4 Malaysia Japan International Institute of Technology (MJIIT), Universiti Teknologi Malaysia, Jalan Sultan Yahya Petra, Kuala Lumpur 54100, Malaysia
5 Department of Chemistry, Faculty of Science, University of Hradec Kralove, 50003 Hradec Kralove, Czech Republic
6 Department of Vegetable Science, Punjab Agricultural University, Ludhiana 141004, Punjab, India; sharmasp@pau.edu
7 Green House Gas Laboratory, Department of Soil Science, Punjab Agricultural University, Ludhiana 141004, Punjab, India; gsdheri@pau.edu
8 School of Agricultural Biotechnology, Punjab Agricultural University, Ludhiana 141004, Punjab, India; yvikal-soab@pau.edu
9 Department of Renewable Energy Engineering, Punjab Agricultural University, Ludhiana 141004, Punjab, India; monicasachdeva@pau.edu
10 Agricultural Research Center, Plant Pathology Research Institute, Giza 12619, Egypt; kamelabdelsalam@gmail.com
11 Department of Mechanical Engineering, Iowa State University, Ames, IA 50011, USA; carmen@iastate.edu
* Correspondence: kaliaanu@pau.edu (A.K.); kamil.kuca@uhk.cz (K.K.); Tel.: +91-161-2401960 (A.K.); +420-603-289-166 (K.K.)

Citation: Kondal, R.; Kalia, A.; Krejcar, O.; Kuca, K.; Sharma, S.P.; Luthra, K.; Dheri, G.S.; Vikal, Y.; Taggar, M.S.; Abd-Elsalam, K.A.; et al. Chitosan-Urea Nanocomposite for Improved Fertilizer Applications: The Effect on the Soil Enzymatic Activities and Microflora Dynamics in N Cycle of Potatoes (*Solanum tuberosum* L.). *Polymers* **2021**, *13*, 2887. http://doi.org/10.3390/polym13172887

Academic Editor: Alina Sionkowska

Received: 4 August 2021
Accepted: 23 August 2021
Published: 27 August 2021

Abstract: The impact of polymer-based slow-release urea formulations on soil microbial N dynamics in potatoes has been sparingly deciphered. The present study investigated the effect of a biodegradable nano-polymer urea formulation on soil enzymatic activities and microflora involved in the N cycling of potato (*Solanum tuberosum* L.). The nano-chitosan-urea composite (NCUC) treatment significantly increased the soil dehydrogenase activity, organic carbon content and available potassium compared to the conventional urea (CU) treatment. The soil ammonical nitrogen (NH_4^+-N) and nitrate nitrogen (NO_3^--N) contents and urease activity were significantly decreased in the NCUC-amended soil. The slow urea hydrolysis rate led to low concentrations of NH_4^+-N and NO_3^--N in the tested potato soil. Furthermore, these results corroborate the low count of ammonia oxidizer and nitrate reducer populations. Quantitative PCR (q-PCR) studies revealed that the relative abundance of eubacterial (AOB) and archaeal ammonia-oxidizing (AOA) populations was reduced in the NCUC-treated soil compared to CU. The abundance of AOA was particularly lower than AOB, probably due to the more neutral and alkaline conditions of the tested soil. Our results suggest that the biodegradable polymer urea composite had a significant effect on the microbiota associated with soil N dynamics. Therefore, the developed NCUC could be used as a slow N-release fertilizer for enhanced growth and crop yields of potato.

Keywords: chitosan; nanocomposite; nanofertilizer; slow release; ammonia oxidase gene; quantitative polymerase chain reaction; microflora N cycle; nutrient use efficiency; soil N content

1. Introduction

Potato (*Solanum tuberosum* L.) is one of the most abundant and widely cultivated vegetable crops in the world. It is the third most important crop after rice and wheat, with an annual production of 376 million tons (FAOSTAT, 2018). The growth and yield of potato strongly depend on the availability of nutrients, especially nitrogen (N) [1]. Urea is widely used as a N source because it can trigger biomass and total N accumulation in potatoes [2]. Potato requires high doses of N fertilizers to achieve maximum yield; however, N recovery in potatoes is often low because of the plant's poorly developed root system [3]. Therefore, the input of N fertilizers has been increased over the years to maximize tuber yields; however, excessive application of N fertilizers may reduce tuber yields [4]. In addition, it also significantly influences N cycling and augments NH_3^+ emissions (volatilization) and NO_3^--N accumulation (leaching), which have serious environmental implications [5]. This calls for the development of effective N management strategies such as controlled-release fertilizers (CRF) to improve N use efficiency and also to combat the environmental impact of fertilizer application [6].

Polymer-coated urea (PCU) is a urea fertilizer formulation that exhibits a more predictable release pattern. It enables the nutrient release to synchronize with the needs of the plant [7,8]. Recently, the use of controlled-release urea (CRU) has become a new trend to enhance crop yields due to its high potential for enhancing nutrient use efficiency (NUE) [9]. A study has shown that PCU in cotton can significantly affect the NH_4^+ and NO_3^- concentrations in the soil by lowering them as compared to conventional nitrogen fertilizers [10]. Another study on PCU reported that NH_3 volatilization was significantly reduced when compared to surface-applied urea on creeping bent grass [11].

The concentrations of NO_3^- and NH_4^+ in soils are controlled by various factors, such as soil temperature, pH, soil microbiota, fertilizer form and moisture conditions [12]. The biological component of soils usually responds to the changing soil conditions more rapidly than either the chemical or physical properties [13]. Soil enzymatic activities have been used as indicators of microbial activity because these activities reflect the total range of oxidative activity in soil microflora [14,15]. Previous studies on dehydrogenase activities in maize revealed that a polyolefin-coated CR urea fertilizer formulation significantly increased the soil microbial activity compared to conventional urea [16]. However, the urease enzyme activity was decreased when different biodegradable urea coating formulations were incorporated in comparison to conventional urea, leading to slower urea hydrolysis rates [17].

Previous studies focused on improving the slow release properties of chitosan have been carried out [18,19]. However, these studies did not focus on nitrogen release properties. Likewise, few other published reports involved evaluation of the effect of controlled-release urea fertilizers in potato, but these studies were on the use of polymer sulfur or polyurethane as a coating material and not chitosan [20]. The benefits of the use of chitosan as a urea-encapsulating/embedding agent are two-fold. The primary benefit is that the chitosan polymer has a biological origin [21] and is bio-safe and biodegradable compared to synthetic polymers which cannot be metabolized by the soil microbes [22]. Further, it has already been utilized for immobilization of enzymes for food [23], and biosensing [24,25] applications. The second benefit compared to inorganic coatings such as sulfur coatings is the absence of cracks and the lowered ability to cause acidification of the soil [26,27]. Furthermore, there are no reports published that discern the effect of different chitosan concentrations on the controlled release properties of nitrogen nanofertilizers for use as N fertilizers in potato. In this study, the hypothesis tested stated that the nano-chitosan-urea composite (NCUC) will have a significant effect on the microflora involved in N cycling.

This formulation may also decrease the ammonical N and nitrate N content in the soil. A pot experiment was conducted using a nano-chitosan-urea composite in comparison to conventional urea at four different N application rates (0, 50, 75 and 100% recommended N dose per pot). The primary objectives of this study were to determine the effect of NCUC on (1) soil microbial viable counts and soil enzymatic activities, (2) soil chemical properties and (3) soil ammonia-oxidizing eubacterial and archaea-bacterial populations involved in the N cycling of potato cv. Kufri Pukhraj.

2. Materials and Methods

2.1. Materials

Chitosan (medium molecular weight: 190 to 310 kilo Daltons, and low DDA percentage (>75%), 100% purity) was purchased from Sigma-Aldrich, St. Louis, USA. Sodium tripolyphosphate (purity: 98.0%) was also procured from Sigma Aldrich, St. Louis, USA. Analytical-grade urea (purity: 99.5 to 100%) was purchased from HiMedia Laboratories Pvt. Ltd., Mumbai, India. HPLC-grade water was utilized for the preparation of all the formulations. The other analytical-grade chemicals utilized for the determination of the various enzyme activities were purchased from Hi-Media Laboratories Pvt. Ltd., Mumbai, India, while the consumables required for the q-PCR studies were purchased from DSS Takara BioIndia Pvt. Ltd., New Delhi, India.

2.2. Synthesis and Characterization of Nano-Polymer Urea Composite

Chitosan (1.5% w/v) was dissolved in acetic acid (1.0% v/v) containing Tween 80 as a surfactant to prevent the agglomeration of the nanoparticles during and after synthesis. Urea (analytical grade, 10% w/v) was dissolved in the chitosan-acetic acid suspension. Under constant stirring conditions, the aqueous tripolyphosphate (TPP) solution was added drop wise (16 drops per minute) to chitosan-urea suspensions under room temperature conditions for gelation. Instantaneously, the synthesis of chitosan-urea nanoparticles started due to the TPP-initiated ionic gelation mechanism [28–31]. The coalescence of the formed nanoparticles was reduced by ultrasonication of the formulation using a probe sonicator (VCX-750, Sonics and Materials Inc., Newtown, CT USA) operated at 35% amplitude for 15 min under ice bath conditions to avoid over-heating of the formulation. The chitosan-urea formulations were subjected to freeze drying in a lyophilizer assembly (REVA commercial lyophilizer, Gujarat, India) at a pressure of 10^{-1} torr (at -40 °C) for 24 h to obtain nano-chitosan-urea composite (NCUC) powdered samples.

2.3. UV-Vis Spectroscopy

The chitosan-urea composite formulation was analyzed for its unique light matter interaction using a double beam UV-Vis spectrophotometer (model SL 218, ELICO, Hyderabad, India) for wavelengths ranging from 190 to 800 nm. The absorbance was plotted against the wavelength to observe the characteristic UV-Vis absorbance peaks for free urea, the chitosan-TPP composite (control) and the chitosan-urea nanocomposite.

2.4. Transmission Electron Microscopy (TEM)

The specimens for TEM analysis were prepared by dispensing a known amount of freeze-dried powdered NCUC in deionized distilled water. These suspensions were sonicated for 15 min in a bath sonicator (model SW-4, Toshcon, India). A known aliquot of the sonicated sample suspension was placed on a 200 mesh size carbon-coated copper grid (Tedpella, Redding, CA, USA). The sample was allowed to adsorb and form a thin film on the carbon coating by incubating for 2 to 3 min. The grids were then allowed to air dry at room temperature overnight. The dried grids were imaged in high-resolution imaging mode using a transmission electron microscope (model Hitachi H-7650, Hitachi High-Technologies Corporation, Tokyo, Japan) at an 80 kV accelerating voltage. The size dimensions of the chitosan and chitosan-urea nanoparticles were measured through

Image J software (version 1.46 r, U. S. National Institute of Health, Bethesda, MD, USA) by manually obtaining the diameters for >50 particles from five TEM micrographs.

2.5. Scanning Electron Microscopy (SEM) and SEM-Energy-Dispersive Spectroscopy (SEM-EDS)

The morphology and cross-sectional topography of the synthesized NCUC were studied by a scanning electron microscope (model Hitachi s-3400N, Hitachi High-Technologies Corporation, Tokyo, Japan) operated at a 15 kV accelerating voltage in secondary electron imaging mode. The freeze-dried NCUC was placed on the stub and sputter coated for 30 s (current: 18–20 mA) with gold plasma in an ion sputter coater (model Hitachi E-1010, Hitachi High-Technologies Corporation, Tokyo, Japan) to make the sample conductive. The elemental composition, the percentage of atoms and the weight of elements present on the sample surface were analyzed by SEM-EDS (model Thermo Noran, Thermo Fisher Scientific Inc., Waltham, Massachusetts, USA).

2.6. Fourier Transform-Infrared Spectroscopy (FT-IR Spectroscopy)

FT-IR spectroscopy for the freeze-dried powdered samples of the NCUC was performed. The samples were mixed uniformly with pre-activated potassium bromide in an appropriate ratio (1:100), and pellets were prepared by using a hydraulic pellet press assembly (Model no. 1701, Maharashtra, India). The prepared pellets were placed in the sample mounting region of the FT-IR spectroscope (model Thermo Nicolet 6700, Thermo Fisher Scientific Inc., Waltham, Massachusetts, USA) and were scanned in the mid-IR range of 400–4000 cm^{-1} to obtain transmission spectra at a 4.0 cm^{-1} spectral resolution using 32 scans.

2.7. Urea Release Profile and Encapsulation Efficiency

The release of the encapsulated urea from the nano-chitosan-urea composite was performed by extracting aliquots (1 mL) at different time intervals (after every five days) from 500 mL autoclaved deionized water containing known concentrations (1, 10 and 100 mg per liter) of nano-chitosan-urea composites (sink conditions). The aliquots drawn were evaluated for urea content (in mg L^{-1}) using the acidic p-dimethyl aminobenzaldehyde method [32] by measuring the absorbance of the solution at 422 nm against a reagent blank in a dual beam UV-Vis spectrophotometer (model SL-218, ELICO, Hyderabad, India). At each time interval, the sample aliquot was filtered using a 0.2 μm nylon membrane syringe filter and then analyzed spectrophotometrically at 422 nm. For the encapsulation efficiency, a known amount of NCUC was crushed and dispersed in distilled water, and then left in suspension for 72 h at 400 rpm at room temperature. The suspensions were passed through a 0.2 μm polypropylene syringe filter (VWR Intl., Radnor, PA, USA) to remove the chitosan. The amount of urea was determined by using the acidic p-dimethyl amino benzaldehyde method [32]. The entrapment efficiency was calculated according to Equation (1).

$$EE(\%) = \frac{amount\ of\ active\ compound\ entrapped}{initial\ active\ compound\ amount} * 100 \quad (1)$$

2.8. In Vitro Biodegradation Assay

The biodegradability of the NCUC was evaluated through an in vitro biodegradation assay by incubating a known pre-weighed quantity of the sample in phosphate-buffered saline (2 mL, pH = 7.4) and a lysozyme mixture (Hen egg white origin, Sigma Aldrich, St. Louis, MO, USA, 1 mg mL^{-1}) followed by weighing of the incubated sample retrieved after 1, 3 and 7 days of incubation. Both the initial and retrieved samples were freeze dried to get rid of the water molecules. The biodegradation was calculated as

$$\text{Biodegradation } (\%) = W_1 - W_2/W_1 * 100$$

where W_1 = initial weight of NCUC, and W_2 = weight of retrieved NCUC.

2.9. Experimental Design

The soil (topsoil 0–20 cm) used in this study was collected from fields of the Vegetable Farm in the Department of Vegetable Science, Punjab Agricultural University, Ludhiana, Punjab, India. The collected sandy loam soil (Dystric Cambisol) was sieved through a 2 mm iron mesh. Then, the plastic pots (30 cm height and 11 cm diameter, 8 kg capacity) were filled with 5 kg of soil. The experiment was carried out from October 2016 to March 2017 in open conditions with average temperatures ranging between 23 and 34.6 °C and relative humidity of 75%. The experimental tuber seeds of *Solanum tuberosum* L. (potato) cv. Kufri Pukhraj were also obtained from the Department of Vegetable Science. Each pot was watered 24 h before sowing the tuber, and sowing was conducted with one tuber per pot. The recommended dose of nitrogen was applied at the rate of 75 kg N or 165 kg urea acre^{-1}. All three types of fertilizer, the nano-chitosan-urea composite (NCUC), conventional urea (CU) and chitosan (CS), were applied at four different concentrations of 0, 50, 75 and 100% of the recommended nitrogen fertilizer dosage (75 kg acre^{-1}). The phosphorus and potassium fertilizers were applied at a rate of 25 kg acre^{-1}. The soil samples were collected at 0, 20, 40, 60 and 90 days after treatment (DAT) and subjected to various analyses (described below).

2.10. Soil Chemical Characterization

The soil samples taken at different time intervals were dried in an oven at 60 °C for 2 days or until a constant weight of the samples was achieved. The dried soil was grinded in a pestle and mortar and sieved through a 2.0 mm sieve to get rid of any organic or root debris. The dried and sieved soil samples were used for chemical property characterization. The pH and electrical conductivity (EC, soil/water ratio of 1:2) of the soil samples were determined by the glass electrode conductivity method on pH (model µpH system 361, Systronics India (Pvt.) Ltd., Ahmedabad, Gujarat, India) and a solubridge meter [33]. Soil organic carbon (OC) was estimated by following the procedure described by Walkley and Black (1934), which involved the use of a soil (2.0 g) sample which is reacted with 1 N potassium dichromate (100 mL)+20 mL sulfuric acid+NaF (0.5 g)+Diphenylamine indicator solution followed by titration of 10 mL of this mixture against 0.5 N ferrous ammonium sulfate solution. The macronutrient status (N, P and K) of the soil was determined using methods described by Kjeldahl [34], and Reed and Scott [35]. The soil P content was measured through the ammonium molybdate technique, while the K content was estimated through flame photometry. The available ammonical nitrogen and nitrate nitrogen in the soil were determined by the Kjeldahl technique and included mixing the soil sample (10.0 g) in KCl (2N, 100 mL) solution. The contents were shaken (1 h), filtered through filter paper (Whatman no. 1 filter paper, Kent, U.K.) and analyzed. For ammonical N estimation, sulfuric acid (N/200, 5.0 mL) and methyl red indicator (50.0 µL) were mixed, and the delivery tube end of the Kjeldahl flask was dipped in it. The soil extract (10.0 mL) was incubated with MgO powder (0.2 g) in a distillation flask, and the collected distillate (30.0 mL) was titrated against NaOH (N/200). The amount of ammonical N was calculated from the volume of sulfuric acid (N/200) used for the absorption of NH_3. For the nitrate N content estimation, the soil extract (10.0 mL) taken in the distillation flask was mixed with 0.2 g each of MgO and Deverda's alloy. The distillation process was performed in a similar manner to that described for ammonical N, and nitrate N was calculated.

2.11. Enumeration of the Culturable Microbial Population

The soil microbial count was enumerated using the spread plate method [36]. Briefly, ten grams of fresh soil samples was transferred to an Erlenmeyer flask (150 mL) containing 90 mL sterile distilled water and shaken at 120 rpm for 15 min to obtain a homogenous suspension. Serial dilutions (up to 10^{-6}) were produced by pipetting 1 mL of the soil suspension into 9 mL of sterile water. Finally, a 0.1 mL aliquot of the diluted soil suspension was uniformly spread with the help of a sterile spreader on solidified Petri plates with nutrient agar for total aerobic bacteria, NH_4 oxidizer agar for ammonia oxidizers, nitrate agar for nitrate-reducing bacteria and Potato Dextrose Agar (PDA)/Sabouroaud's Dextrose Agar (SDA) (HiMedia Laboratories Pvt. Ltd., Mumbai, India) for total fungi counts, at 0, 20, 40, 60 and 90 days after sowing (DAS). Dilutions of 10^{-5} and 10^{-6} were selected for the enumeration of bacteria, and 10^{-3} to 10^{-4} for fungi, ammonia oxidizers and nitrate-reducing bacteria based on their population in the soil. The Petri plates were incubated for 2 to 6 days at 28 ± 2 °C. The total aerobic bacterial count of the soil samples was enumerated by plating the 10^{-5} and 10^{-6} dilutions on nutrient agar, and the number of bacterial colonies appearing after 24 to 48 h of incubation was counted. The nitrate-reducing bacteria were enumerated by counting the microbial colonies that grew on nitrate-reducing media. Two plates per dilution of each soil sample were plated, counted and reported in CFU g^{-1} of soil.

2.12. Soil Enzyme Activity

The effect of different N fertilizer treatments was observed on two specific soil enzyme activities, dehydrogenase and urease. The collected soil was dried and sieved through a 2 mm mesh. The procedures for the dehydrogenase and urease activities were performed following the procedures described by [37,38]. The brief procedure for the dehydrogenase activity that was followed included mixing of $CaCO_3$ (0.1 g) in pre-sieved sample soil (10.0 g). The contents were pulverized to obtain a homogenous mixture. A known amount (3.0 g) of the above mixture was mixed with 2,3,5-Triphenyl tetrazolium chloride solution (3% *w*/*v*, 0.5 mL) and distilled water (1.5 mL). The contents were incubated for 24 h at 37 °C followed by methanol (5.0 mL) extraction under shaking conditions to obtain a pink-colored solution after repeated washing of the soil and filtration. The volume was made up to 25.0 mL with methanol, and the absorbance was measured at 485 nm wavelength on the spectrophotometer using methanol as a blank. The protocol used for the estimation of the urease activity of the soil samples involved addition of urea (2000 mg L^{-1}, 5.0 mL) to grinded and sieved (sieve size: <2 mm) soil samples (5.0 g). After 5 h of incubation at 37 °C, potassium chloride-phenylmercuric acetate solution (2.0 M, 50.0 mL) was added, and the contents were further incubated for 60 min under shaking conditions. After filtering the contents, the extract (1.0 mL) was mixed with extracting (5.0 mL) and coloring reagents (15.0 mL, Diacetyl monoxime (2.5% *w*/*v*, 25.0 mL)+Thiosemicarbazone (0.25% *w*/*v*, 5.0 mL)). The contents were heated in a boiling water bath for 30 min and cooled downed to room temperature, and the volume was made up to 25 mL using distilled water. The contents were then mixed thoroughly, and the intensity of the red color was measured (λ = 527 nm).

2.13. Molecular Characterization of Eubacterial and Archaeal Ammonia-Oxidizing Populations by q-Polymerase Chain Reaction (q-PCR)

Soil samples at 0, 20, 40 and 60 DAT were collected, and isolation of total bacterial community DNA was carried out by HiPurA^{TM} soil DNA purification kit (MB542) purchased from Hi-Media Laboratories (Mumbai, India). DNA quantification was performed on 0.8% gel. Ammonia oxidase gene primers used in this study included the eubacterial ammonia-oxidizing bacteria (AOB) (forward primer: amoA-1F (GGGGTTTCTACTGGTGGT), reverse primer: amoA-2R (CCCCTCTGGAAAGCCTTCTTC)) [39] and archaebacterial ammonia-oxidizing archaea (AOA) (forward primer: amoA310f (TGGATACCGTCAGCAATG), reverse primer: amoA529r (GCAACAGGACTATTGTAGAA)) primers for the ammonia oxidase gene [40], while the 16s rRNA gene universal primer (forward primer: F984 (AACGCGAAGAACCTTAC), reverse primer: R1401 (GGGTCTTGTACACACCG)) was also used. These primers were used for total soil DNA and in all in vitro amplification reactions performed in master cyclers (Eppendorf, Hamburg, Germany and Roche Applied Sciences, Penzberg, Upper Bavaria, Germany) using 14–50 ng genomic DNA of each sample in a final volume of 10 μL of PCR mix.

2.14. Potato Vegetative Growth and Yield Traits

The potato plants were uprooted from pots, and the shoot fresh weight per plant was recorded; afterwards, they were oven dried at 60 °C for 2 days, and their dry weight was recorded using an electronic weighing balance in grams at 90 DAS. The yield-attributing characteristics such as the number of tubers formed per plant, total tuber weight per plant and marketable tuber yield per plant were also determined. The marketable tubers were considered by the number and weight of the potatoes above 35 to 40 g [41].

2.15. Statistical Analysis

The experiment was carried out in a completely randomized block design with a factorial arrangement. The main plots were fertilizer types (nano-chito-urea (NCUC) and conventional urea (CU)), with the level of fertilizer applications as the sub-plot factors. For all the analyses of this study, determinations were performed in triplicate as independent experiments. Data on different variables were subjected to three-way analysis of variance (ANOVA) with SAS software (Version 9.3, Cary, NC, USA) [42,43], and significant differences between treatments were determined by pair-wise comparisons using Fischer's least significant difference test (LSD; $p \leq 0.05$).

3. Results

3.1. Characterization of Nano-Polymer Urea Composite

The UV-Vis absorbance spectra of chitosan and the nano-chitosan-urea composite are shown in Figure 1D. The chitosan-TPP composite exhibited an absorption peak at 220 nm [44]. An earlier study by Abd-Elhady et al. [44] also reported the occurrence of a specific UV absorption peak for the chitosan nanoparticles to exist at 226 nm. The two UV chromogenic functional groups in chitosan, i.e., N-acetyl glucosamine and glucosamine, exhibit absorption of the far UV wavelength(s). Both of these UV chromogenic groups have an additive role for the absorption peak at 201 nm [45]. However, after incorporation of urea (10% w/v) into chitosan (1.5% w/v), dichotomization of the absorption peak at 220 nm occurred with the appearance of a blue-shifted shoulder peak at 200 nm. This indicates a shared occurrence of both the chitosan matrix and chitosan-TPP nanoparticles in the NCUC sample.

These results were further verified through TEM analysis with size dimensions of the CS-urea NPs embedded in NCUC ranging from 5.67 to 55.67 nm (Figure 1(Ab)). While the overall range of the particle size varied from 69.61 to 154.186 nm in the chitosan sample, the average size of the nanoparticles was 33.39 ± 11.84 and 113.55 ± 19.02 nm in the NCUC and chitosan samples, respectively. The SEM analysis of the 1.5% NCUC (10% w/v) exhibited an increase in the networked structure of the NCUC, probably due to the formation of rodlets and rodlet bundles within the matrix of the lyophilized product (Figure 1(Bd)). The elemental composition of the prepared NCUC, as depicted by the EDS spectra, showed the presence of three major elements in percentage weight and percentage atom composition, i.e., C, N and O elements (Figure 1(Cf)). The urea incorporation enhanced the nitrogen percentage, as indicated by the corresponding increase in percentage weight and percentage atom content for the nitrogen element (Figure 1(Cf)).

The FT-IR spectra of chitosan showed the presence of characteristic absorption peaks for NCUC (Figure 1E). The spectra demonstrated the occurrence of specific bands at 1550–1650 cm^{-1}, 1600–1700 cm^{-1} and 2900 cm^{-1}, which corresponded to the individual amide group vibrations [46,47]. The FT-IR analysis (Figure 1E) revealed the occurrence of peaks at 1634 cm^{-1} in chitosan associated with the stretching vibrations of NH-CO and C=O along with bending bonds of N-H and NH_2 [48], whereas incorporation of urea in chitosan showed sharp and narrow peaks at 1620 cm^{-1} and 1461 cm^{-1}, which are due to amide II bending and C–N stretching vibrations [46]. The urea release profiles of NCUC showed a two-step biphasic process (Figure 1F). These results indicate an initial burst effect up to 10–18 days followed by successive slower release up to the 30th day. The 1.5% NCUC, at 10% (w/v) urea nanoparticle loading, showed an increased release rate of urea compared to the 3 and 5% urea-chitosan formulations. The 1.5% NCUC with 10% urea encapsulated exhibited an 85% encapsulation efficiency. The in vitro biodegradation study involved loss of weight to half by the seventh day for the CS sample (Figure 2). However, the enhanced cross-linking by the addition of urea in the NCUC sample led to a lowered degradation rate compared to the CS sample.

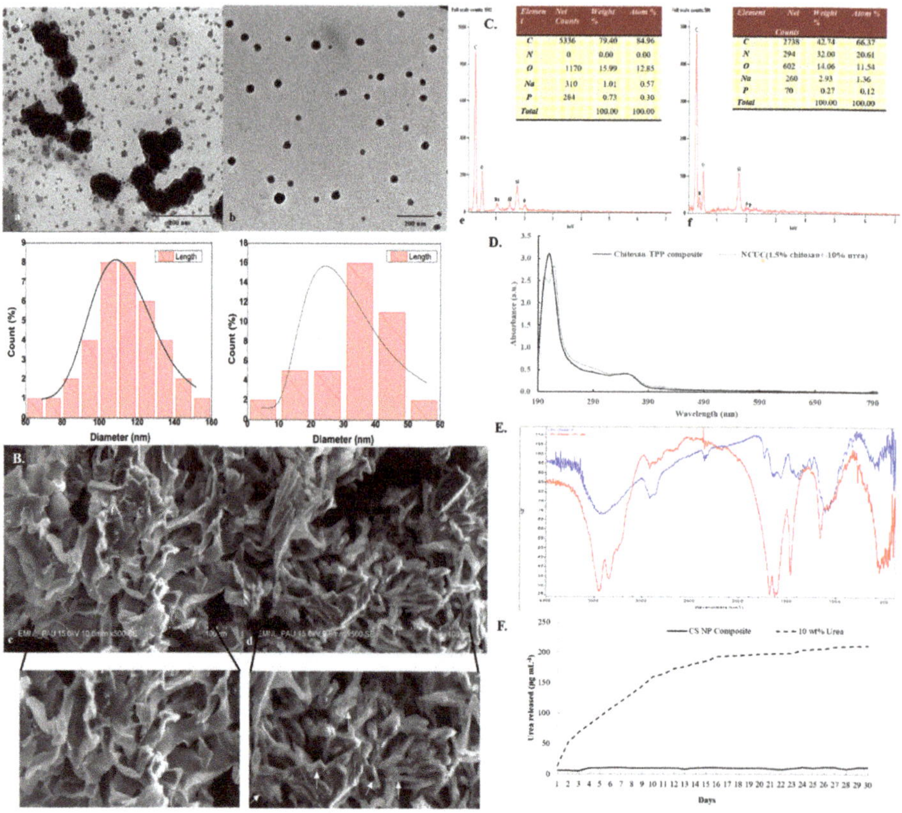

Figure 1. Characterization of chitosan and NCUC. Representative (**A**) TEM micrograph (15.0 K magnification, 80 kV acceleration voltage, high-contrast imaging mode), (**B**) SEM micrograph (500X magnification, 15.0 kV acceleration voltage, secondary electron imaging mode), (**C**) SEM-EDS spectra and elemental composition, (**D**) UV-Vis spectra, (**E**) FT-IR spectra and (**F**) release profile of NCUC. Solid white arrows represent the formation of layered structures in NCUC.

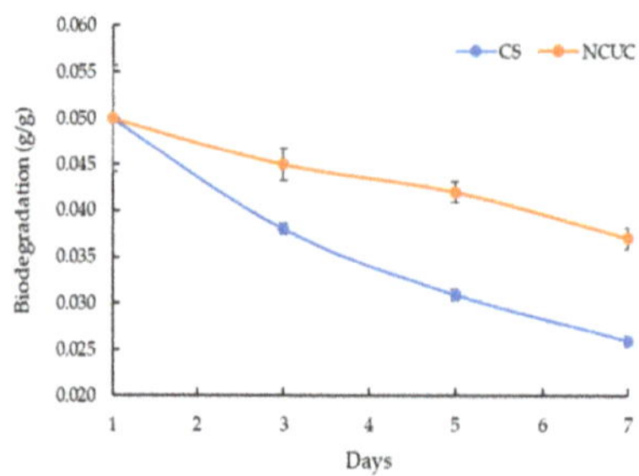

Figure 2. Biodegradation of the chitosan nanoparticles and NCUC samples.

3.2. Soil Chemical and Microbiological Properties

The effect of the slow-release nano-chitosan-urea composite fertilizer on different parameters of soil chemical properties is presented in Tables 1 and 2. The CU recorded the highest ($p \leq 0.05$) soil pH and EC values followed by CS and NCUC. Among the days after treatment, CU showed a significantly higher soil pH at 60 DAT, whereas the highest EC was observed at 90 DAT (Table 2). On the other hand, the OC showed a different trend, with the CS treatment exhibiting the highest OC value (0.35%) at 90 DAT followed by NCUC and CU (Supplementary Table S1).

Table 1. Effect of different types of fertilizer on soil chemical properties and microbial viable cell counts (log cfu g^{-1} soil) during potato cultivation.

Source of Variation	pH	OC (%)	EC (dS/m)	ABC	AOB	NRB	Fungi
DAT							
0	6.50e	0.21e	0.27e	7.45d	4.61e	4.64e	3.83e
20	6.97d	0.27d	0.57d	7.56c	4.83c	4.81c	3.93d
40	7.36b	0.28c	0.59c	7.89b	4.98b	4.92b	4.02c
60	7.62a	0.29a	0.60b	8.02a	5.07a	4.98a	4.14b
90	7.07c	0.28b	0.62a	7.40d	4.78d	4.71d	4.27a
N fertilizer							
CU	7.14a	0.25c	0.54a	7.70b	5.05a	4.95a	4.08b
NCUC	7.12c	0.27b	0.52c	7.81a	4.88b	4.85b	4.16a
CS	7.10b	0.27a	0.53b	7.64c	4.81c	4.76c	4.02c
N level							
0	7.00d	0.23d	0.49d	7.33d	4.64d	4.60d	3.94d
50	7.05c	0.26c	0.51c	7.75c	4.93c	4.86c	4.06c
75	7.16b	0.28b	0.55b	7.85b	5.01b	4.94b	4.14b
100	7.21a	0.29a	0.57a	7.94a	5.08a	5.02a	4.22a

Means within a sub-factor followed by a different letter in a column are significantly different at $p \leq 0.05$ according to pair-wise comparison of least square means.

Table 2. Analysis of variance of days after treatment (DAT), N fertilizer and N level of different types of fertilizers on soil chemical properties during potato cultivation.

SOURCE	DF	pH	OC (%)	EC (dS/m)	ABC	AOB	NRB	Fungi
DAS	4	6.471 ***	0.043 ***	0.788 ***	2.970 ***	0.621 ***	0.501 ***	0.771 ***
Source	2	0.080 ***	0.006 ***	0.006 ***	0.377 ***	0.683 ***	0.397 ***	0.220 ***
DAS*Source	8	0.119 ***	0.005 ***	0.005 ***	0.003 ***	0.003 ***	0.010 ***	0.002ns
N level	3	0.437 ***	0.031 ***	0.054 ***	2.664 ***	1.308 ***	1.218 ***	0.520 ***
DAS*N level	12	0.089 ***	0.002 ***	0.003 ***	0.165 ***	0.027 ***	0.002 ***	0.002ns
Source*N level	6	0.053 ***	0.007 ***	0.001 ***	0.044 ***	0.076 ***	0.051 ***	0.026 ***
DAS*Source*N level	24	0.022 ***	0.001 ***	0.0001 ***	0.002 ***	0.002 ***	0.002 ***	0.001ns

*** = $p \leq 0.001$, ns = non-significant. DF: degree of freedom, OC: organic carbon, EC: electrical conductivity, ABC: aerobic bacterial count, AOB: ammonia-oxidizing bacteria, NRB: nitrate-reducing bacteria.

The soil microbial viable cell populations of total aerobic bacteria, ammonia oxidizers, nitrate-reducing bacteria and fungi in the soil amended with fertilizer showed significant changes among the N fertilizer sources. The count of total aerobic bacteria in the NCUC treatment was significantly higher than that for the CU and CS treatments. On average, the populations of aerobic bacteria in the soil amended with NCUC were 0.11-fold higher than those in the CU treatment. The highest counts of ammonia-oxidizing and nitrate-reducing bacterial populations were observed for CU at 100% concentration-treated soils followed by treatment with NCUC at the same N application level. Contrarily, the CS treatment showed a meager increase in ammonia oxidizers and the nitrate-reducing bacterial count at the 100% N concentration level. The fungal population was the highest in NCUC-treated soils followed by the CU and CS treatments.

The macronutrient status of the soil (i.e., NH_4^+-N, NO_3^--N and K) varied among the different fertilizer sources during potato cultivation (Supplementary Table S2). The CU recorded significantly higher ammonical N (Figure 3a) followed by NCUC and CS over a period of 60 DAS. After 60 DAS, there was a sudden decline in the concentration among all the fertilizers. A similar trend was observed for the nitrate N contents (Figure 3b). Among the DAS, significantly higher nitrate N contents were observed at 60 DAS followed by a gradual decrease for both CU and NCUC treatments, with a greater decrease for the NCUC treatment. Conversely, the highest K (Figure 3c) content in the soil was recorded for the NCUC treatment at 60 DAT followed by CS and CU treatments.

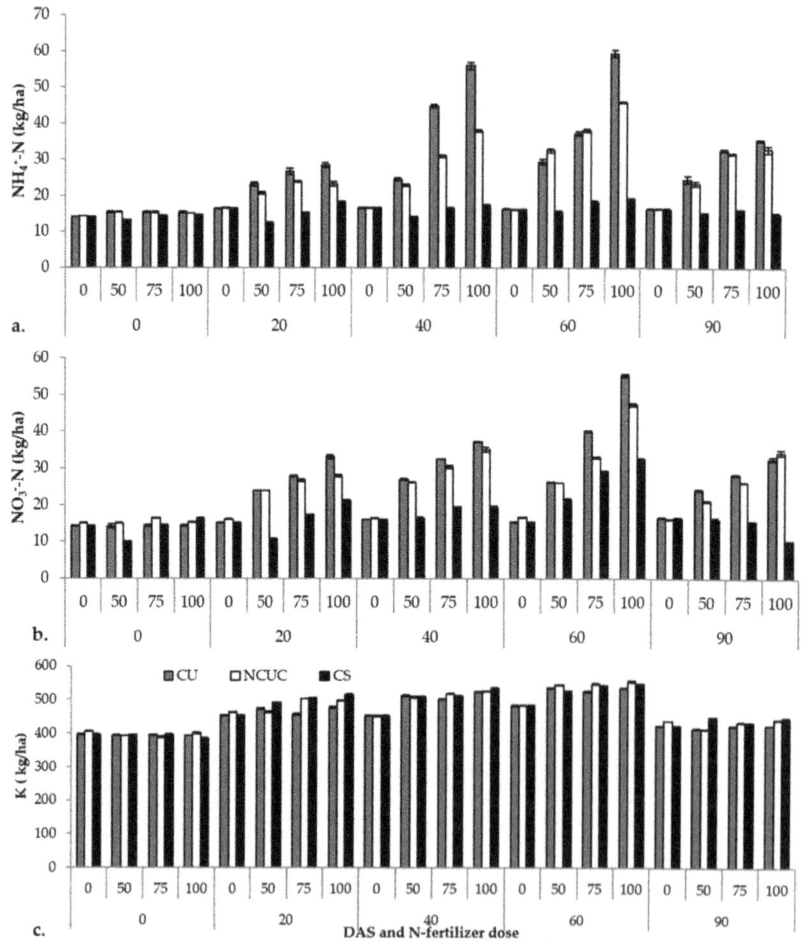

Figure 3. Effect of application of different types of N fertilizers on soil ammonical N (**a**), nitrate N (**b**) and K (**c**) contents of the soil samples under potato cultivation.

3.3. Soil Ammonia-Oxidizing Microbial Population Analyzed through Real-Time PCR (q-PCR)

The relative abundance of the ammonia-oxidizing eubacterial and archaeal gene (amoA) and the 16S rRNA reference gene in response to two fertilizer sources, CU and NCUC, was analyzed by real-time PCR (q-PCR). The results show that ammonia-oxidizing bacteria exhibited a significantly higher abundance in the CU than in the NCUC treatment, and in particular, there was a 2-fold decrease in the amoA gene abundance in NCUC with 75% and 100% of the recommended dose of the fertilizer treatment at 40 and 20 DAT compared to the respective conventional fertilizer treatment (Figure 4). Among the various treatments applied, the 100% N application level showed the highest relative abundance for both archaeal (AOA) and eubacterial (AOB) ammonia-oxidizing bacterial populations. In comparison, the abundance of AOA was found to be relatively lower than that of AOB. Thus, it seems that the AOB population can be considered the predominant N-transforming microbes affecting the nitrification process in the aerobic sandy loam soils.

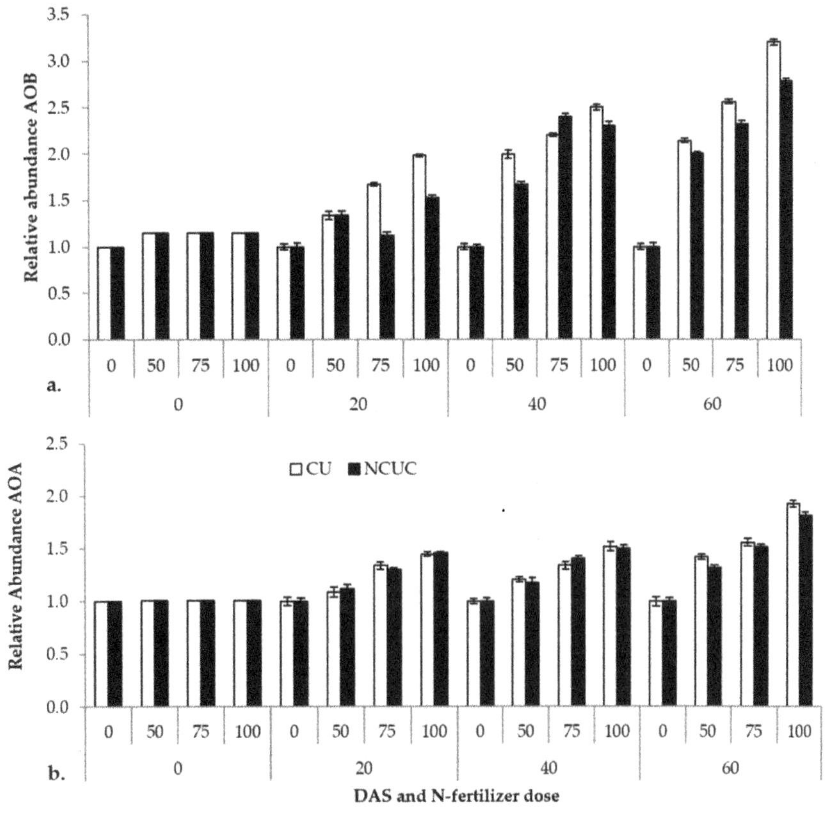

Figure 4. Effect of different types of fertilizer on relative abundance of soil (**a**) ammonia-oxidizing (AOB) and, (**b**) archaeal (AOA) bacteria in potato soil.

3.4. Soil Enzymatic Activities

The various sources of fertilizers were tested for soil enzymatic activities. The days after sowing (DAS), the sources (CU, NCUC and CS) and their concentrations exhibited significant differences in the soil dehydrogenase activity. Among the various sources applied, CSNF showed a sudden ($p \leq 0.05$) increase in dehydrogenase activity followed by NCUC and CU treatments up to 20 DAS, with a slight increase up to 60 DAS (Figure 5a). The highest urease activity was observed in the CU treatment at 60 DAS at the 100% N application level compared to the other treatments (Figure 5b).

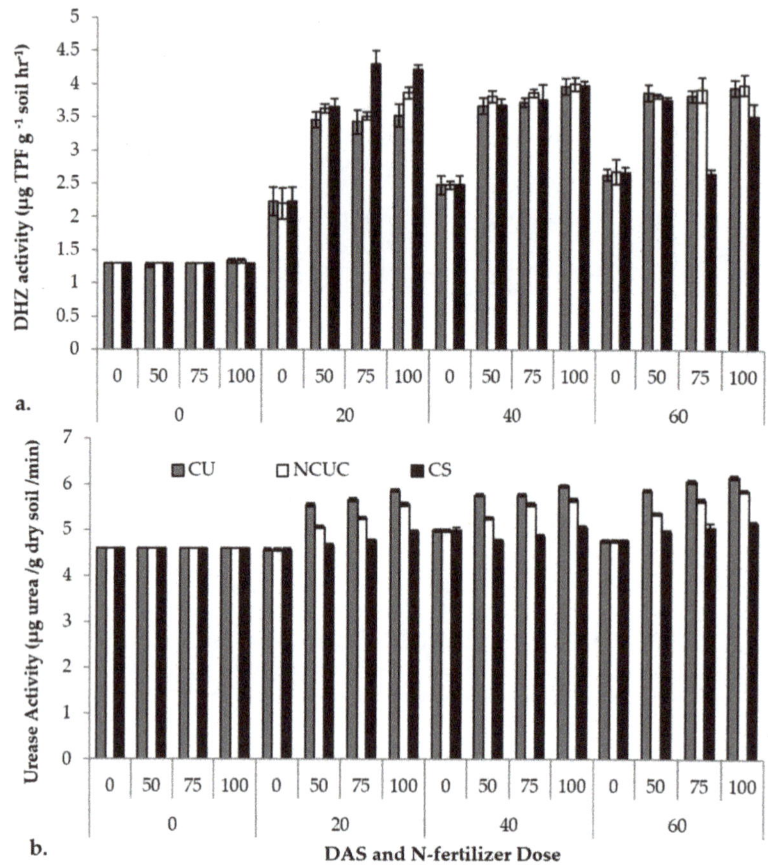

Figure 5. Effect of application of different types of N fertilizers on soil dehydrogenase (**a**) and urease (**b**) activity of the rhizospheric soil of potato cv. Kufri Pukhraj.

3.5. Potato Vegetative Growth and Yield Traits

The shoot growth was significantly enhanced by the NCUC application (Figure 6) compared to the conventional urea treatment at an equivalent dose of application.

Figure 6. Effect of application of conventional and nano-chitosan-urea composite fertilizer on vegetative growth of potato cv. Kufri Pukhraj: (**a**). N level 75% RDF; (**b**). N level 100% RDF.

The shoot fresh and dry weights of the sampled plants were observed to be highest in the NCUC treatment (Table 3). In addition, a dose-dependent effect on the shoot fresh and dry weight parameters was recorded for both the CU and NCUC treatments.

Table 3. Effect of different sources and concentrations of fertilizers on shoot fresh and dry weight (g) parameters.

N Fertilizer Level	Fresh wt.			Dry wt.			Mean
	CU	NCUC	CS	CU	NCUC	CS	
0	18.27 ± 0.10d	18.27 ± 0.10d	18.27 ± 0.10d	15.27 ± 0.14d	15.00 ± 0.25d	15.00 ± 0.25d	16.68
50	23.39 ± 0.36c	28.63 ± 0.12c	23.50 ± 0.21c	19.67 ± 0.19c	24.63 ± 0.17c	18.53 ± 0.19c	23.06
75	26.48 ± 0.03b	31.64 ± 0.21b	24.54 ± 0.23b	23.33 ± 0.06b	28.67 ± 0.12b	19.67 ± 0.15b	25.72
100	27.44 ± 0.41a	35.70 ± 0.08a	26.33 ± 0.38a	24.93 ± 0.19a	32.40 ± 0.12a	23.27 ± 0.20a	28.34
Mean	23.89	28.56	23.16	20.8	25.17	19.12	

Means within a sub-factor followed by a different letter in a column are significantly different at $p \leq 0.05$ according to a pair-wise comparison of least square means.

The tuber yield traits including the tuber number and average tuber weight per plant were also recorded to have improved by application of the nano-chitosan-urea composite. The type of N fertilizer and dose of application exhibited a dual interactive effect (Supplementary Tables S3 and S4) for the above yield traits.

4. Discussion

The TEM studies (Figure 1A) revealed that the addition of urea (10% w/v) in NCUC resulted in a reduction in the nanoparticle sizes, with the size range varying from 11 to 35 nm. This may be due to the pore size of the NCUC matrix, which could have decreased due to the enhanced networking within the matrix. These results were found to be consistent with another study [49] which reported that chitosan particle sizes decreased significantly with increases in the concentrations of bovine serum albumin (entrapped compound). The SEM images further corroborated the TEM results, indicating that the higher concentration of urea (10% w/v) in chitosan resulted in a decrease in the pore size of the NCUC matrix (Figure 1B), which may be due to the enhanced networking followed by the formation of rodlets and rodlet bundles within the matrix. Similar structural changes in ZnO-CTS-Starch and ZnO-CTS-Whey powder NPs have been observed by [50]. The UV-Vis spectroscopy characterization of the NCUC showed that the incorporation of urea in chitosan shifted the absorption peak from 220 to 200 nm [44]. As urea was incorporated, there was shift in the peak to a lower wavelength, which could be attributed to the formation of chito-urea nanoparticles (Figure 1D). The mechanism for the formation of the chito-urea nanoparticles may include step-wise ionic or hydrogen bond interactions among the linear chitosan chains, sodium TPP and urea. Upon drop-wise addition of sodium TPP in a chitosan-acetic acid solution, the NH_3^+ functional group of chitosan exhibited an ionic interaction with the PO_4^- group of tripolyphosphate [51]. The incorporation of urea in a chitosan-acetic acid solution led to its adsorption on the linear chitosan chains through interaction of the amine group(s) in urea with the hydroxyl groups of the chitosan chains [52]. Later, TPP cross-linked the chitosan chains, leading to formation of chito-urea nanoparticles. Previous reports of UV-Vis spectroscopy analysis of chitosan nanoparticles indicated the occurrence of an absorption peak at 217 to 250 nm [53], 226 [54] and 320 [48]. However, a Zn/chitosan NP study provided insights on the blue shift in the absorption peak on increasing the concentration of zinc (Zn) in Zn/chitosan NPs. This possibly indicated the formation of a smaller size of nanoparticles [50], though this may be attributed, partly, to the metallic nature of the ZnO NPs. The major peaks in the FT-IR spectra obtained for the NCUC sample indicated the occurrence of shared peaks for urea and chitosan. The chitosan hydroxyl (O-H) functional group stretching vibrations led to the emergence of peaks at 3450 and 3310 cm^{-1} which have also been observed by Jae-Wook et al. [48]. Likewise, the N-H (amino group) and C-H (alkyl group) bending vibration peaks at 1630 and 1537 cm^{-1} were also identified by them [48].

The release kinetics (Figure 1F) of 1.5% NCUC showed that when the concentration of urea increased, the amount of chitosan might not be sufficient to retain all the urea, due to which some urea molecules might remain on the surface of the chitosan. Likewise, the alpha tocopherol encapsulation study of Luo et al. [55] showed a gradual reduction in the free TOC content released due to the burst effect. It can be inferred that 1.5% NCUC showed the sustainable release of urea within 30 days. The CS and NCUC in vitro biodegradation assay revealed degradation of both the samples, with NCUC exhibiting relatively lowered degradation compared to the CS sample. Likewise, it showed degradation of the chitosan scaffolds [56–58], chitosan fibers [59] and chitosan nanoparticles [60].

The results were found to be significant for the soil chemical properties which is consistent with the hypothesis of the slow release of urea from the chitosan matrix (Table 2). The CU-amended soil showed the highest soil organic carbon compared to NCUC. This might be due to chitosan which acted as a carbon source, and it may have contributed to increasing the organic carbon in the soil. Liu et al. [13] observed high OC in an organic fertilizer treated with 400 kg N ha^{-1} compared to a conventional fertilizer treated with 400 kg N ha^{-1}. On the other hand, EC was recorded to be the highest in CU followed by NCUC and CS. This observation agrees with the findings of Alva et al. [61] in which the EC of the urea-amended soil was 2- to 10-fold greater than the urea-based CRF treatments. The hydrolysis of urea (without coating) and subsequent transformations were fairly rapid. Higher EC values could potentially decrease the water potential of the soil water and thus inhibit plant growth [62]. However, the EC of the sampled soil under different fertilizer treatments varied from 0.27 to 0.68 $dS{\cdot}m^{-1}$. These EC values of the tested soil remained lower than the threshold of a saline soil, i.e., EC = 2.4 $dS{\cdot}m^{-1}$ [63]. Similarly, the pH was recorded to be significantly higher in CU followed by NCUC and CS. The initial increase in the soil pH following the application of urea to the soil was due to hydrolysis of urea into ammonium carbonate, through the action of the urease enzyme [64]. In comparison, the pH values were lower in NCUC, which could be due to the preparation of chitosan-urea nanoformulations in (1% v/v) acetic acid.

The macronutrient profile of the soil also showed significant variations for the three different types of N sources. Among the different sources of fertilizer applied, CU showed an increased propensity for loss of the applied urea due to the poorly developed root system. The concentration of the ammonical N content was substantially lower in the soil treated with NCUC, which could be due to the controlled urea release property of the chitosan polymer (Figure 3a). Similar studies in cotton involving application of controlled-release urea formulations developed by mixing an epoxy resin coating on sulfur-coated urea and polymer-coated urea had significant positive effects on delaying the NH_3 volatilization peaks and reducing the overall NH_3 volatilization in comparison to the urea treatment at the same N application [10]. Likewise, observations on the nitrate contents (Figure 3b) were recorded during the experiment of controlled-release urea, in which the accumulation of nitrate in the soil was significantly lower with polyurethane-coated urea than conventional urea and sulfur-coated urea, which might be due to the controlled release property of the polymer coating [65]. The increase in the concentration of K was supported by Chang et al. [16], who reported an increase in the concentration of the soil K (Figure 3c) content in response to organic fertilizer treatment.

Among the soil microbial populations, the bacterial population at 60 DAT in comparison to 20 and 40 DAT rose significantly (Table 2). This may be due to changes in root secretions that increased the rhizo-deposits and nutrient accessibility to the growing bacteria. These results are in accordance with Daudurand and Knudsen [66], who have reported that bacteria utilize simple amino acids from the rhizosphere of young plants, while they can also utilize complex carbohydrates from the rhizosphere of mature plants, thus indicating that there are modifications in the quality of plant root exudates. Further, the bacterial population can adapt under variable conditions and competition in the niche. The count of ammonia-oxidizing and nitrate-reducing bacteria was lower in NCUC in comparison with CU. The reason behind this can be the slow release of urea from the chitosan polymer, which controlled the N release through the transformation of urea by soil enzymes, besides reducing the emission of N_2O and diminishing the leaching of urea from the root zone. Similar results have been reported by [67], corroborating that the source of ammonia applied to the soils affects the count and activity of the ammonia oxidizers in the applied soils. A high fungal population was observed in soil treated with NCUC as compared to CU since fungi can flourish in environments having a low N content [68], a condition that is exhibited by application of nanofertilizers owing to their slow nutrient release characteristic [69]. Conversely, CU acted as a rapid source of the carbon nutrient for the growing fungal species, which was observed by a rise in the fungal population from 60 to 90 DAT. The opposite was observed for the bacterial population.

Previous studies have shown that the functional groups involved in N cycling will respond differently to environmental changes, but very little information on their relative responsiveness has been discussed [70]. Ammonia-oxidizing bacteria showed a higher abundance in CU- than NCUC-treated soils. This was due to the increased availability of the ammonia concentration for the CU-treated soil (Figure 4a). The increased concentration of ammonia triggered the process of ammonia oxidation [71]. On the other hand, there was slight variation in the abundance of the AOA population in CU- and NCUC-treated soils (Figure 4b). Similar results were reported by Jung et al. [71], who have observed that AOB were more abundant compared to AOA. Thus, in this study, eubacterial ammonia oxidizers were the most dominant bacterial populations in the nitrification process for the NCUC treatment.

For the soil enzyme activity, CU showed the highest dehydrogenase activity at 60 days at the 100% recommended dose of the fertilizer compared to NCUC and CS. The NCUC showed low urease activity compared to CU, which may be due to the decrease in the availability of urea encapsulated in the chitosan polymer and, thus, lower urea transformation rates which may prevent non-target N losses (Figure 5a). Urease activity increased with an increase in the level of urea application [16,72]. The CU treatment showed the highest enzyme activity at 60 DAT in 100% RDF than the NCUC and CS treatments. The NCUC treatment showed low urease activity compared to the conventional urea treatment. This may be attributed to the presence of the polymeric chitosan matrix which decreased the availability of encapsulated urea and, thus, the transformation of urea, thereby preventing N losses. Similarly, Junejo et al. [17] observed a slowdown in urea hydrolysis in different biodegradable urea coatings in comparison to urea, while the dehydrogenase activity was observed to increase in NCUC and CS treatments. The reason for an increase in the dehydrogenase activity in NCUC and CS could probably be due to chitosan, which may have acted as a degradable carbon source. This extra C source must have led to an enhancement in the number of metabolically active bacteria in these treatments. Previous studies on dehydrogenase activities in maize revealed that CRF (polyolefin-coated urea) significantly increased the soil microbial activity compared to conventional urea [16]. Slow- or controlled-release nitrogen fertilizers substantially affect the total marketable tuber yield in potato besides the N use efficiency [73]. In a three-year field study, Ziadi et al. [74] reported an improved tuber yield of two potato cultivars (Chieftain and Goldrush) by application of commercially available CRU at 150 kg ha^{-1}. The researchers argued that this improvement may be attributed to the better availability and uptake of the slowly released N nutrient from the CRF formulation by the root system of the potato plants.

5. Conclusions

In this study, encapsulation of urea with the chitosan polymer was carried out, and the final product was characterized through various microscopy and spectroscopy tools. The NCUC product was applied to potato crops under pot conditions. This formulation helped to obtain a potato yield equivalent to the recommended dose (75 kg N ha^{-1}) by application of 25% of the required level of N fertilizer. Further, NCUC application significantly reduced the viable counts of soil bacteria (3.36 and 2.02% AOB and NRB, respectively) involved in N transformation and caused a numerical decrease in the urease enzyme activity in the test soil. The soil chemical properties showed significant changes in response to different fertilizer treatments as the pH, EC and OC were negatively and positively influenced by the NCUC treatment. The soil ammonical N and nitrate N contents were low in the NCUC compared to the CU treatment, i.e., 10.69 and 4.55%, respectively. The low concentrations of ammonia and nitrate in the soil supported the slow process of nitrification and controlled release of urea. These results are also corroborated by the reduced population of ammonia-oxidizing and nitrate-reducing bacteria in the soil amended with the NCUC fertilizer. Furthermore, the results show that the relative abundance of AOB and AOA was lower in response to the NCUC treatment. This low abundance of nitrogen-cycling microbial populations in the NCUC-treated soil supports the lower accumulation of nitrate in the soil, and thus its effect on the nitrification and denitrification processes. Thus, the abundance of nitrogen-cycling microbial communities decreased markedly due to the efficient controlled release of the N nutrient by the polymer-encapsulated fertilizer.

Supplementary Materials: The following are available online at https://www.mdpi.com/article/10.3390/polym13172887/s1, Table S1: Effect of application of different concentrations of nano-chitosan-urea composite and conventional urea fertilizer on the soil chemical properties, Table S2: Effect of application of different concentrations of nano-chitosan-urea composite and conventional urea fertilizer on the percent nitrogen content in plant root, shoot and tuber tissues, Table S3: Analysis of variance of N fertilizer source and dose of application (N level) on vegetative and yield attributes of potato cv. Kufri Pukhraj. Table S4: Effect of application of different concentrations of nano-chitosan-urea composite and conventional urea fertilizer on the yield attributes of potato.

Author Contributions: Conceptualization, A.K., C.L.G., K.K., K.A.A.-E. and S.P.S.; methodology, A.K., C.L.G., Y.V., G.S.D. and M.S.T.; validation, A.K., R.K. and K.L.; formal analysis, R.K., K.L., A.K. and O.K.; investigation, R.K., K.L., A.K. and S.P.S.; resources, A.K.; data curation, A.K., R.K. and K.L.; writing—original draft preparation, A.K. and R.K.; writing—review and editing, C.L.G. and K.A.A.-E.; visualization, A.K., C.L.G., S.P.S. and O.K.; supervision, A.K.; project administration, A.K.; funding acquisition, A.K., S.P.S., K.K. and K.A.A.-E. All authors have read and agreed to the published version of the manuscript.

Funding: This research was funded by Rashtriya Krishi Vikas Yojna (RKVY), Government of India, and by Excellence project PrF UHK 2011/2021-2022.

Institutional Review Board Statement: Not applicable.

Informed Consent Statement: Not applicable.

Data Availability Statement: The data are included in the article.

Acknowledgments: The authors graciously thank the head of the Department of Soil Science and Department of Vegetable Science for the infrastructural facilities. A sincere thank you goes to the Director of Research for the allocation of the funds in the RKVY scheme.

Conflicts of Interest: The authors declare no conflict of interest.

References

1. Cao, W.; Tibbitts, T.W. Study of various NH4+/NO3-mixtures for enhancing growth of potatoes. *J. Plant Nutr.* **1993**, *16*, 1691–1704. [CrossRef] [PubMed]
2. Silva, J.G.; França, M.G.C.; Gomide, F.T.F.; Magalhaes, J.R. Different Nitrogen Sources Affect Biomass Partitioning and Quality of Potato Production in a Hydroponic System. *Am. J. Potato Res.* **2013**, *90*, 179–185. [CrossRef]
3. Bundy, L.G.; Andraski, T.W. Recovery of Fertilizer Nitrogen in Crop Residues and Cover Crops on an Irrigated Sandy Soil. *Soil Sci. Soc. Am. J.* **2005**, *69*, 640. [CrossRef]
4. Laboski, C.A.M.; Kelling, K.A. Influence of fertilizer management and soil fertility on tuber specific gravity: A review. *Am. J. Potato Res.* **2007**, *84*, 283–290. [CrossRef]
5. Shoji, S.; Delgado, J.; Mosier, A.; Miura, Y. Use of controlled release fertilizers and nitrification inhibitors to increase nitrogen use efficiency and to conserve air and water quality. *Commun. Soil Sci. Plant Anal.* **2001**, *32*, 1051–1070. [CrossRef]
6. Joern, B.C.; Vitosh, M.L. Influence of applied nitrogen on potato Part I: Yield, quality, and nitrogen uptake. *Am. Potato J.* **1995**, *72*, 51–63. [CrossRef]
7. DeRosa, M.C.; Monreal, C.; Schnitzer, M.; Walsh, R.; Sultan, Y. Nanotechnology in fertilizers. *Nat. Nanotechnol.* **2010**, *5*, 91. [CrossRef]
8. Naderi, M.R.; Danesh-Shahraki, A. Nanofertilizers and their roles in sustainable agriculture. *Int. J. Agric. Crop Sci.* **2013**, *IJACS/2013*, 2229–2232.
9. Zheng, W.; Sui, C.; Liu, Z.; Geng, J.; Tian, X.; Yang, X.; Li, C.; Zhang, M. Long-Term effects of controlled-release urea on crop yields and soil fertility under wheat–corn double cropping systems. *Agron. J.* **2016**, *108*, 1703–1716. [CrossRef]

10. Tian, X.; Geng, J.; Guo, Y.; Li, C.; Zhang, M.; Chen, J. Controlled-release urea decreased ammonia volatilization and increased nitrogen use efficiency of cotton. *J. Plant Nutr. Soil Sci.* **2017**, *180*, 667–675. [CrossRef]
11. Knight, E.C.; Guertal, E.A.; Wood, C.W. Mowing and nitrogen source effects on ammonia volatilization from turfgrass. *Crop Sci.* **2007**, *47*, 1628–1634. [CrossRef]
12. Norton, J.M.; Schepers, J.S.; Raun, W.R. Nitrification in Agricultural Soils. In *Nitrogen in Agricultural Soils*; Raun, W.R., Schepers, J.S., Eds.; American Society for Agronomy, CSSA SSSA Inc.: Madison, WI, USA, 2008; pp. 173–199.
13. Liu, C.W.; Sung, Y.; Chen, B.C.; Lai, H.Y. Effects of nitrogen fertilizers on the growth and nitrate content of lettuce (Lactuca sativa L.). *Int. J. Environ. Res. Public Health* **2014**, *11*, 4427–4440. [CrossRef] [PubMed]
14. Meena, A.; Rao, K.S. Assessment of soil microbial and enzyme activity in the rhizosphere zone under different land use/cover of a semiarid region, India. *Ecol. Process.* **2021**, *10*, 1–12. [CrossRef]
15. Bhowmik, A.; Kukal, S.S.; Saha, D.; Sharma, H.; Kalia, A.; Sharma, S. Potential indicators of soil health degradation in different land use-based ecosystems in the shiwaliks of northwestern India. *Sustainability* **2019**, *11*, 3908. [CrossRef]
16. Chang, E.H.; Chung, R.S.; Tsai, Y.H. Effect of different application rates of organic fertilizer on soil enzyme activity and microbial population. *Soil Sci. Plant Nutr.* **2007**, *53*, 132–140. [CrossRef]
17. Junejo, N.; Khanif, M.Y.; Dharejo, K.A.; Abdu, A.; Abdul-Hamid, H. A field evaluation of coated urea with biodegradable materials and selected urease inhibitors. *Afr. J. Biotechnol.* **2011**, *10*, 19729–19736. [CrossRef]
18. Corradini, E.; de Moura, M.R.; Mattoso, L.H.C. A preliminary study of the incorparation of NPK fertilizer into chitosan nanoparticles. *Express Polym. Lett.* **2010**, *4*, 509–515. [CrossRef]
19. Valášková, M.; Rieder, M.; Matějka, V.; Čapková, P.; Slíva, A. Exfoliation/delamination of kaolinite by low-temperature washing of kaolinite-urea intercalates. *Appl. Clay Sci.* **2007**, *35*, 108–118. [CrossRef]
20. Gao, X.; Li, C.; Zhang, M.; Wang, R.; Chen, B. Controlled release urea improved the nitrogen use efficiency, yield and quality of potato (*Solanum tuberosum* L.) on silt loamy soil. *Field Crop. Res.* **2015**, *181*, 60–68. [CrossRef]
21. Garavand, F.; Cacciotti, I.; Vahedikia, N.; Rehman, A.; Tarhan, Ö.; Akbari-Alavijeh, S.; Shaddel, R.; Rashidinejad, A.; Nejatian, M.; Jafarzadeh, S.; et al. A comprehensive review on the nanocomposites loaded with chitosan nanoparticles for food packaging. *Crit. Rev. Food Sci. Nutr.* **2020**, 1–34. [CrossRef] [PubMed]
22. Lawrencia, D.; Wong, S.K.; Low, D.Y.S.; Goh, B.H.; Goh, J.K.; Ruktanonchai, U.R.; Soottitantawat, A.; Lee, L.H.; Tang, S.Y. Controlled release fertilizers: A review on coating materials and mechanism of release. *Plants* **2021**, *10*, 238. [CrossRef] [PubMed]
23. Cacciotti, I.; Lombardelli, C.; Benucci, I.; Esti, M. Clay/chitosan biocomposite systems as novel green carriers for covalent immobilization of food enzymes. *J. Mater. Res. Technol.* **2019**, *8*, 3644–3652. [CrossRef]
24. Talarico, D.; Arduini, F.; Amine, A.; Cacciotti, I.; Moscone, D.; Palleschi, G. Screen-printed electrode modified with carbon black and chitosan: A novel platform for acetylcholinesterase biosensor development. *Anal. Bioanal. Chem.* **2016**, *408*, 7299–7309. [CrossRef]
25. Barbieri, M.; Cellini, F.; Cacciotti, I.; Peterson, S.D.; Porfiri, M. In situ temperature sensing with fluorescent chitosan-coated PNIPAAm/alginate beads. *J. Mater. Sci.* **2017**, *52*, 12506–12512. [CrossRef]
26. Ayub, G.S.E.; Rocha, S.C.S.; Perrucci, A.L.I. Analysis of the surface quality of sulphur-coated urea particles in a two-dimensional spouted bed. *Braz. J. Chem. Eng.* **2001**, *18*, 13–22. [CrossRef]
27. Rajan, M.; Shahena, S.; Chandran, V.; Mathew, L. Controlled release of fertilizers—Concept, reality, and mechanism. In *Controlled Release Fertilizers for Sustainable Agriculture*; Elsevier: Amsterdam, The Netherlands, 2021; pp. 41–56. ISBN 9780128195550.
28. Jarudilokkul, S.; Tongthammachat, A.; Boonamnuayvittaya, V. Preparation of chitosan nanoparticles for encapsulation and release of protein. *Korean J. Chem. Eng.* **2011**, *28*, 1247–1251. [CrossRef]
29. Kamat, V.; Bodas, D.; Paknikar, K. Chitosan nanoparticles synthesis caught in action using microdroplet reactions. *Sci. Rep.* **2016**, *6*, 8–11. [CrossRef]
30. Yan, J.; Guan, Z.Y.; Zhu, W.F.; Zhong, L.Y.; Qiu, Z.Q.; Yue, P.F.; Wu, W.T.; Liu, J.; Huang, X. Preparation of puerarin chitosan oral nanoparticles by ionic gelation method and its related kinetics. *Pharmaceutics* **2020**, *12*, 216. [CrossRef]
31. Vahedikia, N.; Garavand, F.; Tajeddin, B.; Cacciotti, I.; Jafari, S.M.; Omidi, T.; Zahedi, Z. Biodegradable zein film composites reinforced with chitosan nanoparticles and cinnamon essential oil: Physical, mechanical, structural and antimicrobial attributes. *Colloids Surf. B Biointerfaces* **2019**, *177*, 25–32. [CrossRef]
32. Giraldo, J.D.; Rivas, B.L. Determination of urea using p-N,N-dimethylaminobenzaldehyde: Solvent effect and interference of chitosan. *J. Chil. Chem. Soc.* **2017**, *62*, 3538–3542. [CrossRef]
33. Jackson, M.L. *Soil Chemical Analysis*; Prentice Hall, Inc.: Englewood Cliffs, NJ, USA, 1958.
34. Kjeldhal, J. A new method for the estimation of nitrogen in organic compounds. *Ann. Chem.* **1883**, *22*, 366–382.
35. Reed, M.G.; Scott, A.D. Flame photometric methods of determining the potassium in potassium tetraphenylborate. *Anal. Chem.* **1961**, *33*, 773–775.
36. Throndsen, J. Preservation and storage. In *Phytoplankton Manual Monographs on Oceanographic Methodology*; Sournia, A., Ed.; UNESCO: Paris, France, 1978; pp. 70–71.
37. Casida, J.E. Biochemistry of pyrethrins. In *Pyrethrum the natural insecticide*; Casida, J.E., Ed.; Academic Press: Cambridge, MA, USA, 1973; pp. 101–120.
38. Bremner, J.M.; Douglas, L.A. Inhibition of urease activity in soils. *Soil Biol. Biochem.* **1971**, *3*, 297–307. [CrossRef]

39. Hoshino, T.; Noda, N.; Tsuneda, S.; Hirata, A.; Inamori, Y. Direct Detection by In Situ PCR of the amoA Gene in Biofilm Resulting from a Nitrogen Removal Process. *Appl. Environ. Microbiol.* **2001**, *67*, 5261–5266. [CrossRef] [PubMed]
40. Marusenko, Y.; Bates, S.T.; Anderson, I.; Johnson, S.L.; Soule, T.; Garcia-Pichel, F. Ammonia-oxidizing archaea and bacteria are structured by geography in biological soil crusts across North American arid lands. *Ecol. Process.* **2013**, *2*, 1–10. [CrossRef]
41. Das, S.; Mitra, B.; Luthra, S.K.; Saha, A.; Hassan, M.M.; Hossain, A. Study on Morphological, Physiological Characteristics and Yields of Twenty-One Potato (Solanum tuberosum L.) Cultivars Grown in Eastern Sub-Himalayan Plains of India. *Agronomy* **2021**, *11*, 335. [CrossRef]
42. Steel, R.G.D.; Torrie, J.H. *Principles and Procedures of Statistics: A Biometrical Approach*, 2nd ed.; McGraw-Hill Book Co.: New York, NY, USA, 1980; ISBN 9780070609266.
43. Sharma, S.P.; Leskovar, D.I.; Volder, A.; Crosby, K.M.; Ibrahim, A.M.H. Root distribution patterns of reticulatus and inodorus melon (Cucumis melo L.) under subsurface deficit irrigation. *Irrig. Sci.* **2018**, *36*, 301–317. [CrossRef]
44. AbdElhady, M.M. Preparation and Characterization of Chitosan/Zinc Oxide Nanoparticles for Imparting Antimicrobial and UV Protection to Cotton Fabric. *Int. J. Carbohydr. Chem.* **2012**, *2012*, 1–6. [CrossRef]
45. Liu, D.; Wei, Y.; Yao, P.; Jiang, L. Determination of the degree of acetylation of chitosan by UV spectrophotometry using dual standards. *Carbohydr. Res.* **2006**, *341*, 782–785. [CrossRef] [PubMed]
46. Hussain, M.R.; Devi, R.R.; Maji, T.K. Controlled release of urea from chitosan microspheres prepared by emulsification and cross-linking method. *Iran. Polym. J.* **2012**, *21*, 473–479. [CrossRef]
47. Benucci, I.; Liburdi, K.; Cacciotti, I.; Lombardelli, C.; Zappino, M.; Nanni, F.; Esti, M. Chitosan/clay nanocomposite films as supports for enzyme immobilization: An innovative green approach for winemaking applications. *Food Hydrocoll.* **2018**, *74*, 124–131. [CrossRef]
48. Oh, J.W.; Chun, S.C.; Chandrasekaran, M. Preparation and in vitro characterization of chitosan nanoparticles and their broad-spectrum antifungal action compared to antibacterial activities against phytopathogens of tomato. *Agronomy* **2019**, *9*, 21. [CrossRef]
49. Gan, Q.; Wang, T. Chitosan nanoparticle as protein delivery carrier-Systematic examination of fabrication conditions for efficient loading and release. *Colloids Surf. B Biointerfaces* **2007**, *59*, 24–34. [CrossRef]
50. Dhillon, G.S.; Kaur, S.; Brar, S.K. Facile fabrication and characterization of chitosan-based zinc oxide nanoparticles and evaluation of their antimicrobial and antibiofilm activity. *Int. Nano Lett.* **2014**, *4*. [CrossRef]
51. Sreekumar, S.; Goycoolea, F.M.; Moerschbacher, B.M.; Rivera-Rodriguez, G.R. Parameters influencing the size of chitosan-TPP nano- and microparticles. *Sci. Rep.* **2018**, *8*, 4695. [CrossRef]
52. Xue, C.; Wilson, L.D. Kinetic study on urea uptake with chitosan based sorbent materials. *Carbohydr. Polym.* **2016**, *135*, 180–186. [CrossRef]
53. Vijayalakshmi, V.; Hina Kousar, P.A.; Das, S. Optimization and characterization of chitosan based nanocarrier for the application of cancer drug delivery. *J. Crit. Rev.* **2020**, *7*, 762–769. [CrossRef]
54. Agarwal, M.; Agarwal, M.K.; Shrivastav, N.; Pandey, S.; Das, R.; Gaur, P. Preparation of Chitosan Nanoparticles and their In-vitro Characterization. *Int. J. Life-Sci. Sci. Res.* **2018**, *4*, 1713–1720. [CrossRef]
55. Luo, Y.; Zhang, B.; Whent, M.; Yu, L.L.; Wang, Q. Preparation and characterization of zein/chitosan complex for encapsulation of α-tocopherol, and its in vitro controlled release study. *Colloids Surf. B Biointerfaces* **2011**, *85*, 145–152. [CrossRef] [PubMed]
56. Liu, Y.; Zhou, C.; Sun, Y. A biomimetic strategy for controllable degradation of chitosan scaffolds. *J. Mater. Res.* **2012**, *27*, 1859–1868. [CrossRef]
57. Cunha-Reis, C.; Tuzlakoglu, K.; Baas, E.; Yang, Y.; El Haj, A.; Reis, R.L. Influence of porosity and fibre diameter on the degradation of chitosan fibre-mesh scaffolds and cell adhesion. *J. Mater. Sci. Mater. Med.* **2007**, *18*, 195–200. [CrossRef] [PubMed]
58. Thein-Han, W.W.; Kitiyanant, Y. Chitosan Scaffolds for In Vitro Buffalo Embryonic Stem-Like Cell Culture: An Approach to Tissue Engineering. *J. Biomed. Mater. Res.* **2007**, *80B*, 92–101. [CrossRef]
59. Lu, G.; Kong, L.; Sheng, B.; Wang, G.; Gong, Y.; Zhang, X. Degradation of covalently cross-linked carboxymethyl chitosan and its potential application for peripheral nerve regeneration. *Eur. Polym. J.* **2007**, *43*, 3807–3818. [CrossRef]
60. Islam, N.; Dmour, I.; Taha, M.O. Degradability of chitosan micro/nanoparticles for pulmonary drug delivery. *Heliyon* **2019**, *5*, e01684. [CrossRef]
61. Alva, A.K.; Hodges, T.; Boydston, R.A.; Collins, H.P. Effects of irrigation and tillage practices on yield of potato under high production conditions in the pacific northwest. *Commun. Soil Sci. Plant Anal.* **2002**, *33*, 1451–1460. [CrossRef]
62. Ostonen, I.; Püttsepp, Ü.; Biel, C.; Alberton, O.; Bakker, M.R.; Lõhmus, K.; Majdi, H.; Metcalfe, D.; Olsthoorn, A.F.M.; Pronk, A.; et al. Specific root length as an indicator of environmental change. *Plant Biosyst.* **2007**, *141*, 426–442. [CrossRef]
63. Rhoades, J.D.; Miyamoto, S. Testing Soils for Salinity and Sodicity. In *Soil Testing and Plant Analysis*; Westerman, R.L., Ed.; Soil Science Society of America Inc.: Madison, WI, USA, 1990; pp. 299–336.
64. Paramasivam, S.; Alva, A.K. Leaching of nitrogen forms from controlled-release nitrogen fertilizers. *Commun. Soil Sci. Plant Anal.* **1997**, *28*, 1663–1674. [CrossRef]
65. Zhang, K.; Xu, J.; Wang, K.Y.; Cheng, L.; Wang, J.; Liu, B. Preparation and characterization of chitosan nanocomposites with vermiculite of different modification. *Polym. Degrad. Stab.* **2009**, *94*, 2121–2127. [CrossRef]

66. Dandurand, L.M.C.; Knudsen, G.R. Sampling microbes from the rhizosphere and phyllosphere. In *Manual of Environmental Microbiology*; Hurst, C.J., Crawford, R.L., Knudsen, G.R., McInerney, M.J., Stetzenbach, L.D., Eds.; American Society for Microbiology (ASM): Washington, DC, USA, 2002; pp. 516–526.
67. Prosser, J.I.; Nicol, G.W. Archaeal and bacterial ammonia-oxidisers in soil: The quest for niche specialisation and differentiation. *Trends Microbiol.* **2012**, *20*, 523–531. [CrossRef] [PubMed]
68. de Vries, F.T.; Hoffland, E.; van Eekeren, N.; Brussaard, L.; Bloem, J. Fungal/bacterial ratios in grasslands with contrasting nitrogen management. *Soil Biol. Biochem.* **2006**, *38*, 2092–2103. [CrossRef]
69. Pereira, E.I.; Da Cruz, C.C.T.; Solomon, A.; Le, A.; Cavigelli, M.A.; Ribeiro, C. Novel slow-release nanocomposite nitrogen fertilizers: The impact of polymers on nanocomposite properties and function. *Ind. Eng. Chem. Res.* **2015**, *54*, 3717–3725. [CrossRef]
70. Zhang, X.; Liu, W.; Schloter, M.; Zhang, G.; Chen, Q.; Huang, J.; Li, L.; Elser, J.J.; Han, X. Response of the Abundance of Key Soil Microbial Nitrogen-Cycling Genes to Multi-Factorial Global Changes. *PLoS ONE* **2013**, *8*, 2–11. [CrossRef] [PubMed]
71. Jung, J.; Yeom, J.; Kim, J.; Han, J.; Lim, H.S.; Park, H.; Hyun, S.; Park, W. Change in gene abundance in the nitrogen biogeochemical cycle with temperature and nitrogen addition in Antarctic soils. *Res. Microbiol.* **2011**, *162*, 1018–1026. [CrossRef] [PubMed]
72. Dharmakeerthi, R.S.; Thenabadu, M.W. Urease activity in soils: A review. *J. Natl. Sci. Ctries. Sri Lanka* **1996**, *24*, 159–195. [CrossRef]
73. Simonne, E.H.; Hutchinson, C.M. Controlled-release Fertilizers for Vegetable Production in the Era of Best Management Practices: Teaching New Tricks to an Old Dog. *Horttechnology* **2005**, *15*, 36–46. [CrossRef]
74. Ziadi, N.; Grant, C.; Samson, N.; Nyiraneza, J.; Bélanger, G.; Parent, L.É. Efficiency of controlled-release urea for a potato production system in Quebec, Canada. *Agron. J.* **2011**, *103*, 60–66. [CrossRef]

polymers

MDPI

Article

Thermal Stability and Decomposition Mechanism of PLA Nanocomposites with Kraft Lignin and Tannin

Nina Maria Ainali [1,2], Evangelia Tarani [3], Alexandra Zamboulis [1], Klementina Pušnik Črešnar [4], Lidija Fras Zemljič [4], Konstantinos Chrissafis [3], Dimitra A. Lambropoulou [2] and Dimitrios N. Bikiaris [1,*]

1 Laboratory of Chemistry and Technology of Polymers and Dyes, Department of Chemistry, Aristotle University of Thessaloniki, GR54124 Thessaloniki, Greece; nsainali@chem.auth.gr (N.M.A.); azamboulis@gmail.com (A.Z.)
2 Laboratory of Environmental Pollution Control, Department of Chemistry, Aristotle University of Thessaloniki, GR54124 Thessaloniki, Greece; dlambro@chem.auth.gr
3 Department of Physics, Aristotle University of Thessaloniki, GR54124 Thessaloniki, Greece; etarani@physics.auth.gr (E.T.); hrisafis@physics.auth.gr (K.C.)
4 Faculty of Mechanical Engineering, University of Maribor, 2000 Maribor, Slovenia; klementina.pusnik@um.si (K.P.Č.); lidija.fras@um.si (L.F.Z.)
* Correspondence: dbic@chem.auth.gr

Citation: Ainali, N.M.; Tarani, E.; Zamboulis, A.; Črešnar, K.P.; Zemljič, L.F.; Chrissafis, K.; Lambropoulou, D.A.; Bikiaris, D.N. Thermal Stability and Decomposition Mechanism of PLA Nanocomposites with Kraft Lignin and Tannin. *Polymers* **2021**, *13*, 2818. https://doi.org/10.3390/polym13162818

Academic Editors: Cédric Delattre and Beom-Soo Kim

Received: 31 July 2021
Accepted: 20 August 2021
Published: 22 August 2021

Abstract: Packaging applications cover approximately 40% of the total plastics production, whereas food packaging possesses a high proportion within this context. Due to several environmental concerns, petroleum-based polymers have been shifted to their biobased counterparts. Poly(lactic acid) (PLA) has been proved the most dynamic biobased candidate as a substitute of the conventional polymers. Despite its numerous merits, PLA exhibits some limitations, and thus reinforcing agents are commonly investigated as fillers to ameliorate several characteristics. In the present study, two series of PLA-based nanocomposites filled with biobased kraft-lignin (KL) and tannin (T) in different contents were prepared. A melt–extrusion method was pursued for nanocomposites preparation. The thermal stability of the prepared nanocomposites was examined by Thermogravimetric Analysis, while thermal degradation kinetics was applied to deepen this process. Pyrolysis–Gas Chromatography/Mass Spectrometry was employed to provide more details of the degradation process of PLA filled with the two polyphenolic fillers. It was found that the PLA/lignin nanocomposites show better thermostability than neat PLA, while tannin filler has a small catalytic effect that can reduce the thermal stability of PLA. The calculated $E\alpha$ value of PLA-T nanocomposite was lower than that of PLA-KL resulting in a substantially higher decomposition rate constant, which accelerate the thermal degradation.

Keywords: poly(lactic acid); nanocomposites; tannin; lignin; thermal degradation kinetics; decomposition mechanism; pyrolysis

1. Introduction

Poly(lactic acid) (PLA), the so-called "polymer of the 21st century", is one of the most dynamic polymers conforming to a circular and green economy and applied in a wide spectrum of applications due to its exceptional peculiarities [1–3]. Among them, its biodegradability, biocompatibility, bioabsorption, versatility and advanced mechanical properties are outpaced, whereas its origin from biomass including corn, sugarcane and other renewable resources should also be remarked upon. One of the most sufficient pathways to synthesize PLA includes the ring opening polymerization of lactide, which is formed by lactic acid; a product of agricultural compounds fermentation [4]. The extended range of its applications can be outlined by its use from the biomedical and bioengineering field [5–7] to packaging and everyday-life products [1,8–10]. In a more specific context, this aliphatic polyester has been employed for the construction of 3D-printed structures utilized to fulfil several purposes [11–13], whereas its application for the preparation of materials with advanced antimicrobial and antioxidant surfaces should also be highlighted [8,14–16].

However, since every coin has two sides, PLA downsides are mainly centred on its poor hydrophilicity and toughness, inadequate gas barrier properties, slow degradation rate, low thermal conductivity and weak crystallization [14,17]. Since properties such as mechanical and enzymatic hydrolysis are dependent on the crystallinity of the polymer, it can be concluded that crystallinity of PLA-possessing a commonly semicrystalline nature affects significantly its performance. In fact, polymeric chain architecture greatly affects the ability of nucleation, and thus, the value of crystallinity. Concerning the PLA counterpart, the L-/D- lactide isomers ratio and the macromolecular extent also play a crucial role in the degree of crystallinity [18]. To overcome the limitations attributed to crystallinity and tune the relative properties desired for specific applications, scientific research has been directed towards the synthesis of PLA nanocomposites with several types of nanoscale fillers, including metals, metal oxides, carbon nanotubes and several other clays [10,19–28]. In general, nano-sized fillers are proved to be excellent agents to boost several properties of polymers, including thermal stability, degradation efficiency, optical, mechanical and permeation properties [29–32]. In this context, several studies were published concerning the combination of lignin with PLA, since such composite materials exhibit promising characteristics including antioxidant, antifungal and antimicrobial properties; resistance to UV exposure; and fire-retardant capacities [33]. Tannins were also proven to be dynamic polymer additives due to their antioxidant, stabilizing and UV-protective characteristics; nevertheless, tannins were scarcely explored within the context of PLA composites [34]. Nevertheless, some limitations and gaps occur in the field, since the preparation of composite materials with synergistic and advanced properties is a tough aspect.

In detail concerning the relative works, Spiridon et al. [35] and Fereira da Silva et al. [36] have been reported that the addition of lignins into the PLA matrix cause an increase in thermal stability of PLA because of the good adhesion between matrix and filler. However, in both cases, a deterioration in the tensile and impact strength of the composites was demonstrated, especially for fillers loading above 5 wt%, whereas in the second case a decrease in the degree of crystallinity was also observed. Park et al. [37] synthesized PLA-lignin and lignin plasticized composites. They found that the filler decreases the thermal degradation rate against temperature. Zhai et al. [38] demonstrated that the onset of thermal degradation temperature and maximum thermal degradation temperature of PLA composites decrease with increasing filler content. Concerning the PLA-T composites, Anwer et al. [39] and Liao et al. [40] have shown that the onset of thermal degradation of PLA-T composites occurs at a slightly lower temperatures than neat PLA.

Previously, our team reported on the preparation of eight PLA-based nanocomposites with two polyphenolic and natural fillers, namely, kraft-lignin (KL) and tannin (T) at several contents, for the amelioration of their antioxidant character [19]. The detailed preparation, as well as the characterization of the PLA composites with lignin and tannin by several methods, was described previously [19]. Since the results of the aforementioned study revealed the effective addition of the two fillers into the PLA-matrix through the strong suppression of the crystallization time and before the investigation of antioxidant behaviour of the most promising composite materials, the examination of thermal stability, thermal degradation pathway and kinetic parameters has been also considered crucial for the thermal stability control.

The main purpose of the current manuscript was to observe and report thoroughly for the first time the decomposition mechanism of PLA in the presence of KL and T, that could be potential appropriate for packaging applications. To achieve this goal, thermogravimetric analysis (TGA) was initially employed in order to study the effect of the two different polyphenolic fillers on the thermal stability of the PLA matrix and to select the proper temperatures for the pyrolysis tests. Pyrolysis–gas chromatography/mass spectrometry (Py–GC/MS) was also utilized for the analysis of the compounds during the thermal decomposition and the representation of the detailed degradation mechanism of the samples. Additionally, the isoconversional differential method and multivariate non-linear regression method were used in order to determine the effective activation energy and the kinetic triplet (the conversion function f(α), the pre-exponential parameter A and the activation energy) of the degradation reactions. It is noteworthy to mention that the investigation and identification of degradation products are critical for the research of new materials, because they can be formed during treatment processes, and could lead to undesired colouring. A thorough understanding of the degradation mechanism could enhance the selection of thermal stability enhancing agents.

2. Materials and Methods

2.1. Materials

For the preparation of the nanocomposites, PLA of molar mass Mw~75 kg/mol with ~96% of L- and ~4% of D- forms was employed, and was obtained from Plastika Kritis S.A. (Iraklion, Greece). As reported in our previous study [19], for this molecular weight and based on titration measurements, the fractions of –COOH and –OH groups of PLA were calculated as, approximately, 3 meq/kg and 17 meq/kg, respectively. Tannin (T) and Kraft-Lignin (KL) were purchased from Sigma-Aldrich. According to dynamic light scattering (ZetaSizer 5000, Malvern company, Worcestershire, United Kingdom) measurements, the average particle size of the powders, which were measured in water, was found to be 612 nm for kraft-lignin (polydispersity index, PDI~0.64) and 454 nm for tannin (PDI~0.44).

2.2. Preparation of PLA Nanocomposite Materials with Kraft Lignin and Tannin

A melt-mixing via extrusion method was applied for the preparation of the PLA-based nanocomposites filled with tannin and lignin, with a previous overnight step of drying under vacuum, at 110 °C. Dried PLA and tannin or lignin at several wt. fractions were melt mixed in a Haake-Buchler reomixer (model 600) (Haake-Buchler Instruments Ltd., Saddle Brooke, NJ, USA) equipped with roller blades and a mixing head with a volumetric capacity of 69 cm^3, operating at 195 °C and 30 rpm for 10 min. In total, six nanocomposites were prepared, i.e., with four different lignin contents, 1.0, 2.5, 5.0 and 10.0 wt.%, as well as two different concentrations for tannin nanocomposites, 1.0 and 2.5 wt.%.

2.3. Characterization Techniques

2.3.1. Thermogravimetric Analysis (TGA)

Thermogravimetric analysis (TGA) of PLA nanocomposites was performed by a SETARAM SETSYS TG-DTA 16/18 instrument. The samples (6 $\pm$ 0.5 mg) were placed in alumina crucibles, while an empty alumina crucible was used as reference. To eliminate the buoyancy effect, a blank measurement was performed and subsequently was subtracted by the experimental curve. For the kinetic analysis study [41], PLA nanocomposites were heated from 25 °C to 600 °C, in a 50 mL/min flow of N_2 at heating rates of 5, 10, 15 and 20 °C/min. Continuous recordings of sample temperature, sample mass, its first derivative and heat flow were taken. Thermal degradation kinetic analysis of the PLA nanocomposites was performed using NETZSCH Kinetics Neo software (NETZSCH, Selb, Germany). This model-fitting kinetic approach aims to find a kinetic model with minimal adjustable parameters which quantitatively describes the kinetics of the complete degradation reaction.

2.3.2. Pyrolysis-Gas Chromatography/Mass Spectrometry (Py-GC/MS) Study

For Py-GC/MS analysis of the PLA nanocomposites, a very small amount of each material was "dropped" initially into the "Double-Shot" EGA/PY 3030D Pyrolyzer (Frontier Laboratories Ltd., Fukushima, Japan) using a CGS-1050Ex carrier gas selector. For pyrolysis analysis (flash pyrolysis), each sample was placed into the sample cup which afterwards fell into the Pyrolyzer furnace. The pre-selected pyrolysis temperature was differentiated for each nanocomposite sample, while the GC oven temperature was programmed at 50 °C for 2 min, followed by a stepped increase to 200 °C with a heating rate of 5 °C/min, where it was held for 8 min, and then the temperature was increased at 300 °C by a rate 20 °C/min, where it was held for 5 min. Sample vapours generated in the furnace were split (at a ratio of 1/50), a portion moved to the column at a flow rate of 1 mL/min, pressure 53.6 kPa and the remaining portion exited the system via the vent. The pyrolyzates were separated using temperature programmed capillary column of a Shimadzu QP-2010 Ultra Plus (Kyoto, Japan) gas chromatogram and analysed by the mass spectrometer MS-QP2010SE of Shimadzu (Kyoto, Japan). Ultra-ALLOY® metal capillary column from Frontier Laboratories LTD (Fukushima, Japan) was used containing 5% diphenyl and 95% dimethylpolysiloxane stationary phase, column length 30 m and column ID 0.25 mm. For the mass spectrometer, the following conditions were used: ion source heater 200 °C, interface temperature 300 °C, vacuum 10^{-4}–10^{0} Pa, m/z range 45–500 amu (atomic mass unit) and scan speed 10,000. The ion gas chromatograms and spectra retrieved by each experiment were subjected to further interpretation through Shimadzu and Frontier post-run software. The chromatogram and spectra retrieved by each experiment were subject to further interpretation through Shimadzu (NIST11.0) and Frontier (F-Search software 4.3) post-run software. Identification was recognized depending on the similarity percentage (minimum value of 80%) between average mass spectra on the entire chromatogram.

3. Results and Discussion

3.1. Thermogravimetric Analysis (TGA)

The thermal degradation of KL, T and PLA nanocomposites filled with the two different polyphenolic fillers was studied by TGA analysis. Figure 1 shows the TGA thermograms and derivative mass loss (dTG) curves of KL and T. The temperatures corresponding to 2.5% and 5% mass loss were found to be 49.2 °C and 71 °C for KL, as well as 83.1 °C and 106.1 °C for T sample. KL shows a small weight loss of ~12% below 100 °C due to water evaporation and the main degradation step between 200 °C and 500 °C because of the gradual degradation, and emission of phenol molecules [36,42]. Tannin exhibits a three-step degradation process that includes the loss of adsorbed water, the decomposition of linkage between flavonoid units and pyrolytic degradation [34]. Additionally, the residual mass of KL and T fillers reaches ~41% and 50% at 600 °C, indicating that the fillers do not fully degrade at that specific temperature. Based on these thermograms, it is clear that lignin is more stable than tannin. The maximum rate of the main decomposition steep of lignin is recorded at 425 °C while in tannin at 260 °C (Figure 1b).

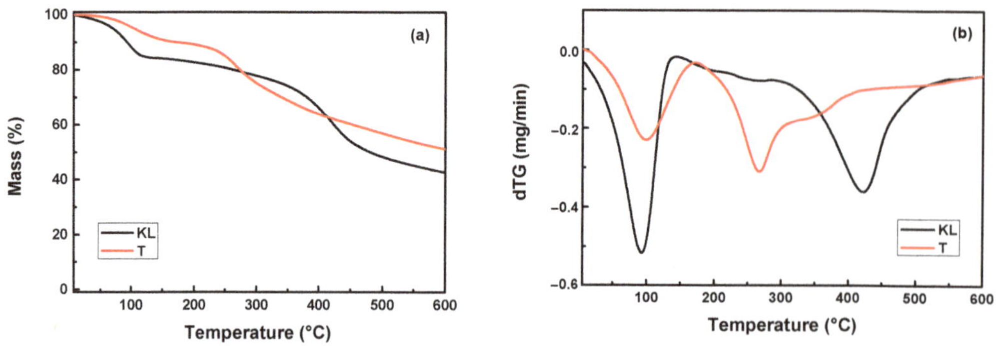

Figure 1. (**a**) TGA thermograms and (**b**) dTG curves of KL and T samples at a heating rate of 20 °C/min under a nitrogen atmosphere.

Figure 2 presents the TGA thermograms and dTG curves of PLA nanocomposites filled with the two different polyphenolic fillers at a heating rate of 20 °C/min, while the

TGA results of all the studied nanocomposites are shown in Table 1. It can be seen from the TGA curves, Figure 2a,b, that PLA/KL nanocomposites show good thermostability since no significant mass loss (<0.5%) occurs until 310 °C. It should be mentioned that the mass loss curves of the PLA nanocomposites appear to be almost identical with a one-step procedure yielding the same curve. Additionally, the residue content of the PLA nanocomposites was found to be close to the added filler content due to the fact that the fillers do not fully decompose until 600 °C (Figure 1). The temperatures corresponding to 0.5%, 2.5% and 5% mass loss, as well as the char residue of the samples at 600 °C, are shown in Table 1. An increase in thermal stability of PLA-lignin nanocomposites was found due to the presence of aromatic phenyl groups and hydroxyl groups [35,42–44]. However, Zhai et al. and Park et al. [37,38] found that both the onset thermal degradation temperature and the maximum thermal degradation temperature of PLA nanocomposites decrease with increasing filler content because of the lower onset thermal degradation temperature and maximum thermal degradation temperature of lignin compared to neat PLA. From the dTG curves, the highest decomposition rate of neat PLA-1%KL, PLA-2.5%KL, PLA-5%KL and PLA-10%KL nanocomposites was found to be 394.4 °C, 396.7 °C, 392.2 °C and 394.4 °C, respectively. Additionally, the PLA-1%T and PLA-2.5% nanocomposites decompose at 283 °C and 266 °C, respectively, with the highest decomposition rate at 387.4 °C and 382.2 °C, which are significantly lower than the previous ones. So, T filler has a strong catalytic effect and can markedly reduce PLA thermal stability because of the presence of highly polarized hydroxyl groups and low polarity PLA matrix, resulting in poor interfacial adhesion and dispersion of T in the PLA matrix [19,34,40]. Concerning the char residue of the nanocomposite materials, it was shown that for all the samples, the addition of both the two polyphenolic fillers tended to rise the char residue, as reported by previous studies [42,45].

As a final point, the difference between the behaviour of KL and T regarding the thermal stability of the PLA composites may be ascribed to their slightly differentiated structures. In fact, T is a smaller compound with a lower molecular weight than KL, while it contains on its structure fewer aromatic groups and higher amount of phenolic -OH groups in contrast to the methoxy- groups of KL. Due to these differences, tannin decomposes at lower temperatures than KL (Figure 1) and this could also affect the thermal stability of PLA/T nanocomposites.

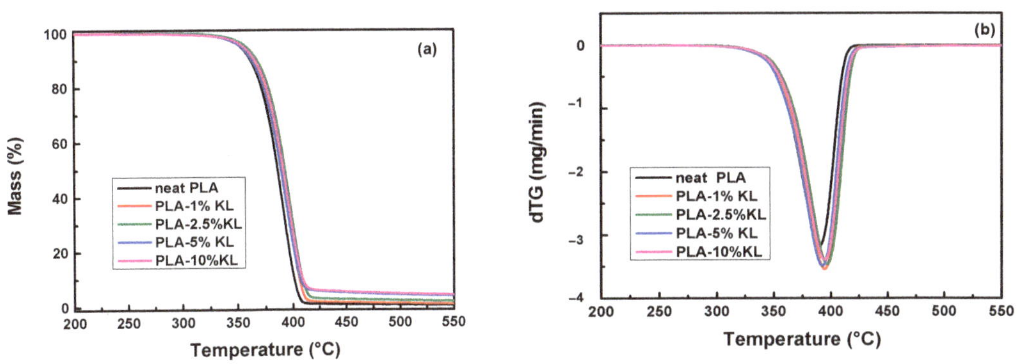

Figure 2. *Cont.*

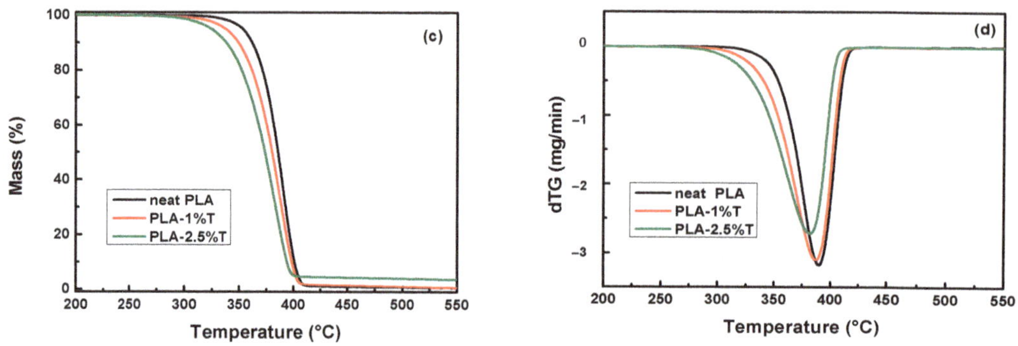

Figure 2. TGA thermograms and dTG curves of PLA nanocomposites filled with (**a**,**b**) KL and (**c**,**d**) T at a heating rate of 20 °C/min under a nitrogen atmosphere.

Table 1. TGA results of all studied samples.

Sample	$T_{0.5}$ (°C)	$T_{2.5}$ (°C)	T_5 (°C)	$T_{d,max}$ (°C)	CR_{600} (%)
Neat PLA	320.9	343.4	353.3	389.7	0.4
PLA-1%KL	320.2	344.1	354.4	394.4	0.6
PLA-2.5%KL	324.6	348.6	358.1	396.7	1.6
PLA-5%KL	310.7	342.6	353.7	392.2	3.4
PLA-10%KL	306.6	345.4	355.9	394.4	3.9
PLA-1%T	290.6	325.7	337.6	387.4	0.4
PLA-2.5%T	282.2	311.2	324.2	382.2	3.1

3.2. *Study of the Degradation Mechanism with Py-GC/MS*

Analytical techniques based on pyrolysis have proven particularly successful to investigate the degradation process of several polymers as well as their blends, since information is provided at a molecular level on these complex polymeric systems by examining their pyrolytic products. In fact, the latter are small volatile molecules formed by selective bond cleavage induced by thermal energy [46]. Once coupled with gas chromatography (GC) and mass spectrometry (MS), thermoanalytical pathway is highly sensitive, requires small masses and a minimum sample preparation while the analysis time is generally short.

Herein, in order to investigate the degradation process of the prepared nanocomposites based on a PLA matrix and the two polyphenolic fillers, namely kraft lignin and tannin, temperatures related to each sample's degradation peak were chosen for Py-GC/MS analysis. This thermoanalytical approach gives an insight into the optimal parameters required for the thermal stability control, and thus, is a useful tool for the preparation of the PLA nanocomposites for further packaging applications. Moreover, in an attempt to investigate how the pyrolytic profile of PLA materials is affected by the incorporation of the two fillers in different contents is also illustrated. All nanocomposites were pyrolyzed at a temperature range of 387–397 °C (Table 2), while the recorded chromatographs of the degradation products are shown in Figure 3. The most characteristic pyrolysis products were identified through their MS spectra and are presented in Table 2, while some indicative mass spectra of the pyrolytic compounds of KL and PLA-composites are depicted in Figure 4.

Generally, thermal degradation of thermoplastic polyesters takes place via heterolytic and homolytic scission of their aliphatic fragments. At small retention times (Rt), more volatile compounds with low molecular weight are found, whereas as retention time increases, larger and more complex compounds are identified, such as dimer, trimer and tetramer structures. The foremost pyrolyzates correspond to characteristic products of β-scission reactions including compounds with vinyl- and carboxyl- end groups. Regarding PLA, the complexity of its thermal decomposition has been extensively explored in literature [47–54]. In brief, at temperatures above 200 °C, intramolecular trans-esterification processes hold an outpacing position with the formation of lactide and cyclic oligomers due to cis-elimination proceeds, while the formation of acrylic acid and other oligomers is due to the β-scission reactions. Further fragmentation leads to the final formation of acetaldehyde and CO_2 [47].

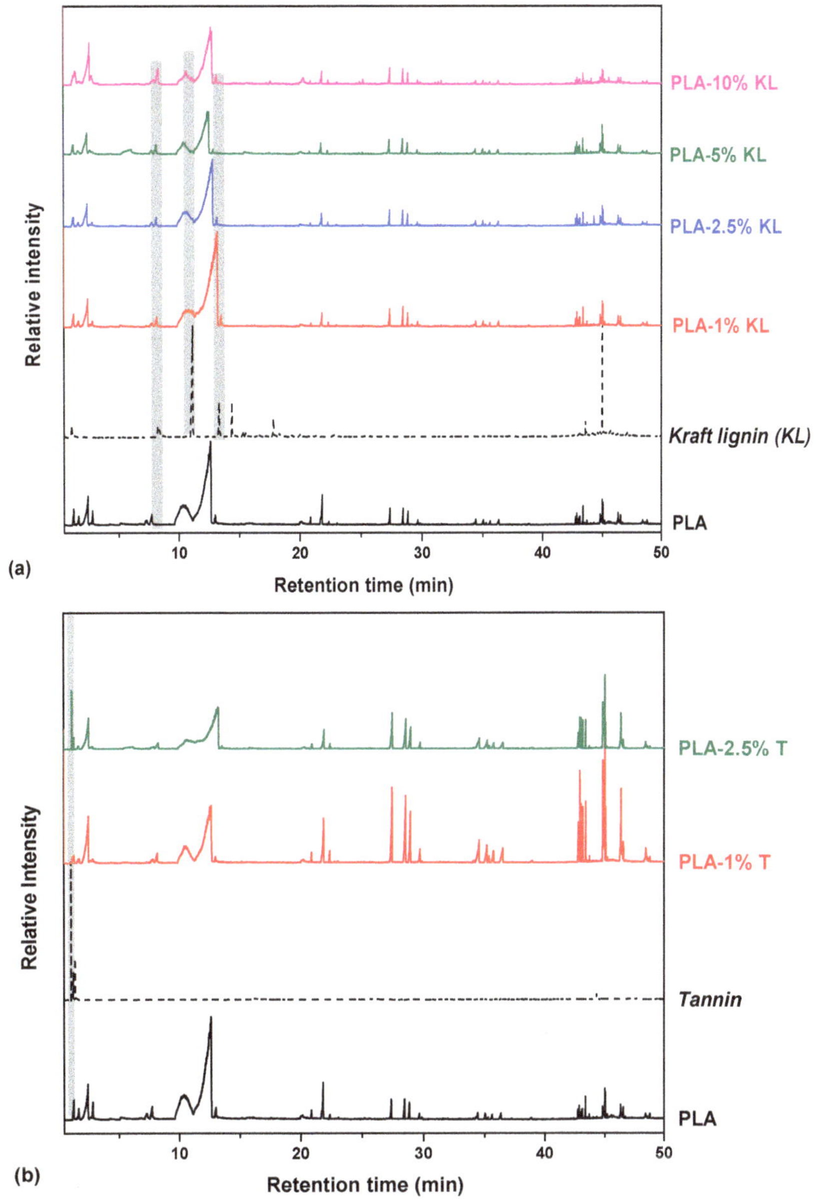

Figure 3. Total ion gas chromatograms of (**a**) neat PLA, KL, PLA nanocomposites with 1, 2.5, 5 and 10% wt. KL content and (**b**) neat PLA, T and PLA nanocomposites with 1 and 2.5% wt. T.

The principal thermal degradation products of PLA are depicted in Table 2. Specifically, according to the chromatogram of PLA (Figure 3), the first identified peak corresponds to acetaldehyde, while 2-propenoic acid (Figure 4d) and 2,3-pentanedione can also be detected due to the β-scission mechanism. Figure 3 exhibits the presence of two peaks (Rt = 10.25 and 12.47 min), assigned to the fragments with m/z of 45, 56 and 144 (Figure 4c). The aforementioned peaks conformed to meso-lactide and D,L-lactide structures, respectively, whereas peaks appearing sporadically at retention times higher than 20 min were ascribed to larger cyclic lactide oligomers [52,53].

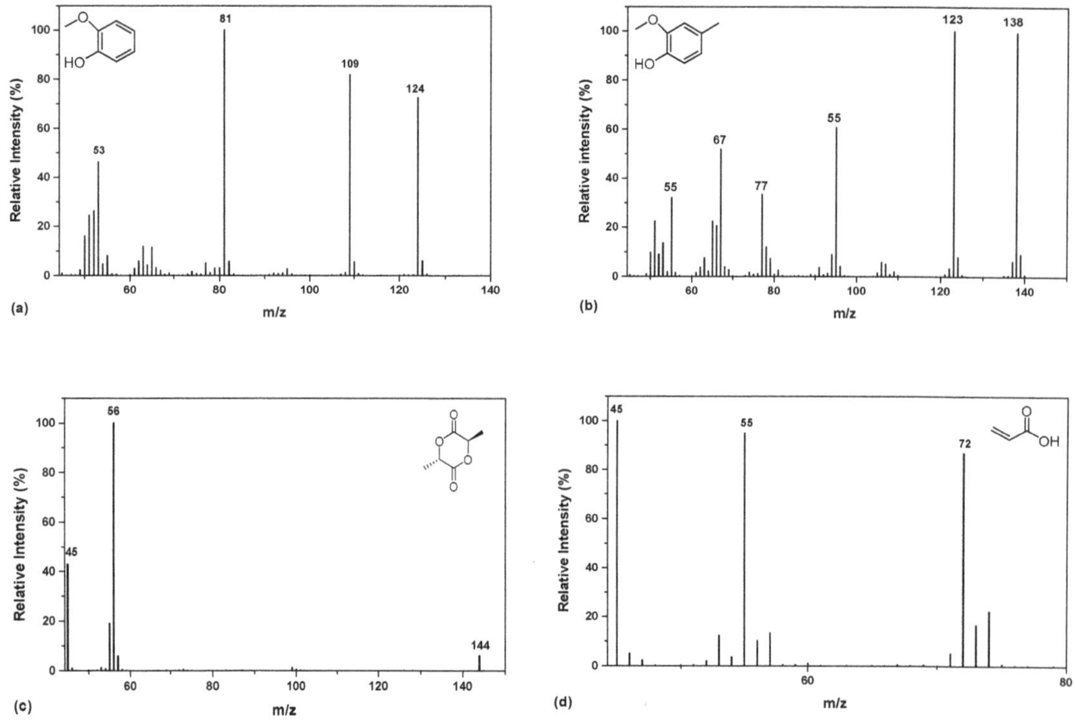

Figure 4. Mass spectra of (**a**) 2-methoxy-phenol, (**b**) 2-methoxy-4-methylphenol, (**c**) D,L-lactide and (**d**) 2-propenoic acid.

Among the products of KL, phenol derivatives and aromatic hydrocarbons were identified, with the most characteristic peak the relative found at Rt = 11.15 min, attributed to 2-methoxy-4-methylphenol (Figure 4b) [55–57]. Concerning the chromatographs of Figure 3a, all the PLA nanocomposites released compounds of low molecular weight at low retention times, including acetaldehyde, 2-propenoic acid and 2,3-pentanedione. In retention times within the range 10.25–12.47 min, the two distinctive peaks of PLA are depicted; meso-lactide and D,L-lactide. A noteworthy remark about the PLA nanocomposites is that as the content of KL increases the morphology of the peaks is slightly modified, whereas the meso-lactide intensity increases from 22.5% (1% KL) to 30% (10% KL) in comparison with the relative of D,L-lactide counterpart. Moreover, a new small peak assigned to 2-methoxy-4-methylphenol is displayed in Figure 3 at Rt = 11.18 min, mainly due to the larger content of the filler in this material. Generally, for the PLA nanocomposites with KL as well as T, it could be noticed that their pyrolysis products identified were identical, presenting that KL and T as fillers do not affect the degradation products of PLA formed at low content. In detail, both fillers possess aromatic structure, with KL containing mainly hydroxy- and methoxy- groups, and T holding plentiful hydroxyl groups (phenolic) on its architecture. From the other hand, PLA has a poor hydrophilic structure which may prevent the interactions between the polymer matrix and the two fillers. Due to this aforementioned fact, the addition of both phenolic compounds seems that do not affect the thermal degradation mechanism of PLA.

Table 2. Main pyrolysis products evolved during thermal decomposition of PLA, KL (KL), T (T) and their nanocomposites, at different temperatures as identified with mass spectroscopy based on the Py-GC chromatograms.

PLA	Kraft Lignin (KL)	PLA-1%KL	PLA-2.5%KL	PLA-5%KL	PLA-10%KL	Tannin (T)	PLA-1%T	PLA-2.5%T	Mw (amu)	Possible Product
Pyrolysis Temperature										
390 °C	400 °C	394 °C	397 °C	392 °C	394 °C	400 °C	387 °C	382 °C		
Rt (min)										
-	1.16	-	-	-	-	-	-	-	64	Sulphur dioxide
-	-	-	-	-	-	1.19	1.21	1.21	58	Acetone 
1.27	-	1.26	1.27	1.24	1.23	-	1.29	1.33	44	Acetaldehyde 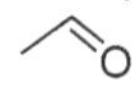
2.45	-	2.46	2.43	2.39	2.64	-	2.49	2.54	72	2-propenoic acid (acrylic acid) 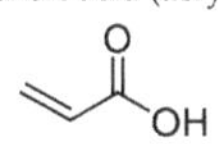
2.83	-	2.87	2.80	2.75	2.79	-	2.71	2.84	100	2,3-pentanedione
-	8.24	-	-	-	8.26	-	-	-	124	2-methoxy-phenol 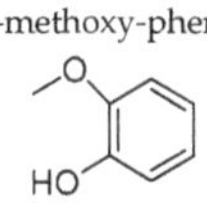
-	11.10	-	-	11.10	11.11	-	-	-	138	2-methoxy-4-methylphenol (creosol) 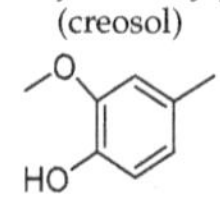
10.25 12.47	-	10.73 13.11	10.63 12.71	10.52 12.77	10.63 12.71	-	10.52 12.61	10.67 13.21	144	meso-lactide or D,L-lactide 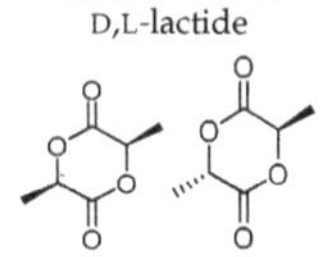
-	13.30	-	-	-	-	-	-	-	164	2-methoxy-3-(2-propenyl)-phenol 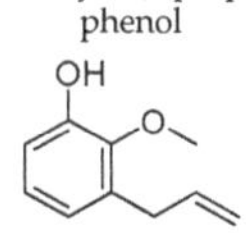
-	14.38	-	-	-	-	-	-	-	180	4-((1E)-3-Hydroxy-1-propenyl)-2-methoxyphenol 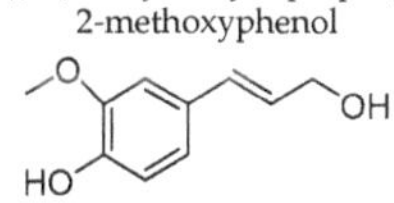
21.74	-	21.82	21.78	21.70	21.80	-	21.90	21.90	202	PLA trimer

3.3. Kinetic Analysis Based on Thermogravimetric Data

The degradation mechanism of PLA-1% nanocomposites filled with the two different polyphenolic fillers was described by calculating the degree of conversion (α) and kinetic parameters. For this reason, the isoconversional methods and the model fitting methods were used. The mass curves were recorded at various heating rates (5, 10, 15 and 20 °C/min) under nitrogen. The reaction rate can be defined using the general equation of solid-state reactions [41]:

$$\frac{d\alpha}{dt} = k(T)f(\alpha) \tag{1}$$

where k(T) is the reaction rate constant, f(α) is the reaction model and α is the degree of conversion. The term degree of conversion a refers to the ratio of the actual mass loss at a given temperature Δm to the total mass loss Δm_{tot} which occurs after the completion of the degradation process [22]:

$$\alpha = \frac{m_0 - m}{m_0 - m_f} = \frac{\Delta m}{\Delta m_{tot}} \tag{2}$$

The kinetic model, f(α) is an algebraic expression that describes the kinetics of the solid-state reaction. The dependence of temperature on the rate of reaction is defined by the Arrhenius equation, $k(T) = Ae^{-E/RT}$, where E is the apparent activation energy (kJ/mol), R the gas constant (8.314 J/mol·K), A the pre-exponential factor (s^{-1}) and T the absolute temperature (K).

The isoconversional methods are divided into differential and integral methods. The most frequently used methods are the differential isoconversional method of Friedman [58], the integral isoconversional method of Vyazovkin [59], and the integral isoconversional method of Ozawa, Flynn and Wall (OFW) [60].

The differential isoconversional Friedman's method can be described [58]:

$$\ln\left[\beta_i\left(\frac{d\alpha}{dt}\right)_{\alpha,i}\right] = \ln[f(\alpha)A_\alpha] - \frac{E_\alpha}{RT_{\alpha,i}} \tag{3}$$

where A is the pre-exponential factor and β is the heating rate. To obtain the activation energy E_α values at a constant conversion function, the slope of the straight lines of the plot ln[βi(dα/dt)a i] vs. $1/T_{\alpha,i}$ should be calculated. Differential methods have the advantage of requiring no approximations and applying them to any temperature program. However, a determination limit on the baseline causes limited accuracy, and they occasionally display numerical instability compared to integral ones.

An isoconversional nonlinear method has been proposed by Vyazovkin in order to calculate the Eα [59]:

$$\Phi(E_\alpha) = \sum_{i=1}^{n}\sum_{j\neq 1}^{n} \frac{J[E_\alpha, T_i(t_\alpha)]}{J[E_\alpha, T_j(t_\alpha)]} \tag{4}$$

where the indexes i and j denote set of experiments performed under different heating rates, n is the total number of experiments and J is evaluated over small intervals of Eα variation:

$$J[E_\alpha, T_i(t_\alpha)] = \int_{t_{\alpha-\Delta\alpha}}^{t_\alpha} \exp\left[\frac{-E_\alpha}{RT_i(t)}\right]dt \tag{5}$$

The value that minimizes the function Φ(Eα), Equation (4), is used for the calculation of Eα. The time tα,i and temperature $T_{\alpha,i}$ of selected α values are found by an exact interpolation utilizing a Lagrangian algorithm for each i-th temperature program.

The Ozawa, Flynn and Wall equation can be presented as [61–63]:

$$\ln(\beta_i) = \text{Const} - 1.052\left(\frac{E_\alpha}{RT_\alpha}\right) \tag{6}$$

where $\beta = dT/dt = \text{const}$ is the heating rate. The index i express the different heating rates that were applied to the experimental data. The activation energy E_α values can be obtained by the slope of the $\ln(\beta i)$ vs. $1/T_\alpha$ plots.

Three iso-conversional methods, the differential isoconversional method of Friedman, the integral isoconversional method of Vyazovkin, and the integral isoconversional method of OFW were used for the calculation of the activation energies [58,59,64–66]. Figure 5 shows the E_α values of PLA nanocomposites filled with the two different polyphenolic fillers versus the degree of conversion α by using the above-mentioned iso-conversional methods. The E_α mean value obtained by using the Vyazovkin method from the TGA curves was found to be in excellent agreement with the values computed by the Friedman method, while the differences in E_α values between the OFW and Friedman methods can be explained by a systematic error owing to poor integration [67]. In all cases, the E_α-dependency suggests that the thermal degradation mechanism is complex, and the degradation kinetics is regulated by different processes at the initial and final stages. The E_α values of PLA-1%KL and PLA-1%T nanocomposites present a rapid increase in the first region, while it remains almost constant for greater values. According to the Friedman method, Vyazovkin analysis, and OFW method, the E_α mean values of PLA-1%KL nanocomposite for $\alpha < 0.1$ were found to be 115.2, 114.5, and 103.7 kJ/mol, respectively, while the values were 135.8, 136.4, and 126.0 kJ/mol for $\alpha > 0.1$. Additionally, the E_α values of PLA-1%T nanocomposite ($\alpha < 0.3$) were 109.3, 109.9 and 96.1 kJ/mol for Friedman method, Vyazovkin analysis, and OFW method, respectively, and 129.6, 130.5, and 115.6 kJ/mol for $\alpha > 0.3$. The PLA-1%T nanocomposite has lower activation energy than PLA-1%KL, requiring less activation energy to undergo thermal degradation.

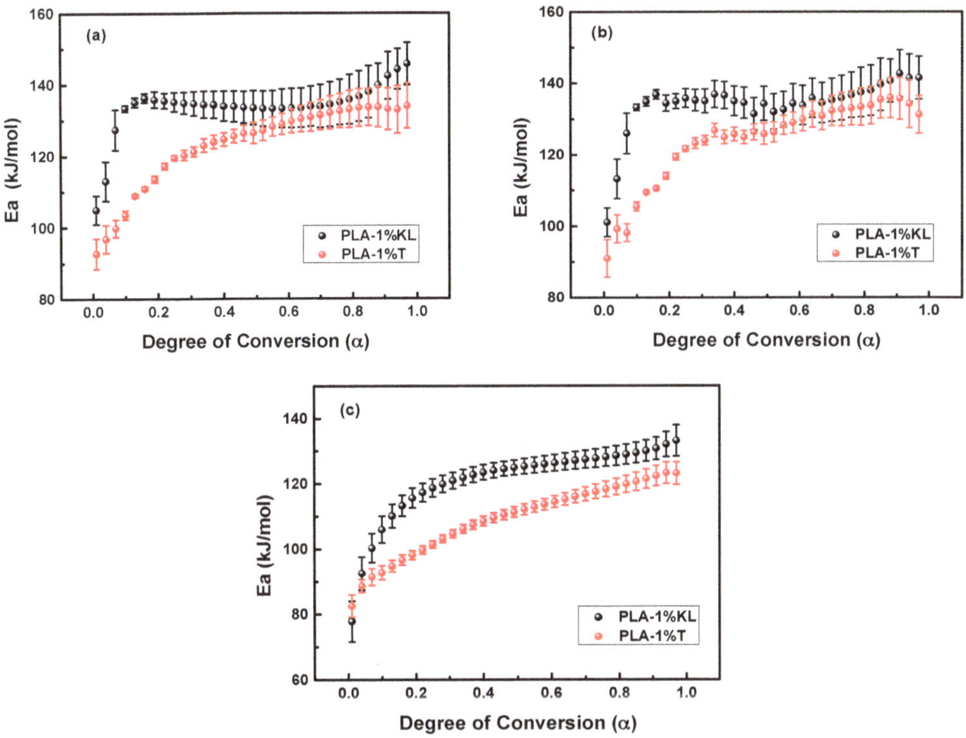

Figure 5. The dependence of activation energy (E_α) on the extent of conversion (α) for the thermal degradation of PLA-1% nanocomposites filled with KL and T as calculated by (**a**) Friedman method, (**b**) Vyazovkin analysis and (**c**) OFW analysis.

The degradation mechanism and the kinetic triplet (E, A, f(α)) of each reaction were determined by using multivariate non-linear regression method or model-fitting method through the comparison of the experimental (four heating rates) and theoretical data. Various reaction models (16 models) were adopted for the fitting to the experimental data at different heating rates. First, a single-step reaction mechanism is assumed to occur, which corresponds to the major mass loss. If the fitting to the experimental data does not yield satisfactory results, two or more mechanisms are combined. In our previous work [14], it was found that the form of the conversion function, which describes well the neat PLA is the mechanism of autocatalysis n-order (Cn):

$$f(\alpha) = (1-\alpha)^n(1+K_{cat}X) \quad (7)$$

where K_{cat} is the autocatalysis rate constant and X is the extent of conversion of the autocatalytic reactions. According to autocatalytic reaction, the thermally induced polymer fragments (reaction products) attack the unreacted parts of the polymer breaking the chemical bonds and causing acceleration of the degradation process. Figure 5 shows that the combination of two consecutive Cn-Cn reaction models fits well the experimental data of PLA-1%KL and PLA-1%T nanocomposites. The calculated parameters (activation energy, pre-exponential factor, and the reaction order) obtained by the tested models are summarized in Table 3. The E_α values calculated by the two-step mechanism model were found to be in good agreement with those estimated by the isoconversional methods (Figure 6). The calculated values suggest that the T filler, added to the PLA matrix, has led to lower E_α values, and resulted in a substantially higher decomposition rate constant. Additionally, the pre-exponential factor A of the PLA-1%T nanocomposite presents lower values in both stages than those of PLA-1%KL following the calculated E_α values. This means that the rate constant of PLA-1%T nanocomposite is significantly larger than that of PLA-1%KL, accelerating the thermal degradation. The values of correlation coefficient (R^2) were found to be 0.99993 and 0.99992 for PLA-1%KL and PLA-1%T nanocomposites, respectively.

Figure 6. Mass (%) curves of (**a**) neat PLA-1%KL and (**b**) PLA-1%T at heating rates of 5, 10, 15 and 20 °C/min in a nitrogen atmosphere (symbols) and the corresponding fitting curves with the combination of Cn-Cn reaction models (continuous black lines).

Table 3. Activation energy, pre-exponential factor and reaction order of PLA nanocomposites.

Sample	Model	Activation Energy, $E/kJmol^{-1}$	Pre-Exponential Factor, $logA_1/s^{-1}$	Reaction Order/n	Log K_{cat}	R^2
PLA-1%KL	Cn	119.02	5.182	0.694	1.081	0.99993
	Cn	134.15	6.277	1.241	1.381	
PLA-1%T	Cn	102.69	4.086	0.635	0.930	0.99992
	Cn	132.71	6.106	1.210	1.324	

4. Conclusions

The effect of two natural polyphenolic fillers, namely kraft lignin and tannin on the thermal stability and degradation of PLA was thoroughly evaluated throughout this manuscript. For this reason, TGA in conjunction with Py-GC/MS was effectively employed to investigate the thermal stability, decomposition kinetic and mechanism of PLA nanocomposites. The isoconversional differential method and multivariate non-linear regression method were used for the determination of the effective activation energy and the kinetic triplet of the degradation reactions of PLA nanocomposites. TGA thermographs depicted that PLA/KL nanocomposites present good thermostability, since no significant mass loss takes place until 310 °C. This fact may be attributed to the additional aromatic groups of KL. On the other hand, T filler displayed a severe catalytic effect on the PLA/T nanocomposites and provoked a reduction of PLA thermal stability. Concerning the investigation of the thermal degradation pathway through Py-GC/MS analysis, it was found that for both PLA and its composites with KL and T, intramolecular *trans*-esterification processes hold a principal role during decomposition with the formation of mainly lactide and cyclic oligomers, while cis-elimination provokes the formation of acrylic acid and oligomers. It seems that the addition of both the fillers does not greatly affect the thermal degradation pathway of PLA. Additionally, a two-step mechanism was detected for the thermal degradation of PLA-1%KL and PLA-1%T nanocomposites, which was best fitted by the Cn–Cn model (n-th order model). It was found that the T filler, added in the polymer matrix, decreased the E_α values, leading to higher decomposition rate constant compared to that of PLA-KL nanocomposite and accelerating the thermal degradation process.

Author Contributions: Conceptualization, D.N.B., A.Z., K.P.Č. and L.F.Z.; methodology, E.T., N.M.A. and A.Z.; software, E.T.; validation, N.M.A. and E.T.; formal analysis, N.M.A., E.T. and A.Z.; investigation, E.T., N.M.A. and A.Z.; resources, D.N.B. and L.F.Z.; writing—original draft preparation, E.T. and N.M.A.; writing—review and editing, D.N.B., L.F.Z., D.A.L., A.Z. and E.T.; supervision, D.N.B., K.C. and L.F.Z. All authors have read and agreed to the published version of the manuscript.

Funding: This work was funded from the European Union's Horizon 2020 Research and Innovation Programme under Grant Agreement No. 952941 (BIOMAC Project).

Institutional Review Board Statement: Not applicable.

Informed Consent Statement: Not applicable.

Data Availability Statement: The data presented in this study are available on request from the corresponding author.

Conflicts of Interest: The authors declare no conflict of interest.

References

1. Maharana, T.; Mohanty, B.; Negi, Y.S. Melt-solid polycondensation of lactic acid and its biodegradability. *Prog. Polym. Sci.* **2009**, *34*, 99–124. [CrossRef]
2. Balla, E.; Daniilidis, V.; Karlioti, G.; Kalamas, T.; Stefanidou, M.; Bikiaris, N.D.; Vlachopoulos, A.; Koumentakou, I.; Bikiaris, D.N. Poly(lactic Acid): A Versatile Biobased Polymer for the Future with Multifunctional Properties—From Monomer Synthesis, Polymerization Techniques and Molecular Weight Increase to PLA Applications. *Polymers* **2021**, *13*, 1822. [CrossRef]
3. Hong, M.; Chen, E.Y.X. Chemically recyclable polymers: A circular economy approach to sustainability. *Green Chem.* **2017**, *19*, 3692–3706. [CrossRef]

4. Armentano, I.; Bitinis, N.; Fortunati, E.; Mattioli, S.; Rescignano, N.; Verdejo, R.; Lopez-Manchado, M.A.; Kenny, J.M. Multifunctional nanostructured PLA materials for packaging and tissue engineering. *Prog. Polym. Sci.* **2013**, *38*, 1720–1747. [CrossRef]
5. Zisopoulou, S.A.; Chatzinikolaou, C.K.; Gallos, J.K.; Ofrydopoulou, A.; Lambropoulou, D.A.; Psochia, E.; Bikiaris, D.N.; Nanaki, S.G. Synthesis of Dacus Pheromone, Controlled Release Devices. *Agronomy* **2020**, *10*, 1053. [CrossRef]
6. Sha, L.; Chen, Z.; Chen, Z.; Zhang, A.; Yang, Z. Polylactic acid based nanocomposites: Promising safe and biodegradable materials in biomedical field. *Int. J. Polym. Sci.* **2016**, *2016*. [CrossRef]
7. Casalini, T.; Rossi, F.; Castrovinci, A.; Perale, G. A Perspective on Polylactic Acid-Based Polymers Use for Nanoparticles Synthesis and Applications. *Front. Bioeng. Biotechnol.* **2019**, *7*, 1–16. [CrossRef]
8. Psochia, E.; Papadopoulos, L.; Gkiliopoulos, D.J.; Francone, A.; Grigora, M.-E.; Tzetzis, D.; de Castro, J.V.; Neves, N.M.; Triantafyllidis, K.S.; Torres, C.M.S.; et al. Bottom-Up Development of Nanoimprinted PLLA Composite Films with Enhanced Antibacterial Properties for Smart Packaging Applications. *Macromol* **2021**, *1*, 49–63. [CrossRef]
9. Ahmed, J.; Varshney, S.K. Polylactides-chemistry, properties and green packaging technology: A review. *Int. J. Food Prop.* **2011**, *14*, 37–58. [CrossRef]
10. Sarfraz, J.; Gulin-Sarfraz, T.; Nilsen-Nygaard, J.; Pettersen, M.K. Nanocomposites for food packaging applications: An overview. *Nanomaterials* **2021**, *11*, 10. [CrossRef]
11. Ngo, T.D.; Kashani, A.; Imbalzano, G.; Nguyen, K.T.Q.; Hui, D. Additive manufacturing (3D printing): A review of materials, methods, applications and challenges. *Compos. Part B Eng.* **2018**, *143*, 172–196. [CrossRef]
12. Costanzo, A.; Spotorno, R.; Candal, M.V.; Fernández, M.M.; Müller, A.J.; Graham, R.S.; Cavallo, D.; McIlroy, C. Residual alignment and its effect on weld strength in material-extrusion 3D-printing of polylactic acid. *Addit. Manuf.* **2020**, *36*, 101415. [CrossRef]
13. Domínguez-Robles, J.; Martin, N.K.; Fong, M.L.; Stewart, S.A.; Irwin, N.J.; Rial-Hermida, M.I.; Donnelly, R.F.; Larrañeta, E. Antioxidant pla composites containing lignin for 3D printing applications: A potential material for healthcare applications. *Pharmaceutics* **2019**, *11*, 165. [CrossRef] [PubMed]
14. Tarani, E.; Pušnik Črešnar, K.; Zemljič, L.F.; Chrissafis, K.; Papageorgiou, G.Z.; Lambropoulou, D.; Zamboulis, A.; Bikiaris, D.N.; Terzopoulou, Z. Cold crystallization kinetics and thermal degradation of pla composites with metal oxide nanofillers. *Appl. Sci.* **2021**, *11*, 3004. [CrossRef]
15. Gumienna, M.; Górna, B. Antimicrobial Food Packaging with Biodegradable Polymers and Bacteriocins. *Molecules* **2021**, *26*, 3735. [CrossRef] [PubMed]
16. Bezerra Lima, E.M.; Middea, A.; Marconcini, J.M.; Corrêa, A.C.; Fernandes Pereira, J.; Vieira Guimarães, A.; Firmino de Lima, J.; Ramos dos Anjos, M.; Miranda de Castro, I.; Nunes Oliveira, R.; et al. Biodegradable PLA based nanocomposites for packaging applications: The effects of organo-modified bentonite concentration. *J. Appl. Polym. Sci.* **2021**, *138*, 1–17. [CrossRef]
17. Fredi, G.; Rigotti, D.; Bikiaris, D.N.; Dorigato, A. Tuning thermo-mechanical properties of poly(lactic acid) films through blending with bioderived poly(alkylene furanoate)s with different alkyl chain length for sustainable packaging. *Polymer (Guildf)* **2021**, *218*, 123527. [CrossRef]
18. Delpouve, N.; Saiter, A.; Dargent, E. Cooperativity length evolution during crystallization of poly(lactic acid). *Eur. Polym. J.* **2011**, *47*, 2414–2423. [CrossRef]
19. Cresnar, K.P.; Klonos, P.A.; Zamboulis, A.; Terzopoulou, Z.; Xanthopoulou, E.; Papadopoulos, L.; Kyritsis, A.; Bikiaris, D.N. Structure-Properties relationships in renewable composites based on polylactide filled with Tannin and Kraft Lignin—Crystallization and molecular mobility. *Thermochim. Acta* **2021**, *703*, 178998. [CrossRef]
20. Sanusi, O.M.; Benelfellah, A.; Bikiaris, D.N.; Aït Hocine, N. Effect of rigid nanoparticles and preparation techniques on the performances of poly(lactic acid) nanocomposites: A review. *Polym. Adv. Technol.* **2021**, *32*, 444–460. [CrossRef]
21. Klonos, P.; Terzopoulou, Z.; Koutsoumpis, S.; Zidropoulos, S.; Kripotou, S.; Papageorgiou, G.Z.; Bikiaris, D.N.; Kyritsis, A.; Pissis, P. Rigid amorphous fraction and segmental dynamics in nanocomposites based on poly(L–lactic acid) and nano-inclusions of 1–3D geometry studied by thermal and dielectric techniques. *Eur. Polym. J.* **2016**, *82*, 16–34. [CrossRef]
22. Nerantzaki, M.; Prokopiou, L.; Bikiaris, D.N.; Patsiaoura, D.; Chrissafis, K.; Klonos, P.; Kyritsis, A.; Pissis, P. In situ prepared poly(DL-lactic acid)/silica nanocomposites: Study of molecular composition, thermal stability, glass transition and molecular dynamics. *Thermochim. Acta* **2018**, *669*, 16–29. [CrossRef]
23. Shen, X.; Zheng, Q.; Kim, J.K. Rational design of two-dimensional nanofillers for polymer nanocomposites toward multifunctional applications. *Prog. Mater. Sci.* **2021**, *115*, 100708. [CrossRef]
24. Garcia, C.V.; Shin, G.H.; Kim, J.T. Metal oxide-based nanocomposites in food packaging: Applications, migration, and regulations. *Trends Food Sci. Technol.* **2018**, *82*, 21–31. [CrossRef]
25. Liao, C.; Li, Y.; Tjong, S.C. Antibacterial activities of aliphatic polyester nanocomposites with silver nanoparticles and/or graphene oxide sheets. *Nanomaterials* **2019**, *9*, 1102. [CrossRef]
26. Zhang, H.; Hortal, M.; Jordá-Beneyto, M.; Rosa, E.; Lara-Lledo, M.; Lorente, I. ZnO-PLA nanocomposite coated paper for antimicrobial packaging application. *LWT Food Sci. Technol.* **2017**, *78*, 250–257. [CrossRef]
27. Papadopoulos, L.; Klonos, P.A.; Terzopoulou, Z.; Psochia, E.; Sanusi, O.M.; Hocine, N.A.; Benelfellah, A.; Giliopoulos, D.; Triantafyllidis, K.; Kyritsis, A.; et al. Comparative study of crystallization, semicrystalline morphology, and molecular mobility in nanocomposites based on polylactide and various inclusions at low filler loadings. *Polymer (Guildf)* **2021**, *217*, 123457. [CrossRef]

28. Bužarovska, A.; Blazevska-Gilev, J.; Pérez-Martnez, B.T.; Balahura, L.R.; Pircalabioru, G.G.; Dinescu, S.; Costache, M. Poly(l-lactic acid)/alkali lignin composites: Properties, biocompatibility, cytotoxicity and antimicrobial behavior. *J. Mater. Sci.* **2021**, *56*, 13785–13800. [CrossRef]
29. Bikiaris, D. Can nanoparticles really enhance thermal stability of polymers? Part II: An overview on thermal decomposition of polycondensation polymers. *Thermochim. Acta* **2011**, *523*, 25–45. [CrossRef]
30. Terzopoulou, Z.; Klonos, P.A.; Kyritsis, A.; Tziolas, A.; Avgeropoulos, A.; Papageorgiou, G.Z.; Bikiaris, D.N. Interfacial interactions, crystallization and molecular mobility in nanocomposites of Poly(lactic acid) filled with new hybrid inclusions based on graphene oxide and silica nanoparticles. *Polymer (Guildf)* **2019**, *166*, 1–12. [CrossRef]
31. Sanusi, O.M.; Benelfellah, A.; Papadopoulos, L.; Terzopoulou, Z.; Malletzidou, L.; Vasileiadis, I.G.; Chrissafis, K.; Bikiaris, D.N.; Aït Hocine, N. Influence of montmorillonite/carbon nanotube hybrid nanofillers on the properties of poly(lactic acid). *Appl. Clay Sci.* **2021**, *201*, 105925. [CrossRef]
32. Saeidlou, S.; Huneault, M.A.; Li, H.; Park, C.B. Poly(lactic acid) crystallization. *Prog. Polym. Sci.* **2012**, *37*, 1657–1677. [CrossRef]
33. Kai, D.; Tan, M.J.; Chee, P.L.; Chua, Y.K.; Yap, Y.L.; Loh, X.J. Towards lignin-based functional materials in a sustainable world. *Green Chem.* **2016**, *18*, 1175–1200. [CrossRef]
34. Liao, J.; Brosse, N.; Hoppe, S.; Du, G.; Zhou, X.; Pizzi, A. One-step compatibilization of poly(lactic acid) and tannin via reactive extrusion. *Mater. Des.* **2020**, *191*, 108603. [CrossRef]
35. Spiridon, I.; Leluk, K.; Resmerita, A.M.; Darie, R.N. Evaluation of PLA-lignin bioplastics properties before and after accelerated weathering. *Compos. Part B Eng.* **2015**, *69*, 342–349. [CrossRef]
36. da Silva, T.F.; Menezes, F.; Montagna, L.S.; Lemes, A.P.; Passador, F.R. Effect of lignin as accelerator of the biodegradation process of poly(lactic acid)/lignin composites. *Mater. Sci. Eng. B Solid-State Mater. Adv. Technol.* **2019**, *251*, 114441. [CrossRef]
37. Park, C.W.; Youe, W.J.; Kim, S.J.; Han, S.Y.; Park, J.S.; Lee, E.A.; Kwon, G.J.; Kim, Y.S.; Kim, N.H.; Lee, S.H. Effect of lignin plasticization on physico-mechanical properties of lignin/poly(lactic acid) composites. *Polymers* **2019**, *11*, 2089. [CrossRef]
38. Zhai, S.; Liu, Q.; Zhao, Y.; Sun, H.; Yang, B.; Weng, Y. A review: Research progress in modification of poly(lactic acid) by lignin and cellulose. *Polymers* **2021**, *13*, 776. [CrossRef]
39. Anwer, M.A.S.; Naguib, H.E.; Celzard, A.; Fierro, V. Comparison of the thermal, dynamic mechanical and morphological properties of PLA-Lignin & PLA-Tannin particulate green composites. *Compos. Part B Eng.* **2015**, *82*, 92–99. [CrossRef]
40. Liao, J.; Brosse, N.; Pizzi, A.; Hoppe, S.; Zhou, X.; Liao, J.; Brosse, N.; Pizzi, A.; Hoppe, S.; Zhou, X. Characterization and 3D printability of poly(lactic acid)/acetylated tannin. *Ind. Crop. Prod.* **2020**, *149*, 112320. [CrossRef]
41. Vyazovkin, S.; Burnham, A.K.; Criado, J.M.; Pérez-Maqueda, L.A.; Popescu, C.; Sbirrazzuoli, N. ICTAC Kinetics Committee recommendations for performing kinetic computations on thermal analysis data. *Thermochim. Acta* **2011**, *520*, 1–19. [CrossRef]
42. Gordobil, O.; Egüés, I.; Llano-Ponte, R.; Labidi, J. Physicochemical properties of PLA lignin blends. *Polym. Degrad. Stab.* **2014**, *108*, 330–338. [CrossRef]
43. Nfc, P. Analysis of PLA Composite Filaments Reinforced with Lignin. *Polymers* **2021**, *13*, 2174.
44. Wang, X.; Jia, Y.; Liu, Z.; Miao, J. Influence of the lignin content on the properties of poly(lactic acid)/lignin-containing cellulose nanofibrils composite films. *Polymers* **2018**, *10*, 1013. [CrossRef]
45. Sahoo, S.; Misra, M.; Mohanty, A.K. Enhanced properties of lignin-based biodegradable polymer composites using injection moulding process. *Compos. Part A Appl. Sci. Manuf.* **2011**, *42*, 1710–1718. [CrossRef]
46. Moldoveanu, S.C. General Information About Pyrolysis. *Pyrolysis Org. Mol.* **2019**, 1–33. [CrossRef]
47. Aoyagi, Y.; Yamashita, K.; Doi, Y. Thermal degradation of poly[(R)-3-hydroxybutyrate], poly[ε-caprolactone], and poly[(S)-lactide]. *Polym. Degrad. Stab.* **2002**, *76*, 53–59. [CrossRef]
48. Kopinke, F.D.; Remmler, M.; Mackenzie, K.; Möder, M.; Wachsen, O. Thermal decomposition of biodegradable polyesters—II. Poly(lactic acid). *Polym. Degrad. Stab.* **1996**, *53*, 329–342. [CrossRef]
49. Sun, C.; Li, C.; Tan, H.; Zhang, Y. Synergistic effects of wood fiber and polylactic acid during co-pyrolysis using TG-FTIR-MS and Py-GC/MS. *Energy Convers. Manag.* **2019**, *202*, 112212. [CrossRef]
50. Ozdemir, E.; Lekesiz, T.O.; Hacaloglu, J. Polylactide/organically modified montmorillonite composites; effects of organic modifier on thermal characteristics. *Polym. Degrad. Stab.* **2016**, *134*, 87–96. [CrossRef]
51. Khabbaz, F.; Karlsson, S.; Albertsson, A.C. Py-GC/MS an effective technique to characterizing of degradation mechanism of poly(L-lactide) in the different environment. *J. Appl. Polym. Sci.* **2000**, *78*, 2369–2378. [CrossRef]
52. Arrieta, M.P.; Parres, F.; López, J.; Jiménez, A. Development of a novel pyrolysis-gas chromatography/mass spectrometry method for the analysis of poly(lactic acid) thermal degradation products. *J. Anal. Appl. Pyrolysis* **2013**, *101*, 150–155. [CrossRef]
53. Kopinke, F.D.; Mackenzie, K. Mechanistic aspects of the thermal degradation of poly(lactic acid) and poly(β-hydroxybutyric acid). *J. Anal. Appl. Pyrolysis* **1997**, *40–41*, 43–53. [CrossRef]
54. Chrysafi, I.; Ainali, N.M.; Bikiaris, D.N. Thermal degradation mechanism and decomposition kinetic studies of poly(Lactic acid) and its copolymers with poly(hexylene succinate). *Polymers* **2021**, *13*, 1365. [CrossRef] [PubMed]
55. Wądrzyk, M.; Janus, R.; Lewandowski, M.; Magdziarz, A. On mechanism of lignin decomposition—Investigation using microscale techniques: Py-GC-MS, Py-FT-IR and TGA. *Renew. Energy* **2021**, *177*, 942–952. [CrossRef]
56. Dai, L.; Liu, R.; Si, C. A novel functional lignin-based filler for pyrolysis and feedstock recycling of poly(l-lactide). *Green Chem.* **2018**, *20*, 1777–1783. [CrossRef]
57. Kawamoto, H. Lignin pyrolysis reactions. *J. Wood Sci.* **2017**, *63*, 117–132. [CrossRef]

58. Friedman, H.L. Kinetics of thermal degradation of char-forming plastics from thermogravimetry. Application to a phenolic plastic. *J. Polym. Sci. Part C Polym. Symp.* **2007**, *6*, 183–195. [CrossRef]
59. Vyazovkin, S. Evaluation of activation energy of thermally stimulated solid-state reactions under arbitrary variation of temperature. *J. Comput. Chem.* **1997**, *18*, 393–402. [CrossRef]
60. Ozawa, T. Kinetics of non-isothermal. *Polymer (Guildf)* **1971**, *12*, 150–158. [CrossRef]
61. Ozawa, T. A New Method of Analyzing Thermogravimetric Data. *Bull. Chem. Soc. Jpn.* **1965**, *38*, 1881–1886. [CrossRef]
62. Flynn, J.H.; Wall, L.A. A Quick, Direct Method for the Determination of Activation Energy from Thermogravimetric Data. *Polym. Lett.* **1966**, *4*, 323–328. [CrossRef]
63. Vyazovkin, S. Model-free kinetics: Staying free of multiplying entities without necessity. *J. Therm. Anal. Calorim.* **2006**, *83*, 45–51. [CrossRef]
64. Budrugeac, P.; Segal, E.; Pérez-Maqueda, L.A.; Criado, J.M. The use of the IKP method for evaluating the kinetic parameters and the conversion function of the thermal dehydrochlorination of PVC from non-isothermal data. *Polym. Degrad. Stab.* **2004**, *84*, 311–320. [CrossRef]
65. Tarani, E.; Papageorgiou, G.Z.; Bikiaris, D.N.; Chrissafis, K. Kinetics of crystallization and thermal degradation of an isotactic polypropylene matrix reinforced with graphene/glass-fiber filler. *Molecules* **2019**, *24*, 1984. [CrossRef] [PubMed]
66. Tarani, E.; Terzopoulou, Z.; Bikiaris, D.N.; Kyratsi, T.; Chrissafis, K.; Vourlias, G. Thermal conductivity and degradation behavior of HDPE/graphene nanocomposites: Pyrolysis, kinetics and mechanism. *J. Therm. Anal. Calorim.* **2017**, *129*, 1715–1726. [CrossRef]
67. Vyazovkin, S. Modification of the Integral Isoconversional Method to Account for Variation in the Activation Energy. *J. Comput. Chem.* **2001**, *22*, 178–183. [CrossRef]

polymers

MDPI

Article

Synthesis, Characterization and Mechanical Properties of Novel Bio-Based Polyurethane Foams Using Cellulose-Derived Polyol for Chain Extension and Cellulose Citrate as a Thickener Additive

Loredana Maiuolo [1,*], Fabrizio Olivito [1,*], Vincenzo Algieri [1], Paola Costanzo [1], Antonio Jiritano [1], Matteo Antonio Tallarida [1], Antonio Tursi [1], Corradino Sposato [2], Andrea Feo [2] and Antonio De Nino [1,*]

1 Department of Chemistry and Chemical Technologies, University of Calabria, 87036 Rende, CS, Italy; vincenzo.algieri@unical.it (V.A.); paola.costanzo@unical.it (P.C.); antonio.jiritano@unical.it (A.J.); matteoa.tallarida@unical.it (M.A.T.); antonio.tursi@unical.it (A.T.)

2 ENEA, Italian National Agency for New Technologies, Energy and Sustainable Economic Development, Trisaia Research Centre, S.S. 106 Ionica, km 419 + 500, 75026 Rotondella, MT, Italy; corradino.sposato@enea.it (C.S.); andrea.feo@enea.it (A.F.)

* Correspondence: maiuolo@unical.it (L.M.); fabrizio.olivito@unical.it (F.O.); denino@unical.it (A.D.N.)

Citation: Maiuolo, L.; Olivito, F.; Algieri, V.; Costanzo, P.; Jiritano, A.; Tallarida, M.A.; Tursi, A.; Sposato, C.; Feo, A.; De Nino, A. Synthesis, Characterization and Mechanical Properties of Novel Bio-Based Polyurethane Foams Using Cellulose-Derived Polyol for Chain Extension and Cellulose Citrate as a Thickener Additive. *Polymers* **2021**, *13*, 2802. https://doi.org/10.3390/polym13162802

Academic Editors: Arn Mignon and Miguel Angel Rodríguez-Pérez

Received: 22 July 2021
Accepted: 16 August 2021
Published: 20 August 2021

Publisher's Note: MDPI stays neutral with regard to jurisdictional claims in published maps and institutional affiliations.

Abstract: A novel series of bio-based polyurethane composite foams was prepared, employing a cellulose-derived polyol for chain extension and cellulose-citrate as a thickener additive. The utilized polyol was obtained from the reduction reaction of cellulose-derived bio-oil through the use of sodium borohydride and iodine. Primarily, we produced both rigid and flexible polyurethane foams through chain extension of the prepolymers. Secondly, we investigated the role of cellulose citrate as a polyurethane additive to improve the mechanical properties of the realized composite materials. The products were characterized by FT-IR spectroscopy and their morphologies were analysed by SEM. Mechanical tests were evaluated to open new perspectives towards different applications.

Keywords: bio-based polyurethanes; prepolymers; cellulose-derived polyol; cellulose-citrate; polyurethane composites

1. Introduction

During the last decade, industrial and academic interest in the conversion of renewable biomass into fuels, chemicals and useful materials grew enormously [1,2]. In contrast to the limited fossil sources, agricultural wastes represent an opportunity because they are widely available, with their annually produced quantity estimated at around 2×10^{11} tons per year [3,4]. In terms of the future prospect of a totally sustainable economy, chemistry plays a central role [5–7]. Cellulose is the major component of natural feedstocks and the most important efforts of the scientific community are focused on the development of new methodologies for its transformation [8–11]. Pyrolysis is one of the oldest chemical methods to convert cellulose into useful platform molecules such as sugars, furans or bio-oil mixtures [12–14]. The harsh conditions related to high temperature and pressure were substituted, over the years, by milder approaches [15–18]. However, under milder pressure and temperature, hydrolysis is usually accompanied by the use of such mineral acids as sulfuric or hydrochloric acid, and this involves damage to process equipment as well as negative environmental impact [19,20]. Unconventional solvents such as ionic liquids or deep eutectic solvents solved many environmental and procedural problems [21,22], but they are still limited by the difficulty of recovering the products, due to their high affinity for these phases [23–26]. In this ambit, novel surfactant-based ionic liquids are receiving attention due to their ease of preparation, low cost, and inherent added-value properties; in fact, their (tuneable) amphiphilic nature can, in principle, help in solving affinity problems, but applications are still far from being achieved [27,28]. Ball milling and ultrasounds are the most used physical methods for the pre-treatment and the conversion of cellulosic biomass, but the scalability of these methods is usually limited [29,30]. Cellulose-derived products are extremely versatile building blocks for the production of materials for engineering applications, medical equipment, food-related purposes, biofuels and many other applications [31–33]. Polyurethanes (PUs) are multifunctional polymers that have been known and produced for a long time, with wide commercial diffusion due to their low cost and significant versatility, having been used, for example, as thermal insulation materials [34,35]. Conventional PUs can be formed by a very exothermic reaction of isocyanates and polyols [36] but, in recent years, industrial and research efforts have been focused on the production of bio-based products with more eco-sustainable methodologies [37–40].

Cellulose can be turned into valuable polyols that can be used as a starting materials or chain extenders for polyurethane synthesis [41–43]. We recently developed a simple and fast method for the contemporary conversion of microcrystalline cellulose into a furan-enriched bio-oil and cellulose citrate [44]. In another recent work, we proved the efficacy of some inorganic salts, such as sodium chloride, in the reaction of PEG 400 and different aliphatic diisocyanates for the production of polyurethanes [45].

The principal goal of this work was the development of a novel, efficient and sustainable approach for the synthesis of polyurethane foams and polyurethane composites using two renewable materials and a green catalyst. More specifically, the furan-enriched bio-oil obtained through our previous procedure [44] was successfully reduced using a simple open flask reaction into a polyol mixture, composed mainly of 2,5-bis(hydroxyl methyl)furan. More precisely, this furan diol was proven, over the years, to have different applications, and the synthetic efforts towards its facile production are growing rapidly [46–48]. For the aforementioned reasons, we used—for the first time, to our knowledge—this cellulose-derived polyol, which was synthesized through our simple and eco-compatible procedure, as a chain extender to synthesize novel bio-based polyurethane foams. Subsequently, the addition of the second renewable product into the same reaction mixture, with cellulose citrate as an additive, allowed us to realize novel polyurethane composite foams with enhanced mechanical properties. The final materials were chemically characterized by FT-IR spectroscopy and their morphologies were analysed by SEM. Their mechanical properties were preliminary evaluated to facilitate new application-based studies using these materials. Finally, a conceptualization regarding the chemical interactions of cellulose citrate inside the urethane chains was carried out.

2. Materials and Methods

Tetrahydrofuran (THF) was purchased from Carlo Erba (Milan, Italy) at analytical grade, and freshly distilled before use after drying over sodium sulfate. Acetone was purchased from Honeywell at a high purity grade and used without further purification. Microsrystalline cellulose type 102 was purchased from Roquette (Lestrem, France) at high purity. Citric acid was purchased from Sigma Aldrich (St. Louis, MO, USA) at 99% purity grade. Sodium borohydride was purchased from Carlo Erba (Milan, Italy) at 95% purity grade. Iodine was purchased from Carlo Erba (Milan, Itay) at the analytical grade. Polyethylene glycol (PEG) 400 was purchased from Thermo Fisher Scientific (Waltham, MA, USA) at 99% purity grade. Isophorone diisocyanate (IPDI) **3**, 4,4′-methylenedicyclohexyl diisocyanate (H_{12}MDI) **4**, and 2,2,4-trimethyl hexamethylene diisocyanate (TMDI) **5** were purchased from EVONIK INDUSTRIES (Essen, Germany), at 95% purity grade. Sodium chloride and sodium sulfate anhydrous was purchased from Sigma Aldrich (St. Louis, MO, USA) at analytical grade. IR spectra of all products are reported in the Supplementary Material (Sections S2, S3 and S5).

2.1. Bio-Oil Reduction into Polyol **1**

A quantity of 50 g of bio-oil was dissolved in 200 mL of THF dry in a 500 mL one-neck round bottom flask. In addition, 12.5 g (25% in weight respect to bio-oil) of $NaBH_4$ was added portion-wise and hydrogen was released from the reaction flask. After one hour, 15 g (30% in weight respect to bio-oil) of iodine was added slowly while the hydrogen was released from the reaction flask. After one hour, the excess $NaBH_4$ was quenched with aqueous HCl 1M. The mixture was filtered through a sintered glass funnel. The solution was dried over sodium sulphate, filtered, and the solvent was removed under vacuum to obtain a polyol in the form of a brown oil, with a yield of 85–90%. The LC-MS and FT-IR spectra of bio-oil and polyol are reported in the Supplementary Material (Sections S4 and S5).

2.2. *Hydroxyl Group Content of Biomass Derived Polyol* **1**

Polyol **1** was initially esterified using phthalic anhydride according to an earlier study [49]. In the same reaction flask, phthalic anhydride (112 g, 0.76 mols) and imidazole (17 g, 0.25 mols) were dissolved in pyridine (700 mL) (this mixture is called phthalation reagent). One gram of polyol and 25 mL of the phthalation reagent were mixed under stirring at 100 °C for 15 min. The mixture was cooled down to room temperature and 50 mL of pyridine and 10 mL of distilled water were added. The tritant solution NaOH (0.5 M) was added until the pink ending point (1 mL of phenol-phthalein solution, 1% *w*/*v* in EtOH, was used as indicator). The experiments were replicated three times and the value expressed as a mean value. The hydroxyl group content was determined using the following equation:

$$Hydroxyl\ number = \frac{(B-S)(56.1)(N)}{W} \tag{1}$$

B = the volume of the blank (mL) at the ending point; S = the volume of the substrate (mL) at the ending point; N = the normality of the NaOH solution; W = the weight of the substrate. The OH content of cellulose-derived polyol **1** was 310 mg KOH/g.

2.3. *Prepolymer Synthesis* **6–8**

The detailed procedure is reported in the literature [45]. Briefly, three different prepolymers **6–8** were synthesized using the relative diisocyanates **3–5** at a molar excess of 2.5:1 with respect to PEG 400 **2**. Water was used as a blowing agent at a weight percentage of 5.6% with respect to PEG 400 **2**. The reagents were added in the following order in a plastic container: PEG 400 **2**, distilled water and NaCl as catalyst. The mixture was stirred using a mechanical apparatus. The appropriate diisocyanate was added and the mixture was vigorously stirred. The blend was warmed up to 70 °C for one hour until diisocyanate consumption. The three prepolymers **6–8** were obtained using the same procedure in the form of a colourless gel. The reaction was monitored by FT-IR spectroscopy with respect to the isocyanate signal. The produced prepolymers were defined as stable in accordance with the standard titration method, ASTM D 2572-97, using di-*n*-butylamine [49].

2.4. *Synthesis of PU Foams* **9–11** *Using Polyol 1 as a Chain Extender*

Polyol **1** at 30% weight with respect to the prepolymer was added to the freshly prepared prepolymer, and the mixture was vigorously and mechanically stirred for a few minutes in the same plastic container. After that, the same mixture was reversed in a steel mold (Supplementary Material, Section S6) and the system was closed under pressure at room temperature. After eight hours, the polyurethane foam was obtained in the final form for further use (see Supplementary Material, Section S7). The disappearance of the isocyanate signal was monitored by FT-IR spectroscopy.

2.5. *Preparation of PU Composite Foams* **12–14** *Using Cellulose Citrate as a Thickener Additive*

Polyol **1** at 30% weight and cellulose citrate at 20% weight, with respect to the prepolymer, were added sequentially in the same plastic container as the prepolymer, and the mixture was vigorously and mechanically stirred for a few minutes. After that, the same mixture was reversed in a steel mold and the system was closed under pressure at room temperature. After eight hours, the polyurethane foam was obtained in the final form for further use. The disappearance of the isocyanate signal was monitored by FT-IR spectroscopy.

2.6. Fourier-Transform Infrared Spectroscopy (FT-IR)

FT-IR spectra were acquired using the Shimadzu IRAffinity-1S spectrometer (Shimadzu Corporation, Kyoto, Japan) in the spectral region of 400 to 4000 cm^{-1} with a resolution of 1 cm^{-1}, with 48 scans undertaken for a single analysis, and the KBr pellets technique used. The KBr pellets were obtained by mixing the substrate with KBr powder (ratio 1:100) and by pressing with a hydraulic press. All spectra are collected in the Supplementary Material (Sections S2, S3 and S5).

2.7. Scanning Electron Microscopy (SEM)

Morphological studies of PU composites were carried out using a LEO 420 scanning electron microscope (SEM, Zeiss, Oberkochen, Germany), operating with vacuum conditions of 9×10^6 Torr at an accelerating voltage of 15 kV. Before analysis, all substrates were gold metallized using an Auto Sputter Coater (Agar, Stansted, UK). Images were taken with 100 SEM micrograph magnications. The cell structure analysis was conducted using the Aphelion^{TM} software version 4.5.0. The program, through the use of SEM images, provided the cell surface and cell number. Furthermore, the anisotropy index was calculated as the ratio of cell height to width in the perpendicular direction of the foam growth.

2.8. Mechanical Test

Mechanical tests were performed in accordance with UNI EN ISO 3386-2:2010, "Flexible cellular polymeric materials, Determination of stress–strain characteristic in compression. Part 2: High density material". The test pieces were cubes with square load-bearing surfaces of 40 mm minimum size.

An Instron 3369 double column universal machine with a load cell of 50 kN, equipped with two circular plates of 160 mm in diameter, was used for the test. The compression plate moved at a uniform low rate of 2 mm/min.

3. Results and Discussion

The conversion of microcrystalline cellulose (MCC) into multi-functional molecules is intricate because of its high ordered crystalline structure and the low accessibility, which limit the interactions with chemicals and the consequent depolymerisation [50,51]. In our previous work, we proved the role of molten citric acid in the contemporary hydrolysis and esterification of MCC [44]. With the aim of producing a renewable polyol, the obtained furan-enriched bio-oil was reduced by a procedure employed for the conversion of carboxylic acids, which makes use of sodium borohydride and iodine, as mild reducing agents, in an open-air system [52]. This reaction is useful for bio-oil reduction because aldehydes and ketones can be reduced by sodium borohydride while the sodium borohydride/iodine system is necessary for the reduction of carboxylic groups. The reaction is shown in Scheme 1.

The newly synthesized polyol **1** was obtained as a dark highly viscous oil following two hours of the aforementioned one-pot reaction, and 2,5-bis(hydroxymethyl)furan was identified as the main compound through the LC-MS technique (see Supplementary Material for FT-IR and LC-MS spectra of polyo **1**, Sections S3 and S4).

It is a common practice for the production of polyurethane foams to use a two-step methodology through which the prepolymers produced in the first step can be extended with short diols in the second step [53]. Initially, we followed our previously reported method for the synthesis of prepolymers using polyethylene glycol (PEG 400) (**2**) with three different aliphatic diisocyanates—isophorone diisocyanate (IPDI) (**3**), 4,4′-methylenedicyclohexyl diisocyanate ($H_{12}MDI$) (**4**) and 2,2,4-trimethyl hexamethylene diisocyanate (TMDI) (**5**)—and sodium chloride as the catalyst [45]. We chose these diisocyanates as source of polyurethanes not only because of their easy commercial availability, but also due to their aliphatic nature. In fact, the aromatic diisocyanates are certainly more commonly used, but they are very toxic. Moreover, few works in the literature that involve the use of aliphatic diisocyanates are reported [54,55], and this fact prompted us to develop an adequate and more eco-friendly synthetic strategy.

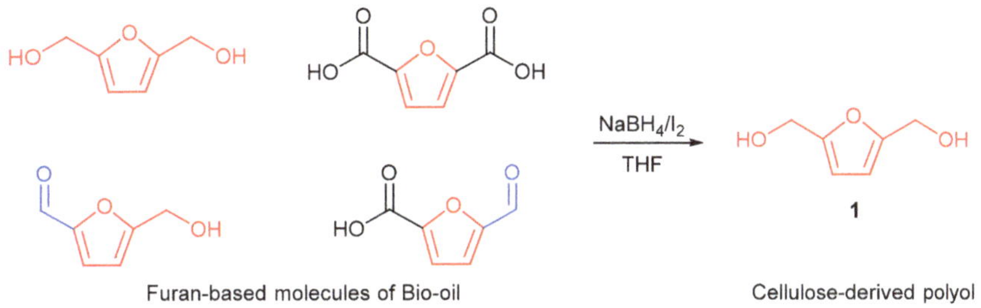

Scheme 1. Bio-oil reduction to cellulose-derived polyol.

The reaction products (**6–8**) are shown in Scheme 2.

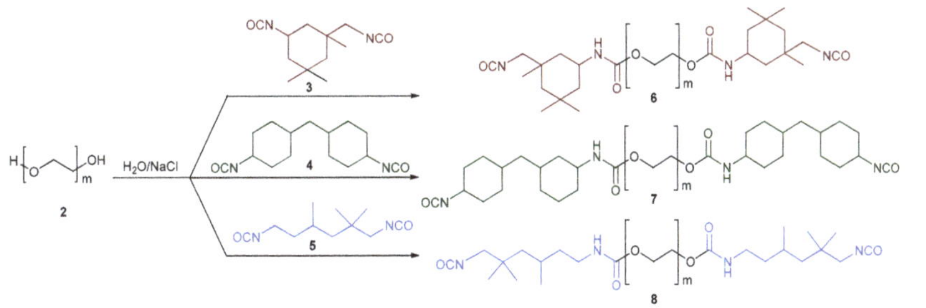

Scheme 2. Reaction scheme for the synthesis of the prepolymers (**6–8**).

Successively, we employed the cellulose-derived polyol (**1**) as a chain extender for the three prepolymers (**6–8**) (Scheme 3). The reaction was conducted, without the addition of any ulterior catalyst, in a closed steel mold, at room temperature, and the products (**9–11**) were obtained in the form of polyurethane foams after 8 h. The quantitative conversion was verified by FT-IR analysis through the complete disappearance of the isocyanate signal (see Supplementary Material, Section S5).

The properties of the realized composite materials were chemically characterized by FT-IR spectroscopy (Figure 1). We marked the characteristic peaks of the polyurethane structure. The vibrational bands near 3330 cm^{-1} are relative to N-H stretching vibrations. The shoulder at its high-frequency side (around 3600 cm^{-1}) can be due to tightly bounded or monomeric residual molecules of water [56]. The IR signal in the region near 3000 cm^{-1} corresponds to both the asymmetrical and symmetrical stretching vibrations of aliphatic C-H. The signal near 1700 cm^{-1} corresponds to C=O stretching vibration. The band above 1500 cm^{-1} is probably relative to N-H bending vibrations and C-N stretching vibrations. The IR signal near 1100 cm^{-1} is relative to O-C-O stretching frequency. The band in the region near 950 cm^{-1} is relative to the =C-H bending vibrations of furan [57,58].

Finally, cellulose citrate was employed in the same polyurethane formulations as an additive and the relative polyurethane composites were obtained (**12–14**) (Table 1, Section S8).

For example, Figure 2 shows the interaction forces between cellulose citrate and the chains of polyurethane **14**. Considering the chemical nature of polyurethane chains and cellulose citrate, we suppose a strong interaction between them, assisted by a hydrogen bond network.

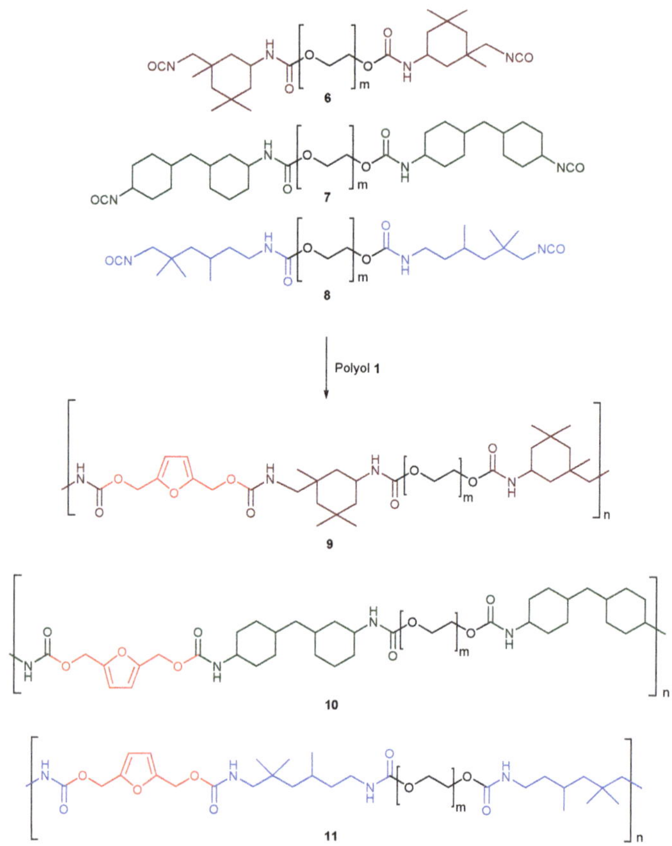

Scheme 3. The final products after chain extension with polyol **1**.

The presence of the hydrogen bonds could be confirmed by FT-IR spectra, as can be observed in the spectrum of materials **12–14**. In fact, the band in the range of 3600–2800 cm^{-1} is very broad and can also encompass the vibration signal of the -OH group (Figure 3).

Moreover, in the literature, a shift to lower frequencies was documented in the carbonyl group when this group took part in hydrogen bonds [59]. The shift was evident for the C=O stretching of the citrate group, which, from its typical vibration signal near 1735 cm^{-1} in cellulose citrate, was evidently shifted to lower frequency of 1700 cm^{-1} in all of the composites due to the fact that this molecule was engaged in a network of hydrogen bonds.

Scanning electron microscopy observations were carried out to monitor the structure of PUs **9–11** and PU composites **12–14** (Figure 4). In all cases, the morphological analysis shows the typical polyurethane structure. In the case of **9** (Figure 4a) and **12** (Figure 4b), two types of pores can be observed, large and small, with dimensions of 500 and 150 μm, respectively. Both types of pores are evenly distributed in the matrix. For **10** (Figure 4c) and **13** (Figure 4d), the structure appears compact and devoid of pores; in particular, for **13** (Figure 4d), a roughness, attributable to the presence of cellulose citrate, is observed on its surface. For **11** (Figure 4e) and **14** (Figure 4f), both still have a pore-rich structure with sizes between 250 and 50 μm, but **14** appears more compact than **11** **[60]**. Ultimately, the morphological analysis indicates that the presence of cellulose citrate generates a decrease in the size of the pores and/or a greater compactness, with a consequent increase in density, as also found in the mechanical characterizations. These findings could be attributable to the interaction forces between the cellulose citrate and the polyurethane chains. Apart from their mutual chemical affinity, this hypothesis could be reinforced by the results of some studies which pointed out that, in cases of blending of differently sized molecules, a spontaneous structuring takes place, driven by the fact that smaller molecules tend to be more likely to occupy the voids made by bigger ones, with a compression of volume and, overall, a more compact structure [61,62].

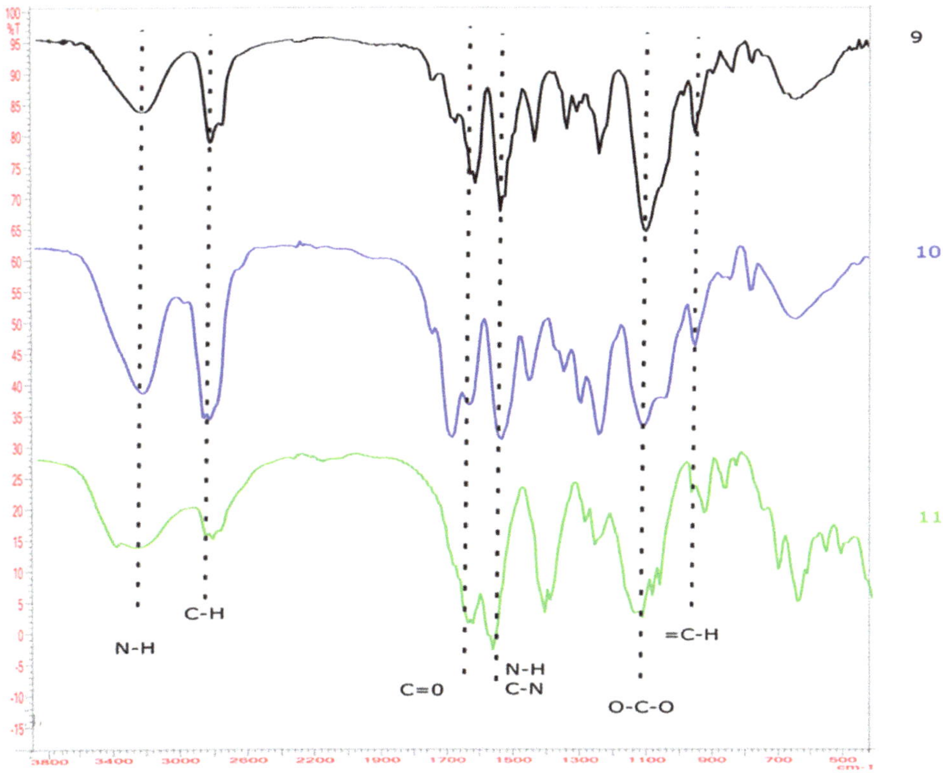

Figure 1. FT-IR spectra of the products (**9–11**).

Table 1. Polyurethane composite materials **12–14**.

PU	Chain Extender	Additive	PU Composite
9	Polyol **1**	Cellulose citrate	**12**
10	Polyol **1**	Cellulose citrate	**13**
11	Polyol **1**	Cellulose citrate	**14**

Structural parameters were determined, and are included in Table 2.

The anisotropy index is defined as the ratio of the length to the width of the cell. When this ratio is close to 1, the shape of the cell approaches a sphere [63]. The cells of **9**, **11**, **12** and **14** in the cross section, which were perpendicular to the direction of the foam growth, were characterized by anisotropy index very close to 1. These values indicate that the cells have almost spherical shapes. Therefore, the introduction of cellulose citrate into the systems does not change the shape of the cells, considering that the values of the anisotropy index are similar in all systems. It was observed that, due to the nucleation effect, the addition of cellulose citrate for PU composites **12** and **14** increased the number of cells in the cross section of the foam, as compared to the corresponding additive-free PU systems

9 and 11, respectively. The effect of increasing the number of cells after adding cellulose citrate or other fillers was reported in the literature [64,65]. For 10 and 13, the evaluation of the characteristics of the cellular structures was not performed as they appeared compact and devoid of pores, as can be seen in Figure 4c,d.

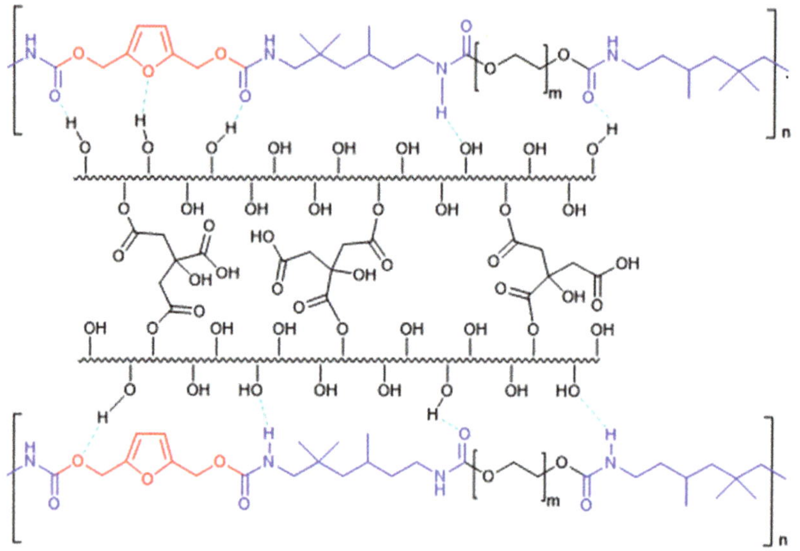

Figure 2. Graphical representation of the composite material **14** derived from cellulose citrate and polyurethane **11**.

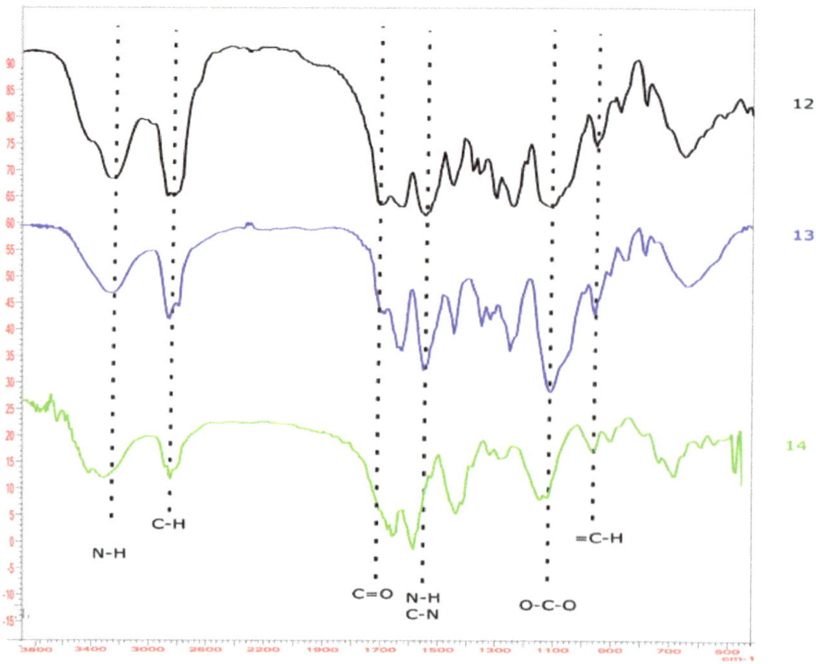

Figure 3. FT-IR spectrum of **14** that can confirm the presence of hydrogen bonds.

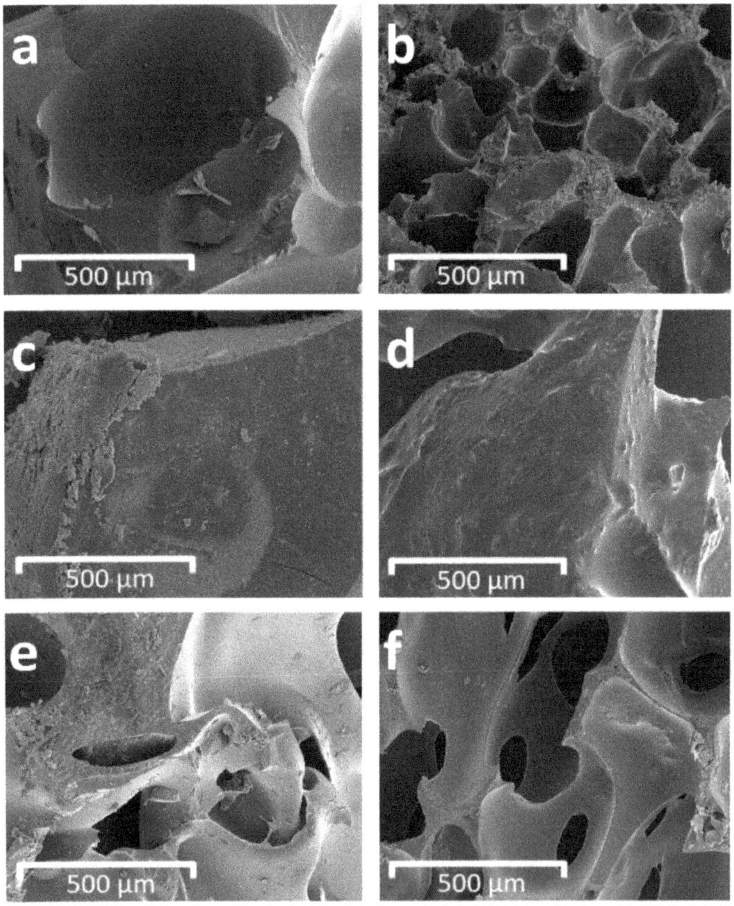

Figure 4. Scanning electron microscopy (SEM) images of structure of PUs and PU composites (×100 magnification): **(a) 9**, **(b) 12**, **(c) 10**, **(d) 13**, **(e) 11**, **(f) 14**.

Table 2. Characteristics of PU cellular structure.

PU	Anisotropy Index	Number of Cell/(mm^2)	Average Cell Area (mm^2)
9	0.97 ± 0.05	21 ± 2	0.242 ± 0.044
11	1.09 ± 0.12	8 ± 2	0.503 ± 0.063
12	1.12 ± 0.06	109 ± 7	0.023 ± 0.008
14	1.16 ± 0.07	39 ± 4	0.152 ± 0.012

Finally, a comparison of the mechanical properties of the PUs **9–11** and the PU composites **12–14** was carried out [37,60]. The results are reported in Table 3.

Products **10** and **13** show a high density, of about 1000 kg/m^3, and a brittle behaviour. The material, during the compression test, fractured without significant plastic deformation. The maximum values of the stress–strain curves are 0.95 MPa and 1.1 MPa for **10** and **13**, respectively. The compounds **9** and **12** show lower density (about 290 kg/m^3), and better compressive strength (compared to compounds **10** and **13**), with a leathery behaviour. In the stress–strain curve, after an initial linear loading regime, the foam specimens show relatively abrupt yielding, followed by a sustained plateau region. Finally, **11** and **14** have a rubbery behaviour and a density comparable to **9** and **12**. For this type of material, the breaking load has not been evaluated, as they are elastomers; in fact, they deform during the compression test and, after the load is released, they return to their initial dimensions.

Table 3. Mechanical properties of PUs **9–11** and PU composites **12–14**.

Substrates	Density (kg/m^3)	Compressive Strength (MPa)
9	260 ± 10	2.03 ± 0.04
10	970 ± 10	0.95 ± 0.05
11	270 ± 10	-
12	290 ± 10	2.41 ± 0.09
13	1000 ± 10	1.10 ± 0.06
14	280 ± 10	-

4. Conclusions

In conclusion, bio-based composite materials were successfully synthesized and characterized. The syntheses were carried out by two-step routes by using, in the second step (elongation stage), a furan-enriched polyol, derived from cellulose, as a chain extender to synthesize eco-sustainable polyurethane foams. In particular, our efforts made it possible to reduce bio-oil in an innovative manner, in an open-air reaction flask and under atmospheric pressure, to obtain a furan-enriched polyol, a privileged renewable platform molecule with a wide range of applications in the present and for the future. We used this product and we proved that is an optimal chain extender for this kind of polymer. For the second series of composites, cellulose citrate, another renewable material, was employed as a thickener for the first time, and it was added into the same reaction mixture to obtain novel composite materials with enhanced mechanical properties. In fact, from preliminary mechanical data, it seems that the use of cellulose citrate as an additive improves the mechanical performance of the materials, increasing their compressive strength. Moreover, the experimental data confirm that the hyperbranched structure of the cellulose citrate introduced into polyurethane chains increases the hydrogen bonds in the PU system (see Figure 2), causing a compression of volume and a more compact structure in the composite materials. In general, prepared bio-based polyurethane foams and their composites show good mechanical properties; therefore, we believe that our innovative method should be considered very attractive. Certainly, ulterior studies will be necessary in order to establish other properties of our materials—such as thermal conductivity and thermal stability—for such applications as insulators.

In the future, we can consider the potential use of our products in the industrial sector, and in agrifood and agroforestry waste processing, as a source of cellulose citrate and furan-enriched polyol, with strong environmental impact due to the recycling of waste material. Therefore, a wider application of this simple, efficient and eco-sustainable methodology may represent a valid alternative to other elaborate processes that are currently in use; moreover, the production of novel bio-based polyurethane foams and polyurethane composites, with their chemical and physical properties, will surely be informative for future applications in this research field.

Supplementary Materials: The following are available online at https://www.mdpi.com/article/10.3390/polym13162802/s1. Section S1: Reagents; Section S2: FT-IR spectra of PEG 400, isocyanates and the relative prepolymers; Section S3: FT-IR spectra of biomass derived polyol (1); Section S4: LC-MS spectrum of polyol (1); Section S5: FT-IR spectra of biomass-based polyurethanes; Section S6: Composite material processing equipment; Section S7: Photographs of bio-based polyurethanes; Section S8: Photographs of bio-based composite materials.

Author Contributions: Conceptualization: L.M. and A.D.N.; Supervision: L.M. and A.D.N.; investigation: L.M., F.O., P.C. and A.D.N.; resources: L.M. and A.D.N.; writing—original draft: L.M., F.O. and A.D.N.; writing—review and editing: L.M.; P.C. and A.D.N.; funding acquisition: L.M. and A.D.N.; methodology: F.O., V.A., A.J. and M.A.T.; formal analysis: F.O., V.A., A.J., M.A.T., A.T., C.S.

and A.F.; data curation: F.O.; V.A., A.J., A.T., C.S. and A.F. All authors have read and agreed to the published version of the manuscript.

Funding: This research was supported by the FOREST-COMP—CUP H56C18000080005—PON "Ricerca e Innovazione 2014–2020" project. The University of Calabria and Calabria Region (PAC CALABRIA 2014–2020—Asse Prioritario 12, Azione B 10.5.12 CUP: H28D19000040006) are acknowledged for their financial support.

Institutional Review Board Statement: Not applicable.

Informed Consent Statement: Not applicable.

Data Availability Statement: Data are contained within the article.

Conflicts of Interest: The authors declare no conflict of interest.

References

1. Alonso, D.M.; Bond, J.Q.; Dumesic, J.A. Catalytic conversion of biomass to biofuels. *Green Chem.* **2010**, *12*, 1493–1513. [CrossRef]
2. Mascal, M. 5-(Chloromethyl) furfural (CMF): A platform for transforming cellulose into commercial products. *ACS Sustain. Chem. Eng.* **2019**, *7*, 5588–5601. [CrossRef]
3. Singh, H.; Sharma, A.; Bhardwaj, S.K.; Arya, S.K.; Bhardwaj, N.; Khatri, M. Recent advances in the applications of nano-agrochemicals for sustainable agricultural development. *Environ. Sci. Process. Impacts* **2021**, *23*, 213–239. [CrossRef]
4. Li, J.; Jiang, Z.; Hu, L.; Hu, C. Selective conversion of cellulose in corncob residue to levulinic acid in an aluminum tri-chloride–sodiumchloride system. *ChemSusChem* **2014**, *7*, 2482–2488. [CrossRef]
5. Collins, T.J. Review of the twenty-three year evolution of the first university course in green chemistry: Teaching future leaders how to create sustainable societies. *J. Clean. Prod.* **2017**, *140*, 93–110. [CrossRef]
6. Horváth, I.T. Introduction: Sustainable chemistry. *Chem. Rev.* **2018**, *118*, 369–371. [CrossRef] [PubMed]
7. Anjali, K.; Venkatesha, N.J.; Christopher, J.; Sakthivel, A. Rhodium porphyrin molecule-based catalysts for the hydrogenation of biomass derived levulinic acid to biofuel additive c-valerolactone. *New J. Chem.* **2020**, *44*, 11064–11075. [CrossRef]
8. Qiu, X.; Hu, S. "Smart" materials based on cellulose: A review of the preparations, properties, and applications. *Materials* **2013**, *6*, 738–781. [CrossRef] [PubMed]
9. Hassan, N.S.; Jalil, A.A.; Hitam, C.N.C.; Vo, D.V.N.; Nabgan, W. Biofuels and renewable chemicals production by catalytic pyrolysis of cellulose: A review. *Environ. Chem. Lett.* **2020**, *18*, 1625–1648. [CrossRef]
10. Maiuolo, L.; Algieri, V.; Olivito, F.; Tallarida, M.A.; Costanzo, P.; Jiritano, A.; De Nino, A. Chronicle of nanocelluloses (NCs) for catalytic applications: Key advances. *Catalysts* **2021**, *11*, 96. [CrossRef]
11. Baghaei, B.; Skrifvars, M. All-cellulose composites: A Review of recent studies on structure, properties and applications. *Molecules* **2020**, *25*, 2836. [CrossRef]
12. Shen, D.; Xiao, R.; Gu, S.; Luo, K. The pyrolytic behavior of cellulose in lignocellulosic biomass: A review. *RSC Adv.* **2011**, *1*, 1641–1660. [CrossRef]
13. Kan, T.; Strezov, V.; Evans, T.J. Lignocellulosic biomass pyrolysis: A review of product properties and effects of pyrolysis parameters. *Renew. Sustain. Energy Rev.* **2016**, *57*, 1126–1140. [CrossRef]
14. Yang, G.; Yang, J.; Huang, D.; Zhou, W.; Yang, L.; Lv, P.; Yi, W.; Sun, Y.; Yan, B. BTX production from rice husk by fast catalytic pyrolysis over a Ga-modified ZSM-5/SBA-15 catalyst. *New J. Chem.* **2021**, *45*, 3809–3816. [CrossRef]
15. Chinnappan, A.; Baskar, C.; Kim, H. Biomass into chemicals: Green chemical conversion of carbohydrates into 5-hydroxymethylfurfural in ionic liquids. *RSC Adv.* **2016**, *6*, 63991–64002. [CrossRef]
16. Pang, J.; Zheng, M.; Li, X.; Sebastian, J.; Jiang, Y.; Zhao, Y.; Wang, A.; Zhang, T. Unlock the compact structure of lignocel-lulosic biomass by mild ball milling for ethylene glycol production. *ACS Sustain. Chem. Eng.* **2019**, *7*, 679–687. [CrossRef]
17. Gaudino, E.C.; Cravotto, G.; Manzoli, M.; Tabasso, S. From waste biomass to chemicals and energy via microwave-assisted processes. *Green Chem.* **2019**, *21*, 1202–1235. [CrossRef]
18. Li, Z.; Su, K.; Ren, J.; Yang, D.; Cheng, B.; Kim, C.K.; Yao, X. Direct catalytic conversion of glucose and cellulose. *Green Chem.* **2018**, *20*, 863–872. [CrossRef]
19. Palkovits, R.; Tajvidi, K.; Procelewska, J.; Rinaldia, R.; Ruppert, A. Hydrogenolysis of cellulose combining mineral acids and hydrogenationcatalysts. *Green Chem.* **2010**, *12*, 972–978. [CrossRef]
20. Chang, J.K.-W.; Duret, X.; Berberi, V.; Zahedi-Niaki, H.; Lavoie, J.-M. Two-step thermochemical cellulose hydrolysis with partial neutralization for glucose production. *Front. Chem.* **2018**, *6*, 117–127. [CrossRef]
21. Calandra, P.; Szerb, E.; Lombardo, D.; Algieri, V.; De Nino, A.; Maiuolo, L. A Presentation of ionic liquids as lubricants: Some critical comments. *Appl. Sci.* **2021**, *11*, 5677. [CrossRef]
22. Maiuolo, L.; Algieri, V.; Olivito, F.; De Nino, A. Recent developments on 1,3-dipolar cycloaddition reactions by catalysis in green solvents. *Catalysts* **2020**, *10*, 65. [CrossRef]

23. Kuroda, K.; Miyamura, K.; Satria, H.; Takada, K.; Ninomiya, K.; Takahashi, K. Hydrolysis of cellulose using an acidic and hydrophobic ionic liquid and subsequent separation of glucose aqueous solution from the ionic liquid and 5-(Hydroxymethyl) furfural. *ACS Sustain. Chem. Eng.* **2016**, *4*, 3352–3356. [CrossRef]
24. Chen, T.; Xiong, C.; Tao, Y. Enhanced hydrolysis of cellulose in ionic liquid using mesoporous ZSM-5. *Molecules* **2018**, *23*, 529. [CrossRef]
25. Loow, Y.-L.; New, E.K.; Yang, G.H.; Ang, L.Y.; Foo, L.Y.W.; Wu, T.Y. Potential use of deep eutectic solvents to facilitate lignocellulosic biomass utilization and conversion. *Cellulose* **2017**, *24*, 3591–3618. [CrossRef]
26. Liu, F.; Xue, Z.; Zhao, X.; Mou, H.; He, J.; Mu, T. Catalytic deep eutectic solvents for highly efficient conversion of cellulose to gluconic acid with gluconic acid self-precipitation separation. *Chem. Commun.* **2018**, *54*, 6140–6143. [CrossRef] [PubMed]
27. Liveri, V.T.; Lombardo, D.; Pochylski, M.; Calandra, P. Molecular association of small amphiphiles: Origin of ionic liquid properties in dibutyl phosphate/propylamine binary mixtures. *J. Mol. Liq.* **2018**, *263*, 274–281. [CrossRef]
28. Calandra, P.; Mandanici, A.; Liveri, V.T. Self-assembly in surfactant-based mixtures driven by acid–base reactions: Bis(2-ethylhexyl) phosphoric acid–n-octylamine systems. *RSC Adv.* **2013**, *3*, 5148–5155. [CrossRef]
29. Barakat, A.; Mayer-Laigle, C.; Solhy, A.; Arancon, R.A.D.; De Vries, H.; Luque, R. Mechanical pretreatments of lignocellulosic biomass: Towards facile and environmentally sound technologies for biofuels production. *RSC Adv.* **2014**, *4*, 48109–48127. [CrossRef]
30. Yang, C.; Yuan, X.; Wang, X.; Wu, K.; Liu, Y.; Liu, C.; Lu, H.; Liang, B. Ball milling promoted direct liquefaction of lignocellulosic biomass in supercritical ethanol. *Front. Chem. Sci. Eng.* **2019**, *14*, 605–613. [CrossRef]
31. Wang, Q.; Sun, J.; Yao, Q.; Ji, C.; Liu, J.; Zhu, Q. 3D printing with cellulose materials. *Cellulose* **2018**, *25*, 4275–4301. [CrossRef]
32. Khan, A.; Jagdale, P.; Rovere, M.; Nogués, M.; Rosso, C.; Tagliaferro, A. Carbon from waste source: An eco-friendly way for strengthening polymer composites. *Compos. Part B Eng.* **2018**, *132*, 87–96. [CrossRef]
33. Golling, F.E.; Pires, R.; Hecking, A.; Weikard, J.; Richter, F.; Danielmeier, K.; Dijkstra, D. Polyurethanes for coatings and ad-hesives–chemistry and applications. *Polym. Int.* **2019**, *68*, 848–855. [CrossRef]
34. De Avila Delucis, R.; Magalhães, W.L.E.; Petzhold, C.L.; Amico, S.C. Forest-based resources as fillers in biobased polyurethane foams. *J. Appl. Polym. Sci.* **2018**, *135*, 45684. [CrossRef]
35. Zhang, H.; Fang, W.-Z.; Li, Y.-M.; Tao, W.-Q. Experimental study of the thermal conductivity of polyurethane foams. *Appl. Therm. Eng.* **2017**, *115*, 528–538. [CrossRef]
36. Gharib, J.; Pang, S.; Holland, D. Synthesis and characterisation of polyurethane made from pyrolysis bio-oil of pine wood. *Eur. Polym. J.* **2020**, *133*, 109725. [CrossRef]
37. Akindoyo, J.O.; Beg, M.D.H.; Ghazali, S.; Islam, M.R.; Jeyaratnam, N.; Yuvaraj, A.R. Polyurethane types, synthesis and applications—A review. *RSC Adv.* **2016**, *6*, 114453–114482. [CrossRef]
38. Rokickia, G.; Parzuchowskia, P.G.; Mazurek, M. Non-isocyanate polyurethanes: Synthesis, properties, and applications. *Polym. Adv. Technol.* **2015**, *26*, 707–761. [CrossRef]
39. Calvo-Correas, T.; Martin, M.D.; Retegi, A.; Gabilondo, N.; Corcuera, M.A.; Eceiza, A. Synthesis and Characterization of Polyurethanes with High Renewable Carbon Content and Tailored Properties. *ACS Sustain. Chem. Eng.* **2016**, *4*, 5684–5692. [CrossRef]
40. Cao, J.; Chen, N.; Chen, Y.; Luo, X. Synthesis of a novel biodegradable polyurethane with phosphatidylcholines. *Int. J. Mol. Sci.* **2010**, *11*, 1870–1877. [CrossRef] [PubMed]
41. Aberasturi, O.G.J.; Gómez, J.R.O. New approaches to producing polyols from biomass. *J. Chem. Technol. Biotechnol.* **2017**, *92*, 705–711. [CrossRef]
42. Arnaud, S.P.; Wu, L.; Chang, M.-A.W.; Comerford, J.W.; Farmer, T.J.; Schmid, M.; Chang, F.; Li, Z.; Mascal, M. New bio-based monomers: Tuneable polyester properties using branched diols from biomass. *Faraday Discuss.* **2017**, *202*, 61–77. [CrossRef] [PubMed]
43. Choi, S.M.; Lee, M.W.; Shin, E.J. One-pot processing of regenerated cellulose nanoparticles/waterborne polyurethane nanocomposite for eco-friendly polyurethane matrix. *Polymers* **2019**, *11*, 356. [CrossRef] [PubMed]
44. Romeo, I.; Olivito, F.; Tursi, A.; Algieri, V.; Beneduci, A.; Chidichimo, G.; Maiuolo, L.; Sicilia, E.; De Nino, A. Totally green cellulose conversion into bio-oil and cellulose citrate using molten citric acid in an open system: Synthesis, characterization and computational investigation of reaction mechanisms. *RSC Adv.* **2020**, *10*, 34738–34751. [CrossRef]
45. Maiuolo, L.; Olivito, F.; Ponte, F.; Algieri, V.; Tallarida, M.A.; Tursi, A.; Chidichimo, G.; Sicilia, E.; De Nino, A. A novel catalytic two-step process for the preparation of rigid polyurethane foams: Synthesis, mechanism and computational studies. *React. Chem. Eng.* **2021**, *6*, 1238–1245. [CrossRef]
46. Yuan, H.; Kusema, B.T.; Yan, Z.; Streiff, S.; Shi, F. Highly selective synthesis of 2,5-bis(aminomethyl)furan via catalytic ami-nation of 5-(hydroxymethyl)furfural with NH_3 over a bifunctional catalyst. *RSC Adv.* **2019**, *9*, 38877–38881. [CrossRef]
47. Zhao, W.; Wu, W.; Li, H.; Fang, C.; Yang, T.; Wang, Z.; He, C.; Yang, S. Quantitative synthesis of 2,5-bis(hydroxymethyl)furan from biomass-derived 5-hydroxymethylfurfural and sugars over reusable solid catalysts at low temperatures. *Fuel* **2018**, *217*, 365–369. [CrossRef]
48. Nguyen, H.; Wang, Y.; Moglia, D.; Fu, J.; Zheng, W.; Orazov, M.; Vlachos, D.G. Production of renewable oleo-furan surfactants by cross-ketonization of biomass-derived furoic acid and fatty acids. *Catal. Sci. Technol.* **2021**, *11*, 2762–2769. [CrossRef]

39. Amran, U.A.; Zakaria, S.; Chia, C.H.; Roslan, R.; Jaafar, S.N.S.; Salleh, K.M. Polyols and rigid polyurethane foams derived from liquefied lignocellulosic and cellulosic biomass. *Cellulose* **2019**, *26*, 3231–3246. [CrossRef]
40. Zoghlami, A.; Paës, G. Lignocellulosic biomass: Understanding recalcitrance and predicting hydrolysis. *Front. Chem.* **2019**, *7*, 874. [CrossRef] [PubMed]
41. Zhou, L.; Gao, D.; Yang, J.; Yang, X.; Su, Y.; Lu, T. Conversion of recalcitrant cellulose to alkyl levulinates and levulinic acid via oxidation pretreatment combined with alcoholysis over $Al_2(SO_4)_3$. *Cellulose* **2019**, *27*, 1451–1463. [CrossRef]
42. Kanth, J.V.B.; Periasamy, M. Selective reduction of carboxylic acids into alcohols using sodium borohydride and iodine. *J. Org. Chem.* **1991**, *56*, 5964–5965. [CrossRef]
43. Sardon, H.; Pascual, A.; Mecerreyes, D.; Taton, D.; Cramail, H.; Hedrick, J.L. Synthesis of polyurethanes using organocatalysis: A perspective. *Macromolecules* **2015**, *48*, 3153–3165. [CrossRef]
44. Motokucho, S.; Nakayama, Y.; Morikawa, H.; Nakatani, H. Environment-friendly chemical recycling of aliphatic polyure-thanes by hydrolysis in a CO_2-water system. *J. Appl. Polym. Sci.* **2018**, *135*, 45897–45903. [CrossRef]
45. Głowińska, E.; Wolak, W.; Datta, J. Eco-friendly route for thermoplastic polyurethane elastomers with bio-based hard seg-ments composed of bio-glycol and mixtures of aromatic–aliphatic and aliphatic–aliphatic diisocyanate. *J. Polym. Environ.* **2021**, *29*, 2140–2149. [CrossRef] [PubMed]
46. Calandra, P.; Caponetti, E.; Chillura Martino, D.; D'Angelo, P.; Minore, A.; Turco Liveri, V. FT-IR and dielectric study of water/AOT liquid crystals. *J. Mol. Struct.* **2000**, *522*, 165–178. [CrossRef]
47. Kowalczuk, D.; Pitucha, M. Application of FTIR method for the assessment of immobilization of active substances in the matrix of biomedical materials. *Materials* **2019**, *12*, 2972. [CrossRef]
48. Shen, X.; Dai, J.; Liu, Y.; Liu, X.; Zhu, J. Synthesis of high performance polybenzoxazine networks from bio-based furfurylamine: Furan vs benzene ring. *Polymers* **2017**, *122*, 258–269. [CrossRef]
49. Nie, B.; Stutzman, J.; Xie, A. A vibrational spectral maker for probing the hydrogen-bonding status of protonated asp and glu residues. *Biophys. J.* **2005**, *88*, 2833–2847. [CrossRef] [PubMed]
50. Ciecierska, E.; Jurczyk-Kowalska, M.; Bazarnik, P.; Gloc, M.; Kulesza, M.; Kowalski, M.; Krauze, S.; Lewandowska, M. Flammability, mechanical properties and structure of rigid polyurethane foams with different types of carbon reinforcing materials. *Compos. Struct.* **2016**, *140*, 67–76. [CrossRef]
51. Calandra, P.; Mandanici, A.; Liveri, V.T.; Pochylski, M.; Aliotta, F. Emerging dynamics in surfactant-based liquid mixtures: Octanoic acid/bis(2-ethylhexyl) amine systems. *J. Chem. Phys.* **2012**, *136*, 064515. [CrossRef]
52. Calandra, P. On the physico-chemical basis of self-nanosegregation giving magnetically-induced birefringence in dibutyl phosphate/bis(2-ethylhexyl) amine systems. *J. Mol. Liq.* **2020**, *310*, 113186. [CrossRef]
53. Uram, K.; Leszczyńska, M.; Prociak, A.; Czajka, A.; Gloc, M.; Leszczyński, M.; Michałowski, S.; Ryszkowska, J. Polyurethane composite foams synthesized using bio-polyols and cellulose filler. *Materials* **2021**, *14*, 3474. [CrossRef] [PubMed]
54. Septevani, A.A.; Evans, D.A.C.; Martin, D.J.; Annamalai, P.K. Hybrid polyether-palm oil polyester polyol based rigid polyurethane foam reinforced with cellulose nanocrystal. *Ind. Crops Prod.* **2018**, *112*, 378–388. [CrossRef]
55. Zhou, X.; Sain, M.M.; Oksman, K. Semi-rigid biopolyurethane foams based on palm-oil polyol and reinforced with cellulose nanocrystals. *Compos. Part A Appl. Sci. Manuf.* **2016**, *83*, 56–62. [CrossRef]

polymers

MDPI

Article

Albumin Acts as a Lubricant on the Surface of Hydrogel and Silicone Hydrogel Contact Lenses

Chen-Ying Su [1,†], Lung-Kun Yeh [2,3,†], Tzu-Wei Fan [1], Chi-Chun Lai [2,3] and Hsu-Wei Fang [1,4,*]

1 Department of Chemical Engineering and Biotechnology, National Taipei University of Technology, 1, Sec. 3, Zhongxiao E. Rd., Taipei 10608, Taiwan; chenying.su@ntut.edu.tw (C.-Y.S.); sdfwer119@gmail.com (T.-W.F.)
2 Department of Ophthalmology, Chang Gung Memorial Hospital, Linkou, No. 5, Fuxing St., Taoyuan 333, Taiwan; yehlungkun@gmail.com (L.-K.Y.); chichun.lai@gmail.com (C.-C.L.)
3 College of Medicine, Chang Gung University, No. 259, Wenhua 1st Rd., Taoyuan 333, Taiwan
4 Institute of Biomedical Engineering and Nanomedicine, National Health Research Institutes, No. 35, Keyan Road, Zhunan Town, Miaoli County 35053, Taiwan
* Correspondence: hwfang@ntut.edu.tw
† These authors contributed equally to this work.

Abstract: Feeling comfortable is the greatest concern for contact lens wearers, and it has been suggested that in vivo comfort could be corresponded to the in vitro friction coefficient of contact lenses. How tear albumin could affect the friction coefficient of silicone hydrogel and hydrogel contact lenses was analyzed by sliding a lens against a quartz glass in normal and extremely high concentration of albumin solution. Albumin deposition testing and surface roughness analysis were also conducted. The results showed that the friction coefficient of tested contact lenses did not correspond to both the albumin deposition amount and surface roughness, but we proposed a model of how albumin might act as a lubricant on the surface of some hydrogel and silicone hydrogel contact lenses. In conclusion, albumin provided lubrication for silicone hydrogel contact lenses regardless of albumin concentrations, while albumin only acted as a lubricant for hydrogel contact under normal concentration.

Keywords: lubricant; tribology; albumin deposition; contact lens; surface roughness

Citation: Su, C.-Y.; Yeh, L.-K.; Fan, T.-W.; Lai, C.-C.; Fang, H.-W. Albumin Acts as a Lubricant on the Surface of Hydrogel and Silicone Hydrogel Contact Lenses. *Polymers* **2021**, *13*, 2051. https://doi.org/10.3390/polym13132051

Academic Editor: Arn Mignon

Received: 9 June 2021
Accepted: 21 June 2021
Published: 23 June 2021

Publisher's Note: MDPI stays neutral with regard to jurisdictional claims in published maps and institutional affiliations.

1. Introduction

Tear film is composed of lipids, proteins, electrolytes, mucins, and water [1]. Once a contact lens is put into the eye, tear proteins and lipids will be instantly deposited on the surface of the lens and will easily accumulate after wearing contact lenses for a long day. Although tear proteins can protect eyes from being infected, protein accumulation caused by incomplete cleaning often can trigger immune reactions and lead to discomfort, red eyes, or contact lens-induced papillary conjunctivitis [2,3]. Many researches about protein deposition on contact lenses have been focused on lysozyme, which is the most plentiful tear protein and the main protein that gets deposited on contact lenses [4]. Lysozyme is not the only tear protein, and other tear proteins such as albumin, a natural polymer, also play a critical role in the interaction with contact lens materials [5]. The tear albumin concentration is much lower (0.02–1.1 mg/mL) than lysozyme (1.9 mg/mL) [4,6]. However, the concentration of albumin increases after wearing contact lenses or orthokeratology lenses [6–8]. It also has been shown that the levels of albumin rose significantly in patients with infections or dry eye, and albumin concentration could be increased to 8.3 mg/mL on average, which was 415 times higher than 0.02 mg/mL [7,9]. Therefore, the change of albumin concentration may be a critical clinical index, making albumin a protein of interest when studying protein deposition on contact lenses.

Tear protein deposition can trigger immune reactions of contact lens wearers, and it also can result in discomfort. Since the in vitro coefficient of friction (COF) of contact lenses has been shown to correspond to degree of comfort in vivo [10,11], some researchers have been focused on investigating the relationship between tear protein deposition and the COF of contact lenses. Ngai et al. showed that COFs were lower when contact lenses were exposed to a mixture of lysozyme and albumin prior to friction testing and suggested that proteins might contribute to lubricating lenses at the early stage of protein deposition [12]. Sterner et al. demonstrated that the COF of most contact lenses increased after contact lenses were repeatedly immersed in tear-like fluid and exposed to air, which may imply that the degree of comfort from wearing contact lenses changes during the course of the day [13]. Although numerous studies have focused on the relationship between protein deposition on the contact lenses and the clinical degree of comfort, there is still no direct evidence indicating a correlation [10]. We have previously shown that the COFs of some hydrogel contact lenses such as Polymacon or Hefilcon-A increased in higher concentrations of lysozyme, and a higher COF corresponded to changes in the lysozyme secondary structure rather than lysozyme deposition amounts [14–16].

In order to understand whether albumin deposition on contact lenses corresponded to COFs, we investigated the COF of two hydrogel (Etafilcon-A and Polymacon) and two silicone hydrogel contact lenses (Somofilcon-A and Senofilcon-A) in solutions with different albumin concentrations by using an in vitro friction testing system we established previously [15]. In addition, contact lenses were immersed in albumin before friction testing to understand the tribological properties of contact lenses if protein depositions were not removed completely. Surface roughness was also tested to provide a qualitative evaluation of the four contact lens types under investigation.

2. Materials and Methods

2.1. Contact Lenses and Reagents

The two hydrogel contact lenses used here were Etafilcon-A contact lenses (1 Day Acuvue Moist, Johnson & Johnson, New Brunswick, NJ, USA), and Polymacon contact lenses (Hydron Eye Secret Aspheric Daily, Yung Sheng Optical Co., Ltd., Taichung City, Taiwan). The two silicone hydrogel contact lenses used were Somofilcon-A contact lenses (Clariti 1 Day, Cooper Vision, Victor, NY, USA), and Senofilcon-A contact lenses with HydraLux (Acuvue Oasys Brand, Johnson & Johnson, New Brunswick, NJ, USA). Bovine serum albumin powder (Sigma-Aldrich, St. Louis, MO, USA) was dissolved in phosphate-buffered saline (PBS) for a final concentration of 0.2 or 50 mg/mL. Then, 0.2 mg/mL albumin was considered as control in this study but was higher than normal tear albumin to simulate the condition after wearing contact lenses [17]. Then, 50 mg/mL of albumin was set extremely high and could not represent the albumin level of dry eye patients, but it was investigated in this study in order to observe the influence of albumin adsorption on the tribological properties of different contact lens materials.

2.2. Coefficient of Friction Testing System

A CETR universal micro-tribometer-2 (UMT-2, Bruker, Campbell, CA, USA) was used for measuring COF for different contact lens materials in PBS or PBS with 0.2 or 50 mg/mL albumin. The testing system has been previously described [15]. The friction testing program used in this study was as follows; force: 0.76 kPa, rotation radius: 8 mm (mm), rotation speed: 1 revolution per minute (rpm) or 50.24 mm/minute, rotation time: 900 s. The friction force was recorded every 0.03 s by the UMT-2 sensor, and the COF was obtained by dividing friction force by normal force. The COF of each cycle (1 min) was averaged. Four independent lenses were tested for each condition.

2.3. Albumin Deposition Analysis

Each contact lens was placed in 3 mL of 50 mg/mL albumin solution at room temperature for 15 min. Then, the lens was taken out of the albumin solution and washed three times with 1 mL of PBS for each wash. These 3 mls of PBS from the wash were combined with the initial albumin solution. The Bio-Rad DC protein assay (Bio-Rad, Hercules, CA,

USA) was used for measuring the amount of albumin in the combined solution. The optical density (OD) value was obtained by an Enzyme-Linked Immuno Sorbent Assay (ELISA) reader with a wavelength of 280 nm. The total albumin amount in the solution was determined by multiplying the concentration (mg/mL) by the total volume (6 mL). In order to determine the deposition amount onto the lens, the amount in the solution was subtracted from the starting mass, which was 150 mg (3 mL × 50 mg/mL). Three independent lenses were tested for each condition.

2.4. Surface Roughness Measurement

Atomic force microscopy (AFM, XE-100, Park) was used for measuring the surface roughness. The tip of AFM was PointProbe ® Plus from Nanosensors (Neuchatel, Switzerland). The shape of the tip was a polygon-based pyramid, the radius of the tip was smaller than 7 nm, and the tip height was 10–15 μm. The four periphery edges of each contact lens were cut in order to create a flat surface, resulting in a 4 × 4 mm square of lens for analysis. The micro-arm was used to sense and amplify the force between the sharp probe on the cantilever and the surface of the contact lens. The probe frequency was set to a range between 0 and 1000 kilohertz, the scan range was 5 × 5 μm, and the *z*-axis range was smaller than 12 μm.

2.5. Statistical Analysis

The 2-tailed *t*-test was assessed in order to compare differences in albumin deposition amounts between two different lens materials. A value of $p < 0.05$ was considered significant.

3. Results

3.1. Friction Coefficient of Contact Lenses in Different Concentrations of Albumin

The COF of four different contact lenses in PBS or PBS with 0.2 or 50 mg/mL albumin were measured and shown in Figure 1. For Somofilcon-A lenses, the COF in PBS was the most stable during the period of fifteen cycles (Figure 1a). The COF was 0.009 and 0.007 during the first cycle when Somofilcon-A lenses were sliding against the glass in 0.2 and 50 mg/mL albumin, respectively, but it dropped to 0.002 and 0.003 starting at cycle two (Figure 1a). A similar phenomenon was observed for Senofilcon-A lenses. The COFs were both 0.004 when Senofilcon-A lenses were sliding in 0.2 and 50 mg/mL albumin but decreased to 0.003 in cycle 2 and 3, respectively (Figure 1b). The COF of Etafilcon-A lenses was 0.007 during the first cycle but dropped below 0.003 afterwards when sliding in 0.2 mg/mL albumin and had stable COFs when sliding in PBS or 50 mg/mL albumin (Figure 1d). In contrast, the COF increased in cycle 3 when Polymacon lenses were sliding in PBS or 50 mg/mL albumin solution. The COF of Polymacon lenses in 0.2 mg/mL albumin was relatively stable (Figure 1c).

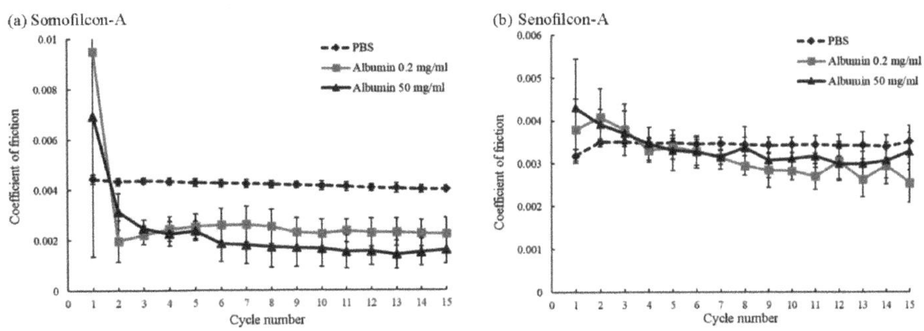

Figure 1. *Cont.*

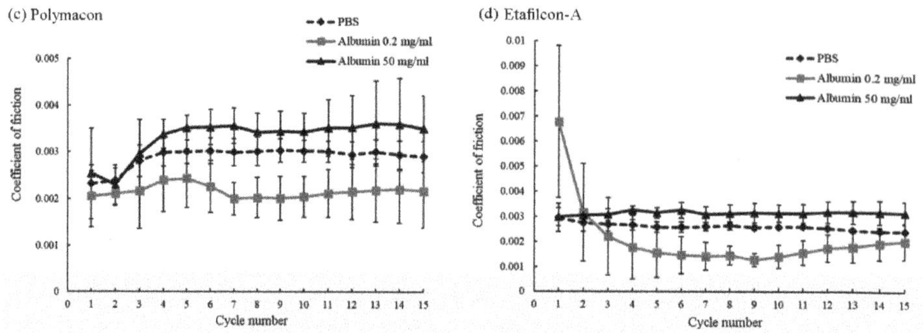

Figure 1. The coefficients of friction of Somofilcon-A (**a**), Senofilcon-A (**b**), Polymacon (**c**), and Etafilcon-A (**d**) lenses when sliding against glass in phosphate-buffered saline (black dashed line), 0.2 mg/mL albumin (gray line), or 50 mg/mL albumin (black line). Error bars represented standard deviation and were calculated from four experiments.

3.2. Coefficients of Friction of Contact Lenses after Being Immersed in a High Concentration of Albumin

All contact lenses were immersed in 50 mg/mL albumin for fifteen minutes and then were slid against the glass in 50 mg/mL albumin for fifteen cycles. The results showed the decrease of COF from cycle 1 to cycle 2 for all the lens materials (Figure 2). The COF of Polymacon lenses was the highest (all above 0.005), while the COF of Somofilcon-A lenses was the lowest (all below 0.003).

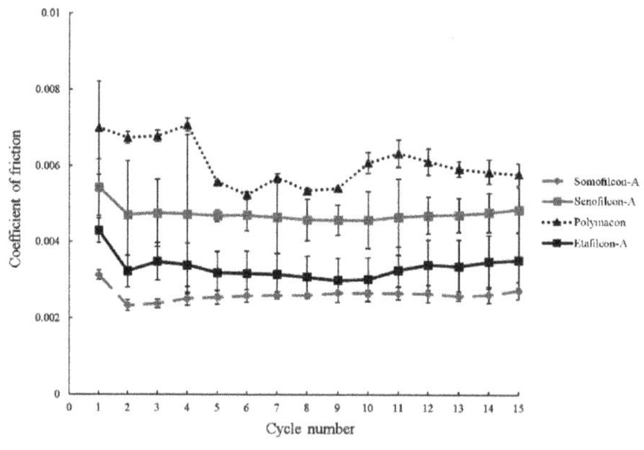

Figure 2. The coefficients of friction of Somofilcon-A (gray dashed line), Senofilcon-A (gray line), Polymacon (black dotted line), and Etafilcon-A (black line) in 50 mg/mL albumin solution after contact lenses are immersed in 50 mg/mL albumin for 15 min. Error bars represented standard deviation, and four experiments were used for calculating error bars.

3.3. Albumin Deposition on the Contact Lenses

To investigate the amount of albumin deposition on contact lenses prior to the friction testing, contact lenses were immersed in 50 mg/mL albumin for 15 min. Somofilcon-A displayed the highest albumin deposition amount, while Etafilcon-A showed the lowest deposition amount, but there was no statistical difference (Figure 3).

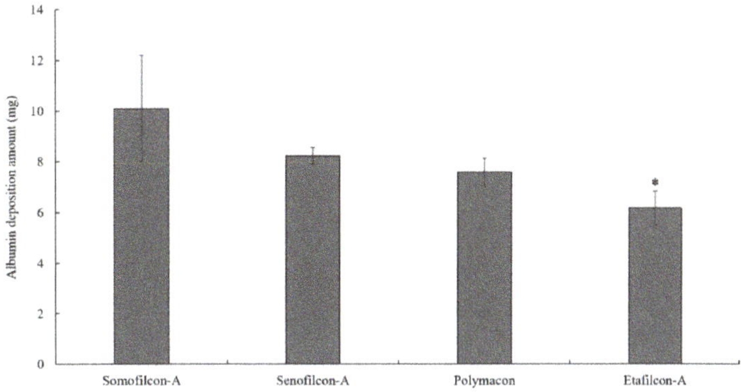

Figure 3. Deposited albumin concentrations are measured after Somofilcon-A, Senofilcon-A, Polymacon, and Etafilcon-A lenses are immersed in 50 mg/mL albumin solution for 15 min. * $p < 0.05$ when comparing albumin deposition amount on Senofilcon-A versus on Etafilcon-A lenses. Error bars represented standard deviation and were obtained from three experiments.

3.4. Surface Roughness

The surface of Etafilcon-A was the smoothest, and the average roughness (Ra) was 0.564 μm (Figure 4d). The Somofilcon-A lens had the roughest surface with an Ra of 4.864 μm (Figure 4a). The Ra values for Senofilcon-A and Polymacon lenses were 3.254 μm and 2.478 μm, respectively (Figure 4b–c).

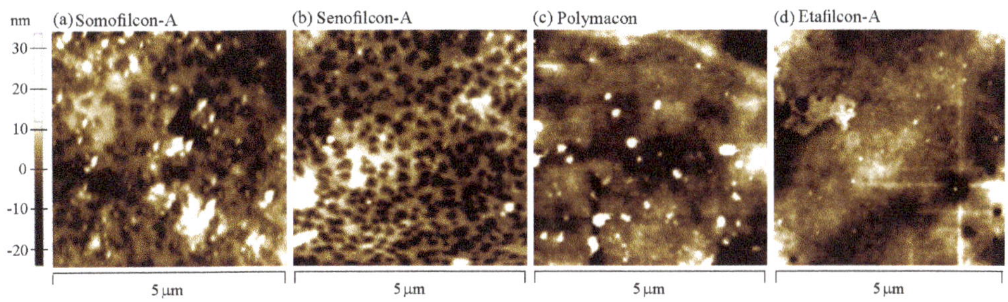

Figure 4. Surface roughness analysis for Somofilcon-A (**a**), Senofilcon-A (**b**), Polymacon (**c**), and Etafilcon-A (**d**) lenses.

4. Discussion

We analyzed the role of albumin deposition on the coefficients of friction (COF) of four contact lens materials in this study. When contact lenses were sliding in albumin solution immediately after being taken out of the packaging, it mimicked the conditions of when contact lenses are initially put into the eyes. The result showed that the COF was relatively high during the first sliding minute for Somofilcon-A and Etafilcon-A lenses in 0.2 mg/mL albumin, which is similar to the concentration of albumin after wearing contact lenses [17]. Since the in vitro COF of contact lenses could correspond to in vivo comfort degree [10,11], it may suggest that wearers would feel some degree of discomfort when they first wear Somofilcon-A or Etafilcon-A lenses. However, the COF was subsequently reduced for both Somofilcon-A and Etafilcon-A lenses, suggesting that wearers would not feel discomfort once their eyes adapted to the contact lenses. In summary, the COFs of all contact lenses in 0.2 mg/mL albumin solution were lower than in PBS. When contact lenses were sliding in 50 mg/mL albumin, the COF of Somofilcon-A was even lower than when sliding in 0.2 mg/mL albumin; thus, we speculate that albumin might act as a lubricant for these four contact lens materials.

The protein conformational change, but not the amount of protein deposition on the contact lens, has been shown to correspond to comfort [18,19]. Indeed, the albumin deposition analysis showed that the amount of albumin deposited on Somofilcon-A was the highest when lenses were immersed in 50 mg/mL albumin for 15 min. However, the COF of Somofilcon-A lenses was lowest when sliding in 50 mg/mL albumin solution after being immersed in 50 mg/mL albumin for 15 min. The COF of Polymacon lenses was the highest in albumin solution, but the amount of albumin deposition was the second lowest. Although the measurement of albumin deposition on the lens in this study was not the most accurate method of protein adsorption, the results demonstrated that the amount of albumin deposition did not appear to correspond to the COF. Therefore, the mechanism of how albumin undergoes a conformational change after deposition during friction needs to be considered.

Proteins are first in solution and move toward the contact lens surface; then, the proteins are adsorbed to the surface followed by structural changes [20]. It has been shown that the contact lens material composition, pore size, water content, hydrophobicity, surface roughness, contact lens or protein charge, protein size, etc. all play a role in protein deposition [5]. Then, we proposed a model that may explain the results (Figure 5). Both Somofilcon-A and Senofilcon-A lenses are silicone hydrogel contact lenses, but Senofilcon-A lenses are coated with vinyl pyrrolidone (PVP) as a wetting agent to reduce hydrophobicity [21]. Albumin has been demonstrated to be denatured on hydrophobic surfaces more easily than on hydrophilic surfaces [22]. However, the COF of Somofilcon-A lenses was lower than that of Senofilcon-A lenses, suggesting that surface hydrophobicity might not be a factor that affects albumin conformational change. Other factors could be the water content and surface roughness of contact lenses. Higher water content leads to larger pore sizes, which may result in protein penetration into the matrix of contact lenses [23,24]. Water content is 56% in Somofilcon-A lenses and 38% in Senofilcon-A lenses [25]; thus, the pore sizes of Somofilcon-A should be larger than those of Senofilcon-A. In addition, the surface roughness results showed that the Ra value of Somofilcon-A was 4.864 μm, while the Ra value of Senofilcon-A was 3.254 μm, suggesting there might be more deposits forming on imperfections of the surface. Taken together, it is possible that albumin was more likely to penetrate into the matrix of Somofilcon-A and get deposited on the imperfections of the surface, resulting in less albumin on the surface against the glass. Therefore, less albumin might undergo conformational changes on the surface, leading to a smaller COF of Somofilcon-A lenses compared to Senofilcon-A lenses (Figure 5a,b). However, higher water contents display lower oxygen permeability for silicone hydrogel contact lenses [26]. Wearing contact lenses with high oxygen permeability can reduce contact lens-induced hypoxia, resulting in better ocular physiology; thus, wearers may choose Senofilcon-A even though Somofilcon-A lenses provide better lubrication than was shown here. Therefore, more clinical investigation will be required to understand whether the in vitro friction coefficient of contact lenses could be directly corresponded to the in vivo comfort.

Both Polymacon and Etafilcon-A are materials used in hydrogel contact lenses, but Polymacon is hydrophobic, while Etafilcon-A has hydrophilic properties [27,28]. Water contents of Etafilcon-A and Polymacon are 58% and 38.6%, respectively [29]. It is possible that the low water content of Polymacon lenses causes albumin to stay on the surface, and albumin may go through conformational changes because of the hydrophobic properties of the material (Figure 5c). It might be the reason why the COF of Polymacon was higher than Etafilcon-A. In addition, Etafilcon-A is negatively charged [28], whereas the pI pH of albumin is 5.16 [5]. Therefore, albumin is not easily attracted or bound to Etafilcon-A lenses. The higher water content and electrical repulsion might cause albumin to either penetrate into the matrix or repel away from the surface, resulting in less albumin on the surface of Etafilcon-A lenses to undergo conformational changes, resulting in a lower COF (Figure 5d).

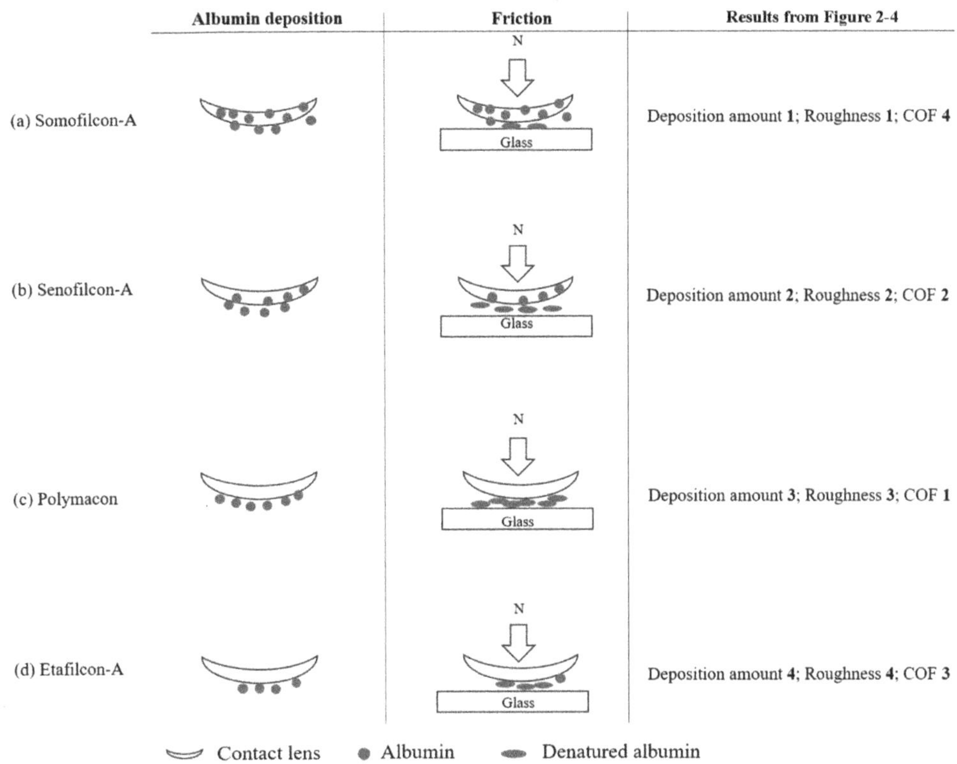

Figure 5. The potential model of how different materials of contact lenses display distinct behavior of albumin deposition and tribological properties. (**a**) Majority of albumin may permeate into the matrix of Somofilcon-A lens, resulting in the highest albumin deposition amount but the lowest COF. (**b**) Although the rough surface of Senofilcon-A results in a large amount of adsorbed albumin, friction still causes albumin on the surface to undergo conformational change, leading to an increased COF. (**c**) Albumin is only adsorbed on the surface and may undergo the conformational change resulting in the highest COF of Polymacon lens even though the amount of albumin deposition is the second lowest. (**d**) A lower amount of albumin deposition on the surface of Etafilcon-A results in a lower COF. In the column of results from Figures 2–4, the number represents ranking. For example, deposition amount 1 represents the highest amount of albumin deposition.

Under control tear albumin concentrations (0.2 mg/mL), albumin was able to provide lubrication for Somofilcon-A, Senofilcon-A, Polymacon, and Etafilcon-A lenses. The result showed that albumin acted distinctly from lysozyme, as we previously demonstrated that lysozyme would increase the COF of some hydrogel contact lenses [15,16]. However, only one protein was investigated here. Whether the effect of albumin alone on the tribological properties of contact lenses is the same as the impact of a mixture of tear proteins and whether the in vitro friction test results demonstrated here could relate to the in vivo bio-tribological property between the contact lens and the eyelid needs further investigation. Albumin has been used as eye drops for treating severe dry eye [30,31]. Since many contact lens wearers feel eyes are dry after a long period of time, albumin might be considered as a lubricating additive in the artificial tears. The dosages of albumin in the artificial tears need to be investigated when albumin is used as a lubricant without affecting ocular physiology.

5. Conclusions

The current study showed that under control tear albumin concentration, albumin acted as a lubricant for both silicone hydrogel and hydrogel lenses investigated here. Many factors may affect the COFs of contact lenses, but the results demonstrated that no correspondence was observed between the amount of deposited albumin or between the surface roughness and the tribological properties of contact lenses. The results suggested that albumin might be applied as a lubricating additive in the artificial tears and can be used for contact lens wearers when eyes feel dry after wearing for a long period of time.

Author Contributions: Conceptualization, H.-W.F.; methodology, T.-W.F. and C.-C.L.; validation, C.-Y.S., L.-K.Y. and T.-W.F.; formal analysis, C.-Y.S. and L.-K.Y.; investigation, T.-W.F. and C.-C.L.; data curation, C.-Y.S., L.-K.Y. and H.-W.F.; writing—original draft preparation, C.-Y.S.; writing—review and editing, L.-K.Y. and H.-W.F.; funding acquisition, C.-Y.S., L.-K.Y. and H.-W.F. All authors have read and agreed to the published version of the manuscript.

Funding: This research was supported by the Ministry of Science and Technology (MOST), Taiwan, under grant number 107-2218-E-027-012-MY3; National Health Research Institute (BN-108-PP-12); National Taipei University of Technology and Chang Gung Memorial Hospital Joint Research Program (NTUT-CGMH-108-01); and Chang Gung Medical Research Project CORPG3I0131.

Institutional Review Board Statement: Not applicable.

Informed Consent Statement: Not applicable.

Data Availability Statement: The data presented in this study are available on request from the corresponding author.

Acknowledgments: The Precision Analysis and Material Research Center at National Taipei University of Technology kindly provided AFM instruments.

Conflicts of Interest: The authors declare no conflict of interest.

References

1. Tiffany, J.M. Tears in health and disease. *Eye* **2003**, *17*, 923–926. [CrossRef]
2. Allansmith, M.R.; Korb, D.R.; Greiner, J.V.; Henriquez, A.S.; Simon, M.A.; Finnemore, V.M. Giant papillary conjunctivitis in contact lens wearers. *Am. J. Ophthalmol.* **1977**, *83*, 697–708. [CrossRef]
3. Skotnitsky, C.; Sankaridurg, P.R.; Sweeney, D.F.; Holden, B.A. General and local contact lens induced papillary conjunctivitis (CLPC). *Clin. Exp. Optom.* **2002**, *85*, 193–197. [CrossRef] [PubMed]
4. Sack, R.A.; Tan, K.O.; Tan, A. Diurnal tear cycle: Evidence for a nocturnal inflammatory constitutive tear fluid. *Investig. Ophthalmol. Vis. Sci.* **1992**, *33*, 626–640.
5. Luensmann, D.; Jones, L. Albumin adsorption to contact lens materials: A review. *Contact Lens Anterior Eye* **2008**, *31*, 179–187. [CrossRef]
6. Choy, C.K.; Cho, P.; Benzie, I.F.; Ng, V. Effect of one overnight wear of orthokeratology lenses on tear composition. *Optom. Vis. Sci.* **2004**, *81*, 414–420. [CrossRef]
7. Barishak, Y.; Zavaro, A.; Samra, Z.; Sompolinsky, D. An immunological study of papillary conjunctivitis due to contact lenses. *Curr. Eye Res.* **1984**, *3*, 1161–1168. [CrossRef]
8. Lundh, R.L.; Liotet, S.; Pouliquen, Y. Study of the human blood-tear barrier and the biochemical changes in the tears of 30 contact lens wearers (50 eyes). *Ophthalmologica* **1984**, *188*, 100–105. [CrossRef] [PubMed]
9. Runstrom, G.; Mann, A.; Tighe, B. The fall and rise of tear albumin levels: A multifactorial phenomenon. *Ocul. Surf.* **2013**, *11*, 165–180. [CrossRef]
10. Jones, L.; Brennan, N.A.; Gonzalez-Meijome, J.; Lally, J.; Maldonado-Codina, C.; Schmidt, T.A.; Subbaraman, L.; Young, G.; Nichols, J.J. The TFOS International Workshop on Contact Lens Discomfort: Report of the contact lens materials, design, and care subcommittee. *Investig. Ophthalmol. Vis. Sci.* **2013**, *54*, TFOS37–TFOS70. [CrossRef]
11. Roba, M.; Duncan, E.G.; Hill, G.A.; Spencer, N.D.; Tosatti, S.G.P. Friction measurements on contact lenses in their operating environment. *Tribol. Lett.* **2011**, *44*, 387–397. [CrossRef]
12. Ngai, V.; Medley, J.; Jones, L.; Forrest, J.; Teichroeb, J. *Friction of Contact Lenses: Silicone Hydrogel versus Conventional Hydrogel*; Elsevier: Amsterdam, The Netherlands, 2005; Volume 48, pp. 371–379.
13. Sterner, O.; Aeschlimann, R.; Zurcher, S.; Osborn Lorenz, K.; Kakkassery, J.; Spencer, N.D.; Tosatti, S.G. Friction Measurements on Contact Lenses in a Physiologically Relevant Environment: Effect of Testing Conditions on Friction. *Investig. Ophthalmol. Vis. Sci.* **2016**, *57*, 5383–5392. [CrossRef] [PubMed]
14. Chang, Y.C.; Su, C.Y.; Chang, C.H.; Fang, H.W.; Wei, Y. Correlation between Tribological Properties and the Quantified Structural Changes of Lysozyme on Poly (2-hydroxyethyl methacrylate) Contact Lens. *Polymers* **2020**, *12*, 1639. [CrossRef] [PubMed]

15. Su, C.Y.; Lai, C.C.; Yeh, L.K.; Li, K.Y.; Shih, B.W.; Tseng, C.L.; Fang, H.W. The characteristics of a preservative-free contact lens care solution on lysozyme adsorption and interfacial friction behavior. *Colloids Surf. B Biointerfaces* **2018**, *171*, 538–543. [CrossRef] [PubMed]
16. Su, C.Y.; Yeh, L.K.; Lai, C.C.; Li, K.Y.; Tseng, C.L.; Fang, H.W. Effects of lysosomal deposition on the friction coefficient of hydrogel contact lenses. *Contact Lens Anterior Eye* **2020**, *43*, 144–148. [CrossRef] [PubMed]
17. Mirejovsky, D.; Patel, A.S.; Rodriguez, D.D.; Hunt, T.J. Lipid adsorption onto hydrogel contact lens materials. Advantages of Nile red over oil red O in visualization of lipids. *Optom. Vis. Sci.* **1991**, *68*, 858–864. [CrossRef]
18. Heuberger, M.P.; Widmer, M.R.; Zobeley, E.; Glockshuber, R.; Spencer, N.D. Protein-mediated boundary lubrication in arthroplasty. *Biomaterials* **2005**, *26*, 1165–1173. [CrossRef]
19. Subbaraman, L.N.; Glasier, M.A.; Varikooty, J.; Srinivasan, S.; Jones, L. Protein deposition and clinical symptoms in daily wear of etafilcon lenses. *Optom. Vis. Sci.* **2012**, *89*, 1450–1459. [CrossRef]
20. Dijt, J.C.; Cohen-Stuart, M.A.; Hofman, J.E.; Fleer, G.J. Kinetics of polymer adsorption in stagnation point flow. *Colloids Surf.* **1990**, *51*, 141–158. [CrossRef]
21. Hoteling, A.J.; Nichols, W.F.; Harmon, P.S.; Conlon, S.M.; Nunez, I.M.; Hoff, J.W.; Cabarcos, O.M.; Steffen, R.B.; Hook, D.J. Characterization and quantitation of PVP content in a silicone hydrogel contact lens produced by dual-phase polymerization processing. *J. Biomed. Mater. Res. B Appl. Biomater.* **2018**, *106*, 1064–1072. [CrossRef]
22. Garrett, Q.; Griesser, H.J.; Milthorpe, B.K.; Garrett, R.W. Irreversible adsorption of human serum albumin to hydrogel contact lenses: A study using electron spin resonance spectroscopy. *Biomaterials* **1999**, *20*, 1345–1356. [CrossRef]
23. Chirila, T.V.; Constable, I.J.; Crawford, G.J.; Vijayasekaran, S.; Thompson, D.E.; Chen, Y.C.; Fletcher, W.A.; Griffin, B.J. Poly(2-hydroxyethyl methacrylate) sponges as implant materials: In vivo and in vitro evaluation of cellular invasion. *Biomaterials* **1993**, *14*, 26–38. [CrossRef]
24. Garrett, Q.; Laycock, B.; Garrett, R.W. Hydrogel lens monomer constituents modulate protein sorption. *Investig. Ophthalmol. Vis. Sci.* **2000**, *41*, 1687–1695.
25. Sulley, A.; Dumbleton, K. Silicone hydrogel daily disposable benefits: The evidence. *Contact Lens Anterior Eye* **2020**, *43*, 298–307. [CrossRef]
26. Stapleton, F.; Stretton, S.; Papas, E.; Skotnitsky, C.; Sweeney, D.F. Silicone hydrogel contact lenses and the ocular surface. *Ocul. Surf.* **2006**, *4*, 24–43. [CrossRef]
27. Ajayi, B.O.; Kio, F.E.; Otajevwo, F.D. Adhesive capabilities of Staphylococcus aureus and Pseudomonas aeruginosa isolated from tears of HIV/AIDS patients to soft contact lenses. *Glob. J. Health Sci.* **2012**, *4*, 140–148. [CrossRef] [PubMed]
28. Carney, F.P.; Keay, L.; Stapleton, F.; Morris, C.A.; Willcox, M.D.P. Hydrogel lens wettability and deposition in vivo. *Clin. Exp. Optom.* **1998**, *81*, 51–55. [CrossRef]
29. Lee, S.E.; Kim, S.R.; Park, M. Influence of Tear Protein Deposition on the Oxygen Permeability of Soft Contact Lenses. *J. Ophthalmol.* **2017**, *2017*, 5131764. [CrossRef]
30. Schargus, M.; Kohlhaas, M.; Unterlauft, J.D. Treatment of severe ocular surface disorders with albumin eye drops. *J. Ocul. Pharmacol. Ther.* **2015**, *31*, 291–295. [CrossRef]
31. Shimmura, S.; Ueno, R.; Matsumoto, Y.; Goto, E.; Higuchi, A.; Shimazaki, J.; Tsubota, K. Albumin as a tear supplement in the treatment of severe dry eye. *Br. J. Ophthalmol.* **2003**, *87*, 1279–1283. [CrossRef]

polymers

Article

Production of Sucrolytic Enzyme by *Bacillus licheniformis* by the Bioconversion of Pomelo Albedo as a Carbon Source

Chien Thang Doan [1,2], Thi Ngoc Tran [1,2], Thi Thanh Nguyen [2], Thi Phuong Hanh Tran [2], Van Bon Nguyen [3], Trung Dung Tran [2], Anh Dzung Nguyen [3] and San-Lang Wang [1,4,*]

1 Department of Chemistry, Tamkang University, New Taipei City 25137, Taiwan; dcthang@ttn.edu.vn (C.T.D.); ttngoc@ttn.edu.vn (T.N.T.)
2 Faculty of Natural Sciences and Technology, Tay Nguyen University, Buon Ma Thuot 630000, Vietnam; ntthanh@ttn.edu.vn (T.T.N.); ttphanh@ttn.edu.vn (T.P.H.T.); ttdung@ttn.edu.vn (T.D.T.)
3 Institute of Biotechnology and Environment, Tay Nguyen University, Buon Ma Thuot 630000, Vietnam; nvbon@ttn.edu.vn (V.B.N.); nadzung@ttn.edu.vn (A.D.N.)
4 Life Science Development Center, Tamkang University, New Taipei City 25137, Taiwan
* Correspondence: sabulo@mail.tku.edu.tw; Tel.: +886-2-2621-5656; Fax: +886-2-2620-9924

Abstract: Recently, there has been increasing use of agro-byproducts in microbial fermentation to produce a variety of value-added products. In this study, among various kinds of agro-byproducts, pomelo albedo powder (PAP) was found to be the most effective carbon source for the production of sucrose hydrolyzing enzyme by *Bacillus licheniformis* TKU004. The optimal medium for sucrolytic enzyme production contained 2% PAP, 0.75% NH_4NO_3, 0.05% $MgSO_4$, and 0.05% NaH_2PO_4 and the optimal culture conditions were pH 6.7, 35 °C, 150 rpm, and 24 h. Accordingly, the highest sucrolytic activity was 1.87 U/mL, 4.79-fold higher than that from standard conditions using sucrose as the carbon source. The purified sucrolytic enzyme (sleTKU004) is a 53 kDa monomeric protein and belongs to the glycoside hydrolase family 68. The optimum temperature and pH of sleTKU004 were 50 °C, and pH = 6, respectively. SleTKU004 could hydrolyze sucrose, raffinose, and stachyose by attacking the glycoside linkage between glucose and fructose molecules of the sucrose unit. The K_m and V_{max} of sleTKU004 were 1.16 M and 5.99 μmol/min, respectively. Finally, sleTKU004 showed strong sucrose tolerance and presented the highest hydrolytic activity at the sucrose concentration of 1.2 M–1.5 M.

Keywords: agro-byproduct; *Bacillus licheniformis*; bioconversion; pomelo albedo; sucrolytic

Citation: Doan, C.T.; Tran, T.N.; Nguyen, T.T.; Tran, T.P.H.; Nguyen, V.B.; Tran, T.D.; Nguyen, A.D.; Wang, S.-L. Production of Sucrolytic Enzyme by *Bacillus licheniformis* by the Bioconversion of Pomelo Albedo as a Carbon Source. *Polymers* **2021**, *13*, 1959. https://doi.org/10.3390/polym13121959

Academic Editor: Arn Mignon

Received: 14 May 2021
Accepted: 10 June 2021
Published: 13 June 2021

Publisher's Note: MDPI stays neutral with regard to jurisdictional claims in published maps and institutional affiliations.

1. Introduction

Sucrose (α-D-glucopyranosyl-1,2-β-D-fructofuranoside), the world's most abundant disaccharide, comprises a glucose unit and a fructose unit linked together by an $\alpha 1$–$\beta 2$ linkage [1]. Sucrose hydrolysis is catalyzed by sucrolytic enzymes to produce a mixture of glucose and fructose (also known as invert sugar). Sucrolytic enzymes are of various types such as invertases (sucrases) [2], levansucrases [3], dextransucrases [4], glucansucrases [5], and inulosucrases [6]. They use sucrose as the donor of fructosyl moiety to be transferred to various kinds of acceptors, such as water (sucrose hydrolysis), sucrose, and fructan [7]. Therefore, these enzymes are also responsible for the production of some compounds such as fructooligosaccharides (FOSs), inulins, and levans [8,9]. To date, sucrolytic enzymes have diverse applications in the food [10–12], beverage [13,14], medicine [1,14], cosmetic [5,15], bioenergy [16], paper [15], and biochemical sectors [15]. Thus, the use of sucrolytic enzymes is highly widespread, with a large number of industrial applications.

The worldwide market for enzymes is anticipated to reach $10.5 billion in 2024 [17]. Among these, enzymes from microbial sources are the most important group, comprising 90% of the global market [18]. Regardless of their high demand, the price of enzymes still remains relatively high, which thus increases the cost of the processes using enzymatic methods. Therefore, for the cost-effective production of enzymes, several strategies are being developed including using cheap and readily available carbon and nitrogen sources such as the wastes from food processing, fishery, and agriculture as the alternative substrates for the microbial fermentation process [19–22]. Besides, the utilization of by-products for enzyme production is also an environmentally friendly technique because it can reduce the amount of waste generated and released into the environment [23]. Accordingly, this green technique has received a great attention, and as a result, various kinds of enzymes have been effectively produced by using by-products as the major nutrient source such as proteases [22,24], xylanases [21,25], chitosanases [26,27], chitinases [28,29], pectinases [23,30], invertases [31,32], etc.

Bacillus licheniformis is a rod-shaped, facultatively anaerobic, Gram-positive, and spore-forming bacterium [33]. *B. licheniformis* is favored because of its outstanding fermentation properties, high protein yield, and completely toxin-free productivity [34]. To date, various applications of *B. licheniformis* have been explored for the production of probiotics [35], industrial enzymes [36,37], and biofuels [37]. *B. licheniformis* has been commonly used for the manufacture of amylase and protease [37–41], however, the production of sucrolytic enzymes by this genus has not been fully explored. In fact, there are only a few reports on the production of sucrose hydrolyzing enzymes by *B. licheniformis*, especially using low-cost materials for the fermentation process [42–44]. In comparison, several *B. licheniformis* strains are known to have a great capacity for processing agricultural byproducts to produce enzymes [39,45]. This leads to the idea of using agro-byproducts as economically effective nutritional supplements for the synthesis of sucrose hydrolytic enzymes by *B. licheniformis*.

The exploration of new enzymes, having a low-cost manufacturing method, and possessing desirable characteristics, has always been regarded as important. Accordingly, in this study, various kinds of agro-byproducts such as pomelo (*Citrus maxima*) albedo powder (PAP), pomelo flavedo powder (PFP), orange (*Citrus sinensis*) peel powder (OPP), banana (*Musa paradisiaca*) peel powder (BPP), rice bran powder (RBP), and wheat bran powder (WBP) were used as carbon sources for *B. licheniformis* TKU004 to produce sucrolytic enzyme. Besides, the medium composition and culture conditions were optimized for maximum sucrolytic enzyme production. In addition, the properties of the obtained enzyme, as well as its purification, were studied.

2. Materials and Methods

2.1. Materials

B. licheniformis TKU004 was isolated and described in an earlier report [46]. Sucrose, glucose, maltose, and fructose were bought from Katayama Chemical Industries Corporation (Osaka, Japan). 3,5-Dinitrosalicylic acid (DNS), and fructooligosaccharide (FOS) were purchased from Sigma Aldrich (Saint Louis, MO, USA). Rice bran and wheat bran were collected from Miaoli (Miaoli City, Taiwan). Orange, banana, and pomelo were bought from the local markets of Tamshui (New Taipei, Taiwan) and then peeled to prepare orange peel, banana peel, pomelo albedo, and pomelo flavedo. Finally, these materials were dried and ground to powder (particle size smaller than 0.297 mm). All other chemicals were of the highest possible quality.

2.2. Sucrolytic Assay

Sucrolytic activity was determined using the DNS method [21] with certain modifications. Briefly, a mixture of 50 µL enzyme and 150 mL sucrose (1%, *w/v*) was incubated at 37 °C for 60 min to allow the hydrolysis of sucrose. Later, 750 µL of DNS reagent was added to the reaction and the obtained mixture was heated at 100 °C for 10 min. Consequently, the absorbance of red color of the mixture was read at 515 nm using an ELISA reader (Bio-Rad, Hercules, CA, USA) and was used to calculate the amount of reducing sugar in the hydrolysis reaction. The amount of enzyme required to release 1 µM of reducing sugar in one minute was described as one unit of sucrolytic activity.

2.3. Agro-Byproducts as the Sole Carbon Source for Sucrolytic Enzymes Production

One percent of each carbon source (PAP, PFP, OPP, BPP, RBP, WBP, and sucrose) was added into a 250 mL flask containing 100 mL of basal medium (1% KNO_3, 0.1% KH_2PO_4, and 0.05% $MgSO_4$). The bacterial seed was prepared by adding a loopful of pure culture into a 250 mL flask containing 100 mL of nutrient broth and then culturing at 37 °C and 150 rpm for 24 h. The cultivation of *B. licheniformis* TKU004 on different carbon source-containing mediums was initiated by adding 1 mL of the bacterial seed solution (OD 660 nm = 1) to each medium and kept at 37 °C and 150 rpm. One milliliter of culture medium was withdrawn each 24 h and the sucrolytic activity was estimated. Different amounts of PAP (0.25 g, 0.5 g, 1.0 g, 1.5 g, 2.0 g, 2.5 g, and 3 g) were used to explore the optimal PAP concentration for sucrolytic enzyme production. Each experiment was performed in triplicate.

2.4. Effect of Other Conditions on Sucrolytic Enzymes Production

To explore the effect of other conditions on sucrolytic enzyme production by *B. licheniformis* TKU004, the one-factor-at-a-time method was used in the order of nitrogen source (KNO_3, NH_4NO_3, $(NH_4)_2SO_4$, casein, peptone, beef extract powder (BEP), squid pens powder (SPP), shrimp shells powder (SSP), and shrimp heads powder (SHP)), NH_4NO_3 concentration (0%, 0.25%, 0.50%, 0.75%, 1%, 1.25%, 1.50%, 1.75%, and 2%), phosphate salts (NaH_2PO_4, Na_2HPO_4, KH_2PO_4, and K_2HPO_4), NaH_2PO_4 concentration (0%, 0.025%, 0.05%, 0.075%, and 0.1%), yeast extract powder concentration (0%, 0.05%, 0.1%, 0.15%, and 0.2%), pH (3.7, 4.7, 5.7, 6.7, 7.7, and 8.7), temperature (25 °C, 30 °C, 35 °C, 37 °C, and 40 °C), shaking speed (100 rpm, 125 rpm, 150 rpm, 175 rpm, and 200 rpm), and cultivation time (0 h, 12 h, 24 h, 36 h, 48 h, 60 h, and 72 h). The original conditions for the experiments were 2% PAP, 1% KNO_3, 0.1% KH_2PO_4, 0.05% $MgSO_4$, pH = 5.7, 37 °C, 100/250 mL, 150 rpm, and 24 h. At a time, only one factor was assessed while other factors were constant and the condition giving the highest sucrolytic activity was chosen for further experiments. Each experiment was performed in triplicate.

2.5. Enzyme Purification

To obtain cell-free supernatant, the culture medium of *B. licheniformis* TKU004 was centrifuged (9000× rpm for 30 min, 320 R, Hettich, Tuttlingen, Germany) and then mixed with $(NH_4)_2SO_4$ (60% *w*/*v*, 4 °C, overnight). The precipitate was conveniently removed from the mixture by centrifugation (10.000× *g* rpm, 30 min) and dissolved in sodium phosphate buffer (50 mM, pH = 7). The residual $(NH_4)_2SO_4$ in the crude enzyme solution was removed using a cellulose membrane (CelluSep T2, Interchim, Montluçon, France). The crude enzyme solution was then loaded onto a DEAE sepharose column pre-equilibrated with sodium phosphate buffer (50 mM, pH = 7). A linear gradient of NaCl (0 M–1.0 M) was applied to elute the target enzymes. Fractions exhibiting sucrolytic activity were then pooled and concentrated by lyophilization. Finally, the obtained enzyme was purified by a high-performance liquid chromatography (HPLC, Hitachi Chromaster HPLC system, Hitachi, Tokyo, Japan) system consisting of a KW-802.5 column. The molecular weight (MW) of the obtained enzyme was determined using sodium dodecyl sulfate-polyacrylamide gel electrophoresis (SDS-PAGE) and matrix-assisted laser desorption ionization-time of flight mass spectrometry (MALDI-TOF MS, Bruker Daltonics, Bremen, Germany) methods, as described in early reports [22,47]. The identification of the obtained enzyme was determined by liquid chromatography-tandem mass spectrometry (LC-MS/MS, performed by Mission Biotech, Taipei, Taiwan) [48].

2.6. Effect of Temperature and pH

The optimum temperature of sleTKU004 activity was examined over a range of 30 °C to 80 °C. The thermal stability of sleTKU004 was accorded to its residual sucrolytic activity after allowing the sleTKU004 solution to stand at 20 °C, 40 °C, 50 °C, and 60 °C for 0.5 h, 1 h, 2 h, and 4 h. The optimum pH and pH stability of sleTKU004 were explored at the pH range of 3 to 10 using a system buffer (sodium carbonate buffer, pH 9–10; sodium phosphate buffer, pH 6–8; sodium acetate buffer, pH = 5; glycine HCl buffer, pH = 3–4). The pH stability was accorded to the residual sucrolytic activity after allowing the sleTKU004 solution to stand at different pH points for 1 h at 20 °C. Each experiment was performed in triplicate.

2.7. Effect of Various Chemicals

Various chemicals including KCl, $BaCl_2$, $CaCl_2$, $CaSO_4$, $FeCl_3$, $FeCl_2$, $CuSO_4$, $ZnSO_4$, $MgSO_4$, $MnSO_4$, ethylenediaminetetraacetic acid (EDTA), Tween 20, Tween 40, Triton X-100, and sodium dodecyl sulfate (SDS) were used to examine their effect on the sucrolytic activity of sleTKU004. KCl, $BaCl_2$, $CaCl_2$, $CaSO_4$, $FeCl_3$, $FeCl_2$, $CuSO_4$, $ZnSO_4$, $MgSO_4$, $MnSO_4$, and EDTA were tested at the final concentration of 1 mM, 5 mM, and 10 mM whereas Tween 20, Tween 40, Triton X-100, and SDS were assessed at 1%, 5%, and 10% concentrations. Each experiment was performed in triplicate.

2.8. Substrate Specificity

The specificity of the sleTKU004 enzyme was tested on substrates including sucrose, raffinose, stachyose, FOS, pectin, dialyzed pectin, dextran starch, gum arabic, and 1,3-β-glucan using the conditions of the sucrolytic activity assay (as described above). All the substrates were prepared at a concentration of 1% (*w*/*v*). Each experiment was performed in triplicate. The HPLC analysis of the sucrose, raffinose, stachyose hydrolysis was conducted under the following conditions: NH_2P-50 4E column; a mobile phase of 75% acetonitrile and 25% water; flowrate of 1.0 mL/min; infrared (IR) detector.

2.9. Effect of Sucrose Concentration

The effect of sucrose concentration on the sucrolytic activity of sleTKU004 was examined by mixing sleTKU004 solution with sucrose at different concentrations to gain the final sucrose concentration of 0.2 M to 1.5 M. Sucrolytic activity was then determined following the assay described above. Each experiment was performed in triplicate.

2.10. Statistical Analysis

The experiments pertaining to the culture parameters and effects of sucrose concentration on the activity of sleTKU004 were analyzed through one-way analysis of variance (ANOVA) and the post hoc Duncan's Multiple Range Test. The experiment on the effects of various chemicals on the activity of sleTKU004 was performed using one-way ANOVA with the post hoc Tukey Honestly Significant Difference (HSD). Statistical significance was achieved when $p < 0.05$.

3. Results and Discussion

3.1. Agro-Byproducts as the Sole Carbon Source for Sucrolytic Enzyme Production

Various carbon sources were examined for the production of sucrolytic enzymes by *B. licheniformis* TKU004. These included PAP, PFP, OPP, BPP, RBP, WBP, and sucrose. As shown in Figure 1a, the highest sucrolytic activity was observed in the culture supernatant containing PAP as the carbon source (0.81 U/mL on the first day of fermentation), followed by OPP (0.56 U/mL on the 1st day and 0.58 U/mL on the 2nd day of fermentation), RBP (0.59 U/mL on the 2nd day of fermentation), WBP (0.56 U/mL on the 2nd day of fermentation), BPP (0.46 U/mL on the 1st day and 0.46 U/mL on the 2nd day of fermentation), sucrose (0.38 U/mL on the 1st day of fermentation), and PFP (0.28 U/mL on the 1st day of fermentation). Pomelo albedo contains 72.62% carbohydrate, 16.13% moisture, 6.27% protein, 3.41% ash and 1.56% fat [49]. Huang et al., (2014) reported that the carbohydrates of pomelo peel include pectin (35.42%), cellulose (16.50%), hemicellulose (6.86%), lignin (3.16%), and soluble sugar (12.62%) [50]. The high carbohydrate ratio, especially the soluble sugar could make the pomelo albedo an attractive carbon source for microbial fermentation. Currently, bioethanol production is the primary commercial operational process that utilizes pomelo peel waste as a raw material [51,52]. The potential of pomelo albedo in the manufacture of value-added products such as enzymes is not yet fully explored. In this study, among the tested byproducts, pomelo albedo exhibited the highest enzyme activity and was thus chosen as the most suitable carbon source for sucrolytic enzyme production by *B. licheniformis* TKU004.

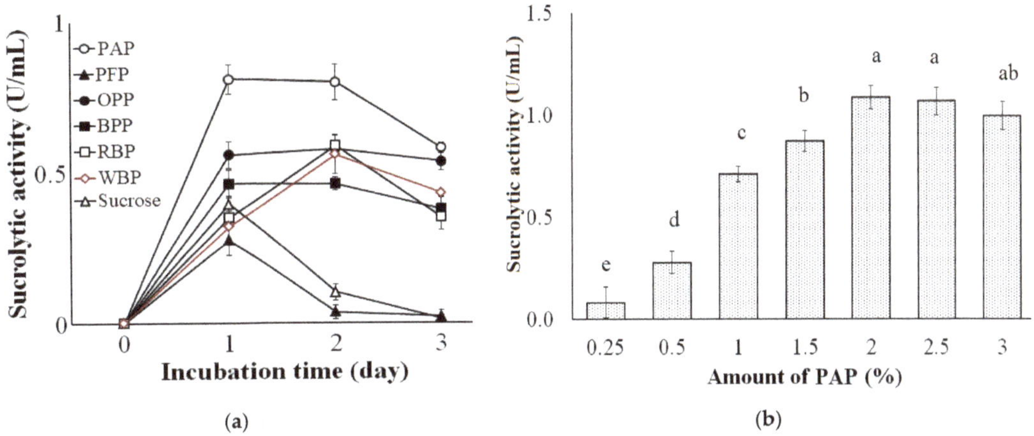

Figure 1. Effect of different carbon sources (**a**) and amount of PAP (**b**) on the sucrolytic enzyme production by *B. licheniformis* TKU004. PAP, pomelo albedo powder; PFP, pomelo flavedo powder; OPP, orange peel powder; BPP, banana peel powder; RBP, rice bran powder; WBP, wheat bran powder. All data points are the mean ± standard deviation. Sucrolytic activity values with different letters (a–e) are significantly different based on Duncan's Multiple Range Test at the level of 5%.

To examine the effect of PAP concentration on the production of sucrolytic enzymes, the amount of PAP was adjusted in a range of 0.25–3% (*w/v*). As shown in Figure 1B, the increase of PAP concentration from 0.25% to 2% could significantly increase the production of sucrolytic enzymes (from 0.08 U/mL to 1.09 U/mL, respectively). Besides, higher PAP concentration (2%, 2.5%, 3%) did not show any significant improvement in the production of sucrolytic enzymes (1.09 U/mL, 1.07 U/mL, and 0.99 U/mL). Due to the lower cost PAP material and the highest sucrolytic enzyme productivity, 2% PAP was chosen as the most suitable concentration for the production of sucrolytic enzymes by *B. licheniformis* TKU004. From the literature, various substances have been used as suitable carbon sources for producing sucrolytic enzymes, such as sucrose [43], glucose [53], fructose [54], orange peel [55], and pineapple crown [56].

3.2. Effect of Nitrogen Sources on Sucrolytic Enzymes Production

Pomelo albedo has a low nitrogen content and comprises only 6.27% protein [49]. Thus, supplementing the nitrogen source to a PAP-containing medium may be a necessity. Accordingly, different nitrogen sources such as KNO_3, NH_4NO_3, $(NH_4)_2SO_4$, casein, peptone, BEP, SPP, SSP, and SHP were supplemented to the PAP-containing medium to assess the most suitable nitrogen source for the production of sucrolytic enzymes by *B. licheniformis* TKU004. As shown in Figure 2a, a higher sucrolytic activity was observed using NH_4NO_3 (1.10 U/mL), followed by KNO_3 (0.94 U/mL), peptone (0.53 U/mL), casein (0.48 U/mL), and BEP (0.41 U/mL). The sucrolytic activity of SSP-, SPP-, and SHP-containing mediums were not significantly different from that of the control medium (0.33 U/mL, 0.31 U/mL, 0.31 U/mL, and 0.31 U/mL respectively), which only contained 2% PAP, 0.1% KH_2PO_4, and 0.05% $MgSO_4$. Besides, the use of 1% $(NH_4)_2SO_4$ could potentially inhibit the production of sucrolytic enzymes by *B. licheniformis* TKU004 (0.03 U/mL). These results indicate that NH_4NO_3 is the best nitrogen source as a supplement to PAP-containing medium for the production of sucrolytic enzymes by *B. licheniformis* TKU004. There are reports of various organic and inorganic nitrogen sources ideal for high sucrolytic enzyme production such as yeast extract, peptone, urea, $NaNO_3$, tryptone, NH_4Cl, and $(NH_4)_2HPO_4$ [15].

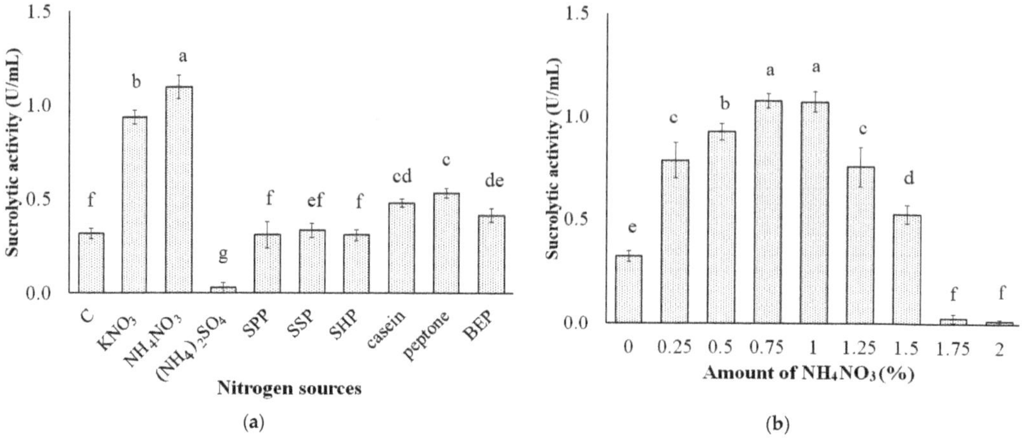

Figure 2. Effect of different nitrogen sources (**a**) and amount of NH_4NO_3 (**b**) on the sucrolytic enzyme production by *B. licheniformis* TKU004. C, control; SPP, squid pens powder; SSP, shrimp shells powder; SHP, shrimp heads powder; BEP, beef extract powder. All data points are the mean ± standard deviation. Sucrolytic activity values with different letters (a–g) are significantly different based on Duncan's Multiple Range Test at the level of 5%.

The effect of NH_4NO_3 concentration on sucrolytic enzyme production was studied by adding 0–2% NH_4NO_3 into the PAP-containing medium. As shown in Figure 2b, the sucrolytic activity of medium with NH_4NO_3 at 0%, 0.25%, 0.50%, 0.75%, 1%, 1.25%, 1.50%, 1.75% and 2% was 0.31 U/mL, 0.79 U/mL, 0.93 U/mL, 1.08 U/mL, 1.07 U/mL, 0.76 U/mL, 0.53 U/mL, 0.03 U/mL, and 0.02 U/mL, respectively. The sucrolytic activity of the medium containing 0.75% NH_4NO_3 and 1% NH_4NO_3 were not significantly different, indicating that 0.75% NH_4NO_3 is the optimum concentration for the production of the sucrolytic enzyme by *B. licheniformis* TKU004.

3.3. *Effect of Phosphate Salts and Yeast Extract Powder Concentration on Sucrolytic Enzyme Production*

Phosphate salts also affect microbial bio-active productivity [57]. Thus, the effect of some phosphate salts such as NaH_2PO_4, Na_2HPO_4, KH_2PO_4, and K_2HPO_4 on the production of sucrolytic enzymes by *B. licheniformis* TKU004 was explored herein. Each of those phosphate salts was added to a medium containing 2% PAP, 0.75% NH_4NO_3, and 0.05% $MgSO_4$. As shown in Figure 3a, NaH_2PO_4 was screened as the most suitable form of phosphate for the optimum sucrolytic activity and was higher than that of the others (1.20 U/mL compared to 0.80 U/mL in Na_2HPO_4-containing medium, 1.08 U/mL in KH_2PO_4-containing medium, and 1.05 U/mL in K_2HPO_4-containing medium). As shown in Figure 3b, the sucrolytic activity of medium with NaH_2PO_4 at 0%, 0.025%, 0.05%, 0.075%, and 0.1% was, 0.16 U/mL, 0.95 U/mL, U/mL, 1.30 U/mL, 1.27 U/mL, and 1.17 U/mL (respectively). The sucrolytic activity of the medium containing 0.05% NaH_2PO_4 and 0.075% NaH_2PO_4 were not significantly different, indicating that 0.05% NaH_2PO_4 is the optimum concentration for the production of the sucrolytic enzyme by *B. licheniformis* TKU004.

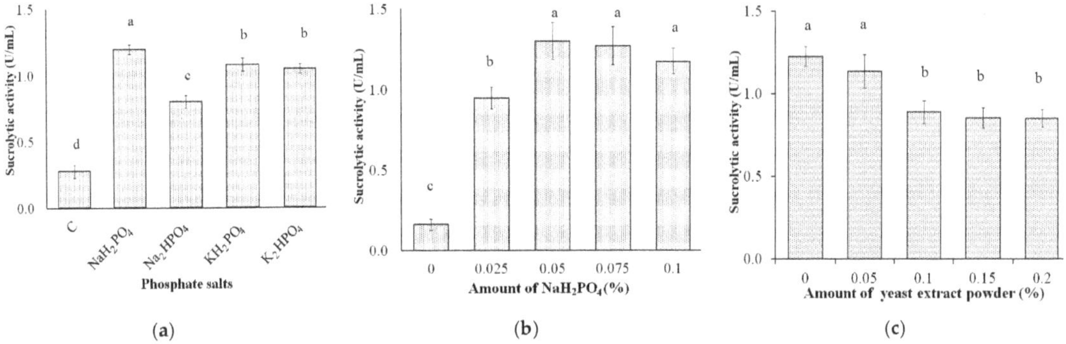

Figure 3. Effect of phosphate salts (**a**), amount of NaH_2PO_4 (**b**), and amount of yeast extract powder (**c**) on the sucrolytic enzyme production by *B. licheniformis* TKU004. C, control. All data points are mean ± standard deviation. Sucrolytic activity values with the different letters (a–d) are significantly different based on Duncan's Multiple Range Test at the level of 5%.

Various microbial strains, for example, *Pichia* sp. [58], *B. cereus* TA-11 [59], *B. macerans* [60], *Lactobacillus brevis* Mm-6 [61], and *Zymomonas mobilis* CDBB-B 603 [53] need yeast extract to enhance the sucrolytic enzyme productivity. In this study, a range of yeast extract powder from 0% to 0.2% was added to a medium containing 2% PAP, 0.75% NH_4NO_3, 0.05% $MgSO_4$, and 0.05% NaH_2PO_4 to explore its effect on the production of sucrolytic enzymes from *B. licheniformis* TKU004. As shown in Figure 3c, yeast extract powder had a negative effect on the enzyme production and the highest sucrolytic activity was obtained from the medium without the addition of yeast extract powder (1.23 U/mL). In fact, yeast extract powder has a high ratio of proteins (48.52%), and carbohydrates (32.92%) [62]. According to the result, the supplement of yeast extract powder can provide nutrients to the bacteria without the necessity to use sucrose or sucrose-based substances, thereby drastically reducing the production of sucrolytic enzymes.

3.4. Effect of Cultural Conditions on Sucrolytic Enzymes Production

The optimum sucrolytic enzyme production by *B. licheniformis* TKU004 using pomelo albedo as the sole carbon source was in the medium containing 2% PAP, 0.75% NH_4NO_3, 0.05% $MgSO_4$, and 0.05% NaH_2PO_4. We then explored the effect of the cultural conditions on the sucrolytic enzyme production by *B. licheniformis* TKU004. These included initial pH of the medium (3.7–8.7), cultivation temperature (25–40 °C), shaking speed (100 rpm, 125 rpm, 150 rpm, 175 rpm, and 200 rpm), and cultivation time (0–72 h). The sucrolytic activity produced in media of different pH (Figure 4a) were in the order of pH = 6.7 (1.29 U/mL), pH = 5.7 (1.09 U/mL), pH = 7.7 (0.89 U/mL), pH = 8.7 (0.89 U/mL), pH = 4.7 (0.23 U/mL), and pH = 3.7 (0.13 U/mL); at different temperatures (Figure 4b) were in the order of 35 °C (1.72 U/mL), 37 °C (1.12 U/mL), 40 °C 0.82 U/mL), 30 °C (0.18 U/mL), and 25 °C (0.02 U/mL); and at different shaking speeds (Figure 4c) were in the order of 150 rpm (1.60 U/mL), 125 rpm (1.09 U/mL), 175 rpm (0.97 U/mL), 200 rpm (0.16 U/mL), and 100 rpm (0.15 U/mL). The time course of cell growth and sucrolytic enzyme production of *B. licheniformis* TKU004 under the optimized conditions is shown in Figure 4d. The number of the cells of *B. licheniformis* TKU004 increased quickly at 0–24 h and then entered the stationary phase in the 24–60 h period. The sucrolytic activity of the culture medium also reached its maximum in the early stationary phase (1.87 U/mL at 24 h) and then decreased over time. Likewise, the time course for sucrolytic enzyme production by bacteria was around 12–72 h [15] for example *Streptomyces* sp. ALKC8 (24 h, 0.35 U/mL) [63], *Brevibacterium divaricatum* (12 h) [64], *Lactobacillus brevis* Mm-6 (72 h, 1399 U/mL) [61], and *Arthrobacter* sp.10137 (22.5 h, 26.69 U/mL) [65]. Another study found that the short fermentation period has more advantages than the long one [21]. This suggested that the time course for the production of sucrolytic enzyme by *B. licheniformis*

TKU004 was acceptable. From the literature, the optimum medium for the production of sucrolytic enzymes may require the supplement of sugar (sucrose [43], fructose [54], and glucose [53]) and organic nitrogen (peptone [60], malt extract [61], and yeast extract [58]). However, the price of those materials may be a significant obstacle in the production of sucrolytic enzymes. Thus, to reduce the cost of the process of sucrolytic enyzme production, the use of agricultural wastes can be an effective alternative. In this study, we could successfully develop a low-cost and simple medium based on pomelo albedo and optimize the culture conditions for sucrolytic enzyme production by *B. licheniformis* TKU004. The composition of the optimal medium was 2% PAP, 0.75% NH_4NO_3, 0.05% $MgSO_4$, and 0.05% NaH_2PO_4, and the optimal culture conditions were observed to be initial medium pH of 6.7, cultivation temperature of 35 °C, shaking speed of 150 rpm, and cultivation time of 24 h. Accordingly, the highest sucrolytic activity was 1.87 U/mL, 4.79-fold higher than that in original conditions using sucrose as the carbon source.

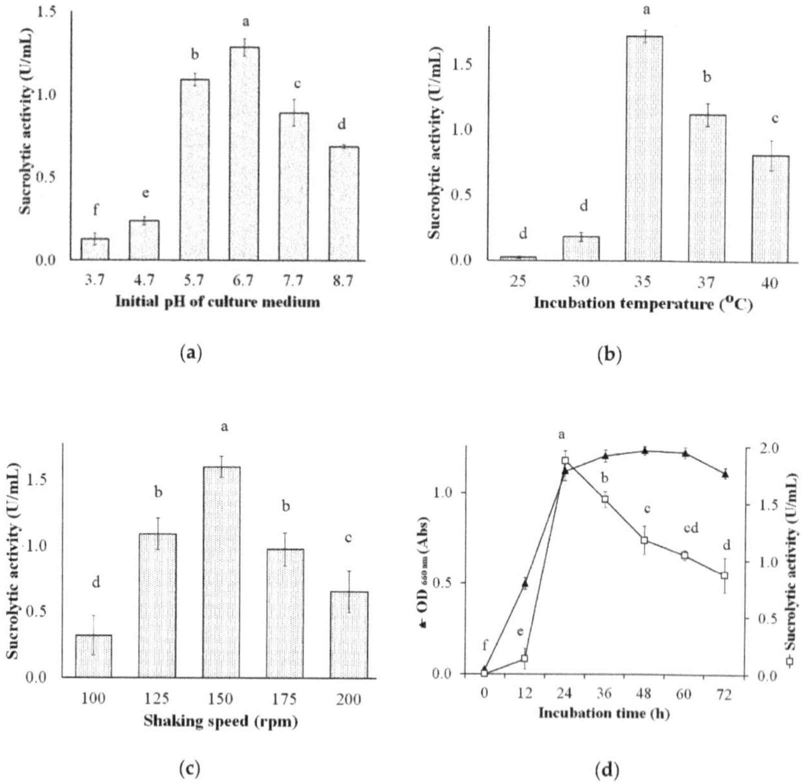

Figure 4. Effect of initial pH (**a**), temperature (**b**), shaking speed (**c**), and incubation time (**d**) on the sucrolytic enzyme production by *B. licheniformis* TKU004. All data points are the mean ± standard deviation. Sucrolytic activity values with different letters (a–f) are significantly different based on Duncan's Multiple Range Test at the level of 5%.

3.5. Purification of the Produced Sucrolytic Enzyme

One-day culture supernatant (0.4 L) was used to purify the produced sucrolytic enzymes under the following condition: DEAE-Sepharose resin, 50 mM sodium phosphate (pH = 7.0), and NaCl gradient from 0 M to 1 M. The fractions showing sucrolytic activity from tube number 45 to 59 (Figure 5) were dialyzed against 50 mM sodium phosphate (pH = 7.0), and concentrated by a freeze-dryer. HPLC with size-exclusion chromatography mode was then used for the purification of the enzyme. After purification, approximately 1.34 mg of *B. licheniformis* TKU004 sucrolytic enzyme (sleTKU004) was obtained. The recovery yield of the obtained enzyme was 1.31% with 11.4-folds of the specific activity (Table 1). As shown in Table 2, the MW of 53 kDa (by SDS-PAGE method) of sleTKU004 is highly similar to sucrase from *B. stearothermophilus* NUB36 (51.519 Da) and levansucrases from *B. licheniformis* RN-01 (52 kDa), *B. licheniformis* 8-37-0-1 (51 kDa), *B. amyloliquefaciens* BH072 (55 kDa), and *B. amyloliquefaciens* KK9 (52974 Da). In general, the sucrolytic enzymes from the *Bacillus* genus have an MW range of 14–105 kDa (Table 2). The non-reducing SDS-PAGE analysis of sleTKU004 also revealed a single band at 53 kDa (Figure 6a), indicating that the enzyme is a monomer in its native form. Other reports show that the sucrose hydrolytic enzymes from the *Bacillus* genus could be monomers [66] or dimers [67]. The native form of sleTKU004 was also analyzed by MALDI-TOF MS and a peak was observed at 53,228 Da (Figure 6b).

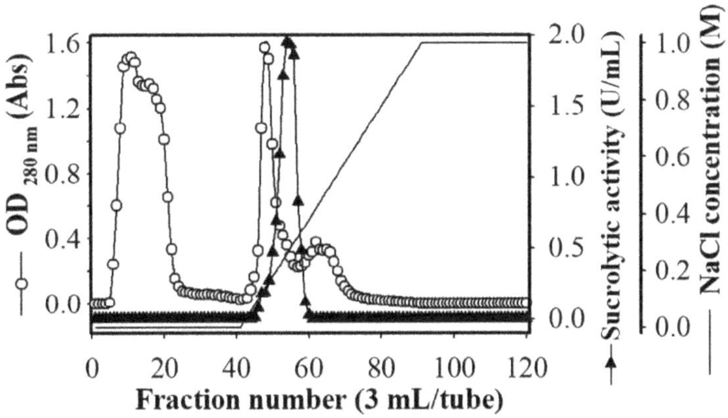

Figure 5. DEAE-Sepharose chromatography profile of the crude sleTKU004 enzyme.

Table 1. A summary of the purification of sleTKU004 from *B. licheniformis* TKU004.

Step	Total Protein (mg)	Total Activity (U)	Specific Activity (U/mg)	Recovery (%)	Purification (Fold)
Cultural supernatant	1170.91	582.31	0.497	100.00	1.00
$(NH_4)_2SO_4$ precipitation	359.71	297.74	0.827	51.13	1.66
Ion-exchange chromatography	79.98	199.47	2.494	34.26	5.02
Size-exclusion chromatography	1.34	7.63	5.670	1.31	11.40

Table 2. A comparison of sleTKU004 with sucrolytic enzyme produced by other *Bacillus* strains.

Enzyme/Strain	Opt. Temp.	Opt. pH	Carbon Source	MW	Ref
Levansucrase *B. licheniformis* TKU004	50 °C	6	PAP	53 kDa	This study
β-d-fructofuranosidase *B. subtilis* LYN12	30–60 °C	4–8	wheat bran and molasses	66 kDa and 64,512.31 Da [1]	[68]
Levansucrase *B. licheniformis* RN-01	50 °C	6	peptone	52 kDa	[42]
Levansucrase *B. licheniformis* 8-37-0-1	45–50 °C	6	sucrose	51 kDa	[43]

Table 2. *Cont.*

Enzyme/Strain	Opt. Temp.	Opt. pH	Carbon Source	MW	Ref
Fructan sucrase *B. subtilis* ZW019	50	5.6	tryptone	58 kDa	[69]
Sucrase and levansucrase *B. subtilis* Marburg				40 kDa	[70]
Invertase *Bacillus* sp. HJ14	30–32.5 °C	8		58 kDa	[71]
Levanase *B. subtilis*			sucrose	73 kDa	[72]
Levansucrase *B. amyloliquefaciens* BH072	40 °C	6	sucrose and corn starch	55 kDa	[73]
Invertase *B. cereus* TA-11	50 °C	7	sucrose	23 kDa and 26 kDa [2]	[59]
Sucrase *B. stearothermophilus* NUB36	55 °C			51,519 Da [3] 105,000 Da [2]	[66]
α-glucosidase *B. subtilis*				66 kDa	[74]
levansucrase *B. amyloliquefaciens* KK9				52,974 Da [3]	[75]
levansucrase *B. subtilis* NRC16	45 °C	8.2	starch	14 kDa	[76]
Levansucrase *B. methylotrophicus* SK 21.002	40 °C	6.5	sucrose	60 kDa	[77]
Levansucrase *B. licheniformis* ANT 179	60 °C	6	sugarcane juice	25 kDa	[44]
Levansucrase *Bacillus* sp.	60 °C	6	starch	56 kDa	[78]

[1], LC/MS; [2], gel filtration; [3], deduced polypeptide.

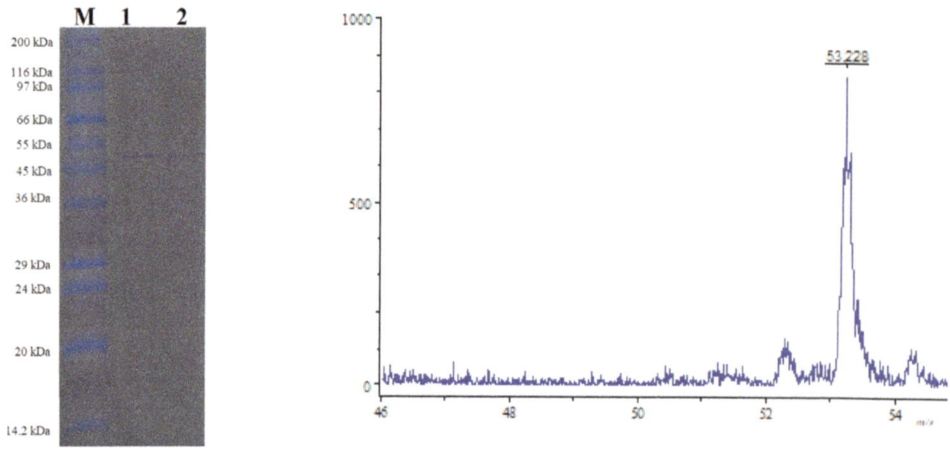

(a) (b)

Figure 6. SDS-PAGE (**a**) and MALDI-TOF MS profile (**b**) of the sleTKU004. M, protein markers; 1, with 2-mercaptoethanol (2-ME); 2, without 2-ME.

3.6. Identification of sleTKU004 by LC-MS/MS Analysis

To identify sleTKU004 that appeared as a prominent 53-kDa band via SDS-PAGE, the band was excised and analyzed after tryptic digestion. The excised band from the gel was subjected to electrospray tandem mass spectrometry analysis. The fragment spectra were subjected to an NCBI non-redundant protein database search. As shown in Table 3, the spectra of sleTKU004 matched nine tryptic peptides that were identical to the levansucrase (sacB) from *B. subtilis* subsp. *subtilis* str. 168 with 30% sequence coverage. The peptide sequences indicate that sleTKU004 belongs to the family 68 glycosyl hydrosylase based on the amino acid sequence similarity of the cited GH-68 enzymes from *B. subtilis*. Besides, the MW of sleTKU004 (53 kDa) was also similar to that of levansucrase from *B. subtilis* subsp. *subtilis* str. 16 (52.938 Da).

Table 3. Identification of sleTKU004 by LC-MS/MS Analysis.

Matched Peptide Sequence	Identified Protein and Coverage Rate	Mass and pI	Strain
^{39}ETYGISHITR^{48} ^{63}YQVPEFDSSTIK^{74} ^{115}NADDTSIYMFYQKVGETSIDSWK^{137} ^{145}DSDKFDANDSILK^{157} ^{177}LFYTDFSGK^{185} ^{216}SIFDGDGKTYQNVQQFIDEGNYSSGDNHTLR^{246} ^{326}KVMKPLIASNTVTDEIER^{343} ^{421}GNNVVITSYMTNR^{433} ^{461}DSILEQGQLTVNK^{473}	Levansucrase 30%	52,938 Da 6.18	*B. subtilis* subsp. *subtilis* strain 168

Superscripts (^{39} and ^{48}; ^{63} and ^{74}; ^{115} and ^{137}; ^{145} and ^{157}; ^{177} and ^{185}; ^{216} and ^{246}; ^{326} and ^{343}; ^{421} and ^{433}; ^{461} and ^{473}) indicate regions in the protein sequence of the levansucrase from *B. subtilis* subsp. *subtilis* strain 168.

3.7. Effects of Temperature and pH on the Activity and Stability of sleTKU004

The optimum temperature for sucrolytic activity of sleTKU004 was examined in a temperature range of 30–80 °C. The sucrolytic activity of the enzyme increased with a rise in temperature from 30 °C to 50 °C, and then sharply decreased at higher temperatures (Figure 7a). The optimum temperature for sucrolytic activity of sleTKU004 was 50 °C. Likewise, sucrolytic enzymes from *Bacillus* genus have also been found to work best in the temperature range of 30 °C to 60 °C. To explore the thermal stability of sleTKU004, the enzyme was treated at 20 °C, 40 °C, 50 °C, and 60 °C for 0.5–4 h. The result showed that the sucrolytic activity of sleTKU004 could be retained to 100% at 20 °C in 4 h, 92% at 40 °C in 3 h, and 73% at 40 °C in 4 h (Figure 7b). At higher temperatures, the half-life of sleTKU004 was estimated to be around 2 h (at 50 °C), and 0.3 h (at 60 °C). The pH activity profile sleTKU004 is shown in Figure 7c,d. The optimum enzyme activity was observed at pH = 6 and it stable at the pH range of pH = 5–9. Likewise, many sucrolytic enzymes from other *Bacilli* expressed the highest activity at pH = 6 (Table 2).

3.8. Effect of Various Chemicals on the Activity of sleTKU004

Generally, metal ions have a great influence on protein folding and catalytic processes [58]. As shown in Figure 8, some metal salts such as KCl, $BaCl_2$, $CaCl_2$, and $CaSO_4$ had no significant effect or only slightly inhibited the sucrolytic activity of sleTKU004. At 1 mM, $FeCl_3$, $FeCl_2$, $FeSO_4$, $CuSO_4$, $ZnSO_4$, and EDTA did not affect sleTKU004, however, at higher concentrations (5 mM, and 10 mM), they could inhibit the sucrolytic activity of the enzyme. On the contrary, $MgSO_4$, and $MnSO_4$ in the concentration range of 1 mM–10 mM slightly promoted the sucrolytic activity of sleTKU004. Tween 20, Tween 40, Triton X-100, and SDS were also tested for their effect on sleTKU004 sucrolytic activity. A slight inhibiting effect was found on tween 20 (10%), tween 40 (10%), and triton X-100 (5% and 10%). At 1%, SDS did not affect sleTKU004, however, at higher concentrations (5% and 10%) it completely inhibited the enzyme activity.

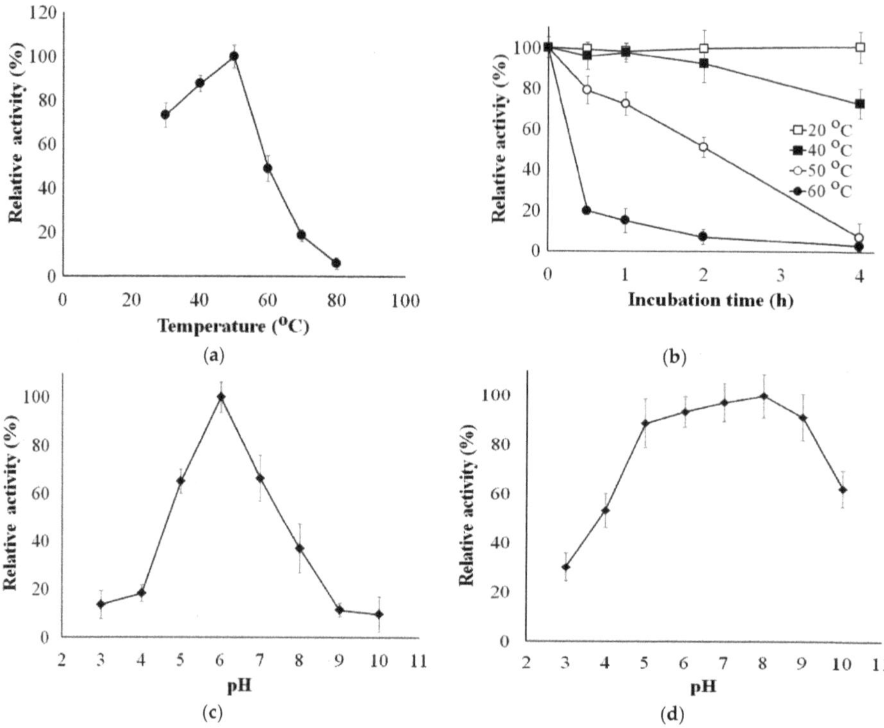

Figure 7. Effects of Temperature and pH on the activity and stability of sleTKU004. (**a**), optimum temperature; (**b**), thermal stability; (**c**), optimum pH; (**d**), pH stability. All data points are the mean ± standard deviation.

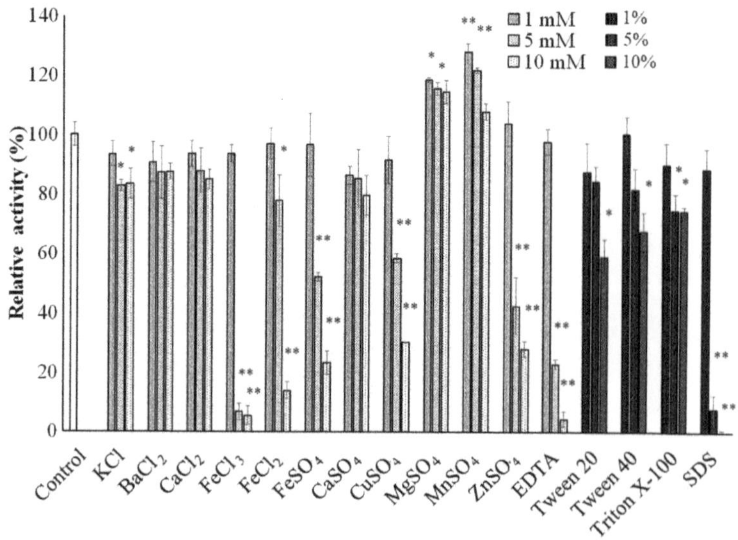

Figure 8. Effect of various chemicals on the activity of sleTKU004. All data points are the mean ± standard deviation. All data points are the mean ± standard deviation. Asterisks were used to highlight the statistical differences between control and treatment groups using the Tukey HSD Test. *, and ** indicate statistical significance at $p < 0.05$ and $p < 0.01$, respectively.

3.9. Substrate Specificity of sleTKU004

Substrate specificity sleTKU004 is summarized in Table 4. In the case of disaccharide and oligosaccharide substrates, sleTKU004 expressed the greatest activity in the order of sucrose (100%) > raffinose (85.74%) > stachyose (73.60%) > FOS (19.15%). No activity was observed against maltose and sucralose. In the case of polysaccharide substrates, sleTKU004 seemingly showed high pectinolytic activity and expressed a relative activity of 81.38% on pectin. Surprisingly, a significant amount of residual sucrose was found in the pectin substrate and was the primary contributor to the production of reducing sugar which was then used to calculate the enzyme's relative activity (data not shown). Thus, 2-days dialyzed pectin was used to test the hydrolysis activity of sleTKU004, and the enzyme only expressed 15.26% relative activity on this type of pectin. Likewise, the relative activity on dextran was 16.94%. These results indicate that sleTKU004 may have a slight pectinolytic, and dextranolytic activity. Besides, the activity of sleTKU004 was insignificant on starch (8.82%), gum arabic (9.96%), and 1,3-β-glucan (3.59%).

Table 4. Substrate specificity of sleTKU004.

Substrate	Relative Activity (%)
Sucrose	100.00 ± 4.44
Raffinose	85.74 ± 3.45
Stachyose	73.60 ± 5.39
FOS	19.15 ± 2.47
Starch	8.82 ± 1.93
Pectin	81.38 ± 2.60
Dialyzed pectin	15.26 ± 3.28
Dextran	16.94 ± 5.04
Gum arabic	9.96 ± 1.65
1,3-β-glucan	3.59 ± 2.24
Maltose	N.A.
Sucralose	N.A.

N.A.: no activity. All data points are the mean ± standard deviation.

The mechanism of high hydrolytic activity of sleTKU004 on sucrose, raffinose, and stachyose was further analyzed by HPLC. The results showed that fructose was one of the major products of the hydrolysis, indicating that sleTKU004 strictly attacked the glycoside linkage between glucose and fructose molecules of the sucrose unit (Figure 9). The other products of sucrose, raffinose, and stachyose hydrolysis were found to be glucose, melibiose, and manninotriose.

3.10. Effect of Sucrose Concentration and Kinetic Characterization

The effect of sucrose concentration on the activity of sleTKU004 was shown in Figure 10a. With an increase in sucrose concentration from 0.15 M to 1.2 M, there was a gradual increase in the sucrolytic activity of sleTKU004 (1.98–8.39 U/mL, respectively). Higher sucrose concentrations (1.2–1.5 M) maintained the high sucrose hydrolytic activity (8.39–8.79 U/mL). It was shown that the enzymatic method is preferable in the production of high-quality invert syrups compared with the acidic method. However, high sucrose concentration, which is regarded as one of the inhibitors that block the preparation of inverted sugar [2], could be a significant obstacle for the application of the enzymatic method. For this reason, candidates producing sucrolytic enzymes, which are tolerant to high-concentration sucrose, are being explored for the production of high-quality invert sugar syrups [2]. Together, the result indicates that sleTKU004 could be a potential candidate for the inverted sugar preparation.

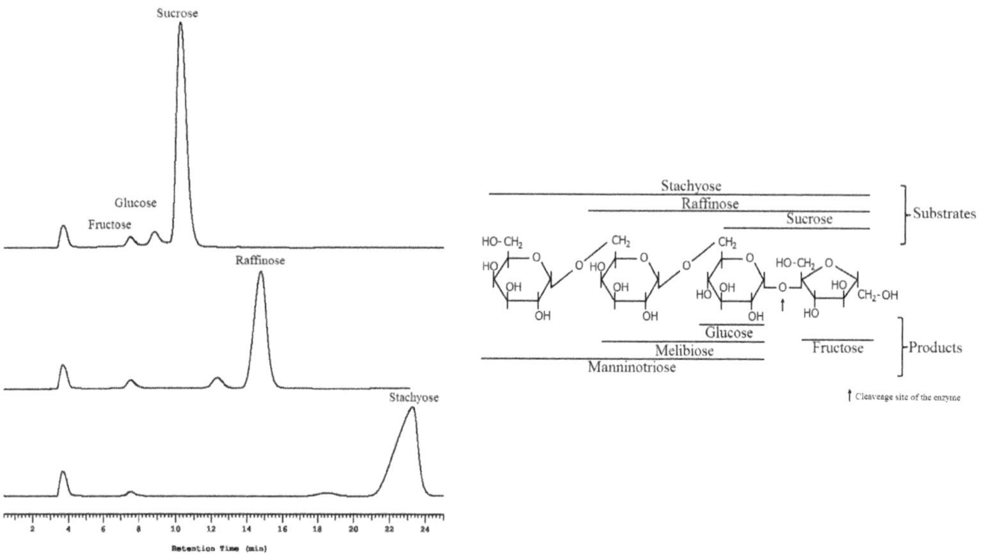

(a) (b)

Figure 9. HPLC profile (**a**) and schematic diagram (**b**) of the sucrose, raffinose, and stachyose hydrolysis catalyzed by sleTKU004.

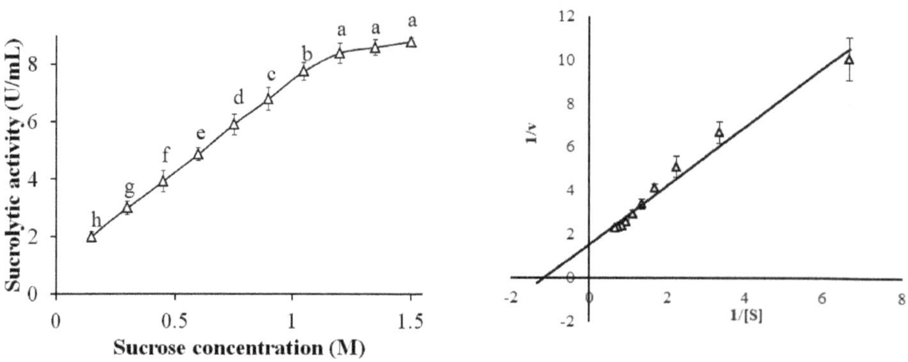

(a) (b)

Figure 10. The effect of sucrose concentration on the activity of sleTKU004 (**a**) and the Lineweaver Burk plot (**b**). All data points are the mean ± standard deviation. Sucrolytic activity values with different letters (a–h) are significantly different based on Duncan's Multiple Range Test at the level of 5%.

K_m and V_{max} of sleTKU004 were estimated to be 1.16 M and 5.99 µmol/min based on a Lineweaver-Burk plot (Figure 10b). K_m value of sleTKU004 was higher than that of sucrolytic enzymes from other *Bacilli* such as *B. methylotrophicus* SK 21.002 (117.2 mM) [77], *B. amyloliquefacien* BH072 (0.71 mM) [73], *B. cereus* TA-11 (370 mM) [59], and *B. subtilis* (33.96 g/L) [69]. A high K_m value indicates that high concentration of sucrose must be present to saturate sleTKU004. This suggests that the affinity of sleTKU004 towards sucrose is low [59].

4. Conclusions

The current study aimed to establish bioprocessing of agro-byproducts for sucrolytic enzyme production by *B. licheniformis* TKU004. A medium containing PAP as the sole carbon source was found to be the most suitable for sucrolytic enzyme production. Then, the culture condition for PAP-containing mediums was successfully optimized with a sucrolytic activity 4.79-fold higher than that from original conditions. A 53 kDa sucrolytic enzyme (sleTKU004) was isolated from the PAP culture supernatant and purified and was then determined to belong to the GH 68 family. Furthermore, the characteristics of sleTKU004 were also determined. Finally, sleTKU004 showed strong sucrose tolerance and presented the highest hydrolytic activity at sucrose concentration of 1.2–1.5 M, indicating that sleTKU004 may be a good candidate for industrial applications requiring sucrolytic enzymes.

Author Contributions: Conceptualization and methodology, C.T.D., T.N.T. and S.-L.W.; software and validation, C.T.D., T.N.T. and S.-L.W.; formal analysis, C.T.D., T.N.T., T.T.N., T.P.H.T., T.D.T., V.B.N., A.D.N. and S.-L.W.; investigation, C.T.D., T.N.T. and S.-L.W.; resources, S.-L.W.; data curation, C.T.D., T.N.T., T.T.N., T.P.H.T., T.D.T., V.B.N., A.D.N., and S.-L.W.; investigation, C.T.D., T.N.T. and S.-L.W.; writing—original draft preparation, C.T.D., T.N.T. and S.-L.W.; writing—review and editing, C.T.D., T.N.T. and S.-L.W.; visualization, supervision, and project administration, S.-L.W. All authors have read and agreed to the published version of the manuscript.

Funding: This study was supported in part by a grant from the Ministry of Science and Technology, Taiwan (MOST 109-2313-B-032-002-; MOST 109-2811-B-032-500; MOST 110-2923-B-032-001).

Institutional Review Board Statement: Not applicable.

Informed Consent Statement: Not applicable.

Data Availability Statement: The data presented in this study are available on request from the corresponding author.

Conflicts of Interest: The authors declare no conflict of interest.

References

1. Xu, W.; Ni, D.; Zhang, W.; Guang, C.; Zhang, T.; Mu, W. Recent advances in levansucrase and inulosucrase: Evolution, characteristics, and application. *Crit. Rev. Food Sci. Nutr.* **2019**, *59*, 3630–3647. [CrossRef] [PubMed]
2. Zhou, G.; Peng, C.; Liu, X.; Chang, F.; Xiao, Y.; Liu, J.; Fang, Z. Identification and immobilization of an invertase with high specific activity and sucrose tolerance ability of *Gongronella* sp. w5 for high fructose syrup preparation. *Front. Microbiol.* **2020**, *11*, 633. [CrossRef] [PubMed]
3. Hill, A.; Karboune, S.; Narwani, T.J.; de Brevern, A.G. Investigating the product profiles and structural relationships of new levansucrases with conventional and non-conventional substrates. *Int. J. Mol. Sci.* **2020**, *21*, 5402. [CrossRef] [PubMed]
4. Koirala, P.; Maina, N.H.; Nihtilä, H.; Katina, K.; Coda, R. Brewers' spent grain as substrate for dextran biosynthesis by *Leuconostoc pseudomesenteroides* DSM20193 and *Weissella confusa* A16. *Microb Cell Fact.* **2021**, *20*, 23. [CrossRef] [PubMed]
5. Li, X.; Wang, X.; Meng, X.; Dijkhuizen, L.; Liu, W. Structures, physico-chemical properties, production and (potential) applications of sucrose-derived α-d-glucans synthesized by glucansucrases. *Carbohydr. Polym.* **2020**, *249*, 116818. [CrossRef] [PubMed]
6. Trollope, K.M.; van Wyk, N.; Kotjomela, M.A.; Volschenk, H. Sequence and structure-based prediction of fructosyltransferase activity for functional subclassification of fungal GH32 enzymes. *FEBS J.* **2015**, *282*, 4782–4796. [CrossRef] [PubMed]
7. Le Roy, K.; Lammens, W.; Verhaest, M.; de Coninck, B.; Rabijns, A.; van Laere, A.; van den Ende, W. Unraveling the difference between invertases and fructan exohydrolases: A single amino acid (Asp-239) substitution transforms Arabidopsis cell wall invertase1 into a fructan 1-exohydrolase. *Plant. Physiol.* **2007**, *145*, 616–625. [CrossRef] [PubMed]
8. Ackerman, D.L.; Craft, K.M.; Townsend, S.D. Infant food applications of complex carbohydrates: Structure, synthesis, and function. *Carbohydr. Res.* **2017**, *437*, 16–27. [CrossRef]
9. González-Garcinuño, A.; Tabernero, A.; Domínguez, A.; Galán, M.A.; del Valle, E.M.M. Levan and levansucrases: Polymer, enzyme, micro-organisms and biomedical applications. *Biocatal. Biotransform.* **2018**, *36*, 233–244. [CrossRef]
10. Taskin, M.; Ortucu, S.; Unver, Y.; Canli, O. Invertase production and molasses decolourization by cold-adapted filamentous fungus *Cladosporium herbarum* ER-25 in non-sterile molasses medium. *Process. Saf. Environ. Protect. Instit. Chem. Eng.* **2016**, *103*, 136–143. [CrossRef]
11. Hoffmann, J.J.; Hövels, M.; Kosciow, K.; Deppenmeier, U. Synthesis of the alternative sweetener 5-ketofructose from sucrose by fructose dehydrogenase and invertase producing *Gluconobacter* strains. *J. Biotechnol.* **2020**, *307*, 164–174. [CrossRef] [PubMed]

12. Rasbold, L.M.; Heinen, P.R.; da Conceição Silva, J.L.; Simão, R.d.C.G.; Kadowaki, M.K.; Maller, A. *Cunninghamella echinulata* PA3S12MM invertase: Biochemical characterization of a promiscuous enzyme. *J. Food Biochem.* **2021**, *45*, e13654. [CrossRef]
13. Contesini, F.J.; de Alencar Figueira, J.; Kawaguti, H.Y.; de Barros Fernandes, P.C.; de Oliveira Carvalho, P.; da Graça Nascimento, M.; Sato, H.H. Potential applications of carbohydrases immobilization in the food industry. *Int. J. Mol. Sci.* **2013**, *14*, 1335–1369. [CrossRef]
14. Manoochehri, H.; Hosseini, N.F.; Saidijam, M.; Taheri, M.; Rezaee, H.; Nouri, F. A review on invertase: Its potentials and applications. *Biocatal. Agric. Biotechnol.* **2020**, *25*, 101599. [CrossRef]
15. Lincoln, L.; More, S.S. Bacterial invertases: Occurrence, production, biochemical characterization, and significance of transfructosylation. *J. Basic Microbiol.* **2017**, *57*, 803–813. [CrossRef]
16. Kulshrestha, S.; Tyagi, P.; Yadavilli, S. Invertase and its application—A brief review. *J. Pharm. Res.* **2013**, *7*, 792–797. [CrossRef]
17. Mariz, B.P.; Carvalho, S.; Batalha, I.L.; Pina, A.S. Artificial enzymes bringing together computational design and directed evolution. *Org. Biomol. Chem.* **2021**, *19*, 1915–1925. [CrossRef] [PubMed]
18. Mehta, P.K.; Sehgal, S. Microbial enzymes in food processing. In *Biocatalysis*; Husain, Q., Ullah, M., Eds.; Springer: Cham, Switzerland, 2019; pp. 255–272.
19. Wang, C.H.; Doan, C.T.; Nguyen, V.B.; Nguyen, A.D.; Wang, S.L. Reclamation of fishery processing waste: A mini-review. *Molecules* **2019**, *24*, 2234. [CrossRef] [PubMed]
20. Doan, C.T.; Tran, T.N.; Nguyen, V.B.; Vo, T.P.K.; Nguyen, A.D.; Wang, S.L. Chitin extraction from shrimp waste by liquid fermentation using an alkaline protease-producing strain, *Brevibacillus parabrevis*. *Int. J. Biol. Macromol.* **2019**, *131*, 706–715. [CrossRef]
21. Tran, T.N.; Doan, C.T.; Wang, S.-L. Conversion of wheat bran to xylanases and dye adsorbent by *Streptomyces thermocarboxydus*. *Polymers* **2021**, *13*, 287. [CrossRef]
22. Doan, C.T.; Tran, T.N.; Nguyen, V.B.; Nguyen, A.D.; Wang, S.L. Utilization of seafood processing by-products for production of proteases by *Paenibacillus* sp. TKU052 and their application in biopeptides' preparation. *Mar. Drugs* **2020**, *18*, 574. [CrossRef] [PubMed]
23. Doan, C.T.; Chen, C.L.; Nguyen, V.B.; Tran, T.N.; Nguyen, A.D.; Wang, S.L. Conversion of wheat bran to pectinases by *Bacillus amyloliquefaciens* and its applications on hydrolyzing banana peels for prebiotics production. *Polymers* **2021**, *13*, 1483. [CrossRef]
24. Hammami, A.; Bayoudh, A.; Abdelhedi, O.; Nasri, M. Low-cost culture medium for the production of proteases by *Bacillus mojavensis* SA and their potential use for the preparation of antioxidant protein hydrolysate from meat sausage by-products. *Ann. Microbiol.* **2018**, *68*, 473–484. [CrossRef]
25. Walia, A.; Guleria, S.; Mehta, P.; Chauhan, A.; Parkash, J. Microbial xylanases and their industrial application in pulp and paper biobleaching: A review. *3 Biotech.* **2017**, *7*, 11. [CrossRef] [PubMed]
26. Doan, C.T.; Tran, T.N.; Nguyen, V.B.; Tran, T.D.; Nguyen, A.D.; Wang, S.L. Bioprocessing of squid pens waste into chitosanase by *Paenibacillus* sp. TKU047 and its application in low-molecular weight chitosan oligosaccharides production. *Polymers* **2020**, *12*, 1163. [CrossRef]
27. Wang, C.L.; Su, J.W.; Liang, T.W.; Nguyen, A.D.; Wang, S.L. Production, purification and characterization of a chitosanase from *Bacillus cereus*. *Res. Chem. Intermed.* **2014**, *40*, 2237–2248. [CrossRef]
28. Tran, T.N.; Doan, C.T.; Nguyen, M.T.; Nguyen, V.B.; Vo, T.P.K.; Nguyen, A.D.; Wang, S.L. An exochitinase with *N*-acetyl-β-glucosaminidase-like activity from shrimp head conversion by *Streptomyces speibonae* and its application in hydrolyzing β-chitin powder to produce *N*-acetyl-D-glucosamine. *Polymers* **2019**, *11*, 1600. [CrossRef]
29. Tran, T.N.; Doan, C.T.; Nguyen, V.B.; Nguyen, A.D.; Wang, S.L. The isolation of chitinase from *Streptomyces thermocarboxydus* and its application in the preparation of chitin oligomers. *Res. Chem. Intermed.* **2019**, *45*, 727–742. [CrossRef]
30. Biz, A.; Finkler, A.T.J.; Pitol, L.O.; Medina, B.S.; Krieger, N.; Mitchell, D.A. Production of pectinases by solid-state fermentation of a mixture of citrus waste and sugarcane bagasse in a pilot-scale packed-bed bioreactor. *Biochem. Eng. J.* **2016**, *111*, 54–62. [CrossRef]
31. Ravindran, R.; Hassan, S.S.; Williams, G.A.; Jaiswal, A.K. A Review on bioconversion of agro-industrial wastes to industrially important enzymes. *Bioengineering* **2018**, *5*, 93. [CrossRef]
32. Ohara, A.; de Castro, R.J.S.; Nisshde, T.G.; Dias, F.F.G.; Bagagli, M.P.; Sato, H.H. Invertase production by *Asspergillus niger* under solid state fermentation: Focus on physical-chemical parameters, synergistic and antagonistic effects using agro-industrial wastes. *Biocatal. Agric. Biotechnol.* **2015**, *4*, 645–652. [CrossRef]
33. Clements, L.D.; Miller, B.S.; Streips, U.N. Comparative growth analysis of the facultative anaerobes *Bacillus subtilis, Bacillus licheniformis,* and *Escherichia coli*. *Syst. Appl. Microbiol.* **2002**, *25*, 284–286. [CrossRef]
34. Pham, J.V.; Yilma, M.A.; Feliz, A.; Majid, M.T.; Maffetone, N.; Walker, J.R.; Kim, E.; Cho, H.J.; Reynolds, J.M.; Song, M.C.; et al. A review of the microbial production of bioactive natural products and biologics. *Front. Microbiol.* **2019**, *10*, 1404. [CrossRef] [PubMed]
35. Deng, W.; Dong, X.F.; Tong, J.M.; Zhang, Q. The probiotic *Bacillus licheniformis* ameliorates heat stress-induced impairment of egg production, gut morphology, and intestinal mucosal immunity in laying hens. *Poult. Sci.* **2012**, *91*, 575–582. [CrossRef]
36. Sellami-Kamoun, A.; Haddar, A.; Ne, H.A.; Ghorbel-Frikha, B.; Kanoun, S.; Nasri, M. Stability of thermostable alkaline protease from *Bacillus licheniformis* RP1 in commercial solid laundry detergent formulations. *Microbiol. Res.* **2008**, *163*, 299–306. [CrossRef] [PubMed]

37. Muras, A.; Romero, M.; Mayer, C.; Otero, A. Biotechnological applications of *Bacillus licheniformis*. *Crit. Rev. Biotechnol.* **2021**, *41*, 609–627. [CrossRef]
38. Schallmey, M.; Singh, A.; Ward, O.P. Developments in the use of *Bacillus* species for industrial production. *Can. J. Microbiol.* **2004**, *50*, 1–17. [CrossRef] [PubMed]
39. Doan, C.T.; Tran, T.N.; Nguyen, M.T.; Nguyen, V.B.; Nguyen, A.D.; Wang, S.L. Anti-α-glucosidase activity by a protease from *Bacillus licheniformis*. *Molecules* **2019**, *24*, 691. [CrossRef] [PubMed]
40. Fincan, S.A.; Özdemir, S.; Karakaya, A.; Enez, B.; Mustafov, S.D.; Ulutaş, M.S.; Şen, F. Purification and characterization of thermostable α-amylase produced from *Bacillus licheniformis* So-B3 and its potential in hydrolyzing raw starch. *Life Sci.* **2021**, *264*, 118639. [CrossRef]
41. De Boer, A.S.; Priest, F.; Diderichsen, B. On the industrial use of *Bacillus licheniformis*: A review. *Appl. Microbiol. Biotechnol.* **1994**, *40*, 595–598. [CrossRef]
42. Nakapong, S.; Pichyangkura, R.; Ito, K.; Iizuka, M.; Pongsawasdi, P. High expression level of levansucrase from *Bacillus licheniformis* RN-01 and synthesis of levan nanoparticles. *Int. J. Biol. Macromol.* **2013**, *54*, 30–36. [CrossRef]
43. Lu, L.; Fu, F.; Zhao, R.; Jin, L.; He, C.; Xu, L.; Xiao, M. A recombinant levansucrase from *Bacillus licheniformis* 8-37-0-1 catalyzes versatile transfructosylation reactions. *Process. Biochem.* **2014**, *49*, 1503–1510. [CrossRef]
44. Xavier, J.R.; Ramana, K.V. Optimization of levan production by cold-active *Bacillus licheniformis* ANT 179 and fructooligosaccharide synthesis by its levansucrase. *Appl. Biochem. Biotechnol.* **2017**, *181*, 986–1006. [CrossRef]
45. Parrado, J.; Rodriguez-Morgado, B.; Tejada, M.; Hernandez, T.; Garcia, C. Proteomic analysis of enzyme production by *Bacillus licheniformis* using different feather wastes as the sole fermentation media. *Enzyme Microb. Technol.* **2014**, *57*, 1–7. [CrossRef]
46. Wang, S.L.; Kao, T.Y.; Wang, C.L.; Yen, Y.H.; Chern, M.K.; Chen, Y.H. A solvent stable metalloprotease produced by *Bacillus* sp. TKU004 and its application in the deproteinization of squid pen for β-chitin preparation. *Enzym. Microb. Technol.* **2006**, *39*, 724–731. [CrossRef]
47. Tran, T.N.; Doan, C.T.; Nguyen, V.B.; Nguyen, A.D.; Wang, S.L. Anti-oxidant and anti-diabetes potential of water-soluble chitosan–glucose derivatives produced by Maillard reaction. *Polymers* **2019**, *11*, 1714. [CrossRef]
48. Liang, T.-W.; Lo, B.-C.; Wang, S.-L. Chitinolytic bacteria-assisted conversion of squid pen and its effect on dyes and pigments adsorption. *Mar. Drugs* **2015**, *13*, 4576–4593. [CrossRef] [PubMed]
49. Zain, N.F.M.; Yusop, S.M.; Ahmad, I. Preparation and characterization of cellulose and nanocellulose from pomelo (*Citrus grandis*) albedo. *J. Nutr. Food Sci.* **2014**, *5*, 1000334.
50. Huang, R.; Cao, M.; Guo, H.; Qi, W.; Su, R.; He, Z. Enhanced ethanol production from pomelo peel waste by integrated hydrothermal treatment, multienzyme formulation, and fed-batch operation. *J. Agric. Food Chem.* **2014**, *62*, 4643–4651. [CrossRef] [PubMed]
51. Tocmo, R.; Pena-Fronteras, J.; Calumba, K.F.; Mendoza, M.; Johnson, J.J. Valorization of pomelo (*Citrus grandis* Osbeck) peel: A review of current utilization, phytochemistry, bioactivities, and mechanisms of action. *Compr. Rev. Food Sci. Food Saf.* **2020**, *19*, 1969–2012. [CrossRef] [PubMed]
52. Multari, S.; Guzzon, R.; Caruso, M.; Licciardello, C.; Martens, S. Alcoholic fermentation of citrus flavedo and albedo with pure and mixed yeast strains: Physicochemical characteristics and phytochemical profiles. *LWT* **2021**, *144*, 111133. [CrossRef]
53. Bahena, J.M.V.; Estrada, J.V.; Hernandez, J.A.S.; Lopez, J.O. Expression and improved production of the soluble extracellular invertase from *Zymomonas mobilisin* in *Escherichia coli*. *Enzyme Microb. Technol.* **2006**, *40*, 61–66. [CrossRef]
54. Warchol, M.; Perrin, S.; Grill, J.P.; Schneider, F. Characterization of a purified β-fructofuranosidase from *Bifidobacterium infantis* ATCC 15697. *Lett. Appl. Microbiol.* **2002**, *35*, 462–467. [CrossRef]
55. Nehad, E.A.; Atalla, S.M. Production and immobilization of invertase from *Penicillium* sp. using orange peel waste as substrate. *Egypt Pharmaceut. J.* **2020**, *19*, 103–112.
56. Do Nascimento, G.C.; Batista, R.D.; Santos, C.C.A.D.A.; da Silva, E.M.; de Paula, F.C.; Mendes, D.B.; de Oliveira, D.P.; de Almeida, A.F. *β*-Fructofuranosidase and *β*-D-Fructosyltransferase from new *Aspergillus carbonarius* PC-4 strain isolated from canned peach syrup: Effect of carbon and nitrogen sources on enzyme production. *Sci. World J.* **2019**, *2019*, 6956202. [CrossRef] [PubMed]
57. Nguyen, V.B.; Nguyen, D.N.; Nguyen, A.D.; Ngo, V.A.; Ton, T.Q.; Doan, C.T.; Pham, T.P.; Tran, T.P.H.; Wang, S.-L. Utilization of crab waste for cost-effective bioproduction of prodigiosin. *Mar. Drugs* **2020**, *18*, 523. [CrossRef]
58. Ghasemi, Y.; Mohkam, M.; Ghasemian, A.; Rasoul-Amini, S. Experimental design of medium optimization for invertase production by *Pichia* sp. *J. Food Sci. Technol.* **2014**, *51*, 267–275. [CrossRef] [PubMed]
59. Yoon, M.; Choi, W.; Kwon, S.; Yi, S.; Lee, D.; Lee, J. Purification and properties of intracellulase invertase from alkalophilic and thermophilic *Bacillus cereus* TA-11. J. *Appl. Biol. Chem.* **2007**, *50*, 196–201.
60. Ahmed, S.A. Invertase production by *Bacillus macerans* immobilized on calcium alginate beads. *J. Appl. Sci. Res.* **2008**, *4*, 1777–1781.
61. Awad, G.E.A.; Amer, H.; Gammal, E.W.E.; Helmy, W.A. Production optimization of invertase by *Lactobacillus brevis* Mm-6 and its immobilization on alginate beads. *Carbohydr. Polym.* **2013**, *93*, 740–746. [CrossRef]
62. Aracri, F.M.; Cavalcanti, R.M.F.; Guimaraes, L.H.S. Extracellular tannase from *Aspergillus ochraceus*: Influence of the culture conditions on biofilm formation, enzyme production, and application. *J. Microbiol. Biotechnol.* **2019**, *29*, 1749–1759. [CrossRef]
63. Kaur, N.; Sharma, A.D. Production, optimization and characterization of extracellular invertase by an actinomycete strain. *J. Sci. Ind. Res.* **2005**, *64*, 515–519.

64. Yamamoto, K.; Kitamoto, Y.; Ohata, N.; Isshiki, S. Purification and properties of invertase from a glutamate-producing bacterium. *J. Ferment. Technol.* **1986**, *64*, 285–291. [CrossRef]
65. Xu, Z.W.; Li, Y.Q.; Wang, Y.H.; Yang, B. Production of β-fructofuranosidase by *Arthrobacter* sp. and its application in the modification of stevioside and rebaudioside A. *Food Technol. Biotechnol.* **2009**, *47*, 137–143.
66. Li, Y.; Ferenci, T. The *Bacillus stearothermophilus* NUB36 surA gene encodes a thermophilic sucrase related to *Bacillus subtilis* SacA. *Microbiology* **1996**, *142*, 1651–1657. [CrossRef] [PubMed]
67. Chambert, R.; Rain-Guion, M.C.; Petit-Glatron, M.F. Readthrough of the *Bacillus subtilis* stop codon produces an extended enzyme displaying a higher polymerase activity. *Biochim. Biophys. Acta* **1992**, *1132*, 145–153. [CrossRef]
68. Lincon, L.; More, S.S.; Reddy, S.V. Purification and biochemical characterization of β-D-fructofuranosidase from *Bacillus subtilis* LYN12. *J. Food Biochem.* **2018**, *42*, e12592. [CrossRef]
69. Wang, J.; Xiao, H.; Zhao, F.; Zhao, B.; Xu, M.; Zhou, Z.; Han, Y. A fructan sucrase secreted extracellular and purified in one-step by gram-positive enhancer matrix particles. *Processes* **2021**, *9*, 95. [CrossRef]
70. Pascal, M.; Kunst, F.; Lepesant, J.A.; Dedonder, R. Characterization of two sucrase activities in *Bacillus subtilis* Marburg. *Biochimie* **1971**, *53*, 1059–1066. [CrossRef]
71. Zhou, J.; He, L.; Gao, Y.; Han, N.; Zhang, R.; Wu, Q.; Li, J.; Tang, X.; Xu, B.; Ding, J.; et al. Characterization of a novel low-temperature-active, alkaline and sucrose-tolerant invertase. *Sci. Rep.* **2016**, *6*, 32081. [CrossRef]
72. Martin, I.; Débarbouillé, M.; Ferrari, E.; Klier, A.; Rapoport, G. Characterization of the levanase gene of *Bacillus subtilis* which shows homology to yeast invertase. *Mol. Gen. Genet.* **1987**, *208*, 177–184. [CrossRef] [PubMed]
73. Mu, D.; Zhou, Y.; Wu, X.; Montalban-Lopez, M.; Wang, L.; Li, X.; Zheng, Z. Secretion of *Bacillus amyloliquefaciens* levansucrase from *Bacillus subtilis* and its application in the enzymatic synthesis of levan. *ACS Food Sci. Technol.* **2021**, *1*, 249–259. [CrossRef]
74. Schönert, S.; Buder, T.; Dahl, M.K. Identification and enzymatic characterization of the maltose-inducible alpha-glucosidase MalL (sucrase-isomaltase-maltase) of *Bacillus subtilis*. *J. Bacteriol.* **1998**, *180*, 2574–2578. [CrossRef] [PubMed]
75. Phengnoi, P.; Charoenwongpaiboon, T.; Wangpaiboon, K.; Klaewkla, M.; Nakapong, S.; Visessanguan, W.; Ito, K.; Pichyangkura, R.; Kuttiyawong, K. Levansucrase from *Bacillus amyloliquefaciens* KK9 and its Y237S variant producing the high bioactive levan-type fructooligosaccharides. *Biomolecules* **2020**, *10*, 692. [CrossRef] [PubMed]
76. Salama, B.M.; Helmy, W.A.; Ragab, T.I.M.; Ali, M.M.; Taie, H.A.A.; Esawy, M.A. Characterization of a new efficient low molecular weight *Bacillus subtilis* NRC 16 levansucrase and its levan. *J. Basic Microbiol.* **2019**, *59*, 1004–1015. [CrossRef] [PubMed]
77. Li, R.; Zhang, T.; Jiang, B.; Mu, W.; Miao, M. Purification and characterization of an intracellular levansucrase derived from *Bacillus methylotrophicus* SK 21.002. *Biotechnol. Appl. Biochem.* **2015**, *62*, 815–822. [CrossRef]
78. Ammar, Y.B.; Matsubara, T.; Ito, K.; Iizuka, M.; Limpaseni, T.; Pongsawasdi, P.; Minamiura, N. Characterization of a thermostable levansucrase from *Bacillus* sp. TH4-2 capable of producing high molecular weight levan at high temperature. *J. Biotechnol.* **2002**, *99*, 111–119. [CrossRef]

polymers

MDPI

Article

Isolation and Characterization of Alpha and Nanocrystalline Cellulose from Date Palm (*Phoenix dactylifera* L.) Trunk Mesh

Hamid M. Shaikh [1,*], Arfat Anis [1], Anesh Manjaly Poulose [1], Saeed M. Al-Zahrani [1], Niyaz Ahamad Madhar [2], Abdullah Alhamidi [1] and Mohammad Asif Alam [3]

1 SABIC Polymer Research Center (SPRC), Department of Chemical Engineering, King Saud University, P.O. Box 800, Riyadh 11421, Saudi Arabia; aarfat@ksu.edu.sa (A.A.); apoulose@ksu.edu.sa (A.M.P.); szahrani@ksu.edu.sa (S.M.A.-Z.); AKFHK90@hotmail.com (A.A.)

2 Department of Physics and Astronomy, College of Sciences, King Saud University, Riyadh 11451, Saudi Arabia; nmadhar@ksu.edu.sa

3 Center of Excellence for Research in Engineering Materials (CEREM), King Saud University, P.O. Box 800, Riyadh 11421, Saudi Arabia; moalam@ksu.edu.sa

* Correspondence: hamshaikh@ksu.edu.sa; Tel.: +966-41-1176747

Abstract: Highly pure cellulosic polymers obtained from waste lignocellulose offer great potential for designing novel materials in the concept of biorefinery. In this work, alpha-cellulose and nanocrystalline cellulose were isolated from the date palm trunk mesh (DPTM) through a series of physicochemical treatments. Supercritical carbon dioxide treatment was used to remove soluble extractives, and concentrated alkali pretreatment was used to eliminate the lignin portion selectively to obtain alpha-cellulose in approximately 94% yield. Further treatments of this cellulose yielded nanocrystalline cellulose. The structure–property relationship studies were carried out by characterizing the obtained polymers by various standard methods and analytical techniques such as Fourier transform infrared spectroscopy-attenuated total reflection (FTIR-ATR), thermogravimetric analysis (TGA), differential scanning calorimetry (DSC), energy dispersive X-ray diffraction (EDX-XRD), scanning electron microscopy (SEM) and transmission electron microscopy (TEM). Almost 65% yield of pure cellulose was achieved, out of which 94% is the alpha-cellulose. This cellulose shows good thermal stability and crystallinity. The microscopic analysis of the nanocellulose showed a heterogeneous mix of irregular-shaped particles with a size range of 20–60 nm. The percentage crystallinity of alpha-cellulose and nanocellulose was found to be 68.9 and 71.8, respectively. Thus, this study shows that, this DPTM-based low-cost waste biomass can be a potential source to obtain cellulose and nano-cellulose.

Keywords: date palm trunk mesh; cellulose; lignocellulosic waste; alpha cellulose; nanocellulose

Citation: Shaikh, H.M.; Anis, A.; Poulose, A.M.; Al-Zahrani, S.M.; Madhar, N.A.; Alhamidi, A.; Alam, M.A. Isolation and Characterization of Alpha and Nanocrystalline Cellulose from Date Palm (*Phoenix dactylifera* L.) Trunk Mesh. *Polymers* **2021**, *13*, 1893. https://doi.org/10.3390/polym13111893

Academic Editor: Arn Mignon

Received: 4 May 2021
Accepted: 4 June 2021
Published: 7 June 2021

Publisher's Note: MDPI stays neutral with regard to jurisdictional claims in published maps and institutional affiliations.

1. Introduction

The current global initiative to promote research and innovation in the area of renewable agri-resources has led to the development of novel and high-value products based on lignocellulose. For the sustainable development of non-food consumer goods and various industrial products, cellulose is a cornerstone resource available to human beings. Cellulose, a linear polymer composed of glucose units, is the most abundant of all naturally occurring organic materials, with annual production exceeding 10^{11}–10^{12} tons [1]. To convert this natural polymer into advanced materials, many studies pertaining to various sources of this polymer, particularly waste agri-byproducts, different isolation strategies and its physicochemical properties have been carried out.

For many centuries, in arid regions, the date palm has been one of the most widely cultivated plants, because of its sweet edible fruits. Additionally, it has been utilized for shelter, various handicraft products and for many other applications. The Kingdom of Saudi Arabia is the second largest date producer globally. According to the 2018 statistical database of the United Nations Food and Agricultural Organization, the date production in Saudi Arabia reached 1,302,859 tonnes over 116,125 ha of harvesting areas [2]. Moreover, according to reports by Saudi Arabia's Ministry of Environment, Water, and Agriculture, over 30 million palm trees were cultivated in 2018 [3]. As a result, substantial quantities of waste biomass are generated during the seasonal pruning and refinement of the palms. It is estimated that about 20 to 35 kg of biomass is generated per palm tree, and annually, approximately 1 million metric tons of the biomass waste is generated [4–6]. This vast amount of date-palm waste is either buried in landfills or burned directly in open fields, causing a severe threat to the environment. Therefore, innovative ways of valorizing this renewable resource should be analyzed for value additions [7,8]. This lignocellulosic waste biomass can be viewed as a sustainable source of cellulosic materials for various applications. Moreover, this will help address the issue of solid waste disposal to a certain extent and generate additional income sources for farmers.

Recently, a few studies have been conducted to obtain pure cellulose from selected parts of the date palm. For example, microcrystalline cellulose was extracted from date seeds by sequential treatment with alkali, and then bleaching with sodium hypochlorite, followed by acid hydrolysis. Cellulose of type I with a crystallinity index of 62% was obtained and was further enhanced to 72% after acid hydrolysis [9]. In a similar study, microcrystalline cellulose was isolated from a date palm fruit bunch stalk by treating with a bleaching agent at first, followed by alkali and then finally, subjecting it to acid hydrolysis. It was observed that cellulose fibers having diameters of 21–96 μm and lengths of more than 200 μm were obtained. Also, the cellulose so obtained was reported to have a 79.4% crystallinity index [10].

In another work, cellulose was isolated from waste date rachis, leaflets and fibers by treatment with dilute acid, alkali, followed by bleaching with acetic acid, hydrogen peroxide and sulfuric acid. A more than 70% yield of alpha-cellulose was obtained, with an average crystallinity index of 52.27% [11]. In yet another study, microfibrillated cellulose and cellulose nanowhiskers were isolated from date rachis and leaflets. Cellulose was isolated from these parts through a series of reactions, including dewaxing with two wt.% of alkali (NaOH) and bleaching with sodium chlorite($NaClO_2$). Furthermore, a high-pressure microfluidizer was used to obtain microfibrillated cellulose, while sulfuric acid treatment was used to get nanosized whiskers. Rod-shaped nanoparticles having an average diameter of 6.1 nm and length of 260 nm were also reported [12].

Besides these, very few studies have been performed to obtain nano-sized cellulosic fibers/crystals from the various parts of the date tree. Patrik Sobolčiak et al. isolated cellulose nanofibers from date leaves by treating them with 1 M ammonium persulphate solution at 60 °C with vigorous overnight stirring followed by probe sonication. Cellulose fibers with an average thickness of 18 ± 5 nm and a length of approximately 300 ± 24 nm were obtained [13]. Similarly, nanocrystalline cellulose was obtained from date tree fibers using a combination of sulfuric acid and acetic acid for hydrolysis. Cellulose nanocrystals having a good aspect ratio were obtained [14].

Similarly, Sami Boufi et al. reported the isolation of cellulose from the date rachis; first, biomass was treated at higher temperature with alkali and then bleached, followed by oxidation with 2,2,6,6-tetramethylpiperidin-1-yl) oxyl (TEMPO) in the second stage. This was followed by high-pressure homogenization to obtain cellulose nanofibers with 20–50 nm width and 200–1000 nm length. Further hydrolysis of the cellulose with high content of sulfuric acid (~65 wt.%) yielded cellulose nanocrystals with 15–25 nm average width and 150–250 nm length [15]. Likewise, date rachis was used to extract nanocellulose fibers by sequential oxidation to obtain shape-controlled fibers. Depending on the time of oxidation and specific reaction conditions, cellulose fibers with widths of 20–30 nm were obtained [16].

In all of the previous works, a specific portion of the date tree, i.e., rachis, leaflets, seeds, etc., was used to obtain cellulose and/or nanocellulose. Comparatively, fewer studies have been carried out to obtain cellulose and/or nanocellulose from DPTM, a fibrous waste surrounding the trunk that can be obtained after cutting the fronds (Figure 1). Each date palm tree generates approximately 1.25 kg of mesh annually [17], contributing significant amounts of waste that can be valorized. Nonetheless, every feedstock remains an essential factor in extracting cellulose and other constituents under the biorefinery principle.

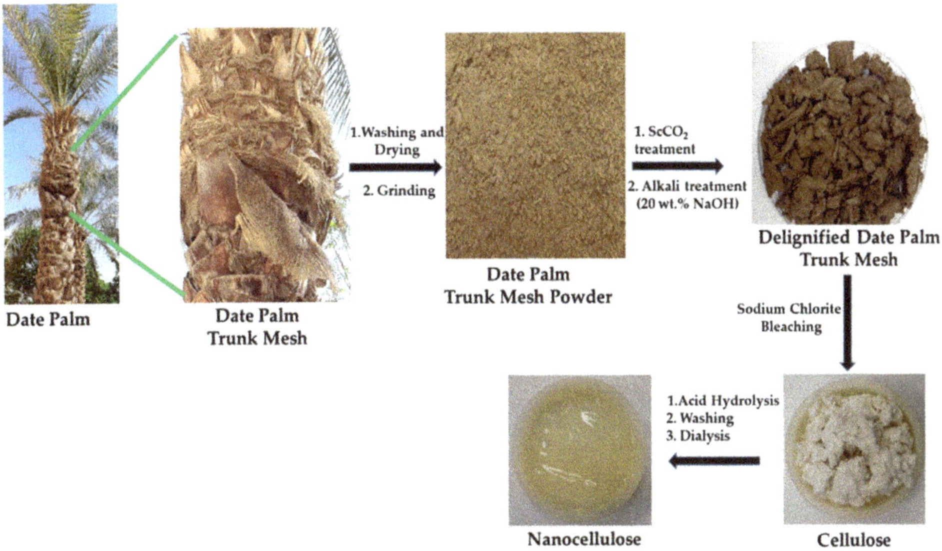

Figure 1. Photographs of DPTM fibers and products at various experimental stages.

Over the last century, research to understand various facets of cellulose, including its isolation/extraction from various sources, gained momentum due to the increasing demand for pure cellulose and its derivatives. Additionally, according to the US Department of Energy, renewable resources will supply almost 50% of essential chemicals by 2050 [18]. Furthermore, the gross domestic product demand for nanocellulose alone is expected to increase to USD 1176 million by 2025 [19]. Recently, studies to overcome the challenges arising due to matrix–filler interaction and adhesion in the area of composites have led to an emerging concept of 'all-cellulose composites' [20]. In general, lignocellulosic biomass is mainly composed of cellulose, hemicellulose, lignin and, to a minor extent, pectin, waxes and inorganic minerals. The final properties and performance of the cellulose depend not only on plant locations, age of the plants, defects, and even various parts of the same plant from which it is derived, but also on several other factors such as cell dimensions, fiber structure, microfibril angles, etc. In addition, pretreatment approaches and chemicals used during extraction play a crucial role in determining the final composition of the cellulose [21].

This study focuses on the isolation and characterization of alpha- and nanocellulose from the DPTM. Alpha cellulose is a high molecular-weight portion of the pure cellulose that can withstand treatments with concentrated alkali (~17.5 wt.% of NaOH), while beta and gamma celluloses are soluble fractions [22]. For the isolation of pure cellulose, few pre-treatments were carried out, including treatments with supercritical carbon dioxide, alkali and sodium chlorite. Furthermore, delignified DPTM and cellulose were characterized and used to prepare nanocellulose. The properties of the nanocellulose were thoroughly investigated for its further utilization.

2. Materials and Methods

2.1. Isolation of Cellulose

Waste fibers from a palm tree trunk were collected from the King Saud University campus. This biomass was washed several times to remove dirt, dried at room temperature and ground into fine powder. It was then treated with supercritical carbon dioxide ($ScCO_2$) by using a pilot unit from PID Eng & Tech (Colmenar Viejo, Madrid, Spain). This pilot unit is composed of a L lit capacity extractor vessel that can be operated to maximum pressures up to 340 bar and a temperature up to 90 °C. Initially, the raw powder was filled into the extractor and treated with liquid CO_2 through a pump that was maintained at a temperature below 10 °C to keep CO_2 in the liquid state. The flow rate was kept at 1.5 L min^{-1}. This pretreatment was carried out only to increase the surface area of the biomass and to remove unwanted minor components, such as waxes and other soluble extractives. The extractant was discarded without further analysis, and the solid powder was used for alkali treatments. A known amount of the above powder was treated with 20% (w/v) of NaOH solution at 90 °C for about 6 h, filtered through nylon cloth, the extractant discarded and solid materials washed thoroughly until it became neutral. It was finally dried at 100 °C and subjected to bleaching with a sodium chlorite solution using 1:0.5 weight ratio of powder to sodium chlorite. The pH of this solution was brought down to 3.7 through the addition of acetic acid, and the reaction was carried out at 70 °C for 4 h. The pure cellulose was collected by filtration, washed repeatedly until it is free from acid, and finally dried to obtain constant weight. The schematic experimental procedures and images of the products obtained are depicted in Figure 1.

2.2. Extraction of Nanocellulose (NC)

The extraction of cellulose nanocrystals was performed, first by twin-screw defibrillation using a procedure reported earlier [23], followed by hydrolysis with sulfuric acid. In general, mechanical treatments, such as grinding, ultrasonication, high-pressure homogenization and microfluidization, are commonly used to produce cellulose nanofibrils from the cellulosic fibers. However, these are very energy-intensive processes. Therefore, various pre- and/or post-treatment methods, such as acid hydrolysis, 2,2,6,6-tetramethylpiperidine-1-oxyl radical (TEMPO)-mediated oxidation and enzymatic hydrolysis have been explored to reduce energy consumption. This combination of chemical processes is less energy-intensive because it reduces the negative or positive charges on fiber surfaces, which in turn helps in improving the colloidal stability of the final nanocellulose [24]. On the other hand, twin-screw compounding is a very simple and continuous extrusion process and is viewed as a feasible solution for cellulose defibrillation even at nanoscale. With this method, higher yields can be achieved within a short period and at room temperature. It is estimated that the energy consumption required for this process is about 4.1 kWh/kg which is much lower than what is required in high pressure homogenization methods, i.e., approximately 30–50 kWh/kg [25].

Similarly, acid hydrolysis was carried out to convert these defibrillated fibers into nanocrystals. In this process, the amorphous or paracrystalline portion of cellulose is hydrolyzed and dissolved in acidic solution. The crystalline part, on the other hand, remains intact because it is chemically resistant to the acid. As a result, the cellulose fibrils are transversely cleaved, resulting in short cellulose nanocrystals with high crystallinities [26].

Considering all these factors, a combination of mechanical disintegration and chemical treatment with sulfuric acid was used to obtain nanocellulose in this work. For twin-screw extrusion/defibrillation, a laboratory-scale microcompounder (15 cm^3 Xplore® Sittard, Sittard, The Netherland) was used. It has a compact barrel with a conical-shaped co-rotating twin-screw and the provision of a channel for the recirculation of materials. Homogenous cellulose suspension having a solid content of ~5 wt.% was fed into the barrel and continuously recirculated within the barrel for about 30 min at a constant screw speed of 250 rpm. After this treatment, the solids were collected and subjected to sulfuric acid hydrolysis as described elsewhere [26]. In short, 10 g of treated cellulose samples were hydrolyzed in

100 mL 50 wt.% H_2SO_4 solution. The reaction was performed at 45 °C with continuous stirring for about 60 min. Finally, the hydrolysis reaction was quenched by the addition of a large amount of distilled water. This suspension was centrifuged several times and the supernatant fluid was discarded until it became neutral. This cloudy suspension was then dialyzed against distilled water until the pH of the suspension reached a constant value. This portion of the nanocellulose suspension was stored in a refrigerator at 4 °C; the other was freeze-dried and used for further characterization.

2.3. Assessment of Chemical Composition

For the determination of the chemical composition and yield, different standard methods of the Technical Association of the Pulp and Paper Industry (TAPPI) were used. Table 1 depicts the data and the detailed procedure is provided in the Supplementary Information.

Table 1. Chemical composition of date palm trunk mesh (DPTM).

Samples	α-Cellulose %	Kappa No.	Lignin %	Moisture %	Ash%	Yield %	Crystallinity %
DPTM	33.70 (±3.17)	82.06 (±3.93)	22.53 (±2.91)	6.40 (±1.01)	4.26 (±1.20)	-	51.33 (±2.06)
Delignified DPTM	65.91 (±2.50)	54.36 (±1.96)	2.16 (±0.30)	5.16 (±0.65)	2.33 (±0.51)	71.76 (±2.32)	57.66 (±2.15)
Cellulose	94.50 (±1.70)	0.39 (±0.06)	0.32 (±0.08)	4.45 (±0.50)	0.23 (±0.07)	66.53 (±3.27)	68.60 (±1.68)
Nanocellulose (NC)	92.26 (±1.35)	0.25 (±0.03)	0.13 (±0.05)	4.20 (±0.30)	-	35.26 (±4.28)	89.61 (±1.13)

2.4. Functional Group Analysis

Attenuated total reflectance—Fourier transform infrared spectroscopy (ATR-FTIR) analysis was carried out using a (Thermo Scientific, Winsford, UK) Nicolet iN10 FTIR microscope having a germanium microtip. The analysis was carried out in the wavenumber range of 650–4000 at a resolution of 6 cm^{-1} over 16 scans.

2.5. Thermal Analysis

A thermogravimetric analyzer (DTG 60H, Shimadzu., Kyoto, Japan) was used for the thermogravimetric analysis. A sample size of 10–15 mg was analyzed. The heating rate was kept at 20 °C/min with a temperature range from 30 °C to 800 °C under a continuous nitrogen flow of 50 mL/min. Similarly, differential scanning calorimetry (DSC, Shimadzu DSC-60, Kyoto, Japan) was used for thermal analysis. About 10–15 mg was sealed in the DSC aluminum pan and heating was carried out at a rate of 20 °C/min from 30 °C to 600 °C.

2.6. Study of Crystallinity

The crystallinities of the celluloses were analyzed using Wwde-angle X-ray diffraction (WAXRD, D8 Advance, Bruker, Berlin, Germany). An automated wide-angle goniometer coupled to a sealed-tube with Cu-Kα source radiation (λ = 1.54056 Å) was used. In reflection mode, a range of 2θ was scanned from 5° to 50° at 5°/min. Peak deconvolution was carried out using Origin software, and the Voigt function was used for the distribution of the complex pattern into the components for attaining the best fit shape. The degree of crystallinity (Xc) was assessed per the Segal method, using the following equation [26];

$$CrI(\%) = \frac{I_{002} - I_{Am}}{I_{002}} \times 100 \quad (1)$$

where, CrI is the percentage crystallinity index; I_{002} is the maximum intensity of the peak at 22°, which is a crystalline part, and I_{Am} is the intensity of diffraction due to the amorphous part at 18°.

2.7. Morphological Analysis

Morphological studies were carried out at different stages of treatments by scanning electron microscopy (SEM, JSM-6360A, JEOL, Tokyo, Japan) equipped with energy-dispersive X-ray spectroscopy (EDS). The analysis was carried by placing the samples on conducting carbon tape. The accelerating voltage was kept at 6 kV for surface morphology determination, while elemental analysis by EDS was conducted at an accelerating voltage of 20 kV and with a working distance of 10 mm. All the samples were gold-sputtered by using a JEOL, JFC-1600 auto fine coater before observation to avoid samples being overcharged. The operating condition was maintained at 20 mA for 30 s. Transmission electron microscopy (TEM) imaging was performed to study the nanostructure by using a JEM-1400 (JEOL, Tokyo, Japan) field-emission electron microscope operating at an accelerating voltage of 120 kV.

3. Results and Discussion

3.1. Compositional Analysis

To obtain high-purity cellulose, effective pretreatment processes are required to remove other components, such as lignin, hemicellulose and other extractives from the recalcitrant biomass. The delignification (alkali treatment) processes removed a major amount of lignin, as seen from Table 1. The chemical functionalization of solid lignin in the alkaline delignification process is known to transform solid lignin of biomass into a soluble liquid phase [27]. Furthermore, the removal of lignin is influenced by the concentration of alkali, reaction time and temperature. It was observed that alkali treatment removed ~90% of the lignin and other soluble fractions like hemicelluloses, while bleaching with sodium chlorite removed the remainder. The maximum amount of lignin removal by alkali treatment was also reported by N Y Abu Thabit et al [9]. The reduction in kappa number indicates a lack of lignin in the biomass and can also be correlated with the amount and purity of the cellulosic components.

Similarly, alpha-cellulose represents undegraded higher-molecular-weight cellulose as mentioned earlier. The alpha-cellulose content in pure cellulose rose from 33% to 94% when compared to the DPTM. This is the highest quantity of alpha-cellulose obtained so far and is further used to make high-quality nanocellulose materials.

Moreover, according to EDX analysis (details in Supplementary Information), the raw biomass primarily contained carbon and oxygen elements with traces of inorganic elements, such as magnesium, silica, calcium, iron and chlorine. In general, lignocellulose biomass contains more than half of its compounds in the form of carbon. However, the individual components of biomass, such as cellulose, hemicellulose and lignin, have different carbon contents and properties. Lignin alone has almost 60% carbon content, followed by cellulose (42%) and hemicellulose (40%) [28]. As pristine biomass contains all of these components, including lignin, it contributes to an increase in the carbon content of the raw biomass. Furthermore, some inorganic carbons from carbonates and other inorganic sources may contribute to the increase in total carbon content in the raw biomass, as evidenced by the EDX analysis.

During alkali treatment, these inorganic elements were extracted, due to which traces of sodium could be found in the samples treated with sodium hydroxide. The cellulose and nanocellulose samples were mostly composed of carbon and oxygen, suggesting the high purity of the obtained cellulosic materials.

3.2. Functional Group Analysis by ATR-FTIR

The FTIR spectra of all the samples are presented in Figure 2. A broad peak at around 3320 cm^{-1} due to stretching vibration of the O-H bond was observed in all the samples, while peaks due to C-H bond stretching frequency were detected at 2896 cm^{-1} [29]. Owing to the increased cellulose content and decreased lignin and hemicellulose portions of the raw biomass, these peaks appeared sharper for these samples. A diffused peak at 1640 cm^{-1} was due to the adsorbed moisture and indicated cellulose–water interaction [30].

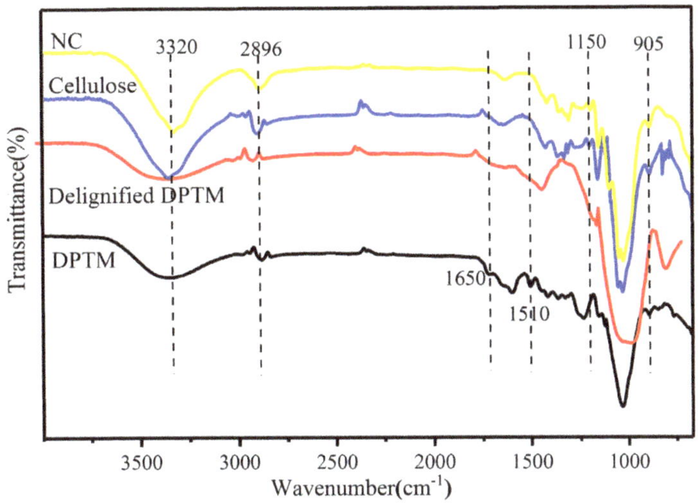

Figure 2. FTIR spectra of DPTM, delignified DPTM, cellulose and nanocellulose samples.

Similarly, peaks at 1650 cm^{-1} and 1510 cm^{-1} indicated the presence of lignin in the DPTM sample. These are due to the stretching frequencies of C=C present in the aromatic structure of lignin and that of C=O present in the ester groups of the hemicellulose. These peaks disappeared in the spectra of the other samples, suggesting the complete removal of the lignin and thereby indicating the purity of the cellulose. In cellulose samples, the C-H vibration of the glycosidic bond was observed at 1365 cm^{-1} and 1150 cm^{-1}, while the peak due to C-O-C glycosidic bond vibration (pyranose C-O-C ring stretching) appeared at 1060 cm^{-1} [31]. The peak at 905 cm^{-1} was due to C-H rocking vibrations and could be correlated to the purity of the cellulose structure [32].

3.3. Thermal Analysis

Figure 3 represents the relative thermogravimetric curves of the samples and Table 2 summarized the results of the analysis. The decomposition or stability can be evaluated by determining the temperature at the onset of decomposition.

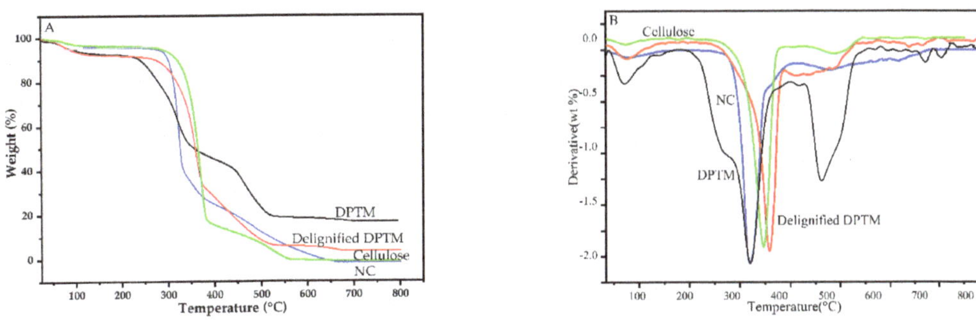

Figure 3. Thermogravimetric curve (**A**) and derivative thermogravimetric curve (**B**) of DPTM, delignified DPTM, cellulose and nanocellulose samples.

Table 2. Summarized results of TGA analysis of samples.

Sample	T_{onset} (°C)	T_{10} (°C)	T_{max} (°C)	Residue (W%)
DPTM	221.16	241.80	321.56 (Cellulose) 461.10 (Lignin)	17.65
Delignified DPTM	240.98	275.66	358.72	6.02
Cellulose	280.50	321.09	368.50	0.62
NC	292.58	302.28	319.00	0.01

The first small decomposition curve (Figure 3B) at around 100 °C represents the elimination of adsorbed water or humidity, which is nearly 5.49%, 6.20%, 2.71% and 3.5% for DPTM, alkali-treated DPTM, cellulose and nanocellulose respectively. Then samples displayed thermal stability with negligible mass loss that depends on the nature of the samples. For example, raw DPTM showed the onset of decomposition at about 221 °C, while for alkali-treated samples it was at 240 °C. This difference could be attributed to the presence of different amounts of hemicellulose (a mixture of low-molecular-weight pentose and hexose sugars) and lignin. The cellulose sample showed a relatively high onset-of-degradation temperature, 280 °C. This was found to be more than the onset-of-degradation temperature range of cellulose (150–210 °C) obtained from bleached eucalyptus pulp [33]. Similarly, it is in close agreement with the onset-of-degradation temperature of cellulose obtained from viscose wood pulp, which is reported to be 272 °C [34]. The thermal stability of cellulose depends upon its degree of crystallinity and the type and/or source of cellulose [35]. Moreover, degradation is believed to take place initially through the cellulose activation process owing, to the scission of glycosidic bonds that leads to the reduction in the degree of polymerization without losing mass [36]. This activated cellulose then releases gases like CO_2, CO and H_2O, which cause the partial crosslinking of cellulose, which finally contributes to the formation of solid char or tar [37].

The onset-of-degradation temperature of nanocellulose was 292 °C, considerably higher than that of cellulose. The onset-of-degradation temperature ranges for cellulose nanofibers were reported to be 184–207 °C by Y. Peng et al., 2013 [38]. According to the DTG curve, the maximum thermal decomposition of these samples occurred until 500 °C. The DPTM exhibited a typical lignocellulosic pattern with two distinct thermal decomposition phases. The first thermal decomposition occurred between 195 °C and 420 °C, corresponding to the hemicellulose and cellulose portions. Hemicellulose degrades at around 265 °C, while cellulose decomposes at temperatures ranging from 320 °C to 370 °C. The mass loss beyond 417 °C was due to the decomposition of lignin and biochar components, which took place over a wide temperature range, up to 800 °C [39].

Figure 4 shows a DSC analysis of the samples, performed to evaluate the energy consumption properties of these samples. The first broad endothermic peak was observed in the range of 80–140 °C for all samples, due to the evaporation of moisture and other volatile components. Two broad peaks appeared at 323.78 °C and 420 °C during the analysis of the DPTM sample, which corresponded to the presence of the cellulosic component and lignin, respectively [30]. The disappearance of these peaks in all other samples and the increase in the intensity of the peak due to cellulose suggested the purity of the sample. The second endothermic peak of cellulose appeared in the temperature range of 370–375 °C which could be due to the depolymerization of cellulose and decarboxylation reactions [40]. The sharp intensities of these peaks could be attributed to uniform heat absorption by the pure cellulosic components [41]. Furthermore, as a result of various treatment methods, cellulose crystallites can undergo reorientation and rearrangement, leading to a more compact crystal structure and hence, improved thermal stability [42].

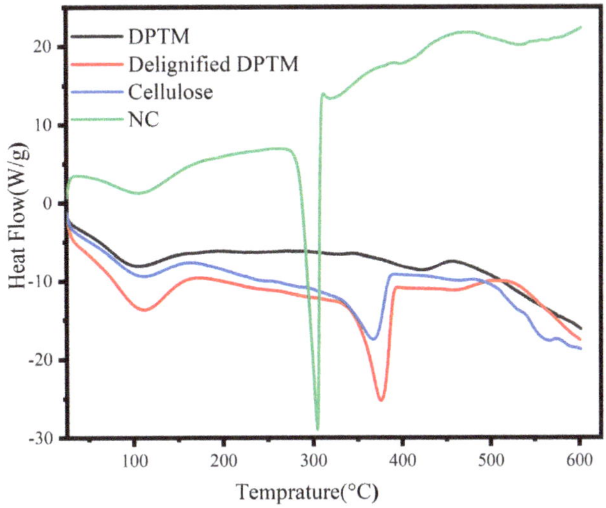

Figure 4. DSC curve of DPTM, delignified mesh, cellulose and nanocellulose samples.

An endothermic peak occurred much earlier in the case of nanocellulose due to the formation of sulfate groups (-OSO_3H), formed due to the reaction of sulfuric acid with the hydroxyl groups during hydrolysis. This acid hydrolysis also decreased the amorphous regions, molecular weight and degree of polymerization of cellulose.

3.4. Wide-Angle X-ray Diffraction (WAXRD) Analysis

Figure 5 displays XRD spectra of all the samples. As can be seen in diffractograms, in all the samples, peaks were found to be superimposed as a wide halo, due to the presence of an amorphous phase. In semicrystalline lignocellulosic polymers from softwood and vegetable fibers, this type of pattern is commonly observed [43].

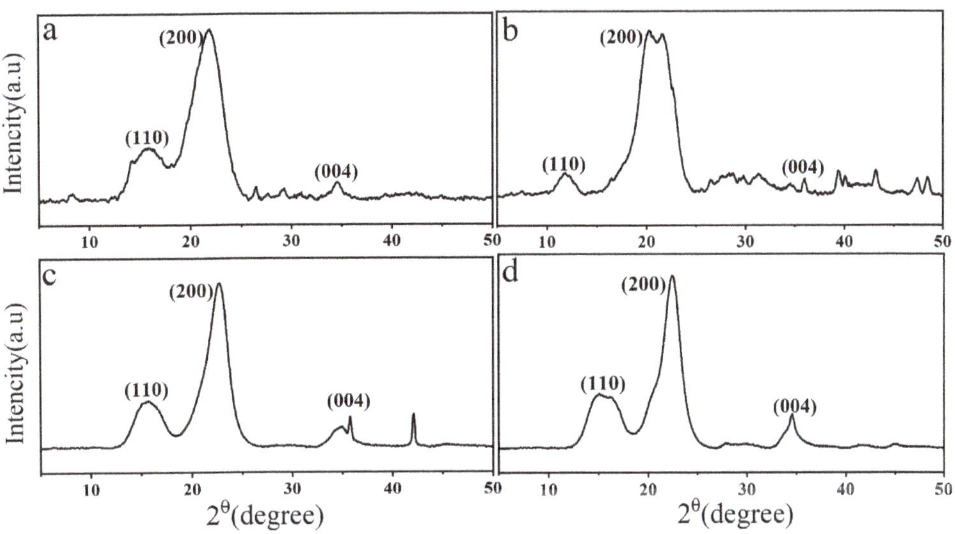

Figure 5. XRD diffractograms of (**a**) DPTM, (**b**) delignified DPTM, (**c**) cellulose and (**d**) NC.

Moreover, the main peaks of lignocellulosic materials were observed at 2θ = 15.9°, 21.8° and 34.4° for each sample, corresponding to the (110), (002) and (004) crystal planes, respectively. This also indicated the presence of Iβ cellulose structures in cellulose and cellulose nanoparticles [44]. The sharper peak at 22.7° and 22.4° in cellulose and nanocellulose, respectively, confirmed the formation of highly ordered crystalline cellulose domains. The more prominent peaks reflected in these samples could be due to the increased crystallinity s because of the improved molecular hydrogen bonding [45]. In alkali-treated samples, the doublet peaks were observed at 20.3° and 21.5° exhibiting two types of polymorphs of cellulose, i.e., cellulose I and cellulose II. Alkali treatment by sodium hydroxide is known to partially convert cellulose I into cellulose II [46]. However, this polymorph again rearranged into cellulose I structure in the cellulose and nanocellulose after bleaching and acid hydrolysis treatments. The degree of crystallinity was estimated to be 51.3% for the DPTM samples and 56.5% for fibers treated with alkali. This is significantly more than reported by Abdal-hay et al. [47]. Additionally, an increase in the crystallinity of pure cellulose was also observed, and it was found to be approximately 68.06%. This increase in crystallinity could be attributed to the decrease in the amorphous domain due to the cleavage of glycosidic bonds present in the disordered paracrystalline section. Moreover, a bleaching reaction eventually leads to the removal of lignin and the rearrangement of cellulose molecules into more crystalline order [48]. It is also in close agreement with the crystallinity of the cellulose obtained from different parts of date fibers [11]. Similarly, the crystallinity of nanocellulose was found to be 89.61% which was higher than that of nanocellulose obtained with an oxidizing agent and homogenization treatment of palm biomass [16]. It was also found to be greater than that of nanocellulose obtained from other wood-based bioresidues such as cotton waste and oil palm leaves via sulfuric acid treatments [49,50].

3.5. Morphological Analysis

In general, lignocellulosic biomass is composed of three layers of the cell wall in which semi-crystalline cellulose is reinforced with lignin and hemicellulose fractions. Morphological analysis showed that the raw date fiber bundle (Figure 6A) had a rigid structure, as evidenced by broken thick filaments. The individual cellulose fibers were packed together and displayed a compact and inflexible structure. This rigidity could arise in the raw biomass due to the rigid nature of the lignin and other cementitious components that bind these polymeric components (cellulose, hemicellulose and lignin) together as a composite structure. Significant changes to the surface of the fibers could be seen after alkali treatment with partial cellulose microfibril separation (Figure 6B). Alkali treatment is known to eliminate lignin and hemicellulose from biomass as these components are soluble in hot aqueous NaOH solution, while the cellulose fraction remains intact [51]. This treatment also removes the inorganic impurities and waxes and reduces the size of the cellulose fibers/fibrils, which eventually results in increasing the surface area of the fibers [52].

The surface of the cellulose fiber became more uniform and smooth after bleaching treatment, owing to the complete removal of the non-cellulosic components (Figure 6C). The complete defibrillation of the microfibrils into the individual fibrils resulted in a substantial reduction of the diameter of the fibrils. After exposure to air, the nanoparticles tend to agglomerate due to strong hydrogen bonding between surface hydroxyl groups of the cellulose. It is difficult to separate these fibrils (Figure 6D). Moreover, EDX analysis confirmed that the cellulose and nanocellulose are free from alkali or other residual elements and thereby also confirmed the purity of the samples (Supplementary Information).

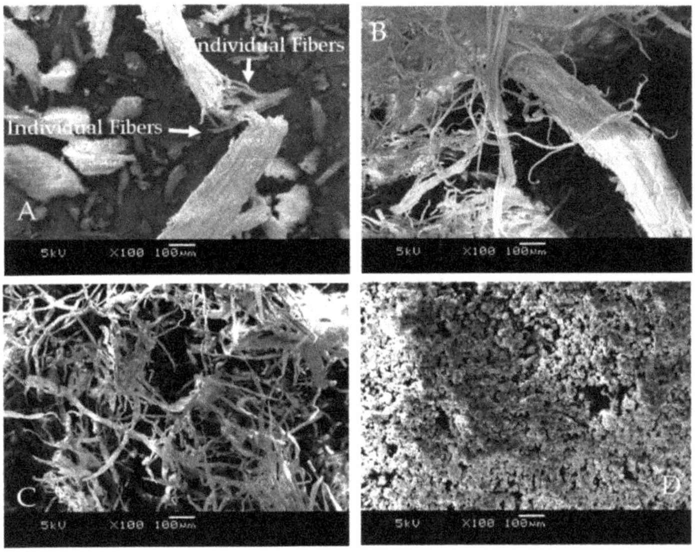

Figure 6. SEM images of (**A**) DPTM, (**B**) delignified DPTM, (**C**) cellulose and (**D**) NC.

3.6. Characterization of Nanocellulose

From Table 1, it can be seen that the acid hydrolysis of the cellulose had a negligible effect on increasing the alpha-cellulose content of nanocellulose. Cellulose chain scission via acid hydrolysis yielded nearly 89.6% of the nanocellulose. This hydrolysis also decreased the particle size of nanocellulose while further smoothing the surface morphology (Figure 5D) by eliminating the amorphous content. The TEM analysis of the particles in Figure 7 showed that the nanoparticles were agglomerated, and the native characteristics of cellulose were retained. Irregular-shaped nanoparticles with sizes ranging from 26 nm to 61 nm were observed.

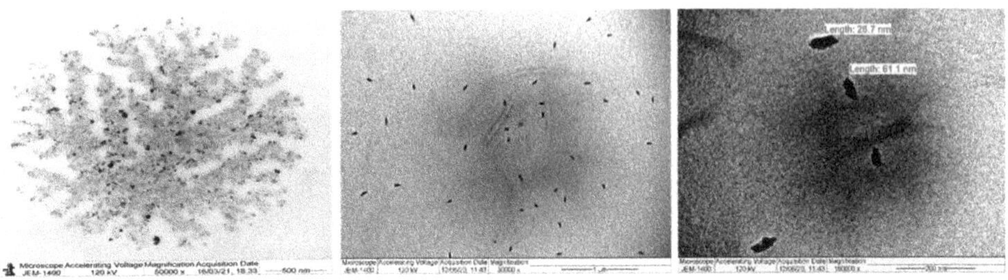

Figure 7. TEM images of Nanocellulose.

Individual nanoparticles and smaller nanocrystallites were formed as a result of a reduction in hydrogen bonding caused by cellulose chain fragmentation [53]. The EDX analysis indicated that the carbon content of the nanocellulose was slightly higher than that of the cellulose, while oxygen content was found to be lower, owing to the applied hydrolysis conditions. Furthermore, the aqueous suspension of nanocellulose had a high negative zeta potential value of −35 mV. This could be due to the repulsive forces arising from anionic sulfate moieties, which contribute to the formation of a stable dispersion system. However, further research is required to determine the overall efficiency of the isolation method in terms of yield and qualities of nanocellulose.

4. Conclusions

In conclusion, a series of pretreatments were carried out to obtain pure cellulosic components from the waste mesh of the date palm trunk, and their structure–property relationship was analyzed in detail. The isolated celluloses exhibited similar characteristics to those reported in the literature and were free of lignin and hemicellulose components. Nearly 94% of pure alpha-cellulose, with a total yield of about 66% compared to the starting material, was obtained from this waste feedstock. The low Kappa number suggested its purity, which was also confirmed by FTIR and DSC analysis. These celluloses showed high thermal stability with onset-of-degradation temperatures of ~280 and 290 °C for the cellulose and nanocellulose, respectively. XRD analysis showed that this cellulose had the cellulose I type of polymorph, having a crystallinity of around 68% for the cellulose and 89% for the nanocellulose. According to morphological studies, the resulting cellulose had fibers with smooth surfaces, whereas nanocellulose had irregular-shaped rod-like structures with nanoscale features. Cellulose and nanocellulose are widely used in food, pharmaceutical, paint and fillers for polymer composites. Therefore, this waste agricultural biomass can be valorized as a potential source for these materials and used in such applications.

Supplementary Materials: The following are available online at https://www.mdpi.com/article/10.3390/polym13111893/s1, Figure S1: EXD analysis of raw date palm mesh, Figure S2: EXD analysis of delignified mesh, Figure S3: EXD analysis of cellulose, Figure S4: EXD analysis of nanocellulose.

Author Contributions: Conceptualization, H.M.S.; Methodology, H.M.S., A.A. (Arfat Anis); Formal analysis, A.A. (Abdullah Alhamidi), A.M.P., A.A. (Arfat Anis); A.A. (Abdullah Alhamidi), Investigation, H.M.S., A.M.P., N.A.M., A.A. (Arfat Anis), M.A.A.; Resources, H.M.S., S.M.A.-Z.; Data curation, H.M.S., A.A. (Abdullah Alhamidi); Writing—original draft preparation, H.M.S., A.A. (Arfat Anis); Writing—review and editing, H.M.S., N.A.M.; Supervision, H.M.S.; Project administration, S.M.A.-Z., and H.M.S.; Funding acquisition, H.M.S. All authors have read and agreed to the published version of the manuscript.

Funding: This project was funded by the National Plan for Science, Technology and Innovation (MAARIFAH), King Abdulaziz City for Science and Technology, Kingdom of Saudi Arabia, Award Number (13-NAN1120-02).

Institutional Review Board Statement: Not Applicable.

Informed Consent Statement: Not Applicable.

Data Availability Statement: The data presented in this study are available on request from the corresponding author.

Acknowledgments: This project was funded by the National Plan for Science, Technology and Innovation (MAARIFAH), King Abdulaziz City for Science and Technology, Kingdom of Saudi Arabia, Award Number (13-NAN1120-02).

Conflicts of Interest: The authors declare no conflict of interest.

References

1. Klemm, D.; Philipp, B.; Heinze, T.; Heinze, U.; Wagenknecht, W. (Eds.) Front Matter. In *Comprehensive Cellulose Chemistry*; WILEY-VCH Verlag GmbH: Weinheim, Germany, 1998; pp. i–xxii. [CrossRef]
2. FAOSTAT. Food and Agriculture Organization of the United Nations, Statistics Division. Forestry Production and Trade, Dataset, Date Production In Saudi Arabia. Available online: http://www.fao.org/faostat/en/ (accessed on 28 May 2021).
3. Ministry of Environment, Water and Agriculture (MEWA). *Annual Statistical Book*; Ministry of Environment, Water and Agriculture (MEWA): Riyadh, Saudi Arabia, 2019.
4. Hussain, A.; Farooq, A.; Bassyouni, M.I.; Sait, H.H.; El-Wafa, M.A.; Hasan, S.W.U.; Ani, F.N. Pyrolysis Of Saudi Arabian Date Palm Waste: A Viable Option For Converting Waste Into Wealth. *Life Sci. J.* **2014**, *11*, 667–671.
5. El-Juhany, L.I. *Surveying of Lignocellulosic Agricultural Residues in Some Major Cities of Saudi Arabia*; Research Bulletin No. 1-Agricultural Research Center, College of Agriculture, King Saud University: Riyadh, Saudi Arabia, 2001.
6. Nasser, R.A.; Salem, M.Z.M.; Hiziroglu, S.; Al-Mefarrej, H.A.; Mohareb, A.S.; Alam, M.; Aref, I.M. Chemical Analysis of Different Parts of Date Palm (*Phoenix dactylifera* L.) Using Ultimate, Proximate and Thermo-Gravimetric Techniques for Energy Production. *Energies* **2016**, *9*, 374. [CrossRef]

7. Hakeem, K.R.; Jawaid, M.; Rashid, U. *Biomass and Bioenergy*; Springer: Berlin/Heidelberg, Germany, 2014.
8. Chandrasekaran, M.; Bahkali, A.H. Valorization of date palm (*Phoenix dactylifera*) fruit processing by-products and wastes using bioprocess technology—Review. *Saudi J. Biol. Sci.* **2013**, *20*, 105–120. [CrossRef]
9. Abu-Thabit, N.Y.; Judeh, A.A.; Hakeem, A.S.; Ul-Hamid, A.; Umar, Y.; Ahmad, A. Isolation and characterization of microcrystalline cellulose from date seeds (*Phoenix dactylifera* L.). *Int. J. Biol. Macromol.* **2020**, *155*, 730–739. [CrossRef]
10. Alotabi, M.D.; Alshammari, B.A.; Saba, N.; Alothman, O.Y.; Kian, L.K.; Khan, A.; Jawaid, M. Microcrystalline Cellulose from Fruit Bunch Stalk of Date Palm: Isolation and Characterization. *J. Polym. Environ.* **2020**, *28*, 1766–1775. [CrossRef]
11. Galiwango, E.; Abdel Rahman, N.S.; Al-Marzouqi, A.H.; Abu-Omar, M.M.; Khaleel, A.A. Isolation and characterization of cellulose and α-cellulose from date palm biomass waste. *Heliyon* **2019**, *5*, e02937. [CrossRef]
12. Bendahou, A.; Kaddami, H.; Dufresne, A. Investigation on the effect of cellulosic "nanoparticles' morphology on the properties of natural rubber based nanocomposites. *Eur. Polym. J.* **2010**, *46*, 609–620. [CrossRef]
13. Sobolčiak, P.; Tanvir, A.; Popelka, A.; Moffat, J.; Mahmoud, K.A.; Krupa, I. The preparation, properties and applications of electrospun co-polyamide 6,12 membranes modified by cellulose nanocrystals. *Mater. Des.* **2017**, *132*, 314–323. [CrossRef]
14. Alothman, O.Y.; Kian, L.K.; Saba, N.; Jawaid, M.; Khiari, R. Cellulose nanocrystal extracted from date palm fibre: Morphological, structural and thermal properties. *Ind. Crop. Prod.* **2021**, *159*, 113075. [CrossRef]
15. Boufi, S.; Kaddami, H.; Dufresne, A. Mechanical performance and transparency of nanocellulose reinforced polymer nanocomposites. *Macromol. Mater. Eng.* **2014**, *299*, 560–568. [CrossRef]
16. Benhamou, K.; Dufresne, A.; Magnin, A.; Mortha, G.; Kaddami, H. Control of size and viscoelastic properties of nanofibrillated cellulose from palm tree by varying the TEMPO-mediated oxidation time. *Carbohydr. Polym.* **2014**, *99*, 74–83. [CrossRef] [PubMed]
17. El-Mously, H. The palm fibers for the reinforcement of polymer composites: Prospects and challenge. In Proceedings of the First Ain Shams Conference on Environmental Engineering, Aim Shams University, Cairo, Egypt, 14 April 2009.
18. Mohanty, A.K.; Misra, M.; Drzal, L.T. Sustainable Bio-Composites from Renewable Resources: Opportunities and Challenges in the Green Materials World. *J. Polym. Environ.* **2002**, *10*, 19–26. [CrossRef]
19. Global Nanocellulose Market by Type (Nanocrystalline Cellulose, N.C., Others), Application, Region, Global Industry Analysis, Market Size, Share, Growth, Trends, and Forecast 2018 to 2025. Available online: https://www.fiormarkets.com/report/global-nanocellulose-market-by-type-nanocrystalline-cellulose-nanofibrillated-367085.html (accessed on 30 May 2021).
20. Foroughi, F.; Rezvani Ghomi, E.; Morshedi Dehaghi, F.; Borayek, R.; Ramakrishna, S. A Review on the Life Cycle Assessment of Cellulose: From Properties to the Potential of Making It a Low Carbon Material. *Materials* **2021**, *14*, 714. [CrossRef]
21. Fakirov, S.; Bhattacharyya, D. (Eds.) *Engineering Biopolymers: Homopolymers, Blends, and Composites*; Carl Hanser Verlag GmbH Co. KG: Munich, Germany, 2015.
22. TAPPI, T203 om-99. *Alpha-, Beta- and Gamma-Cellulose in Pulp*; Technical Association of the Pulp and Paper Industry: Atlanta, GA, USA, 1999.
23. Rol, F.; Vergnes, B.; El Kissi, N.; Bras, J. Nanocellulose Production by Twin-Screw Extrusion: Simulation of the Screw Profile To Increase the Productivity. *ACS Sustain. Chem. Eng.* **2020**, *8*, 50–59. [CrossRef]
24. Peng, B.L.; Dhar, N.; Liu, H.L.; Tam, K.C. Chemistry and applications of nanocrystalline cellulose and its derivatives: A nanotechnology perspective. *Can. J. Chem. Eng.* **2011**, *89*, 1191–1206. [CrossRef]
25. Chaker, A.; Mutjé, P.; Vilar, M.R.; Boufi, S. Agriculture crop residues as a source for the production of nanofibrillated cellulose with low energy demand. *Cellulose* **2014**, *21*, 4247–4259. [CrossRef]
26. Zheng, D.; Zhang, Y.; Guo, Y.; Yue, J. Isolation and Characterization of Nanocellulose with a Novel Shape from Walnut (*Juglans Regia* L.) Shell Agricultural Waste. *Polymers* **2019**, *11*, 1130. [CrossRef] [PubMed]
27. Evstigneyev, E.I.; Shevchenko, S.M. Structure, chemical reactivity and solubility of lignin: A fresh look. *Wood Sci. Technol.* **2019**, *53*, 7–47. [CrossRef]
28. Darmawan, S.; Wistara, N.J.; Pari, G.; Maddu, A.; Syafii, W. Characterization of Lignocellulosic Biomass as Raw Material for the Production of Porous Carbon-Based Materials. *Bioresources* **2016**, *1*, 3561–3574. [CrossRef]
29. Sánchez, R.; Espinosa, E.; Domínguez-Robles, J.; Loaiza, J.M.; Rodríguez, A. Isolation and characterization of lignocellulose nanofibers from different wheat straw pulps. *Int. J. Biol. Macromol.* **2016**, *92*, 1025–1033. [CrossRef]
30. Trilokesh, C.; Uppuluri, K.B. Isolation and characterization of cellulose nanocrystals from jackfruit peel. *Sci. Rep.* **2019**, *9*, 16709. [CrossRef]
31. Xiang, Z.; Gao, W.; Chen, L.; Lan, W.; Zhu, J.Y.; Runge, T. A comparison of cellulose nanofibrils produced from Cladophora glomerata algae and bleached eucalyptus pulp. *Cellulose* **2016**, *23*, 493–503. [CrossRef]
32. Oliveira, J.P.; Bruni, G.P.; Lima, K.O.; Halal, S.; Rosa, G.S.D.; Dias, A.R.G.; Zavareze, E.D.R. Cellulose fibers extracted from rice and oat husks and their application in hydrogel. *Food Chem.* **2017**, *221*, 153–160. [CrossRef] [PubMed]
33. Calahorra, M.E.; Cortázar, M.; Eguiazábal, J.I.; Guzmán, G.M. Thermogravimetric analysis of cellulose: Effect of the molecular weight on thermal decomposition. *J. Appl. Polym. Sci.* **1989**, *37*, 3305–3314. [CrossRef]
34. El-Saied, H.; El-Diwany, A.I.; Basta, A.H.; Atwa, N.A.; El-Ghwas, D.E. Production and Characterization of Economical Bacterial Cellulose. *BioResources* **2008**, *3*, 22.
35. Santmartí, A.; Lee, K.-Y. Crystallinity and thermal stability of nanocellulose. In *Nanocellulose and Sustainability*, 1st ed.; CRC Press: Boca Raton, FL, USA, 2018.

36. Mamleev, V.; Bourbigot, S.; Yvon, J. Kinetic analysis of the thermal decomposition of cellulose: The main step of mass loss. *J. Anal. Appl. Pyrolysis* **2007**, *80*, 151–165. [CrossRef]
37. Shen, D.K.; Gu, S. The mechanism for thermal decomposition of cellulose and its main products. *Bioresour. Technol.* **2009**, *100*, 6496–6504. [CrossRef]
38. Peng, Y.; Gardner, D.J.; Han, Y.; Kiziltas, A.; Cai, Z.; Tshabalala, M.A. Influence of drying method on the material properties of nanocellulose I: Thermostability and crystallinity. *Cellulose* **2013**, *20*, 2379–2392. [CrossRef]
39. Makkawi, Y.; El Sayed, Y.; Salih, M.; Nancarrow, P.; Banks, S.; Bridgwater, T. Fast pyrolysis of date palm (Phoenix dactylifera) waste in a bubbling fluidized bed reactor. *Renew. Energy* **2019**, *143*, 719–730. [CrossRef]
40. Taflick, T.; Schwendler, L.A.; Rosa, S.M.L.; Bica, C.I.D.; Nachtigall, S.M.B. Cellulose nanocrystals from acacia bark–Influence of solvent extraction. *Int. J. Biol. Macromol.* **2017**, *101*, 553–561. [CrossRef]
41. Ye, D.; Yang, J. Ion-responsive liquid crystals of cellulose nanowhiskers grafted with acrylamide. *Carbohydr. Polym.* **2015**, *134*, 458–466. [CrossRef]
42. Mandal, A.; Chakrabarty, D. Isolation of nanocellulose from waste sugarcane bagasse (SCB) and its characterization. *Carbohydr. Polym.* **2011**, *86*, 1291–1299. [CrossRef]
43. Da Silva, C.G.; Kano, F.S.; Rosa, D.S. Lignocellulosic Nanofiber from Eucalyptus Waste by a Green Process and Their Influence in Bionanocomposites. *Waste Biomass Valoriz.* **2020**, *11*, 3761–3774. [CrossRef]
44. French, A.D. Idealized powder diffraction patterns for cellulose polymorphs. *Cellulose* **2014**, *21*, 885–896. [CrossRef]
45. Wang, L.-F.; Shankar, S.; Rhim, J.-W. Properties of alginate-based films reinforced with cellulose fibers and cellulose nanowhiskers isolated from mulberry pulp. *Food Hydrocoll.* **2017**, *63*, 201–208. [CrossRef]
46. Borchani, K.E.; Carrot, C.; Jaziri, M. Untreated and alkali treated fibers from Alfa stem: Effect of alkali treatment on structural, morphological and thermal features. *Cellulose* **2017**, *22*, 1577–1589. [CrossRef]
47. Abdal-hay, A.; Suardana, N.P.G.; Jung, D.Y.; Choi, K.-S.; Lim, J.K. Effect of diameters and alkali treatment on the tensile properties of date palm fiber reinforced epoxy composites. *Int. J. Precis. Eng. Manuf.* **2012**, *13*, 1199–1206. [CrossRef]
48. Krishnan, J.; Sunil Kumar, S.; Krishna Prasad, R. Characterization of kraft pulp delignification using sodium dithionite as bleaching agent. *Chem. Eng. Commun.* **2020**, *207*, 837–846. [CrossRef]
49. Jordan, J.H.; Easson, M.W.; Dien, B.; Thompson, S.; Condon, B.D. Extraction and characterization of nanocellulose crystals from cotton gin motes and cotton gin waste. *Cellulose* **2019**, *26*, 5959–5979. [CrossRef]
50. Hussin, F.N.N.M.; Attan, N.; Wahab, R.A. Extraction and Characterization of Nanocellulose from Raw Oil Palm Leaves (Elaeis guineensis). *Arab. J. Sci. Eng.* **2020**, *45*, 175–186. [CrossRef]
51. Han, Y.; Bai, Y.; Zhang, J.; Liu, D.; Zhao, X. A comparison of different oxidative pretreatments on polysaccharide hydrolyzability and cell wall structure for interpreting the greatly improved enzymatic digestibility of sugarcane bagasse by delignification. *Bioresour. Bioprocess.* **2020**, *7*, 24. [CrossRef]
52. Mohamed Sutan, N.; Mazlan, S.M.; Taib, S.N.L.; Lee, D.T.C.; Hassan, A.; Sahari, S.K.; Mohamad Said, K.A.; Sobuz, H.R. Biomass Morphology Subjected to Different Chemical Treatment. *E3S Web Conf.* **2018**, *34*, 02051. [CrossRef]
53. Shahrousvand, E.; Shahrousvand, M.; Ghollasi, M.; Seyedjafari, E.; Jouibari, I.S.; Babaei, A.; Salimi, A. Preparation and evaluation of polyurethane/cellulose nanowhisker bimodal foam nanocomposites for osteogenic differentiation of hMSCs. *Carbohydr. Polym.* **2017**, *171*, 281–291. [CrossRef] [PubMed]

polymers

MDPI

Article

Double-Cross-Linked Networks Based on Methacryloyl Mucin

Elena Olăreț [1], Brîndușa Bălănucă [1,2], Andra Mihaela Onaș [1], Jana Ghițman [1], Horia Iovu [1,3], Izabela-Cristina Stancu [1,4] and Andrada Serafim [1,*]

1 Advanced Polymer Materials Group, University Politehnica of Bucharest, 1–7 Ghe. Polizu Street, 011061 Bucharest, Romania; olaretelena@gmail.com (E.O.); brindusa.balanuca@yahoo.com (B.B.); andra_mihaela.onas@upb.ro (A.M.O.); jana.ghitman@upb.ro (J.G.); horia.iovu@upb.ro (H.I.); izabela.stancu@upb.ro (I.-C.S.)
2 Department of Organic Chemistry Costin Nenitescu, University Politehnica of Bucharest, 1–7 Ghe. Polizu Street, 011061 Bucharest, Romania
3 Academy of Romanian Scientists, 54 Splaiul Independentei, 050094 Bucharest, Romania
4 Faculty of Medical Engineering, University Politehnica of Bucharest, 1–7 Ghe. Polizu Street, 011061 Bucharest, Romania
* Correspondence: andrada.serafim@gmail.com or andrada.serafim0810@upb.ro

Abstract: Mucin is a glycoprotein with proven potential in the biomaterials field, but its use is still underexploited for such applications. The present work aims to produce a synthesis of methacryloyl mucin single-network (SN) hydrogels and their double-cross-linked-network (DCN) counterparts. Following the synthesis of the mucin methacryloyl derivative, various SN hydrogels are prepared through the photopolymerization of methacrylate bonds, using reaction media with different pH values. The SN hydrogels are converted into DCN systems via supplementary cross-linking in tannic acid aqueous solution. The chemical modification of mucin is described, and the obtained product is characterized; the structural modification of mucin is assessed through FTIR spectroscopy, and the circular dichroism and the isoelectric point of methacryloyl mucin is evaluated. The affinity for aqueous media of both SN and DCN hydrogels is estimated, and the mechanical properties of the systems are assessed, both at macroscale through uniaxial compression and rheology tests and also at microscale through nanoindentation tests.

Keywords: porcine gastric mucin; methacryloyl mucin; double-cross-linked networks; circular dichroism; mechanical characterization

Citation: Olăreț, E.; Bălănucă, B.; Onaș, A.M.; Ghițman, J.; Iovu, H.; Stancu, I.-C.; Serafim, A. Double-Cross-Linked Networks Based on Methacryloyl Mucin. *Polymers* **2021**, *13*, 1706. https://doi.org/10.3390/polym13111706

Academic Editor: Arn Mignon

Received: 23 April 2021
Accepted: 20 May 2021
Published: 23 May 2021

1. Introduction

Mucin represents a major component of mucus, which covers several tissues (e.g., nose, eyes, gastrointestinal and respiratory tract, articular joins, etc.) with the aim of protecting them from the adsorption of unwanted molecules, reducing friction, preventing wear [1,2] or even autodigestion, in the case of gastric mucin [3]. Regardless of the covered tissue, mucus is secreted and degraded continuously, and most nanospecies are unable to penetrate it, being ultimately covered in a thin mucus layer before reaching the epithelial tissue [4]. Generally speaking, mucins contain a long polypeptide backbone rich in proline, threonine, and serine (PTS domain), on which dense brushes of carbohydrate chains are grafted [3,5,6]. Mucins are large compounds with a molecular weight of up to 20 MDa, and while the protein represents only about 20% of their molecular weight, the rest is made up of *O*-linked and *N*-linked oligosaccharides [1,6,7]. The composition and structure of mucins depend on their origin and are mirrored in slightly different behavior. For example, Madsen et al. demonstrated that while both bovine submaxillary mucin (BSM) and porcine gastric mucin (PGM) form hydrated films on hydrophobic surfaces, the adsorbed layers have different mechanical properties (BSM leads to elastic films, while those formed by the PGM are viscous), and the authors attributed this behavior to the structural differences between the two macromolecules [8]. Although mucin is pricier than other natural polymers such as collagen, gelatin, chitosan, and alginate, it is considered a material of interest for a variety of applications in the biomedical field. Since commercial mucins do not exhibit the required features for the aimed applications (e.g., virus inhibition, hydrogel formation, etc.) [7], these products are often purified through chromatography [5,9,10], thus increasing the actual price of the starting material. Nevertheless, the complexity of their structure proved to be appealing, and mucins have been studied for their use as carriers for bioactive species [11,12], coatings with improved tribological performance [13–15], and antifouling properties [16–19]. Furthermore, to obtain chemically cross-linked stable hydrogels, protocols describing the modification of other natural macromolecules (such as gelatin [20,21] or chitosan [22,23]) have been used to modify mucin, and they allow for its subsequent polymerization in a manner similar to that performed for synthetic polymers. The direct reaction of mucin with methacrylic anhydride has been previously exploited in order to obtain systems with controlled drugs release properties [24,25]. Mucin's methacryloyl derivative was also used as coating on polypropylene meshes for ventral hernia repair, and its potential as a bioactivator was proven through cellular tests using murine fibroblast cells [26]. However, as far as the authors of the present study are aware, the ability of methacryloyl mucin (MuMA) to form double-cross-linked network (DCN) systems has not yet been explored.

Typically, double-network hydrogels contain simultaneously a tough but brittle network formed through covalent bonds and a secondary weak and ductile network obtained through physical interactions [27–29]. This combination results in a soft hydrogel with superior mechanical properties when compared to their single-network (SN) system counterparts [30] and is increasingly used in developing biomaterials with applications in substitutes for load-bearing soft tissues such as cartilage or muscles [27].

This work presents a two-step procedure of obtaining DCN systems based on methacryloyl-modified PGM by (1) covalent bonding of MuMA methacrylate groups through radical photopolymerization, leading to SN hydrogels; and (2) additional hydrogen bonds (H-bonds) formed between the functional groups of MuMA and tannic acid (TA).

Considering the well-researched influence of a medium's pH on mucin behavior [31–33], in this work, hydrogels are prepared using media with different pH values. The synthesized SN systems are further incubated in a TA solution, and the properties of the obtained DCN hydrogels are compared with those of the SN counterparts. The efficiency of the photo-induced polymerization leading to SN hydrogels is assessed, and the affinity for aqueous media of all synthesized systems is evaluated. The mechanical properties of the obtained systems are assessed at both macro- and microscales. In addition, the study presents a characterization of the modified protein, using for the first time—in addition to Fourier-transformed infrared—the methods of circular dichroism spectroscopy and dynamic light scattering. This work sustains the idea that commercial PGM modified through a relatively easy and cost-efficient protocol may be used as a primary constituent in the synthesis of DCN hydrogels with applications in the field of biomaterials for load-bearing soft tissue.

2. Materials and Methods

All materials were purchased from Sigma-Aldrich (St. Louis, MI, USA) and used without any purification. Porcine gastric mucin type III (PGM), methacrylic anhydride (MA), carbonate-bicarbonate buffer (CB), phosphate buffer saline (PBS), 2-(N-Morpholino), ethanesulfonic acid hydrate (MES), and tannic acid (TA) were used in this experimental procedure. Here, 1-[4-(2-Hydroxyethoxy)-phenyl]-2-hydroxy-2-methyl-1-propane-1-one (Irgacure 2959) was used as a 10% ethanolic solution. HCl and NaOH were used as 1 mol/L solutions as pH adjusters. Double-distilled water (DDW) was used throughout the experiments.

2.1. Modification and Characterization of Natural Macromolecules

2.1.1. Modification of Commercial Porcine Gastric Mucin

Mucin was modified through direct reaction with methacrylic anhydride (MA) by adapting the protocol reported in [24,25]. In brief, 1% *w*/*v* porcine gastric mucin (PGM) solution in a carbonate-bicarbonate (CB) buffer (0.05 mol/L) (pH 8) was cooled in an ice bath, followed by an addition of MA to a final concentration of 0.5% *w*/*v*. The reaction was kept under continuous stirring, at 4 °C overnight. The reaction mixture was dialyzed against double-distilled water (DDW), using MWCO 12–14 kDa dialysis membranes (Spectrum™ Labs Spectra/Por) for 24 h with 3 changes of water, at room temperature. The obtained mixture was freeze-dried (48 h, −80 °C) and kept at −20 °C until later use for the hydrogel synthesis.

2.1.2. Fourier Transform Infrared (FTIR) Spectroscopy

FTIR spectroscopy was employed to assess the efficiency of the PGM modification. To this end, analyses were performed using a Vertex 70 Brucker FTIR spectrometer, equipped with an attenuated total reflectance (ATR) accessory. All FTIR spectra were recorded at room temperature, using 32 scans in the 600−4000 cm^{-1} wavenumber region, with a resolution of 4 cm^{-1}. Fourier self-deconvolution was done using Omnic software in the spectral region of 1500–1750 cm^{-1} and also in 850–1200 cm^{-1}. This computational tool was used to elucidate these wavelength intervals of interest, where the overlapped bands were detected in a separate manner and the important changes in the chemical structure of the modified mucin could be identified. This method was also employed to evaluate the formation of DCN systems, which is later presented.

2.1.3. Circular Dichroism (CD) Spectroscopy

CD spectroscopy was performed to characterize the secondary and tertiary structures of MuMA. The spectral results were registered using the JASCO J-1500 circular dichroism spectrophotometer equipped with a Peltier cell holder unit for precise temperature control. Data were collected in the wavelength range 176–350 nm using a 100 nm/min scanning speed, at a 1 nm bandwidth. A 0.025 nm data pitch and 3 accumulations for each analysis were used. The samples were analyzed as 0.5 mg/mL solutions in a 1 mm path-length quartz rectangular cell.

2.1.4. Zeta Potential

The zeta potential (ζ) of PMG and MuMA was determined in order to evaluate the modification of the isoelectric point (IP) of the modified protein when compared to the pristine commercial product. The IP is defined as the pH value at which the zeta potential value is zero, implying no electric charge on the surface of molecule [34]. The zeta potential of PMG and MuMA was measured using a Zetasizer Nano-ZS (Malvern Instruments Ltd., United Kingdom) equipped with a He–Ne linear polarized laser operating at a wavelength of λ = 632.8 nm and an angle of 13°. The equipment determines the electrophoretic mobility of colloids, employing the principle of laser Doppler velocimetry which is converted to a ζ-potential using the Helmholtz–Smoluchowski equation [35]. For these measurements, 1 mg/mL solutions of PMG and of MuMA, with pH values ranging from 1 to 6, were prepared in DDW, loaded in transparent rectangle cuvette with electrodes at each end and subsequently placed in the equipment where an electric field was applied to the cell. The pH of the tested samples was adjusted using 1 mol/L HCl solution and NaOH 1 mol/L solution, both prepared using DDW. For each sample, 3 measurements with 20 successive cycles at 25 °C were performed.

2.2. Synthesis of MuMA-Based Networks

Hydrogels consisting of 10% *w*/*v* MuMA in three different media (i.e., PBS, MES, and CB) were synthetized through photopolymerization. The corresponding amount of MuMA was gradually added in each of the previously mentioned media, followed by 2 h 30 min of vigorous stirring at 4 °C. 0.1% w/w (with respect to MuMA content), after which Irgacure 2959 was added. After homogenization, the reaction mixtures were poured into Petri dishes, followed by UV exposure (30 min at 312 nm wavelength using UV transilluminator ECX-F26). As-obtained samples of each composition were dried and preserved for the evaluation of polymerization efficiency through gel fraction assessment. The rest of the synthesized hydrogels were rinsed with DDW and cut into suitable geometries for further evaluation. Additionally, samples from each hydrogel composition were immersed in 10% TA solution in DDW for 24 h to form the secondary network, while their corresponding control samples were incubated 24 h in DDW. Hydrogels composition and codes are presented in Table 1.

Table 1. Hydrogel compositions and codes.

Type of System	Sample Code	Dispersion Medium	pH Value	Postsynthesis Incubation
Single network (SN)	MuMA I MES	MES	4.0	DDW, 24 h
	MuMA I PBS	PBS	7.4	
	MuMA I CB	CB	10.0	
Double cross-linked network (DCN)	MuMA I MES_TA	MES	4.0	10% TA aqueous solution, 24 h
	MuMA I PBS_TA	PBS	7.4	
	MuMA I CB_TA	CB	10.0	

To avoid any interference from the dispersion media, the hydrogels subjected to FTIR measurements were synthesized in DDW following the same procedure as previously described.

2.3. Gel Fraction Evaluation

The efficiency of the network-forming polymerization was correlated with the polymerization reaction media through a gel fraction (GF, %) assessment. Briefly, samples of each synthesized system were dried (at 37 °C for 24 h) immediately after synthesis. Following the complete evaporation of the solvent, the samples were weighted (w_0) and subsequently washed thoroughly in DDW for 24 h at 37 °C, then were dried and weighted again (w_f). GF was assessed through Equation (1) [36]. The measurements were performed in triplicate.

$$GF,\ \% = \frac{w_f}{w_0} \times 100 \quad (1)$$

The GF analysis aimed exclusively at the assessment of the polymerization efficiency and was performed only on the SCN systems. The DCNs were obtained through the immersion of the synthesized hydrogels in 10% TA aqueous solution for a period of 24 h, and it was expected that any unbounded protein chains would be dispersed in the incubation medium, and thus the GF analysis would result in inaccurate data.

2.4. Swelling Ability

The affinity for aqueous media of the MuMA and MuMA_TA systems was investigated through a conventional gravimetric method. In this respect, dried samples of each composition were weighed (w_0), immersed in DDW at 37 °C, removed from the incubation medium, blotted with filter paper, and weighted at pre-established time points (w_t). All measurements were performed in triplicate, and the swelling content (SC) was determined, as in Equation (2). A SC–time graph was plotted, and the swelling profile of the synthesized systems was constructed [37].

$$SC, g/g = \frac{w_t - w_0}{w_0} \quad (2)$$

In addition, the equilibrium water content (EWC, %) was computed as reported previously [36] for both SN and DCN systems, using Equation (3).

$$EWC, \% = \frac{w_{max} - w_0}{w_{max}} \times 100 \quad (3)$$

where w_{max} represents the weight of the hydrogel at swelling equilibrium.

2.5. Mechanical Investigations

Mechanical assessments were performed to characterize the synthesized hydrogels at both macro- and microscales. To this end, uniaxial compressions were carried out to estimate the reinforcing role of TA in the DCN hydrogels, and rheology tests offered information about the storage and loss moduli of the bulk samples, while nanoindentation provided insights regarding the correlation local microstructure–mechanical behavior.

2.5.1. Uniaxial Compressions

The synthesized systems were subjected to compression tests using a CT3 Texture Analyzer (Brookfield) equipped with a 4500 g cell. Cylindrical samples (d = 5 mm, h = 2 mm) were compressed up to 50% deformation, and the strain–stress curves of the compositions were plotted.

2.5.2. Rheological Tests

The investigation of the rheologic behavior was performed through dynamic oscillatory measurements using a Kinexus Pro rheometer equipped with a Peltier element for precise temperature control, performed at 25 °C. Samples of hydrogel hydrated at equilibrium (d = 20 mm, h = 1.5 mm) were placed on the bottom plate of the rheometer, and parallel-plate geometry was used. Sand paper was used on the plates to avoid slipping. Dehydration was prevented using a water lock. The linear viscous region was established through amplitude sweep measurements, and subsequently a frequency sweep analysis was performed in the frequency range of 0.01–100 Hz at a constant stress value of 10 Pa. The storage and loss moduli (G′ and G″, respectively) were plotted in a logarithm graph.

2.5.3. Nanoindentation

Local mechanical characterization was carried out through nanoindentation using the Nano Indenter G200 system supplied with a DCM II head and CSM option (KLA Instruments). The shear storage modulus (G′) and loss modulus (G″) were obtained using a cylindrical flat punch diamond tip with a face diameter of 100 μm. Four different sites were tested for each hydrogel sample, while ensuring a minimum distance of 500 μm between them. The testing frequency was set to 10 Hz. Poisson's ratio was set to 0.5 (as it is typical for gels), and the oscillation amplitude (in sample) was set to 500 nm. Pretest compression was kept at 5% of the punch diameter.

3. Results and Discussions

3.1. Synthesis and Characterization of PGM Methacryloyl Derivative

The reaction of mucin with methacrylic anhydride was first described by Duffy et al., who confirmed the modification of bovine submaxillary mucin by FTIR spectroscopy and the ability of the macromolecule to form stable hydrogels [25]. Following the same modification protocol, our work group previously reported the use of porcine gastric mucin (PGM) methacryloyl derivative (MuMA) in hybrid systems with applications in wound treatment [24] and as coating for the bioactivation of polypropylene meshes for ventral hernia repair [26]. Herein, we used a slightly modified version of the previous protocols, namely, a carbonate buffer (CB) was used as the reaction medium for the modification. The pH was monitored throughout the reaction, and no additional measures were required to maintain the pH at 8. The circular dichroism (CD) spectra registered for PGM in the CB confirm that at pH 10, the PGM molecule was less folded when compared to the spectra registered for PGM in DDW (pH 6) (Figure S1). The formation of a small shoulder-like peak at approximately 185 nm for PGM in CB was noticed. This is characteristic for a better exposure of lateral groups to the surrounding medium and therefore a better unfolding of the PGM structure. Moreover, a small decrease in the intensity of the peak at approximately 190 nm suggests the unfolding of the polypeptide chain for higher pH medium. As a consequence, it is expected that the NH_2 groups are more exposed, allowing for the reaction with the anhydride.

Following the synthesis and purification of MuMA, spectral analyses were employed to characterize the freeze-dried product. The registered ATR-FTIR spectra for native PGM, MA, and MuMA are presented in Figure 1A. The FTIR spectrum of MuMA shows all characteristic signals of a protein [36]: a broad peak in the region 3100–3700 cm^{-1} attributed to overlapped N–H and O–H stretching vibration, an amide I typical signal at 1642 cm^{-1}, and a peak attributed to amide II that slightly shifted from 1554 cm^{-1} in PGM to 1550 cm^{-1} in MuMA. In order to gain a better visualization of the specific vibration bands from MA, a deconvolution process of the MuMA spectrum was performed in the 1500–1750 cm^{-1} and 850–1200 cm^{-1} regions (Figure 1A, inset A.1 and A.2, respectively). The outcome reveals the presence of two peaks, i.e., around 1633 cm^{-1} and 945 cm^{-1}, which were assigned to the stretching vibration of –C=C– bonds and the deformation of =CH_2 bonds, respectively, and were specific bands for the new grafted methacrylate groups [38,39].

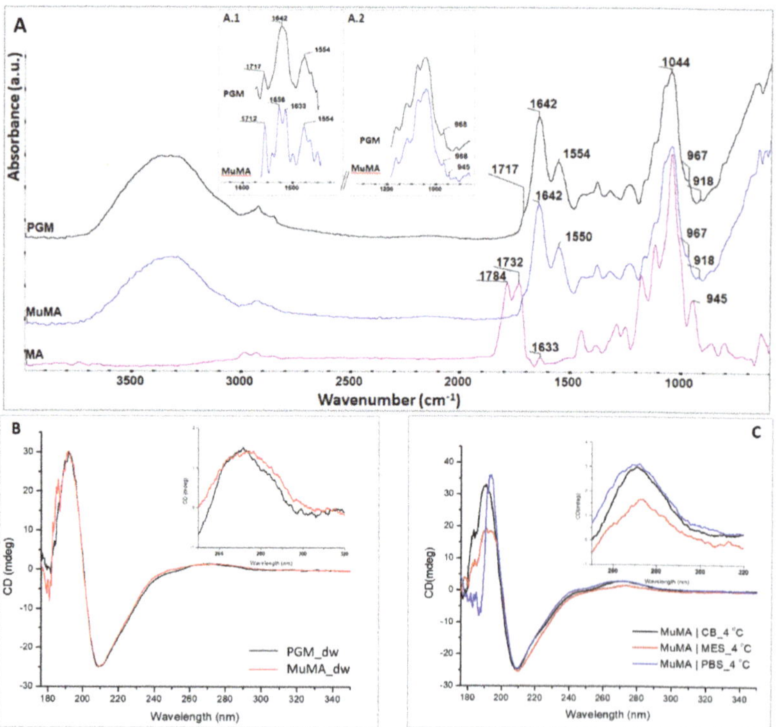

Figure 1. (**A**) FTIR spectra of PGM (**black**), MuMA (**blue**), and MA (**purple**); inset A.1: deconvoluted spectra of PGM and MuMA in the wavenumber region 1500–1750 cm^{-1}; inset A.2: deconvoluted spectra of PGM and MuMA in the wavenumber region 850–1200 cm^{-1}; (**B**) CD spectra of PGM and MuMA; inset: magnification in the 250–320 nm region; (**C**) CD spectra of MuMA | MES, MuMA | PBS, and MuMA | CB at 4 °C; inset: magnification in the 250–320 nm region.

The successful methacrylation was also confirmed through CD spectroscopy. As depicted in Figure 1B, the spectrum registered for MuMA presents a prominent shoulder around 182 nm attributed to the n–σ* transition of the C=O groups introduced by the methacrylation process. Additionally, the CD results indicate that the methacrylation process does not lead to the denaturation of the polypeptidic chains, as the registered MuMA spectra indicate an unaltered tertiary structure of the methacryloyl derivative, similar to the one of the pristine protein. Moreover, CD spectroscopy indicated that the MA-modified product maintains the pH dependence of the pristine PGM. The spectra registered for MuMA I MES, MuMA I PBS, and MuMA I CB indicated that the methacryloyl derivative is most unfolded in an acidic medium. Figure 1C depicts the MuMA spectrum in each medium at 4 °C.

For the MuMA I MES CD spectra, a decrease of the positive contribution—as reflected by the magnitude of the peak—at around 190 nm may be noticed, which is associated to the π–π^* perpendicular polarized transitions of helix-containing polypeptides, suggesting a less folded structure at pH 4. From the 250–320 nm region, we observed that the MuMA I CB sample exhibits the most folded tertiary structure of the polypeptidic chains, while for the MuMA I MES sample, a slight deviation of the 270 nm peak suggests a possible modification of the folded architecture, and its obvious decrease in intensity could be assigned to an unfolding of the macromolecular chains.

The CD spectra for MuMA samples in all three media were registered both at 4 and 25 °C (Figure S2). The results indicate that in MES, the polypeptide chains were more unfolded at 4 °C. There was a clear reduction in the positive contribution at approximately 190 nm for the MES spectra at low temperature when compared to ambient temperature (25 °C) caused by the unfolding of the molecule. Conversely, the spectra registered for MuMA in PBS showed that at 4 °C, the positive contribution around 195 nm increased, leading to a more tightly folded structure of the macromolecule when compared to ambient temperature spectral results. At high pH values, the arrangement of the macromolecular chains was not influenced by this parameter, as the spectra registered for MuMA in the CB is similar for the two investigated temperatures.

Zeta potential measurements were also performed to characterize the methacryloyl derivative. Generally, it is known that zeta potential is a function of the surface charge of a colloid in solution as well as of the pH and ionic strength of the solution. In the pH range of 5–2.5, the zeta potential of both PMG and MuMA was negative due to the negatively charged carboxylic functionalities (–COO–) from the protein chains. The decrease of pH value below 2.5 led to a positive reversion of the zeta potential, exhibiting a cross point (IP value) at pH $\approx$ 2.1 for PMG and pH $\approx$1.1 for MuMA (Figure S3). In this pH range, the amino groups from protein chains were positively charged ($-NH_3^+$) and the zeta potential became positive. The results regarding the IP of the unmodified protein are in agreement with the findings of other research groups that have reported an IP value in the range of 2–3 [6,40].

The IP value of MuMA shifted to a lower pH value when compared to that of PMG. In the methacrylation reaction of PMG, the primary amino groups ($-NH_2$) were mostly involved, covalently binding the methacrylic functionality on the protein chain. This led to the formation of MuMA and methacrylic acid as a byproduct that is easily eliminated from the system by dialysis. Thus, this pH shift might have originated in the increase of the carboxyl/amino groups' ratio from the MuMA structure (considering that a part of the $-NH_2$ functionalities are consumed in the methacrylation reaction). These results are in good agreement with the previously discussed FTIR data, indicating that some of the $-NH_2$ groups were consumed during the reaction with the methacrylic anhydride.

3.2. Synthesis of the SN and DCN Systems

Considering the pH dependency of MuMA, as indicated by the CD spectroscopy, the polymerization precursors were prepared using reaction media with different pH values, all above the IP of MuMA (approx. 1.1 as indicated by zeta potential investigations). As a consequence of this pH dependency, corroborated with the amphoteric character of the polymer due to the abundance of both positively and negatively charged groups (unmodified $-NH_3^+$ groups form the protein core, and –O– and –COO– form the carbohydrates side chains), it was expected that the reaction media of the polymerization process would imprint a certain orientation of the MuMA's chains.

The DCN systems were synthesized through a two-step procedure consisting of (1) the formation of SN systems through covalent bonds established through the bulk polymerization of the methacrylic groups of MuMA, followed by (2) the formation of an additional physical network based on H-bonds established through TA diffusion into the MuMA SN. While the first step of the process led to opaque, easy-to-handle, soft hydrogels, following the incubation in TA the hydrogels changed to a brownish color and exhibited a slight shrinkage with a more robust nature. A schematic representation of the synthesis procedure is shown in Figure 2A.

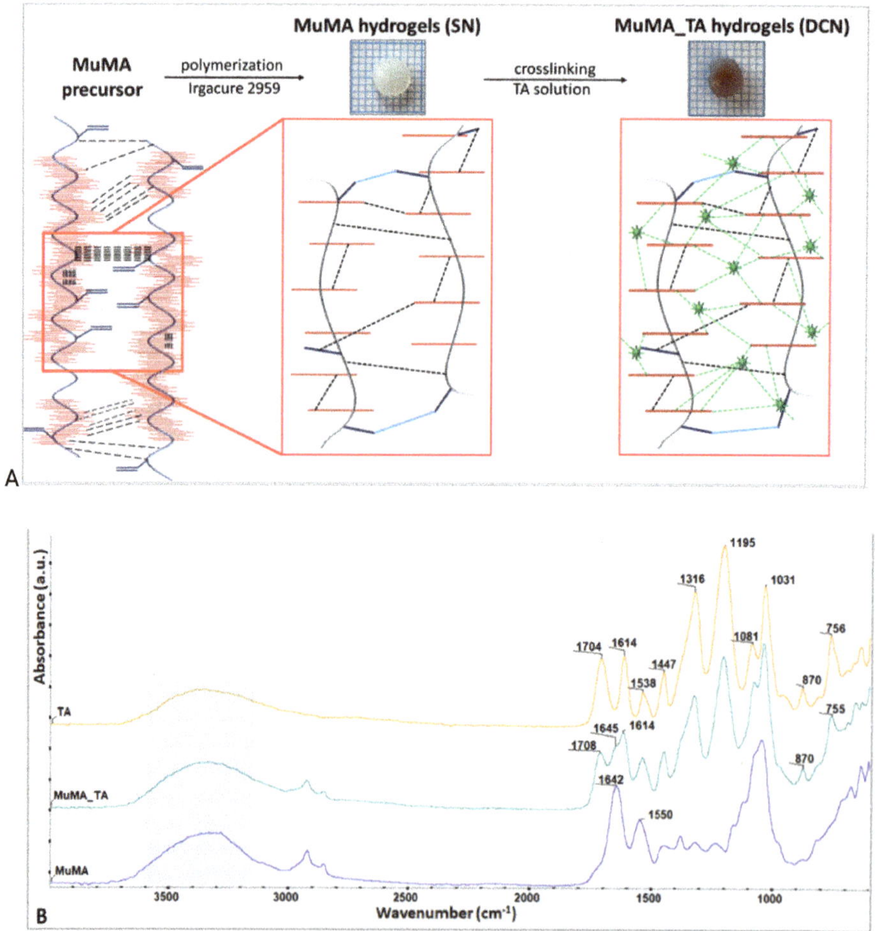

Figure 2. (**A**) Schematic representation of the synthesized SN and DCN systems (blue: "C–C" covalent bonds formed through the polymerization of "C=C" groups in MuMA; dashed black: intra- and intermolecular H-bonds in MuMA; dashed green: TA-MuMA H-bonds); (**B**) FTIR spectra of TA (**orange**), MuMA_TA (**teal**), and MuMA (**blue**).

Given the numerous functional groups of MuMA, intra- and intermolecular H-bonds were formed in the hydrogel precursor prior to the photopolymerization of "C=C" bonds. SN formation led to new "C–C" covalent bonds, while TA treatment additionally strengthened the hydrogel with TA–MuMA H-bonds, resulting in DCN.

3.3. Assessment of the Polymerization Process

Successful polymerization was confirmed through the GF analysis, based on gravimetric measurements and using Equation (1). The overall results indicate that the polymerization was successful for all compositions. The GF values that were computed for all three systems are in the range of 68–78%, indicating a successful polymerization process that led to insoluble, stable hydrogels regardless of the pH of the reaction media. However, slight differences between the analyzed SN systems were noticed (Table 2).

Table 2. Gel fraction values registered for the MuMA-based SN systems.

Single-Network System	Gel Fraction (GF, %)
MuMA I MES	78.63 ± 1.04
MuMA I PBS	68.31 ± 0.72
MuMA I CB	75.62 ± 0.82

Considering that all systems incorporated the same solid content and were synthesized following the same protocol, the different values obtained for GF are attributed to the different pH values of the polymerization media, and they are in good agreement with the insights provided by CD spectroscopy, which indicated the highest GF for the system obtained in MES (in which MuMA exhibited the most unfolded structure at 4 °C). This behavior led to a better exposure of the methacrylic groups on the polypeptidic chains and thus the formation of covalent bonds in the presence of the photoinitiator. Conversely, the macromolecular chains are more folded, exposing fewer reaction centers in the systems in which PBS and CB were used, i.e., the dense "brush" of carbohydrates hindering the methacrylic groups and therefore leading to the formation of systems with lower GF values. The lowest GF value was registered for the systems obtained in PBS, indicating that the macromolecular chains were neither unfolded enough to sufficiently expose the methacrylic groups (as they were in MES) nor properly folded to form additional physical bonds (as this is probably the case with the systems synthesized in a CB). These findings are in agreement with the CD results, which show that the MuMA I PBS spectrum exhibited in the region of 190–210 nm a narrower peak of higher intensity when compared to the other registered spectra, confirming the distinct folding of the modified protein at pH 7.4 and 4 °C.

The GF analysis was not considered relevant for the DCN systems due to the relatively long period of incubation (24 h) in 10% TA aqueous solution. In these conditions, it was expected that any unbounded modified protein would be lost in the incubation medium, thus leading to incorrect results.

The formation of DCN systems was firstly evaluated through FTIR spectroscopy; the corresponding spectra for TA, MuMA, and MuMA_TA are presented in Figure 2B. The MuMA_TA spectrum shows specific signals attributed to TA and also the MuMA matrix in 1600–1710 cm^{-1} region as follows: 1708 cm^{-1}, $\nu_{C=O}$ and aromatic $\nu_{C=C}$ from TA; 1614 cm^{-1}, $\nu_{C\text{-}C}$ and $\delta_{C\text{-}H}$ from the TA aromatic structure; and 1645 cm^{-1}, amide I from long protein backbone of PGM. All other absorption bands in between the 600–1538 cm^{-1} wavenumber interval could be assigned to the specific bond vibrations from PGM and TA [14]. Intermolecular H-bonds formed between the –OH groups from TA and –COOH moieties from the MuMA-based polymerized structure during the incubation process led to the shape modification of the broad band at 3000–3600 cm^{-1}. The TA spectrum indicates a more intense and broader O–H band at higher wavenumber values, due to the many hydroxyl groups from its structure while the O–H stretch band for MuMA slightly shifted to lower wavenumber values due to the great number of –COOH moieties from the protein region of macromolecule.

3.4. Investigation of the Water Uptake Potential of the Synthesized Systems

One of the most exploited features of hydrogels is their water affinity, as these materials are able to absorb large quantities of aqueous media without losing their dimensional integrity [41]. In addition, the time required to reach swelling equilibrium is also extremely important. This characteristic is pivotal for several precise applications, such as drug delivery and the design of scaffolds for tissue regeneration and repair [42–44]. Our study established not only the amount of water absorbed by the two type of systems but also their swelling profile, as in the computed parameters, i.e., equilibrium water content, (EWC, %) and swelling content (SC, g/g), providing important insights regarding the time required for the synthesized systems to reach the maximum hydration and also the maximum amount of aqueous media they are able to uptake.

The swelling profile was represented as a function of time, and it indicated that there are major differences between the synthesized systems. As depicted in Figure 3A, the highest SC was registered for MuMA I PBS, which is in accordance with the GF study and thus confirming the formation of networks that allow for the uptake of large amounts of water. Conversely, the more tightly packed networks of MuMA I MES and MuMA I CB, for which a higher GF was obtained, led to slightly lower SC values. These differences are less pronounced in the case of the DCN systems, indicating that the formation of the supplementary network through incubation in TA is efficient for all systems, leading to more compact matrices. Moreover, the considerably lower values of SC registered for the DCN when compared to the SN counterparts confirm the formation of a second dense network, leading to systems with a decreased water uptake ability. Furthermore, the DCN systems require a longer period of time to reach swelling equilibrium (250 min) when compared to the SN systems (180 min) (Figure 3A).

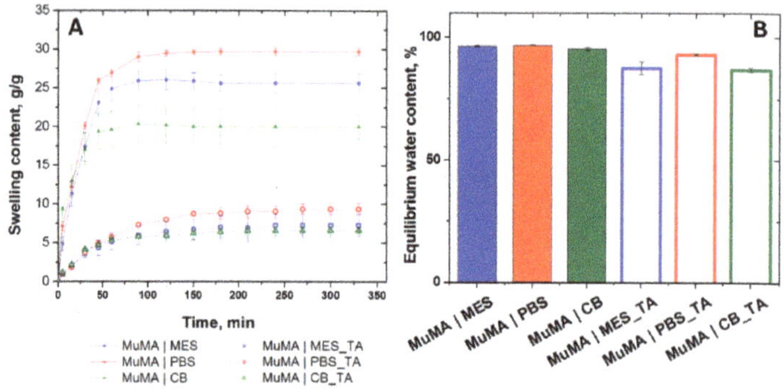

Figure 3. (**A**) Swelling kinetics representing the SC increase during hydration of SN and DCN systems; (**B**) equilibrium water content (EWC, %) computed for the SN and DCN hydrogels.

After swelling equilibrium was reached, EWC values were computed, which are represented in Figure 3B. Although there are no significant differences between the EWC of the SN systems, the data show that the formation of the secondary network had a different impact on the obtained DCN systems. The EWC computed for the systems that were synthesized in MES presented the highest difference between SN (96.24 $\pm$ 0.19) and DCN (87.47 $\pm$ 2.5), while the MuMA I PBS registered a value that is only 3.79% higher than MuMA I PBS_TA. This behavior is also a confirmation of the different orientation of the functional groups in MuMA structure during the photopolymerization process due to the pH value of the dispersion media, which ultimately leads to a more unfolded (in the case of MES) or folded (in the case of PBS and CB) structure.

3.5. Evaluation of the Mechanical Behavior of the SN and DCN Systems

The hydrogels' mechanical properties are generally considered weak, and thus strategies to enhance mechanical properties while preserving their unique characteristics are in continuous development [45]. Among these, the sequential chemical and physical cross-linking of natural hydrogels proved to be a valid method for obtaining materials with superior mechanical behavior [46]. The results registered following conventional uniaxial compression tests of MuMA_TA hydrogels demonstrated that the DCN exhibited an improved resistance at break when compared to their SN counterparts. The strain–stress curves (Figure S4) show that the SN systems broke around a strain of 30–40% but were able to undergo higher deformation. Oppositely, the DCN systems required a significantly higher stress to be deformed at the same strain value, and this behavior correlated to the lower EWC and SC values. Furthermore, these data confirm that the interactions between the MuMA-based SN and TA significantly improved the materials' resistance to deformation.

The storage modulus and the loss modulus of the composition were evaluated at macroscale through rheology tests performed on equilibrium-swollen samples. While the storage modulus (G′) provides information regarding the elasticity of the material due to the ability to store energy, the loss modulus provides insights into the ability to dissipate energy, and it is a measure of the hydrogel's viscous behavior [47].

The results show that elastic behavior is predominant in both SN and DCN hydrogels, with G′ higher than G″. In addition, G′ is independent of the frequency value in the studied interval, which shows good stability. As presented in Figure 4, regardless of the pH value of the polymerization medium, the formation of the supplementary network with TA led to a decrease of elasticity. Although the G′ value increased slightly, it is interesting that the G″ value remained unchanged in the DCN systems, when compared with the SN ones. This detail suggests that the presence of the supplementary cross-linking of the hydrogel brought by the presence of H-bonds was not accompanied by the immobilization of any chain parts. The formation of the first network through covalent bonds led to soft yet brittle materials, while further incubating these materials in TA aqueous solution led to more robust systems that were able to undergo higher deformations without breaking under stress.

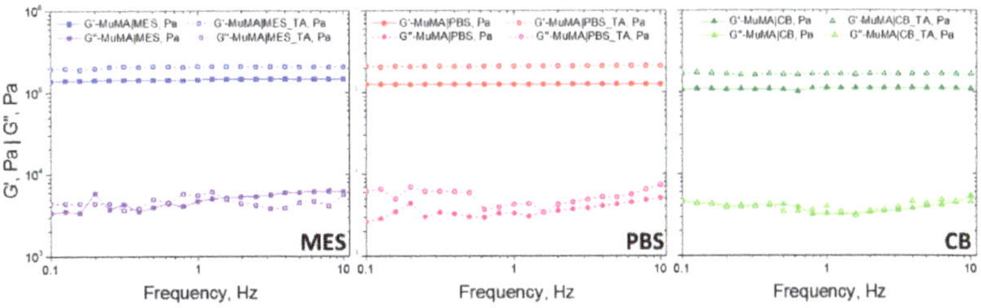

Figure 4. The rheological behavior of control samples and TA-treated samples.

In addition to the macroscale mechanical properties of the synthesized materials, the study also offers important insights regarding their properties at microscale, as obtained through nanoindentation tests. Surface mechanical properties are of great interest when considering biomedical applications since they influence the cellular behavior [48]. As mentioned by Youssefian et al. [49], H-bonds act as springs between atoms and have a significant influence on polymer chains' arrangement and on their elastic behavior. Thus, the storage and the loss moduli of equilibrium-swollen hydrogels were evaluated also at microscale through dynamic instrumented indentation tests (Figure 5). Overall, the storage moduli increased after SN incubation in TA, suggesting the surface reinforcing of the hydrogels. Such behavior may be assigned to MuMA–TA H-bonds formation, resulting in DCN. Furthermore, the surface elasticity placed the SN hydrogels in the order of MuMA | PBS < MuMA | MES < MuMA | CB, and TA treatment changed this pattern to MuMA | MES_TA < MuMA | CB_TA < MuMA | PBS_TA. TA exhibited a lower influence on MuMA | PBS (29.58% increase). For MuMA | CB, a doubling of the storage modulus was observed after TA incubation (97.85% increase). The highest increase in storage modulus after TA treatment was recorded for MuMA | MES_TA (114.45% increase). This is consistent with the CD results, where MuMA | MES presented the most unfolded structure, and therefore it was expected that this composition provides the most available sites to H-bond formation and hence a more pronounced TA-cross-linking effect

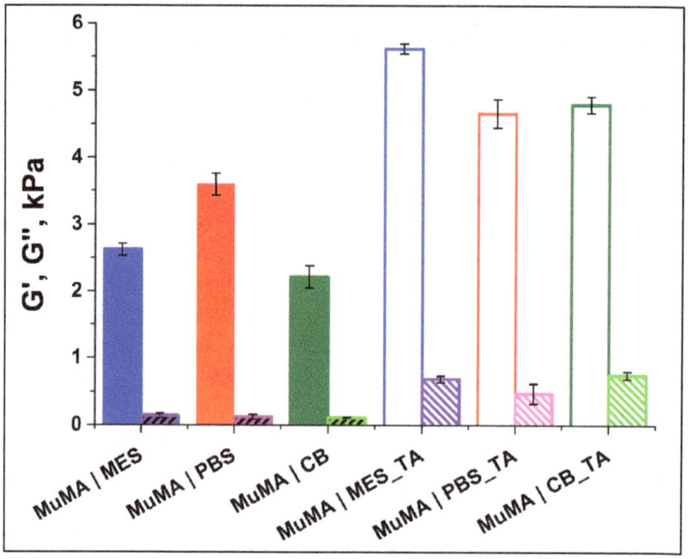

Figure 5. Microscale storage modulus (G′, simple bars) and loss modulus (G″, patterned bars) as obtained from instrumented indentation tests.

4. Conclusions

The successful synthesis of DCNs based on methacryloyl-modified porcine gastric mucin was demonstrated in three media with pH values representative for acid, neutral, and alkaline conditions. The study highlighted the importance of the pH value of the initial dispersion media on the folding of MuMA, as mirrored on the water uptake ability and the mechanical properties of the SN systems. Furthermore, our findings showed that the supplementary cross-linking with polyphenol allows an additional adjustment of the materials' properties, both at the macro- and microscales. The conventional mechanical tests showed that TA cross-linking led to more robust systems that require additional effort to be deformed under stress. The surface mechanical properties indicated that the most significant influence of the supplementary cross-linking was exhibited on the MuMA | MES hydrogels, which doubled their storage modulus. Overall, the results demonstrated the capacity of MuMA to form DCN systems that have the ability to uptake large amounts of aqueous media without losing their integrity. The high water uptake ability and the macro- and microscale mechanical properties indicate that these systems may find applications in the field of tissue reconstruction and repair.

Supplementary Materials: The following are available online at https://www.mdpi.com/article/10.3390/polym13111706/s1; Figure S1: CD spectra registered for PGM in DDW and CB; the additional peak at 185 nm confirms a better exposure of the lateral groups of PGM at pH 10 (in CB) when compared to pH 6 (in DDW); Figure S2: CD spectra for MuMA in media with different pH values (MES at pH 4, PBS at pH 7.4, and CB at pH 10) registered at 25 °C and 4 °C; the results indicate that at 4 °C MES has the most unfolded structure, while PBS presents the most tightly packed structure; Figure S3: Graphical representation of the zeta potential measurements performed for the native mucin (PGM) and the methacryloyl derivative (MuMA). The results indicate that the modification with methacrylic anhydride decreases the isoelectric point from ≈2.1 (PGM) to ≈1.1 (MuMA); Figure S4: Stress–strain curves for both SN (solid line) and DCN (dash line) systems.

Author Contributions: E.O.: Conceptualization, methodology, validation, formal analysis, investigation, writing—original draft preparation; B.B.: Methodology, validation, formal analysis, investigation, writing—original draft preparation; A.M.O.: Methodology, validation, formal analysis, investigation; J.G.: Methodology, validation, formal analysis, investigation; H.I.: writing—review and editing, project administration, funding acquisition; I.-C.S.: writing—review and editing, supervision, project administration; A.S.: Conceptualization, methodology, validation, writing—original draft preparation, writing—review and editing, visualization, supervision, project administration, funding acquisition. All authors have read and agreed to the published version of the manuscript.

Funding: This work was supported by a grant of the Romanian Ministry of Education and Research, CNS-UEFISCDI, project number PNIII-P1-1.1-TE-2019-1161, within PNCDI III. The work of I.-C. Stancu was supported by a grant of the Romanian Ministry of Research and Innovation, CCCDI-UEFISCDI, grant number PN-III-P1-1.2-PCCDI-2017-0782/REGMED—project 2 NERVE, within PNCDI III. The nanoindentation tests, dynamic light scattering measurements, and circular dichroism spectroscopy were possible due to the European Regional Development Fund through the Competitiveness Operational Program 2014–2020, priority axis 1, project number P_36_611, MySMIS code 107066: Innovative Technologies for Materials Quality Assurance in Health, Energy and Environmental—Center for Innovative Manufacturing Solutions of Smart Biomaterials and Biomedical Surfaces (INOVABIOMED).

Institutional Review Board Statement: Not applicable.

Informed Consent Statement: Not applicable.

Data Availability Statement: The data presented in this study are available on request from the corresponding author.

Acknowledgments: The help of Livia Butac and Claudiu Patrascu is also greatly acknowledged. Their remarks and advice helped in building a better research paper.

Conflicts of Interest: The authors declare no conflict of interest.

References

1. Bansil, R.; Turner, B.S. Mucin structure, aggregation, physiological functions and biomedical applications. *Curr. Opin. Colloid Interface Sci.* **2006**, *11*, 164–170. [CrossRef]
2. Perez-Vilar, J.; Hill, R.L. The structure and assembly of secreted mucins. *J. Biol. Chem.* **1999**, *274*, 31751–31754. [CrossRef]
3. Barz, B.; Turner, B.S.; Bansil, R.; Urbanc, B. Folding of pig gastric mucin non-glycosylated domains: A discrete molecular dynamics study. *J. Biol. Phys.* **2012**, *38*, 681–703. [CrossRef] [PubMed]
4. Cone, R.A. Barrier properties of mucus. *Adv. Drug Deliv. Rev.* **2009**, *61*, 75–85. [CrossRef] [PubMed]
5. Zappone, B.; Patil, N.J.; Madsen, J.B.; Pakkanen, K.I.; Lee, S. Molecular structure and equilibrium forces of bovine submaxillary mucin adsorbed at a solid-liquid interface. *Langmuir* **2015**, *31*, 4524–4533. [CrossRef] [PubMed]
6. Lee, S.; Müller, M.; Rezwan, K.; Spencer, N.D. Porcine gastric mucin (PGM) at the water/poly(dimethylsiloxane) (PDMS) interface: Influence of pH and ionic strength on its conformation, adsorption, and aqueous lubrication properties. *Langmuir* **2005**, *21*, 8344–8353. [CrossRef] [PubMed]
7. Schömig, V.J.; Käsdorf, B.T.; Scholz, C.; Bidmon, K.; Lieleg, O.; Berensmeier, S. An optimized purification process for porcine gastric mucin with preservation of its native functional properties. *RSC Adv.* **2016**, *6*, 44932–44943. [CrossRef]
8. Madsen, J.B.; Sotres, J.; Pakkanen, K.I.; Efler, P.; Svensson, B.; Abou Hachem, M.; Arnebrant, T.; Lee, S. Structural and mechanical properties of thin films of bovine submaxillary mucin versus porcine gastric mucin on a hydrophobic surface in aqueous solutions. *Langmuir* **2016**, *32*, 9687–9696. [CrossRef] [PubMed]
9. Madsen, J.B.; Pakkanen, K.I.; Duelund, L.; Svensson, B.; Hachem, M.A.; Lee, S. A Simplified Chromatographic Approach to Purify Commercially Available Bovine Submaxillary Mucins (BSM). *Prep. Biochem. Biotechnol.* **2015**, *45*, 84–99. [CrossRef] [PubMed]
10. Celli, J.; Gregor, B.; Turner, B.; Afdhal, N.H.; Bansil, R.; Erramilli, S. Viscoelastic Properties and Dynamics of Porcine Gastric Mucin. *Biomacromolecules* **2005**, *6*, 1329–1333. [CrossRef]
11. Li, C.; Liu, Z.; Yan, X.; Lu, W.; Liu, Y. Mucin-controlled drug release from mucoadhesive phenylboronic acid-rich nanoparticles. *Int. J. Pharm.* **2015**, *479*, 261–264. [CrossRef] [PubMed]

12. Mumuni, M.A.; Kenechukwu, F.C.; Ofokansi, K.C.; Attama, A.A.; Díaz, D.D. Insulin-loaded mucoadhesive nanoparticles based on mucin-chitosan complexes for oral delivery and diabetes treatment. *Carbohydr. Polym.* **2020**, *229*, 115506. [CrossRef] [PubMed]
13. Rickert, C.A.; Wittmann, B.; Fromme, R.; Lieleg, O. Highly Transparent Covalent Mucin Coatings Improve the Wettability and Tribology of Hydrophobic Contact Lenses. *ACS Appl. Mater. Interfaces* **2020**, *12*, 28024–28033. [CrossRef]
14. Olaret, E.; Ghitman, J.; Iovu, H.; Serafim, A.; Stancu, I.C. Coatings based on mucin-tannic acid assembled multilayers. Influence of pH. *Polym. Adv. Technol.* **2020**, *31*, 645–653. [CrossRef]
15. Song, J.; Winkeljann, B.; Lieleg, O. The Lubricity of Mucin Solutions Is Robust toward Changes in Physiological Conditions. *ACS Appl. Bio Mater.* **2019**, *2*, 3448–3457. [CrossRef]
16. Janairo, R.R.R.; Zhu, Y.; Chen, T.; Li, S. Mucin covalently bonded to microfibers improves the patency of vascular grafts. *Tissue Eng. Part A* **2014**, *20*, 285–293. [CrossRef] [PubMed]
17. Co, J.Y.; Crouzier, T.; Ribbeck, K. Probing the Role of Mucin-Bound Glycans in Bacterial Repulsion by Mucin Coatings. *Adv. Mater. Interfaces* **2015**, *2*, 1–6. [CrossRef]
18. Winkeljann, B.; Bauer, M.G.; Marczynski, M.; Rauh, T.; Sieber, S.A.; Lieleg, O. Covalent Mucin Coatings Form Stable Anti-Biofouling Layers on a Broad Range of Medical Polymer Materials. *Adv. Mater. Interfaces* **2020**, *7*. [CrossRef]
19. Shi, L.; Ardehali, R.; Caldwell, K.D.; Valint, P. Mucin coating on polymeric material surfaces to suppress bacterial adhesion. *Colloids Surf. B Biointerfaces* **2000**, *17*, 229–239. [CrossRef]
20. Zhu, M.; Wang, Y.; Ferracci, G.; Zheng, J.; Cho, N.-J.; Lee, B.H. Gelatin methacryloyl and its hydrogels with an exceptional degree of controllability and batch-to-batch consistency. *Sci. Rep.* **2019**, *9*, 6863. [CrossRef]
21. Van Den Bulcke, A.I.; Bogdanov, B.; De Rooze, N.; Schacht, E.H.; Cornelissen, M.; Berghmans, H. Structural and rheological properties of methacrylamide modified gelatin hydrogels. *Biomacromolecules* **2000**, *1*, 31–38. [CrossRef] [PubMed]
22. Kolawole, O.M.; Lau, W.M.; Khutoryanskiy, V.V. Methacrylated chitosan as a polymer with enhanced mucoadhesive properties for transmucosal drug delivery. *Int. J. Pharm.* **2018**, *550*, 123–129. [CrossRef] [PubMed]
23. Yu, L.M.Y.; Kazazian, K.; Shoichet, M.S. Peptide surface modification of methacrylamide chitosan for neural tissue engineering applications. *J. Biomed. Mater. Res. A* **2007**, *82*, 243–255. [CrossRef] [PubMed]
24. Serafim, A.; Olaret, E.; Cecoltan, S.; Butac, L.M.; Balanuca, B.; Vasile, E. Bicomponent Hydrogels Based on Methacryloyl Derivatives of Gelatin and Mucin with Potential Wound Dressing Applications. *Mater. Plast.* **2018**, *55*, 68. [CrossRef]
25. Duffy, C.V.; David, L.; Crouzier, T. Covalently-crosslinked mucin biopolymer hydrogels for sustained drug delivery. *Acta Biomater.* **2015**, *20*, 51–59. [CrossRef]
26. Serafim, A.; Cecoltan, S.; Olaret, E.; Dragusin, D.-M.; Vasile, E.; Popescu, V.; Mastalier, B.S.M.; Iovu, H.; Stancu, I.-C. Bioinspired Hydrogel Coating Based on Methacryloyl Gelatin Bioactivates Polypropylene Meshes for Abdominal Wall Repair. *Polymers* **2020**, *12*, 1677. [CrossRef]
27. Gong, J.P. Why are double network hydrogels so tough? *Soft Matter* **2010**, *6*, 2583–2590. [CrossRef]
28. Gong, J.P. Materials science. Materials both tough and soft. *Science* **2014**, *344*, 161–162. [CrossRef]
29. Wang, J.J.; Wang, Y.; Wang, Q.; Yang, J.; Hu, S.-Q.; Chen, L. Mechanically Strong and Highly Tough Prolamin Protein Hydrogels Designed from Double-Cross-Linked Assembled Networks. *ACS Appl. Polym. Mater.* **2019**, *1*, 1272–1279. [CrossRef]
30. Panteli, P.A.; Patrickios, C.S. Multiply interpenetrating polymer networks: Preparation, mechanical properties, and applications. *Gels* **2019**, *5*, 36. [CrossRef]
31. Maleki, A.; Lafitte, G.; Kjøniksen, A.-L.; Thuresson, K.; Nyström, B. Effect of pH on the association behavior in aqueous solutions of pig gastric mucin. *Carbohydr. Res.* **2008**, *343*, 328–340. [CrossRef] [PubMed]
32. Curnutt, A.; Smith, K.; Darrow, E.; Walters, K.B. Chemical and Microstructural Characterization of pH and [Ca^{2+}] Dependent Sol-Gel Transitions in Mucin Biopolymer. *Sci. Rep.* **2020**, *10*, 8760. [CrossRef] [PubMed]
33. Celli, J.P.; Turner, B.S.; Afdhal, N.H.; Ewoldt, R.H.; McKinley, G.H.; Bansil, R.; Erramilli, S. Rheology of Gastric Mucin Exhibits a pH-Dependent Sol−Gel Transition. *Biomacromolecules* **2007**, *8*, 1580–1586. [CrossRef] [PubMed]
34. Salis, A.; Boström, M.; Medda, L.; Cugia, F.; Barse, B.; Parsons, D.F.; Ninham, B.W.; Monduzzi, M. Measurements and Theoretical Interpretation of Points of Zero Charge/Potential of BSA Protein. *Langmuir* **2011**, *27*, 11597–11604. [CrossRef] [PubMed]
35. Cho, D.; Lee, S.G.; Frey, M.W. Characterizing zeta potential of functional nanofibers in a microfluidic device. *J. Colloid Interface Sci.* **2012**, *372*, 252–260. [CrossRef] [PubMed]
36. Serafim, A.; Tucureanu, C.; Petre, D.-G.; Dragusin, D.-M.; Salageanu, A.; Van Vlierberghe, S.; Dubruel, P.; Stancu, I.-C. One-pot synthesis of superabsorbent hybrid hydrogels based on methacrylamide gelatin and polyacrylamide. Effortless control of hydrogel properties through composition design. *New J. Chem.* **2014**, *38*, 3112–3126. [CrossRef]
37. Kipcak, A.S.; Ismail, O.; Doymaz, I.; Piskin, S. Modeling and investigation of the swelling kinetics of acrylamide-sodium acrylate hydrogel. *J. Chem.* **2014**, *2014*. [CrossRef]
38. Balanuca, B.; Lungu, A.; Hanganu, A.M.; Stan, L.R.; Vasile, E.; Iovu, H. Hybrid nanocomposites based on POSS and networks of methacrylated camelina oil and various PEG derivatives. *Eur. J. Lipid Sci. Technol.* **2014**, *116*, 458–469. [CrossRef]
39. Balanuca, B.; Stan, R.; Hanganu, A.; Lungu, A.; Iovu, H. Design of new camelina oil-based hydrophilic monomers for novel polymeric materials. *JAOCS J. Am. Oil Chem. Soc.* **2015**, *92*, 881–891. [CrossRef]
40. Argenis Caicedo, J.; Perilla, J.E. Effect of pH on the rheological response of reconstituted gastric mucin. *Ing. Investig.* **2015**, *35*, 43–48. [CrossRef]

41. Hoffman, A.S.; Ratner, B.D. Synthetic Hydrogels for Biomedical Applications—A Review. *Am. Chem. Soc. Polym. Prepr. Div. Polym. Chem.* **1975**, *16*, 272–275.
42. Nisar, S.; Pandit, A.H.; Wang, L.-F.; Rattan, S. Strategy to design a smart photocleavable and pH sensitive chitosan based hydrogel through a novel crosslinker: A potential vehicle for controlled drug delivery. *RSC Adv.* **2020**, *10*, 14694–14704. [CrossRef]
43. Jayaramudu, T.; Ko, H.-U.; Kim, H.C.; Kim, J.W.; Kim, J. Swelling Behavior of Polyacrylamide–Cellulose Nanocrystal Hydrogels: Swelling Kinetics, Temperature, and pH Effects. *Materials* **2019**, *12*, 2080. [CrossRef] [PubMed]
44. Saidi, M.; Dabbaghi, A.; Rahmani, S. Swelling and drug delivery kinetics of click-synthesized hydrogels based on various combinations of PEG and star-shaped PCL: Influence of network parameters on swelling and release behavior. *Polym. Bull.* **2020**, *77*, 3989–4010. [CrossRef]
45. Li, X.; Sun, Q.; Li, Q.; Kawazoe, N.; Chen, G. Functional hydrogels with tunable structures and properties for tissue engineering applications. *Front. Chem.* **2018**, *6*. [CrossRef] [PubMed]
46. Zhao, D.; Huang, J.; Zhong, Y.; Li, K.; Zhang, L.; Cai, J. High-Strength and High-Toughness Double-Cross-Linked Cellulose Hydrogels: A New Strategy Using Sequential Chemical and Physical Cross-Linking. *Adv. Funct. Mater.* **2016**, *26*, 6279–6287. [CrossRef]
47. Mandal, B.B.; Kapoor, S.; Kundu, S.C. Silk fibroin/polyacrylamide semi-interpenetrating network hydrogels for controlled drug release. *Biomaterials* **2009**, *30*, 2826–2836. [CrossRef] [PubMed]
48. Amani, H.; Arzaghi, H.; Bayandori, M.; Dezfuli, A.S.; Pazoki-Toroudi, H.; Shafiee, A.; Moradi, L. Controlling Cell Behavior through the Design of Biomaterial Surfaces: A Focus on Surface Modification Techniques. *Adv. Mater. Interfaces* **2019**, *6*, 1900572. [CrossRef]
49. Youssefian, S.; Jakes, J.E.; Rahbar, N. Variation of Nanostructures, Molecular Interactions, and Anisotropic Elastic Moduli of Lignocellulosic Cell Walls with Moisture. *Sci. Rep.* **2017**, *7*, 2054. [CrossRef]

polymers

MDPI

Article

Extraction and Physicochemical Characterization of Dried Powder Mucilage from *Opuntia ficus-indica* Cladodes and Aloe Vera Leaves: A Comparative Study

María Carolina Otálora [1,*], Andrea Wilches-Torres [1] and Jovanny A. Gómez Castaño [2,*]

[1] Grupo de Investigación en Ciencias Básicas (NÚCLEO), Facultad de Ciencias e Ingeniería, Universidad de Boyacá, Tunja, Boyacá 050030, Colombia; andreawilches@uniboyaca.edu.co

[2] Grupo Química-Física Molecular y Modelamiento Computacional (QUIMOL®), Facultad de Ciencias, Universidad Pedagógica y Tecnológica de Colombia (UPTC), Avenida Central del Norte, Tunja, Boyacá 050030, Colombia

* Correspondence: marotalora@uniboyaca.edu.co (M.C.O.); jovanny.gomez@uptc.edu.co (J.A.G.C.)

Abstract: Cactaceae and Asphodelaceae are native desert plants known for their high mucilage content, which is a polysaccharide of growing interest in the food, cosmetic, and pharmaceutical industries. In this study, powdered mucilage was obtained from cladodes of *Opuntia ficus-indica* (OFI) and aloe vera (AV) leaves, and their molecular, morphological, and thermal properties were investigated and compared. Additionally, their dietary fiber content was determined. Three-dimensional molecular models were calculated for both mucilages using ab initio methods. Vibrational spectra (FTIR and Raman) revealed intramolecular interactions and functional groups that were specified with the help of theoretical ab initio and semi-empirical calculations. SEM micrographs measured at magnifications of 500× and 2000× demonstrated significantly different superficial and internal morphologies between these two mucilages. Thermal analysis using DSC/TGA demonstrated superior thermal stability for the OFI mucilage. The dietary fiber content in OFI mucilage was more than double that of AV mucilage. Our results show that both dehydrated mucilages present adequate thermal and nutritional properties to be used as functional ingredients in industrial formulations; however, OFI mucilage exhibited better physicochemical and functional characteristics than AV mucilage as a raw material.

Keywords: mucilage; pectin polysaccharide; *Opuntia ficus-indica*; aloe vera; acemannan; Cactaceae; Asphodelaceae

Citation: Otálora, M.C.; Wilches-Torres, A.; Castaño, J.A.G. Extraction and Physicochemical Characterization of Dried Powder Mucilage from *Opuntia ficus-indica* Cladodes and Aloe Vera Leaves: A Comparative Study. *Polymers* **2021**, *13*, 1689. https://doi.org/10.3390/polym13111689

Academic Editor: Arn Mignon

Received: 26 April 2021
Accepted: 18 May 2021
Published: 22 May 2021

1. Introduction

Mucilages are complex polymeric substances that are carbohydrates in nature and have garnered considerable interest in the food, pharmaceutical, and cosmetic industries for their functional, health, and nutritional benefits. Their composition is mainly based on branched polysaccharide macromolecules that form intricate molecular networks capable of retaining large amounts of water, making them a potential source of natural hydrocolloids that provide a substantial thickening effect that is desirable for the chemical and cosmetic industries [1,2]. They have medicinal attributes such us wound, burn, and ulcer healing, as well as antidiabetic and antiglycation effects [3,4]. Other mucilage applications include its use in foods as a stabilizer, flavoring agent, fat substitute [5], and edible coating to extend the useful life of fruit [6], among others [7,8].

For centuries the plants belonging to the Cactaceae and Asphodelaceae botanical families have been recognized for their abundant content of mucilage (commonly called gel) [9]. Among the most representative members of these families are the nopal cactus (*Opuntia ficus-indica* (L.) Mill) and the aloe vera plant (*Aloe barbadensis* Miller), the mucilages of which are widely renowned because of their varied nutritional, cosmetic, and medicinal properties. Mucilage obtained from *O. ficus-indica* (OFI mucilage) has been reported to be able to lower cholesterol levels [10,11] and exhibit hypoglycemic effects [12] as well as antiulcer [13], hepatoprotective, antigenotoxic, and cardioprotective pharmacological effects [14]. OFI mucilage is also considered a potent antioxidant due to its carbohydrate and polyphenol content [15,16]. Additionally, a high total dietary fiber content, between 57 and 72 g/100 g dry matter, has been reported for OFI mucilage [17–19].

On the other hand, numerous biological activities, including antiviral, antibacterial, laxative, radiation protection, wound healing, antioxidant, anti-inflammatory, anticancer, antidiabetic, antiallergic, and immunostimulatory effects, among others, have been attributed to the aloe vera mucilage (AV mucilage) [20–23].

The mucilage extracted from the cladodes of *Opuntia ficus-indica* is a heteropolysaccharide of an anionic polyelectrolyte nature, with a molar mass of 3×10^6 g/mol, containing L-arabinose (24.6–42.0%), D-galactose (21.0–40.1%), D-xylose (22.0–22.2%), L-rhamnose (7.0–13.1%), and α-D-(1 $\rightarrow$ 4) galacturonic acid (8.0–12.7%), as well as insoluble (lignin and polysaccharides) and soluble (neutral sugars and uronic acid) fiber [5]. This natural hydrocolloid has garnered growing industrial interest due to its edible, biodegradable, non-toxic characteristics, and its cost effectiveness [24]. Its potential applications in the food, pharmaceutical, and cosmetic industries include its use as thickening, binding, emulsifying, stabilizing, and gelling agent [25]; film-former [26]; and as wall material for encapsulation processes [18,19,27]. Previous studies have shown that OFI mucilage is considered a rich source in polyunsaturated fatty acid, especially α-linolenic (C18: 3 ω-3) and linoleic (C18: 2 ω-6), which is essential for human nutrition [16].

On the other hand, the mucilage extracted from the hydroparenchyma of the succulent leaves of aloe vera is a biopolymer mainly constituted by acemannan and pectic substances [28]. The acemannan polysaccharide, with a molecular weight of around 40–50 kDa, contains partially acetylated mannose units (>60%), glucose ($\sim$20%), and galactose (<10%) [29,30]. Previous studies have shown that the AV mucilage is considered a rich source in dietary fiber (35.5%) [28] as well as a potent antioxidant due to its aloe-emodin content [31]. This hydrocolloid also has potential use in the food industry as an edible coating [32–34], biodegradable film [35], and encapsulating agent [36].

Most of the reports on OFI and AV mucilages have focused on the biological properties and possible applications of these biopolymers in their fresh form or in hydrated preparations, while much less is known about their chemical and physical characteristics, as well as the possible uses of their concentrated powder. This study focuses on comparatively evaluating the structural, morphological, and thermal properties, as well as the total dietary fiber content of hot air-dried mucilages extracted from OFI cladodes and AV leaves. These two mucilages have been particularly selected for the present study given their potential usefulness as functional ingredients in formulations for nutritional, health, medicinal, and cosmetic purposes [37], as well as the scalability of the cultivation of OFI and AV species in arid and semi-arid areas [1,2]. The results of this study seek to contribute to the determination of the structural and thermal behavior of these powdered mucilages, which constitute a fundamental factor in guaranteeing their chemical stability within a food, cosmetic, or pharmacological matrix until their consumption or final application.

2. Materials and Methods

2.1. Vegetal Materials

Cladodes of *Opuntia ficus-indica* were collected in the rural jurisdiction of the city of Duitama in the Department of Boyacá, Colombia. Aloe vera leaves were purchased at a local supermarket in the city of Tunja in the Department of Boyacá, Colombia. The size of the leaves ranged between 30 and 40 cm in length.

2.2. Mucilage Extraction

For the extraction of mucilage from cladodes of *Opuntia ficus-indica*, the methodology reported by Quinzio et al. [38] was used with some modifications, as explained below. The cladodes were washed with distilled water, transversally cut with a Teflon knife, and manually peeled. The cladode outer layer was removed, leaving only the white and inner part (i.e., the medulla). The medulla was cut into small pieces and placed in a 1000 mL beaker. Distilled water at room temperature was added to the beaker in a 1:2 *v/v* (medulla:water) ratio, and the mixture was left for 24 h. According to the literature [1,39–41], these extraction parameters ensure maximum viscosity and high yield of the extracted mucilage. The hydrated medulla pieces were manually squeezed through a nylon cloth. The obtained gel was centrifuged at 10,000 rpm for 30 min, and then 95% ethanol (provided by Merck, Darmstadt, Germany) was added in a ratio of 3:1 (ethanol/centrifuged gel) at room temperature for the precipitation of the mucilage. Extraction with ethanol was used to dissolve the chlorophyll present in the pads and produce a white mucilage powder, that is, with a presentation similar to the other gums used in the food industry [5]. The precipitate was dried in Petri dishes at 105 °C in an oven (UM 400, Memmert, Schwabach, Germany) for 24 h. The dried material was manually macerated in a porcelain mortar until a fine powder was obtained. The OFI powdered mucilage was placed in high-density polyethylene bags and stored in a desiccator at room temperature with a relative humidity of 30% until use.

The aloe vera leaves were washed with distilled water at room temperature, and then the epidermis was carefully separated from the inner pulp using a Teflon knife. The pulp filets were cut into small pieces and manually squeezed to extract the leaf gel. The gel obtained was filtered through a nylon-cloth. Then 95% ethanol was added at room temperature to the filtered gel in an ethanol:gel ratio of 3:1 *v/v*, and then the mixture was stirred manually with a glass rod until the appearance of a white-milky gel (mucilage) was present. The mucilage gel was placed in Petri dishes and dried at 60 °C in an oven (UM 400, Memmert, Schwabach, Germany) for 24 h. According to literature [42], at a drying temperature of 60–70 °C a high quality aloe vera gel is obtained, that is, with minor alterations in its physicochemical and nutritional properties. The dried material was manually macerated in a porcelain mortar until a fine powder was obtained. The AV powdered mucilage was placed in high-density polyethylene bags and stored in a desiccator at room temperature with a relative humidity of 30% until use.

2.3. Structural Characterization

2.3.1. Fourier-Transform Infrared (FTIR) Spectroscopy

The infrared spectra of the powdered mucilages of *Opuntia ficus-indica* and aloe vera were recorded on a Shimadzu Prestigie 21 spectrophotometer (Duisburg, Germany) equipped with a Michelson-type interferometer, a KBr/Ge beam-splitter, a ceramic lamp, and a DLATGS detector. The FTIR spectra were measured in the range of 4500–4520/cm with a resolution of 3.0/cm and 30 cumulative scans using the attenuated total reflectance/reflection (ATR) technique.

2.3.2. Raman Spectroscopy

The surface composition of the *Opuntia ficus-indica* and aloe vera powdered mucilages were analyzed using a Raman spectrophotometer (DXR™ Smart Raman, Thermo Scientific, Waltham, MA, USA) equipped with a 785 nm excitation diode laser. Spectra were measured with an average scan time of 1.0 s using a laser power of 20.0 mW. A total of 20 scans per spectra was performed in order to improve the signal-to-noise ratio.

2.3.3. Scanning Electron Microscopy (SEM)

The microscopic morphology of the powdered mucilages of *Opuntia ficus-indica* and aloe vera was evaluated by scanning electron microscopy (SEM) using EVO MA 10-Carl Zeiss equipment (Oberkochen, Germany) operating at 20 kV. All samples were coated by gold–palladium sputtering before their examination.

2.4. Thermal Properties

Thermogravimetric analysis (TGA)/differential scanning calorimetry (DSC) was performed on a TA Instrument (SDT Q600 V20.9 Build 20, New Castle, DE, USA). Argon was used as a purge gas (100 mL/min). The dried samples of OFI and AV mucilages were placed in aluminum pans and heated from 20 to 600 °C at a heating rate of 10 °C/min.

2.5. Dietary Fiber Content

Total dietary fiber (TDF) content in *Opuntia ficus-indica* and aloe vera powdered mucilage samples was determined using a total dietary fiber test kit (TDF-100A), provided by Sigma Aldrich (St. Louis, MO, USA), which is based on the enzymatic–gravimetric method AOAC 985.29 [43]. The dry, fat-free mucilage powder was gelatinized with the thermostable α-amylase and then enzymatically digested with protease and amyloglucosidase to remove protein and starch. The soluble fiber was precipitated with ethanol; then the residues were washed with ethanol and acetone, filtered, dried, and weighed. The dry material was determined for protein and ash content. Total dietary fiber (TDF) was calculated as the residue minus the weight of protein and ash and was expressed as g/100 g of powder.

2.6. Molecular Models and Quantum Chemical Calculations

The 3D molecular models of the mucilages were initially built in the Chem3D Pro 12.0 program [44] taking into consideration the three-dimensional conformations of the monosaccharide structures as reported in the PubChem database of the National Center for Biotechnology Information (NCBI) [45] and their preliminary minimized structures using the force field calculation method MM2 [46] as implemented in Chem3D Pro 12.0.

All quantum chemical calculations were performed using the Gaussian 09 software package (Rev. A.02) [47], while molecular models and simulated IR and Raman spectra were visualized through the GaussView 6.0.16 graphical user interface [48]. The minima structures and vibrational harmonic frequencies for acemannan and OFI mucilage polysaccharides were calculated in the gas phase using the HF/6-31(d) and the HF/6-31(d)//PM6 approximation levels of theory, respectively, assuming a temperature of 298.15 K and 1 atmosphere of pressure. The equilibrium structures were confirmed as minima in the potential energy surface by the absence of imaginary frequencies.

3. Results and Discussion

3.1. Molecular Modelling of the Isolated Dehydrated OFI and AV Mucilages

To achieve a representative isolated molecular model for the OFI mucilage, a branched biopolymeric structure was built and made up of a total of 25 sugar residues that were linked according to the type and arrangement of the glycosidic bonds reported for this heteropolysaccharide [49]. The sugar residues used were β-D-galactopyranose (Gal-β), β-L-rhamnopyranose (Rha-β), and α-D-galactopyranosyl uronic acid (Gal-Ac-α) in a proportion of 60, 20, and 20%, respectively. The central chain consisted of 10 residues with intercalated Gal-Ac-α(1→2)Rha-β(1→4) glycosidic bonds, while five branched chains, each one made up of Gal-β(1→6)Gal-β(1→6)Gal-β glycosidic bonds, were connected through Gal-β(1→4)Rha-β bonds to the central chain. A two-dimensional representation of the molecular model for the OFI mucilage is presented in Figure 1.

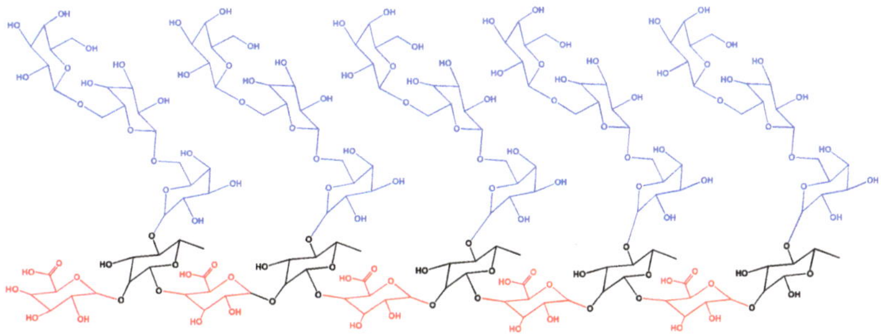

Figure 1. Two-dimensional representation used for the construction of the molecular model of the isolated OFI mucilage. Gal-β (in blue), Rha-β (in black), and Gal-Ac-α (in red).

To approximate the molecular model for the dry AV mucilage, an isolated structure of acemannan was built considering the type and disposition of the glycosidic bonds reported for this polysaccharide [30]. In particular, the base structure chosen for building this model corresponded to the carrisyn (acemannan) molecule (MF: $C_{66}H_{100}NO_{49}$, MW: 1691.5 g/mol), as reported in the PubChem database under the compound CID number 72041 [45]. This molecule is a representative eight-sugar fragment of the acemannan typically found in the AV leaf mucilage, which is comprised of α-D-galacto mannan, α-D-galacto mannan acetate, and N-acetyl-D-glucosamine residues linked by α-1,4-glycosidic bonds. A two-dimensional representation used for the molecular modelling of acemannan is presented in Figure 2.

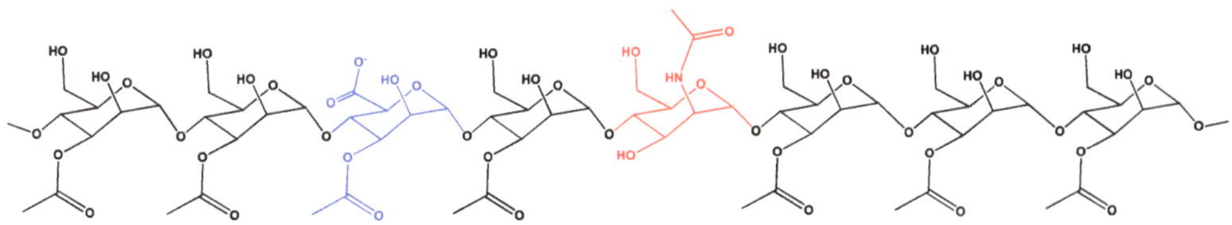

Figure 2. Two-dimensional representation used for the construction of the molecular model of the isolated acemannan (carrisyn). Sugar residues: α-D-galacto mannan (in black), α-D-galacto mannan acetate (in blue), and N-acetyl-D-glucosamine (in red).

Figure 3 shows the 3D structures of the isolated molecules for the OFI mucilage and acemannan, optimized at the HF/6-31(d) approximation level. The stationary structure of the isolated OFI mucilage (Figure 3a) adopted a three-dimensional arrangement that resembles the shape of a "starfish". Each of the five "extremities" in this structure corresponds to one of the Gal-β(1→6)Gal-β(1→6)Gal-β branched chains, which fold into the inner part, while the central chain forms a circle that is located in the outermost non-folded part. The folding on this structure is largely favored by the formation of intramolecular hydrogen bonds (HBs) between the carbonyl oxygen of the carboxyl group in L-rhamnopyranose central residues and hydroxyl oxygens (>*C*=*O*···*H*–*O*, $2.164 \leq d \leq 2.328$ Å) or aliphatic hydrogens (>*C*=*O*···*H*–*C*, $2.278 \leq d \leq 2.377$ Å) in D-galactopyranose branched-chain residues. A cavernous form, added to a high concentration of hydroxyl (OH) and ether (ROR') groups in the folded part of this structure, together with the presence of some carbonyl groups (*C*=*O*), constitute a reticular system capable of coordinating a high number of water molecules, which explains the reservoir properties of the OFI mucilage.

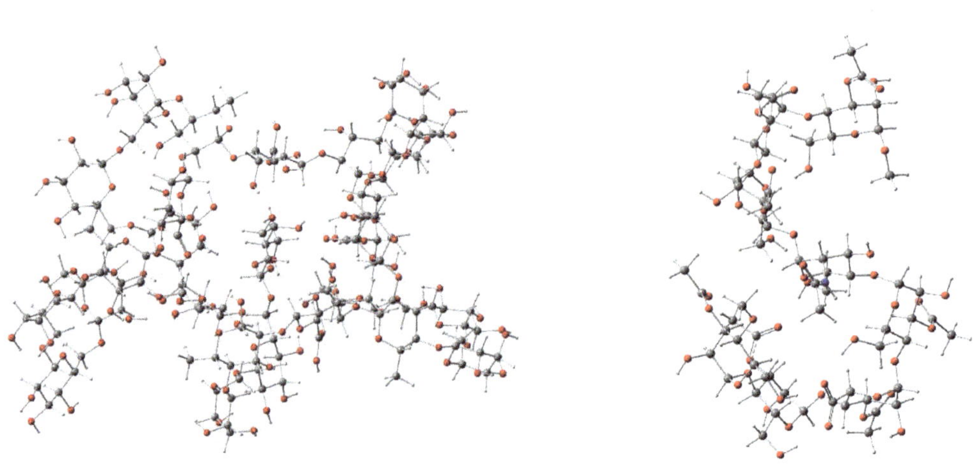

(a) (b)

Figure 3. Stationary structures for the isolated molecules of OFI mucilage (**a**) and acemannan (**b**) minimized using the HF/6-31(d) level of approximation.

On the other hand, the optimized isolated structure of acemannan acquired a wavy spatial orientation in the shape of the letter "s", as shown in Figure 3b. The folding in this structure is governed by the formation of intramolecular HBs between the carbonyl oxygen and methyl hydrogens in acetyl groups (*>C=O···H–CH*$_2$*–*, $2.402 \leq d \leq 2.843$ Å) or with hydroxyl oxygens (*>C=O···H–O*, $d = 1.931$ Å), as well as a sequence of HBs between hydroxyl groups (*–(H)O···H–O*, $1.998 \leq d \leq 2.101$ Å). The oxygens atoms of the carboxylate group in particular form a set of short HBs ($1.884 \leq d \leq 1.970$ Å) with hydroxyl hydrogens (*OC=O···H–O*) of the vicinal D-galacto mannan residues, whereas the amide hydrogen forms an HB with the hydroxyl group of another D-galacto mannan residue (*N–H···O(H)–*, $d = 2.319$ Å). As in the case of OFI mucilage, the optimized structure of acemannan exhibits a high capacity to retain water molecules.

3.2. Fourier-Transform Infrared (FTIR) Spectroscopy

The experimental FTIR spectra of OFI and AV mucilages measured in powder, along with the respective theoretical gas-phase IR spectra, are shown in Figure 4. The OFI mucilage FTIR spectrum (Figure 4a) showed a broad band centered at 3350/cm that was attributed to a combined contribution of OH stretching modes from both alcohol (*R–OH*) and carboxylic acid (*–C(O)–OH*) moieties involved in the intramolecular OH bonding, while the absorption observed at 2895/cm was assigned to a sum of *C–H* stretches from both pyranose *CH* and glycoside $-OCH_2$ moieties [26,50]. The absorptions observed at 1728/cm (shoulder) and 1609/cm, as well as the band at 1402/cm, were assigned to the carbonyl (υ(*C=O*)) and *COO–* stretching modes, respectively, of the D-galactopyranosyl uronic acid residues [51]. The signals at 1315 and 1246/cm were attributed to characteristic stretching vibrations of the pyranose ring, whereas the intense band at 1040/cm was related to the polysaccharide backbone [52]. As shown in Figure 4a, the comparison between the experimental (in powder) FTIR spectrum and the theoretical (in gas-phase) IR spectra of OFI mucilage demonstrated positive vibrational shifting values ($+\Delta\upsilon$) between 275 and 161 cm^{-1} for signals in the fingerprint region (2000 to 500 cm^{-1}) and negative $\Delta\upsilon$ values for absorptions located at higher frequencies. In particular, the semi-empirical PM6 method revealed a very poor estimation ($\Delta\upsilon = -770$ cm^{-1}) for the hydroxyl (*O–H*) harmonic frequencies, while the CH vibrations were the best approximate ($\Delta = -99$ cm^{-1}) in this macromolecule.

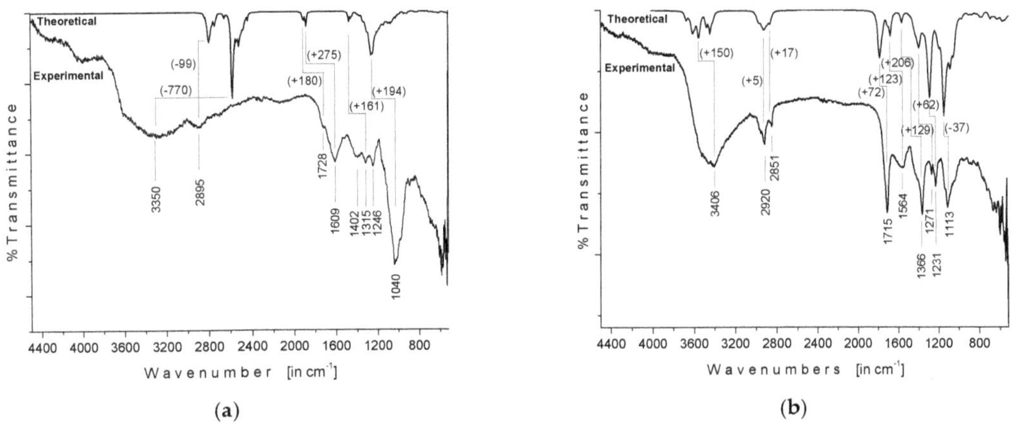

Figure 4. Experimental (in powder) and theoretical (in gas phase) IR spectra of OFI and AV mucilages: (**a**) FTIR spectrum of OFI mucilage versus the respective theoretical (unscaled) IR spectrum calculated at the HF/6-31(d)//PM6 level of approximation; (**b**) FTIR spectrum of AV mucilage versus the theoretical IR spectrum (scaled at factor 0.8953 [51]) of acemannan calculated at the HF/6-31-G(d) level of approximation. Theoretical vibrational shifting ($\Delta\upsilon$ in cm^{-1}) are presented in parenthesis.

The FTIR spectrum of AV mucilage in powder (Figure 4b) showed a broad band centered at 3406/cm that was attributed to the combined presence of multiple hydroxyl groups. According to our ab initio calculations (HF/6-31(d)), the greatest contribution to this band can be attributed to the OH stretching modes belonging to the sequential chain of intramolecular (H)*O*···*H–O* bonds in acemannan, as explained above in Section 3.1. The absorptions observed at 2920 and 2851/cm were assigned to the aliphatic *C–H* stretching. The signals at 1715 and 1564/cm were attributed to *C=O* stretching and *COO–* asymmetric stretching, respectively, indicating the presence of carbonyl compounds (acemannan content). The bands at 1366, 1271, and 1231/cm were related to the asymmetric CH_3 scissoring, the bending (*C–O–C*) glycosidic symmetric stretching vibration, and the *C–O–C* stretching of acetyl groups, respectively, with the latter related to the bioactive acetylated polysaccharide acemannan. Finally, the absorption observed at 1113/cm was assigned to the *C–O* stretching [53,54]. As can be seen in Figure 4b, the comparison between the IR spectrum of the experimental AV mucilage and the theoretical spectrum of the isolated acemannan revealed a close correlation between them, which showed a high composition of acemannan in the dehydrated sample. Likewise, the theoretical spectrum (scaled at factor 0.8953, [50]) presented an overestimation of less than 200 cm^{-1} for all absorptions, which indicates a good vibrational prediction for this molecule using the approximation level HF/6-31(d).

3.3. Raman Spectroscopy

The Raman spectra of OFI and AV-powdered mucilages, along with the calculated (HF/6-31G(d)) Raman spectrum for isolated acemannan in the 100–3400/cm spectral region are shown in Figure 5a–c. Both OFI and AV mucilage's spectra presented a similar well-resolved pattern with strong signals in the 2800–3000/cm range and medium-intensity signals in the 1000–1500/cm range. As is also noted in Figure 5, an excellent agreement was obtained when comparing the experimental Raman spectrum of the AV mucilage in powder (Figure 5b) and the Raman spectrum calculated for the acemannan molecule (Figure 5c), thus reaffirming the high content of acemannan in the sample of dry AV mucilage, and allowing the possibility of making more specific assignments of the signals in the Raman spectra. Consequently, the three narrow signals observed in both mucilage spectra in the υ(*C–H*) region around 2940, 2890, and 2850/cm were associated with the asymmetric mode of the hydrogen attached to the anomeric carbons, the asymmetric mode of the hydrogens of the pyranose ring, and the symmetric mode of the methyl hydrogens, respectively. The two sharp signals observed close to 1560 and 1540/cm in both mucilage's Raman spectra were attributed to the symmetric deformations, ρ(CH_2), in $-CH_2OH$ moieties, and ρ(CH_3) in acetyl groups, respectively. The acute signal around 1300/cm was assigned to the υ(C–C) stretching mode in pyranose rings, whereas the two narrow signals around 1120 and 1050/cm were assigned to a complex motion involving the polysaccharide backbone and the υ(C–O) mode in the pyranose ring, respectively.

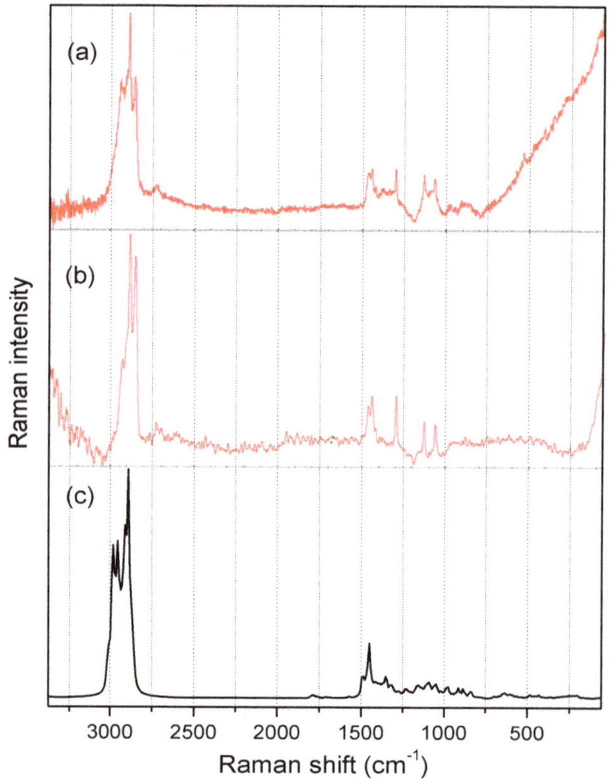

Figure 5. Raman spectra of powdered OFI (**a**) and AV (**b**) mucilage in the range of 3400 to 100/cm. The theoretical Raman spectrum for the isolated acemannan molecule (**c**), calculated at the HF/6-31G(d) approximation level and scaled at factor 0.8953 [50], is included for comparative purposes.

3.4. Microscopic Morphology

Figure 6 shows SEM micrographs of the surface and the internal structures of the powered samples for OFI and AV mucilages obtained by a forced air convection drying process. The OFI mucilage sample observed with a magnification of 500× showed an irregular, compact, dense, and cracked surface that coincides with the morphological descriptions given previously for this kind of mucilages [27]. In contrast, the AV mucilage surface observed at 500× exhibited a rough morphology characterized by the presence of scales and pores.

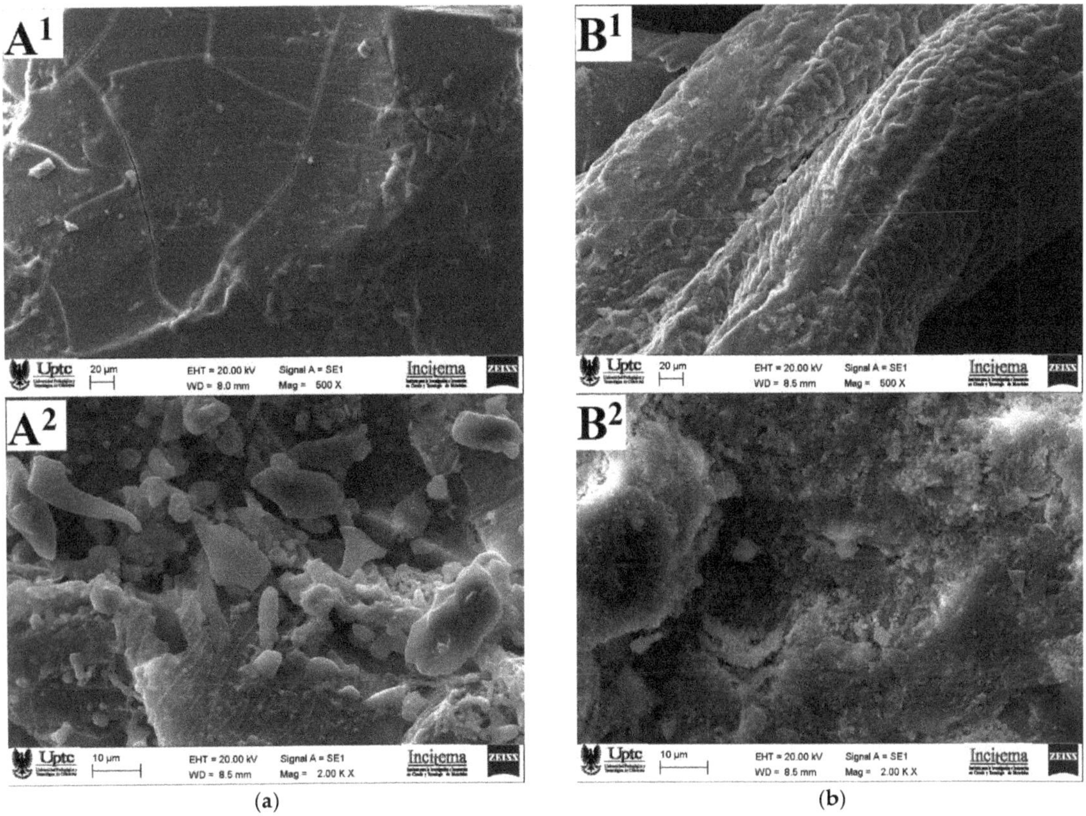

Figure 6. SEM micrograph images of the surface at 500× (1) and of the internal structure at 2000× (2) for mucilage of OFI (**a**) and AV (**b**) in powder.

Regarding the internal mucilage structures observed using a magnification of 2000×, the OFI sample showed the presence of small particles with diameters ranging between 2 and 10 μm that possibly correspond to protein aggregates adhered to the carbohydrate blocks, whereas the AV sample showed a heterogeneous slightly rough internal morphology with the presence of cavities. This last microstructure resembled the morphology recently described for mucilage samples extracted from the rhizomes of Taro (*Colocasia esculente*) by means of an ethanol precipitation [55]. The morphological differences observed between OFI and AV mucilages may be associated with the different conditions used during extraction processes, flocculation during ethanol precipitation, and sample dryness [56].

3.5. Thermal Properties

The thermal behavior of the OFI and AV mucilage powder samples was analyzed by DSC/TGA, resulting in the thermograms shown in Figure 7. The thermogram for OFI mucilage (Figure 7a) was characterized by two principal thermal events. First, an endothermic event occurred between 50 and 250 °C (peak 108.11 °C) with a related mass loss of 11.94%. This event was attributed to water loss followed by gelatinization. This moisture content was attributed to the hygroscopicity of the powdered mucilage. The second event corresponded to an exothermic process that occurred between 250 and 325 °C (peak 300 °C), with a mass loss of 65.59%. This event was attributed to the degradation of the polysaccharide structure and the subsequent decomposition/volatilization of material [18]. Similar thermal behavior has been previously reported for OFI mucilage samples [27].

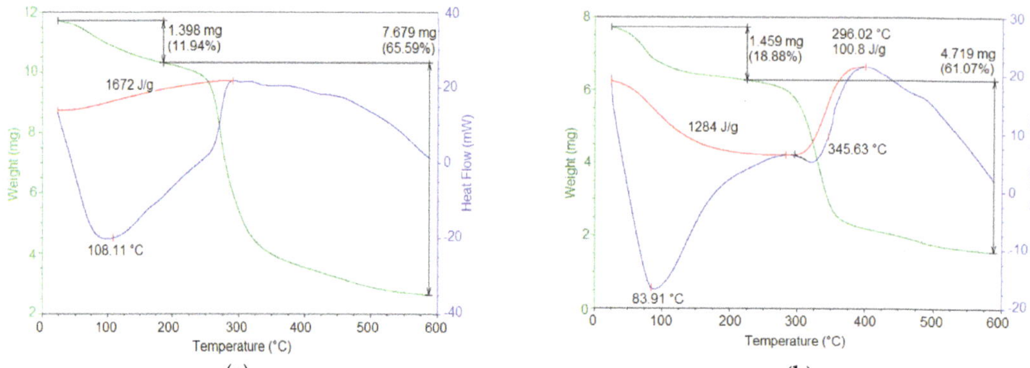

Figure 7. Representative TGA/DSC thermograms of OFI (**a**) and AV (**b**) mucilage samples in powder.

On the other hand, the AV mucilage thermogram (Figure 7b) showed two endothermic events and one exothermic event. The first endothermic event occurred between 60 and 200 °C (peak 82 °C) and was attributed to the glass transition temperature. This event was accompanied by a related mass loss of 18.88% due to the loss of water (absorbed by hygroscopicity) and the decomposition of structural units. The second endothermic event was observed between 210 and 400 °C (peak 310 °C), which was assigned to the melting temperature. This event was accompanied by a related mass loss of 61.07% due to the thermal polymer degradation. The exothermic event occurred between 375 and 425 °C (peak 400 °C), which was attributed to a new crystallinity. This behavior was similar to that reported by Agha Mohammadi et al. for AV gel-dehydrated powder [57].

As described above, the AV mucilage showed lower thermal stability, as well as a lower melting and degradation temperature, compared to the OFI mucilage. This difference in thermal behavior between these two mucilages is related to the lower molecular weight and the hydrophilic nature of the functional groups of the AV mucilage biopolymer. These results revealed that OFI mucilage has a potential use in foods processed at high temperatures, while AV mucilage must be added to industrial processes at milder temperatures.

3.6. Dietary Fiber Content

The total dietary fiber content in the dry samples of OFI and AV mucilage was 73.9 g/100 g of powder and 37.0 g/100 g of powder, respectively. The greater dietary fiber content in OFI mucilage, resulted in its higher thermal stability and provides this biopolymer the advantage of being used in nutritional preparations with more potent prebiotic potential and better positive effects on the immune system [58]. Therefore, OFI mucilage should be considered as an optimal raw biomaterial to reduce the sugar and fat content in certain products.

4. Conclusions

Powder samples of mucilage from cladodes of *Opuntia ficus-indica* (OFI) and aloe vera (AV) leaves were extracted using precipitation in ethanol and dried by means of forced air convection techniques. Molecular characterization of these solid samples was performed using FTIR and Raman spectroscopy in combination with the execution of quantum-chemical calculations based on ab initio and semi-empirical theoretical methods. Computational modeling at the HF/6-31(d) approximation level, built using characteristic sugar residues and glycosidic connections found in this type of polysaccharide, led to representative three-dimensional stationary molecular structures for both mucilages. The isolated molecular models for OFI and AV mucilages adopted intricate structures, resembling the shape of a starfish and a wavy ribbon, respectively, both with an intrinsic capacity to coordinate a high number of water molecules. Such structures were rationalized based on different types of intramolecular hydrogen bonds, the formation of which was supported with the assignment of vibrational signals obtained by FTIR and Raman spectroscopies. The study of the microscopic morphology of the powdered mucilage samples using SEM revealed significant differences in the surface and internal structures of the OFI and AV samples. Such morphological changes were attributed to the different experimental conditions used during the extraction and drying processes of the mucilages. The OFI and AV mucilages obtained have a nutritional value due to the presence of total dietary fiber and polysaccharides, as revealed by enzymatic–gravimetric method and the infrared spectroscopy, respectively. According to the DSC/TGA and fiber content analysis, solid OFI mucilage had a greater thermal stability and dietary content, which makes this heteropolysaccharide a more convenient ingredient in medium–high temperature industrial preparations and with a greater nutritional benefit than powder AV mucilage. Nevertheless, it is worth mentioning that both OFI and AV powdered mucilages can offer high gelling and thickening capacities as well as various nutritional and health benefits. These benefits make them extremely desirable natural ingredients for the cosmetic, pharmaceutical, and food industries. Consequently, the powdered mucilage extracted from both OFI cladodes and AV leaf has the potential to replace starch in the formulation of beverages, soups, bakery products, and meat derivatives. Likewise, the incorporation of these mucilages rich in dietary fiber in food matrices will allow their consideration in products with the "healthy label", which will benefit the health of the consumer.

Author Contributions: Data curation, M.C.O. and J.A.G.C.; formal analysis, J.A.G.C. and M.C.O.; investigation, A.W.-T. and M.C.O.; project administration, M.C.O.; writing (original draft preparation), M.C.O.; writing (review and editing), J.A.G.C. and M.C.O. All authors have read and agreed to the published version of the manuscript.

Funding: This work was funded by the Universidad de Boyacá and the Universidad Pedagógica y Tecnológica de Colombia through the interinstitutional project SGI 2384 of the Vicerrectoría de Investigaciones of the Universidad Pedagógica y Tecnológica de Colombia.

Institutional Review Board Statement: Not applicable.

Informed Consent Statement: Not applicable.

Data Availability Statement: The data presented in this study are available on request from the corresponding author.

Acknowledgments: The authors greatly acknowledge the financial support provided by the Universidad de Boyacá and the Universidad Pedagógica y Tecnológica de Colombia.

Conflicts of Interest: The authors declare that there is no conflict of interest.

References

1. Sepulveda, E.; Sáenz, C.; Aliaga, E.; Aceituno, C. Extraction and characterization of mucilage in *Opuntia* spp. *J. Arid Environ.* **2007**, *68*, 534–545. [CrossRef]
2. Monrroy, M.; García, E.; Ríos, K.; García, J.R. Extraction and Physicochemical Characterization of Mucilage from *Opuntia cochenillifera* (L.) Miller. *J. Chem.* **2017**, *2017*, 4301901. [CrossRef]
3. El-Mostafa, K.; El Kharrassi, Y.; Badreddine, A.; Andreoletti, P.; Vamecq, J.; El Kebbaj, M.H.; Latruffe, N.; Lizard, G.; Nasser, B.; Cherkaoui-Malki, M. Nopal Cactus (*Opuntia ficus-indica*) as a Source of Bioactive Compounds for Nutrition, Health and Disease. *Molecules* **2014**, *19*, 14879–14901. [CrossRef] [PubMed]
4. Hamman, J.H. Composition and Applications of *Aloe vera* Leaf Gel. *Molecules* **2008**, *13*, 1599–1616. [CrossRef]
5. Sáenz, C.; Sepulveda, E.; Matsuhirob, B. Opuntia spp mucilage's: A functional component with industrial perspectives. *J. Arid Environ.* **2004**, *57*, 275–290. [CrossRef]
6. Del-Valle, V.; Hernandez-Muñoz, P.; Guarda, A.; Galotto, M.J. Development of a cactus-mucilage edible coating (*Opuntia ficus indica*) and its application to extend strawberry (*Fragaria ananassa*) shelf-life. *Food Chem.* **2005**, *91*, 751–756. [CrossRef]

7. Rachdi, R.; Srarfi, F.; Slim Shimi, N. Cactus Opuntia as natural flocculant for urban wastewater treatment. *Water Sci. Technol.* **2017**, *76*, 1875–1883. [CrossRef] [PubMed]
8. Pichler, T.; Young, K.; Alcantar, N. Eliminating turbidity in drinking water using the mucilage of a common cactus. *Water Supply* **2012**, *12*, 179–186. [CrossRef]
9. Gebresamuel, N.; Gebre-Mariam, T. Comparative Physico-Chemical Characterization of the Mucilages of Two Cactus Pears (*Opuntia* spp.) Obtained from Mekelle, Northern Ethiopia. *J. Biomater. Nanobiotechnol.* **2012**, *3*, 79–86. [CrossRef]
10. Gharras, H.E. Cactus stems (*Opuntia* spp.): Anti-hyperlipidemic, cholesterol lowering and other positive health benefit effects. In *Low and High-Fat Diets: Myths vs. Reality*, 1st ed.; Ferreira, J.E., Muniz, N., Eds.; Nova Biomedical: Runcorn, UK, 2013; pp. 25–47.
11. Cardenas Medellin, M.L.; Serna Saldivar, S.O.; Velazco de la Garza, J. Effect of raw and cooked nopal (*Opuntia ficus indica*) ingestion on growth and profile of total cholesterol, lipoproteins, and blood glucose in rats. *Arch. Lat. Nutr.* **1998**, *48*, 316–323.
12. Andrade-Cetto, A.; Heinrich, M. Mexican plants with hypoglycaemic effect used in the treatment of diabetes. *J. Ethnopharmacol.* **2005**, *99*, 325–348. [CrossRef]
13. Galati, E.M.; Monforte, M.T.; Tripodo, M.M.; D'Aquino, A.; Mondello, M.R. Antiulcer activity of *Opuntia ficus indica* (L.) Mill. (Cactaceae): Ultrastructural study. *J. Ethnopharmacol.* **2001**, *76*, 1–9. [CrossRef]
14. Das, G.; Jik Lim, K.; Tantengco, O.A.G.; Carag, H.M.; Gonçalves, S.; Romano, A.; Das, S.K.; Coy-Barrera, E.; Shin, H.-S.; Gutiérrez-Grijalva, E.P.; et al. Cactus: Chemical, nutraceutical composition and potential bio-pharmacological properties. *Phytother Res.* **2020**, *35*, 1248–1283. [CrossRef] [PubMed]
15. Ben Saad, A.; Dalel, B.; Rjeibi, I.; Smida, A.; Ncib, S.; Zouari, N.; Zourgui, L. Phytochemical, antioxidant and protective effect of cactus cladodes extract against lithium-induced liver injury in rats. *Pharm. Biol.* **2017**, *55*, 516–525. [CrossRef] [PubMed]
16. Messina, C.M.; Arena, R.; Morghese, M.; Santulli, A.; Liguori, G.; Paolo Inglese, P. Seasonal characterization of nutritional and antioxidant properties of *Opuntia ficus-indica* [(L.) Mill.] mucilage. *Food Hydrocoll.* **2021**, *111*, 106398. [CrossRef]
17. Rodríguez-González, S.; Martínez-Flores, H.E.; Chávez-Moreno, C.K.; Macías-Rodríguez, L.I.; Zavala-Mendoza, E.; Garnica-Romo, M.G.; Chacón-García, L. Extraction and characterization of mucilage from wild species of Opuntia. *J. Food Process Eng.* **2014**, *37*, 285–292. [CrossRef]
18. Otálora, M.C.; Carriazo, J.G.; Iturriaga, L.; Nazareno, M.A.; Osorio, C. Microencapsulation of betalains obtained from cactus fruit (*Opuntia ficus-indica*) by spray drying using cactus cladode mucilage and maltodextrin as encapsulating agents. *Food Chem.* **2015**, *187*, 174–181. [CrossRef]
19. Camelo Caballero, L.R.; Wilches-Torres, A.; Cárdenas-Chaparro, A.; Gómez Castaño, J.A.; Otálora, M.C. Preparation and physicochemical characterization of softgels cross-linked with cactus mucilage extracted from cladodes of *Opuntia Ficus–indica*. *Molecules* **2019**, *24*, 2531. [CrossRef]
20. Kumar, M.; Rakesh, S.; Nagpal, R.; Hemalatha, R.; Ramakrishna, A.; Sudarshan, V.; Ramagoni, R.; Shujauddin, M.; Verma, V.; Kumar, A.; et al. Probiotic Lactobacillus rhamnosus GG and *Aloe vera* gel improve lipid profiles in hypercholesterolemic rats. *Nutrition* **2013**, *29*, 574–579. [CrossRef]
21. Ray, A.; Ghosh, S. *Aloe vera* L. Gel: Biochemical Composition, Processing and Nutraceutical Applications. In *Recent Progress in Medicinal Plants*, 41st ed.; Pathak, M., Govil, J.N., Eds.; Studium Press LLC: Houston, TX, USA, 2014; pp. 1–22.
22. Añibarro-Ortega, M.; Pinela, J.; Barros, L.; Ćirić, A.; Silva, S.P.; Coelho, E.; Mocan, A.; Calhelha, R.C.; Soković, M.; Coimbra, M.A.; et al. Compositional Features and Bioactive Properties of *Aloe vera* Leaf (Fillet, Mucilage, and Rind) and Flower. *Antioxidants* **2019**, *8*, 444. [CrossRef]
23. Sánchez-Machado, D.I.; López-Cervantes, J.; Sendón, R.; Sanches-Silva, A. *Aloe vera*: Ancient knowledge with new frontiers. *Trends Food Sci. Technol.* **2017**, *61*, 94–102. [CrossRef]
24. Kaur, M.; Kaur, A.; Sharma, R. Pharmacological actions of Opuntia ficus indica: A review. *J. Appl. Pharm. Sci.* **2012**, *2*, 15–18. [CrossRef]
25. Malviya, P. Extraction characterization and evaluation of selected mucilage as pharmaceutical excipient. *Polim. Med.* **2011**, *41*, 39–44. [PubMed]
26. Gheribi, R.; Puchot, L.; Verge, P.; Jaoued-Grayaa, N.; Mezni, M.; Habibi, Y.; Khwaldia, K. Development of plasticized edible films from Opuntia ficus-indica mucilage: A comparative study of various polyol plasticizers. *Carbohydr. Polym.* **2018**, *190*, 204–211. [CrossRef] [PubMed]
27. Otálora, M.C.; Gómez Castaño, J.A.; Wilches-Torres, A. Preparation, study and characterization of complex coacervates formed between gelatin and cactus mucilage extracted from cladodes of Opuntia ficus-indica. *Food Sci. Technol.* **2019**, *112*, 108234. [CrossRef]
28. Femenia, A.; Sánchez, E.S.; Simal, S.; Rosselló, C. Compositional features of polysaccharides from *Aloe vera* (*Aloe barbadensis Miller*) plant tissues. *Carbohydr. Polym.* **1999**, *39*, 109–117. [CrossRef]
29. Liu, C.; Cui, Y.; Pi, F.; Cheng, Y.; Guo, Y.; Qian, H. Extraction, Purification, Structural Characteristics, Biological Activities and Pharmacological Applications of Acemannan, a Polysaccharide from *Aloe vera*: A Review. *Molecules* **2019**, *24*, 1554. [CrossRef] [PubMed]
30. Minjares-Fuentes, J.R.; Femenia, A. Effect of processing on the bioactive polysaccharides and phenolic compounds from *Aloe vera* (*Aloe barbadensis* miller). In *Dietary Fiber Functionality in Food and Nutraceuticals*; Hosseinian, F., Oomah, B.D., Campos-Vega, R., Eds.; John Wiley & Sons, Ltd: Chichester, UK, 2016; pp. 263–287.

31. Ni, Y.; Turner, D.; Yates, K.M.; Tizard, I. Isolation and characterization of structural components of *Aloe vera* L. Leaf pulp. *Int. Immunopharmacol.* **2004**, *4*, 1745–1755. [CrossRef]
32. Marpudi, S.L.; Ramachandran, P.; Srividya, N. *Aloe vera* gel coating for post-harvest quality maintenance of fresh fig fruits. *Res. J. Pharm. Biol. Chem. Sci.* **2013**, *4*, 878–887.
33. Sogvar, O.B.; Saba, M.K.; Emamifar, A. *Aloe vera* and ascorbic acid coatings maintain postharvest quality and reduce microbial load of strawberry fruit. *Postharvest Biol. Technol.* **2016**, *114*, 29–35. [CrossRef]
34. Khaliq, G.; Ramzan, M.; Baloch, A.H. Effect of *Aloe vera* gel coating enriched with Fagonia indica plant extract on physicochemical and antioxidant activity of sapodilla fruit during postharvest storage. *Food Chem.* **2019**, *286*, 346–353. [CrossRef] [PubMed]
35. Chin, S.S.; Lyn, F.H.; Hanani, Z.A.N. Effect of *Aloe vera* (*Aloe barbadensis* Miller) gel on the physical and functional properties of fish gelatin films as active packaging. *Food Packag. Shelf Life* **2017**, *12*, 128–134. [CrossRef]
36. Medina-Torres, L.; Núñez-Ramírez, D.M.; Calderas, F.; González-Laredo, R.F.; Minjares-Fuentes, R.; Valadez-García, M.A.; Bernad-Bernad, M.J.; Manero, O. Microencapsulation of gallic acid by spray drying with *Aloe vera* mucilage (*Aloe barbadensis* miller) as wall material. *Ind. Crops Prod.* **2019**, *1385*, 111461. [CrossRef]
37. Sonawane, S.K.; Gokhale, J.S.; Mulla, M.Z.; Kandu, V.R.; Patil, S. A comprehensive overview of functional and rheological properties of *Aloe vera* and its application in foods. *J. Food Sci. Technol.* **2021**, *58*, 1217–1226. [CrossRef]
38. Quinzio, C.; Corvalán, M.; López, B.; Iturriaga, L. Studying stability against coalescence in tuna mucilage emulsions. *Acta Hortic.* **2009**, *811*, 427–431. [CrossRef]
39. Koocheki, A.; Mortazavi, S.A.; Shahidi, F.; Razavi, S.M.A.; Taherian, A.R. Rheological properties of mucilage extracted from Alyssum homolocarpum seed as a new source of thickening agent. *J. Food Eng.* **2009**, *91*, 490–496. [CrossRef]
40. He, L.; Yan, X.; Liang, J.; Li, S.; He, H.; Xiong, Q.; Lai, X.; Hou, S.; Huang, S. Comparison of different extraction methods for polysaccharides from *Dendrobium officinale* stem. *Carbohydr. Polym.* **2018**, *198*, 101–108. [CrossRef]
41. Luo, Q.; Wu, M.; Sun, Y.; Lv, J.; Zhang, Y.; Cao, H.; Wu, D.; Lin, D.; Zhang, Q.; Liu, Y.; et al. Optimizing the Extraction and Encapsulation of Mucilage from *Brasenia schreberi*. *Polymers* **2019**, *11*, 822. [CrossRef] [PubMed]
42. Miranda, M.; Maureira, H.; Rodríguez, K.; Vega-Gálveza, A. Influence of temperature on the drying kinetics, physicochemical properties, and antioxidant capacity of *Aloe vera* (*Aloe barbadensis* Miller) gel. *J. Food Eng.* **2009**, *91*, 297–304. [CrossRef]
43. Cunniff, P. Enzymatic-gravimetric method. In *Official Methods of Analysis of AOAC International*, 16th ed.; AOAC: Gaithersburg, MD, USA, 1997.
44. Chem3D Pro 12.0, Cambridgesoft, PerkinElmer Informatics. Available online: http://www.cambridgesoft.com/ (accessed on 1 May 2021).
45. National Library of Medicine. Available online: https://pubchem.ncbi.nlm.nih.gov/ (accessed on 1 May 2021).
46. Burkert, U.; Allinger, N.L. Molecular Mechanics. In *ACS Monograph 177*; American Chemical Society: Washington, DC, USA, 1982.
47. Frisch, M.J.; Trucks, G.W.; Schlegel, H.B.; Scuseria, G.E.; Robb, M.A.; Cheeseman, J.R.; Scalmani, G.; Barone, V.; Mennucci, B.; Petersson, G.A.; et al. *Gaussian 09, Revision D.01*; Gaussian, Inc.: Wallingford, CT, USA, 2009.
48. Dennington, R.; Keith, T.; Millam, J. *GaussView 6.0.16*; Semichem, Inc.: Shawnee Mission, KS, USA, 2016.
49. Cano-Barrita, P.F.J.; León-Martínez, F.M. Biopolymers with viscosity-enhancing properties for concrete. In *Biopolymers and Biotech Admixtures for Eco-Efficient Construction*, 1st ed.; Pacheco-Torgal, F., Ivanov, V., Karak, N., Jonkers, H., Eds.; Woodhead Publishing: Duxford, UK, 2016; pp. 221–252.
50. Scott, A.P.; Radom, L. Harmonic Vibrational Frequencies: An Evaluation of Hartree−Fock, Møller−Plesset, Quadratic Configuration Interaction, Density Functional Theory, and Semiempirical Scale Factors. *J. Phys. Chem.* **1996**, *100*, 16502–16513. [CrossRef]
51. Contreras-Padilla, M.; Rodríguez-García, M.E.; Gutiérrez-Cortez, E.; Valderrama-Bravoc, M.C.; Rojas-Molina, J.I.; Rivera-Muñoz, E.M. Physicochemical and rheological characterization of *Opuntia ficus* mucilage at three different maturity stages of cladode. *Eur. Polym. J.* **2016**, *78*, 226–234. [CrossRef]
52. González-Martínez, D.A.; Carrillo-Navas, H.; Barrera-Díaz, C.E.; Martínez-Vargas, S.L.; Alvarez-Ramírez, J.; Perez-Alonso, C. Characterization of a novel complex coacervate based on whey protein isolate-tamarind seed mucilage. *Food Hydrocoll.* **2017**, *72*, 115–126. [CrossRef]
53. Minjares-Fuentes, J.R.; Femenia, A.; Comas-Serra, F.C.; Rossello, C.; Rodríguez-Gonzalez, V.M.; Gonzalez-Laredo, R.F.; Gallegos-Infante, J.A.; Medina-Torres, L. Effect of different drying procedures on physicochemical properties and flow behavior of *Aloe vera* (*Aloe barbadensis* Miller) gel. *Food Sci. Technol.* **2016**, *74*, 378–386. [CrossRef]
54. Ray, A.; Ghosh, S. Chemometrics for Functional Group Distribution, and UV Absorption Potential of *Aloe vera* L. Gel at Different Growth Periods. *Mater. Today Proc.* **2018**, *5*, 22245–22253. [CrossRef]
55. Andrade, L.A.; Aparecida de Oliveira Silva, D.; Nunes, C.A.; Pereira, J. Experimental techniques for the extraction of taro mucilage with enhanced emulsifier properties using chemical characterization. *Food Chem.* **2020**, *327*, 127095. [CrossRef] [PubMed]
56. Jiang, Y.; Du, J.; Zhang, L.; Li, W. Properties of pectin extracted from fermented and steeped hawthorn wine pomace: A comparison. *Carbohydr. Polym.* **2018**, *1971*, 174–182. [CrossRef] [PubMed]
57. Aghamohamadi, N.; Sanjani, N.S.; Majidi, R.F.; Nasrollahi, S.A. Preparation and characterization of *Aloe vera* acetate and electrospinning fibers as promising antibacterial properties materials. *Mater. Sci. Eng. C* **2019**, *94*, 445–452. [CrossRef] [PubMed]
58. Soukoulis, C.; Gaiani, C.; Hoffmann, L. Plant seed mucilage as emerging biopolymer in food industry applications. *Curr. Opin. Food Sci.* **2018**, *22*, 28–42. [CrossRef]

polymers

MDPI

Article

Conversion of Pectin-Containing By-Products to Pectinases by *Bacillus amyloliquefaciens* and Its Applications on Hydrolyzing Banana Peels for Prebiotics Production

Chien Thang Doan [1,2], Chien-Lin Chen [1], Van Bon Nguyen [3], Thi Ngoc Tran [1,2], Anh Dzung Nguyen [3] and San-Lang Wang [1,4,*]

1 Department of Chemistry, Tamkang University, New Taipei City 25137, Taiwan; dcthang@ttn.edu.vn (C.T.D.); vic900065@gmail.com (C.-L.C.); ttngoc@ttn.edu.vn (T.N.T.)
2 Faculty of Natural Sciences and Technology, Tay Nguyen University, Buon Ma Thuot 630000, Vietnam
3 Institute of Biotechnology and Environment, Tay Nguyen University, Buon Ma Thuot 630000, Vietnam; nvbon@ttn.edu.vn (V.B.N.); nadzung@ttn.edu.vn (A.D.N.)
4 Life Science Development Center, Tamkang University, New Taipei City 25137, Taiwan
* Correspondence: sabulo@mail.tku.edu.tw; Tel.: +886-2-2621-5656; Fax: +886-2-2620-9924

Abstract: The utilization of pectin-containing by-products may be useful in a variety of fields. This study aims to establish the processing of pectin-containing by-products to produce pectinases using *Bacillus amyloliquefaciens* TKU050 strain. In this study, several kinds of agricultural pectin-containing by-products from banana (banana peel), rice (rice bran), orange (orange peel), coffee (spent coffee grounds), and wheat (wheat bran) were utilized to provide carbon sources for the production of a pectinase by *B. amyloliquefaciens* TKU050. *B. amyloliquefaciens* TKU050 expressed the highest pectinase productivity (0.76 U/mL) on 0.5% wheat bran-containing medium at 37°C for four days. A 58 kDa pectinase was purified from the four-day cultured medium fermented under optimized culture conditions with 7.24% of a recovery ratio and 0.51 U/mg of specific activity, respectively. The optimum temperature, optimum pH, thermal stability, and pH stability of the TKU050 pectinase were 50 °C, pH 6, <50 °C, and pH 6–9, respectively. The TKU050 pectinase was inhibited by sodium dodecyl sulfate and Cu^{2+}. The reducing sugar obtained by hydrolyzing banana peel with TKU050 pectinase showed the growth-enhancing effect on the growth of four tested lactic acid bacteria.

Keywords: pectin; pectinase; wheat bran; banana peel; *Bacillus amyloliquefaciens*; prebiotics

Citation: Doan, C.T.; Chen, C.-L.; Nguyen, V.B.; Tran, T.N.; Nguyen, A.D.; Wang, S.-L. Conversion of Pectin-Containing By-Products to Pectinases by *Bacillus amyloliquefaciens* and Its Applications on Hydrolyzing Banana Peels for Prebiotics Production. *Polymers* **2021**, *13*, 1483. https://doi.org/10.3390/polym13091483

Academic Editor: Arn Mignon

Received: 20 April 2021
Accepted: 3 May 2021
Published: 4 May 2021

1. Introduction

Pectins are a wide category of polysaccharides that consist mainly of galacturonic acid as well as some L-arabinose, D-galactose, and L-rhamnose units that can be split into three major groups, including homogalacturonan, rhamnogalacturonan-I, and rhamnogalacturonan-II [1]. In nature, pectins are found abundantly in higher plants' primary cell walls and middle lamella [2,3]. The main enzymes responsible for pectin degradation are pectinases [4]. There are three categories of pectinases based on their action mode and enzymatic substrate, including polygalacturonases, lyases, and pectin methylesterases [2].

According to studies, microbial pectinase accounts for 25% of global food and industrial enzyme revenues, and the demand is constantly expanding [5]. *Penicillium* spp., *Aspergillus* spp., *Erwinia* spp., and *Bacillus* spp. are among the microbial strains that have been commonly used in the commercial development of pectinases. [2]. However, many of those researches used commercial pectin as the carbon source for pectinase production [2,6]. Consequently, the high cost of pectin could raise the price of enzyme production, thereby limiting pectinase's applications. This problem can be partially solved by using agro-waste sources, such as citrus peel [7], wheat bran [8], apple pomace [9], grape pomace [10], papaya peel [9], or banana peel [11], etc., as alternative substrates for the microbial fermentation process. Additionally, bacterial pectinase is favored over fungal pectinase due to its ease of fermentation and modern techniques for increasing production yield [12].

Pectinases are essential in a variety of industries. In the food and beverage industries, they are a useful tool for making jams and jellies [2], clarifying wine [13], extracting juice [14] and vegetable oil [15], removing mucilage from coffee beans [16], concentrating and fermenting tea, cocoa, and coffee, and pickling [17]. In the paper and textile industries, pectinases are used for bleaching pulp, bio-scouring cotton, retting, and degumming plant fibers [2]. Pectinases are also useful in the processing of bioenergy, poultry feed, wastewater treatment, and protoplast fusion technology [2,17].

The search for novel enzymes with a low-cost production process and appropriate properties is considered critical research. Thus, this study aims to produce pectinases by *B. licheniformis* TKU050 using pectin-containing by-products as the carbon source. This can simultaneously achieve two objectives: lowers the cost of enzyme production and reducing the amount of waste generated in the environment. Accordingly, various kinds of pectin-containing byproducts, such as wheat bran, rice bran, banana peel, pomelo peel, spent coffee grounds, and orange peel, have been used as carbon sources for *B. amyloliquefaciens* TKU050 to produce pectinase. The properties of pectinase, as well as its purification, have been studied. Furthermore, among several kinds of pectin-containing by-products, banana peel was observed as the most suitable substrate for *B. amyloliquefaciens* TKU050 pectinase. Banana peel is a well-known agricultural waste that contains a range of nutrients and bioactivity, including antioxidants [18,19], antimicrobial [20], anti-cancer [21], and enhance α-glucosidase inhibitory in yogurt [22]. Banana peel makes up about 40% of the fruit weight, and the recovery of banana peel for its bioactivity compounds is an essential step toward agricultural sustainability. However, there is no relevant report on its prebiotic effects after being hydrolyzed by pectinases. Therefore, the growth-enhancing effects of the reducing sugar obtained by hydrolyzing with TKU050 pectinase on lactic acid bacteria were also investigated.

2. Materials and Methods

2.1. Materials

Bacillus amyloliquefaciens TKU050 was isolated from the soil using pectin as the sole carbon source. Pectin was bought from Acros Organics (Fisher Scientific, Göteborg, Sweden). *Lactococcus lactis* subsp. *lactis* BCRC10791, *Lactobacillus rhamnosus* BCRC104940, *Lactobacillus paracasei* subsp. *paracasei* BCRC14023, and *Lactobacillus rhamnosus* BCRC16000 were obtained from the Bioresource Collection and Research Center (Hsinchu, Taiwan). The other chemicals used were of the highest possible quality.

2.2. Screening, Selection, and Identification of the Pectinase-Producing Bacterium

A pectin-containing medium (1% (*w*/*v*) pectin, 0.2% K_2HPO_4, 0.2% KH_2PO_4, 0.2% KNO_3, and 0.3% yeast extract) was used to isolate the pectinase-producing bacteria. Bacterial strains collected from Tamkang University (New Taipei, Taiwan) soil were incubated in a pectin-containing agar medium for three days at 37 °C and a 150-rpm shaking speed, after which the supernatant was withdrawn to examine its pectinase activity. Later on, the selected strain was characterized, and its scientific name was identified by morphological, API 50 CHB/E kit, and 16S rDNA sequences analysis.

2.3. Pectinase Assay

The dinitrosalicylic acid (DNS) method [23] was used to determine the amount of reducing sugars produced from the pectin hydrolysis reaction catalyzed by pectinase. A pectin hydrolysis reaction was carried out with an equivalent volume of substrate (0.2 mL of 1% *w*/*v* pectin) and diluted sample (0.2 mL) with 30 min of reaction time and an incubation temperature of 37 °C. Next, 1.5 mL of DNS reagent was introduced to the mixture, which was then heated for 10 min at 100 °C. After centrifugation (4000× *g*, 10 min), the color of the reaction was measure by an ELISA plate reader to calculate the pectinase of the sample.

A pectinase unit is the volume of enzyme needed to produce 1 μmol of D-galacturonic acid in one minute.

2.4. By-Products of Agriculture Processing as the Sole C/N Source for Pectinase Production

To prepare a medium containing agriculture by-products (ABP), 1% of each ABP (wheat bran, rice bran, banana peel powder, orange peel powder, coffee grounds, and pectin) was added into 100 mL of a basal medium containing 0.2% K_2HPO_4, 0.2% KH_2PO_4, 0.2% KNO_3, and 0.3% yeast extract. The fermentation of *B. amyloliquefaciens* TKU050 on different ABP-containing mediums was initiated by adding 1 mL of bacterial seed solution to 100 mL of each medium and then incubating the solution at 37 °C and a 150-rpm shaking speed. During the cultivation period, the pectinase activity of the growth media was tested every 24 h. Some wheat bran powder (WBP) amounts (0.5, 1.0, 1.5, and 2.0 g) were added to 100 mL of basal medium to explore the optimal wheat bran concentration for *B. amyloliquefaciens* TKU050 pectinase synthesis. The fermentation conditions have already been mentioned in this study.

2.5. Enzyme Purification

B. amyloliquefaciens TKU050 was incubated as per the optimal conditions for pectinase production. The culture broth was centrifuged at 4000× *g* for 10 min to remove the residual solids and bacterial cells. The liquid supernatant was then mixed with ammonium sulfate (80% *w*/*v*) and kept at 4 °C for one night. The precipitate that appeared in the culture supernatant was conveniently obtained by centrifugation (10,000× *g*, 20 min). A small volume of 50 mM NPB at pH 7 was used to dissolve the obtained precipitate. The residual ammonium sulfate in the crude supernatant was removed by dialyzing the solution against NPB for one night using a cellulose membrane (cellusep T2, Interchim, Montluçon, France). For the enzyme purification, the crude enzyme was loaded onto a Macro Prep High Q column (Bio-rad, Hercules, California, USA) pre-equilibrated with NPB (50 mM, pH 7). The washing step took place until the A280 nm reached a stable value. After the washing step, a linear gradient of NaCl (0 M–1.0 M) was conducted to elute the target enzymes. Fractions exhibiting pectinase activity were then pooled and concentrated by the lyophilization method. Finally, an HPLC system consisting of a KW-802.5 column was used to purify the obtained enzyme (a flowrate of 0.6 mL/min, an A280 nm detector, and a column temperature of 25 °C).

2.6. Sodium Dodecyl Sulfate-Polyacrylamide Gel Electrophoresis (SDS-PAGE) Analysis

SDS-PAGE was conducted in a 10% resolving gel according to the method described previously [24]. The sample was prepared in a sample buffer containing 2-mercaptoethanol and SDS, and the mixture was heated at 100 °C for two min. Then, a 10 μL sample/well was loaded for electrophoresis at 25 mA, after which the gel was stained by a protein assay dye reagent (Bio-Rad, Berkeley, CA, USA) and then de-stained by methanol/acetic acid/water solution (1/1/8, *v*/*v*/*v*). The visual band of the enzyme was compared with the bands of the protein markers to determine its MW.

2.7. Effect of pH and Temperature

By altering the pH of the reaction solution, the optimal pH of *B. amyloliquefaciens* TKU050 pectinase was determined in the same way as the pectinase activity test described above. Sodium carbonate (pH 9–11), sodium phosphate (pH 6–8), sodium acetate (pH 5), and glycine HCl (pH 2–4) made up the pH change buffer system. The pH stability of *B. amyloliquefaciens* TKU050 pectinase was calculated by measuring its residual activity after incubating the enzyme for one hour at different pH levels ranging from 2 to 11. The reaction of the enzyme and 1% pectin solution was held at various temperatures to determine the optimal temperature of the *B. amyloliquefaciens* TKU050 pectinase. The remaining activity of *B. amyloliquefaciens* TKU050 pectinase after one hour of incubation at different temperatures (10–100 °C) was the basis for determining its thermal stability.

The remaining activity of *B. amyloliquefaciens* TKU050 pectinase was tested following the procedure mentioned above.

2.8. Substrate Specificity

The substrate specificity of *B. amyloliquefaciens* TKU050 pectinase was tested on pectin, starch, dextran, gum arabic, carboxymethyl cellulose (CMC), banana peel powder (BPP), orange peel powder, wheat bran powder, rice bran powder (RBP), pomelo peel powder (PPP), coffee pulp powder (CPP), and spent coffee grounds (SCG) using the conditions for the pectinase activity test as described above. The pectinase activity in pectin was used as the control.

2.9. Effect of Metal Ions, Inhibitors, and Surfactants

The chemicals tested included metal ions (Zn^{2+}, Fe^{2+}, Mn^{2+}, Cu^{2+}, Ba^{2+}, and Ca^{2+}), EDTA, and surfactants (SDS, triton X-100, tween 20, and tween 40). Initially, a *B. amyloliquefaciens* TKU050 pectinase solution was added to each of those chemicals in the same proportion for 30 min, and then, the residual activity of the pectinase was determined by the methods described above.

2.10. Growth Enhancing Effect of BPP Oligosaccharides on Lactic Acid Bacteria Test

Lactobacillus paracasei subsp. *paracasei* BCRC14023, *L. rhamnosus* BCRC16000, *L. rhamnosus* BCRC10494, and *Lactobacillus lactis* BCRC10791 were used to evaluate the growth-enhancing effect of BPP hydrolysate, pectin hydrolysate, and BPP. The bacteria were cultured in an MRS medium containing 0.1% (w/v) BPP hydrolysate, pectin hydrolysate, BPP, or glucose for 48 h at 37 °C. The 655 nm optical density of the culture supernatant was used to measure the cell growth of the bacteria.

3. Results and Discussion

3.1. Screening and Identification of Strain TKU050

Over 300 bacterial strains isolated from the soils of Tamsui (New Taipei City, Taiwan) were cultivated at 37 °C in a medium containing 1% pectin. Among them, strain TKU050 exhibited the highest pectinase activity and was selected for further investigations. To characterize strain TKU050, 16S rDNA sequencing and phylogenetic analyses were performed. Based on the evaluation of the 16S rDNA gene sequence, TKU050 was most closely aligned to *Bacillus amyloliquefaciens*, with 95% similarity. To further characterize strain TKU050, standard morphological, physiological, and biochemical plate analysis showed that strain TKU050 was a Gram-positive and endospore-forming bacillus that grows in both aerobic and anaerobic environments. According to the API 50 CHB/E identification, TKU50 was most closely related to *B. amyloliquefaciens*. The phylogenetic identification and API 50 CHB/E analysis indicated that strain TKU050 belonged to the species *B. amyloliquefaciens*.

3.2. Effect of the Carbon (C) Source on Pectinase Production

Six types of pectin sources, including banana peel powder (BPP), orange peel powder (OPP), spent coffee grounds (SCG), wheat bran powder (WBP), rice bran powder (RBP), and pectin powder was used as the sole C source to investigate their effect on pectinase production by *B. amyloliquefaciens* TKU050. One gram of each pectin source was added to 100 mL of a basal medium containing 0.2% K_2HPO_4, 0.2% KH_2PO_4, 0.2%, KNO_3, and 0.3% yeast extract, which was then shaken at 150 rpm for five days.

As shown in Figure 1, all five tested agricultural by-products received higher pectinase productivity than that of commercial pectin products. *B. amyloliquefaciens* TKU050 produced the highest pectinase activity in the medium containing WBP (0.612 U/mL on the fourth day of the fermentation period) and then decreased in the order of BPP (0.608 U/mL on the fourth day), OPP (0.322 U/mL on the fourth day), RBP (0.158 U/mL on the third day), SCG (0.150 U/mL on the fourth day), and pectin (0.118 U/mL on the fourth day). As such, WBP showed promise as a potential C source for pectinase production by *B. amyloliquefaciens* TKU050. Further investigation of the pectinase production of *B. amyloliquefaciens* TKU050 using the WBP-containing medium indicated that a lower concentration of WBP would be more appropriate for higher pectinase productivity. Accordingly, the highest pectinase activity was observed in the four-day medium containing 0.5% WBP (0.515 U/mL), followed by the 1% WBP (0.495 U/mL), 1.5% WBP (0.448 U/mL), and 2% WBP (0.260 U/mL). As such, 0.5% (*w*/*v*) was selected as the optimum WBP concentration for pectinase production by *B. amyloliquefaciens* TKU050.

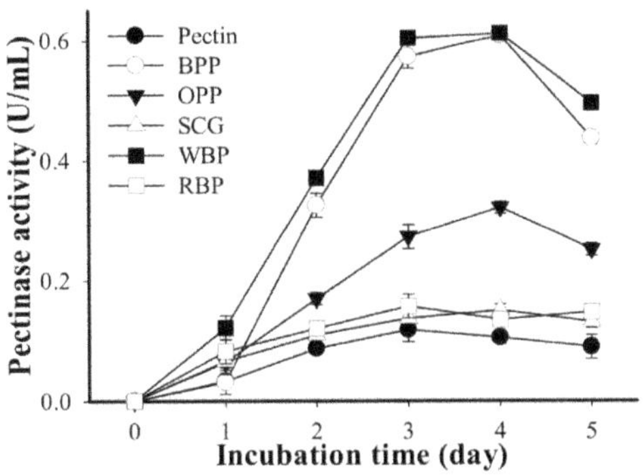

Figure 1. Screening of a 1% (*w*/*v*) suitable C source for pectinase production by *B. amyloliquefaciens* TKU050, including BPP (banana peel powder), OPP (orange peel powder), SCG (spent coffee grounds), WBP (wheat bran powder), and RBP (rice bran powder). All error bars are the ±SD (standard deviation) of three different experiments.

3.3. Optimization of Culture Conditions for Pectinase Production

WBP (0.5%, *w*/*v*) was confirmed as a potential source of carbon and was chosen for an optimization study of a number of parameters, including cultivation pH (3–9), cultivation temperature (30, 37, 40, 45, and 50 °C), and shaking speed (50, 100, 150, and 200 rpm). The effect of cultivation pH on pectinase production was performed using a 100-rpm shaking speed, a medium/flask volume ratio of 100/250 mL (0.2% K_2HPO_4, 0.2% KH_2PO_4, 0.2% KNO_3, and 0.3% yeast extract), and 0.5% WBP (C source). The pectinase activities produced in different pH media were all highest on the fourth day in the order of pH 6 (0.764 U/mL), pH 7 (0.722 U/mL), pH 5 (0.714 U/mL), pH 8 (0.274 U/mL), pH 3 (0.136 U/mL), and pH 4 (0.104 U/mL). The original pH of the media before pH adjustment and autoclaving was 6.3. The higher pectinase activity produced was found in the media at pH 5–7. The effect of cultivation temperature on pectinase production was performed at pH 6, a shaking speed of 100 rpm, 0.5% WBP (C source), and a medium/flask volume ratio of 100/250 mL. The pectinase activities produced at different temperatures were all highest on the fourth day in the order of 37 °C (0.754 U/mL), 40 °C (0.734 U/mL), 45 °C (0.668 U/mL), 30 °C (0.486 U/mL), and 50 °C (0.035 U/mL). The effect of shaking speed on pectinase production was performed at pH 6 and 37 °C using 0.5% WBP (C source). The pectinase activities produced in different shaking speeds were all highest on the fourth day in the order of 100 rpm (0.715 U/mL), 150 rpm (0.706 U/mL), 200 rpm (0.680 U/mL), and 50 rpm (0.303 U/mL). The culture conditions for pectinase production by *B. amyloliquefaciens* TKU050 before and after the optimized study are summarized in Table 1. The pectinase productivity was increased from 0.11 U/mL to 0.76 U/mL after the optimized investigation. From the above study, it could be deduced that *B. amyloliquefaciens* TKU050 was able to produce an appreciable amount of pectinase using wheat bran and could be considered as a potential candidate for industrial applications such as valorization of wastes [25], clarification of wine and fruit juices [13,14], preparation of prebiotic [26], degradation of biomass to produce biofuel [27], etc.

Table 1. The optimized culture conditions for the production of pectinase by *B. amyloliquefaciens* TKU050.

Compared Factors	Before Optimization	After Optimization
Culture time (day)	3	4
C concentration (*w*/*v*)	1% pectin	0.5% WBP
Cultivation temperature (°C)	37	37
pH value	6.6	6.0
Shaking speed (rpm)	150	100
Pectinase activity (U/mL)	0.11	0.76

3.4. Purification of the Pectinase

To date, there have been few reports of the pectinase being produced by *B. amyloliquefaciens* strains. The pectinase production of *B. amyloliquefaciens* SW106 has been studied; however, its purification process was not performed [28]. To compare the characteristics of the pectinase produced by *B. amyloliquefaciens* TKU050 against other reported strains, the pectinase of this strain was purified from the four-day culture supernatant by a series of steps. The protein was precipitated from the culture supernatant by ammonium sulfate (80% *w*/*v*) and eluted by ion-exchange chromatography using a Macro-Prep High Q column with a linear gradient of 0–1 M NaCl (Figure 2). The eluted peaks of the pectinase fractions were pooled and desalted for further purification by high-performance liquid chromatography (HPLC) with a KW-802.5 column.

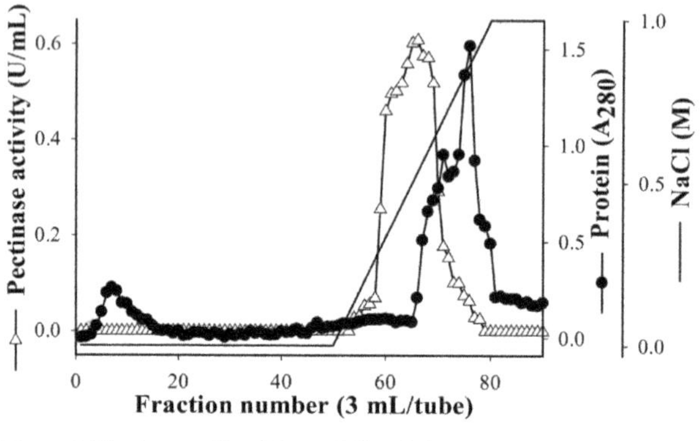

Figure 2. Elution profile of *B. amyloliquefaciens* TKU050 pectinase in a High Q column.

After isolation and purification, approximately 54.9 mg of *B. amyloliquefaciens* TKU050 pectinase was obtained. The recovery yield of the obtained TKU050 pectinase was 7.2% with 19.8-folds of specific activity (Table 2). The molecular weight of the TKU050 pectinase was approximately 58 kDa, as determined by the SDS-PAGE method (Figure 3). *Bacillus* pectinases have molecular weights ranging from 37 to 66 kDa except for *Bacillus* sp. DT7 (Table 3).

Table 2. Purification of *B. amyloliquefaciens* TKU050 pectinase.

Step	Total Protein (mg)	Total Activity (U)	Specific Activity (U/mg)	Recovery (%)	Purification (fold)
Cultural supernatant	15,003.63	689.06	0.046	100.00	1.00
$(NH_4)_2SO_4$ precipitation	2136.57	566.83	0.265	82.26	5.76
Ion-exchange chromatography	396.50	319.87	0.807	46.42	17.54
Gel filtration	54.90	49.89	0.909	7.24	19.79

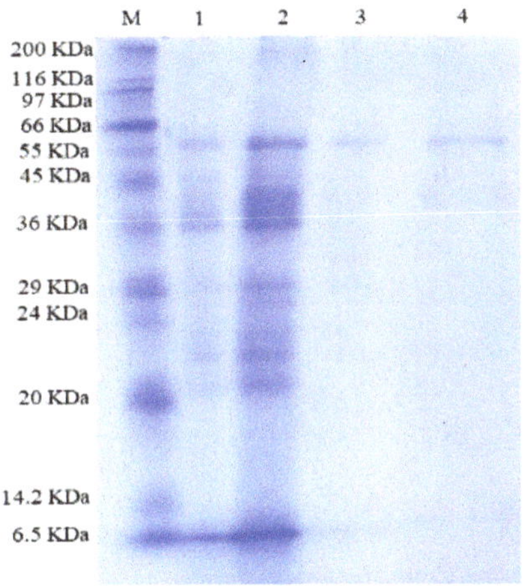

Figure 3. SDS-PAGE profiles of *B. amyloliquefaciens* TKU050 pectinase: (1) culture supernatant; (2) crude enzyme after $(NH_4)_2SO_4$ precipitation; (3) pectinase fraction after running FPLC with a Macro-prep High Q column; (4) purified pectinase after running HPLC with a KW802.5 column. M represents the protein markers.

Table 3. Comparison of pectinase produced by *Bacillus* strains.

Strain	MW (KDa)	Opt. Temp. (°C)	Opt. pH	Carbon Source	Reference
B. amyloliquefaciens TKU050	58	50	6	wheat bran	This study
Bacillus sp. DT7	106	60	8	pectin	[29]
B. paralicheniformis CBS3	53	60	9.1	pectin	[30]
Bacillus sp. MBRL576	66	-	-	sucrose	[31]
B. subtilis Btk27	-	50	7.5	wheat bran	[16]
B. subtilis	38	-	-	pectin	[4]
B. subtilis CM5	56	50	7	cassava bagasse	[32]
Bacillus sp. AD 1	-	37	7	pectin	[33]
B. tequilensis CAS-MEI-2-33	45.4	40	10	tobacco stalk	[12]
B. sonorensis MPTD1	-	-	-	pectin	[34]
Bacillus sp. MFW7	37	-	-	cassava waste	[35]
B. subtilis TYg4-3	-	-	-	lactose	[28]
B. amyloliquefaciens SW106	-	-	-	maltose	[28]
B. subtilis SS	-	-	-	wheat bran	[36]
Bacillus sp. Y1	-	-	-	starch, wheat bran, and sucrose	[37]
B. subtilis ZGL14	65	50	8.6	starch	[38]
B. licheniformis KIBGE-IB21	-	-	-	wheat bran	[39]
B. pumilus dcsr1				sesame oilseed cake	[40]
B. mojavensis I4	-	60	8	carrot peel	[41]
B. licheniformis KIBE-IB3	-	-	-	wheat bran	[42]
B. safensis M35	-	-	-	wheat bran and citrus peel	[7]
B. altitudinis J208	-	-	-	wheat bran and citrus peel	[7]

3.5. Properties of TKU050 Pectinase

3.5.1. Effects of Temperature and pH

The thermal stability of *B. amyloliquefaciens* TKU050 pectinase was tested by treating the enzyme solution at different temperatures for one h. The optimal temperature for *B. amyloliquefaciens* TKU050 pectinase was found to be 50 °C (Figure 4a). *B. amyloliquefaciens* TKU050 pectinase retained above 80% of the initial activity until 50 °C and then gradually lost its activity at higher temperatures. Pectinase from *Bacillus* strains has also been found to work best at temperatures ranging from 40 °C to 60 °C [16,29,32,33].

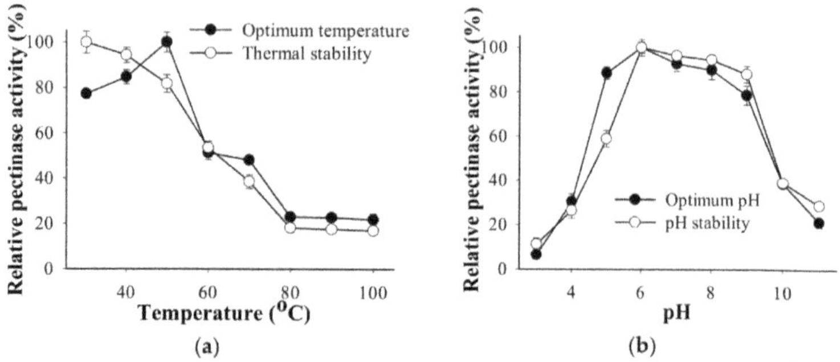

Figure 4. Effects of temperature and pH on the activity and stability of *B. amyloliquefaciens* TKU050 pectinase: (**a**) temperature; (**b**) pH. All data points are the mean and standard deviation.

The pH activity profile of *B. amyloliquefaciens* TKU050 pectinase is shown in Figure 4b. The optimum enzyme activity was observed at pH 6 and was stable at the pH range of pH 6 to pH 8. In earlier studies, the optimum pH of pectinase from *Bacillus* strains was commonly reported at neutral or alkaline values, such as *B. subtilis* CM5 (pH 7) [32], *Bacillus* sp. AD1 (pH 7) [33], *B. subtilis* Btk27 (pH 7.5) [16], and *Bacillus* sp. DT7 (pH 8) [29].

3.5.2. Effects of Divalent Metal Ions, EDTA, and Surfactants

Table 4 illustrates the activity of *B. amyloliquefaciens* TKU050 pectinase incubated with various chemicals. EDTA and majority of the tested divalent metal ions (Zn^{2+}, Fe^{2+}, Mn^{2+}, Mg^{2+}, and Ba^{2+}) had no impact on pectinase activity except for Ca^{2+}, and Cu^{2+}. Surfactants (Triton X-100, Tween 20, Tween 40, and SDS) were also tested for their effect on *B. amyloliquefaciens* pectinase activity. A slight enhancing effect was found using 0.5% of the nonionic surfactants Triton X-100 (117.48%) and 0.5% Tween 40 (131.94%), whereas the ionic surfactant SDS remarkably inhibited the pectinase activity (retained 20.56%). The activating effect of Triton X-100 and Tween 40 may be related to the capacity of surface-active reagents to enhance the frequency of contact between the pectin and the pectinase active site [43]. The suppressing effect of SDS on the pectinase activity could be attributed to its capacity to alter the pectinase secondary structure [44].

Table 4. Effects of various chemicals on the activity of *B. amyloliquefaciens* TKU050 pectinase. All error bars are the ±SD (standard deviation) of three different experiments.

Metal Ion/Surfactant	Concentration	Relative Pectinase Activity (%)
None		100 ± 2.29
Zn^{2+}	2 mM	80.83 ± 3.38
Fe^{2+}	2 mM	96.75 ± 6.03
Mn^{2+}	2 mM	97.47 ± 1.21
Cu^{2+}	2 mM	26.35 ± 0.24
Mg^{2+}	2 mM	102.29 ± 1.93
Ba^{2+}	2 mM	87.46 ± 4.94
Ca^{2+}	2 mM	114.71 ± 2.53
EDTA	2 mM	106.99 ± 3.74
Triton X-100	0.5%	117.48 ± 2.41
Tween 20	0.5%	96.75 ± 2.41
Tween 40	0.5%	131.94 ± 8.2
SDS	2 mM	20.56 ± 0.72

3.5.3. Substrate Specificity

As shown in Figure 5, *B. amyloliquefaciens* TKU050 pectinase expressed the greatest activity in the order of pectin > banana peel powder (BPP) > pomelo peel powder (PPP) > coffee pulp powder > wheat bran powder (WBP) > spent coffee grounds (SCG) > rice bran powder (RBP). No activity was found against starch, dextran, carboxymethyl cellulose (CMC), xylan, or gum arabic. This result suggested that *B. amyloliquefaciens* TKU050 pectinase could exhibit pectolytic ability compared to other enzyme activities, such as amylases, xylanases, and cellulases.

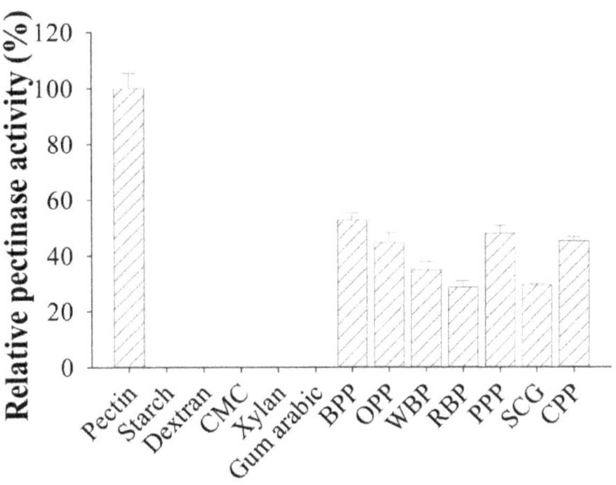

Figure 5. Substrate specificity of *B. amyloliquefaciens* TKU050 pectinase, including CMC (Carboxymethyl Cellulose), BPP (banana peel powder), OPP (orange peel powder), WBP (wheat bran powder), RBP (rice bran powder), PPP (pomelo peel powder), SCG (spent coffee grounds), and CPP (coffee pulp powder). All error bars are the ±SD (standard deviation) of three different experiments.

Based on the results shown in Figure 5, banana peel powder (BPP) was observed to be the most suitable substrate next to pectin for the TKU050 pectinases; therefore, their hydrolysis was conducted and analyzed via the amount of reducing sugars released during the reaction period. The pectinases from the culture supernatant (60 mL) were used to catalyze the hydrolysis of BPP (1% w/v) and pectin (1% w/v) in a 50 mM sodium phosphate buffer (pH 7). As shown in Figure 6a, the rate of BPP hydrolysis was high within the first hour (from 1.25 mg/mL at 0 h to 2.2 mg/mL at 1 h) and remained stationary thereafter. A similar hydrolysis pattern was observed in pectin, as the hydrolysis rate increased dramatically with the first hour (from 0.1 mg/mL at 0 h to 2.3 mg/mL at 1 h), reached 2.4 mg/mL at the second hour, and remained the same thereafter. In a further experiment, the BPP-hydrolyzing ability of *B. amyloliquefaciens* TKU050 pectinase was also compared to a commercial pectinase from *Aspergillus niger*. The result showed that the amount of reducing sugar released from the BPP-hydrolysis showed no significant difference between *B. amyloliquefaciens* pectinase (2.15 mg/mL) and *A. niger* pectinase (2.16 mg/mL), indicating that *B. amyloliquefaciens* TKU050 pectinase could be a potential enzyme for hydrolyzing BPP (Figure 6b).

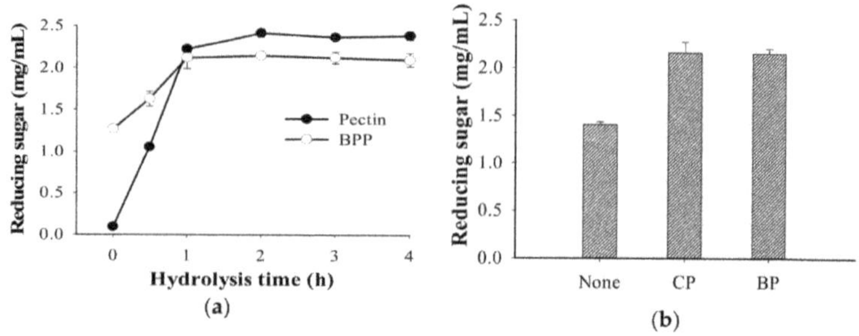

Figure 6. The hydrolysis of BPP and pectin catalyzed by TKU050 pectinase (**a**); BPP catalyzed by commercial pectinase and TKU050 pectinase (**b**). BPP, banana peel powder; CP, commercial pectinase from *A. niger*; and BP, *B. amyloliquefaciens* TKU050 pectinase. All error bars are the ±SD (standard deviation) of three different experiments.

3.6. Evaluation of the Growth Enhancing Effect of BPP Hydrolysates on Lactic Acid Bacteria

The effect of the hydrolysates obtained from the hydrolysis of BPP and the pectin generated by TKU050 pectinases were studied for their prebiotic effects on four lactic acid bacteria: *L. paracasei* subsp. *paracasei* BCRC14023 (Figure 7a), *L. rhamnosus* BCRC16000 (Figure 7b), *L. rhamnosus* BCRC10494 (Figure 7c), and *L. lactis* BCRC10791 (Figure 7d). For comparison, BPP and glucose were also investigated. As shown in Figure 7, the four tested supplemented samples all showed enhancing effect on the tested four lactic acid bacteria. Among them, the BPP hydrolysates, pectin hydrolysates, and BPP had better prebiotic effects than that of glucose. The BPP hydrolysates gained a higher enhancing effect than that of BPB on the growth of *L. rhamnosus* BCRC16000 (Figure 7b). These results suggested that BPP and its hydrolysates produced by TKU050 pectinases may have the potential to become prebiotic candidates.

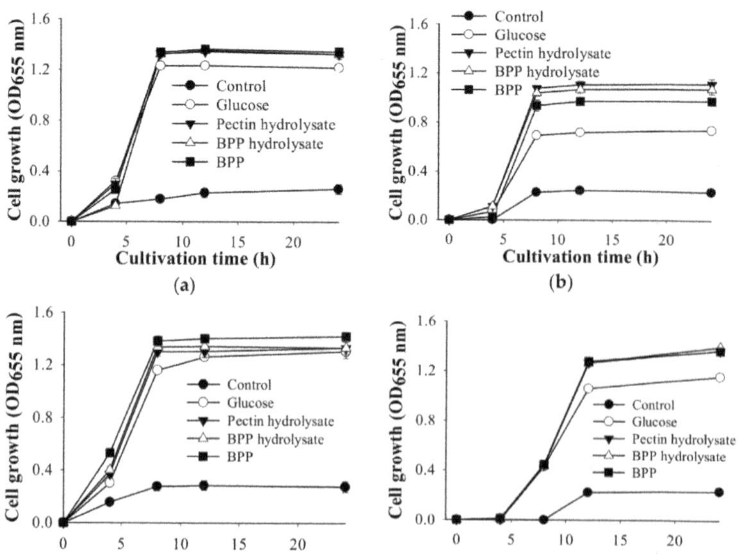

Figure 7. Effect of BPB hydrolysate, pectin hydrolysate, BPB, and glucose on the growth of lactic acid bacteria strains: (**a**) *L. paracasei* subsp. *paracasei* BCRC14023; (**b**) *L. rhamnosus* BCRC16000; (**c**) *L. rhamnosus* BCRC10494; (**d**) *L. lactis* BCRC10791. The error bars represent standard deviations (n = 3).

4. Conclusions

Pectinase is considered to be an effective tool for preparing pectin oligosaccharides, but the enzyme's cost is a possible stumbling block for this application. Thus, the objective of the current study was to post a cost-effective and environmentally friendly bioprocessing of pectin-containing by-products for pectinase production by *B. amyloliquefaciens* TKU050. Following that, a 58 kDa pectinase was purified from the wheat bran culture medium. Additionally, the hydrolysis product of banana peel powder catalyzed by this enzyme showed prebiotic effects on the four tested lactic acid bacteria, indicating that *B. amyloliquefaciens* TKU050 pectinase could be a good candidate for producing prebiotics.

Author Contributions: Conceptualization, methodology, S.-L.W., C.-L.C., T.N.T., C.T.D., V.B.N. and A.D.N.; software, validation, formal analysis, investigation, resources, data curation, C.T.D., T.N.T. and C.-L.C.; writing—original draft preparation, S.-L.W. and C.T.D.; writing—review and editing, S.-L.W. and C.T.D.; visualization S.-L.W. and C.T.D.; supervision, project administration, funding acquisition, S.-L.W. All authors have read and agreed to the published version of the manuscript.

Funding: This study was supported in part by a grant from the Ministry of Science and Technology, Taiwan (MOST 109-2313-B-032-002; MOST 109-2811-B-032-500).

Institutional Review Board Statement: Not applicable.

Informed Consent Statement: Not applicable.

Data Availability Statement: The data presented in this study are available on request from the corresponding author.

Conflicts of Interest: The authors declare no conflict of interest.

References

1. Amid, M.; Manap, Y.; Zohdi, K. Purification and characterisation of thermo-alkaline pectinase enzyme from *Hylocereus polyrhizus*. *Eur. Food Res. Technol.* **2014**, *239*, 21–29. [CrossRef]
2. Kc, S.; Upadhyaya, J.; Joshi, D.R.; Lekhak, B.; Chaudhary, D.K.; Pant, B.R.; Bajgai, T.R.; Dhital, R.; Khanal, S.; Koirala, N.; et al. Production, Characterization, and Industrial Application of Pectinase Enzyme Isolated from Fungal Strains. *Fermentation* **2020**, *6*, 59. [CrossRef]
3. Zdunek, A.; Pieczywek, P.M.; Cybulska, J. The primary, secondary, and structures of higher levels of pectin polysaccharides. *Compr. Rev. Food Sci. Food Saf.* **2021**, *20*, 1101–1117. [CrossRef] [PubMed]
4. Saharan, R.; Sharma, K.P. Production, purification and characterization of pectin lyase from *Bacillus subtilis* isolated from moong beans leaves (*Vigna radiata*). *Biocatal. Agric. Biotechnol.* **2019**, *21*, 101306. [CrossRef]
5. Dranca, F.; Oroian, M. Optimization of pectin enzymatic extraction from Malus domestica 'Falticeni' apple pomace with celluclast 1.5L. *Molecules* **2019**, *24*, 2518. [CrossRef]
6. Abdollahzadeh, R.; Pazhang, M.; Najavand, S.; Fallahzadeh-Mamaghani, V.; Amani-Ghadim, A.R. Screening of pectinase-producing bacteria from farmlands and optimization of enzyme production from selected strain by RSM. *Folia Microbiol.* **2020**, *65*, 705–719. [CrossRef]
7. Thite, V.S.; Nerurkar, A.S.; Baxi, N.N. Optimization of concurrent production of xylanolytic and pectinolytic enzymes by *Bacillus safensis* M35 and *Bacillus altitudinis* J208 using agro-industrial biomass through Response Surface Methodology. *Sci. Rep.* **2020**, *10*, 1–12. [CrossRef]
8. Katileviciute, A.; Plakys, G.; Budreviciute, A.; Onder, K.; Damiati, S.; Kodzius, R. A Sight to Wheat Bran: High Value-Added Products. *Biomolecules* **2019**, *9*, 887. [CrossRef]
9. Patidar, M.K.; Nighojkar, S.; Kumar, A.; Nighojkar, A. Pectinolytic enzymes-solid state fermentation, assay methods and applications in fruit juice industries: A review. *3 Biotech* **2018**, *8*, 199. [CrossRef]
10. Botella, C.; Diaz, A.; Ory, I.; Webb, C.; Blandino, A. Xylanase and pectinase production by Aspegillus awamori on grape pomace in solid state fermentation. *Process Biochem.* **2007**, *42*, 98–101. [CrossRef]
11. Sethi, B.K.; Nanda, P.K.; Sahoo, S. Enhanced production of pectinase by Aspergillus terreus NCFT 4269.10 using banana peels as substrate. *3 Biotech* **2016**, *6*, 1–15. [CrossRef]
12. Zhang, G.; Li, S.; Xu, Y.; Wang, J.; Wang, F.; Xin, Y.; Shen, Z.; Zhang, H.; Ma, M.; Liu, H. Production of alkaline pectinase: A case study investigating the use of tobacco stalk with the newly isolated strain Bacillus tequilensis CAS-MEI-2-33. *BMC Biotechnol.* **2019**, *19*, 1–11. [CrossRef]
13. Claus, H.; Mojsov, K. Enzymes for Wine Fermentation: Current and Perspective Applications. *Fermentation* **2018**, *4*, 52. [CrossRef]
14. Mahmoodi, M.; Najafpour, G.D.; Mohammadi, M. Production of pectinases for quality apple juice through fermentation of orange pomace. *J. Food Sci. Technol.* **2017**, *54*, 4123–4128. [CrossRef]

15. Perez, E.E.; Fernández, M.B.; Nolasco, S.M.; Crapiste, G.H. Effect of pectinase on the oil solvent extraction from different genotypes of sunflower (*Helianthus annuus* L.). *J. Food Eng.* **2013**, *117*, 393–398. [CrossRef]
16. Oumer, O.J.; Abate, D. Characterization of Pectinase from Bacillus subtilis Strain Btk 27 and Its Potential Application in Removal of Mucilage from Coffee Beans. *Enzym. Res.* **2017**, *2017*, 1–7. [CrossRef]
17. Kubra, K.T.; Ali, S.; Walait, M.; Sundus, H. Potential applications of pectinases in food, agricultural, and environmental sectors. *Int. J. Pharm. Chem. Biol. Sci.* **2018**, *6*, 23–34.
18. Singh, B.; Singh, J.P.; Kaur, A.; Singh, N. Bioactive compounds in banana and their associated health benefits—A review. *Food Chem.* **2016**, *206*, 1–11. [CrossRef]
19. González-Montelongo, R.; Lobo, M.G.; González, M. Antioxidant activity in banana peel extracts: Testing extraction conditions and related bioactive compounds. *Food Chem.* **2010**, *119*, 1030–1039. [CrossRef]
20. Kapadia, S.P.; Pudakalkatti, P.S.; Shivanaikar, S. Detection of antimicrobial activity of banana peel (*Musa paradisiaca* L.) on Porphyromonas gingivalis and Aggregatibacter actinomycetemcomitans: An in vitro study. *Contemp. Clin. Dent.* **2015**, *6*, 496–499. [CrossRef]
21. Gurumallesh, P.; Ramakrishnan, B.; Dhurai, B. A novel metalloprotease from banana peel and its biochemical characterization. *Int. J. Biol. Macromol.* **2019**, *134*, 527–535. [CrossRef] [PubMed]
22. Kabir, R.; Hasan, M.; Islam, R.; Haque, A.R.; Hasan, S.M.K. Formulation of yogurt with banana peel extracts to enhance storability and bioactive properties. *J. Food Process. Preserv.* **2021**, *45*, 15191. [CrossRef]
23. Tran, T.N.; Doan, C.T.; Wang, S.-L. Conversion of Wheat Bran to Xylanases and Dye Adsorbent by *Streptomyces thermocarboxydus*. *Polymers* **2021**, *13*, 287. [CrossRef] [PubMed]
24. Doan, C.T.; Tran, T.N.; Nguyen, V.B.; Nguyen, A.D.; Wang, S.L. Utilization of seafood processing by-products for production of proteases by *Paenibacillus* sp. TKU052 and their application in biopeptides' preparation. *Mar. Drugs* **2020**, *18*, 574. [CrossRef]
25. Amin, F.; Bhatti, H.N.; Bilal, M. Recent advances in the production strategies of microbial pectinases—A review. *Int. J. Biol. Macromol.* **2019**, *122*, 1017–1026. [CrossRef]
26. Chung, W.S.F.; Meijerink, M.; Zeuner, B.; Holck, J.; Louis, P.; Meyer, A.S.; Wells, J.M.; Flint, H.J.; Duncan, S.H. Prebiotic potential of pectin and pectic oligosaccharides to promote anti-inflammatory commensal bacteria in the human colon. *FEMS Microbiol. Ecol.* **2017**, *93*, 93. [CrossRef]
27. Mihajlovski, K.; Buntić, A.; Milić, M.; Rajilić-Stojanović, M.; Dimitrijević-Branković, S. From Agricultural Waste to Biofuel: Enzymatic Potential of a Bacterial Isolate Streptomyces fulvissimus CKS7 for Bioethanol Production. *Waste Biomass Valorization* **2021**, *12*, 165–174. [CrossRef]
28. Arekemase, O.M.; Omotosho, I.O.; Agbabiaka, T.O.; Ajide-Bamigboye, N.T.; Lawal, A.K.; Ahmed, T.; Orogu, J.O. Optimization of bacteria pectinolytic enzyme production using banana peel as substrate under submerged fermentation. *Sci. World J.* **2020**, *15*, 56–63.
29. Kashyap, D.; Chandra, S.; Kaul, A.; Tewari, R. Production, purification and characterization of pectinase from a Bacillus sp. DT7. *World J. Microbiol. Biotechnol.* **2000**, *16*, 277–282. [CrossRef]
30. Khan, M.; Choi, Y.S.; Kim, Y.K.; Yoo, J.C. Immobilization of an alkaline endopolygalacturonase purified from *Bacillus paralicheniformis* exhibits bioscouring of cotton fabrics. *Bioprocess Biosyst. Eng.* **2018**, *41*, 1425–1436. [CrossRef]
31. Bhardwaj, V.; Garg, N. Production, purification of pectinase from Bacillus sp. MBRL567 isolate and its application in extraction of juice. *Int. J. Sci. Res.* **2014**, *3*, 648–652.
32. Swain, M.R.; Ray, R.C. Production, Characterization and Application of a Thermostable Exo-polygalacturonase by *Bacillus subtilis* CM5. *Food Biotechnol.* **2010**, *24*, 37–50. [CrossRef]
33. Dey, A.; Karmakar, M.; Ray, R.R. Extracellular thermostable polygalacturonase from Bacillus sp. AD 1. *Der Pharm. Lett.* **2011**, *3*, 358–367.
34. Mohandas, A.; Raveendran, S.; Parameswaran, B.; Abraham, A.; Athira, R.S.R.; Mathew, A.K.; Pandey, A. Production of Pectinase from Bacillus sonorensis MPTD1. *Food Technol. Biotechnol.* **2018**, *56*, 110–116. [CrossRef]
35. Mukesh kumar, D.J.; Saranya, G.M.; Suresh, K.; Andal Priyadharshini, D.; Rajakumar, R.; Kalaichelvan, P.T. Production and optimization of pectinase from Bacillus sp. MFW7 using Cassava waste. *Asian J. Plant Sci. Res.* **2012**, *2*, 369–375.
36. Ahlawat, S.; Mandhan, R.P.; Dhiman, S.S.; Kumar, R.; Sharma, J. Potential Application of Alkaline Pectinase from Bacillus subtilis SS in Pulp and Paper Industry. *Appl. Biochem. Biotechnol.* **2007**, *149*, 287–293. [CrossRef]
37. Guo, F.; Li, X.; Zhao, J.; Li, G.; Gao, P.; Han, X. Optimizing Culture Conditions by Statistical Approach to Enhance Production of Pectinase from Bacillus sp. Y1. *BioMed Res. Int.* **2019**, *2019*, 1–10. [CrossRef]
38. Yu, P.; Zhang, Y.; Gu, D. Production optimization of a heat-tolerant alkaline pectinase fromBacillus subtilisZGL14 and its purification and characterization. *Bioengineered* **2017**, *8*, 613–623. [CrossRef]
39. Rehman, H.U.; Siddique, N.N.; Aman, A.; Nawaz, M.A.; Baloch, A.H.; Qader, S.A.U. Morphological and molecular based identification of pectinase producing Bacillus licheniformis from rotten vegetable. *J. Genet. Eng. Biotechnol.* **2015**, *13*, 139–144. [CrossRef]
40. Sharma, D.C.; Satyanarayana, T. Biotechnological Potential of Agro Residues for Economical Production of Thermoalkali-Stable Pectinase by *Bacillus pumilus* dcsr1 by Solid-State Fermentation and Its Efficacy in the Treatment of Ramie Fibres. *Enzym. Res.* **2012**, *2012*, 1–7. [CrossRef]

41. Ghazala, I.; Sayari, N.; Ben Romdhane, M.; Ellouz-Chaabouni, S.; Haddar, A. Assessment of pectinase production by *Bacillus mojavensis* I4 using an economical substrate and its potential application in oil sesame extraction. *J. Food Sci. Technol.* **2015**, *52*, 7710–7722. [CrossRef]
42. Jahan, N.; Shahid, F.; Aman, A.; Mujahid, T.Y.; Qader, S.A.U. Utilization of agro waste pectin for the production of industrially important polygalacturonase. *Heliyon* **2017**, *3*, e00330. [CrossRef]
43. Doan, C.T.; Tran, T.N.; Nguyen, V.B.; Tran, T.D.; Nguyen, A.D.; Wang, S.-L. Bioprocessing of Squid Pens Waste into Chitosanase by *Paenibacillus* sp. TKU047 and Its Application in Low-Molecular Weight Chitosan Oligosaccharides Production. *Polymers* **2020**, *12*, 1163. [CrossRef]
44. Liang, T.-W.; Lo, B.-C.; Wang, S.-L. Chitinolytic Bacteria-Assisted Conversion of Squid Pen and Its Effect on Dyes and Pigments Adsorption. *Mar. Drugs* **2015**, *13*, 4576–4593. [CrossRef]

polymers

MDPI

Article

Combination of Computational Techniques to Obtain High-Quality Gelatin-Base Gels from Chicken Feet

José C. C. Santana [1,2,*], Poliana F. Almeida [3], Nykael Costa [4], Isabella Vasconcelos [4], Flavio Guerhardt [4], Dimitria T. Boukouvalas [5], Wonder A. L. Alves [5], Pedro C. Mendoza [6], Felix M. C. Gamarra [7], Segundo A. V. Llanos [8], Sidnei A. Araujo [5], Ada P. B. Quispe [8], Rosangela M. Vanalle [4] and Fernando T. Berssaneti [1,*]

1 Department of Production Engineering, Polytechnic School, University of São Paulo, Av. Prof. Luciano Gualberto, 1380, Butantã, São Paulo 05508-010, Brazil
2 Department of Management Engineering, Federal University of ABC, University Mall, São Bernardo do Campo 09606-045, Brazil
3 Federal Institute of Mato Grosso, São Vicente Campus, São Vicente da Serra 78106-000, Brazil; poliana.almeida@svc.ifmt.edu.br
4 Industrial Engineering Post Graduation Program, Nine July University, Vergueiro Avenue, 235/249, Liberdade, São Paulo 01504-000, Brazil; nykaelcosta@gmail.com (N.C.); belinnh1@gmail.com (I.V.); flavioguerhardt@gmail.com (F.G.); rvanalle@uni9.pro.br (R.M.V.)
5 Informatics and Knowledge Management Post Graduate Program, Nove de Julho University, Vergueiro Avenue, 235/249, Liberdade, São Paulo 01504-000, Brazil; dtbouk@outlook.com (D.T.B.); wonder@uni9.pro.br (W.A.L.A.); saraujo@uni9.pro.br (S.A.A.)
6 Facultad de Ingeniería Ambiental y Sanitaria, Universidad Nacional San Luis Gonzaga de Ica, Ciudad Universitaria, Km 305, Ica 11000, Peru; pedro.cordova@unica.edu.pe
7 Energy Engineering, University of Brasilia, FGA-UnB, St. Leste Projeção A—Gama Leste, Brasília 72444-240, Brazil; fcarbajal@unb.br
8 Chemical Engineering Department and CYMAIDS, Universidad Nacional Pedro Ruiz Gallo, Juan XXIII 391 Street, Lambayeque 14000, Peru; svasquezll@unprg.edu.pe (S.A.V.L.); abarturen@unprg.edu.pe (A.P.B.Q.)
* Correspondence: jose.curvelo@ufabc.edu.br (J.C.C.S.); fernando.berssaneti@usp.br (F.T.B.)

Citation: Santana, J.C.C.; Almeida, P.F.; Costa, N.; Vasconcelos, I.; Guerhardt, F.; Boukouvalas, D.T.; Alves, W.A.L.; Mendoza, P.C.; Gamarra, F.M.C.; Llanos, S.A.V.; et al. Combination of Computational Techniques to Obtain High-Quality Gelatin-Base Gels from Chicken Feet. *Polymers* **2021**, *13*, 1289. https://doi.org/10.3390/polym13081289

Academic Editor: Arn Mignon

Received: 30 March 2021
Accepted: 12 April 2021
Published: 15 April 2021

Abstract: With the increasing global population, it has become necessary to explore new alternative food sources to meet the increasing demand. However, these alternatives sources should not only be nutritive and suitable for large scale production at low cost, but also present good sensory characteristics. Therefore, this situation has influenced some industries to develop new food sources with competitive advantages, which require continuous innovation by generating and utilising new technologies and tools to create opportunities for new products, services, and industrial processes. Thus, this study aimed to optimise the production of gelatin-base gels from chicken feet by response surface methodology (RSM) and facilitate its sensorial classification by Kohonen's self-organising maps (SOM). Herein, a 2^2 experimental design was developed by varying sugar and powdered collagen contents to obtain grape flavoured gelatin from chicken feet. The colour, flavour, aroma, and texture attributes of gelatines were evaluated by consumers according to a hedonic scale of 1–9 points. Least squares method was used to develop models relating the gelatin attributes with the sugar content and collagen mass, and their sensorial qualities were analysed and classified using the SOM algorithm. Results showed that all gelatin samples had an average above six hedonic points, implying that they had good consumer acceptance and can be marketed. Furthermore, gelatin D, with 3.65–3.80% (*w/w*) powdered collagen and 26.5–28.6% (*w/w*) sugar, was determined as the best. Thus, the SOM algorithm proved to be a useful computational tool for comparing sensory samples and identifying the best gelatin product.

Keywords: neural network; chicken feet; sensorial quality; food quality; gelatine

1. Introduction

1.1. Poultry Production in Brazil

Many countries have been economically influenced by the trade of chicken and despite the international crisis, the production of chicken increased in 2019 with a global poultry production of approximately 98.5 million tons. The three major world producers are USA, China, and Brazil, with 20.6%, 15.1% and 14.2% production, respectively [1]. Brazil was the second largest producer of chicken in 2015, at 13.5 million tons, but in 2019 it was the third largest producer, at 13.3 million tons and it was estimated to reach 14 million tons in 2020. Moreover, Brazil is also the largest global exporter of chicken meat with 4.2 million tons, with an increase of 2% per year. The people of Brazil have been estimated to consume approximately 97.6 kg per capita of meat, comprising 47.3 kg chicken, 36.3 kg beef, and 14.1 kg pork (accounting for 48.5%, 37.2%, and 14.4% of the total, respectively) [1]. However, the Brazilians usually do not consume the chicken by-products (carcasses, chicken feet, viscera, etc.); therefore, these are commonly discarded as industrial waste [2,3].

Hence, with the increased meat production, a significant amount of organic waste is generated in the different stages of the production chain of chicken and beef, in particular, and the residue leads to economic and environmental problems. It is a serious concern as the high organic matter content in these wastes can serve as sources for proliferation of microorganisms [2].

Despite the commercial significance of the waste generated during the slaughter of chickens, previous studies have been limited in identifying other treatment and disposal technologies, which are of concern to the poultry industry in Brazil. Some companies have gradually developed technologies to solve these bottlenecks considering the environmental effects of the production process [4].

1.2. Gelatin as a Solution for the Poultry Industry Waste

Gelatin is a denatured fibrous protein derived from collagen by partial thermal hydrolysis. It is an important functional biopolymer that has wide applications in food, material, pharmaceutical, and photography industries [5,6].

The global market for gelatin is estimated at 300 thousand tons per year accounting for more than $2 billion per year of the economy worldwide. Brazil exports approximately 25 tons per year which accounts for 80% of the total production, considering growth in the gelatin market, because of its notable competitive advantages in domestic animal production [7]. Thus, food and pharmaceutical industries worldwide have faced an increasing demand for collagen and gelatin. The most popular and commonly used is the gelatin of mammals (pigs and cattle), which are subjected to greater restrictions and scepticism among consumers due to socio-cultural and health concerns [4,6,8]. In this context, Prosekov and Voroshilin [9] stated that the optimisation of gelatin production technologies and investigation of new raw material sources for its production will contribute to solving the problem of substituting the import of this product and stimulating its export.

The increasing demand for new gelling agents to replace the gelatin of mammals has guided several studies on different raw materials, such as the gelatin of marine origin (fish skin, bone, and fins) [5]. Moreover, other research focussing on the extraction and classification of gelatin from fish have been conducted by [5,10–12]. However, it is an underused source, and with the increasing world population, it is necessary to explore new suitable alternatives to meet the increasing demand. Furthermore, these alternative sources should not only have nutritive value and ability to be produced at a large scale with low cost, but also present good sensory characteristics [4,13].

The materials considered as waste in some regions may be the raw materials for traditional products with high benefits in other regions. For example, in some Asian countries, chicken feet are a delicacy, but in Brazil, these are not generally consumed. The cost of chicken feet is below 0.5 US$/ton [14]. These characteristics of national market are crucial in defining its low sale price. Thus, the development of a gelatin with good sensorial qualities from this raw material will add value to the production chain of the meat industry [2,15], for example, a Nile tilapia hamburger developed by incorporating whey, collagen from chicken feet, and taro flour [7]; or gelatines, jellies, and biofilms obtained from the collagen of chicken feet [2,4,15,16]. Chicken feet gelatin has also been used to improve the texture of chocolate spread [16].

1.3. Application of Algorithms in Sampling

To ensure that the evaluation of samples by the consumers have significance, a large amount of data is required; however, a part of these data is not required for knowledge extraction. Moreover, data are pre-processed to reduce the amount of information and select more relevant attributes. Therefore, various techniques have been developed; for example, Pereira and Sassi [17] used a Hough sets data clustering technique for the data provided by an insurance company, Chen et al. [18] used a fuzzy neural network on a consumer advertising data set, and Pourahmad et al. [19] used a hybrid fuzzy neural network to assess the service quality of an academic library.

De Pelsmaeker et al. [20] used an algorithm based on the fuzzy set theory using the house of quality (HOQ) in a quality function deployment (QFD) study applied to the food industry and evaluated its possibilities and limitations for improving food products. A case study was conducted on filled chocolates, where consumer preferences, processing parameters, and sensory attributes were evaluated. The results revealed limitations in the application and interpretation of HOQ within the food industry and that by using the fuzzy set theory, chain information was incorporated in the HOQ, which established good communication between departments. Thus, the company was able to produce products with high consumer preference and had a high success rate.

Almeida et al. [21] standardised the preparation of Barbados cherry wines by simulated annealing technique. It is a probabilistic search technique that simulates the annealing process of metals, wherein the metal is heated to high temperatures and then systematically cooled in the same order to achieve an equilibrium, characterised by a uniform and stable microstructure. The optimisation results showed that the best conditions were observed at a mass ratio of 1/7.5–1/6 and 28.6–29.0° Brix total soluble solids, from which a sensory acceptance between 7–9 hedonic points were obtained for colour, aroma, and flavour.

According to Santana et al. [22], the statistical techniques of F-test and t-Student test based on six Sigma, data envelopment analysis (DEA), and Taguchi procedures are commonly used in food technology [21–24]. However, in the last decade artificial neural networks (ANN) have been increasingly applied in the sensorial quality evaluation of food. Among the ANN models, the Multilayer Perceptron (MLP) and Kohonen's self-organising maps (SOM) are noteworthy. The MLP is a supervised learning model, commonly used for solving nonlinear problems, that learns a specific function by training on a dataset [25]. In addition to the input and output layers, it has one or more hidden layers that allow the network to map input patterns with similar structures for different outputs. The SOM is an unsupervised learning model to produce a low-dimensional (typically two-dimensional) discrete representation of the input space of the training samples, in the form of a topological map [26]. This model comprises of two layers (input and output), and due to its characteristics, SOM can be used as a dimensionality reduction method.

Santana et al. [22] have applied SOM for clustering sensorial data from Barbados cherry wines, demonstrating that this algorithm is a promising assessment technique for the sensorial quality of foods. Alves et al. [27] employed the SOM to classify wine samples by their sensorial attributes. They demonstrated the potential of this ANN type in the sensory classification of food and to replace known statistical techniques.

Liu et al. [28] compared the Arrhenius model and MLP for the quality prediction of rainbow trout (*Oncorhynchus mykiss*) fillets during storage at different temperatures. Based on the obtained results, they concluded that MLP was more effective, and was a potential tool for modelling quality changes of rainbow trout fillets within the temperature range of 270–282 K.

Ouyang et al. [29] proposed a method that combines MLP and adaptive boosting (AdaBoost) algorithm for estimating Chinese rice wine quality. They observed that the proposed method showed superior results when compared with other algorithms, and therefore, appropriate for predicting the sensory quality in Chinese rice wine.

Yu et al. [30] developed a hybrid model combining linear partial least squares (PLS) regression and MLP for prediction of consumer approval scores for ready-to-drink green tea beverages, which obtained a coefficient of determination (R^2) of 0.875 and a root-mean-square error (RMSE) of 2.41%.

Lu et al. [31] proposed an approach that combines the principal component analysis (PCA) technique with MLP for modelling the effect of vibration on the quality of stirred yogurt during transportation and determined the optimal transportation distances.

Sarkar et al. [32] compared the classification efficiency of SOM, PCA, and hierarchical cluster analysis (HCA) in an analysis of the nutritional value, X-ray diffraction properties, texture, and sensory characteristics for normal "rasgulla" samples and pineapple with different drying process. The authors observed approximately similar trends in the results obtained by the three compared algorithms.

Notably, there are several applications of computational techniques in the optimisation of production processes and in analysing the quality of food products. However, after an extensive review of literature from international scientific document bases, no reports were found on studies that applied software to assess the sensory quality of gelatines.

This study aimed to optimise the production process of gelatin-base gels from chicken feet by response surface methodology (RSM) and facilitate the assessment of their sensory quality using Kohonen's SOM. Factors such as collagen and sugar content were varied in a 2^2 experimental design (DOE) to assess their effects on the sensorial attributes of colour, flavour, aroma, and texture of these gelatines. A Hough assessment of the sensory characteristics was conducted by applying the ANN of SOM on a set with 2000 data.

2. Material and Methods

2.1. Preparation of Gelatines

The physicochemical analysis of collagen and gelatin processing was developed by a Federal Institute of Mato Grosso, Brazil and chicken feet were acquired from a partner food company. Chicken feet collagen powder was prepared according to the method described by Santana et al. (2020) [4]. For the production of gelatines, the Good Manufacturing Practices (GMP) were adhered to during the entire process until the storage of the final product, to ensure safety and to avoid contamination of the end product. Eight gelatin samples were used in this study, seven freshly prepared from chicken feet and a commercial gelatin (produced from ox hide) to act as a reference to compare the experimental results. The chicken feet were obtained from markets in the city of São Paulo, Brazil. For preparation, the chicken feet were washed, the nails were removed, and washed again with cold water to remove any residue of dirt, cleaned with chlorine (2 ppm active Cl_2) water, and then cut and stored in a refrigerator. Then, 200 g of chicken feet was placed in a thermal bath of 4% acetic acid solution at 60 °C for 4 h, to extract the collagen [2,4,5,11,14].

2.2. Experimental Design

A 2^2 experimental design was adopted wherein the collagen powder and sugar contents were used as factors that highly influenced the sensorial responses of consumer to each attribute (colour, aroma, flavour, texture, and general aspect), as showed in Table 1 [22, 33–36]. The gelatines were prepared to a volume of 700 mL, in grape flavour. Commercial gelatin-base gel was prepared by the procedure provided by its manufacturer, Dr. Oetker® (São Paulo, Brazil), and was referred to as Gelatin H [2,4].

Table 1. Level of variables of experimental design.

Gelatin-Base Gel Samples	x_1	x_2	Gelatin (g/100 g)	Sugar (g/100 g)
Gelatin A	−1	−1	1.903	7.1
Gelatin B	1	−1	3.806	7.1
Gelatin C	−1	1	1.903	28.6
Gelatin D	1	1	3.806	28.6
Gelatin E	0	0	2.857	17.9
Gelatin F	0	0	2.857	17.9
Gelatin G	0	0	2.857	17.9

After conducting the tests, a model was obtained by the least squares method and its fitting was verified using the analysis of variance (ANOVA). Models were tested for the following attributes as responses: colour, flavour, aroma, and texture with sugar content and collagen mass as factors. Optimisation was performed using the RSM in Software Statistica 10.0 for Windows®, São Paulo, Brazil, based on the concept proposed by [22,34–37]. The extracted material was transferred to Petri dishes and placed in a vacuum oven at 55 °C for 12 h. The dried material was then ground to obtain a powder and characterised according to the standards provided by the Association of Official Analytical Chemists (AOAC) [38].

2.3. Sensory Analyses

The acceptability of the gelatin samples was analysed by using sensory affective tests and compared with the sensorial qualities of commercial gelatin. Fifty (20 mL each) gelatin samples were served to the 50 consumers in codified plastic cups covered with a thin layer of plastic film, using a monadic presentation and rated using a hedonic nine-point scale. The consumers also registered their purchasing intentions for each sample on the same score sheet, using a nine-point scale. Sensorial characteristics such as flavour, colour, and aroma of gelatines were evaluated. The quantitative experimental research was conducted wherein the standard sample was used for sensorial analysis and random sampling was applied for each of the above attributes using a hedonic scale (1–9 points). This methodology is described in [15,21,24,30,39].

The numerical values in the hedonic scale represented the sensorial responses of the consumer as: 1—disliked very extremely, 2—disliked extremely, 3—disliked regularly, 4—disliked, 5—no perceived difference, 6—liked slightly, 7—liked regularly, 8—liked extremely, and 9—liked very extremely [15,21,24,30]. Furthermore, based on the frequency of responses, the sensorial data were compared by their *t*-values [24,30,39].

2.4. Determination of Chemical Properties

2.4.1. Percentage Composition

All measurements were conducted according to the standard methods of the AOAC [15,23,34,38] and results were expressed as the percent weight loss during the drying process. The moisture content of gelatine and chicken feet were analysed by drying at 105 °C for 8 h. The protein and lipid contents in gelatine and chicken feet were determined by the Kjeldahl method and the Soxhlet method, respectively. The total ash content of previously dried samples was evaluated by calcining at 500–600 °C for 4 h. The average of three batches under optimal extraction conditions was considered. To measure the energy contained in the gelatines, it was assumed that one gram of protein and one gram of lipid had 4 kcal of energy each and one gram of carbohydrate had 9 kcal of energy. A texture analyser TA-XT2, Stable Micro System (Surrey, UK) was used for determination of gel strength at 25 °C and 6.67% gelatin content (*w*/*w*) [4,19].

2.4.2. Spectroscopic Analyses

The elements Mn, Fe, Al, K, Ca, Ti, Mg, Na, Cr, V, Ni, Zn, Pb, Li, Cu, La, Ce, Th, U, Sr, and Be were determined in collagen ash samples by inductively coupled mass spectroscopy (ICP-MS) using the method by [4,11,12,40,41]. Gelatin content was measured by FTIR spectroscopy using a Nicolet iS5 FTIR spectrometer equipped with an ATR/iD3 with an argon horizontal cell (Thermo Fisher Scientific®, Waltham, MA, USA) at 16 °C. The spectra were analysed in the range of 400–4000 cm^{-1} and the automatic signals were obtained in 32 scans at a resolution of 4 cm^{-1} against a background spectrum recorded from a clean empty cell at 16 °C [4,11,12,40,41]. The average of 32 batch samples under optimal extraction conditions was considered as the result.

2.5. Gelatin-Base Gel Classification by Self-Organisation Maps

An initial prototype was implemented using the ANN technology and the Visual Basic ® software from Microsoft Co., for evaluating the sensorial qualities of seven chicken feet gelatin-base gel samples. The similar samples were classified into groups by SOM algorithm. The sensorial values attributed by consumers to each gelatin-base gel sample were catalogued in a spreadsheet. Then, the SOM algorithm was applied for clustering these sensorial values, creating a map where each region represented a group [22].

The working of SOM algorithm can be summarised as follows: when a pattern (a set of sensorial values in this study) is presented to the input layer, a neuron of the output layer (topological map) is selected to represent this pattern through a competitive process. During the training phase, the network increases the similarity of chosen neuron in the output layer and their neighbours to the pattern presented in the input layer. Thus, a topological map is developed wherein the neurons that are topologically close respond similarly to input patterns with similar characteristics.

Sensorial data set have been represented by the expression: $\zeta = \{(c_i, p_i) \in Z^4 \times \{A, B, \ldots, H\} : i = 1, 2, \ldots, 50\}$ so that $X_i = (c_i, p_i) \in \zeta$ to combine a sequence of sensorial evaluations, c_i (for aroma, flavour, appearance, and texture), from the product, p_i by consumer, i. Their similarities are presented in sensorial ζ in a Kohonen network. Furthermore, the neural network has the following configuration: a 1 × 18 enter vector, 5 × 5 topological structure, 500 iterations, and 0.0005 learning rate [22,27].

3. Results and Discussion

3.1. Gelatin Composition

Figure 1 presents photographs of the gelatin-base gel samples obtained from each of the formulations as shown in Table 1. There are negligible visual differences between the samples with respect to colour quality. There is a slight visual difference of gelatin F compared to the other samples. However, as this is one sample among 50, therefore, this analysis does not reveal anything definite about the sample quality.

Table 2 shows the analysis results of gelatin from chicken feet. The lipid content was found to be less than 7%, indicating a low-fat content (15% of total daily consumption). Protein content in chicken feet was higher than 78%, and 90% of it is gelatin (70.90% of total composition), revealing that this product was very rich in gelatin. Chicken feet gelatin has almost two and half times more gelatin content than cowhide (~35%) [4]. Approximately 99% of the mineral composition comprised the elements Na, K, Ca, Mg, P, and S, which are of great importance for human health. The gel strength of chicken feet gelatin powder was 295 bloom, which was five times higher than that of cow leather gelatin. Additionally, the amount of calories from this gelatin was 376 kcal, which corresponds to 18.8% of the daily consumption requirement (2000 kcal/day). All quality attributes of collagen from chicken feet were superior to those presented by cow leather gelatin (commercial), as already reported in [2,4,15]. This can be further corroborated with [7,9,12,42], which stated that gelatin should be available to trade from other alternative sources, as pork and beef products are not consumed by everyone, such as Islam and Jewish. The low molecular weight peptides formed during prolonged extraction processes possibly form covalent cross-links during the freeze-drying process [12,42–44]. This affected the gelatin content from commercial gelatin and reduced one of the primary qualities expected in gelatin. The process used in this study did not have the same effect on chicken gelatin.

Figure 1. Photographs of the different formulations of gelatin samples produced from chicken feet. (**A–G**) gelatin-base gels obtained of according to formulation from experimental design.

Table 2. Composition analysis of powdered gelatin from chicken feet.

Composition *	(g/100 g)	Sub-Composition (g/100 g)
Moisture	9.749	
Minerals	4.807	
	Na	0.0285 ± 1.10^{-4} *
	Mg	0.0088 ± 3.10^{-5}
	Ca	0.0081 ± 4.10^{-5}
	S	0.0072 ± 4.10^{-5}
	K	0.0041 ± 1.10^{-6}
	P	0.0008 ± 8.10^{-6}
	Si	0.0001 ± 8.10^{-6}
Lipids	6.919	
Proteins	78.525	
	Gelatin	70.90 ± 0.52 **
Gel strength, bloom	294.79 ± 0.50 ***	
Calories (kcal)	376.37	

Obtained by ICP-MS analysis *; FTIR analysis **, and texturometrical analysis *** [4].

3.2. Classification of Sensorial Attributes

Table 3 presents the averages of 50 tests obtained after compiling all sensory data derived from responses provided by the consumers. It was seen that the formulation of the gelatin "A", with lower concentrations of gelatin and sugar obtained the lowest values, and the mean was less than four points on the hedonic scale (disliked slightly). The formulation with higher concentrations of factors, that is gelatin "D", had the highest average for their sensory qualities, reaching values close to six points on the hedonic scale (liked slightly).

Table 3. Results of statistical analysis of gelatin qualities.

Sample	Colour [a]	Aroma [b]	Flavour [c]	Texture [d]	General Appearance [e]
Gelatin A	3.76	2.68	2.60	2.32	2.80
Gelatin B	4.60	3.10	2.70 [c]	5.88	3.72
Gelatin C	4.38	4.16	4.18 [c]	3.64	3.64
Gelatin D	5.92	5.30	5.76 [c]	6.76	6.06
Gelatin E	4.76	4.08	4.28 [c]	5.30	4.9
Gelatin F	4.72	4.54	4.14 [c]	4.40	4.3
Gelatin G	5.90	4.90	5.16 [c]	6.10	5.32
Gelatin H	7.32	7.36	7.34 [c]	7.2	7.26

Same letters means that there exist significant differences at 95% of level confidence: $t_{0.05,(98)} = 1.66$.

Samples of gelatin-base gel formulation "G" had sensory values represented by over five points on the hedonic scale following those of Gelatin "D". However, as it is one of the central point triplicates (Gelatines E, F, and G) and samples E and F did not exhibit similar performance, gelatin G cannot be indicated as one of the best. Furthermore, on adding the 150 analyses of the three gelatin samples, the average would be lower than the five points hedonic scale. Nevertheless, the *t*-test statistical evaluation showed no significant differences among all the gelatin formulations. Table 3 also shows the statistical analysis results. Notably, the lack of significant differences among gelatin samples according to the statistical analyses may demonstrate limitations of the statistical technique. This was because the samples had a high variation in their sensory results, including several outliers that were not eliminated from the original data sets of the samples.

Figures 2–5 show the clustering results of the sensorial data from Table 3. They show that the samples were perfectly classified into distinct groups by the hedonic values. It may be noted that the groups obtained by SOM algorithm were different from those of the statistical method. This demonstrated the efficiency of SOM algorithm on the sample clustering, as observed in the significant differences among means of samples presented in this table. Figure 2 presents the results obtained for clustering of aroma. After comparison, sensorial values for some gelatin samples were grouped based on their similarity by the SOM neural network and their average values. Three groups were obtained, with commercial gelatin (gelatin H) in the first group with an average of eight hedonic points; gelatines C, D, E, F, and G in the second group with approximately six hedonic points; and gelatines A and B in the third group with a low average of approximately two hedonic points.

Figure 3 shows the results obtained for clustering of the sensorial quality, flavour. Three groups are also shown wherein commercial gelatin (gelatin H) was in the first group with an average more than eight hedonic points; gelatines D, E, F, and G were in the second group with approximately six hedonic points and gelatines A, B, and C were clustered in the third group with approximately two hedonic points.

Figure 4 shows three groups in which gelatines have been classified according to their attribute, appearance. Here, the commercial gelatin (gelatin H) and gelatin D were clustered in the first group with an average higher than eight hedonic points; gelatines C, E, F, and G comprised the second group with approximately six hedonic points; and gelatines A and B were clustered to the third group with an average of approximately two hedonic points.

For the attribute texture, as shown in Figure 5, the gelatin D was grouped with the commercial gelatin H and their average is higher than eight hedonic points; gelatines B, C, E, F and G comprised the second group with approximately six hedonic points; and gelatin A was in a third group with a low average of approximately two hedonic points.

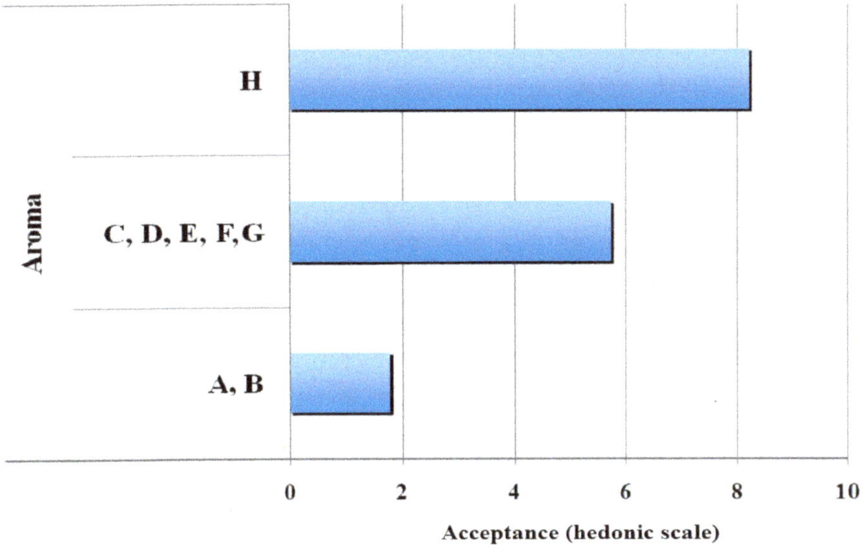

Figure 2. Results of clustering of gelatin-base gel samples for the attribute, aroma.

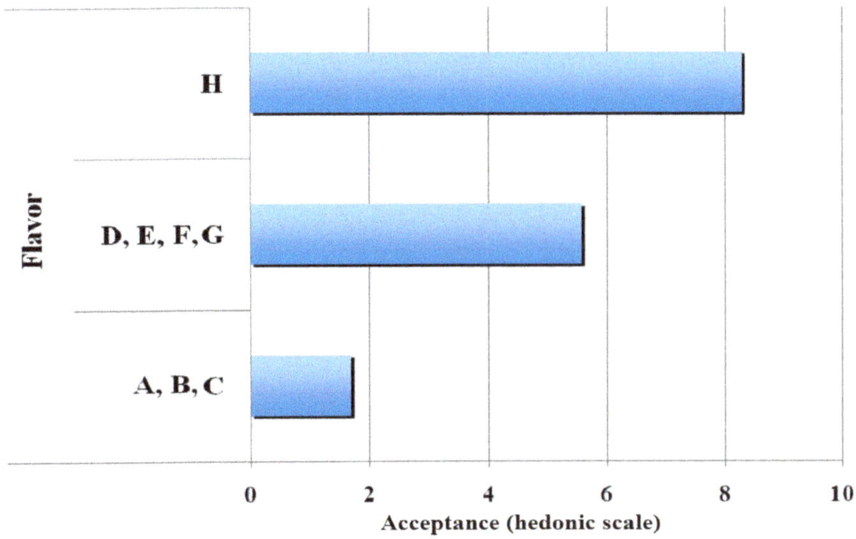

Figure 3. Results of clustering of gelatin-base gel samples for the attribute, flavour.

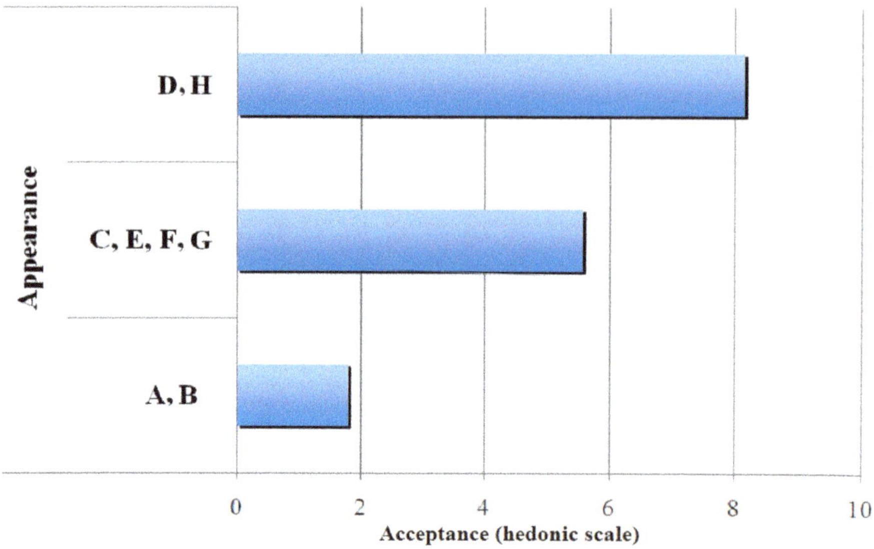

Figure 4. Results of clustering of gelatin-gal base samples for the attribute, appearance.

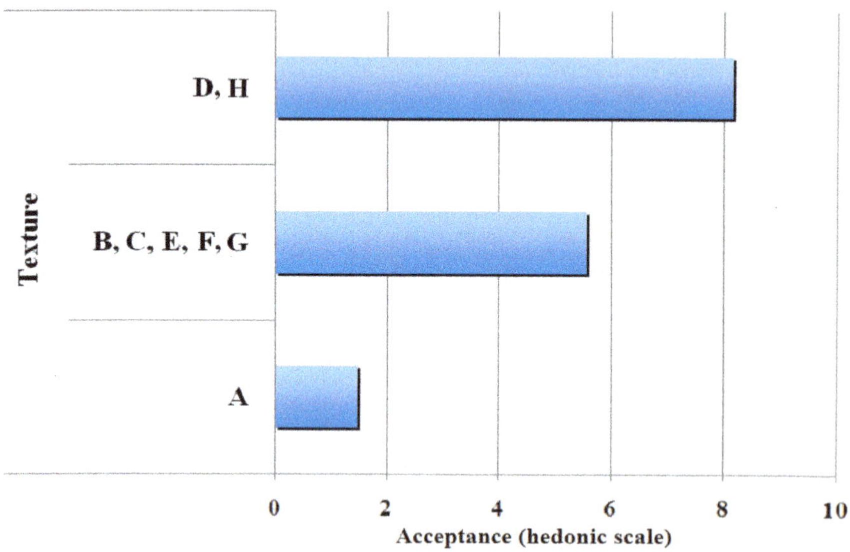

Figure 5. Results of clustering of gelatine samples for the attribute, texture.

3.3. *Optimisation of the Gelatin Production*

After regression by the least squares method, the models presented by Equations (1)–(4) were obtained. The values of the correlations were approximately close to unity, and though these are not the best as indicated by previous literature, this does not disprove the validity of the models. As noted, all the models showed a hyperbolic dependence of the factors with the responses. Similar models were obtained by Almeida et al. [15,16], when evaluating the dependence of appearance, flavour, and aroma on sugar concentration as factor, and mass of acerola pulp for the wines of the same fruit. The models in their study also did not have a good fit, but their correlation values were approximately close to unity as those obtained in this study.

Equation (1) shows that the influence of concentrations of the parameters (factors) are similar to the responses obtained for the sensory attribute, colour, as it was solely associated with the addition of the dye; all the gelatin samples had similar concentrations of dyes. As shown by Equations (2) and (3), the influence of sugar concentration was more significant than that of the collagen concentration on the sensory attributes of flavour and aroma. Almeida et al. [15,16] showed that sugar has a major influence in the responses for the aroma. Equation (4) shows that there was a higher influence of the concentration of gelatin than that of sugar in the sensory evaluation of the texture of gelatin, which is due to the fact that the higher amount of gelatin (collagen) provided a texture to the more rigid gels, leading to good consumer acceptance. These behaviours have also been observed by previous studies [5,6,11,12].

$$Color = 4.86 + 0.60 \cdot x_1 + 0.48 \cdot x_2 + 0.18 \cdot x_1 . x_2 \quad (\mathrm{R} = 0.8139) \quad (1)$$

$$Flavor = 4.11 + 0.39 \cdot x_1 + 0.92 \cdot x_2 + 0.18 \cdot x_1 . x_2 \quad (\mathrm{R} = 0.8827) \quad (2)$$

$$Aroma = 4.12 + 0.42 \cdot x_1 + 1.16 \cdot x_2 + 0.37 \cdot x_1 . x_2 \quad (\mathrm{R} = 0.9036) \quad (3)$$

$$Texture = 4.91 + 1.67 \cdot x_1 + 0.55 \cdot x_2 + 0.11 \cdot x_1 . x_2 \quad (\mathrm{R} = 0.9249) \quad (4)$$

Figure 6a–d show the response surfaces generated after optimising the production process of gelatine obtained from chicken feet, under the conditions considered in this study. An overall analysis showed that increasing the amount of sugar and collagen leads to a better product acceptance for all sensory attributes.

Based on the gelatin that presented the best set of sensory qualities, obtained by the SOM algorithm combined with the result obtained by RSM optimisation, a range of sugar and collagen contents can be used to obtain high quality gelatines, for example Gelatin D. Thus, the optimisation indicated that the qualities of gelatin will also be maintained if the gelatin is produced with levels between 0.8 and 1 for both the factors, implying the production of gelatines with sugar content ranging 26.5–28.5 g/100 g and collagen content ranging 3.62–3.80 g/100 g [23,33,35].

From the results obtained in this study the following advantages may be inferred for the gelatines produced from chicken feet over other gelatines: (1) it uses a very low-cost raw material; chicken feet are available for sale at approximately 1.0 R$/ton [14]; (2) it shows good nutritional and sensorial qualities; (3) the product is acceptable for consumption by everyone, irrespective of their socio-cultural beliefs [4,5,11].

Gelatin obtained from a tailing reduces their disposal in landfills, thereby decreasing their environmental impacts, such as the release of wastewaters and soil contamination. This further reduces the release of foul odours and prevents disease caused by pests generated due to its addition to landfills. Additionally, animal skin is very poorly regarded by consumers, which favours the shift from gelatines obtained from these raw materials to that obtained from chicken feet. The adoption of this technology by the meat industry will result in an increase in their profits, making it economically advantageous for industries [44–47].

According to Maceta, P.R.M., Berssaneti [48] and Cardoso et al. [49], considering the increased demand for natural foods and the variety of food choices, producers aim to differentiate their products to gain consumers. Therefore, producers may invest in new products, such as gelatine or collagen from chicken feet, which is considered a nutritive food, leading to increased consumption of the healthy food products, and increasing the quality of life of consumers. Further studies may be conducted to show that the investment in the sensory analysis software based on SOM neural network will result in high economic returns, because it results in innovative products with unique competitive advantages over other products in the market.

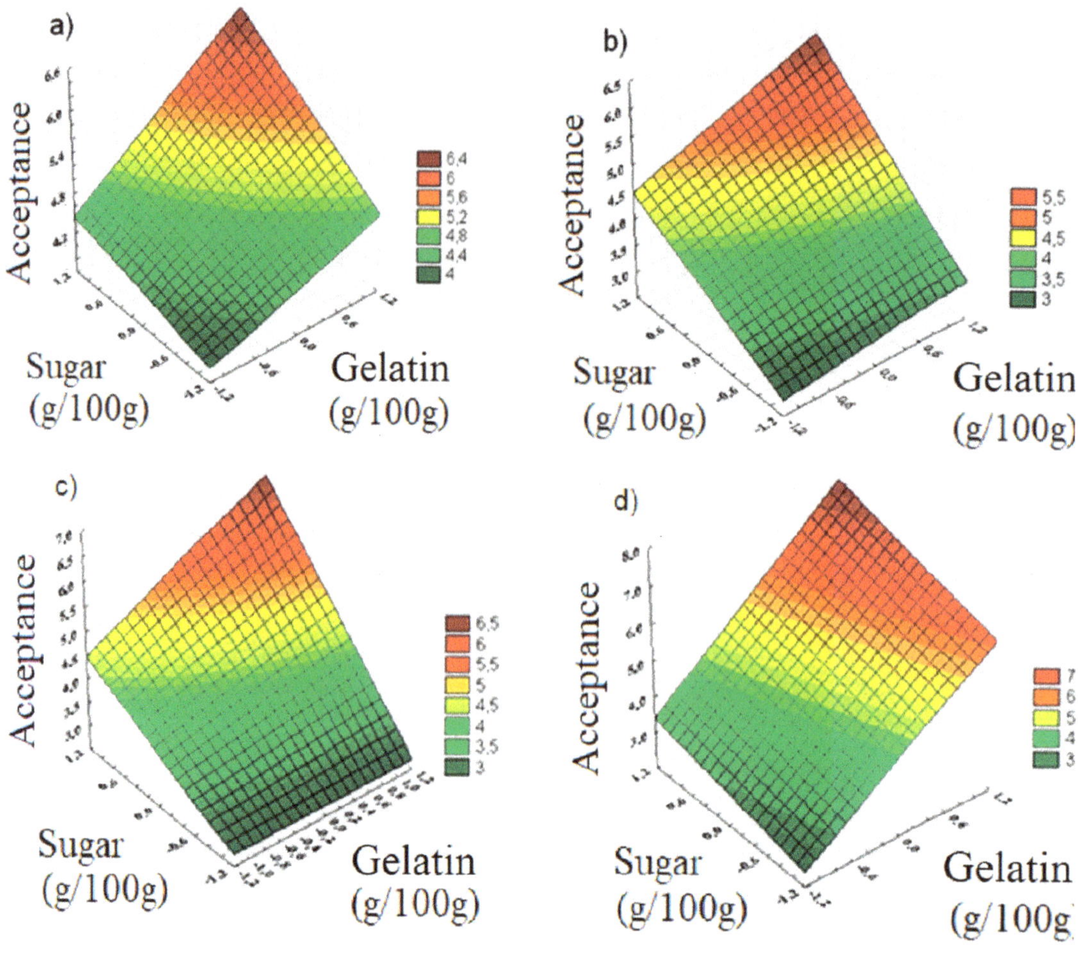

Figure 6. Response surfaces to optimise the gelatin-base gel attributes: (**a**) colour, (**b**) flavour, (**c**) aroma, and (**d**) texture.

4. Conclusions

The results showed that SOM is a suitable computational tool to study sensory characteristics of the samples and identify the best gelatin addition to prepare consumer acceptable gelatin-based gels. The developed models show that colour, flavour, aroma, and texture vary hyperbolically with the sugar and collagen contents. Gelatin samples D, E, F, and G had an average of approximately six hedonic points implying that they had good consumer acceptance and were marketable. Gelatin D, with 3.65–3.80% (*w/w*) powdered collagen and 26.5–28.6% (*w/w*) sugar, was found to be the best. In addition, this sample also showed the best sensorial qualities, varying between six and eight hedonic points. The results obtained may have significant contribution to the poultry production chain, because this study explored the utilisation of by-products. The findings of this research may be expanded to the supply chain to particularly include feet gelatin produced from the meat industry, so as to utilise the entire product life cycle. Furthermore, the use of chicken feet to produce gelatin may contribute to sustainability by minimising the waste from the poultry industry and providing an effective alternative through the development of a new product which also adds value to the supply chain of the meat industry.

Author Contributions: All authors contributed according with the fallowing distribution: Conceptualization, D.T.B., J.C.C.S., P.F.A., W.A.L.A. and S.A.A.; methodology, P.F.A., W.A.L.A., S.A.A., N.C., I.V. and J.C.C.S.; formal analysis, D.T.B., P.F.A., I.V., N.C., F.G. and A.P.B.Q.; resources, A.P.B.Q., S.A.V.L., F.M.C.G., F.T.B. and J.C.C.S.; writing—original draft preparation, F.T.B., P.C.M., J.C.C.S., F.G. and F.M.C.G.; writing—review and editing, A.P.B.Q., S.A.V.L., P.C.M., F.M.C.G., F.T.B., and R.M.V.; supervision, S.A.A., R.M.V., F.T.B. and J.C.C.S.; project administration, P.F.A., F.M.C.G., F.T.B., R.M.V., S.A.A. and J.C.C.S. All authors have read and agreed to the published version of the manuscript.

Funding: This study was financed by Coordination for the Improvement of Post-Graduation Level, Brazil, on funded number 01 and National Council for Scientific and Technological Development, CNPq, Brazil on funded number 305987/2018-6.

Institutional Review Board Statement: This research was developed under the Ethics Committee Protocol: P32432814.8.0000.5511, which was generated by the Brazil Platform of the Ministry of Health of the Federative Republic of Brazil in 06/16/2014.

Informed Consent Statement: Informed consent was obtained from all subjects involved in the study.

Data Availability Statement: http://www.saraujo.pro.br/sas/, accessed on 14 April 2021.

Acknowledgments: The authors thank CNPq, CAPES, FAPIC/CNPq (UNINOVE), Polytechnic School of USP and Carlos Alberto Vanzonini Fundation (FCAV) for financial supports.

Conflicts of Interest: All authors declare have no conflict of interest, and we warrant that this article is the authors' original work.

References

1. AviNews. Statistics and Prices. Available online: https://avicultura.info/pt-br/carne-de-frango-producao-exportacao-2020/ (accessed on 25 March 2021).
2. Almeida, P.F.; Araújo, M.G.O.; Santana, J.C.C. Collagen extraction from chicken feet for jelly production. *Acta Scientiarum. Technol.* **2012**, *34*, 345–351. [CrossRef]
3. Lopes, I.R.V.; Zapata, J.F.F.; Freitas, E.R.; Souza, D.V.; Viana Neto, J.L.; Lima, J.R. Meat quality and color of abdominal fat of broilers fed diets containing cashew nut meal treated with antioxidant. *Acta Sci. Technol.* **2013**, *35*, 163–174. [CrossRef]
4. Santana, J.C.C.; Gardim, R.B.; Almeida, P.F.; Borini, G.B.; Quispe, A.P.B.; Llanos, S.A.V.; Heredia, J.A.; Zamuner, S.; Gamarra, F.M.C.; Farias, T.M.B.; et al. Valorization of Chicken Feet By-Product of the Poultry Industry: High Qualities of Gelatin and Biofilm from Extraction of Collagen. *Polymers* **2020**, *12*, 529. [CrossRef] [PubMed]
5. Haug, I.J.; Draget, K.I.; Smidsrod, O. Physical and rheological properties of fish gelatin compared to mammalian gelatin. *Food Hydrocoll.* **2004**, *18*, 203–213. [CrossRef]
6. Karim, A.A.; Bhat, R. Extraction and characterisation of gelatin from Atlantic salmon (*Salmo salar*) skin. *Food Hydrocoll.* **2009**, *23*, 563–576. [CrossRef]
7. Maciel, C.N.; Seller, L.F.F.; Souza, A.B.; Almeida, P.F. Formulation of fishburgers with the addition of different protein sources and taro flour. *Ciência Rural* **2021**, *51*, e20200380. [CrossRef]
8. Eaqub Ali, M.d.; Sultana, S.; Abd Hamid, S.B.; Hossain, H.; Yehya, W.A.; Kader, A.; Bhargava, S.K. Gelatin controversies in food, pharmaceuticals, and personal care products: Authentication methods, current status, and future challenges. *Crit Rev Food Sci. Nutr.* **2018**, *58*, 1495–1511. [CrossRef]
9. Prosekov, A.Y.; Voroshilin, R.A. Gelatin production—Market status and prospects, alternative sources, production technologies. *BCË O MЯCE* **2020**, *5S*, 265–268. [CrossRef]
10. Badii, F.; Howell, N.K. Fish gelatin: Structure, gelling properties and interaction with egg albumen proteins. *Food Hydrocoll.* **2006**, *20*, 630–640. [CrossRef]
11. Gómez-Estaca, J.; Montero, P.; Fernández-Mantín, F.; Gómez-Guillén, M.C. Physico-chemical and film-forming properties of bovine-hide and tuna-skin gelatin: A comparative study. *J. Food Eng.* **2009**, *90*, 480–486. [CrossRef]
12. Oliveira, V.M.; Assis, C.R.A.; Costa, B.A.M.; Neri, R.C.A.; Monte, F.T.D.; Freitas, H.M.S.C.V.; França, R.C.P.; Santos, J.F.; Bezerra, R.S.; Porto, A.L.F. Physical, biochemical, densitometric and spectroscopic techniques for characterization collagen from alternative sources: A review based on the sustainable valorization of aquatic by-products. *J. Mol. Struct.* **2021**, *1224*, 129023. [CrossRef]
13. Acevedo, C.A.; Olguín, Y.; Orellana, N.; Sánchez, E.; Pepczynska, M.; Enrione, J. Anatase Incorporation to Bioactive Scaffolds Based on Salmon Gelatin and Its Effects on Muscle Cell Growth. *Polymers* **2020**, *12*, 1943. [CrossRef] [PubMed]
14. Almeida, P.F.; Silva, J.R.; Lannes, S.C.S.; Farias, T.M.B.; Santana, J.C.C. Quality Assurance and Economical Feasibility of an Innovative Product Obtained from a Byproduct of the Meat Industry in Brazil. *Afr. J. Bus. Manag.* **2013**, *7*, 2745–2756.
15. Almeida, P.F.; Calarge, F.A.; Santana, J.C.C. Production of a product similar to gelatin from chicken feet collagen. *Eng. Agrícola* **2013**, *33*, 1289–1300. [CrossRef]

16. Almeida, P.F.; Lannes, S.C.S. Effects of chicken by-product gelatin on the physicochemical properties and texture of chocolate spread. *J. Texture Stud.* **2017**, *48*, 392–402. [CrossRef]
17. Pereira, F.H.; Sassi, R.J. Rough sets and principal components analysis: A comparative study on customer database attributes selection. *Afr. J. Bus. Manag.* **2012**, *6*, 3822–3828.
18. Chen, T.-Y.; Wang, H.-P.; Wang, J.-C. Fuzzy automata based on Atanassov fuzzy sets and applications on consumers' advertising involvement. *Afr. J. Bus. Manag.* **2012**, *6*, 865–880.
19. Pourahmad, A.A.; Neshat, M.; Baghi, A. Service quality assessment in the academic library: Use of hybrid fuzzy expert system. *Afr. J. Bus. Manag.* **2012**, *6*, 11511–11529.
20. De Pelsmaeker, S.; Gellynck, X.; Delbaere, C.; Declercq, N.; Dewettinck, K. Consumer-driven product development and improvement combined with sensory analysis: A case-study for European filled chocolates. *Food Qual. Prefer.* **2015**, *41*, 20–29. [CrossRef]
21. Almeida, S.S.; Alves, W.A.L.; Araújo, S.A.; Santana, J.C.C.; Narain, N.; Souza, R.R. Use of simulated annealing in standardization and optimization of the *acerola* wine production. *Food Sci. Technol.* **2014**, *34*, 292–297. [CrossRef]
22. Santana, J.C.C.; Dias, C.G.; Souza, R.R.; Tambourgi, E.B. Applying of neural network on the wine sensorial analysis from Barbados cherry. *J. Food Process Eng.* **2010**, *33*, 365–378. [CrossRef]
23. Curvelo-Santana, J.C.; Ehrhardt, D.D.; Tambourgi, E.B. Otimização da produção de álcool de mandioca. *Ciência E Tecnol. De Aliment.* **2010**, *30*, 613–617. [CrossRef]
24. Severo Júnior, J.B.; Almeida, S.S.; Narain, N.; Souza, R.R.; Santana, J.C.C.; Tambourgi, E.B. Wine clarification from *Spondias mombin L.* pulp by hollow fiber membrane system. *Process Biochem.* **2007**, *42*, 1516–1520. [CrossRef]
25. Haykin, S. *Neural Networks: A Comprehensive Foundation*, 3rd ed.; Prentice-Hall, Inc.: Hoboken, NJ, USA, 2007.
26. Kohonen, T. Self-Organized Formation of Topologically Correct Feature Maps. *Biol. Cybern.* **1982**, *43*, 59–69. [CrossRef]
27. Alves, W.A.L.; Araújo, S.A.D.; Pessota, J.H.; Santos, R.A.B.O.D. A Methodology for Sensory Evaluation of Food Products Using Self-Organizing Maps and K-Means Algorithm. *Appl. Mech. Mater.* **2013**, *263–266*, 2191–2194. [CrossRef]
28. Liu, X.; Jiang, Y.; Shen, S.; Luo, Y.; Gao, L. Comparison of Arrhenius model and artificial neuronal network for the quality prediction of rainbow trout (*Oncorhynchus mykiss*) fillets during storage at different temperatures. *LWT Food Sci. Technol.* **2015**, *60*, 142–147. [CrossRef]
29. Ouyang, Q.; Chen, Q.; Zhao, J. Intelligent sensing sensory quality of Chinese rice wine using near infrared spectroscopy and nonlinear tools. *Spectrochim. Acta Part A Mol. Biomol. Spectrosc.* **2016**, *154*, 42–46. [CrossRef] [PubMed]
30. Yu, P.; Low, M.Y.; Zhou, W. Development of a partial least squares-artificial neural network (PLS-ANN) hybrid model for the prediction of consumer liking scores of ready-to-drink green tea beverages. *Food Res. Int.* **2018**, *103*, 68–75. [CrossRef] [PubMed]
31. Lu, A.; Wei, X.; Cai, R.; Xiao, S.; Yuan, H.; Gong, J.; Chu, B.; Xiao, G. Modeling the effect of vibration on the quality of stirred yogurt during transportation. *Food Sci. Biotechnol.* **2020**, *29*, 889–896. [CrossRef] [PubMed]
32. Sarkar, T.; Salauddin, M.; Hazra, S.K.; Chakraborty, R. Comparative classification efficiency of self-organizing map, principal component analysis, and hierarchical cluster Analysis for normal dairy and differently pineapple fortified rasgulla. *Es Food Agrofor.* **2020**, *1*, 94–105. [CrossRef]
33. Benvenga, M.A.C.; Librantz, A.F.H.; Santana, J.C.C.; Tambourgi, E.B. Genetic algorithm applied to study of the economic viability of alcohol production from Cassava root from 2002 to 2013. *J. Clean. Prod.* **2016**, *113*, 483–494. [CrossRef]
34. Severo, J.B., Jr.; De Sá Oliveira, L.S.; Sardeiro, F.S.; De Souza, R.R.; Lopes, F.L.G.; Santana, J.C.C.; Tambourgi, E.B. Response surface methodology to evaluation the recovery of amylases by hollow fiber membrane. *Braz. Arch. Biol. Technol.* **2007**, *50*, 713–718. [CrossRef]
35. Leite, N.S.; Lima, A.A.B.; Santana, J.C.C.; Lopes, F.L.G.; Lédo, A.S.; Tambourgi, E.B.; Souza, R.R. Determination of optimal condition to obtain the bromelain from pineapple plants produced by micropropagation. *Braz. Arch. Biol. Technol.* **2012**, *55*, 647–652. [CrossRef]
36. Vilanova Neta, J.L.; Lédo, A.S.; Lima, A.A.B.; Santana, J.C.C.; Leite, N.S.; Ruzende, D.S.; Silva, D.P.; Souza, R.R. Bromelain enzyme from pineapple: In vitro activity study under different micropropagation conditions. *Appl. Biochem. Biotechnol.* **2012**, *168*, 234–246. [CrossRef]
37. Rosa, J.M.; Tambourgi, E.B.; Santana, J.C.C.; Araujo, M.C.; Ming, W.C.; Trindade, N. Development of colors with sustainability: A comparative study between dyeing of cotton with reactive and vat dyestuffs. *Text. Res. J.* **2014**, *84*, 1009–1017. [CrossRef]
38. AOAC-Association of Official Analytical Chemists. *Official Methods of Analysis*, 15th ed.; AOAC: Washington, DC, USA, 1996.
39. Ehab, K.; Abouldahab, O.; Hassan, A.; El-Sayed, K.M.F. Alvogyl and absorbable gelatin sponge as palatal wound dressings following epithelialized free gingival graft harvest: A randomized clinical trial. *Clin. Oral. Investig.* **2020**, *24*, 1517–1525. [CrossRef]
40. Muyonga, J.H.; Cole, C.G.B.; Duodu, K.G. Fourier transform infrared (FTIR) spectroscopic study of acid soluble collagen and gelatin from skins and bones of young and adult Nile perch (*Lates niloticus*). *Food Chem.* **2004**, *86*, 325–333. [CrossRef]
41. Miranda, A.C.; Klepa, R.B.; Farias, T.M.B.; Santana, J.C.C. Quantification study of Azithromycin drugs in soil, by the infrared technique with Fourier Transform (IFTR). *Revista Ambiente e Agua* **2019**, *14*, e2268. [CrossRef]
42. Abedinia, A.; Nafchi, A.M.; Sharifi, M.; Ghalambor, P.; Oladzadabbasabadi, N.; Ariffin, F.; Huda, N. Poultry gelatin: Characteristics, developments, challenges, and future outlooks as a sustainable alternative for mammalian gelatin. *Trends Food Sci. Technol.* **2020**, *104*, 14–26. [CrossRef]

43. Nizar, N.N.A.; Hossain, M.; Sultana, S.; Ahamad, M.N.; Johan, M.R.; Eaqub Ali, M. Quantitative duplex real-time polymerase chain reaction assay with TaqMan probe detects and quantifies *Crocodylus porosus* in food chain and traditional medicines. *Food Addit. Contam. Part A* **2019**, *36*, 825–835. [CrossRef]
44. Lucato, W.C.; Vieira Jr, M.; Vanalle, R.M.; Salles, J.A.A. Model to measure the degree of competitiveness for auto parts manufacturing companies. *Int. J. Prod. Res.* **2012**, *50*, 5508–5522. [CrossRef]
45. Berssaneti, F.T.; Assumpção, A.; Nakao, O.S. Engineering, procurement and construction (EPC): What are the variables that impact the success of the projects currently running in Brazil? *Gest. Prod.* **2014**, *21*, 95–109. [CrossRef]
46. Miranda, A.C.; Silva Filho, S.C.; Tambourgi, E.B.; Santana, J.C.C.; Vanalle, R.M.; Guerhardt, F. Analysis of the costs and logistics of biodiesel production from used cooking oil in the metropolitan region of Campinas (Brazil). *Renew. Sustain. Energy Rev.* **2018**, *88*, 373–379. [CrossRef]
47. Santana, J.C.C.; Miranda, A.C.; Yamamura, C.L.K.; Filho, S.C.S.; Tambourgi, E.B.; Ho, L.L.; Berssaneti, F.T. Effects of air pollution on human health and costs: Current situation in São Paulo, Brazil. *Sustainability* **2020**, *12*, 4875. [CrossRef]
48. Maceta, P.R.M.; Berssaneti, F.T. Comparison of project portfolio management practices in the public and private sectors in Brazil: Characteristics, similarities, and differences. *Int. J. Manag. Proj. Bus.* **2019**, *13*, 1405–1422. [CrossRef]
49. Cardoso, J.F.; Casarotto Filho, N.; Miguel, P.A.C. Application of quality function deployment for the development of an organic product. *Food Qual. Prefer.* **2015**, *40*, 180–190. [CrossRef]

polymers

MDPI

Article

New Hyaluronic Acid/Polyethylene Oxide-Based Electrospun Nanofibers: Design, Characterization and In Vitro Biological Evaluation

Oana Maria Ionescu [1], Arn Mignon [2,*], Andreea Teodora Iacob [1], Natalia Simionescu [3,4], Luminita Georgeta Confederat [5], Cristina Tuchilus [5] and Lenuța Profire [1,*]

1 Department of Pharmaceutical Chemistry, Faculty of Pharmacy, Grigore T. Popa University of Medicine and Pharmacy of Iași, 16 University Street, 700028 Iasi, Romania; oana-maria.dc.ionescu@d.umfiasi.ro (O.M.I.); andreea.panzariu@umfiasi.ro (A.T.I.)
2 Smart Polymeric Biomaterials, Surface and Interface Engineered Materials, Campus Group T, KU Leuven, Andreas Vesaliusstraat 13, 3000 Leuven, Belgium
3 Centre of Advanced Research in Bionanoconjugates and Biopolymers, Petru Poni Institute of Macromolecular Chemistry, 41A Grigore Ghica Voda Alley, 700487 Iasi, Romania; natalia.simionescu@icmpp.ro
4 Emergency Clinical Hospital "Prof. Dr. Nicolae Oblu", 2 Ateneului Street, 700309 Iasi, Romania
5 Department of Microbiology, Grigore T. Popa University of Medicine and Pharmacy of Iasi, 700115 Iasi, Romania; georgeta-luminita.confederat@umfiasi.ro (L.G.C.); cristina.tuchilus@umfiasi.ro (C.T.)
* Correspondence: arn.mignon@kuleuven.be (A.M.); lenuta.profire@umfiasi.ro (L.P.)

Citation: Ionescu, O.M.; Mignon, A.; Iacob, A.T.; Simionescu, N.; Confederat, L.G.; Tuchilus, C.; Profire, L. New Hyaluronic Acid/Polyethylene Oxide-Based Electrospun Nanofibers: Design, Characterization and In Vitro Biological Evaluation. *Polymers* **2021**, *13*, 1291. https://doi.org/10.3390/polym13081291

Academic Editor: Andrea Ehrmann

Received: 17 March 2021
Accepted: 10 April 2021
Published: 15 April 2021

Abstract: Natural compounds have been used as wound-healing promoters and are also present in today's clinical proceedings. In this research, different natural active components such as propolis, Manuka honey, insulin, L-arginine, and *Calendula officinalis* infusion were included into hyaluronic acid/poly(ethylene)oxide-based electrospun nanofiber membranes to design innovative wound-dressing biomaterials. Morphology and average fiber diameter were analyzed by scanning electron microscopy. Chemical composition was proved by Fourier transform infrared spectroscopy, which indicated successful incorporation of the active components. The nanofiber membranes with propolis and *Calendula officinalis* showed best antioxidant activity, cytocompatibility, and antimicrobial properties against pathogen strains *Staphylococcus aureus*, *Escherichia coli*, and *Pseudomonas aeruginosa* and had an average diameter of 217 ± 19 nm with smooth surface aspect. Water vapor transmission rate was in agreement with the range suitable for preventing infections or wound dehydration (~5000 g/m^2 24 h). Therefore, the developed hyaluronic acid/poly(ethylene)oxide nanofibers with additional natural components showed favorable features for clinical use as wound dressings.

Keywords: hyaluronic acid; polyethylene oxide; electrospinning; nanofibers; wound dressings

1. Introduction

The acute and chronic wound care management represents a major health problem in the entire world, and biomedical research is constantly seeking new options with respect to promoting the healing process and to reducing the therapy cost. In 2021, the wound therapy cost is estimated to exceed 20 billion dollars [1]. Therefore, a cost-effective therapy is needed to aid the patient overcome this health issue in a shorter timeframe to prevent further complications that can debilitate someone's social, economic, and mental wellbeing [1].

The dynamic process of wound healing needs the support of a proper dressing that can mimic and promote the natural process of healing [2]. While it is important to keep a moist environment that can aid healing, the designing of a dressing that would meet the demanding qualities for maintaining cell growth whilst ensuring their differentiation, still represents an important aspect of biomaterial engineering. Different dressing biomaterials such as films, hydrogels, sponges, and electrospun membranes are studied for their properties as a support in wound healing [3]. These materials are characterized for their surface properties and morphology that are highlighted using scanning electronic microscopy, transmission electronic microscopy, or fractal model for porous membranes [4–6].

Nanofibers (NFs) are relatively new porous systems currently investigated [7,8] for their multiple purposes. Based on their unique properties (large surface and high porosity), NFs act as excellent extracellular matrices [9] which enhance the tissue formation. Mass transport through porous nanofibers is also an important characteristic and using a mathematical model based on fractal character of micropores, the imbibition height and the imbibition mass could be calculated [5]. Designing NFs delivers a large range of possibilities in the matter of selecting the polymers and active components (ACs) [10,11].

In the case of wound healing, excessive production of reactive oxygen species may lead to slow regeneration of the tissue. Vitamins, catechol-based products, and glutathione proteins have been studied for their antioxidant activity and presented benefits when added to wound dressings [12,13]. An antimicrobial effect is a desirable property of a wound-dressing material since it can diminish the inflammatory response to local infection and help with the ever-growing problem of antibiotic resistance [14].

The rationale of the wound dressings developed in this study starts from having a polymeric matrix, based on hyaluronic acid (HA), that not only protects the wound from external damaging agents yet adds value through biological effects, with HA promoting the cellular proliferation. The active components chosen to be incorporated were taken into consideration for their benefits in speeding the healing process, because of the multitude of risks associated with wounds. Therefore, re-epithelization agents (Manuka honey and insulin), antimicrobial agents (propolis, Manuka honey), an angiogenetic compound (L-arginine), and an antioxidant/re-epithelization agent (*Calendula officinalis* infusion) were selected for this study.

Hyaluronic acid (HA) is a glycosaminoglycan found in high concentrations in connective tissue and skin. Chemically, it is a linear polysaccharide made of repeating units of the following disaccharides: 1-4-d-glucuronic acid and 1-2-*N*-acetyl-D-glucosamine [15]. Shortly after the wound has occurred, HA may act as a temporary supporting structure, as a result of its high molecular weight and viscoelasticity [16]. Therefore, it permits nutrients and cells (keratinocytes) to diffuse easily. In the first phase of wound healing (haemostasis), HA combines with fibrin and forms a compound that accelerates cell's migration. It can also stimulate cutaneous adnexa (follicles, apocrine or sebaceous glands). Unfortunately, HA has low electrospinning properties by itself, and various successful attempts have been made using additional compounds to convert it into NFs [6,17,18]. In water, the HA chains form an expanded coil and low concentration solutions deliver polymer chain entanglements. The higher the concentration conducts, the higher the viscosity, which is not beneficial for the electrospinning process. To provide lower viscosity and successful electrospinning of HA solutions, sodium chloride and PEO was used as viscosity modifiers. This blended HA solution is less viscous, has a pH level above 4, and provides successful NFs [18].

Polyethylene oxide (PEO) is a synthetic, biocompatible, biodegradable, hydrophilic polymer and is easily soluble in water. It is often used in the fabrication of NFs through the electrospinning technique and mostly plays the role of a copolymer in the formulation.

Although different HA-PEO nanofibers have been previously prepared [6,19] into blend or core-shell structures, none have yet incorporated Manuka honey, propolis, *Calendula officinalis* infusion, and L-arginine to our knowledge.

The main component of Manuka honey is methylglyoxal (MGO). Its immunomodulatory action during healing is associated with the release of cytokines from leucocytes, thus regulating proliferation of fibroblasts. It is in fact a natural derivative with re-epithelization and antibacterial effects at the wound site. By reducing inflammation (via decreasing the reactive oxygen species scavenging activity) and absorbing wound exudate through its osmotic mechanism [20], it prevents maceration while promoting angiogenesis and granulation tissue production. The final effects are accelerating wound contraction and epithelialization [21,22].

Propolis is a resinous product made by bees. Through its complex chemical composition, it displays numerous biological effects. Local flora, climate condition, weather, and other components such as wax or pollen may influence the chemical composition. Of high interest from a biological point of view are the flavonoids, phenolic acids, and their ester derivatives. The literature is abundant of examples of applications for propolis based on its antioxidant, anti-inflammatory, and antimicrobial properties. Propolis was previously incorporated into NFs for the treatment of leg ulcers [23,24].

L-arginineis a basic α-amino acid that plays several key roles in the physiology of the cell. Its alkaline guanidine group (carbon-nitrogen double bond) has reductive effects that can enhance antioxidant activity in wound-dressing materials [13]. It may also be considered as a nitric oxide precursor [25] that is responsible for local responses at the wound site [26] and is involved in granulation tissue formation and also epithelialization [27]. It was noticed that low concentrations of L-arginine are present at the wound site because of the arginase enzyme which is responsible for the hydrolysis of L-arginine to degradation compounds (urea and L-ornithine) [28].

Insulinis a hormone which acts as a growth factor and promotes healing at the wound site [29]. When applied to a wound, insulin promotes the production of granulation tissue, speeds epithelialization, and wound contraction [30].

Calendula officinalis is often used in therapy as an infusion, tincture, or ointment when treating inflammation of the skin, burns, insect bites, and leg ulcers. It displays curative properties due to its flavonoids and saponins [31–33].

Although propolis has already been successfully incorporated into polyurethane-HA [34] and L-arginine into polyvinylalcohol-HA [35] scaffolds, no combination of them with other ACs, such as insulin, Manuka honey, and *Calendula officinalis* has been yet reported. Many benefits may derive from combination of ACs that can enhance the biological properties of HA with respect to regenerative medicine. In the emerging concept of biomimetism, which is adapting principles from nature into science and medicine, we report the preparation, physicochemical characterization and in vitrobiological evaluation of HA-PEO-NFs, with biologically ACs, such as Manuka honey, propolis, *Calendula officinalis* infusion, insulin, and L-arginine, blended into the polymer matrix in different combinations, as possible candidate materials for wound applications.

2. Materials and Methods

2.1. Materials

Hyaluronic acid (HA) sodium salt from *Streptococcus equi*, with a molecular weight of 1.5×10^6 Da (which is classified as high molecular weight) and poly(ethylene oxide) (PEO), with a molecular weight of 200 kDa, L-arginine·HCl, DPPH (2,2-Diphenyl-1-picrylhydrazyl), α-bromo-naphtalene (97%), ABTS (2,2′-azino-bis(3-ethyl-benzthiazoline-6-sulfonic acid), disodium hydrogen phosphate, sodium chloride, and sodium dihydrogen phosphate, were purchased from Sigma-Aldrich (Merck Group, Schnelldorf, Germany). Insulin (Novo Rapid Penfill 100 IU/mL, (Novo Nordisk, Bagsvaerd, Dennmark), Manuka honey 550+ (Manuka HealthLtd., Te Awamutu, New Zealand), *Calendula officinalis flos* (Fares, Orăştie, Romania), and propolis (Apiland, Baia Mare, Romania) were incorporated into the nanofibers. Gram-negative (*Escherichia coli* ATCC 25922, *Pseudomonas aeruginosa* ATCC 27853), and Gram-positive (*Staphylococcus aureus* ATCC 25923) bacterial strains as well as pathogenic yeasts (*Candida albicans* ATCC 10231) were provided by the Department of Microbiology of Grigore T. Popa University of Medicine and Pharmacy of Iasi, Romania. All materials were used without further purification. A sterile, isotonic, and aqueous solution of sodium chloride (0.9%) (B. Braun®, Melsungen, Germany) was purchased. Alpha-MEM medium (Lonza, Basel, Switzerland), fetal bovine serum (FBS, Gibco, Thermo Fisher Scientific, Waltham, MA, USA), and 1% penicillin-streptomycin-amphotericin B mixture (10 K/10 K/25 μg, Lonza, Basel, Switzerland) were also used.

2.2. Preparation of HA-PEO-ACs Solutions

Firstly, the HA was dissolved in normal saline solution, and then PEO was added, and a homogeneous polymeric solution was obtained after 24 h of stirring (EcoStir, DLAB Scientific Inc., Ontario, CA, USA). The molecular weight of PEO is a key factor in obtaining HA-PEO-NFs. Previous attempts which have included PEO 1000 kDa proved unsuccessful in obtaining NFs. When PEO 200 kDa was used, the electrospinnability of the polymeric solutions was optimal, and successful HA-PEO-NFs were obtained. The total polymer concentration was kept at 12.5% (wt/wt). Afterwards, different ACs, such as Manuka honey (M), propolis (P), insulin (I), and L-arginine (L), were added. The electrospinning solutions were blended on a tabletop shaker (MaxQ HP Tabletop Shaker, ThermoFisher Scientific, Waltham, MA, USA) until a homogenous composition was reached. Afterwards, the solutions were left to rest for 1 h for degassing at room temperature. Detailed data on the electrospinning solutions are listed in Table 1. To prepare the HA_PEO@PC solution, slight modifications were applied to the protocol: an infusion of *Calendula officinalis flos* (C) was firstly prepared according to the EMEA monograph from 1.5 g of *flos* and 150 mL water (70 °C). Then, the solution was filtered and kept at room temperature and used in the first 24 h as a solvent for HA, PEO, and propolis.

Table 1. Parameters of hyaluronic acid-polyethylene oxide-active components (HA-PEO-ACs) blending electrospinning solutions.

HA-PEO-NFs	Solvent	ACs
HA_PEO@P	normal saline	P: 7% (wt/V) [1]
HA_PEO@IP	normal saline	I: 100 IU/mL P: 7% (wt/V)
HA_PEO@PC	*Calendula officinalis* infusion	P: 7% (wt/V)
HA_PEO@ML	normal saline	M: 7% (wt/V) [1] L: 7%(wt/V) [1]

[1] Higher concentrations led to dripping and bead formation.

2.3. The Viscosity Measurements of HA-PEO-ACs Solutions

The rheological properties of the HA-PEO-ACs solutions were studied using the Ostwald model to show the pseudoplastic behavior and the Carreau–Yassuda model to determine the zero-shear viscosity. A rheometer from Anton Paar (Physica MCR, Graz, Austria) with a 50 mm upper plate set in a parallel-plate system was used. The flow curves were carried out with shear-rates ranging from 0.01 to 500 s^{-1}. Before each determination, the solutions were firstly equilibrated at 25 °C. The apparent viscosity of the polymeric solutions was determined at 100 s^{-1}.

2.4. Preparation of HA-PEO-NFs

Referring to HA-PEO blending solution, HA is a negatively charged polymer, whose carboxyl groups are ionized in carboxylate ions in the normal saline, which serves as an aqueous solvent and endows the HA electrical properties. Under the electrostatic force, the HA molecules move along the opposite direction of the electric field. Electrospinning was carried out using the Nanospinner Inovenso (Inovenso Ltd., Istanbul, Turkey) equipment, which is composed of an infusion pump, a generator, and a negative electrode under the form of a plate collector. Electrospinning parameters were kept constant during the experiments, with a 13–15 cm distance from the needle tip to the collector, a flow rate of 0.5–0.7 mL/h, and a voltage below 20 kV.

2.5. Characterization of the HA-PEO-NFs

2.5.1. Surface Morphology and Fiber Diameter

Fiber morphology was carried out using scanning electron microscopy (SEM). The micrographs were performed with a Phenom Desktop SEM-FEI (Hillsboro, OR, USA). Dry samples were set to a sample holder with a double-sided carbon tape, and then gold-sputter-coated (20 mA, 60 s, vacuum) by an automatic Au sputter coater EmiTech K550X

(EmiTech Ltd., Ashford, UK) with a RV3 two-stage rotary vane pump. Fiber diameter and distribution were analyzed using the phenom fiber metric software (Phenom Pro, Eindhoven, the Netherlands). At least three images per sample and 150 fibers per image were measured. Data are shown as average diameter ± standard deviation (SD).

2.5.2. Fourier Transform Infrared Spectroscopy (FTIR)

The infrared spectra of the developed HA-PEO-NFs were recorded using the IR ABB MB300 spectrophotometer (ABB, Zurich, Switzerland). All spectra were collected in the wave number range 4000–600 cm^{-1} at a resolution of $4 cm^{-1}$.

2.5.3. Water-Vapor Transmission Rate (WVTR)

In order to measure the HA-PEO-NFs moisture permeability, the evaluation of WVTR was carried out according to the standard water method assayed by the American Society for Testing and Materials (ASTM 95-96) and European Pharmacopoeia, with slightly modifications [36]. All samples were dried before the test, then cut (3 cm diameter samples), and set on the mouth of glass vials (of 2 cm diameter) which contained 10 mL of distilled water each. The NFs were secured with paraffin tape, alongside the vials' edges. Afterward, the vials were placed in an incubator (Roth M, Karlsruhe, Germany) and kept at constant temperature and humidity (37 °C, to mimic physiological temperature). The rate of change of mass in the water from the vials over which the NFs were mounted was recorded. The weight change was measured, and the WVTR was calculated according to the following equation:

$$\text{WVTR}\ (\text{g/m}^2 \cdot 24\ \text{h}) = (W_i - W_f)/A \times t \qquad (1)$$

where W_i and W_f represent the initial and final weight of bottles containing water and A is the exposure area. The experiment was performed in triplicate, and the results are expressed as mean ±SD.

2.5.4. In Vitro Antioxidant Assays

The radical scavenging activity of HA-PEO-NFs toward DPPH and $ABTS^{\bullet+}$ was measured. Then, the total antioxidant capacity and also ferric-reducing antioxidant power were determined. Samples of 60 mg, corresponding to the HA-PEO-NFs, were immersed in 5 mL ethanol, and a set volume of the ethanolic extract was used for experiments. A GBC Cintral UV–Vis spectrophotometer (GBC Scientific Equipment, Braeside, Victoria, Australia) was used to perform measurements. All experiments were performed in triplicate, and the results are expressed as mean ± SD.

DPPH Radical Scavenging Assay

The DPPH assay is one of the most used methods to determine the radical scavenging effects. The violet color of DPPH in ethanolic solution turns into a yellow in the presence of a proton-donating agent [37], indicating the reduction of the DPPH (Figure 1).

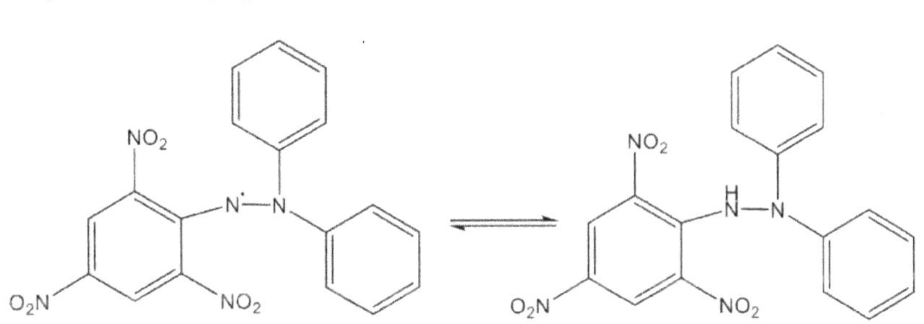

Figure 1. The reduction reaction of DPPH.

In brief, a volume of 0.25 mL of the ethanolic extract was mixed with 1 mL of 100 μM DPPH ethanolic solution. The obtained solution was kept in the dark at room temperature. After 60 min, the absorbance was measured at 517 nm, and the inhibition (%) was calculated according to the following equation:

$$\text{Inhibition (Scavenging activity) \%} = [(A_{DPPH} - A_s)/A_{DPPH}] \times 100 \quad (2)$$

where A_s represents the absorbance of the HA-PEO-NFs sample and A_{DPPH} is the absorbance of the DPPH ethanolic solution.

$ABTS^{\bullet+}$ Radical Scavenging Assay

The ABTS (7 mM) aqueous solution was prepared with ammonium persulfate (2.45 mM) and was kept in the dark for 16 h in order to initiate the formation of the $ABTS^{\bullet+}$. The literature protocols indicate that the solution be diluted with ethanol to obtain an absorbance of 0.700 ± 0.02 at 734 nm [38]. The blue chromophore $ABTS^{\bullet+}$is reduced in the presence of hydrogen donating agents (Figure 2). The result is the decrease of the absorbance [39].

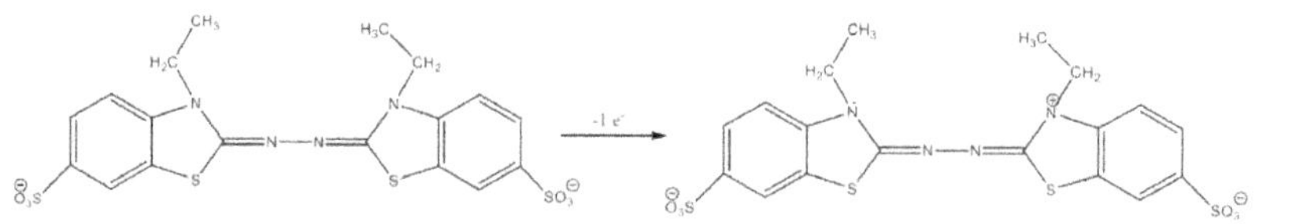

Figure 2. The generation reaction of $ABTS^{\bullet+}$.

A volume of 1 mL ABTS solution was added to 0.25 mL of HA-PEO-NFs ethanolic extract. After 6 min, the absorbance of the samples was measured at 734 nm against a blank (ethanol was used instead of the set volume of HA-PEO-NFs ethanolic extract). The inhibition (%) representing the $ABTS^{\bullet+}$radical-inhibiting capacity was calculated according to the following equation [38,39]:

$$\text{Inhibition (Scavenging activity) \%} = [(A_{ABTS} - A_s)/A_{ABTS}] \times 100 \quad (3)$$

where A_{ABTS} is absorbance of the $ABTS^{\bullet+}$ solution; A_s is absorbance of the HA-PEO-NFs sample.

Phosphomolybdenum-Reducing Antioxidant Power (PRAP) Assay

The total antioxidant capacity of HA-PEO-NFs was assayed using the modified phosphomolibdenum method, as was previously described in literature [40]. This spectrophotometric determination is based on the reducing of Mo(VI) to Mo(V) in the presence of electron donating agents in acidic medium [41]. The resulting phosphomolibdenum complex is of green color. Briefly, a volume of 2 mL reagent solution (consisting of 0.6 M sulfuric acid, 4 mM ammonium molibdate, and 28 mM disodium hydrogen phosphate) was mixed with 0.2 mL of HA-PEO-NFs ethanolic extract. All samples were then incubated at 95 °C for 90 min. After cooling (reaching room temperature), the absorbance was measured at 695 nm against the blank (where ethanol was used instead of the set volume of HA-PEO-NFs extract). The increase of optical density implies superior antioxidant effects.

Ferric-Reducing Antioxidant Power (FRAP) Assay

The ferric-reducing antioxidant power method is based on the reduction of ferricyanide into ferrocyanide (blue), in the presence of electron donating agents [31]. To 0.5 mL HA-PEO-NFs extract, 0.5 mL of 0.2 M phosphate buffer (pH = 6.6) were added. To initiate the reaction, 0.5 mL of potassium ferricyanide 1% (wt/*V*) were added, and the samples were incubated at 50 °C for 20 min. To complete the reaction, 0.5 mL of trichloroacetic acid 10% (wt/*V*) were added. In addition, 1 mL of double distilled water and 0.2 mL of ferric chloride 0.1% (wt/*V*) were added to 1 mL of the previously resulting solution. The samples were left to rest for 10 min at room temperature, and their absorbance was measured at 700 nm against a blank (consisting of ethanol instead of the set volume of HA-PEO-NFs ethanolic extract). As the case of the PRAP assay, the increase of optical density implies superior antioxidant effects.

2.5.5. In Vitro Cytotoxicity (MTS) Assay

The cytotoxicity degree of HA-PEO-NFs was determined by MTS assay, by using the CellTiter 96®aqueous one-solution cell proliferation assay (Promega, Madison, WI, USA). The method was according to the manufacturer's set of instructions and a procedure adapted from ISO 10993-5 [42]. Normal dermal fibroblast cells were grown in alpha-MEM medium supplemented with 10% fetal bovine serum and 1% penicillin-streptomycin-amphotericin B mixture and seeded at a density of 0.5×10^5 cells/mL into 96-well plates treated with tissue culture and were allowed to adhere for 24 h. The cells were afterwards incubated for 72 h with 100 μL of different concentrations of HA-PEO-NFs ethanolic extract (12.5, 25, 50, and 100 μg/mL) or with fresh complete medium (which also served as control). A volume of 20 μL of the MTS reagent (tetrazolium inner salt) was added to the samples, and after 3 h, the absorbance was read at 490 nm on a FLUOstar® Omega microplate reader (from BMG LABTECH, Ortenberg, Germany). The viability of the cells was expressed as a percentage (%) of the control cells' viability. The experiment was performed in triplicate, and the results are expressed as mean ± SD.

2.5.6. Antimicrobial Assay

The antimicrobial assay consisted of determination of antibacterial and antifungal activity, by measuring the diameter of the inhibition area, according to the agar disc diffusion method [43]. The Gram-negative (*Escherichia coli* ATCC 25922, *Pseudomonas aeruginosa* ATCC 27853) and Gram-positive (*Staphylococcus aureus* ATCC 25923) bacteria were used as bacterial strains and *Candida albicans* ATCC 10231as pathogenic yeast. The discs containing Ciprofloxacin (5 μg/disc) and Voriconazole (1 μg/disc) were used as controls (Whatman plc, Buckinghamshire, UK) for bacterial and yeast strains, respectively. Suspensions of the bacterial and fungal strains (10^6 CFU/mL in NaCl 0.9%) were inoculated onto the surface of the culture medium spread at a volume of 25 mL/sterile Petri plate (Mueller–Hinton for bacterial strains and Sabouraud for fungal strain, Merck, Darmstadt, Germany). A volume of 200 μL of HA-PEO-NFs aqueous extract was added into sterile stainless-steel cylinders of 5 mm internal diameter and 10 mm height and placed on the agar surface. The plates were incubated at 37 °C for 24 h (antibacterial activity assessment) and at 35 °C for 48 h (antifungal activity assessment), after which the diameters of the inhibition area (mm, including the disc size) were measured.

2.5.7. Data Analysis

All the experiments were performed in triplicate. The statistical significance was calculated with a t-Test and a one-way ANOVA, where a p-value < 0.05 was considered statistically significant.

3. Results and Discussion

3.1. Viscosity Measurements of HA-PEO-ACs Solutions

The rheological characterization of HA-PEO-ACs solutions, based on Carreau–Yassuda model, is shown in Table 2. Adding Manuka honey to the polymeric solution was partly responsible for the higher zero-shear viscosity value (HA_PEO@ML, 1.7 Pa·s) when compared to the solutions containing insulin, propolis, or *Calendula officinalis* infusion (HA_PEO@IP; 1.5 Pa·s; HA_PEO@P, 1.2 Pa·s; HA_PEO@PC, 0.9 Pa·s). Viscosity values are directly related to the composition of polymeric solution (content, molecular weight of polymer, and solvent type).

Table 2. Viscosity parameters of HA-PEO-ACs blending electrospinning solutions.

HA-PEO-ACs	Ostwald I Model		Carreau–Yassuda Model Zero-Shear Viscosity, η_0 (Pa·s)	Apparent Viscosity, at 100 s^{-1} ($\eta_{a,100}$, Pa·s)
	Flow Behavior Index, a (K)	Consistency Coefficient, b (n)		
HA_PEO@IP	2.5	0.56	1.5	0.4
HA_PEO@ML	3.8	0.59	1.7	1.5
HA_PEO@P	2.3	0.57	1.2	0.4
HA_PEO@PC	1.9	0.58	0.9	0.3

3.2. *Physicochemical Characterization of HA-PEO-NFs*

3.2.1. SEM Morphology and Fiber Diameter

The relative humidity (RH) and concentration of HA have been previously reported to affect the electrospinning of PEO-HA solutions [44,45]. At a lower RH, the aqueous solvent evaporates at a faster rate, and the jet is solidified very quickly during the electrospinning process. In addition, it was noticed that high HA content in PEO solutions and low RH leads to the increase of the viscosity of the polymer solution, which decreases the stretching of the jet and forms NFs with a larger diameter [19].Therefore, the assuring constant RH is a key factor to uniform results. The experiments were carried out at 35–40% RH. As depicted in Figure 3, the developed HA-PEO-NFs displayed a bead-free nanostructure, a homogenous network with smooth surfaces, and uniform diameters. The HA_PEO matrix presented an average diameter of 166 ± 11 nm. The adding of Manuka honey has as result increasing of the diameter of the NFs to 201 ± 12 nm (HA_PEO@ML). This is possibly due to an increase in the charge density caused by the highly hydrophilic Manuka honey, which in turn leads to lesser extension of the polymer jet. In the presence of propolis and *Calendula officinalis* infusion, the average diameter of the NFs increased to 217 ± 19 nm (HA_PEO@PC). Generally, the diameter of developed NFs was less than other NFs reported in the literature [43]. This could be due to higher molecular weight of HA which was used in combination with PEO low molecular weight.

3.2.2. Fourier Transform Infrared Spectroscopy (FTIR)

The composition of the HA-PEO-ACs NFs was proved by the IR spectra (Figure 4), and the collected data are in agreement with the literature [46,47]. The characteristic amides I and II groups of HA were attributed to the absorption bands at 1500 to 1670 cm^{-1}. Characteristic bands for carbonyl (C=O) groups appear in the HA spectrum at 1638 cm^{-1} (correspondingly to the amide) and at 1420 cm^{-1} (attributed to the –COO stretching). The free OH groups and hydrogen-bonded OH were identified at 3000–3600 cm^{-1}. In addition, the pyranose ring vibrations (at 950 and 11,250 cm^{-1}) and C–O–C due to the glycosidic bonds (at 1103 cm^{-1}) were also identified. The absorption band at 3447 cm^{-1} was attributed to OH and NH stretching while the absorption band at 2925 cm^{-1} to stretching vibrations of C–H. The characteristic bands that indicate the presence of PEO are around at 2885 cm^{-1} corresponding to the methylene stretching and at around 1100 cm^{-1} as a result of combination of methylene group and ether group stretching, also referred to in the literature [48] as the –C–O–C– absorption complex.

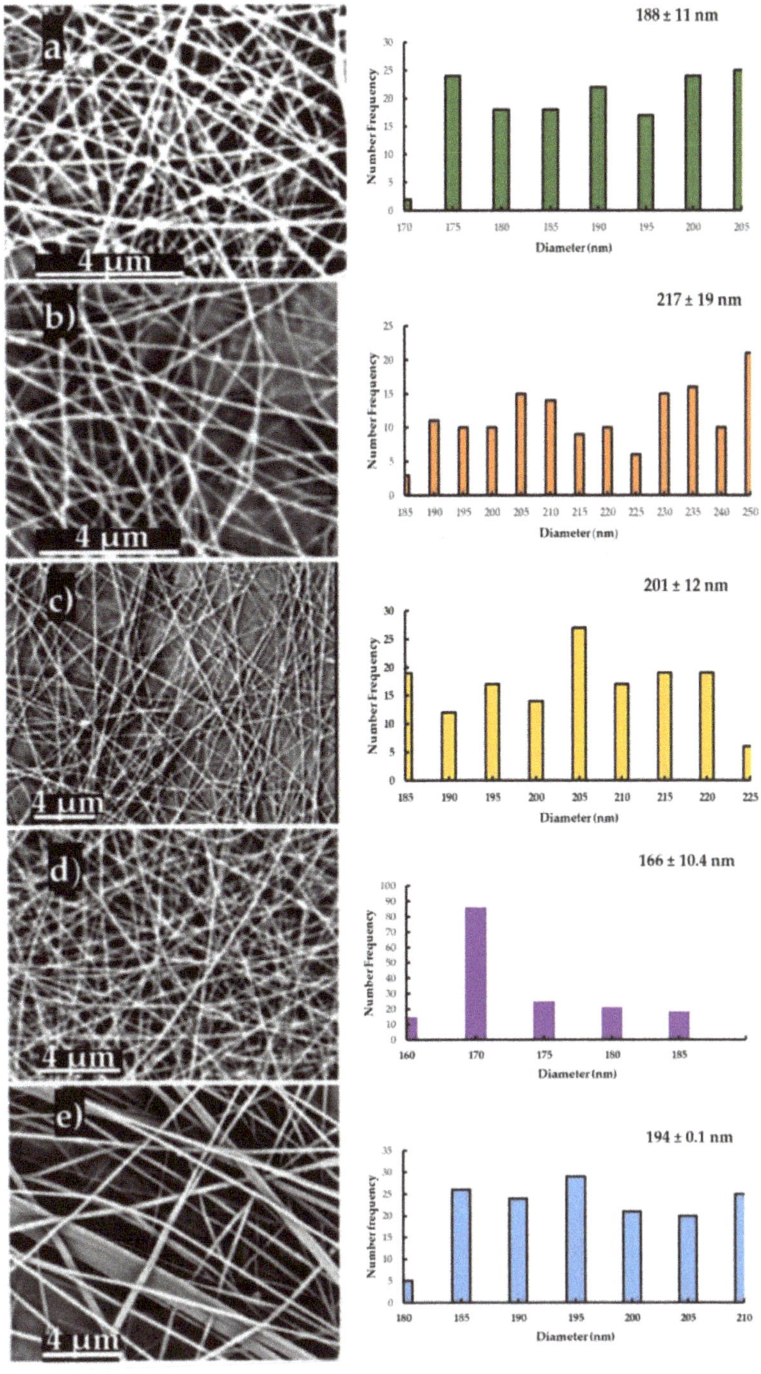

Figure 3. SEM micrographs, fiber distribution (number frequency) and average diameter (nm) of HA_PEO-NFs: HA_PEO@P (**a**), HA_PEO@IP (**b**), HA_PEO@PC (**c**), HA_PEO@ML (**d**), HA_PEO (**e**).

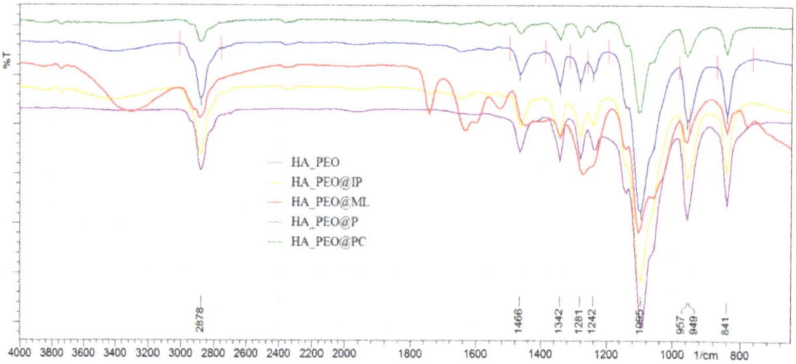

Figure 4. Fourier transform infrared spectroscopy (FTIR) assay of HA-PEO-ACs nanofibers (NFs).

Upon the addition of ACs, characteristic bands were noticed for Manuka honey: at 1050 cm^{-1}(C–O), 1636 cm^{-1}(C=O), according to the literature data [22]. Specific bands for propolis were present at 1610 cm^{-1}, 1490 cm^{-1}, and 1450 cm^{-1} (due to C=C stretches of aromatic rings) [34]. Absorption bands at 1640 cm^{-1} and at 1530 cm^{-1} are characteristic of two amide groups belonging to insulin [49]. L-arginine presents characteristic –C=N stretching vibration at 1630 cm^{-1} (due to the guanidine group) [23]. *Calendula officinalis* presented characteristic bands at 1050–1030 cm^{-1} in agreement with literature data on its flavonoid content and ether groups [33].

3.2.3. Water-Vapor Transmission Rate (WVTR)

WVTR is a notable factor in the wound-healing process by promoting gas exchange through the material. Wound dehydration may occur with a high WVTR value and lead to formation of a scar. Moreover, low WVTR acts as wound-healing moisture reservoir because of exudates present at the wound bed which may lead to a possible infection [44]. The results of our study show that by adding propolis to the HA_PEO matrix, the WVTR was significantly improved (5119.6 $\pm$ 1 g/m^2 24 h for HA_PEO@P, 4959.0 $\pm$ 1 g/m^2 24 h for HA_PEO@IP and 5122.6 $\pm$ 2 g/m^2 24 h for HA_PEO@PC) when compared with HA_PEO, unlike HA_PEO@ML which presented similar results with HA_PEO (4636.33 $\pm$ 4 g/m^2 24 h and 4646.0 $\pm$ 5 g/m^2 24 h, respectively) (Figure 5).

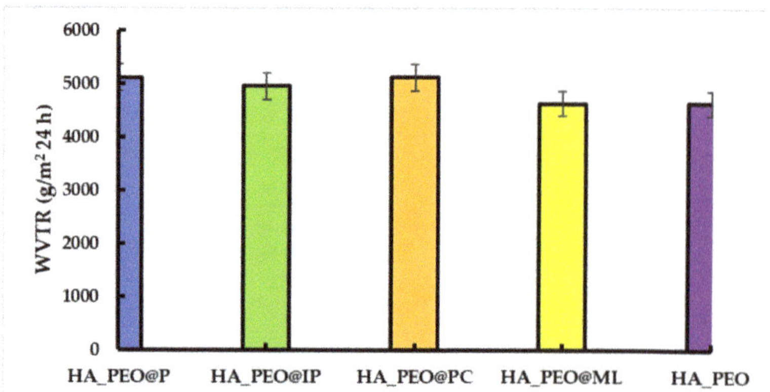

Figure 5. Water-vapor transmission rate (WVTR) assay of HA-PEO-NFs (expressed as g/m^2 24 h).

3.2.4. In Vitro Antioxidant Assays

Since the reactive oxygen species (ROS) are often produced at the wound site, an antioxidant effect is a desirable property for a wound dressing that promotes the healing process [4,50]. The analysis of the data presented in Figure 6 revealed that the best antioxidant effect was obtained for HA_PEO@PC.

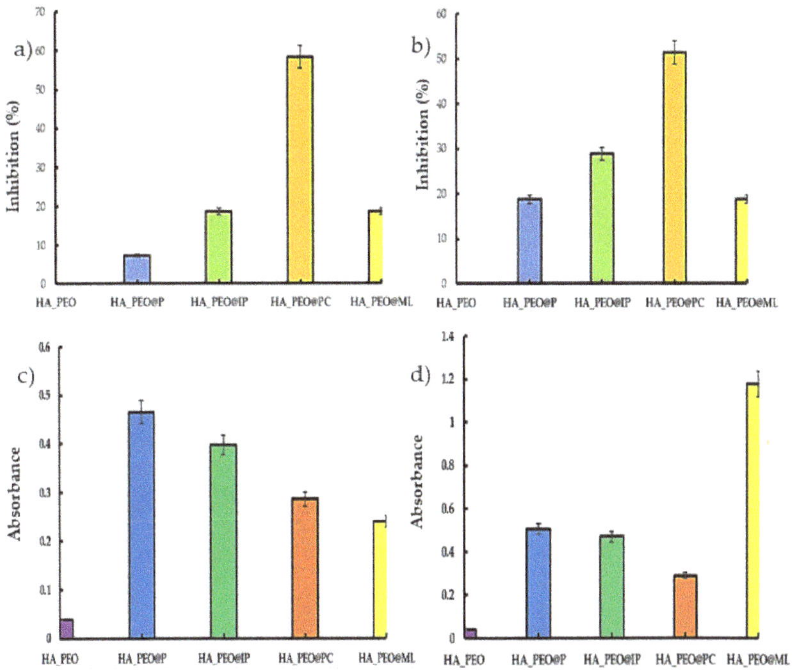

Figure 6. The antioxidant effects of HA-PEO-NFs: DPPH (**a**), ABTS (**b**), ferric-reducing antioxidant power (FRAP) (**c**), and phosphomolybdenum-reducing antioxidant power (PRAP) (**d**).

For this sample, the value of scavenging ability of the DPPH radical was 58.35 ± 6%, and for the ABTS, 51.43 ± 9%. This property may be the result of *Calendula officinalis* infusion's high flavonoid content, as the propolis containing samples presented significantly less activity against DPPH and ABTS (18.60 ± 8% for the HA_PEO@IP against DPPH and 28.79 ± 2% for HA_PEO@IP against ABTS). HA_PEO@ML presented the best results in the case of the PRAP assay, a fact also supported by literature data [51] and based on the phenolic compounds content.

3.2.5. In Vitro Cytotoxicity Assay

In vitro biocompatibility of HA_PEO, HA_PEO@ML, HA_PEO@PC, and HA_PEO@IP was assessed by MTS assay after 72 h incubation with different extract concentrations (500, 250, 125, and 62.5 µg/mL) or fresh complete medium. The results showed that NFs extracts were not cytotoxic at concentrations up to 500 µg/mL (Figure 7) and are therefore in agreement with ISO 10993-5 guidelines [52]. Furthermore, there was no significant difference in cell viability between NFs extracts at 125 µg/mL. Additionally, HA_PEO@PC stimulates normal fibroblasts' proliferation by 21% at concentrations of 250 µg/mL and by 37% at 500 µg/mL (Figure 7).

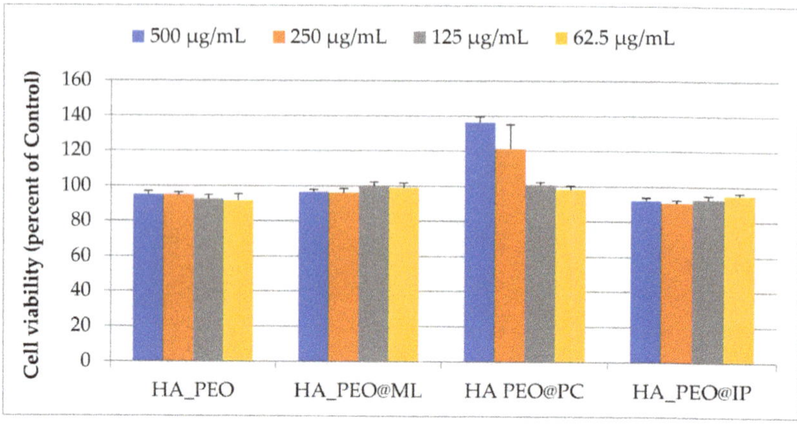

Figure 7. Cell viability (%) of HA-PEO-NFs, at different concentrations tested on normal human dermal fibroblasts.

3.2.6. Antimicrobial Assay

The antibacterial and antifungal evaluation of HA-PEO-NFs assayed the area of inhibition against Gram-negative (*Escherichia coli, Pseudomonas aeruginosa*), Gram-positive (*Staphylococcus aureus*), and *Candida albicans* strains. These strains are a frequent source of infection and may afterwards interfere with wound healing. A high percentage of clinically acquired infections are due to *Staphylococcus aureus*. *Escherichia coli* is the main cause of infection in burn wounds. Manuka honey was studied for its benefits in nanofibrous materials with the help of inhibition-zone determination [53], and its effect is largely attributed to its methylglyoxal constituent. Different activity against the bacterial strains used may be the result of the different cell wall structure of the Gram-positive and Gram-negative bacteria. The antimicrobial effect of HA-PEO-NFs expressed as diameters of the inhibition area (mm) are presented in Table 3. The samples displayed good antibacterial effects, especially against *Staphylococcus aureus*. The most active were HA_PEO@PC and HA_PEO@ML with their effect being half of Ciprofloxacin, used as control. Concerning the antifungal assessment, no sample displayed any effect. Considering the infection risk of wounds, the dressing materials which possess antibacterial activity are of high interest given their beneficial effects toward wound healing.

Table 3. Antimicrobial effects of the HA-PEO-NFs tested on bacterial and fungal strains.

Sample	Diameter of Inhibition Area (mm)			
	S. aureus **ATCC 25923**	***E. coli*** **ATCC 25922**	***P. aeruginosa*** **ATCC 27853**	***C. albicans*** **ATCC 10231**
HA_PEO@P	11	0	0	0
HA_PEO@IP	11	0	0	0
HA_PEO@PC	17	16	14	0
HA_PEO@ML	12	15	10	0
HA_PEO	0	0	0	0
CIP (5 µg/disc)	26	28	30	*Nt*
VRC (1 µg/disc)	*nt*	*nt*	*nt*	27

All designed HA-PEO-ACs NFs display promising results referring to in vitro cell toxicity, antimicrobial effect, and antioxidant potential. This can be explained as a cumulative effect of the polymeric matrix and the incorporated ACs. Taken separately, the ACs have proven their wound-healing effects [25,54–56]. The novelty of this work is based on combination of different ACs in order have a synergetic effect and so to improve the healing effects of HA. The results proved this hypothesis which so support the wound-dressing potential of HA-PEO-ACs NFs. The antimicrobial effects recorded for HA_PEO@PC and HA_PEO@ML against *Staphylococcus aureus* are even improved compared with other previously reported NFs. Our results are in agreement with other reported data for Manuka honey NFs based on cellulose acetate [53], polyvinylalcohol [57], and silk fibroin [22], which had a favorable effect for fibroblasts development but were detrimental to pathogenic bacteria proliferation. Except for the Manuka honey–polyvinylalcohol NFs which were tested for wound-healing effects in combination with pomegranate extract, no other reported Manuka honey NFs were tested as dressing materials. In addition, HA_PEO@PC and HA_PEO@ML were proved to be active also on *Pseudomonas aeruginosa* strain, while other research did not study the effect on this strain. This is important novelty element because *Pseudomonas aeruginosa* is a common strain causing nosocomial infection [58,59], which can aggravate and delay the wound healing. The presence of L-arginine together with Manuka honey into the polymer matrix further promotes the possibility of proline synthesis at the wound site and late collagen synthesis with the prospects of a speedier wound recovery [27]. Insulin is a signaling molecule with key role in cell migration, proliferation, and differentiation which translates into a beneficial effect on the fibroblasts [60]. Moreover, the association with propolis enhances insulin's properties regarding proliferation, a content of less that 10% propolis being favorable in this matter [61], which was proved also by our results. A correlation between the active components of *Calendula officinalis* has been previously reported in the literature, based on its use in inflammatory cutaneous pathologies, burns, or sun exposure-caused erythema [31,32]. Referring to the composition of *Calendula officinalis* extract is reported that the essential oil has antibacterial and antifungal effects, the flavonoids present anti-inflammatory properties, and the saponins displayed anti-inflammatory, antibacterial, and antifungal properties, while the carotenoids are the components responsible for cell-interactive properties [62]. These constituents form a phyto-complex which acts synergic for its biological effects, proving antioxidant, cell-compatible, and antimicrobial effects.

4. Conclusions

The new HA-PEO-NFs, which have inglobated different ACs such as Manuka honey, propolis, *Calendula officinalis*, insulin, and L-arginine, have been prepared and characterized in terms of physicochemical and structural features, including morphology, fiber diameters, FTIR spectra, and WVTR. The thin NFs (ranging from 156 to 230 nm) exhibited suitable WVTR values (ranging from 4634 ± 4 g/m^2 24 h to 5122.6 ± 2 g/m^2 24 h) and could maintain a moist environment favorable for wound healing. That means that obtained NFs form vapor-permeable and microbe-preventive membranes. The biological evaluation, based on in vitroassays, proved that the antimicrobial and antioxidant effects of the developed NFs are due to the ACs, since HA_PEO matrix does not exhibit any effect. Moreover, the addition of the ACs maintained the cytocompatibility of the polymeric membranes. In this respect, the obtained NFs may be considered safe nanocarriers for natural active components. A future step in improving the characteristics of the obtained NFs will be to assure the support as proper wound dressings. Taking into consideration that chemical or physical crosslinking of porous materials improves stability in aqueous media, the crosslinking of HA-based NFs with EDC/NHS will be performed. Even more, layer-by-layer dressings can be formulated based on the results obtained. In addition, in vitro degradation assay, using phosphate buffered solution, pH = 7.4 (PBS), and pseudo extracellular fluid solution (PECF), will be performed. Based on in vitro degradation assay the molecular weight distribution of the HA in the corresponding experimental conditions will be also determined.

The results of our study suggest that the developed formulations had a synergistic effect on behalf of the ACs, especially for HA_PEO@PC, that promoted cellular proliferation. This formulation may be considered as an effective and safe candidate for wound healing, and the results encourage us to include it in anin vivo study using an acute wound model (surgical excision) induced to rats.

Author Contributions: Conceptualization, L.P. and O.M.I.; methodology, O.M.I., A.M., A.T.I., L.G.C. and N.S.; validation, L.P. and C.T.; investigation, O.M.I., A.T.I., L.G.C. and N.S.; writing—original draft preparation, O.M.I. and A.T.I.; writing—review and editing, L.P. and A.M.; supervision, L.P.; funding acquisition, O.M.I., A.T.I., L.P. and N.S. All authors have read and agreed to the published version of the manuscript.

Funding: This work was financially supported by UEFISCDI grants PN-III-P1-1.1-MC-2017-2462, PN-III-P1-1.1-MC-2018-2451, COST-CA19140, AUF-IFA 2019-2020 (contract no. 28/2019), University of Medicine and Pharmacy Grigore T. Popa of Iasi (contract no. 27496/20.12.2018) and by the European Social Fund for Regional Development, Competitiveness Operational Programme Axis 1, InoMatPol, ID P_36_570 (contract no. 142/10.10.2016).

Institutional Review Board Statement: Not applicable.

Informed Consent Statement: Not applicable.

Data Availability Statement: The data presented in this study are available on request from the corresponding author.

Conflicts of Interest: The authors declare no conflict of interest. The funders had no role in the design of the study; in the collection, analyses, or interpretation of data; in the writing of the manuscript, or in the decision to publish the results.

References

1. Veith, A.P.; Henderson, K.; Spencer, A.; Sligar, A.D.; Baker, A.B. Therapeutic strategies for enhancing angiogenesis in wound healing. *Adv. Drug Deliv. Rev.* **2019**, *146*, 97–125. [CrossRef] [PubMed]
2. Halstead, F.D.; Rauf, M.; Bamford, A.; Wearn, C.M.; Bishop, J.R.B.; Burt, R.; Fraise, A.P.; Moiemen, N.S.; Oppenheim, B.A.; Webber, M.A. Antimicrobial dressings: Comparison of the ability of a panel of dressings to prevent biofilm formation by key burn wound pathogens. *Burns* **2015**, *41*, 1683–1694. [CrossRef] [PubMed]
3. Bhardwaj, N.; Kundu, S.C. Electrospinning: A fascinating fiber fabrication technique. *Biotechnol. Adv.* **2010**, *28*, 325–347. [CrossRef] [PubMed]
4. Tang, Y.; Lan, X.; Liang, C.; Zhong, Z.; Xie, R.; Zhou, Y.; Miao, X.; Wang, H.; Wang, W. Honey loaded alginate/PVA nanofibrous membrane as potential bioactive wound dressing. *Carbohydr. Polym.* **2019**, *219*, 113–120. [CrossRef]
5. Xiao, B.; Huang, Q.; Chen, H.; Chen, X.; Long, G. A Fractal model for capillary flow through a single tortuous capillary with roughened surfaces in fibrous porous media. *Fractals* **2021**, *29*, 2150017. [CrossRef]
6. Petrova, V.A.; Chernyakov, D.D.; Poshina, D.N.; Gofman, I.V.; Romanov, D.P.; Mishanin, A.I.; Golovkin, A.S.; Skorik, Y.A. Electrospun bilayer chitosan/hyaluronan material and its compatibility with mesenchymal stem cells. *Materials* **2019**, *12*, 2016. [CrossRef]
7. Wang, Y.; Beekman, J.; Hew, J.; Jackson, S.; Issler-Fisher, A.C.; Parungao, R.; Lajevardi, S.S.; Li, Z.; Maitz, P.K.M. Burn injury: Challenges and advances in burn wound healing, infection, pain and scarring. *Adv. Drug Deliv. Rev.* **2018**, *123*, 3–17. [CrossRef]
8. Boateng, J.; Catanzano, O. Advanced Therapeutic Dressings for Effective Wound Healing—A Review. *J. Pharm. Sci.* **2015**, *104*, 3653–3680. [CrossRef]
9. Wade, R.J.; Bassin, E.J.; Rodell, C.B.; Burdick, J.A. Protease-degradable electrospun fibrous hydrogels. *Nat. Commun.* **2015**, *6*, 6639. [CrossRef]
10. Weng, L.; Xie, J. Smart Electrospun Nanofibers for Controlled Drug Release: Recent Advances and New Perspectives. *Curr. Pharm. Des.* **2015**, *21*, 1944–1959. [CrossRef]
11. Hassiba, A.J.; El Zowalaty, M.E.; Webster, T.J.; Abdullah, A.M.; Nasrallah, G.K.; Khalil, K.A.; Luyt, A.S.; Elzatahry, A.A. Synthesis, characterization, and antimicrobial properties of novel double layer nanocomposite electrospun fibers for wound dressing applications. *Int. J. Nanomed.* **2017**, *12*, 2205–2213. [CrossRef]
12. Li, M.; Chen, J.; Shi, M.; Zhang, H.; Ma, P.X.; Guo, B. Electroactive anti-oxidant polyurethane elastomers with shape memory property as non-adherent wound dressing to enhance wound healing. *Chem. Eng. J.* **2019**, *375*, 121999. [CrossRef]
13. Zhang, S.; Hou, J.; Yuan, Q.; Xin, P.; Cheng, H.; Gu, Z.; Wu, J. Arginine derivatives assist dopamine-hyaluronic acid hybrid hydrogels to have enhanced antioxidant activity for wound healing. *Chem. Eng. J.* **2020**, *392*, 123775. [CrossRef]
14. Zhao, X.; Wu, H.; Guo, B.; Dong, R.; Qiu, Y.; Ma, P.X. Antibacterial anti-oxidant electroactive injectable hydrogel as self-healing wound dressing with hemostasis and adhesiveness for cutaneous wound healing. *Biomaterials* **2017**, *122*, 34–47. [CrossRef]
15. Aduba, D.C.; Yang, H. Polysaccharide fabrication platforms and biocompatibility assessment as candidate wound dressing materials. *Bioengineering* **2017**, *4*, 1. [CrossRef]

16. Zhu, J.; Tang, X.; Jia, Y.; Ho, C.T.; Huang, Q. Applications and delivery mechanisms of hyaluronic acid used for topical/transdermal delivery—A review. *Int. J. Pharm.* **2020**, *578*, 119127. [CrossRef]
17. El-Aassar, M.R.; Ibrahim, O.M.; Fouda, M.M.G.; El-Beheri, N.G.; Agwa, M.M. Wound healing of nanofiber comprising Polygalacturonic/Hyaluronic acid embedded silver nanoparticles: In-vitro and in-vivo studies. *Carbohydr. Polym.* **2020**, *238*, 116175. [CrossRef]
18. Brenner, E.K.; Schiffman, J.D.; Toth, L.J.; Szewczyk, J.C.; Schauer, C.L. Phosphate salts facilitate the electrospinning of hyaluronic acid fiber mats. *J. Mater. Sci.* **2013**, *48*, 7805–7811. [CrossRef]
19. Ahire, J.J.; Robertson, D.D.; van Reenen, A.J.; Dicks, L.M.T. Polyethylene oxide (PEO)-hyaluronic acid (HA) nanofibers with kanamycin inhibits the growth of Listeria monocytogenes. *Biomed. Pharmacother.* **2017**, *86*, 143–148. [CrossRef]
20. Sarkar, S.; Chaudhary, A.; Saha, T.K.; Das, A.K.; Chatterjee, J. Modulation of collagen population under honey assisted wound healing in diabetic rat model. *Wound Med.* **2018**, *20*, 7–17. [CrossRef]
21. Rückriemen, J.; Klemm, O.; Henle, T. Manuka honey (*Leptospermum scoparium*) inhibits jack bean urease activity due to methylglyoxal and dihydroxyacetone. *Food Chem.* **2017**, *230*, 540–546. [CrossRef]
22. Yang, X.; Fan, L.; Ma, L.; Wang, Y.; Lin, S.; Yu, F.; Pan, X.; Luo, G.; Zhang, D.; Wang, H. Green electrospun Manuka honey/silk fibroin fibrous matrices as potential wound dressing. *Mater. Des.* **2017**, *119*, 76–84. [CrossRef]
23. Liu, Y.; Zhou, S.; Gao, Y.; Zhai, Y. Electrospun nanofibers as a wound dressing for treating diabetic foot ulcer. *Asian J. Pharm. Sci.* **2019**, *14*, 130–143. [CrossRef]
24. Patil, S.; Desai, N.; Mahadik, K.; Paradkar, A. Can green synthesized propolis loaded silver nanoparticulate gel enhance wound healing caused by burns? *Eur. J. Integr. Med.* **2015**, *7*, 243–250. [CrossRef]
25. Debats, I.B.J.G.; Wolfs, T.G.A.M.; Gotoh, T.; Cleutjens, J.P.M.; Peutz-Kootstra, C.J.; van der Hulst, R.R.W.J. Role of arginine in superficial wound healing in man. *Nitric OxideBiol. Chem.* **2009**, *21*, 175–183. [CrossRef]
26. Iacob, A.T.; Drăgan, M.; Ghețu, N.; Pieptu, D.; Vasile, C.; Buron, F.; Routier, S.; Giusca, S.E.; Caruntu, I.D.; Profire, L. Preparation, characterization and wound healing effects of new membranes based on chitosan, hyaluronic acid and arginine derivatives. *Polymers* **2018**, *8*, 607. [CrossRef]
27. Bennacef-Heffar, N.; Laraba-Djebari, F. Beneficial effects of Heparin and L Arginine on dermonecrosis effect induced by Vipera lebetina venom: Involvement of NO in skin regeneration. *Acta Trop.* **2017**, *171*, 226–232. [CrossRef]
28. Rieger, K.A.; Birch, N.P.; Schiffman, J.D. Electrospinning chitosan/poly(ethylene oxide) solutions with essential oils: Correlating solution rheology to nanofiber formation. *Carbohydr. Polym.* **2016**, *139*, 131–138. [CrossRef] [PubMed]
29. Oryan, A.; Alemzadeh, E. Effects of insulin on wound healing: A review of animal and human evidences. *Life Sci.* **2017**, *174*, 59–67. [CrossRef]
30. Emanuelli, T.; Burgeiro, A.; Carvalho, E. Effects of insulin on the skin: Possible healing benefits for diabetic foot ulcers. *Arch. Dermatol. Res.* **2016**, *308*, 677–694. [CrossRef]
31. Agatonovic-Kustrin, S.; Babazadeh Ortakand, D.; Morton, D.W.; Yusof, A.P. Rapid evaluation and comparison of natural products and antioxidant activity in calendula, feverfew, and German chamomile extracts. *J. Chromatogr. A* **2015**, *1385*, 103–110. [CrossRef] [PubMed]
32. Nicolaus, C.; Junghanns, S.; Hartmann, A.; Murillo, R.; Ganzera, M.; Merfort, I. In vitro studies to evaluate the wound healing properties of Calendula officinalis extracts. *J. Ethnopharmacol.* **2017**, *196*, 94–103. [CrossRef] [PubMed]
33. Rad, Z.P.; Mokhtari, J.; Abbasi, M. Calendula officinalis extract/PCL/Zein/Gum arabic nanofibrous bio-composite scaffolds via suspension, two-nozzle and multilayer electrospinning for skin tissue engineering. *Int. J. Biol. Macromol.* **2019**, *135*, 530–543.
34. Khodabakhshi, D.; Eskandarinia, A.; Kefayat, A.; Rafienia, M.; Navid, S.; Karbasi, S.; Moshtaghian, J. In vitro and in vivo performance of a propolis-coated polyurethane wound dressing with high porosity and antibacterial efficacy. *Colloids Surf. B Biointerfaces* **2019**, *178*, 177–184. [CrossRef]
35. Hussein, Y.; El-Fakharany, E.M.; Kamoun, E.A.; Loutfy, S.A.; Amin, R.; Taha, T.H.; Salim, S.A.; Amer, M. Electrospun PVA/hyaluronic acid/L-arginine nanofibers for wound healing applications: Nanofibers optimization and in vitro bioevaluation. *Int. J. Biol. Macromol.* **2020**, *164*, 667–676. [CrossRef]
36. Adeli, H.; Khorasani, M.T.; Parvazinia, M. Wound dressing based on electrospun PVA/chitosan/starch nanofibrous mats: Fabrication, antibacterial and cytocompatibility evaluation and in vitro healing assay. *Int. J. Biol. Macromol.* **2019**, *122*, 238–254. [CrossRef]
37. Tomadoni, B.; Ponce, A.; Pereda, M.; Ansorena, M.R. Vanillin as a natural cross-linking agent in chitosan-based films: Optimizing formulation by response surface methodology. *Polym. Test.* **2019**, *78*, 105935. [CrossRef]
38. Sadiq, A.; Mahmood, F.; Ullah, F.; Ayaz, M.; Ahmad, S.; Haq, F.U.; Khan, G.; Jan, M.S. Synthesis, anticholinesterase and antioxidant potentials of ketoesters derivatives of succinimides: A possible role in the management of alzheimer's. *Chem. Cent. J.* **2015**, *9*, 31. [CrossRef]
39. Xue, M.; Wang, Z.; Sun, S.F.; Huang, Z.S.; Zhang, X.X.; Ma, J.; Dong, X.L. Mechanism investigation on the formation of high valent iron intermediate in Fe(VI) oxidation using ABTS as a probe: Effect of excess Fe(VI). *Chem. Eng. J.* **2020**, *387*, 124123. [CrossRef]
40. Orhan, I.E.; Senol, F.S.; Yılmaz, B.S.; Altun, M.L.; Özbilgin, S.; Yazgan, A.N.; Yüksel, E.; İşcan, G.S. Neuroprotective potential of Viburnum orientale Pallas through enzyme inhibition and antioxidant activity assays. *S.Afr. J. Bot.* **2018**, *114*, 126–131. [CrossRef]

41. Ajmal, G.; Bonde, G.V.; Mittal, P.; Khan, G.; Pandey, V.K.; Bakade, B.V.; Mishra, B. Biomimetic PCL-gelatin based nanofibers loaded with ciprofloxacin hydrochloride and quercetin: A potential antibacterial and anti-oxidant dressing material for accelerated healing of a full thickness wound. *Int. J. Pharm.* **2019**, *567*, 118480. [CrossRef]
42. ISO. *ISO 10993-5:2009(E): Biological Evaluation of Medical Devices—Part 5: Tests for In Vitro Cytotoxicity*; ISO: Geneva, Switzerland, 2009.
43. ISO. *ISO 10993-4. Biological Evaluation of Medical Devices—Part 4: Selection of Tests for Interactions with Blood*; ISO: Geneva, Switzerland, 2017.
44. Brenner, E.K.; Schiffman, J.D.; Thompson, E.A.; Toth, L.J.; Schauer, C.L. Electrospinning of hyaluronic acid nanofibers from aqueous ammonium solutions. *Carbohydr. Polym.* **2012**, *87*, 926–929. [CrossRef]
45. Séon-Lutz, M.; Couffin, A.C.; Vignoud, S.; Schlatter, G.; Hébraud, A. Electrospinning in water and in situ crosslinking of hyaluronic acid/cyclodextrin nanofibers: Towards wound dressing with controlled drug release. *Carbohydr. Polym.* **2019**, *207*, 276–287. [CrossRef]
46. El-Aassar, M.R.; El-Beheri, N.G.; Agwa, M.M.; Eltaher, H.M.; Alseqely, M.; Sadik, W.S.; El-Khordagui, L. Antibiotic-free combinational hyaluronic acid blend nanofibers for wound healing enhancement. *Int. J. Biol. Macromol.* **2021**, *167*, 1552–1563. [CrossRef]
47. De Oliveira, S.A.; da Silva, B.C.; Riegel-Vidotti, I.C.; Urbano, A.; de Sousa Faria-Tischer, P.C.; Tischer, C.A. Production and characterization of bacterial cellulose membranes with hyaluronic acid from chicken comb. *Int. J. Biol. Macromol.* **2017**, *97*, 642–653. [CrossRef]
48. Pucić, I.; Jurkin, T. FTIR assessment of poly(ethylene oxide) irradiated in solid state, melt and aqeuous solution. *Radiat. Phys. Chem.* **2012**, *81*, 1426–1429. [CrossRef]
49. Sarmento, B.; Ferreira, D.; Veiga, F.; Ribeiro, A. Characterization of insulin-loaded alginate nanoparticles produced by ionotropic pre-gelation through DSC and FTIR studies. *Carbohydr. Polym.* **2006**, *66*, 1–7. [CrossRef]
50. Hossen, M.S.; Ali, M.Y.; Jahurul, M.H.A.; Abdel-Daim, M.M.; Gan, S.H.; Khalil, M.I. Beneficial roles of honey polyphenols against some human degenerative diseases: A review. *Pharmacol. Rep.* **2017**, *69*, 1194–1205. [CrossRef]
51. Bolanos De La Torre, A.A.S.; Henderson, T.; Nigam, P.S.; Owusu-Apenten, R.K. A universally calibrated microplate ferric reducing antioxidant power (FRAP) assay for foods and applications to Manuka honey. *Food Chem.* **2015**, *174*, 119–123. [CrossRef]
52. Weber, M.; Steinle, H.; Golombek, S.; Hann, L.; Schlensak, C.; Wendel, H.P.; Avci-Adali, M. Blood-Contacting Biomaterials: In Vitro Evaluation of the Hemocompatibility. *Front. Bioeng. Biotechnol.* **2018**, *6*, 99. [CrossRef]
53. Ullah, A.; Ullah, S.; Khan, M.Q.; Hashmi, M.; Nam, P.D.; Kato, Y.; Tamada, Y.; Kim, I.S. Manuka honey incorporated cellulose acetate nanofibrous mats: Fabrication and in vitro evaluation as a potential wound dressing. *Int. J. Biol. Macromol.* **2020**, *155*, 479–489. [CrossRef]
54. Oryan, A.; Alemzadeh, E.; Moshiri, A. Potential role of propolis in wound healing: Biological properties and therapeutic activities. *Biomed. Pharmacother.* **2018**, *98*, 469–483. [CrossRef]
55. Rathinamoorthy, R.; Sasikala, L. In vivo—Wound healing studies of *Leptospermum scoparium* honey loaded chitosan bioactive wound dressing. *Wound Med.* **2019**, *26*, 100162. [CrossRef]
56. Ebrahimifard, F.; Nooraei, N.; Fathi, M.; Dehghani, M. Effect of Insulin on Healing of Pressure Sore. *Arch. Crit. Care Med.* **2015**, *1*, e527. [CrossRef]
57. Abou Zekry, S.S.; Abdellatif, A.; Azzazy, H.M.E. Fabrication of pomegranate/honey nanofibers for use as antibacterial wound dressings. *Wound Med.* **2020**, *28*, 100181. [CrossRef]
58. Fayemi, O.E.; Ekennia, A.C.; Katata-Seru, L.; Ebokaiwe, A.P.; Ijomone, O.M.; Onwudiwe, D.C.; Ebenso, E.E. Antimicrobial and Wound Healing Properties of Polyacrylonitrile-Moringa Extract Nanofibers. *ACS Omega* **2018**, *3*, 4791–4797. [CrossRef]
59. Motealleh, B.; Zahedi, P.; Rezaeian, I.; Moghimi, M.; Abdolghaffari, A.H.; Zarandi, M.A. Morphology, drug release, antibacterial, cell proliferation, and histology studies of chamomile-loaded wound dressing mats based on electrospun nanofibrous poly(ε-caprolactone)/polystyrene blends. *J. Biomed. Mater. Res. Part B Appl. Biomater.* **2014**, *102*, 977–987. [CrossRef]
60. Hrynyk, M.; Martins-Green, M.; Barron, A.E.; Neufeld, R.J. Sustained prolonged topical delivery of bioactive human insulin for potential treatment of cutaneous wounds. *Int. J. Pharm.* **2010**, *398*, 146–154. [CrossRef]
61. Unnithan, A.R.; Gnanasekaran, G.; Sathishkumar, Y.; Lee, Y.S.; Kim, C.S. Electrospun antibacterial polyurethane-cellulose acetate-zein composite mats for wound dressing. *Carbohydr. Polym.* **2014**, *102*, 884–892. [CrossRef] [PubMed]
62. Tyagi, V.; Singh, V.K.; Sharma, P.K.; Singh, V. Essential oil-based nanostructures for inflammation and rheumatoid arthritis. *J. Drug Deliv. Sci. Technol.* **2020**, *60*, 101983. [CrossRef]

polymers

MDPI

Review

A Comprehensive Review on Plant-Derived Mucilage: Characterization, Functional Properties, Applications, and Its Utilization for Nanocarrier Fabrication

Mansuri M. Tosif [1], Agnieszka Najda [2,*], Aarti Bains [3], Ravinder Kaushik [4], Sanju Bala Dhull [5], Prince Chawla [1,*] and Magdalena Walasek-Janusz [2]

1 Department of Food Technology and Nutrition, Lovely Professional University, Phagwara, Punjab 144411, India; tosifmansuri444@gmail.com
2 Department of Vegetable Crops and Medicinal Plants, University of Life Sciences in Lublin, 20-280 Lublin, Poland; magdalena.walasek@up.lublin.pl
3 Department of Biotechnology, Chandigarh Group of Colleges Landran, Mohali, Punjab 140307, India; aarti05888@gmail.com
4 Department of Food Technology, School of Health Sciences, University of Petroleum and Energy Studies, Dehradun, Uttarakhand 248007, India; ravinderfoodtech2007@rediffmail.com
5 Department of Food Science and Technology, Chaudhary Devi Lal University, Sirsa, Haryana 125055, India; sanjudhull@gmail.com
* Correspondence: agnieszka.najda@up.lublin.pl (A.N.); princefoodtech@gmail.com (P.C.)

Citation: Tosif, M.M.; Najda, A.; Bains, A.; Kaushik, R.; Dhull, S.B.; Chawla, P.; Walasek-Janusz, M. A Comprehensive Review on Plant-Derived Mucilage: Characterization, Functional Properties, Applications, and Its Utilization for Nanocarrier Fabrication. *Polymers* **2021**, *13*, 1066. https://doi.org/10.3390/polym13071066

Academic Editor: Arn Mignon

Received: 6 March 2021
Accepted: 25 March 2021
Published: 28 March 2021

Publisher's Note: MDPI stays neutral with regard to jurisdictional claims in published maps and institutional affiliations.

Abstract: Easily sourced mucus from various plant parts is an odorless, colorless and tasteless substance with emerging commercial potential in agriculture, food, cosmetics and pharmaceuticals due to its non-toxic and biodegradable properties. It has been found that plant-derived mucilage can be used as a natural thickener or emulsifier and an alternative to synthetic polymers and additives. Because it is an invisible barrier that separates the surface from the surrounding atmosphere, it is used as edible coatings to extend the shelf life of fresh vegetables and fruits as well as many food products. In addition to its functional properties, mucilage can also be used for the production of nanocarriers. In this review, we focus on mucus extraction methods and its use as a natural preservative for fresh produce. We detailed the key properties related to the extraction and preservation of food, the mechanism of the effect of mucus on the sensory properties of products, coating methods when using mucus and its recipe for preserving fruit and vegetables. Understanding the ecological, economic and scientific factors of production and the efficiency of mucus as a multi-directional agent will open up its practical application in many industries.

Keywords: nanohydrogel; food applications; biopolymers; polysaccharide

1. Introduction

Plant-derived polymers have attained high demand in food and other industries due to their diverse industrial applications such as film coating, emulsifier, binder, and gelling agents, therefore they are excessively used in the textile industry, paper industry, and cosmetic industry [1,2]. Nowadays, due to the hazardous effect of synthetic polymers on human health, people showed major interest in plant-based naturally derived biopolymers (gums, mucilage, cellulose, and glucans) as an effective ingredient for the formulation of eco-friendly, sustainable, cost-effective products [3]. Moreover, a large number of polysaccharides can also be biosynthetically fabricated by several living organisms including plants, algae, animals, bacteria, and fungi [4]. Also, natural polysaccharides are used in the food industry as they are regarded as safe for human consumption [5]. Among various polysaccharides, plant-originated mucilage is widely used in various food industries due to its valuable broad-spectrum applications [6]. Generally, mucilage can be obtained from several plants or their different parts such as *Aloe vera*, *Salvia hispanica* seeds, *Cordia dichotoma, Basella alba, Plantago psyllium, Cyamopsis tetragonoloba, Cactaceae, Abelmoschus esculentus, Trigonella foenum-graecum, Moringa Oleifera,* and *Linum usitatissimum.* Plant-derived mucilage, due to its distinctive health (anticancer, angiotensin-converting enzyme inhibition extends to diabetes, and immunity stimulation) and food properties, is widely used as an active ingredient for the formulation of pharmaceutics, functional, and nutraceutical products [7]. Structurally, mucilage (a complex of polymeric polysaccharide) is mainly composed of carbohydrates with highly branched structures that consist of monomer units of L-arabinose, D-xylose, D-galactose, L-rhamnose, and galacturonic acid. They also contain glycoproteins and different bioactive components such as tannins, alkaloids, and steroids [8–10]. Also, mucilage produces an indefinite number of monosaccharides on hydrolysis, depending on the type of hydrolysis products obtained due to the nature of the polysaccharide. It can also further classify into pentose sugars (xylan) and hexose sugars (cellulose and starch) and can be considered as gum like components due to their similar physiological properties. However, both mucilage and gum are mostly related to hemicelluloses in composition, except the sugars produced by hemicelluloses such as xylose, glucose, and mannose instead of sugars produced by the gums such as galactose and arabinose [11,12]. Moreover, that can be utilized in several applications such as edible coating, wound healing, tablet formation, encapsulation, water purification, and various nanocarriers. Mucilage exhibits an excellent functional property, however, due to the hydrogen bonding in between different functional and other polar groups, they also have an important role in film, emulsion, coated metal nanoparticles, and gel formation [13]. In recent years, nanostructured hydrogels and mucilage coated metal nanoparticles are intensively used as a significant delivery vehicle for various hydrophilic and hydrophobic components [14]. For the formulation of nanohydrogel, different types of biopolymers and cross-linking polymers can be used and mucilage can act as either a primary biopolymer or a cross-linking component for the formulation of nanohydrogel [15]. Several reports have been published on the formulation of stable nanohydrogels using mucilage as an active component and researchers revealed various therapeutic and food applications of the formulated nanohydrogels [11–17]. Furthermore, nanohydrogels formulated with mucilage exhibit higher stability than that of other conventional plant-based biopolymers. Furthermore, metal nanoparticles coated with polymeric carbohydrates such as starch, dextran, chitosan, and mucilage are the most abundant nanocarriers used for targeted drug delivery. Because, in addition to increasing blood circulation time by hiding them from the immune system, their polymeric shells enable them to transfer and release the drug during biodegradation [16,17]. However, only a few reports are published on the comprehensive knowledge of plant-derived mucilage, therefore, the present review is emphasized on the physicochemical properties, characterization, health, and functional attributes of the mucilage. Also, the application of mucilage crosslinked nanohydrogels and mucilage coated metal nanoparticles are discussed with mechanisms and schematic diagrams.

2. Origin of Mucilage in Different Plant Parts

The Mucilage is a water-soluble edible adhesive material that constitutes carbohydrates and uranic acids units present in different parts of plants including the mucous epidermis of the outer layer of seeds, bark, leaves, and buds [18]. The majority of plants produce mucilage from the seed coat and this process of producing mucilage is termed Myxospermy and some plant species produce it from the fruit epicarp which is known as Myxocarpy. Plants producing mucilage from seed coat belong to the family *Plantaginaceae, Acanthaceae, Linaceae,* and *Brassicaceae,* while Myxocarpy (fruit mucilage) is commonly found in families like *Poaceae, Asteraceae,* and *Lamiaceae* [19]. The presence of mucilage on the seed coat prevents the plant from early seedling development and drought stress during the germination. depending upon its origin it is characterized into many groups including hair secretion, intracellular mucilage, and cell membrane mucilage [11]. The mucilage obtained from the seed coat is classified into three classes that are endosperm non-starch polysaccharide (galactomannans), cell wall material of the endosperm (soybean hemicelluloses and xyloglucans), and mucilaginous constituents of the seed coat (flaxseed, Chia seed, and yellow mustard) [20,21]. Mucilage develops a jelly-like structure around fruit and maintains moisture and prevents seeds from completely drying out and therefore act as a hydrating agent and also acts as an energy reservoir [22,23]. Mucilage also plays a significant role in the control of germination, the promotion of dispersal, and soil adhesion, root mucilage is usually exhibited from the outer layers of the root cap, consisting of mostly root border cells and polysaccharides, which produce various chemical substances such as flavonoids, phenolics acids, amino acids, galactosidase, antibiotics, sugars, peroxidase, proteins, and anthocyanins [24]. Moreover, root mucilage plays a very important role for plant growth, such as for the maintenance of root-to-soil touch, root tip lubrication, soil microaggregate stabilization, water storage ability, selective storage, and the absorption of ions (Na^+, Cd^{2+}, Pb^{2+}, and Al^{3+}) through root cells. Furthermore, it is primarily secreted by the secretory vesicles of hypersecretory root cap cells as a coagulated polysaccharide (poly-galacturonic acid) and is subsequently passed during root extension to older root areas, but epidermal cells are also effective in secreting mucilage [25–27]. Mucilage is also produced in the leaves and buds of several plant species; it may allow the leaves to retain water capacity when soil water deficits emerge; therefore, it helps in the storage of food and water. Various mucilage containing plants and their origins are highlighted in Table 1.

Table 1. Origin of mucilage in different parts of a plant.

Source of Mucilage	Part	Yield of Mucilage	Extraction Method	References
Linum usitatissimum L.	Seed	7.3%	Extracted by centrifugation	[28]
Spinacia oleracea L.	Leaves	12.62%	Extracted by acetone	[29]
Salvia hispanica	Seed	8.3% 6.4%	Non-thermal extraction Thermal extraction	[30]
Plantago major	Seed	15.18%	Thermal extraction	[31]
Basella alba	Leaves Stem	4.83% 6.20%	Hot water extraction	[32]
Psyllium seed	Husk	37–52%	Ultrasonic bath extraction	[33]

3. Extraction of Plant-Derived Mucilage

Mucilage can be extracted from any part of the plant is considered a valuable natural polysaccharides source with excellent potential in pharmaceutical and food applications. several studies said that the yield, functional, and rheological properties of mucilage are highly dependent on the extraction method and extraction conditions [28]. Generally, all the extraction method of mucilage comprises of two successive procedures which are maceration and precipitation. Usually, the maceration extraction method of mucilage is simple and valuable, although the disadvantage of maceration is low efficiency and long extraction time [33]. This method consists of soaking the raw material in the chosen solvent at room temperature with regular agitation. maceration for extraction of mucilage is typically done using the low ratio of solid-liquid and hot water treatment. acid solutions, ammonium oxalate, and EDTA are also used to improve mucilage extraction [26]. According to several studies, the water-extracted mucilage showed high viscosity than alkali and acid-extracted mucilage due to marked differences in the mucilage structure of monosaccharides. However, the limitations of these methods are higher levels of protein with low yields of mucilage and subsequent denaturation. Apart from its nutritional value, protein is known as an impurity in the final product which affects its purity and restricts its industrial use due to the instability and undesirable taste caused by microbial spoilage. while protein presence in the mucilage has been identified regardless of the extraction methods used and an emulsifying action has been taken to demonstrate that protein is desirable to improve the texture and consistency of emulsions and beverages [34]. The general outline revealed an increase in protein levels as time and temperature of maceration increased, resulting in a higher molecular weight of mucilage. Also, higher maceration temperature and extended stirring period contributed to highly colored mucilage, which is unacceptable for industrial usage. Acid pre-treatment should be considered in order to protect against this color effect [35]. Solvent treatment is a conventional extraction technique of mucilage. generally, the aqueous procedure included the mucilage extracted from the dry parts of the plant (seeds, leaves, roots, stems) by using hot distilled water. The procedure occurs under continuous shaking and stirring of solution. The solution is then filtered, these stages can be repeated often. Consequently, the mucilage is precipitated by the addition of alcohol to the filtrate. Then, precipitated mucilage is dried via freeze-drying or in an oven to obtaining the final mucilage powder [29] whereas ultrasonication treatment is a emerging non-thermal technique for mucilage extraction with several applications in the pharmaceutical and food industry. Furthermore, in some plant parts, two layers of mucilage (Arabidopsis seed coating) make the extraction process very difficult; therefore, ultrasonication treatment can be much helpful in such cases. The outer layer of the seed coat can be extracted through shaking, and an inner layer of the seed coat is composed of rhamnogalacturonan I, which is a very difficult challenge to extract. in this case, ultrasonic treatment for 20 s showed excellent results. ultrasound with low-frequency has been used for mucilage extraction by disrupting the biological cell wall through the formation of pores. The formation of cavitation bubbles and their resulting collapse produce high spots with higher pressure and temperature, capable of breaking the bonds between the mucilage and seed coat. As a result, the amplitude of the ultrasounds, extraction temperature, and time must all be selected carefully to prevent the extraction of undesirable compounds and mechanical disturbance of the seeds [36]. Likewise, microwave is a thermal emerging extraction method that can be used for the extraction of mucilage from several parts of the plant. Moreover, microwave-assisted extraction is a potential alternative method providing advantages in terms of improved extraction efficiency, reduced solvent consumption, and time reduction. Conventional aqueous extraction methods include the effects of solvent, pH, and temperature, which change the nutritional value, functional and structural property of mucilage [37]. Using hot water extraction with a long extraction period is considered to be cost-effective, although it can reduce the consistency of the mucilage. Several studies said that conventional hot water extraction methods result in the loss of heat liable compounds of mucilage, to overcome this major drawback, microwave-assisted extraction with extraction at 300 to 400 W for 120 to180 s was used, resulting in a high mucilage yield. Consequently, enzymes also play a very crucial role in the extraction process due to their wide range of applications in the food industry. During the enzymatic hydrolysis for reduction of viscosity and molecular weight of mucilage, several enzymes can be used such as rhamnase, arabinase, xylanse, and mannonase [13]. Furthermore, cold extraction can be applied in order to produce more viscous mucilage, which is more natural but with a lower yield compared to the hot extraction method [38]. Moreover, an increase in the ratio of solid-liquid is proportional to more yield due to more availability of solvents, which improves the driving force of mucilage from raw plant material. However, the high ratio of solid-liquid cannot significantly affect mucilage yield, also consisting of high process costs. In this context, Elboutachfaiti et al. [37] extracted mucilage from flaxseed mucilage after maceration in hot water, six rhamnogalacturonan fractions of flaxseed mucilage were extracted and purified by using ion-exchange chromatography. As regards the pH effect on the extraction yield of mucilage, a significant increase in the yield of mucilage was observed with an increase in pH due to the separation of the acidic groups such as uranic acids, and due to the attraction between the negatively charged ones, which increased the solubility of the mucilage, although further decreased after a certain pH value. Moreover, lower pH values of mucilage are likely to improve protein recovery due to protein solubilization. Although, below 3 pH, the action of acid can result in a lower protein yield due to its hydrolysis [39]. In order to obtain the quality polysaccharides (mucilage) and highest yield, it is necessary to improve the extraction procedure (Figure 1). Deionized water can also be used for the extraction of mucilage [39,40].

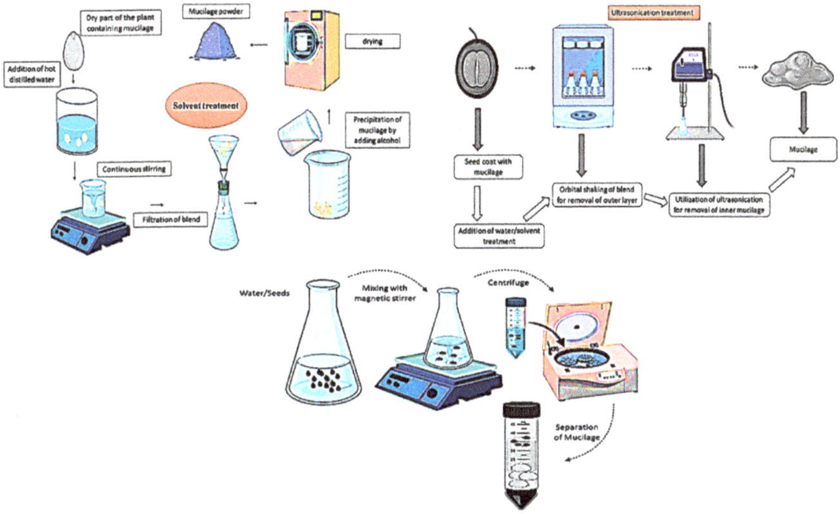

Figure 1. Extraction of mucilage from plant parts using solvent extraction, microwave assisted, and centrifugation process.

4. Structural Chemistry of Mucilage

The mucilage is a water-soluble component constituting different functional chemical components with potential human health benefits [41]. Mucilage and gum are a subgroup of hydrocolloids containing monosaccharides linked with organic acids and are close to each other due to the hydrophilic and hydrocolloid components that create a sticky solution or gel in the presence of water [2]. Plant hydrocolloids (Gum and mucilage) contain pentose, galactose and methyl pentose sugar joined to uranic acid residues by glycosidic linkages. The terminal atom of carbon (at the opposite side of the carbonyl chain) of the monosaccharide unit may occur in an oxidized (carboxylic acid) form. Six carbon with aldohexose in the form of a carboxylic acid is termed uranic acid [24]. Furthermore, Monosaccharides are the most common carbohydrate molecules that cannot be broken down into simpler sugar molecules by hydrolysis and are occasionally referred to as simple sugars. Mucilage present in plant consists of two main polysaccharides pectin and hemicellulose each of which is comprised of rhamnogalacturonan, and arabinoxylans respectively [42]. Generally, mucilage arabinoxylans encompass by β-1,4-linked xylose backbones, they are mostly replaced by 1-3 sugar residues at O-2 or/and O-3 positions. While, mucilage rhamnogalacturonan I (RG-I) comprise a backbone of the repeating disaccharide of α-(1,2)-rhamnose and α-d-(1,4)-galacturonic acid [43]. Moreover, there is some evidence that rhamnogalacturonan I side chains can be covalently bound to hemicelluloses, creating a super-macromolecular polymeric network. The neutral sugars are mainly D-xylose, L-arabinose, and D-galactose, with the proportions and types of neutral sugars varying with the origin of mucilage. In mucilage, both carboxyl and hydroxyl groups are present as functional groups due to the presence of these functional groups it acts as a polyelectrolyte [44]. The different chemical structures of polysaccharides are illustrated in Figure 2.

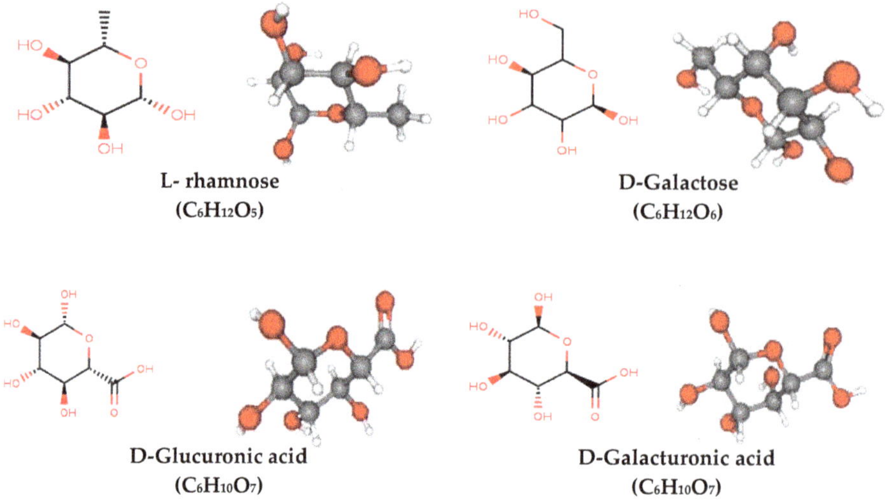

Figure 2. Chemical structure of different mucilage (L- rhamnose ($C_6H_{12}O_5$) 6-deoxy-hexose or methyl-pentose, D-Galactose ($C_6H_{12}O_6$) C-4 epimer of glucose, D-Glucuronic acid ($C_6H_{10}O_7$) sixth carbon atom oxidized to a carboxylic acid, D-Glucuronic acid ($C_6H_{10}O_7$) an aldehyde group at C1 and a carboxylic acid group at C6).

In a study done by Hosseini-Parvar et al. [45], Beikzadeh et al. [9], Timilsena et al. [46], and Mirhosseini et al. [47] observed that basil seed mucilage contains xylan (24.29%), glucan (2.31%), and glucomannan (43%), chia seed contains glucose (19.6%), galactose (6.1%), arabinose (9.6%), xylose (38.5%), galacturonic acid (5.3%), and glucuronic acid (18.7%), cress seed mucilage contains glucose (1%), fructose (6.8%), arabinose (19.4%) rhamnose (1.9%), glucuronic acid (6.7%), and galactose (4.7%), and flaxseed contains xylose (21.1–37.4%), fructose (5–7.1%), galactose (20–28%), and arabinose (9.2–13.5%) [17]. All these studies conclude that mucilage present in the seeds coat of plants of different species exhibits different forms of carbohydrates and uranic acid units.

5. Characterization of the Mucilage

This technique is used to determine the chain configuration, i.e., the microstructure of the polymer when present in either a solid state or a liquid state. This technique can also be applied to any form of the sample containing a spin-filled nucleus. In this context, Singh et al. [2] observed nuclear magnetic resonance spectra (^{1}H and ^{13}C) of the mucilage of the seed/fruit of *Diospyros melonoxylon Roxb* at 400 MHz, and revealed that many sugar compositions are consist of CH and OH groups of mannose (δ 3065 to δ 3060 ppm), CH group of rhamnose (δ 72.2 ppm), CH group of arabinose (δ 70.1–δ 71.8 ppm), the CH_2 group of arabinose (δ 3.81 to δ 3.55 ppm), and the CH group of mannose (δ 72.3 ppm), respectively. Likewise, Deore et al. [48] and Dehghani et al. [49] studied the composition of the mucilage of *Cassia obtusifolia* and chia seeds and concluded that the mucilage of *Cassia obtusifolia* contains the CH group of arabinose (δ 69.61–δ 71.25 ppm), glucose (δ 4.15 and δ 3.84 ppm), OH and CH groups of mannose (d 3.62 and d 3.41 ppm), methyl group (1.23 ppm), non-anomeric protons (3.1 and 4.1 ppm), OH and CH_2 groups of arabinose (δ 3.55 and δ 3.39 ppm), while chia seed mucilage contains OH and CH groups of mannose (3.6 and 3.65 ppm), OH and CH groups of arabinose (3.55 and 3.81 ppm), and the bond between methyl and protons with C_6 and C_4 of galactose. In another study done by Devi et al. [50], it was observed that flaxseed mucilage contains methylene and thio portions of thioglycolate resonance at the peak value of 4.14 and 5.56 ppm. FTIR analysis is performed to determine chemical structure and functional groups of mucilage, FTIR spectroscopy can be used in wavelength regions between 4000 and 400 cm^{-1} at a resolution of 2 cm^{-1} or 4 cm^{-1}. The area between 800 and 1200 cm^{-1} is known as the carbohydrate fingerprinting area. Mucilage contains polymers such as CH_2, O-H, C-H, C-O-C, and the carboxylate group was observed in different studies. It was also observed that the spectra obtained from mucilage showed a huge peak value at the range of 3500–3300 cm^{-1}, the absorption band at around 3000–2800 cm^{-1}, 1270–1080 cm^{-1}, at 1600 cm^{-1}, and 1400 cm^{-1} confirming the vibrational stretching of the polymeric O-H group, CH_2 and C-H group, C-O group, Carboxylate asymmetric stretching, and symmetric stretching respectively [51–53]. Naji-Tabasi et al. [54] conducted a study on basil seed mucilage, the FTIR analysis of the study indicated the existence of uronic acids, absorptions at wavelengths 1600 and 1400 cm^{-1} confirmed the presence of uronic acid assigned to C-OO asymmetrical and symmetrical stretching, respectively. Consequently, Pratik and Shadique [55] extracted mucilage from the fruits of Tilkor (*Mamoradica monadelpha*), the mucilage sample was prepared in powder form by using freeze-drying. FTIR result confirmed the presence of complex carbohydrate (starch), moreover, FTIR of Tilkor mucilage showed the vibrational stretching of C-H bending of Alkynes, C-H bending in aromatic rings, C-C Stretching vibrations, free O-H groups Vibrations, and C=N bond showed Aminoacids/ proteins and the absorption band wavelength at around 685 cm^{-1}–665 cm^{-1}, 900 cm^{-1}–625cm^{-1}, 822 cm^{-1}, 3710–3513 cm^{-1}, and 1623 cm^{-1} respectively.

Rheological characterization is used to determine the shear rate-dependent flow behavior of the mucilage solutions, and is generally examined over a shear rate range of 0 to 100 Hz. The rheological characterization of the mucilage of seeds from different plants was performed by several scientists. In this context, Punia et al. [56] measured the rheological characteristics of the mucilage of chia seeds by an oscillatory shear (range of frequency stress sweep is about 1 Hz to 10 Hz) and revealed the highest correlation coefficient of chia seed ($R^2 > 98.58$) and with the increase in shear rate, they observed a rapid decrease in the viscosity of the solution. In another study by Capitain et al. [57] it was observed that the chia seed dispersions viscosity increases by increasing the concentration from a range of 0.25 to 1.00 (*w/v*). Likewise, Abbastabar et al. [58] characterized quince seed mucilage and observed curve turned from the linear range at strain 11.4%. Increases of storage modulus and decrease of viscoelastic linear range (7.7%) were observed in the presence of 0.2 M NaCl as NaCl shows its dual behavior on the mucilage or gum. Quince seed mucilage has high activation energy which is about 6988.74 J mol^1. activation energy is usually used to determine the chain flexibility of biopolymers. Keshani-Dokht et al. [59] in their study observed a decrease in the magnitude of *Cordia myxa* mucilage and an increase in solution concentration from 0.99 to 0.89 and 0.2–2% respectively. In their study, they observed that the mucilage solution has a tendency of more shear thinning at a high concentration level. The activation energy of *Cordia myxa* mucilage was calculated at about 446.23 KJ. The thermogravimetry analysis (TGA) technique is used for measuring the mass variation in a particular mucilage-containing sample as a function of temperature in a stable atmosphere [53]. The vibration of mass can be negative when the part of the sample is transformed in vapor and that can be positive when the sample is subjected to corrosion or oxidation [60]. The thermal stability of mucilage can be measured in two conditions such as isothermal (temperature is kept constant) and dynamic (temperature is increased at a linear rate) [6]. Thermal stability and thermal behavior of *Manilkara zapota* (Linn.) Seed mucilage data showed that heating at 10 °C /min from 0 °C to a maximum of 900 °C resulted in two mass loss events according to the derivative thermograms and primary thermograms. Furthermore, there was 41.17% weight loss at 178.6–359.7 °C temperature range in the first Decomposition stage and 30.06% weight loss at 359.7–600.6 °C in the second Decomposition stage. Besides, at the same temperature range the enthalpy (315.8729 and 3624.787 J/g), DTG peak (350.3 °C and 614.4 °C), and heat change (138.4354 and 1082.215 μVs/mg) were observed for both first and second decomposition stages [61]. Differential scanning calorimetry (DSC) has emerged as an excellent physical technique to investigated physical and chemical or exothermic and endothermic changes that occur in the mucilage during thermal processing [62]. The high thermo-stable property of mucilage can be utilized in paints for stabilizing suspensions or emulsions, improve bake-stability in cakes, preventing crystal growth, and improving freeze-thaw stability [6]. The outcome of the DSC analysis of *Diospyros melonoxylon* Roxb seed mucilage discovered the glass transition temperature is 78 °C. the high intense peak was observed in Differential scanning calorimetry thermograms at around 200 °C which was an endothermic transition. The large endothermic peak correlates to the hydrophilic nature of the functional groups of chia seed mucilage and can be due to in regular packing structure of the gums [63]. The endothermic peak transition temperatures (Te, To, Tp) of chia seed mucilage were 215 °C, 52.8 °C, and 107.9 °C, respectively. 233.9 J/g was enthalpy change value of chia seed mucilage. In the case of exothermic peak transition temperatures (Te, To, Tp) of chia seed mucilage were 354.9 °C, 277.7 °C, and 316.8 °C, respectively, and the enthalpy value of exothermic peak (101.9 J/g) was lower than endothermic. The high enthalpy value of mucilage indicates high energy needed in releasing water-related with the loss of crystallization as well as hydrogel bounded [56].

6. Functional Properties of Mucilage

Hydrocolloid polymers in the food industry and pharmaceutical industry are generally made up of large molecular biopolymers. A biopolymer contains a hydroxyl group which leading to viscous dispersions, increases the water attraction. Also, those hydrocolloids are widely used as a stabilizer, thickener, dietary fiber, whipping agent, and fat replacer. Moreover, they also have applications in crystallization inhibition, edible coatings, edible films, and encapsulating flavors [13,64]. The different functional properties of mucilage are shown in Table 2.

Table 2. Functional properties of various plant-based mucilage.

Mucilage Source	Functional Property	References
Chia seed mucilage	Stabilizing agent, thickening agent, fat replacer, and emulsifying agent.	[21,56]
Okra seed mucilage	Oil absorption, water absorption, emulsion stability, foaming capacity, and emulsifying capacity.	[65,66]
Tamarind seed mucilage	Water holding capacity, oil holding capacity, and solubility.	[13,67]
Flaxseed mucilage	Water holding capacity, thickening agent, gelling agent, emulsifying agent, and foaming agent.	[40,68,69]
Cress seed mucilage	water absorption, Film-forming agent, and gelling agent.	[70]

6.1. Water Holding Capacity of Mucilage

All the scientific literature has shown that high water holding capacity is due to the presence of hydroxyl groups and protein substituents in the gum and mucilage structure. In 1996, the Food and Agriculture Organization of the united nation (FAO) described it as a potential source of polysaccharides due to its excellent property in aqueous solutions at low concentrations [63]. Water holding capacity is the capacity of a moist polymer or sample to hold water when exposed to an external centrifugal gravity force or compression. It comprises the sum of linked water, physically trapped water, and hydrodynamic water, the latter of which contributes most to this capacity [13]. The low amount of water holding capacity can be attributed to the high solubility of mucilage even at a high concentration which leads to an inability to form a gel. Furthermore, mucilage is widely selected in the food system (fruits, vegetables, poultry, and fish) and contributed to improved emulsifying and water binding capacity of food [60]. According to a study the water holding capacity of *Cordia myxa* mucilage (14.94 ± 2.44 g/g), *Opuntia dilleniid* mucilage (4.00 ± 0.10 g/g), sweet basil seed mucilage (97.5 ± 2.4 g/g), Tamarind seed at 25 °C (0.18 ± 0.014 g/g), at 45 °C (0.22 ± 0.167 g/g), and at 65 °C (1.07 ± 0.025 g/g), which is shown that the water holding capacity of tamarind seed is increasing as increasing of temperature [13,26,59,71,72].

6.2. Oil Holding Capacity of Mucilage

Oil holding capacity is the most important functional property of hydrocolloids, which indicates oil absorption capacity. According to the reports, due to an increase in the accessibility of nonpolar chains with increasing molecular mobility of hydrophobic hydrocarbon domains, plant-derived mucilage show increased oil binding capacity with increases in temperature. The oil holding capacity of tamarind seed mucilage at 25 °C (0.068 ± 0.014 g/g), at 45 °C (0.104 ± 0.010 g/g), at 65 °C (0.133 ± 0.004 g/g); chia seed mucilage (12.97 ± 1.90 g/g); guar gum (0.87 ± 0.06 g/g); and basil seed mucilage (8.37 ± 1.02 g/g) [13,26,27,73,74]. Good oil holding values indicate that mucilage could improve the texture of food products. Fruit or seed extracted mucilage has good oil binding capacity due to presence of monopolar molecules, therefore, it can trap higher amounts of oil particles and also prevent oil and the loss of flavor from food systems. Therefore, fruit or seed-derived mucilage can be used as a good functional ingredient in formulated foods [75].

6.3. Emulsifying Property of Mucilage

Plant-derived polysaccharides such as gum, mucilage, and starch as well as Carboxymethyl cellulose (CMC) are widely used in emulsion products [76,77]. Polysaccharide mucilage and gum are attracting attention towards commercial use in the pharmaceutical and food industry due to their excellent emulsifying properties, where they are used for suspension of particulates, stabilization of emulsions, control of crystallization, the thickening, and formation of films, encapsulation [30]. The emulsion products prepared from different plants are checked for their stability by several scientists to use them for different purposes in different food and pharmaceutical industries. Sangeethapriya and Siddhuraju [78] prepared the emulsion products from quince seed mucilage and they observed that the stability of the resulting product was 94.89% likewise Behrouzian et al. [79] observed the high emulsifying capacity (92%) of cress seed mucilage while in another study by Jouki et al. [80] it was observed that emulsifying capacity of *Zizyphus mauritiana* fruit mucilage is about 54.12% and emulsifying stability is 42.14%.

6.4. Foaming Property of Mucilage

Good foaming property is associated with the elastic structure of the mucilage, foaming property of mucilage is depending upon many factors, such as the presence of further compounds in the hydrocolloid, molecular weight, protein, structure, and carbohydrates. The good foaming capacity of mucilage is highly associated with the flexible structure of mucilage that can reduce the surface tension [81]. The foaming capacity and foaming stability of *Abelmoschus esculentus* mucilage is 50–51% to 50–62% and 36.04%, 1% *w/v* respectively, foaming capacity flaxseed mucilage reported about 75%, and Plantago seed mucilage has foaming stability (88.4%) [13,66,82].

6.5. Antioxidant Property of Mucilage

Plant-derived polymers contain different phenolic compounds including flavonoids and polyphenols as bioactive compounds. These bioactive compounds prevent the chain propagation reactions originated by free radical reactions and prevent disease-related oxidative damage [83–86]. Sangeethapriya and Siddhuraju [78] studied the antioxidant activity of the polyphenols isolated from *Zizyphus mauritiana* fruit mucilage against DPPH-2,2-diphenyl-1-picrylhydrazyl is 5.27 g mucilage/g DPPH, superoxide is 85.12%, ABTS-2,2-azino-bis-3-ethylbenzothiazoline-6-sulphonic acid is 16587.32 mmol Trolox Eq./g, and hydroxyl is 76.13%. Sardarodiyan et al. [87] also studied the antioxidant activity and total phenolic compounds of *Lallemantia royleana* seed mucilage and in their study, they observed antioxidant activity of 528.54 μg/mL of DPPH and total phenolic content of 82.56 ± 1.6 μg/GAE. In two different studies done by Jouki et al. [80] and Silveira Coelho et al. [88] the antioxidant property of Quince seed and phenolic content of chia seeds were 29.88% and 641.71 μg/GAE, respectively.

6.6. Antimicrobial Activity of Mucilage

Nowadays, food safety and food quality are major challenges for all countries, it is estimated that there are around 30% of people are suffering annually from food illness disease in developed countries [84,85]. Many natural polymers such as proteins and carbohydrates have antibacterial and antifungal properties, which prevent or reduce the growth of microorganisms in food. Infectious microorganisms are extensively spread everywhere on the earth which causes a serious problem for society. Furthermore, natural polymers are more effective in the treatment of those diseases and also prevent several side effects that can be led by antimicrobial agents [86]. While mucilage has an antibacterial and antifungal property which can also prevent or reduce the risk of food-borne illness disease and food spoilage. Much seed-based mucilage has an antimicrobial property such as Chia seed [86], Cress seed [78], okra seed [65], and quince seed [80] Tantiwatcharothai and Prachayawarakorn [89] prepared Basil seed-based sponges hydrogel for measuring the antibacterial activity (against *S. aureus* and *E. coli*) in wound dressing. ZnO nanocomposite and borax crosslinking were also incorporated with basil seed mucilage. The result showed that there was no effect of borax in the antibacterial activity of Basil seed mucilage sponge. Whereas, ZnO nanocomposite loaded resulted in good antimicrobial activity against the gram-positive (*Staphylococcus aureus*) and gram-negative (*Escherichia coli*) bacteria. Furthermore, better antibacterial activity was observed against gram-positive (*S. aureus*) than that of gram-negative (*Escherichia coli*) because of the structural difference of the cell wall. Gram-negative bacteria possess a complex cell wall, while gram-positive bacteria contain a thick cell wall of peptidoglycan, therefore, it is more resistant to attack by the microbial agent. Jouki et al. [80] prepared quince seed mucilage-based film with the incorporation of thyme essential oil in different concentrations ranging from 0 to 2.0%. The film was separated against eleven food associated with bacterial strains by agar disc-diffusion assay. And observed that 1% thyme essential oil containing film was more effective against all microorganisms including gram-positive and gram-negative bacteria such as *Staphylococcus aureus*, *Listeria monocytogenes*, and *Shewanella putrefaciens*. Mujtaba et al. [90] also prepared chia seed mucilage composite films with the incorporation of cellulose nanofibrils at various concentrations by using the solution casting method. Microorganisms such as *Escherichia coli*, *Pseudomonas aeruginosa*, *Staphylococcus mutans*, *Salmonella typhimurium*, *Bacillus thuringiensis*, and *Staphylococcus aureus* were incubated in LB agar at 37 °C for 24 h. the result indicated that chia seed-based cellulose nanofibrils composite film improved antimicrobial activity with the increasing of cellulose nanofibrils concentrations (3–6%) and observed that mucilage films containing 6% cellulose nanofibrils had the highest antimicrobial capacity against all the exanimated microorganisms. The inhibition zone range of all the exanimated microorganisms was *Escherichia coli* (20.17 mm), *Pseudomonas aeruginosa* (18.25 mm), *Staphylococcus mutans* (17.79 mm), *Salmonella typhimurium* (6.12 mm), *Bacillus thuringiensis* (17.19 mm), and *Staphylococcus aureus* (18.89 mm). The antimicrobial activity of different mucilage-containing plants is represented in Table 3.

Seed and fruit mucilage show excellent activity against pathogenic gram-positive and gram-negative bacteria. The mode of action of their mechanism is proposed by several authors in various theories as shown in Figure 3, mucilage with incorporated substance exhibit antibacterial ability by binding to bacterial membranes and enters inside the cell, thus resulting in the damage of the Nucleus: DNA damage, Degradation of Protein, Cell membrane damage and Mitochondrial damage [37].

Table 3. Antimicrobial activity of different mucilage-containing plants.

Mucilage	Incorporated Agent	Gram-Positive Bacteria	Gram-Negative Bacteria	References
Basil seed mucilage	Zinc oxide nanocomposites crosslinking with borax	*Staphylococcus aureus*	*Escherichia coli*	[89]
Quince seed mucilage	Thyme essential oil	*Bacillus cereus, Listeria monocytogenes*	*Salmonella typhimurium, Escherichia coli, Vibrio cholera, Pseudomonas aeruginosa, Yersinia enterocolitica*	[80]
Chia seed mucilage	Oregano essential oils, cellulose nanofibers	*Staphylococcus aureus, Staphylococcus mutans, Bacillus thuringiensis*	*Escherichia coli, Pseudomonas aeruginosa, Salmonella typhmurium*	[37,90]
Okra mucilage	Zinc oxide nanoparticles and carboxymethylcellulose (CMC)	*Staphylococcus aureus*	*Escherichia coli*	[91]
Qodume shirazi seed mucilage	Lavender essential oil	*Staphylococcus aureus*	*Escherichia coli*	[92]
Lepidium sativum seed mucilage	Heracleum lasiopetalum essential oil	*Staphylococcus aureus*	*Escherichia coli*	[93]

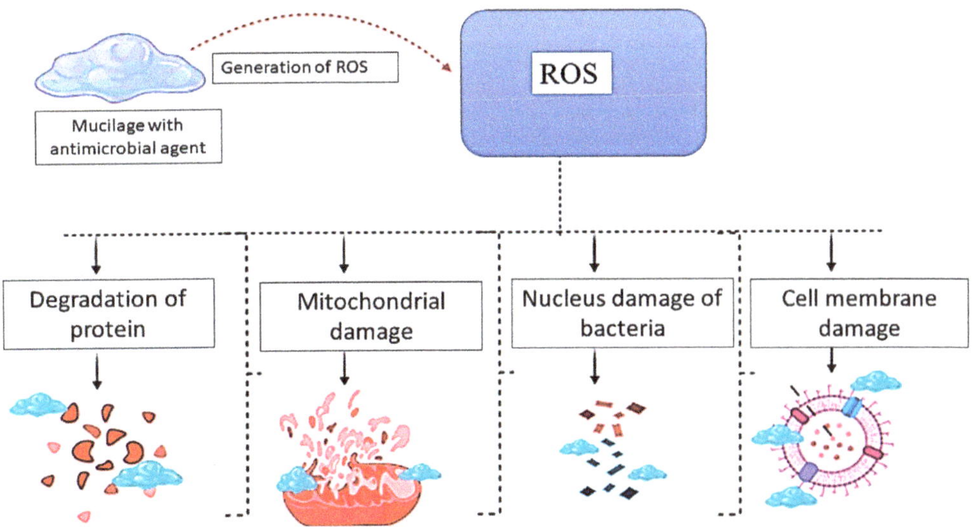

Figure 3. Possible mechanism of antimicrobial efficacy of plant-based mucilage.

The mucilage is helpful in the development of free radicals by the membrane, which leads to porosity and the death of the cell. Moreover, mucilage generates ROS (reactive oxygen species), which are free radicals that easily react with other molecules in the cell due to the presence of oxygen, which causes the damaging of proteins, RNA, and DNA [91]. Oxidative stress of ROS is a major cause of damage to the cell membrane due to lipid peroxidation. The mucilage can attach to the bacterial cell wall through several interactions in the cell membrane; therefore, the reducing the viability of the bacterial cell by inhibiting DNA replication. The bacterial cell comprises DNA bases enriched in sulfur and phosphorus, these biomaterials associate with the mucilage. Moreover, mucilage can break down the cell membrane permeability of pathogenic bacteria such as *E. coli*, *S.*

aureus, *Bacillus* spp., and *Pseudomonas* spp., and also reduce the amount of ATP (adenosine triphosphate) by regulating the production of ATPase enzymes [92,93].

7. Application of Mucilage

The plant-derived additives or polymers due to their positive escape gaining importance in recent days and can be utilized as natural thickeners or emulsifiers for the human diet, therefore, act as an alternative to synthetic polymers or additives [19]. Moreover, mucilage derived from plants due to their property of elasticity can form a large number of networks molecules, therefore, can be widely used as an edible film or edible coating in food packaging applications in food industries, disintegrants in tablets, application of tablet binders, and several other applications in the pharmaceutical industry [11,94].

7.1. Mucilage as a Coating Material

Mucilage derived from plants can enhance the shelf life, storage capacity and also reduce the loss of moisture, therefore, extensively used as an edible coating in food packaging industries [95]. The mucilage has high tensile strength and barrier properties against different gases which reduce the firmness and weight loss of coated products [96]. Several studies have been conducted to check the effect of mucilage obtained from seeds of plants by different scientists [97]. In a study by Alizadeh Behbahani et al. [39] edible coating of Shahri Balangu seed mucilage (SBM) incorporated with cumin essential oil (CEO) on beef slices was done and a sensory study of beef slices during the storage period of 9 days was carried out. It was observed that mucilage coated beef slices reduced the loss of lipid oxidation, microbial counts (total viable count TVC), *Escherichia coli*, psychotropic count, yeast, and molds during the storage period. There was no adverse effect of SBM with CEO on sensory characteristics of beef and it conferred an excellent texture to the product as compared to the control sample. The different food applications of mucilage are shown in Table 4.

Table 4. Application of various plant-based Mucilage as a coating material in the food industry.

Source of Mucilage	Coating on Food	Reference
Aloe vera	Application on tomatoes	[98]
Aloe vera and Basil mucilage	Application on Apricots	[99]
Barbery fig mucilage	Application coating on kiwi slices	[100]
Cress mucilage	Application on fresh beef	[93]
Shahri Balangu seed mucilage with cumin essential oil	Application on beef slices	[39]
Hibiscus mucilage	Application on tomato	[101]
Flaxseed mucilage and xanthan gum	Application on cheddar cheese	[102]
Aloe vera gel	Application on apple slices	[103]

7.2. Application of Mucilage as Encapsulation Agents

Encapsulation of food ingredients, pharmaceutical ingredients, and nutraceutical ingredients is an effective technique to improve the stability of targeted compounds, apart from providing benefits on delivery characteristics [99]. The technique is used in the flavor industry to conceal undesirable taste/flavor notes such as the characteristic pungency and bitterness of polyphenols. This can be applied to a decline in the formation of rancidity in oil powders [84]. The preparation of the wall or filler material for the encapsulation of the bioactive substance is an important step because it influences the properties of the emulsion, compound preservation, and final product stabilization. several food-grade polymers have been discovered for nano and microencapsulation applications such as lipids, proteins, and carbohydrates. Furthermore, mucilage has a good ability to form a dense network after drying and their high molecular weight carbohydrate polymers make them suitable for nano and microencapsulation of food and bioactive compounds [102]. High molecular weights can aid in the retention of more key substances and increase stability. Plant-based mucilages are currently being investigated as wall materials for various encapsulation tests, in addition to their other applications.

7.3. Application of Mucilage in Wound Healing

The skin comprises the three layers of the dermis, epidermis, and hypodermis, these are known as the largest organ in the body. A wound is a mechanical or physical injury that results in breaking or opening of the skin [104]. Dressing a wound to prevent it from bacterial infection is most important to allow the wound to heal quickly. Generally, dressing absorbs the exudates from the wounded skin and prevents the pathogenic microorganisms by producing moisture at the border of the wound [89]. Plant-derived polymer due to hydrogen bond-forming groups are more extensively used for advanced and traditional wound healing and drug delivery system [105]. Mucilage-based hydrogels are non-toxic, biocompatible, biodegradable, and easily available, therefore, used for the wound healing process. These hydrogels can be naturally removed or applied with wound beds without any significant interference [106]. Likewise, Tantiwatcharothai and Prachayawarakorn [89] studied the antimicrobial effect of basil seed mucilage-based hydrogel and they observed good antimicrobial activity against pathogenic bacteria *E. coli* and *S. aureus*, isolated from the wounds. The application of mucilage in the wound healing process is explained in Figure 4.

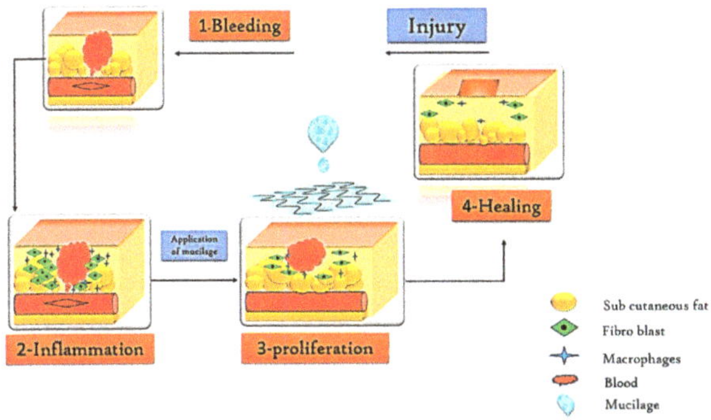

Figure 4. Application of plant-based mucilage in wound healing.

7.4. Application of Mucilage as an Emulsifying or Suspending Agent

Mucilage can act as a good emulsifying and suspending agent. An emulsion is a thermodynamically unstable, biphasic system, containing two non-miscible liquids that can be classified based on their droplet size into several categories, such as nano-emulsions (20–200 nm), microemulsions (100 Å–100 nm), and macroemulsions (1–100 μm); therefore, they are extensively used in various industrial applications such as in the food and pharmaceutical industries [107]. Moreover, plant-derived natural emulsifying agents have effective emulsifying properties by imparting a charge to dispersed droplets, forming multimolecular sheaths around emulsion droplets by increasing the viscosity of a material [49]. The emulsifying properties of *Cucumis sativus* Linnaeus mucilage were also compared with commercially available tragacanth and gelatin with an additional valuation of their combined effect on the emulsion properties. They observed a high emulsifying capacity of *Cucumis sativus Linnaeus* mucilage as a primary emulsifying agent for oil–water emulsions. Likewise, Campos et al. [108] used chia seed mucilage as a replacer emulsifier for the formation of ice cream. Chia seed mucilage maintained the quality of the product, but its sensory properties were affected due to the dark color of mucilage. Moreover, plant-derived mucilage can increase the tensile strength of the hydration layer formed across the suspended solid particles by molecular interactions and hydrogen bonding. Therefore, these suspending agents perform well in the presence of moistening agents, as they do not reduce the interfacial surface tension. Thus, natural polysaccharides (mucilage) are commonly used as protective thickeners or colloids

7.5. Application of Mucilage in Tablet Formations

Mucilage is widely used for the preparation of tablets due to its high adhesive property, good swelling, and water absorption ability, therefore it can be used as a disintegrant or binding agent. Moreover, mucilage obtained from the seeds and fruits is used to sustain or modify drug release [109,110]. Mucilage is a binding agent that is pharmaceutical excipients used in the manufacturing of tablets to affect the coherence and aggregation of the mixed powder for the enhanced flow properties of particles and to impart physical strength to the tablet. Chaudhary et al. [111] manufactured tablets by using *Grewia asiatica* mucilage as a binding agent. During the in vitro dissolution study, released more than 80% of the drug within 30 min. They observed that percentage of friability and hardness of tablets was decreased with the increasing concentration of mucilage.

7.6. Application of Mucilage for Removal of Contaminants from Water

Nowadays, plant-derived polymers (gums and mucilage) have attained a high demand in water purification due to their high water absorption capacities. Many components are responsible for water impurities, such as inorganic impurities (dyes), organic impurities (microorganisms), organic impurities (microorganisms), dissolved salts, heavy metals, and oils. moreover, seed or fruit mucilage is most effectively used towards such kind of water contaminators [112,113]. Furthermore, a Large concentration of contaminants in groundwater continues to be recorded as threats to the wellbeing of millions of people worldwide. Heavy metals are typically found in trace quantities in natural waters, but most are toxic even at low concentrations. Metals such as arsenic, lead, cadmium, nickel, mercury, and limited concentrations of zinc are highly dangerous for human health [114]. Heavy metals may reach the human body through food, water, air, or absorption through the skin when they come into contact with humans in agriculture, manufacturing, medicinal, industrial, or residential conditions [115]. Jones et al. [116] removed the selected heavy metal ions from the aqueous solution by using mucilage of *Dicerocaryum eriocarpum* (act as a biosorption). Mucilage was modified with potassium chlorides and sodium. Also Compared the potassium chlorides mucilage and sodium mucilage with non-modified deionized water mucilage. They observed that potassium chlorides mucilage increased the intensity of mucilage but did not produce new functional groups. The result of the study showed that *Dicerocaryum eriocarpum* mucilage has excellent binding attractions with heavy metal ions such as Fe, Zn, Cr, and Cd in the aqueous solution. In another study, Fox et al. [117] had worked on removing heavy metal from water using *cactus* mucilage. Intially, they extracted mucilage from (cactus) for the removal of arsenic. and prepared two types of the extracts (i) gelling extract and (ii) non-gelling extract. Their work mainly revealed an interaction of cactus mucilage with arsenic. -CO groups (carboxyl and carbonyl groups) and -OH (hydroxyl) functional groups interaction with the arsenate. The outcome from their work comes is that the gelling extract shows a better average than the non-gelling extract. Mucilage extracted from *Hibiscus esculentus* (okra) and *Malvasylvestris* (mallow) and have been successful in eliminating turbidity from wastewater due to their substantial flocculation activity, however, there has been an increased (COD) chemical oxygen demand [118]. consequently, *Viciafaba* and *Ficus-indica* mucilage are environmentally friendly bio-flocculants in the treatment of COD as well as in the decoloration and turbidity from textile wastewater and have shown the capacity for more contamination removal relative to industrial flocculants such as Polyacrylamide A100PWG and EPENWATE EXP31/1, and are more likely to be environmentally safe [119].

8. Therapeutic Importance of Mucilage

For many decades medicinal herbs and plants are used as traditional medicine for medical treatments. It is estimated that in developing countries around 80% of the people are still using traditional medicines for the prevention of diseases and among them, around 25% of drugs are derived from plants [7]. Many studies revealed that plant-derived mucilage such as (fenugreek gum, yellow mustard mucilage, and flaxseed) has many health beneficial properties due to their potential of reducing or preventing the risk of type 2 diabetes [67]. Mucilage is conventionally used in the medical field via topical or oral routes for urinary, respiratory, gastrointestinal, reproductive, and musculoskeletal systems and is also used for skin disorders. Moreover, they are also extending to diabetes, cancer, immunity stimulation, and have antioxidant, and antimicrobial property. Furthermore, mucilage containing plants such as *Trigonella foenum* graecum has antioxidative, antimicrobial, and anti-inflammatory properties [67].

9. Mucilage Based Nanocarriers and Their Application

Nowadays, synthetic and non-synthetic polymers have been successfully used for the formation of a hydrogel, but plant-derived (synthetic) polymers such as proteins, polysaccharides, and polypeptides are the most preferable choice, because of their extensive use of applications. Mucilage has an excellent potential to synthesize hydrogels because of its hydrophilicity, safety, and biodegradability. Hydrogels are hydrophilic and polymeric 3D material, which retains diffusive transport of liquids as well as also retains cohesive property of solids. They have attained high demand for technologists and researchers due to their extensive range of applications. The first synthetic hydrogel was prepared in 1960. Moreover, hydrogels from the plant-derived polymers are in high demand due to the presence of functional groups such as sulfate, amide, hydroxyl, and carboxylic which increases their swelling and water holding capacity, they are also interconnected with elasticity, and microscopic pores. There are many stimuli factors such as (pH, temperature, and electric field) [120]. There are generally two methods (physical and chemical crosslinking) that are used for the formation of hydrogel along with the principles of crosslinking of a polymer chain. The chemical crosslinking method includes the creation of new covalent bonds with the hydrogel's polymer chain, while physical interaction can be also present between the polymer chain of the hydrogel. Both chemical and physical crosslinking methods can be applied for the synthesize of hydrogel from the plant-derived polymers (gum and mucilage) [121]. The formation of a nanohydrogel is explained in Figure 5. These characteristics increase the value of hydrogel as an applicant in food, pharma, and several industries [122].

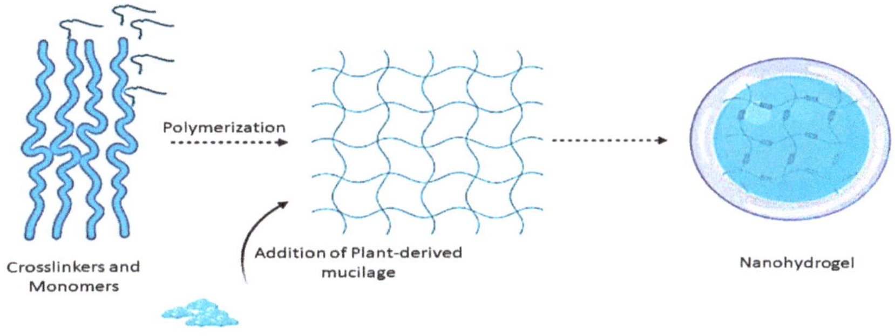

Figure 5. Synthesis of nanohydrogel using plant-based mucilage as an effective biopolymer.

The characterization of hydrogels is dependent upon the cross-linking (physical or chemical) measures during the formulation of the gel. Nanohydrogel is mostly similar to a normal hydrogel, which can be defined as a three-dimensional network of hydrophilic material (e.g., polysaccharide) with a diameter of less than 100 nm. Nanoparticulates have many benefits when compared to micro and macrocategorization in food and several other industries. The term nanohydrogel was first introduced to describe the cross-linking and networking of poly-anions [123]. They are used in several applications such as wound healing, drug delivery, vaccine delivery, the enhancement of film properties, and enzyme immobilization [124,125]. Mucilage-based hydrogels containing nanocomposites form a 3D network of extreme porosity, which allows a large absorption of food or drugs in water [126]. Nanocomposites are divided into three classes: ceramic matrix nanocomposites, polymer matrix nanocomposites, and metal matrix nanocomposites. They are chosen related to macro and microcomposites due to their excellent potential properties such as mechanical, barrier, and optical characteristics. The characterization of mucilage-based nanohydrogel can be performed through different methods such as field emission scanning electron microscope (FESEM), Fourier-transform infrared spectroscopy (FTIR), X-ray diffraction (XRD), X-ray photoelectron spectroscopy (XPS), and high-resolution transmission electron microscopy (HRTEM) [127]. The mucilage-based hydrogels can act as a protector, which prevents active ingredients from degradation, oxidation, and destruction, and also has several applications in water purification, drug delivery, the food industry, tissue engineering, and agriculture. Nanomaterials improve the barrier and mechanical aspects of food packages, and other developments for intelligent and active applications in the food industry [91]. Polymers containing hydrophilic groups such as -COOH, -OH, -CONH-, -SO_3H, and -$CONH_2$ interact with each other. Nanohydrogels are responsible for several stimuli such as temperature, electromagnetic field, pH, ionic strength, and light. Moreover, mucilage-based nanohydrogel is mostly utilized for the preparation of the edible coatings of edible films, and it is estimated that high mucilage-containing seeds or fruits are a good source of edible gum and can be used for various applications [122]. Nanohydrogels combine the great characteristics of hydrogels, such as absorption capacity, hydrophilicity, flexibility, great water holding capacity, with the advantages of nanoparticles, allowing for obtaining better dispersion in food packaging material and decreasing the number of bioactive compounds to be applied [128]. Plant-derived mucilage-based nanohydrogels are in great demand due to their unique properties, such as biocompatibility, biodegradability, stimuli-responsive properties, and biological characteristics, making them a good material for selection in diverse applications. Furthermore, nanohydrogels have potential applications such as controlled drug delivery, biomimetic materials, and biological or chemical sensors. Nowadays, nanotechnology plays a very important role in drug delivery systems, food applications, and water purification [129]. Therefore, nanoparticles (magnetic and non-magnetic), nanofibers, nanocomposites, and nanoencapsulation are widely used as nanocarriers in various industrial applications, such as for the controlled delivery of drugs, the removal of dye, and the development of film, which are highlighted in Table 5.

Table 5. Application of seed mucilage with various nanocarriers.

Seed Mucilage	Nanocarrier	Applications	References
Basil seed mucilage	Magnetic nanoparticles (Fe_3O_4)	Application for the controlled delivery of antibiotic (Cephalexin)	[130]
Cress seed mucilage	Nanofibers	Application for the delivery of vitamin A	[18]
Quince seed mucilage	Zinc oxide nanoparticles	Application for photocatalytic dye degradation	[131]
Quince seed mucilage	Magnetic nanocomposites	Application for removal of cationic dyes from the aqueous solutions	[132]
Basil seed mucilage	Zinc based magnetic bio nanocomposites	Application for removal of azo anionic and cationic dyes from the aqueous solutions	[133]
Okra seed mucilage	Zinc oxide nanoparticles	Application for nanocomposites-based films	[91]
Basil seed mucilage	ZnO nanocomposites	Application for wound healing	[68]
Chia seed mucilage	Nanoencapsulation	Application as wall material	[115]

Moreover, Rayegan et al. [130] synthesized magnetic Fe_3O_4 nanoparticles coated with basil seed mucilage for the application of the controlled drug delivery of an antibiotic (cephalexin). The sample was characterized using XRD, FTIR, TEM, FESEM, and VSM. One-hundred and fifty magnetic nanoparticles were randomly selected for FESEM, which showed that the mean size of the nanoparticles was 6 nm and 12 nm, with 0.25 and 0.28

PDI values, respectively. Moreover, the antibacterial efficacy was evaluated by the disk diffusion method, and it was observed that there were no negative effects on the performance of drugs or on the structure by the loading of cephalexin onto the basil seed mucilage-coated magnetic nanoparticles. Moreover, it also increased the antibacterial properties of cephalexin. Consequently, Mohammadi et al. [91] prepared nanocomposite films based on okra mucilage (OM), carboxymethylcellulose (CMC), and ZnO nanoparticles, and evaluated their antibacterial and physicomechanical properties. In their study, they used different proportions of okra mucilage and carboxymethylcellulose (0/100, 30/70, 40/60, and 50/50, respectively). Colored films were observed with high levels of ZnO nanoparticles and okra mucilage. Moreover, due to the addition of mucilage, tensile strength was increased and elongation at the break value was decreased by the incorporation of ZnO nanoparticles into carboxy methylcellulose film.

Synthesis of Nanoparticles with Mucilage

Nanoparticles have attained widespread attention for the development of green, facile, and sustainable synthesis of nanoparticles. Physicochemical methods such as direct precipitation, microwave, and hydrothermal methods are widely used [132,133]. Therefore, plant-derived polymers such as gums and mucilages are effectively used for the preparation and functionalization of nanoparticles. Mucilage used for the formulation of microparticles was also exploited in nano-formulations using vitamin D loaded nanoparticles on cress seed mucilage and gelatin. The prepared particles enhanced encapsulation efficiency (around 70%), with better in vitro release percentages of vitamin D in intestinal fluids and simulated gastroitestinal conditions [130]. Consequently, a study conducted by Pathak et al. [134] explained the achievement of using mucilage as an anionic ocular polymer for the synthesis of nanoparticles with a cationic polysaccharide polymer (chitosan). The antibacterial activity of the prepared sample was investigated against Gram-positive (*Bacillus cereus* and *Staphylococcus aureus*) and Gram-negative bacteria (*Salmonella typhimurium* and *Escherichia coli*). This type of nanoparticle can be useful in the medicinal industry for sustaining the anti-bacterial effect of antibiotics. Furthermore, the negative charge of the mucilage was additionally subjugated through the preparation of cephalexin-loaded/basil seed mucilage-coated iron oxide (Fe_3O_4) magnetic nanoparticles. The in vitro release of antibiotic (cephalexin) from the drug-loaded coated nanoparticles at pH 7.4 displayed biphasic form, starting with an initial quick release, then a continuous release phase. Apart from abusing the negative charge of the mucilage, phase inversion techniques and anti-solvent supercritical gas using supercritical carbon dioxide as an anti-solvent were applied for the preparation of mucilage nanoparticles loaded with the cytotoxic drug paclitaxel. The nanoparticles showed high drug loading above 75, and small particle size (around 200 nm) [135]. Mucilage contains flavonoids and polyphenolic contents. These compounds can efficiently bind zinc ions from an aqueous medium, and they function as stabilizing agents and natural reducers during the synthesis of nanoparticles. During the oxidation process, several carbohydrates and the counter acetate ions are also decomposed to produce high amounts of carbonate ions (Ag, Zn, Cu, and Co) and carbon dioxide; therefore, the binders can be removed [136–140].

10. Market Outlook of Mucilage

Mucilage is a gelatinous component that contains a good amount of proteins and polysaccharides, which are produced by almost all plants and a few microorganisms. It is abundantly present in agar obtained from red algae, and also in flaxseeds, fenugreek, chia seeds, marshmallow roots, psyllium, nopales, and aloe. However, a significant amount of mucilage that is used in industries is not available, and statistical consumption of mucilage is not reported. Earlier, it was primarily used for therapeutic purposes; however, it now has a variety of applications in the food, cosmetics, and health care industries [141]. The American and Asian markets have excessive utilization of mucilage in the ink, glue, and adhesive industries. The global market for cosmetics is growing, and so the use of natural products is as well. Therefore, the demand for mucilage for cosmetics will also increase in the upcoming years. Mucilage is useful as a binding agent in medicine, and, due to its healing properties, is used in ointments, and its cryoprotective property is increasing demands in the pharmaceutical industry [142]. The American and European mucilage market is in demand and is expected to grow positively during the forecast period. In the food industry, there is a growing demand for processed foods that contain mucilage as a thickening, gel-forming, and emulsifying agent. It is also used as a stabilizer in the milk industry, such as in ice cream, yogurt, and, flavored milk. Moreover, mucilage-derived hydrocolloids have been used in the food industry to provide textural functionality, such as in fruit fillings, water binding in meat products, dairy-based beverages, desserts, and jams [143]. Mucilages contain different kinds of antioxidants that slow the aging process, and thus can also be used in cosmetics products. Furthermore, mucilage is used for its health advantages, including bolstering the immune system, soothing the gastrointestinal tract, the ability to lower low-density lipids, and lowering blood pressure, which makes it applicable in pharmaceuticals and medicine. In the forecast period, there will be significant market growth in the pharmaceutical, health care, and medicine markets.

Nowadays, intensive research is underway on plant-derived mucilage-based edible coatings and films to enhance the shelf life of fruits, vegetables, meats, and fish products. Similarly, mucilage is being used to develop nanofibers, which creates a large market for the textile industry in fabrication. The hydrophilic nature of mucus can act as a barrier to water penetration, delaying water loss and prolonging firmness [144]. This property makes it a promising component for cosmetic industry applications. Therefore, it is to be expected that, in the coming years, the slime market will include a wide range of applications and new market development.

11. Conclusions, Future Research Perspectives, and Challenges

In recent years, plant-derived polymeric carbohydrates have acquired high demand from food industries due to their diverse applications, such as as film coatings, emulsifiers, binders, and gelling agents. Among all carbohydrate polymers, mucilage has been abundantly used in the field of modern science due to its diverse applications. Plant-derived mucilage can be obtained from the special mucilage cells of different plant parts. Moreover, mucilage is mainly composed of complex carbohydrate polymers with highly branched structures that consist of L-arabinose, D-xylose, D-galactose, L-rhamnose, and galacturonic acid. They also contain glycoproteins and different bioactive components such as tannins, alkaloids, and steroids. Due to these properties, plant-derived mucilages can be used as an active functional ingredient, emulsifier, surfactant, stabilizer, encapsulating material, or cross-linker; therefore, they could be used for the fabrication of different types of nanocarrier. Moreover, researchers revealed broad-spectrum applications of biopolymer fabricated nanocarriers, especially nanohydrogels and metal nanoparticles, in the fields of medical and food sciences. Mucilage is a plant-derived polymer, their availability varies based on seasonal and environmental conditions. Apart from agronomical variations, these variations may affect the quality and production of mucilages; however, the extraction and purification processes are very complicated. Inadequate removal of mucilage, physical damage to the seed, and morphological features of the plant parts containing mucilage may all affect mucilage yield and consistency, posing a serious challenge to associated costs and the potential for mass production. The toxicity of mucilages is determined by their chemical composition and mode of action in food systems. As a result, it is important to determine the toxicity level of mucilages, particularly the presence of heavy metals in different plant sources. The toxic effects of mucilages can be determined using the fixed-dose procedure, which is recommended by guideline number 425 of the Organization for Economic Cooperation and Development (OECD). Mucilage contains around 10% moisture content, therefore the risk of microbial contamination during any stage of their processing may be high. This is due to the presence of biological substances that promote the growth and production of microorganisms in favorable conditions. Furthermore, the storage period is also a key factor in the contamination of mucilage. Studies have reported that variation in storage conditions leads to changes in the quality of mucilage. This often requires proper control of the different handling methods used at various stages of the supply chain. Moreover, modifications can also be made to address the disadvantages of uncontrolled biodegradability, shear instability, thermal decomposition, pH dependence, thickening, and uncontrolled hydration.

Author Contributions: Conceptualization, P.C., A.N. and M.M.T.; Resources, A.B., R.K. and S.B.D.; Writing—original draft preparation, M.M.T., A.B., R.K. and S.B.D.; writing—review and editing, A.N.; visualization, A.B. and M.W.-J.; supervision, A.N. and P.C. All authors have read and agreed to the published version of the manuscript.

Funding: This research received no external funding.

Institutional Review Board Statement: Not applicable.

Informed Consent Statement: Not applicable.

Data Availability Statement: Data sharing is not applicable to this article.

Conflicts of Interest: The authors declare no conflict of interest.

References

1. Ma, F.; Wang, R.; Li, X.; Kang, W.; Bell, A.E.; Zhao, D.; Liu, X.; Chen, W. Physical properties of mucilage polysaccharides from dioscorea *Opposita* Thunb. *Food Chem.* **2020**, *311*. [CrossRef]
2. Singh, R.; Barreca, D.N. Analysis of gums and mucilages. In *Recent Advances in Natural Products Analysis*; Silva, A.S., Nabavi, S.F., Saeedi, M., Nabavi, S.M., Eds.; Elsevier: Amsterdam, The Netherlands, 2020; pp. 663–676. [CrossRef]
3. Freitas, T.K.F.S.; Oliveira, V.M.; de Souza, M.T.F.; Geraldino, H.C.L.; Almeida, V.C.; Fávaro, S.L.; Garcia, J.C. Optimization of coagulation-flocculation process for treatment of industrial textile wastewater using okra (*A. esculentus*) mucilage as natural coagulant. *Ind. Crop. Prod.* **2015**, *76*, 538–544. [CrossRef]
4. Gasperini, L.; Mano, J.F.; Reis, R.L. Natural polymers for the microencapsulation of cells. *J. R. Soc. Interface* **2014**, *11*. [CrossRef]
5. Chawla, P.; Kumar, N.; Bains, A.; Dhull, S.B.; Kumar, M.; Kaushik, R.; Punia, S. Gum arabic capped copper nanoparticles: Synthesis, characterization, and applications. *Int. J. Biol. Macromol.* **2020**, *146*, 232–242. [CrossRef] [PubMed]
6. Archana, G.; Sabina, K.; Babuskin, S.; Radhakrishnan, K.; Fayidh, M.A.; Azhagu Saravana Babu, P.; Sivarajan, M.; Sukumar, M. Preparation and characterization of mucilage polysaccharide for biomedical applications. *Carbohydr. Polym.* **2013**, *98*, 89–94. [CrossRef]
7. Ameri, A.; Heydarirad, G.; Jafari, J.M.; Ghobadi, A.; Rezaeizadeh, H.; Choopani, R. Medicinal plants contain mucilage used in traditional Persian medicine (TPM). *Pharm. Biol.* **2015**, *53*, 615–623. [CrossRef]
8. Fernandes, S.S.; de las Mercedes Salas-Mellado, M. Addition of chia seed mucilage for reduction of fat content in bread and cakes. *Food Chem.* **2017**, *227*, 237–244. [CrossRef]
9. Beikzadeh, S.; Khezerlou, A.; Jafari, S.M.; Pilevar, Z.; Mortazavian, A.M. Seed Mucilages as the functional ingredients for biodegradable films and edible coatings in the food industry. *Adv. Colloid Interface Sci.* **2020**, *280*, 102164. [CrossRef]
10. Gebresamuel, N.; Gebre-Mariam, T. Comparative physico-chemical characterization of the mucilages of two cactus pears (*Opuntia* Spp.) obtained from Mekelle, Northern Ethiopia. *J. Biomater. Nanobiotechnol.* **2012**, *3*, 79–86. [CrossRef]
11. Prajapati, V.D.; Jani, G.K.; Moradiya, N.G.; Randeria, N.P. Pharmaceutical applications of various natural gums, mucilages and their modified forms. *Carbohydr. Polym.* **2013**, *92*, 1685–1699. [CrossRef] [PubMed]
12. Petera, B.; Delattre, C.; Pierre, G.; Wadouachi, A.; Elboutachfaiti, R.; Engel, E.; Poughon, L.; Michaud, P.; Fenoradosoa, T.A. Characterization of arabinogalactan-rich mucilage from cereus triangularis cladodes. *Carbohydr. Polym.* **2015**, *127*, 372–380. [CrossRef]
13. Alpizar-Reyes, E.; Carrillo-Navas, H.; Gallardo-Rivera, R.; Varela-Guerrero, V.; Alvarez-Ramirez, J.; Pérez-Alonso, C. Functional properties and physicochemical characteristics of tamarind (*Tamarindus indica* L.) seed mucilage powder as a novel hydrocolloid. *J. Food Eng.* **2017**, *209*, 68–75. [CrossRef]
14. Hosseini, S.M.; Hemmati, K.; Ghaemy, M. Synthesis of nanohydrogels based on tragacanth gum biopolymer and investigation of swelling and drug delivery. *Int. J. Biol. Macromol.* **2016**, *82*, 806–815. [CrossRef]
15. Sharma, G.; Kumar, A.; Devi, K.; Sharma, S.; Naushad, M.; Ghfar, A.A.; Ahamad, T.; Stadler, F.J. Guar Gum-Crosslinked-Soya lecithin nanohydrogel sheets as effective adsorbent for the removal of thiophanate methyl fungicide. *Int. J. Biol. Macromol.* **2018**, *114*, 295–305. [CrossRef] [PubMed]
16. Oh, J.K.; Lee, D.I.; Park, J.M. Biopolymer-based microgels/nanogels for drug delivery applications. *Prog. Polym. Sci.* **2009**, *34*, 1261–1282. [CrossRef]
17. Suner, S.S.; Sahiner, M.; Sengel, S.B.; Rees, D.J.; Reed, W.F.; Sahiner, N. Responsive biopolymer-based microgels/nanogels for drug delivery applications. *Stimuli Responsive Polym. Nanocarriers Drug Deliv. Appl.* **2018**, *1*, 453–500. [CrossRef]

18. Mukherjee, T.; Lerma-Reyes, R.; Thompson, K.A.; Schrick, K. Making glue from seeds and gums: Working with plant-based polymers to introduce students to plant biochemistry. *Biochem. Mol. Biol. Educ.* **2019**, *47*, 468–475. [CrossRef]
19. Viudes, S.; Burlat, V.; Dunand, C. Seed mucilage evolution: Diverse molecular mechanisms generate versatile ecological functions for particular environments. *Plant Cell Environ.* **2020**, *43*, 2857–2870. [CrossRef]
20. Ferreira, B.; Montesinos, D.; Sales, F. Mucilage in Portuguese Lamiaceae. *Bot. Lett.* **2020**, *167*, 430–438. [CrossRef]
21. Cuomo, F.; Iacovino, S.; Cinelli, G.; Messia, M.C.; Marconi, E.; Lopez, F. Effect of additives on chia mucilage suspensions: A rheological approach. *Food Hydrocoll.* **2020**, *109*, 106118. [CrossRef]
22. Zhang, K.; Zhang, Y.; Ji, Y.; Walck, J.L.; Tao, J. Seed Biology of *Lepidium apetalum* (Brassicaceae), with particular reference to dormancy and mucilage development. *Plants* **2020**, *9*, 333. [CrossRef] [PubMed]
23. Yang, X.; Baskin, J.M.; Baskin, C.C.; Huang, Z. More than just a coating: Ecological importance, taxonomic occurrence and phylogenetic relationships of seed coat mucilage. *Perspect. Plant Ecol. Evol. Syst.* **2012**, *14*, 434–442. [CrossRef]
24. Cai, M.; Wang, N.; Xing, C.; Wang, F.; Wu, K.; Du, X. Immobilization of aluminum with mucilage secreted by root cap and root border cells is related to aluminum resistance in *Glycine max* L. *Environ. Sci. Pollut. Res.* **2013**, *20*, 8924–8933. [CrossRef] [PubMed]
25. Hawes, M.C.; Curlango-Rivera, G.; Xiong, Z.; Kessler, J.O. Roles of root border cells in plant defense and regulation of rhizosphere microbial populations by extracellular DNA 'trappin'. *Plant Soil* **2012**, *355*, 1–16. [CrossRef]
26. Rashid, F.; Ahmed, Z.; Hussain, S.; Huang, J.Y.; Ahmad, A. *Linum usitatissimum* L. seeds: Flax gum extraction, physicochemical and functional characterization. *Carbohydr. Polym.* **2019**, *215*, 29–38. [CrossRef]
27. Coorey, R.; Tjoe, A.; Jayasena, V. Gelling properties of chia seed and flour. *J. Food Sci.* **2014**, *79*, E859–E866. [CrossRef]
28. Nayak, A.K.; Pal, D.; Pradhan, J.; Hasnain, M.S. Fenugreek seed mucilage-alginate mucoadhesive beads of metformin HCl: Design, optimization and evaluation. *Int. J. Biol. Macromol.* **2013**, *54*, 144–154. [CrossRef]
29. Nayak, A.K.; Pal, D.; Pany, D.R.; Mohanty, B. Evaluation of *Spinacia oleracea* L. leaves mucilage as an innovative suspending agent. *J. Adv. Pharm. Technol. Res.* **2010**, *1*, 338–341. [CrossRef]
30. Tavares, L.S.; Junqueira, L.A.; de Oliveira Guimarães, Í.C.; de Resende, J.V. Cold extraction method of chia seed mucilage (*Salvia hispanica* L.): Effect on yield and rheological behavior. *J. Food Sci. Technol.* **2018**, *55*, 457–466. [CrossRef]
31. Jindal, M.; Kumar, V.; Rana, V.; Tiwary, A.K. Exploring potential new gum source aegle marmelos for food and pharmaceuticals: Physical, chemical and functional performance. *Ind. Crop. Prod.* **2013**, *45*, 312–318. [CrossRef]
32. Hung, P.Y.; Lai, L.S. Structural characterization and rheological properties of the water extracted mucilage of basella alba and the starch/aqueous mucilage blends. *Food Hydrocoll.* **2019**, *93*, 413–421. [CrossRef]
33. Souza, G.; Siqueira dos Santos, S.; Bergamasco, R.; Antigo, J.; Madrona, G.S. Antioxidant activity, extraction and application of psyllium mucilage in chocolate drink. *Nutr. Food Sci.* **2020**, *50*, 1175–1185. [CrossRef]
34. Andrade, L.A.; de Oliveira Silva, D.A.; Nunes, C.A.; Pereira, J. Experimental techniques for the extraction of taro mucilage with enhanced emulsifier properties using chemical characterization. *Food Chem.* **2020**, *327*, 127095. [CrossRef] [PubMed]
35. Kamel, R.; Afifi, S.M.; Kassem, I.A.A.; Elkasabgy, N.A.; Farag, M.A. Arabinoxylan and rhamnogalacturonan mucilage: Outgoing and potential trends of pharmaceutical, environmental, and medicinal merits. *Int. J. Biol. Macromol.* **2020**, *165*, 2550–2564. [CrossRef]
36. Andrade, L.A.; Nunes, C.A.; Pereira, J. Relationship between the chemical components of taro rhizome mucilage and its emulsifying property. *Food Chem.* **2015**, *178*, 331–338. [CrossRef] [PubMed]
37. Elboutachfaiti, R.; Delattre, C.; Quéro, A.; Roulard, R.; Duchêne, J.; Mesnard, F.; Petit, E. Fractionation and structural characterization of six purified rhamnogalacturonans type i from flaxseed mucilage. *Food Hydrocoll.* **2017**, *62*, 273–279. [CrossRef]
38. Luo, M.; Cao, Y.; Wang, W.; Chen, X.; Cai, J.; Wang, L.; Xiao, J. Sustained-release antimicrobial gelatin film: Effect of chia mucilage on physicochemical and antimicrobial properties. *Food Hydrocoll.* **2019**, *87*, 783–791. [CrossRef]
39. Behbahani, B.A.; Noshad, M.; Jooyandeh, H. Improving oxidative and microbial stability of beef using Shahri Balangu seed mucilage loaded with Cumin essential oil as a bioactive edible coating. *Biocatal. Agric. Biotechnol.* **2020**, *24*, 101563. [CrossRef]
40. Pereira, G.A.; Silva, E.K.; Peixoto Araujo, N.M.; Arruda, H.S.; Meireles, M.A.A.; Pastore, G.M. Mutamba seed mucilage as a novel emulsifier: Stabilization mechanisms, kinetic stability and volatile compounds retention. *Food Hydrocoll.* **2019**, *97*, 105190. [CrossRef]
41. Soukoulis, C.; Gaiani, C.; Hoffmann, L. Plant seed mucilage as emerging biopolymer in food industry applications. *Curr. Opin. Food Sci.* **2018**, *22*, 28–42. [CrossRef]
42. Vardhanabhuti, B.; Ikeda, S. Isolation and characterization of hydrocolloids from monoi (*Cissampelos pareira*) leaves. *Food Hydrocoll.* **2006**, *20*, 885–891. [CrossRef]
43. Naji-Tabasi, S.; Razavi, S.M.A. Functional properties and applications of basil seed gum: An overview. *Food Hydrocoll.* **2017**, *73*, 313–325. [CrossRef]
44. Hesarinejad, M.A.; Sami Jokandan, M.; Mohammadifar, M.A.; Koocheki, A.; Razavi, S.M.A.; Ale, M.T.; Attar, F.R. The effects of concentration and heating-cooling rate on rheological properties of plantago lanceolata seed mucilage. *Int. J. Biol. Macromol.* **2018**, *115*, 1260–1266. [CrossRef]
45. Hosseini-Parvar, S.H.; Matia-Merino, L.; Goh, K.K.T.; Razavi, S.M.A.; Mortazavi, S.A. Steady shear flow behavior of gum extracted from *Ocimum basilicum* L. seed: Effect of concentration and temperature. *J. Food Eng.* **2010**, *101*, 236–243. [CrossRef]
46. Timilsena, Y.P.; Adhikari, R.; Barrow, C.J.; Adhikari, B. Physicochemical and functional properties of protein isolate produced from Australian chia seeds. *Food Chem.* **2016**, *212*, 648–656. [CrossRef] [PubMed]

47. Mirhosseini, H.; Amid, B.T. A Review study on chemical composition and molecular structure of newly plant gum exudates and seed gums. *Food Res. Int.* **2012**, *46*, 387–398. [CrossRef]
48. Deore, U.V.; Mahajan, H.S. Isolation and characterization of natural polysaccharide from cassia obtustifolia seed mucilage as film forming material for drug delivery. *Int. J. Biol. Macromol.* **2018**, *115*, 1071–1078. [CrossRef] [PubMed]
49. Dehghani, S.; Noshad, M.; Rastegarzadeh, S.; Hojjati, M.; Fazlara, A. Electrospun chia seed mucilage/pva encapsulated with green cardamonmum essential oils: Antioxidant and antibacterial property. *Int. J. Biol. Macromol.* **2020**, *161*, 1–9. [CrossRef]
50. Devi, R.; Bhatia, M. Thiol Functionalization of Flaxseed Mucilage: Preparation, characterization and evaluation as mucoadhesive polymer. *Int. J. Biol. Macromol.* **2019**, *126*, 101–106. [CrossRef] [PubMed]
51. Thessrimuang, N.; Prachayawarakorn, J. Development, modification and characterization of new biodegradable film from basil seed (*Ocimum basilicum* L.) mucilage. *J. Sci. Food Agric.* **2019**, *99*, 5508–5515. [CrossRef]
52. Fahami, A.; Fathi, M. Development of cress seed mucilage/PVA nanofibers as a novel carrier for vitamin a delivery. *Food Hydrocoll.* **2018**, *81*, 31–38. [CrossRef]
53. Kurd, F.; Fathi, M.; Shekarchizadeh, H. Basil seed mucilage as a new source for electrospinning: Production and physicochemical characterization. *Int. J. Biol. Macromol.* **2017**, *95*, 689–695. [CrossRef]
54. Naji-Tabasi, S.; Razavi, S.M.A.; Mohebbi, M.; Malaekeh-Nikouei, B. New studies on basil (*Ocimum bacilicum* L.) seed gum: Part i—Fractionation, physicochemical and surface activity characterization. *Food Hydrocoll.* **2016**, *52*, 350–358. [CrossRef]
55. Pratik, P.; Shadique, A. Antimicrobial examination of mucilage obtained from fruits of tilkor [*Mamoradica monadelpha*]: A potential medicinal plant. *J. Biol. Chem. Chron.* **2020**, *6*, 5–9. [CrossRef]
56. Punia, S.; Dhull, S.B. Chia seed (*Salvia hispanica* L.) mucilage (a heteropolysaccharide): Functional, thermal, rheological behaviour and its utilization. *Int. J. Biol. Macromol.* **2019**, *140*, 1084–1090. [CrossRef] [PubMed]
57. Capitani, M.I.; Ixtaina, V.Y.; Nolasco, S.M.; Tomás, M.C. Microstructure, chemical composition and mucilage exudation of chia (*Salvia hispanica* L.) nutlets from Argentina. *J. Sci. Food Agric.* **2013**, *93*, 3856–3862. [CrossRef]
58. Abbastabar, B.; Azizi, M.H.; Adnani, A.; Abbasi, S. Determining and modeling rheological characteristics of quince seed gum. *Food Hydrocoll.* **2015**, *43*, 259–264. [CrossRef]
59. Keshani-Dokht, S.; Emam-Djomeh, Z.; Yarmand, M.S.; Fathi, M. Extraction, chemical composition, rheological behavior, antioxidant activity and functional properties of *Cordia myxa* mucilage. *Int. J. Biol. Macromol.* **2018**, *118*, 485–493. [CrossRef] [PubMed]
60. Hadad, S.; Goli, S.A.H. Fabrication and characterization of electrospun nanofibers using flaxseed (*Linum usitatissimum*) Mucilage. *Int. J. Biol. Macromol.* **2018**, *114*, 408–414. [CrossRef] [PubMed]
61. Singh, S.; Bothara, S.B. *Manilkara zapota* (Linn.) seeds: A potential source of natural gum. *ISRN Pharm.* **2014**, *2014*, 647174. [CrossRef] [PubMed]
62. de Alvarenga Pinto Cotrim, M.; Mottin, A.C.; Ayres, E. Preparation and characterization of Okra Mucilage (*Abelmoschus esculentus*) edible films. *Macromol. Symp.* **2016**, *367*, 90–100. [CrossRef]
63. Pawar, H.A.; Lalitha, K.G. Isolation, purification and characterization of galactomannans as an excipient from *Senna tora* seeds. *Int. J. Biol. Macromol.* **2014**, *65*, 167–175. [CrossRef]
64. Viebke, C.; Al-Assaf, S.; Phillips, G.O. Food hydrocolloids and health claims. *Bioact. Carbohydr. Diet. Fibre* **2014**, *4*, 101–114. [CrossRef]
65. Kang, S.; Wang, H.; Xia, L.; Chen, M.; Li, L.; Cheng, J.; Li, X.; Jiang, S. Colorimetric film based on polyvinyl alcohol/okra mucilage polysaccharide incorporated with rose anthocyanins for shrimp freshness monitoring. *Carbohydr. Polym.* **2020**, *229*, 115402. [CrossRef]
66. Gemede, H.F.; Haki, G.D.; Beyene, F.; Rakshit, S.K.; Woldegiorgis, A.Z. Indigenous Ethiopian Okra (Abelmoschus Esculentus) Mucilage: A Novel Ingredient with Functional and Antioxidant Properties. *Food Sci. Nutr.* **2018**, *6*, 563–571. [CrossRef] [PubMed]
67. Ubeyitogullari, A.; Ciftci, O.N. Fabrication of bioaerogels from camelina seed mucilage for food applications. *Food Hydrocoll.* **2020**, *102*, 105597. [CrossRef]
68. Kaur, M.; Kaur, R.; Punia, S. Characterization of mucilages extracted from different flaxseed (*Linum usitatissiumum* L.) cultivars: A heteropolysaccharide with desirable functional and rheological properties. *Int. J. Biol. Macromol.* **2018**, *117*, 919–927. [CrossRef]
69. Lan, Y.; Ohm, J.B.; Chen, B.; Rao, J. Physicochemical properties and aroma profiles of flaxseed proteins extracted from whole flaxseed and flaxseed meal. *Food Hydrocoll.* **2020**, *104*, 105731. [CrossRef]
70. Hassan, L.K.; Haggag, H.F.; ElKalyoubi, M.H.; Abd EL-Aziz, M.; El-Sayed, M.M.; Sayed, A.F. Physico-chemical properties of yoghurt containing cress seed mucilage or guar gum. *Ann. Agric. Sci.* **2015**, *60*, 21–28. [CrossRef]
71. Dhull, S.B.; Sandhu, K.S.; Punia, S.; Kaur, M.; Chawla, P.; Malik, A. Functional, thermal and rheological behavior of fenugreek (*Trigonella foenum–graecum* L.) gums from different cultivars: A comparative study. *Int. J. Biol. Macromol.* **2020**, *159*, 406–414. [CrossRef] [PubMed]
72. Galla, N.R.; Dubasi, G.R. Chemical and functional characterization of Gum Karaya (*Sterculia urens* L.) seed meal. *Food Hydrocoll.* **2010**, *24*, 479–485. [CrossRef]
73. Felisberto, M.H.F.; Wahanik, A.L.; Gomes-Ruffi, C.R.; Clerici, M.T.P.S.; Chang, Y.K.; Steel, C.J. Use of chia (*Salvia hispanica* L.) mucilage gel to reduce fat in pound cakes. *LWT Food Sci. Technol.* **2015**, *63*, 1049–1055. [CrossRef]
74. Monrroy, M.; García, E.; Ríos, K.; García, J.R. Extraction and physicochemical characterization of mucilage from *Opuntia cochenillifera* (L.) Miller. *J. Chem.* **2017**, *2017*. [CrossRef]

75. Kalegowda, P.; Chauhan, A.S.; Nanjaraj Urs, S.M. *Opuntia dillenii* (Ker-Gawl) haw cladode mucilage: Physico-chemical, rheological and functional behavior. *Carbohydr. Polym.* **2017**, *157*, 1057–1064. [CrossRef] [PubMed]
76. Hayati, I.N.; Ching, C.W.; Rozaini, M.Z.H. Flow properties of o/w emulsions as affected by xanthan gum, guar gum and carboxymethyl cellulose interactions studied by a mixture regression modelling. *Food Hydrocoll.* **2016**, *53*, 199–208. [CrossRef]
77. Quinzio, C.; Ayunta, C.; López de Mishima, B.; Iturriaga, L. Stability and rheology properties of oil-in-water emulsions prepared with mucilage extracted from opuntia *Ficus-indica* (L). Miller. *Food Hydrocoll.* **2018**, *84*, 154–165. [CrossRef]
78. Sangeethapriya, M.; Siddhuraju, P. Health related functional characteristics and antioxidant potential of mucilage (dietary fiber) from *Zizyphus mauritiana* fruits. *Food Sci. Hum. Wellness* **2014**, *3*, 79–88. [CrossRef]
79. Behrouzian, F.; Razavi, S.M.A.; Phillips, G.O. Cress seed (*Lepidium sativum*) mucilage, an overview. *Bioact. Carbohydr. Diet. Fibre* **2014**, *3*, 17–28. [CrossRef]
80. Jouki, M.; Mortazavi, S.A.; Yazdi, F.T.; Koocheki, A. Characterization of antioxidant-antibacterial quince seed mucilage films containing thyme essential oil. *Carbohydr. Polym.* **2014**, *99*, 537–546. [CrossRef]
81. Mahfoudhi, N.; Sessa, M.; Chouaibi, M.; Ferrari, G.; Donsì, F.; Hamdi, S. assessment of emulsifying ability of almond gum in comparison with gum arabic using response surface methodology. *Food Hydrocoll.* **2014**, *37*, 49–59. [CrossRef]
82. Alizadeh Behbahani, B.; Tabatabaei Yazdi, F.; Shahidi, F.; Hesarinejad, M.A.; Mortazavi, S.A.; Mohebbi, M. *Plantago major* seed mucilage: Optimization of extraction and some physicochemical and rheological aspects. *Carbohydr. Polym.* **2017**, *155*, 68–77. [CrossRef]
83. Safaei-Ghomi, J.; Ebrahimabadi, A.H.; Djafari-Bidgoli, Z.; Batooli, H. GC/MS analysis and in vitro antioxidant activity of essential oil and methanol extracts of *Thymus caramanicus* Jalas and its main constituent carvacrol. *Food Chem.* **2009**, *115*, 1524–1528. [CrossRef]
84. Roche, A.; Ross, E.; Walsh, N.; O'Donnell, K.; Williams, A.; Klapp, M.; Fullard, N.; Edelstein, S. Representative literature on the phytonutrients category: Phenolic acids. *Crit. Rev. Food Sci. Nutr.* **2017**, *57*, 1089–1096. [CrossRef]
85. Wang, J.; Hu, S.; Nie, S.; Yu, Q.; Xie, M. Reviews on mechanisms of in vitro antioxidant activity of polysaccharides. *Oxidative Med. Cell. Longev.* **2016**, *64*, 1–13. [CrossRef]
86. Zhao, Q.; Dong, B.; Chen, J.; Zhao, B.; Wang, X.; Wang, L.; Zha, S.; Wang, Y.; Zhang, J.; Wang, Y. Effect of drying methods on physicochemical properties and antioxidant activities of wolfberry (*Lycium barbarum*) polysaccharide. *Carbohydr. Polym.* **2015**, *127*, 176–181. [CrossRef]
87. Liguori, G.; Gentile, C.; Gaglio, R.; Perrone, A.; Guarcello, R.; Francesca, N.; Fretto, S.; Inglese, P.; Settanni, L. Effect of addition of Opuntia ficus-indica mucilage on the biological leavening, physical, nutritional, antioxidant and sensory aspects of bread. *J. Biosci. Bioeng.* **2020**, *129*, 184–191. [CrossRef] [PubMed]
88. Silveira Coelho, M.; de las Mercedes Salas-Mellado, M. Chemical characterization of chia (*Salvia Hispanica* L.) for use in food products. *J. Food Nutr. Res.* **2014**, *2*, 263–269. [CrossRef]
89. Tantiwatcharothai, S.; Prachayawarakorn, J. Property improvement of antibacterial wound dressing from basil seed (*O. basilicum* L.) mucilage-ZnO nanocomposite by borax crosslinking. *Carbohydr. Polym.* **2020**, *227*, 115360–115367. [CrossRef] [PubMed]
90. Mujtaba, M.; Akyuz, L.; Koc, B.; Kaya, M.; Ilk, S.; Cansaran-Duman, D.; Martinez, A.S.; Cakmak, Y.S.; Labidi, J.; Boufi, S. Novel, multifunctional mucilage composite films incorporated with cellulose nanofibers. *Food Hydrocoll.* **2019**, *89*, 20–28. [CrossRef]
91. Mohammadi, H.; Kamkar, A.; Misaghi, A. Nanocomposite films based on cmc, okra mucilage and ZnO nanoparticles: Physico mechanical and antibacterial properties. *Carbohydr. Polym.* **2018**, *181*, 351–357. [CrossRef] [PubMed]
92. Heydari, S.; Jooyandeh, H.; Alizadeh Behbahani, B.; Noshad, M. The Impact of Qodume Shirazi seed mucilage-based edible coating containing lavender essential oil on the quality enhancement and shelf life improvement of fresh ostrich meat: An experimental and modeling study. *Food Sci. Nutr.* **2020**, *8*, 6497–6512. [CrossRef] [PubMed]
93. Barzegar, H.; Alizadeh Behbahani, B.; Mehrnia, M.A. Quality retention and shelf life extension of fresh beef using *Lepidium sativum* seed mucilage-based edible coating containing *Heracleum lasiopetalum* essential oil: An experimental and modeling study. *Food Sci. Biotechnol.* **2020**, *29*, 717–728. [CrossRef] [PubMed]
94. Gheribi, R.; Puchot, L.; Verge, P.; Jaoued-Grayaa, N.; Mezni, M.; Habibi, Y.; Khwaldia, K. Development of plasticized edible films from *Opuntia ficus-indica* mucilage: A comparative study of various polyol plasticizers. *Carbohydr. Polym.* **2018**, *190*, 204–211. [CrossRef]
95. Mir, S.A.; Dar, B.N.; Wani, A.A.; Shah, M.A. Effect of plant extracts on the techno-functional properties of biodegradable packaging films. *Trends Food Sci. Technol.* **2018**, *80*, 141–154. [CrossRef]
96. Espino-Díaz, M.; De Jesús Ornelas-Paz, J.; Martínez-Téllez, M.A.; Santillán, C.; Barbosa-Cánovas, G.V.; Zamudio-Flores, P.B.; Olivas, G.I. Development and characterization of edible films based on mucilage of *Opuntia ficus-indica* (L.). *J. Food Sci.* **2010**, *75*, E347–E352. [CrossRef] [PubMed]
97. Kozlu, A.; Elmacı, Y. Quince seed mucilage as edible coating for mandarin fruit; determination of the quality characteristics during storage. *J. Food Process. Preserv.* **2020**, *44*, e14854. [CrossRef]
98. Chrysargyris, A.; Nikou, A.; Tzortzakis, N. Effectiveness of *Aloe vera* gel coating for maintaining tomato fruit quality. *N. Z. J. Crop Hortic. Sci.* **2016**, *44*, 203–217. [CrossRef]
99. Nourozi, F.; Sayyari, M. Enrichment of *Aloe vera* gel with basil seed mucilage preserve bioactive compounds and postharvest quality of apricot fruits. *Sci. Hortic.* **2020**, *262*, 109041. [CrossRef]

100. Allegra, A.; Inglese, P.; Sortino, G.; Settanni, L.; Todaro, A.; Liguori, G. The influence of *Opuntia ficus-indica* mucilage edible coating on the quality of 'Hayward' kiwifruit slices. *Postharvest Biol. Technol.* **2016**, *120*, 45–51. [CrossRef]
101. Vignesh, R.M.; Nair, B.R. Improvement of shelf life quality of tomatoes using a novel edible coating formulation. *Plant Sci. Today* **2019**, *2*, 84–90. [CrossRef]
102. Soleimani-Rambod, A.; Zomorodi, S.; Naghizadeh Raeisi, S.; Khosrowshahi Asl, A.; Shahidi, S.-A. The Effect of Xanthan Gum and Flaxseed Mucilage as Edible Coatings in Cheddar Cheese during Ripening. *Coatings* **2018**, *8*, 80. [CrossRef]
103. Song, H.Y.; Jo, W.S.; Song, N.B.; Min, S.C.; Song, K.B. Quality change of apple slices coated with *Aloe vera* gel during storage. *J. Food Sci.* **2013**, *78*, 817–822. [CrossRef]
104. Tamri, P.; Hemmati, A.; Boroujerdnia, M.G. wound healing properties of quince seed mucilage: In vivo evaluation in rabbit full-thickness wound model. *Int. J. Surg.* **2014**, *12*, 843–847. [CrossRef] [PubMed]
105. George, B.; Suchithra, T.V. Plant-derived bioadhesives for wound dressing and drug delivery system. *Fitoterapia* **2019**, *137*, 104241. [CrossRef]
106. Ahmad, S.; Ahmad, M.; Manzoor, K.; Purwar, R.; Ikram, S. A review on latest innovations in natural gums based hydrogels: Preparations & applications. *Int. J. Biol. Macromol.* **2019**, *136*, 870–890. [CrossRef] [PubMed]
107. Bouyer, E.; Mekhloufi, G.; Rosilio, V.; Grossiord, J.L.; Agnely, F. Proteins, polysaccharides, and their complexes used as stabilizers for emulsions: Alternatives to synthetic surfactants in the pharmaceutical field? *Int. J. Pharm.* **2012**, *436*, 359–378. [CrossRef] [PubMed]
108. Campos, B.E.; Dias Ruivo, T.; da Silva Scapim, M.R.; Madrona, G.S.; Bergamasco, R.D. Optimization of the mucilage extraction process from chia seeds and application in ice cream as a stabilizer and emulsifier. *LWT Food Sci. Technol.* **2016**, *65*, 874–883. [CrossRef]
109. Ahuja, M.; Kumar, A.; Yadav, P.; Singh, K. Mimosa pudica seed mucilage: Isolation; characterization and evaluation as tablet disintegrant and binder. *Int. J. Biol. Macromol.* **2013**, *57*, 105–110. [CrossRef]
110. Kaleemullah, M.; Jiyauddin, K.; Thiban, E.; Rasha, S.; Al-Dhalli, S.; Budiasih, S.; Gamal, O.E.; Fadli, A.; Eddy, Y. Development and evaluation of Ketoprofen sustained release matrix tablet using *Hibiscus rosa-sinensis* leaves mucilage. *Saudi Pharm. J.* **2017**, *25*, 770–779. [CrossRef]
111. Chaudhary, A.; Kulkarnı, G.T.; Awasthı, R.; Kumar, P. Investigation on binding properties of *Grewi asiatica* mucilage in tablet formulations. *Marmara Pharm. J.* **2016**, *20*, 353–367. [CrossRef]
112. Raj, V.; Shim, J.J.; Lee, J. Grafting modification of okra mucilage: Recent findings, applications, and future directions. *Carbohydr. Polym.* **2020**, 116653. [CrossRef]
113. Braghiroli, F.L.; Bouafif, H.; Neculita, C.M.; Koubaa, A. Activated biochar as an effective sorbent for organic and inorganic contaminants in water. *Water Air Soil Pollut.* **2018**, 229–230. [CrossRef]
114. Hanjra, M.A.; Blackwell, J.; Carr, G.; Zhang, F.; Jackson, T.M. wastewater irrigation and environmental health: Implications for water governance and public policy. *Int. J. Hyg. Environ. Health* **2012**, *215*, 255–269. [CrossRef]
115. Rahman, Z.; Singh, V.P. The relative impact of toxic heavy metals (THMs) (arsenic (As), cadmium (Cd), chromium (Cr)(VI), mercury (Hg), and lead (Pb)) on the total environment: An overview. *Environ. Monit. Assess.* **2019**, *191*, 419. [CrossRef] [PubMed]
116. Jones, B.O.; John, O.O.; Luke, C.; Ochieng, A.; Bassey, B.J. Application of mucilage from dicerocaryum eriocarpum plant as biosorption medium in the removal of selected heavy metal ions. *J. Environ. Manag.* **2016**, *177*, 365–372. [CrossRef]
117. Fox, D.I.; Pichler, T.; Yeh, D.H.; Alcantar, N.A. Removing heavy metals in water: The interaction of cactus mucilage and arsenate (As (V)). *Environ. Sci. Technol.* **2012**, *46*, 4553–4559. [CrossRef]
118. Anastasakis, K.; Kalderis, D.; Diamadopoulos, E. Flocculation behavior of mallow and okra mucilage in treating wastewater. *Desalination* **2009**, *249*, 786–791. [CrossRef]
119. Bouatay, F.; Eljebsi, N.; Dridi-Dhaouadi, S.; Mhenni, F. Valorization of the vicia faba mucilage on textile wastewater treatment as a bio-flocculant: Process development and optimization using response surface methodology (RSM). *Water Sci. Technol.* **2017**, *75*, 629–642. [CrossRef]
120. Sindhu, G.; Ratheesh, M.; Shyni, G.L.; Nambisan, B.; Helen, A. Anti-inflammatory and antioxidative effects of mucilage of *Trigonella Foenum Graecum* (fenugreek) on adjuvant induced arthritic rats. *Int. Immunopharmacol.* **2012**, *12*, 205–211. [CrossRef]
121. Ahmed, E.M. Hydrogel: Preparation, characterization, and applications: A review. *J. Adv. Res.* **2015**, *6*, 105–121. [CrossRef]
122. Sharma, G.; Naushad, M.; Kumar, A.; Rana, S.; Sharma, S.; Bhatnagar, A.; Stadler, F.J.; Ghfar, A.A.; Khan, M.R. Efficient removal of coomassie brilliant blue r-250 dye using starch/poly(alginic acid-cl-acrylamide) nanohydrogel. *Process Saf. Environ. Prot.* **2017**, *109*, 301–310. [CrossRef]
123. Lodhi, B.A.; Hussain, M.A.; Sher, M.; Haseeb, M.T.; Ashraf, M.U.; Hussain, S.Z.; Hussain, I.; Bukhari, S.N.A. Polysaccharide-based superporous, superabsorbent, and stimuli responsive hydrogel from sweet basil: A novel material for sustained drug release. *Adv. Polym. Technol.* **2019**, *2119*, 9583516. [CrossRef]
124. Setia, A.; Ahuja, P. Nanohydrogels. In *Organic Materials as Smart Nanocarriers for Drug Delivery*; Elsevier: Amsterdam, The Netherlands, 2018; pp. 293–368. [CrossRef]
125. Okutan, N.; Terzi, P.; Altay, F. Affecting parameters on electrospinning process and characterization of electrospun gelatin nanofibers. *Food Hydrocoll.* **2014**, *39*, 19–26. [CrossRef]
126. Fathi, M.; Martín, Á.; McClements, D.J. Nanoencapsulation of food ingredients using carbohydrate based delivery systems. *Trends Food Sci. Technol.* **2014**, *39*, 18–39. [CrossRef]

127. Prusty, K.; Swain, S.K. Release of ciprofloxacin drugs by nano gold embedded cellulose grafted polyacrylamide hybrid nanocomposite hydrogels. *Int. J. Biol. Macromol.* **2019**, *126*, 765–775. [CrossRef]
128. Mohammadinejad, R.; Kumar, A.; Ranjbar-Mohammadi, M.; Ashrafizadeh, M.; Han, S.S.; Khang, G.; Roveimiab, Z. Recent advances in natural gum-based biomaterials for tissue engineering and regenerative medicine: A review. *Polymers* **2020**, *12*, 176. [CrossRef]
129. Thakur, V.K.; Thakur, M.K. Recent Trends in Hydrogels Based on Psyllium Polysaccharide: A Review. *J. Clean. Prod.* **2014**, *82*, 1–15. [CrossRef]
130. Rayegan, A.; Allafchian, A.; Abdolhosseini Sarsari, I.; Kameli, P. Synthesis and characterization of basil seed mucilage coated Fe_3O_4 magnetic nanoparticles as a drug carrier for the controlled delivery of cephalexin. *Int. J. Biol. Macromol.* **2018**, *113*, 317–328. [CrossRef] [PubMed]
131. Seyyed, M.; Tabrizi, H.M.; Behrouz, E.; Vahid, J. Biosynthesis of pure zinc oxide nanoparticles using Quince seed mucilage for photocatalytic dye degradation. *J. Alloy. Compd.* **2020**, *821*, 153519. [CrossRef]
132. Prasad, A.R.; Garvasis, J.; Oruvil, S.K.; Joseph, A. Improving oxidative and microbial stability of beef using Shahri Balangu seed mucilage loaded with Cumin essential oil as a bioactive edible coating. *J. Phys. Chem. Solids* **2019**, *127*, 265–274. [CrossRef]
133. Mahmoodi, M.; Javanbakht, V. Fabrication of Zn-based magnetic zeolitic imidazolate framework bionanocomposite using basil seed mucilage for removal of azo cationic and anionic dyes from aqueous solution. *Int. J. Biol. Macromol.* **2021**, *167*, 1076–1090. [CrossRef] [PubMed]
134. Pathak, D.; Kumar, P.; Kuppusamy, G.; Gupta, A.; Kamble, B.; Wadhwani, A. Physicochemical characterization and toxicological evaluation of plant-based anionic polymers and their nanoparticulated system for ocular delivery. *Nanotoxicology* **2014**, *8*, 843–855. [CrossRef] [PubMed]
135. Akbari, I.; Ghoreishi, S.M.; Habibi, N. Generation and precipitation of paclitaxel nanoparticles in basil seed mucilage via combination of supercritical gas antisolvent and phase inversion techniques. *J. Supercrit. Fluids* **2014**, *94*, 182–188. [CrossRef]
136. Haseeb, M.T.; Khaliq, N.U.; Yuk, S.H.; Hussain, M.A.; Bashir, S. Linseed polysaccharides based nanoparticles for controlled delivery of docetaxel: Design, in vitro drug release and cellular uptake. *J. Drug Deliv. Sci. Technol.* **2019**, *49*, 143–151. [CrossRef]
137. Chawla, P.; Kumar, N.; Kaushik, R.; Dhull, S.B. Synthesis, characterization and cellular mineral absorption of nanoemulsions of Rhododendron arboreum flower extracts stabilized with gum Arabic. *J. Food Sci. Technol.* **2019**, *56*, 5194–5203. [CrossRef]
138. Carmona, J.C.; Robert, P.; Vergara, C.; Sáenz, C. Microparticles of yellow-orange cactus pear pulp (Opuntia ficus-indica) with cladode mucilage and maltodextrin as a food coloring in yogurt. *LWT* **2021**, *138*, 110672. [CrossRef]
139. Miart, F.; Fournet, F.; Dubrulle, N.; Petit, E.; Demailly, H.; Dupont, L.; Zabijak, L.; Marcelo, P.; Boudaoud, A.; Pineau, C.; et al. Cytological approaches combined with chemical analysis reveals the layered nature of flax mucilage. *Front. Plant Sci.* **2019**, *10*, 684. [CrossRef]
140. Waghmare, R.; Moses, J.A.; Anandharamakrishnan, C. Mucilages: Sources, extraction methods, and characteristics for their use as encapsulation agents. *Crit. Rev. Food Sci. Nutr.* **2021**, 1–22. [CrossRef]
141. Soto-Cerda, B.J.; Cloutier, S.; Quian, R.; Gajardo, H.A.; Olivos, M.; You, F.M. Genome-wide association analysis of mucilage and hull content in flax (Linum usitatissimum L.) seeds. *Int. J. Mol. Sci.* **2018**, *19*, 2870. [CrossRef]
142. Orrego, D.; Zapata-Zapata, A.D.; Kim, D. Optimization and scale-up of coffee mucilage fermentation for ethanol production. *Energies* **2018**, *11*, 786. [CrossRef]
143. Añibarro-Ortega, M.; Pinela, J.; Barros, L.; Ćirić, A.; Silva, S.P.; Coelho, E.; Mocan, A.; Calhelha, R.C.; Soković, M.; Coimbra, M.A.; et al. Compositional features and bioactive properties of aloe vera leaf (Fillet, mucilage, and rind) and flower. *Antioxidants* **2019**, *8*, 444. [CrossRef] [PubMed]
144. Allegra, A.; Sortino, G.; Inglese, P.; Settanni, L.; Todaro, A.; Gallotta, A. The effectiveness of Opuntia ficus-indica mucilage edible coating on post-harvest maintenance of 'Dottato'fig (Ficus carica L.) fruit. *Food Packag. Shelf Life* **2017**, *12*, 135–141. [CrossRef]

polymers

MDPI

Article

Green Natural Rubber Composites Reinforced with Black/White Rice Husk Ashes: Effects of Reinforcing Agent on Film's Mechanical and Dielectric Properties

Praewpakun Sintharm and Muenduen Phisalaphong *

Bio-Circular-Green-Economy Technology & Engineering Center, BCGeTEC, Department of Chemical Engineering, Faculty of Engineering, Chulalongkorn University, Bangkok 10330, Thailand; praewpakun.s@gmail.com
* Correspondence: muenduen.p@chula.ac.th; Tel.: +66-2-218-6875

Abstract: Green natural rubber (NR) composites reinforced with black rice husk ash (BRHA)/white rice husk ash (WRHA), using alginate as a thickening and dispersing agent and crosslinking by $CaCl_2$, was developed to improve mechanical, chemical and dielectric properties of NR-based films by using a latex aqueous microdispersion process. A maximum of 100 per hundred rubbers (phr) of rice husk ashes (RHAs) could be integrated in NR matrix without phase separation. Mechanical properties of the composite films were considerably enhanced, compared to the neat NR film. The composite films reinforced with WRHA demonstrated relatively better mechanical properties than those reinforced with BRHA, whereas the composites filled with BRHA demonstrated higher elongation at break. The crosslinking by $CaCl_2$ improved the film tensile strength but lowered the film elasticity. The reinforcement strongly improved chemical resistance of the composite films in toluene. The films are biodegradable in soil, with weight loss of 7.6–18.3% of the initial dry weight after 3 months. Dielectric constant and dielectric loss factors of the composite films were enhanced with RHAs loading. According to the obtained properties, the composites offer potential for further development as stretchable conductive substrate or semiconducting polymer films for electronic applications.

Keywords: natural rubber; rice husk ash; alginate; mechanical properties; dielectric properties

Citation: Sintharm, P.; Phisalaphong, M. Green Natural Rubber Composites Reinforced with Black/White Rice Husk Ashes: Effects of Reinforcing Agent on Film's Mechanical and Dielectric Properties. *Polymers* **2021**, *13*, 882. https://doi.org/10.3390/polym13060882

Academic Editor: Arn Mignon

Received: 14 February 2021
Accepted: 8 March 2021
Published: 13 March 2021

1. Introduction

In recent years, biopolymers have attracted considerable interest for their potential to reduce consumption of fossil resources and nonbiodegradable polymers in order to reduce plastic pollution. Biopolymers are usually biodegradable, meaning that they can be decomposed and degraded into CO_2 and water by microorganisms in the environment. Recent research has focused on improving properties of biopolymer composites, including starch-based materials, cellulose-derived polymers, bacterial polyesters and natural rubber composites. The blending process, which involves adding biofillers (such as agricultural waste, cellulose and ash), can be used to modify the properties of biopolymer composites [1]. Biodegradable polymers can be used in many electronics applications; for instance, they can be used as insulators, (semi)conductors, dielectric material or electronic packaging [2].

Thailand is the world's largest producer and exporter of natural rubber (NR). In 2017, over 4.5 million tons of NR were produced in the country (about 37% of global production) [3]. NR is exported in several basic forms, including ribbed smoked sheets, concentrated latex and compound rubber. However, the NR prices have been unstable for several years, because of oversupply from major rubber producing countries, a weak global economy and cheaper synthetic rubbers for replacement natural rubber [3]. In order to increase the value of NR, technologies for improving NR properties or transforming NR latex into higher value products have to be developed. NR is one of the important elastomers and is widely utilized to prepare many rubber products such as vehicle tires, belts, condoms, gloves, health equipment and accessories, coatings and floor covering [4]. NR is nonsaturated rubber, because its structural unit contains a double bond. Thus, it has very low aging resistance, because ozone and oxygen can react with the double bond very easily. Consequently, NR often has its chemical structure modified by a vulcanization process and reinforced by adding fillers to improve its properties [5]. Carbon black and silica are synthetic fillers that are commonly used as reinforcing fillers in the rubber industry. Both carbon black and silica are added into an NR matrix in order to improve the mechanical and thermal properties of NR composites. However, the major problem of this type of reinforcement is the poor dispersion and agglomeration of fillers in the NR matrix, which limits the loading content of fillers.

Natural fillers are becoming more popular as alternative materials for NR reinforcement, because they are biodegradable, environmentally friendly, cheap to purchase and help to reduce domestic waste [6]. In agricultural countries, rice husk is one of the major agricultural residues generated from the rice milling process and is usually used as a biomass fuel in steam boilers to generate steam in the power plant and rice milling plant. Burning rice husk as fuel results in the production of rice husk ash (RHA) as a waste material [7]. There are two types of RHA, white rice husk ash (WRHA) and black rice husk ash (BRHA). WRHA is obtained from the combustion of rice husk in the atmosphere, whereas BRHA is from the pyrolysis of rice husk in a poor oxygen atmosphere [8]. Usually, rice husk ash waste is released into the environment without any commercial return. Because rice husk ash is usually composed of silica (SiO_2) (over 80–90%) and carbon black, it is attractive for use as reinforcing filler for NR composites. Previous studies have reported on NR composites produced from NR and mixed rice husk ash in solid state using a two-roll mill (the conventional process) [9–11]; however, the limitations of this method include poor NR matrix–filler interaction and low dispersion of fillers in NR composite [11–15]. A new latex compounding method was developed to solve the problem of incompatibility between NR and fillers and improve the dispersion of fillers in the NR matrix [16,17]. Moreover, the latex compounding method consumes less energy during mixing when compared to the conventional process [18].

The aim of this study is to improve the mechanical, chemical and dielectric properties of uncured NR composite films prepared via a latex aqueous microdispersion process through the addition of different types of rice husk ashes (RHAs). In order to obtain good dispersion of polar fillers in a nonpolar NR matrix, sodium alginate, one of the most commonly used natural polysaccharides, was employed as a thickening and dispersing agent in the mixture. Through the use of this technique, high concentrations of fillers can be added into the NR matrix without phase separation. The effects of reinforcing NR composites with RHAs and crosslinking of the dispersing agent (alginate) by $CaCl_2$ on the chemical, mechanical, biodegradation and dielectric properties were investigated for further development of the composite films.

2. Materials and Methods

2.1. Materials

Rice husk and rice husk ash were obtained from a rice milling plant in Phra nakhon si Ayutthaya province, Thailand. NR latex with a 60% dry rubber content was purchased from the Rubber Research Institute of Thailand, Bangkok, Thailand. Hydrochloric acid, sodium alginate and calcium chloride were purchased from Sigma-Aldrich (Thailand) Co Ltd., Bangkok, Thailand.

2.2. Methods

2.2.1. Preparation of Black Rice Husk Ash (BRHA) and White Rice Husk Ash (WRHA).

BRHA used in this study was kindly provided by Nakhon Luang Rice Mill (Phra Nakhon Si Ayutthaya, Thailand), in which the combustion process was under oxygen-depleted atmosphere due to the limited air supply. BRHA from a rice milling plant was dried in an electric oven at 105 °C for 12 h to remove water and stored in a glass bottle container at room temperature (~30 °C). WRHA was prepared in our laboratory (Chulalongkorn University, Bangkok, Thailand). Initially, rice husks were washed with distilled water to remove dirt and impurities and then were dried in an oven at 105 °C for 24 h. The dried rice husks were submitted to the heat treatment in ceramic crucibles, which was carried out in an atmosphere of air in an electric furnace at the temperature of 500 °C for 2 h [19]. The particle sizes of both types of rice husk ashes were reduced using ball milling (PM 100, Haan, Germany) at revolution speed 400 rpm for 15 min and sieved (Test sieve ASTM, 203 × 50 mm, 106 μm, 140 mesh) in order to obtain th2e powder with a size of less than 106 μm.

2.2.2. Preparation of NR Composite Films.

The NR composite films were reinforced by using BRHA and WRHA as filler. The filler loading used in this work was varied at 0, 20, 60 and 100 per hundred rubbers (phr). The microdispersion process [16] was applied for the preparation of composite films. Initially, the filler was added into the aqueous solution of 1% *w*/*v* alginate (which was found to be a suitable solution for good dispersion of BRHA and WRHA). Then, the slurry was thoroughly mixed under mechanical stirring at room temperature for 30 min. Then, 5 g of 60 phr NR latex was slowly added into 30 mL of the slurry under continuous mixing by high-frequency mechanical stirring until the mixture was homogenous. After that, the mixture was poured into a plastic tray and dried overnight (~12 h) in an oven at 40 °C to obtain NR composite films. The composite films of NR combined with BRHA and WRHA are referred to as NR–B and NR–W, respectively.

2.2.3. Crosslinking with $CaCl_2$.

The aqueous solution of $CaCl_2$ at 1% *w*/*v* was used as crosslinking agent. The solution was prepared by dispersing $CaCl_2$ in distilled water and stirring at room temperature for 30 min. The NR composite films were soaked in 1% *w*/*v* $CaCl_2$ solution for crosslinking of alginate for ~1 h, and then the films were rinsed with distilled water to remove excess chloride. After that, the composite films were dried in an oven overnight at 40 °C [20]. After crosslinking with Ca^{2+}, the crosslinked composite films of NR combined with BRHA and WRHA are referred to as NR–BC and NR–WC, respectively.

2.3. *Characterization*

Morphologies of rice husk ashes and composite films were observed by scanning electron microscope and energy dispersive X-ray spectrometer (SEM–EDS) (IT-500HR) using a JEOL, JSM-IT-500HR and JEOL, JED-2300 (JEOL, Tokyo, Japan). The specimens were frozen in liquid nitrogen and vacuum dried. After that, the specimens were sputtered with gold. The SEM–EDS performed at an accelerating voltage of 10 kV.

The overall components and particle size distribution of each dried sample were determined by X-ray fluorescence spectrometer analysis (Bruker model S8 Tiger, Karlsruhe, Germany) and laser particle size distribution analyzer (Mastersizer 3000, Malvern Panalytical, Malvern, UK), respectively. The surface area, pore volume and average pore diameter of samples were determined by nitrogen (N_2) physisorption-desorption using a surface area and pore size analyzer (Autosorb-iQ-MP, Quantachrome, Boynton Beach, FL, USA).

The structural information and crystallinity of fillers and NR composite films were characterized using X-ray diffractometer (XRD, Bruker AXS Model D8 Discover, Karlsruhe, Germany) with Cu-Kα radiation in the 2θ range of 5–40°. The operation conditions were at the accelerating voltage of 40 kV and electric current of 30 mA.

The functional groups and possible interaction between fillers were determined by Fourier transform infrared (FTIR) spectroscopy (PerkinElmer, Waltham, MA, USA) in the ranges of 4000–650 cm^{-1} with a resolution of 4 cm^{-1}.

For mechanical properties tests of dry film of NR, NR composites (Young's modulus, tensile strength and elongation at break) were performed using Universal Testing Machine

(Instron, Norwood, MA, USA). The test conditions were according to ASTM D882. At least five specimens for each different blend composition were tested.

The measurement of water absorption capacity (WAC) was performed by using the specimen films of 2 × 2 cm^2, with a thickness of 0.5–0.6 mm. The specimens were immersed in distilled water at room temperature for 0–18 days. The specimens were removed from water every 2 days, and excess water at the surface of the samples was blotted by Kimwipes® paper. The weight of wet sample was measured. All testing was carried out in triplicate. Water absorption capacity was calculated by using the formula:

$$\text{Water absorption } (\%) = \frac{W_h - W_d}{W_d} \times 100$$

where, W_h and W_d are the weights of the hydrated and dried specimens, respectively.

For the test of toluene uptake, dried sample films (2 × 2 cm^2), with a thickness of 0.5–0.6 mm, were immersed in toluene at room temperature. The weight change was monitored at 1 h intervals for 8 h. All testing was carried out in triplicate. Toluene uptake was calculated by using the formula:

$$\text{Toluene uptake } (\%) = \frac{W_t - W_d}{W_d} \times 100$$

where, W_d and W_t are the weights of the specimens before and after immersion in toluene at a time (t), respectively.

For the preliminary test of biodegradation in soil, each test specimen, having 5 × 5 cm^2 and a thickness of 0.5–0.6 mm, was used for the measurement of biodegradation in soil. Potting soil purchased from a garden center (Bangkok, Thailand) was used for the experiment. The main composition of potting soil was loam soil, compost manure and coconut coir. The soil temperature was around 28–30 °C. The moisture content of the soil was around 50–60%. The specimens were weighed and buried in soil at a depth of 10 cm for 3 months under ambient conditions, where the temperature range was 24 to 35 °C. After 1, 2 and 3 months, the samples were removed from soil, washed with deionized water (DI)water, dried at 40 °C for 12 h and recorded for their weights. The degradation was evaluated by measuring the weight loss by using the formula:

$$\text{Biodegradation } (\%) = \frac{W_0 - W_i}{W_0} \times 100$$

where, W_0 and W_i are the weights of the specimens before and after being buried in soil, respectively.

The dielectric constant and dielectric loss factors were measured directly by using impedance analyzer with a precision impedance analyzer (4294A, Agilent, Santa Clara, CA, USA) at room temperature. The measurements were done at varying frequencies ranging from 103–106 Hz. The samples were coated by silver paint as electrode on both sides before measurement.

3. Results and Discussion

3.1. *Characterization of BRHA and WRHA Particles*

Rice husk ash is a major by-product of the combustion of rice husk to generate heat for boilers in rice milling plants, in which the burning temperature is usually above 800 °C. BRHA results from the pyrolysis of rice husk in poor oxygen atmosphere, whereas WRHA results from the combustion of rice husk in atmospheric air. The results of XRF analysis for chemical compositions of BRHA and WRHA are shown in Table 1. The main components of BRHA and WRHA were found to be silica as silicon dioxide (SiO_2), at 87.0% and 95.3%, respectively. The minor components were alumina oxide (Al_2O_3) and potassium oxide (K_2O) and small amounts of CaO, P_2O_5, MgO, Fe_2O_3, SO_3, MnO, ZnO and Rb_2O. In addition, the unburned carbon or loss of ignition value of BRHA and WRHA were found to be about 7.8% and 0.1%, respectively. The color of the rice husk depends on amount of unburned carbon in the ash [21]; the reported unburned carbon contents of gray-RHA and WRHA were 2.4% and 1.4%, respectively. The product of WRHA from combustion treatment contains higher silica content than BRHA, but the unburned carbon content of WRHA was lower than that of BRHA.

Table 1. Chemical compositions of black rice husk ash (BRHA) and white rice husk ash (WRHA).

Chemical Compositions	BRHA (%wt)	WRHA(%wt)
SiO_2	87.00	95.30
Al_2O_3	2.51	1.76
K_2O	1.07	0.89
CaO	0.52	0.71
P_2O_5	0.63	0.57
MgO	0.32	0.36
Fe_2O_3	0.16	0.13
SO_3	-	0.17
MnO	719 ppm	743 ppm
ZnO	64.1 ppm	113 ppm
Rb_2O	44.4 ppm	-
Loss on ignition	7.783	0.105

The SEM images of BRHA and WRHA particles are shown in Figure 1a,b, respectively. BRHA formed black platelets of partially crystalline oxides of silicon and others, whereas WRHA was in roughly spherical form of aggregated white powders. Both ash particles were produced in a wide range of sizes, from 1 to 10 μm. However, the average size of BRHA was relatively smaller, compared to WRHA.

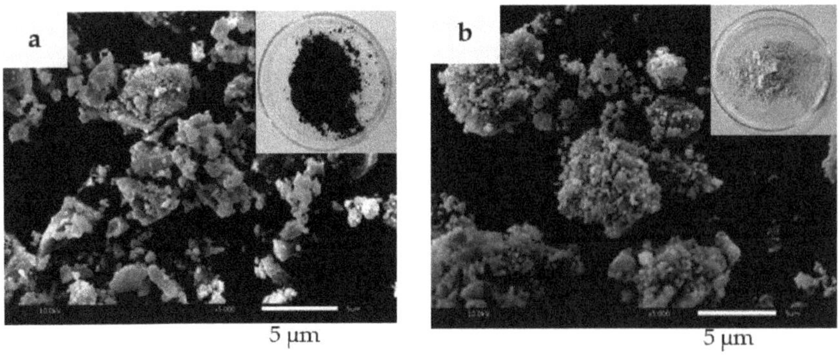

Figure 1. Scanning electron microscope (SEM) images of BRHA (**a**) and WRHA (**b**) particles.

The XRD analysis was conducted on BRHA and WRHA particles, as shown in Figure 2a,b. According to the XRD pattern, BRHA was in the form of partially crystalline oxides of cristobalite, corundum and quartz alpha. The observation of cristobalite and quartz alpha in BRHA indicates that BRHA should be treated at temperatures greater than 900 °C during the combustion process [22]. On the other hand, the XRD pattern of WRHA exhibited a very small peak of quartz alpha and broadening of the cristobalite peak (at 2θ = 22.5°), which indicates the nature of amorphous silica [23]. The degrees of crystallinity of BRHA and WRHA were 16.9% and 2.6%, respectively. The SEM micrographs and XRD results confirm that BRHA is a partially crystalline silica oxide, whereas WRHA is an amorphous silica oxide. These results agree with those reported by Xu et al. and Osman et al. [24,25], where WRHA with silica in a mainly amorphous form was produced at a controlled temperature below 800 °C, and BRHA with silica in partial crystalline phases was generated at temperatures above 800 °C.

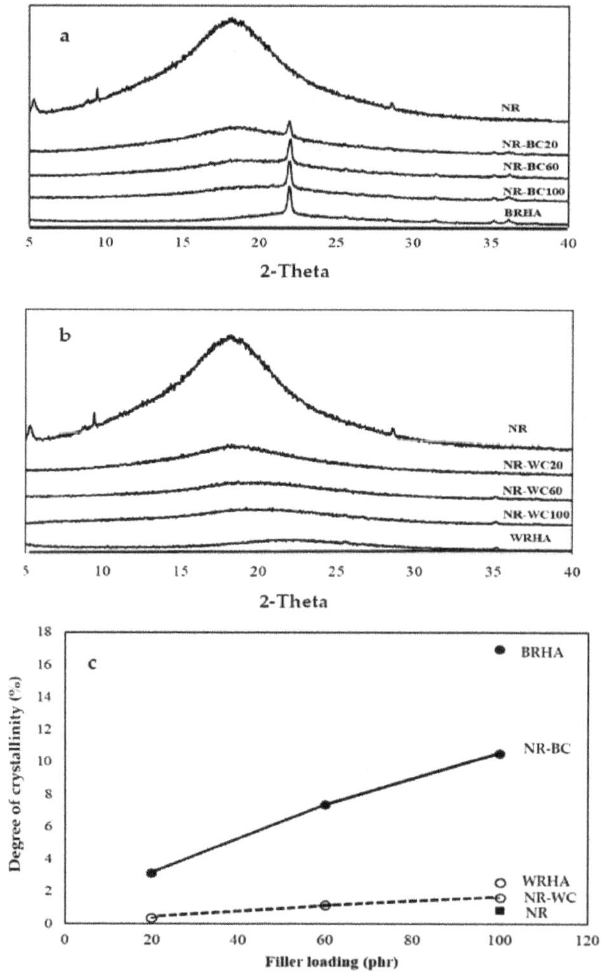

Figure 2. X-ray diffraction of the composite films of crosslinked composite films of natural rubber (NR) combined with BRHA (NR–BC) (**a**) and with WRHA (NR–WC) (**b**) and degree of crystallinity (**c**) with loaded content of BRHA and WRHA at 0, 20, 60 and 100 per hundred rubbers (phr).

The particle size distributions of the BRHA and WRHA are shown in Figure 3. The BRHA particles ranged in size from 0.4 to ~100 μm, with an average size ≈ 6 μm. The d10, d50 and d90 values based on the volume distribution were 1.5, 6.2 and 28.5 μm, respectively. The WRHA particles ranged in size from 0.4 to 144 μm, with an average size ≈ 40 μm. The d10, d50 and d90 of WRHA were higher than those of BRHA (1.9, 17.8 and 70.4 μm, respectively). The results from the SEM observation (Figure 1) and the particle size distribution (Figure 3) indicated that an average particle size of WRHA was larger than that of BRHA. This result can be attributed to hydroxyl groups of silanol (Si–OH) on the surface of rice husk ash particles. The hydroxyl group had very strong intermolecular hydrogen bonds with another hydroxyl of adjacent silica particle. These hydrogen bonds could easily cause the formation of agglomeration of particles. WRHA particles have larger particle sizes, as compared to BRHA particles, due to the higher degree of agglomeration of the WRHA powders, because of their higher silica content.

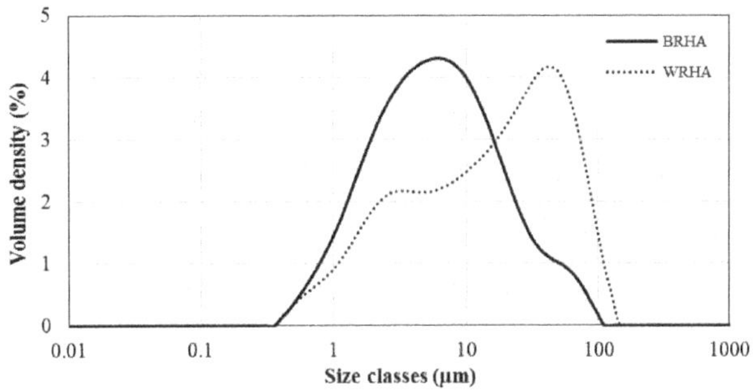

Figure 3. Particle size distributions of BRHA and WRHA particles.

Pore size distributions of BRHA and WRHA are shown in Figure 4. Because WRHA formed agglomerated particles, WRHA presents bimodal pore structure of two groups of pores that are considerably different in size. The small pores inside of fine particle are less than 2.5 nm, and the pores between particles are 4–30 nm. On the other hand, BRHA shows a monodispersed mesoporous structure. Table 2 shows surface area, pore volume and pore size of BRHA and WRHA particles. Owing to the smaller particle size of BRHA, the specific surface area of BRHA (51.57 m^2/g) was greater than that of WRHA (40.06 m^2/g), as the decrease of particle size resulted in an increase in surface area. The average pore volume and average pore diameter of BRHA were lower than those of WRHA. The pore volumes of BRHA and WRHA were 0.24 and 0.48 cm^3/g, respectively, with average pore diameter of 9.22 and 23.92 nm, respectively. These results can be attributed to larger pore size in WRHA agglomerated particles.

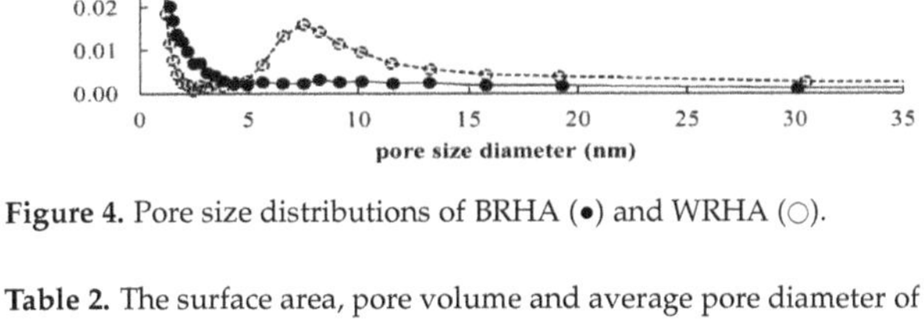

Figure 4. Pore size distributions of BRHA (•) and WRHA (○).

Table 2. The surface area, pore volume and average pore diameter of BRHA and WRHA.

Rice Husk Ash	BET Surface Area (m^2/g)	Pore Volume (cm^3/g)	Average Pore Diameter (nm)
BRHA	51.57	0.24	9.22
WRHA	40.06	0.48	23.92

3.2. Morphology of NR and NR Composite Films

The morphologies of the NR, NR–BC and NR–WC specimens are illustrated in Figure 5. From the outlook (Figure 5a), both NR–BC and NR–WC composite films had a rather smooth surface. NR–BC films were black in color, whereas NR–WC films were light brown. The surface and cross-section areas of the neat NR film were smooth, whereas NR–BC and NR–WC had relatively rough surfaces. The degree of surface roughness increased with increases in BRHA and WRHA loading content. In addition, the cross-section area images show that both BRHA and WRHA demonstrated homogenous dispersion in NR matrix without phase separation. The results reveal that alginate is a good dispersing agent for the dispersion of rice husk ashes in NR matrix up to 100 phr.

Figure 5. The outlook (**a**) and SEM micrographs (**b**) of surface morphologies (on the top) and cross-section (on the bottom) of NR, NR–BC20, NR–BC100, NR–WC20 and NR–WC100.

3.3. X-ray Diffraction (XRD) of NR and NR Composite Films

As shown in Figure 2, the XRD pattern of pure NR film demonstrates a broad peak with very low degree of crystallinity (0.8%), which indicates an amorphous structure for the polymer phase. NR–BC and NR–WC composite films revealed diffraction peaks corresponding to BRHA and WRHA structures, respectively. In addition, the diffraction peaks of both ash composites increased linearly with increasing filler loading content. The degree of crystallinity of composite films also linearly increased along with the filler loading content (Figure 2c).

3.4. Fourier Transform Infrared (FTIR) Spectroscopy

Figure 6 demonstrates the FTIR spectrum of NR, BRHA, WRHA, NR–B, NR–BC, NR–W and NR–WC composites. Pure NR consists mainly of cis-polyisoprene. The functional groups for identifying cis-polyisoprene were the asymmetric stretching vibration of methyl groups (–CH3) at 2960 cm^{-1} and C–H deformation at 1446 cm^{-1} [26]. The peaks observed at 2917 and 1663 cm^{-1} are assigned to symmetric stretching vibration of methylene (–CH2) and C=C stretching [17]. The BRHA and WRHA have similar FTIR spectrum. The strong sharp peaks at 1080 and 1083 cm^{-1} are attributed to Si–O–Si asymmetric stretching, along with peaks at 796 and 800 cm^{-1}, which are due to symmetric Si–O–Si stretching and Si–O quartz [27,28]. The main broad peak between 1000 and 1200 cm^{-1} of WRHA indicates characteristics of amorphous silica [21]. The FTIR spectrum clearly indicates that functional groups found on NR–B, NR–W, NR–BC and NR–WC are the same as those found on NR, BRHA and WRHA. The spectra of the composites contain peaks at 2961, 2915, 1661 and 1067 cm^{-1}, which are assigned to ($-CH_3$), ($-CH_2$), C=C stretching and Si–O–Si asymmetric stretching, respectively. The composite films reveal a broad absorption band between 3400 and 3100 cm^{-1}, referring to the stretching of the O–H group. The hydroxyl regions of NR–BC and NR–WC are more dominant than those of the composite films without $CaCl_2$ crosslinking, which might indicate that the $CaCl_2$ crosslinking had an effect on the O–H interaction of sodium alginate (absorption band between 3500 and 3100 cm^{-1}) [29]. Daemi and Barikani (2012) had previously reported that the absorption regions of O–H group in calcium alginate were narrower than the normal peak of sodium alginate, because the calcium ion decreased in hydrogen bonding between hydroxyl functional groups, resulting in narrower bands of calcium alginate [30]. The position peaks of functional groups of composite films are slightly shifted, as compared to the original spectra of NR, BRHA and WRHA, which indicates the formation of interactions of BRHA and WRHA in the NR matrix without a chemical reaction [16,17,20].

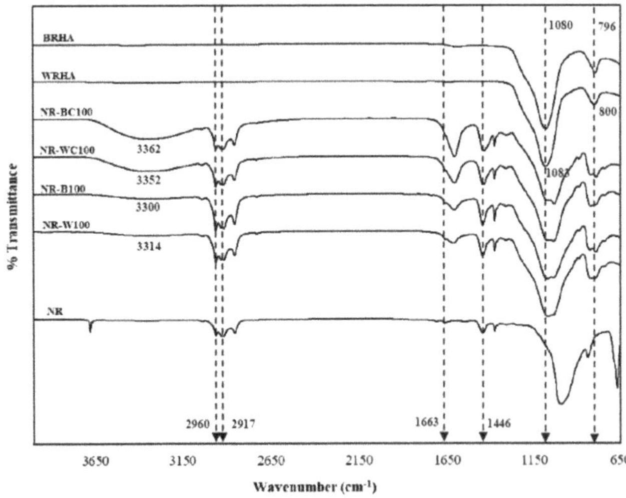

Figure 6. Fourier transform infrared (FTIR) of the WRHA; BRHA; NR; composite films of NR combined with BRHA (NR–B), NR–B100; composite films of NR combined with WRHA (NR–W), NR–W100; NR–BC100 and NR–WC100.

3.5. Mechanical Properties

The mechanical properties of BRHA and WRHA composites were examined in terms of tensile strength, Young's modulus and elongation at break, as shown in Figure 7. Uncured NR film was high elastic elongation, but it demonstrated low tensile strength. The tensile strength, Young's modulus and elongation at break of the uncured NR film were 1 MPa, 2 MPa and 113.3%, respectively. Figure 7a clearly shows that the addition of BRHA and WRHA to the NR matrix resulted in improvement among all observed mechanical properties of the composite films. The tensile strength was relatively enhanced through the increase of BRHA and WRHA loading content in all cases. The maximum tensile strength values of the NR–B, NR–W, NR–BC and NR–WC were obtained by the filler loading at 100 phr, at 6.5, 10.4, 3.2 and 13.1 MPa, respectively. The results indicate that WRHA-filled NR films exhibit significantly higher tensile strength than those reinforced by BRHA, especially at filler loading content of 60 and 100 phr. The treatment of $CaCl_2$ crosslinking tended to improve tensile strength of the composite films reinforced with WRHA but reduced the strength of the films reinforced with BRHA.

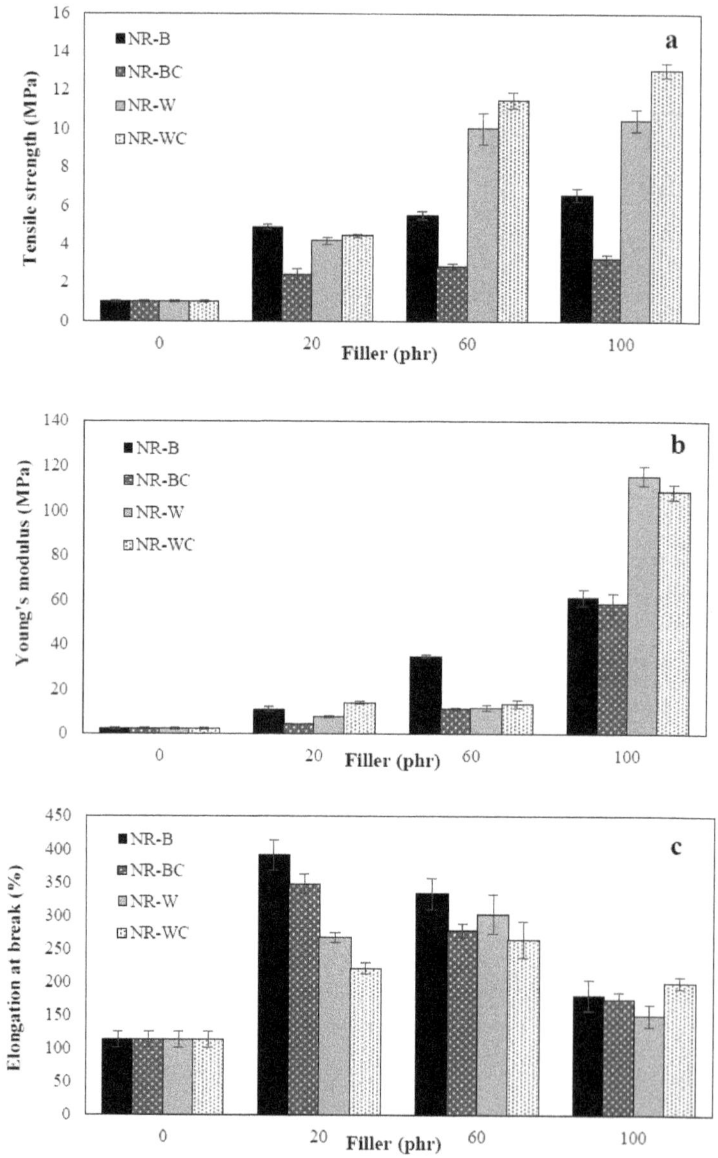

Figure 7. Mechanical properties, tensile strength (**a**), Young's modulus (**b**) and elongation at break (**c**) of NR, NR–B, NR–BC, NR–W and NR–WC composite with filler loading content at 0, 20, 60 and 100 phr.

The reinforcement effects on Young's modulus of composites filled with BRHA and WRHA were quite similar to those on tensile strength. The values of Young's modulus were improved significantly, as compared to that of pure NR film, and the values increased with increasing filler content, as shown in Figure 7b. The maximum values of Young's modulus of NR–B, NR–W, NR–BC and NR–WC films were obtained by BRHA and WRHA loadings at 100 phr, which were 61.1, 115.6, 58.4 and 108.6 MPa, respectively. As compared to BRHA, the reinforcement by WRHA was more effective in improving strength and modulus, which is likely due to the higher content of SiO_2 in WRHA.

As compared to the result of the NR film shown in Figure 7c, it is demonstrated that reinforcement with BRHA and WRHA can also improve elongation at break of the composite films. The maximum values of elongation at break of NR–B at 392.0% and NR–W at 302.7% were obtained at filler loading of 20 phr and 60 phr, which were 3.5-fold and 2.7-fold increases of that of the NR film, respectively. However, the values relatively decreased with further increases in filler loading beyond the optimal point, because the elasticity of the rubber chains was reduced when filler particles exceeded a certain amount in the rubber matrix [31,32].

In the case of a small amount of filler in NR composite (20 phr), the tensile strength, Young's modulus and elongations at break of composites filled with BRHA were better than those filled with WRHA. These results might be attributable to the smaller particle size of BRHA, since the smaller particle with higher specific surface area should facilitate a better filler—NR interaction [9]. However, when increasing filler loading in the NR composites to 60 and 100 phr, the tensile strength and Young's modulus of NR–WRHA were greater than those of NR–BRHA. Thus, factors, such as surface activity, chemical composition of fillers, concentration, shape and interactions between fillers and NR matrix play important roles in the reinforcement of NR composites. As compared to BRHA, WRHA contains a higher concentration of silica but lower concentration of carbon. Higher concentration of silica and metal oxides in fillers could promote reinforcement in NR composites. Additionally, in this study, the solid dispersion of SiO_2 in NR matrix was successfully improved by the addition of alginate as dispersion agent, resulting in no phase separation at high filler loading in NR matrix. The results agree with those of previous works on vulcanized rubber composites [32,33]; it was reported that the mechanical properties of rubber vulcanizates filled with WRHA were better than composites filled with BRHA.

Further crosslinking of the composites by $CaCl_2$ significantly improved the tensile strength of NR–WRHA, especially at high WRHA loading content. After the immersion of NR composites in $CaCl_2$ solution, calcium ion created gelation and ionic crosslinking with specific and strong interactions between Ca^{2+} with G blocks of alginate, forming the net structure [34]. This ionic crosslinking can improve the structural stability of the composite films and reinforce filler–filler and filler–rubber interactions, resulting in enhancement of the tensile strength of the composite film. After crosslinking sodium alginate/natural rubber/coconut composite by $CaCl_2$, the composites were highly stable [35]. However, it was found that the mechanical properties of NR–BC films were reduced by crosslinking the composites by $CaCl_2$. The different effects of $CaCl_2$ crosslinking on NR–BRHA and NR–WRHA might be attributable to differences in the porous structure of those fillers. WRHA presents bimodal porous structure with higher porosity and higher pore sizes, resulting in better dispersion of $CaCl_2$ into the mixtures and more interactions of fillers–alginate–NR of NR–W films. Consequently, alginate crosslinking with Ca^{2+} could enhance the tensile strength of NR–WC. A similar observation was also noted by Costa et al. [36]. The efficiency and rate constant of vulcanization of NR–BRHA was found to be lower than that of NR–WRHA vulcanization, because the pores of BRHA were too small for polymer chains to enter. Moreover, large internal surface areas might be a disadvantage, because a certain proportion of accelerator might become immobilized and inactivated. The adsorbed Ca^{2+} on surface might result in weakened filler–rubber interactions. Thus, mechanical properties of the NR–BC composite films were reduced by $CaCl_2$ crosslinking.

The previous studies demonstrated the limitations of polar filler loading in NR composites. They found the tensile strength of the NR composite increased with increasing the filler content up to a maximum value (20–60 phr) and then decreased [12,13,31,32,37]. Higher filler loading in NR composite could lead to formation of filler agglomeration and undispersed filler in NR matrix, resulting in a weak filler–rubber interaction. RHA has high surface polarity and strong filler-filler interaction, which leads to filler agglomerates and poor RHA dispersion in NR matrix, resulting in low mechanical properties of NR composite. The present study demonstrates that the tensile strength of uncured NR composites increases with increasing BRHA and WRHA loading, and that the maximum tensile strength was obtained at BRHA/WRHA loading of 100 phr. The results indicate that the latex aqueous microdispersion process with dispersing agent is suitable for composite preparation to achieve homogeneous dispersion of RHA fillers in NR matrix. Sodium alginate is suitable for the use as a dispersing agent in this system, because it can improve RHA dispersion and reduce filler-filler interactions. Previously, sodium alginate as a dispersing agent was proven to improve the stability, filler dispersion and viscosity of composites of NR–coal fly ash [16], NR–sago starch [38] and NR–microfibrillated cellulose [39].

3.6. Water Absorption Capacity

In the water sorption and toluene sorption tests, the length and width of the film specimen were 2 cm and 2 cm. The thickness was 0.5–0.6 mm. The film thickness slightly increased with increasing filler content but was not significantly changed by $CaCl_2$ crosslinking. The results of the water absorption capacity (WAC) tests of NR and NR composites filled with BRHA/WRHA are presented in Figure 8. The WAC of NR film was lower than NR–B and NR–W films, due to the hydrophobic nature of NR. The water adsorption of NR film was nonlinearly increased during a period of 0–6 days for water uptake of 0–31% and then was slightly increased during the immersion in water for 6–18 days. The water absorption behavior of filled polymer composites depended on the properties and nature of fillers such as functionality, polarity, specific surface area, filler loading and time in water [40]. The results demonstrate that the water absorption of NR–B, NR–BC and NR–W increase with increasing filler loading content, due to the hydrophilic nature of silica and other metal oxides, which are the main components in both ashes. The water uptake rates of the composite NR films filled with BRHA/WRHA were higher, as compared to NR film, and the saturation point was reached after the immersion in water for 2 days. The fillers, both BRHA and WRHA, are polar compounds and contain hydroxyl groups from silica and other metal oxides. The surface hydroxyl groups on metal oxides are sites for the adsorption of water molecules by hydrogen bonding. Thus, the increased RHA loading content promoted WAC of the composite films. The superior water absorption of NR composite filled with RHA was previously reported [40,41]. Without the $CaCl_2$ crosslinking process, the composite film filled with WRHA demonstrated higher WAC than that filled with BRHA, which is likely due to the higher amount of SiO_2 in WRHA, as compared to BRHA. The maximum WAC at around 78% was obtained from NR–W films at WRHA loading content of 100 phr.

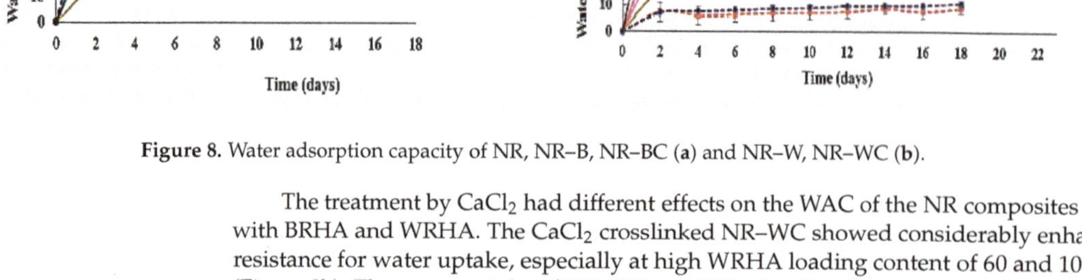

Figure 8. Water adsorption capacity of NR, NR–B, NR–BC (**a**) and NR–W, NR–WC (**b**).

The treatment by $CaCl_2$ had different effects on the WAC of the NR composites filled with BRHA and WRHA. The $CaCl_2$ crosslinked NR–WC showed considerably enhanced resistance for water uptake, especially at high WRHA loading content of 60 and 100 phr (Figure 9b). The water uptake of NR–WC at WRHA loading 100 phr was reduced to 11%, which was only ≈1/7 that of NR–W100 and was only ≈1/4 that of NR film. Previously, a significant reduction in water solubility of crosslinked alginate-based films by 1% $CaCl_2$ was reported [34]. It was suggested that the crosslinking alginate by $CaCl_2$ generated an "egg box" network. The COO– in the alginate binds to the Ca^{2+}, resulting in reduced available alginate chains to bind with H_2O molecules. The crosslinking NR–W composites by $CaCl_2$ could also increase filler–rubber interactions and limited water diffusion into the NR–W composites. Accordingly, NR–WC presented lower WAC after crosslinking by $CaCl_2$. On the other hand, the WAC of NR composite films filled with BRHA was increased after crosslinking by $CaCl_2$ (Figure 8a). As previously discussed in the noneffective alginate crosslinking by Ca^{2+} of NR–BC composite films, in this case, the addition of $CaCl_2$ was found to reduce filler–rubber interaction in NR–BC composites. The available COO– in alginate molecule and OH– in silica oxides could interact with water molecules, leading to higher water absorption of NR–BC. The addition of $CaCl_2$ might also increase hydrophilicity of the films and, therefore, increase the water diffusion into NR–BC films.

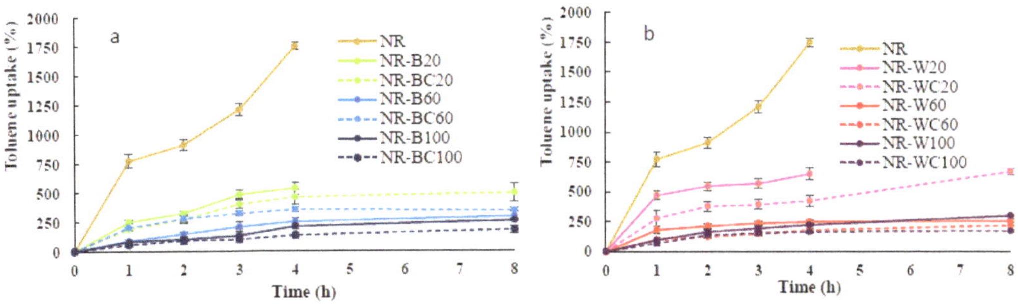

Figure 9. Toluene uptake of NR, NR–B, NR–BC (**a**) and NR–W, NR–WC (**b**).

3.7. Toluene Uptake

The results of the toluene uptake of NR and NR composite films soaked in toluene for 8 h are shown in Figure 9. The maximum toluene uptake of NR film was 1746% at the immersion in toluene for 4 h; after that, the film was decomposed in toluene. Toluene and NR are nonpolar; thus, this solvent can be highly absorbed in NR film. The high uptake rate and poor chemical resistance to nonpolar solvents of uncured NR were previously reported [16,17,42]. Uncrosslinked NR chains can dissolve in toluene, and the decrease in the solvent uptake of NR composites may reveal good filler—rubber interaction [42]. The toluene uptake rate of NR–B and NR–W composites at RHA loading of 20 phr was much less than that of NR film; however, after the immersion in toluene for 4 h, both NR–B and NR–W were, to some extent, decomposed in toluene. Uncured NR composites might be dissolved in nonpolar solvents, such as toluene [43]. More resistance to toluene uptake was observed in the NR composites with higher BRHA/WRHA content. The toluene uptake of the composite films decreased with increased filler loading and was considerably lower, as compared to that of pure NR film [16,39]. The low toluene uptake of NR–B and NR–W films was exhibited at RHA loading of 60–100 phr, which represented about 260–300% (or 0.15–0.17 of that of the NR film), due to polar nature of silica oxides, the major constituent in both RHAs. The integration of polar fillers into NR matrix can reduce the toluene uptake rate and adsorption capacity. In addition, the NR composites filled with BRHA/WRHA at 60–100 phr maintained their structural stability during immersion in toluene for 8 h. This observation indicates that the composite films had good dispersion of RHA in the NR matrix.

After crosslinking by $CaCl_2$, the NR composites filled with WRHA/BRHA demonstrated slightly improved resistance to toluene. The toluene uptakes of NR–BC and NR–WC were reduced to 170–180%. The crosslinked films had better stability in toluene, as no decomposition of the films was observed during immersion in toluene for 8 h. Thus, the

$CaCl_2$ treatment seems to improve the solvent resistance of composite films. The addition of calcium ions into NR matrix possibly makes the surface more hydrophilic, resulting in lower diffusion of nonpolar solvent into NR matrix. The lowering of toluene uptake might also be attributable to the good interaction between filler and NR matrix [44].

3.8. Biodegradation in Soil

The preliminary study of biodegradation in soil of pure NR film and NR composite films filled with RHAs was determined in uncontrolled conditions (with an average temperature of ≈30 °C). The percentage weight loss and the visual analysis of the biodegraded samples after 1, 2 and 3 months are shown in Figures 10 and 11, respectively. The weight loss of the pure NR film was lower than all NR composite films at any degradation time, due to the slow degradation of NR in natural environments by microorganisms [45]. After 3 months, the weight loss of NR, NR–BRHA and NR–WRHA were 5.3%, 7.6–13.0% and 7.6–18.4%, respectively. Overall, the NR–BRHA composite presented comparable weight loss to NR–WRHA composites at similar RHA loading contents, except for the NR composites with 100 phr RHA loadings, of which, NR–W100 demonstrated relatively higher biodegradability in soil. The weight loss of composites increased with increased filler loading in all conditions. These results agree with those of Ramasamy et al. [46], who reported that the weight loss percentage of NRL foam increased with increased rice husk powder, because, at lower rice husk powder loading in foam composite, microbial activity was not high due to lack of microbial growth supports.

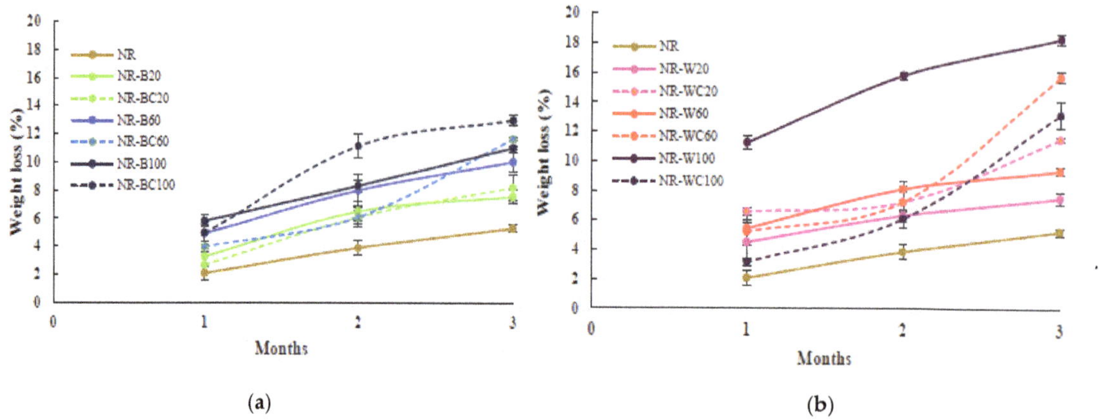

Figure 10. Preliminary test for biodegradation in soil of NR, NR–B, NR–BC (**a**) and NR–W, NR–WC (**b**).

The effect of the $CaCl_2$ treatment of the NR composites was observed from the composites at filler loading of 100 phr in both cases but in opposite directions. The weight loss of NR–BC100 was higher than that of NR–B100, whereas the weight loss of NR–WC100 was lower than NR–W100. These results are in accord with those on the water absorption capacity, which can be explained by the high number of fillers in NR matrix generating hydrophilic surfaces, resulting in enhanced water absorption of the composites. The higher water content in NR matrix could promote microbial activities that increase the rate of degradation [31,47]. The weight loss of NR–WC100 was lower than NR–W100, because the $CaCl_2$ crosslinked film displayed a higher tensile strength with lower WAC, as compared to NR–W100. On the other hand, the weight loss of NR–BC 100 was higher than NR–B100, due to the higher WAC of the composite after $CaCl_2$ crosslinking, as discussed earlier. It was also suggested that the use of sodium alginate as a dispersing agent in NR composites could increase the rate of biodegradation of the composites [38]. Figure 11 presents visual analysis of the biodegradation of NR and NR composite films during a period of 1–3 months. The enhanced degradation of NR composite films filled with higher RHA loading was clearly observed, especially at 3 months of biodegradation in soil. Black spots were found in the composite films reinforced with WRHA, while white spots, as well as fogging on film surfaces, were observed after 1 month in soil, indicating soil microflora activities in the composites, which is a good indicator of biodegradation [48]. Moreover, after 3 months in soil, some parts of the composite films disappeared. The results revealed that NR–BRHA and NR–WRHA could be biodegradable in soil.

Figure 11. The visual analysis of biodegradation in soil of NR–BC and NR–WC for 3 months.

3.9. Dielectric Properties

Dielectric properties of materials are important characteristics for predicting behavior with respect to application in electronic materials, such as print circuit boards, power transformers, microelectronics, radio propagation, remote sensing and actuator applications [15]. The dielectric properties of polymers depend on cations, dipole moment and space charge polarizations [49]. On the other hand, dielectric behavior of composite materials depends on dielectric properties of the matrix and filler, chemical composition, structure, shape and morphology of particles and filler dispersion in the composite [50]. The dielectric constant is the amount of energy from an external electrical field stored in materials. The dielectric constant of NR–BRHA and NR–WRHA composites at different filler loadings are presented in Figure 12a,b. The frequency dependence of the dielectric constant at ambient temperature was analyzed. The NR film had a low dielectric constant, in the range of 1.8–2, attributable to the nonpolar nature of NR, which has only instantaneous electronic and atomic polarization [51]. It was demonstrated that the dielectric constant of the NR composite films filled with RHAs increased with increased BRHA/WRHA loading at all frequencies. BRHA and WRHA contained 87 and 95% silica oxides with polar hydroxyl groups of silanol (Si–OH) on surface. Silica as a dielectric metal oxide semiconductor has a dielectric constant of about 3–9, and it has been used in micro- and nano-electronic industries [52]. Thus, the increase of RHA in NR composites resulted in higher dipole or orientation polarization; thus, the dielectric constant was enhanced [51]. In addition, the increase in RHA loading resulted in an increased space charge, leading to polarization in the composite matrix [53]. It was shown that the NR composite films presented a higher dielectric constant at lower frequencies. At high frequency, molecular movements arrested or decreased orientation polarization that leads to the decreased value of the dielectric constant [53,54]. The dielectric constants of NR–BRHA were relatively higher than those of NR–WRHA, which should be due to the higher degree of crystallinity of BHRA, as compared to that of WRHA. Correlations between crystal structure and dielectric properties of metal oxides were reported [55]. Metal and carbon-based filler could be acting as charge centers and increasing segmental mobility in the polymer matrix, which would enhance the dielectric constant of composites [56,57]. Oxides are suggested as promising materials in high dielectric materials, as they possess both high dielectric constants and large band gaps [55].

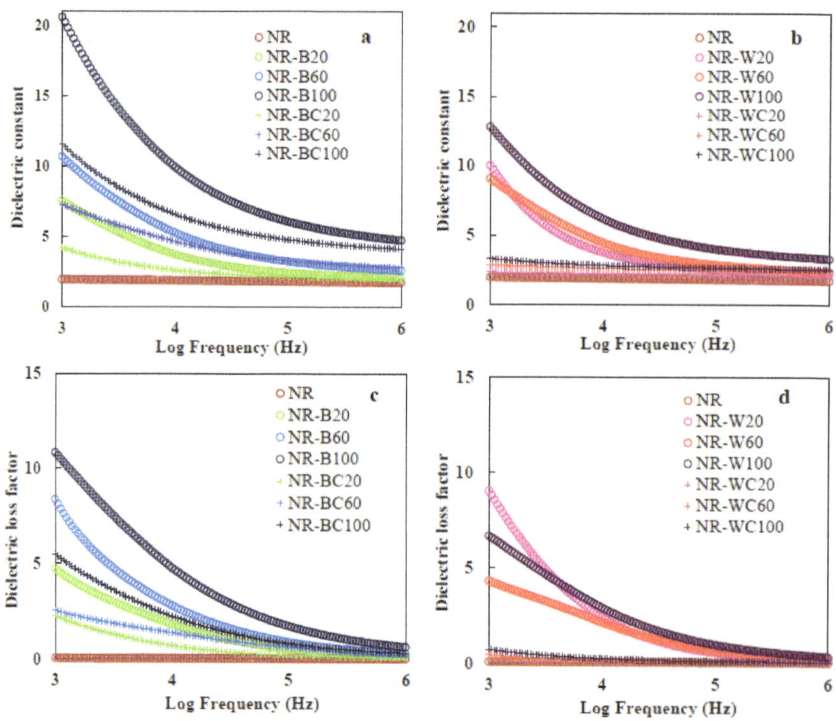

Figure 12. Dielectric constant of (**a**) NR, NR–B, NR–BC and (**b**) NR, NR–W, NR–WC and dielectric loss of (**c**) NR, NR–B, NR–BC and (**d**) NR, NR–W, NR–WC.

The crosslinking of the composite films by $CaCl_2$ caused a considerable drop in dielectric properties in all cases. The addition of $CaCl_2$ might reduce the amount of electron charges, including COO– (from alginate) and OH– (from silica), in the composite films. The higher number of free protons suggests more localization of charge carriers along with mobile ions, resulting in higher dipole polarization and a higher dielectric constant [54]. For this reason, the value of the dielectric constant of composite film after crosslinking alginate was much lower than that of films without $CaCl_2$ treatment. The obtained values for the dielectric constants of the films from the lowest value to the highest value were NR < NR–WC < NR–BC < NR–W < NR–B. These results could also imply that the stronger interactions between WRHA filler and NR matrix could lower the hydrophilicity and polarity of the composite films. A similar effect from the reinforcement that caused the reduction of orientation polarization and the dielectric constant was previously reported [51].

The dielectric loss factor presents the amount of energy loss from a material due to an external electric field [58]. The frequency dependence of dielectric loss of composites is shown in Figure 12c,d. The values of dielectric loss demonstrate similar trends as those of the dielectric constant, which increased with increased filler loading and declined with increases in frequency. The increase of RHA loading enhanced the polar groups in the composite films, which led to enlargement of orientation polarization and relaxation in composite, resulting in relatively high energy loss [53]. In addition, increasing amorphous phase to the composite enhanced the flow of current through the amorphous region and generated amorphous phase relaxation, resulting in higher dielectric loss [51,59]. At low frequency, dielectric loss was high, because of interfacial polarization enhanced by the difference between the conductivity of various phases [53]. This result agrees with the previous work by Jamal et al. [57], in which the dielectric loss of nickel–rubber nanocomposites was high at frequency lower than 105 Hz, due to relaxation time of pure rubber enhanced when nickel nanoparticles were incorporated into the matrix. Moreover, the results indicate dielectric loss factor of NR–WC composites was lower than that of other composites at all WRHA loadings (~0.7–2.1 at frequency 103 Hz). The effect of crosslinking composite by $CaCl_2$ on the reduction of the dielectric loss factor was the same as that on the dielectric constant for similar reasons. Less free space in the matrix resulted in a low dielectric loss factor [54].

4. Conclusions

BRHA/WRHA-filled uncured NR composites were successfully prepared via a latex aqueous microdispersion process with the use of sodium alginate as a dispersing agent. SEM images presented good dispersion of RHAs in composite matrices and no phase separation with RHA loading in the range of 20–100 phr. The filler–rubber interactions of composites were observed via FTIR analysis. The effective crosslinking of alginate in the composites by $CaCl_2$ was achieved on NR–WC films. The mechanical properties of NR composite films were improved by BRHA/WRHA reinforcement. NR–WC at filler loading of 100 phr presented the highest tensile strength and the lowest water and toluene uptake. The results indicate that crosslinking NR–W composite by $CaCl_2$ promotes filler–alginate–NR interaction, mechanical properties and structural stability of NR–WC films. In addition, the preliminary biodegradation in soil for 3 months revealed relatively improved biodegradability with RHA loading. The weight losses of NR–Bs and NR–Ws after 3 months in soil were about 7.6–13.0% and 7.6–18.4%, respectively. Moreover, the reinforcement considerably improved dielectric properties of the composite films. Since the NR composites filled with RHAs exhibited improved properties in terms of mechanical properties, chemical resistance, biodegradability and dielectric properties, they have potential for further development as stretchable conductive substrate or semiconducting polymer films for electronic applications.

Author Contributions: Conceptualization, M.P. and P.S.; methodology, M.P. and P.S.; validation, M.P.; formal analysis, P.S.; investigation, M.P. and P.S.; resources, M.P.; data curation, P.S.; writing—original draft preparation, P.S.; writing—review and editing, M.P.; supervision, M.P.; project administration, M.P.; funding acquisition, M.P. All authors have read and agreed to the published version of the manuscript.

Funding: This research was funded by the 90th Anniversary of Chulalongkorn University Fund (Ratchadaphiseksomphot Endowment Fund). The authors also would like to thank the International Research Network Program, Thailand Science Research and Innovation (IRN62W0001) for supporting this research.

Institutional Review Board Statement: Not applicable.

Informed Consent Statement: Not applicable.

Data Availability Statement: The data presented in this study are available on request from the corresponding author.

Conflicts of Interest: The authors declare no conflict of interest.

References

1. Katarzyna, L.; Lewandowicz, G. Polymer Biodegradation and Biodegradable Polymers—A Review. *Pol. J. Environ. Stud.* **2010**, *19*, 255–266.
2. Liu, H.; Jian, R.; Chen, H.; Tian, X.; Sun, C.; Zhu, J.; Yang, Z.; Sun, J.; Wang, C. Application of Biodegradable and Biocompatible Nanocomposites in Electronics: Current Status and Future Directions. *Nanomaterials* **2019**, *9*, 950. [CrossRef]
3. Chetchuda, C. Natural Rubber Processing. Available online: https://www.krungsri.com/bank/getmedia/5fc8fc4b-5f60-48ad-a306-5533ed9df767/IO_Rubber_2018_EN.aspx (accessed on 2 April 2019).
4. Rippel, M.M.; Galembeck, F. Nanostructures and adhesion in natural rubber: New era for a classic. *J. Braz. Chem. Soc.* **2009**, *20*, 1024–1030. [CrossRef]
5. Imran, K.; Bhat, A.H. Micro and Nano Calcium Carbonate Filled Natural Rubber Composites and Nanocomposites. In *Natural Rubber Materials, Volume 2: Composites and Nanocomposites*; Maria, H.J., Thomas, S., Joy, J.P., Chan, C.H., Pothen, L.A., Eds.; The Royal Society of Chemistry: London, UK, 2013; pp. 467–487.
6. Raja, S. Effect of Natural Filler on the Characterization of Natural Rubber. Bachelor's Thesis, Chemical Engineering, Universiti Malaysia Pahang, Pekan Pahang, Malaysia, 2010.
7. Pode, R. Potential applications of rice husk ash waste from rice husk biomass power plant. *Renew. Sustain. Energy Rev.* **2016**, *53*, 1468–1485. [CrossRef]
8. Stefani, P.M.; Cyras, V.; Barchi, A.T.; Vazquez, A. Mechanical properties and thermal stability of rice husk ash filled epoxy foams. *J. Appl. Polym. Sci.* **2006**, *99*, 2957–2965. [CrossRef]
9. Boonmee, A.; Sabsiriroht, P.; Jarukumjorn, K. Preparation and characterization of rice husk ash for using as a filler in natural rubber. *Mater. Today Proc.* **2019**, *17*, 2097–2103. [CrossRef]
10. Da Costa, H.M.; Visconte, L.L.Y.; Nunes, R.C.R.; Furtado, C.R.G. Mechanical and dynamic mechanical properties of rice husk ash-filled natural rubber compounds. *J. Appl. Polym. Sci.* **2002**, *83*, 2331–2346. [CrossRef]
11. Ismail, H.; Nasaruddin, M.; Ishiaku, U. White rice husk ash filled natural rubber compounds: The effect of multifunctional additive and silane coupling agents. *Polym. Test.* **1999**, *18*, 287–298. [CrossRef]
12. Ishak, Z.; Bakar, A. An investigation on the potential of rice husk ash as fillers for epoxidized natural rubber (ENR). *Eur. Polym. J.* **1995**, *31*, 259–269. [CrossRef]
13. Sae-Oui, P.; Rakdee, C.; Thanmathorn, P. Use of rice husk ash as filler in natural rubber vulcanizates: In comparison with other commercial fillers. *J. Appl. Polym. Sci.* **2002**, *83*, 2485–2493. [CrossRef]
14. Sarkawi, S.S.; Aziz, Y. Ground Rice Husk as Filler in Rubber Compounding. *J. Teknol.* **2003**, *39*, 135–148. [CrossRef]
15. Arayapranee, W.; Na-Ranong, N.; Rempel, G.L. Application of rice husk ash as fillers in the natural rubber industry. *J. Appl. Polym. Sci.* **2005**, *98*, 34–41. [CrossRef]
16. Panitchakarn, P.; Wikranvanich, J.; Phisalaphong, M. Synthesis and characterization of natural rubber/coal fly ash composites via latex aqueous microdispersion. *J. Mater. Cycles Waste Manag.* **2018**, *21*, 134–144. [CrossRef]
17. Phomrak, S.; Phisalaphong, M. Reinforcement of Natural Rubber with Bacterial Cellulose via a Latex Aqueous Microdispersion Process. *J. Nanomater.* **2017**, *2017*, 1–9. [CrossRef]
18. Ranjan, G.; Shelukar, S.; Kar, S.; Urumise, A. Carbon Black Latex Masterbatch Composition. U.S. Patent No.WO 2017/093805 Al, 8 June 2017.
19. Muthadhi, A.; Kothandaraman, S. Optimum production conditions for reactive rice husk ash. *Mater. Struct.* **2010**, *43*, 1303–1315. [CrossRef]
20. Chiaoprakobkij, N.; Seetabhawang, S.; Sanchavanakit, N.; Phisalaphong, M. Fabrication and characterization of novel bacterial cellulose/alginate/gelatin biocomposite film. *J. Biomater. Sci. Polym. Ed.* **2019**, *30*, 961–982. [CrossRef] [PubMed]
21. Jaya, R.P.; Nor, M.A.A.M.; Ahmad, Z.A.; Amin, Z.M. Properties of Mortar Containing Rice Husk Ash at Different Temperature and Exposed to Aggressive Environment. *Adv. Mater. Res.* **2012**, *620*, 87–93. [CrossRef]
22. Venkatanarayanan, H.K.; Rangaraju, P.R. Material Characterization Studies on Low- and High-Carbon Rice Husk Ash and Their Performance in Portland Cement Mixtures. *Adv. Civ. Eng. Mater.* **2013**, *2*. [CrossRef]
23. Sreekumar, V.M.; Pillai, R.M.; Pai, B.C.; Chakraborty, M. Microstructural development in Al/MgAl2O4 in situ metal matrix composite using value-added silica sources. *Sci. Technol. Adv. Mater.* **2008**, *9*, 015004. [CrossRef] [PubMed]
24. Xu, W.; Lo, T.Y.; Memon, S.A. Microstructure and reactivity of rich husk ash. *Constr. Build. Mater.* **2012**, *29*, 541–547. [CrossRef]
25. Osman, R.; Abdullah, N.H.; Matori, K.A.; Hamidon, M.N.; Ismail, I.; Mustaffa, S. Effect of Temperature Towards Rice Husk Silica Characterization with Different Preparation Methods. *Int. J. Basic Appl. Sci.* **2017**, *17*, 15–20.
26. Rahmah, M.; Khusyairi, A.R.A.; Khairunnisya, H.N.; Nurbalqis, M.S. Oven Ageing versus UV Ageing Properties of Natural Rubber Cup Lump/EPDM Rubber Blend with Mangosteen Powder (MPP) as Natural Antioxidant. In Proceedings of the IOP Conference Series: Materials Science and Engineering; IOP Publishing: Aceh, Indonesia, 2019; Volume 548, p. 012014. Available online: https://iopscience.iop.org/article/10.1088/1757-899X/548/1/012014 (accessed on 20 April 2019).
27. Deshmukh, P.; Peshwe, D.; Pathak, S. FTIR and TGA Analysis in Relation with the % Crystallinity of the SiO2 Obtained by Burning Rice Husk at Various Temperatures. *Adv. Mater. Res.* **2012**, *585*, 77–81. [CrossRef]
28. Saikia, B.J.; Parthasarathy, G. Fourier Transform Infrared Spectroscopic Characterization of Kaolinite from Assam and Meghalaya, Northeastern India. *J. Mod. Phys.* **2010**, *1*, 206–210. [CrossRef]

29. Andreia Grossi Santos de, L.; de Souza Costa Junior, E.; de Souza Costa, H. A Study of Sodium Alginate and Calcium Chloride Interaction through Films for Intervertebral Disc Regeneration Uses. In Proceedings of the 21 CBECIMAT: Brazilian Congress of Engineering and Materials Science, Cuiabá, MT, Brazil, 9–13 November 2014; pp. 7341–7348.
30. Daemi, H.; Barikani, M. Synthesis and characterization of calcium alginate nanoparticles, sodium homopolymannuronate salt and its calcium nanoparticles. *Sci. Iran.* **2012**, *19*, 2023–2028. [CrossRef]
31. Jayathilaka, L.P.I.; Ariyadasa, T.U.; Egodage, S.M. Development of biodegradable natural rubber latex composites by employing corn derivative bio-fillers. *J. Appl. Polym. Sci.* **2020**, *137*. [CrossRef]
32. Da Costa, H.M.; Nunes, R.C.R.; Visconte, L.L.Y.; Furtado, C.R.G. Physical Properties and Swelling of Natural Rubber Compounds Containing Rice Husk Ash. *Raw Mater. Appl.* **2001**, *54*, 242–249.
33. Costa, H.M.; Visconte, L.L.Y.; Nunes, R.C.R.; Furtado, C.R.G. Rice husk ash filled natural rubber compounds – the use of rheometric data to qualitatively estimate optimum filler loading. *Int. J. Polym. Mater.* **2004**, *53*, 475–497. [CrossRef]
34. Costa, M.J.; Marques, A.M.; Pastrana, L.M.; Teixeira, J.A.; Sillankorva, S.M.; Cerqueira, M.A. Physicochemical properties of alginate-based films: Effect of ionic crosslinking and mannuronic and guluronic acid ratio. *Food Hydrocoll.* **2018**, *81*, 442–448. [CrossRef]
35. Riyajan, S.-A.; Tangboriboonrat, P. Novel composite biopolymers of sodium alginate/natural rubber/coconut waste for adsorption of Pb(II) ions. *Polym. Compos.* **2013**, *35*, 1013–1021. [CrossRef]
36. Da Costa, H.M.; Visconte, L.L.Y.; Nunes, R.C.R.; Furtado, C.R.G. Rice husk ash filled natural rubber. I. Overall rate constant determination for the vulcanization process from rheometric data. *J. Appl. Polym. Sci.* **2003**, *87*, 1194–1203. [CrossRef]
37. Khed, V.C.; Mohammed, B.S.; Liew, M.; Zawawi, N.A.W.A. Development of response surface models for self-compacting hybrid fibre reinforced rubberized cementitious composite. *Constr. Build. Mater.* **2020**, *232*, 117191. [CrossRef]
38. Ab Rahman, M.; Norfaizal, N.; Azura, A. The Influence of Sago Starch Dispersion on Mechanical Properties of Biodegradable Natural Rubber Latex Films. *Mater. Today: Proc.* **2019**, *17*, 1040–1046. [CrossRef]
39. Supanakorn, G.; Varatkowpairote, N.; Taokaew, S.; Phisalaphong, M. Alginate as Dispersing Agent for Compounding Natural Rubber with High Loading Microfibrillated Cellulose. *Polymers* **2021**, *13*, 468. [CrossRef] [PubMed]
40. Turmanova, S.; Genieva, S.; Vlaev, L. Obtaining Some Polymer Composites Filled with Rice Husks Ash-A Review. *Int. J. Chem.* **2012**, *4*. [CrossRef]
41. Ismail, H.; Mega, L.; Khalil, H.P.S.A. Effect of a silane coupling agent on the properties of white rice husk ash-polypropylene/natural rubber composites. *Polym. Int.* **2001**, *50*, 606–611. [CrossRef]
42. Visakh, P.M.; Thomas, S.; Oksman, K.; Mathew, A.P. Effect of cellulose nanofibers isolated from bamboo pulp residue on vulcanized natural rubber. *Bioresources* **2012**, *7*, 2156–2168. [CrossRef]
43. Abraham, E.; Thomas, M.S.; John, C.; Pothen, L.; Shoseyov, O.; Thomas, S. Green nanocomposites of natural rubber/nanocellulose: Membrane transport, rheological and thermal degradation characterisations. *Ind. Crop. Prod.* **2013**, *51*, 415–424. [CrossRef]
44. Roy, K.; Debnath, S.C.; Tzounis, L.; Pongwisuthiruchte, A.; Potiyaraj, P. Effect of Various Surface Treatments on the Performance of Jute Fibers Filled Natural Rubber (NR) Composites. *Polymers* **2020**, *12*, 369. [CrossRef]
45. Bhatt, R.; Shah, D.; Patel, K.; Trivedi, U. PHA–rubber blends: Synthesis, characterization and biodegradation. *Bioresour. Technol.* **2008**, *99*, 4615–4620. [CrossRef] [PubMed]
46. Ramasamy, S.; Ismail, H.; Munusamy, Y. Soil burial, tensile properties, morphology, and biodegradability of (rice husk powder)-filled natural rubber latex foam. *J. Vinyl Addit. Technol.* **2015**, *21*, 128–133. [CrossRef]
47. Yahya, S.R.; Rashid, A.A.; Azahari, B. Soil Burial Studies for Biodegradation of Natural Rubber Latex Films. *Adv. Mater. Res.* **2013**, *844*, 406–409. [CrossRef]
48. Zaborski, M.; Piotrowska, M.; Zakowska, Z.; Piotrowska, M. Hydrophilic-hydrophobic rubber composites with increased susceptibility to biodegradation. *Polymers* **2006**, *51*, 534–538. [CrossRef]
49. Kambale, R.C.; Shaikh, P.A.; Bhosale, C.H.; Rajpure, K.Y.; Kolekar, Y.D. The effect of Mn substitution on the magnetic and dielectric properties of cobalt ferrite synthesized by an autocombustion route. *Smart Mater. Struct.* **2009**, *18*. [CrossRef]
50. Pal, R. *Electromagnetic, Mechanical, and Transport Properties of Composite Materials*; CRC Press: Boca Raton, FL, USA, 2014; Volume 158.
51. Jacob, M.; Varughese, K.T.; Thomas, S. Dielectric characteristics of sisal–oil palm hybrid biofibre reinforced natural rubber biocomposites. *J. Mater. Sci.* **2006**, *41*, 5538–5547. [CrossRef]
52. Iriani, Q.; Kusumandari, Y.; Khoirum, S. Effect of Sintering Temperature on Dielectric Constant of Silica Prepared from Rice Husk Ash. In Proceedings of the ASEAN Conference on Science and Technology 2014, Bogor, Indonesia, 18–19 August 2014.
53. Louis, N.; Maria, S.; Banu, A.; Sudha, S. Effect of Rice Husk Ash and Hydrated Silica on the Dielectric Proper-Ties of Ldpe Composites. *Mater. Sci. Indian J.* **2014**, *11*, 0974–7486.
54. Juntaro, J.; Ummartyotin, S.; Sain, M.; Manuspiya, H. Bacterial cellulose reinforced polyurethane-based resin nanocomposite: A study of how ethanol and processing pressure affect physical, mechanical and dielectric properties. *Carbohydr. Polym.* **2012**, *87*, 2464–2469. [CrossRef]
55. Wu, H.; Kim, E.S. Correlations between crystal structure and dielectric properties of high-Q materials in rock-salt structure Li_2O–MgO–BO_2 (B = Ti, Sn, Zr) systems at microwave frequency. *RSC Adv.* **2016**, *6*, 47443–47453. [CrossRef]
56. Samanta, B.; Kumar, P.; Nanda, D.; Sahu, R. Dielectric properties of Epoxy-Al composites for embedded capacitor applications. *Results Phys.* **2019**, *14*, 102384. [CrossRef]

57. Jamal, E.M.A.; Joy, P.; Kurian, P.; Anantharaman, M. Synthesis of nickel–rubber nanocomposites and evaluation of their dielectric properties. *Mater. Sci. Eng. B* **2009**, *156*, 24–31. [CrossRef]
58. Khor, S.F. Dielectric spectroscopy on mixture of rice husk, rice husk ash and rice bran from 4 Hz to 1 MHz. *Int. J. Geomate* **2016**, *11*. [CrossRef]
59. Sheykhnazari, S.; Tabarsa, T.; Mashkour, M.; Khazaeian, A.; Ghanbari, A. Multilayer bacterial cellulose/resole nanocomposites: Relationship between structural and electro-thermo-mechanical properties. *Int. J. Biol. Macromol.* **2018**, *120*, 2115–2122. [CrossRef] [PubMed]

polymers

MDPI

Article

Effects of Lyophilization on the Release Profiles of 3D Printed Delivery Systems Fabricated with Carboxymethyl Cellulose Hydrogel

Xuepeng Jiang [1], Yanhua Huang [1], Yiliang Cheng [2], Zhan Zhang [3], Xiaolei Shi [2,*] and Hantang Qin [1,3,*]

1 Department of Industrial and Manufacturing Systems Engineering, Iowa State University, Ames, IA 50011, USA; xuepengj@iastate.edu (X.J.); yanhuah@iastate.edu (Y.H.)
2 Department of Food Science and Human Nutrition, Iowa State University, Ames, IA 50011, USA; ycheng8@iastate.edu
3 Center for Nondestructive Evaluation, Iowa State University, Ames, IA 50014, USA; zhan@iastate.edu
* Correspondence: xshi@iastate.edu (X.S.); qin@iastate.edu (H.Q.)

Abstract: Recently, increasing numbers of researchers are becoming interested in 3D bioprinting because it provides customizability and structural complexity, which is difficult for traditional subtractive manufacturing to achieve. One of the most critical factors in bioprinting is the material. Depending on the bio-applications, materials should be bio-inert or bio-active, non-toxic, and along with those characteristics, mechanical properties should also meet the applicational or manufacturing requirement. As previously validated for bioprinting, carboxymethyl cellulose (CMC) hydrogel is focused on the printability and release control test in this study. With a differentiated weight percentage of CMC hydrogels were used to 3D print capsules filled with food degradable colorant at designated voids to mimic capsules manufactured for oral delivery. Standard USP (United States Pharmacopeia) dissolution apparatus II (Paddle) evaluations were performed both on lyophilized and non-lyophilized printed capsules. The first-order model was selected due to high linear fitting regression. Upon 24 h dissolution, non-lyophilized capsules showed a different release efficiency when the CMC percentage varied, while lyophilized capsules showed no significant difference. This study signifies the possibility of customizing oral drug delivery by printing capsules with CMC hydrogel. The improved delivery efficiency demonstrated by capsules with post-process lyophilizing proposed potential optimization options for pharmaceutical manufacturing industries.

Keywords: 3D printing; carboxymethyl cellulose; hydrogel; lyophilization; dissolution; release model; customization

Citation: Jiang, X.; Huang, Y.; Cheng, Y.; Zhang, Z.; Shi, X.; Qin, H. Effects of Lyophilization on the Release Profiles of 3D Printed Delivery Systems Fabricated with Carboxymethyl Cellulose Hydrogel. *Polymers* **2021**, *13*, 749. https://doi.org/10.3390/polym13050749

Academic Editor: Arn Mignon

Received: 7 January 2021
Accepted: 2 February 2021
Published: 28 February 2021

1. Introduction

3D printing is an additive manufacturing process that fabricates three-dimensional CAD (computer-aided design) objects in a layer-by-layer buildup subsequence. Although the 3D printing idea came out in the 1980s [1], researchers in the engineering area still pay close attention to the 3D printing technique. Its advantage of high precision, controlling deposition, cost-effectiveness, simple processing, and fast prototyping compare favorably with traditional subtractive manufacturing in recent years [2]. Nowadays, the 3D printing technique involves interdisciplinary areas that include not only manufacturing but also civil construction, fashion and design, tissue engineering, pharmaceutical, and food production [2–6]. Material jetting (including drop on demand (DOD), nanoparticle jetting (NPJ) and material jetting (MJ)), powder bed fusion (including selective laser sintering (SLS), and selective laser melting (SLM)), directed energy deposition (DED) and material extrusion (ME) or semi-solid extrusion (SSE) are four major branches made up of 3D printing technology [7]. Among all additive manufacturing technologies, ME or SSE, FDM (fused deposition modeling), inkjet, and polyjet printing methods are commonly used in the bioprinting field [8,9].

3D bioprinting is of tremendous interest to researchers due to the need for high customizability for organ and tissue engineering [10] and wound therapy [11,12]. The biomaterial is the most crucial carrier to achieve an additive manufacturing 3D structure for bioprinting. Materials that meet the requirement of not inducing inflammation and allergy symptoms, non-toxic, biocompatibility or biodegradation, biofunctional, bioactive, bioinert, and sterilizable could be called biomaterials [13,14]. The most common biomaterials used in pharmaceutical and medical areas include metals, polymers, ceramics, and composite [15]. However, biopolymers have the largest applications because of their properties. In addition to the properties that biomaterials have, biopolymers also have some distinctive properties: flexibility (composition and shape of the form), resistance to biochemical attack, and light weight [15]. Among all biopolymers, carboxymethyl cellulose (CMC) is the most abundant native polymer on the earth, a renewable resource. CMC can easily form an aqueous hydrogel below 37 °C [16]; its properties brought tremendous interest to researchers for 3D bioprinting. Habib et al. studied the printability of alginate-carboxymethyl cellulose hybrid hydrogel and 3D printed scaffolds to investigate the viability of pancreatic cancer cell culturing [17]. Janarthanan et al. investigated the printability of hyaluronic acid–carboxymethyl cellulose hybrid hydrogel for soft tissue and organ regeneration with three different compositions by 50-layer height in complex structures without support materials or any post-processing. In their study, texture profile analysis, morphological analysis, in vitro cytotoxicity, and mice studies were conducted to indicate that the self-crosslinking capacity, composition, and other factors will have huge impacts on the generation of multilayered constructions for soft tissue engineering [16]. Kageyama et al. validated the feasibility of fabricating perfusable vasculatures using in situ cross-linkable gelatin–CMC hybrid hydrogels with cell embedding [18]. Calcagnile et al. composited polydimethylsiloxane (PDMS) and CMC to improve its hybrid hydrogel tactile properties and simulated the slimy effect of organic in the human body [19]. Pasqui et al. researched bone tissue scaffolds using hydroxyapatite–CMC hybrid hydrogel [10]. Ahlfeld et al. investigated the printing fidelity of alginate–CMC hybrid hydrogel as well with two different concentrations. Moreover, cell viability was tested with immortalized human Mesenchymal Stromal Cells (hMSCs) over 21 days of in vitro culturing [20]. Maver et al. used an polyethylene oxide (PEO) –CMC hybrid hydrogel with a spin-assisted layer-by-layer coating process to develop the wound dressing materials [11]. With different composites, carboxymethyl cellulose hybrid hydrogels could present specific features to meet bioprinting's requirements.

Most researchers studied a 3D printing CMC hydrogel composite (such as composite with alginate, hyaluronic acid, gelatin, PDMS, and hydroxyapatite) in tissue engineering and wound dressing area [10–12]. Only a few researchers did investigate the release profile for their wound dressing materials, such as Ahlfeld et al. who studied the incorporation of synthetic nano-silicate clays with alginate–CMC hybrid hydrogel, which further increased the ability of the alginate–CMC samples to release loaded drugs in a more sustained manner [20]. Maver et al. tested PEO–CMC hybrid hydrogel 3D printed scaffold with a pain reliever, prolonging its efficacy for two days upon changing the dressing [11]. Few studies have investigated the food colorant acting as a dye for the drug delivery system during the dissolution test. Erythrosine (red-3) and Allura Red AC (red-40) were commonly used around the world. Red-3 can be used in colored food and ingested drugs; however, red-3 cannot be used in cosmetics and external drugs [21]. Thus, a red food degradable colorant (consists of red-3 and red-40) was chosen in this study for the in-fill material not only because of easily eye-observing during the printing and dissolution test but also because red-3 and red-40 can be used in the food and ingested drugs according to FDA approval as an edible dye.

In the pharmaceutical field, lyophilization has typically been used for anti-infectives, biotechnology-derived products and in-vitro diagnostics. Lyophilization or freeze-drying is a low-temperature dehydration process without passing through a liquid phase that involves freezing the product, lowering the pressure, then removing the ice by sublimation. However, the lyophilization post-processing costs time and resources to achieve long-term room temperature storage compared with non-lyophilization. Hence, this study focused on the effect of lyophilization on the dissolution test. A pre-study was conducted which found that carboxymethyl cellulose aqueous (mixed with DI water) hydrogel could be the potential printing material by adjusting the CMC concentration. This study used degradable food colorant and lyophilization post-processing to verify the release profile with three different CMC aqueous hydrogel concentrations.

2. Materials and Methods

2.1. Materials and Preparation

Carboxymethyl cellulose hydrogel was formulated with CMC 6000 Fine Powder (curtesy of Ticalose, White Marsh, MD, USA). The CMC to total aqueous hydrogel ratio was narrowed down from 8% to 12% (*w*/*w*) based on our preliminary experiment with a broader concentration range of 4–16% (*w*/*w*). CMC hydrogel below 8% showed insufficient mechanical integrity withholding the printing dimension, while above 12% demonstrated difficulties during the extrusion when it printed with a tip nozzle of 0.672 mm in diameter. The aqueous CMC hydrogels with concentrations of 8%, 10%, and 12% (*w*/*w*) were prepared as follows: 4, 5, and 6 g of CMC powder were mixed and homogenized with 44, 45, and 46 g of deionized water to formulate of 8%, 10%, and 12% (*w*/*w* in total solids) aqueous hydrogels at room temperature (24 °C), respectively; then, all samples were centrifuged at 3000 rpm for 10 min for air bubble removal. Food degradable colorant was purchased from Wilton (Naperville, IL, USA). The red colorants combined with red-3 (Erythrosine, $C_{20}H_6I_4Na_2O_5$) and red-40 (Allura Red AC, $C_{18}H_{14}N_2Na_2I_8S_2$) was chosen as the in-fill material. Due to the pressure of inert thickener glycerol, in the red colorant formulation, the stock colorant was diluted ten times with DI water. All other chemicals were purchased from Fisher Scientific (Hampton, NH, USA) and used as they were received, unless specified otherwise.

2.2. Rheological Properties

The rheological tests were carried out using a Discovery HR-2 Rheometer (TA Instruments, New Castle, DE, USA) with parallel plate geometry at a gap size of 1 mm at 23 °C. The flow ramp test was conducted to determine the apparent viscosity under the increased shear rate (0.1 to 50 1/s). The oscillatory frequency sweep test was performed to characterize the materials' dynamic modulus with 0.1–600 rad/s angular frequency increases. The rheological properties, i.e., shear rate, apparent viscosity, angular frequency, storage modulus (G'), loss modulus (G''), and loss tangent ($\tan\delta = G''/G'$) were recorded. All rheological properties tests were conducted in triplicate.

2.3. 3D Printing

Based on an FDM printer, K8200 (Velleman, Inc., Gavere, Belgium), a syringe holder add-on kit K8205 (Velleman, Inc., Gavere, Belgium) was used to allow the FDM printer to print semi-solids. For quality control, the capsule model was designed with ten layers in total, meaning the layer height was equal to the nozzle size. The first and last three layers were the enclosure. A circular void section with a diameter of 6.72 mm was printed from layer 4 to layer 7 to form a cylindrical hollow section for filling the colorant. Diluted red food degradable colorant was added to the capsule's void section during the printing process. To be more specific, 20 μL of 10-fold diluted colorant was injected into the void section after the printer finished the 7th layer of the CMC aqueous hydrogel (See Figure 1a) and before the 8th layer enclosed the void section (See Figure 1b). The Cura-Engine was used for the 3D model slicing and G-code generation. 3D printing parameters were set as follows: the printing speed was 2.4 mm/s for the outer wall and 4 mm/s for the in-fill of 4% with the extrusion multiplier of 0.125.

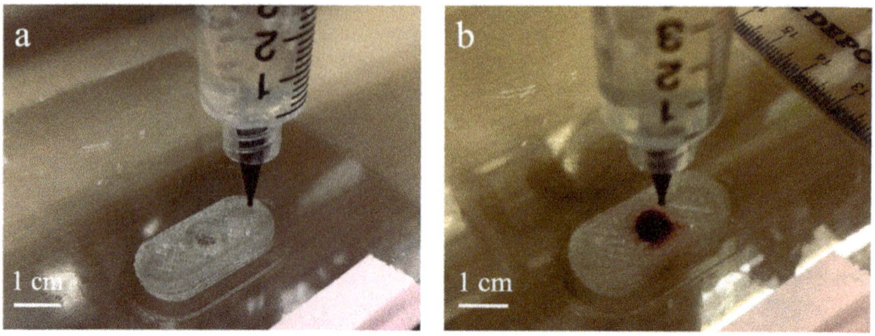

Figure 1. Printing process: (**a**) 7th layer; (**b**) colorant injected after the 7th layer.

2.4. Lyophilization

Lyophilization (also known as freeze-drying) post-processing was performed by freeze-dryer (Virtis Genesis SQ freeze-dryer, Gardiner, NY, USA) for 12 h at −45 °C in the condensation chamber under a low vacuum (below 2.6×10^4 Pa). After 3D printing, the lyophilized samples were placed in the freezer at −20 °C with at least 4 h to wait for the samples to be fully frosted before lyophilization. The non-lyophilized samples were kept in the fridge at −20 °C until the lyophilized samples were taken out from the freeze-dryer. All samples were kept at room temperature (24 °C) 30 min before the dissolution test for a stable temperature.

2.5. Dissolution Test

The dissolution test was conducted following the USP apparatus II (Paddle) procedure (37 °C, 100 rpm) using a DT 126 light dissolution tester (Erweka, Hessen, Germany) in a time as mannerly as the 1, 2, 3, 4, 6, 8, 12, 24 h time points. For the first two hours, the samples were dissolved in the pH = 1 environment with 500 mL of 0.1 M HCl and switched to a pH = 7 environment with 500 mL phosphate buffer (K_2HPO_4 and KH_2PO_4). At each time point, four 100 µL samples were acquired and stored in a 96-well plate with a lid in the dark 4 °C fridge waiting for plate reading. The amount withdrawn was not replaced by a fresh medium. This experiment was performed in triplicate, and the mean value was taken. Blank and standard curves of the colorant were also stored on the same plate for each set.

2.6. Data Analysis

In this study, the zero-order equation, first-order equation, and Higuchi equation were selected to verify the release kinetic as these release models were used to determine the release kinetic, which is prominent nowadays [22].

Zero-order equation:

$$Q_t = Q_0 + K \cdot t, \quad (1)$$

based on the linear fit [22–24], here was simplified with:

$$Y = A \cdot X + B, \quad (2)$$

where Y is the cumulative release level, X is time in hour. First-order equation:

$$\frac{dC}{dt} = -K \cdot c, \quad (3)$$

$$\log(C) = \log(C_0) - \frac{K \cdot t}{2.303} \quad (4)$$

based on logarithm fit [22–24]. Here was simplified it with:

$$Y = A - e^{(-B \cdot X)}, \quad (5)$$

Higuchi equation is the most prominent model used in nowadays release profile. It was defined as

$$f_t = Q = A \cdot \sqrt{D \cdot (2 \cdot C - C_s) \cdot C_s \cdot t}, \quad (6)$$

and which can be simplified in most cases as [22–24],

$$Y = A \cdot X^{0.5}. \quad (7)$$

Nelder–Mead Simplex algorithm was chosen for use in SciDAVis for fitting the release model, and Origin 2019 (Northampton, MA, USA) was used for analyzing the cumulative release level. Standard curves for red-3 and red-40 were statically evaluated by linear fit with a concentration weight factor. All experiments were performed in triplicate.

3. Results

3.1. Rheological Properties

As shown in Figure 2, CMC hydrogels' apparent viscosity curves were plotted at three different concentrations. Regardless of the CMC concentration, the hydrogels demonstrated shear-thinning behavior, i.e., as the shear rate increased, the apparent viscosity decreased. The shear-thinning behavior of the CMC hydrogel is beneficial for semi-solid extrusion-based 3D printing. The shear rate-dependent viscosity could make material flowable during extrusion and hold the printed shape once the stress is removed [25]. In addition, the CMC hydrogels' apparent viscosity increased as the concentration increased, which could be explained by the increased content of CMC provide more hydroxyl groups and results in more hydrogen bonding [26]. Compared with low concentration, a high CMC hydrogel concentration was associated with stronger mechanical integrity and shape retention ability due to the relatively high apparent viscosity. However, the low viscosity would make the low-concentration hydrogel easier to be extrude through the nozzle [17,26].

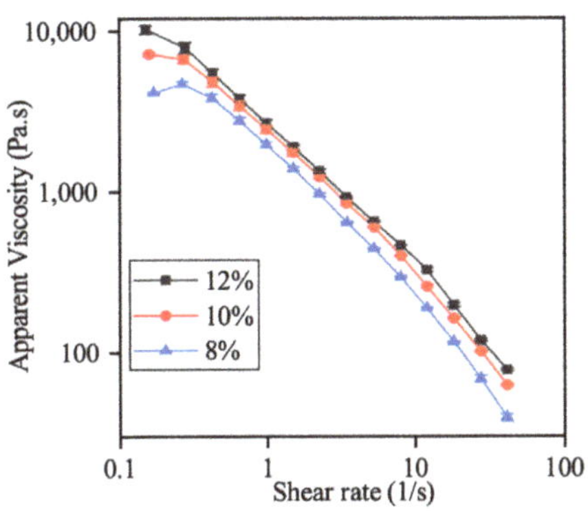

Figure 2. Apparent viscosity versus shear rate profiles of carboxymethyl cellulose (CMC) at different concentrations (8%, 10%, 12% w/w). Values are presented as the mean ± SD (n = 3).

Figure 3 shows the storage modulus (G′), elastic modulus (G″) and phase angle (tan δ) as a function of angular frequency for the oscillatory frequency sweep test. The elastic modulus G′ and G″ represent the ability of energy storage or dissipation via the polymer, respectively [27]. Tan δ is the ratio between G″ and G′, which are less than one no matter the concentration of CMC hydrogel. This indicated that the G′ is always higher than G″ during the frequency sweep test. Since the value of tan δ is less than one and G′ is higher than G″, the CMC hydrogel showed viscoelastic solid-like characteristics rather than liquid-like properties [27]. The highest concentration of CMC hydrogel, 12% *w*/*w*, showed the highest G′ and lowest tan δ at the same angular frequency when compared with the lower concentration. The higher G′ indicated that the material has better shape retention ability after extrusion, and the lower tan δ revealed more solid-like behavior and poor fluidity [28,29].

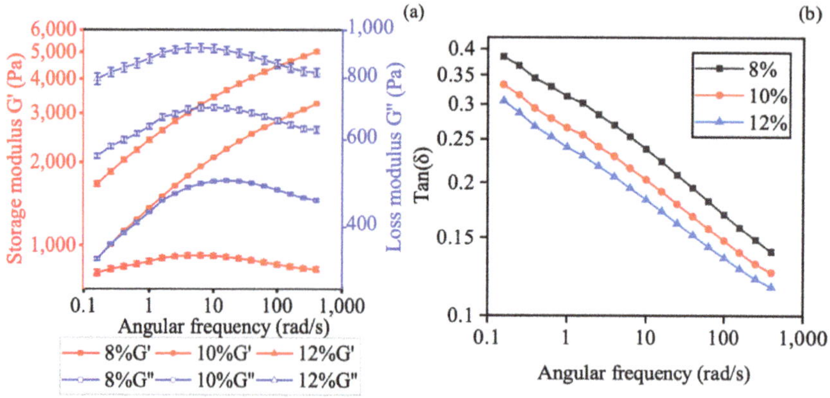

Figure 3. Oscillatory frequency sweep profile of CMC at different concentrations (8%, 10%, 12% *w/w*): (**a**) storage modulus (G′) and loss modulus (G″); and (**b**) tan δ. Values are presented as the mean ± SD (n = 3).

3.2. 3D Printing and Lyophilization of CMC Aqueous Hydrogel

In this study, CMC aqueous hydrogel's printability was defined into two categories: (1) hydrogel ink should be continuously and uniformly extruded out through a nozzle tip with the desired diameter; (2) the hydrogel ink should possess sufficient mechanical strength to support its own weight and minimize deformation after printing. However, high mechanical strength usually requires high viscosity and yield stress; high viscosity materials usually block the nozzle tip when extruding, thus resulting in a printing failure [2]. Our preliminary study found that the high-concentration (14%, 16%) CMC hydrogels resulted in blocking the 0.672 mm nozzle tip. The low concentration (4%, 6%) CMC hydrogels resulted in deformation during printing because of a lack of mechanical strength given the printing dimensions. Furthermore, liquid-like samples (printed with 4% and 6% CMC hydrogel) could not be easily peeled off from the glass substrate due to low viscosity, which increased the manufacturing failure rate.

Based on the preliminary study result, 8%, 10%, and 12% CMC hydrogels with printed samples (lyophilized and non-lyophilized) were chosen to be evaluated by the dissolution tests. Moreover, 8%, 10%, and 12% samples (Figure 4a–c, respectively) showed excellent retention after SSE printing. Moreover, they could be easily peeled off from the glass substrate without using any tools. Among the three concentrations, 8% CMC aqueous hydrogel presented the highest transparency. The higher the concentration is, the hydrogel became opaquer due to the higher polymer content. CMC shows amphiphilic characteristics as containing a hydrophobic polysaccharide backbone and many hydrophilic carboxyl groups [30]. Hence, as red-3 and red-40 are hydrophilic, the red colorants started diffusing when added to the capsule's void section. The red colorants showed different diffusion degrees from Figures 4 and 5 due to the colorant being hydrophilic and the different degrees of water content in each capsule; it started diffusing inside the capsule when the colorant was deposited. When the concentration increased, the diffusion level became less (as can be seen from both non-lyophilized samples and lyophilized samples).

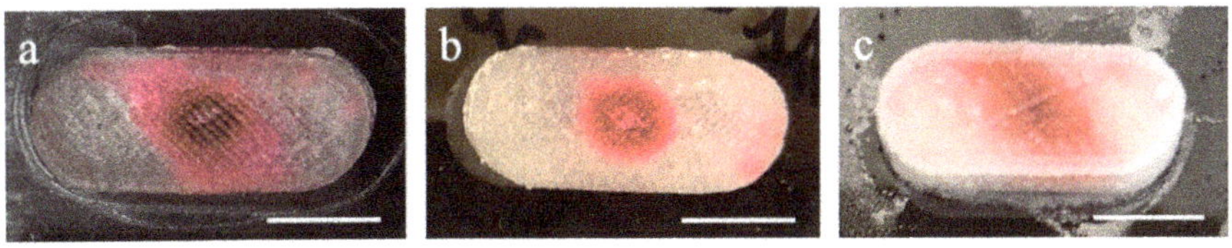

Figure 4. Capsule printed using: (**a**) 8%; (**b**) 10%; and (**c**) 12% CMC hydrogels (non-lyophilized, scale bar in 1 cm).

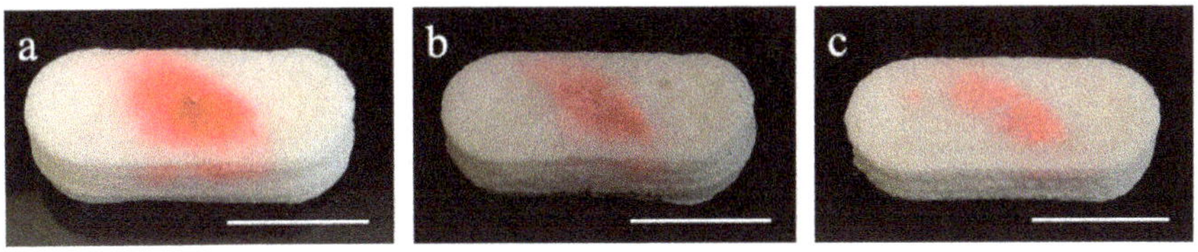

Figure 5. Capsule printed using: (**a**) CMC 8%; (**b**) CMC 10%; and (**c**) CMC 12% (lyophilized, scale bar in 1 cm).

Prior to the lyophilization post-process, the samples were frozen in a −20 °C freezer for 24 h. Subsequently, the samples were lyophilized at −40 °C for 24 h for dehydration. Compared to the non-lyophilized samples, the lyophilized samples consist of different amounts of CMC and the same amount of colorant (red-3 and red-40) in the capsule without moisture. Compared with samples without lyophilization, lyophilized samples require additional water to swell when placed into the solution. As shown in Figure 5, SSE-printed CMC hydrogel layers did not split or separate by hand, which means the attachment between layers by layers is stickiness or adhesiveness to make the capsules still retain their shape after lyophilization post-processing. Moreover, the lyophilization post-processing retains the capsule's shape without significant dimension reduction (less than 5%). The red colorant showed a parallelogram pattern on the top surface because the colorant solution showed the capillarity following the tool path. The trivial gap between the filaments formed a tube-like structure provides the capillarity. The adhesion from the tube (trivial gap here) to the water (colorant solution here) let the solution keep going until the surface tension cannot hold its weight. In addition, the printing speed and feed rate did not match with each other—the feed rate is faster than the printing speed as a solid filament was wanted so the gap between the filaments can be kept under minimum air bubble—so the extruded filament was whirled in the void section.

4. Discussion

Release kinetics is an essential parameter of a drug delivery system as it indicates its efficacy and stability. An early burst release is sometimes preferable as giving the maximum relief instantly, followed by a sustained release to prevent repeated administration. Such as practical anti-cancer treatment demands the sustained release of the drugs [31]. Release mechanisms include a dissolution-controlled system (encapsulation), diffusion-controlled system (reservoir), dissolution and diffusion mixed controlled system (matrix), water penetration-controlled system (swelling, osmotically), chemically controlled system (erodible, pendent), and hydrogel system (diffusion-controlled systems, swelling-controlled system, chemically controlled systems, and environmentally responsive systems) [32]. The different release models represented different release mechanisms. A release model prediction will help researchers find the right track to investigate or modify their samples to a desired controlled release system. The release kinetics models include zero-order release model, which refers to the process of constant drug delivery or matrix tablets with low soluble drugs; first-order release model, which refers to the process of no change in the shape of the solid during the dissolution test or porous matrices with water-soluble drugs; Hixson–Crowell release model, which refers to the process of a change in surface area and diameter or proportionally diminishing of tablets; Higuchi release model, which refers to the process to the simple laws of diffusion or the combination of the zero-order release model and the first-order release model; and Korsmeyer–Peppas release model, which refers to the process of a polymeric system [22,24].

In order to locate the absorption peak of the degradable food colorant, Beckman DU520 UV/Vis spectrophotometer (Beckman Coulter, Brea, CA) was used to perform a 1 nm wavelength sweep. The absorption peaks for the red food degradable colorant ten times diluted aqueous solutions were 504 and 530 nm by red-40 and red-3. In order to evaluate the release percentage, Synergy Neo2 microplate reader (BioTek, Winoski, VT, USA) was used to access the acquired sampled at each time point with absorption at 504 and 530 nm. Within each plate, standard samples (triplicate) were prepared for plate reading. Based on Beer–Lambert's law, the standard curves were calculated individually for each plate to ensure the consistency of the sampling condition within the plates.

From Table 1, compared with the zero-order equation and the Higuchi equation, the first-order equation showed the highest R-square value within non-lyophilized samples and lyophilized samples' dissolution behavior, indicating that the dissolution process was dominated by a process with no change in the solid or porous matrices with soluble drugs. All the samples showed the lowest R square value in the Higuchi equation and excluded the Higuchi release profile in this study. However, most of the R square value was under 0.9. The highest one in the non-lyophilized samples and overall samples was the non-lyophilized 8%, which showed 0.96 under the first-order equation in 504 nm (red-40) wavelength reading (Figure 6a). The average R square value between the reading of red-40 and red-3 of the non-lyophilized also indicated that 8% of the CMC aqueous hydrogel was capable of a hydrophilic material release profile. The lyophilized samples' highest R square value was the lyophilized at 12%, which showed that it was at a rate of 0.92 under the first-order equation in 530 nm (red-3) wavelength reading (Figure 6b). The average R square value between the reading of red-40 and red-3 of the lyophilized also indicated 12% CMC aqueous hydrogel was capable for the hydrophilic material release profile. As expected, the results were with a lower CMC concentration (8%) with a non-lyophilization feature that made the saturated capsule more solid-like, which made no-change during the dissolution test. With a higher CMC concentration (12%), the lyophilization post-processing could cause the capsule to be more solid-like as the first-order release model refers to the process of no change in the solid shape during the dissolution test. Above all, the results showed that the 8% CMC aqueous hydrogel without lyophilization could achieve the same function as the lyophilized 12% CMC aqueous hydrogel; both can be used as a capsule for drug delivery use under the first-order release model.

Table 1. R-square value results of release model fitting.

CMC %	Wavelength	Zero-Order Model		First-Order Model		Higuchi Model	
		Non-Lyophilization	Lyophilization	Non-Lyophilization	Lyophilization	Non-Lyophilization	Lyophilization
8	504	0.39034905	0.48813441	0.9592987	0.85242592	0.36952187	0.5348747
10	504	0.58273114	0.49129531	0.86269848	0.81052361	0.45650565	0.34067544
12	504	0.5272996	0.41533232	0.88521804	0.89801676	0.24596937	−0.0187975
8	530	0.21204131	0.50842099	0.75179267	0.84067689	−0.5735051	0.46366171
10	530	0.51079008	0.23943026	0.77361972	0.75222391	0.38693773	−1.3676172
12	530	0.35421321	0.46153722	0.83338572	0.92251983	−1.4281945	0.39575727

Compared to the non-lyophilized samples, due to the loss of water, and lyophilized samples swelled water in the solution in the beginning, which is why the results showed a more delayed release compared to the non-lyophilized samples. Results showed that within a pH = 1 environment, the colorant release was suppressed due to the integrity preservation under the acidic environment. Upon switching to a pH = 7 environment, the cumulative release can reach a maximum plateau 2 h after environmental changes, which fitted the first-order release model as expected. As shown in Figure 7 the conclusion can be made that with lyophilization post-processing which suppresses the dissolution impact caused by the differentiation of polymer content, the colorants release level which reached 100% at the 24 h time point regardless of the CMC percentage when formulating the hydrogel. However, the non-lyophilized samples showed that a different percentage of CMC hydrogels presented different abilities releasing the colorants in 24 h (Figure 8).

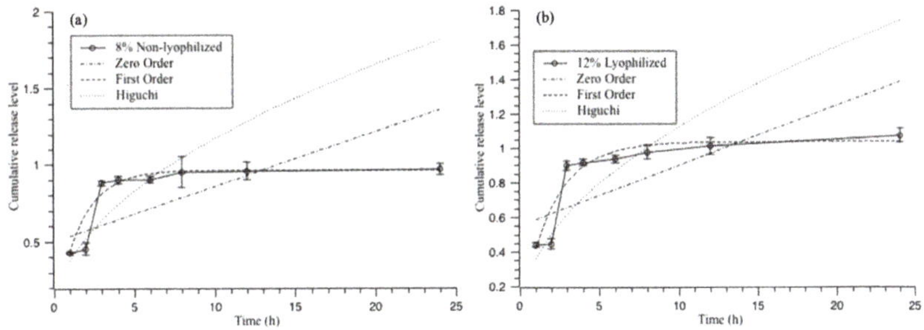

Figure 6. Release model fitting for a standard dissolution test (pH switching at the second hour): (**a**) 8% non-lyophilized CMC hydrogel under 504 nm; and (**b**) 12% lyophilized CMC hydrogel under 530 nm.

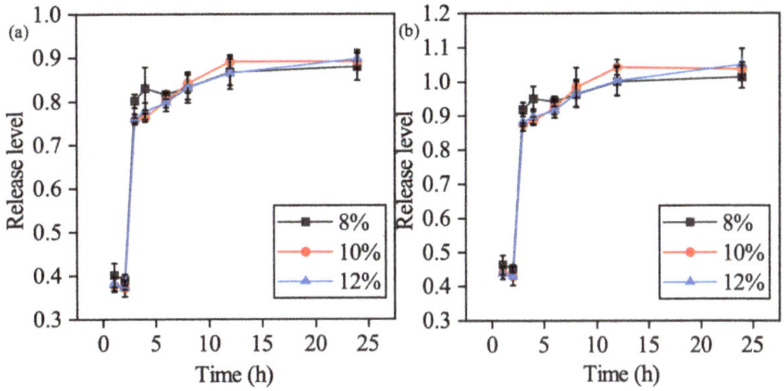

Figure 7. Cumulative release level for the standard dissolution test (pH switching at the second hour) of the lyophilized sample: (**a**) 504 nm and (**b**) 530 nm.

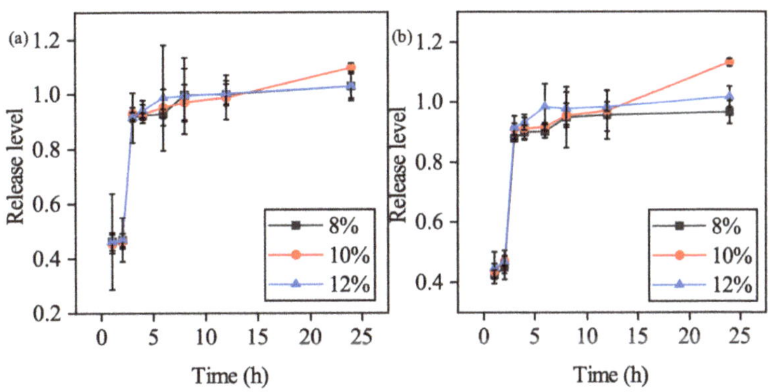

Figure 8. Cumulative release level for the standard dissolution test (pH switching at the second hour) of the non-lyophilized sample: (**a**) 504 nm, and (**b**) 530 nm.

5. Conclusions

This study demonstrated the preparation of CMC aqueous hydrogel with lyophilization post-processing. The samples were successfully 3D printed without deformation. The cumulative release profile showed that an 8% non-lyophilized CMC aqueous hydrogel had the same capability as a 12% lyophilized CMC aqueous hydrogel to achieve a no shape change of the solid during the dissolution test with the food colorant. In addition, the lyophilized CMC aqueous hydrogel capsules showed full dissolvability in 24 h. However, non-lyophilized CMC aqueous hydrogel capsules showed different dissolvability under 24 h, compared with lyophilized CMC aqueous hydrogel capsules. The standard dissolution test showed that the CMC aqueous hydrogel had potential application in the pharmaceutical field as a targeted drug delivery system for intestinal delivery. This study could pave the path for the future work of SSE cost-effectively printing drug delivery systems.

6. Patents

"Cellulose Derivative Based Biodegradable Support Structures for 3D Printing", Approved at Iowa State University Office of Intellectual Property and Technology Transfer Office (ISURF 04987), provisional application filed and approved for US Patent.

Author Contributions: Conceptualization, X.J., X.S., and H.Q.; methodology, X.J., X.S., Y.H., and H.Q.; software, X.J.; validation, X.J., and Y.H.; formal analysis, X.J., and Y.H.; investigation, X.J., and Y.H.; resources, X.J.; data curation, X.J.; writing—original draft preparation, X.J.; writing—review and editing, Y.H., Y.C., X.S., Z.Z., and H.Q.; visualization, X.J.; supervision, X.S., and H.Q.; project administration, Z.Z.; funding acquisition, H.Q. All authors have read and agreed to the published version of the manuscript.

Funding: This research was funded by Regents Innovation Fund by Iowa Economic Development to Qin's group.

Institutional Review Board Statement: All procedures performed in studies, where applicable, were in accordance with the ethical standards of Iowa State University and/or national research committee and with the 1964 Helsinki declaration and its later amendments or comparable ethical standards.

Informed Consent Statement: Informed consent, where applicable, was obtained from all individual participants included in the study, and to the submission of the research to the journal.

Data Availability Statement: The authors confirm that the data supporting the findings of this study are available within the article.

Acknowledgments: This project is supported by the Regents Innovation Fund by Iowa Economic Development to Qin's group and the Undergraduate Research Assistant Program from the Department of Industrial and Manufacturing Systems Engineering (IMSE_URA) at Iowa State University.

Conflicts of Interest: Cyber Metrology and Manufacturing Solutions LLC. is one of the award recipients of Regents Innovation Fund by Iowa Economic Development. Qin is the lead PI at Iowa State University on the project, and Professor Shi is the co-founder of the company. The study in the research is conducted at Iowa State University and has no oversight relationship with the company-sponsored research project. There is no other potential conflict of interest.

References

1. Holzmann, P.; Breitenecker, R.J.; Soomro, A.A.; Schwarz, E.J. User entrepreneur business models in 3D printing. *J. Manuf. Technol. Manag.* **2017**, *28*, 75–94. [CrossRef]
2. Kuo, C.C.; Qin, H.; Cheng, Y.; Jiang, X.; Shi, X. An integrated manufacturing strategy to fabricate delivery system using gelatin/alginate hybrid hydrogels: 3D printing and freeze-drying. *Food Hydrocoll.* **2021**, *111*, 106262. [CrossRef]
3. Cheng, Y.; Qin, H.; Acevedo, N.C.; Jiang, X.; Shi, X. 3D printing of extended-release tablets of theophylline using hydroxypropyl methylcellulose (HPMC) hydrogels. *Int. J. Pharm.* **2020**, *591*, 119983. [CrossRef] [PubMed]
4. Cheng, Y.; Shi, X.; Jiang, X.; Wang, X.; Qin, H. Printability of a Cellulose Derivative for Extrusion-Based 3D Printing: The Application on a Biodegradable Support Material. *Front. Mater. Sci.* **2020**, *7*, 86. [CrossRef]

5. Polamaplly, P.; Cheng, Y.L.; Shi, X.L.; Manikandan, K.; Zhang, X.; Kremer, G.E.; Qin, H.T. 3D printing and characterization of hydroxypropyl methylcellulose and methylcellulose for biodegradable support structures. *Polymer* **2019**, *173*, 119–126. [CrossRef]
6. Manikandan, K.; Jiang, X.; Singh, A.A.; Li, B.; Qin, H. Effects of nozzle geometries on 3D printing of clay constructs: Quantifying contour deviation and mechanical properties. *Procedia Manuf.* **2020**, *48*, 678–683. [CrossRef]
7. Vithani, K.; Goyanes, A.; Jannin, V.; Basit, A.W.; Gaisford, S.; Boyd, B.J. An Overview of 3D Printing Technologies for Soft Materials and Potential Opportunities for Lipid-based Drug Delivery Systems. *Pharm. Res.* **2019**, *36*, 4. [CrossRef] [PubMed]
8. Ahangar, P.; Cooke, M.E.; Weber, M.H.; Rosenzweig, D.H. Current biomedical applications of 3D printing and additive manufacturing. *Appl. Sci.* **2019**, *9*, 1713. [CrossRef]
9. Kyle, S.; Jessop, Z.M.; Al-Sabah, A.; Whitaker, I.S. "'Printability" of Candidate Biomaterials for Extrusion Based 3D Printing: State-of-the-Art. *Adv. Healthc. Mater.* **2017**, *6*, 1700264. [CrossRef]
10. Pasqui, D.; Torricelli, P.; de Cagna, M.; Fini, M.; Barbucci, R. Carboxymethyl cellulose-Hydroxyapatite hybrid hydrogel as a composite material for bone tissue engineering applications. *J. Biomed. Mater. Res. A* **2014**, *102*, 1568–1579. [CrossRef]
11. Maver, T.; Kurečič, M.; Smrke, D.M.; Kleinschek, K.S.; Maver, U. Electrospun nanofibrous CMC/PEO as a part of an effective pain-relieving wound dressing J. Sol-Gel Sci. Technol. **2016**, *79*, 475–486. [CrossRef]
12. Wang, M.; Xu, L.; Hu, H.; Zhai, M.; Peng, J.; Nho, Y.; Li, J.; Wei, G. Radiation synthesis of PVP/CMC hydrogels as wound dressing. *Nucl. Instrum. Methods Phys. Res. B* **2007**, *265*, 385–389. [CrossRef]
13. Domanska, A.; Boczkowska, A. Biodegradable polyurethanes from crystalline prepolymers. *Polym. Degrad. Stab.* **2014**, *108*, 175–181. [CrossRef]
14. Bauer, S.; Schmuki, P.; von der Mark, K.; Park, J. Engineering biocompatible implant surfaces: Part I: Materials and surfaces. *Prog. Mater. Sci.* **2013**, *58*, 261–326. [CrossRef]
15. Patel, N.; Gohil, P. A review on biomaterials: Scope, applications & human anatomy significance. *Int. J. Emerg. Technol. Learn.* **2012**, *2*, 91–101.
16. Gopinathan, J.; Noh, I. Recent trends in bioinks for 3D printing. *Biomater. Res.* **2018**, *22*, 11. [CrossRef] [PubMed]
17. Habib, A.; Sathish, V.; Mallik, S.; Khoda, B. 3D printability of alginate-carboxymethyl cellulose hydrogel. *Materials* **2018**, *11*, 454. [CrossRef]
18. Kageyama, T.; Osaki, T.; Enomoto, J.; Myasnikova, D.; Nittami, T.; Hozumi, T.; Ito, T.; Fukuda, J. In Situ Cross-Linkable Gelatin-CMC Hydrogels Designed for Rapid Engineering of Perfusable Vasculatures. *ACS Biomater. Sci. Eng.* **2016**, *2*, 1059–1066. [CrossRef]
19. Calcagnile, P.; Cacciatore, G.; Demitri, C.; Montagna, F.; Corcione, C.E. A feasibility study of processing polydimethylsiloxane-sodium carboxymethylcellulose composites by a low-cost fused deposition modeling 3D printer. *Materials* **2018**, *11*, 1578. [CrossRef]
20. Ahlfeld, T.; Cidonio, G.; Kilian, D.; Duin, S.; Akkineni, A.R.; Dawson, J.I.; Yang, S.; Lode, A.; Oreffo, R.O.C.; Gelinsky, M. Development of a clay based bioink for 3D cell printing for skeletal application. *Biofabrication* **2017**, *9*, 034103. [CrossRef] [PubMed]
21. Sharma, V.; McKone, H.T.; Markow, P.G. A global perspective on the history, use, and identification of synthetic food dyes. *J. Chem. Educ.* **2011**, *88*, 24–28. [CrossRef]
22. Dash, S.; Murthy, P.N.; Nath, L.; Chowdhury, P. Kinetic modeling on drug release from controlled drug delivery systems. *Acta Pol. Pharm.* **2010**, *67*, 217–223.
23. Costa, P.; Sousa Lobo, J.M. Modeling and comparison of dissolution profiles. *Eur. J. Pharm. Sci.* **2001**, *13*, 123–133. [CrossRef]
24. Singhvi, G.; Singh, M. In-vitro drug release characterization models. *Int. J. Pharm. Stud. Res.* **2011**, *2*, 77–84.
25. Liu, J.; Sun, L.; Xu, W.; Wang, Q.; Yu, S.; Sun, J. Current advances and future perspectives of 3D printing natural-derived biopolymers. *Carbohydr. Polym.* **2019**, *207*, 297–316. [CrossRef]
26. Mulakkal, M.C.; Trask, R.S.; Ting, V.P.; Seddon, A.M. Responsive cellulose-hydrogel composite ink for 4D printing. *Mater. Des.* **2018**, *160*, 108–118. [CrossRef]
27. Gunasekaran, S.; Ak, M.M. Dynamic oscillatory shear testing of foods-Selected applications. *Trends Food Sci. Technol.* **2000**, *11*, 115–127. [CrossRef]
28. Wilson, S.A.; Cross, L.M.; Peak, C.W.; Gaharwar, A.K. Shear-Thinning and Thermo-Reversible Nanoengineered Inks for 3D Bioprinting. *ACS Appl. Mater. Interfaces* **2017**, *9*, 43449–43458. [CrossRef] [PubMed]
29. Yang, F.; Zhang, M.; Bhandari, B.; Liu, Y. Investigation on lemon juice gel as food material for 3D printing and optimization of printing parameters. *LWT* **2018**, *87*, 67–76. [CrossRef]
30. Hu, S.; Lu, Q.; Xu, Y. Biosensors based on direct electron transfer of protein. In *Electrochemical Sensors, Biosensors and Their Biomedical Applications*; Elsevier Inc.: Amsterdam, The Netherlands, 2008; pp. 531–581.
31. Allen, T.M.; Cullis, P.R. Drug Delivery Systems: Entering the Mainstream. *Science* **2004**, *303*, 1818–1822. [CrossRef]
32. Siepmann, J.; Peppas, N.A. Modeling of drug release from delivery systems based on hydroxypropyl methylcellulose (HPMC). *Adv. Drug Deliv. Rev.* **2012**, *64*, 163–174. [CrossRef]

polymers

MDPI

Article

X-ray Photoelectron Spectroscopy Analysis of Chitosan–Graphene Oxide-Based Composite Thin Films for Potential Optical Sensing Applications

Wan Mohd Ebtisyam Mustaqim Mohd Daniyal [1], Yap Wing Fen [1,2,*], Silvan Saleviter [1], Narong Chanlek [3], Hideki Nakajima [3], Jaafar Abdullah [4] and Nor Azah Yusof [4]

1 Institute of Advanced Technology, Universiti Putra Malaysia, UPM Serdang, Selangor 43400, Malaysia; wanmdsyam@gmail.com (W.M.E.M.M.D.); silvansaleviter94@gmail.com (S.S.)
2 Department of Physics, Faculty of Science, Universiti Putra Malaysia, UPM Serdang, Selangor 43400, Malaysia
3 Synchrotron Light Research Institute, Maung, Nakhon Ratchasima 30000, Thailand; n.chanlek@gmail.com (N.C.); hideki@slri.or.th (H.N.)
4 Department of Chemistry, Faculty of Science, Universiti Putra Malaysia, UPM Serdang, Selangor 43400, Malaysia; jafar@upm.edu.my (J.A.); azahy@upm.edu.my (N.A.Y.)
* Correspondence: yapwingfen@gmail.com or yapwingfen@upm.edu.my

Abstract: In this study, X-ray photoelectron spectroscopy (XPS) was used to study chitosan–graphene oxide (chitosan–GO) incorporated with 4-(2-pyridylazo)resorcinol (PAR) and cadmium sulfide quantum dot (CdS QD) composite thin films for the potential optical sensing of cobalt ions (Co^{2+}). From the XPS results, it was confirmed that carbon, oxygen, and nitrogen elements existed on the PAR–chitosan–GO thin film, while for CdS QD–chitosan–GO, the existence of carbon, oxygen, cadmium, nitrogen, and sulfur were confirmed. Further deconvolution of each element using the Gaussian–Lorentzian curve fitting program revealed the sub-peak component of each element and hence the corresponding functional group was identified. Next, investigation using surface plasmon resonance (SPR) optical sensor proved that both chitosan–GO-based thin films were able to detect Co^{2+} as low as 0.01 ppm for both composite thin films, while the PAR had the higher binding affinity. The interaction of the Co^{2+} with the thin films was characterized again using XPS to confirm the functional group involved during the reaction. The XPS results proved that primary amino in the PAR–chitosan–GO thin film contributed more important role for the reaction with Co^{2+}, as in agreement with the SPR results.

Keywords: X-ray photoelectron spectroscopy; surface plasmon resonance; thin film; quantum dot; 4-(2-pyridylazo)resorcinol; chitosan; graphene oxide

Citation: Daniyal, W.M.E.M.M.; Fen, Y.W.; Saleviter, S.; Chanlek, N.; Nakajima, H.; Abdullah, J.; Yusof, N.A. X-ray Photoelectron Spectroscopy Analysis of Chitosan–Graphene Oxide-Based Composite Thin Films for Potential Optical Sensing Applications. *Polymers* **2021**, *13*, 478. https://doi.org/10.3390/polym13030478

Academic Editor: Arn Mignon
Received: 12 January 2021
Accepted: 26 January 2021
Published: 2 February 2021

1. Introduction

Chitosan is a semicrystalline polymer material that is derived from chitin by deacetylation under alkaline condition. Chitosan is said to be the most important chitin derivative, much easier to process, and has a good mechanical and optical properties. Owing to its excellent advantages that include non-toxic, biodegradable, and biocompatible properties, as well as possessing excellent film-forming ability, chitosan has attracted many researchers to be used in various research field. For instance, chitosan has been applied in the biomedical applications including tissue engineering and drug delivery [1–3]. However, the stability of chitosan is low because of their hydrophilic character and also pH sensitivity [4]. Therefore, a number of techniques have been used to improve the mechanical and chemical properties of chitosan. One of the technique is by crosslinking with some reagent such as ionophore, glutaraldehyde or formaldehyde [5]. Other than that, chitosan can be reinforced by blending it with other novel material such as graphene-based substances. Graphene is the most interesting allotrope of carbon, given by its one-atom-thick layer of sp^2-bonded carbons atom arranged into a 2D honeycomb lattice. Due to its unique 2D crystal structure, graphene can be controlled to have unlimited dimension, such as carbon materials with 0, 1, 2 and 3 dimensions [6]. In addition, there are lots of studies that have proven graphene to have remarkable strength, excellent electrical and thermal conductivity, large surface area, and biocompatibility [7,8]. Graphene can be chemically derived into graphene oxide (GO). These graphene derivatives can be easily obtained from inexpensive graphite, therefore are cost effective and also highly hydrophilic (thus, are stable in aqueous solution) which makes it easy to facilitate for the assembly of macroscopic structure [9]. Moreover, GO is known to have potential binding sites that drove to its reputation for the past decades [10–21]. Owing to the metal ion adsorption properties of both chitosan and GO, chitosan–GO based materials have received continuous attention from the scientific community. A significant amount of research has been conducted for the development of this material to match with the intended application [22–29]. In this research, 4-(2-pyridilazo)resorcinol (PAR) and cadmium sulfide quantum dots (CdS QDs) were incorporated with chitosan–graphene oxide composite material for the potential detection of cobalt ions (Co^{2+}).

The compound 4-(2-pyridylazo)resorcinol is a well-known chromogenic reagent, a type of reagent that gives or changes color in a reaction [30]. This compound is a popular reagent applied in spectrophotometric, chelatometric, and colorimetric analysis due to its ability to correlate with many different metals [31]. The versatility of this compound may be contributed by its several reactive sites, such as a pyridyl nitrogen atom, azo group, and o-hydroxyl group [32]. Pyridyl is a group derived from pyridine (C_5H_5N) by removal of a hydrogen atom from a ring carbon atom. The removal of the hydrogen atom opens room for a bonding with the azo group (–N=N–) at the second carbon. Resorcinol ($C_6H_6O_2$), on the other hand, is an organic compound synthesized from the destructive distillation of a natural resin. In its compound form, it appears as a white crystalline compound with a weak odor and a bittersweet taste. The reaction of resorcinol with 2-pyridylazo led to the synthesis of PAR for the first time in 1959 [33]. Since then, PAR has been widely used as a chromogenic reagent for the detection of mainly metal ions. Besides PAR, other materials such as quantum dots (QDs) are also reported to have excellent properties for potential metal ion sensing [34–37]. QDs are made up from atoms of groups II–VI, III–V, or IV–VI elements in the periodic table. In the past few decades, QDs have attracted considerable attention due to their special properties, and CdS QDs are one of the most studied QDs which are composed of semiconductors of atoms from groups II–VI [38]. There is much research information which can be searched and obtained from various sources regarding the preparation, properties, and applications of QDs [39–44]. Cadmium sulfide-based QDs (CdS QDs) are promising materials for optics, optoelectronics, medicine, and sensor technology development [45–48]. Owing to the excellent properties of these materials, both PAR and CdS QDs have been explored for their incorporation with chitosan–graphene oxide (chitosan–GO) for a potential sensing layer of cobalt ions (Co^{2+}) using surface plasmon resonance (SPR) optical sensor.

Cobalt (Co^{2+}) is a metal ion which is essential to human biological systems. It is one of the important components of vitamin B_{12}, and is also needed as a coenzyme for cell mitosis [49]. It is present in amounts of around 1–2 mg in the human body, and can be found in the kidney, liver, heart, pancreas, and brain [50,51]. Excess intake of Co^{2+} can lead to various health effects such as dizziness, headaches, increased heart rate, asthma, and fibrosis in the lungs [52]. Therefore, it is crucial to detect Co^{2+} for continuous monitoring of this metal ion at trace amounts. Several sensors have been developed for sensing Co^{2+}, which include X-ray fluorescence spectrometry (XRF), anodic stripping voltammetry (ASV), and atomic absorption spectroscopy (AAS). However, all these sensors have limitations, such as long measuring periods, expensive, and complicated sample treatments. This can be overcome using SPR sensors as an alternative, owing to its advantages such as fast measurements, simple sample preparation, cost-effectiveness, and no reference solution is necessary.

SPR is a well-known optical sensor, which works by measuring the changes of refractive index near the thin metal film surface [53]. Any changes of the optical properties on the thin film surface will change the resonance angle. Hence, SPR can be used to detect any binding interaction, and as a sensor [54]. Interaction between molecules, such as metal ions with active layer thin films for sensing applications, have been studied extensively in the past few decades [55–91]. However, most of the studies only focused on the metal ion adsorption uptake performance [92]. There are limited studies on the chemical interaction and adsorption mechanism between the thin film active layer and metal ion, that become one of the main motivations in this research. X-ray photoelectron spectroscopy (XPS) is a sensitive surface analysis technique employed to explore the chemical composition on the surface of a sample. To the best of our knowledge, there was still a lack of studies on the chemical analysis of the PAR–chitosan–GO and CdS QD–chitosan–GO thin film using XPS. Therefore, in this research, XPS was used to study the surface chemistry and the interaction between PAR–chitosan–GO and CdS QD–chitosan–GO thin film with Co^{2+} as evidence for SPR sensing.

2. Materials and Methods

2.1. Prepartion of Chemicals

Medium molecular weight (MMW) chitosan with an MW of 190,000–310,000, degree of deacetylation of 75–85%, acetic acid (assay $\geq$ 99.7%), and 4-(2-pyridylazo)resorcinol were purchased from Sigma Aldrich (St. Louis, MO, USA); cadmium chloride decahydrate ($CdCl_2 \cdot 10H_2O$), mercaptoacetic acid (MPA) (HS–CH_2–COOH) and disodium sulfide nonahydrate ($Na_2S{\cdot}9H_2O$) were purchased from R&M chemicals. The graphene oxide (GO) (4 mg/mL) was purchased from Graphenea (Cambridge, MA, USA).

PAR (4-(2-pyridylazo)resorcinol) with a concentration of 1.5 mg/mL was prepared by diluting 0.15 g of powdered PAR with 100 mL of deionized water. To prepare the PAR–chitosan–GO composite solution, 5 mL of 1.5 mg/mL PAR was added into the mixture of chitosan–GO solution. Then, in room temperature, the mixture was stirred constantly for 24 h or until the solution mixed completely.

On the other hand, CdS QDs were synthesized by a wet method. The process began by dissolving 0.5 mmol of MPA and 0.5 mmol of $CdCl_2 \cdot 10H_2O$ by adding 250 mL of sterile ultra-pure water (ddH_2O) in a 500 mL beaker. Then, the pH of the solution was adjusted to 6.0 by adding NaOH solution (1M) dropwise with constant stirring. Subsequently, the solution was purged with nitrogen gas for at least 60 min under vigorous stirring to remove excess oxygen in the solution. Sodium sulfide ($Na_2S{\cdot}9H_2O$) (0.5 mmol) was then added dropwise into the stirred solution until the clear yellowish suspension of CdS QDs was obtained. The obtained aqueous CdS QDs were then quenched at 0 °C in the freezer (45 min) and stored in a refrigerator at 4 °C. CdS QD–chitosan–GO composite solution was prepared by mixing the synthesized CdS QDs with chitosan–GO solution with a 1:1:1 volume ratio.

2.2. Prepartion of Thin Films

Firstly, the glass slip was cleaned by using acetone or water to remove any fingerprint marks or dirt on the surface of the glass. Before the composite solution was coated onto the glass slip, the glass slip was first deposited with gold layer. The glass cover slip was deposited with a gold layer by using an SC7640 Sputter Coater (Quorum Technologies, West Sussex, UK) with the duration of 67 s for the sputtering process to obtain a 50 nm thick gold layer, which is the optimum thickness for an SPR sensor [93].

PAR–chitosan–GO and CdS QD–chitosan–GO active layers were produced by spin coating method using a P-6708D spin Coater (Inc. Medical Devices, Indianapolis, USA). Approximately 0.55 mL of the composite solution were placed on the gold layered glass film, which was spun at 6000 rpm for 30 s.

2.3. X-ray Photoelectron Spectroscopy

The XPS analysis was separated in two sections. The first section was carried out to investigate the existing functional groups of PAR–chitosan–GO and CdS QD–chitosan–GO thin films. The second part was to investigate the chemical changes of the PAR–chitosan–GO and CdS QD–chitosan–GO thin films after it came into contact with Co^{2+}. The XPS analysis was performed in the SUT-NANOTEC-SLRI joint research facility located at the beamline in the Synchrotron Light Research Institute (SLRI), Nakhon-Ratchasima, Thailand. The samples were mounted on the stainless plate sample holder before being transferred into the vacuum chamber where the base pressure was controlled at 4×10^{-9} mbar. A scanning XPS Microprobe (PHI5000 Versa Probe II, ULVAC-PHI) equipped at the beamline generated X-rays from an aluminium Kα radiation at 1486.6 eV, monochromatized and focused on the sample surface at a 100 µm size spot by the quartz crystal. The energy of electrons emitted from the sample was analyzed by the concentric hemi-spherical analyzer at 45° from surface normal. The total energy resolution was typically about 0.85 eV. The charging effect was neutralized by the low-energy electrons on the sample surface. The binding energy was calibrated at the C–C peak of C 1s at 284.8 eV. The fitting procedure, was performed after the subtraction of the Shirley's background method. The fitting was optimized in the least-square method under the conditions of constraints such as the full-width half-maximum and peak position from the reference literature [94,95]. The XPS data were fitted using Fityk fitting and data analysis software (version 0.9.8, Marcin Wojdyr, Warsaw, Poland) by Gaussian–Lorentzian functions [96].

2.4. Surface Plasmon Resonance Spectroscopy

SPR is an optical sensor that measured the changes of the optical properties of the thin film. Any binding interaction that occurred on the thin film surface changed the optical properties; thus, altering the resonance angle. Hence, SPR can be used to monitor any binding event optically, and also for sensing applications. SPR was previously reported to have numerous advantages, such as fast measurement, no necessity for a reference solution, cost-effective, and simple sample preparation [97]. However, SPR has low sensitivity in sensing metal ions below 100 ppm. Therefore, PAR–chitosan–GO and CdS QD–chitosan–GO have been introduced on top of the gold layer as an attempt to increase the SPR sensitivity.

To test the ability of the thin film in sensing Co^{2+}, the SPR experiment setup was designed in the laboratory. The thin film was placed between the dielectric medium and the prism, as shown in Figure 1. Then, the Co^{2+} solutions of 0.01, 0.1, 1, 5, 10, 20, 40, 60, 80, and 100 ppm concentrations were inserted into the dielectric medium one at a time before the reflected beam was recorded [98]. The resonance angle was determined from the lowest peak of SPR curves for each concentration.

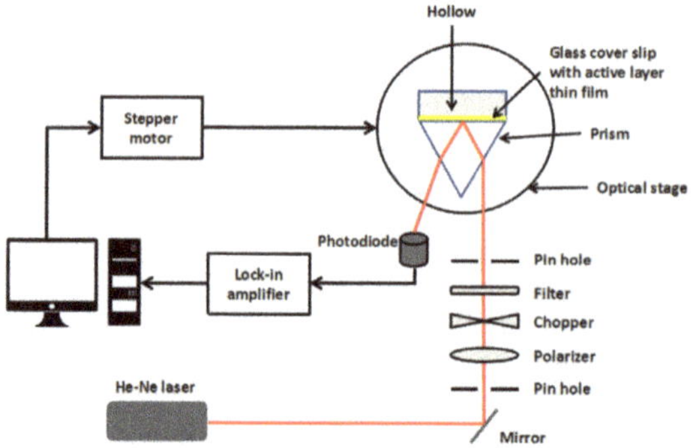

Figure 1. Optical setup of surface plasmon resonance spectroscopy.

3. Results and Discussion

3.1. XPS Characterization of Thin Films

XPS analysis was conducted for the PAR–chitosan–GO, and CdS QD–chitosan–GO thin films in order to study the functional group that exist on the surface of each thin films. After the XPS analysis, the existence of C, O, and N elements on the PAR–chitosan–GO thin film was revealed and the ratio between elements was obtained by evaluation of the peak area. The peak area of each elements was first normalized with the corrected relative sensitivity factor (RSF) value (which includes the mean free path and the transmission function of the system) before the ratio was calculated. The ratio of C, O, and N that existed on the surface of the PAR–chitosan–GO thin film was 66.41%, 27.96%, and 5.63%, respectively, as shown in Table 1. The deconvolution of the peak for each elements that existed on the surface of the thin films was also carried out to investigate the sub-peak component using the Gaussian–Lorentzian function based on theoretical prediction and previously reported works from the High Resolution XPS of Organic Polymers: The Scienta ESCA300 Database [99].

The C 1s peak for PAR–chitosan–GO thin film was mainly resolved into C–C bonds and C–O or C–O–C bonds at 284.7 eV and 286.6 eV, respectively, as shown in Figure 2a. A lower intensity of C=O or O–C=O bonds was also deconvoluted at 288.1 eV. The existence of C–O or C–O–C and C=O or O–C=O from the deconvolution of the C 1s peak indicated the existence of the oxygen-containing group in GO's structure, and has a good agreement with the theoretical structure of GO [100]. Moreover, the deconvolution of the O 1s peak by Gaussian–Lorentzian function was carried out to confirm the deconvolution of oxygen-containing functional groups of the C 1s peak, as shown in Figure 2b. The O 1s spectra for the gold/GO thin film were deconvoluted into two sub-peaks at 530.9 eV and 532.9 eV that were assigned to C=O and C–O or C–O–C bonds, respectively. For the N 1s peak, the peak was deconvoluted into three sub-peaks at 397.9 eV, 399.7 eV, and 400.9 eV, which were assigned for pyridinic, amine, and pyrrolic N, respectively, as shown in Figure 2c. The deconvoluted N 1s peak had a good agreement with the N functional groups that exist on PAR and chitosan [101].

On the other hand, the XPS analysis revealed the existence of C, O, Cd, S, and N on the surface of the CdS QD–chitosan–GO thin film with ratios of 67.53%, 28.63%, 0.24%, 1.65%, and 1.95%, respectively [102]. The quantitative result is summarized in Table 2. Each elements peak that existed on the surface of the CdS QD–chitosan–GO thin film was also deconvoluted using the Gaussian–Lorentzian function to obtain the information on the sub-peak component. The C 1s peak for the CdS QD–chitosan–GO thin film can be resolved into three sub-peak components. The first and second peaks at 284.6 eV and 286.8 eV were assigned to a C–C bond and C–O or C–O–C bonds, respectively, while the third peak at 288.1 eV could be assigned to C=O or O–C=O bonds, as shown in Figure 3a. For the O 1s, the peak at 531.1 eV and 533.1 eV could be resolved into C=O and C–O or C–O–C, as shown in Figure 3b. Moreover, the Cd 3d peak could be deconvoluted into two components. The first component at 405 eV was assigned to $Cd_{5/2}$, while the second peak at 412 eV could be assigned to $Cd_{3/2}$, as shown in Figure 3c. For the deconvolution of the S 2p peak, two pairs of sub-peaks were revealed, as shown in Figure 3d. The first pair at 163.3 eV and 164.6 eV, assigned to $S_{3/2}$ and $S_{1/2}$, respectively, corresponds to the S–Cd bond [103]. The second pair also contained $S_{3/2}$ and $S_{1/2}$ sub-peaks: 167.8 eV and 169.1 eV, respectively. The existence of the second pair was related to the presence of residual sulfate on the surface of the CdS QD–chitosan–GO thin film that might not have completely been removed during preparation [104]. The existence of Cd and S elements on the surface of thin film proved that the combination of CdS QDs with the chitosan–GO was successful. For the last element, the N 1s could be resolved into two sub-peaks, which were the amine groups and C–N or NH_2 bonds at 399.2 eV and 400.2 eV, respectively, as shown in Figure 3e [92].

Table 1. Elemental composition of the PAR–chitosan–GO thin film. RSF, relative sensitivity factor.

Sample	Element	Corrected RSF	Area	Corrected Area	Ratio (%)
PAR–chitosan–GO	C	1.2070	15,701	13,008.3	66.41
	O	3.1048	17,005	5477.0	27.96
	N	2.0520	2264	1103.3	5.63

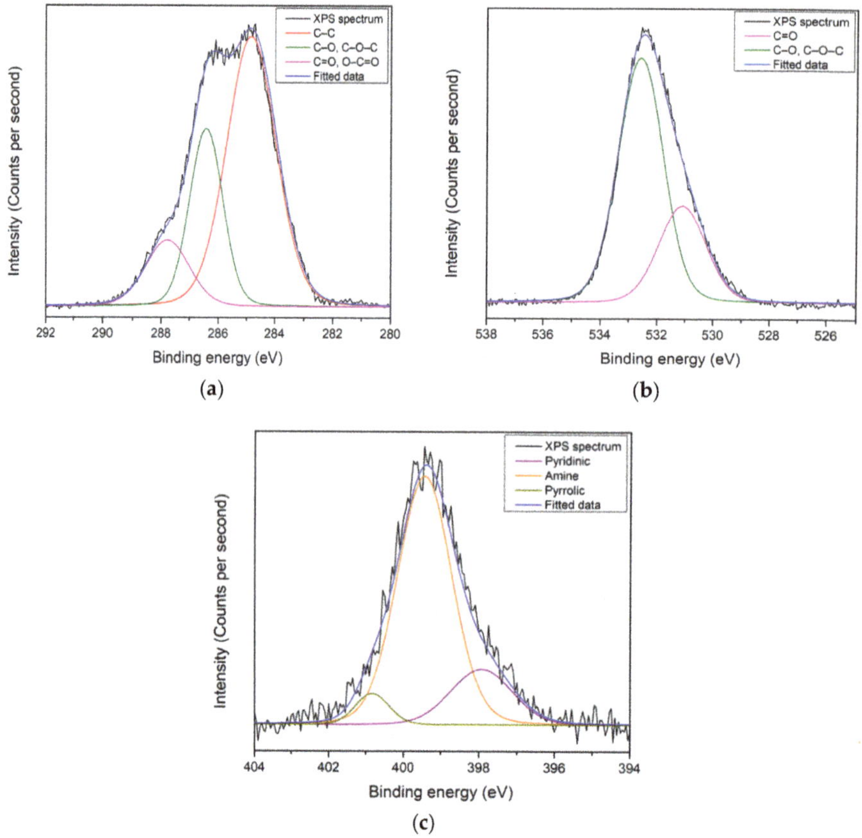

Figure 2. XPS narrow scan spectra of PAR–chitosan–GO for (**a**) C 1s; (**b**) O 1s; and (**c**) N 1s.

Table 2. Elemental composition of the CdS QD–chitosan–GO thin film.

Sample	Element	Corrected RSF	Area	Corrected Area	Ratio (%)
CdS QD–chitosan–GO	C	1.2070	15,033	12,454.8	67.53
	O	3.1048	16,396	5280.9	28.63
	Cd	25.9923	1147	44.1	0.24
	S	2.4319	742	305.1	1.65
	N	2.0520	738	359.6	1.95

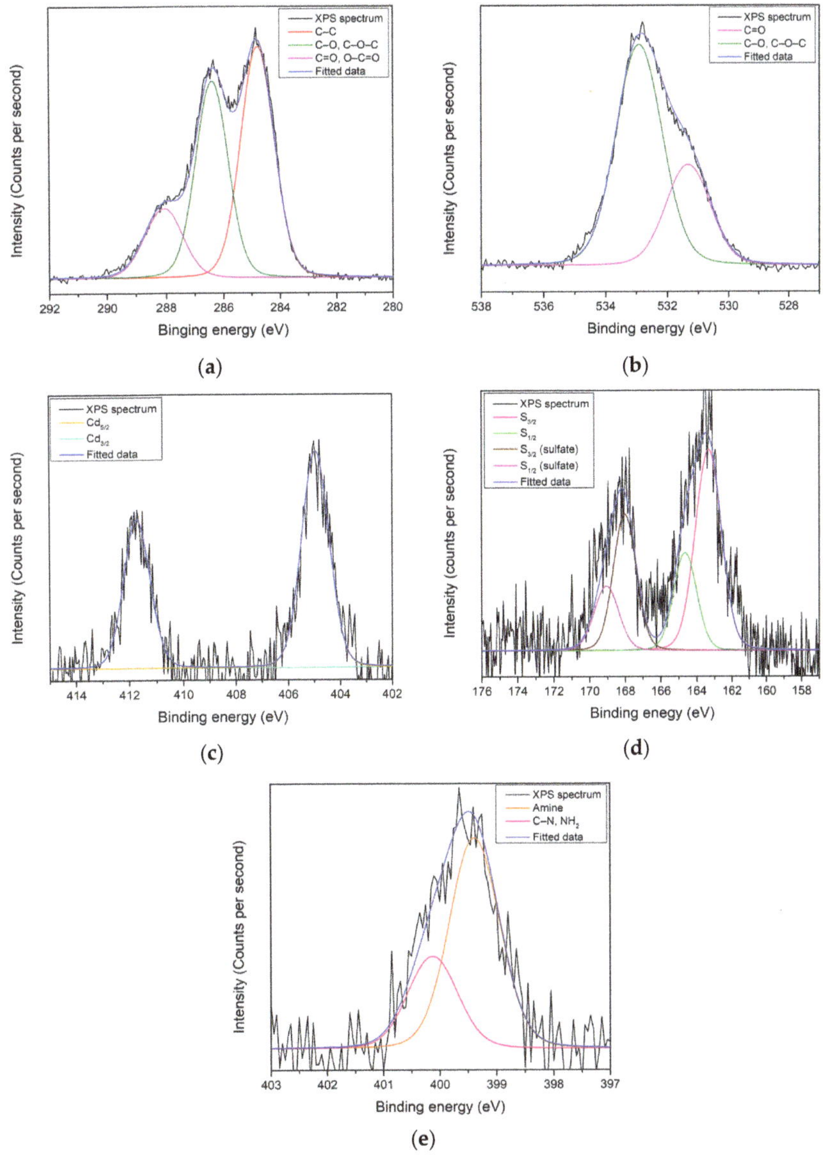

Figure 3. XPS narrow scan spectra of CdS QD–chitosan–GO for (**a**) C 1s; (**b**) O 1s; (**c**) Cd; (**d**) S 2p; and (**e**) N 1s.

3.2. SPR Response of Thin Films With Co^{2+}

The SPR analysis began by finding the SPR curve of the thin films when in contact with Co^{2+} solution at 0.01, 0.1, 1, 5, 10, 20, 40, 60, 80, and 100 ppm concentrations. The resonance angle then was determined from each of the SPR curves, as shown in Figure 4a [105]. The SPR curves for both PAR–chitosan–GO and CdS QD–chitosan–GO thin films were shifted to the right when Co^{2+} solution was used, with the lowest detection of 0.01 ppm for both thin films. This result proved that the binding event did occur between the Co^{2+} and the modified gold thin films [106].

The shift of the resonance angle has been applied as a parameter to measure the sensitivity of the SPR sensor [107]. The comparison of the resonance angle shift of gold, PAR–chitosan–GO, and CdS QD–chitosan–GO thin films towards various concentration of Co^{2+} is shown in Figure 4b. From the figure, it can be observed that there are no changes in the resonance angle when the gold layer is introduced with different concentrations of Co^{2+}. The figure clearly demonstrates the sensitivity enhancement of the active layers compared to only the gold layer film. To be exact, the linear regression analysis revealed that the sensitivity of the active layers towards Co^{2+} are different; the PAR–chitosan–GO thin film exhibited a greater slope compared to the CdS QD–chitosan–GO thin film with slopes of 0.2370° ppm^{-1} and 0.1188° ppm^{-1}, respectively, for lower concentrations (0 ppm to 1 ppm). At high concentrations (10 ppm to 100 ppm), PAR–chitosan–GO thin film exhibited a greater slope at 0.00069° ppm^{-1} compared to the CdS QD–chitosan–GO thin film with a slope of 0.00026° ppm^{-1}. These results confirmed that the PAR–chitosan–GO thin film attracts more Co^{2+} compared to the CdS QD–chitosan–GO thin film [108].

Furthermore, the shift of resonance angle data was fitted by Langmuir–Freundlich models to obtain the binding affinity of Co^{2+} ions with the thin films [109–112]. The Langmuir–Freundlich models can be expressed as $\Delta\theta = \frac{\Delta\theta_{\max}(K_aC)^n}{1+(K_aC)^n}$, where $\Delta\theta_{\max}$ is the maximum of the resonance angle shift, K_a is the affinity constant, C is the concentration of the Co^{2+} ions, and n is the heterogeneity index. As shown in Figure 5, it can be observed that the SPR angle shift of the PAR–chitosan–GO thin film produced a greater maximum angle shift compared to the CdS QD–chitosan–GO thin film. The maximum resonance angle shifts for the PAR–chitosan–GO and CdS QD–chitosan–GO thin films were 0.334° and 0.213°, respectively. Resonance angle shift was caused by the changes in the material composition, and the higher the angle shift, the greater the changes of the material composition [113,114]. Other than that, the binding affinity constant of PAR–chitosan–GO thin films were higher as compared to the CdS QD–chitosan–GO thin films, i.e., 1.649 ppm^{-1} and 0.939 ppm^{-1}, respectively, with R^2 of 0.96. The calculated fitting parameters are summarized in Table 3. Overall, the results obtained from the binding model fitting also show a great response for the adsorption of Co^{2+} ions with the composite sensor layer.

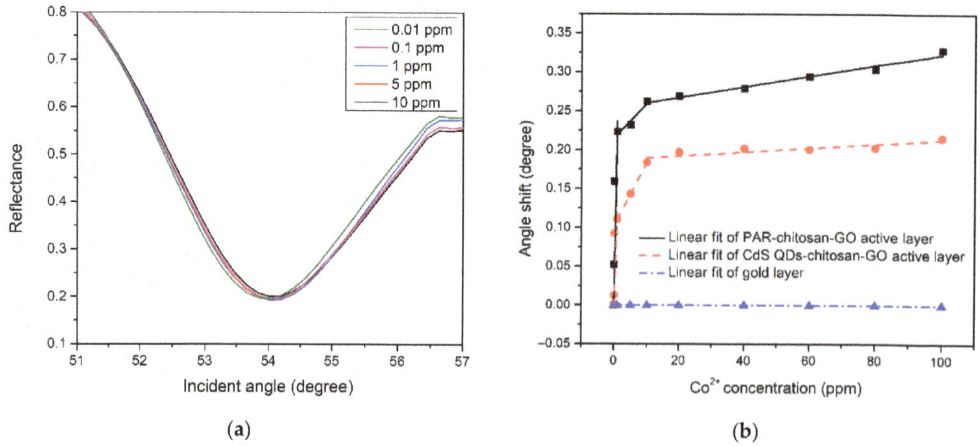

Figure 4. (**a**) SPR response of thin films in contact with Co^{2+} at 0.01–10 ppm concentration, and (**b**) the comparison of the linear regression analysis for various concentration of Co^{2+} in contact with gold layer and modified gold layer thin films.

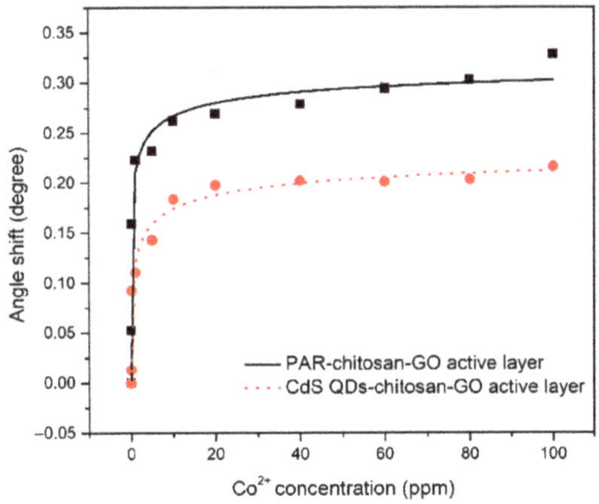

Figure 5. SPR angle shift comparison (fitted with Sips model) between PAR–chitosan–GO and CdS QD–chitosan–GO thin films in contact with different concentration of Co^{2+} from 0 to 100 ppm.

Table 3. The calculated fitting parameters using Langmuir–Freundlich models.

Sample	K (ppm^{-1})	R^2	n	$\Delta\theta_{max}$ (°)
PAR–chitosan–GO	1.649	0.96	0.400	0.334
CdS QD–chitosan–GO	0.939	0.96	0.400	0.213

3.3. Evidence of Co^{2+} Interaction with the Thin Films

After the SPR analysis, XPS analysis was carried out to study the changes of the surface chemistry on the thin films. The PAR–chitosan–GO thin film that was in contact with Co^{2+} (1 ppm) was analyzed by XPS, and the results revealed that the same elements were detected: C, O, N, and an extra peak for Co at 63.51%, 29.32%, 6.94%, and 0.22% ratio, respectively [92]. The ratio for each element is recorded in Table 4. Each element peak was then deconvoluted using the Gaussian–Lorentzian function to investigate the changes of functional groups on the thin film surface.

Identical to the C 1s before it was in contact with Co^{2+}, the C 1s peak for the PAR–chitosan–GO thin film was resolved into three sub-peaks at 284.7 eV, 286.8 eV, and 287.9 eV, which were assigned to a C–C bond, C–O or C–O–C bonds, and C=O or O–C=O bonds, respectively, as shown in Figure 6a. The ratios of the chemical composition were also calculated to provide a better description on the changes observed for the XPS results. When compared with the C 1s of the PAR–chitosan–GO thin film before it was in contact with Co^{2+}, there was decrease in the sub-peak intensity for the C 1s spectra after the interaction. The ratio for the sub-peak at 287.9 eV that was assigned to the O–C=O bond decreased from 13.31% to 12.61% after it was in contact with Co^{2+}. Decreasing of this sub-peak indicated that the chemical state of the O–C=O functional groups undergo changes after the interaction. Moreover, the intensity for the sub-peak at 286.8 eV that was assigned to the C–O or C–O–C bond increased due to the decrease in the sub-peak intensity for the C=O or O–C=O bond. As the result, the calculated ratio for the C–C bond decreased from 60.28% to 48.68%.

To confirm the changes of the sub-peak component, O 1s after the interaction with Co^{2+} was also fitted using the Gaussian–Lorentzian function. For the O 1s peak, the deconvolution of the spectra also comprised two sub-peaks at 530.9 eV and 532.9 eV, which were assigned to C=O and C–O or C–O–C bonds, respectively. As shown in Figure 6b, the sub-peak at 532.9 eV that was assigned to C=O decreased, which was also proved by calculating the ratio. The ratio of C=O decreased from 29.5% to 23.15%. This result is in good agreement with the changes of the sub-peak of C 1s after it was in contact with Co^{2+}.

The O–C=O can lose a proton and form a negatively charged ion (COO^-). The existence of these negatively charged functional groups may attract the positively charge Co^{2+} to form electrostatic interactions; thus, changing the optical properties of the thin film surface [106]. Moreover, the XPS results also proved that Co^{2+} may have interacted electrostatically with the amino group that existed on the surface of the PAR–chitosan–GO thin film [92]. The N 1s peak was deconvoluted into three sub-peaks at 397.9 eV, 399.3 eV, and 400.8 eV, which were assigned to pyridinic, amine, and pyrrolic N, respectively, as shown in Figure 6c. From the plotted graph, there was decrease in the amine functional group that existed on the surface of the PAR–chitosan–GO thin film. The amine group decreased from 74.81% to 65.01% after contact with Co^{2+}. The XPS analysis also revealed an extra peak that was assigned to Co 2p, that proved the existence of Co on the surface of the PAR–chitosan–GO thin film, as shown in Figure 6d. The Co 2p peak consisted of two sub-peaks that were assigned to $Co_{3/2}$ at 781.7 eV, associated with the Co^{2+} oxidation state, while the second peak at 796.4 eV was assigned to $Co_{1/2}$. The quantitative results are summarized in Table 5.

Table 4. Elements composition of the PAR–chitosan–GO thin film after it was in contact with Co^{2+}.

Sample	Element	Corrected RSF	Area	Corrected Area	Ratio (%)
PAR–chitosan–GO	C	1.2070	16,139	13,371.2	63.51
	O	3.1048	19,169	6174	29.32
	N	2.0520	2998	1461	6.94
	Co	18.5764	858	46.2	0.22

Table 5. Chemical composition ratio for C 1s, O 1s, and N 1s for the PAR–chitosan–GO thin film before and after contact with Co^{2+}.

Sample	Element	Corrected RSF	Sub-Peak	Area	Corrected Area	Ratio (%)
PAR–chitosan–GO before contact with Co^{2+}	C	1.2070	C–C	9491.2	7863.5	60.28
			C–O, C–O–C	4159.1	3445.8	26.41
			C=O, O–C=O	2095.1	1735.8	13.31
	O	3.1048	O–C, C–O–C	11,196.8	3606.3	70.50
			O=C	4685.2	1509.0	29.50
	N	2.0520	Pyridinic	442.4	214.6	19.41
			Amine	1705.1	830.9	74.79
			Pyrrolic	132.3	64.5	5.80
PAR–chitosan–GO after contact with Co^{2+}	C	1.2070	C–C	7526.9	6236.0	48.68
			C–O, C–O–C	5985.3	4958.8	38.71
			C=O, O–C=O	1949.8	1615.4	12.61
	O	3.1048	O–C, C–O–C	14,541.3	4683.5	76.80
			O=C	4392.7	1414.8	23.15
	N	2.0520	Pyridinic	772	376.21	25.87
			Amine	1939.9	946.29	65.01
			Pyrrolic	272.1	132.60	9.12

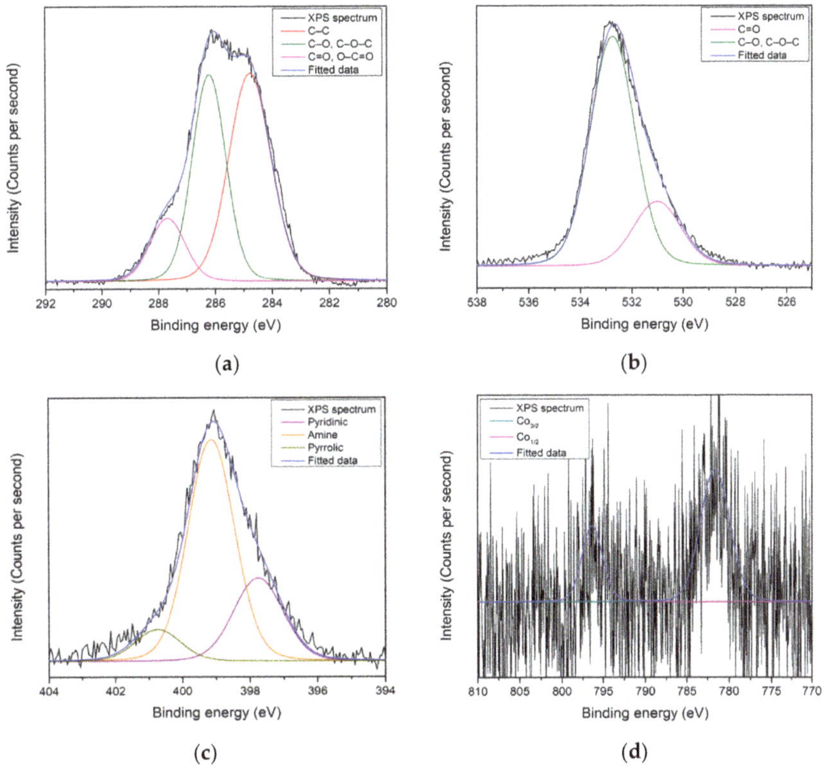

Figure 6. XPS narrow scan spectra of PAR–chitosan–GO after contact with Co^{2+} for (**a**) C 1s; (**b**) O 1s, (**c**) N 1s; and (**d**) Co 2p.

Moving on to the XPS results of the CdS QD–chitosan–GO thin film after it came into contact with Co^{2+}: the surface of the CdS QD–chitosan–GO thin film was identified with C, O, Cd, S, and N at 66.6%, 28.05%, 0.27%, 2.25%, and 2.59% ratios, respectively, as shown in Table 6. An extra peak of Co at a 0.29% ratio was also detected, proving the existence of Co at the surface of the thin film. The first element, C 1s, was deconvoluted into the three sub-peak components. The C 1s of the CdS QD–chitosan–GO thin film comprised C–C, C–O or C–O–C, and C=O or O–C=O bonds first at 284.7 eV, 286.8 eV, and 288.1 eV, respectively, as shown in Figure 7a. Interestingly, after the CdS QD–chitosan–GO thin film was in contact with Co^{2+}, the sub-peak component at 288.1 eV that was assigned to C=O or O–C=O decreased. The ratio for this sub-peak decreased from 15.24% to 12.12% after it came into contact with Co^{2+}. Moreover, the O 1s peak of CdS QD–chitosan–GO thin film after contact with Co^{2+} could also be resolved into two sub-peaks at 531.0 eV and 532.9 eV, which were assigned to C=O and C–O or C–O–C, respectively, as shown in Figure 7b. From the deconvolution of the O 1s peak after contact with Co^{2+}, it was confirmed that the peak intensity at 532.9 eV decreased when the ratio changed from 30.43% to 26.71%. Furthermore, no changes occurred after the Cd 3d peak came into contact with Co^{2+}. The Cd 3d peak could be resolved into two sub-peaks at 405 eV and 412 eV, which were assigned to $Cd_{5/2}$ and $Cd_{3/2}$, respectively, as shown in Figure 7c. For S 2p, two pairs of sub-peak components were identified. The first pair at 162.4 eV and 163.5 eV was assigned to $S_{3/2}$ and $S_{1/2}$, respectively. The second pair that corresponds to the presence of residual sulfate on the surface was also assigned to $S_{3/2}$ and $S_{1/2}$, at 167.8 eV and 168.5 eV, respectively, as shown in Figure 7d. For N 1s, two sub-peaks were deconvoluted at 399.2 eV and 400.2 eV, which were assigned to the amine groups and C–N or NH_2 bonds, respectively, as shown in Figure 7e. The sub-peak ratio of N 1s that was assigned to the amine functional group also decreased from 70.01% to 67.32%. Furthermore, one spin-orbit pair corresponding to the Co^{2+} oxidation state was identified at 781. 5 eV and 797.3 eV, which was assigned to $Co_{3/2}$ and $Co_{1/2}$ sub-peaks, respectively, as shown in Figure 7f.

Similarly to the PAR–chitosan–GO thin film, it is believed that Co^{2+} interacts with negatively charged functional groups on the surface of the CdS QD–chitosan–GO thin film, i.e., COO^- and the amine functional groups through electrostatic interaction. Moreover, it is also suggested that PAR played more important role during the interaction, based on the XPS and SPR results. The surface of the PAR–chitosan–GO thin film contained more active sites for interaction with Co^{2+} owing to the presence of PAR; it provides more amine functional groups. As a result, more Co^{2+} could be absorbed on the surface of the PAR–chitosan–GO thin film. The existence of more active sites makes PAR–chitosan–GO thin film more advantageous in sensing Co^{2+} compared to CdS QD–chitosan–GO thin film; this was proven using SPR where the PAR–chitosan–GO thin film had a higher binding affinity towards Co^{2+}. The quantitative results are summarized in Table 7.

Table 6. Elemental composition of the CdS QD–chitosan–GO thin film after contact with Co^{2+}.

Sample	Element	Corrected RSF	Area	Corrected Area	Ratio (%)
CdS QD–chitosan–GO	C	1.2070	18,606	15,415.1	66.60
	O	3.1048	20,161	6493.5	28.05
	Cd	25.9923	1608	61.9	0.27
	S	2.4319	1267	521	2.25
	N	2.0520	1230	599.4	2.59
	Co	18.5764	1070	57.6	0.29

Table 7. Chemical composition ratios for C 1s, O 1s, and N 1s for the CdS QD–chitosan–GO thin film before and after contact with Co^{2+}.

Sample	Element	Corrected RSF	Sub-Peak	Area	Corrected Area	Ratio (%)
CdS QD–chitosan–GO before contact with Co^{2+}	C	1.2070	C–C	6594.5	5463.5	46.45
			C–O, C–O–C	5438.9	4506.1	38.31
			C=O, O–C=O	2163.6	1792.5	15.24
	O	3.1048	O–C, C–O–C	10,782.7	3472.9	69.57
			O=C	4716.3	1519.0	30.43
	N	2.0520	Amine	515.9	251.4	70.01
			C–N	221.1	107.7	29.91
CdS QD–chitosan–GO after contact with Co^{2+}	C	1.2070	C–C	7542.7	6249.1	44.56
			C–O, C–O–C	7332.8	6075.2	43.32
			C=O, O–C=O	2051.6	1699.8	12.12
	O	3.1048	O–C, C–O–C	14,528.3	4679.3	73.29
			O=C	5294.7	1705.3	26.71
	N	2.0520	Amine	935.1	455.7	67.32
			C–N	453.9	221.2	32.68

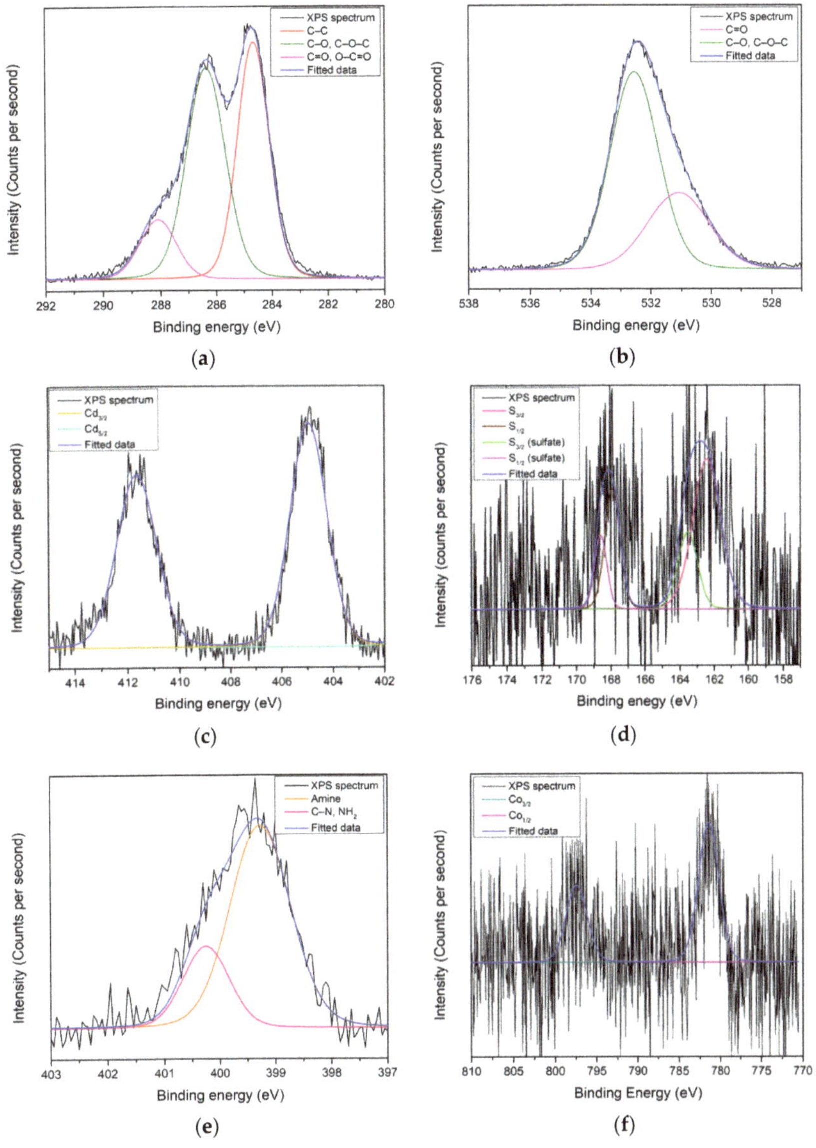

Figure 7. XPS narrow scan spectra of CdS QD–chitosan–GO after contact with Co^{2+} for (**a**) C 1s; (**b**) O 1s; (**c**) Cd; (**d**) N 1s; (**e**) S 2p; and (**f**) Co 2p.

4. Conclusions

In this study, PAR–chitosan–GO and CdS QD–chitosan–GO thin films have been characterized using X-ray photoelectron spectroscopy (XPS), before and after interaction with Co^{2+}. The XPS results revealed the existence of C, O, and N on the surface of the PAR–chitosan–GO thin film, while C, O, Cd, S, and N were detected on the surface of the CdS QD–chitosan–GO thin film. Each element peak has been deconvoluted by a Gaussian–Lorentzian curve fitting program in order to determine the sub-peak components. Further analysis of the PAR–chitosan–GO and CdS QD–chitosan–GO thin films using SPR has also been carried out; both thin films were able to detect Co^{2+} as low as 0.01 ppm. The binding affinity constant of PAR–chitosan–GO thin film was higher compared to the CdS QD–chitosan–GO thin film, i.e., 1.649 ppm^{-1} and 0.939 ppm^{-1}, respectively, with R^2 of 0.96. After interaction with Co^{2+}, the thin films were characterized again by XPS to confirm the chemical interactions involved between the Co^{2+} with the existing functional groups on the thin film surfaces. From the deconvolution of each element, Co^{2+} may have interacted electrostatically with negatively charged functional groups on the surface of the both chitosan–GO thin films, i.e., COO^- and amine groups. Moreover, it is suggested that PAR played more important role during the interaction, based on the XPS and SPR results. The surface of the PAR–chitosan–GO thin film contained more active sites due to the presence of PAR, which provided more amine functional groups. The existence of more active sites promote the use of PAR–chitosan–GO thin film in sensing Co^{2+} compared to CdS QD–chitosan–GO thin film.

Author Contributions: Conceptualization, Y.W.F.; methodology, Y.W.F., W.M.E.M.M.D. and S.S.; validation, Y.W.F., N.C. and H.N.; formal analysis, W.M.E.M.M.D. and S.S.; investigation, W.M.E.M.M.D. and S.S.; data curation, W.M.E.M.M.D. and S.S.; writing—original draft preparation, W.M.E.M.M.D.; writing—review and editing, Y.W.F.; supervision, Y.W.F., J.A. and N.A.Y. All authors have read and agreed to the published version of the manuscript.

Funding: This research work was funded by the Ministry of Education, Malaysia, through the Fundamental Research Grant Scheme (FRGS/1/2019/STG02/UPM/02/1).

Institutional Review Board Statement: Not applicable.

Informed Consent Statement: Not applicable.

Data Availability Statement: Not applicable.

Conflicts of Interest: The authors declare no conflict of interest.

References

1. Banerjee, A.; Ganguly, S. Alginate–chitosan composite hydrogel film with macrovoids in the inner layer for biomedical applications. *J. Appl. Polym. Sci.* **2019**, *136*, 30–33. [CrossRef]
2. Lisuzzo, L.; Cavallaro, G.; Milioto, S.; Lazzara, G. Halloysite nanotubes coated by chitosan for the controlled release of khellin. *Polymers* **2020**, *12*, 1766. [CrossRef]
3. Bertolino, V.; Cavallaro, G.; Milioto, S.; Lazzara, G. Polysaccharides/Halloysite nanotubes for smart bionanocomposite materials. *Carbohydr. Polym.* **2020**, *245*, 116502–116513. [CrossRef] [PubMed]
4. Rinaudo, M. Chitin and chitosan: Properties and applications. *Prog. Polym. Sci.* **2006**, *31*, 603–632. [CrossRef]
5. Maitra, J.; Shukla, V.K. Cross-linking in hydrogels—A review. *Am. J. Polym. Sci.* **2014**, *4*, 25–31.
6. Geim, A.K.; Novoselov, K.S. The rise of graphene. *Nat. Mater.* **2007**, *6*, 183–191. [CrossRef]
7. Avouris, P. Graphene: Electronic and photonic properties and devices. *Nano Lett.* **2010**, *10*, 4285–4294. [CrossRef]
8. Geim, A.K. Graphene: Status and Prospects. *Science* **2009**, *324*, 1530–1535. [CrossRef]
9. Pei, S.; Cheng, H.M. The reduction of graphene oxide. *Carbon* **2012**, *50*, 3210–3228. [CrossRef]
10. Zainudin, A.A.; Fen, Y.W.; Yusof, N.A.; Omar, N.A.S. Structural, optical and sensing properties of ionophore doped graphene based bionanocomposite thin film. *Optik* **2017**, *144*, 308–315.
11. Hunt, A.; Kurmaev, E.Z.; Moewes, A. Band gap engineering of graphene oxide by chemical modification. *Carbon* **2014**, *75*, 366–371. [CrossRef]
12. Chen, P.; Cao, Z.-F.; Wen, X.; Wang, J.; Yang, F.; Wang, S.; Zhong, H. In situ nano-silicate functionalized graphene oxide composites to improve MB removal. *J. Taiwan Inst. Chem. Eng.* **2017**, *81*, 87–94. [CrossRef]
13. Wang, W.L.; Jang, J.; Nguyen, V.H.; Auxilia, F.M.; Song, H.; Jang, K.; Jin, E.M.; Lee, G.Y.; Gu, H.B.; Ham, M.H. Cerium vanadate and reduced graphene oxide composites for lithium-ion batteries. *J. Alloy. Compd.* **2017**, *724*, 1075–1082. [CrossRef]
14. Yang, X.; Tu, Y.; Li, L.; Shang, S.; Tao, X. Well-dispersed chitosan/graphene oxide nanocomposites. *ACS Appl. Mater. Interfaces* **2010**, *2*, 1707–1713. [CrossRef]
15. Zhu, Y.; Murali, S.; Cai, W.; Li, X.; Suk, J.W.; Potts, J.R.; Ruoff, R.S. Graphene and graphene oxide: Synthesis, properties, and applications. *Adv. Mater.* **2010**, *22*, 3906–3924. [CrossRef] [PubMed]
16. Daniyal, W.M.E.M.M.; Fen, Y.W.; Abdullah, J.; Saleviter, S.; Omar, N.A.S. Preparation and characterization of hexadecyltrimethylammonium bromide modified nanocrystalline cellulose/graphene oxide composite thin film and its potential in sensing copper ion using surface plasmon resonance technique. *Optik* **2018**, *173*, 71–77. [CrossRef]
17. Joshi, R.K.; Carbone, P.; Wang, F.C.; Kravets, V.G.; Su, Y.; Grigorieva, I.V.; Wu, H.A.; Geim, A.K.; Nair, R.R. Precise and ultrafast molecular sieving through graphene oxide membranes. *Science* **2014**, *343*, 752–754. [CrossRef]

18. He, H.; Klinowski, J.; Forster, M.; Lerf, A. A new structural model for graphite oxide. *Chem. Phys. Lett.* **1998**, *287*, 53–56. [CrossRef]
19. Eda, G.; Chhowalla, M. Chemically derived graphene oxide: Towards large-area thin-film electronics and optoelectronics. *Adv. Mater.* **2010**, *22*, 2392–2415. [CrossRef]
20. Jiang, J.; Feng, C.; Qian, W.; Zhu, L.; Han, S.; Lin, H. Effect of graphene oxide nanosheets and ultrasonic electrodeposition technique on Ni–Mo/graphene oxide composite coatings. *Mater. Chem. Phys.* **2017**, *199*, 239–248. [CrossRef]
21. Cheng, Z.-L.; Li, Y.; Liu, Z. Fabrication of graphene oxide/silicalite-1 composites with hierarchical porous structure and investigation on their adsorption performance for rhodamine B. *J. Ind. Eng. Chem.* **2017**, *55*, 234–243. [CrossRef]
22. Yu, B.; Xu, J.; Liu, J.H.; Yang, S.T.; Luo, J.; Zhou, Q.; Wan, J.; Liao, R.; Wang, H.; Liu, Y. Adsorption behavior of copper ions on graphene oxide-chitosan aerogel. *J. Environ. Chem. Eng.* **2013**, *1*, 1044–1050. [CrossRef]
23. Anandhavelu, S.; Thambidurai, S. Single step synthesis of chitin/chitosan-based graphene oxide-ZnO hybrid composites for better electrical conductivity and optical properties. *Electrochim. Acta* **2013**, *90*, 194–202. [CrossRef]
24. Kamaruddin, N.H.; Bakar, A.A.A.; Yaacob, M.H.; Mahdi, M.A.; Zan, M.S.D.; Shaari, S. Enhancement of chitosan-graphene oxide SPR sensor with a multi-metallic layers of Au-Ag-Au nanostructure for lead (II) ion detection. *Appl. Surf. Sci.* **2016**, *361*, 177–184. [CrossRef]
25. Pan, Y.; Wu, T.; Bao, H.; Li, L. Green fabrication of chitosan films reinforced with parallel aligned graphene oxide. *Carbohydr. Polym.* **2011**, *83*, 1908–1915. [CrossRef]
26. Kamaruddin, N.; Bakar, A.A.; Mobarak, N.; Zan, M.S.; Arsad, N. Binding affinity of a highly sensitive Au/Ag/Au/chitosan-graphene oxide sensor based on direct detection of Pb^{2+} and Hg^{2+} ions. *Sensors* **2017**, *17*, 2277. [CrossRef]
27. Zainudin, A.A.; Fen, Y.W.; Yusof, N.A.; Al-Rekabi, S.H.; Mahdi, M.A.; Omar, N.A.S. Incorporation of surface plasmon resonance with novel valinomycin doped chitosan-graphene oxide thin film for sensing potassium ion. *Spectrochim. Acta Mol. Biomol. Spectrosc.* **2018**, *191*, 111–115. [CrossRef]
28. Daniyal, W.M.E.M.M.; Fen, Y.W.; Fauzi, N.I.M.; Hashim, H.S.; Ramdzan, N.S.M.; Omar, N.A.S. Recent advances in surface plasmon resonance optical sensors for potential application in environmental monitoring. *Sens. Mater.* **2020**, *32*, 4191–4200.
29. Daniyal, W.M.E.M.M.; Fen, Y.W.; Abdullah, J.; Hashim, H.S.; Fauzi, N.I.M.; Chanlek, N.; Mahdi, M.A. X-ray photoelectron study on gold/nanocrystalline cellulose-graphene oxide thin film as surface plasmon resonance active layer for metal ion detection. *Thin Solid Film.* **2020**, *713*, 138340–138350. [CrossRef]
30. Florence, T.M.; Farrar, Y. Spectrophotometric determination of uranium with 4-(2-pyridylazo) resorcinol. *Anal. Chem.* **1963**, *35*, 1613–1616. [CrossRef]
31. Starosta, R.; Nitek, W.; Komarnicka, U.K.; Stochel, G. Synthesis and characterization of copper (I) coordination compounds with (1-(2-pyridylazo)-2-naphthol) and (4-(2-pyridylazo) resorcinol). *Polyhedron* **2014**, *68*, 357–364.
32. Karmakar, A.; Singh, B. Charge-transfer interaction of 4-(2-pyridylazo) resorcinol with nitroaromatics: Insights from experimental and theoretical results. *J. Mol. Liq.* **2017**, *236*, 135–143. [CrossRef]
33. Pollard, F.H.; Hanson, P.; Geary, W.J. 4-(2-pyridylazo) resorcinol as a possible analytical reagent for the colorimetric estimation of cobalt, lead, and uranium. *Anal. Chim. Acta* **1959**, *20*, 26–31. [CrossRef]
34. Zi, L.; Huang, Y.; Yan, Z.; Liao, S. Thioglycolic acid-capped $CuInS_2$/ZnS quantum dots as fluorescent probe for cobalt ion detection. *J. Lumin.* **2014**, *148*, 359–363. [CrossRef]
35. Amjadi, M.; Shokri, R.; Hallaj, T. A new turn-off fluorescence probe based on graphene quantum dots for detection of Au (III) ion. *Spectrochim. Acta Mol. Biomol. Spectrosc.* **2016**, *153*, 619–624. [CrossRef]
36. Anas, N.A.A.; Fen, Y.W.; Yusof, N.A.; Omar, N.A.S.; Daniyal, W.M.E.M.M.; Ramdzan, N.S.M. Highly sensitive surface plasmon resonance optical detection of ferric ion using CTAB/hydroxylated graphene quantum dots thin film. *J. Appl. Phys.* **2020**, *128*, 1–10. [CrossRef]
37. Manan, F.A.A.; Hong, W.W.; Abdullah, J.; Yusof, N.A.; Ahmad, I. Nanocrystalline cellulose decorated quantum dots based tyrosinase biosensor for phenol determination. *Mater. Sci. Eng.* **2019**, *99*, 37–46. [CrossRef] [PubMed]
38. Tyagi, C.; Sharma, A.; Kurchania, R. Synthesis of cds quantum dots using wet chemical co-precipitation method. *J. NonOxide Glas.* **2014**, *6*, 23–26.
39. Liu, S.; Na, W.; Pang, S.; Su, X. Fluorescence detection of Pb^{2+} based on the DNA sequence functionalized CdS quantum dots. *Biosens. Bioelectron.* **2014**, *58*, 17–21. [CrossRef]
40. Uppa, Y.; Kulchat, S.; Ngamdee, K.; Pradublai, K. Silver ion modulated CdS quantum dots for highly selective detection of trace Hg^{2+}. *J. Lumin.* **2016**, *178*, 437–445. [CrossRef]
41. Wang, J.; Wang, X.; Tang, H. Ultrasensitive electrochemical detection of tumor cells based on multiple layer CdS Quantum dots-functionalized polystyrene microspheres and graphene oxide–polyaniline composite. *Biosens. Bioelectron.* **2018**, *100*, 1–7. [CrossRef]
42. Li, Z.; Du, Y.; Zhang, Z.; Pang, D. Preparation and characterization of CdS quantum dots chitosan biocomposite. *React. Funct. Polym.* **2003**, *55*, 35–43. [CrossRef]
43. Tshemese, Z.; Mlowe, S.; Revaprasadu, N.; Deenadayalu, N. Synthesis of CdS quantum dots in an imidazolium based ionic liquid. *Mater. Sci. Semicond. Process.* **2017**, *71*, 258–262. [CrossRef]
44. Koneswaran, M.; Narayanaswamy, R. Mercaptoacetic acid capped CdS quantum dots as fluorescence single shot probe for mercury (II). *Sens. Actuators Chem.* **2009**, *139*, 91–96. [CrossRef]

45. Veerathangam, K.; Pandian, M.S.; Ramasamy, P. Photovoltaic performance of Ag-doped CdS quantum dots for solar cell application. *Mater. Res. Bull.* **2017**, *94*, 371–377.
46. Choi, H.-I.; Hong, J.A.; Kim, M.-S.; Lee, S.E.; Jung, S.-H.; Yoon, P.W.; Song, J.S.; Kim, J.-J. Severe cardiomyopathy due to arthroprosthetic cobaltism: Report of two cases with different outcomes. *Cardiovasc. Toxicol.* **2018**, *19*, 82–89. [CrossRef] [PubMed]
47. Omar, N.A.S.; Fen, Y.W.; Saleviter, S.; Daniyal, W.M.E.M.M.; Anas, N.A.A.; Ramdzan, N.S.M.; Roshidi, M.D.A. Development of a graphene-based surface plasmon resonance optical sensor chip for potential biomedical application. *Materials* **2019**, *12*, 1928. [CrossRef] [PubMed]
48. Omar, N.A.S.; Fen, Y.W.; Abdullah, J.; Zaid, M.H.M.; Daniyal, W.M.E.M.M.; Mahdi, M.A. Sensitive surface plasmon resonance performance of cadmium sulfide quantum dots-amine functionalized graphene oxide based thin film towards dengue virus E-protein. *Opt. Laser Technol.* **2019**, *114*, 204–208. [CrossRef]
49. Ortega, R.; Bresson, C.; Fraysse, A.; Sandre, C.; Devès, G.; Gombert, C.; Tabarant, M.; Bleuet, P.; Seznec, H.; Simionovici, A.; et al. Cobalt distribution in keratinocyte cells indicates nuclear and perinuclear accumulation and interaction with magnesium and zinc homeostasis. *Toxicol. Lett.* **2009**, *188*, 26–32. [CrossRef]
50. Battaglia, V.; Compagnone, A.; Bandino, A.; Bragadin, M.; Rossi, C.A.; Zanetti, F.; Colombatto, S.; Grillo, M.A.; Toninello, A. Cobalt induces oxidative stress in isolated liver mitochondria responsible for permeability transition and intrinsic apoptosis in hepatocyte primary cultures. *Int. J. Biochem. Cell Biol.* **2009**, *41*, 586–594. [CrossRef]
51. Karovic, O.; Tonazzini, I.; Rebola, N.; Edström, E.; Lövdahl, C.; Fredholm, B.B.; Daré, E. Toxic effects of cobalt in primary cultures of mouse astrocytes. Similarities with hypoxia and role of HIF-1α. *Biochem. Pharmacol.* **2007**, *73*, 694–708. [CrossRef]
52. Lombaert, N.; Lison, D.; Van Hummelen, P.; Kirsch-Volders, M. In vitro expression of hard metal dust (WC-Co)-responsive genes in human peripheral blood mononucleated cells. *Toxicol. Appl. Pharmacol.* **2008**, *227*, 299–312. [CrossRef] [PubMed]
53. Daniyal, W.M.E.M.M.; Saleviter, S.; Fen, Y.W. Development of surface plasmon resonance spectroscopy for metal ion detection. *Sens. Mater.* **2018**, *30*, 2023–2038. [CrossRef]
54. Roshidi, M.D.A.; Fen, Y.W.; Omar, N.A.S.; Saleviter, S.; Daniyal, W.M.E.M.M. Optical studies of graphene oxide/poly amidoamine) dendrimer composite thin film and its potential for sensing Hg^{2+} using surface plasmon resonance spectroscopy. *Sens. Mater.* **2019**, *31*, 1147–1156. [CrossRef]
55. Ock, K.; Jang, G.; Roh, Y.; Kim, S.; Kim, J.; Koh, K. Optical detection of Cu^{2+} ion using a SQ-dye containing polymeric thin-film on Au surface. *Microchem. J.* **2001**, *70*, 301–305. [CrossRef]
56. Lee, S.M.; Kang, S.W.; Kim, D.U.; Cui, J.Z.; Kim, S.H. Effect of metal ions on the absorption spectra and surface plasmon resonance of an azacrown indoaniline dye. *Dye. Pigment.* **2001**, *49*, 109–115. [CrossRef]
57. Ramdzan, N.S.M.; Fen, Y.W.; Omar, N.A.S.; Anas, N.A.A.; Daniyal, W.M.E.M.M.; Saleviter, S.; Zainudin, A.A. Optical and surface plasmon resonance sensing properties for chitosan/carboxyl-functionalized graphene quantum dots thin film. *Optik* **2019**, *178*, 802–812. [CrossRef]
58. Zhang, Y.; Xu, M.; Wang, Y.; Toledo, F.; Zhou, F. Studies of metal ion binding by apo-metallothioneins attached onto preformed self-assembled monolayers using a highly sensitive surface plasmon resonance spectrometer. *Sens. Actuators Chem.* **2007**, *123*, 784–792. [CrossRef]
59. Forzani, E.S.; Foley, K.; Westerhoff, P.; Tao, N. Detection of arsenic in groundwater using a surface plasmon resonance sensor. *Sens. Actuators Chem.* **2007**, *123*, 82–88. [CrossRef]
60. Hong, S.; Kang, T.; Moon, J.; Oh, S.; Yi, J. Surface plasmon resonance analysis of aqueous copper ions with amino-terminated self-assembled monolayers. *Colloids Surf. Physicochem. Eng. Asp.* **2007**, *292*, 264–270. [CrossRef]
61. Fen, Y.W.; Yunus, W.M.M. Surface plasmon resonance spectroscopy as an alternative for sensing heavy metal ions: A review. *Sens. Rev.* **2013**, *33*, 305–314.
62. Chen, H.; Gal, Y.-S.; Kim, S.-H.; Choi, H.-J.; Oh, M.-C.; Lee, J.; Koh, K. Potassium ion sensing using a self-assembled calix[4]crown monolayer by surface plasmon resonance. *Sens. Actuators Chem.* **2008**, *133*, 577–581. [CrossRef]
63. Fahnestock, K.J.; Manesse, M.; McIlwee, H.A.; Schauer, C.L.; Boukherroub, R.; Szunerits, S. Selective detection of hexachromium ions by localized surface plasmon resonance measurements using gold nanoparticles/chitosan composite interfaces. *Analyst* **2009**, *134*, 881–886. [CrossRef] [PubMed]
64. Wang, L.; Li, T.; Du, Y.; Chen, C.; Li, B.; Zhou, M.; Dong, S. Au NPs-enhanced surface plasmon resonance for sensitive detection of mercury (II) ions. *Biosens. Bioelectron.* **2010**, *25*, 2622–2626. [CrossRef]
65. Chang, C.C.; Lin, S.; Wei, S.C.; Chen, C.Y.; Lin, C.W. An amplified surface plasmon resonance "turn-on" sensor for mercury ion using gold nanoparticles. *Biosens. Bioelectron.* **2011**, *30*, 235–240. [CrossRef] [PubMed]
66. Fen, Y.W.; Yunus, W.M.M.; Yusof, N.A. Detection of mercury and copper ions using surface plasmon resonance optical sensor. *Sens. Mater.* **2011**, *23*, 325–334.
67. Fen, Y.W.; Yunus, W.M.M.; Yusof, N.A. Surface plasmon resonance optical sensor for detection of essential heavy metal ions with potential for toxicity: Copper, zinc and manganese ions. *Sens. Lett.* **2011**, *9*, 1704–1711. [CrossRef]
68. Fen, Y.W.; Yunus, W.M.M.; Yusof, N.A.; Ishak, N.S.; Omar, N.A.S.; Zainudin, A.A. Preparation, characterization and optical properties of ionophore doped chitosan biopolymer thin film and its potential application for sensing metal ion. *Optik* **2015**, *126*, 4688–4692. [CrossRef]
69. Hur, Y.; Ock, K.; Kim, K.; Jin, S.; Gal, Y.; Kim, J.; Kim, S.; Koh, K. Surface plasmon resonance study on enhanced refractive index change of an Ag^{+} ion-sensing membrane containing dithiosquarylium dye. *Anal. Chim. Acta* **2002**, *460*, 133–139. [CrossRef]

70. Sadrolhosseini, A.R.; Noor, A.S.M.; Moksin, M.M.; Abdi, M.M.; Mohammadi, A. Application of polypyrrole-chitosan layer for detection of Zn (II) and Ni (II) in aqueous solutions using surface plasmon resonance. *Int. J. Polym. Mater.* **2013**, *62*, 284–287. [CrossRef]
71. Fen, Y.W.; Yunus, W.M.M.; Yusof, N.A. Surface plasmon resonance optical sensor for detection of Pb^{2+} based on immobilized p-tert-butylcalix[4]arene-tetrakis in chitosan thin film as an active layer. *Sens. Actuators Chem.* **2012**, *171*, 287–293. [CrossRef]
72. Pelossof, G.; Tel-Vered, R.; Willner, I. Amplified surface plasmon resonance and electrochemical detection of Pb^{2+} ions using the Pb^{2+}-dependent DNAzyme and hemin/G-quadruplex as a label. *Anal. Chem.* **2012**, *84*, 3703–3709. [CrossRef] [PubMed]
73. Kim, E.J.; Chung, B.H.; Lee, H.J. Parts per trillion detection of Ni (II) ions by nanoparticle-enhanced surface plasmon resonance. *Anal. Chem.* **2012**, *84*, 10091–10096. [CrossRef] [PubMed]
74. Fen, Y.W.; Yunus, W.M.M.; Moksin, M.M.; Talib, Z.A.; Yusof, N.A. Surface plasmon resonance optical sensor for mercury ion detection by crosslinked chitosan thin film. *J. Optoelectron. Adv. Mater.* **2011**, *13*, 279–285.
75. Verma, R.; Gupta, B.D. Detection of heavy metal ions in contaminated water by surface plasmon resonance based optical fibre sensor using conducting polymer and chitosan. *Food Chem.* **2015**, *166*, 568–575. [CrossRef] [PubMed]
76. Peralta-Domínguez, D.; Rodriguez, M.; Ramos-Ortiz, G.; Maldonado, J.L.; Luna-Moreno, D.; Ortiz-Gutierrez, M.; Barba, V. A Schiff base derivative used as sensor of copper through colorimetric and surface plasmon resonance techniques. *Sens. Actuators Chem.* **2016**, *225*, 221–227. [CrossRef]
77. May, L.M.; Russell, D.A. Novel determination of cadmium ions using an enzyme self-assembled monolayer with surface plasmon resonance. *Anal. Chim. Acta* **2003**, *500*, 119–125. [CrossRef]
78. Sadrolhosseini, A.R.; Naseri, M.; Rashid, S.A. Polypyrrole-chitosan/nickel-ferrite nanoparticle composite layer for detecting heavy metal ions using surface plasmon resonance technique. *Opt. Laser Technol.* **2017**, *93*, 216–223. [CrossRef]
79. Saleviter, S.; Fen, Y.W.; Sheh Omar, N.A.; Daniyal, W.M.E.M.M.; Abdullah, J.; Mahdi, M.A. Label-free binding analysis of 4-(2-Pyridylazo)-resorcinol-based composite layer with cobalt ion using surface plasmon resonance optical sensor. *Sens. Mater.* **2020**, *32*, 2877–2889. [CrossRef]
80. Saleviter, S.; Fen, Y.W.; Omar, N.A.S.; Daniyal, W.M.E.M.M.; Abdullah, J.; Zaid, M.H.M. Structural and optical studies of cadmium sulfide quantum dot-graphene oxide-chitosan nanocomposite thin film as a novel SPR spectroscopy active layer. *J. Nanomater.* **2018**, *2018*, 1–8. [CrossRef]
81. Zhao, Y.; Gan, S.; Wu, L.; Zhu, J.; Xiang, Y.; Dai, X. GeSe nanosheets modified surface plasmon resonance sensors for enhancing sensitivity. *Nanophotonics* **2020**, *9*, 327–336. [CrossRef]
82. Bakhshpour, M.; Denizli, A. Highly sensitive detection of Cd (II) ions using ion-imprinted surface plasmon resonance sensors. *Microchem. J.* **2020**, *159*, 105572–105583. [CrossRef]
83. Saleviter, S.; Fen, Y.W.; Alia, N.; Omar, S.; Zainudin, A.A.; Yusof, N.A. Development of optical sensor for determination of Co (II) based on surface plasmon resonance phenomenon. *Sens. Lett.* **2017**, *15*, 1–6. [CrossRef]
84. Chah, S.; Yi, J.; Zare, R.N. Surface plasmon resonance analysis of aqueous mercuric ions. *Sens. Actuators Chem.* **2004**, *99*, 216–222. [CrossRef]
85. Daniyal, W.M.E.M.M.; Fen, Y.W.; Abdullah, J.; Sadrolhosseini, A.R.; Saleviter, S.; Omar, N.A.S. Exploration of surface plasmon resonance for sensing copper ion based on nanocrystalline cellulose-modified thin film. *Opt. Express* **2018**, *26*, 34880–34893. [CrossRef] [PubMed]
86. Yu, J.C.C.; Lai, E.P.C.; Sadeghi, S. Surface plasmon resonance sensor for Hg (II) detection by binding interactions with polypyrole and 2-mercaptobenzothiazole. *Sens. Actuators Chem.* **2004**, *101*, 236–241. [CrossRef]
87. Wu, C.M.; Lin, L.Y. Immobilization of metallothionein as a sensitive biosensor chip for the detection of metal ions by surface plasmon resonance. *Biosens. Bioelectron.* **2004**, *20*, 863–870. [CrossRef] [PubMed]
88. Wu, C.M.; Lin, L.Y. Utilization of albumin-based sensor chips for the detection of metal content and characterization of metal-protein interaction by surface plasmon resonance. *Sens. Actuators Chem.* **2005**, *110*, 231–238. [CrossRef]
89. Anas, N.A.A.; Fen, Y.W.; Omar, N.A.S.; Ramdzan, N.S.M.; Daniyal, W.M.E.M.M.; Saleviter, S.; Zainudin, A.A. Optical properties of chitosan/hydroxyl-functionalized graphene quantum dots thin film for potential optical detection of ferric (III) ion. *Opt. Laser Technol.* **2019**, *120*, 105724–105732. [CrossRef]
90. Forzani, E.S.; Zhang, H.; Chen, W.; Tao, N. Detection of heavy metal ions in drinking water using a high-resolution differential surface plasmon resonance sensor. *Environ. Sci. Technol.* **2005**, *39*, 1257–1262. [CrossRef]
91. Moon, J.; Kang, T.; Oh, S.; Hong, S.; Yi, J. In situ sensing of metal ion adsorption to a thiolated surface using surface plasmon resonance spectroscopy. *J. Colloid Interface Sci.* **2006**, *298*, 543–549. [CrossRef] [PubMed]
92. Fen, Y.W.; Yunus, W.M.M.; Talib, Z.A. Analysis of Pb (II) ion sensing by crosslinked chitosan thin film using surface plasmon resonance spectroscopy. *Optik* **2013**, *124*, 126–133. [CrossRef]
93. Hashim, H.S.; Fen, Y.W.; Omar, N.A.S.; Daniyal, W.M.E.M.M.; Saleviter, S.; Abdullah, J. Structural, optical and potential sensing properties of tyrosinase immobilized graphene oxide thin film on gold surface. *Optik* **2020**, *212*, 164786–164797. [CrossRef]
94. Crist, B.V. XPS in industry-problems with binding energies in journals and binding energy databases. *J. Electron Spectros. Relat. Phenom.* **2019**, *231*, 75–87. [CrossRef]
95. Shinotsuka, H.; Yoshikawa, H.; Murakami, R.; Nakamura, K.; Tanaka, H. Automated information compression of XPS spectrum using information criteria. *J. Electron Spectros. Relat. Phenom.* **2020**, *239*, 146903–146910. [CrossRef]
96. Wojdyr, M. Fityk: A general-purpose peak fitting program. *J. Appl. Crystallogr.* **2010**, *43*, 1126–1128. [CrossRef]

97. Fauzi, N.I.M.; Fen, Y.W.; Omar, N.A.S.; Saleviter, S.; Daniyal, W.M.E.M.M.; Hashim, H.S.; Nasrullah, M. Nanostructured chitosan/maghemite composites thin film for potential optical detection of mercury ion by surface plasmon resonance investigation. *Polymers* **2020**, *12*, 1497. [CrossRef]
98. Fen, Y.W.; Yunus, W.M.M.; Talib, Z.A.; Yusof, N.A. Development of surface plasmon resonance sensor for determining zinc ion using novel active nanolayers as probe. *Spectrochim. Acta Mol. Biomol. Spectrosc.* **2015**, *134*, 48–52. [CrossRef]
99. Bennettand, H.; Ouver, G.J. High Resolution XPS of Organic Polymers: The Scienta ESCA300 Database. (Beamson, G.; Briggs, D.). *J. Chem. Educ.* **1993**, *70*, A25.
100. De Moraes, A.C.M.; Lima, B.A.; De Faria, A.F.; Brocchi, M.; Alves, O.L. Graphene oxide-silver nanocomposite as a promising biocidal agent against methicillin- resistant Staphylococcus aureus. *Int. J. Nanomed.* **2015**, *10*, 6847–6861. [CrossRef]
101. Liu, Q.; Li, D.; Zhu, Z.; Yu, S.; Zhang, Y.; Yu, D.; Jiang, Y. N-doped carbon dots from phenol derivatives for excellent colour rendering WLEDs. *RSC Adv.* **2018**, *8*, 4850–4856. [CrossRef]
102. Anas, N.A.A.; Fen, Y.W.; Omar, N.A.S.; Daniyal, W.M.E.M.M.; Ramdzan, N.S.M.; Saleviter, S. Development of graphene quantum dots-based optical sensor for toxic metal ion detection. *Sensors* **2019**, *19*, 3850. [CrossRef]
103. Gu, Y.; Tang, L.; Guo, X.; Xiang, J.; Seng Teng, K.; Ping Lau, S. Preparation and photoelectric properties of cadmium sulfide quantum dots. *Chin. Phys.* **2019**, *28*, 47803. [CrossRef]
104. Fantauzzi, M.; Elsener, B.; Atzei, D.; Rigoldi, A.; Rossi, A. Exploiting XPS for the identification of sulfides and polysulfides. *RSC Adv.* **2015**, *5*, 75953–75963. [CrossRef]
105. Fen, Y.W.; Yunus, W.M.M.; Yusof, N.A. Optical properties of crosslinked chitosan thin film as copper ion detection using surface plasmon resonance technique. *Opt. Appl.* **2011**, *41*, 999–1013.
106. Daniyal, W.M.E.M.M.; Fen, Y.W.; Anas, N.A.A.; Omar, N.A.S.; Ramdzan, N.S.M.; Nakajima, H.; Mahdi, M.A. Enhancing the sensitivity of a surface plasmon resonance-based optical sensor for zinc ion detection by the modification of a gold thin film. *RSC Adv.* **2019**, *9*, 41729–41736. [CrossRef]
107. Lokman, N.F.; Bakar, A.A.A.; Suja, F.; Abdullah, H.; Rahman, W.B.W.A.; Huang, N.M.; Yaacob, M.H. Highly sensitive SPR response of Au/chitosan/graphene oxide nanostructured thin films toward Pb (II) ions. *Sens. Actuators Chem.* **2014**, *195*, 459–466. [CrossRef]
108. Ho, Y.S.; Porter, J.F.; McKay, G. Equilibrium isotherm studies for the sorption of divalent metal ions onto peat: Copper, nickel and lead single component systems. *Water Air Soil Pollut.* **2002**, *141*, 1–33. [CrossRef]
109. Daniyal, W.M.E.M.M.; Fen, Y.W.; Abdullah, J.; Sadrolhosseini, A.R.; Saleviter, S.; Omar, A.S. Lable-free optical spectroscopy for characterizing binding properties of highly sensitive nanocrystalline cellulose-graphene oxide based nanocomposite towards nickel ion. *Spectrochim. Acta Mol. Biomol. Spectrosc.* **2018**, *212*, 25–31. [CrossRef]
110. Rosddi, N.N.M.; Fen, Y.W.; Anas, N.A.A.; Omar, N.A.S.; Ramdzan, N.S.M.; Daniyal, W.M.E.M.M. Cationically modified nanocrystalline cellulose/carboxyl-functionalized graphene quantum dots nanocomposite thin film: Characterization and potential sensing application. *Crystals* **2020**, *10*, 875. [CrossRef]
111. Omar, N.A.S.; Fen, Y.W.; Abdullah, J.; Chik, C.E.N.C.E.; Mahdi, M.A. Development of an optical sensor based on surface plasmon resonance phenomenon for diagnosis of dengue virus E-protein. *Sens. Biosensing Res.* **2018**, *20*, 16–21. [CrossRef]
112. Saleviter, S.; Fen, Y.W.; Daniyal, W.M.E.M.M.; Abdullah, J.; Sadrolhosseini, A.R.; Omar, N.A.S. Design and analysis of surface plasmon resonance optical sensor for determining cobalt ion based on chitosan-graphene oxide decorated quantum dots-modified gold active layer. *Opt. Express* **2019**, *27*, 32294–32307. [CrossRef] [PubMed]
113. Omar, N.A.S.; Fen, Y.W.; Abdullah, J.; Kamil, Y.M.; Daniyal, W.M.E.M.M.; Sadrolhosseini, A.R.; Mahdi, M.A. Sensitive detection of dengue virus type 2 E-proteins signals using self-assembled monolayers/reduced graphene oxide-pamam dendrimer thin film-SPR optical sensor. *Sci. Rep.* **2020**, *10*, 1–15. [CrossRef]
114. Sadrolhosseini, A.R.; Shafie, S.; Rashid, S.A.; Mahdi, M.A. Surface plasmon resonance measurement of arsenic in low concentration using polypyrrole-graphene quantum dots layer. *Measurement* **2020**, *10*, 108546–108557. [CrossRef]

polymers MDPI

Article

One-Step Synthesis of Eu^{3+}-Modified Cellulose Acetate Film and Light Conversion Mechanism

Zhihui Zhang [1,2,3], Zhengdong Zhao [1,2,3], Yujia Lu [1,2,3], Di Wang [1,2,3,*], Chengyu Wang [1,2,3] and Jian Li [1,2,3]

1 Engineering Research Center of Advanced Wooden Materials (Ministry of Education), Northeast Forestry University, Harbin 150040, China; zzh13251613906@nefu.edu.cn (Z.Z.); Zhaozd6866@163.com (Z.Z.); luyujia1999@163.com (Y.L.); wangcy@nefu.edu.cn (C.W.); nefujianli@163.com (J.L.)

2 Key Laboratory of Bio-based Material Science and Technology (Ministry of Education), Northeast Forestry University, Harbin 150040, China

3 Collage of Material Science & Engineering, Northeast Forestry University, Harbin 150040, China

* Correspondence: diwang1030@nefu.edu.cn; Tel.: +86-1864-506-5081

Abstract: A CA-Eu(III) complex was synthesized by the coordination reaction of cellulose acetate (CA) and Eu^{3+} to obtain a CA-Eu light conversion film. This product was then doped with Tb(III) to sensitize the luminescence of Eu^{3+}, which could functionalize the CA film. FTIR and XPS showed that the oxygen atoms in C=O, C–O (O=C–O), and O–H were involved in the complexation with Eu^{3+} and formed a Eu–O bond. SEM revealed that Eu^{3+} filled in the pores of the CA film. By changing the experimental conditions, the best fluorescence performance was obtained at the CA: Eu^{3+} ratio of 3:1 with a reaction time of 65 min. The energy transfer between Tb^{3+}–Eu^{3+} could be realized by doping Tb^{3+} to enhance the luminescence of Eu^{3+}. The best fluorescence performance of the CA-Eu-Tb light conversion film was at a Eu^{3+}:Tb^{3+} ratio of 3:1. Compared with the CA film, the light conversion film has high transparency, high tensile strength, and good flexibility. It can convert the ultraviolet light harmful to plants into red light that is beneficial to photosynthesis. This offers high efficiency and environmental protection in the field of agricultural films.

Keywords: light conversion film; cellulose acetate; europium; sensitization

Citation: Zhang, Z.; Zhao, Z.; Lu, Y.; Wang, D.; Wang, C.; Li, J. One-Step Synthesis of Eu^{3+}-Modified Cellulose Acetate Film and Light Conversion Mechanism. *Polymers* **2021**, *13*, 113. https://doi.org/10.3390/polym13010113

Received: 4 December 2020
Accepted: 24 December 2020
Published: 30 December 2020

1. Introduction

Agricultural light conversion films achieve light conversion by adding light conversion agents, thereby changing the light quality. The absorption spectrum of chlorophyll is in the blue-violet light region (400–500 nm) and the red-orange light region (600–680 nm). The agricultural light conversion film blocked the harmful ultraviolet light and converted it into red-orange light, which promoted the photosynthesis of plants and improved the use of light, thereby increasing the yield of crops [1].

The light conversion agents used in agricultural films can be divided into rare earth organic ligand light conversion agents, rare earth inorganic light conversion agents, and fluorescent dyes according to the light conversion principles and the general category of elements. To explore highly efficient yellow-green light conversion agents used for agricultural film, Yu et al. [2] prepared six bay-substituted perylene diimides and their polyvinyl chloride-doped films. Qi et al. [3] explored a new type of light conversion film based on aggregation-induced emission (AIE). Gong et al. [4] obtained a series of multicolor Si NPs powders with fluorescence colors from blue to red by a facile high temperature calcination strategy for the first time and the carboxymethyl cellulose (CMC)/Si NPs powders obtained by calcining under air atmosphere for 2 min (SSF-2) (CMC/SSF-2) films with strong UV absorption, good transparency, and high tensile strength have been obtained with great potential in the field of agricultural planting. However, most mulching films were made of polyethylene, causing white pollution when they remained in the soil after use [5]. Therefore, the use of degradable mulch is an excellent way to solve the pollution problem [6].

Yu [7] et al. researched and synthesized a polylactic acid/polybutylene terephthalate/rare earth complex biodegradable light conversion agricultural film. Wang et al. [8] synthesized a rare earth europium(III) complex with α-thenoyltrifluoroacetone and triphenylphosphine oxide ($Eu(TTA)_3(TPPO)_2$) and then blended this product with polylactide (PLA) and poly(butylene adipate-co-terephthalate) (PBAT) to prepare a biodegradable and efficient blue-violet light conversion agricultural film. They further explored this for environmental pollution. In addition, some light conversion agents had poor dispersibility and compatibility with the base resin of the greenhouse film, which may deteriorate the processing performance and even affect the mechanical properties of the agricultural film during the production of the greenhouse film. To solve these problems, research into rare earth polymer complex fluorescent materials is needed.

The polymer ligands of this type of material had good compatibility with the base resin, and the central ions usually included Eu^{3+}, Sm^{3+}, and Er^{3+} [9]. Cellulose is the most abundant renewable resource on the earth, and its functionalization has always been a hot research topic. Cellulose is renewable with large reserves and easy biodegradability. It can be used to construct various functional polymer materials, and cellulose acetate (CA) is one of the most widely used cellulose derivative materials. This is a film-forming material with good performance—the molecular chain contained a large number of polar groups such as hydroxyl and acetyl groups, and the intermolecular forces are large as well [10,11].

Based on reference with the light conversion agricultural film of the predecessors, we used a rare earth ion Eu(III) with higher light conversion efficiency to modify CA. Here, CA-Eu(III) complexes were synthesized by the reaction of CA and Eu^{3+} to prepare a CA film modified with rare earth ions, and doped with Tb(III) to sensitize the luminescence of Eu(III) and functionalize the CA film. The luminescence performance was mainly studied to determine the best production conditions to improve the utilization rate of rare earths to reduce costs; this can be used in a wide range of agricultural films. The organic ligand CA in the films had a good absorption coefficient in the ultraviolet wavelength range. As long as the ligand could meet the requirements of energy transfer, it could solve the f-f prohibition of rare earth ions through ultraviolet absorption and transfer after coordination with the rare earth ions to enhance the luminous efficiency of rare earth (Antenna effect) [12]. Due to its high transparency, high tensile strength, good flexibility, and absorption of ultraviolet light to enhance visible light and biodegradable characteristics, it offers high efficiency and environmental protection in the field of agricultural films.

In addition, the product (casting liquid) of this experiment was the organic rare earth materials, which has the characteristics of low toxicity, high fluorescence chroma, and good oil solubility. Therefore, it could also be used for anti-counterfeiting identification, highlighter oils, and other directions.

2. Materials and Methods

2.1. Preparation of Ion-Modified Cellulose Acetate Light Conversion Film

First, cellulose acetate (CA, the degree of substitution (DS) = 2.45, acetyl 39.8 wt%, hydroxyl 3.5 wt%, Aladdin Bio-Chem Technology Co., Ltd., Shanghai, China) (2.6667 g, 1.0 mol) and europium nitrate hexahydrate (99.9%, Huawei Ruike Chemical Co., Ltd., Beijing, China) (2.2453 g, 5.03×10^{-3} mol) were dissolved separately in acetone (analytical grade, Fuyu Fine Chemical Co., Ltd., Tianjin, China) (20 mL for CA (in a single-mouthed flask) and 10 mL for Eu (in a beaker)) until they were completely dissolved. The sample was then poured into a single-mouthed flask and mixed evenly. CA in the mixed solution was 10 wt%, and the molar ratio of CA (structural units) to rare earth ions was 2:1. The mixing solution was stirred with magnetic stirrer at constant temperature (80 °C) under reflux for 35 min and cooled to room temperature, leading to a rare earth metal europium-modified cellulose acetate solution.

Control variables and change of other reaction conditions. Here, CA: Eu^{3+} ratios of 1:1, 2:1, 3:1, 4:1, and 5:1 were used. The doping molar ratio variable (Eu^{3+}: Tb^{3+} ratio of 1:1, 2:1, 3:1, 4:1, 5:1 and 1:1, 1:2, 1:3, 1:4, 1:5) and the reaction time variable (35, 45, 65, and 75 min) were also optimized. A series of ion-modified cellulose acetate solutions were then prepared under different reaction conditions.

The appropriate amount of ion-modified cellulose acetate solution was cast on a dry and clean glass plate, and the film was scraped with a SZQ type four-sided preparation device (AI Testing Instrument Co., Ltd., Taizhou, China); the thickness of the film was set to 1000 μm.

2.2. Structural Characterization and Performance Testing

The chemical groups of the films were detected by the Fourier-transform infrared (FTIR) spectroscopy (FTIR-650) (Gangdong Technology Co. Ltd., Tianjin, China): 4000–600 cm^{-1}, 32 scans, and resolution of 4 cm^{-1}. X-ray photoelectron spectroscopy (XPS, PHI5700 ULVCA-PHI, Chigasaki, Japan) was used to characterize the chemical structure of the films, and Casa XSP software was used to analyze the spectra.

Surface and cross-sectional morphology of the films were obtained by scanning electron microscope (SEM) (Apreo s, Thermo Fisher Scientific, Waltham, MA, USA). The films were immersed in liquid nitrogen for two minutes, and the cross-section samples were prepared by the brittle fracture method. The surface and the cross-section samples were sprayed with gold for observation on the experimental bench. The acceleration voltage was 5 kV.

The UV absorption and transmission spectra of the films were detected by TU-1950 UV-visible spectrophotometer (General Analysis General Instrument Co., Ltd., Beijing, China) from 200 to 700 nm.

The fluorescence spectroscopy (FL) was detected by HORIBA highly sensitive integrated FluoroMax-4 fluorescence spectrometer (HORIBA Scientific Instruments Division, Sunnyvale, CA, USA), and the scanning speed was 600 $nm \cdot min^{-1}$.

The mechanical properties were detected by CMT5504 electronic universal testing machine (New Sansi Metrology Technology Co., Ltd., Shenzhen, China). The test method was based on GB/T 1040.3-2006, and three sets of samples were tested in parallel. The flat, clean, and defect-free films were cut into 12 × 100 mm^2 strips for testing. The test speed was set to 2.0 mm/min, and the nominal distance was 45 mm. The load displacement curve was obtained through tensile testing, and the tensile performance parameters of the films could then be obtained: tensile strength (σ_M) and elongation at break (ε_B).

The thermal stability of the films was obtained by thermogravimetric analyzer (TG 209 F1, Netzsch instruments Co. Ltd., Bavarian, Germany) under a nitrogen atmosphere. The test temperature range was 30 to 800 °C at a heating rate of 10 °C/min.

3. Results

3.1. Structure and Morphology Characterization of Films

The infrared spectra of CA, CA–Eu (2:1), and CA–Eu–Tb (Eu:Tb = 1:1) films were shown in Figure 1. Of these, there was a wide peak at 3480 cm^{-1} in the pure CA film, which is attributed to the stretching vibration peak of O–H. However, the O–H group in the light conversion films red-shifted (the O–H had moved to 3396 cm^{-1}) and the absorption peak was obviously broadened and stronger. This is because the hydroxyl group can be freed from the hydrogen bonds formed with other groups after the hydroxyl group on the CA molecular chain coordinated with Eu^{3+}/Tb^{3+} [13]. The band observed at 2966 cm^{-1} belongs to the C–H stretching vibration peak [14]. The peaks at 1737 cm^{-1} and 1431 cm^{-1} correspond to the symmetric and antisymmetric stretching vibrations of $-COO^-$ in CA, whereas both ν_s ($-COO^-$) and ν_{as} ($-COO^-$) had shifted in the light conversion films with a broadened peak shape, which initially indicated that Eu^{3+}/Tb^{3+} coordinated with the O atom in the carbonyl group. The absorption peaks at 1031 cm^{-1} and 1215 cm^{-1} are attributed to the symmetrical stretching peak of C–O–C (there was basically no movement in the light conversion films) and the asymmetric stretching of C–O–C (blue shift to the position near 1225 cm^{-1} in the light conversion films) in the ester group [15], respectively. The peaks observed at 1367 cm^{-1} and 900 cm^{-1} are due to the characteristic absorption peaks of $-CH_3$ [16] and the β-linked glucan structure [17]. The new vibration peaks appeared in the light conversion films at 810 cm^{-1} and 737 cm^{-1}, which correspond to the infrared characteristic peaks of Eu–O and Tb–O [18] indicating the formation of coordination structures. Table S1 (see Supporting Information) summarizes the positions and changes of some absorption peaks of the films.

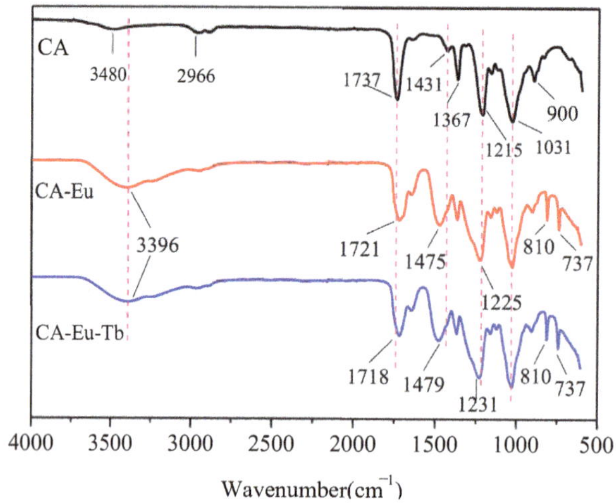

Figure 1. FT-IR spectra pure CA, CA-Eu (2:1), and CA-Eu-Tb (Eu: Tb = 1:1) films.

To further explore the elements and structural composition of the light conversion films, XPS was performed to characterize the functional groups on the surface. An XPS survey spectrum was shown in Figure 2a. The main constituent elements of the pure CA film were C and O, and their contents were C (68.21 At%), and O (31.79 At%), respectively. In addition to the C and O elements contained in the light-converting films, Eu or Tb also appeared further indicating that Eu (III)/Tb(III) reacted with CA successfully and doped into the light-converting films.

The C1s and O1s chemical bonds in the pure CA film, CA-Eu, and CA–Eu–Tb light conversion films were analyzed by peak fitting method. Figure 2b is the C1s spectrum of CA, containing three peaks positioned at 284.8, 286.774, and 289.002 eV, which were respectively assigned to C–C or C–H in the molecular chain skeleton, C–OH and C–O–C bond in the pyranose ring, and O=C–O in the acetyl group, respectively [19–21]. As shown in Figure 2c,d, the spectral shapes changed and the O–C=O bond of CA–Eu and CA–Eu–Tb light-converting films had moved 0.163 eV and 0.269 eV to the lower binding energy position compared with CA film, respectively. This is because the electron density around the C atom increased because of the coordination reaction accompanied by electron transfer; thus, its inner electron binding energy decreased inevitably. The O1s spectrum of CA was de-convoluted into three peaks positioned at 531.493, 531.907, and 532.693 eV; they belonged to the –OH, C–O, and carbonyl C=O bonds, respectively [22]. The three peaks in the CA–Eu film were located at 531.945, 532.393, and 533.131 eV, and these peaks in the CA–Eu–Tb film were located at 531.74, 532.297, and 533.06 eV. Compared with the CA film, the binding energy of O1s in the light conversion films all shifted to a higher position because of the coordination between Eu(III)/Tb(III). The O atoms in C=O, C–O (O=C–O), and –OH bonds transferred some electrons to the rare earth metals and formed Eu-O/Tb-O bond. Therefore, the electron density decreased and the binding energy of O1s increased in XPS spectra. The binding energy of C1s and O1s is summarized in Table S2 (see Supporting Information). The internal coordination structure of the light conversion films could be obtained combined with the FTIR analysis; two schematic diagrams of the CA-Eu structure are shown in Figure 3.

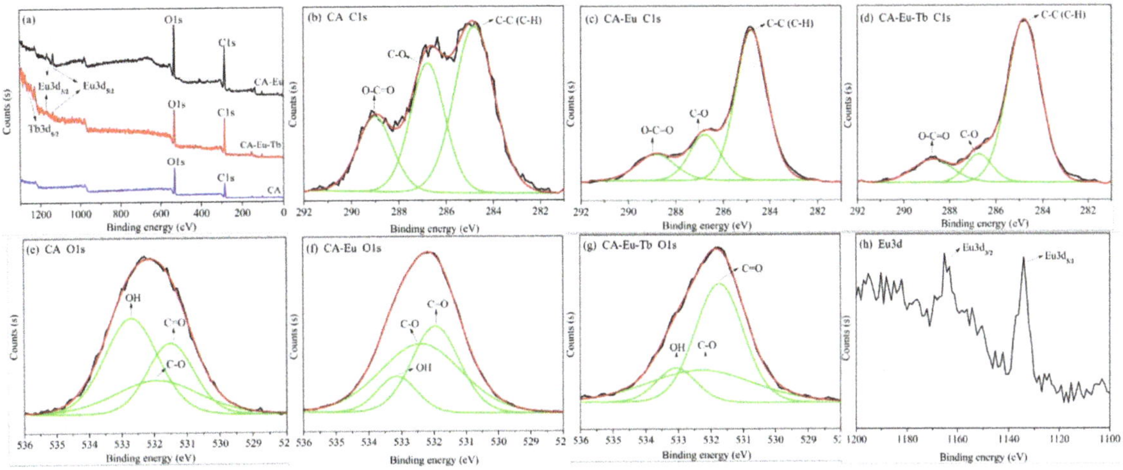

Figure 2. The XPS spectra of CA, CA–Eu and CA–Eu–Tb films. (**a**)—the survey spectrum, (**b**)—the C1s spectra for CA, (**c**)—CA–Eu, (**d**)—CA–Eu–Tb; (**e**)—the O1s spectra for CA, (**f**)—CA–Eu, (**g**)—CA–Eu–Tb, (**h**)—the Eu3d spectra.

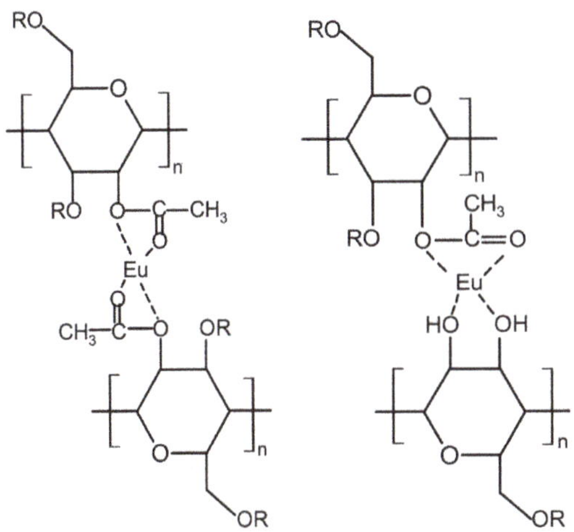

Figure 3. Schematic diagrams of the coordination structure between CA and Eu (III). (R = H or CH_3CO–).

The XPS spectrum of CA-Eu film in Figure 2h revealed the characteristic peaks of Eu3d (1165.08 eV for $Eu3d_{3/2}$ and 1134.08 eV for $Eu3d_{5/2}$), which indicated the presence of Eu(III) in CA-Eu light conversion film. These two spin orbitals made Eu^{3+} emit red and orange light, while the integration area for $Eu3d_{5/2}$ orbitals was larger so that the CA–Eu light conversion film mainly had red emission [23], which is consistent with the results.

Figure 4 shows SEM images of the surface and section of CA film and CA–Eu film. Figure 4a,b (magnification of 10,000×) shows that the surface of the pure CA film had the porous structure of different sizes [24–26], however, the pores in CA–Eu light conversion film were filled. And these tiny pores at the surface are a result of the phase separation [27]. During the process of scraping the film, bubbles were generated in the film due to the rapid volatilization of acetone (act as a pore creating agent). Due to the short film formation time (about 2–3 min), the cellulose acetate solution did not have enough time to gather and formed the pores of different sizes in the bubbles. Similarly, bubbles were generated as well during preparation process of the light conversion film. However, the distance between molecules increased after doping Eu^{3+}, which could reduce the entanglement between long molecular chains and accelerated the movement of molecules, thereby reducing the viscosity. The conditional viscosity was measured using the QNO-4 viscometer (CA-22.9s, CA–Eu-11.0s, it is confirmed that the viscosity reduced after doping Eu^{3+} and measurement method can be seen in Supporting Information.). Low viscosity made the cellulose acetate solution had enough time to gather after rapid volatilization of acetone, which shows that the pores were filled in CA–Eu film. The surface of the CA–Eu film was uniform and smooth, indicating that CA and Eu^{3+} had good compatibilities so that the high transparency of the light conversion film could be observed. Section electron micrographs (magnification of 40,000×) in Figure 4c,d show that the longitudinal distribution of the films was uniform and dense. Uniformly distributed spherical particles could be observed in the CA-Eu light conversion film. There was no obvious particle agglomeration phenomenon, further indicating that Eu^{3+} had good compatibility with CA.

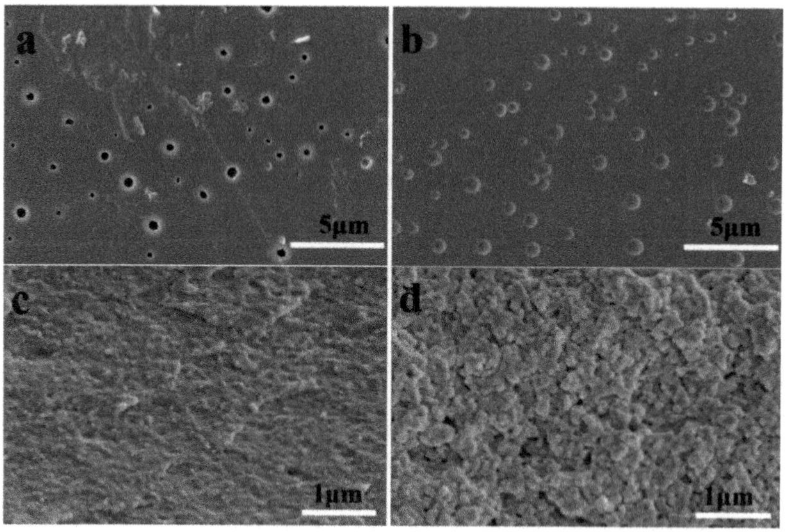

Figure 4. Scanning electron micrographs of CA film and CA-Eu light conversion film. (**a**) the surface of CA film, (**b**) the surface of CA-Eu film, (**c**) the section of CA film, (**d**) the section of CA-Eu film.

Ultraviolet absorption spectra of the pure CA and light-conversion films are shown in Figure 5. The UV absorption of the light conversion films was stronger than that of the pure CA film at 200–300 nm and converted ultraviolet light into red light. The CA film formed a broad absorption band at about 208 nm, which originated from the π–π^* transition absorption peak of the carbonyl group. After doping with rare earth ions, the electronic environment around the ligand CA changed due to the presence of the central ion, which resulted in a red shift of the carbonyl absorption wavelength. Additionally, in the light-conversion films, the shape of the carbonyl absorption peak became sharper, and new absorption peaks appeared. This is mainly the result of the reaction of rare earth ions with CA to form a coordination structure. In addition, the UV absorption (300–400 nm) of the films was basically equal to zero, indicating that the light-conversion films played an important role in avoiding harmful ultraviolet radiation [28].

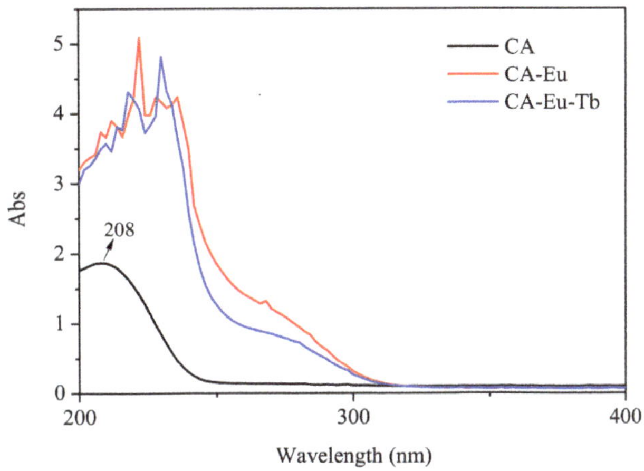

Figure 5. UV absorption spectra of CA, CA-Eu and CA-Eu-Tb films.

3.2. *Fluorescent Characterization of the Light Conversion Films*

The emission spectra (fluorescence spectroscopy) of CA-Eu light conversion film with different proportions (λ_{ex} = 395 nm) are shown in Figure 6a. The emission light purity of CA-Eu film was high, and different ratios had no effect on the position of the emission peak. This was mainly composed of four emission peaks located at 578, 591, 615, and 649 nm, corresponding to $^5D_0 \rightarrow {}^7F_0$, $^5D_0 \rightarrow {}^7F_1$, $^5D_0 \rightarrow {}^7F_2$, and $^5D_0 \rightarrow {}^7F_3$ energy level transition of Eu^{3+}. Of these, the electric dipole transition ($^5D_0 \rightarrow {}^7F_2$) at 615 nm made CA-Eu light conversion film emit high-purity red light; the intensity of the emission peak was the largest, indicating that the CA-Eu film could efficiently convert harmful ultraviolet light into red light beneficial to plant growth. The intensity of the magnetic dipole transition (red-orange light) at 591 nm was the second highest. Therefore, the emission peak intensity at 615 nm was used to compare the fluorescence intensity of CA-Eu light-converting films prepared under different conditions. Figure S1a (in Supporting Information) shows that the fluorescence intensity first increased and then decreased with increasing Eu^{3+}, i.e., concentration quenching [29,30]. The initial increase in Eu^{3+} concentration implied that the number of complexes would increase accordingly; the fluorescence intensity gradually increased as well, and the coordination reaction was complete when the doping amount of Eu^{3+} exceeded a certain range; thus, the ion distance was shortened to generate energy resonance, which caused the luminescence intensity to decrease. Therefore, the fluorescence intensity of the light conversion film for a CA:Eu^{3+} ratio of 3:1 was optimal when other conditions were the same.

CA, CA-Eu and CA-Eu-Tb light conversion films were put under the ZF-20D dark-box ultraviolet analyzer (Yuhua Instrument Co., Ltd., Gongyi, China) and appeared different fluorescent color. The optical photograph was shown in Figure 6d.

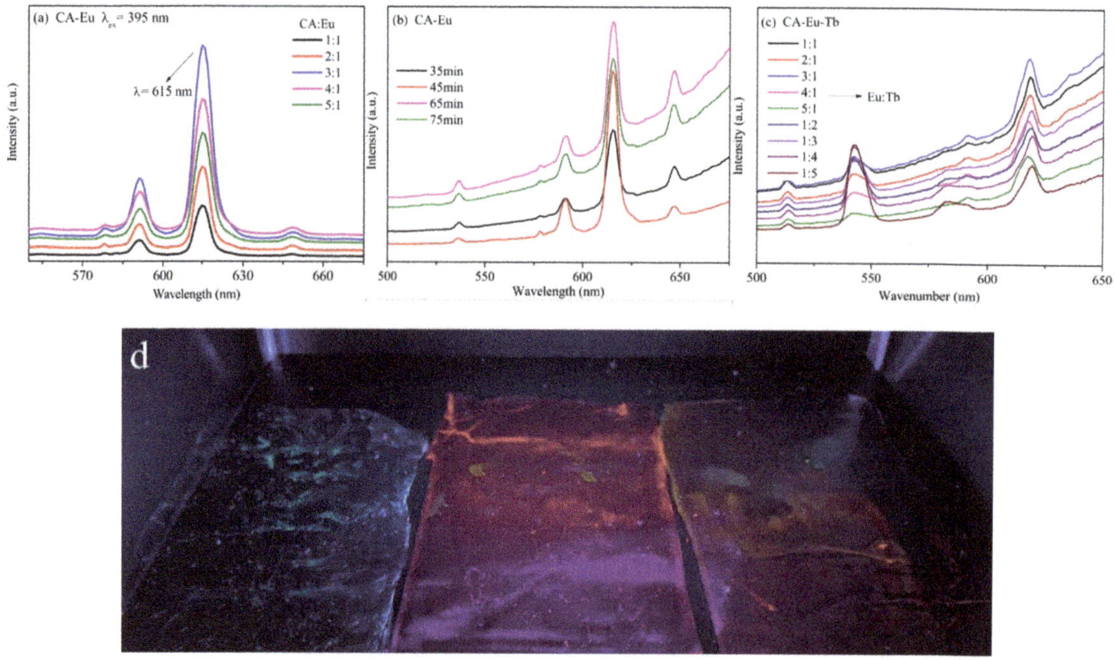

Figure 6. Emission spectra of light conversion films. (**a**) CA–Eu film with different proportions, (**b**) CA–Eu film with different reaction time. (**c**) CA–Eu–Tb film with different Eu^{3+}:Tb^{3+} ratios, (**d**) the optical photograph of CA and the light conversion films. (From left to right are CA, CA–Eu and CA–Eu–Tb).

Figure 6b shows the emission spectra of CA-Eu light conversion films prepared under different reaction time conditions. The intensity comparison of the characteristic emission peak of Eu^{3+} at 615 nm (red light) is shown in Figure S1b; it was obvious that the fluorescence intensity increased first and then decreased with increasing reaction time. This is because the coordination reaction was not sufficient at short times and fewer complexes were formed; thus, the fluorescence intensity was relatively low. As the reaction time continued, the coordination between CA and Eu^{3+} became increasingly complete, and more complexes were formed accordingly. As the reaction time reached 65 min, the UV absorption of the ligand was the strongest, and CA absorbed the most energy and transferred the energy to Eu^{3+} effectively at this moment; thus, optimal fluorescence properties were obtained [31]. With longer reaction times, heat would destroy the balance that established between the intra molecular and intermolecular reactions so that the length of the Eu–O bond became larger. This caused a decline in the fluorescence intensity and proved that the Eu–O bond distance had an important effect on the emission intensity: A shorter bond length leads to a higher luminescence intensity [26]. Therefore, the fluorescence intensity of the light conversion film for the reaction time at 65 min was optimal when other conditions were the same.

The emission spectra of CA–Eu–Tb light conversion films with different Eu/Tb doping ratios (the total amount of rare earth ions was fixed) are shown in Figure 6c. In addition to the emission peak of Eu^{3+} in the fluorescence spectrum, the characteristic emission peaks of Tb^{3+} could also be monitored under excitation light at 375 nm (the characteristic excitation wavelength of Tb^{3+}) located at 488, 542, 583, and 620 nm corresponding to $^5D_4 \rightarrow {}^7F_6$, $^5D_4 \rightarrow {}^7F_5$, $^5D_4 \rightarrow {}^7F_4$ and $^5D_4 \rightarrow {}^7F_3$ energy level transitions in Tb^{3+}, respectively. Figure S1c compares the intensity of the characteristic emission peak at 615 nm. The fluorescence intensity of CA-Eu-Tb light-conversion film was stronger than that of CA–Eu light conversion film (Eu:Tb = 1:0), which indicated that Tb^{3+} could sensitize Eu^{3+} to emit light. After the CA–Eu–Tb film absorbed the ultraviolet light with a wavelength at 375 nm, part of the energy was transferred to the $^5D_4$ energy level of Tb^{3+}; the other part of the energy was absorbed by Eu^{3+}, making it transition from the ground state ($^7F_0$) to the excited state ($^5L_6$). Part of Tb^{3+} at the $^5D_4$ energy level transitioned to the $^7F_5$ excited state to generate visible light at 545 nm—most of that was transferred to the $^5D_1$ energy level of Eu^{3+} and then transitioned to the $^5D_0$ energy level through cross relaxation. Eu^{3+} at the $^5L_6$ energy level reached the $^5D_0$ energy level after ($^5D_3$, $^5D_2$, $^5D_1$) continuous cross relaxation, and then transitioned to ($^7F_1$, $^7F_2$, $^7F_3$) energy level to generate the light with different wavelengths and realize the energy transfer between Tb^{3+} and Eu^{3+}. This process increased the luminescence intensity of Eu^{3+} and played a sensitization role [32]. In addition, the fluorescence intensity increased first and then decreased with increasing Tb^{3+} concentration, which led to concentration quenching. The fluorescence performance of the light conversion film was optimal at a Eu^{3+}:Tb^{3+} ratio of 3:1 (0.75:0.25).

3.3. Performance Characterization of the Films

Figure 7 shows the UV-visible transmission spectra of the films at 200–700 nm. The pure CA film had good UV-Vis transmittance (80%). The light transmittance from 200 to 240 nm was almost 0%, but it reached ~80% after 240 nm; CA–Eu–Tb film had the strongest light transmittance and reached 86%. The pores probably scatter the light, and the film with greater porosity makes higher scattering intensity [33]. SEM images show that CA film had the porous structure and the pores were filled in CA–Eu film, so the light transmittance of the light conversion films was higher than that of the CA film. On the other hand, the section electron micrograph of CA film shows that the longitudinal distribution was uniform and dense and the molecular was the stretched state, which indicated the light transmission probably reduced. After doping Eu^{3+}, the stretched molecular changed into uniformly distributed spherical particles structure in CA–Eu film, which may leave the empty space and conducive to light transmission. Therefore, the light transmittance of the light conversion films was higher than that of the CA film. The average light transmittance of the films at 200–350 nm and 350–700 nm is summarized in Table 1. Compared with the pure CA film, the light conversion films had a significant effect on UV absorption and could maintain a high transmittance of the CA film as well.

Table 1. Mean of the pure CA and light-conversion film for transmittance.

Sample	T% Mean(200–350 nm)	T% Mean(350–700 nm)
CA	54.3625	81.20739
CA–Eu	24.0450	83.23807
CA–Eu–Tb	30.5563	85.94716

The above data is calculated by origin 8.5. (OriginLab, Northampton, MA, USA)

Adjusting the size (250–1000 μm) of the preparation device could lead to different thicknesses of light conversion films. A stainless steel electronic digital caliper (Shenhan Measuring Tool Co., Ltd., Shanghai, China) was used to accurately measure the thickness of the films. Five points were selected on the films to measure and take the average values, and the actual thicknesses could be obtained: 1000–72 μm, 750–24 μm, 500–20 μm, and 250–10 μm. Figure 7b shows the UV-visible transmission spectra of CA-Eu light conversion films for different thicknesses and the average light transmittance in the wavelength range of 200–700 nm was calculated in Table 2. Obviously, as the film thickness decreased, the transmittance of CA–Eu film in the visible light region gradually increased. The average light transmittance of CA–Eu light conversion film with a thickness of 10 μm could reach the maximum exceeding 90%. Compared with the previously work, the light conversion films present the high transmittance in the visible range [34–37].

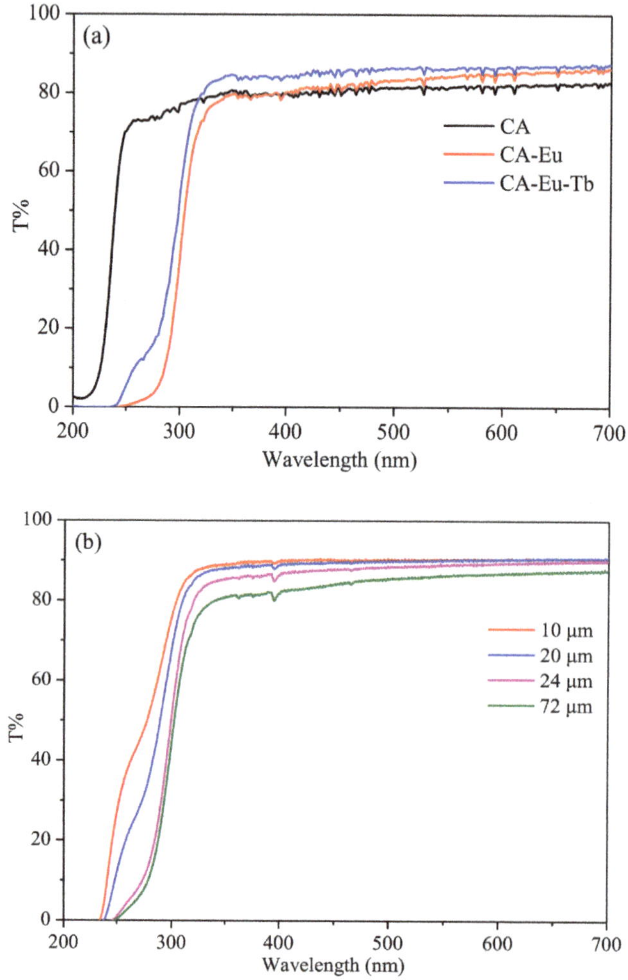

Figure 7. UV-visible transmittance spectra of CA, CA–Eu and CA–Eu–Tb films. (**a**) the transmittance spectra, (**b**) the transmittance spectra of CA–Eu film of different thicknesses.

Table 2. Mean of light-conversion film (CA–Eu) of different thicknesses for transmittance.

Sample (CA–Eu)	T% Mean(200–350 nm)	T% Mean(350–700 nm)
10 μm	43.78562	90.17846
20 μm	37.51844	89.81298
24 μm	29.26031	88.35093
72 μm	26.37656	85.15506

Mechanical properties include tensile strength that represents the strength, and elongation at breaking that reflects flexibility and elasticity of the films. The tensile strength could be calculated by the formula:

$$\sigma_M = \frac{F}{b \times d} \quad (1)$$

where σ_M represents the tensile strength (MPa), F is the maximum tensile force (N), b is the width (mm), and d is the thickness (mm) of the films. The elongation at breaking could be calculated by the formula:

$$\varepsilon_B = \frac{L - L_0}{L_0} \times 100\% \tag{2}$$

where ε_B is the elongation at breaking (%), L_0 is the distance between the clamps before the film was stretched (mm), and L is the distance between the clamps when the film was broken (mm). Figure 8 shows the load-displacement curves of CA and CA–Eu films; the mechanical properties are summarized in Table 3. The tensile strength and elongation at breaking of CA films were 11.4 MPa, and 2.7%, respectively. The mechanical properties of CA–Eu light conversion film were higher than those of CA film, which were 13.3 MPa, and 4.8%, respectively. These results indicated that the tensile strength of CA film could be enhanced by adding Eu^{3+} because of the interaction between Eu^{3+} and CA molecules to form a stable coordination structure. The interaction also enhanced the fluidity of the molecular chain to enhance the flexibility and elasticity of the light transfer films on the macroscale [27,38]. And CA–Eu light conversion film presents good mechanical properties compare with the previously reported CA based films, such as cellulose acetate–polyurethane (CA–PU) film [39], the CA-based composite membranes incorporated with bamboo-based lignocellulose nanofibrils (LCNFs) [40] and et al.

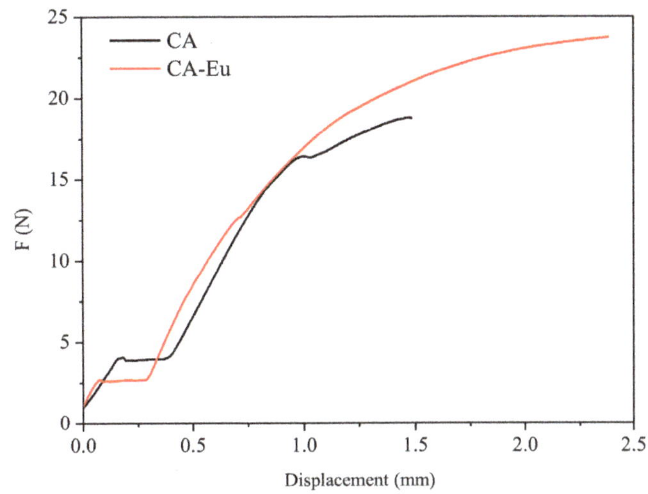

Figure 8. The load-displacement curves of CA and CA–Eu films.

Table 3. The mechanical properties of CA film and CA-Eu light conversion film.

Sample	σ_M (MPa)					Mean/SD	ε_B (%)					Mean/SD
	1	2	3	4	5		1	2	3	4	5	
CA	6.3744	13.0921	13.8606	12.2708	11.5697	11.4/3.0	2.8119	3.3088	1.8889	2.3977	3.2536	2.7/0.6
CA-Eu	16.2126	13.1485	9.6287	14.9523	12.7006	13.3/2.5	5.9627	5.309	4.9064	4.1741	3.8461	4.8/0.9

SD represents the standard deviation.

The thermogravimetric (TG) and derivative thermogravimetric (DTG) curves of CA and CA–Eu films are shown in Figure 9. The TG curve showed that the quality of the pure CA film remained basically unchanged before 250 °C, and the acetyl group on the CA molecule chains began to undergo pyrolysis when the temperature rose to 250 °C. The acidity generated by the dissociation of the acetyl group would accelerate the degradation of CA molecule chains so that the weight loss rate gradually increased. The DTG curve indicates that the largest weight loss rate of CA film was at 356 °C. The weight loss rate then gradually decreased, and the quality was basically stable above 410 °C. Finally, the residual mass of the CA film was 29.94%. The initial weight loss temperature and residual mass were relatively lower for the CA–Eu film, indicating that its thermal stability was poor. There were three obvious weight loss stages: (1) Evaporation of trace residual solvent and adsorbed water between 40 °C and 143 °C [41], where the weight loss was less about 5.94%; (2) 143–194 °C due to the pyrolysis of Eu–O coordination structure and heat loss of 9.63%; and (3) main mass loss above 194 °C. The final stage was mainly the degradation and carbonization of the CA molecular backbone [42]; the weight loss rate reached a maximum at 356 °C and then gradually decreased. Until 544 °C, the weight of CA-Eu film basically no longer changed and finally the residual mass was 26.02%.

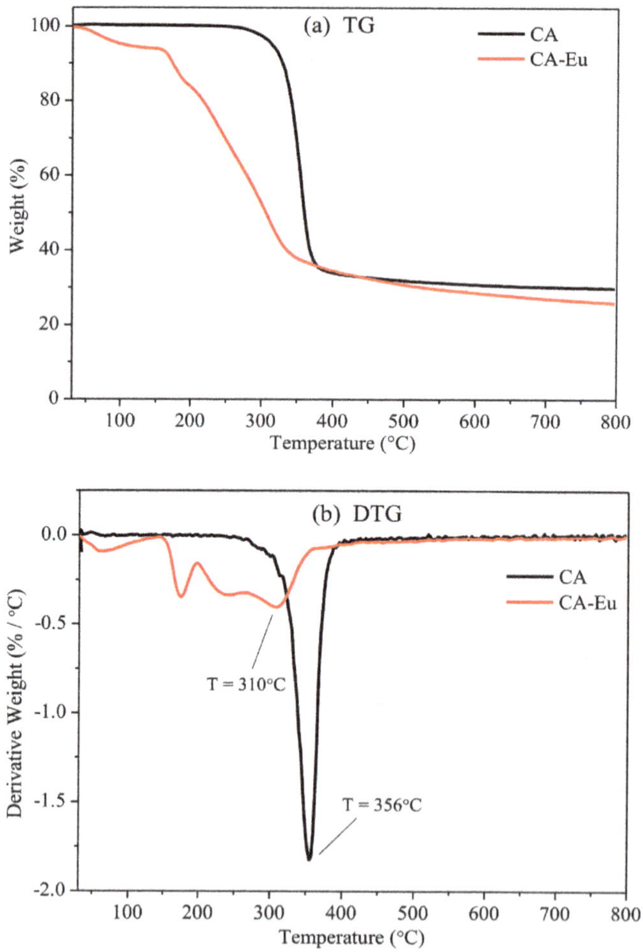

Figure 9. Thermogravimetric (TG) and derivative thermogravimetric (DTG) curves of CA and CA-Eu films. (**a**) TG, (**b**) DTG.

Cabbage seeds were grown in the soil. The seedlings were then covered with CA film and the CA–Eu light conversion film and placed in an environment with the same temperature, light, and moisture conditions for 20 days [43]. Figure 10 recorded the growth of seedlings from day 1 to day 20. The maximum leaf width changes are shown in Figure 11. The growth rate of the two seedlings in the first 10 days was almost the same. The growth rate of the seedlings increased, and it was obvious that the growth of the seedling covered by the CA–Eu light conversion film was better than that of CA film. The seedlings covered by a CA–Eu film had a larger leaf width (CA film: 2.5 cm; CA–Eu film: 3.5 cm). They also had more leaves and were higher, which proved that CA–Eu light conversion film could indeed convert ultraviolet light into red light for a positive effect on the photosynthetic growth of plants.

Figure 10. The growth of cabbage seedlings from day 1 to day 20. (**left**—CA film, **right**—CA-Eu film).

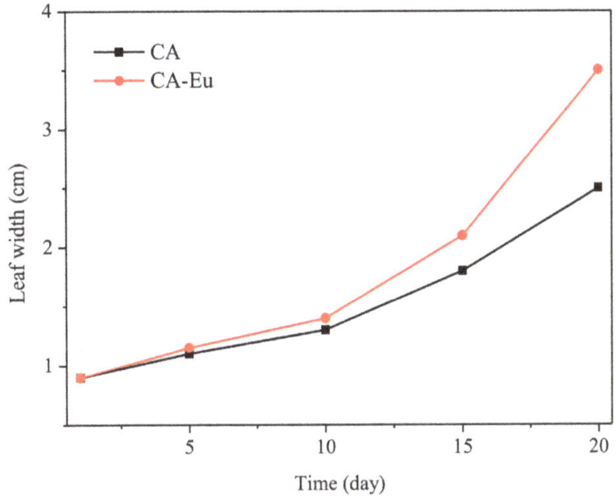

Figure 11. Change of maximum leaf width from day 1 to day 20.

4. Conclusions

(1) Eu^{3+} reacted with CA and formed an Eu-O coordination bond with the O atoms in C=O, C–O (O=C–O), and O–H of CA structure. The organic ligand CA has a good absorption coefficient at UV wavelength range so that it could effectively transfer energy to Eu^{3+} through ultraviolet absorption and make CA–Eu light conversion

film emit high-purity red light through $^5D_0 \rightarrow ^7F_2$ electric dipole transition. Due to the concentration quenching effect, the optimal fluorescence intensity was obtained at a CA: Eu^{3+} ratio of 3:1. In addition, the influence of the reaction time was also explored and the optimal fluorescence performance was obtained for the reaction time at 65 min.

(2) After doping Tb^{3+} in the light-conversion films, the Tb^{3+} could effectively increase the fluorescence intensity of Eu^{3+} to play a sensitization role in the process of energy transfer.

(3) The prepared light conversion film had a simple preparation process, high transparency, high tensile strength, and good flexibility. The material could absorb ultraviolet light and convert it into red light beneficial for the growth of plant photosynthesis. The material was also biodegradable, which is useful for efficiency and environmental protection.

Supplementary Materials: The following are available online at https://www.mdpi.com/2073-4360/13/1/113/s1, Figure S1: Fluorescence intensity of the light conversion films at the peak of 615 nm, Table S1: The characteristic FTIR data of the pure CA and light conversion films, Table S2: Binding energy of C1s and O1s for CA and light-conversion films, Table S3: The conditional viscosity of CA and CA-Eu solutions.

Author Contributions: Conceptualization, Z.Z. (Zhihui Zhang); writing—original draft preparation, Z.Z. (Zhihui Zhang); writing—review and editing, D.W. and Z.Z. (Zhihui Zhang); supervision, D.W., Z.Z. (Zhihui Zhang), Z.Z. (Zhengdong Zhao), Y.L., C.W. and J.L.; project administration, D.W.; funding acquisition, D.W. All authors have read and agreed to the published version of this manuscript.

Funding: This research was funded by [the National Natural Science Foundation of China] grant number [31770593], [China Postdoctoral Science Foundation] grant number [160750], [National College Student Innovation and Entrepreneurship Training Program, Northeast Forestry University, China] grant number [201910225481].

Informed Consent Statement: Informed consent was obtained from all subjects involved in the study.

Data Availability Statement: The data presented in this study are available in the insert article.

Conflicts of Interest: The authors declare no conflict of interest.

References

1. Long, S.P.; Zhu, X.G.; Naidu, S.L.; Ort, D.R. Can improvement in photosynthesis increase crop yields? *Plantcell Environ.* **2006**, *29*, 315–330. [CrossRef]
2. Yu, Y.; Wang, Y.; Liu, W.; Jia, X.; Ma, L.; Ren, L.; Xue, M.; Liu, X. Exploration of highly efficient light conversion agents for agricultural film based on the bay-substituted perylene diimides derivatives. *Dye. Pigment.* **2018**, *159*, 483–490. [CrossRef]
3. Qi, Y.; Wang, Y.; Yu, Y.; Liu, Z.; Zhang, Y.; Qi, Y.; Zhou, C. Exploring highly efficient light conversion agents for agricultural film based on aggregation induced emission effects. *J. Mater. Chem. C* **2016**, *4*, 11291–11297. [CrossRef]
4. Gong, T.; Li, Y.; Lei, B.; Zhang, X.; Liu, Y.; Zhang, H. Solid-state silicon nanoparticles with color-tunable photoluminescence and multifunctional applications. *J. Mater. Chem. C* **2019**, *7*, 5962–5969. [CrossRef]
5. Hou, L.; Xi, J.; Chen, X.; Li, X.; Ma, W.; Lu, J.; Xu, J.; Lin, Y.B. Biodegradability and ecological impacts of polyethylene-based mulching film at agricultural environment. *J. Hazard. Mater.* **2019**, *378*, 120774. [CrossRef] [PubMed]
6. Yang, N.; Sun, Z.-X.; Feng, L.-S.; Zheng, M.-Z.; Chi, D.-C.; Meng, W.-Z.; Hou, Z.-Y.; Bai, W.; Li, K.-Y. Plastic Film Mulching for Water-Efficient Agricultural Applications and Degradable Films Materials Development Research. *Mater. Manuf. Process.* **2015**, *30*, 143–154. [CrossRef]
7. Yu, Y.; Xu, P.; Jia, S.; Pan, H.; Zhang, H.; Wang, D.; Dong, L. Exploring polylactide/poly(butylene adipate-co-terephthalate)/rare earth complexes biodegradable light conversion agricultural films. *Int. J. Biol. Macromol.* **2019**, *127*, 210–221. [CrossRef]
8. Wang, D.; Yu, Y.; Ai, X.; Pan, H.; Zhang, H.; Dong, L. Polylactide/poly(butylene adipate-co-terephthalate)/rare earth complexes as biodegradable light conversion agricultural films. *Polym. Adv. Technol.* **2019**, *30*, 203–211. [CrossRef]
9. Zeshan, H.U.; Lanhua, L.I.; Chengbin, S.; Qiang, C.A.I. Light Conversion Film and Its Investigation Status. *Mater. Rev.* **2008**, *22*, 290–293.
10. Zhao, X.; Huang, C.; Zhang, S.; Wang, C. Cellulose Acetate/Activated Carbon Composite Membrane with Effective Dye Adsorption Performance. *J. Macromol. Sci. Part B-Phys.* **2019**, *58*, 909–920. [CrossRef]

11. Li, Y.; He, Y.; Fan, Y.; Shi, H.; Wang, Y.; Ma, J.; Li, H. Novel dual superlyophobic cellulose membrane for multiple oil/water separation. *Chemosphere* **2020**, *241*, 125067. [CrossRef] [PubMed]
12. Zhang, J.; Wang, L.-X.; Zhang, L.; Chen, Y.; Zhang, Q.-T. Co-luminescence properties of terbium ions-benzoic acid-phen complexes doped with europium ions. *Rare Met.* **2013**, *32*, 599–604. [CrossRef]
13. Ye, J.; Wang, B.; Xiong, J. Effect of Eu(3+) Concentration on the Structure and Fluorescence Quenching of Carboxymethyl Cellulose/Eu(III) Nanoparticles. *Polym. Mater. Sci. Eng.* **2016**, *32*, 32–37.
14. Chen, M.; Li, R.-M.; Runge, T.; Feng, J.; Feng, J.; Hu, S.; Shi, Q.-S. Solvent-Free Acetylation of Cellulose by 1-Ethyl-3-methylimidazolium Acetate-Catalyzed Transesterification. *ACS Sustain. Chem. Eng.* **2019**, *7*, 16971–16978. [CrossRef]
15. Liu, L.; Gong, D.; Bratasz, L.; Zhu, Z.; Wang, C. Degradation markers and plasticizer loss of cellulose acetate films during ageing. *Polym. Degrad. Stab.* **2019**, *168*, 108952. [CrossRef]
16. Jiang, L.; Li, K.; Yang, H.; Liu, X.; Li, W.; Xu, W.; Deng, B. Improving mechanical properties of electrospun cellulose acetate nanofiber membranes by cellulose nanocrystals with and without polyvinylpyrrolidone. *Cellulose* **2020**, *27*, 955–967. [CrossRef]
17. Kaur, H.; Bulasara, V.K.; Gupta, R.K. Influence of pH and temperature of dip-coating solution on the properties of cellulose acetate-ceramic composite membrane for ultrafiltration. *Carbohydr. Polym.* **2018**, *195*, 613–621. [CrossRef]
18. Wei, Y.; Fu, Z.; Zhao, H.; Liang, R.; Wang, C.; Wang, D.; Li, J. Preparation of PVA Fluorescent Gel and Luminescence of Europium Sensitized by Terbium(III). *Polymers* **2020**, *12*, 893. [CrossRef]
19. Xiang, L.; Yang, J.; Luo, D.; Su, X.; Qin, S. Construction of efficient desalting layer on a cellulose acetate membrane by acetalized surface crosslinking treatment. *Polym. Eng. Sci.* **2019**, *59*, 913–918. [CrossRef]
20. Song, H.-m.; Zhu, L.-j.; Zeng, Z.-x.; Xue, Q.-j. High performance forward osmosis cellulose acetate (CA) membrane modified by polyvinyl alcohol and polydopamine. *J. Polym. Res.* **2018**, *25*, 159. [CrossRef]
21. Serbanescu, O.S.; Pandele, A.M.; Miculescu, F.; Voicu, S.I. Synthesis and Characterization of Cellulose Acetate Membranes with Self-Indicating Properties by Changing the Membrane Surface Color for Separation of Gd(III). *Coatings* **2020**, *10*, 468. [CrossRef]
22. Huang, Z.; Huang, Z.; Feng, L.; Luo, X.; Wu, P.; Cui, L.; Mao, X. Modified cellulose by polyethyleneimine and ethylenediamine with induced Cu(II) and Pb(II) adsorption potentialities. *Carbohydr. Polym.* **2018**, *202*, 470–478. [CrossRef] [PubMed]
23. Ye, J.; Wang, B.; Xiong, J.; Sun, R. Enhanced fluorescence and structural characteristics of carboxymethyl cellulose/Eu(III) nano-complex: Influence of reaction time. *Carbohydr. Polym.* **2016**, *135*, 57–63. [CrossRef] [PubMed]
24. Asghar, M.R.; Zhang, Y.; Wu, A.; Yan, X.; Shen, S.; Ke, C.; Zhang, J. Preparation of microporous Cellulose/Poly(vinylidene fluoride-hexafluoropropylene) membrane for lithium ion batteries by phase inversion method. *J. Power Sources* **2018**, *379*, 197–205. [CrossRef]
25. Cobo, F.N.; Faria-Tisher, P.C.S.; Duarte, J.L.; Carvalho, G.M. Preparation and characterization of microporous cellulose acetate films using breath figure method by spin coating technique. *Cellulose* **2017**, *24*, 4981–4995. [CrossRef]
26. Mafirad, S.; Mehrnia, M.R.; Zahedi, P.; Hosseini, S.-N. Chitosan-based nanocomposite membranes with improved properties: Effect of cellulose acetate blending and TiO2 nanoparticles incorporation. *Polym. Compos.* **2018**, *39*, 4452–4466. [CrossRef]
27. Sajjan, P.; Nayak, V.; Padaki, M.; Zadorozhnyy, V.Y.; Klyamkin, S.N.; Konik, P.A. Fabrication of Cellulose Acetate Film through Blending Technique with Palladium Acetate for Hydrogen Gas Separation. *Energy Fuels* **2020**, *34*, 11699–11707. [CrossRef]
28. Zhang, Y.; Luo, B.; Sun, Y. Properties of Ultraviolet-Shielding Composite Film Prepared from Cellulose Acetate with Eu(III) Complex. *Chemistryselect* **2020**, *5*, 1688–1693. [CrossRef]
29. Wang, Z.; Guo, S.; Li, Q.; Zhang, X.; Li, T.; Li, P.; Yang, Z.; Guo, Q. Luminescent properties of Ba2SiO4:Eu3+ for white light emitting diodes. *Phys. B Condens. Matter* **2013**, *411*, 110–113. [CrossRef]
30. Raju, G.S.R.; Jung, H.C.; Park, J.Y.; Moon, B.K.; Balakrishnaiah, R.; Jeong, J.H.; Kim, J.H. The influence of sintering temperature on the photoluminescence properties of oxyapatite Eu3+:Ca2Gd8Si6O26 nanophosphors. *Sens. Actuators B Chem.* **2010**, *146*, 395–402. [CrossRef]
31. Chen, L.M.; Liu, Y.N.; Li, Y.D. Preparation and characterization of ZrO2: Eu3+ phosphors. *J. Alloys Compd.* **2004**, *381*, 266–271. [CrossRef]
32. Du, H.; Zhou, W.; Niu, J.; Xu, D.; Li, S.; Zhang, Z. Energy transfer and tunable photoluminescence of Sr6Gd2Na2(PO4)(6)F-2:Tb3+, Eu3+ phosphors for near-UV white LEDs. *J. Mater. Sci. Mater. Electron.* **2019**, *30*, 18575–18583. [CrossRef]
33. Wu, J.; Hu, D. Synthesis of PAM membranes with different porous structure and their light-scattering behavior. In Proceedings of the 14th Colloid and Interface Chemistry Conference of the Chinese Chemical Society, Changchun, China, 21–25 July 2013; 2.
34. Chen, H.; Li, R.; Xu, X.; Zhao, P.; Wong, D.S.H.; Chen, X.; Chen, S.; Yan, X. Citrate-based fluorophores in polymeric matrix by easy and green in situ synthesis for full-band UV shielding and emissive transparent display. *J. Mater. Sci.* **2019**, *54*, 1236–1247. [CrossRef]
35. Guo, Y.; Tian, D.; Shen, F.; Yang, G.; Long, L.; He, J.; Song, C.; Zhang, J.; Zhu, Y.; Huang, C.; et al. Transparent Cellulose/Technical Lignin Composite Films for Advanced Packaging. *Polymers* **2019**, *11*, 1455. [CrossRef] [PubMed]
36. Marmol, G.; Gauss, C.; Fangueiro, R. Potential of Cellulose Microfibers for PHA and PLA Biopolymers Reinforcement. *Molecules* **2020**, *25*, 4653. [CrossRef] [PubMed]
37. Rincon-Iglesias, M.; Lizundia, E.; Lanceros-Mendez, S. Water-Soluble Cellulose Derivatives as Suitable Matrices for Multifunctional Materials. *Biomacromolecules* **2019**, *20*, 2786–2795. [CrossRef] [PubMed]
38. Fischer, S.; Thümmler, K.; Volkert, B.; Hettrich, K.; Schmidt, I.; Fischer, K. Properties and Applications of Cellulose Acetate. *J. Macromol. Symp.* **2008**, *262*, 89–96. [CrossRef]

39. Iqhrammullah, M.; Marlina, M.; Khalil, H.P.S.A.; Kurniawan, K.H.; Suyanto, H.; Hedwig, R.; Karnadi, I.; Olaiya, N.G.; Abdullah, C.K.; Abdulmadjid, S.N. Characterization and Performance Evaluation of Cellulose Acetate-Polyurethane Film for Lead II Ion Removal. *Polymers* **2020**, *12*, 1317. [CrossRef]
40. Yang, S.; Wang, T.; Tang, R.; Yan, Q.; Tian, W.; Zhang, L. Enhanced permeability, mechanical and antibacterial properties of cellulose acetate ultrafiltration membranes incorporated with lignocellulose nanofibrils. *Int. J. Biol. Macromol.* **2020**, *151*, 159–167. [CrossRef]
41. Rajeswari, A.; Vismaiya, S.; Pius, A. Preparation, characterization of nano ZnO-blended cellulose acetate-polyurethane membrane for photocatalytic degradation of dyes from water. *Chem. Eng. J.* **2017**, *313*, 928–937. [CrossRef]
42. Xu, P.; Cen, C.; Chen, N.; Lin, H.; Wang, Q.; Xu, N.; Tang, J.; Teng, Z. Facile fabrication of silver nanoparticles deposited cellulose microfiber nanocomposites for catalytic application. *J. Colloid Interface Sci.* **2018**, *526*, 194–200. [CrossRef] [PubMed]
43. Wu, W.; Zhang, Z.; Dong, R.; Xie, G.; Zhou, J.; Wu, K.; Zhang, H.; Cai, Q.; Lei, B. Characterization and properties of a Sr2Si5N8:Eu2+-based light-conversion agricultural film. *J. Rare Earths* **2020**, *38*, 539–545. [CrossRef]

MDPI
St. Alban-Anlage 66
4052 Basel
Switzerland
Tel. +41 61 683 77 34
Fax +41 61 302 89 18
www.mdpi.com

Polymers Editorial Office
E-mail: polymers@mdpi.com
www.mdpi.com/journal/polymers

www.ingramcontent.com/pod-product-compliance
Lightning Source LLC
Chambersburg PA
CBHW041014270926
51850CB00024B/12
* 9 7 8 3 0 3 6 5 5 2 3 6 1 *